Physical Geology

EXPLORING THE EARTH

Shiprock, New Mexico; Jim Wark, Airphoto

Physical Geology

EXPLORING THE EARTH SIXTH EDITION

James S. Monroe
Professor Emeritus
Central Michigan University

Reed Wicander
Central Michigan University

Richard Hazlett
Pomona College

Australia • Brazil • Japan •Korea • Mexico • Singapore • Spain • United Kingdom • United States

Physical Geology: Exploring the Earth, Sixth Edition
James S. Monroe, Reed Wicander, Richard Hazlett

Earth Sciences Editor: Keith Dodson

Development Editor: Alyssa White

Assistant Editor: Carol Benedict

Editorial Assistants: Brandi Kirksey, Anna Jarzab

Technology Project Manager: Ericka Yeoman-Saler

Marketing Manager: Mark Santee

Marketing Assistant: Sylvia Krick

Marketing Communications Manager: Kelley McAllister

Art Director: Vernon T. Boes

Print Buyer: Barbara Britton

Permissions Editor: Kiely Sisk

Production Service: Graphic World Inc.

Cover and Text Designer: Patrick Devine

Frontmatter and Endmatter Designer: John Walker

Photo Researcher: Kathleen Olson

Copy Editor: Carol Reitz

Illustrator: Precision Graphics

Cover and Frontispiece Image: Jim Wark, Airphoto

Compositor: Graphic World Inc

© 2007 Brooks/Cole, Cengage Learning

ALL RIGHTS RESERVED. No part of this work covered by the copyright herein may be reproduced, transmitted, stored, or used in any form or by any means graphic, electronic, or mechanical, including but not limited to photocopying, recording, scanning, digitizing, taping, Web distribution, information networks, or information storage and retrieval systems, except as permitted under Section 107 or 108 of the 1976 United States Copyright Act, without the prior written permission of the publisher.

> For product information and technology assistance, contact us at
> **Cengage Learning Customer & Sales Support, 1-800-354-9706**
> For permission to use material from this text or product, submit all requests online at **cengage.com/permissions**
> Further permissions questions can be emailed to
> **permissionrequest@cengage.com**

ExamView® and *ExamView Pro*® are registered trademarks of FSCreations, Inc. Windows is a registered trademark of the Microsoft Corporation used herein under license. Macintosh and Power Macintosh are registered trademarks of Apple Computer, Inc. Used herein under license.

Library of Congress Control Number: 2006920086

ISBN-13: 978-0-495-01148-4

ISBN-10: 0-495-01148-7

Brooks/Cole
10 Davis Drive
Belmont, CA 94002-3098
USA

Cengage Learning is a leading provider of customized learning solutions with office locations around the globe, including Singapore, the United Kingdom, Australia, Mexico, Brazil, and Japan. Locate your local office at: **international.cengage.com/region**

Cengage Learning products are represented in Canada by Nelson Education, Ltd.

For your course and learning solutions, visit **academic.cengage.com**

Purchase any of our products at your local college store or at our preferred online store **www.ichapters.com**

Printed in China by China Translation & Printing Services Limited
5 6 7 8 9 11 10 09

ABOUT THE AUTHORS

James S. Monroe is professor emeritus of geology at Central Michigan University where he taught physical geology, historical geology, prehistoric life, and stratigraphy and sedimentology from 1975 until he retired in 1997. He is the co-author of several textbooks with Reed Wicander and has interests in Cenozoic geology and geologic education.

Reed Wicander is a geology professor at Central Michigan University where he teaches physical geology, historical geology, prehistoric life, and invertebrate paleontology. He is the co-author of several geology textbooks with James S. Monroe. His main research interests involve various aspects of Pale-ozoic palynology, specifically the study of acritarchs, on which he has published many papers. He is past president of the American Association of Stratigraphic Palynologists and a former councillor of the International Federation of Palynological Societies. He is the current chairman of the Acritarch Subcommission of the Commission Internationale de Microflore du Paléozoique.

Richard W. Hazlett is the winner of the 1996 and 2001 Wig Awards for teaching excellence at Pomona College. He is the first Stephen M. Pauley Chair in Environmental Analysis at the college, receiving this appointment in 2001. His main research interests involve volcanic stratigraphy, igneous petrology and resources, but most recently has been exploring land use issues focusing on the American West. Dr. Hazlett styles himself as transitioning from volcanology to environmental studies with a land use/natural resource emphasis. He is a co-author with Bernard Pipkin and Dee D. Trent on *Geology and the Environment* and was a script author and animation specialist for the highly successful *Earth Revealed* telecourse series.

ABOUT THE COVER

SHIPROCK, NEW MEXICO: Rising 600 meters (2000 feet) above the surrounding desert of northwestern New Mexico, Shiprock formed about 30 million years ago when an uncommon basaltic magma called *minette* worked its way through the crust to erupt, forming a maar, which is a small, explosive type of volcano built up from large amounts of ash and blocky rubble. The volcanic debris hardened into stone as it settled, compacted, welded, and cemented together through groundwater infiltration. Geologists call such transformed material *volcanic breccia*.

Shiprock volcano itself has long since eroded away. Presently, the land surface stands 750–1000 meters below the level that it did at the time of eruption. Shiprock is merely the minette breccia filling the throat, or pipe, that fed the volcano. It is more resistant to erosion than the surrounding landscape, hence it stands out, like an old-time sailing ship at sea visible to travelers as far as a hundred kilometers away.

Radiating from the base of Shiprock, like the spokes of a wheel, are several rib-like bands of minette, called dikes, that formed when magma entered cracks as the crust stretched open in response to the upwelling of molten rock through the central volcanic pipe. The early Navajo, unaware of "old-time sailing ships," had their own name for Shiprock, based upon the radiating set of dikes. They called it "Bitai," or "Winged Rock." Can you see why?

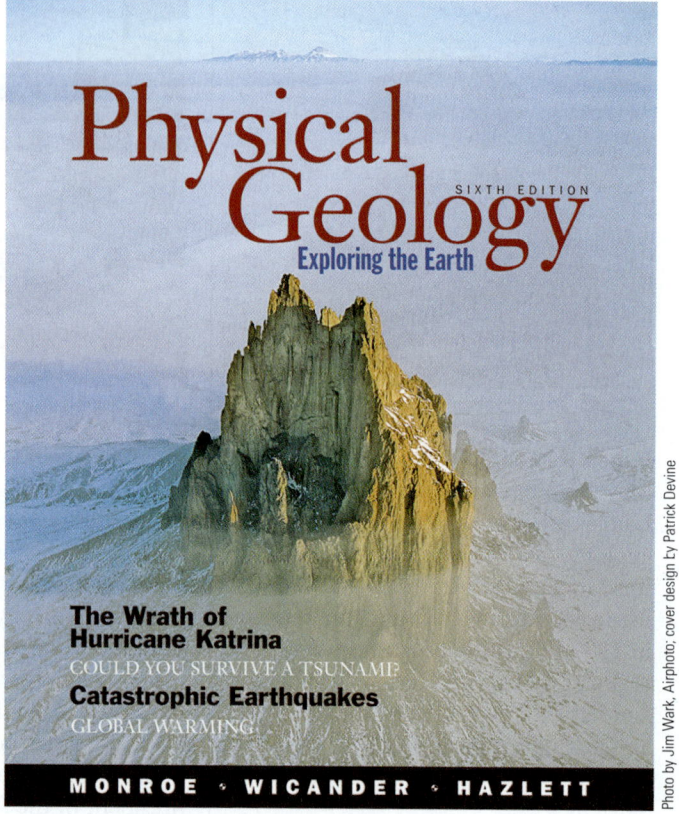

BRIEF CONTENTS

1. Understanding Earth: A Dynamic and Evolving Planet   2
2. Plate Tectonics: A Unifying Theory   32
3. Minerals—The Building Blocks of Rocks   70
4. Igneous Rocks and Intrusive Igneous Activity   100
5. Volcanoes and Volcanism   132
6. Weathering, Erosion, and Soil   168
7. Sediment and Sedimentary Rocks   198
8. Metamorphism and Metamorphic Rocks   230
9. Geologic Time: Concepts and Principles   260
10. Earthquakes   298
11. Earth's Interior   334
12. The Seafloor   358
13. Deformation, Mountain Building, and the Evolution of Continents   388
14. Mass Wasting   424
15. Running Water   456
16. Groundwater   496
17. Glaciers and Glaciation   530
18. The Work of Wind and Deserts   566
19. Shorelines and Shoreline Processes   596
20. Geology and Humanity   630

CONTENTS

1 Understanding Earth: A Dynamic and Evolving Planet 2

1.1 Introduction 4
1.2 Geology in Our Everyday Lives 6
1.3 Global Geologic and Environmental Issues Facing Humankind 8
1.4 Origin of the Universe and Solar System, and Earth's Place in Them 10
1.5 Earth as a Dynamic and Evolving Planet 13
  ■ *The Terrestrial and Jovian Planets* 14
1.6 Geology and the Formulation of Theories 16
1.7 Plate Tectonic Theory 17
1.8 The Rock Cycle 19
1.9 Geologic Time and Uniformitarianism 21

REVIEW WORKBOOK 25

GEOLOGY IN UNEXPECTED PLACES:
A Bit of Egypt in Central Park, New York City, and London, England 28

GEOLOGY IN FOCUS:
The Kyoto Protocol—An Update 29

GEOLOGY AT WORK:
Forensic Geology and Palynology 30

GEOLOGY IN YOUR LIFE:
Global Warming and Climate Change, and How They Affect You 31

2 Plate Tectonics: A Unifying Theory 32

2.1 Introduction 34
2.2 Continental Drift 34
2.3 Evidence for Continental Drift 36
2.4 Paleomagnetism and Polar Wandering 40
2.5 Magnetic Reversals and Seafloor Spreading 41
2.6 Plate Tectonics: A Unifying Theory 46
2.7 The Three Types of Plate Boundaries 47
  ■ *Tectonics of the Terrestrial Planets* 50
2.8 Hot Spots: An Intraplate Feature 57
2.9 Plate Movement and Motion 58
2.10 The Driving Mechanism of Plate Tectonics 59
2.11 Plate Tectonics and the Distribution of Natural Resources 60

REVIEW WORKBOOK 62

GEOLOGY IN FOCUS:
Oil, Plate Tectonics, and Politics 66

GEOLOGY IN UNEXPECTED PLACES:
A Man's Home Is His Castle 67

GEOLOGY AND CULTURAL CONNECTIONS:
The Struggle toward Scientific Progress 68

GEOLOGY AT WORK:
The Geologic Exploration of Mars 69

3 Minerals—The Building Blocks of Rocks 70

3.1 Introduction 72
3.2 Matter, Atoms, Elements, and Bonding 73
3.3 Explore the World of Minerals 77
3.4 Mineral Groups Recognized by Geologists 80
3.5 Mineral Identification 85
  ■ *The Precious Metals* 86
3.6 The Significance of Rock-Forming Minerals 90
3.7 The Origin of Minerals 91
3.8 Natural Resources and Reserves 92

REVIEW WORKBOOK 94

GEOLOGY IN FOCUS:
Mineral Crystals 96

GEOLOGY IN UNEXPECTED PLACES:
The Queen's Jewels 97

GEOLOGY IN YOUR LIFE:
Welcome to the Wonderful World of Micas 98

GEOLOGY AT WORK:
Terry S. Mollo: Sculpture Out of Stone 99

4 Igneous Rocks and Intrusive Igneous Activity 100

4.1 Introduction 102
4.2 The Properties and Behavior of Magma and Lava 103
  ■ *From Pluton to Volcano* 104
4.3 Igneous Rocks—Their Characteristics and Classification 115
4.4 Plutons—Their Characteristics and Origins 122

Contents

REVIEW WORKBOOK 125
GEOLOGY IN YOUR LIFE:
Of Mineral Water and Meteorites 128
GEOLOGY IN FOCUS:
Some Remarkable Volcanic Necks 129
GEOLOGY IN UNEXPECTED PLACES:
Little Rock, Big Story 130
GEOLOGY AND CULTURAL CONNECTIONS:
Hell Helped Early Earth Science 131

5 Volcanoes and Volcanism 132

5.1 Introduction 134
5.2 Volcanism and Volcanoes 136
5.3 Types of Volcanoes 141
5.4 Other Volcanic Landforms 149
5.5 The Distribution of Volcanoes 151
5.6 North America's Active Volcanoes 152
5.7 Plate Tectonics, Volcanoes, and Plutons 153
   ■ *Cascade Range Volcanos* 154
5.8 Volcanic Hazards, Volcano Monitoring, and Forecasting Eruptions 156

REVIEW WORKBOOK 160
GEOLOGY IN FOCUS:
The Bronze Age Eruption of Santorini 164
GEOLOGY IN UNEXPECTED PLACES:
Oldoinyo Lengai Volcano 165
GEOLOGY IN YOUR LIFE:
Do Volcanic Gases Cause Ozone Depletion? 166
GEOLOGY AT WORK:
Volcano Observatories and Volcano Monitoring 167

6 Weathering, Erosion, and Soil 168

6.1 Introduction 170
6.2 Alteration of Minerals and Rocks 171
6.3 Mechanical Weathering—Disaggregation of Earth Materials 171
6.4 Chemical Weathering—Decomposition of Earth Materials 175
   ■ *Arches National Park, Utah* 176
6.5 Soil and Its Origin 181
6.6 Expansive Soils and Soil Degradation 188
6.7 Weathering and Natural Resources 191

REVIEW WORKBOOK 192
GEOLOGY IN FOCUS:
The Dust Bowl 194

GEOLOGY IN UNEXPECTED PLACES:
Gravestones and Geology 195
GEOLOGY IN YOUR LIFE:
Industrialization and Acid Rain 196
GEOLOGY AND CULTURAL CONNECTIONS:
The Agricultural Revolution 197

7 Sediment and Sedimentary Rocks 198

7.1 Introduction 200
7.2 Sediment Sources, Transport, and Deposition 201
7.3 Lithification: Converting Sediment into Sedimentary Rock 203
7.4 The Types of Sedimentary Rocks 204
7.5 Sedimentary Facies 210
7.6 Read the Story Told by Sedimentary Rocks 212
   ■ *Fossilization* 216
7.7 Important Resources in Sediments and Sedimentary Rocks 219

REVIEW WORKBOOK 222
GEOLOGY IN FOCUS:
Fossils and Fossilization 226
GEOLOGY IN UNEXPECTED PLACES:
Sandstone Lion 227
GEOLOGY AT WORK:
Exploring for Oil and Natural Gas, Susan M. Landon 228
GEOLOGY AND CULTURAL CONNECTIONS:
Deep Time 229

8 Metamorphism and Metamorphic Rocks 230

8.1 Introduction 232
8.2 Equilibrium and the Causes of Metamorphism 233
   ■ *Marble, Plaster, and Art* 234
8.3 The Main Types of Metamorphism 237
8.4 Classification of Metamorphic Rocks 243
8.5 Metamorphic Zones and Facies 248
8.6 Plate Tectonics and Metamorphism 250
8.7 Metamorphism and Global Climate Change 253
8.8 Some Economic Uses of Metamorphic Materials 253

REVIEW WORKBOOK 254
GEOLOGY IN UNEXPECTED PLACES:
Starting Off with a Clean Slate 258

GEOLOGY AT WORK:
What Do Metamorphic Petrologists Do? 258
GEOLOGY IN FOCUS:
Asbestos: Good or Bad? 259

9 Geologic Time: Concepts and Principles 260

9.1 Introduction 262
9.2 Early Concepts of Geologic Time and the Age of Earth 263
9.3 James Hutton and the Recognition of Geologic Time 264
9.4 Relative Dating Methods 265
  ■ *Uluru and Kata Tjuta* 276
9.5 Correlating Rock Units 268
9.6 Absolute Dating Methods 281
9.7 Development of the Geologic Time Scale 287
9.8 Geologic Time and Climate Change 288

REVIEW WORKBOOK 291
GEOLOGY IN FOCUS:
Denver's Weather—280 Million Years Ago! 295
GEOLOGY IN UNEXPECTED PLACES:
Time Marches On—The Great Wall of China 296
GEOLOGY AT WORK:
Geology in Hollywood, John R. Horner 297

10 Earthquakes 298

10.1 Introduction 300
10.2 Elastic Rebound Theory 301
10.3 Seismology 303
10.4 The Frequency and Distribution of Earthquakes 304
10.5 Seismic Waves 307
10.6 Locating an Earthquake 308
10.7 Measuring the Strength of an Earthquake 311
10.8 The Destructive Effects of Earthquakes 314
  ■ *The San Andreas Fault* 316
10.9 Earthquake Prediction 323
10.10 Earthquake Control 326

REVIEW WORKBOOK 327
GEOLOGY IN FOCUS:
Paleoseismology 330
GEOLOGY IN FOCUS:
Designing Earthquake-Resistant Structures 331
GEOLOGY IN UNEXPECTED PLACES:
It's Not My Fault 332
GEOLOGY AT WORK:
Pacific Tsunami Warning System 333

11 Earth's Interior 334

11.1 Introduction 336
11.2 Earth's Size, Density, and Internal Structure 337
  ■ *Earth's Place in the Cosmos* 338
11.3 Earth's Crust—Its Outermost Part 342
11.4 Earth's Mantle—The Layer Below the Crust 343
11.5 The Core 344
11.6 Earth's Internal Heat 346
11.7 Gravity and How Its Force Is Determined 347
11.8 Floating Continents—The Principle of Isostasy 349
11.9 Earth's Magnetic Field 352

REVIEW WORKBOOK 354
GEOLOGY IN FOCUS:
Planetary Alignments, Gravity, and Catastrophes 356
GEOLOGY IN UNEXPECTED PLACES:
Diamonds and Earth's Interior 357

12 The Seafloor 358

12.1 Introduction 360
12.2 Methods Used to Study the Seafloor 360
12.3 Oceanic Crust—Its Structure and Composition 364
12.4 The Continental Margins 364
12.5 Features Found in the Deep-Ocean Basins 369
12.6 Sediments on the Deep Seafloor 375
12.7 Reefs—Rocks Made by Organisms 376
12.8 Resources from the Oceans 378
  ■ *Reefs: Rocks Made by Organisms* 380

REVIEW WORKBOOK 382
GEOLOGY IN FOCUS:
Oceanic Circulation and Resources from the Sea 385
GEOLOGY IN UNEXPECTED PLACES:
Ancient Seafloor in San Francisco 386
GEOLOGY AT WORK:
The Future beneath the Sea 387

13 Deformation, Mountain Building, and the Evolution of Continents 388

13.1 Introduction 390
13.2 Rock Deformation 391
13.3 Folded Rock Layers 394
13.4 Joints and Faults—Deformation by Fracturing 398

- 13.5 Deformation and the Origin of Mountains 404
  - *Types of Faults* 410
- 13.6 The Formation and Evolution of Continents 413

REVIEW WORKBOOK 418

GEOLOGY IN UNEXPECTED PLACES:
Ancient Ruins and Geology 421

GEOLOGY IN FOCUS:
Geologic Maps—Their Construction and Uses 422

GEOLOGY AT WORK:
Engineering and Geology 423

14 Mass Wasting 424

- 14.1 Introduction 426
- 14.2 Factors That Influence Mass Wasting 427
- 14.3 Types of Mass Wasting 431
  - *Point Fermin—Slip Sliding Away* 436
- 14.4 Recognizing and Minimizing the Effects of Mass Wasting 446

REVIEW WORKBOOK 451

GEOLOGY IN FOCUS:
The Vaiont Dam Disaster 453

GEOLOGY IN UNEXPECTED PLACES:
New Hampshire Say Good-Bye to the "Old Man" 454

GEOLOGY IN YOUR LIFE:
Southern California Landslides 455

15 Running Water 456

- 15.1 Introduction 458
- 15.2 Water on Earth 459
- 15.3 Running Water 461
- 15.4 Running Water, Erosion, and Sediment Transport 464
- 15.5 Deposition by Running Water 467
- 15.6 Predicting and Controlling Floods 474
  - *The Flood of '93* 476
- 15.7 Drainage Systems 479
- 15.8 The Evolution of Valleys 484

REVIEW WORKBOOK 490

GEOLOGY IN FOCUS:
The River Nile and the History of Egypt 493

GEOLOGY AT WORK:
Dams, Reservoirs, and Hydroelectric Power Plants 494

GEOLOGY IN UNEXPECTED PLACES:
Floating Burial Chambers and the Mississippi River Delta 495

16 Groundwater 496

- 16.1 Introduction 498
- 16.2 Groundwater and the Hydrologic Cycle 498
- 16.3 Porosity and Permeability 498
- 16.4 The Water Table 500
- 16.5 Groundwater Movement 501
- 16.6 Springs, Water Wells, and Artesian Systems 502
- 16.7 Groundwater Erosion and Deposition 505
  - *The Burren Area of Ireland* 508
- 16.8 Modifications of the Groundwater System and Their Effects 512
- 16.9 Hydrothermal Activity 518

REVIEW WORKBOOK 524

GEOLOGY IN FOCUS:
Arsenic and Old Lace 527

GEOLOGY IN UNEXPECTED PLACES:
Water-Treatment Plants 528

GEOLOGY AND CULTURAL CONNECTIONS:
Dowsing 529

17 Glaciers and Glaciation 530

- 17.1 Introduction 532
- 17.2 The Kinds of Glaciers 533
- 17.3 Glaciers—Moving Bodies of Ice on Land 534
- 17.4 The Glacial Budget—Accumulation and Wastage 538
- 17.5 Erosion and Transport by Glaciers 541
  - *Valley Glaciers and Erosion* 544
- 17.6 Deposits of Glaciers 547
- 17.7 The Ice Age (The Pleistocene Epoch) 553
- 17.8 Causes of the Ice Age 557

REVIEW WORKBOOK 559

GEOLOGY IN UNEXPECTED PLACES:
Evidence of Glaciation in New York City 562

GEOLOGY IN YOUR LIFE:
A Visit to Cape Cod, Massachusetts 563

GEOLOGY IN FOCUS:
Glaciers and Global Warming 564

GEOLOGY AND CULTURAL CONNECTIONS:
The Ice Man 565

18 The Work of Wind and Deserts 566

- 18.1 Introduction 568
- 18.2 Sediment Transport by Wind 569
- 18.3 Wind Erosion 570
- 18.4 Wind Deposits 574
- 18.5 Air-Pressure Belts and Global Wind Patterns 580
- 18.6 The Distribution of Deserts 581
- 18.7 Characteristics of Deserts 582
  - ■ *Rock Art for the Ages* 584
- 18.8 Desert Landforms 587

REVIEW WORKBOOK 591

GEOLOGY IN FOCUS:
Radioactive Waste Disposal—Safe or Sorry? 593

GEOLOGY IN UNEXPECTED PLACES:
Spiral Jetty 594

GEOLOGY IN YOUR LIFE:
Windmills and Wind Power 595

19 Shorelines and Shoreline Processes 596

- 19.1 Introduction 598
- 19.2 Shoreline Processes 599
- 19.3 Shoreline Erosion 606
- 19.4 Deposition along Shorelines 608
  - ■ *Shoreline Processes and Beaches* 610
- 19.5 The Nearshore Sediment Budget 615
- 19.6 The Classification of Coasts 617
- 19.7 Storm Waves and Coastal Flooding 618
- 19.8 Managing Coastal Areas as Sea Level Rises 620

REVIEW WORKBOOK 623

GEOLOGY IN UNEXPECTED PLACES:
Rising Sea Level and the Fate of Venice, Italy 626

GEOLOGY IN YOUR LIFE:
Giant Killer Waves 627

GEOLOGY AT WORK:
Erosion and the Cape Hatteras Lighthouse 628

GEOLOGY IN FOCUS:
Energy from the Oceans 629

20 Geology and Humanity 630

- 20.1 Introduction 632
- 20.2 Geology and History 633
- 20.3 Humans as Geologic Agents 637
- 20.4 Climate Change 639

REVIEW WORKBOOK 644

GEOLOGY IN FOCUS:
Atomic Testing and the Earth System 646

GEOLOGY AT WORK:
Environmental Geology 647

GEOLOGY IN YOUR LIFE:
Organic Farming 648

GEOLOGY IN UNEXPECTED PLACES:
Mines and Flooding 649

Appendix A: English–Metric Conversion Chart 651
Appendix B: Periodic Table of the Elements 652
Appendix C: Mineral Identification Tables 654
Appendix D: Topographic Maps 657
Answers 661
Glossary 666
Index 677

PREFACE

Earth is a dynamic planet that has changed continuously during its 4.6 billion years of existence. The size, shape, and geographic distribution of the continents and ocean basins have changed through time, as have the atmosphere and biota. As scientists and concerned citizens, we have become increasingly aware of how fragile our planet is and, more importantly, how interdependent all of its various systems and subsystems are.

We have also learned that we cannot continually pollute our environment and that our natural resources are limited and, in most cases, nonrenewable. Furthermore, we are coming to realize how central geology is to our everyday lives. For example, a 9.0-magnitude earthquake in the Indian Ocean on December 26, 2004 generated a tsunami that killed more than 220,000 people in Indonesia, Sri Lanka, India, Thailand, Somalia, Myanmar, Malaysia, and the Maldives, and caused billions of dollars in damage. Two hurricanes in 2005, Katrina and Rita, damaged many offshore oil platforms and oil refineries in Texas and Louisiana, demonstrating how fragile our energy network, from production to the finished refined product, is and how dependent we are on petroleum to run our economy. For these and other reasons, geology is one of the most important college or university courses a student can take.

Physical Geology: Exploring the Earth is designed for a one-semester introductory course in geology that serves both majors and nonmajors in geology and the Earth sciences. One of the problems with any introductory science course is that students are overwhelmed by the amount of material that must be learned. Furthermore, most of the material does not seem to be linked by any unifying theme and does not always appear to be relevant to their lives. This book, however, is written to address that problem in that it shows, in its easy-to-read style, that geology is an exciting and ever-changing science, and one in which new discoveries and insights are continually being made.

The goals of this book are to provide students with a basic understanding of geology and its processes and, most importantly, with an understanding of how geology relates to the human experience: that is, how geology affects not only individuals, but society in general. With these goals in mind, we introduce the major themes of the book in the first chapter to provide students with an overview of the subject and to enable them to see how the various systems and subsystems of Earth are interrelated. We then cover the unifying theme of geology, plate tectonics, in the second chapter. Plate tectonic theory is central to the study of geology, because it links many aspects of geology together. It is a theme that is woven throughout this edition.

We also discuss the economic and environmental aspects of geology throughout the book rather than treating these topics in separate chapters. In this way, students can see, through relevant and interesting examples, how geology impacts our lives.

New Features in the Sixth Edition

Just as Earth is a dynamic and evolving planet, so too is *Physical Geology: Exploring the Earth*. The sixth edition has undergone a major revision that has resulted in a textbook that is still not only easy to read, but has a high level of current information, many new photographs, a completely revamped art program, and various new features to help students maximize their learning and understanding of Earth and its systems. Drawing on the comments and suggestions of reviewers, we have incorporated many new features into this edition, including a new coauthor, Dr. Richard W. Hazlett, from Pomona College, California. Perhaps the most noticeable change is that the chapter Plate Tectonics: A Unifying Theory has been moved to Chapter 2, so that students can see how plate tectonic theory ties many aspects of geology together. In addition, the final chapter, Geology and Humanity, has been completely rewritten to show how geology relates to, and impacts, such diverse topics as history, human affairs, the environment, and climate change.

The beginning of each chapter has been redesigned to make it easier for the student to see what the important and relevant topics are in the chapter. Each topic heading is followed by a series of second-order headings asking relevant questions about the topic that the student should learn. The chapter-opening photograph relates to the human aspects of the chapter topic in a way that is new and different than many textbooks. The photographs were chosen to engage students to think about the topic in a different way.

The format of the chapters has also been completely redone to make it more learning-friendly for the student. Each topic covered in a chapter has subheadings that ask essential questions about the topic students should know. That question is answered in the first paragraph following the question heading and elaborated upon, in subsequent paragraphs. At the end of each major topic, the material is summarized in a table, to make sure students know what the relevant and salient points were in that section.

The end-of-chapter summary has also been redone to emphasize and reinforce the chapter material. Instead of a recitation of the topics, the chapter summary provides the essential questions asked throughout the chapter and follows with the answer. In this way, students see what the important questions were and how they are answered. We have replaced the multiple-choice questions with Review Questions that require an understanding of the material to answer. We also have added Apply Your Knowledge questions that involve solving geologic problems based on the chapter material, and added Field Questions that engage the student in real-life geologic problems based on sketches or photographs of geologic outcrops or situations.

A new section has been added at the end of each chapter and is titled Geology Matters. Written in a magazine-style format, this section contains some of the previous *Geology in Unexpected Places*, *Geo-Focus*, and *Geo-Profiles* chapter inserts that were found throughout the chapter in earlier editions. By incorporating them together in one place, the flow of the chapter is not interrupted. The *Geo-Focus* boxes are now called *Geology in Focus* and contain either new topics or updated previous ones. *Geology at Work* is a new feature that covers the different types of jobs geologists do or profiles of people in careers students may not associate with geology, such as sculpting or movie-making. Some of these profiles are from the previous edition's *Geo-Profiles*, but most are new. Another new feature is *Geology in Your Life*, which looks at different ways geology affects or influences your life. Lastly, in several chapters, are articles titled *Geology and Cultural Connections*, which look at the link between geology and society.

In addition to text changes, the art program has been completely revamped. The figure captions have been expanded to better explain what the student is seeing in the figure, and numerous captions contain questions about the figure that are also answered for the student's benefit. Many of the figures have been completely redone to make them easier to understand and to better illustrate the material described in the text.

Many of the features that were so successful in the fifth edition have been kept and redone or updated. These include the *Geology in Unexpected Places* boxes which are now at the end of the chapter in the Geology Matters section, the *What Would You Do?* boxes in the text, and the Physical GeologyNow interactive media program, which has been seamlessly integrated with the text, enhancing students' understanding of important geological processes. This interactive resource brings geology alive with animated figures, media-enhanced activities, tutorials, and personalized learning plans. And like other features in our new edition, it encourages students to be curious, to think about geology in new ways, and to connect their newfound knowledge of the world around them to their own lives.

We feel the numerous changes made here greatly improve the sixth edition by making it more relevant, easier to read and comprehend, as well as being a more effective teaching tool. Additionally, improvements have been made in the ancillary package that accompanies the book.

Text Organization

Plate tectonic theory is the unifying theme of geology and this book. This theory has revolutionized geology because it provides a global perspective of Earth and allows geologists to treat many seemingly unrelated geologic phenomena as part of a total planetary system. Because plate tectonic theory is so important, it has been moved to Chapter 2 and is discussed in most subsequent chapters in terms of the subject matter of that chapter.

Another theme of this book is that Earth is a complex, dynamic planet that has changed continually since its origin some 4.6 billion years ago. We can better understand this complexity by using a systems approach in the study of Earth and emphasizing this approach through the book.

We have organized *Physical Geology: Exploring the Earth* into several informal sections. Chapter 1 is an introduction to geology and Earth systems, its relevance to the human experience, the origin of the solar system and Earth's place in it, an introduction to plate tectonic theory, the rock cycle, and geologic time and uniformitarianism. Chapter 2 covers plate tectonic theory and its relevance and importance in all aspects of geology. Chapters 3–8 examine Earth's materials (minerals and igneous, sedimentary, and metamorphic rocks) and the geologic processes associated with them, including the role of plate tectonics in their origin and distribution. Chapter 9 discusses geologic time, introduces several dating methods, and explains how geologists correlate rocks.

Chapters 10–13 deal with the related topics of earthquakes, Earth's interior, the seafloor, and deformation and mountain building. Chapters 14–19 cover Earth's surface processes. Chapter 20 provides an overview of the relationship between geology and humanity.

We have found that presenting the material in this order works well for most students. We know, however, that many instructors prefer an entirely different order of topics, depending on the emphasis in their course. We have therefore written this book so instructors can present the chapters in any order that suits the needs of their course.

Chapter Organization

Each chapter has a photograph relating to the chapter material, an outline of the topics covered, using an "Essential Questions to Ask" format so that the student knows what the major topics are and what he or she should know about the topics. This is followed by an *Introduction* that is intended to stimulate interest in the chapter by discussing some aspect of the material and showing students how the chapter material fits into the larger geologic perspective.

The text is written in a clear, informal style, making it easy for students to comprehend. Numerous newly rendered color diagrams and photographs complement the text, providing a visual representation of the concepts and information presented. In addition, many figures ask the student to describe what they are seeing.

Each chapter contains at least one, and usually two *What Would You Do?* boxes, designed to encourage thinking by students as they attempt to solve a hypothetical problem or issue relating to the chapter material. Mineral and energy resources are discussed in the final sections of a number of chapters to provide interesting, relevant information in the context of the chapter topics. The previous *Geology in Unexpected Places*, *Geo-Focus*, and *Geo-Profile* boxes are now found in the Geology Matters at the end of each chapter.

The end-of-chapter materials begin with an *Essential Questions Summary* in the Review Workbook. The *Essential Terms to Know*, which are printed in boldface type in the chapter text, are listed at the end of each chapter for easy review as well as the page number where they are first defined. A full *Glossary* of important terms appears at the end of the text. The *Review Questions* are another important feature of this book, and include new short answer review questions that test the student's understanding of the material presented, as well as thought-provoking and quantitative questions in the *Apply Your Knowledge* section. A third feature are the *Field Questions*. These photographs or drawings, ask the student to describe the geologic outcrop or feature illustrated as either a sketch or a photograph.

Ancillary Materials

Class Preparation

Online Instructor's Manual with Test Bank The Instructor's Manual contains chapter by chapter outlines, objectives, summaries, lecture suggestions, enrichment topics, answers to chapter review questions and a complete test bank with answers. Available exclusively online at our password-protected instructor's Web site. ISBN: 0495011517

Lecture Tools

JoinIn™ on TurningPoint® for Response Systems Using an audience response system? Brooks/Cole, a part of Cengage Learning, is now pleased to offer you book-specific JoinIn™ content for Response Systems tailored to *Physical Geology: Exploring the Earth*, allowing you to transform your classroom and assess your students' progress with instant in-class quizzes and polls. Our exclusive agreement to offer TurningPoint® software lets you pose book-specific questions and dis-play students' answers seamlessly within the Microsoft® PowerPoint® slides of your own lecture, in conjunction with the infrared or radio frequency "clicker" hardware of your choice. Enhance how your students interact with you, your lecture, and each other. For college and university adopters only. Contact your local Cengage Learning representative to learn more. ISBN: 0495017485

Multimedia Manager with Living Lecture™ Tools This one-stop lecture tool makes it easy for you to assemble, edit, publish, and present custom lectures for your course, using Microsoft PowerPoint™. Enhanced with Living Lecture Tools (including all of the text illustrations, photos, and Active Figure animations), the Multimedia Manager quickly and easily lets you build media-enhanced lecture presentations to suit your course. ISBN: 0495011533

Transparency Acetates This set of 100 transparency acetates contains the most popular photos and images from the text. ISBN: 0495126438

Slides This set of slides contains many photos and images from the text. ISBN: 0495126799

Testing/Assessment

ExamView® Computerized Testing Create, deliver, and customize tests and study guides (both print and online) in minutes with this easy-to-use assessment and tutorial system. ExamView offers both a Quick Test Wizard and an Online Test Wizard that guide you step by step through the process of creating tests, while its "what you see is what you get" interface allows you to see the test you are crating on the screen exactly as it will print or display online. You can build tests of up to 250 questions using up to 12 question types. Using ExamView's complete word processing capabilities, you can enter an unlimited number of new questions or edit existing questions from our test bank. ISBN: 0495011525

Course Management

WebCT/Now Integration Integrate the conceptual testing and multimedia tutorial features of Physical GeologyNow™ within your familiar WebCT environment by packaging the text with this special access code. You can assign the Physical GeologyNow™ materials and have the results flow automatically to your WebCT grade book, creating a robust online course. Students access Physical GeologyNow™ via your WebCT course, without using a separate user name or password. For college and university adopters only. Contact your local Cengage Learning representative to learn more.

Blackboard/Now Integration Integrate the conceptual testing and multimedia tutorial features of Physical GeologyNow™ within your familiar Blackboard environment by packaging the text with this special access code. You can assign the Physical GeologyNow™ materials and have the results flow automatically to your Blackboard grade book, creating a robust online course. Students access Physical GeologyNow™ via your Blackboard course, without using a separate user name or password. For college and university adopters only. Contact your local Cengage Learning representative to learn more.

Concept Mastery: Homework/Tutorial/Remediation

Physical GeologyNow™ academic.cengage.com

Instructor Empower your students with the first assessment-centered student tutorial system for a Physical Geology course. This powerful and interactive online resource uses a series of chapter-specific diagnostics to gauge students' unique study needs, then provides a Personalized Learning Plan that focuses their study time on the concepts they need to review most by directing them to specific sections of the text, Active Figures (animated text figures), Geo-focus Figures (figures you can zoom in on), and more! By providing students with a better understanding of exactly what they need to focus on, Physical GeologyNow™ helps students maximize their study time to bring them closer to success!

Student What do you need to learn NOW? Physical GeologyNow™ is a powerful, personalized online learning companion designed to help you gauge your own learning needs and identify the concepts on which you most need to focus your study time. How does it work? After you read the chapter, take the diagnostic Pre-Test ("What Do I Know?") for a quick assessment of how well you already understand the material. Your Personalized Learning Plan ("What Do I Need to Learn?") outlines the interactive animations, tutorials, and exercises you need to review. The Post-Test ("What Have I Learned?") helps you measure how well you have mastered the core concepts from the chapter. Study smarter—and make every minute count!

Study Guide This student study guide contains Chapter Overviews, Important Terms, Study Questions and Answers, Activities, and Internet Connections. ISBN: 0495011509

Labs/Applications/Courseware/Other Resources

Earth Lab: Exploring the Earth Sciences, 2e This lab manual, by Claudia Owen (University of Oregon), Diane Pirie, and Grenville Draper (both from Florida International University) features a variety of engaging, hands-on activities designed to help students develop data gathering and interpretive skills to create their own models of Earth's natural processes. ISBN: 0495013285

Investigations into Physical Geology, 2e This full-color lab manual, by Jim Mazzulo of Texas A&M University, provides a wide-range of activities about rock and mineral specimens, topographic maps, and aerial photography. ISBN: 0030202949

GIS Investigations for the Earth Sciences Learn by doing with this innovative series by the Saguaro Project team. Each of the guides in the series leads students to explore, analyze and elaborate from a robust GIS dataset using ArcView® GIS software. Each guide comes free with a CD-based, 120-day time-locked version of the ArcView software. An extensive Instructor's Manual provides assistance on incorporating GIS activities into the classroom.

1. **Exploring the Dynamic Earth: GIS Investigations for the Earth Sciences**, ISBN: 0534391389
2. **Exploring the Ocean Environment: GIS Investigations for the Earth Sciences**, ISBN: 0534423507
3. **Exploring Water Resources: GIS Investigations for the Earth Sciences**, ISBN: 0534391567
4. **Exploring Tropical Cyclones: GIS Investigations for the Earth Sciences**, ISBN: 0534391478

Website

URL academic.cengage.com/earthscience/monroe
Description: When you adopt Physical Geology: Exploring the Earth, you and your students will have access to a rich array of teaching and learning resources that you won't find anywhere else. This outstanding site features chapter-by-chapter online tutorial quizzes, a final exam, chapter outlines, chapter review, chapter-by-chapter web links, flash cards, and more!

Acknowledgments

As authors, we are, of course, responsible for the organization, style, and accuracy of the text, and any mistakes, omissions, or errors are our responsibility. The finished product is the culmination of many years of work during which we received numerous comments and advice from many geologists who reviewed parts of the text. We wish to express our sincere appreciation to the reviewers who reviewed the fifth edition and made many helpful and useful comments that led to the improvements in this sixth edition:

David Black, *University of Akron*
Kathleen Bower, *Eastern Illinois University*
Beth Christensen, *Georgia State University*
Katherine Clancy, *University of Maryland, College Park*
Mark Everett, *Texas A&M University*
Andrew M. Goodliffe, *University of Alabama*
Nathan Green, *University of Alabama*
Roger Hoggan, *Brigham Young University–Idaho*
Curtis Hollabaugh, *University of West Georgia*
Kent S. Murray, *University of Michigan–Dearborn*
Debra Simpson, *Delaware County Community College*
David Steffy, *Jacksonville State University*
Matthew M. Uliana, *Texas State University–San Marcos*

We would also like to thank the reviewers of the fourth edition for their many comments and helpful suggestions. They are:

William J. Frazier, *Columbus State University*
Thomas J. Leonard, *William Paterson University*
Donald Lovejoy, *Palm Beach Atlantic University*
Peter Bower, *Barnard College/Columbia University*
Tom Shoberg, *Pittsburg State University*
Bethany D. Rinard, *Tarleton State University*
Thom Wilch, *Albion College*
Glen Merrill, *University of Houston–Downtown*
Jon C. Crawley, *Roanoke College*
Neil Johnson, *Appalachian State University*
Richard. L. Mauger, *East Carolina University*
Eddie B. Robertson, *Reinhardt College*
John Tacinelli, *Rochester Community and Technical College*
Ntungwa Maasha, *Coastal Georgia Community College*
Cathy Baker, *Arkansas Tech University*
David P. Lawrence, *East Carolina University*

Nicholas A. Gioppo, *Mohawk Valley Community College*
Stan P. Dunagan, *University of Tennessee at Martin*
Dave Thomas, *Washtenaw Community College*
Ronald A. Johnston, *Fayetteville State University*
Steve Mattox, *Grand Valley State University*
John Leland, *Glendale Community College*
Jim Van Alstine, *University of Minnesota*
Ed van Hees, *Wayne State University*
Ed Wehling, *Anoka-Ramsey Community College*

Our thanks also to the third edition reviewers, whose comments improved the fourth edition:

Steven R. Dent, *Northern Kentucky University*
Michael R. Forrest, *Rio Hondo Community College*
René De Hon, *Northeast Louisiana University*
Robert B. Jorstad, *Eastern Illinois University*
David T. King, *Auburn University*
Peter L. Kresan, *University of Arizona*
Gary D. Rosenberg, *Indiana University–Purdue University at Indianapolis*
Darrel W. Schmitz, *Mississippi State University*
David G. Towell, *Indiana University*

We would also like to thank the reviewers of the second edition whose comments and suggestions greatly improved the third edition:

Gary Allen, *University of New Orleans*
Richard Beck, *Miami University*
Roger Bilham, *University of Colorado, Boulder*
John P. Buchanan, *Eastern Washington University*
Jeffrey B. Connelly, *University of Arkansas at Little Rock*
Peter Copeland, *University of Houston*
Rachael Craig, *Kent State University*
P. Johnathan Patchett, *University of Arizona*
Gary Rosenberg, *Indiana University–Purdue University at Indianapolis*
Chris Sanders, *Southeast Missouri State University*
Paul B. Tomascak, *University of Maryland at College Park*
Harve S. Waff, *University of Oregon*

In addition, we thank those individuals who reviewed the first edition and made many useful and insightful comments that were incorporated into the second edition of this book:

Theodore G. Benitt, *Nassau Community College*
Bruce A. Blackerby, *California State University–Fresno*
Gerald F. Brem, *California State University–Fullerton*
Ronald E. Davenport, *Louisiana Tech University*
Isabella M. Drew, *Ramapo College of New Jersey*
Jeremy Dunning, *Indiana University*
Norman K. Grant, *Miami University*
Norris W. Jones, *University of Wisconsin–Oshkosh*
Patricia M. Kenyon, *University of Alabama*
Peter L. Kresan, *University of Arizona*
Dave B. Loope, *University of Nebraska–Lincoln*
David N. Lumsden, *Memphis State University*

Steven K. Reid, *Morehead State University*
M.J. Richardson, *Texas A&M University*
Charles J. Ritter, *University of Dayton*
Gary D. Rosenberg, *Indiana University–Purdue University at Indianapolis*
J. Alexander Speer, *North Carolina State University*
James C. Walters, *University of Northern Iowa*

Lastly, we would like to provide thanks to the individuals who reviewed the first edition of this book in manuscript form. Their comments helped make the book a success for students and instructors alike.

Gary C. Allen, *University of New Orleans*
R. Scott Babcock, *Western Washington University*
Kennard Bork, *Denison University*
Thomas W. Broadhead, *University of Tennessee at Knoxville*
Anna Buising, *California State University at Hayward*
F. Howard Campbell III, *James Madison University*
Larry E. Davis, *Washington State University*
Noel Eberz, *California State University at San Jose*
Allan A. Ekdale, *University of Utah*
Stewart S. Farrar, *Eastern Kentucky University*
Richard H. Fluegeman, Jr., *Ball State University*
William J. Fritz, *Georgia State University*
Kazuya Fujita, *Michigan State University*
Norman Gray, *University of Connecticut*
Jack Green, *California State University at Long Beach*
David R. Hickey, *Lansing Community College*
R. W. Hodder, *University of Western Ontario*
Cornelis Klein, *University of New Mexico*
Lawrence W. Knight, *William Rainey Harper College*
Martin B. Lagoe, *University of Texas at Austin*
Richard H. Lefevre, *Grand Valley State University*
I. P. Martini, *University of Guelph, Ontario*
Michael McKinney, *University of Tennessee at Knoxville*
Robert Merrill, *California State University at Fresno*
Carleton Moore, *Arizona State University*
Alan P. Morris, *University of Texas at San Antonio*
Harold Pelton, *Seattle Central Community College*
James F. Petersen, *Southwest Texas State University*
Katherine H. Price, *DePauw University*
William D. Romey, *St. Lawrence University*
Gary Rosenberg, *Indiana University–Purdue University at Indianapolis*
David B. Slavsky, *Loyola University of Chicago*
Edward F. Stoddard, *North Carolina State University*
Charles P. Thornton, *Pennsylvania State University*
Samuel B. Upchurch, *University of South Florida*
John R. Wagner, *Clemson University*

We also wish to thank Kathy Benison, R. V. Dietrich (Professor Emeritus), David J. Matty, Jane M. Matty, Wayne E. Moore (Professor Emeritus), and Sven Morgan of the Geology Department, and Bruce M. C. Pape (Emeritus) of the Geography Department of Central Michigan University, as well as Eric Johnson (Hartwick College, New

York) and Stephen D. Stahl (St. Bonaventure, New York) for providing us with photographs and answering our questions concerning various topics. We are also grateful for the generosity of the various agencies and individuals from many countries who provided photographs.

Special thanks must go to Keith Dodson, Earth Sciences Editor at Brooks/Cole, who initiated this sixth edition and saw it through to completion, and to Alyssa White, Associate Development Editor, Physical Sciences, who took a special interest in the new features of this edition as well as the art program to ensure that the figures would not only convey the geologic information, but be visually pleasing and interesting as well. We are equally indebted to our production product manager, Hal Humprey, for all his help. We would also like to thank Carol Reitz for her copyediting skills. We appreciate her help in improving our manuscript. We thank Kathleen Olson for her invaluable help in locating appropriate photos and checking on photo permissions, and Kiely Sisk, permissions editor. Special thanks also to Brandi Kirksey, Editorial Assistant; Mark Santee, Sr. Marketing Manager; Ericka Yeoman-Saler, Technology Project Manager; and E. Kirsten Peters for her work on "Geology and Cultural Connections." Because geology is largely a visual science, we extend special thanks to the artists at Precision Graphics and Graphic World Inc., who were responsible for rendering and updating much of the art program. They all did an excellent job, and we enjoyed working with them.

As always, our families were very patient and encouraging when much of our spare time and energy were devoted to this book. We again thank them for their continued support and understanding.

JAMES S. MONROE

REED WICANDER

RICK HAZLETT

DEVELOPING CRITICAL THINKING AND STUDY SKILLS

Introduction

College is a demanding and important time, a time when your values will be challenged, and you will try out new ideas and philosophies. You will make personal and career decisions that will affect your entire life. One of the most important lessons you can learn in college is how to balance your time among work, study, and recreation. If you develop good time management and study skills early in your college career, you will find that your college years will be successful and rewarding.

This section offers some suggestions to help you maximize your study time and develop critical thinking and study skills that will benefit you, not only in college, but throughout your life. While mastering the content of a course is obviously important, learning how to study and to think critically is, in many ways, far more important. Like most things in life, learning to think critically and study efficiently will initially require additional time and effort, but once mastered, these skills will save you time in the long run.

You may already be familiar with many of the suggestions and may find that others do not directly apply to you. Nevertheless, if you take the time to read this section and apply the appropriate suggestions to your own situation, we are confident that you will become a better and more efficient student, find your classes more rewarding, have more time for yourself, and get better grades. We have found that the better students are usually also the busiest. Because these students are busy with work or extracurricular activities, they have had to learn to study efficiently and manage their time effectively.

One of the keys to success in college is avoiding procrastination. While procrastination provides temporary satisfaction because you have avoided doing something you did not want to do, in the long run it leads to stress. While a small amount of stress can be beneficial, waiting until the last minute usually leads to mistakes and a subpar performance. By setting clear, specific goals and working toward them on a regular basis, you can greatly reduce the temptation to procrastinate. It is better to work efficiently for short periods of time than to put in long, unproductive hours on a task, which is usually what happens when you procrastinate.

Another key to success in college is staying physically fit. It is easy to fall into the habit of eating junk food and never exercising. To be mentally alert, you must be physically fit. Try to develop a program of regular exercise. You will find that you have more energy, feel better, and study more efficiently.

General Study Skills

Most courses, and geology in particular, build upon previous material, so it is extremely important to keep up with the coursework and set aside regular time for study in each of your courses. Try to follow these hints, and you will find you do better in school and have more time for yourself:

- Develop the habit of studying on a daily basis.
- Set aside a specific time each day to study. Some people are day people, and others are night people. Determine when you are most alert and use that time for study.
- Have an area dedicated for study. It should include a well-lighted space with a desk and the study materials you need, such as a dictionary, thesaurus, paper, pens, and pencils, and a computer if you have one.
- Study for short periods and take frequent breaks, usually after an hour of study. Get up and move around and do something completely different. This will help you stay alert, and you'll return to your studies with renewed vigor.
- Try to review each subject every day or at least the day of the class. Develop the habit of reviewing lecture material from a class the same day.
- Become familiar with the vocabulary of the course. Look up any unfamiliar words in the glossary of your textbook or in a dictionary. Learning the language of the discipline will help you learn the material.

Getting the Most from Your Notes

If you are to get the most out of a course and do well on exams, you must learn to take good notes. Taking good notes does not mean you should try to write down every word your professor says. Part of being a good note taker is knowing what is important and what you can safely leave out.

Early in the semester, try to determine whether the lecture will follow the textbook or be predominantly new material. If much of the material is covered in the textbook, your notes do not have to be as extensive or detailed as when the material is new. In any case, the following suggestions should make you a better note taker and enable you to derive the maximum amount of information from a lecture:

- Regardless of whether the lecture discusses the same material as the textbook or supplements the reading assignment, read or scan the chapter the lecture will cover *before* class. This way you will be somewhat familiar with the concepts and can listen critically to what is being said rather than trying to write down everything. Later a few key words or phrases will jog your memory about what was said.
- Before each lecture, briefly review your notes from the previous lecture. Doing this will refresh your memory and provide a context for the new material.
- Develop your own style of note taking. Do not try to write down every word. These are notes you're taking, not a transcript. Learn to abbreviate and develop your own set of abbreviations and symbols for common words and phrases: for example, w/o (without), w (with), = (equals), ^ (above or increases), v (below or decreases), < (less than), > (greater than), & (and), u (you).
- Geology lends itself to many abbreviations that can increase your note-taking capability: for example, pt (plate tectonics), ig (igneous), meta (metamorphic), sed (sedimentary), rx (rock or rocks), ss (sandstone), my (million years), and gts (geologic time scale).
- Rewrite your notes soon after the lecture. Rewriting your notes helps reinforce what you heard and gives you an opportunity to determine whether you understand the material.
- By learning the vocabulary of the discipline before the lecture, you can cut down on the amount you have to write—you won't have to write down a definition if you already know the word.
- Learn the mannerisms of the professor. If he or she says something is important or repeats a point, be sure to write it down and highlight it in some way. Students have told me (RW) that when I stated something twice during a lecture, they knew it was important and probably would appear on a test. (They were usually right!)
- Check any unclear points in your notes with a classmate or look them up in your textbook. Pay particular attention to the professor's examples, which usually elucidate and clarify an important point and are easier to remember than an abstract concept.
- Go to class regularly and sit near the front of the class if possible. It is easier to hear and see what is written on the board or projected onto the screen, and there are fewer distractions.
- If the professor allows it, tape record the lecture, but don't use the recording as a substitute for notes. Listen carefully to the lecture and write down the important points; then fill in any gaps when you replay the tape.
- If your school allows it, and if they are available, buy class lecture notes. These are usually taken by a graduate student who is familiar with the material; typically they are quite comprehensive. Again use these notes to supplement your own.
- Ask questions. If you don't understand something, ask the professor. Many students are reluctant to do this, especially in a large lecture hall, but if you don't understand a point, other people are probably confused as well. If you can't ask questions during a lecture, talk to the professor after the lecture or during office hours.

Getting the Most Out of What You Read

The old adage that "you get out of something what you put into it" is true when it comes to reading textbooks. By carefully reading your text and following these suggestions, you can greatly increase your understanding of the subject:

- Look over the chapter outline to see what the material is about and how it flows from topic to topic. If you have time, skim through the chapter before you start to read in depth.
- Pay particular attention to the tables, charts, and figures. They contain a wealth of information in abbreviated form and illustrate important concepts and ideas. Geology, in particular, is a visual science, and the figures and photographs will help you visualize what is being discussed in the text and provide actual examples of features such as faults or unconformities.

- As you read your textbook, highlight or underline key concepts or sentences, but make sure you don't highlight everything. Make notes in the margins. If you don't understand a term or concept, look it up in the glossary.
- Read the chapter summary carefully. Be sure you understand all the key terms, especially those in boldface or italic type. Because geology builds on previous material, it is imperative that you understand the terminology.
- Go over the end-of-chapter questions. Write your answers as if you were taking a test. Only when you see your answer in writing will you know if you really understood the material.
- Knowing how to search the Internet is an essential skill.

Developing Critical Thinking Skills

Few things in life are black and white, and it is important to be able to examine an issue from all sides and come to a logical conclusion. One of the most important things you will learn in college is to think critically and not accept everything you read and hear at face value. Thinking critically is particularly important in learning new material and relating it to what you already know. Although you can't know everything, you can learn to question effectively and arrive at conclusions consistent with the facts. Thus, these suggestions for critical thinking can help you in all your courses:

- Whenever you encounter new facts, ideas, or concepts, be sure you understand and can define all of the terms used in the discussion.
- Determine how the facts or information was derived. If the facts were derived from experiments, were the experiments well executed and free of bias? Can they be repeated? The controversy over cold fusion is an excellent example. Two scientists claimed to have produced cold fusion reactions using simple experimental laboratory apparatus, yet other scientists have never been able to achieve the same reaction by repeating the experiments.
- Do not accept any statement at face value. What is the source of the information? How reliable is the source?
- Consider whether the conclusions follow from the facts. If the facts do not appear to support the conclusions, ask questions and try to determine why they don't. Is the argument logical or is it somehow flawed?
- Be open to new ideas. After all, the underlying principles of plate tectonic theory were known early in this century yet were not accepted until the 1970s despite overwhelming evidence.

- Look at the big picture to determine how various elements are related. For example, how will constructing a dam across a river that flows to the sea affect the stream's profile? What will be the consequences to the beaches that will be deprived of sediment from the river? One of the most important lessons you can learn from your geology course is how interrelated the various systems of Earth are. When you alter one feature, you affect numerous other features as well.

Improving Your Memory

Why do you remember some things and not others? The reason is that the brain stores information in different ways and forms, making it easy to remember some things and difficult to remember others. Because college requires that you learn a vast amount of information, any suggestions that can help you retain more material will help you in your studies:

- Pay attention to what you read or hear. Focus on the task at hand and avoid daydreaming. Repetition of any sort will help you remember material. Review the previous lecture before going to class, or look over the last chapter before beginning the next. Ask yourself questions as you read.
- Use mnemonic devices to help you learn unfamiliar material. For example, the order of the Paleozoic periods (Cambrian, Ordovician, Silurian, Devonian, Mississippian, Pennsylvanian, and Permian) of the geologic time scale can be remembered by the phrase, **C**ampbell's **O**nion **S**oup **D**oes **M**ake **P**eter **P**ale, or the order of the Cenozoic Epochs (Paleocene, Eocene, Oligocene, Miocene, Pliocene, and Pleistocene) can be remembered by the phrase, **P**ut **E**ggs **O**n **M**y **P**late **P**lease. Using rhymes can also be helpful.
- Look up the roots of important terms. If you understand where a word comes from, its meaning will be easier to remember. For example, *pyroclastic* comes from *pyro*, meaning "fire," and *clastic*, meaning "broken pieces." Hence a pyroclastic rock is one formed by volcanism and composed of pieces of other rocks. We have provided the roots of many important terms throughout this text to help you remember their definitions.
- Outline the material you are studying. This practice will help you see how the various components are interrelated. Learning a body of related material is much easier than learning unconnected and discrete facts. Looking for relationships is particularly helpful in geology because so many things are interrelated. For example, plate tectonics explains how mountain building, volcanism, and earthquakes are all related.

The rock cycle relates the three major groups of rocks to each other and to subsurface and surface processes (Chapter 1).

- Use deductive reasoning to tie concepts together. Remember that geology builds on what you learned previously. Use that material as your foundation and see how the new material relates to it.
- Draw a picture. If you can draw a picture and label its parts, you probably understand the material. Geology lends itself very well to this type of memory device because so much is visual. For example, instead of memorizing a long list of glacial terms, draw a picture of a glacier and label its parts and the type of topography it forms.
- Focus on what is important. You can't remember everything, so focus on the important points of the lecture or the chapter. Try to visualize the big picture and use the facts to fill in the details.

Preparing for Exams

For most students, tests are the critical part of a course. To do well on an exam, you must be prepared. These suggestions will help you focus on preparing for examinations:

- The most important advice is to study regularly rather than try to cram everything into one massive study session. Get plenty of rest the night before an exam, and stay physically fit to avoid becoming susceptible to minor illnesses that sap your strength and lessen your ability to concentrate on the subject at hand.
- Set up a schedule so that you cover small parts of the material on a regular basis. Learning some concrete examples will help you understand and remember the material.
- Review the chapter summaries. Construct an outline to make sure you understand how everything fits together. Drawing diagrams will help you remember key points. Make flash cards to help you remember terms and concepts.
- Form a study group, but make sure your group focuses on the task at hand, not on socializing. Quiz each other and compare notes to be sure you have covered all the material. We have found that students dramatically improved their grades after forming or joining a study group.
- Write the answers to all the Review Questions. Before doing so, however, become thoroughly familiar with the subject matter by reviewing your lecture notes and reading the chapter. Otherwise, you will spend an inordinate amount of time looking up answers.
- If you have any questions, visit the professor or teaching assistant. If review sessions are offered, be sure to attend. If you are having problems with the material, ask for help as soon as you have difficulty. Don't wait until the end of the semester.
- If old exams are available, look at them to see what is emphasized and what types of questions are asked. Find out whether the exam will be all objective or all essay or a combination. If you have trouble with a particular type of question (such as multiple choice or essay), practice answering questions of that type—your study group or a classmate may be able to help.

Taking Exams

The most important thing to remember when taking an exam is not to panic. This, of course, is easier said than done. Almost everyone suffers from test anxiety to some degree. Usually, it passes as soon as the exam begins, but in some cases, it is so debilitating that an individual does not perform as well as he or she could. If you are one of those people, get help as soon as possible. Most colleges and universities have a program to help students overcome test anxiety or at least keep it in check. Don't be afraid to seek help if you suffer test anxiety. Your success in college depends to a large extent on how well you perform on exams, so by not seeking help, you are only hurting yourself. In addition, the following suggestions may be helpful:

- First of all, relax. Then look over the exam briefly to see its format and determine which questions are worth the most points. If it helps, quickly jot down any information you are afraid you might forget or particularly want to remember for a question.
- Answer the questions that you know the best first. Make sure, however, that you don't spend too much time on any one question or on one that is worth only a few points.
- If the exam is a combination of multiple choice and essay, answer the multiple-choice questions first. If you are not sure of an answer, go on to the next one. Sometimes the answer to one question can be found in another question. Furthermore, the multiple-choice questions may contain many of the facts needed to answer some of the essay questions.
- Read the question carefully and answer only what it asks. Save time by not repeating the question as your opening sentence to the answer. Get right to the point. Jot down a quick outline for longer essay questions to make sure you cover everything.

- If you don't understand a question, ask the examiner. Don't assume anything. After all, it is your grade that will suffer if you misinterpret the question.
- If you have time, review your exam to make sure you covered all the important points and answered all the questions.
- If you have followed our suggestions, by the time you finish the exam, you should feel confident that you did well and will have cause for celebration.

Concluding Comments

We hope that the suggestions we have offered will be of benefit to you, not only in this course but throughout your college career. Though it is difficult to break old habits and change a familiar routine, we are confident that following these suggestions will make you a better student. Furthermore, many of the suggestions will help you work more efficiently, not only in college, but also throughout your career. Learning is a lifelong process that does not end when you graduate. The critical thinking skills that you learn now will be invaluable throughout your life, both in your career and as an informed citizen.

Physical Geology

EXPLORING THE EARTH

Chapter 1

Understanding Earth: A Dynamic and Evolving Planet

The Great Wave of Kanagawa, by Katsusika Hokusai (1760–1849), color woodblock print.

Tsunami
Fingers of the sea
Pull back their waves, then release
Cerulean blue
—Alyssa White

ESSENTIAL QUESTIONS TO ASK

1.1 Introduction
- *What do natural disasters have to do with geology?*
- *What is a system?*

1.2 Geology in Our Everyday Lives
- *What is geology and what do geologists do?*
- *How does geology relate to the human experience?*
- *How does geology affect our everyday lives?*

1.3 Global Geologic and Environmental Issues Facing Humankind
- *What is considered the greatest environmental problem facing the world today?*
- *How is the greenhouse effect related to global warming?*

1.4 Origin of the Universe and Solar System, and Earth's Place in Them
- *How did the universe begin?*
- *What is our solar system composed of, and what is its origin and history?*
- *How did Earth form?*

1.5 Earth as a Dynamic and Evolving Planet
- *Why is Earth a dynamic planet?*
- *What are Earth's three concentric layers, and what are their characteristics?*

1.6 Geology and the Formulation of Theories
- *What is a theory?*

1.7 Plate Tectonic Theory
- *What is plate tectonic theory?*
- *Why is plate tectonic theory important in geology?*

1.8 The Rock Cycle
- *What is a rock?*
- *What is the rock cycle?*
- *What are the characteristics of each of the three major rock groups?*
- *How are the rock cycle and plate tectonics related?*

1.9 Geologic Time and Uniformitarianism
- *Why is it important to understand how geologic time differs from the human perspective of time?*
- *What is the principle of uniformitarianism, and why is it important in geology?*

GEOLOGY MATTERS

GEOLOGY IN UNEXPECTED PLACES:
A Bit of Egypt in Central Park, New York City, and London, England

GEOLOGY IN FOCUS:
The Kyoto Protocol—An Update

GEOLOGY AT WORK:
Forensic Geology and Palynology

GEOLOGY IN YOUR LIFE:
Global Warming and Climate Change, and How They Affect You

This icon, which appears throughout the book, indicates an opportunity to explore interactive tutorials, animations, or practice problems available on the Physical GeologyNow website at **http://now.brookscole.com/phygeo6**.

1.1 Introduction

Would you know what to do if suddenly the ground beneath you started shaking violently, or you were at the beach and the sea unexpectedly retreated to expose a large area that was formerly underwater, or you were on vacation somewhere near a volcano when it erupted? These are no longer hypothetical questions that you can dismiss because you think they don't apply to you. In this day and age of global business and travel, your knowledge of what to do in such situations may mean the difference between life and death. Every day we read about natural disasters and the toll they take in human suffering. The December 26, 2004, magnitude-9.0 earthquake in the Indian Ocean produced a tsunami that killed, by conservative estimates, more than 220,000 people is just the latest in a long line of natural disasters.

In a recent report published by the Earth Institute at Columbia University and the World Bank, more than half of the world's population is at risk from floods, droughts, cyclones and hurricanes, landslides, volcanoes, or earthquakes. According to the report, "Natural Disaster Hotspots: A Global Risk Analysis," more than a third of the United States population lives in hazard-prone areas, and 82% of the world's population lives in areas subject to flooding! These natural disasters not only result in human suffering but also exact an economic toll on nations. They hit particularly hard the poorest and least developed countries, which also have the fewest resources for recovery.

■ *What do natural disasters have to do with geology?*

Geology is an integral part of our lives. Natural disasters such as earthquakes, volcanic eruptions, tsunami, landslides, and floods are all components of geology that affect everyone. The more we understand how these phenomena work, the better prepared we can be to lessen their destruction and minimize the loss of life resulting from them.

Our standard of living depends directly on our consumption of natural resources, resources that formed millions and billions of years ago. However, the way we consume natural resources and interact with the environment, both as individuals and as a society, also determines our ability to pass on this standard of living to the next generation.

The major theme of this book is that Earth is a complex, dynamic planet that has changed continuously since its origin some 4.6 billion years ago. These changes and the present-day features we observe result from the interactions among Earth's various internal and external systems, subsystems, and cycles. Earth is unique among the planets of our solar system in that it supports life and has oceans of water, a hospitable atmosphere, and a variety of climates. It is ideally suited for life as we know it because of a combination of factors, including its distance from the Sun and the evolution of its interior, crust, oceans, and atmosphere. Life processes have, over time, influenced the evolution of Earth's atmosphere, oceans, and to some extent its crust. In turn, these physical changes have affected the evolution of life.

By viewing Earth as a whole—that is, thinking of it as a system—we not only see how its various components are interconnected but also better appreciate its complex and dynamic nature. The system concept makes it easier for us to study a complex subject such as Earth because it divides the whole into smaller components we can easily understand without losing sight of how the components all fit together as a whole.

■ *What is a system?*

A **system** is a combination of related parts that interact in an organized fashion. An automobile is a good example of a system. Its various components or subsystems, such as the engine, transmission, steering, and brakes, are all interconnected in such a way that a change in any one of them affects the others.

We can examine Earth in the same way we view an automobile—that is, as a system of interconnected components that interact and affect each other in many ways. The principal subsystems of Earth are the *atmosphere, biosphere, hydrosphere, lithosphere, mantle,* and *core* (▶ Figure 1.1). The complex interactions among these subsystems result in a dynamically changing planet in which matter and energy are continuously recycled into different forms (Table 1.1).

Table 1.1

Interactions among Earth's Principal Subsystems

	Atmosphere	Hydrosphere	Biosphere	Lithosphere
Atmosphere	Interaction among various air masses	Surface currents driven by wind Evaporation	Gases for respiration Dispersal of spores, pollen, and seed by wind	Weathering by wind erosion Transport of water vapor for precipitation of rain and snow
Hydrosphere	Input of water vapor and stored solar heat	Hydrologic cycle	Water for life	Precipitation Weathering and erosion
Biosphere	Gases from respiration	Removal of dissolved materials by organisms	Global ecosystems Food cycles	Modification of weathering and erosion processes Formation of soil
Lithosphere	Input of stored solar heat Landscapes affect air movements	Source of solid and dissolved materials	Source of mineral nutrients Modification of ecosystems by plate movements	Plate tectonics

▶ **Figure 1.1 Subsystems of Earth** The atmosphere, hydrosphere, biosphere, lithosphere, mantle, and core are all subsystems of Earth. This simplified diagram shows how these subsystems interact, with some examples of how materials and energy are cycled throughout the Earth system. The interactions between these subsystems make Earth a dynamic planet that has evolved and changed since its origin 4.6 billion years ago.

Atmosphere

Atmospheric gases and precipitation contribute to weathering of rocks.

Evaporation, condensation, and precipitation transfer water between atmosphere and hydrosphere, influencing weather and climate and distribution of water.

Plant, animal, and human activity affect composition of atmospheric gases. Atmospheric temperature and precipitation helps to determine distribution of Earth's biota.

Hydrosphere

Plants absorb and transpire water. Water is used by people for domestic, agricultural, and industrial uses.

Water helps determine abundance, diversity, and distribution of organisms.

Biosphere

Plate movement affects size, shape, and distribution of ocean basins. Running water and glaciers erode rock and sculpt landscapes.

Organisms break down rock into soil. People alter the landscape. Plate movement affects evolution and distribution of Earth's biota.

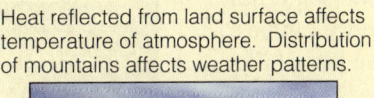

Heat reflected from land surface affects temperature of atmosphere. Distribution of mountains affects weather patterns.

Lithosphere (plates)

Convection cells within mantle contribute to movement of plates (lithosphere) and recycling of lithospheric material.

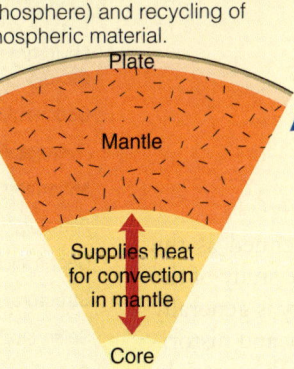

Plate

Mantle

Supplies heat for convection in mantle

Core

For example, the movement of plates has profoundly affected the formation of landscapes, the distribution of mineral resources, and atmospheric and oceanic circulation patterns, which in turn have affected global climate changes.

We must also not forget that humans are part of the Earth system, and our activities can produce changes with potentially wide-ranging consequences. When people discuss and debate such environmental issues as acid rain, the greenhouse effect and global warming, and the depleted ozone layer, it is important to remember that these are not isolated events but part of the larger Earth system. Furthermore, remember that Earth goes through time cycles that are much longer than humans are used to. Although they may have disastrous short-term effects on the human species, global warming and cooling are part of a longer-term cycle that has resulted in many glacial advances and retreats during the past 1.8 million years.

As you study the various topics covered in this book, keep in mind the themes discussed in this chapter and how, like the parts of a system, they are interrelated. By relating each topic to its place in the entire Earth system, you will gain a greater appreciation of why geology is so integral to our lives.

Section 1.1 Summary

- More than half of the world's population is at risk from various natural disasters. The more we understand how these phenomena work, the better prepared we can be to lessen their destructive effects and minimize loss of life.

- Earth can be viewed as a system of interconnected components that interact and affect each other, resulting in a dynamically changing planet.

- Humans are part of the Earth system, and our activities can produce changes that have potentially wide-ranging consequences.

1.2 Geology in Our Everyday Lives

■ *What is geology and what do geologists do?*

Geology, from the Greek *geo* and *logos*, is defined as the study of Earth, but now it must also include the study of the planets and moons in our solar system. Geology is generally divided into two broad areas—physical geology and historical geology. *Physical geology* is the study of Earth materials, such as minerals and rocks, as well as the processes operating within Earth and on its surface. *Historical geology* examines the origin and evolution of Earth, its continents, oceans, atmosphere, and life.

The discipline of geology is so broad that it is subdivided into numerous fields or specialties. Table 1.2 lists many of the diverse fields of geology and their relationship to the sciences of astronomy, biology, chemistry, and physics.

Nearly every aspect of geology has some economic or environmental relevance. Many geologists are involved in exploration for mineral and energy resources, using their specialized knowledge to locate the natural resources on which our industrialized society is based. As the demand for these nonrenewable resources increases, geologists apply the basic principles of geology in increasingly sophisticated ways to help focus their attention on areas with a high potential for economic success.

Whereas some geologists work on locating mineral and energy resources, other geologists use their expertise in solving environmental problems. Finding adequate sources of groundwater for the ever-burgeoning needs of communities and industries is becoming increasingly important, as is monitoring surface and underground water pollution and cleaning it up. Geologic engineers help find safe locations for dams, waste-disposal sites, and power plants and also design earthquake-resistant buildings.

Geologists are also engaged in making short- and long-range predictions about earthquakes and volcanic eruptions and the potential destruction that may result. Following the tragic events in Indonesia at the end of 2004, geologists are now more involved than ever in working with governmental agencies and civil defense planners to ensure that timely

Table 1.2

Specialties of Geology and Their Broad Relationship to the Other Sciences

Specialty	Area of Study	Related Science
Geochronology	Time and history of Earth	Astronomy
Planetary geology	Geology of the planets	
Paleontology	Fossils	Biology
Economic geology	Mineral and energy resources	
Environmental geology	Environment	Chemistry, Biology
Geochemistry	Chemistry of Earth	Chemistry
Hydrogeology	Water resources	
Mineralogy	Minerals	
Petrology	Rocks	
Geophysics	Earth's interior	Physics
Structural geology	Rock deformation	
Seismology	Earthquakes	
Geomorphology	Landforms	
Oceanography	Oceans	
Paleogeography	Ancient geographic features and locations	
Stratigraphy/ sedimentology	Layered rocks and sediments	

warnings are given to potentially affected regions and contingency plans are in place when such natural disasters as tsunami occur.

■ *How does geology relate to the human experience?*

You would probably be surprised at the extent to which geology pervades our everyday lives and at the numerous references to geology in the arts, music, and literature. Many sketches and paintings depict rocks and landscapes realistically. Leonardo da Vinci's *Virgin of the Rocks* and *Virgin and Child with Saint Anne*, Giovanni Bellini's *Saint Francis in Ecstasy* and *Saint Jerome*, and Asher Brown Durand's *Kindred Spirits* are just a few examples by famous painters (▶ Figure 1.2). In addition, we know that glaciers extended farther than they do today during Europe's Little Ice Age (from about 1500 to the middle or late 1800s) from such landscape paintings as Samuel Birmann's *Unterer Grindelwald* painted in 1826 (see Figure 17.1a).

In the field of music, Ferde Grofé's *Grand Canyon Suite* was no doubt inspired by the grandeur and timelessness of Arizona's Grand Canyon and its vast rock exposures. The rocks on the island of Staffa in the Inner Hebrides provided the inspiration for Felix Mendelssohn's famous *Hebrides Overture*.

References to geology abound in *The German Legends of the Brothers Grimm*. Jules Verne's *Journey to the Center of the Earth* describes an expedition into Earth's interior. There is even a forensic geology mystery series featuring fictional geologist Em Hansen, who uses her knowledge of geology to solve crimes. On one level, the poem "Ozymandias" by Percy B. Shelley deals with the fact that nothing lasts forever and even solid rock eventually disintegrates under the ravages of time and weathering. And popular comic strips contain references to geology; two of the best known are *B.C.* by Johnny Hart and *The Far Side* by Gary Larson.

Geology has also played an important role in the history and culture of humankind. Empires throughout history have risen and fallen on the distribution and exploitation of natural resources. Wars have been fought for the control of such natural resources as oil, gas, gold, silver, diamonds, and other valuable minerals. The Earth's surface, or its *topography*, which is shaped by geologic agents, has been a crucial factor in military strategy. Napoleon included two geologists in his expeditionary forces when he invaded Egypt in 1798, and the Russians used geologists as advisors in selecting sites for fortifications during the Russo-Japanese War of 1904–1905. Natural barriers such as mountain ranges and rivers frequently serve as political boundaries, and the shifting of river channels has sparked numerous border disputes. Deserts, which most people think of as inhospitable areas, have been the home to many people, such as the Bedouin, and have shaped their cultures over the centuries.

■ *How does geology affect our everyday lives?*

We observe the most obvious connection between geology and our everyday lives when natural disasters strike. Less apparent, but equally significant, are the connections between geology and economic, social, and political issues. Although most readers of this book will not become professional geologists, everyone should have a basic understanding of the geologic processes that ultimately affect all of us.

Natural Events Events such as destructive volcanic eruptions, devastating earthquakes, disastrous landslides, tsunami, floods, and droughts make headlines and affect people in obvious ways. Although we cannot prevent most of these natural disasters from happening, the more we learn about what causes them, the better we will be able to predict them and mitigate the severity of their impact.

Economics and Politics Equally important, but not always as well understood or appreciated, is the connection between geology and economic and political power. Mineral and energy resources are not equally distributed and no country is self-sufficient in all of them. Throughout history,

▶ **Figure 1.2 Layered Rocks around Gorges** *Kindred Spirits* by Asher Brown Durand (1849) realistically depicts the layered rocks along gorges in the Catskill Mountains of New York State. Durand was one of numerous artists of the 19th-century Hudson River School, which was known for realistic landscapes. This painting shows Durand conversing with the recently deceased Thomas Cole, the original founding force of the Hudson River School.

people have fought wars to secure these resources. The United States was involved in the 1990–1991 Gulf War largely because it needed to protect its oil interests in that region. Many foreign policies and treaties develop from the need to acquire and maintain adequate supplies of mineral and energy resources.

Our Role as Decision Makers You may become involved in geologic decisions in various ways—for instance, as a member of a planning board or as a property owner with mineral rights. In such cases, you must have a basic knowledge of geology to make informed decisions. Many professionals must also deal with geologic issues as part of their jobs. Lawyers, for example, are becoming more involved in issues ranging from ownership of natural resources to how development activities affect the environment. As government plays a greater role in environmental issues and regulations, members of Congress have increased the number of staff devoted to studying issues related to the environment and geology.

Consumers and Citizens If issues like nonrenewable energy resources, waste disposal, and pollution seem simply too far removed or too complex to be fully appreciated, consider for a moment just how dependent we are on geology in our daily routines (▶ Figure 1.3).

Much of the electricity for our appliances comes from the burning of coal, oil, or natural gas or from uranium consumed in nuclear-generating plants. It is geologists who locate the coal, petroleum, natural gas, and uranium. The copper or other metal wires through which electricity travels are manufactured from materials found as the result of mineral exploration. The buildings we live and work in owe their existence to geologic resources. The concrete in the foundations is a mixture of clay, sand, or gravel, and limestone; drywall is made largely from the mineral gypsum; and windows have the mineral quartz as the principal ingredient in their manufacture.

When we go to work, the car or public transportation we use is powered and lubricated by some type of petroleum by-product and is constructed of metal alloys and plastics. And the roads or rails we ride over come from geologic materials, such as gravel, asphalt, concrete, or steel. All these items are the result of processing geologic resources.

As individuals and societies, we enjoy a standard of living that is obviously directly dependent on the consumption of geologic materials. We therefore need to be aware of how our use and misuse of geologic resources may affect the environment. We must develop policies that not only encourage the wise management of our natural resources but also allow for continuing economic development among all the world's nations. Geologists will continue to play an important role in meeting these demands by locating the needed resources and ensuring that the environment is protected for the benefit of future generations.

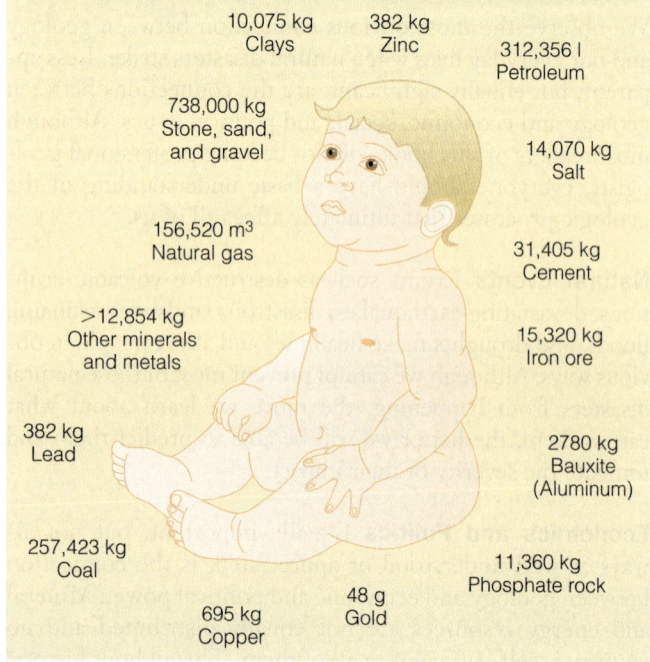

▶ **Figure 1.3 Life Expectancy** According to the Mineral Information Institute in Golden, Colorado, the average American born in 2004 has a life expectancy of 77.3 years and will need 1,620,000 kg of minerals, metals, and fuels to sustain his or her standard of living over a lifetime. That is an average of 20,886 kg of mineral and energy resources per year for every man, woman, and child in the United States.

10,075 kg Clays
382 kg Zinc
312,356 l Petroleum
738,000 kg Stone, sand, and gravel
14,070 kg Salt
156,520 m³ Natural gas
31,405 kg Cement
>12,854 kg Other minerals and metals
15,320 kg Iron ore
382 kg Lead
2780 kg Bauxite (Aluminum)
257,423 kg Coal
11,360 kg Phosphate rock
695 kg Copper
48 g Gold

Section 1.2 Summary

• Geology is the study of Earth. It is divided into two broad areas—physical geology (the study of Earth materials) and historical geology (the study of the origin and evolution of Earth and life).

• Geology pervades the human experience in ways most people are not even aware of, such as in art, music, and literature. Geology has also played an important role in the history and culture of humankind.

• Geology affects all of us as it relates to natural disasters, our economic well-being, politics, and our responsibility as global citizens.

1.3 Global Geologic and Environmental Issues Facing Humankind

■ *What is considered the greatest environmental problem facing the world today?*

Most scientists would argue that overpopulation is the greatest environmental problem facing the world today (▶ Figure 1.4). The world's population reached 6.4 billion in 2004, and projections indicate that this number will grow by at least another billion during the next two decades, bringing Earth's human population to more than 7.5 billion. Although this may not seem to be a geologic problem, remember that these people must be fed, housed, and clothed, and all with a minimal impact on the environment. Much of this population growth will be in areas that are already at risk from such natural hazards as earthquakes, tsunami, volcanic eruptions, and floods. Adequate water supplies must be found and kept from being polluted. Additional energy resources will be needed to help fuel the economies of nations with ever-increasing populations. New techniques must be developed to reduce the use of our dwindling nonrenewable resource base and to increase our recycling efforts so that we can decrease our dependence on new sources of these materials.

▶ **Figure 1.4 Overpopulation** Overpopulation is the greatest environmental problem facing the world today. Until the world's increasing population is brought under control, people will continue to strain Earth's limited resources. Shown here is the Oshodi Market in Lagos, Nigeria. It is located at the crossroads of several bus, rail, and highway junctions in one of the world's most populous cities. Upwards of 4 million people buy and sell goods in this large market daily.

The problems of overpopulation and how it affects the global ecosystem vary from country to country. For many poor and nonindustrialized countries, the problem is too many people and not enough food. For the more developed and industrialized countries, it is too many people rapidly depleting both the nonrenewable and renewable natural resource base. And in the most industrially developed countries, it is people producing more pollutants than the environment can safely recycle on a human time scale. The common thread tying these varied situations together is an environmental imbalance caused by a human population that is exceeding Earth's short-term carrying capacity.

■ *How is the greenhouse effect related to global warming?*

An excellent example of how Earth's various subsystems are interrelated is the relationship between the greenhouse effect and global warming. As a by-product of respiration and the burning of organic material, carbon dioxide is a component of the global ecosystem and is constantly being recycled as part of the carbon cycle. The concern in recent years over the increase in atmospheric carbon dioxide levels relates to its role in the greenhouse effect.

The recycling of carbon dioxide between Earth's crust and the atmosphere is an important climate regulator because carbon dioxide as well as other gases, such as methane, nitrous oxide, chlorofluorocarbons, and water vapor, allow sunlight to pass through them but trap the heat reflected back from Earth's surface. This retention of heat is called the *greenhouse effect*. It results in an increase in the temperature of Earth's surface and, more important, its atmosphere, thus producing global warming (▶ Figure 1.5).

With industrialization and its accompanying burning of tremendous amounts of fossil fuels, carbon dioxide levels in the atmosphere have been steadily increasing since about 1880, causing many scientists to conclude that a global warming trend has already begun and will result in severe global climate shifts. Most computer models based on the current rate of increase in greenhouse gases show Earth warming as a whole by as much as 5°C during the next hundred years. Such a temperature change will be uneven, however, with the greatest warming occurring in the higher latitudes. As a consequence of this warming, rainfall patterns will shift dramatically, which will have a major effect on the largest grain-producing areas of the world, such as the American Midwest. Drier and hotter conditions will intensify the severity and frequency of droughts, leading to increased crop failure and higher food prices. With such shifts in climate, Earth's deserts may expand, with a resulting decrease in the amount of valuable crop and grazing lands.

Continued global warming will result in a rise in mean sea level as ice caps and glaciers melt and contribute their water to the world's oceans. It is predicted that at the current rate of glacial melting, sea level will rise 21 cm by the 2050s, thus increasing the number of people at risk from flooding in coastal areas by approximately 20 million!

We would be remiss, however, if we did not point out that many other scientists are not convinced that the global warming trend is the direct result of increased human activity related to industrialization. They point out that although the amount of greenhouse gases has increased, we are still uncertain about their rate of generation and rate of removal, and whether the rise in global temperature during the past century resulted from normal climate variations through time or from human activity. Furthermore, these scientists point out that even if there is a general global warming during the next

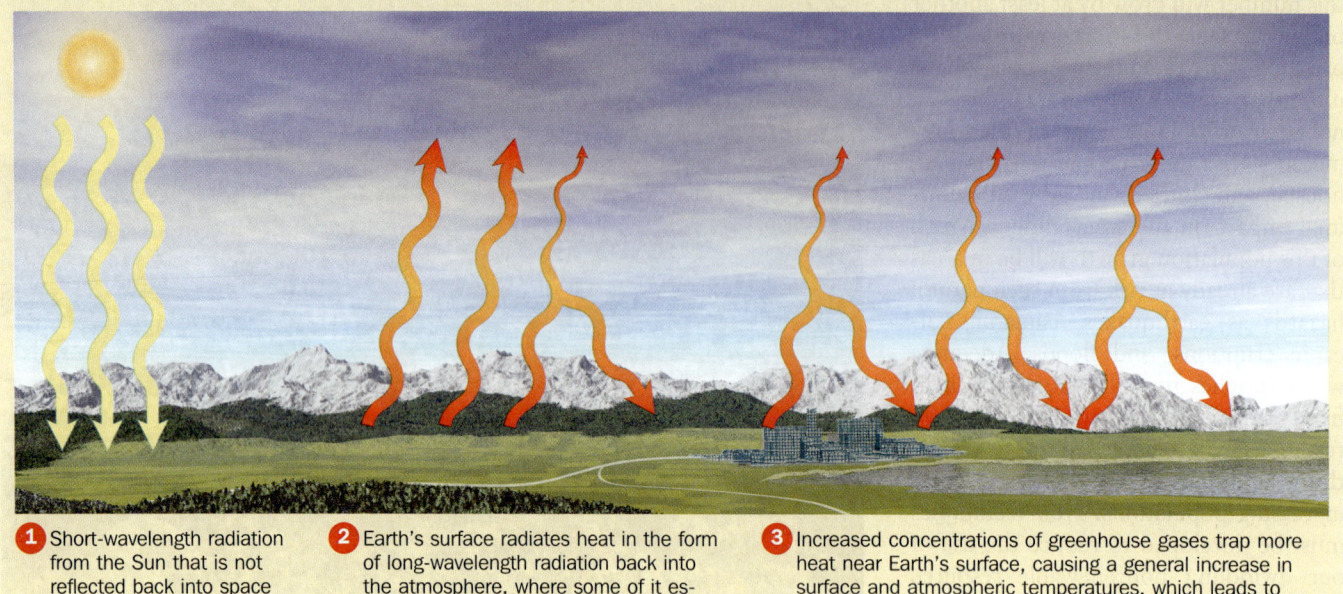

▶ **Figure 1.5 The Greenhouse Effect and Global Warming**

1 Short-wavelength radiation from the Sun that is not reflected back into space penetrates the atmosphere and warms Earth's surface.

2 Earth's surface radiates heat in the form of long-wavelength radiation back into the atmosphere, where some of it escapes into space. The rest is absorbed by greenhouse gases and water vapor and reradiated back toward Earth.

3 Increased concentrations of greenhouse gases trap more heat near Earth's surface, causing a general increase in surface and atmospheric temperatures, which leads to global warming.

hundred years, it is not certain that the dire predictions made by proponents of global warming will come true.

Earth, as we know, is a remarkably complex system, with many feedback mechanisms and interconnections throughout its various subsystems and cycles. It is very difficult to predict all of the consequences that global warming would have for atmospheric and oceanic circulation patterns and its ultimate effect on Earth's biota.

Section 1.3 Summary

• Most scientists would argue that the greatest environmental problem facing the world today is overpopulation. This is because a way must be found to feed, clothe, and house the world's increasing population with a minimal negative effect on the environment.

• Carbon dioxide is a by-product of respiration and the burning of organic materials, and it is constantly being recycled as part of the global carbon cycle. As such, carbon dioxide is an important climate regulator.

• The greenhouse effect is the phenomenon in which carbon dioxide and other gases allow sunlight to pass through them in the atmosphere but trap the heat reflected back from Earth's surface, thus increasing the temperature of Earth's surface and atmosphere.

• Global warming is an increase in the temperature of the atmosphere over time due to the greenhouse effect. As a consequence, global weather patterns will shift and mean sea level will rise, resulting in environmental changes that affect the world's biota.

1.4 Origin of the Universe and Solar System, and Earth's Place in Them

■ *How did the universe begin?*

Most scientists think that the universe originated about 15 billion years ago in what is popularly called the **Big Bang**. The Big Bang is a model for the evolution of the universe in which a dense, hot state was followed by expansion, cooling, and a less dense state. In a region infinitely smaller than an atom, both time and space were set at zero. Therefore there is no "before the Big Bang," only what occurred after it. The reason is that space and time are unalterably linked to form a space–time continuum demonstrated by Einstein's theory of relativity. Without space, there can be no time.

How do we know the Big Bang took place approximately 15 billion years ago? Why couldn't the universe have always existed as we know it today? Two fundamental phenomena

What Would You Do?

An important environmental issue facing the world today is global warming. How can this problem be approached from a global systems perspective? What are the possible consequences of global warming, and can we really do anything about it? Are there ways to tell whether global warming occurred in the geologic past?

indicate that the Big Bang occurred. First, the universe is expanding. When astronomers look beyond our own solar system, they observe that everywhere in the universe galaxies are moving away from each other at tremendous speeds. By measuring this expansion rate, astronomers can calculate how long ago the galaxies were all together at a single point. Second, everywhere in the universe there is a pervasive background radiation of 2.7 Kelvin (K) above absolute zero (absolute zero equals −273°C; 2.7 K = −270.3°C). This background radiation is thought to be the faint afterglow of the Big Bang.

According to the currently accepted theory, matter as we know it did not exist at the moment of the Big Bang and the universe consisted of pure energy. During the first second following the Big Bang, the four basic forces—*gravity* (the attraction of one body toward another), *electromagnetic force* (combines electricity and magnetism into one force and binds atoms into molecules), *strong nuclear force* (binds protons and neutrons together), and *weak nuclear force* (responsible for the breakdown of an atom's nucleus, producing radioactive decay)—separated and the universe experienced enormous expansion. About 300,000 years later, the universe was cool enough for complete atoms of hydrogen and helium to form; photons (the energetic particles of light) separated from matter and light burst forth for the first time.

During the next 200 million years, as the universe continued expanding and cooling, stars and galaxies began to form and the chemical makeup of the universe changed. Initially the universe was 100% hydrogen and helium, whereas today it is 98% hydrogen and helium and 2% all other elements by weight. How did such a change in the universe's composition occur? Throughout their life cycle, stars undergo many nuclear reactions in which lighter elements are converted into heavier elements by nuclear fusion. When a star dies, often explosively, the heavier elements that were formed in its core are returned to interstellar space and are available for inclusion in new stars. In this way, the composition of the universe is gradually enhanced in heavier elements.

■ *What is our solar system composed of, and what is its origin and history?*

Our solar system, which is part of the Milky Way Galaxy, consists of the Sun, 9 planets, 101 known moons or satellites (although this number keeps changing with the discovery of new moons and satellites surrounding the Jovian planets), a tremendous number of asteroids—most of which orbit the Sun in a zone between Mars and Jupiter—and millions of comets and meteorites as well as interplanetary dust and gases (▶ Figure 1.6). Any theory formulated to explain the origin and evolution of our solar system must therefore take into account its various features and characteristics.

Many scientific theories for the origin of the solar system have been proposed, modified, and discarded since the French scientist and philosopher René Descartes first proposed, in 1644, that the solar system formed from a gigantic whirlpool within a universal fluid. Today the **solar nebula theory** for the origin of our solar system involves the condensation and collapse of interstellar material in a spiral arm of the Milky Way Galaxy.

The collapse of this cloud of gases and small grains into a counterclockwise-rotating disk concentrated about 90% of the material in the central part of the disk and formed an embryonic Sun, around which swirled a rotating cloud of material called a *solar nebula*. Within this solar nebula were localized eddies in which gases and solid particles condensed. During the condensation process, gaseous, liquid, and solid particles began to accrete into ever-larger masses

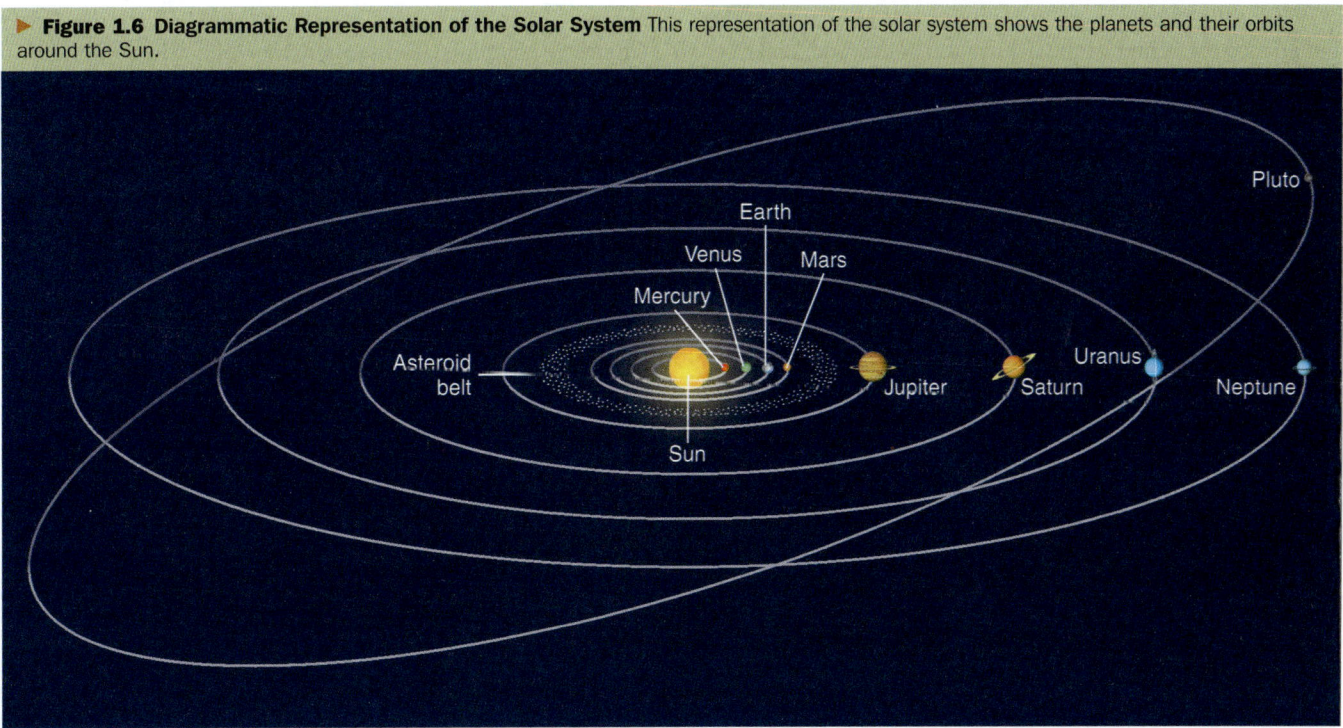

▶ **Figure 1.6 Diagrammatic Representation of the Solar System** This representation of the solar system shows the planets and their orbits around the Sun.

▶ **Figure 1.7 Planetesimals** At the stage of development shown here, planetesimals have formed in the inner solar system, and large eddies of gas and dust remain at great distances from the embryonic Sun.

called *planetesimals* (▶ Figure 1.7), which collided and grew in size and mass until they eventually became planets.

The composition and evolutionary history of the planets are a consequence, in part, of their distance from the Sun (see "The Terrestrial and Jovian Planets" on pages 14 and 15). The **terrestrial planets**—Mercury, Venus, Earth, and Mars—so named because they are similar to *terra*, Latin for "earth," are all small and composed of rock and metallic elements that condensed at the high temperatures of the inner nebula. The **Jovian planets**—Jupiter, Saturn, Uranus, and Neptune—so named because they resemble Jupiter (the Roman god was also named Jove), all have small rocky cores compared to their overall size, and are composed mostly of hydrogen, helium, ammonia, and methane, which condense at low temperatures.

While the planets were accreting, material that had been pulled into the center of the nebula also condensed, collapsed, and was heated to several million degrees by gravitational compression. The result was the birth of a star, our Sun.

During the early accretionary phase of the solar system's history, collisions between various bodies were common, as indicated by the craters on many planets and moons. Asteroids probably formed as planetesimals in a localized eddy between what eventually became Mars and Jupiter in much the same way that other planetesimals formed the terrestrial planets. The tremendous gravitational field of Jupiter, however, prevented this material from ever accreting into a planet. Comets, which are interplanetary bodies composed of loosely bound rocky and icy material, are thought to have condensed near the orbits of Uranus and Neptune.

The solar nebula theory of the formation of the solar system thus accounts for most of the characteristics of the planets and their moons, the differences in composition between the terrestrial and Jovian planets, and the presence of the asteroid belt. Based on the available data, the solar nebula theory best explains the features of the solar system and provides a logical explanation for its evolutionary history.

■ *How did Earth form?*

Some 4.6 billion years ago, various planetesimals in our solar system gathered enough material together to form Earth and eight other planets. Scientists think that this early Earth was probably cool, of generally uniform composition and density throughout, and composed mostly of silicates (compounds of silicon and oxygen), iron and magnesium oxides, and small amounts of all the other chemical elements. Subsequently, when the combination

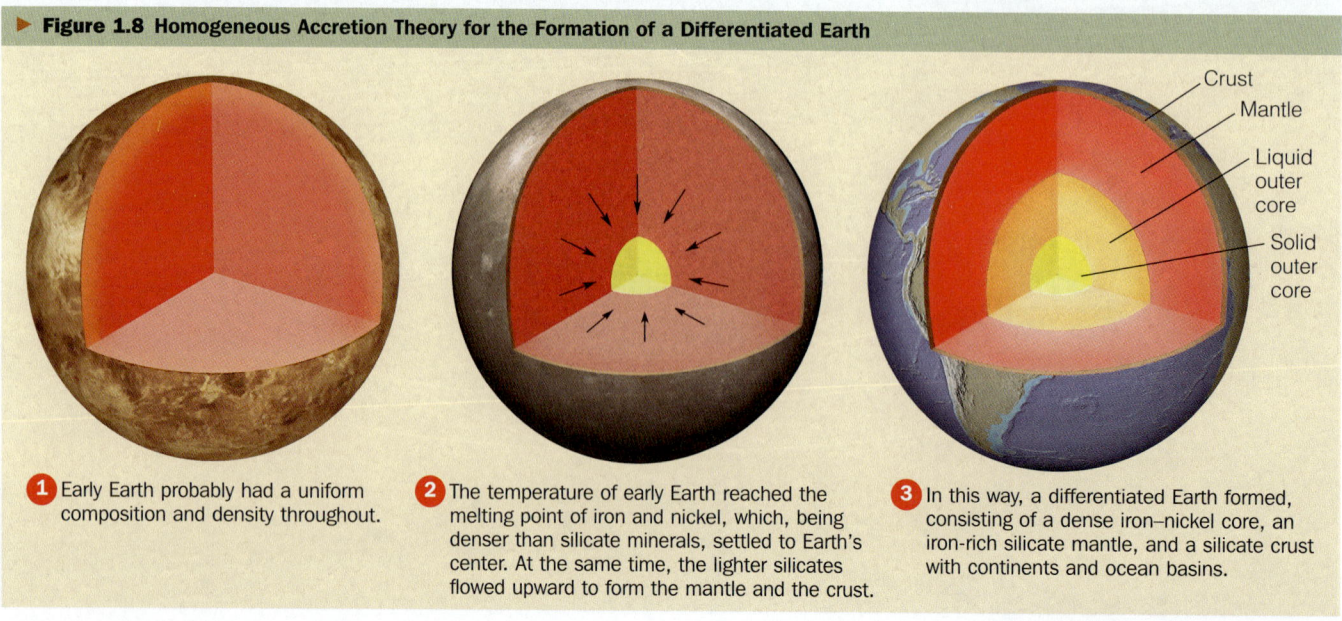

▶ **Figure 1.8 Homogeneous Accretion Theory for the Formation of a Differentiated Earth**

① Early Earth probably had a uniform composition and density throughout.

② The temperature of early Earth reached the melting point of iron and nickel, which, being denser than silicate minerals, settled to Earth's center. At the same time, the lighter silicates flowed upward to form the mantle and the crust.

③ In this way, a differentiated Earth formed, consisting of a dense iron–nickel core, an iron-rich silicate mantle, and a silicate crust with continents and ocean basins.

of meteorite impacts, gravitational compression, and heat from radioactive decay increased the temperature of Earth enough to melt iron and nickel, this homogeneous composition disappeared and was replaced by a series of concentric layers of differing composition and density, resulting in a differentiated planet (▶ Figure 1.8).

This differentiation into a layered planet is probably the most significant event in Earth history. Not only did it lead to the formation of a crust and eventually continents, but it also was probably responsible for the emission of gases from the interior that eventually led to the formation of the oceans and atmosphere.

Section 1.4 Summary

- The universe originated some 15 billion years ago in what is popularly called the Big Bang. The evidence for the Big Bang is that the universe is expanding and has a pervasive background radiation of 2.7 K above absolute zero (absolute zero equals −273°C; 2.7 K = −270.3°C).

- Our solar system began with the condensation and collapse of interstellar matter in a spiral arm of the Milky Way Galaxy. The collapse formed a counterclockwise-rotating disk in which a rotating cloud of gases and particles swirled around an embryonic Sun. This material eventually condensed into larger bodies and formed the nine planets of our solar system.

- The planets are classified as either terrestrial (Mercury, Venus, Earth, and Mars), which are small and composed of rocky and metallic elements that condense at high temperatures, or Jovian (Jupiter, Saturn, Uranus, and Neptune), which have small rocky cores compared to their overall size and are composed mostly of hydrogen, helium, ammonia, and methane, which condense at low temperatures.

- When Earth formed, it was probably cool and of generally uniform composition and density throughout. As a result of meteorite impacts, gravitational compression, and heat from radioactive decay, the temperature of Earth increased, which caused its various elements and compounds to form a series of concentric layers of differing composition and density, resulting in a differentiated planet.

- This differentiation not only led to the formation of a crust and continents but also contributed to the eventual formation of Earth's oceans and atmosphere.

1.5 Earth as a Dynamic and Evolving Planet

■ *Why is Earth a dynamic planet?*

Earth is a dynamic planet that has continuously changed during its 4.6-billion-year existence. The size, shape, and geographic distribution of continents and ocean basins have changed through time, the composition of the atmosphere has evolved, and life-forms existing today differ from those that lived during the past. Mountains and hills have been worn away by erosion, and the forces of wind, water, and ice have sculpted a diversity of landscapes. Volcanic eruptions and earthquakes reveal an active interior, and folded and fractured rocks are testimony to the tremendous power of Earth's internal forces.

■ *What are Earth's three concentric layers, and what are their characteristics?*

Earth consists of three concentric layers: the core, the mantle, and the crust (▶ Figure 1.9). This orderly division results from density differences between the layers as a function of variations in composition, temperature, and pressure.

The **core** has a calculated density of 10–13 grams per cubic centimeter (g/cm^3) and occupies about 16% of Earth's total volume. Seismic (earthquake) data indicate that the core consists of a small, solid inner region and a larger, apparently liquid, outer portion. Both are thought to consist largely of iron and a small amount of nickel.

The **mantle** surrounds the core and comprises about 83% of Earth's volume. It is less dense than the core (3.3–5.7 g/cm^3) and is thought to be composed largely of *peridotite*, a dark, dense igneous rock containing abundant iron and magnesium. The mantle can be divided into three distinct zones based on physical characteristics. The lower mantle is solid and forms most of the volume of Earth's interior. The **asthenosphere**

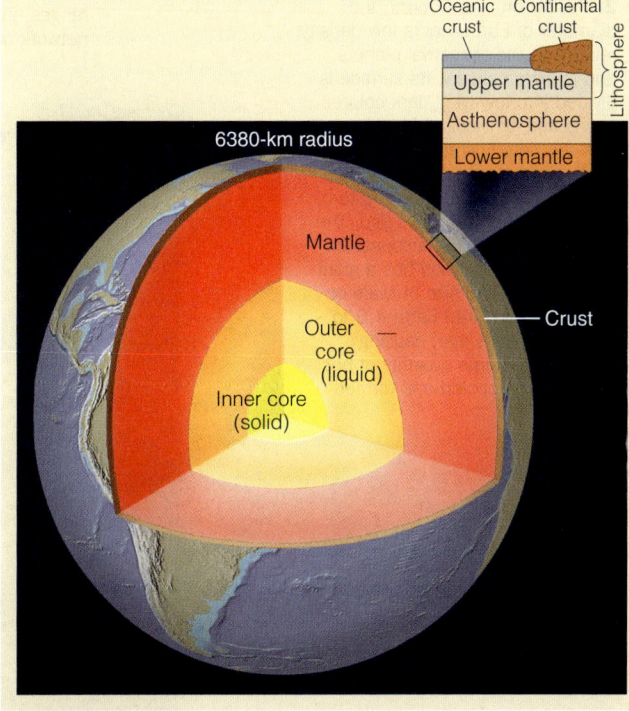

▶ **Figure 1.9 Cross Section of Earth Illustrating the Core, Mantle, and Crust** The enlarged portion shows the relationship between the lithosphere (composed of the continental crust, oceanic crust, and solid upper mantle) and the underlying asthenosphere and lower mantle.

The Terrestrial and Jovian Planets

The planets of our solar system are divided into two major groups that are quite different, indicating that the two underwent very different evolutionary histories. The four inner planets—Mercury, Venus, Earth, and Mars—are the terrestrial planets; they are small and dense (composed of a metallic core and silicate mantle-crust), ranging from no atmosphere (Mercury) to an oppressively thick one (Venus). The outer four planets (excluding Pluto, which some astronomers don't regard as a planet at all)—Jupiter, Saturn, Uranus, and Neptune—are the Jovian planets; they are large, ringed, low-density planets with rocky cores surrounded by thick atmospheres.

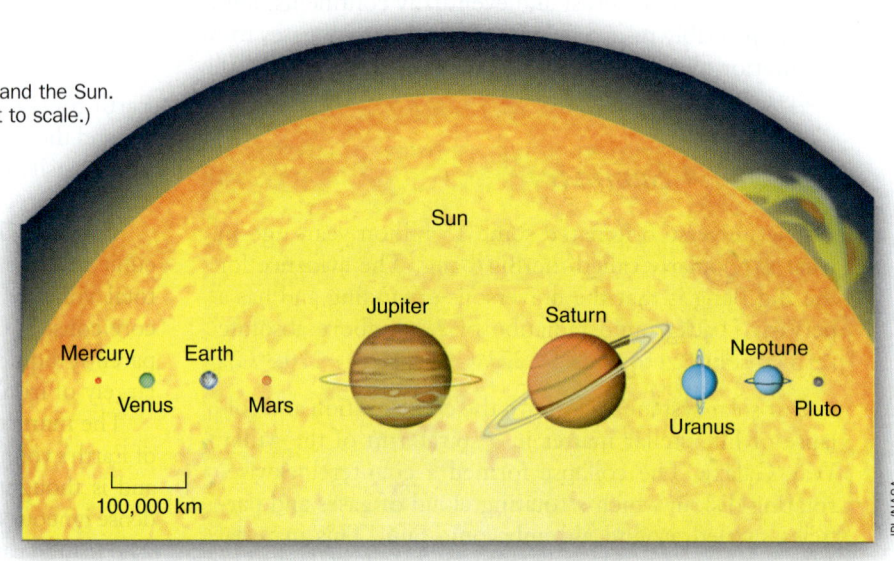

▶ **1.** The relative sizes of the planets and the Sun. (Distances between planets are not to scale.)

▲ **2.** The **Moon** is one-fourth the diameter of Earth, has a low density relative to the terrestrial planets, and is extremely dry. Its surface is divided into low-lying dark colored plains and light-colored highlands that are heavily cratered, attesting to a period of massive meteorite bombardment in our solar system more than 4 billion years ago. The hypothesis that best accounts for the origin of the Moon has a giant planetesimal, the size of Mars or larger, crashing into Earth 4.6 to 4.4 billion years ago, causing ejection of a large quantity of hot material that cooled and formed the Moon.

▶ **3. Venus** is surrounded by an oppressively thick atmosphere that completely obscures its surface. However, radar images from orbiting spacecraft reveal a wide variety of terrains, including volcanic features, folded mountain ranges, and a complex network of faults.

◀ **4. Mercury** has a heavily cratered surface that has changed very little since its early history. Because Mercury is so small, its gravitation attraction is insufficient to retain atmospheric gases; any atmosphere that it may have held when it formed probably escaped into space quickly.

▶ **5. Earth** is unique among our solar system's planets in that is has a hospitable atmosphere, oceans of water, a variety of climates, and it supports life.

▼ **6. Mars** has a thin atmosphere, little water, and distinct seasons. Its southern hemisphere is heavily cratered like the surfaces of Mercury and the Moon. The northern hemisphere has large, smooth plains, fewer craters, and evidence of extensive volcanism. The largest volcano in the solar system is found in the northern hemisphere, as are huge canyons, the largest of which, if present on Earth, would stretch from San Francisco to New York!

◀ **7. Jupiter** is the largest of the Jovian planets. With its moons, rings, strong magnetic field, and intense radiation belts, Jupiter is the most complex and varied planet in our solar system. Jupiter's cloudy and violent atmosphere is divided into a series of different colored bands and a variety of spots (the Great Red Spot) that interact in incredibly complex motions.

◀ **8. Saturn's** most conspicuous feature is its ring system, consisting of thousands of rippling, spiraling bands of countless particles. The width of Saturn's rings would just reach from Earth to the Moon.

▲ **9. Uranus** is the only planet that lies on its side, that is, its axis of rotation nearly parallels the plane of the ecliptic. Some scientists think that a collision with an Earth-sized body early in its history may have knocked Uranus on its side. Like the other Jovian planets, Uranus has a ring system, albeit a faint one.

▶ **10. Neptune** is a dynamic stormy planet with an atmosphere similar to those of the other Jovian planets. Winds up to 2000 km/h blow over the planet, creating tremendous storms, the largest of which, the Great Dark Spot, seen in the center, is nearly as big as Earth and is similar to the Great Red Spot on Jupiter.

surrounds the lower mantle. It has the same composition as the lower mantle but behaves plastically and flows slowly. Partial melting within the asthenosphere generates *magma* (molten material), some of which rises to the surface because it is less dense than the rock from which it was derived. The upper mantle surrounds the asthenosphere. The solid upper mantle and the overlying crust constitute the **lithosphere,** which is broken into numerous individual pieces called **plates** that move over the asthenosphere, partially as a result of underlying *convection cells* (▶ Figure 1.10). Interactions of these plates are responsible for such phenomena as earthquakes, volcanic eruptions, and the formation of mountain ranges and ocean basins.

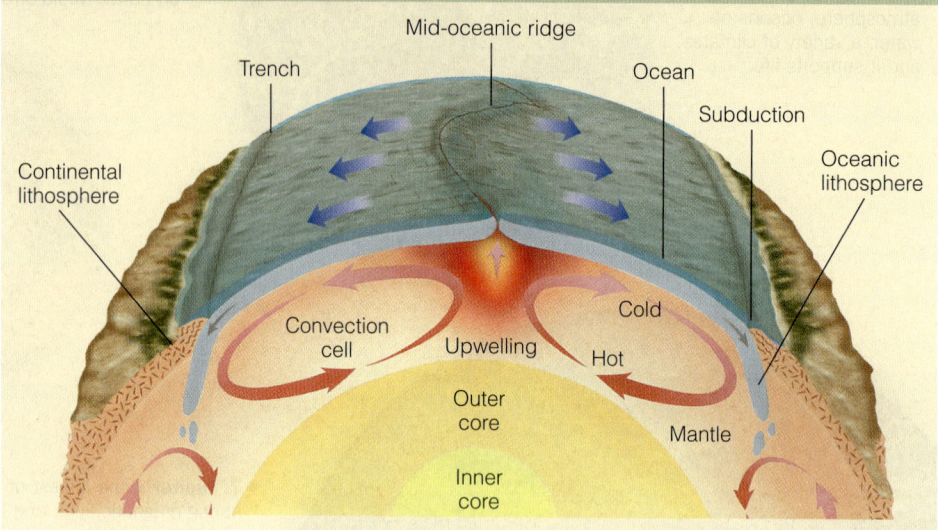

▶ **Active Figure 1.10 Movement of Earth's Plates** Earth's plates are thought to move partially as a result of underlying mantle convection cells in which warm material from deep within Earth rises toward the surface, cools, and then upon losing heat descends back into the interior, as shown in this diagrammatic cross section.

The **crust,** Earth's outermost layer, consists of two types. *Continental crust* is thick (20–90 km), has an average density of 2.7 g/cm³, and contains considerable silicon and aluminum. *Oceanic crust* is thin (5–10 km), denser than continental crust (3.0 g/cm³), and is composed of the dark igneous rock *basalt* and *gabbro*.

Log into GeologyNow and select this chapter to work through a **Geology Interactive** activity on "Core Studies" (click Earth's Layers→Core Studies).

Section 1.5 Summary

- Earth is a dynamic planet that has changed continuously during its 4.6-billion-year history.

- Earth is differentiated into three concentric layers: core, mantle, and crust.

- The core occupies about 16% of Earth's volume, has a calculated density of 10–13 g/cm³, is composed largely of iron with some nickel, and is divided into a small solid portion surrounded by an apparently liquid larger portion.

- The mantle surrounds the core, occupies about 83% of Earth's volume, has a density of 3.3–5.7 g/cm³, and is composed largely of peridotite. It is divided into a solid lower portion, followed by a plastically behaving asthenosphere, and overlain by a solid upper portion.

- Earth's outermost layer is its crust. It is divided into thick continental crust that has an average density of 2.7 g/cm³ and contains considerable silicon and aluminum, and thin oceanic crust that has a density of 3.0 g/cm³ and is composed of basalt and gabbro.

1.6 Geology and the Formulation of Theories

■ *What is a theory?*

The term **theory** has various meanings. In colloquial usage, it means a speculative or conjectural view of something—hence, the widespread belief that scientific theories are little more than unsubstantiated wild guesses. In scientific usage, however, a theory is a coherent explanation for one or several related natural phenomena supported by a large body of objective evidence. From a theory scientists derive predictive statements that can be tested by observations and/or experiments so that their validity can be assessed. The law of universal gravitation is an example of a theory that describes the attraction between masses (an apple and Earth in the popularized account of Newton and his discovery).

Theories are formulated through the process known as the **scientific method.** This method is an orderly, logical approach that involves gathering and analyzing facts or data about the problem under consideration. Tentative explanations, or **hypotheses,** are then formulated to explain the observed phenomena. Next the hypotheses are tested to see whether what was predicted actually occurs in a given situation. Finally, if one of the hypotheses is found, after repeated tests, to explain the phenomena, then the hypothesis is proposed as a theory. Remember, however, that in science even a theory is still subject to further testing and refinement as new data become available.

The fact that a scientific theory can be tested and is subject to such testing separates science from other forms of human inquiry. Because scientific theories can be tested, they have the potential of being supported or even proved wrong. Accordingly, science must proceed without any appeal to beliefs or supernatural explanations, not because such beliefs or explanations are necessarily untrue but because we have no way to investigate them. For this reason, science makes no claim about the existence or nonexistence of a supernatural or spiritual realm.

Each scientific discipline has certain theories that are of particular importance. In geology, plate tectonic theory has changed the way geologists view Earth. Geologists now look at Earth from a global perspective in which all its subsystems and cycles are interconnected, and Earth history is seen as a continuum of interrelated events that are part of a global pattern of change.

Section 1.6 Summary

- The scientific method is a logical, orderly approach that involves gathering data, formulating and testing hypotheses, and proposing theories.

- A hypothesis is a provisional explanation for observations that is subject to repeated testing.

- A theory is a coherent explanation for natural phenomena that is supported by a large body of objective evidence.

1.7 Plate Tectonic Theory

■ *What is plate tectonic theory?*

The recognition that the lithosphere is divided into rigid plates that move over the asthenosphere forms the foundation of **plate tectonic theory** (▶ Figure 1.11). Zones of volcanic activity, earthquakes, or both mark most plate boundaries. Along these boundaries plates separate (diverge), collide (converge), or slide sideways past each other (▶ Figure 1.12).

The acceptance of plate tectonic theory is recognized as a major milestone in the geologic sciences, comparable to the revolution Darwin's theory of evolution caused in biology. Plate tectonics has provided a framework for interpreting the composition, structure, and internal processes of Earth on a global scale. It has led to the realization that the continents and ocean basins are part of a lithosphere–atmosphere–hydrosphere system that evolved together with Earth's interior (Table 1.3).

■ *Why is plate tectonic theory important in geology?*

A revolutionary concept when it was proposed in the 1960s, plate tectonic theory has had far-reaching consequences in all fields of geology because it provides the basis for relating many seemingly unrelated phenomena. Besides being responsible for the major features of Earth's crust, plate

▶ **Figure 1.11 Earth's Plates** Earth's lithosphere is divided into rigid plates of various sizes that move over the asthenosphere.

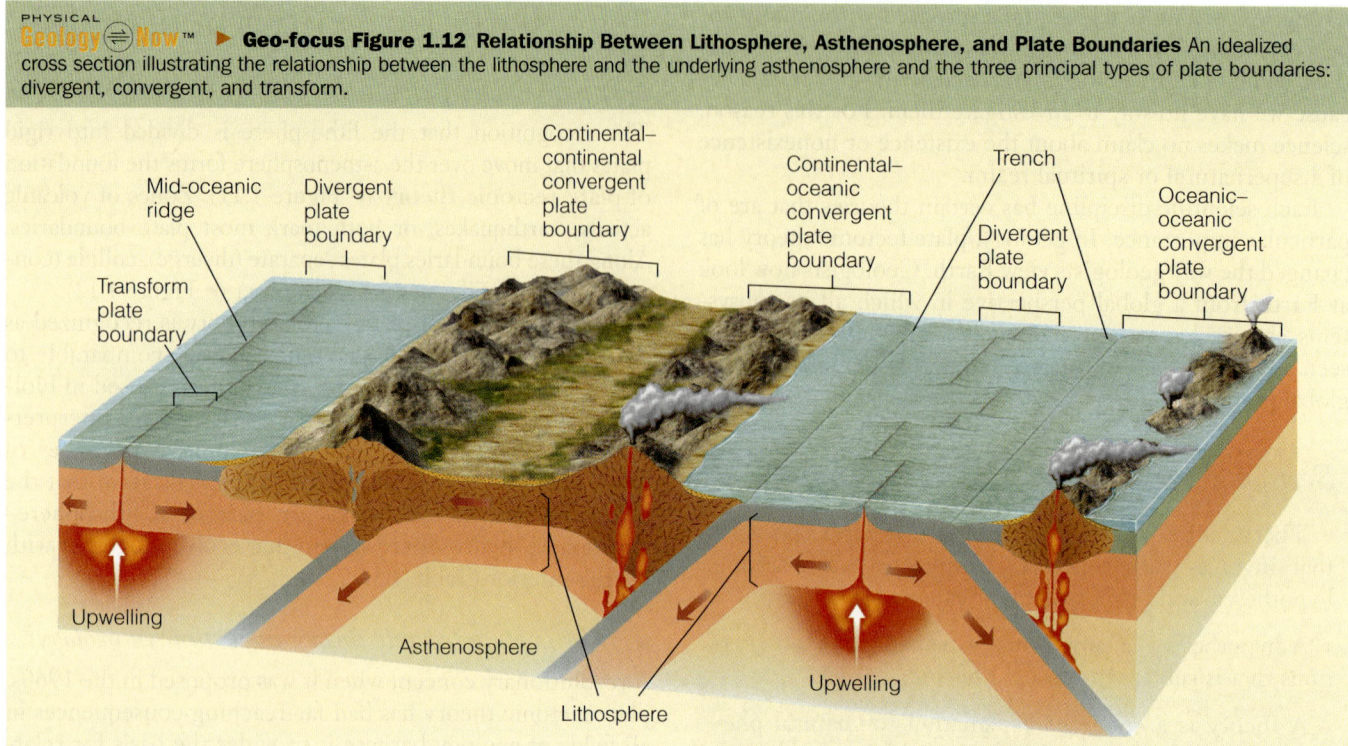

Geo-focus Figure 1.12 Relationship Between Lithosphere, Asthenosphere, and Plate Boundaries An idealized cross section illustrating the relationship between the lithosphere and the underlying asthenosphere and the three principal types of plate boundaries: divergent, convergent, and transform.

movements also affect the formation and occurrence of Earth's natural resources as well as the distribution and evolution of the world's biota.

The impact of plate tectonic theory has been particularly notable in the interpretation of Earth's history. For example, the Appalachian Mountains in eastern North America and the mountain ranges of Greenland, Scotland, Norway, and Sweden are not the result of unrelated mountain-building episodes but, rather, are part of a larger mountain-building event that involved the closing of an ancient "Atlantic Ocean" and the formation of the supercontinent Pangaea about 251 million years ago.

 Log into GeologyNow and select this chapter to work through a **Geology Interactive** activity on "Plate Locations" (click Plate Tectonics→Plate Locations).

Table 1.3
Plate Tectonics and Earth Systems

Solid Earth
Plate tectonics is driven by convection in the mantle and in turn drives mountain-building and associated igneous and metamorphic activity.

Atmosphere
Arrangement of continents affects solar heating and cooling, and thus winds and weather systems. Rapid plate spreading and hot-spot activity may release volcanic carbon dioxide and affect global climate.

Hydrosphere
Continental arrangement affects ocean currents. Rate of spreading affects volume of mid-oceanic ridges and hence sea level. Placement of continents may contribute to onset of ice ages.

Biosphere
Movement of continents creates corridors or barriers to migration, the creation of ecological niches, and transport of habitats into more or less favorable climates.

Extraterrestrial
Arrangement of continents affects free circulation of ocean tides and influences tidal slowing of Earth's rotation.

Source: Adapted by permission from Stephen Dutch, James S. Monroe, and Joseph Moran, *Earth Science* (Minneapolis/St. Paul: West Publishing Co., 1997).

> **Section 1.7 Summary**
>
> - Plate tectonic theory holds that the Earth's lithosphere is divided into rigid plates that move over the asthenosphere and diverge, converge, and slide past each other.
>
> - Plate movement results in volcanic activity, earthquakes, and mountain building. It also affects the formation and distribution of natural resources and influences the distribution and evolution of Earth's biota.

1.8 The Rock Cycle

■ *What is a rock?*

A **rock** is an aggregate of **minerals,** which are naturally occurring, inorganic, crystalline solids that have definite physical and chemical properties. Minerals are composed of elements such as oxygen, silicon, and aluminum, and elements are made up of atoms, the smallest particles of matter that retain the characteristics of an element. More than 3500 minerals have been identified and described, but only about a dozen make up the bulk of the rocks in Earth's crust (see Table 3.3).

Geologists recognize three major groups of rocks—*igneous*, *sedimentary*, and *metamorphic*—each of which is characterized by its mode of formation. Each group contains a variety of individual rock types that differ from one another on the basis of their composition or texture (the size, shape, and arrangement of mineral grains).

■ *What is the rock cycle?*

The **rock cycle** provides a way of viewing the interrelationships between Earth's internal and external processes (▶ Figure 1.13). It relates the three rock groups to each other; to surficial processes such as weathering, transportation, and deposition; and to internal processes such as magma generation and metamorphism.

■ *What are the characteristics of each of the three major rock groups?*

Igneous rocks result when magma crystallizes or volcanic ejecta such as ash accumulate and consolidate. As magma cools, minerals crystallize, and the resulting rock is characterized by interlocking mineral grains. Magma that cools slowly beneath the surface produces *intrusive igneous rocks* (▶ Figure 1.14a); magma that cools at the surface produces *extrusive igneous rocks* (▶ Figure 1.14b).

Rocks exposed at Earth's surface are broken into particles and dissolved by various weathering processes. The particles and dissolved materials may be transported by wind, water, or ice and eventually deposited as *sediment*. This sediment may then be compacted or cemented (lithified) into sedimentary rock.

Sedimentary rocks form in one of three ways: consolidation of rock fragments, precipitation of mineral matter from solution, or compaction of plant or animal remains (▶ Figure 1.14c, d). Because sedimentary rocks form at or near Earth's surface, geologists can make inferences about the environment in which they were deposited, the transporting agent, and perhaps even something about the source from which the sediments were derived (see Chapter 7). Accordingly, sedimentary rocks are especially useful for interpreting Earth history.

Metamorphic rocks result from the alteration of other rocks, usually beneath the surface, by heat, pressure, and the chemical activity of fluids. For example, marble, a rock preferred by many sculptors and builders, is a metamorphic rock produced when the agents of metamorphism are applied to the sedimentary rocks limestone or dolostone. Metamorphic rocks are either *foliated* (▶ Figure 1.14e) or *nonfoliated* (▶ Figure 1.14f). Foliation, the parallel alignment of minerals due to pressure, gives the rock a layered or banded appearance.

■ *How are the rock cycle and plate tectonics related?*

Interactions between plates determine, to some extent, which of the three rock groups will form (▶ Figure 1.15). For example, when plates converge, heat and pressure generated along the plate boundary may lead to igneous activity and metamorphism within the descending oceanic plate, thus producing various igneous and metamorphic rocks.

Some of the sediments and sedimentary rocks on the descending plate are melted, whereas other sediments and sedimentary rocks along the boundary of the nondescending plate are metamorphosed by the heat and pressure generated along the converging plate boundary. Later, the mountain range or chain of volcanic islands formed along the convergent plate boundary will be weathered and eroded, and the new sediments will be transported to the ocean to begin yet another cycle.

The interrelationship between the rock cycle and plate tectonics is just one example of how Earth's subsystems and cycles are all interrelated. Heating within Earth's interior results in convection cells that power the movement of plates, and also in magma, which forms intrusive and extrusive igneous rocks. Movement along plate boundaries may result in volcanic activity, earthquakes, and in some cases mountain building. The interaction between the atmosphere, hydrosphere, and biosphere contributes to the weathering of rocks exposed on Earth's surface. Plates descending back into Earth's interior are subjected to increasing heat and pressure, which may lead to metamorphism as well as the generation of magma and yet another recycling of materials.

PHYSICAL Geology⇌Now™ Log into GeologyNow and select this chapter to work through a **Geology Interactive** activity on "The Rock Cycle" (click Rocks and the Rock Cycle→Rock Cycle).

Figure 1.13 The Rock Cycle This cycle shows the interrelationships between Earth's internal and external processes and how the three major rock groups are related.

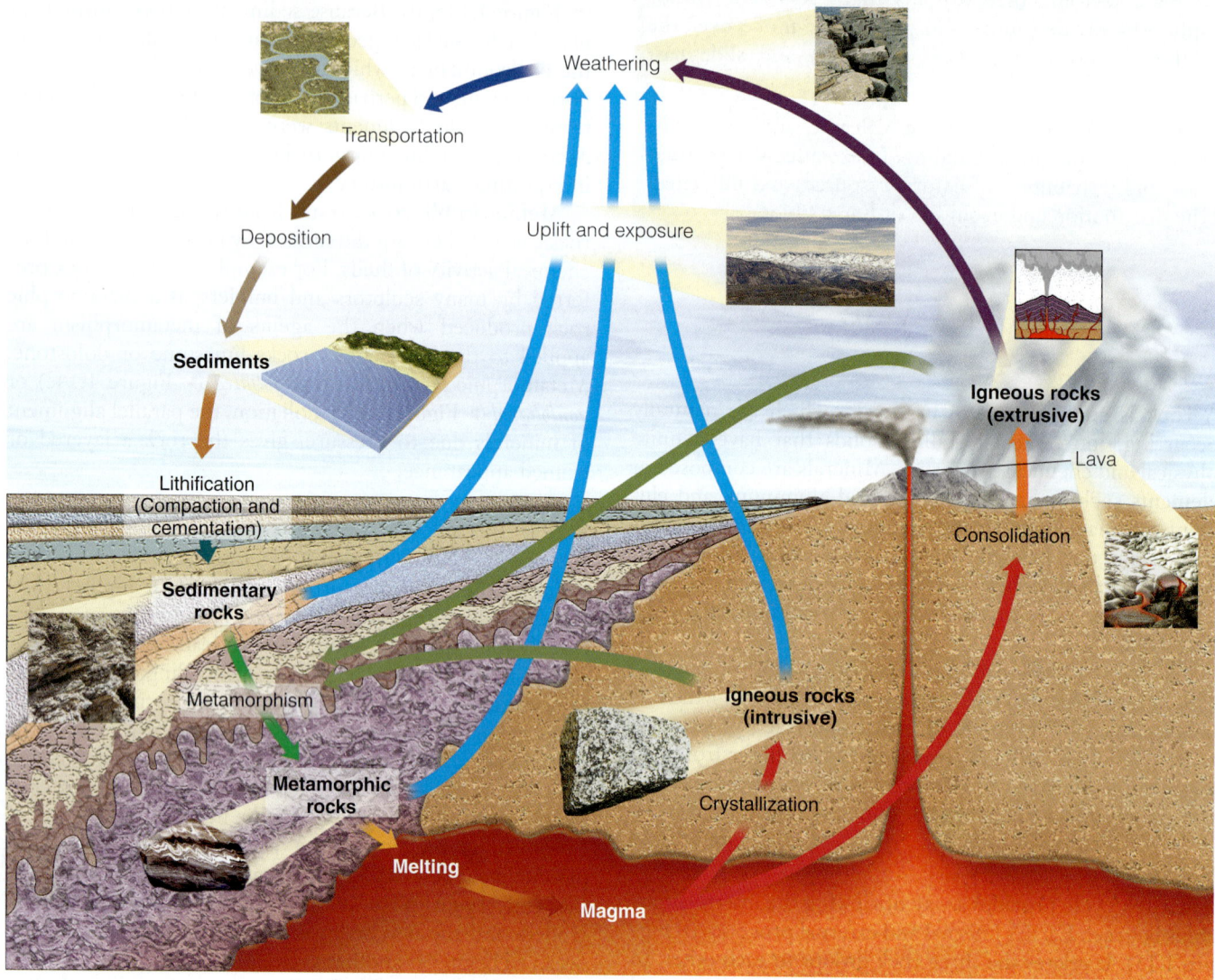

Section 1.8 Summary

• A rock is an aggregate of minerals.

• There are three major groups of rocks, each characterized by its mode of formation.

• Igneous rocks result from the cooling and crystallization of magma or the accumulation and consolidation of volcanic ejecta such as ash.

• Sedimentary rocks form at Earth's surface from the consolidation of rock fragments, precipitation of mineral matter from solution, or compaction of plant or animal remains.

• Metamorphic rocks result from the alteration of other rocks, typically beneath Earth's surface, by heat, pressure, and the chemical activity of fluids.

• The rock cycle relates the three rock groups to each other; to Earth's surficial processes such as weathering, transportation, and deposition; and to its internal processes such as magma generation and metamorphism.

• Plate tectonics is the mechanism that drives the rock cycle and recycles the three rock groups between Earth's interior and its surface.

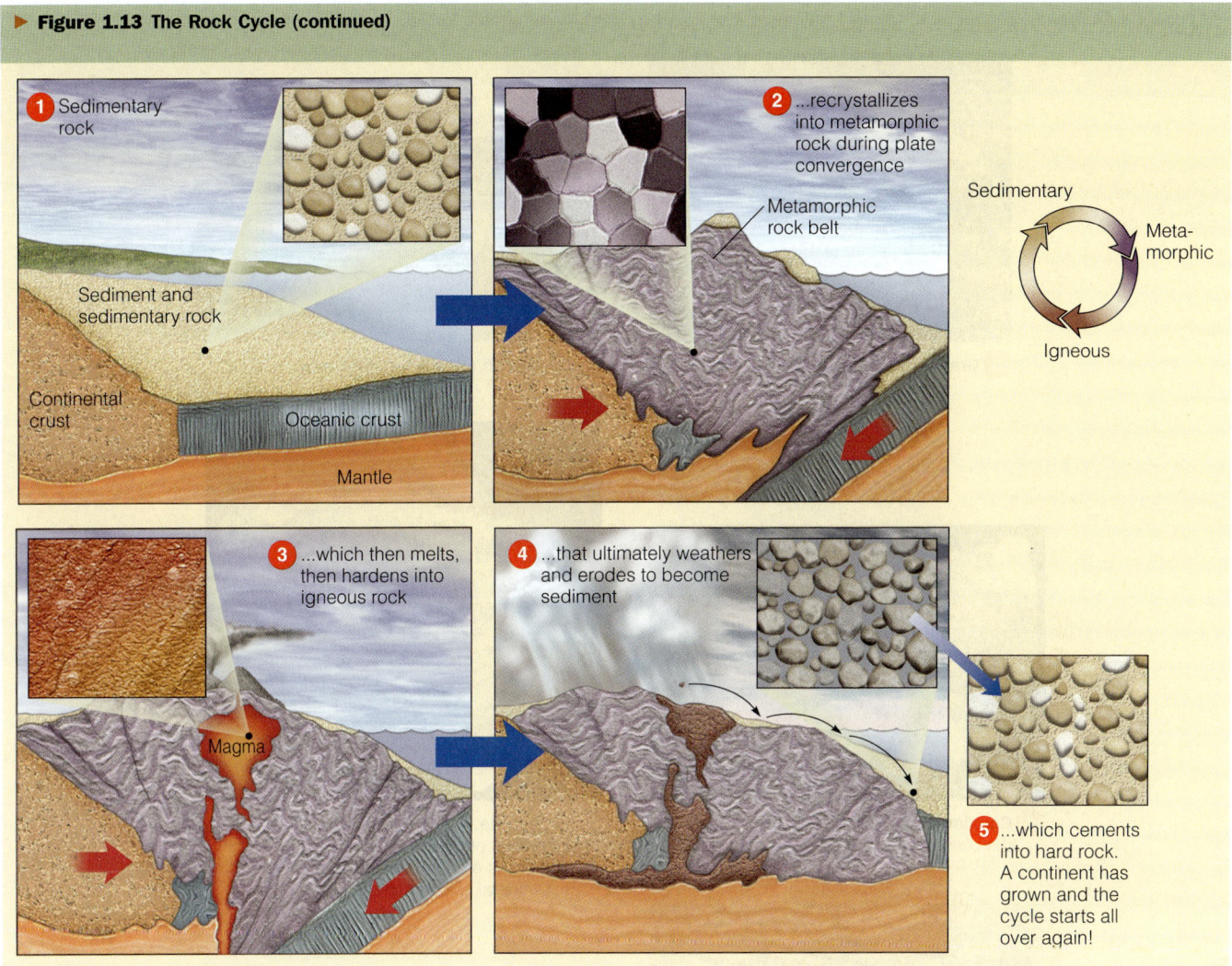

▶ Figure 1.13 The Rock Cycle (continued)

1.9 Geologic Time and Uniformitarianism

■ *Why is it important to understand how geologic time differs from the human perspective of time?*

An appreciation of the immensity of geologic time is central to understanding the evolution of Earth and its biota. Indeed, time is one of the main aspects that sets geology apart from the other sciences, except astronomy. Most people have difficulty comprehending geologic time because they tend to think in terms of the human perspective—seconds, hours, days, and years. Ancient history is what occurred hundreds or even thousands of years ago. When geologists talk of ancient geologic history, however, they are referring to events that happened hundreds of millions or even billions of years ago. To a geologist, recent geologic events are those that occurred within the last million years or so.

It is also important to remember that Earth goes through cycles of much longer duration than the human perspective of time. Although they may have disastrous effects on the human species, global warming and cooling are part of a larger cycle that has resulted in numerous glacial advances and retreats during the past 1.8 million years. Because of their geologic perspective on time and how the various Earth subsystems and cycles are interrelated, geologists can make valuable contributions to many of the current environmental debates, such as those involving global warming and sea-level changes.

The **geologic time scale** subdivides geologic time into a hierarchy of increasingly shorter time intervals; each time subdivision has a specific name. The geologic time scale resulted from the work of many 19th-century geologists who pieced together information from numerous rock exposures and constructed a chronology based on changes in Earth's biota through time. Subsequently, with the discovery of radioactivity in 1895 and the development of various

▶ **Figure 1.14** Hand Specimens of Common Igneous, Sedimentary, and Metamorphic Rocks

a **Granite,** an intrusive igneous rock.

b **Basalt,** an extrusive igneous rock.

c **Conglomerate,** a sedimentary rock formed by the consolidation of rounded rock fragments.

d **Limestone,** a sedimentary rock formed by the extraction of mineral matter from seawater by organisms or by the inorganic precipitation of the mineral calcite from seawater.

e **Gneiss,** a foliated metamorphic rock.

f **Quartzite,** a nonfoliated metamorphic rock.

radiometric dating techniques, geologists have been able to assign numerical ages (also known as absolute ages) in years to the subdivisions of the geologic time scale (▶ Figure 1.16).

- *What is the principle of uniformitarianism, and why is it important in geology?*

One of the cornerstones of geology is the **principle of uniformitarianism,** which is based on the premise that present-day processes have operated throughout geologic time.

Therefore, to understand and interpret geologic events from evidence preserved in rocks, we must first understand present-day processes and their results. In fact, uniformitarianism fits in completely with the system approach we are following for the study of Earth.

Uniformitarianism is a powerful principle that allows us to use present-day processes as the basis for interpreting the past and for predicting potential future events. We should keep in mind, however, that uniformitarianism does not exclude sudden or catastrophic events such as volcanic erup-

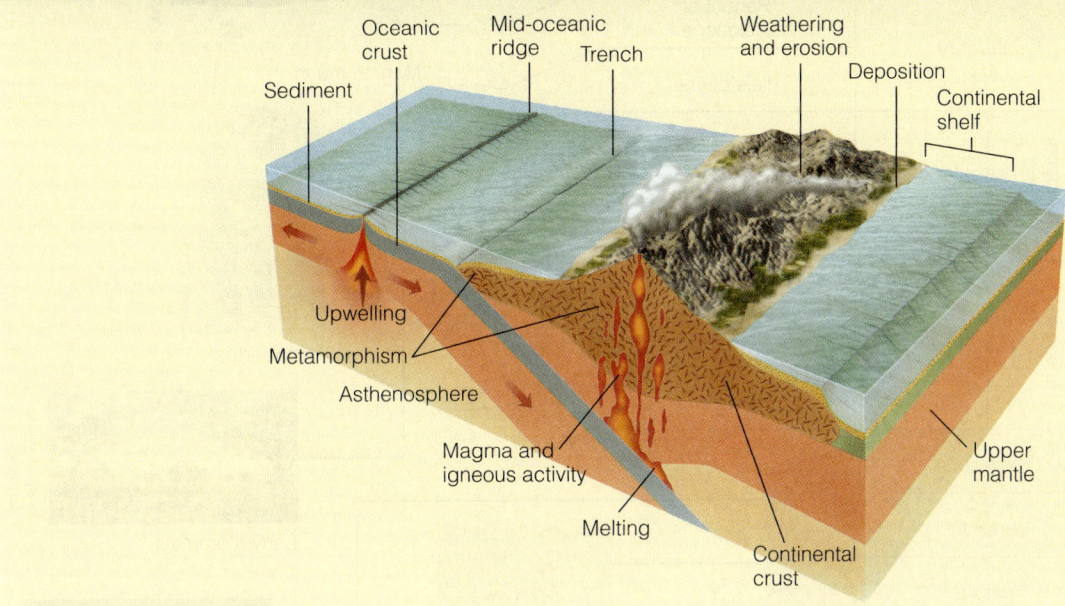

PHYSICAL Geology Now™ ▶ Geo-focus Figure 1.15 Plate Tectonics and the Rock Cycle Plate movement provides the driving mechanism that recycles Earth materials. The cross section shows how the three major rock groups—igneous, metamorphic, and sedimentary—are recycled through both the continental and oceanic regions. Subducting plates are partially melted to produce magma, which rises and either crystallizes beneath Earth's surface as intrusive igneous rock or spills out on the surface, solidifying as extrusive igneous rock. Rocks exposed at the surface are weathered and eroded to produce sediments that are transported and eventually lithified into sedimentary rocks. Metamorphic rocks result from pressure generated along converging plates or adjacent to rising magma.

tions, earthquakes, tsunami, landslides, or floods. These are processes that shape our modern world, and some geologists view Earth history as a series of such short term or punctuated events. This view is certainly in keeping with the modern principle of uniformitarianism.

Furthermore, uniformitarianism does not require that the rates and intensities of geologic processes be constant through time. We know that volcanic activity was more intense in North America 5 to 10 million years ago than it is today, and that glaciation has been more prevalent during the last several million years than in the preceding 300 million years.

What uniformitarianism means is that even though the rates and intensities of geologic processes have varied during the past, the physical and chemical laws of nature have remained the same. Although Earth is in a dynamic state of change and has been ever since it formed, the processes that have shaped it during the past are the same ones operating today.

What Would You Do?

Because of budget shortfalls and in an effort to save money, your local school board is considering eliminating the required geology course all students must pass to graduate from high school. As a concerned parent and citizen, you have rallied community support to keep geology in the curriculum, and you will be making your presentation to the school board next week. What will be your arguments to keep geology as part of the basic body of knowledge all students should have when they graduate from high school? Why should students have a working knowledge of geology? What arguments would you expect from the school board in favor of eliminating geology from the curriculum, and how will you counter their reasons?

Section 1.9 Summary

• Time is one aspect that sets geology apart from most other sciences, and an appreciation of the immensity of geologic time is crucial to understanding the evolution of Earth and its biota.

• The geologic time scale subdivides geologic time into intervals of various duration, with names and numerical age dates in years assigned to each interval.

• The principle of uniformitarianism states that present-day processes have operated throughout geologic time. It is one of the cornerstones of geology because it provides us with the basis for interpreting the past and predicting possible future events.

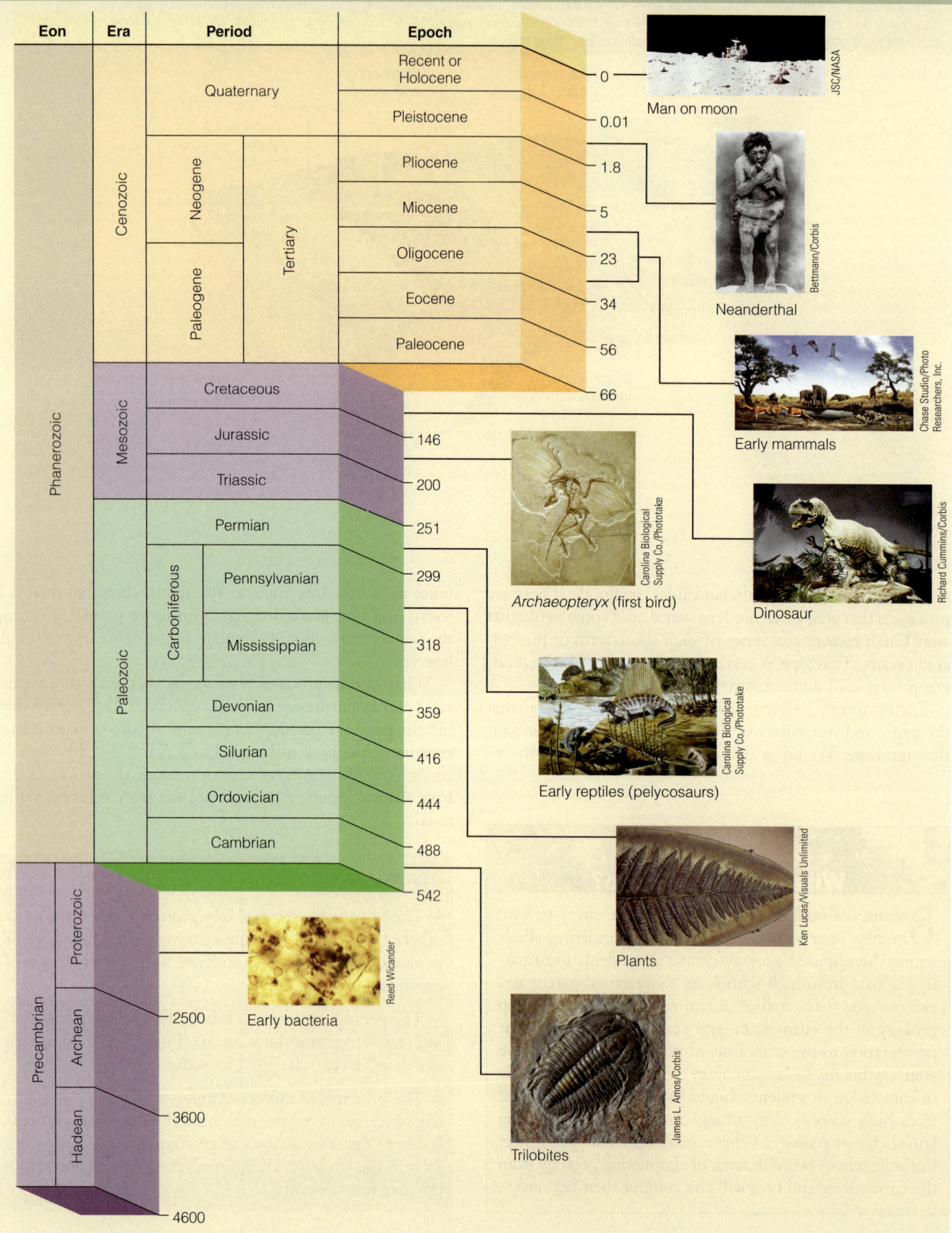

Geo-focus Figure 1.16 **Geologic Time Scale** Numbers to the right of the columns are ages in millions of years before the present. Photographs show characteristic life-forms for the indicated time interval. Dates are from Gradstein, F., Ogg, J., and Smith, A., *A Geologic Time Scale 2004* (Cambridge, UK: Cambridge University Press, 2005), Figure 1.2.

Review Workbook

ESSENTIAL QUESTIONS SUMMARY

1.1 Introduction
■ *What do natural disasters have to do with geology?*
Geology is an integral part of our lives, and natural disasters such as earthquakes, volcanic eruptions, tsunami, landslides, and floods are geologic phenomena that affect everyone. The more we understand how these phenomena work, the better prepared we can be to minimize their destructive effects and lessen the loss of life.

■ *What is a system?*
A system is a combination of related parts, components, or subsystems that interact in an organized fashion. Earth can be viewed as a complex system whose principal subsystems are the atmosphere, biosphere, hydrosphere, lithosphere, mantle, and core. The interaction of these subsystems has resulted in a dynamically changing planet in which matter and energy are continuously recycled into different forms.

1.2 Geology in Our Everyday Lives
■ *What is geology and what do geologists do?*
Geology, the study of Earth, is divided into two broad areas—physical geology (the study of Earth materials) and historical geology (the study of the origin and evolution of Earth). Geologists engage in a variety of occupations, such as exploration for mineral and energy resources, to solve various environmental problems.

■ *How does geology relate to the human experience?*
Geology pervades our everyday lives, with references to it in the arts (Asher Brown Durand's painting *Kindred Spirits*), music (Ferde Grofé's *Grand Canyon Suite*), and literature (Percy B. Shelley's poem "Ozymandias"). Earth's surface features, or topography (such as mountain ranges and rivers), are a result of the interplay among its various subsystems and have dictated military tactics as well as political boundaries.

■ *How does geology affect our everyday lives?*
Geology affects all of us when natural disasters occur. Our economic and political well-being depends, in part, on the distribution and concentration of natural resources. And government plays an increasingly important role in environmental issues related to geology.

1.3 Global Geologic and Environmental Issues Facing Humankind
■ *What is considered the greatest environmental problem facing the world today?*
Most scientists would argue that overpopulation is the greatest problem facing the world today. Increasingly large numbers of people must be fed, housed, and clothed, with a minimal impact on the environment.

■ *How is the greenhouse effect related to global warming?*
The greenhouse effect is the retention of heat in the atmosphere, which results in an increase in the temperature of Earth's surface and atmosphere, thus producing global warming.

1.4 Origin of the Universe and Solar System, and Earth's Place in Them
■ *How did the universe begin?*
The universe began with a Big Bang approximately 15 billion years ago. Evidence for the Big Bang is that the universe is expanding and that the entire universe has a pervasive background radiation of 2.7 K above absolute zero (2.7 K = −270.3°C), which is thought to be the faint afterglow of the Big Bang.

■ *What is our solar system composed of, and what is its origin and history?*
Our solar system consists of the Sun, 9 planets, 101 known moons, asteroids, comets, meteorites, and interplanetary dust and gases. It formed from a rotating cloud of interstellar matter that condensed, collapsed under the influence of gravity, and flattened into a rotating disk from which the Sun, planets, and moons formed.

■ *How did Earth form?*
Earth formed from a swirling eddy of nebular material 4.6 billion years ago, accreting as a solid body and soon thereafter differentiating into a layered planet during a period of internal heating.

1.5 Earth as a Dynamic and Evolving Planet
■ *Why is Earth a dynamic planet?*
Earth has continuously changed during its 4.6-billion-year existence as a result of the interactions between its various subsystems and cycles.

■ *What are Earth's three concentric layers, and what are their characteristics?*
Earth's three concentric layers are the core, mantle, and crust. The core consists of a small, solid inner region and a larger, seemingly liquid, outer portion, both of which are thought to be composed of iron and a small amount of nickel. The mantle, composed largely of peridotite, surrounds the core and is divided into a solid lower mantle, an asthenosphere that behaves plastically and flows slowly, and a solid upper mantle. The outermost layer, the crust, is divided into thick continental crust and thin oceanic crust.

1.6 Geology and the Formulation of Theories
■ *What is a theory?*
A theory is a testable explanation for some natural phenomenon that has a large body of supporting evidence. It is arrived at through the scientific method, which involves gathering and analyzing facts, formulating hypotheses to explain the phenomenon, testing the hypotheses, and finally proposing a theory.

1.7 Plate Tectonic Theory
■ *What is plate tectonic theory?*
Plate tectonic theory states that the lithosphere is divided into rigid plates that diverge, converge, or slide past each other as they move over the asthenosphere.

■ *Why is plate tectonic theory important in geology?*
Plate tectonic theory is a unifying explanation for many geologic features and events, and provides a framework for interpreting the composition, structure, and internal processes of Earth on a global scale.

1.8 The Rock Cycle
■ *What is a rock?*
A rock is an aggregate of minerals.

■ *What is the rock cycle?*
The rock cycle illustrates the interactions between Earth's internal and external processes and how the three rock groups are interrelated.

■ *What are the characteristics of each of the three major rock groups?*
Igneous rocks result from the crystallization of magma or the consolidation of volcanic ejecta. Sedimentary rocks are typically formed by the consolidation of rock fragments, precipitation of mineral matter from solution, or compaction of plant or animal remains. Metamorphic rocks result from the alteration of other rocks, usually beneath Earth's surface, by heat, pressure, and chemically active fluids.

■ *How are the rock cycle and plate tectonics related?*
Plate movement is the driving mechanism of the rock cycle. Plate interaction determines, to some extent, which of the three rock groups will form.

1.9 Geologic Time and Uniformitarianism
■ *Why is it important to understand how geologic time differs from the human perspective of time?*
An appreciation of geologic time is essential to understanding Earth's evolution. Earth goes through cycles of much longer duration than the human perspective of time, and thus we must adjust our concept of time to incorporate the fact that Earth is more than 20,000 times as old as the human species.

■ *What is the principle of uniformitarianism, and why is it important in geology?*
This principle holds that the laws of nature have been constant through time and that the same processes operating today have operated in the past, though at different rates. Therefore, to understand and interpret geologic events from evidence preserved in rocks, geologists must first understand present-day processes and their results.

ESSENTIAL TERMS TO KNOW

asthenosphere (p. 13)
Big Bang (p. 10)
core (p. 13)
crust (p. 16)
geologic time scale (p. 21)
geology (p. 6)
hypothesis (p. 16)
igneous rock (p. 19)
Jovian planets (p. 12)
lithosphere (p. 16)
mantle (p. 13)
metamorphic rock (p. 19)
mineral (p. 19)
plate (p. 16)
plate tectonic theory (p. 17)
principle of uniformitarianism (p. 22)
rock (p. 19)
rock cycle (p. 19)
scientific method (p. 16)
sedimentary rock (p. 19)
solar nebula theory (p. 11)
system (p. 4)
terrestrial planets (p. 12)
theory (p. 16)

REVIEW QUESTIONS

1. Why is viewing the Earth as a system a good way to study Earth? Are humans a part of the Earth system? If so, what role, if any, do we play in Earth's evolution?
2. What is geology? Why is it important that everyone have a basic understanding of geology, even if they aren't going to become geologists?
3. In what ways does geology affect our everyday lives at the individual, local, and nation-state levels?
4. Why do most scientists think overpopulation is the greatest environmental problem facing the world today? Do you agree? If not, what do you think is the greatest threat to our existence as a species?
5. What is the Big Bang? What evidence do we have that the universe began approximately 15 billion years ago?
6. How does the solar nebula theory account for the formation of our solar system, its features, and evolutionary history?
7. Why is Earth considered a dynamic planet? What are its three concentric layers and their characteristics?
8. Why is plate tectonic theory so important to geology? How does it fit into a systems approach to the study of Earth?
9. Using plate movement as the driving mechanism of the rock cycle, explain how the three rock groups are related and how each rock group can be converted into a different rock group.
10. Explain how the principle of uniformitarianism allows for catastrophic events.

APPLY YOUR KNOWLEDGE

1. Discuss why an accurate geologic time scale is particularly important for geologists in examining global temperature changes during the past, and how an understanding of geologic time is crucial to the current debate on global warming and its consequences.

2. Based on what you know about landforms and topography, do you think an understanding of geology would be helpful in planning a military campaign against another country? What geologic factors or information would be useful to know? Can you think of any examples from your knowledge of history where geology played a role in the planning or execution of a military campaign? After you have completed your geology course, answer this question again and see whether your answer is any different from what it is now.

GEOLOGY MATTERS

GEOLOGY
IN UNEXPECTED PLACES

A Bit of Egypt in Central Park, New York City, and London, England

Most tourists visiting Central Park in New York City probably don't expect to see a relic of Egypt's past rising above the trees. But that is exactly what you will find if you visit Cleopatra's Needle, located behind the Metropolitan Museum of Art (▶ Figure 1).

The ruler of Egypt, the Khedive Ismail Pasha, promised the United States in 1869 an obelisk in appreciation for its aid in the construction of the Suez Canal. The obelisk and its companion, which is now in London, were originally erected in front of the great temple of Heliopolis during the reign of Pharaoh Tuthmosis III (1504–1450 B.C.). They were later moved to Alexandria, where they remained until one was transported to London in 1878 and erected along the Thames on the Victoria Embankment (▶ Figure 2) and the other was transported to New York in 1880 and erected at its present site in 1881. Both are dubbed "Cleopatra's Needle," even though they have nothing to do with Cleopatra and their inscriptions celebrate the pharaohs of ancient Egypt.

Henry Gorridge, a lieutenant commander of the U.S. Navy, was assigned the task of transporting the second of the twin obelisks and its pedestal to New York in 1880. The obelisk arrived at the Quarantine Station in New York in July 1880, and then it took 112 days to transport it to its final destination in Central Park, where it was raised in early 1881 in front of more than 10,000 cheering New Yorkers.

What of the geology of the obelisks? Both are made from the red granite of Syene and have suffered the effects of weathering. The New York obelisk is 21.4 m tall and weighs 193 tons, whereas the London obelisk is 21.1 m high and weighs 187 tons. Because of air pollution, the New York obelisk is not as well preserved as the one in London, and the inscriptions on the New York "Needle" are readable on only two sides.

Figure 1
Cleopatra's needle is located in New York's Central Park behind the Metropolitan Museum of Art. This obelisk, a gift from the Khedive Ismail, shows columns of hieroglyphic inscriptions on its sides. Chemical weathering, resulting from air pollution, has visibly damaged the obelisk. Also note more weathering and damage to the dower portion of the obelisk.

Figure 2
The London obelisk was presented to Britain by the Viceroy of Egypt in 1819 and erected at its present location on the Victoria Embankment along the Thames River in London in 1878.

GEOLOGY MATTERS

GEOLOGY IN FOCUS The Kyoto Protocol—An Update

At the 1997 Climate Treaty Meeting in Kyoto, Japan, a group of nations took the first steps to reduce greenhouse gas emissions by negotiating a treaty calling for the United States and 37 other industrialized nations to reduce their emissions by an average of 5.2% below their 1990 levels between 2008 and 2012. Although this might not seem like a significant decrease, it should be kept in mind that the industrialized nations, which make up only 20% of the world's population, have to date emitted 90% of all human-produced carbon emissions.

The Kyoto Protocol requires the largest polluters to make the greatest sacrifices. The United States must cut its greenhouse gas emissions to 7% below 1990 levels, and the European Union and Japan must reduce their emissions by 8% and 6%, respectively. The greenhouse gases that must be reduced are carbon dioxide, methane, nitrous oxide, hydrofluorocarbons, perfluorocarbons, and sulfur hexafluoride. It was also agreed that further reductions will be phased in after 2012.

Developing countries are not required to reduce their emissions of greenhouse gases unless they want to, and countries with extensive forests and jungles can receive credit toward their emission-reduction goals. Research indicates that the deforestation of large areas, particularly in the tropics, is also a cause of increased levels of carbon dioxide. This is because trees and plants use carbon dioxide in photosynthesis and thus act as carbon dioxide "sinks" by removing it from the atmosphere. With a decrease in the global vegetation cover, less carbon dioxide is removed from the atmosphere by photosynthesis, and carbon dioxide levels can increase.

Whereas a reduction in greenhouse gas emissions is certainly desirable, the economic impact on industrialized countries must also be considered. It is estimated that meeting the Kyoto Protocol reductions could cause as much as a 2% drop in the U.S. gross domestic production, which is equivalent to about $227 billion. Furthermore, the economic impact would not be spread evenly throughout the population, and there could be devastating effects on many industries, particularly those that produce and use large quantities of energy. Thus the possible benefits to be gained by reducing greenhouse gas emissions must be weighed against the possible economic hardships that meeting such reductions would entail (▶ Figure 1).

On February 16, 2005, the Kyoto Protocol went into effect. One hundred forty nations that produce 55% of the world's greenhouse gas emissions have ratified the Kyoto Protocol and pledged to reduce overall emissions by an average of 5.2% by 2012. Although the protocol officially ends in 2012, most observers think that it is the first step toward controlling global warming and that it will be extended with modifications, based on how successful it is in reducing overall greenhouse gas emissions.

Even though all the European Community members and Japan ratified the agreement, the world's largest polluter, the United States, refused to ratify the protocol, citing that it could cost millions of U.S. jobs. According to the United States, the Kyoto Protocol is flawed because it exempts large, developing countries, such as China (which also refused to ratify the protocol and is the world's second largest greenhouse gas producer), from having to reduce their pollution. Many of these developing and advanced developing countries, such as South Korea, Brazil, and India, are already concerned that if the protocol is extended beyond 2012, their economies may suffer when they must comply with emissions reductions.

Only time will tell how successful the Kyoto Protocol is in reducing greenhouse gas emissions, and what effect it will have on global warming and the global economy. It is a start, however, and as we know, Earth's climate is a complex system with many variables contributing to its evolution over a long period of time.

Figure 1
The scene of smokestacks at an oil refinery serves to focus on some of the issues relating to the recently ratified Kyoto Protocol and the debate concerning global warming. All industrial activity pollutes the environment to some extent, and the debate is, in part, how much human pollution is contributing to global warming. By requiring nations to reduce their greenhouse gas emissions, particularly those that produce and use large quantities of energy, it is hoped that such reductions will lessen the rate of global warming and its consequences. However, several industrialized as well as developing countries have refused to ratify the Kyoto Protocol because of the perceived potentially crippling effects it may have on their economies. Only time will tell whether the Kyoto Protocol is successful in stemming the increasing levels of greenhouse gas emissions resulting from industrialization.

GEOLOGY MATTERS
Geology AT WORK

Forensic Geology and Palynology

Although Sherlock Holmes, created by Sir Arthur Conan Doyle, is probably the best-known detective in the world, many people would be surprised to know that he also was a forensic geologist of sorts. In Doyle's novel *A Study in Scarlet*, we are introduced to Sherlock Holmes by his future biographer and companion, Dr. John H. Watson. Soon after moving into 221B Baker Street, Dr. Watson was still trying to figure out just what Sherlock Holmes did for a living. He knew Holmes was very knowledgeable in such subjects as botany, chemistry, and anatomy, and yet "of contemporary literature, philosophy and politics he appeared to know next to nothing." According to Dr. Watson, Sherlock Holmes's "knowledge of Geology [is]—Practical, but limited. Tells at a glance different soils from each other. After walks has shown me splashes upon his trousers, and told me by their colour and consistence in what part of London he had received them."

About a century later, Sarah Andrews, a geologist and author, has written 10 mysteries featuring fictional forensic geologist Em Hansen, who uses her knowledge of geology to solve mysteries and crimes. Do real-life forensic geologists exist, or are they only fictional characters? The answer is yes; they exist and are employed in a number of public and private laboratories in the United States as well as many other countries. In fact, the Federal Bureau of Investigation laboratory in Quantico, Virginia, employs several geologist/forensic scientists. In its Materials Analysis Unit, specialists perform a variety of geologic/mineralogic, metallurgic, and elemental analyses on evidence associated with various crimes and disasters. In the Mineralogy Subunit, tests can be done on virtually all naturally occurring geologic materials such as minerals, rocks, soil, and gemstones. In addition, human-made products like glass, abrasives, and building materials are analyzed because these products are derived from geologic materials and have characteristic signatures that may reveal their origin.

What do forensic geologists do? A forensic geologist identifies, analyzes, and compares such earth materials as soil, minerals, rocks, and fossils from a crime scene to determine where a particular material came from, when an incident might have taken place, and even what the cause of the incident was or who might be responsible. A solid grounding in the geologic sciences is essential, and a forensic geologist must know how to identify minerals, rocks, soils, and fossils; how to read and interpret geologic and topographic maps; and how to put all this information together so that it can be used in a court of law.

Just as Sherlock Holmes could tell where a person had been by the soil adhering to his shoes or clothing, the identification of soil found on a victim or suspect is probably the most common type of forensic geology practiced. There are numerous instances in the crime annals of soil or rock chips being used to help solve a case.

A good example involved the apparent accidental shooting death of John Dodson in the Uncompahgre Mountains of western Colorado on October 15, 1995. According to her testimony, his wife, Jane, had returned to their hunting camp to find that John had been shot three times. Jane's ex-husband, J. C. Lee, was also camped near the Dodson hunting camp at the time. He naturally became a prime suspect, but he reported that he had been hunting with his boss far from the camp when the shooting occurred, and furthermore someone had stolen his .308-caliber rifle and a box of cartridges from his tent while he was hunting.

Figure 1
Bentonite, a clay brought in to stop water from seeping out of this cattle pond in western Colorado's Uncompahgre Mountains, provided critical geologic evidence that led to the conviction of Janice Dodson for the murder of her husband.

A cattle pond near the campsite of J. C. Lee provided the evidence that eventually convicted Jane Dodson in the shooting death of her husband (▶ Figure 1). The pond was lined with bentonite, a clay not found in the area but brought in to stop water from seeping out of the bottom of the pond. In her initial statement to the police, Jane reported that she had stepped into a mud bog near her camp and had removed her coveralls, which were caked with mud. Investigators tested the mud from the cattle pond near J. C. Lee's camp, the mud in the bog near the Dodson's camp, and the mud on Jane's clothing. The forensic evidence indicated the mud on Jane's coveralls was consistent with mud from the cattle pond, and not consistent with mud near the Dodson's camp. This meant that Jane Dodson was at her ex-husband's camp around the time his rifle (which was the same caliber as a shell casing and bullet found near the body) was stolen. A jury later convicted Jane Dodson of the murder of her husband, and she is serving a life sentence in Colorado's state prison for women.

Another intriguing story, but this time involving fraud, is the collapse of Bre-X, a Canadian mining company that supposedly discovered the world's largest gold deposit at Busang, Indonesia. According to company releases in 1995, the reserves were estimated to be 200 million ounces of gold (approximately 8% of the world's gold), worth more than $70 billion at that time! Unfortunately, very little gold actually existed at Busang, and with that revelation, Bre-X collapsed in 1997. Instead of millions of ounces of gold, the core samples on which projections were based had been "salted" with outside gold.

How did such a massive fraud happen? From the beginning, no outside analyses of core samples at the mine were allowed. Only the analyses performed by the company hired by Bre-X were released to the public. Furthermore, standard practices of core analysis were not followed. Typically, a core is cut lengthwise and half the sample is analyzed, leaving the other half as a backup to verify the initial results should that be necessary. In the case of Bre-X, the entire core was pulverized and handled only by company geologists and the company hired to do the analysis. Thus, no material was available for independent analysis. Only when another company brought in at the insistence of the Indonesian government to run the mine drilled its own test holes was the fraud discovered. Test results showed that the original samples had been salted with gold dust from elsewhere. If standard procedures had been followed in splitting the original samples, the fraud might never have happened. The collapse of Bre-X is one of the largest stock scandals in Canadian history, and it resulted in numerous lawsuits from investors who lost billions of dollars.

A field related to forensic geology is forensic palynology, which is the study of pollen and spores to help solve criminal cases. Pollen is the microscopic grains produced by the male element of a flowering plant; it is what causes allergy sufferers so much discomfort. Pollen also can help crime investigators determine where someone or something has been; trace the origin of items like packages, letters, antiques, and drugs; and corroborate or discredit a suspect's alibi. Forensic palynologists generally work as independent consultants or for private laboratories.

Despite its potential, forensic palynology is still underutilized by law enforcement. An example of how palynology can be useful is a case worked on by Dr. Lynne Milne, who teaches at the Centre for Forensic Science at the University of Western Australia. During a period of several years, a number of rapes and assaults had occurred in the Perth area of Western Australia, and the police believed the crimes were committed by the same person, possibly an itinerant worker. To help the police narrow down their search area, Dr. Milne collected pollen and other samples from slits in the shoe soles the rapist had allegedly left near the scene of an assault in 1995. In addition to pollen that was common to the area, pollen grains produced by cereals were also found, leading police to theorize that the suspect had been walking on the stubble left from cutting hay. Hay of the type that produces the rare pollen grains found in Dr. Milne's samples is very common in the chaff-cutting businesses located near a particular town in Western Australia. Because this industry employs itinerant workers, police concentrated their efforts in this area and several weeks later arrested a suspect who confessed to the rapes and assaults. Though not directly linking the suspect to the crime, the analysis of pollen from the crime scene enabled police to narrow their search for the perpetrator.

GEOLOGY MATTERS

GEOLOGY IN YOUR LIFE

Global Warming and Climate Change, and How They Affect You

The greenhouse effect, global warming, climate change: These headlines and topics are in the news all the time, global issues that affect us all and the planet we live on. But just how will global warming and the resultant climate change personally affect you? Are they something that you should really be concerned about? After all, there are exams to worry about, graduation, finding a job, and that doesn't even include the everyday issues we all must deal with, not to mention your personal life. Yet part of the college experience is examining and debating the "big picture" and issues facing society today. So what about global warming and you?

You may recall that we talked about the greenhouse effect and its relationship to global warming. The greenhouse effect helps regulate Earth's temperature because as sunlight passes through the atmosphere, some of the heat is trapped in the lower atmosphere and not radiated back into space, thus effectively warming Earth's surface and atmosphere. The issue is not whether we have a greenhouse effect, because we do, but the degree to which human activity, such as the burning of fossil fuels, is increasing the greenhouse effect and thus causing global warming.

Based on many studies using a variety of techniques, it is clear that carbon dioxide (one of the greenhouse gases that allows short-wavelength solar radiation to pass through it but traps some of the long-wavelength radiation that is reflected back from Earth's surface) levels have increased since the Industrial Revolution in the 19th century. Furthermore, global surface temperatures have increased about 0.6°C since the late 1800s and about 0.4°C during the past 25 years. However, this warming trend has not been uniform around the globe, and some areas, such as the southwestern part of the United States, have actually cooled during this time period.

One thing we must be careful about is mistaking regional trends for global trends. For example, there is compelling evidence of climate variability or extremes on a regional scale, but on a global scale, there is currently little evidence of a sustained trend in climate variability or extremes. That doesn't mean we can ignore the overall increase in average global surface temperatures because, if left unchecked, such changes can have significant environmental, ecological, and economic effects.

So what are some of the effects we should be worried about? For starters, there is the problem of rising sea level. During the past 100 years, global sea level has been rising at an average rate of 1 to 2 mm per year, and the projected rate of increase by 2100 is anywhere from 9 to 88 cm, depending on which climate model is used. What this means is that low-lying coastal areas will experience flooding and increased erosion along the coastline, endangering housing and communities. For instance, about 17 million people live less than 1 m above sea level in Bangladesh, and they are certainly at risk due to rising sea level. Furthermore, many major cities are just above sea level and could also suffer from rising sea level. A rise of sea level at the upper end could completely submerge such island nations as the Maldives in the Indian Ocean (▶ Figure 1).

Based on various climate computer models and taking into account the complexity and variability of the atmospheric–oceanic system, most predictions show Earth's average surface temperature increasing by 1.4° to 5.8°C during the period 1990–2100. This increase will result in widely varying regional responses; land areas will warm more and faster than ocean areas, particularly in the high latitudes of the Northern Hemisphere. Expect to see more hot days and heat waves over nearly all the land areas, with more droughts in all continental interiors during the summer (▶ Figure 2). There will also be increased precipitation during the 21st century, particularly in the northern middle to high latitudes. And the glaciers and ice caps will continue to retreat, with a decrease in the Northern Hemisphere's snow cover and sea ice.

What do all these predictions mean to you? With increasingly hot summers and more frequent droughts, expect to see higher food prices as crop yields decrease. There will also be an increased risk of wildfires. Deadly heat waves will result in more heat-related deaths, such as occurred in Europe during the summer of 2003. As climates change, diseases such as malaria will spread easily to areas with warmer, wetter climates. Disease-carrying mosquitoes will extend their reach as climate changes allow them to survive in formerly inhospitable regions. Higher temperatures will affect regional water supplies, creating potential water crises in the western United States within the next 20 years as well as in other areas such as Peru and western China. Just as many regions will experience longer and hotter summers, other areas will suffer from intense and increased rainfall, which will result in severe floods and landslides.

Everyone is vulnerable to weather-related disasters, but large-scale changes brought about by climate change will affect people in poor countries more than those in the more industrialized countries. However, whether these climate changes are part of a natural global cycle taking place over thousands or hundreds of thousands of years—that is, on a geologic time scale—or are driven, in part, by human activities is immaterial. The bottom line is that we already are, or eventually will be, affected in some way, be it economic or social, by the climate changes that are taking place.

Figure 1
(a) Map showing the location of the Maldives Islands in the Indian Ocean. (b) Aerial view of the Maldives Islands. A rise in sea level due to global warming could easily submerge these islands.

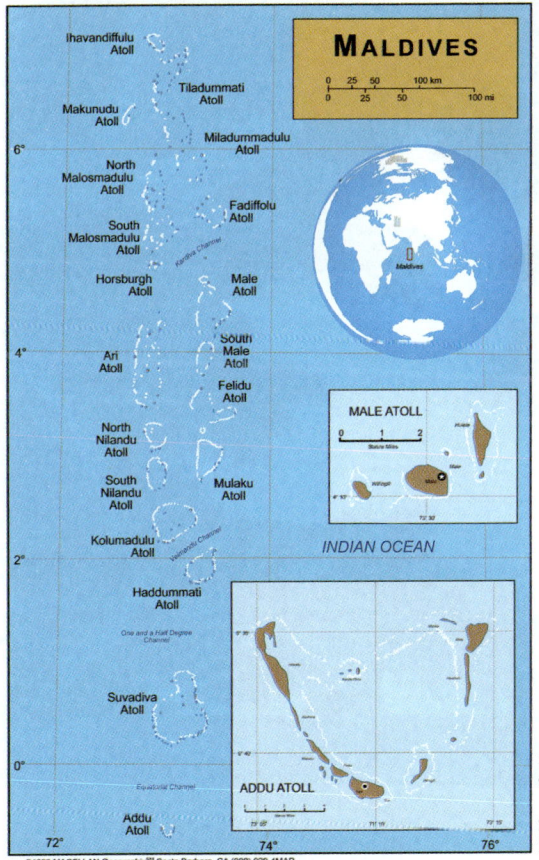

(a)

(b)

Figure 2
Withered corn crop as a result of drought conditions.

Chapter 2

Plate Tectonics: A Unifying Theory

The Himalayas, Southwestern China
Ruby and sapphire veins spread over white like fossils of ferns on rock. This satellite image of the Himalayas reveals an ever-changing tapestry of peaks, ridges, and rivers woven by Earth millions of years ago when India collided with Asia.
—A. W.

ESSENTIAL QUESTIONS TO ASK

2.1 Introduction
- Why should you know about plate tectonics?

2.2 Continental Drift
- What were some early ideas about Earth's past geography?
- What is the continental drift hypothesis and who proposed it?

2.3 Evidence for Continental Drift
- What is the evidence for continental drift?

2.4 Paleomagnetism and Polar Wandering
- What is paleomagnetism?
- What is the Curie point and why is it important?
- How can the apparent wandering of the magnetic poles be best explained?

2.5 Magnetic Reversals and Seafloor Spreading
- What evidence is there that Earth's magnetic field has reversed in the past?
- What is the theory of seafloor spreading, and how does it validate continental drift?
- How was the theory of seafloor spreading confirmed?

2.6 Plate Tectonics: A Unifying Theory
- What are the main tenets of plate tectonic theory?
- Why is plate tectonics a unifying theory of geology?
- What is the supercontinent cycle?

2.7 The Three Types of Plate Boundaries
- What are the three types of plate boundaries?
- What are divergent boundaries?
- What features in the geologic record indicate ancient rifting?
- What are convergent boundaries?
- How can ancient subduction zones be recognized in the geologic record?
- What are transform boundaries?

2.8 Hot Spots: An Intraplate Feature
- What are hot spots and what do they tell us about plate movement?

2.9 Plate Movement and Motion
- How can the rate and direction of plate movement be determined?

2.10 The Driving Mechanism of Plate Tectonics
- What drives plates?
- How do thermal convection cells move plates?
- Can plate movement be gravity driven?

2.11 Plate Tectonics and the Distribution of Natural Resources
- How does plate tectonic theory relate to the origin and distribution of natural resources?
- What is the relationship between plate boundaries and various metallic mineral deposits?

GEOLOGY MATTERS

GEOLOGY IN FOCUS:
Oil, Plate Tectonics, and Politics

GEOLOGY IN UNEXPECTED PLACES:
A Man's Home Is His Castle

GEOLOGY AND CULTURAL CONNECTIONS:
The Struggle toward Scientific Progress

GEOLOGY AT WORK:
The Geologic Exploration of Mars

 This icon, which appears throughout the book, indicates an opportunity to explore interactive tutorials, animations, or practice problems available on the Physical GeologyNow website at **http://now.brookscole.com/phygeo6**.

2.1 Introduction

Imagine it is the day after Christmas, December 26, 2004, and you are vacationing on a beautiful beach in Thailand. You look up from the book you're reading to see the sea suddenly retreat from the shoreline, exposing a vast expanse of seafloor that had moments before been underwater and teeming with exotic and colorful fish. It is hard to believe that within minutes of this unusual event, a powerful tsunami will sweep over your resort and everything in its path for several kilometers inland. Within hours, the coasts of Indonesia, Sri Lanka, India, Thailand, Somalia, Myanmar, Malaysia, and the Maldives will be inundated by the deadliest tsunami in history. More than 220,000 people will die, and billions of dollars in damage will be wreaked on the region.

One year earlier, on December 26, 2003, violent shaking from an earthquake awakened hundreds of thousands of people in the Bam area of southeastern Iran. When the magnitude-6.6 earthquake was over, an estimated 43,000 people were dead, at least 30,000 were injured, and approximately 75,000 survivors were left homeless. At least 85% of the structures in the Bam area were destroyed or damaged. Collapsed buildings were everywhere, streets were strewn with rubble, and all communications were knocked out.

Now go back another 12½ years to June 15, 1991, when Mount Pinatubo in the Philippines erupted violently, discharging huge quantities of ash and gases into the atmosphere. Fortunately, in this case, warnings of an impending eruption were broadcast and heeded, resulting in the evacuation of 200,000 people from areas around the volcano. Unfortunately, the eruption still caused at least 364 deaths not only from the eruption but also from ensuing mudflows.

What do these three recent tragic events have in common? They are part of the dynamic interactions involving Earth's plates. When two plates come together, one plate is pushed or pulled under the other plate, triggering large earthquakes such as the one that shook India in 2001, Iran in 2003, and Pakistan in 2005. If conditions are right, earthquakes can produce a tsunami such as the one in 2004 or the 1998 Papua New Guinea tsunami that killed more than 2200 people.

As the descending plate moves downward and is assimilated into Earth's interior, magma is generated. Being less dense than the surrounding material, the magma rises toward the surface, where it may erupt as a volcano such as Mount Pinatubo did in 1991 and others have since. It therefore should not be surprising that the distribution of volcanoes and earthquakes closely follows plate boundaries.

As we stated in Chapter 1, **plate tectonic theory** has had significant and far-reaching consequences in all fields of geology because it provides the basis for relating many seemingly unrelated phenomena. The interactions between moving plates determine the locations of continents, ocean basins, and mountain systems, which in turn affect atmospheric and oceanic circulation patterns that ultimately determine global climate (see Table 1.3). Plate movements have also profoundly influenced the geographic distribution, evolution, and extinction of plants and animals. Furthermore, the formation and distribution of many geologic resources, such as metal ores, are related to plate tectonic processes, so geologists incorporate plate tectonic theory into their prospecting efforts.

■ *Why should you know about plate tectonics?*

If you're like most people, you probably have no idea or only a vague notion of what plate tectonic theory is. Yet plate tectonics affects all of us. Volcanic eruptions, earthquakes, and tsunami are the result of interactions between plates. Global weather patterns and oceanic currents are caused, in part, by the configuration of the continents and ocean basins. The formation and distribution of many natural resources are related to plate movement and thus have an impact on the economic well-being and political decisions of nations. It is therefore important to understand this unifying theory, not only because it affects us as individuals and as citizens of nation-states but also because it ties together many aspects of the geology you will be studying.

Section 2.1 Summary

- Plate tectonic theory is the unifying theory of geology. It affects all of us because it explains where and why such natural disasters as earthquakes, volcanic eruptions, and tsunami occur as well as the formation and distribution of many economically valuable natural resources.

2.2 Continental Drift

■ *What were some early ideas about Earth's past geography?*

The idea that Earth's past geography was different from today is not new. The earliest maps showing the east coast of South America and the west coast of Africa probably provided people with the first evidence that continents may have once been joined together, then broken apart and moved to their present locations.

During the late 19th century, the Austrian geologist Edward Suess noted the similarities between the Late Paleozoic plant fossils of India, Australia, South Africa, and South America as well as evidence of glaciation in the rock sequences of these southern continents. The plant fossils make up a unique flora in the coal layers just above the glacial deposits of these southern continents. This flora is very different from the contemporaneous coal swamp flora of the northern continents and is collectively known as the *Glossopteris* **flora,** after its most conspicuous genus (▶ Figure 2.1).

In his book, *The Face of the Earth*, published in 1885, Suess proposed the name *Gondwanaland* (or **Gondwana** as we will use here) for a supercontinent composed of the

Figure 2.1 **Fossil *Glossopteris* Leaves** Plant fossils, such as these *Glossopteris* leaves from the Upper Permian Dunedoo Formation in Australia, are found on all five Gondwana continents. The presence of these fossil plants on continents with widely varying climates today is evidence that the continents were at one time connected. The distribution of the plants at that time was in the same climatic latitudinal belt.

aforementioned southern continents. Abundant fossils of the *Glossopteris* flora are found in coal beds in Gondwana, a province in India. Suess thought these southern continents were at one time connected by land bridges over which plants and animals migrated. Thus, in his view, the similarities of fossils on these continents were due to the appearance and disappearance of the connecting land bridges.

The American geologist Frank Taylor published a pamphlet in 1910 presenting his own theory of continental drift. He explained the formation of mountain ranges as a result of the lateral movement of continents. He also envisioned the present-day continents as parts of larger polar continents that eventually broke apart and migrated toward the equator after Earth's rotation was supposedly slowed by gigantic tidal forces. According to Taylor, these tidal forces were generated when Earth captured the Moon about 100 million years ago.

Although we now know that Taylor's mechanism is incorrect, one of his most significant contributions was his suggestion that the Mid-Atlantic Ridge, discovered by the 1872–1876 British HMS *Challenger* expeditions, might mark the site along which an ancient continent broke apart to form the present-day Atlantic Ocean.

■ *What is the continental drift hypothesis and who proposed it?*

Alfred Wegener, a German meteorologist (▶ Figure 2.2), is generally credited with developing the hypothesis of **continental drift.** In his monumental book, *The Origin of Continents and Oceans* (first published in 1915), Wegener proposed that all landmasses were originally united in a single supercontinent that he named **Pangaea,** from the Greek meaning "all land." Wegener portrayed his grand concept of continental movement in a series of maps showing the breakup of Pangaea and the movement of the various continents to their present-day locations. Wegener amassed a tremendous amount of geologic, paleontologic, and climatologic evidence in support of continental drift, but the initial reaction of scientists to his then-heretical ideas can best be described as mixed.

Opposition to Wegener's ideas became particularly widespread in North America after 1928, when the American Association of Petroleum Geologists held an international symposium to review the hypothesis of continental drift. After each side had presented its arguments, the opponents of continental drift were clearly in the majority, even though the evidence in support of continental drift, most of which came from the Southern Hemisphere, was impressive and difficult to refute. The main problem with the hypothesis was its lack of a mechanism to explain how continents, composed of granitic rocks, could seemingly move through the denser basaltic oceanic crust.

Nevertheless, the eminent South African geologist Alexander du Toit further developed Wegener's arguments and gathered more geologic and paleontologic evidence in support of continental drift. In 1937 du Toit published *Our Wandering Continents*, in which he contrasted the glacial deposits of Gondwana with coal deposits of the same age found in the continents of the Northern Hemisphere. To resolve this apparent

Figure 2.2 **Alfred Wegener** Alfred Wegener, a German meteorologist, proposed the continental drift hypothesis in 1912 based on a tremendous amount of geologic, paleontologic, and climatologic evidence. He is shown here waiting out the Arctic winter in an expedition hut in Greenland.

climatologic paradox, du Toit moved the Gondwana continents to the South Pole and brought the northern continents together such that the coal deposits were located at the equator. He named this northern landmass **Laurasia.** It consisted of present-day North America, Greenland, Europe, and Asia (except for India).

Despite what seemed to be overwhelming evidence, most geologists still refused to accept the idea that the continents moved. Not until the 1960s, when oceanographic research provided convincing evidence that the continents had once been joined together and subsequently separated, did the hypothesis of continental drift finally become widely accepted.

Section 2.2 Summary

- The idea that continents have moved in the past is not new and probably goes back to the first maps, in which one could see that the east coast of South America looks like it fits into the west coast of Africa.

- The continental drift hypothesis was first articulated by Alfred Wegener in 1912. He proposed that a single supercontinent, Pangaea, consisting of a northern landmass (later named Laurasia) and a southern landmass (previously named Gondwana), broke apart into what would become Earth's current continents, which then moved across Earth's surface to their present locations.

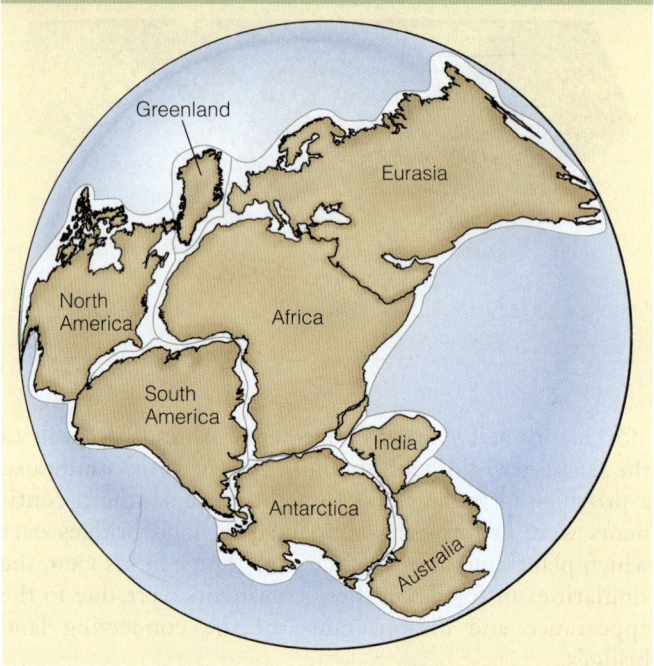

▶ **Figure 2.3 Continental Fit** When continents are placed together based on their outlines, the best fit isn't along their present-day coastlines, but rather along the continental slope at a depth of about 2000 m. **Why is this?** *Because the coastlines are continuously being modified by erosional and depositional processes, and thus one would not expect them to be the same today as they were at any time in the geologic past.*

2.3 Evidence for Continental Drift

■ *What is the evidence for continental drift?*

What, then, was the evidence Wegener, du Toit, and others used to support the hypothesis of continental drift? It includes the fit of the shorelines of continents, the appearance of the same rock sequences and mountain ranges of the same age on continents now widely separated, the matching of glacial deposits and paleoclimatic zones, and the similarities of many extinct plant and animal groups whose fossil remains are found today on widely separated continents. Wegener and his supporters argued that this vast amount of evidence from a variety of sources surely indicated that the continents must have been close together in the past.

Continental Fit

Wegener, like some before him, was impressed by the close resemblance between the coastlines of continents on opposite sides of the Atlantic Ocean, particularly South America and Africa. He cited these similarities as partial evidence that the continents were at one time joined together as a supercontinent that subsequently split apart. As his critics pointed out, though, the configuration of coastlines results from erosional and depositional processes and therefore is continuously being modified. So, even if the continents had separated during the Mesozoic Era, as Wegener proposed, it is not likely that the coastlines would fit exactly.

A more realistic approach is to fit the continents together along the continental slope, where erosion would be minimal. In 1965 Sir Edward Bullard, an English geophysicist, and two associates showed that the best fit between the continents occurs at a depth of about 2000 m (▶ Figure 2.3). Since then, other reconstructions using the latest ocean basin data have confirmed the close fit between continents when they are reassembled to form Pangaea.

Similarity of Rock Sequences and Mountain Ranges

If the continents were at one time joined, then the rocks and mountain ranges of the same age in adjoining locations on the opposite continents should closely match. Such is the case for the Gondwana continents (▶ Figure 2.4). Marine, nonmarine, and glacial rock sequences of Pennsylvanian to Jurassic age are almost identical on all five Gondwana continents, strongly indicating that they were joined at one time.

The trends of several major mountain ranges also support the hypothesis of continental drift. These mountain ranges seemingly end at the coastline of one continent only to apparently continue on another continent across the ocean. The folded Appalachian Mountains of North America, for example, trend northeastward through the eastern United

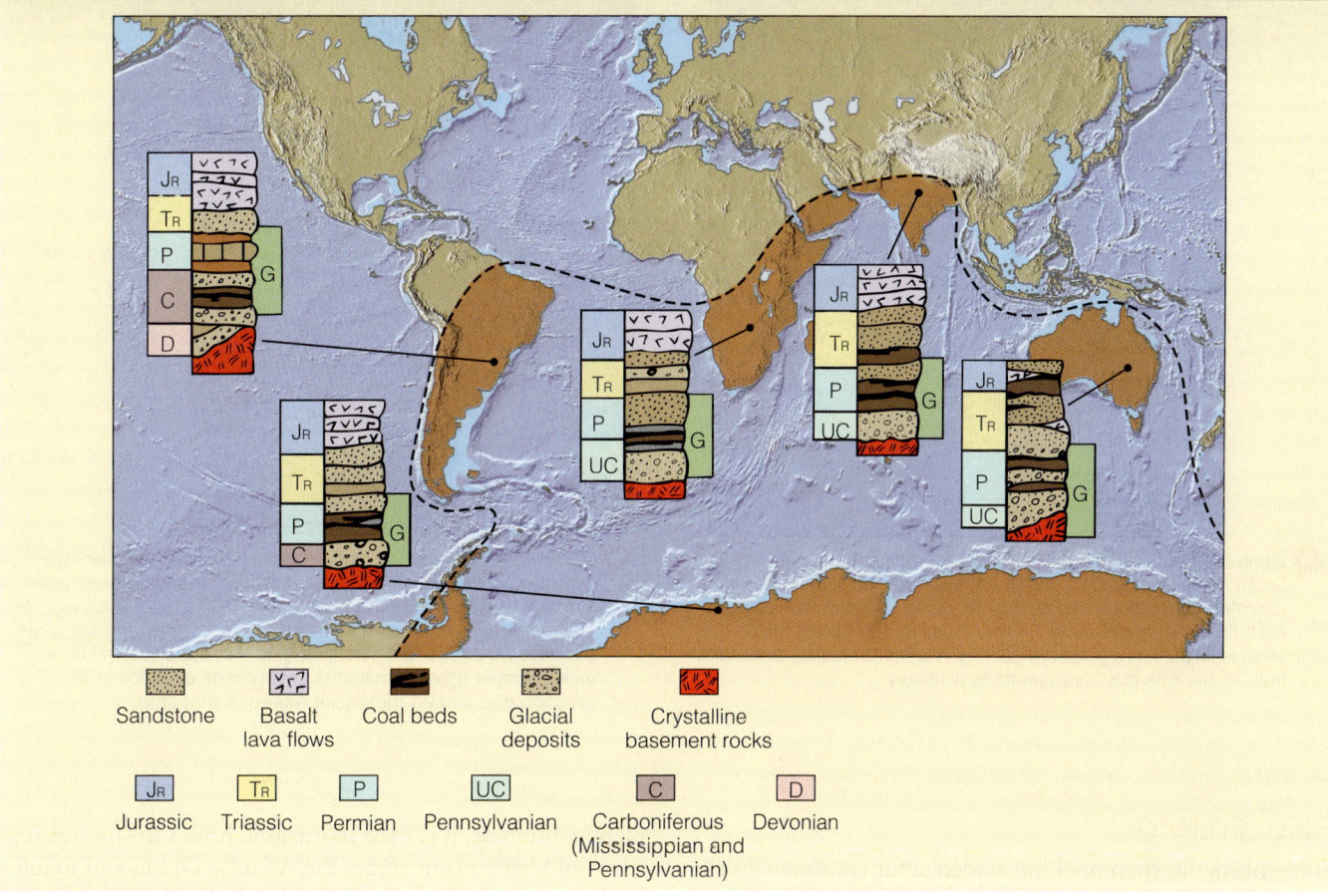

Figure 2.4 Similarity of Rock Sequences on the Gondwana Continents Sequences of marine, nonmarine, and glacial rocks of Pennsylvanian (UC) to Jurassic (JR) age are nearly the same on all five Gondwana continents (South America, Africa, India, Australia, and Antarctica). These continents are widely separated today and have different environments and climates ranging from tropical to polar. Thus the rocks forming on each continent are very different. When the continents were all joined together in the past, however, the environments of adjacent continents were similar and the rocks forming in those areas were similar. The range indicated by G in each column is the age range (Carboniferous–Permian) of the *Glossopteris* flora.

States and Canada and terminate abruptly at the Newfoundland coastline. Mountain ranges of the same age and deformational style are found in eastern Greenland, Ireland, Great Britain, and Norway. In fact, the same red sandstones used in the construction of many English and Scottish castles are used in various buildings throughout New York. So, even though the Appalachian Mountains and their equivalent-age mountain ranges in Great Britain are currently separated by the Atlantic Ocean, they form an essentially continuous mountain range when the continents are positioned next to each other as they were during the Paleozoic Era.

Glacial Evidence

During the Late Paleozoic Era, massive glaciers covered large continental areas of the Southern Hemisphere. Evidence for this glaciation includes layers of till (sediments deposited by glaciers) and striations (scratch marks) in the bedrock beneath the till. Fossils and sedimentary rocks of the same age from the Northern Hemisphere, however, give no indication of glaciation. Fossil plants found in coals indicate that the Northern Hemisphere had a tropical climate during the time the Southern Hemisphere was glaciated.

All the Gondwana continents except Antarctica are currently located near the equator in subtropical to tropical climates. Mapping of glacial striations in bedrock in Australia, India, and South America indicates that the glaciers moved from the areas of the present-day oceans onto land. This would be highly unlikely because large continental glaciers (such as occurred on the Gondwana continents during the Late Paleozoic Era) flow outward from their central area of accumulation toward the sea.

If the continents did not move during the past, one would have to explain how glaciers moved from the oceans onto land and how large-scale continental glaciers formed near the equator. But if the continents are reassembled as a single landmass with South Africa located at the South Pole, the direction of movement of Late Paleozoic continental glaciers makes sense (▶ Figure 2.5). Furthermore, this geographic arrangement places the northern continents nearer the tropics, which is consistent with the fossil and climatologic evidence from Laurasia.

Geo-focus Figure 2.5 Glacial Evidence Indicating Continental Drift

a When the Gondwana continents are placed together so that South Africa is located at the South Pole, the glacial movements indicated by striations (red arrows) found on rock outcrops on each continent make sense. In this situation, the glacier (white area) is located in a polar climate and has moved radially outward from its thick central area toward its periphery.

b Glacial striations (scratch marks) on an outcrop of Permian-age bedrock exposed at Hallet's Cove, Australia, indicate the general direction of glacial movement more than 200 million years ago. As a glacier moves over a continent's surface, it grinds and scratches the underlying rock. The scratch marks that are preserved on a rock's surface (glacial striations) thus provide evidence of the direction (red arrows) the glacier moved at that time.

Fossil Evidence

Some of the most compelling evidence for continental drift comes from the fossil record. Fossils of the *Glossopteris* flora are found in equivalent Pennsylvanian- and Permian-aged coal deposits on all five Gondwana continents. The **Glossopteris flora** is characterized by the seed fern *Glossopteris* (▶ Figure 2.1) as well as by many other distinctive and easily identifiable plants. Pollen and spores of plants can be dispersed over great distances by wind, but *Glossopteris*-type plants produced seeds that are too large to have been carried by winds. Even if the seeds had floated across the ocean, they probably would not have remained viable for any length of time in saltwater.

The present-day climates of South America, Africa, India, Australia, and Antarctica range from tropical to polar and are much too diverse to support the type of plants in the *Glossopteris* flora. Wegener therefore reasoned that these continents must once have been joined so that these widely separated localities were all in the same latitudinal climatic belt (▶ Figure 2.6).

The fossil remains of animals also provide strong evidence for continental drift. One of the best examples is *Mesosaurus*, a freshwater reptile whose fossils are found in Permian-aged rocks in certain regions of Brazil and South Africa and nowhere else in the world (▶ Figure 2.6). Because the physiologies of freshwater and marine animals are completely different, it is hard to imagine how a freshwater reptile could have swum across the Atlantic Ocean and found a freshwater environment nearly identical to its former habitat. Moreover, if *Mesosaurus* could have swum across the ocean, its fossil remains should be widely dispersed. It is more logical to assume that *Mesosaurus* lived in lakes in what are now adjacent areas of South America and Africa but were then united into a single continent.

Lystrosaurus and *Cynognathus* are both land-dwelling reptiles that lived during the Triassic Period; their fossils are found only on the present-day continental fragments of Gondwana (▶ Figure 2.6). Because they are both land animals, they certainly could not have swum across the oceans currently separating the Gondwana continents. Therefore, it is logical to assume that the continents must once have been connected. Recent discoveries of dinosaur fossils in Gondwana continents further solidifies the argument that these landmasses were close to each other during the Early Mesozoic Era.

Notwithstanding all of the empirical evidence presented by Wegener and later by du Toit and others, most geologists simply refused to entertain the idea that continents might have moved during the past. The geologists were not necessarily being obstinate about accepting new ideas; rather, they found the evidence for continental drift inadequate and unconvincing. In part, this was because no one could provide a

▶ **Figure 2.6 Fossil Evidence Supporting Continental Drift** Some of the plants and animals whose fossils are found today on the widely separated continents of South America, Africa, India, Australia, and Antarctica. During the Late Paleozoic Era, these continents were joined together to form Gondwana, the southern landmass of Pangaea. Plants of the *Glossopteris* flora are found on all five continents, which today have widely different climates, but during the Pennsylvanian and Permian periods, they were all located in the same general climatic belt. *Mesosaurus* is a freshwater reptile whose fossils are found only in similar nonmarine Permian-age rocks in Brazil and South Africa. *Cynognathus* and *Lystrosaurus* are land reptiles that lived during the Early Triassic Period. Fossils of *Cynognathus* are found in South America and Africa, whereas fossils of *Lystrosaurus* have been recovered from Africa, India, and Antarctica. It is hard to imagine how a freshwater reptile and land-dwelling reptiles could have swum across the wide oceans that presently separate these continents. It is more logical to assume that the continents were at one time connected.

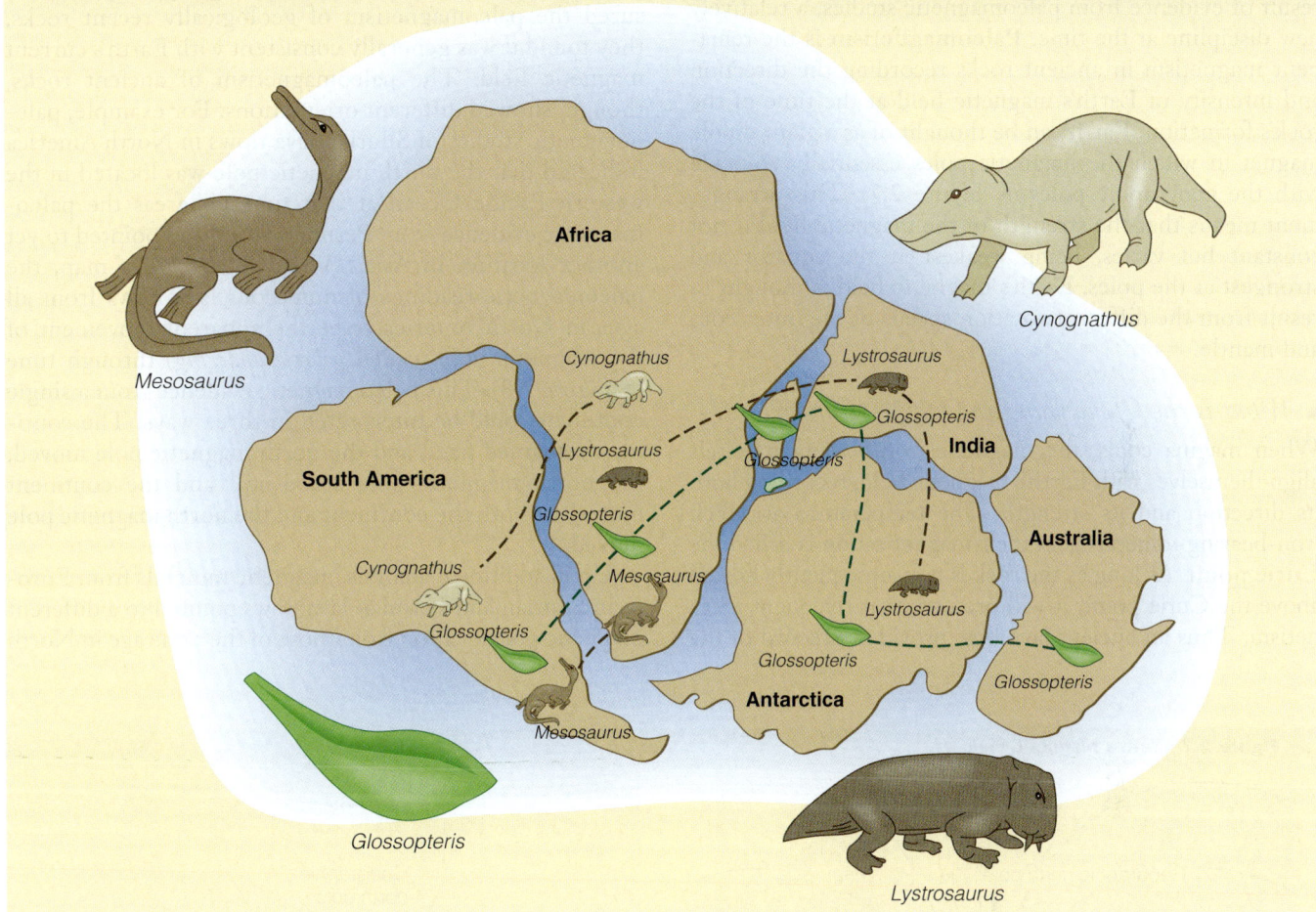

suitable mechanism to explain how continents could move over Earth's surface. Interest in continental drift waned until new evidence from oceanographic research and studies of Earth's magnetic field showed that the present-day ocean basins were not as old as the continents but were geologically young features that resulted from the breakup of Pangaea.

Section 2.3 Summary

- The evidence for continental drift is impressive. It includes the fact that the continents show a close fit along the continental slope at a depth of about 2000 m.

- Furthermore, several major mountain ranges that currently end at the coastline form a continuous range when the present-day continents are assembled into a single landmass.

- Marine, nonmarine, and glacial rock sequences of Pennsylvanian to Jurassic age are nearly identical on the five Gondwana continents, suggesting that these continents were joined together during this time interval.

- Glacial deposits and striations indicate that massive glaciers covered large areas of the Gondwana continents during the Late Paleozoic Era. Placing these continents together with South Africa located at the South Pole shows that the glaciers moved in a radial pattern from a thick central area toward their periphery, as would be expected in such a configuration.

- The distribution of plant and animal fossils also provides strong evidence for continental drift because it is hard to imagine land animals being able to swim across the Atlantic Ocean and the same plants occupying tropical, moderate, and polar environments.

2.4 Paleomagnetism and Polar Wandering

■ *What is paleomagnetism?*

Interest in continental drift revived during the 1950s as a result of evidence from paleomagnetic studies, a relatively new discipline at the time. **Paleomagnetism** is the remanent magnetism in ancient rocks recording the direction and intensity of Earth's magnetic field at the time of the rock's formation. Earth can be thought of as a giant dipole magnet in which the magnetic poles essentially coincide with the geographic poles (▶ Figure 2.7). This arrangement means that the strength of the magnetic field is not constant but varies, being weakest at the equator and strongest at the poles. Earth's magnetic field is thought to result from the different rotation speeds of the outer core and mantle.

■ *What is the Curie point and why is it important?*

When magma cools, the magnetic iron-bearing minerals align themselves with Earth's magnetic field, recording both its direction and its strength. The temperature at which iron-bearing minerals gain their magnetization is called the **Curie point.** As long as the rock is not subsequently heated above the Curie point, it will preserve that remanent magnetism. Thus an ancient lava flow provides a record of the orientation and strength of Earth's magnetic field at the time the lava flow cooled.

■ *How can the apparent wandering of the magnetic poles be best explained?*

As paleomagnetic research progressed during the 1950s, some unexpected results emerged. When geologists measured the paleomagnetism of geologically recent rocks, they found it was generally consistent with Earth's current magnetic field. The paleomagnetism of ancient rocks, though, showed different orientations. For example, paleomagnetic studies of Silurian lava flows in North America indicated that the north magnetic pole was located in the western Pacific Ocean at that time, whereas the paleomagnetic evidence from Permian lava flows pointed to yet another location in Asia. When plotted on a map, the paleomagnetic readings of numerous lava flows from all ages in North America trace the apparent movement of the magnetic pole (called *polar wandering*) through time (▶ Figure 2.8). This paleomagnetic evidence from a single continent could be interpreted in three ways: The continent remained fixed and the north magnetic pole moved; the north magnetic pole stood still and the continent moved; or both the continent and the north magnetic pole moved.

Upon additional analysis, magnetic minerals from European Silurian and Permian lava flows pointed to a different magnetic pole location from those of the same age in North

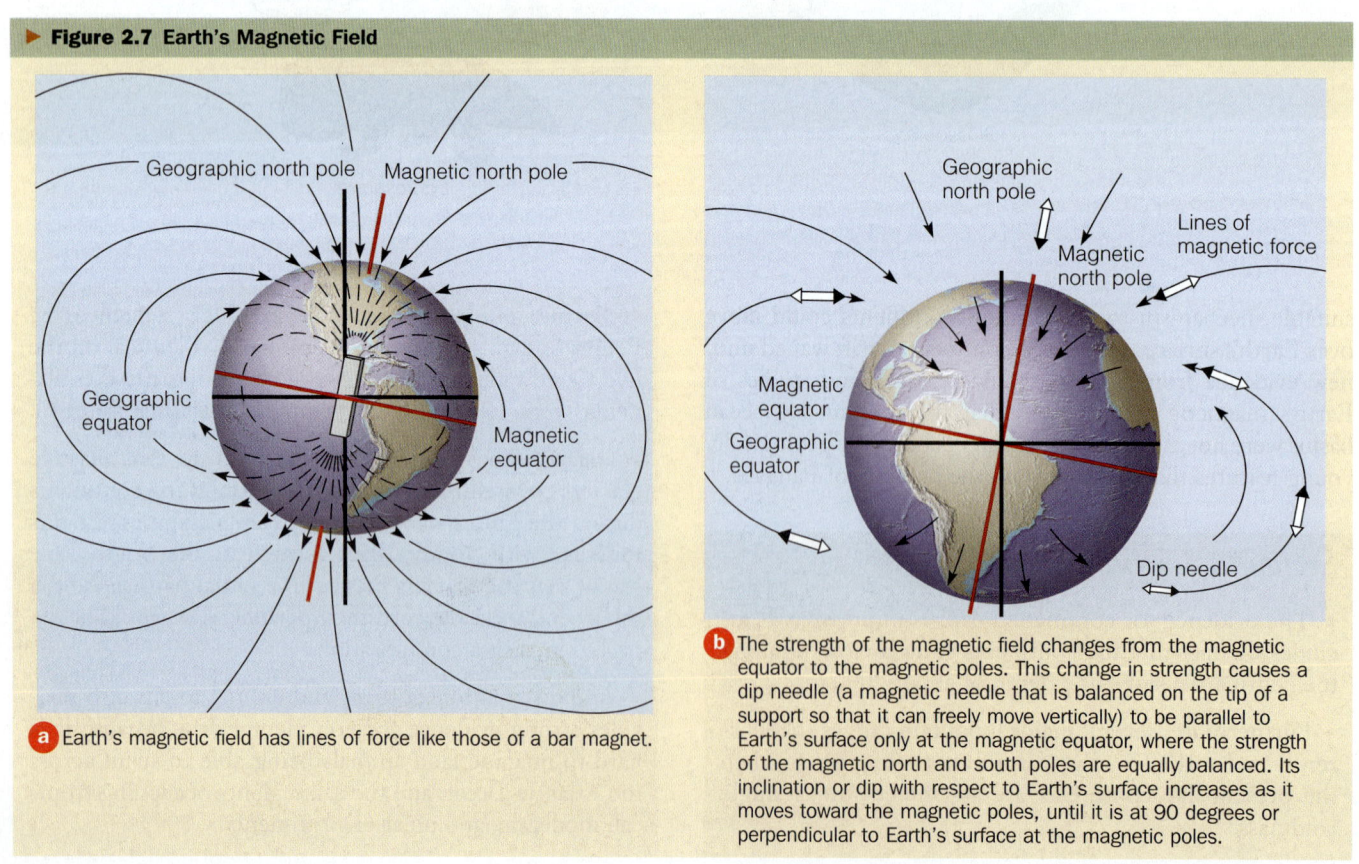

▶ **Figure 2.7 Earth's Magnetic Field**

a Earth's magnetic field has lines of force like those of a bar magnet.

b The strength of the magnetic field changes from the magnetic equator to the magnetic poles. This change in strength causes a dip needle (a magnetic needle that is balanced on the tip of a support so that it can freely move vertically) to be parallel to Earth's surface only at the magnetic equator, where the strength of the magnetic north and south poles are equally balanced. Its inclination or dip with respect to Earth's surface increases as it moves toward the magnetic poles, until it is at 90 degrees or perpendicular to Earth's surface at the magnetic poles.

▶ **Figure 2.8 Polar Wandering** The apparent paths of polar wandering for North America and Europe. The apparent location of the north magnetic pole is shown for different time periods on each continent's polar wandering path. If the continents have not moved through time, and because Earth has only one magnetic pole, the paleomagnetic readings for the same time in the past taken on different continents should all point to the same location. However, the north magnetic pole has different locations for the same time in the past when measured on different continents, indicating multiple north magnetic poles. The logical explanation for this dilemma is that the magnetic north pole has remained at the same approximate geographic location during the past, and the continents have moved.

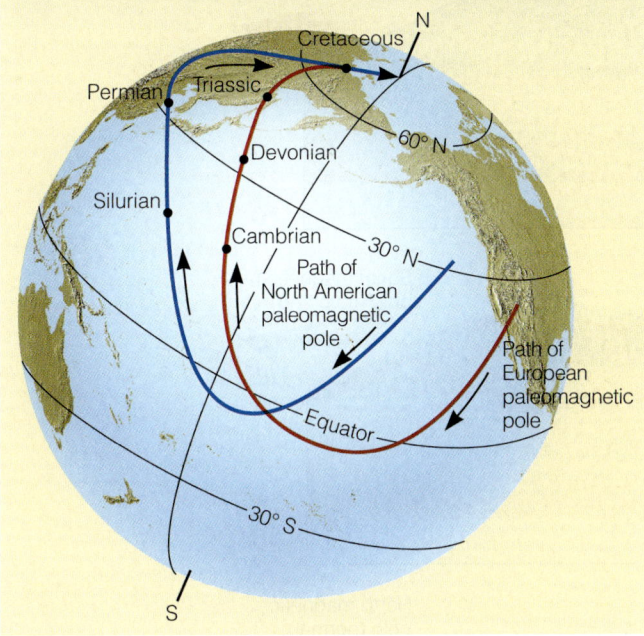

America (▶ Figure 2.8). Furthermore, analysis of lava flows from all continents indicated that each continent seemingly had its own series of magnetic poles. Does this really mean there were different north magnetic poles for each continent? That would be highly unlikely and difficult to reconcile with the theory accounting for Earth's magnetic field.

The best explanation for such data is that the magnetic poles have remained near their present locations at the geographic north and south poles and the continents have moved. When the continental margins are fitted together so that the paleomagnetic data point to only one magnetic pole, we find, just as Wegener did, that the rock sequences and glacial deposits match and that the fossil evidence is consistent with the reconstructed paleogeography.

Section 2.4 Summary

- Paleomagnetism is the remanent magnetism in ancient rocks recording the direction and intensity of Earth's magnetic field at the time of the rock's formation.

- Earth's magnetic field is not constant. Being strongest at the poles and weakest at the equator, Earth's magnetic field is thought to result from the different rotation speeds of the outer core and mantle. Earth's magnetic poles closely coincide with its geographic poles.

- The Curie point is the temperature at which iron-bearing minerals gain their magnetism and align themselves with Earth's magnetic field.

- Polar wandering is the apparent movement of the magnetic poles through time. The best explanation for such apparent movement is that the magnetic poles have remained near their present polar locations and the continents have moved.

2.5 Magnetic Reversals and Seafloor Spreading

■ *What evidence is there that Earth's magnetic field has reversed in the past?*

Geologists refer to Earth's present magnetic field as being normal—that is, with the north and south magnetic poles located approximately at the north and south geographic poles. At various times in the geologic past, however, Earth's magnetic field has completely reversed. The existence of such **magnetic reversals** was discovered by dating and determining the orientation of the remanent magnetism in lava flows on land (▶ Figure 2.9).

Once magnetic reversals were well established for continental lava flows, magnetic reversals were also discovered in igneous rocks in the oceanic crust as part of the large scale mapping of the ocean basins during the 1960s. Although the cause of magnetic reversals is still uncertain, their occurrence in the geologic record is well documented.

■ *What is the theory of seafloor spreading, and how does it validate continental drift?*

A renewed interest in oceanographic research led to extensive mapping of the ocean basins during the 1960s. Such mapping revealed an oceanic ridge system more than 65,000 km long, constituting the most extensive mountain range in the world. Perhaps the best-known part of the ridge system is the Mid-Atlantic Ridge, which divides the Atlantic Ocean basin into two nearly equal parts (▶ Figure 2.10).

As a result of the oceanographic research conducted during the 1950s, Harry Hess of Princeton University proposed the theory of **seafloor spreading** in 1962 to account for continental movement. He suggested that continents do not move across oceanic crust, but rather the continents and oceanic crust move together. Thus the theory of seafloor spreading answered a major objection of the opponents of continental drift—namely, how could continents move through oceanic crust? In fact, the continents moved with the oceanic crust as part of a lithospheric system.

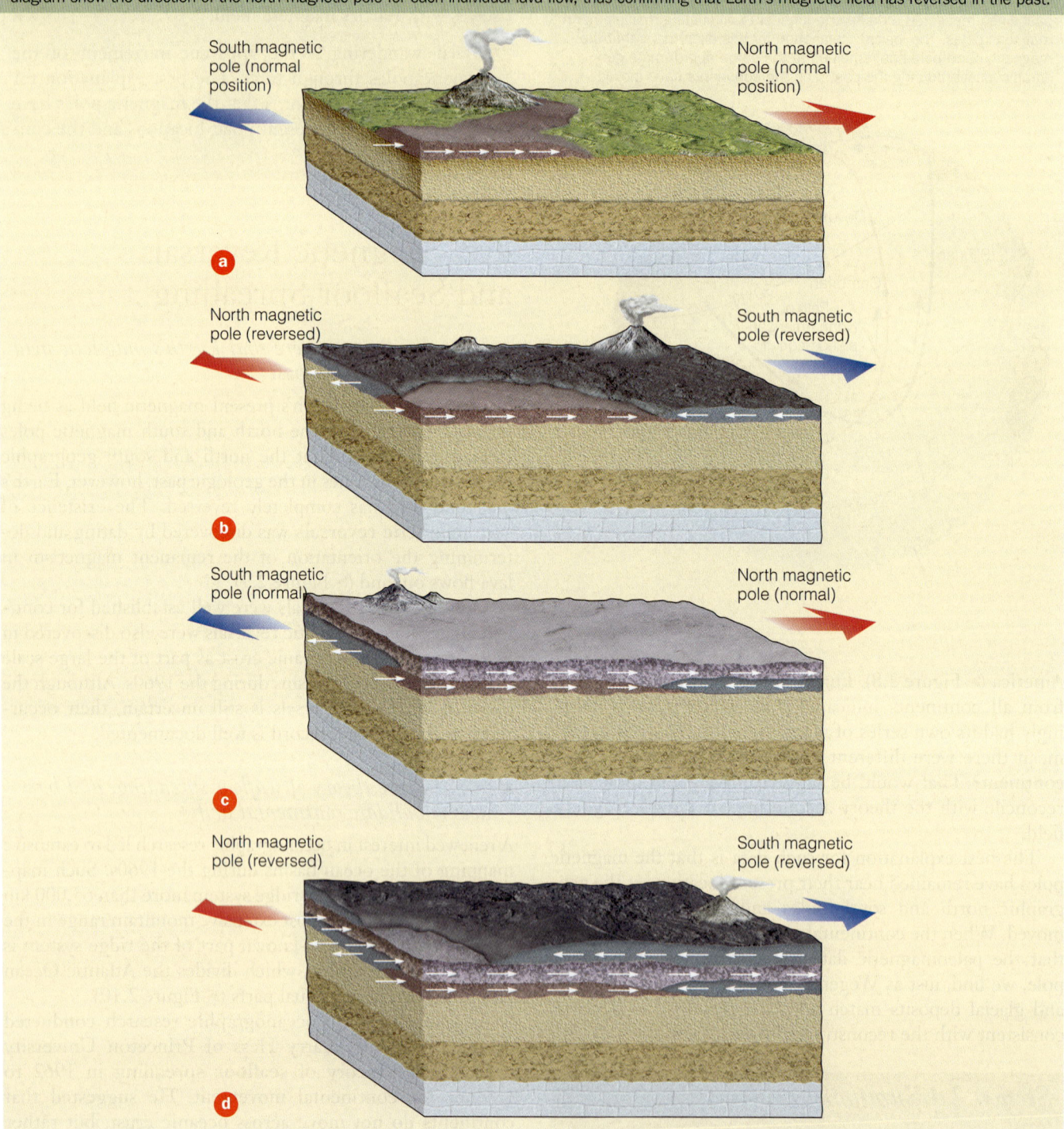

Geo-focus Figure 2.9 Magnetic Reversals During the time period shown **(a–d)**, volcanic eruptions produced a succession of overlapping lava flows. At the time of these volcanic eruptions, Earth's magnetic field completely reversed; that is, the magnetic north pole moved to the geographic south pole, and the magnetic south pole moved to the geographic north pole. Thus the end of the needle on a magnetic compass that today would point to the North Pole would point to the South Pole if the magnetic field should again suddenly reverse. We know that Earth's magnetic field has reversed numerous times in the past because when lava flows cool below the Curie point, magnetic minerals within the flow orient themselves parallel to the magnetic field at the time. They thus record whether the magnetic field was normal or reversed at that time. The white arrows in this diagram show the direction of the north magnetic pole for each individual lava flow, thus confirming that Earth's magnetic field has reversed in the past.

Hess postulated that the seafloor separates at oceanic ridges, where new crust is formed by upwelling magma. As the magma cools, the newly formed oceanic crust moves laterally away from the ridge.

As a mechanism to drive this system, Hess revived the idea (proposed in the 1930s and 1940s by Arthur Holmes and others) of **thermal convection cells** in the mantle; that is, hot magma rises from the mantle, intrudes along frac-

▶ **Figure 2.10 Topography of the Atlantic Ocean Basin** Artistic view of what the Atlantic Ocean basin would look like without water. The major feature is the Mid-Atlantic Ridge, an oceanic ridge system that is longer than 65,000 km and divides the Atlantic Ocean basin in half. It is along such oceanic ridges that the seafloor is separating and new oceanic crust is forming from upwelling magma in Earth's interior.

tures defining oceanic ridges, and thus forms new crust. Cold crust is subducted back into the mantle at oceanic trenches, where it is heated and recycled, thus completing a thermal convection cell (see Figure 1.10).

Paleomagnetic Data

■ *How was the theory of seafloor spreading confirmed?*

Magnetic surveys of the oceanic crust revealed striped **magnetic anomalies** (deviations from the average strength of Earth's magnetic field) in the rocks that are both parallel to and symmetric around the oceanic ridges (▶ Figure 2.11). Furthermore, the pattern of oceanic magnetic anomalies matches the pattern of magnetic reversals already known from studies of continental lava flows (▶ Figure 2.9). When magma wells up and cools along a ridge summit, it records Earth's magnetic field at that time as either normal or reversed. As new crust forms at the summit, the previously formed crust moves laterally away from the ridge. These magnetic stripes represent times of normal and reversed polarity at oceanic ridges (where upwelling magma forms new oceanic crust), conclusively confirming Hess's theory of seafloor spreading.

The seafloor spreading theory also confirms that ocean basins are geologically young features whose openings and closings are partially responsible for continental movement (▶ Figure 2.12). Radiometric dating reveals that the oldest oceanic crust is somewhat less than 180 million years old, whereas the oldest continental crust is 3.96 billion years old. Although geologists do not universally accept the idea of thermal convection cells as a driving mechanism for plate movement, most accept that plates are created at oceanic ridges and destroyed at deep-sea trenches, regardless of the driving mechanism involved.

Deep-Sea Drilling Project Results

For many geologists, the paleomagnetic data amassed in support of continental drift and seafloor spreading were convincing. Results from the Deep-Sea Drilling Project (see Chapter 12) confirmed the interpretations made from earlier paleomagnetic studies. Cores of deep-sea sediments and seismic profiles obtained by the *Glomar Challenger* and other research vessels have provided much of the data that support the seafloor spreading theory.

According to this theory, oceanic crust is continuously forming at mid-oceanic ridges, moves away from these

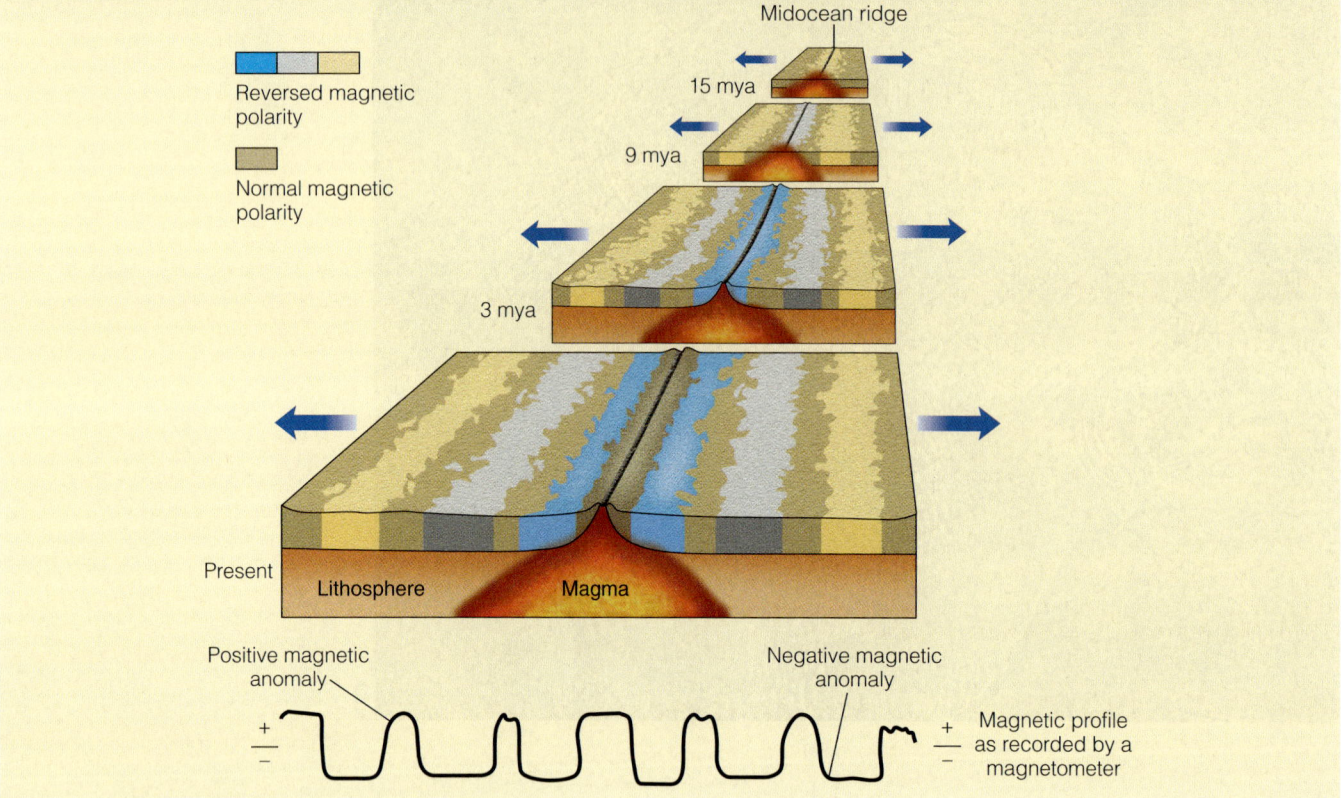

PHYSICAL Geology Now™ ▶ Active Figure 2.11 Magnetic Anomalies and Seafloor Spreading The sequence of magnetic anomalies preserved within the oceanic crust is both parallel to and symmetric around oceanic ridges. Basaltic lava intruding into an oceanic ridge today and spreading laterally away from the ridge records Earth's current magnetic field or polarity (considered by convention to be normal). Basaltic intrusions 3, 9, and 15 million years ago record Earth's reversed magnetic field at that time. This schematic diagram shows how the solidified basalt moves away from the oceanic ridge (or spreading center), carrying with it the magnetic anomalies that are preserved in the oceanic crust. Magnetic anomalies are magnetic readings that are either higher (positive magnetic anomalies) or lower (negative magnetic anomalies) than Earth's current magnetic field strength. The magnetic anomalies are recorded by a magnetometer, which measures the strength of the magnetic field. Modified from Kious and Tilling, USGS and Hyndman & Hyndman *Natural Hazards and Disasters*, Brooks/Cole, 2006, p. 15, Fig. 2.6b.

ridges by seafloor spreading, and is consumed at subduction zones. If this is the case, then oceanic crust should be youngest at the ridges and become progressively older with increasing distance away from them. Moreover, the age of the oceanic crust should be symmetrically distributed about the ridges. As we have just noted, paleomagnetic data confirm these statements. Furthermore, fossils from sediments overlying the oceanic crust and radiometric dating of rocks found on oceanic islands both substantiate this predicted age distribution.

Sediments in the open ocean accumulate, on average, at a rate of less than 0.3 cm in 1000 years. If the ocean basins were as old as the continents, we would expect deep-sea sediments to be several kilometers thick. However, data from numerous drill holes indicate that deep-sea sediments are at most only a few hundred meters thick and are thin or absent at oceanic ridges. Their near-absence at the ridges should come as no surprise because these are the areas where new crust is continuously produced by volcanism and seafloor spreading. Accordingly, sediments have had little time to accumulate at or very close to spreading ridges where the oceanic crust is young, but their thickness increases with distance away from the ridges (▶ Figure 2.13).

Section 2.5 Summary

• According to the theory of seafloor spreading, seafloor separates at oceanic ridges where new crust is formed by upwelling magma generated by thermal convection cells within the mantle. As the magma cools, the newly formed oceanic crust moves laterally away from the ridge.

• Earth's magnetic field has periodically reversed during the past.

▶ **Figure 2.12 Age of the World's Ocean Basins** The age of the world's ocean basins have been determined from magnetic anomalies preserved in oceanic crust. The red colors adjacent to the oceanic ridges are the youngest oceanic crust. Moving laterally away from the ridges, the red colors grade to yellow at 48 million years, to green at 68 million years ago, and to dark blue some 155 million years ago. The darkest blue color is adjacent to the continental margins and is just somewhat less than 180 million years old. **How does the age of the oceanic crust confirm the seafloor spreading theory?** *Based on magnetic anomalies, the age of the oceanic crust gets progressively older away from the oceanic ridges where it is being formed. This means it is moving away from the oceanic ridges; that is, the seafloor is spreading.*

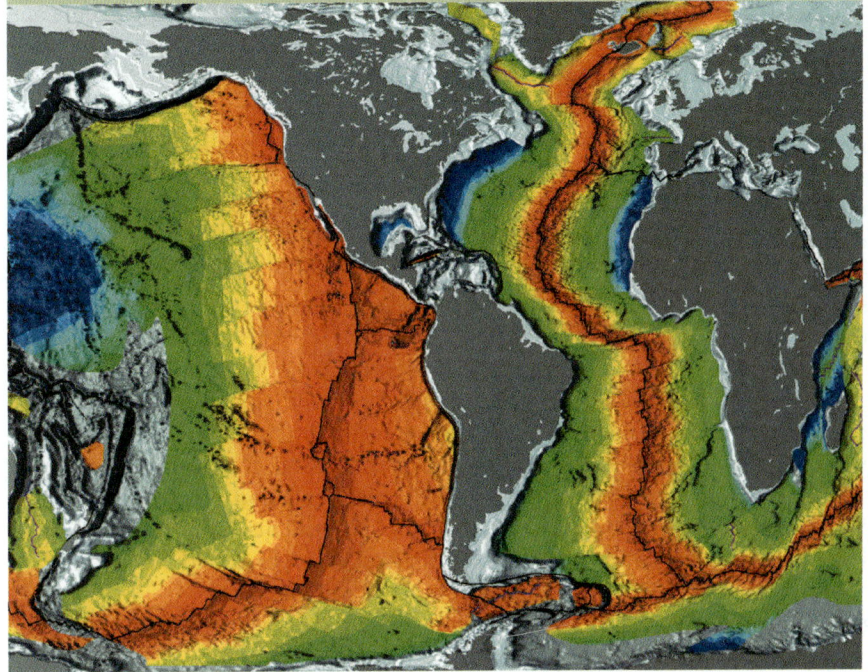

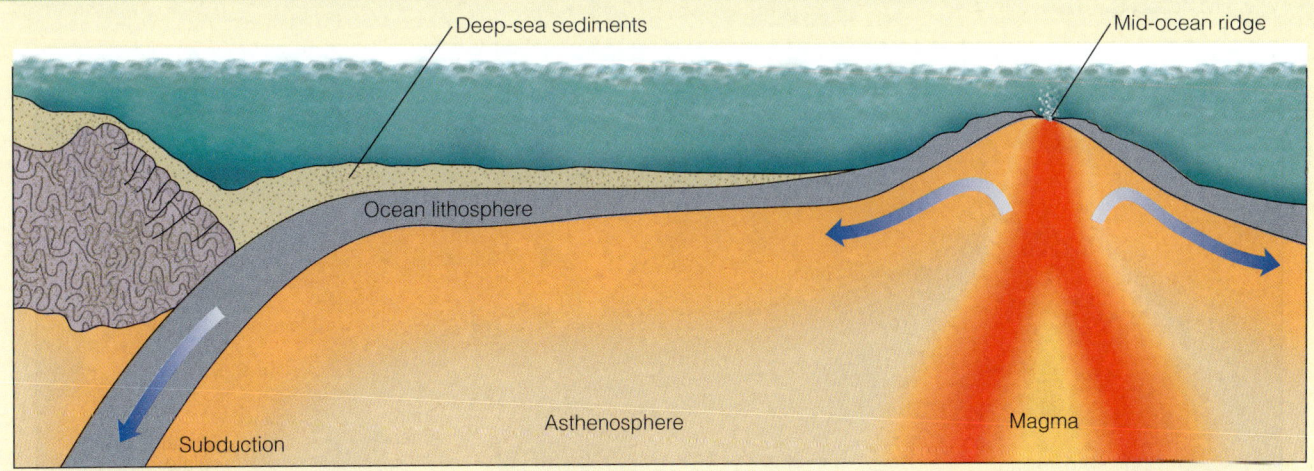

▶ **Figure 2.13 Deep-Sea Sediments and Seafloor Spreading** The total thickness of deep-sea sediments increases away from oceanic ridges. This is because oceanic crust becomes older away from oceanic ridges, and there has been more time for sediment to accumulate.

- Magnetic anomalies in oceanic crust, which are parallel to and symmetric around oceanic ridges, match the pattern of magnetic reversals seen in continental lava flows, thus confirming that new oceanic crust is forming along oceanic ridges and moving the seafloor laterally away from them.

- Cores of deep-sea sediments also confirm the theory of seafloor spreading in that the sediments directly overlying the oceanic crust get older with increasing distance from oceanic ridges, and the sediments become thicker moving away from the ridge.

2.6 Plate Tectonics: A Unifying Theory

■ *What are the main tenets of plate tectonic theory?*

Plate tectonic theory is based on a simple model of Earth. The rigid lithosphere, composed of both oceanic and continental crust as well as the underlying upper mantle, consists of many variable-sized pieces called **plates** (▶ Figure 2.14). The plates vary in thickness; those composed of upper mantle and continental crust are as much as 250 km thick, whereas those of upper mantle and oceanic crust are up to 100 km thick.

The lithosphere overlies the hotter and weaker semiplastic asthenosphere. It is thought that movement resulting from some type of heat-transfer system within the asthenosphere causes the overlying plates to move. As plates move over the asthenosphere, they separate, mostly at oceanic ridges; in other areas, such as at oceanic trenches, they collide and are subducted back into the mantle.

An easy way to visualize plate movement is to think of a conveyor belt moving luggage from an airplane's cargo hold to a baggage cart. The conveyor belt represents convection currents within the mantle, and the luggage represents Earth's lithospheric plates. The luggage is moved along by the conveyor belt until it is dumped into the baggage cart in the same way plates are moved by convection cells until they are subducted into Earth's interior. Although this analogy allows you to visualize how the mechanism of plate movement takes place, remember that this analogy is limited. The major limitation is that, unlike the luggage, plates consist of continental and oceanic crust, which have different densities; only oceanic crust, because it is denser than continental crust, is subducted into Earth's interior.

■ *Why is plate tectonics a unifying theory of geology?*

Most geologists accept plate tectonic theory, in part because the evidence for it is overwhelming and it ties together many seemingly unrelated geologic features and events and shows how they are interrelated. Consequently, geologists now view such geologic processes as mountain building, earthquake activity, and volcanism from the perspective of plate tectonics. Furthermore, because all the inner planets have had a similar origin and early history, geologists are interested in determining whether plate tectonics is unique to Earth or whether it operates in the same way on other planets (see "Tectonics of the Terrestrial Planets" on pp. 50–51).

■ *What is the supercontinent cycle?*

As a result of plate movement, all of the continents came together to form the supercontinent Pangaea by the end of the Paleozoic Era. Pangaea began fragmenting during the Triassic Period and continues to do so, thus accounting for the present distribution of continents and ocean basins. It has been proposed that supercontinents, consisting of all or most of Earth's landmasses, form, break up, and re-form in a cycle spanning about 500 million years.

The *supercontinent cycle hypothesis* is an expansion on the ideas of the Canadian geologist J. Tuzo Wilson. During the early 1970s, Wilson proposed a cycle (now known as the Wilson cycle) that includes continental fragmentation, the

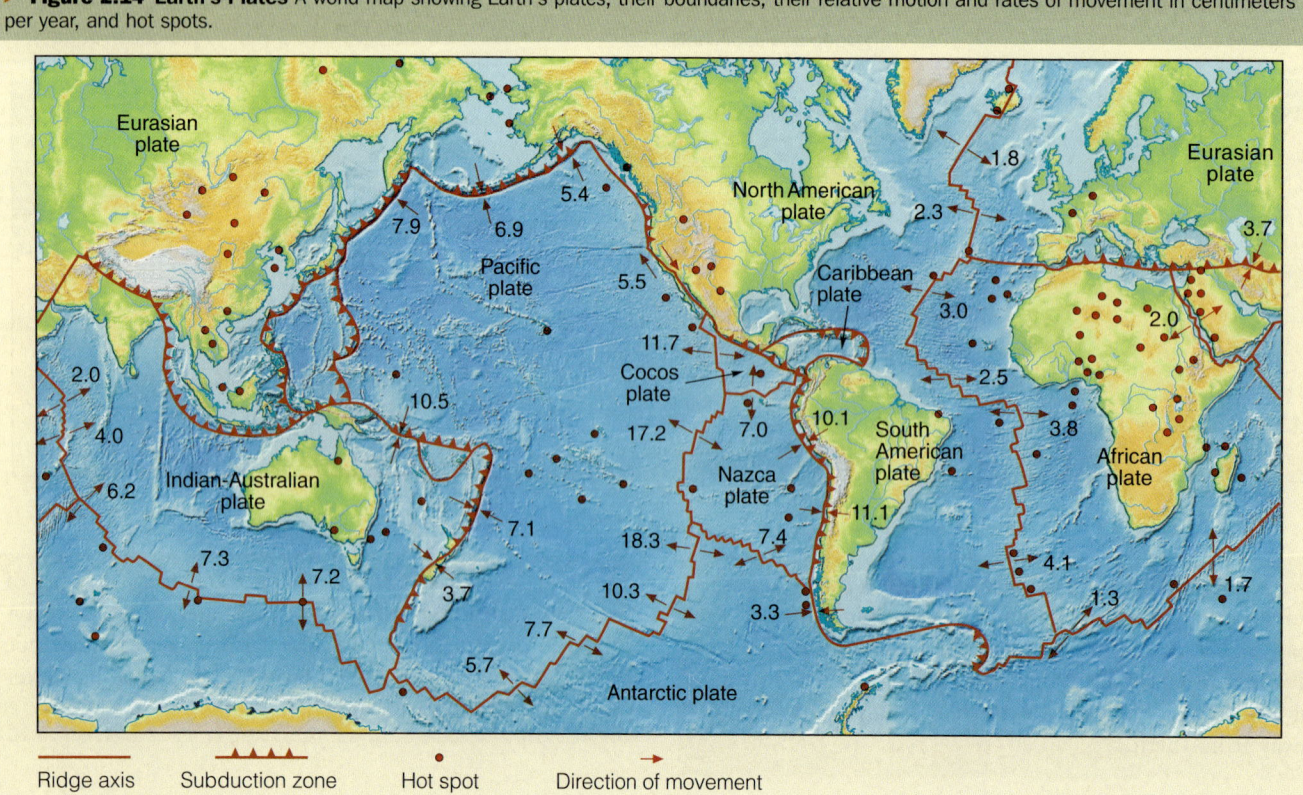

▶ **Figure 2.14 Earth's Plates** A world map showing Earth's plates, their boundaries, their relative motion and rates of movement in centimeters per year, and hot spots.

opening and closing of an ocean basin, and re-assembly of the continent. According to the supercontinent cycle hypothesis, heat accumulates beneath a supercontinent because the rocks of continents are poor conductors of heat. As a result of the heat accumulation, the supercontinent domes upward and fractures. Basaltic magma rising from below fills the fractures. As a basalt-filled fracture widens, it begins subsiding and forms a long, narrow ocean such as the present-day Red Sea. Continued rifting eventually forms an expansive ocean basin such as the Atlantic.

One of the most convincing arguments for the proponents of the supercontinent cycle hypothesis is the "surprising regularity" of mountain building caused by compression during continental collisions. These mountain-building episodes occur about every 400 to 500 million years and are followed by an episode of rifting about 100 million years later. In other words, a supercontinent fragments and its individual plates disperse following a rifting episode, an interior ocean forms, and then the dispersed fragments re-assemble to form another supercontinent.

The supercontinent cycle is yet another example of how interrelated the various systems and subsystems of Earth are and how they operate over vast periods of geologic time.

Section 2.6 Summary

- The main tenets of plate tectonic theory are that the rigid lithosphere consists of numerous variable-sized pieces called plates. These plates move over the hotter and weaker semiplastic asthenosphere as a result of some type of heat-transfer system within the asthenosphere. Plates separate, mostly at oceanic ridges, and collide, usually at oceanic trenches, where they are subducted back into the mantle.

- Plate tectonics is considered a unifying theory of geology because it explains how many geologic features, processes, and events are interrelated.

- In the supercontinent cycle, all or most of Earth's landmasses come together to form a supercontinent, such as Pangaea, then break up, producing ocean basins, and then re-form in a cycle spanning about 500 million years.

2.7 The Three Types of Plate Boundaries

■ *What are the three types of plate boundaries?*

Because it appears that plate tectonics have operated since at least the Proterozoic Eon, it is important that we understand how plates move and interact with each other and how ancient plate boundaries are recognized. After all, the movement of plates has profoundly affected the geologic and biologic history of this planet.

Geologists recognize three major types of plate boundaries: *divergent*, *convergent*, and *transform* (Table 2.1). Along these boundaries new plates are formed, are consumed, or slide laterally past each other. Interaction of plates at their boundaries accounts for most of Earth's volcanic eruptions and earthquakes as well as the formation and evolution of its mountain systems.

■ *What are divergent boundaries?*

Divergent plate boundaries or *spreading ridges* occur where plates are separating and new oceanic lithosphere is forming. Divergent boundaries are places where the crust is extended, thinned, and fractured as magma, derived from the partial melting of the mantle, rises to the surface. The magma is almost entirely basaltic and intrudes into vertical fractures to form dikes and pillow lava flows (see Figure 5.7). As successive injections of magma cool and solidify, they form new oceanic crust and record the intensity and orientation of Earth's magnetic field (▶ Figure 2.11). Divergent boundaries most commonly occur along the crests of oceanic ridges—the Mid-Atlantic Ridge, for example. Oceanic ridges are thus characterized by rugged topography with high relief resulting from the displacement of rocks along large fractures, shallow-depth earthquakes, high heat flow, and basaltic flows or pillow lavas.

Divergent boundaries are also present under continents during the early stages of continental breakup. When magma wells up beneath a continent, the crust is initially elevated, stretched, and thinned, producing fractures, faults, rift valleys, and volcanic activity (▶ Figure 2.15). As magma intrudes into faults and fractures, it

Table 2.1

Types of Plate Boundaries

Type	Example	Landforms	Volcanism
Divergent			
Oceanic	Mid-Atlantic Ridge	Mid-oceanic ridge with axial rift valley	Basalt
Continental	East African Rift Valley	Rift valley	Basalt and rhyolite, no andesite
Convergent			
Oceanic–oceanic	Aleutian Islands	Volcanic island arc, offshore oceanic trench	Andesite
Oceanic–continental	Andes	Offshore oceanic trench, volcanic mountain chain, mountain belt	Andesite
Continental–continental	Himalayas	Mountain belt	Minor
Transform	San Andreas fault	Fault valley	Minor

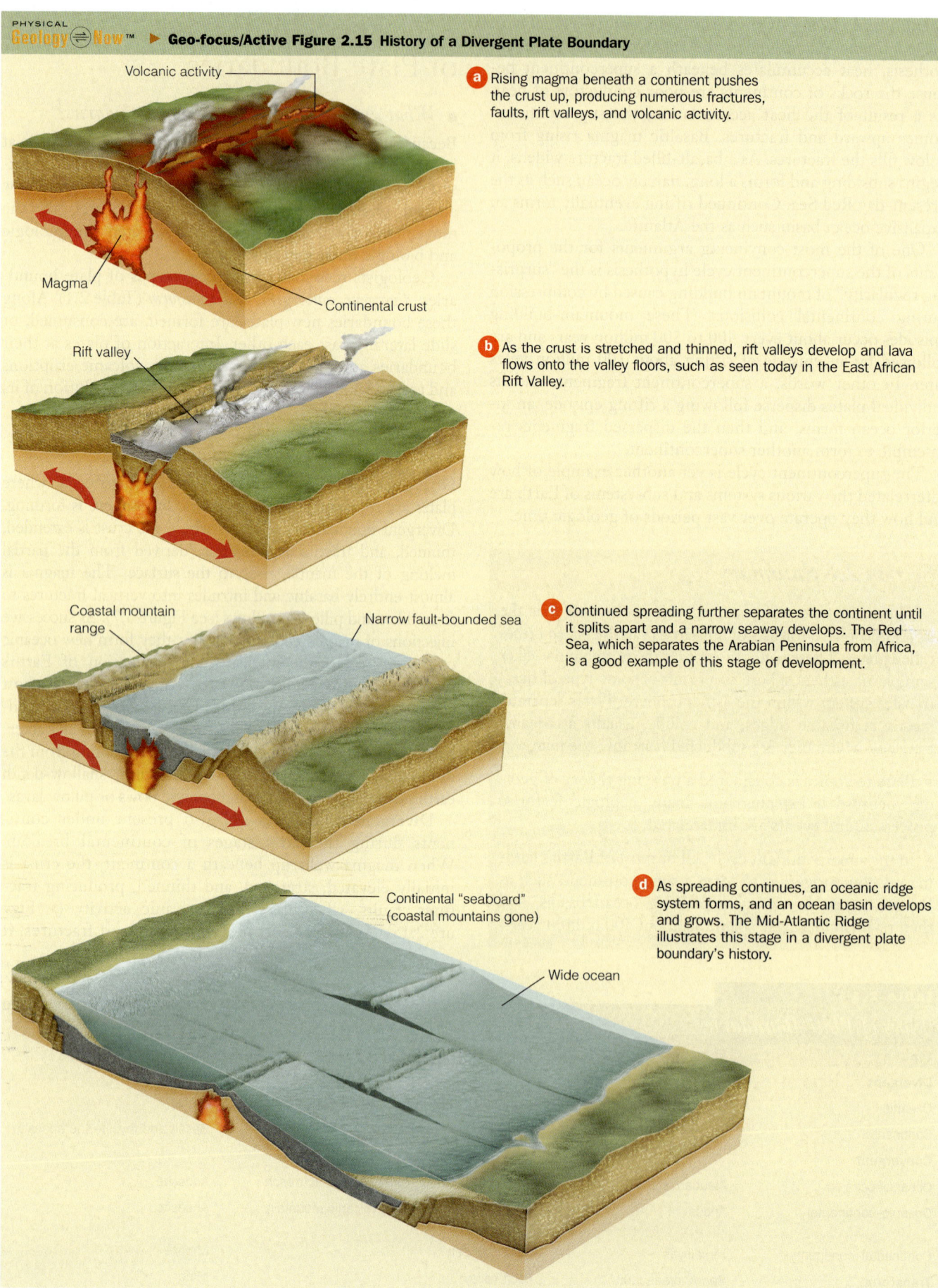

Geo-focus/Active Figure 2.15 History of a Divergent Plate Boundary

a Rising magma beneath a continent pushes the crust up, producing numerous fractures, faults, rift valleys, and volcanic activity.

b As the crust is stretched and thinned, rift valleys develop and lava flows onto the valley floors, such as seen today in the East African Rift Valley.

c Continued spreading further separates the continent until it splits apart and a narrow seaway develops. The Red Sea, which separates the Arabian Peninsula from Africa, is a good example of this stage of development.

d As spreading continues, an oceanic ridge system forms, and an ocean basin develops and grows. The Mid-Atlantic Ridge illustrates this stage in a divergent plate boundary's history.

2.7 The Three Types of Plate Boundaries 49

Figure 2.16 East African Rift Valley and the Red Sea—Present-Day Examples of Divergent Plate Boundaries The East African Rift Valley and the Red Sea represent different stages in the history of a divergent plate boundary.

a The East African Rift Valley is being formed by the separation of eastern Africa from the rest of the continent along a divergent plate boundary.

b The Red Sea represents a more advanced stage of rifting, in which two continental blocks (Africa and the Arabian Peninsula) are separated by a narrow sea.

solidifies or flows out onto the surface as lava flows; the latter often covering the rift valley floor (▶ Figure 2.15b). The East African Rift Valley is an excellent example of continental breakup at this stage (▶ Figure 2.16a).

As spreading proceeds, some rift valleys continue to lengthen and deepen until the continental crust eventually breaks and a narrow linear sea is formed, separating two continental blocks (▶ Figure 2.15c). The Red Sea separating the Arabian Peninsula from Africa (▶ Figure 2.16b) and the Gulf of California, which separates Baja California from mainland Mexico, are good examples of this more advanced stage of rifting.

As a newly created narrow sea continues to enlarge, it may eventually become an expansive ocean basin such as the Atlantic Ocean basin is today, separating North and South America from Europe and Africa by thousands of kilometers (▶ Figure 2.15d). The Mid-Atlantic Ridge is the boundary between these diverging plates (▶ Figure 2.10); the American plates are moving westward, and the Eurasian and African plates are moving eastward.

Tectonics of the Terrestrial Planets

The four inner, or terrestrial, planets—Mercury, Venus, Earth, and Mars—all had a similar early history involving accretion, differentiation into a metallic core and silicate mantle and crust, and formation of an early atmosphere by outgassing. Their early history was also marked by widespread volcanism and meteorite impacts, both of which helped modify their surfaces.

Whereas the other three terrestrial planets as well as some of the Jovian moons display internal activity, Earth appears to be unique in that its surface is broken into a series of plates.

1. Images of **Mercury** sent back by *Mariner 10* show a heavily cratered surface with the largest impact basins filled with what appear to be lava flows similar to the lava plains on Earth's moon. The lava plains are not deformed, however, indicating that there has been little or no tectonic activity.

Another feature of Mercury's surface is a large number of scarps, a feature usually associated with earthquake activity. Yet some scientists think that these scarps formed when Mercury cooled and contracted.

2. A color-enhanced photomosaic of Mercury shows its heavily cratered surface, which has changed very little since its early history.

3. Seven scarps (indicated by arrows) can clearly be seen in this image. These scarps might have formed when Mercury cooled and contracted early in its history.

4. Of all the planets, **Venus** is the most similar in size and mass to Earth, but it differs in most other respects. Whereas Earth is dominated by plate tectonics, volcanism seems to have been the dominant force in the evolution of the Venusian surface. Even though no active volcanism has been observed on Venus, the various-sized volcanic features and what appear to be folded mountains indicate a once-active planetary interior. All of these structures appear to be the products of rising convection currents of magma pushing up under the crust and then sinking back into the Venusian interior.

5. A color-enhanced photomosaic of Venus based on radar images beamed back to Earth by the *Magellan* spacecraft. This image shows impact craters and volcanic features characteristic of the planet.

6. Venus' Aine Corona, about 200 km in diameter, is ringed by concentric faults, suggesting that it was pushed up by rising magma. A network of fractures is visible in the upper right of this image, as well as a recent lava flow at the center of the corona, several volcanic domes in the lower portion of the image, and a large volcanic pancake dome in the upper left of the image.

7. Arrows point to a 600-km segment of Venus' 6800-km long Baltis Vallis, the longest known lava flow channel in our solar system.

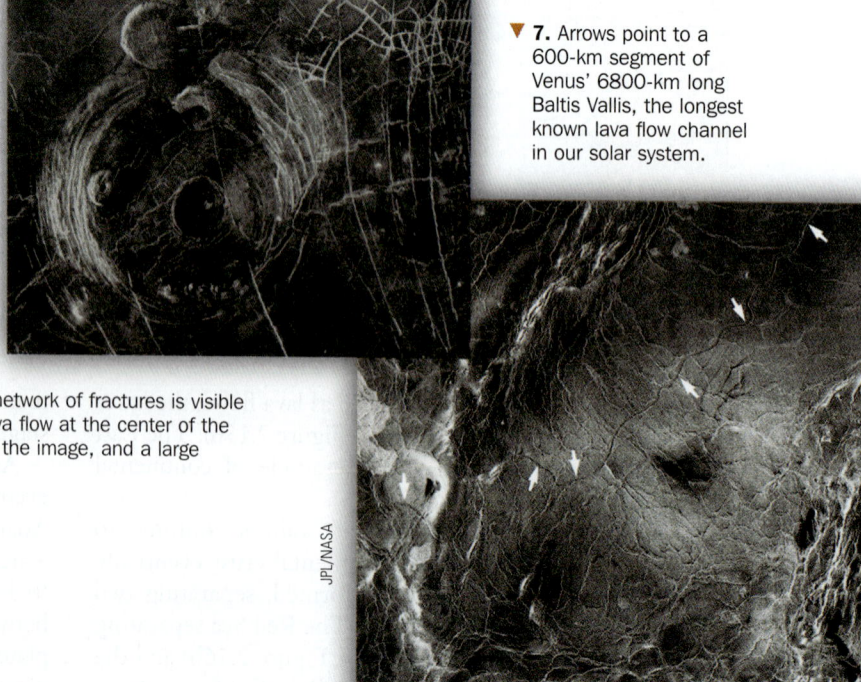

8. Volcano Sapas Mons contains two lava-filled calderas and is flanked by lava flows, attesting to the volcanic activity that was once common on Venus.

9. **Mars,** the Red Planet, has numerous features that indicate an extensive early period of volcanism. These include Olympus Mons, the solar system's largest volcano, lava flows, and uplifted regions thought to have resulted from mantle convection. In addition to volcanic features, Mars displays abundant evidence of tensional tectonics, including numerous faults and large fault-produced valley structures. Although Mars was tectonically active during the past, no evidence indicates that plate tectonics comparable to those on Earth have ever occurred there.

10. A photomosaic of Mars shows a variety of geologic structures, including the southern polar ice cap.

11. A vertical view of Olympus Mons, a shield volcano and the largest volcano known in our solar system. The edge of the Olympus Mons caldera is marked by a cliff several kilometers high. This huge summit crater is large enough to contain the greater New York City metropolitan area.

12. Although not a terrestrial planet, **Io,** the innermost of Jupiter's Galilean moons, must be mentioned. Images from the *Voyager* and *Galileo* spacecrafts show that Io has no impact craters. In fact, more than a hundred active volcanoes are visible on the moon's surface, and the sulfurous gas and ash erupted by these volcanoes bury any newly formed meteorite impact craters. Because of its proximity to Jupiter, the heat source of Io is probably tidal heating, in which the resulting friction is enough to at least partially melt Io's interior and drive its volcanoes.

13. Volcanic features of Io, the innermost moon of Jupiter. As shown in these digitally enhanced color images, Io is a very volcanically active moon.

■ What features in the geologic record indicate ancient rifting?

Associated with regions of continental rifting are faults, dikes (vertical intrusive igneous bodies), sills (horizontal intrusive igneous bodies), lava flows, and thick sedimentary sequences within rift valleys, all features that are preserved in the geologic record. The Triassic fault basins of the eastern United States are a good example of ancient continental rifting (▶ Figure 2.17a). These fault basins mark the zone of rifting that occurred when North America split apart from Africa. The basins contain thousands of meters of continental sediment and are riddled with dikes and sills.

Pillow lavas, in association with deep-sea sediment, are also evidence of ancient rifting. The presence of pillow lavas marks the formation of a spreading ridge in a narrow linear sea. A narrow linear sea forms when the continental crust in the rift valley finally breaks apart, and the area is flooded by seawater. Magma, intruding into the sea along this newly formed spreading ridge, solidifies as pillow lavas, which are preserved in the geologic record, along with the sediment being deposited on them.

■ What are convergent boundaries?

Whereas new crust forms at divergent plate boundaries, older crust must be destroyed and recycled in order for the entire surface area of Earth to remain the same. Otherwise, we would have an expanding Earth. Such plate destruction takes place at **convergent plate boundaries,** where two plates collide and the leading edge of one plate is subducted beneath the margin of the other plate and eventually is incorporated into the asthenosphere. A dipping plane of earthquake foci, called a *Benioff* (or sometimes *Benioff-Wadati*) *zone*, defines a subduction zone (see Figure 10.5). Most of these planes dip from oceanic trenches beneath adjacent island arcs or continents, marking the surface of slippage between the converging plates.

Deformation, volcanism, mountain building, metamorphism, earthquake activity, and deposits of valuable mineral ores characterize convergent boundaries. Three types of convergent plate boundaries are recognized: *oceanic–oceanic*, *oceanic–continental*, and *continental–continental*.

Oceanic–Oceanic Boundaries

When two oceanic plates converge, one is subducted beneath the other along an **oceanic–oceanic plate boundary** (▶ Figure 2.18a). The subducting plate bends downward to form the outer wall of an oceanic trench. A *subduction complex*, composed of wedge-shaped slices of highly folded and faulted marine sediments and oceanic lithosphere scraped off the descending plate, forms along the inner wall of the oceanic trench. As the subducting plate descends into the mantle, it is heated and partially melted, generating magma, commonly of andesitic composition (see Chapter 4). This

▶ **Figure 2.17 Triassic Fault-Block Basins of Eastern North America—An Example of Ancient Rifting**

a Triassic fault-block basin deposits appear in numerous locations throughout eastern North America. These fault-block basins are good examples of ancient continental rifting, and during the Triassic Period they looked like today's fault-block basins (rift valleys) of the East African Rift Valley.

b Palisades of the Hudson River. This sill (tabular-shaped horizontal igneous intrusion) was one of many that were intruded into the fault-block basin sediments during the Late Triassic rifting that marked the initial separation of North America from Africa.

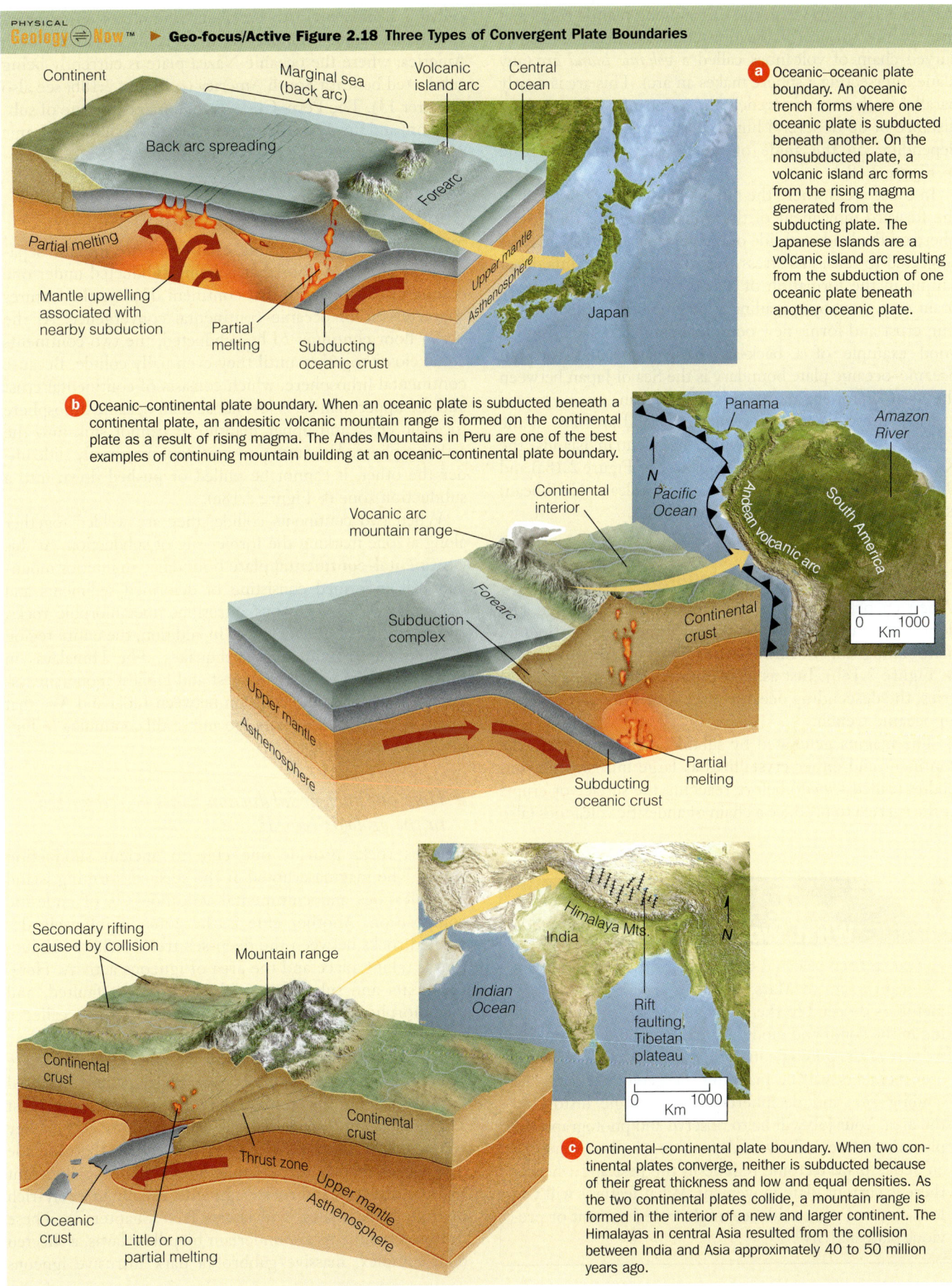

Geo-focus/Active Figure 2.18 Three Types of Convergent Plate Boundaries

a. Oceanic–oceanic plate boundary. An oceanic trench forms where one oceanic plate is subducted beneath another. On the nonsubducted plate, a volcanic island arc forms from the rising magma generated from the subducting plate. The Japanese Islands are a volcanic island arc resulting from the subduction of one oceanic plate beneath another oceanic plate.

b. Oceanic–continental plate boundary. When an oceanic plate is subducted beneath a continental plate, an andesitic volcanic mountain range is formed on the continental plate as a result of rising magma. The Andes Mountains in Peru are one of the best examples of continuing mountain building at an oceanic–continental plate boundary.

c. Continental–continental plate boundary. When two continental plates converge, neither is subducted because of their great thickness and low and equal densities. As the two continental plates collide, a mountain range is formed in the interior of a new and larger continent. The Himalayas in central Asia resulted from the collision between India and Asia approximately 40 to 50 million years ago.

magma is less dense than the surrounding mantle rocks and rises to the surface of the nonsubducted plate to form a curved chain of volcanoes called a *volcanic island arc* (any plane intersecting a sphere makes an arc). This arc is nearly parallel to the oceanic trench and is separated from it by a distance of up to several hundred kilometers—the distance depending on the angle of dip of the subducting plate (▶ Figure 2.18a).

In those areas where the rate of subduction is faster than the forward movement of the overriding plate, the lithosphere on the landward side of the volcanic island arc may be subjected to tensional stress and stretched and thinned, resulting in the formation of a *back-arc basin*. This back-arc basin may grow by spreading if magma breaks through the thin crust and forms new oceanic crust (▶ Figure 2.18a). A good example of a back-arc basin associated with an oceanic–oceanic plate boundary is the Sea of Japan between the Asian continent and the islands of Japan.

Most present-day active volcanic island arcs are in the Pacific Ocean basin and include the Aleutian Islands, the Kermadec–Tonga arc, and the Japanese (▶ Figure 2.18a) and Philippine Islands. The Scotia and Antillean (Caribbean) island arcs are in the Atlantic Ocean basin.

Oceanic–Continental Boundaries

When an oceanic and a continental plate converge, the denser oceanic plate is subducted under the continental plate along an **oceanic–continental plate boundary** (▶ Figure 2.18b). Just as at oceanic–oceanic plate boundaries, the descending oceanic plate forms the outer wall of an oceanic trench.

The magma generated by subduction rises beneath the continent and either crystallizes as large intrusive igneous bodies (called *plutons*) before reaching the surface or erupts at the surface to produce a chain of andesitic volcanoes (also called a *volcanic arc*). An excellent example of an oceanic–continental plate boundary is the Pacific coast of South America, where the oceanic Nazca plate is currently being subducted beneath South America (▶ Figure 2.18b; see also Chapter 13). The Peru–Chile Trench marks the site of subduction, and the Andes Mountains are the resulting volcanic mountain chain on the nonsubducting plate.

Continental–Continental Boundaries

Two continents approaching each other are initially separated by an ocean floor that is being subducted under one continent. The edge of that continent displays the features characteristic of oceanic–continental convergence. As the ocean floor continues to be subducted, the two continents come closer together until they eventually collide. Because continental lithosphere, which consists of continental crust and the upper mantle, is less dense than oceanic lithosphere (oceanic crust and upper mantle), it cannot sink into the asthenosphere. Although one continent may partly slide under the other, it cannot be pulled or pushed down into a subduction zone (▶ Figure 2.18c).

When two continents collide, they are welded together along a zone marking the former site of subduction. At this **continental–continental plate boundary,** an interior mountain belt is formed consisting of deformed sediments and sedimentary rocks, igneous intrusions, metamorphic rocks, and fragments of oceanic crust. In addition, the entire region is subjected to numerous earthquakes. The Himalayas in central Asia, the world's youngest and highest mountain system, resulted from the collision between India and Asia that began 40 to 50 million years ago and is still continuing (▶ Figure 2.18c; see Chapter 13).

■ *How can ancient subduction zones be recognized in the geologic record?*

Igneous rocks provide one clue to ancient subduction zones. The magma erupted at the surface, forming island arc volcanoes and continental volcanoes, is of andesitic composition. Another clue is the zone of intensely deformed rocks between the deep-sea trench where subduction is taking place and the area of igneous activity. Here, sediments and submarine rocks are folded, faulted, and metamorphosed into a chaotic mixture of rocks called a *mélange*.

During subduction, pieces of oceanic lithosphere are sometimes incorporated into the mélange and accreted onto the edge of the continent. Such slices of oceanic crust and upper mantle are called *ophiolites* (▶ Figure 2.19). They consist of a layer of deep-sea sediments that include graywackes (poorly sorted sandstones containing abundant feldspar minerals and rock fragments, usually in a clay-rich matrix), black shales, and cherts (see Chapter 7). These deep-sea sediments are underlain by pillow lavas, a sheeted dike complex, massive gabbro (a dark intrusive igneous

What Would You Do?

You've been selected to be part of the first astronaut team to go to Mars. While your two fellow crewmembers descend to the Martian surface, you'll be staying in the command module and circling the Red Planet. As part of the geologic investigation of Mars, one of the crewmembers will be mapping the geology around the landing site and deciphering the geologic history of the area. Your job will be to observe and photograph the planet's surface and try to determine whether Mars had an active plate tectonic regime in the past and whether there is current plate movement. What features will you look for, and what evidence might reveal current or previous plate activity?

rock), and layered gabbro, all of which form the oceanic crust. Beneath the gabbro is peridotite, which probably represents the upper mantle. The presence of ophiolite in an outcrop or drilling core is a key indication of plate convergence along a subduction zone.

Elongate belts of folded and faulted marine sedimentary rocks, andesites, and ophiolites are found in the Appalachians, Alps, Himalayas, and Andes mountains. The combination of such features is good evidence that these mountain ranges resulted from deformation along convergent plate boundaries.

■ *What are transform boundaries?*

The third type of plate boundary is a **transform plate boundary.** These mostly occur along fractures in the seafloor, known as *transform faults*, where plates slide laterally past each other roughly parallel to the direction of plate movement. Although lithosphere is neither created nor destroyed along a transform boundary, the movement between plates results in a zone of intensely shattered rock and numerous shallow-depth earthquakes.

Transform faults "transform" or change one type of motion between plates into another type of motion. Most commonly, transform faults connect two oceanic ridge segments, but they can also connect ridges to trenches and trenches to trenches (▶ Figure 2.20). Although the majority of transform faults are in oceanic crust and are marked by distinct fracture zones, they may also extend into continents.

One of the best-known transform faults is the San Andreas fault in California. It separates the Pacific plate from the North American plate and connects spreading ridges in the Gulf of California with the Juan de Fuca and Pacific plates off the coast of northern California (▶ Figure 2.21). Many of the earthquakes affecting California are the result of movement along this fault (see Chapter 10).

Unfortunately, transform faults generally do not leave any characteristic or diagnostic features except the obvious displacement of the rocks with which they are associated. This displacement is usually large, on the order of tens to hundreds of kilometers. Such large displacements in ancient rocks can sometimes be related to transform fault systems.

PHYSICAL Geology⇌Now™ Log into GeologyNow and select this chapter to work through **Geology Interactive** activities on "Plate Boundaries" (click Plate Tectonics→Plate Boundaries) and "Triple Junctions and Seafloor Studies" (click Plate Tectonics→Triple Junctions and Seafloor Studies).

Section 2.7 Summary

• The three major types of plate boundaries are divergent, convergent, and transform.

• Divergent plate boundaries occur where plates are separating and new oceanic lithosphere is forming. They are characterized by thinning and fracturing of the crust, formation of rift valleys, intrusion of magma, and shallow-depth earthquakes.

• Zones of ancient continental rifting can be recognized by faults, dikes, sills, lava flows, and thick sedimentary sequences, whereas pillow lavas and associated deep-sea sediments are evidence of ancient spreading ridges.

• Convergent plate boundaries occur where two plates collide, and the leading edge of one plate is subducted beneath the margin of the other plate. They are characterized by metamorphism, mountain building, volcanic and earthquake activity, and the formation of various mineral deposits. Three types of convergent plate boundaries are recognized: oceanic–oceanic, where two oceanic plates converge, oceanic–continental, where an oceanic plate is subducted beneath the continental plate, and continental–continental, where an interior mountain belt is formed.

• Intensely deformed rocks, andesite lavas, and ophiolites are all evidence of ancient subduction zones, marking former convergent plate boundaries.

• Transform plate boundaries occur along fractures in the seafloor, called transform faults, where plates slide laterally past each other.

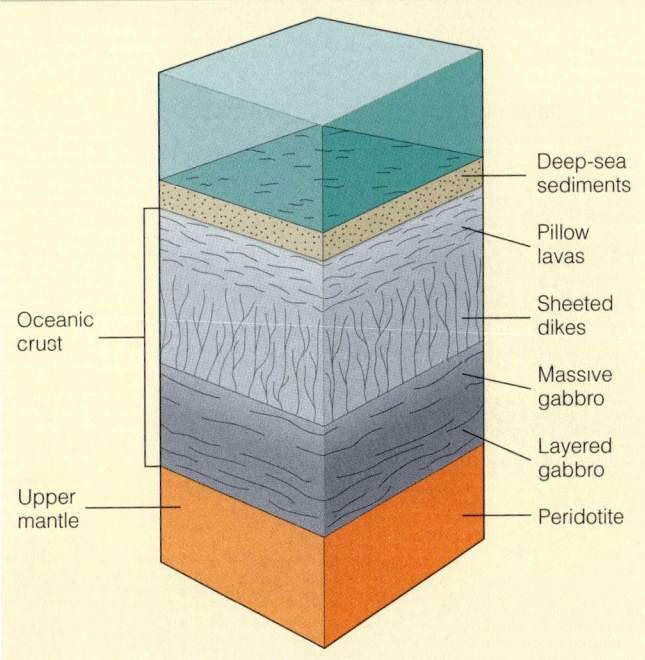

▶ **Figure 2.19 Ophiolites** Ophiolites are sequences of rock on land consisting of deep-sea sediments, oceanic crust, and upper mantle. Ophiolites are one feature used to recognize ancient convergent plate boundaries.

Geo-focus Figure 2.20 Transform Plate Boundaries Horizontal movement between plates occurs along transform faults. Extensions of transform faults on the seafloor form fracture zones.

a Most transform faults connect two oceanic ridge segments.

b A transform fault can connect a ridge and a trench.

c A transform fault can also link two trenches.

2.8 Hot Spots: An Intraplate Feature

■ *What are hot spots and what do they tell us about plate movement?*

Before leaving the topic of plate boundaries, we should mention an intraplate feature found beneath both oceanic and continental plates. A **hot spot** is the location on Earth's surface where a stationary column of magma, originating deep within the mantle (*mantle plume*), has slowly risen to the surface and formed a volcano (▶ Figure 2.22). Because mantle plumes apparently remain stationary (although some evidence suggests that they might not) within the mantle while plates move over them, the resulting hot spots leave a trail of extinct and progressively older volcanoes called *aseismic ridges* that record the movement of the plate.

One of the best examples of aseismic ridges and hot spots is the Emperor Seamount–Hawaiian Island chain (▶ Figure 2.22). This chain of islands and seamounts (structures of volcanic origin rising higher than 1 km above the seafloor) extends from the island of Hawaii to the Aleutian Trench off Alaska, a distance of some 6000 km, and consists of more than 80 volcanic structures.

Currently, the only active volcanoes in this island chain are on the islands of Hawaii and Maui and the Loihi Seamount. The rest of the islands are extinct volcanic structures that become progressively older toward the north and northwest. This means that the Emperor Seamount–Hawaiian Island chain records the direction that the Pacific plate traveled as it moved over an apparently stationary mantle plume. In this case, the Pacific plate first moved in a north-northwesterly direction and then, as indicated by the sharp bend in the chain, changed to a west-northwesterly direction about 43 million years ago. The reason the Pacific plate changed directions is not known, but the shift might be related to the collision of India with the Asian continent at around the same time (see Figure 13.22).

Mantle plumes and hot spots help geologists explain some of the geologic activity occurring within plates as opposed to activity occurring at or near plate boundaries. In addition, if mantle plumes are essentially fixed with respect to Earth's rotational axis, they can be used to determine not only the direction of plate movement but also the rate of movement. They can also provide reference points for determining paleolatitude, an important tool when reconstructing the location of continents in the geologic past.

Section 2.8 Summary

● Hot spots are locations where stationary columns of magma from the mantle (mantle plumes) have risen to the surface and formed volcanoes.

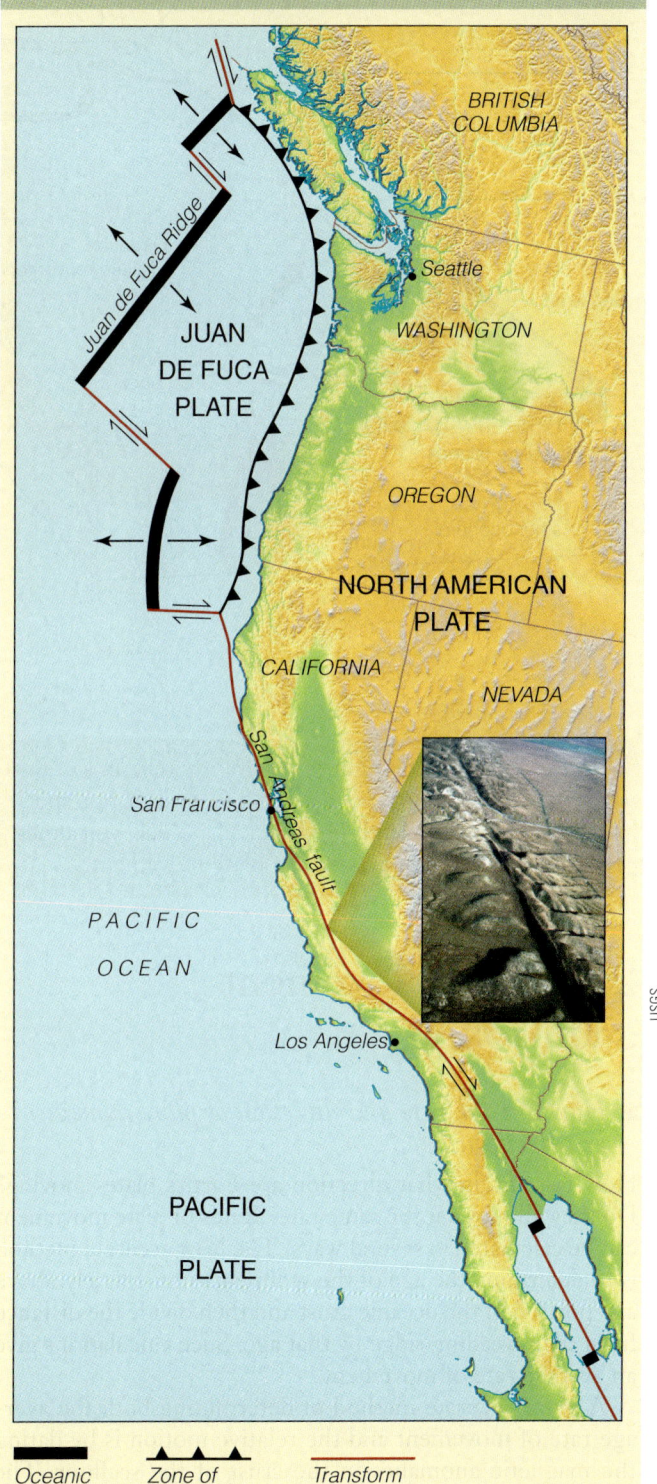

▶ **Figure 2.21 The San Andreas Fault—A Transform Plate Boundary** The San Andreas fault is a transform fault separating the Pacific plate from the North American plate. It connects the spreading ridges in the Gulf of California with the Juan de Fuca and Pacific plates off the coast of northern California. Movement along the San Andreas fault has caused numerous earthquakes. The insert photograph shows a segment of the San Andreas fault as it cuts through the Carrizo Plain, California.

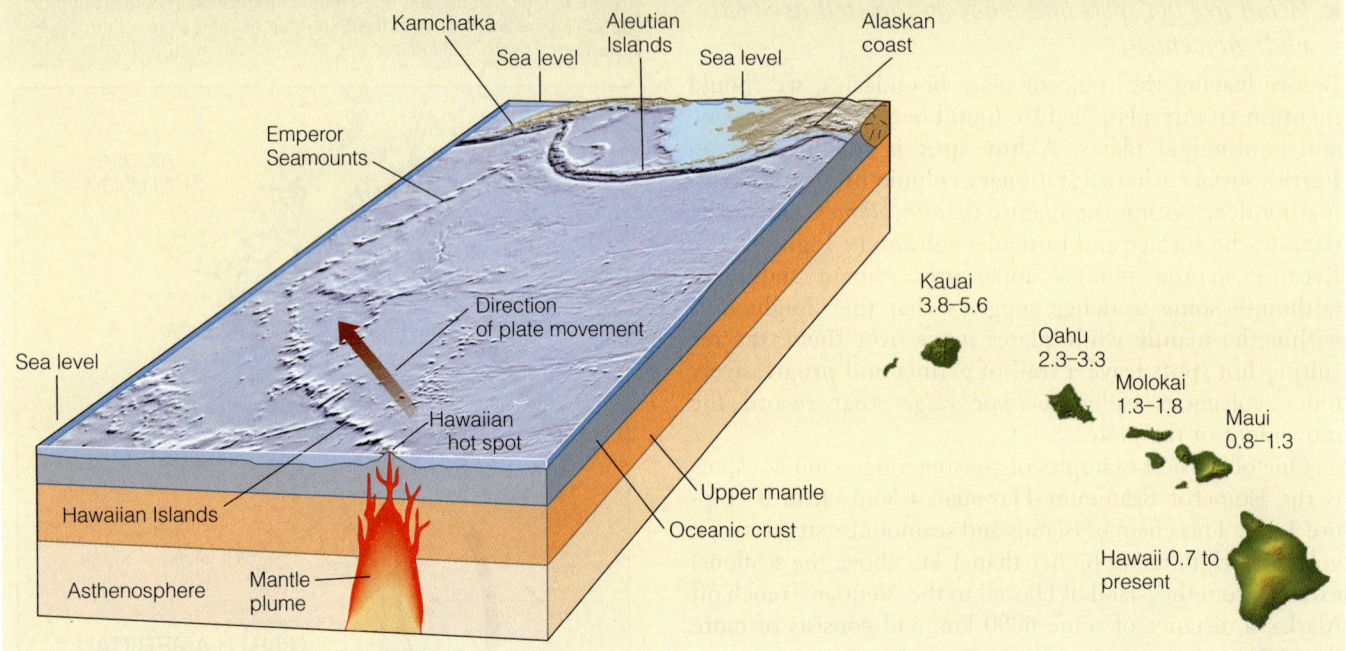

Active Figure 2.22 Hot Spots A hot spot is the location where a stationary mantle plume has risen to the surface and formed a volcano. The Emperor Seamount–Hawaiian Island chain formed as a result of the Pacific plate moving over a mantle plume, and the line of volcanic islands in this chain traces the direction of plate movement. The Hawaiian hot spot currently underlies the southern half of the island of Hawaii and adjoining offshore area. The numbers indicate the ages of the Hawaiian Islands in millions of years.

- Because the mantle plumes apparently remain stationary while plates move over them, the trail of hot spots, marked by extinct and progressively older volcanoes, records the direction and rate of plate movement.

2.9 Plate Movement and Motion

■ *How can the rate and direction of plate movement be determined?*

How fast and in what direction are Earth's plates moving? Do they all move at the same rate? Rates of plate movement can be calculated in several ways. The least accurate method is to determine the age of the sediments immediately above any portion of the oceanic crust and then divide the distance from the spreading ridge by that age. Such calculations give an average rate of movement.

A more accurate method of determining both the average rate of movement and the relative motion is by dating the magnetic anomalies in the crust of the seafloor. The distance from an oceanic ridge axis to any magnetic anomaly indicates the width of new seafloor that formed during that time interval. Thus, for a given interval of time, the wider the strip of seafloor, the faster the plate has moved.

In this way, not only can the present average rate of movement and relative motion be determined (▶ Figure 2.14), but the average rate of movement during the past can also be calculated by dividing the distance between anomalies by the amount of time elapsed between anomalies.

Geologists use magnetic anomalies not only to calculate the average rate of plate movement but also to determine plate positions at various times in the past. Because magnetic anomalies are parallel and symmetric with respect to spreading ridges, all one must do to determine the positions of continents when particular anomalies formed is to move the anomalies back to the spreading ridge, which will also move the continents with them (▶ Figure 2.23). Unfortunately, subduction destroys oceanic crust and the magnetic record it carries. Thus we have an excellent record of plate movements since the breakup of Pangaea, but not as good an understanding of plate movement before that time.

The average rate of movement as well as the relative motion between any two plates can also be determined by satellite–laser ranging techniques. Laser beams from a station on one plate are bounced off a satellite (in geosynchronous orbit) and returned to a station on a different plate. As the plates move away from each other, the laser beam takes more time to go from the sending station to the stationary satellite and back to the receiving station. This difference in elapsed time is used to calculate the rate of movement and the relative motion between plates.

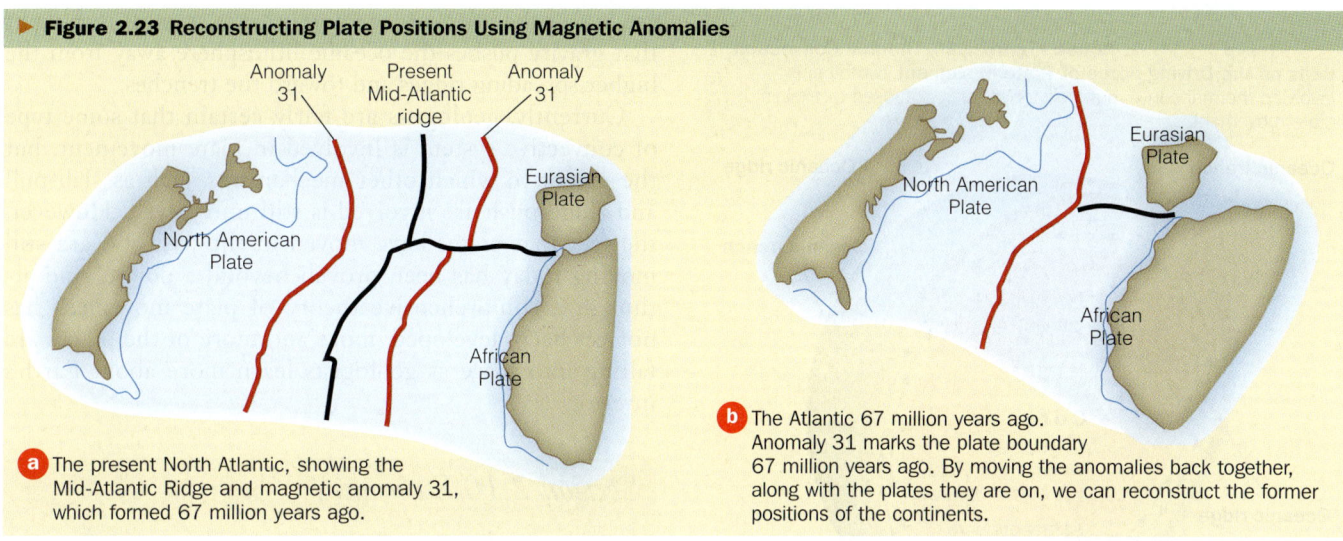

▶ **Figure 2.23 Reconstructing Plate Positions Using Magnetic Anomalies**

a The present North Atlantic, showing the Mid-Atlantic Ridge and magnetic anomaly 31, which formed 67 million years ago.

b The Atlantic 67 million years ago. Anomaly 31 marks the plate boundary 67 million years ago. By moving the anomalies back together, along with the plates they are on, we can reconstruct the former positions of the continents.

Plate motions derived from magnetic reversals and satellite–laser ranging techniques give only the relative motion of one plate with respect to another. Hot spots enable geologists to determine absolute motion because they provide an apparently fixed reference point from which the rate and direction of plate movement can be measured. The previously mentioned Emperor Seamount–Hawaiian Island chain formed as a result of movement over a hot spot. Thus the line of the volcanic islands traces the direction of plate movement, and dating the volcanoes enables geologists to determine the rate of movement.

Section 2.9 Summary

• One technique used to calculate the average rate of plate movement is to divide the distance from an oceanic ridge axis to any magnetic anomaly by the age of that anomaly. The average rate of movement during the past can also be calculated by dividing the distance between anomalies by the amount of time elapsed between anomalies.

• The relative motion of one plate with respect to another can be derived from magnetic reversals and satellite–laser ranging techniques. Hot spots are evidence for absolute motion because they provide an apparently fixed reference point from which the rate and direction of plate movement can be measured.

2.10 The Driving Mechanism of Plate Tectonics

■ *What drives plates?*

A major obstacle to the acceptance of the continental drift hypothesis was the lack of a driving mechanism to explain continental movement. When it was shown that continents and ocean floors moved together, not separately, and that new crust formed at spreading ridges by rising magma, most geologists accepted some type of convective heat system as the basic process responsible for plate motion. The question still remains, however: What exactly drives the plates?

■ *How do thermal convection cells move plates?*

Two models involving thermal convection cells have been proposed to explain plate movement (▶ Figure 2.24). In one model, thermal convection cells are restricted to the asthenosphere; in the second model, the entire mantle is involved. In both models, spreading ridges mark the ascending limbs of adjacent convection cells, and trenches are present where convection cells descend back into Earth's interior. The convection cells therefore determine the location of spreading ridges and trenches, with the lithosphere lying above the thermal convection cells. Each plate thus corresponds to a single convection cell and moves as a result of the convective movement of the cell itself.

Although most geologists agree that Earth's internal heat plays an important role in plate movement, there are problems with both models. The major problem associated with the first model is the difficulty in explaining the source of heat for the convection cells and why they are restricted to the asthenosphere. In the second model, the heat comes from the outer core, but it is still not known how heat is transferred from the outer core to the mantle. Nor is it clear how convection can involve both the lower mantle and the asthenosphere.

■ *Can plate movement be gravity driven?*

In addition to some type of thermal convection system driving plate movement, some geologists think plate movement occurs because of a mechanism involving

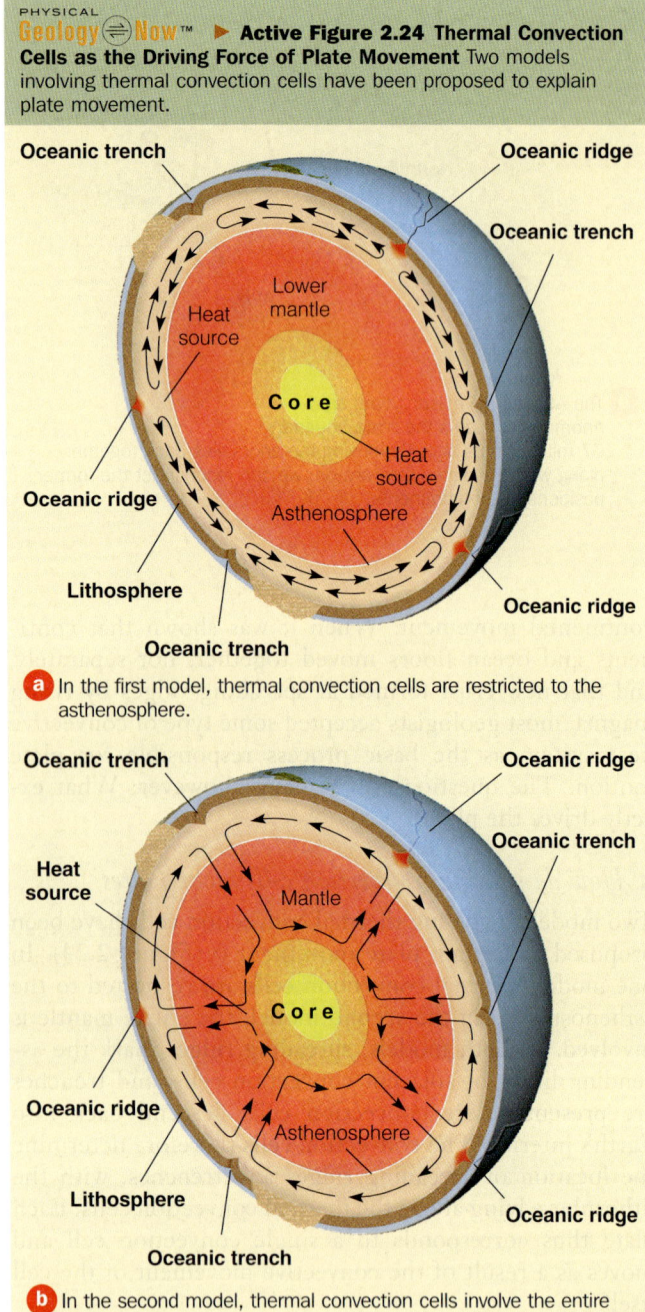

Active Figure 2.24 Thermal Convection Cells as the Driving Force of Plate Movement Two models involving thermal convection cells have been proposed to explain plate movement.

(a) In the first model, thermal convection cells are restricted to the asthenosphere.

(b) In the second model, thermal convection cells involve the entire mantle.

"slab-pull" or "ridge-push," both of which are gravity driven but still dependent on thermal differences within Earth (▶ Figure 2.25). In slab-pull, the subducting cold slab of lithosphere, being denser than the surrounding warmer asthenosphere, pulls the rest of the plate along as it descends into the asthenosphere. As the lithosphere moves downward, there is a corresponding upward flow back into the spreading ridge.

Operating in conjunction with slab-pull is the ridge-push mechanism. As a result of rising magma, the oceanic ridges are higher than the surrounding oceanic crust. It is thought that gravity pushes the oceanic lithosphere away from the higher spreading ridges and toward the trenches.

Currently, geologists are fairly certain that some type of convective system is involved in plate movement, but the extent to which other mechanisms such as slab-pull and ridge-push are involved is still unresolved. However, the fact that plates have moved in the past and are still moving today has been proven beyond a doubt. And although a comprehensive theory of plate movement has not yet been developed, more and more of the pieces are falling into place as geologists learn more about Earth's interior.

Section 2.10 Summary

- Plates are thought to move by some type of thermal convection system. In one model, thermal convection cells are restricted to the asthenosphere, and in the second model, the entire mantle is involved. Both models have problems associated with heat transfer and the source of heat.

- Some geologists think that, in addition to thermal convection, plate movement is primarily gravity driven, by either "slab-pull" or "ridge-push." In slab-pull, the subducting, cold, dense lithosphere pulls the rest of the plate along as it descends into Earth's interior. In ridge-push, gravity pushes the lithosphere away from the higher spreading ridges and toward the subduction trenches.

2.11 Plate Tectonics and the Distribution of Natural Resources

■ *How does plate tectonic theory relate to the origin and distribution of natural resources?*

Besides being responsible for the major features of Earth's crust and influencing the distribution and evolution of the world's biota, plate movements also affect the formation and distribution of some natural resources. The formation of many natural resources results from the interaction between plates, and economically valuable concentrations of such deposits are found associated with current and ancient plate boundaries. Consequently, geologists are using plate tectonic theory in their search for petroleum and mineral deposits and in explaining the occurrence of these natural resources.

It is becoming increasingly clear that if we are to keep up with the continuing demands of a global industrialized society, the application of plate tectonic theory to the origin and distribution of natural resources is essential.

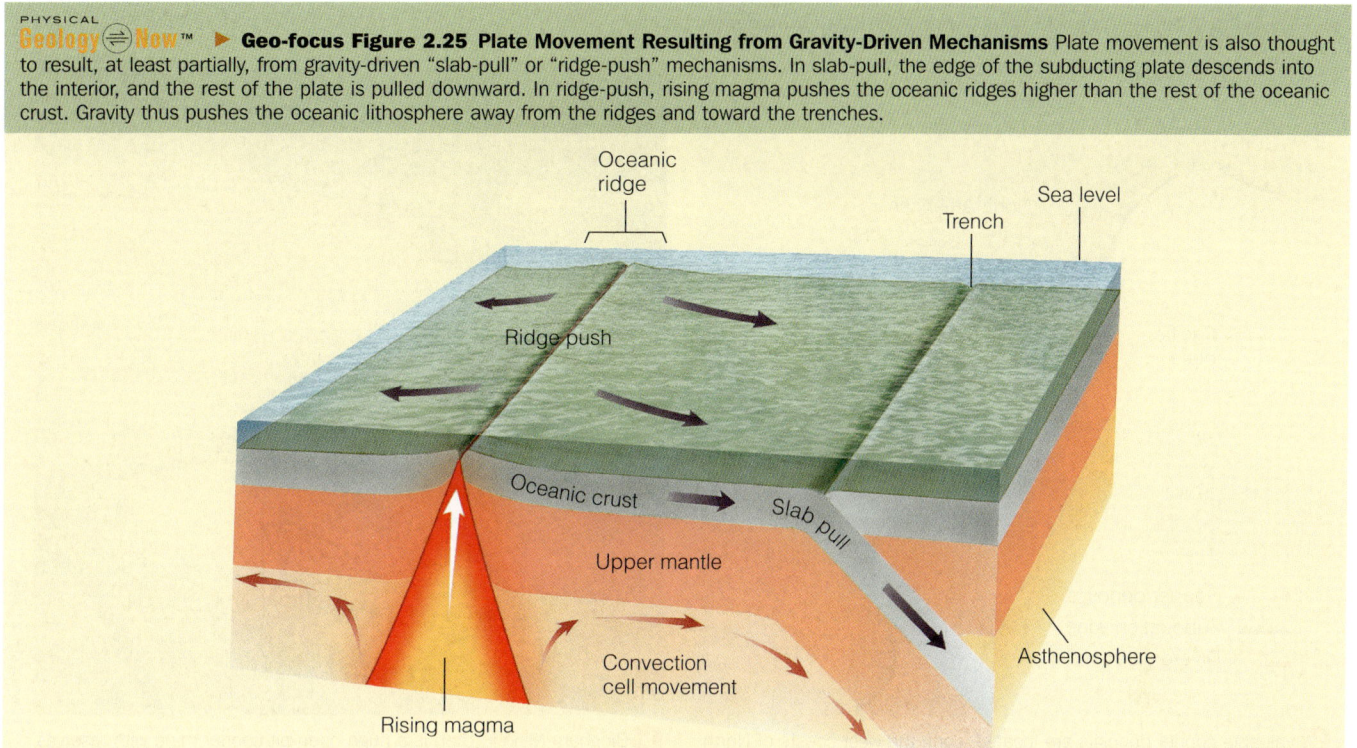

Geo-focus Figure 2.25 Plate Movement Resulting from Gravity-Driven Mechanisms Plate movement is also thought to result, at least partially, from gravity-driven "slab-pull" or "ridge-push" mechanisms. In slab-pull, the edge of the subducting plate descends into the interior, and the rest of the plate is pulled downward. In ridge-push, rising magma pushes the oceanic ridges higher than the rest of the oceanic crust. Gravity thus pushes the oceanic lithosphere away from the ridges and toward the trenches.

■ *What is the relationship between plate boundaries and various metallic mineral deposits?*

Many metallic mineral deposits such as copper, gold, lead, silver, tin, and zinc are related to igneous and associated hydrothermal (hot water) activity, so it is not surprising that a close relationship exists between plate boundaries and the occurrence of these valuable deposits.

The magma generated by partial melting of a subducting plate rises toward the surface, and as it cools, it precipitates and concentrates various metallic ores. Many of the world's major metallic ore deposits are associated with convergent plate boundaries, including those in the Andes of South America, the Coast Ranges and Rockies of North America, Japan, the Philippines, Russia, and a zone extending from the eastern Mediterranean region to Pakistan. In addition, the majority of the world's gold is associated with sulfide deposits located at ancient convergent plate boundaries in such areas as South Africa, Canada, California, Alaska, Venezuela, Brazil, southern India, Russia, and western Australia.

The copper deposits of western North and South America are an excellent example of the relationship between convergent plate boundaries and the distribution, concentration, and exploitation of valuable metallic ores (▶ Figure 2.26a). The world's largest copper deposits are found along this belt. The majority of the copper deposits in the Andes and the southwestern United States formed less than 60 million years ago when oceanic plates were subducted under the North and South American plates. The rising magma and associated hydrothermal fluids carried minute amounts of copper, which was originally widely disseminated but eventually became concentrated in the cracks and fractures of the surrounding andesites. These low-grade copper deposits contain from 0.2 to 2% copper and are extracted from large open-pit mines (▶ Figure 2.26b).

Divergent plate boundaries also yield valuable ore deposits. The island of Cyprus in the Mediterranean is rich in copper and has been supplying all or part of the world's needs for the last 3000 years. The concentration of copper on Cyprus formed as a result of precipitation adjacent to hydrothermal vents along a divergent plate boundary. This

What Would You Do?

You are part of a mining exploration team that is exploring a promising and remote area of central Asia. You know that former convergent and divergent plate boundaries are frequently sites of ore deposits. What evidence would you look for to determine whether the area you're exploring might be an ancient convergent or divergent plate boundary? Is there anything you can do before visiting the area that might help you to determine the geology of the area?

▶ **Figure 2.26 Copper Deposits and Convergent Plate Boundaries**

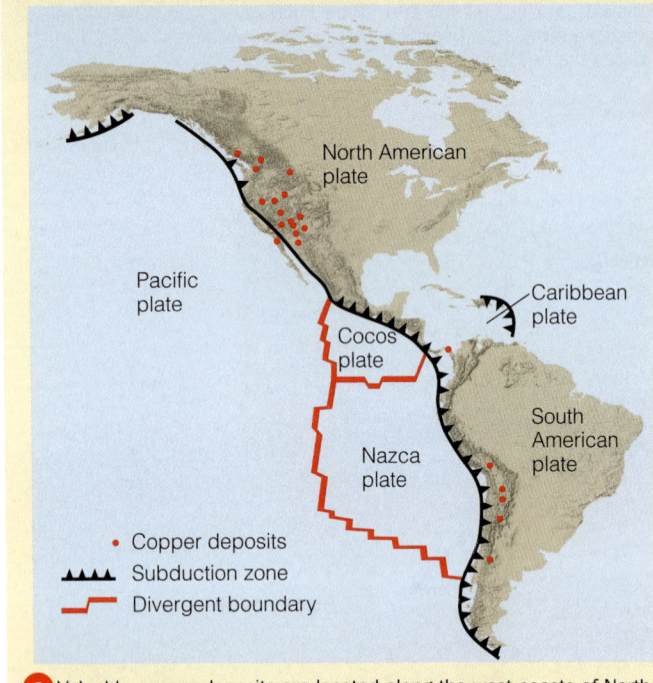

a Valuable copper deposits are located along the west coasts of North and South America in association with convergent plate boundaries. The rising magma and associated hydrothermal activity resulting from subduction carried small amounts of copper, which became trapped and concentrated in the surrounding rocks through time.

b Bingham Mine in Utah is a huge open-pit copper mine with reserves estimated at 1.7 billion tons. More than 400,000 tons of rock are removed for processing each day.

deposit was brought to the surface when the copper-rich seafloor collided with the European plate, warping the seafloor and forming Cyprus.

Studies indicate that minerals of such metals as copper, gold, iron, lead, silver, and zinc are currently forming as sulfides in the Red Sea. The Red Sea is opening as a result of plate divergence and represents the earliest stage in the growth of an ocean basin (▶ Figures 2.15c and 2.16b).

Section 2.11 Summary

- The origin and distribution of many natural resources are related to the interaction between plates. Many metallic ores form as a result of igneous and hydrothermal activity related to the formation of magma along divergent and convergent plate boundaries.

Review Workbook

ESSENTIAL QUESTIONS SUMMARY

2.1 Introduction
■ *Why should you know about plate tectonics?*
Plate tectonics affects all of us, whether in relation to the destruction caused by volcanic eruptions and earthquakes, or politically and economically due to the formation and distribution of valuable natural resources. Furthermore, plate tectonics is the unifying theory of geology, tying together many seemingly unrelated geologic phenomena and illustrating why Earth is a dynamic planet of interacting subsystems and cycles.

2.2 Continental Drift
■ *What were some early idea's about Earth's past geography?*
The idea that continents have moved in the past is not new and probably goes back to the first maps, in which one could see that the east coast of South America looks like it fits into the west coast of Africa.

■ *What is the continental drift hypothesis and who proposed it?*
Alfred Wegener originally proposed the continental drift hypothesis in 1912. He postulated that all landmasses were originally united into a supercontinent he named Pangaea. Pangaea con-

sisted of a northern landmass called Laurasia and a southern landmass called Gondwana. As Pangaea broke up, the various continents moved to their present-day locations.

2.3 Evidence for Continental Drift
■ *What is the evidence for continental drift?*
Wegener and others amassed a large amount of evidence in support of continental drift. There is a close fit between continents off the coasts at a depth of about 2000 m. Marine, nonmarine, and glacial rock sequences of Pennsylvanian to Jurassic age are nearly identical on all the Gondwana continents, and the trend of several major mountain ranges produces a continuous mountain range when the continents are positioned next to each other as they were during the formation of Pangaea. Glacial tills and striations on the bedrock beneath the till provide evidence of glaciation at the same time on all the Gondwana continents, with South Africa located at the South Pole. Lastly, the distribution of fossil plants (*Glossopteris* flora) and animals (the nonmarine reptile *Mesosaurus* in particular) provides convincing evidence that the southern continents were united during the Late Paleozoic Era.

2.4 Paleomagnetism and Polar Wandering
■ *What is paleomagnetism?*
Paleomagnetism is the remanent magnetism in ancient rocks recording the direction and intensity of Earth's magnetic field at the time of the rock's formation.

■ *What is the Curie point and why is it important?*
The Curie point is the temperature at which iron-bearing minerals gain their magnetism. It is important because as long as the rock is not subsequently heated above the Curie point, it will preserve that remanent magnetism.

■ *How can the apparent wandering of the magnetic poles be best explained?*
The best explanation for polar wandering, which is the apparent movement of the magnetic poles through time, is that the magnetic poles have remained near their present locations at the geographic north and south poles and the continents have moved. When the continents are fitted together, the paleomagnetic data point to only one magnetic pole.

2.5 Magnetic Reversals and Seafloor Spreading
■ *What evidence is there that Earth's magnetic field has reversed in the past?*
Earth's present magnetic field is considered normal, that is, with the north and south magnetic poles located approximately at the north and south geographic poles. At various times in the geologic past, Earth's magnetic field has completely reversed. The existence of such magnetic reversals was discovered by dating and determining the orientation of the remanent magnetism in continental lava flows.

■ *What is the theory of seafloor spreading, and how does it validate continental drift?*
Harry Hess proposed the theory of seafloor spreading in 1962. He suggested that the seafloor separates at oceanic ridges, where new crust is formed by upwelling magma. As the magma cools, the newly formed oceanic crust moves laterally away from the ridge. Thus continents and oceanic crust move together, negating the need to explain how continents could plow through oceanic crust.

■ *How was the theory of seafloor spreading confirmed?*
Seafloor spreading was confirmed by the discovery of magnetic anomalies in the ocean crust that were both parallel to and symmetric around the ocean ridges. The pattern of oceanic magnetic anomalies matched the pattern of magnetic reversals already known from continental lava flows. Further evidence confirming seafloor spreading came from the Deep Sea Drilling Project and the age and thickness of the sediments overlying the oceanic crust.

2.6 Plate Tectonics: A Unifying Theory
■ *What are the main tenets of plate tectonic theory?*
According to plate tectonic theory, the rigid lithosphere, composed of oceanic and continental crust as well as the underlying upper mantle, is divided into different-sized plates. The lithosphere overlies the asthenosphere, and through some type of heat-transfer system within the asthenosphere, moves the plates. As the plates move over the asthenosphere, they separate mostly at oceanic ridges and collide and are subducted into Earth's interior at oceanic trenches.

■ *Why is plate tectonics a unifying theory of geology?*
The theory ties together many seemingly unrelated features and events and shows how they are interrelated. Furthermore, it illustrates the dynamic interactions between Earth's various subsystems and cycles.

■ *What is the supercontinent cycle?*
The hypothesis put forth by J. Tuzo Wilson in the early 1970s posits a large-scale global cycle in which a supercontinent fragments to form various ocean basins that widen and then close, thus reassembling another supercontinent.

2.7 The Three Types of Plate Boundaries
■ *What are the three types of plate boundaries?*
The three major types of plate boundaries are divergent, convergent, and transform.

■ *What are divergent boundaries?*
Divergent boundaries, also called spreading ridges, occur where plates are separating and new oceanic lithosphere is forming. Whereas most divergent boundaries occur along the crests of oceanic ridges, they are also present under continents during the early stages of continental breakup.

■ *What features in the geologic record indicate ancient rifting?*
Characteristic features of ancient continental rifting include faulting, dikes, sills, lava flows, and thick sedimentary sequences within rift valleys. Pillow lavas and associated deep-sea sediments are evidence of ancient spreading ridges.

■ *What are convergent boundaries?*
Convergent boundaries are places where two plates collide, and the leading edge of one plate is subducted beneath the margin of the other plate. There are three types of convergent boundaries. Oceanic–oceanic boundaries are where two oceanic plates collide, with one plate subducted beneath the other and a volcanic island arc forming on the nonsubducted plate, parallel to the oceanic trench where subduction is taking place. The volcanoes result from rising magma produced by the partial melting of the subducting plate. An oceanic–continental boundary is where an oceanic plate and a continental plate converge, with the denser oceanic plate being subducted under the continental plate. Just as with an oceanic–oceanic boundary, a chain of volcanoes forms

on the nonsubducted plate. A continental–continental boundary occurs when two continents approach each other and the ocean floor separating them is eventually subducted, resulting in a collision between the two continents. When the two continents collide, they are welded together to form an interior mountain chain along a zone marking the former site of subduction.

■ *How can ancient subduction zones be recognized in the geologic record?*
Intensely deformed rocks, andesite lavas, and ophiolites are all evidence of ancient subduction zones, marking former convergent plate boundaries.

■ *What are transform boundaries?*
These are boundaries along which plates slide laterally past each other along transform faults, which change one type of motion between plates into another type of motion.

2.8 Hot Spots: An Intraplate Feature
■ *What are hot spots and what do they tell us about plate movement?*
A hot spot is the location on Earth's surface where a stationary column of magma, originating deep within the mantle, has slowly risen to the surface and formed a volcano. Because mantle plumes apparently remain stationary within the mantle while plates move over them, the resulting hot spots leave a trail of extinct and progressively older volcanoes that record the movement of the plate.

2.9 Plate Movement and Motion
■ *How can the rate and direction of plate movement be determined?*
The average rate of plate movement is most commonly determined by dividing the distance from an oceanic ridge axis to any magnetic anomaly in the crust of the seafloor by the age of that anomaly. Because magnetic anomalies are parallel and symmetric with respect to spreading ridges, the relative direction of movement of a plate is perpendicular to the spreading ridge. Satellite–laser ranging techniques are also used to determine the rate of movement and relative motion of one plate with respect to another. Hot spots enable geologists to determine absolute motion because they provide an apparently fixed reference point from which the rate and direction of plate movement can be measured.

2.10 The Driving Mechanism of Plate Tectonics
■ *What drives plates?*
Most geologists agree that some type of convective heat system is the basic process responsible for plate motion.

■ *How do thermal convection cells move plates?*
Two models involving thermal convection cells have been proposed to explain plate movement. In one model, thermal cells are restricted to the asthenosphere, whereas in the second model, the entire mantle is involved. Problems with both models involve the source of heat for the convection cells and how heat is transferred from the outer core to the mantle.

■ *Can plate movement be gravity driven?*
Although they accept that some type of thermal convection system is involved in driving plate movement, some geologists think a gravity-driven mechanism such as "slab-pull" or "ridge-push" plays a major role. Both mechanisms still depend on thermal differences within Earth, but slab-pull involves pulling the plate behind a subducting cold slab of lithosphere, and ridge-push involves gravity pushing the oceanic lithosphere away from the higher spreading ridges and toward the subduction trenches.

2.11 Plate Tectonics and the Distribution of Natural Resources
■ *How does plate tectonic theory relate to the origin and distribution of natural resources?*
The formation of many natural resources results from the interaction between plates, and economically valuable concentrations of such deposits are found associated with current and ancient plate boundaries.

■ *What is the relationship between plate boundaries and various metallic mineral deposits?*
Many metallic mineral deposits are related to igneous and associated hydrothermal activity, so it is not surprising that a close relationship exists between plate boundaries and the occurrence of these valuable deposits. Many of the world's major metallic ore deposits are associated with convergent plate boundaries. Divergent plate boundaries also yield valuable ore deposits.

ESSENTIAL TERMS TO KNOW

continental–continental plate boundary (p. 54)
continental drift (p. 35)
convergent plate boundary (p. 52)
Curie point (p. 40)
divergent plate boundary (p. 47)
Glossopteris flora (p. 38)
Gondwana (p. 34)

hot spot (p. 57)
Laurasia (p. 36)
magnetic anomaly (p. 43)
magnetic reversal (p. 41)
oceanic–continental plate boundary (p. 54)
oceanic–oceanic plate boundary (p. 52)
paleomagnetism (p. 40)

Pangaea (p. 35)
plate (p. 46)
plate tectonic theory (p. 34)
seafloor spreading (p. 41)
thermal convection cell (p. 42)
transform fault (p. 55)
transform plate boundary (p. 55)

REVIEW QUESTIONS

1. What evidence convinced Wegener and others that continents must have moved in the past and at one time formed a supercontinent?

2. Why was the continental drift hypothesis proposed by Wegener rejected by so many geologists for so long?

3. How did the theory of seafloor spreading, proposed by Harry Hess in 1962, overcome the objections of those opposed to continental drift?

4. Explain how magnetic anomalies recorded in oceanic crust as well as the sediments deposited on oceanic crust confirm the seafloor spreading theory.

5. Plate tectonic theory builds on the continental drift hypothesis and the theory of seafloor spreading. As such, it is a unifying theory of geology. Explain why it is a unifying theory.

6. Explain why such natural disasters as volcanic eruptions and earthquakes are associated with divergent and convergent plate boundaries.

7. What is the supercontinent cycle? What elements of continental drift and seafloor spreading are embodied in the cycle?

8. Why is some type of thermal convection system thought to be the major force driving plate movement? How have "slab-pull" and "ridge-push," both mainly gravity driven, modified a purely thermal convection model for plate movement?

9. What can hot spots tell us about the absolute direction of plate movement?

10. In addition to the volcanic eruptions and earthquakes associated with convergent and divergent plate boundaries, why are these boundaries also associated with the formation and accumulation of various metallic ore deposits?

APPLY YOUR KNOWLEDGE

1. Using the age for each of the Hawaiian Islands in Figure 2.22 and an atlas in which you can measure the distance between islands, calculate the average rate of movement per year for the Pacific plate since each island formed. Is the average rate of movement the same for each island? Would you expect it to be? Explain why it may not be.

2. Estimate the age of the seafloor crust and the age and thickness of the oldest sediment off the East Coast of North America (e.g., Virginia). In so doing, refer to Figure 2.12 for the ages and to the deep-sea sediment accumulation rate stated in this chapter.

3. If the movement along the San Andreas fault, which separates the Pacific plate from the North American plate, averages 5.5 cm per year, how long will it take before Los Angeles is opposite San Francisco?

4. Based on your knowledge of biology and the distribution of organisms throughout the world, how do you think plate tectonics has affected this distribution both on land and in the oceans?

GEOLOGY MATTERS

GEOLOGY IN FOCUS Oil, Plate Tectonics, and Politics

It is certainly not surprising that oil and politics are closely linked. The Iran–Iraq War of 1980–1989 and the Gulf War of 1990–1991 were both fought over oil (▶ Figure 1). Indeed, many of the conflicts in the Middle East have had as their underlying cause control of the vast deposits of petroleum in the region. Most people, however, are not aware of why there is so much oil in this part of the world.

Although large concentrations of petroleum occur in many areas of the world, more than 50% of all proven reserves are in the Persian Gulf region. It is interesting, however, that this region did not become a significant petroleum-producing area until the economic recovery following World War II (1939–1945). After the war, Western Europe and Japan in particular became dependent on Persian Gulf oil, and they still rely heavily on this region for most of their supply. The United States is also dependent on imports from the Persian Gulf but receives significant quantities of petroleum from other sources such as Mexico and Venezuela.

Why is so much oil in the Persian Gulf region? The answer lies in the ancient geography and plate movements of this region during the Mesozoic and Cenozoic eras. During the Mesozoic Era, and particularly the Cretaceous Period when most of the petroleum formed, the Persian Gulf area was a broad marine shelf extending eastward from Africa. This continental margin lay near the equator where countless microorganisms lived in the surface waters. The remains of these organisms accumulated with the bottom sediments and were buried, beginning the complex process of petroleum generation and the formation of source beds in which the petroleum forms.

As a consequence of rifting in the Red Sea and Gulf of Aden during the Cenozoic Era, the Arabian plate is moving northeast away from Africa and subducting beneath Iran. As the sediments of the continental margin were initially subducted, during the early stages of collision between Arabia and Iran, the heating broke down the organic molecules and led to the formation of petroleum. The tilting of the Arabian block to the northeast allowed the newly formed petroleum to migrate upward into the interior of the Arabian plate. The continued subduction and collision with Iran folded the rocks, creating traps for petroleum to accumulate, such that the vast area south of the collision zone (know as the Zagros suture) is a major oil-producing region.

Figure 1
The Kuwaiti night skies were illuminated by 700 blazing oil wells set on fire by Iraqi troops during the 1991 Gulf War. The fires continued for nine months.

GEOLOGY MATTERS

Geology in Unexpected Places

A Man's Home Is His Castle

Imagine visiting a castle in Devonshire, England, or Glasgow, Scotland, and noticing that the same rocks used in the construction of those castles can be seen in the Catskill Mountains of New York (▶ Figure 1). Whereas most people who visit castles in Great Britain and elsewhere in Europe are learning about the history of the castles—when they were built, why they were built, who lived in them, and other historical facts—geologists frequently are looking at the rocks that make up the walls of a castle and trying to determine what type of rock it is, how old it is, and anything else they can learn about it.

During the Devonian Period (416 to 359 million years ago), the North American continent and what is now Europe were moving toward each other along an oceanic–continental convergent plate boundary. As movement along this boundary continued, the ocean basin separating these landmasses shrunk until the two continents collided in what is known as the Acadian Orogeny. This mountain-building episode formed a large mountain range. Just as is happening today, those mountains began weathering and their eroded sediments were carried by streams and deposited as large deltas in the shallow seas adjacent to the mountains. The sediments deposited in what is now New York are referred to by geologists as the Catskill Delta. The counterpart in Great Britain is known as the Old Red Sandstone. These sediments were deposited in the same environment and reflect the conditions at the time of deposition.

Later during the Mesozoic Era (251 to 66 million years ago), as the supercontinent Pangaea broke apart along divergent plate boundaries, the Atlantic Ocean basin formed, separating North America from Europe. Even though the Devonian rocks of the present-day Catskill Mountains in New York are separated by several thousand kilometers from the Old Red Sandstone rocks of Great Britain, they were formed at the same time and deposited in the same environment hundreds of millions of years ago. That is why they have the same red color and composition and contain many of the same fossils.

So, the next time you see someone closely inspecting the rocks of a castle wall, there is a good chance that person is a geologist or someone, like yourself, who took a geology course.

Figure 1
Remains of Goodrich Castle in Herefordshire, England, dated to the period between 1160 and 1270. Goodrich Castle is one of many castles built from rocks quarried from the Old Red Sandstone, a Devonian-age formation. Identical red sandstone is found in the Catskill Mountains of New York. It too has been used as a building stone for many structures in the New York area.

GEOLOGY MATTERS

Geology and Cultural Connections

The Struggle toward Scientific Progress

Progress in science isn't made easily. It would be nice if a scientist could come up with a good idea and then quickly convince other scientists of its merits, but in fact science is more like one long argument.

When Alfred Wegener proposed his idea of continental drift, it was met by derision from most geologists. They pointed out that Wegener had no convincing explanation for *how* continents moved across entire ocean basins. That was indeed a great weakness in his story, and Wegener's critics capitalized on it. But it wasn't the only issue that they focused on in their attack on Wegener's ideas.

As you read in this chapter, Edward Suess pointed to fossil evidence of similar ancient animals and plants apparently separated by great distances across oceans. Suess thought that land bridges had linked what are now different continents. Wegener, however, took the fossils of Suess and others and argued that the rocks in which the fossils were embedded must have been physically together in one ancient supercontinent (▶ Figure 1). Unfortunately, Wegener was a meteorologist by training, not a paleontologist, and he made mistakes about the details of the fossil evidence he relied on in his argument. Not surprisingly, fossil experts denounced his work because of those errors.

Figure 1
Mesosaurus, a Permian-aged freshwater reptile whose fossil remains are found in Brazil and South Africa, indicating these two continents were joined at the end of the Paleozoic Era.

Wegener also thought that continental movement was quite rapid, much faster than anything we understand today in plate tectonic theory. Indeed, Wegener thought that Greenland was moving away from Europe at speeds up to meters per year—and he died while in Greenland trying to measure that amazing rate of movement. When news of Wegener's disappearance and presumed death in Greenland reached the civilized world, few geologists mourned his passing. The arguments Wegener had spawned had been vociferous even by the standards of scientific disputes.

Progress in science isn't easy, and the arguments that ultimately drive science forward are often unpleasant. In the case of Alfred Wegener, it is now clear that his insights outweighed his errors.

Geology at Work

The Geologic Exploration of Mars

How would you like to have an extra 39 minutes to work or play each day? Well, you would if you lived on Mars, or were part of NASA's (National Aeronautics and Space Administration) Mars Exploration Rover project. With the successful landing on Mars of the two mobile robots *Spirit* and *Opportunity* in January 2004, members of the Mars Exploration Rover team started working on Mars time, or "sols," as the 24-hour, 39-minute Martian day-night cycle became known. They continued to do so for the first few months of the mission, as the two rovers explored the Martian surface. The downside of living on Martian time is that each day begins 39 minutes later than the previous day, which means that, after awhile, your Earth day will be starting in the evening instead of the morning. But most team members would probably agree that is a small price to pay for working on such an exciting voyage of discovery and exploration.

Launched on June 10 and July 7, 2003, respectively, the Mars Exploration Rovers *Spirit* and *Opportunity* successfully landed on opposite sides of Mars on January 3 and 24, 2004 (▶ Figures 1 and 2), and began their primary goal of searching for evidence that Mars once harbored an environment suitable for sustaining life. To achieve this goal, the rovers each carried panoramic cameras that provide 360-degree, stereoscopic, humanlike views of the terrain, as well as being able to focus up close on surface features of the local terrain; a miniature thermal emission spectrometer for identifying minerals at the site; a Moessbauer spectrometer for close-up investigations of the composition of iron-bearing rocks and soils; an alpha particle X-ray spectrometer for close-up analysis of the abundances of elements that make up the Martian rocks and soils; and a microscopic imager for looking at the fine-scale features of the rocks and soils encountered. In addition, they carried a rock abrasion tool for removing the weathered surfaces of rocks to expose the fresh interiors for examination, and magnetic targets for collecting magnetic dust particles.

Shortly after landing both rovers found strong and convincing evidence that Mars harbored abundant quantities of salty water during its early history, several billion years ago. How long this water persisted is still not known, but it does offer the tantalizing prospect that an ancient warmer and wetter Mars might possibly have harbored life. Evidence from *Opportunity* that Mars was awash in water, at least at the Meridiani Planum landing site, was soon forthcoming. Meridiani Planum, a broad, flat, equatorial plain, had been picked as *Opportunity*'s landing site because orbital surveys indicated the presence of gray hematite on the ground, a mineral that usually forms only in the presence of liquid water.

Shortly after landing, *Opportunity* transmitted images of what appeared to be outcrops of layered bedrock, which you would expect to have been deposited in an aqueous environment. However, the most exciting discovery was the presence of what soon came to be called "blueberries," grey BB-size spheres embedded in the rock layers or scattered across the surface as a result of having weathered out from the bedrock. Their resemblance to blueberries in a muffin is what gave them their popular name. These hematite-rich spheres were what the orbital surveys were picking up as the grey layers, making Meridiani Planum such a compelling site. In addition, the spectrometers on *Opportunity* indicated the outcrops were rich in water-deposited sulfate-salt minerals and iron-rich compounds. And lastly, a 0.3-m-long outcrop showed distinct cross-bedding, indicating the sediments were deposited in flowing water.

As exciting as these discoveries are, we still don't know how long water may have existed on Mars, how extensive its "seas" were, and why and how it disappeared so quickly. Nor do we know unequivocally whether Mars ever supported life. *Opportunity* and *Spirit* were not designed to answer such questions, and it will be up to the next generation of robotic rovers or even a manned landing to hopefully answer that question.

Figure 1
Artistic rendition of the Mars Exploration Rover *Opportunity* on the Martian surface at Meridiani Planum. Since landing in January 2004, *Opportunity* has transmitted images of what appear to be outcrops of layered sedimentary rocks, and it has analyzed hematite-rich spheres scattered across the surface that indicate the presence of water during their formation.

Figure 2
Self-portrait of the Mars Exploration Rover *Spirit* taken by its panoramic camera during its 329th and 330th Martian days. The images combined into this mosaic show the Martian surface with numerous pebbles, some of which indicate evidence of wind erosion. Martian dust accumulations are also visible on *Spirit*'s solar panel.

Chapter 3

Minerals—The Building Blocks of Rocks

Blue Agate
*Rings of blue surrounded by a wreath of glistening crystal. Agate forms by filling a hollow inside a host rock. As a result it is often found as a knot-like protuberance with concentric bands that resemble tree rings.
What does the maze-like pattern of blue bands in this slice of agate remind you of?*
—A. W.

ESSENTIAL QUESTIONS TO ASK

3.1 Introduction
- Why do geologists define ice as a mineral but not liquid water or water vapor?
- Are amber and pearls minerals?

3.2 Matter, Atoms, Elements, and Bonding
- What are the atomic number and atomic mass number of an atom?
- What types of chemical bonds are common in minerals, and how do they form?

3.3 Explore the World of Minerals
- Why is it that not all mineral specimens show well-developed crystals but all are crystalline solids?
- How do native elements differ from other minerals?

3.4 Mineral Groups Recognized by Geologists
- Why are there so few common minerals?
- What are the two basic types of silicate minerals, and how do they compare?
- What are carbonate minerals and in what kinds of rocks are they found?

3.5 Mineral Identification
- How does mineral luster differ from color?
- Why is crystal form of limited use in mineral identification?
- What is mineral cleavage, and how is it used in mineral identification?

3.6 The Significance of Rock-Forming Minerals
- How do rock-forming minerals differ from accessory minerals?

3.7 The Origin of Minerals
- What accounts for the origin of minerals from magma?

3.8 Natural Resources and Reserves
- How does a resource differ from a reserve?
- What factors affect the status of a resource?

GEOLOGY MATTERS

GEOLOGY IN FOCUS:
Mineral Crystals

GEOLOGY IN UNEXPECTED PLACES:
The Queen's Jewels

GEOLOGY IN YOUR LIFE:
Welcome to the Wonderful World of Micas

GEOLOGY AT WORK:
Terry S. Mollo—Sculpture Out of Stone

PHYSICAL Geology Now™ — This icon, which appears throughout the book, indicates an opportunity to explore interactive tutorials, animations, or practice problems available on the Physical GeologyNow website at **http://now.brookscole.com/phygeo6**.

3.1 Introduction

■ *Why do geologists define ice as a mineral but not liquid water or water vapor?*

Ice is a mineral because it meets all the criteria in the definition for the term. That is, it is a naturally occurring *crystalline solid*, meaning that its atoms are arranged in a specific three-dimensional pattern, as opposed to liquids and gases which have no such orderly arrangement of atoms. Furthermore, ice has a specific chemical composition (H_2O), and just as all other minerals, it has characteristic physical properties such as hardness, density, and color. To summarize then, a **mineral** is a naturally occurring, crystalline solid, with a narrowly defined chemical composition, and distinctive physical properties. However, the term mineral also brings to mind substances we need for good nutrition, such as calcium, iron, and magnesium. But these are actually chemical elements, not minerals, at least in the geologic sense.

The importance of minerals in many human endeavors cannot be overstated. We rely on minerals and rocks (natural resources) to sustain our industrialized societies and our standard of living. Iron ore, minerals for abrasives, glass, and cement, as well as minerals and rocks used in animal feed supplements and fertilizers are essential for our economic well-being. Much of the ink used in the brilliantly colored medieval manuscripts came from various minerals, and the luster of lipstick, glitter, eye shadow, and paints on appliances and automobiles comes from minerals.

The United States and Canada owe much of their economic success to the availability of abundant natural resources, although both countries must import some essential commodities, thus accounting for political and economic ties with other nations. Indeed, the distribution of various resources such as copper, gold, cobalt, manganese, petroleum, and aluminum ore is one essential consideration in foreign policy decisions.

An important reason for you to study minerals is that they are the building blocks of rocks; so rocks, with few exceptions, are combinations of one or more minerals. And some minerals are attractive and eagerly sought by private collectors, or for museum displays, or for collections of crown jewels. A *gemstone* is any precious or semiprecious mineral or rock used for decoration, especially for jewelry. Many people own small precious gemstones, such as diamonds, sapphires, and emeralds, and perhaps some semiprecious ones, such as garnet, peridot, and turquoise (▶ Figure 3.1). The lore asso-

▶ **Figure 3.1 Minerals in Museum Displays**

a A spectacular specimen of tourmaline (the elongate minerals) and quartz (colorless) from the Himalaya Mine in San Diego County, California. Notice the change in color along the lengths of the tourmaline crystals.

b Turquoise is a sky-blue, blue-green, or light green semiprecious gemstone used in jewelry and as a decorative stone.

ciated with gemstones, such as relating them to one's birth month, makes them appealing to many people.

■ *Are amber and pearls minerals?*

Amber and pearl are included among the semiprecious gemstones, but are they actually minerals? Amber is hardened resin (sap), which is not crystalline. It comes from coniferous trees and is thus an organic substance and not a mineral, but it is nevertheless prized as a decorative "stone" (▶ Figure 3.2a). Amber is best known from the Baltic Sea region of Europe, where sun-worshipping cultures, noting its golden translucence resembling the Sun's rays, believed it had mystical powers.

Pearls form when mollusks, such as clams or oysters, deposit successive layers of tiny mineral crystals around some irritant, perhaps a sand grain. Most pearls are lustrous white, but some are silver gray, green, or black (▶ Figure 3.2b). Unlike other gemstones, pearls need no shaping or polishing before they are used in jewelry; that is, they are essentially ready to use when found.

Section 3.1 Summary

- Minerals are essential constituents of rocks. Many minerals are important natural resources or valued as gemstones.

▶ **Figure 3.2 Semiprecious Gemstones**

a An insect preserved in amber, which is hardened resin (sap) from coniferous trees.

b These black pearls, valued at about $13,000, are on display at Maui Pearls on the island of Roratonga, which is part of the Cook Islands in the South Pacific.

3.2 Matter, Atoms, Elements, and Bonding

Matter is anything that has mass and occupies space and accordingly includes air, water, animals, plants, minerals, and rocks. Physicists recognize four states of matter: *liquids, gases, solids,* and *plasma.* The latter is an ionized gas as in fluorescent and neon lights and the matter in the Sun and stars. Liquids, including groundwater and surface water, as well as atmospheric gases will be important in our discussions of surface processes such as running water and wind, but here our main concern is solids because by definition minerals are solids. So the next question is: What is matter made of?

Atoms and Elements

Matter is made up of chemical *elements*, which in turn are composed of tiny particles known as atoms (▶ Figure 3.3). **Atoms** are the smallest units of matter that retain the characteristics of a particular element. That is, they cannot be split or converted into different substances except in radioactive decay (discussed in Chapter 9). Thus an **element** is made up of atoms that all have the same properties. Scientists have discovered 92 naturally occurring elements, and several others have been made in laboratories. All elements have a name and a symbol—for example, oxygen (O), aluminum (Al), and iron (Fe) (▶ Figure 3.4).

At the center of an atom is a tiny **nucleus** made up of one or more particles known as **protons,** which have a positive electrical charge, and **neutrons,** which are electrically neutral (▶ Figure 3.3). The nucleus is only about 1/100,000 of the diameter of an atom, yet it contains nearly all of the atom's mass. Particles called **electrons** with a negative electrical charge orbit rapidly around the nucleus at specific distances in one or more **electronic shells.** The electrons determine how an atom interacts with other atoms, but the nucleus determines how many electrons an atom has because the positively charged protons attract and hold the negatively charged electrons in their orbits.

▶ **Figure 3.3 Shell Models for Common Atoms** The shell model for several atoms and their electron configurations. A blue circle represents the nucleus of each atom, but remember that atomic nuclei are made up of protons and neutrons, as shown in Figure 3.5.

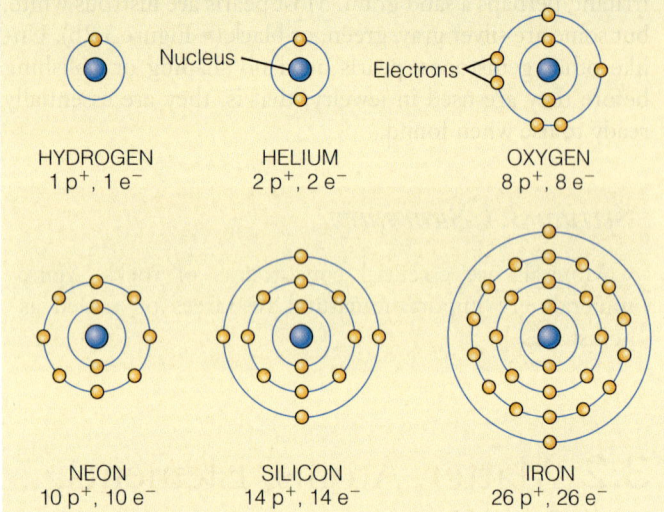

Element	Symbol	Atomic Number	Distribution of Electrons: First Shell	Second Shell	Third Shell	Fourth Shell
Hydrogen	H	1	1	—	—	—
Helium	He	2	2	—	—	—
Carbon	C	6	2	4	—	—
Oxygen	O	8	2	6	—	—
Neon	Ne	10	2	8	—	—
Sodium	Na	11	2	8	1	—
Magnesium	Mg	12	2	8	2	—
Aluminum	Al	13	2	8	3	—
Silicon	Si	14	2	8	4	—
Phosphorus	P	15	2	8	5	—
Sulfur	S	16	2	8	6	—
Chlorine	Cl	17	2	8	7	—
Potassium	K	19	2	8	8	1
Calcium	Ca	20	2	8	8	2
Iron	Fe	26	2	8	14	2

■ *What are the atomic number and atomic mass number of an atom?*

The number of protons in its nucleus determines an atom's identity and its **atomic number.** Hydrogen (H), for instance, has 1 proton in its nucleus and thus has an atomic number of 1. The nuclei of helium (He) atoms possess 2 protons, whereas those of carbon (C) have 6, and uranium (U) have 92, so their atomic numbers are 2, 6, and 92, respectively. Atoms also have an **atomic mass number,** which is the sum of protons and neutrons in the nucleus (electrons contribute negli-

▶ **Figure 3.4 The Periodic Table of Elements** Only about a dozen elements are common in minerals and rocks, but many uncommon ones are important sources of natural resources. For example, lead (Pb) is not found in many minerals, but it is present in the mineral galena, the main ore of lead. Silicon (Si) and oxygen (O), in contrast, are important elements in most of the minerals in Earth's crust.

MAIN GROUP METALS
TRANSITION METALS
METALLOIDS
NONMETALS

Uranium
92 — Atomic number
U — Symbol
238.0289 — Atomic mass number

	1A (1)	2A (2)	3B (3)	4B (4)	5B (5)	6B (6)	7B (7)	8B (8)	8B (9)	8B (10)	1B (11)	2B (12)	3A (13)	4A (14)	5A (15)	6A (16)	7A (17)	8A (18)
1	Hydrogen 1 H 1.0079																	Helium 2 He 4.0026
2	Lithium 3 Li 6.941	Beryllium 4 Be 9.0122											Boron 5 B 10.811	Carbon 6 C 12.011	Nitrogen 7 N 14.0067	Oxygen 8 O 15.9994	Fluorine 9 F 18.9984	Neon 10 Ne 20.1797
3	Sodium 11 Na 22.9898	Magnesium 12 Mg 24.3050											Aluminum 13 Al 26.9815	Silicon 14 Si 28.0855	Phosphorus 15 P 30.9738	Sulfur 16 S 32.066	Chlorine 17 Cl 35.4527	Argon 18 Ar 39.948
4	Potassium 19 K 39.0983	Calcium 20 Ca 40.078	Scandium 21 Sc 44.9559	Titanium 22 Ti 47.867	Vanadium 23 V 50.9415	Chromium 24 Cr 51.9961	Manganese 25 Mn 54.9380	Iron 26 Fe 55.845	Cobalt 27 Co 58.9332	Nickel 28 Ni 58.6934	Copper 29 Cu 63.546	Zinc 30 Zn 65.39	Gallium 31 Ga 69.723	Germanium 32 Ge 72.61	Arsenic 33 As 74.9216	Selenium 34 Se 78.96	Bromine 35 Br 79.904	Krypton 36 Kr 83.80
5	Rubidium 37 Rb 85.4678	Strontium 38 Sr 87.62	Yttrium 39 Y 88.9059	Zirconium 40 Zr 91.224	Niobium 41 Nb 92.9064	Molybdenum 42 Mo 95.94	Technetium 43 Tc (97.907)	Ruthenium 44 Ru 101.07	Rhodium 45 Rh 102.9055	Palladium 46 Pd 106.42	Silver 47 Ag 107.8682	Cadmium 48 Cd 112.411	Indium 49 In 114.818	Tin 50 Sn 118.710	Antimony 51 Sb 121.760	Tellurium 52 Te 127.60	Iodine 53 I 126.9045	Xenon 54 Xe 131.29
6	Cesium 55 Cs 132.9054	Barium 56 Ba 137.327	Lanthanum 57 La 138.9055	Hafnium 72 Hf 178.49	Tantalum 73 Ta 180.9479	Tungsten 74 W 183.84	Rhenium 75 Re 186.207	Osmium 76 Os 190.2	Iridium 77 Ir 192.22	Platinum 78 Pt 195.08	Gold 79 Au 196.9665	Mercury 80 Hg 200.59	Thallium 81 Tl 204.3833	Lead 82 Pb 207.2	Bismuth 83 Bi 208.9804	Polonium 84 Po (208.98)	Astatine 85 At (209.99)	Radon 86 Rn (222.02)
7	Francium 87 Fr (223.02)	Radium 88 Ra (226.0254)	Actinium 89 Ac (227.0278)	Rutherfordium 104 Rf (261.11)	Dubnium 105 Db (262.11)	Seaborgium 106 Sg (263.12)	Bohrium 107 Bh (262.12)	Hassium 108 Hs (265)	Meitnerium 109 Mt (266)	Darmstadtium 110 Ds (271)	Roentgenium 111 Rg (272)	— 112 — Discovered 1996	— 113 — Discovered 2004	— 114 — Discovered 1999	— 115 — Discovered 2004	— 116 — Discovered 1999		

Lanthanides	Cerium 58 Ce 140.115	Praseodymium 59 Pr 140.9076	Neodymium 60 Nd 144.24	Promethium 61 Pm (144.91)	Samarium 62 Sm 150.36	Europium 63 Eu 151.965	Gadolinium 64 Gd 157.25	Terbium 65 Tb 158.9253	Dysprosium 66 Dy 162.50	Holmium 67 Ho 164.9303	Erbium 68 Er 167.26	Thulium 69 Tm 168.9342	Ytterbium 70 Yb 173.04	Lutetium 71 Lu 174.967
Actinides	Thorium 90 Th 232.0381	Protactinium 91 Pa 231.0388	Uranium 92 U 238.0289	Neptunium 93 Np (237.0482)	Plutonium 94 Pu (244.664)	Americium 95 Am (243.061)	Curium 96 Cm (247.07)	Berkelium 97 Bk (247.07)	Californium 98 Cf (251.08)	Einsteinium 99 Es (252.08)	Fermium 100 Fm (257.10)	Mendelevium 101 Md (258.10)	Nobelium 102 No (259.10)	Lawrencium 103 Lr (262.11)

Note: Atomic masses are 1993 IUPAC values (up to four decimal places). Numbers in parentheses are atomic masses or mass numbers of the most stable isotope of an element.

gible mass to atoms). However, atoms of the same chemical element might have different atomic mass numbers because the number of neutrons can vary. All carbon (C) atoms have 6 protons—otherwise they would not be carbon—but the number of neutrons may be 12, 13, or 14. Thus, we recognize three types of carbon, each with a different atomic mass number, or what are known as *isotopes* (▶ Figure 3.5).

These isotopes of carbon, or those of any other element, behave the same chemically; carbon 12 and carbon 14 are both present in carbon dioxide (CO_2), for example. However, isotopes of some elements are radioactive, meaning that they spontaneously decay or change to other stable elements. Carbon 14 is radioactive, whereas both carbon 12 and carbon 13 are stable. Radioactive isotopes are important for determining the absolute ages of rocks (see Chapter 9).

Bonding and Compounds

Interactions among electrons around atoms can result in two or more atoms joining together, a process known as **bonding**. If atoms of two or more elements bond, the resulting substance is a **compound**. Gaseous oxygen consists of only oxygen atoms and is thus an element, whereas the mineral quartz, consisting of silicon and oxygen atoms, is a compound. Most minerals are compounds, although gold, platinum, and several others are important exceptions.

To understand bonding, we must delve deeper into the structure of atoms. Recall that negatively charged electrons orbit the nuclei of atoms in electron shells (▶ Figure 3.3). With the exception of hydrogen, which has only one proton and one electron, the innermost electron shell of an atom contains only two electrons. The other shells contain various numbers of electrons, but the outermost shell never has more than eight (▶ Figure 3.3). The electrons in the outermost shell are those that are usually involved in chemical bonding.

■ *What types of chemical bonds are common in minerals and how do they form?*

Two types of chemical bonds, *ionic* and *covalent*, are particularly important in minerals, and many minerals contain both types of bonds. Two other types of chemical bonds, *metallic* and *van der Waals*, are much less common but extremely important in determining the properties of some useful minerals.

Ionic Bonding Notice in Figure 3.3 that most atoms have fewer than eight electrons in their outermost electron shell. However, some elements, including neon and argon, have complete outer shells containing eight electrons; because of this electron configuration, these elements, known as the *noble gases*, do not react readily with other elements to form compounds. Interactions among atoms tend to produce electron configurations similar to those of the noble gases. That is, atoms interact so that their outermost electron shell is filled with eight electrons, unless the first shell (with two electrons) is also the outermost electron shell, as in helium.

One way for an atom to attain the noble gas configuration is by the transfer of one or more electrons from one atom to another. Common salt is composed of the elements sodium (Na) and chlorine (Cl), each of which is poisonous, but when combined chemically they form sodium chloride (NaCl), the mineral halite, better known as ordinary salt. Notice in Figure 3.6a that sodium has 11 protons and 11 electrons; thus the positive electrical charges of the protons are exactly balanced by the negative charges of the electrons, and the atom is electrically neutral. Likewise, chlorine with 17 protons and 17 electrons is electrically neutral (▶ Figure 3.6a). But neither sodium nor chlorine has 8 electrons in its outermost electron shell; sodium has only 1, whereas chlorine has 7. To attain a stable configuration, sodium loses the electron in its outermost electron shell, leaving its next shell with 8 electrons as the outermost one (▶ Figure 3.6a). Sodium now has one fewer electron (negative charge) than it has protons (positive charge), so it is an electrically charged **ion** and is symbolized Na^{+1}.

The electron lost by sodium is transferred to the outermost electron shell of chlorine, which had 7 electrons to begin with. The addition of one more electron gives chlorine an outermost electron shell of 8 electrons, the configuration of a noble gas. But its total number of electrons is now 18, which exceeds by 1 the number of protons. Accordingly, chlorine also becomes an ion, but it is negatively charged (Cl^{-1}). An **ionic bond** forms between sodium and chlorine because of the attractive force between the positively charged sodium ion and the negatively charged chlorine ion (▶ Figure 3.6a).

In ionic compounds, such as sodium chloride (the mineral halite), the ions are arranged in a three-dimensional framework that results in overall electrical neutrality. In halite, sodium ions are bonded to chlorine ions on all sides, and chlorine ions are surrounded by sodium ions (▶ Figure 3.6b).

▶ **Figure 3.5 Isotopes of Carbon** Schematic representation of the isotopes of carbon. Carbon has an atomic number of 6 and an atomic mass number of 12, 13, or 14, depending on the number of neutrons (n) in its nucleus.

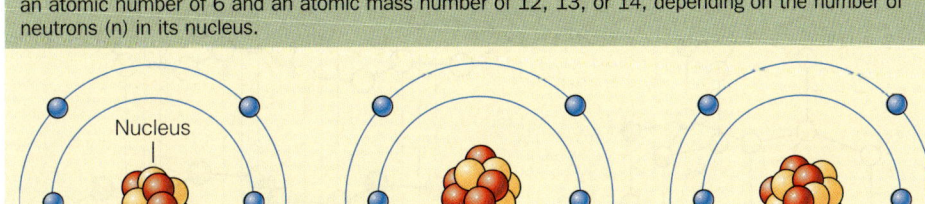

^{12}C (Carbon 12) ^{13}C (Carbon 13) ^{14}C (Carbon 14)

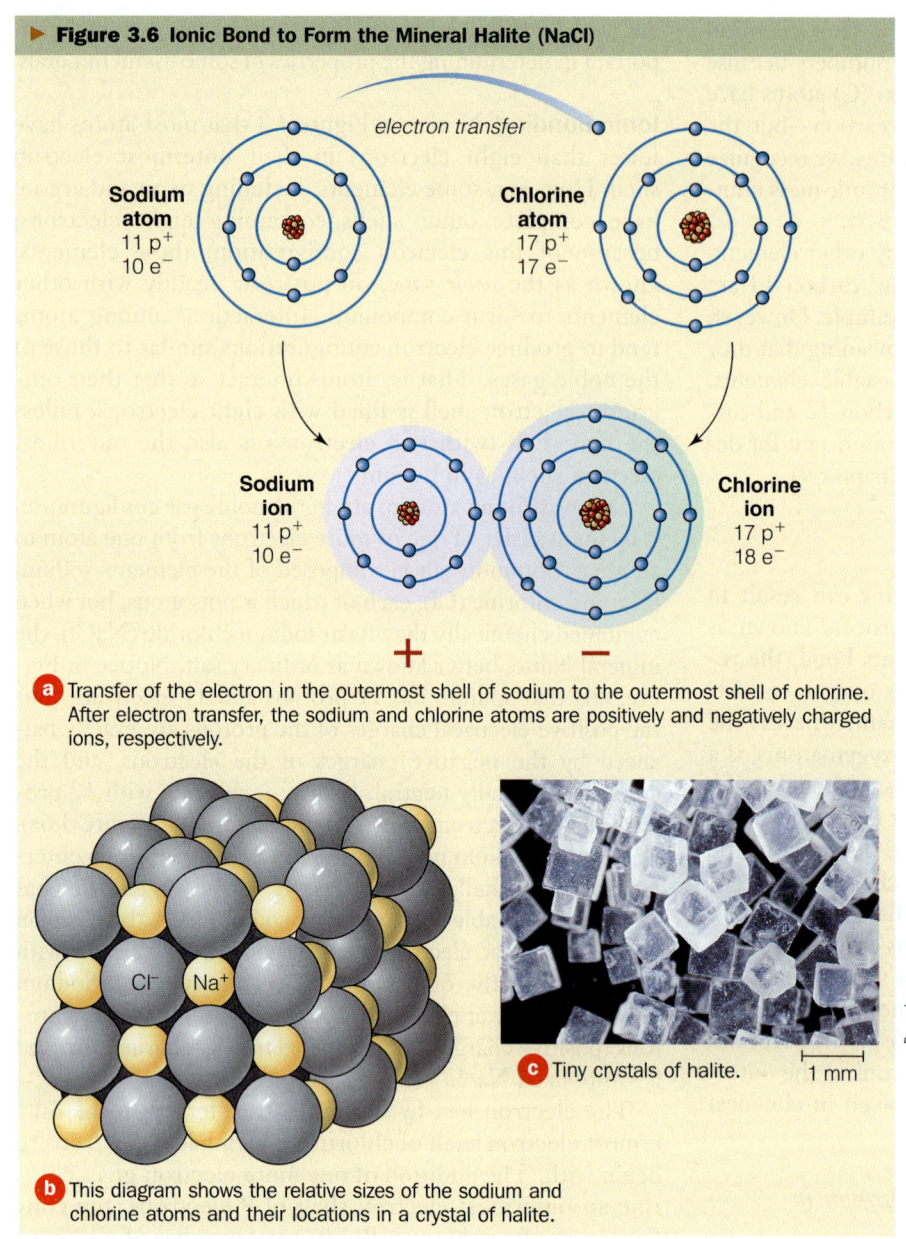

Figure 3.6 Ionic Bond to Form the Mineral Halite (NaCl)

a Transfer of the electron in the outermost shell of sodium to the outermost shell of chlorine. After electron transfer, the sodium and chlorine atoms are positively and negatively charged ions, respectively.

c Tiny crystals of halite.

b This diagram shows the relative sizes of the sodium and chlorine atoms and their locations in a crystal of halite.

Covalent Bonding Covalent bonds form between atoms when their electron shells overlap and they share electrons. For example, atoms of the same element cannot bond by transferring electrons from one atom to another. Carbon (C), which forms the minerals graphite and diamond, has four electrons in its outermost electron shell (▶ Figure 3.7a). If these four electrons were transferred to another carbon atom, the atom receiving the electrons would have the noble gas configuration of eight electrons in its outermost electron shell, but the atom contributing the electrons would not.

In these situations, adjacent atoms share electrons by overlapping their electron shells. A carbon atom in diamond shares all four of its outermost electrons with a neighbor to produce a stable noble gas configuration (▶ Figure 3.7a).

Covalent bonds are not restricted to substances composed of atoms of a single kind. Among the most common minerals, the silicates (discussed later in this chapter), the element silicon forms partly covalent and partly ionic bonds with oxygen.

Metallic and van der Waals Bonds
Metallic bonding results from an extreme type of electron sharing. The electrons of the outermost electron shell of metals such as gold, silver, and copper readily move about from one atom to another. This electron mobil-

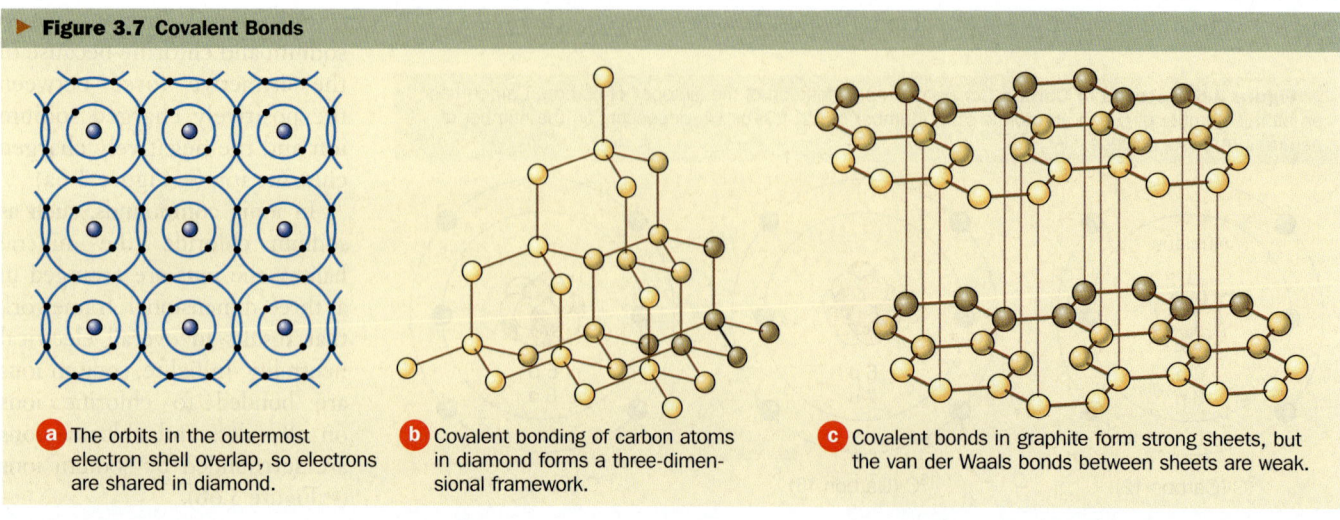

Figure 3.7 Covalent Bonds

a The orbits in the outermost electron shell overlap, so electrons are shared in diamond.

b Covalent bonding of carbon atoms in diamond forms a three-dimensional framework.

c Covalent bonds in graphite form strong sheets, but the van der Waals bonds between sheets are weak.

ity accounts for the fact that metals have a metallic luster (their appearance in reflected light), provide good electrical and thermal conductivity, and can be easily reshaped. Only a few minerals possess metallic bonds, but those that do are very useful; copper, for example, is used for electrical wiring because of its high electrical conductivity.

Some electrically neutral atoms and molecules* have no electrons available for ionic, covalent, or metallic bonding. Nevertheless, when in proximity they have a weak attractive force between them known as a *van der Waals bond* or *residual bond*. The carbon atoms in the mineral graphite are covalently bonded to form sheets, but the sheets are weakly held together by van der Waals bonds (▶ Figure 3.7c). This type of bonding makes graphite useful for pencil leads; when a pencil is moved across a piece of paper, small pieces of graphite flake off along the planes held together by van der Waals bonds and adhere to the paper.

Section 3.2 Summary

- Matter is composed of chemical elements, each of which consists of atoms with protons and neutrons in their nuclei and electrons orbiting the nuclei in electron shells.

- The number of protons in an atom's nucleus determines its atomic number. Its atomic mass number is the number of protons and neutrons in its nucleus.

- Ionic and covalent bonds form as electrons in the outermost electron shells of atoms of different elements interact to form compounds. Most minerals are compounds. Metallic and van der Waals bonds are important in a few minerals.

PHYSICAL Geology⇋Now™ Log into GeologyNow and select this chapter to work through a **Geology Interactive** activity on "Atomic Behavior" (click Atoms and Crystals→Atomic Behavior).

3.3 Explore the World of Minerals

In Section 3.1 we defined a *mineral* as an inorganic, naturally occurring crystalline solid with a narrowly defined chemical composition and characteristic physical properties. Furthermore, we know from Section 3.2 on bonding and compounds that most minerals are compounds of two or more chemically bonded elements, as in quartz (SiO_2). In the following sections we examine each part of the formal definition of the term *mineral*.

---

*A molecule is the smallest unit of a substance that has the properties of that substance. A water molecule (H_2O), for example, possesses two hydrogen atoms and one oxygen atom.

Naturally Occurring Inorganic Substances

The criterion *naturally occurring* excludes from minerals all manufactured substances. Accordingly, most geologists do not regard synthetic diamonds and rubies and other artificially synthesized substances as minerals. This criterion is particularly important to those who buy and sell gemstones, most of which are minerals, because some human-made substances are very difficult to distinguish from natural gem minerals.

Some geologists think the term *inorganic* in the mineral definition is superfluous. It does remind us that animal matter and vegetable matter are not minerals. Nevertheless, some organisms, including corals, clams, and a number of other animals and plants, construct their shells of calcium carbonate ($CaCO_3$), which is either the mineral aragonite or calcite, or their shells are made of silicon dioxide (SiO_2), as in the mineral quartz.

Mineral Crystals

■ *Why is it that not all mineral specimens show well-developed crystals but all are crystalline solids?*

By definition minerals are **crystalline solids** in which the constituent atoms are arranged in a regular, three-dimensional framework (▶ Figure 3.6b). Under ideal conditions, such as in a cavity, mineral crystals grow and form perfect crystals with planar surfaces (crystal faces), sharp corners, and straight edges (▶ Figure 3.8). In other words, the regular geometric shape of a well-formed mineral crystal is the exterior manifestation of an ordered internal atomic arrangement. Not all rigid substances are crystalline solids; natural and manufactured glass lacks the ordered arrangement of atoms and is said to be *amorphous*, meaning "without form."

In the preceding paragraph we used the terms *crystalline* and *crystal*. Keep in mind that *crystalline* refers to a solid with a regular three-dimensional internal framework of atoms, whereas a **crystal** is a geometric shape with planar faces (crystal faces), sharp corners, and straight edges. Minerals are by definition crystalline solids, but crystalline solids do not necessarily always yield well-formed crystals. The reason is that when crystals form, they may grow in proximity and form an interlocking mosaic in which individual crystals are not apparent or easily discerned (▶ Figure 3.9).

So how do we know that the mass of minerals in Figure 3.9b is actually crystalline? X-ray beams and light transmitted through mineral crystals or crystalline masses behave in a predictable manner, providing compelling evidence for an internal orderly structure. Another way we can determine that minerals with no obvious crystals are actually crystalline is by their **cleavage,** the property of breaking or splitting repeatedly along smooth, closely spaced planes. Not all minerals have cleavage planes, but many do, and such regularity certainly indicates that splitting is controlled by internal structure (see Figure 3.17).

As early as 1669, the Danish scientist Nicolas Steno determined that the angles of intersection of equivalent crystal faces on different specimens of quartz are identical. Since

Figure 3.8 A Variety of Mineral Crystal Shapes

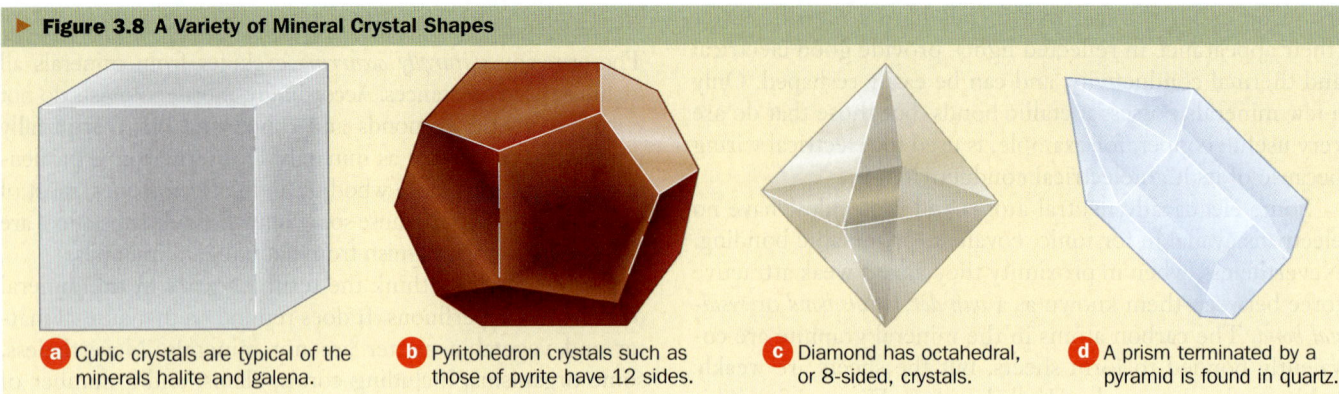

a Cubic crystals are typical of the minerals halite and galena.

b Pyritohedron crystals such as those of pyrite have 12 sides.

c Diamond has octahedral, or 8-sided, crystals.

d A prism terminated by a pyramid is found in quartz.

then, this *constancy of interfacial angles* has been demonstrated for many other minerals, regardless of their size, shape, age, or geographic occurrence. (▶ Figure 3.9c). Steno postulated that mineral crystals are made up of very small, identical building blocks, and that the arrangement of these building blocks determines the external form of mineral crystals. In short, he proposed that external form results from internal structure, a proposal that has since been verified.

Chemical Composition of Minerals

■ *How do native elements differ from other minerals?*

Unlike most minerals, **native elements** are made up of a single chemical element as in gold (Au), silver (Ag), and diamond and graphite, both of which are composed of carbon (C). Whether composed of a single chemical ele-

Figure 3.9 Quartz

a Well-shaped crystal of smoky quartz.

b Specimen of rose quartz in which no obvious crystals can be discerned.

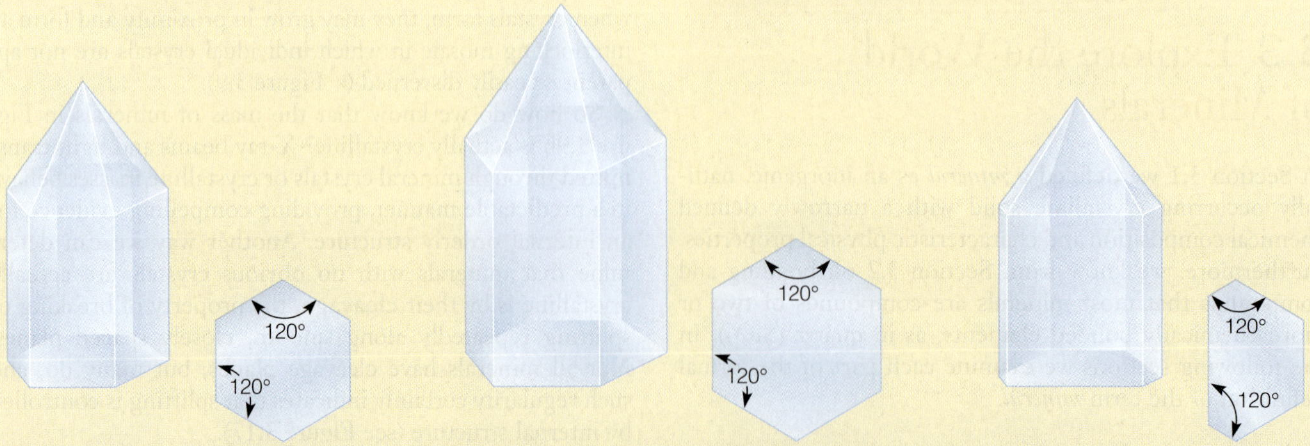

c Side views and cross sections of quartz crystals showing the constancy of interfacial angles. A well-shaped crystal (left), a larger well-shaped crystal (middle), and a poorly shaped crystal (right). The angles formed between equivalent crystal faces on different specimens of the same mineral are the same regardless of size, shape, age, or geographic occurrence of the specimens.

ment or several, mineral composition is shown by a chemical formula, which is a shorthand way of indicating the numbers of atoms of different elements present. For instance, the mineral quartz consists of one silicon (Si) atom for every two oxygen (O) atoms and thus has the formula SiO_2; the subscript number indicates the number of atoms. Orthoclase is composed of one potassium, one aluminum, three silicon, and eight oxygen atoms, so its formula is $KAlSi_3O_8$.

The definition of a mineral includes the phrase *a narrowly defined chemical composition* because some minerals actually have a range of compositions. For many minerals, the chemical composition does not vary. Quartz is always composed of silicon and oxygen (SiO_2), and halite contains only sodium and chlorine (NaCl). Other minerals have a range of compositions because one element can substitute for another if the atoms of two or more elements are nearly the same size and the same charge. Notice in ▶ Figure 3.10 that iron and magnesium atoms are about the same size; therefore they can substitute for each other. The chemical formula for the mineral olivine is $(Mg,Fe)_2SiO_4$, meaning that, in addition to silicon and oxygen, it may contain only magnesium, only iron, or a combination of both. As a matter of fact, the term *olivine* is usually applied to minerals that contain both iron and magnesium, whereas forsterite is olivine with only magnesium (Mg_2SiO_4) and olivine with only iron is fayalite (Fe_2SiO_4). A number of other minerals also have ranges of compositions, so these are actually mineral groups with several members.

Physical Properties of Minerals

The last criterion in our definition of a mineral, *characteristic physical properties*, refers to such properties as hardness, color, and crystal form. These properties are controlled by composition and structure. We have more to say about physical properties of minerals later in this chapter.

Section 3.3 Summary

- Minerals are crystalline solids, meaning they have an ordered internal arrangement of atoms. They are also naturally occurring and inorganic, and have a narrowly defined chemical composition and characteristic physical properties.

- A mineral's composition is indicated by a chemical formula, such as $KAlSi_3O_8$ for orthoclase. A few minerals including gold and silver are made up of a single element and are known as native elements.

- Some minerals have a range of compositions because one element can substitute for another if their atoms are about the same size and have the same electrical charge.

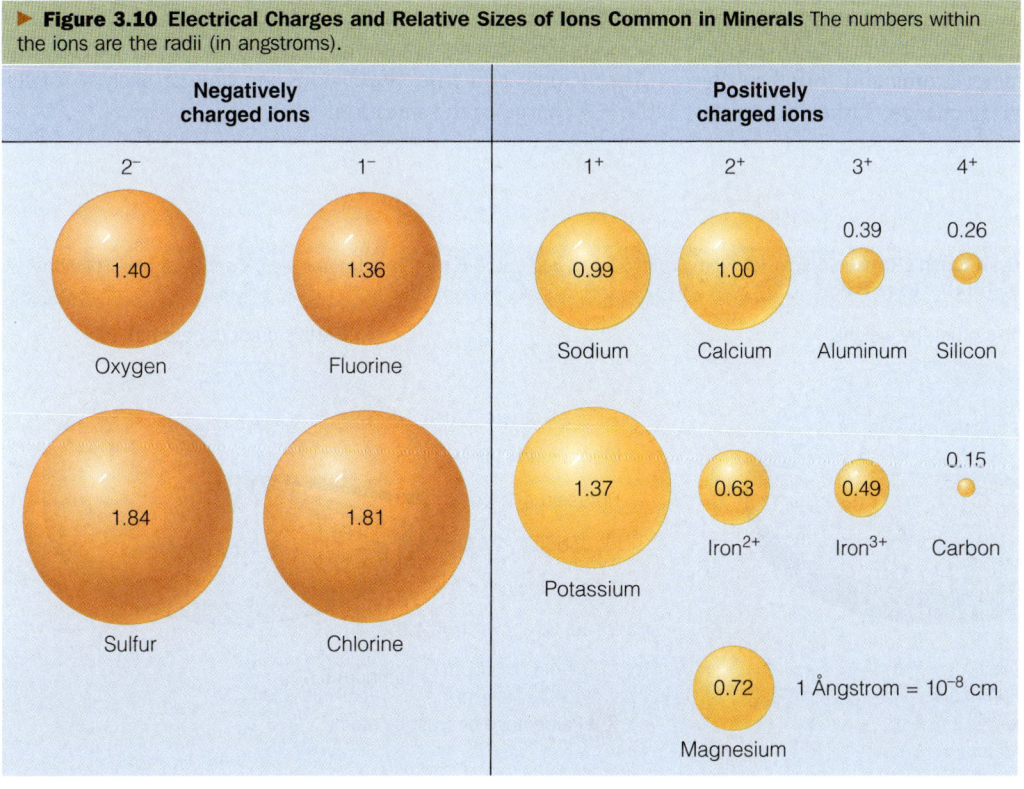

▶ **Figure 3.10 Electrical Charges and Relative Sizes of Ions Common in Minerals** The numbers within the ions are the radii (in angstroms).

3.4 Mineral Groups Recognized by Geologists

■ *Why are there so few common minerals?*

Geologists have identified and described more than 3500 minerals, but only a few—perhaps two dozen—are very common. Why are so few minerals common when the number of possible combinations of 92 naturally occurring elements is enormous? Two factors limit the number of possible minerals. First, many combinations of elements do not occur; no compounds are made up of only positively charged ions or only negatively charged ions, so there are no minerals composed of only potassium and sodium or silicon and iron. Second, the bulk of Earth's crust consists of only eight chemical elements (▶ Figure 3.11). Indeed, oxygen and silicon make up more than 74% (by weight) of the crust and nearly 84% of the atoms available to form minerals. From these percentages you might think that most minerals consist of oxygen and silicon combined with one or more of the other elements listed in Figure 3.11, and you would be correct.

Now you know that more than 3500 minerals are known and you no doubt suspect that they are categorized in some fashion. So what criteria do geologists use to place minerals into categories or groups? The most obvious mineral properties—size, shape, and color—are not adequate for this purpose, but composition is, particularly for minerals that share the same negatively charged ion or ion group (Table 3.1).

Remember that ions are atoms with either a positive or negative electrical charge that results from the loss or gain of electrons in their outermost electron shell. In addition to ions, some minerals contain tightly bonded, complex groups of ions known as *radicals* that act like single units. A good example is the carbonate radical, which consists of a carbon atom bonded to three oxygen atoms and thus has the formula CO_3 and a -2 electrical charge. Other common radicals and their charges are sulfate (SO_4, -2), hydroxyl (OH, -1), and silicate (SiO_4, -4) (▶ Figure 3.12).

Silicate Minerals

Because silicon and oxygen are the two most abundant elements in Earth's crust, it is not surprising that many minerals contain these elements. A combination of silicon and oxygen is known as **silica,** and minerals that contain silica are **silicates.** Quartz (SiO_2) is pure silica because it is composed entirely of silicon and oxygen. But most silicates have one or more additional elements, as in orthoclase ($KAlSi_3O_8$) and olivine [$(Fe,Mg)_2SiO_4$]. Silicate minerals include about one-third of all known minerals, but their abundance is even more impressive when you consider that they make up perhaps 95% of Earth's crust.

The basic building block of silicate minerals is the **silica tetrahedron,** consisting of one silicon atom and four oxygen atoms (▶ Figure 3.13a). These atoms are arranged so that the four oxygen atoms surround a silicon atom that occupies the space between the oxygen atoms, thus forming a four-faced pyramidal structure. The silicon atom has a positive charge of 4, and each of the four oxygen atoms has a negative charge of 2, resulting in a radical with a total negative charge of 4 $(SiO_4)^{-4}$ (▶ Figure 3.12).

Because the silica tetrahedron has a negative charge, it does not exist in nature as an isolated ion group; rather, it combines with positively charged ions or shares its oxygen atoms with other silica tetrahedra. In the simplest silicate minerals, the silica tetrahedra exist as single units bonded to positively charged ions. In minerals containing isolated tetrahedra, the silicon-to-oxygen ratio is 1:4, and the negative charge of the silica ion is balanced by positive ions (▶ Figure 3.13c). Olivine [$(Fe,Mg)_2SiO_4$], for example, has either two magnesium (Mg^{+2}) ions, two iron (Fe^{-2}) ions, or one of each to offset the -4 charge of the silica ion.

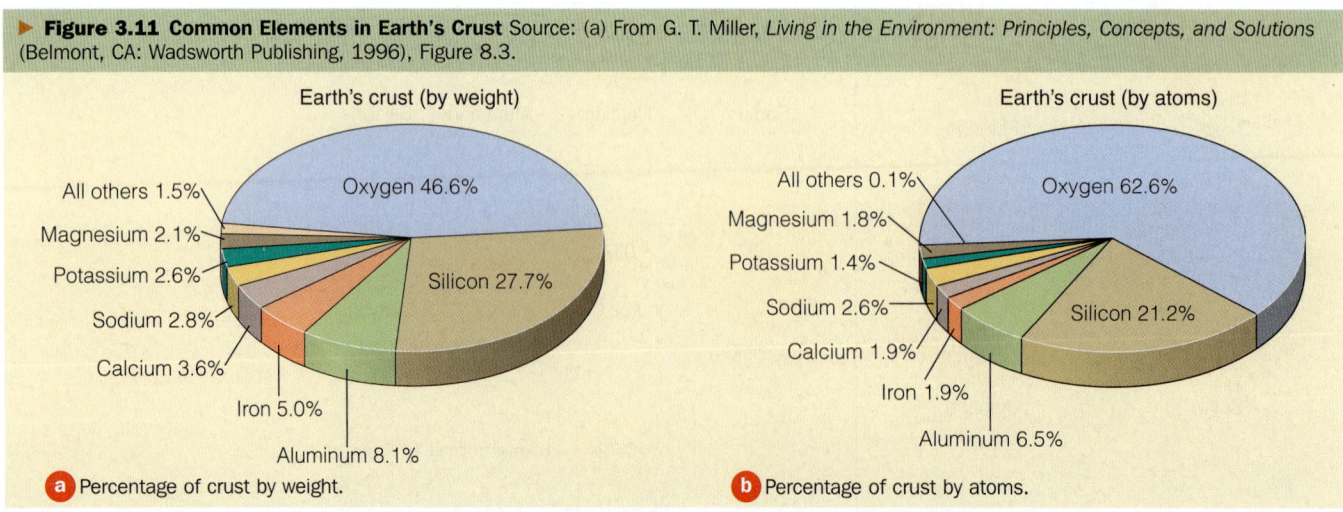

▶ **Figure 3.11 Common Elements in Earth's Crust** Source: (a) From G. T. Miller, *Living in the Environment: Principles, Concepts, and Solutions* (Belmont, CA: Wadsworth Publishing, 1996), Figure 8.3.

a Percentage of crust by weight.

b Percentage of crust by atoms.

Table 3.1
Mineral Groups Recognized by Geologists

Mineral Group	Negatively Charged Ion or Radical	Examples	Composition
Carbonate	$(CO_3)^{-2}$	Calcite	$CaCO_3$
		Dolomite	$CaMg(CO_3)_2$
Halide	Cl^{-1}, F^{-1}	Halite	$NaCl$
		Fluorite	CaF_2
Hydroxide	$(OH)^{-1}$	Limonite	$FeO(OH) \cdot nH_2O$
Native element	—	Gold	Au
		Silver	Ag*
		Diamond	C
Phosphate	$(PO_4)^{-3}$	Apatite	$Ca_5(PO_4)_3(F,Cl)$
Oxide	O^{-2}	Hematite	Fe_2O_3
		Magnetite	Fe_3O_4
Silicate	$(SiO_4)^{-4}$	Quartz	SiO_2
		Potassium feldspar	$KAlSi_3O_8$
		Olivine	$[(Fe,Mg)_2SiO_4]$
Sulfate	$(SO_4)^{-2}$	Anhydrite	$CaSO_4$
		Gypsum	$CaSO_4 \cdot 2H_2O$
Sulfide	S^{-2}	Galena	PbS
		Pyrite	FeS_2
		Argentite	Ag_2S*

*Note that silver is found as a native element and as a sulfide mineral.

Silica tetrahedra may also join together to form chains of indefinite length (▶ Figure 3.13d). Single chains, as in the pyroxene minerals, form when each tetrahedron shares two of its oxygens with an adjacent tetrahedron, resulting in a silicon-to-oxygen ratio of 1:3. Enstatite, a pyroxene-group mineral, reflects this ratio in its chemical formula, $MgSiO_3$. Individual chains, however, possess a net −2 electrical charge, so they are balanced by positive ions, such as Mg^{+2}, that link parallel chains together (▶ Figure 3.13d).

A double-chain structure characterizes the amphibole group of minerals, in which alternate tetrahedra in two parallel rows are cross-linked (▶ Figure 3.13d). The formation of double chains results in a silicon-to-oxygen ratio of 4:11, so each double chain possesses a −6 electrical charge. Mg^{+2}, Fe^{+2}, and Al^{+2} are usually involved in linking the double chains together.

In sheet structure silicates, three oxygens of each tetrahedron are shared by adjacent tetrahedral (▶ Figure 3.13e). Such structures result in continuous sheets of silica tetrahedra with silicon-to-oxygen ratios of 2:5. Continuous sheets also possess a negative electrical charge satisfied by positive ions located between the sheets. This particular structure accounts for the characteristic sheet structure of the *micas*, such as biotite and muscovite, and the *clay minerals*.

Three-dimensional networks of silica tetrahedra form when all four oxygens of the silica tetrahedron are shared by adjacent tetrahedra (▶ Figure 3.13f). Such sharing of oxygen atoms results in a silicon-to-oxygen ratio of 1:2, which is electrically neutral. Quartz is a common framework silicate.

■ *What are the two basic types of silicate minerals, and how do they compare?*

Ferromagnesian Silicates The **ferromagnesian silicates** are silicates that contain iron, magnesium, or both, as in olivine [$(Fe,Mg)_2SiO_4$]. Other minerals in this category include the pyroxenes and the amphiboles, both of which are mineral groups with several members, and biotite (▶ Figure 3.14). Olivine is olive green, but the other ferromagnesian silicates are dark green or black, and all are commonly denser than nonferromagnesian silicates. Olivine is common in some igneous rocks but uncommon in most

▶ **Figure 3.12 Radicals in Minerals** Many minerals contain radicals, which are complex groups of atoms tightly bonded together. The silica and carbonate radicals are particularly common in many minerals, such as quartz (SiO_2) and calcite ($CaCO_3$).

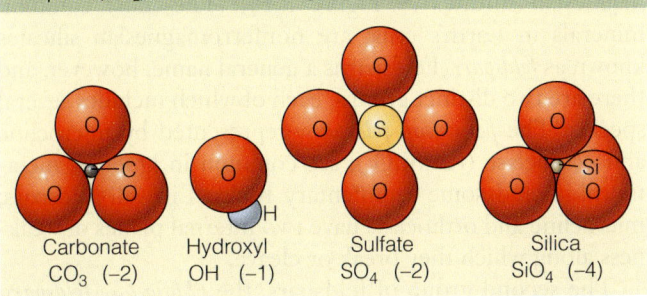

82   Chapter 3   Minerals—The Building Blocks of Rocks

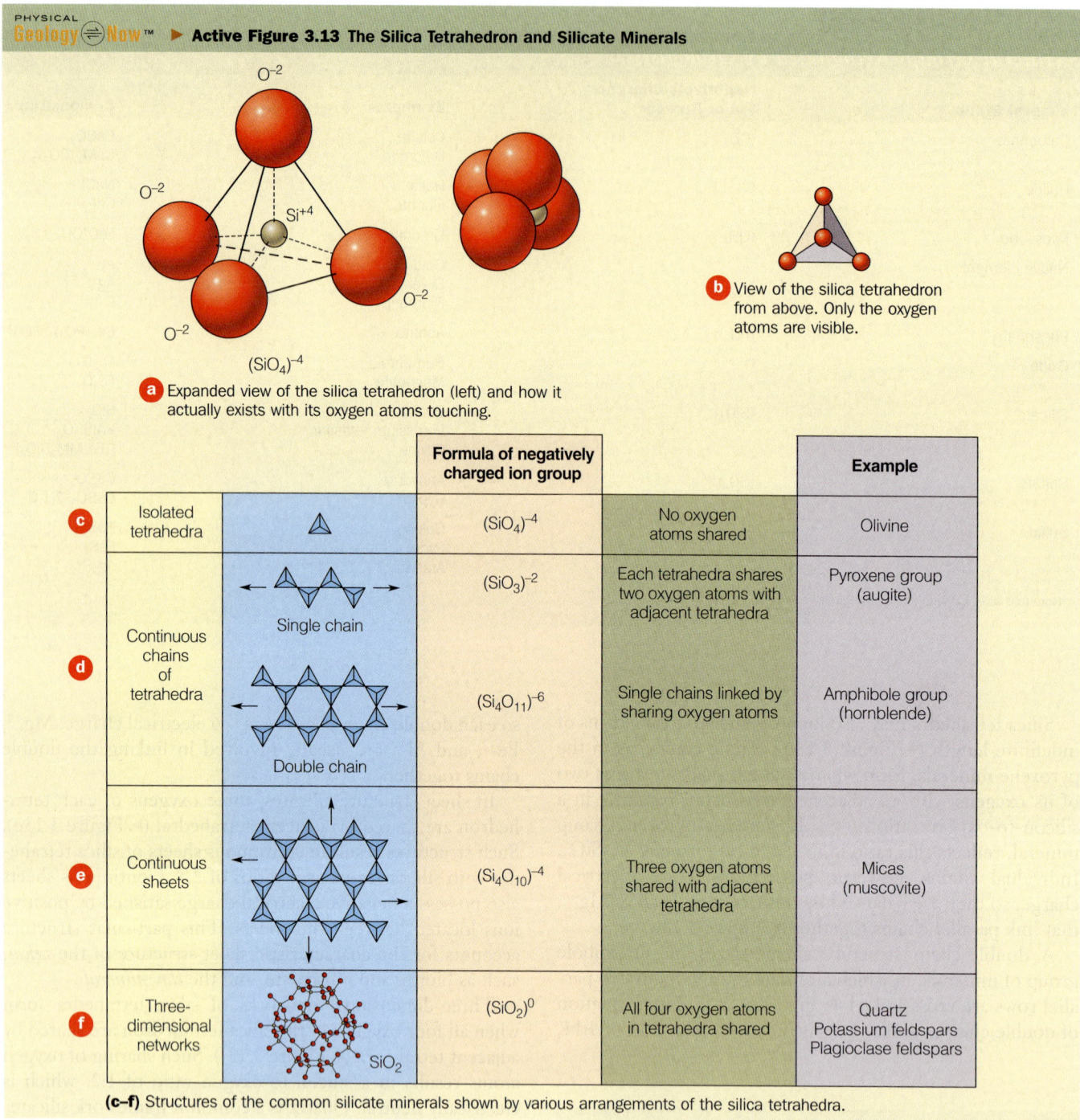

PHYSICAL Geology Now™ ▶ Active Figure 3.13 The Silica Tetrahedron and Silicate Minerals

(a) Expanded view of the silica tetrahedron (left) and how it actually exists with its oxygen atoms touching.

(b) View of the silica tetrahedron from above. Only the oxygen atoms are visible.

(c–f) Structures of the common silicate minerals shown by various arrangements of the silica tetrahedra.

other rock types. Among the pyroxenes, augite is the most common variety, whereas hornblende is the most abundant amphibole; both are found in igneous rocks as well as in some metamorphic rocks. Biotite is a mica, a mineral with a distinctive sheet structure (▶ Figure 3.14). It is common in several igneous and metamorphic rocks, and it is also found in small quantities in some sedimentary rocks.

Nonferromagnesian Silicates The **nonferromagnesian silicates,** as their name implies, lack iron and magnesium, are generally light colored, and are less dense than ferromagnesian silicates (▶ Figure 3.14). The most common minerals in Earth's crust are nonferromagnesian silicates known as *feldspars*. Feldspar is a general name, however, and there are two distinct groups, each of which includes several species. The *potassium feldspars*, represented by microcline and orthoclase ($KAlSi_3O_8$), are common in igneous, metamorphic, and some sedimentary rocks. Like all feldspars, microcline and orthoclase have two internal planes of weakness along which they break or cleave.

The second group of feldspars, the *plagioclase feldspars*, range from calcium-rich ($CaAl_2Si_2O_8$) to sodium-rich

Geo-focus Figure 3.14 Common Rock-Forming Silicate Minerals

a. The ferromagnesian silicates.

b. The nonferromagnesian silicates.

(NaAlSi$_3$O$_8$) varieties. They possess the characteristic feldspar cleavage and typically are white or cream to medium gray. Plagioclase cleavage surfaces commonly show numerous distinctive, closely spaced, parallel lines called *striations*.

Quartz (SiO$_2$), a very abundant nonferromagnesian silicate, is common in the three major rock groups, especially in such rocks as granite, gneiss, and sandstone. A framework silicate, it can usually be recognized by its glassy appearance and hardness (▶ Figure 3.14).

Another fairly common nonferromagnesian silicate is muscovite, which is a mica. Like biotite it is a sheet silicate, but muscovite is typically nearly colorless (▶ Figure 3.14), whereas biotite is black. Various clay minerals also possess the sheet structure typical of the micas, but their crystals are so small that they can be seen only with extremely high magnification. These clay minerals are important constituents of several types of rocks and are essential components of soils (see Chapter 6).

Carbonate Minerals

■ *What are carbonate minerals and in what kinds of rocks are they found?*

Carbonate minerals, those containing the negatively charged carbonate radical (CO$_3$)$^{-2}$, include calcium carbonate (CaCO$_3$), such as the minerals *aragonite* and *calcite*. Aragonite is unstable and commonly changes to calcite, the main

PHYSICAL Geology Now™ ▶ **Geo-focus Figure 3.15 Representative Specimens from Three Mineral Groups**

a The sulfide mineral galena (PbS) is the ore of lead.

b Gypsum (CaSO$_4$·2H$_2$O) is a common sulfate mineral. It is used in plaster of Paris, wallboard, and cement.

c Halite (NaCl) is a good example of a halide mineral. This mineral is a source of chlorine, food seasoning, and hydrochloric acid.

constituent of the sedimentary rock *limestone*. A number of other carbonate minerals are known, but only one of these need concern us: *Dolomite* [CaMg(CO$_3$)$_2$] forms by the chemical alteration of calcite by the addition of magnesium. Sedimentary rock composed of the mineral dolomite is *dolostone* (see Chapter 7).

Other Mineral Groups

In addition to silicates and carbonates, geologists recognize several other mineral groups (Table 3.1). And even though minerals from these groups are less common than silicates and carbonates, many are found in rocks in small quantities, and others are very important resources. In the oxides, an element combines with oxygen, as in hematite (Fe$_2$O$_3$) and magnetite (Fe$_3$O$_4$). Rocks with high concentrations of these minerals in the Lake Superior region of Canada and the United States are sources of iron ore for the manufacture of steel. The related hydroxides form mostly by the chemical alteration of other minerals.

We have already noted that the *native elements* are minerals composed of a single element. Examples are diamond and graphite (C) and the precious metals gold (Au), silver (Ag), and platinum (Pt), two of which are featured in "The Precious Metals" on pages 86 and 87 Some elements such as silver and copper are found both as native elements and as compounds and are thus also included in other mineral groups—the silver sulfide argentite (Ag$_2$S), for example.

Several minerals and rocks containing the phosphate radical (PO$_4$)$^{-3}$ are important sources of phosphorus for fertilizers. The sulfides such as the mineral galena (PbS), the ore of lead, have a positively charged ion combined with sulfur (S^{-2}) (▶ Figure 3.15a), whereas the sulfates have an element combined with the complex radical (SO$_4$)$^{-2}$ as in gypsum (CaSO$_4$·2H$_2$O) (▶ Figure 3.15b). The halides contain the halogen elements, fluorine (F^{-1}), and chlorine (Cl^{-1}): examples are the minerals fluorite (CaF$_2$) and halite (NaCl) (▶ Figure 3.15c).

Section 3.4 Summary

- More than 3500 minerals are known but only a few are very common. Most minerals are composed of silicon and oxygen and other chemical elements common in Earth's crust.

- The silica tetrahedron, consisting of one silicon atom and four oxygen atoms, is the basic building block of all nonferromagnesian and ferromagnesian silicates, which are by far the most common minerals.

- Geologists also recognize several other mineral groups, including carbonates, oxides, halides, and sulfates, which are defined by their negatively charged ions or radicals.

3.5 Mineral Identification

Although geologists may use sophisticated techniques such as chemical analyses, X-rays, and polarizing microscopes to identify minerals, most of the common ones are identified by physical properties such as color, hardness, density, cleavage and fracture, and crystal form. Atomic structure and chemical composition determine all of these properties. Some mineral properties are variable, especially color, whereas many others are remarkably constant for a given type of mineral.

Luster and Color

■ *How does mineral luster differ from color?*

Luster (not to be confused with *color*) is the quality and intensity of light reflected from a mineral's surface. Geologists define two basic types of luster: metallic, or having the appearance of a metal, and nonmetallic. Notice that of the three minerals shown in Figure 3.15 only galena has a metallic luster. Among the several types of nonmetallic luster are glassy or vitreous (as in quartz), dull or earthy, waxy, greasy, and brilliant (as in diamond).

Beginning students are distressed by the fact that the color of some minerals varies considerably, making the most obvious physical property of little use for their identification. Geologists know that color or lack of color in minerals is caused by how the various wavelengths of visible light are absorbed or transmitted. In any case, we can make some generalizations about color that are helpful in mineral identification. Ferromagnesian silicates are typically black, brown, or dark green, although olivine is olive green (▶ Figure 3.14). Nonferromagnesian silicates, on the other hand, vary considerably in color but are rarely very dark. White, cream, colorless, and shades of pink and pale green are more typical (▶ Figure 3.14).

Another helpful generalization is that the color of minerals that have a metallic luster is more consistent than it is for nonmetallic minerals. For example, galena is always lead-gray (▶ Figure 3.15a), whereas pyrite is invariably brassy yellow. In contrast, quartz, a nonmetallic mineral, may be colorless, smoky brown to almost black, rose, yellow-brown, milky white, blue, or violet to purple (▶ Figure 3.9a, b).

Crystal Form

■ *Why is crystal form of limited use in mineral identification?*

We have already mentioned that crystals may grow in proximity and fail to develop smooth crystal faces and straight edges and corners. In short, many minerals you encounter do not show their typical crystal form. Remember, though, that even if crystals are not apparent, all minerals are crystalline and that under ideal circumstances well-formed crystals would have developed.

Some minerals do typically occur as crystals (▶ Figures 3.8 and 3.16). For example, 12-sided crystals of garnet are

PHYSICAL Geology Now™ ▶ **Geo-focus Figure 3.16 Minerals with the Same Kind of Crystals** Mineral crystals are found in a variety of shapes (see Figure 3.8) but different minerals may have the same kinds of crystals as shown by **(a)** pyrite (FeS_2), **(b)** fluorite (CaF_2), and **(c)** halite (NaCl). Differentiating one from the other is easy. Pyrite is brassy yellow (the silvery color results from the way light is reflected), and it is much denser and harder than the two other minerals. You can identify fluorite and halite by their cleavage and taste—halite tastes like salt whereas fluorite has no taste.

a Pyrite (FeS_2)

b Fluorite (CaF_2)

c Halite (NaCl)

The Precious Metals

The discovery of gold by James Marshall at Sutter's Mill near Coloma in 1848 sparked the California gold rush (1849–1853), during which $200 million in gold was recovered.

▲ 1. Specimen of gold from Grass Valley, California. Gold is too heavy and too soft for tools and weapons, so it has been prized for jewelry and as a symbol of wealth, but it is also used in glass making, electrical circuitry, gold plating, the chemical industry, and dentistry.

▼ 2. A miner pans for gold (foreground) by swirling water, sand, and gravel in a broad, shallow pan. The heavier gold sinks to the bottom. At the far left a miner washes sediment in a cradle. As in panning, the cradle separates heavier gold from other materials.

▲ 3. Gold miners on the American River near Sacramento, California. Most of the gold came from placer deposits in which running water separated and concentrated mineral and rock fragments by their density.

◀ **4.** Hydraulic mining in California in which strong jets of water washed gold-bearing sand and gravel into sluices. In this image taken in 1905 at Junction City, California, water is directed through a monitor onto a hillside. Hydraulic mining was efficient from the mining point of view but caused considerable environmental damage.

▶ **5.** Reports in 1876 of gold in the Black Hills of South Dakota resulted in a flood of miners that led to hostilities with the Sioux Indians and the annihilation of Lt. Col. George Armstrong Custer and 260 of his men at the Battle of the Little Big Horn in Montana. This view shows the headworks (upper right) of the Homestake Mine at Lead, South Dakota, in 1900. The headworks is the cluster of buildings near the opening to a mine.

◀ **6.** Like gold, silver is found as a native element, as in this specimen, but it also occurs as a compound in the sulfide mineral argentite (Ag_2S). Silver is used in North America for silver halide film, jewelry, flatware, surgical instruments, and backing for mirrors.

▲ **7.** This image shows the headworks of the Yellowjacket Mine at Gold Hill, Nevada, and the inset shows silver-bearing quartz (white) in volcanic rock. This largest silver discovery in North America, called the Comstock Lode, was responsible for bringing Nevada into the Union in 1864 during the Civil War, even though it had too few people to qualify for statehood. The Comstock Lode was mined for silver and gold from 1859 until 1898.

common, as are 6- and 12-sided crystals of pyrite. Minerals that grow in cavities or are precipitated from circulating hot water (hydrothermal solutions) in cracks and crevices in rocks also commonly occur as crystals.

Crystal form can be a useful characteristic for mineral identification, but a number of minerals have the same crystal form. Pyrite (FeS_2), fluorite (CaF_2), and halite (NaCl) all occur as cubic crystals, but they can be easily identified by other properties such as color, luster, hardness, and density (▶ Figure 3.16).

Cleavage and Fracture

■ *What is mineral cleavage, and how is it used in mineral identification?*

Not all minerals possess cleavage, but those that do tend to break, or split, along a smooth plane or planes of weakness determined by the strength of the bonds within a mineral crystal. Cleavage is characterized in terms of quality (perfect, good, poor), direction, and angles of intersection of cleavage planes. Biotite, a common ferromagnesian silicate, has perfect cleavage in one direction (▶ Figure 3.17a). The fact that biotite preferentially cleaves along a number of closely spaced, parallel planes is related to its structure; it is a sheet silicate with the sheets of silica tetrahedra weakly bonded to one another by iron and magnesium ions (▶ Figure 3.13e).

Feldspars possess two directions of cleavage that intersect at right angles (▶ Figure 3.17b), and the mineral halite has three directions of cleavage, all of which intersect at right angles (▶ Figure 3.17c). Calcite also possesses three directions of cleavage, but none of the intersection angles is a right angle, so cleavage fragments of calcite are rhombohedrons (▶ Figure 3.17d). Minerals with four directions of cleavage include fluorite and diamond (▶ Figure 3.17e). Ironically, diamond, the hardest mineral, can be easily cleaved. A few minerals such as sphalerite, an ore of zinc, have six directions of cleavage (▶ Figure 3.17f).

Cleavage is an important diagnostic property of minerals, and its recognition is essential in distinguishing between some minerals. The pyroxene mineral augite and the amphibole mineral hornblende, for example, look much alike; both are dark green to black, have the same hardness, and possess two directions of cleavage. But the cleavage planes of augite intersect at about 90 degrees, whereas the cleavage

▶ **Figure 3.17 Several Types of Mineral Cleavage**

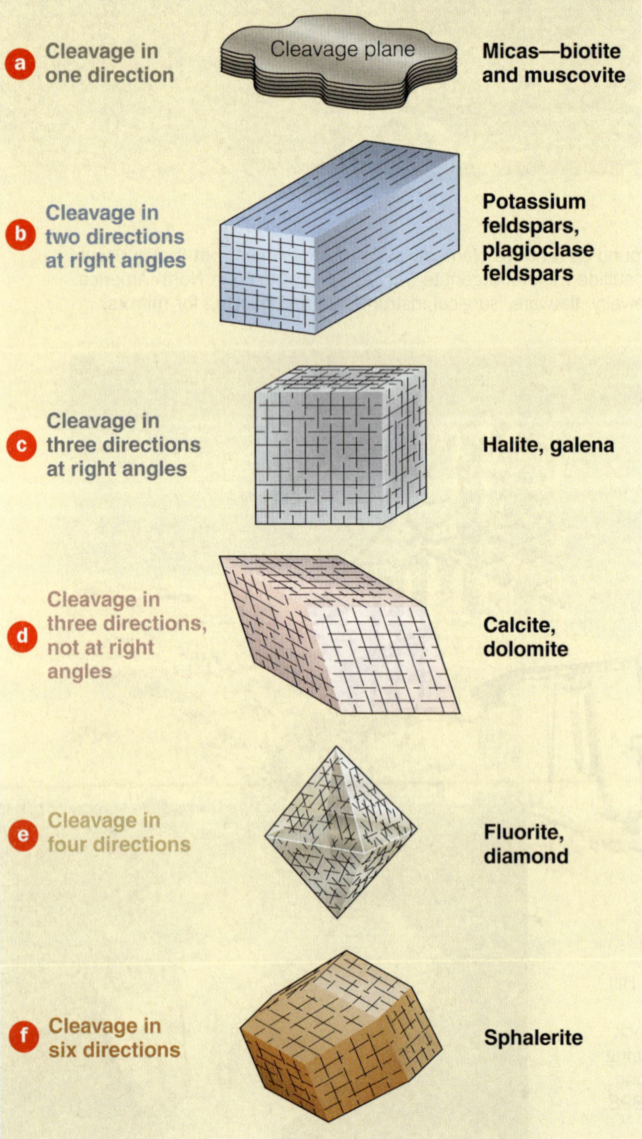

a Cleavage in one direction — Micas—biotite and muscovite

b Cleavage in two directions at right angles — Potassium feldspars, plagioclase feldspars

c Cleavage in three directions at right angles — Halite, galena

d Cleavage in three directions, not at right angles — Calcite, dolomite

e Cleavage in four directions — Fluorite, diamond

f Cleavage in six directions — Sphalerite

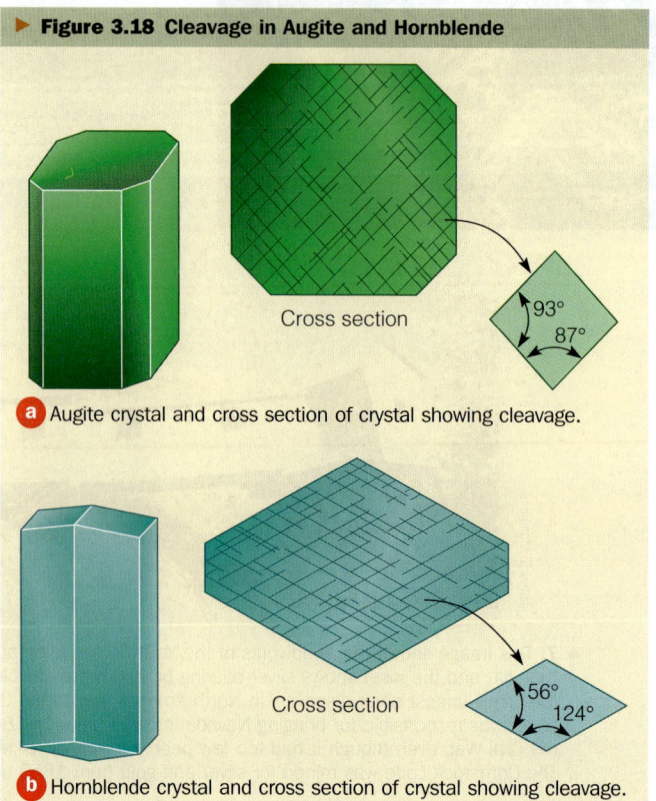

▶ **Figure 3.18 Cleavage in Augite and Hornblende**

a Augite crystal and cross section of crystal showing cleavage.

b Hornblende crystal and cross section of crystal showing cleavage.

planes of hornblende intersect at angles of 56 degrees and 124 degrees (▶ Figure 3.18).

In contrast to cleavage, *fracture* is mineral breakage along irregular surfaces. Any mineral can be fractured if enough force is applied, but the fracture surfaces are uneven or conchoidal (curved) rather than smooth.

Hardness

An Austrian geologist, Friedrich Mohs, devised a relative hardness scale for 10 minerals. He arbitrarily assigned a hardness value of 10 to diamond, the hardest mineral known, and lower values to the other minerals. Relative hardness is easily determined by the use of Mohs hardness scale (Table 3.2). Quartz will scratch fluorite but cannot be scratched by fluorite, gypsum can be scratched by a fingernail, and so on. So **hardness** is defined as a mineral's resistance to abrasion and is controlled mostly by internal structure. For example, both graphite and diamond are composed of carbon, but the former has a hardness of 1 to 2, whereas the latter has a hardness of 10.

Specific Gravity (Density)

Specific gravity and density are two separate concepts, but here we will use them more or less as synonyms. A mineral's **specific gravity** is the ratio of its weight to the weight of an equal volume of pure water. Thus a mineral with a specific gravity of 3.0 is three times as heavy as water. Like all ratios, specific gravity is not expressed in units such as grams per cubic centimeters, it is a dimensionless number. **Density,** in contrast, is a mineral's mass (weight) per unit of volume expressed in grams per cubic centimeters. So the specific gravity of galena (▶ Figure 3.15a) is 7.58 and its density is 7.58 g/cm³. In most instances we will refer to a mineral's density, and in some of the following chapters we will mention the density of various rocks.

Structure and composition control a mineral's specific gravity and density. Because ferromagnesian silicates contain iron, magnesium, or both, they tend to be denser than nonferromagnesian silicates. In general, the metallic minerals, such as galena and hematite, are denser than nonmetals. Pure gold with a density of 19.3 g/cm³ is about two and one half times as dense as lead. Diamond and graphite, both of which are composed of carbon (C), illustrate how structure controls specific gravity or density. The specific gravity of diamond is 3.5, whereas that of graphite varies from 2.09 to 2.33.

Other Useful Mineral Properties

Other physical properties are important for identifying some minerals. Talc has a distinctive soapy feel, graphite writes on paper, halite tastes salty, and magnetite is magnetic (▶ Figure 3.19). Calcite possesses the property of *double refraction*, meaning that an object, when viewed through a

Table 3.2
Mohs Hardness Scale

Hardness	Mineral	Hardness of Some Common Objects
10	Diamond	
9	Corundum	
8	Topaz	
7	Quartz	
		Steel file (6½)
6	Orthoclase	
		Glass (5½–6)
5	Apatite	
4	Fluorite	
3	Calcite	Copper penny (3)
		Fingernail (2½)
2	Gypsum	
1	Talc	

▶ **Figure 3.19 Graphite and Magnetite**

a Graphite is the mineral used to make pencil "lead" write on paper.

b Magnetite is magnetic. Magnetite is an ore of iron and used in toys like "Etch a Sketch."

What Would You Do?

Expensive models are available to illustrate atoms, compounds, and crystalline solids, but you teach at a school that has a severe shortage of funds for science classes. Describe the kinds of models you might construct from inexpensive materials that would adequately illustrate atoms and other structures. What would you use to represent mineral crystals?

Table 3.3

Important Rock-Forming Minerals

Mineral	Primary Occurrence
Ferromagnesian silicates	
Olivine	Igneous, metamorphic rocks
Pyroxene group	
Augite most common	Igneous, metamorphic rocks
Amphibole group	
Hornblende most common	Igneous, metamorphic rocks
Biotite	All rock types
Nonferromagnesian silicates	
Quartz	All rock types
Potassium feldspar group	
Orthoclase, microcline	All rock types
Plagioclase feldspar group	All rock types
Muscovite	All rock types
Clay mineral group	Soils, sedimentary rocks, some metamorphic rocks
Carbonates	
Calcite	Sedimentary rocks
Dolomite	Sedimentary rocks
Sulfates	
Anhydrite	Sedimentary rocks
Gypsum	Sedimentary rocks
Halides	
Halite	Sedimentary rocks

transparent piece of calcite, will have a double image. Some sheet silicates are plastic and, when bent into a new shape, will retain that shape; others are flexible and, if bent, will return to their original position when the forces that bent them are removed.

A simple chemical test to identify the minerals calcite and dolomite involves applying a drop of dilute hydrochloric acid to the mineral specimen. If the mineral is calcite, it will react vigorously with the acid and release carbon dioxide, which causes the acid to bubble or effervesce. Dolomite, in contrast, will not react with hydrochloric acid unless it is powdered.

Section 3.5 Summary

- Although some physical properties of minerals vary, especially color, others such as cleavage, density, luster, and hardness are remarkably constant and provide the necessary information for mineral identification.

3.6 The Significance of Rock-Forming Minerals

■ *How do rock-forming minerals differ from accessory minerals?*

Of the more than 3500 known minerals only a few, designated **rock-forming minerals,** are sufficiently common in rocks to be essential for rock identification and classification (Table 3.3). Geologists use the term **rock** for any solid aggregate of one or more minerals, such as granite, which consists of certain percentages of quartz, potassium feldspars, and plagioclase feldspars (▶ Figure 3.20); and limestone, which is made up of only calcite. However, the term *rock* also refers to masses of mineral-like matter, as in the natural glass obsidian (see Chapter 4), and masses of altered organic matter, as in coal (see Chapter 7). In addition to quartz, potassium, and plagioclase feldspars, granite may contain several *accessory minerals*, those in such small amounts that they can be disregarded for rock classification.

We have emphasized that silicate minerals are by far the most common minerals in Earth's crust, so it follows that most rocks are composed of these minerals. Indeed, feldspar minerals (plagioclase feldspars and potassium feldspars) and quartz make up more than 60% of Earth's crust. But even among the hundreds of silicates only a few are particularly common in rocks, although many others are present as accessory minerals.

The most common nonsilicate rock-forming minerals are the carbonates calcite ($CaCO_3$) and dolomite [$CaMg(CO_3)_2$], the main constituents of the sedimentary rocks limestone and dolostone, respectively (see Chapter 7). Among the sulfates and halides, gypsum ($CaSO_4 \cdot 2H_2O$) in rock gypsum and halite (NaCl) in rock salt (see Chapter 7) are common enough to qualify as rock-forming minerals. But even though these minerals, and their corresponding rocks, might be common in some areas, their overall abundance is limited compared to the silicate and carbonate rock-forming minerals.

Section 3.6 Summary

- Rock-forming minerals are essential for the classification of rocks, whereas accessory minerals can be ignored in this endeavor. Silicates are the most common rock-forming minerals, but some carbonates are also abundant.

Geo-focus Figure 3.20 **Minerals in Granite** The igneous rock granite (see Chapter 4) is made up of mostly three minerals; quartz, potassium feldspar, and plagioclase feldspar, but it may also contain small amounts of biotite, muscovite, and hornblende.

3.7 The Origin of Minerals

■ *What accounts for the origin of minerals from magma?*

Thus far we have discussed the composition, structure, and physical properties of minerals but have not addressed how they originate. One phenomenon that accounts for the origin of minerals is the cooling of molten rock material known as *magma* (magma that flows out onto the surface is called *lava*). As magma or lava cools, minerals crystallize and grow, thereby determining the mineral composition of various igneous rocks such as basalt (dominated by ferromagnesian silicates) and granite (dominated by nonferromagnesian silicates) (see Chapter 4). Hot water solutions derived from magma commonly invade cracks and crevasses in adjacent rocks, and from these solutions a variety of minerals crystallize, some of economic importance. Minerals also originate when water in hot springs cools (see Chapter 16), and when hot, mineral-rich water discharges onto the seafloor at hot springs known as black smokers (see Chapter 12).

Dissolved materials in seawater, and more rarely lake water, combine to form minerals such as halite (NaCl), gypsum ($CaSO_4 \cdot 2H_2O$), and several others when the water evaporates. Aragonite and/or calcite, both varieties of calcium carbonate ($CaCO_3$), might also form from evaporating water, but most originate when organisms such as clams, oysters, corals, and floating microorganisms use this compound to construct their shells. A few plants and animals also use silicon dioxide (SiO_2) for their skeletons, which accumulate as mineral matter on the seafloor when the organisms die (see Chapter 7).

Some clay minerals form when chemical processes alter other minerals compositionally and structurally, such as feldspars (see Chapter 6), and others originate when rocks are changed during metamorphism (see Chapter 8). In fact, the agents that cause metamorphism—heat, pressure, and chemically active fluids—are responsible for the origin of many minerals. A few minerals even originate when gases such as hydrogen sulfide (H_2S) and sulfur dioxide (SO_2) react at volcanic vents to produce sulfur.

Section 3.7 Summary

● Several processes such as cooling magma and lava, evaporation of seawater, the activities of organisms, metamorphism, and inorganic chemical processes account for the origin of minerals.

Log into GeologyNow and select this chapter to work through a **Geology Interactive** activity on the "Mineral Lab" (click Atoms and Crystals→ Mineral Lab).

3.8 Natural Resources and Reserves

■ *How does a resource differ from a reserve?*

Geologists at the U.S. Geological Survey define a **resource** as "a concentration of naturally occurring solid, liquid, or gaseous material in or on Earth's crust in such form and amount that economic extraction of a commodity from the concentration is currently or potentially feasible." So a resource is the total amount of a commodity whether discovered or undiscovered, but a **reserve** is only that part of the resource base that is known and can be recovered economically. The technology exists to extract aluminum from aluminum-rich igneous rocks and sedimentary rocks, but at present it cannot be done economically.

The United States and Canada, both highly industrialized nations, have enjoyed considerable economic success because they have abundant resources, also called natural resources. Most natural resources are concentrations of minerals, rocks, or both, but liquid petroleum and natural gas are also included. In fact, we refer to *metallic resources* (copper, tin, iron ore, etc.), *nonmetallic resources* (sand and gravel, crushed stone, salt, sulfur, etc.), and *energy resources* (petroleum, natural gas, coal, and uranium).

■ *What factors affect the status of a resource?*

We have made the distinction between a resource and a reserve, and whereas the distinction is simple enough in principle, in practice it depends on several factors, not all of which remain constant. For instance, a resource in a remote region might not be mined because transportation costs are too high, and what might be deemed a resource rather than a reserve in the United States and Canada might be mined in a developing country where labor costs are low. The commodity in question is also important. Gold and diamonds in sufficient quantity can be mined profitably almost anywhere, whereas most sand and gravel deposits must be close to their market areas.

Obviously the market price is important in evaluating any resource. From 1935 until 1968, the U.S. government maintained the price of gold at $35 per troy ounce (1 troy ounce = 31.1 g). When this restriction was removed, demand determined the market price and gold prices rose, reaching an all-time high of $843 per troy ounce in 1980. As a result, many marginal deposits became reserves and a number of abandoned mines were reopened.

The status of a resource is also affected by changes in technology. By the time of World War II (1939-1945), the richest iron ores of the Great Lakes region in the United States and Canada had been mostly depleted. But the development of a method for separating the iron from unusable rock and shaping it into pellets ideal for use in blast furnaces made it profitable to mine rocks with less iron. As a matter of fact, a large part of the mineral revenue of Newfoundland and Quebec, Canada, and Minnesota and Michigan comes from mining iron ore.

Most people know that industrialized societies depend on a variety of natural resources but have little knowledge about their occurrence, methods of recovery, and economics. Geologists are, of course, essential in finding and evaluating deposits, but extraction involves engineers and chemists, not to mention many people in support industries that supply mining equipment. Ultimately, though, the decision about whether a deposit should be mined or not is made by people trained in business and economics. In short, extraction must yield a profit. The extraction of natural resources, other than oil, natural gas, and coal, amounted to more than $37 billion during 2003 in the United States, and in Canada the extraction of nonfuel resources during the same year was nearly $20 billion (Canadian dollars). We will have much more to say about energy resources in later chapters, especially Chapter 7.

Everyone is aware of the importance of resources such as petroleum, gold, and ores of iron, copper, and lead. However, some quite common minerals are also essential. For example, pure quartz sand is used to manufacture glass and optical instruments as well as sandpaper and steel alloys. Clay minerals are needed to make ceramics and paper, and feldspars are used for porcelain, ceramics, enamel, and glass. Micas are used in a variety of products including lipstick, glitter, and eye shadow as well as lustrous paints. Phosphate-bearing rock used in fertilizers mined in Florida accounts for a large part of that state's mineral production.

Access to many resources is essential for industrialization and the high standard of living enjoyed in many countries. The United States and Canada are fortunate to be resource-rich nations, but resources are used much faster than they form, so they are *nonrenewable*, meaning that once a resource has been depleted, new deposits or suitable substitutes, if available, must either be found or be imported from elsewhere. For some essential resources the United States is totally dependent on imports. No cobalt or columbium was mined in this country during 2003, and yet cobalt is an essential metal in gas turbine engines, magnets, and corrosion and wear-resistant alloys, and columbium is used in jet engine components and superalloys. All cobalt and columbium is imported, as is all manganese, an element needed for making high-quality steel.

What Would You Do?

Let's say that some reputable businesspeople tell you of opportunities to invest in natural resources. Two ventures look promising: a gold mine and a sand and gravel deposit. If gold sells for about $550 per troy ounce (1 troy ounce = 31.1 g) whereas sand and gravel are worth $6 or $7 per ton, would it be more prudent to invest in the gold mine? Explain not only how market price would influence your decision but also what other factors you might need to consider.

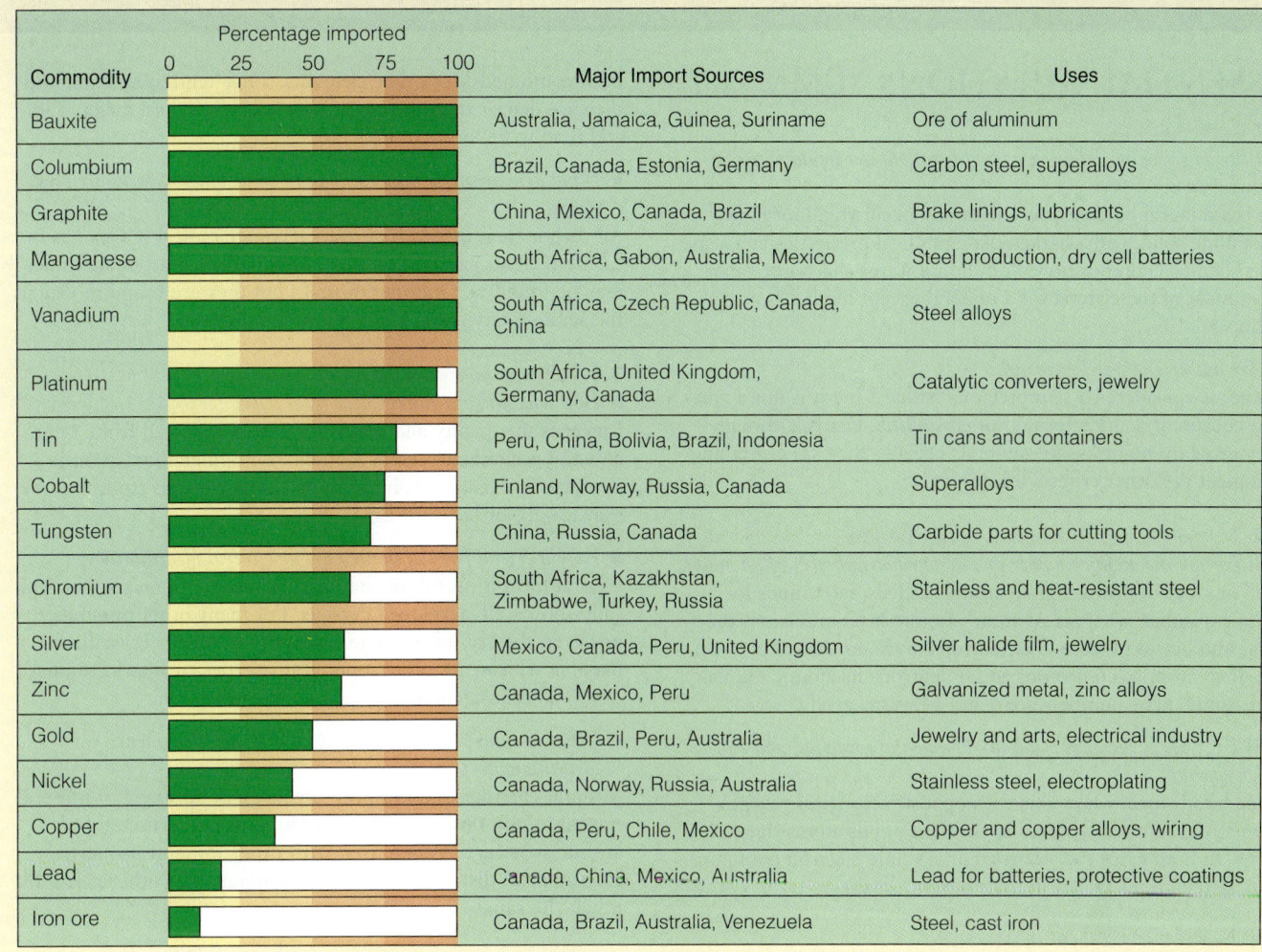

▶ Figure 3.21 Mineral Commodities The dependence of the United States on imports of various mineral commodities is apparent from this chart. The lengths of the green bars correspond to the amounts of the resources imported.

Sources: USGS Minerals Information: http://minerals.usgs.gov/minerals/
USGS Mineral Commodity Summaries 2004: http://usgs.gov/minerals/pubs/mcs/2004.pdf

In addition to cobalt, columbium, and manganese, the United States imports all the aluminum ore it uses as well as all or some of many other resources (▶ Figure 3.21). Canada, in contrast, is more self-reliant, meeting most of its domestic mineral and energy needs. Nevertheless, it must import phosphate, chromium, manganese, and aluminum ore. Canada also produces more crude oil and natural gas than it uses, and it is among the world leaders in producing and exporting uranium.

To ensure continued supplies of essential minerals and energy resources, geologists as well as other scientists, government agencies, and leaders in business and industry continually assess the status of resources in view of changing economic and political conditions and changes in science and technology. The U.S. Geological Survey, for instance, keeps detailed statistical records of mine production, imports, and exports, and regularly publishes reports on the status of numerous commodities. Similar reports appear regularly in the *Canadian Minerals Yearbook*. In several of the following chapters we will discuss the geologic occurrence of several natural resources.

Section 3.8 Summary

- Concentrations of minerals and rocks of economic importance are characterized as metallic resources, nonmetallic resources, and energy resources.

- Reserves are that part of the resource base that can be extracted profitably. The status of a resource versus a reserve depends on market price, labor costs, location, and developments in science and technology.

- The United States must import many resources to maintain its industrial capacity. Canada is more self-reliant, but it too must import some commodities.

Review Workbook

ESSENTIAL QUESTIONS SUMMARY

3.1 Introduction
■ *Why do geologists define ice as a mineral but not liquid water or water vapor?*
Ice is a mineral because it is a naturally occurring, inorganic, crystalline solid, with characteristic physical properties and a specific chemical composition. Both liquid water and water vapor meet most of the criteria for a mineral except neither is a crystalline solid.

■ *Are amber and pearls minerals?*
Amber is considered a semiprecious "stone," but it is not a mineral because it is organic and not crystalline. Pearls, although they grow in mollusks, are minerals because they are crystalline and meet the other criteria for minerals.

3.2 Matter, Atoms, Elements, and Bonding
■ *What are the atomic number and atomic mass number of an atom?*
The number of protons in an atom's nucleus determines its atomic number, whereas an atom's atomic mass number is the total number of protons and neutrons in the nucleus. For example, if an atom has 6 protons and 8 neutrons its atomic number and atomic mass numbers are 6 and 14, respectively.

■ *What types of chemical bonds are common in minerals, and how do they form?*
Ionic bonds form when ions with opposite electrical charges attract one another, whereas in covalent bonds atoms share electrons. In both cases the atoms attain a stable electron configuration with eight electrons in their outermost shell.

3.3 Explore the World of Minerals
■ *Why is it that not all mineral specimens show well-developed crystals but all are crystalline solids?*
As minerals form and grow, they may merge with one another to form a mosaic of interlocking crystalline solids that show no obvious crystals. Nevertheless, individual minerals within this mosaic have their atoms arranged in a specific three-dimensional framework.

■ *How do native elements differ from other minerals?*
By definition native elements are made up of only one chemical element, such as gold (Ag) and diamond (C), whereas most minerals are composed of two of more chemical elements, such as quartz (SiO_2).

3.4 Mineral Groups Recognized by Geologists
■ *Why are there so few common minerals?*
Even though there are 92 naturally occurring elements, only 8 of them are very common in Earth's crust. And even among these 8, oxygen and silicon are by far the most common. As a result, most common minerals are made up of oxygen, silicon, and one or more other elements.

■ *What are the two basic types of silicate minerals, and how do they compare?*
Geologists recognize ferromagnesian silicates and nonferromagnesian silicates. The former are made up of iron, magnesium, or both combined with other elements. They tend to be dark, and they are denser than the latter, which lack iron and magnesium and tend to be light colored.

■ *What are carbonate minerals and in what kinds of rocks are they found?*
All carbonate minerals have the carbonate radical $(CO_3)^{-2}$ as in calcite ($CaCO_3$) and dolomite [$CaMg(CO_3)_2$]. Carbonate minerals may be found in a number of rocks but they are found mostly in the sedimentary rocks limestone and dolostone.

3.5 Mineral Identification
■ *How does mineral luster differ from color?*
Luster is the quality and intensity of light reflected from a mineral and is characterized as metallic or nonmetallic. Color, in contrast, is a visual quality of minerals that results from the way they absorb or transmit light.

■ *Why is crystal form of limited use in mineral identification?*
Crystal form is useful for the identification of minerals that typically show well-developed crystals. For many other minerals, however, the crystals grow in proximity to form a crystalline mass, in which case other mineral properties such as luster, hardness, and cleavage must be used for identification.

■ *What is mineral cleavage, and how is it used in mineral identification?*
Mineral cleavage refers to the breakage or splitting of mineral crystals along one or more smooth planes determined by atomic structure. When used with other mineral properties, cleavage can distinguish between minerals that otherwise look much the same.

3.6 The Significance of Rock-Forming Minerals
■ *How do rock-forming minerals differ from accessory minerals?*
Rock-forming minerals are common enough in rocks to be essential for naming and classifying rocks, whereas accessory minerals can be ignored in this endeavor because they are present in minor quantities.

3.7 The Origin of Minerals
■ *What accounts for the origin of minerals from magma?*
As molten rock material known as magma (lava at the surface) cools, minerals begin to crystallize and grow, thus determining the composition of various igneous rocks.

3.8 Natural Resources and Reserves
■ *How does a resource differ from a reserve?*
A resource is any solid, liquid, or gaseous substance in rocks whose profitable extraction is potentially feasible. In contrast, a reserve is only that part of the resource base that can be extracted economically.

■ *What factors affect the status of a resource?*
Market price is the most obvious determinate of whether a commodity is classified as a resource or a reserve, but other factors are geographic location, labor costs, and developments in science and technology.

ESSENTIAL TERMS TO KNOW

atom (p. 73)
atomic mass number (p. 74)
atomic number (p. 74)
bonding (p. 75)
carbonate mineral (p. 83)
cleavage (p. 77)
compound (p. 75)
covalent bond (p. 76)
crystal (p. 77)
crystalline solid (p. 77)
density (p. 89)
electron (p. 73)

electron shell (p. 73)
element (p. 73)
ferromagnesian silicate (p. 81)
hardness (p. 89)
ion (p. 75)
ionic bond (p. 75)
luster (p. 85)
mineral (p. 72)
native element (p. 78)
neutron (p. 73)
nonferromagnesian silicate (p. 81)

nucleus (p. 73)
proton (p. 73)
reserve (p. 92)
resource (p. 92)
rock (p. 90)
rock-forming mineral (p. 90)
silica (p. 80)
silica tetrahedron (p. 80)
silicate (p. 80)
specific gravity (p. 89)

REVIEW QUESTIONS

1. How can you account for the very different properties of diamond and graphite considering that they are both composed of carbon?

2. What are some of the problems created by the necessity to import resources from other nations?

3. Why does the definition of the term *mineral* include "a narrowly defined chemical composition" rather than a statement such as "a specific chemical composition"?

4. Discuss three ways in which minerals form.

5. What is meant by the terms *crystal* and *crystalline*?

6. Briefly discuss three physical properties of minerals that are useful for their identification.

7. How does a rock made up mostly of ferromagnesian silicates differ from one composed primarily of nonferromagnesian silicates?

8. How does the constancy of interfacial angles indicate that minerals are crystalline?

9. How can you distinguish cubic crystals of pyrite, fluorite, and halite from one another?

10. What are micas, why do they have such good cleavage, and what are their uses?

APPLY YOUR KNOWLEDGE

1. You must clearly explain to an interested audience the distinction between minerals and rocks. How would you do so, and can you think of any analogies that might clarify the points you make?

2. Why do you think the United States and Canada import bauxite (the ore of aluminum) when both countries have vast exposures of igneous rocks and clay-rich rocks from which we can extract aluminum? What factor(s) may change the status of these rocks so that we extract aluminum from them?

3. Refer to the rock cycle in Figure 1.13 and explain at what points in this cycle you would expect minerals to form.

FIELD QUESTION

1. If diamond is perfectly cleaved, it would yield geometric figures like the ones shown here, but this mineral is fluorite. From the image alone you should be able to tell that these specimens are not diamond. How? Look up other properties of these two minerals that you can use to differentiate them.

Geo-focus Figure

GEOLOGY MATTERS

GEOLOGY IN FOCUS Mineral Crystals

Most mineral crystals measure only a few millimeters to centimeters across but a few are gigantic. Spodumene crystals up to 14 m long were mined in South Dakota for their lithium content, quartz crystals weighing several metric tons have been found in Russia, and mines in Ontario, Canada, have yielded sheets of muscovite up to 2.4 m across. Invariably, giant crystals grow in cavities where growth is unrestricted or they are found in pegmatite, an igneous rock similar to granite but with especially large minerals (see Chapter 4).

The most remarkable recent find of giant crystals was in April 2000 in a silver and lead mine in Mexico. A cavity there is lined with hundreds of gypsum crystals more than 1 m long and what one author called "crystal moonbeams," which are gypsum crystals 1.2 m in diameter and up to 15.2 m long (▶ Figure 1). These are probably the largest mineral crystals anywhere. Fearing vandalism, the company that owned the mine kept the crystals secret for some time, but the 65°C temperature and 100% humidity in the crystal-filled cavern would keep out all but the most determined vandals.

For many centuries, mineral crystals as well as several types of rocks and fossils were desired for their alleged healing powers and mystical properties. Indeed, some have served as religious symbols and talismans or have been carried, worn, applied externally, or ingested for their presumed mystical or curative powers. Diamond, according to one myth, wards off evil spirits, sickness, and floods, whereas topaz was thought to avert mental disorders, and ruby was believed to preserve its owner's health.

Even today ads touting the healing qualities of crystals and claims that they enhance emotional stability and clear thinking are seen in magazines and tabloids. Unfortunately for those buying crystals for these purposes, they provide no more benefit than artificial ones. In short, wishful thinking and the placebo effect are responsible for any perceived beneficial results.

One reason some people think that crystals have favorable attributes is the curious property called the piezoelectric effect. When some crystals are compressed or an electrical current is applied, they produce an electrical charge that is quite useful. For example, the electrical current from a watch's battery causes a thin wafer from a quartz crystal to expand and contract about 100,000 times per second. Quartz clocks were first developed in 1928, and now quartz watches and clocks are commonplace (▶ Figure 2). Even inexpensive ones are very accurate, and precision-manufactured quartz clocks used in astronomy do not gain or lose more than 1 second in 10 years.

An interesting historical note is that during World War II (1939–1945) the United States had difficulty obtaining quartz crystals needed for making radios. This shortage prompted the development of artificially synthesized quartz, and now most quartz in watches and clocks is synthetic. So even though the piezoelectric effect imparts no healing or protective powers to crystals, it is essential in applications in which precise measurements of time, pressure, and acceleration are needed. And of course many people are intrigued by crystals simply because they are so attractive.

Figure 1
Some of these gypsum crystals in a cavern in Chicuahua, Mexico, are as long as 15.2 m and may be the largest crystals anywhere. They were discovered in April 2000.

Figure 2
Even inexpensively produced quartz watches and clocks are very accurate.

GEOLOGY MATTERS

GEOLOGY IN UNEXPECTED PLACES

The Queen's Jewels

Geology probably does not immediately come to mind when you visit the Tower of London, a formidable stone structure on the banks of the Thames River in London, England. Construction of the Tower of London began during the reign of William the Conqueror (1066–1087), but it was enlarged and modified until about 1300, and since then has remained much the same. The Tower has served as a fortification, the residence for kings and queens, and a prison for such notable people as Sir Walter Raleigh, who was incarcerated there from 1603 until 1616.

The Tower of London is an impressive stone edifice, but visitors are equally impressed with the Crown Jewels, which have been housed in the Tower in the Waterloo Barracks since the beginning of the 14th century (▶ Figure 1). *Crown jewels* is a collective term for the regalia and vestments worn by the Queen or King of England at her or his coronation and during other important state functions. Only during World War II (1939–1945) were the Crown Jewels removed to a secret location for safekeeping and then later returned to the Waterloo Barracks.

Among the Crown Jewels is the crown made for the coronation of George VI in 1937 and later modified for Queen Elizabeth II's coronation in 1953. It is set with 2868 diamonds, 17 sapphires, 11 emeralds, 5 rubies, and 273 pearls (▶ Figure 2). Notice also in Figure 2 that Queen Elizabeth is holding the Scepter with Cross, which is a symbol of the monarch's power. This 0.9-m-long scepter was originally designed in 1661 and modified in 1905, and it now has the Great Star of Africa diamond in its head, which at 530 carats is the largest cut diamond in the world. Queen Elizabeth is also holding the Sovereign's Orb (▶ Figure 2). The orb measures about 16.5 cm in diameter and is set with various gemstones. It symbolizes the monarch's position as defender of the faith and as head of the Church of England. In addition to crowns, scepters, and orbs, the Crown Jewels include gold plates, christening fonts, and swords. They are one of the most impressive collections of gemstones in the world.

Figure 1
View of the interior of the Tower of London in London, England, from beneath an arch. The building on the left is the Waterloo Barracks, where the Crown Jewels have been kept since the 14th century.

Figure 2
At her coronation in 1953, Queen Elizabeth II wore the Imperial State Crown and held the Scepter with Cross and Sovereign's Orb.

GEOLOGY MATTERS

GEOLOGY IN YOUR LIFE

Welcome to the Wonderful World of Micas

What makes the paint on cars and appliances so lustrous? Why are lipstick, eyeliner, and glitter so attractive? Do you enjoy the amber glow seen through the isinglass window of a wood stove? Some of you may remember the line in the song "The Surrey With the Fringe on Top" from the musical *Oklahoma:* "isinglass curtains that can roll right down in case there's a change in the weather." What is this remarkable substance in paint, lipstick, and isinglass? The answer—micas.

Figure 1
Muscovite mica is colorless, white, pale red, or green. It is a sheet silicate that has many industrial uses in paint, wallboard compound, eyeliner, lipstick, and nail polish.

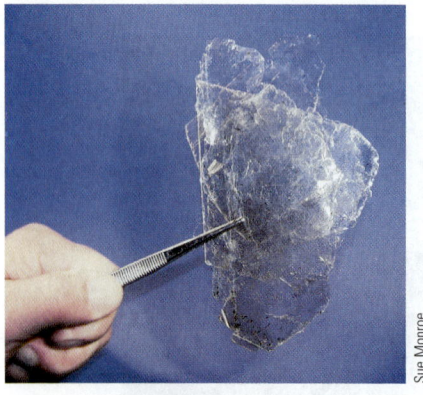

Mica is the name given to a group of 37 sheet silicate minerals that have similar physical properties, particularly they way they split or cleave. Remember that mineral cleavage is breakage along a plane or planes of weakness determined by atomic structure. Some minerals have no cleavage planes, others have two, three, four, or six; but micas have only one. Nevertheless, their cleavage is perfect, meaning that when cleaved they yield very smooth planes. Indeed, the micas split into thin, flexible sheets. The name *mica* probably comes from the Latin term *micare,* "to shine," a reference to their shiny luster.

Even though there are more than three dozen varieties of mica, only a few are common in rocks. One of the most common, *biotite* (black mica) (Figure 3.14), has no commercial uses, although geologists use it in potassium–argon dating (see Chapter 9). *Muscovite* (colorless, white, or pale red or green) mica is also common (▶ Figure 1); it was named for *Moskva* (Moscow), where much of Europe's mica was mined. Isinglass, mentioned above, consists of thin, transparent sheets of muscovite. Muscovite and another mica called *phlogophite* (from Greek *phologopos,* "fiery") have commercial value.

Micas used for products are scrap and flake micas, which either occur naturally or are ground into small pieces, and sheet mica, which is cut into various shapes and sizes for use in the electronics and electrical industries. Scrap and flake micas are produced in many countries and U.S. states, but about half the U.S. production comes from North Carolina; the United States imports micas from several other countries. Micas are found in many types of rocks, but our main concern here is its uses.

When mica is ground up dry, it loses much of its luster but retains its platy nature and is ideal for wallboard joint compound and an additive to paint. It is an essential component of the joint compound because it makes the compound smoother and easier to work with and it prevents cracking. In fact, wallboard compounds and paints account for about 80% of all mica used. In addition, ground dry mica is used in plastics, roofing, rubber, and welding rods.

When mica is ground up wet, however, it retains its sparkling shine and is used in many cosmetics (▶ Figure 2). Mica body powder is brushed onto the skin for an overall sheen. Fortunately, it is chemically inert and poses no risk when applied to the skin. Eye shadow, eyeliner, lipstick, blush, and nail polish have mica added to give them a resinous sheen, or mica-based powder may be added to lipstick. The brilliant sheen of some paints applied to automobiles (▶ Figure 3) and the changing color depending on viewing angles come from micas. Perhaps the statement that *mica is the most amazing stuff on Earth* is an exaggeration, but mica certainly enhances the visual appeal of many products.

Figure 2
Muscovite is used in several kinds of makeup as well as glitter.

Figure 3
Muscovite in paint on automobiles and appliances gives them their lustrous sheen.

GEOLOGY AT WORK

Terry S. Mollo
Sculpture Out of Stone

Terry S. Mollo is a full-time sculptor living in New York. She regularly exhibits her work in New York's Hudson Valley, New York City, and New Jersey. She has a B.A. in Journalism and Communication Arts from Pace University and currently studies sculpture at the Sculpture/Fine Arts Studio in Pomona, New York, under Martin Glick and at the Art Students League of New York with Gary Sussman.

It wasn't until 1997 that I started working with various stones, giving them new form, but I recall being drawn to the beauty of these natural formations at an early age. In 1954, when my grade school class studied the California gold rush, my friends and I formed a geology exploration club, which involved "borrowing" hammers, small chisels, and various other tools from our family garages to break apart pretty "rocks" we collected around the neighborhood. Black and silver mica was one of our favorite specimens; it was easy to find and lots of fun to crumble. We pulverized it with our hammers and chisels, poured it through makeshift strainers into jars, then weighed it, marked it "gem-dust," and stashed it away—a secret, sparkling treasure.

Now, many years later, as a full-time sculptor, I have returned to these remarkable stones that I sought out in childhood, but the tools I use today are specifically designed to file, chip, and carve stone into a new work of art. In my work, I use a method called the "subtractive" method, which is to say that I remove something to create something else. By removing stone and creating areas of depth and shadow, I strive to create an aesthetic work of art. Hundreds of tools are now available for stone carving (manual, pneumatic, electric, and industrial). Despite recent innovations and variations in tools, many sculptors still depend on the same basic handful of tools that Michelangelo used—hammers and chisels. In fact, popular hand tools like files, rasps, and rifflers, which are hand-forged so no two are identical, come from Milan, Italy, where great masters like Michelangelo worked.

When an artist is blocking out a piece in its early stages, large amounts of excess stone can be removed quickly with pneumatic chisels as well as a flexible shaft used for cutting, grinding, and contouring. For fine details, a sculptor often depends on that small traditional group of manual files, rasps, and rifflers to finish a piece. At the end, sanding and polishing are done with silicon carbide, diamond wet/dry sandpapers, or electric machines.

Great strides have been made technologically during the last 10 to 20 years, and now it is actually possible to start with a clay or plaster maquette and use a digitally based carving process called a CNC (Computer Numerically Controlled) carving machine. CNC uses a 3-D scanned image of the original work, digital technology, and an enormous five-axis milling machine. Even with these modern inventions, sculpting stone is a slow and arduous art, and unfortunately a dying one. Those of us who are still active sculptors are often told that we must teach all we know and pass it down, so it doesn't go away.

I have worked with several types of stone, including marble, alabaster, agate, and limestone. There is a large variety of material to choose from, and within each category the colors, shades, shapes, and veining are very exciting to an artist, each lending itself to a different kind of piece, be it abstract, realistic, allegorical—figurative or nonfigurative. Many sculptors today prefer the alabasters, since they offer the widest variety of colors and interesting veining. Alabaster (as well as steatite or soapstone) is a softer stone and is more easily carved as opposed to marble, which is metamorphic, or granite, which is igneous and extremely hard. Marble and granite are both dense and difficult to carve; nonetheless, most sculptors at some point in their career seek out a beautiful piece of Italian marble to carve.

Stones are available from quarries or through suppliers both in the United States and abroad. Domestic alabasters, limestone, and good marbles come from many quarries in the United States in Vermont, Georgia, and Colorado. Italy has amazing marble, alabaster, and travertine. There are also Portuguese pink marble, Belgian black marble, Persian travertine, and African wonderstone, which is a beautiful deep gray and black, although it is very difficult to carve and is one of the most expensive stones. In the New York area current stone prices range from about $0.75 per pound to $4 or $5 per pound. Indiana limestone is quite inexpensive. A piece of Italian white alabaster a little bigger than a football may cost $60.

The most common question people ask me about my work as a sculptor is whether or not I see a piece of stone I admire and bring it to the studio where I decide what to do with it, or whether I have a particular idea and then go out after the perfect piece of stone to create that vision. Well, the truth is that I have experienced the process both ways. Many of the ideas I have with regard to stone involve organic shapes such as leaves, flowers, and trees. For example, one morning I had an idea to create a big flower in an upright position and went out in search of the perfect stone to carve into the flower I imagined (▶ Figure 1). I walked around a gigantic supply basement filled with thousands of pieces that varied in size and color from all over the world. Finally, I chose a white/gold alabaster that weighed about 350 pounds. It had a very nice shape and size for the vision floating in my head. I could see my flower deep within the stone. Then, as I waited for the piece to be loaded into my car, I saw the most beautiful piece of honey-brown-colored Italian agate with beige and pale yellow veining, so I bought it. I had no idea what I would do with it, however, so it sat in my basement for well over a year waiting for my inspiration. Then one evening I came across a framed black and white Ansel Adams photograph of a wave . . . and that was it. I took my Italian brown agate into the studio and carved away. The beige at the tips became sea froth, the opaque, dark chocolate section the undercurrent, and the veins running diagonally my shoreline.

Figure 1
This flower sculpture by Terry Mollo is titled Hu Yi Fang (Breathe Comfortable Fragrance) and is made from white/gold alabaster.

Chapter 4

Igneous Rocks and Intrusive Igneous Activity

Nature: The Architect
A slender spill of water like a shoot of ivory light. Columns of rock, dark and glassy dangle over the cliffs like crystals from a chandelier. The waterfall in this image is Svartifoss, "the black waterfall." The name refers to the backdrop of black, hexagonal, basalt columns. Svartifoss is part of the Skaftafjell National Park in Iceland. Nature, the architect of this design, inspired the architecture of the National Theatre in Reykjavik.
—A. W.

ESSENTIAL QUESTIONS TO ASK

4.1 Introduction
- What are common plutonic and volcanic rocks?
- How are plutonic and volcanic rocks related?

4.2 The Properties and Behavior of Magma and Lava
- What role does magma play in helping heat escape Earth?
- What special conditions are required to make magma?
- What is unique about the origin of granitic magma?
- Why and how does magma rise?
- What are volatiles and why are they important?
- What is the chemical composition of magma?
- How hot is magma?
- What is viscosity and what controls it?
- What is Bowen's reaction series?
- How does magma change as crystallization occurs?
- What happens when two magmas come together?
- What are zoned plutons?

4.3 Igneous Rocks—Their Characteristics and Classification
- What does the term *texture* mean with respect to rocks, and what are some examples of igneous textures?
- Why is mineral content important for classifying igneous rocks?

4.4 Plutons—Their Characteristics and Origins
- What are the largest igneous intrusions, and what do their contacts look like?
- What are the major types of sheetlike igneous intrusions?

GEOLOGY MATTERS

GEOLOGY IN YOUR LIFE:
Of Mineral Water and Meteorites

GEOLOGY IN FOCUS:
Some Remarkable Volcanic Necks

GEOLOGY IN UNEXPECTED PLACES:
Little Rock, Big Story

GEOLOGY AND CULTURAL CONNECTIONS:
Hell Helped Early Earth Science

Physical Geology Now™ This icon, which appears throughout the book, indicates an opportunity to explore interactive tutorials, animations, or practice problems available on the Physical GeologyNow website at **http://now.brookscole.com/phygeo6**.

101

4.1 Introduction

Recall that the term *rock* applies to a solid aggregate of one or more minerals. Less commonly it refers to mineral-like inorganic matter such as natural glass and solid masses of organic matter like coal. Also, remember that there are three main families of rocks: igneous, sedimentary, and metamorphic.

Our exploration of the three main kinds of rocks begins with **igneous rocks,** which in a literal as well as a figurative sense provide a foundation for the other rock types. Recall that igneous rocks form by the cooling of molten matter, or "melt." The melt can take several forms. In its underground state it is called **magma.** Magma can either issue from the ground as **lava** or blast out in roiling clouds of fragmental debris of many sizes and shapes called **pyroclastic materials** (from the Greek, meaning "fire-broken"). Examples of pyroclastic materials include ash, pumice, and cinder.

We are most familiar with the igneous processes of flowing lava and eruptions of volcanic ash because we can see these phenomena in action. We call this kind of igneous activity **volcanic**, or extrusive igneous, in reference to the volcanoes that produce it (see Chapter 5). **Volcanic rocks** form quickly: In a matter of minutes to hours, the blood red and yellow color of molten lava turns gray as the lava hardens into rock (▶ Figure 4.1). But this near-instantaneous geologic change is exceptional. Most igneous rocks are not of volcanic origin, but instead form from melt cooling and crystallizing kilometers beneath Earth's surface in a process that can take several million years. We know about such deep-forming igneous rocks, which are major "building blocks" of the crust, primarily because of the uplift and deep erosion of ancient mountain ranges. In imaginative reference to Pluto, the ancient Roman god of the underworld, we call them **plutonic rocks,** and we call the larger masses that they make in the crust **plutons.**

- *What are common plutonic and volcanic rocks?*

You are probably familiar with one of the most commonly seen and used plutonic rocks, *granite*. Other plutonic rocks, including the less well-known syenite and monzonite, superficially look like granite, and for convenience geologists call them all "granitic."

▶ Figure 4.1 **A Geologist Takes the Temperature of an Active Lava Flow in Hawaii**

▶ Figure 4.2 **Mount Rushmore and Crazy Horse Memorial**

a The presidents' images at Mount Rushmore, South Dakota, were carved in the Harney Peak Granite. The 18-m-high images were carved between 1927 and 1941 and are now the primary attraction at Mount Rushmore National Memorial.

b The nearby Crazy Horse Memorial, also carved in the Harney Peak Granite, is still under construction.

▶ Figure 4.3 Basalt Flow and Cinder Cone in the Medicine Lake Highland, California

Some granitic rocks are quite attractive, especially when sawed and polished. They are used for tombstones, mantlepieces, kitchen counters, facing stones on buildings, pedestals for statues, and statuary itself. The images of the presidents at Mount Rushmore National Memorial in South Dakota as well as part of the nearby Crazy Horse Memorial (under construction) are carved in the 1.7-billion-year-old Harney Peak Granite (▶ Figure 4.2).

Unless you live in places like Iceland, the Hawaiian Islands, or southern Idaho, you may be less familiar with the most common kind of volcanic rock, the dense, dark lava called *basalt* (▶ Figure 4.3). Basalt makes up the bedrock floor of the oceans worldwide, although in many places it is covered by a thin mantle of sediment. Therefore you can think of it as being the most common rock on Earth's surface. It is also the principal lunar lava, making up the dark "eyes" and "mouth" that you see when you look at a full Moon. Basalt is less familiar as a construction material, although it is a great insulator. Many home chimneys at cooler, higher elevations in the Hawaiian Islands are made of mortared basalt blocks.

Granite and basalt are near opposite ends of the compositional and textural spectrum of ordinary igneous rocks. So perhaps it is not surprising that these two rock types result from very different melting processes and rarely occur together.

■ *How are plutonic and volcanic rocks related?*

Although there are many kinds of igneous rocks, there are some simple patterns for classifying and understanding them. Foremost is the fact that each type of volcanic rock has an equivalent plutonic rock of identical composition. For example, granite is the plutonic equivalent of rhyolite, and gabbro is the plutonic equivalent of basalt. To learn the family of igneous rocks, it is a good idea to memorize rock names in plutonic–volcanic pairs—for example, granite/rhyolite, diorite/andesite, and gabbro/basalt (Table 4.1).

In this chapter our main concerns are (1) the origin, composition, textures, and classification of igneous rocks, and (2) the significance and types of plutonic bodies. In Chapter 5 we will consider volcanism, volcanoes, and associated phenomena that result from magma reaching Earth's surface. Keep in mind that the origins of plutons and volcanic rocks are intrinsically related topics (see "From Pluton to Volcano" on pp. 104–105).

Section 4.1 Summary

• Igneous rocks form from the cooling and solidification of magma, molten rock. If this takes place underground, crystalline plutonic rocks develop. If the magma erupts as lava or fragmental (pyroclastic) materials, then the resulting rocks show few if any crystals and are called volcanic. Granite and basalt are two common types of igneous rocks. Each igneous rock type has a plutonic and volcanic equivalent. For example, granite, which is a plutonic rock, is chemically identical to rhyolite, a volcanic rock, even though these two rocks have very different physical appearances.

4.2 The Properties and Behavior of Magma and Lava

■ *What role does magma play in helping heat escape Earth?*

Planets like Earth get hot for several reasons. One of the most important is the slow radioactive decay of various unstable elements, including uranium and potassium. This heat seeks to escape from the deep interior and can do so in only a few ways. One way is convection, discussed in Chapter 1. Another is simple conduction, as shown by the way heat travels along a metal spoon that has one end in a pot of hot soup. A third, very effective way is through the production and ascent of magma. Magma can carry heat toward the

TABLE 4.1
The Major Igneous Rocks

Decreasing Iron and Magnesium, Increasing Silica	Plutonic Rock Names	Volcanic Rock Names	General Compositional Group
↓	Peridotite	n/a	Ultramafic
	Gabbro	Basalt	Mafic
	Diorite	Andesite	Intermediate
	Granodiorite	Dacite	Intermediate
	Granite	Rhyolite	Felsic

From Pluton to Volcano

Every volcano is underlain by igneous intrusions, but rarely do we find preserved the actual *transitions* between intrusive rocks and their associated volcanic deposits. In other words, it is difficult to find ancient volcanic vents preserved in cross-section. But thanks to the quirkiness of erosion, some spectacular examples do exist.

▲ **1.** An intrusion got close enough to the surface to cause the ground to steam on the flank of Bolshoi Semiachik Volcano, Kamchatka.

▲ **2.** A dike slices across pale yellow-brown pyroclastic layers and feeds a lava flow in this road cut on West Maui, Hawaii.

▲ **3.** A sheeted dike swarm marks the position of the Pliocene Mid-Atlantic Ridge, near Blonduos, Iceland. Many of these dikes fed lava flows, some of which are preserved in the adjacent volcanic strata.

◀ **4.** An Icelandic dike filled with explosion breccia. The breccia represents a volcanic eruption extending down into the throat of a fissure—the uppermost part of the dike, in other words.

▲ 5. Detail of dike-filling explosion breccia.

▲ 6. Devil's Backbone, a dike exposed in the wall of Crater Lake Caldera, Oregon. (A caldera is a giant volcanic collapse crater.) The dike extends from the lake shore halfway to the caldera rim, where it spreads into a thick lava flow subsequently buried by younger layers. The dike plus flow record an ancient eruption in the flank of Mt. Mazama, the huge volcano whose later collapse 7700 years ago created the caldera and exposed the dike.

▼ 7. Stages in the emplacement of a shallow dike.

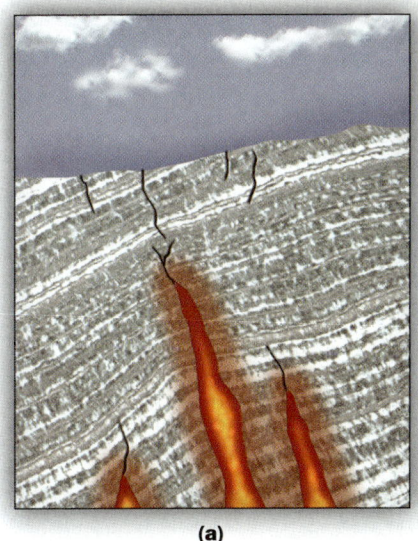
(a)

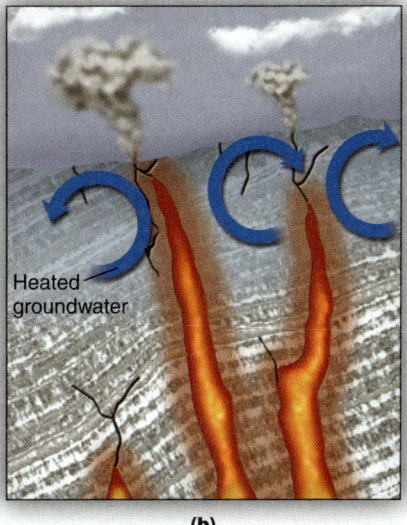
(b)

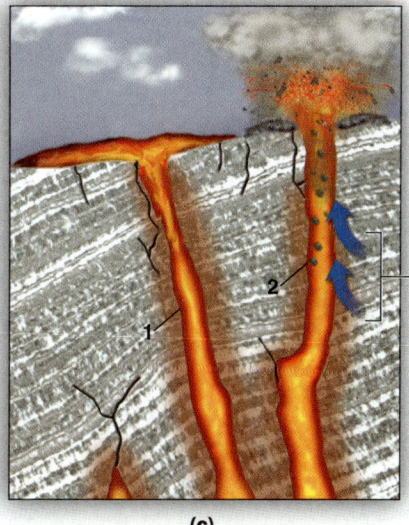
(c)

(a) Magma rises, but too deep to create visible surface effects.

(b) Magma close enough to create steam vents and hot springs.

(c) Volatile-poor magma erupts as lava (1), while volatile-rich magma erupts explosively (2), creating breccia and other pyroclastic deposits.

surface much faster than the surrounding rock can conduct it, and magma can rise through the crust where solid-state convection cannot occur.

■ *What special conditions are required to make magma?*

Magma does not form randomly beneath the surface. Melting in Earth requires very specific conditions. The pressure must be just *low* enough and the temperature just *high* enough for rock to melt. For example, the center of Earth's core, 6300 km beneath your feet, is the hottest point in the planet, yet extremely high pressure prevents the solid metal of the inner core from melting. The atoms simply cannot break their bonds to move randomly in a fluid state. Closer to the surface, though, the pressure drops and the still very hot outer portion of the core melts. In fact, the outer core is the largest **magma chamber** or reservoir in the solar system. Fortunately for us, it is contained in the thick, strong shell of Earth's mantle and crust.

Earth's silicate-rich crust and mantle have different melting temperatures than the core, but the pressure and temperature conditions would still keep the mantle solid if it were not for the convective upwelling of hot mantle rock into the asthenosphere, particularly beneath divergent plate margins (▶ Figure 4.4). The convecting rock cannot cool fast enough to keep from melting as the pressure drops during its ascent. This process generates most of the world's gabbroic magma (Table 4.1). The areas of magma production range from 40 km underneath the Hawaiian hot spot to 10–25 km below divergent plate boundaries. Much smaller pockets of magma form as deep as 100–200 km, where pressure on the mantle drops because of stretching in the overlying lithosphere.

■ *What is unique about the origin of granitic magma?*

The origin of granitic magma is much different. It is a product of melting *within the crust*, not the mantle, typically developing at "mid-crustal" levels 15–25 km deep. Most of this molten material originates from the melting of water-rich sedimentary and metamorphic rocks at convergent plate margins. Water trapped in rock pores has the property of lowering the melting point of ordinary sandstone by as much as 300°C. All that is needed is heat, and this is readily supplied wherever mantle-derived magma ascends into continental crust. In other words, one kind of magma, gabbroic, causes the formation of another, granitic, rather like a flame that melts wax (▶ Figure 4.5).

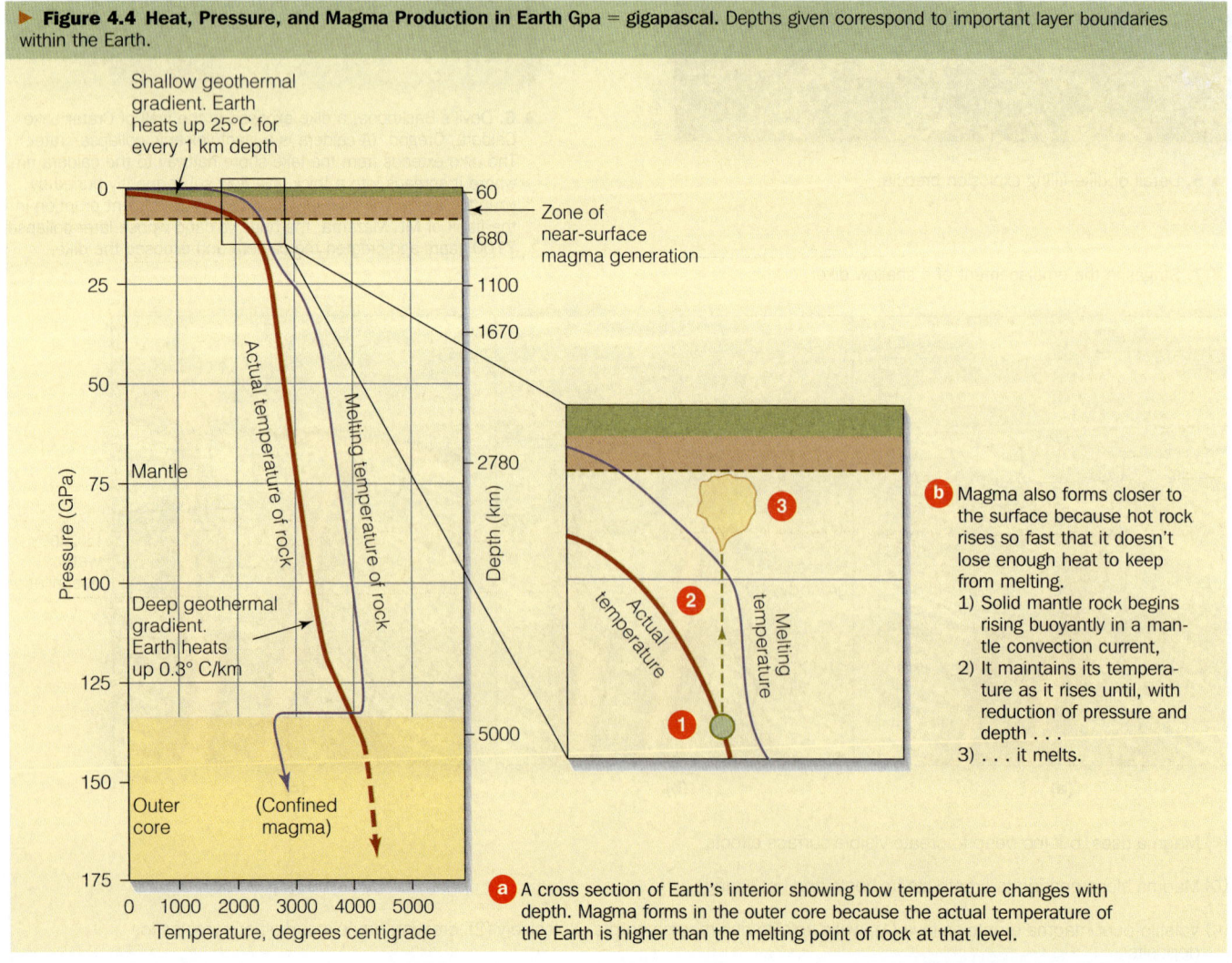

▶ **Figure 4.4 Heat, Pressure, and Magma Production in Earth** Gpa = gigapascal. Depths given correspond to important layer boundaries within the Earth.

a A cross section of Earth's interior showing how temperature changes with depth. Magma forms in the outer core because the actual temperature of the Earth is higher than the melting point of rock at that level.

b Magma also forms closer to the surface because hot rock rises so fast that it doesn't lose enough heat to keep from melting.
1) Solid mantle rock begins rising buoyantly in a mantle convection current,
2) It maintains its temperature as it rises until, with reduction of pressure and depth . . .
3) . . . it melts.

■ Why and how does magma rise?

All magma rises because it is less dense than the surrounding solid rock, which is called **country rock.** The loss of density when rock melts is easy to visualize. In solid mineral structures, atoms are bonded together tightly and efficiently. During melting the bonds break and atoms separate with increased randomness of motion. The volume occupied by the liquid exceeds the volume occupied by a crystalline solid of equivalent mass. In other words, it becomes buoyant.

Below the level of the shallow crust, *density contrast* is the primary factor that controls the rate of ascent; the greater the difference in density between melt and country rock, the faster magma rises. The initial melt filters along crystal grain boundaries and fractures in the solid country rock, eventually accumulating enough mass and buoyancy to push aside the overlying crust as it rises (▶ Figure 4.6a). Closer to the surface, however, the contrast in density drops to the point where equal volumes of magma and enclosing rock may weigh the same. The magma cannot rise farther in this so-called *neutral buoyancy* position. Instead, it begins to build up its volume, stretching and inflating the crust in some areas and in other places passively filling the gaps opened by the fault movements that occur commonly in tectonically active regions (▶ Figure 4.6b). As the volume of magma increases, great slabs of overlying country rock may detach and settle into the molten chamber, a process called **stoping** (▶ Figure 4.6c). Small stoped pieces called *xenoliths*, meaning "foreign rocks," may be smaller than a coin, whereas larger blocks may be the size of buildings. Some stoped blocks can drop hundreds of meters and partly or wholly melt within the magma, altering the magma chemistry in noticeable ways. **Assimilation** is the digestion of stoped blocks and surrounding country rock by magma (▶ Figure 4.7). Chemical studies show that only a small fraction of any typical magma body consists of assimilated material. The continuously cooling magma simply does not have enough heat to incorporate much new material after it stagnates in the shallow crust.

■ What are volatiles and why are they important?

Typical neutral buoyancy levels range from 1.5 to 10 km deep, depending on the magma type and country rock compositions, temperature, and pressure. Magmas stalled at neutral buoyancy can still erupt if sufficiently shallow, due to the buildup of certain dissolved substances, called *volatiles*, within the melt. Volatiles are elements and compounds that do not ordinarily become part of the solid crystalline structure of minerals and that exist as gases or liquids at Earth's surface. The most concentrated volatile by far is carbon dioxide (CO_2). Indeed, much of our atmosphere's carbon dioxide content originated and is maintained by exhalation of CO_2 from erupting volcanoes. After carbon dioxide, hydrogen sulfide (H_2S), a deadly gas with the odor of rotten eggs, is most abundant. Less abundant are chlorine and fluorine (also quite deadly), argon, helium, hydrogen, and water vapor.

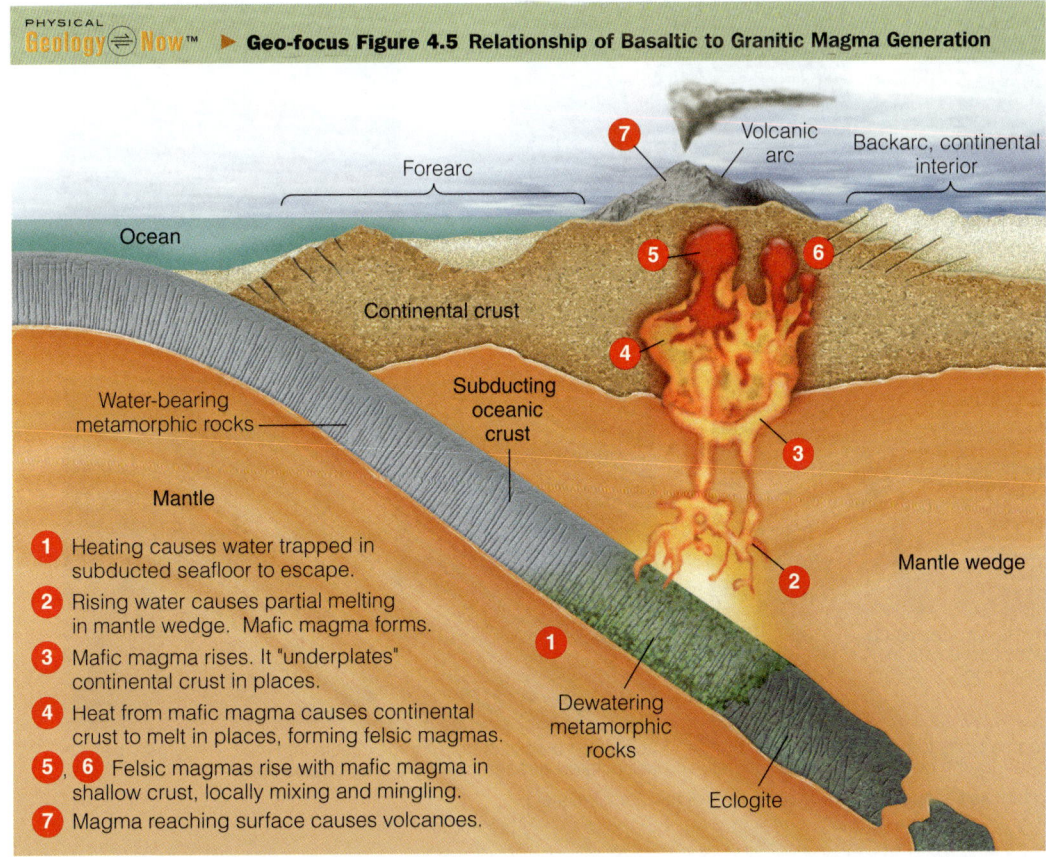

Geo-focus Figure 4.5 Relationship of Basaltic to Granitic Magma Generation

1. Heating causes water trapped in subducted seafloor to escape.
2. Rising water causes partial melting in mantle wedge. Mafic magma forms.
3. Mafic magma rises. It "underplates" continental crust in places.
4. Heat from mafic magma causes continental crust to melt in places, forming felsic magmas.
5, 6 Felsic magmas rise with mafic magma in shallow crust, locally mixing and mingling.
7. Magma reaching surface causes volcanoes.

Figure 4.6 Some Ways in Which Magma Bodies Grow Bigger

What's happening underground

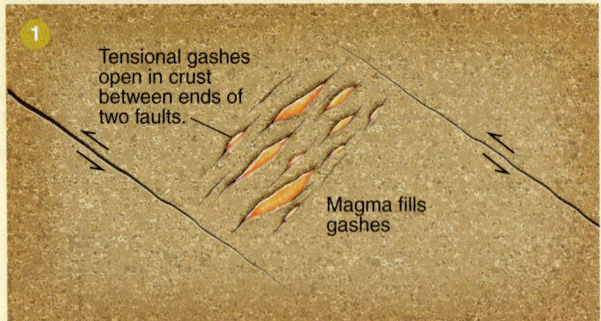

What's seen at the surface

a Magma fills gaps opened by fault movements in the crust.

Rock melts partly deep in crust. Melt is widely dispersed

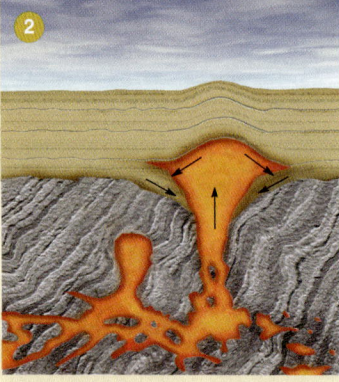

Molten batches grow large enough to start rising, shouldering aside overlying rock.

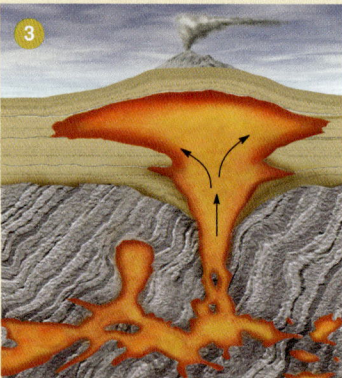

Magma spreads out and swells incrementally in shallow crust, forming giant, lens-shaped pluton.

b Magma bouyantly displaces overlying rock and inflates in heated shallow crust

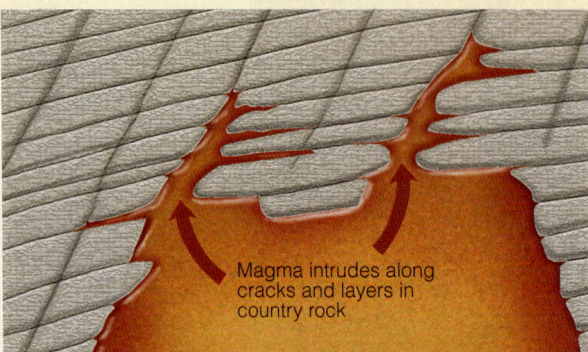

Note: Stoping can assist a magma body to become shallower, but is not a major process in making magma bodies grow larger overall.

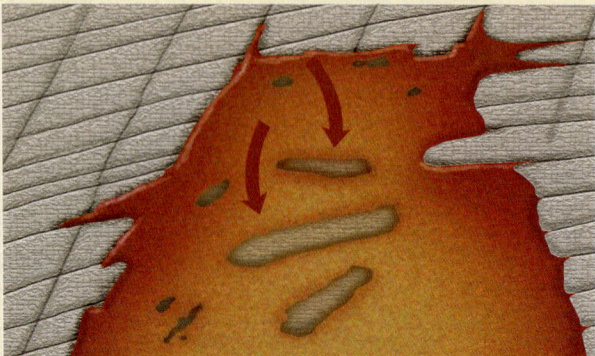

Loosened blocks of country rock settle into melt. They may be assimilated deeper inside the magma chamber.

c Magma detaches ceiling blocks ("stopes" them), which it may digest ("assimilate").

▶ **Figure 4.7 Xenoliths of Pale Country Rock in Darker Lava near Cleetwood Cove, Crater Lake, Oregon** The rounded shapes and darkened margins of the xenoliths show that the lava partly assimilated them as it cooled.

Quite often the volatile composition of magma changes as it approaches the surface. The dissolved H_2O content of the melt increases considerably in the shallow crust because of the assimilation of ordinary groundwater. In fact, H_2O commonly exceeds CO_2 as the most abundant volatile in shallow magma bodies. Given the readiness with which dissolved water expands into steam with the accompanying reduction in pressure, and the power and scale of this expansion, it is easy to see why water, ironically, is a major propellant of explosive volcanic eruptions (▶ Figure 4.8).

As much as 3% or even 6% of the total weight of a granitic magma body may consist of dissolved volatiles and bubbles at neutral buoyancy. Only 0.5–3% of gabbroic melt may be volatiles, however, primarily because mantle magmas derive from the melting of "dry" rocks.

As a magma reservoir grows larger, the pressure on the overlying rocks increases. Fault movement and fracturing in the ceiling rock can suddenly reduce this pressure, causing volatiles to exsolve much like foam atop a quickly uncapped soda bottle. The pressure of gas bubbles may be sufficient to weaken a reservoir roof further. At a depth of 1.5 km, 5 megapascals of bubble pressure can trigger an eruption. (One megapascal is equivalent to 10 times atmospheric pressure.) The escape of volatiles is sometimes enormously violent, blasting out as much as 10% of a magma reservoir in a matter of hours or a few days. In the most powerful volcanic explosions, magma can escape at speeds up to 600 m/sec, accompanied by hypersonic bursts of gas.

Where volatiles cannot escape, they play an important role in altering the heated country rock to create highly valuable ore deposits. Gold, tungsten, molybdenum, lithium, and many other metals such as the europium that we use to give the red pigment in television sets are derived from magma-related volatile releases. Chlorine is an especially important volatile component because it bonds with metals dissolved in magma to create ore-forming solutions. Many of the copper ores that we mine for pipes, chromium for car fenders, and platinum for catalytic converters and satellites accumulated from the slow, incremental crystallization of certain metal-rich magmas.

■ *What is the chemical composition of magma?*

You have already learned that the vast majority of minerals in Earth's crust are composed of tetrahedra of silicon and oxygen atoms and symmetric arrangements of other less abundant elements. It should come as no surprise that the same composition holds true for magmas; silicon and oxygen dominate, with lesser amounts of other elements present in solution. In magmas, of course, these atoms lack the ordered packing of crystalline solids.

In the mantle, where only a small percentage of any given patch of rock may melt, magmas tend to be highly enriched in magnesium (Mg) and iron (Fe) relative to melts formed in the crust. This is shown by the olivine- and pyroxene-rich mineral contents and the very dark and heavy character of gabbro and basalt. We call Mg-Fe–rich melts and their resulting rocks **mafic** (a term derived by combining "<u>ma</u>gnesium" and "<u>fer</u>ric," for iron). In contrast, continental crust is enriched with aluminum, sodium, potassium, and water. This contributes to magmas that crystallize abundant feldspar and hydrous minerals, such as mica and hornblende; they are said to be **felsic** ("fels" refers to feldspar).

▶ **Figure 4.8 A Student Geologist Watches an Eruptive Fissure Open Near the Summit of Kilauea Volcano, Hawaii** Note the clouds of gas and dust issuing with the lava. They are blue and brown-tinged and slightly darker than the ordinary clouds in the background sky.

There is a general correlation between the total silica content and the degree to which an igneous rock is mafic or felsic. If the amount of total silica (SiO_2) in a rock is less than 52%, it is mafic. If greater than about 65%, it is felsic. A gradation of compositions belonging to the **intermediate** igneous clan (53–65% SiO_2) lies between (Table 4.2). This includes such important rock types as andesite, which erupts as lava and ash from many coastal continental volcanoes, and diorite, a salt-and-peppery plutonic rock that is noticeably darker than granite.

The terms *mafic*, *intermediate*, and *felsic* provide a first-order way of classifying igneous rocks and their magmas on the basis of chemical composition.

■ *How hot is magma?*

Whether or not you have witnessed a lava flow, you know that lava is very hot. But how hot? Erupting lavas generally have temperatures in the range of 1000° to 1200°C, although a temperature of 1350°C was once recorded above a Hawaiian lava lake where volcanic gases reacted with the atmosphere. In contrast to mafic lava flows such as those issuing from the Hawaiian volcanoes, erupting felsic lava tends to be somewhat cooler. The temperatures of some erupting lava domes have been measured at a distance with an instrument called an optical pyrometer. The surfaces of these domes are as hot as 900°C. We can only infer that magma must be hotter still.

When Mount St. Helens erupted in 1980, it ejected felsic magma as a ground-hugging hurricane of pyroclastic material that spread over an area of 600 km^2. Two weeks later, beds of this debris still had temperatures between 300°C and 420°C, and a steam explosion took place more than a year later when water encountered some of the still-hot deposits. The reason magma and lava retain heat so well is that rock conducts heat so poorly. Accordingly, the interiors of thick lava flows may remain hot for months or years, whereas plutons, depending on their size and depth, may not completely cool for thousands to millions of years.

■ *What is viscosity and what controls it?*

The term **viscosity** refers to the resistance of a fluid or fluidlike material to flowing; it is widely used to describe the eruptive potential of magma. A high-viscosity material, such as tar or glacial ice, does not flow very easily. Low-viscosity fluids, such as water or oil, flow readily. Magma spans a range of viscosities from high in the case of granitic compositions to low in the case of molten basalt.

Several factors control the viscosity of magma; the most important is temperature. When substances such as tar or wax are heated, their viscosities drop significantly; heated tar flows like syrup. We can generalize and say that hot magma or lava moves more easily than cooler magma or lava, but we must qualify this statement by noting that temperature is not the only determinant of viscosity. Silica content is also important. With increasing concentrations of dissolved silica in magma, networks of silica tetrahedra grow and retard flowage because of their strong silicon–oxygen bonds. Thus mafic magma is generally more fluid than felsic magma because it is hotter and not as rich in silica. In 1783 a mafic flow in Iceland flowed about 80 km, and geologists have traced some ancient flows in Washington State for more than 500 km, from near the Idaho border all the way to the Pacific Ocean. Felsic magma, in contrast, because of its higher viscosity, does not reach the surface as often as mafic magma. And when felsic lava flows do occur, they tend to be slow moving and thick and to move only short distances. A thick, pasty lava flow that erupted in 1915 from Lassen Peak in California flowed only about 300 m before it ceased moving.

Other factors that influence the viscosity of magma are the volatile content, bubble content, crystallinity, and shear stress. A high content of dissolved gases lubricates a liquid, enabling it to flow more easily. A high crystal or gas bubble content, however, retards motion. Likewise, lava flows over rough ground more viscously than over smooth ground because there is a loss of energy owing to friction, or shear, across the irregular surface.

■ *What is Bowen's reaction series?*

During the early part of the last century, a laboratory geologist working at the Carnegie Institute in Washington, DC, Norman L. Bowen, hypothesized that mafic, intermediate, and felsic magmas could all be derived from a parent mafic magma. He knew that minerals do not all crystallize simultaneously from cooling magma, but rather crystallize in a predictable sequence. Based on his observations and experiments with a rock-melting furnace, Bowen proposed a mechanism, now called **Bowen's reaction series,** to account for the derivation of intermediate and felsic magmas from mafic magma. Bowen's reaction series consists of two branches: a *discontinuous branch* and a *continuous branch* (▶ Figure 4.9). As the temperature of magma decreases, minerals crystallize along both branches simultaneously, but for convenience we will discuss them separately.

In the discontinuous branch, which contains only ferromagnesian silicates, one mineral changes to another over specific temperature ranges (▶ Figure 4.9). As the temperature decreases, a temperature range is reached in which a given mineral begins to crystallize. A previously formed mineral reacts with the remaining melt so that it

TABLE 4.2
The Most Common Types of Magma and Their Characteristics

Type of Magma	Silica Content (%)	Sodium, Potassium, and Aluminum	Calcium, Iron, and Magnesium
Ultramafic	<45		Increase ↑
Mafic	45–52	↓	↑
Intermediate	53–65	↓	↑
Felsic	>65	Increase	

forms the next mineral in the sequence. For instance, olivine [$(Mg,Fe)_2SiO_4$] is the earliest ferromagnesian silicate to crystallize. As the magma continues to cool, it reaches the temperature range at which pyroxene is stable; a reaction occurs between the olivine and the remaining melt, and pyroxene forms.

With continued cooling, a similar reaction takes place between pyroxene and the melt, and the pyroxene structure is rearranged to form amphibole. Further cooling causes a reaction between the amphibole and the melt, and its structure is rearranged so that the sheet structure of biotite mica forms. Although the reactions just described tend to convert one mineral to the next in the series, the reactions are not always complete. Olivine, for example, might have a rim of pyroxene, indicating an incomplete reaction. If magma cools rapidly enough, the early-formed minerals do not have time to react with the melt, and thus all the ferromagnesian silicates in the discontinuous branch can be in one rock. In any case, by the time biotite has crystallized, essentially all magnesium and iron present in the original magma have been used.

Plagioclase feldspars, which are nonferromagnesian silicates, are the only minerals in the continuous branch of Bowen's reaction series (▶ Figure 4.9). Calcium-rich plagioclase crystallizes first. As the magma continues to cool, calcium-rich plagioclase reacts with the melt, and plagioclase containing proportionately more sodium crystallizes until all of the calcium and sodium are used up. In many cases cooling is too rapid for a complete transformation from calcium-rich to sodium-rich plagioclase. Plagioclase forming under these conditions is *zoned*, meaning that it has a calcium-rich core surrounded by zones progressively richer in sodium.

As minerals crystallize simultaneously along the two branches of Bowen's reaction series, iron and magnesium are depleted because they are used in ferromagnesian silicates, whereas calcium and sodium are used up in plagioclase feldspars. At this point any leftover magma is enriched in potassium, aluminum, and silicon, which combine to form orthoclase ($KAlSi_3O_8$), a potassium feldspar, and if water pressure is high, the sheet silicate muscovite forms. Finally, any remaining magma is enriched in silicon and oxygen (silica) and forms the mineral quartz (SiO_2). The crystallization of orthoclase and quartz is not a true reaction series as is the crystallization of ferromagnesian silicates and plagioclase feldspars, because they form independently rather than by a reaction of orthoclase with the melt.

■ *How does magma change as crystallization occurs?*

As minerals crystallize sequentially with falling temperature, the composition of the remaining melt becomes progressively more felsic. Bowen thought that if the early-forming Mg-Fe–rich minerals settled to the bottom of a magma reservoir as soon as they crystallized, a process called *gravitational differentiation*, then a single parent melt perhaps of intermediate composition would generate three distinctly different rock types: a mafic layer at the bottom, an intermediate layer in the middle, and a felsic layer at the top of a reservoir (▶ Figure 4.10). In other words, a magma reservoir should behave like a giant natural distillery. The crystal layering of large mafic plutons, such as the Skaergaard Intrusion in Greenland, shows that Bowen's theory is substantially true. But we now know that there is not enough mafic

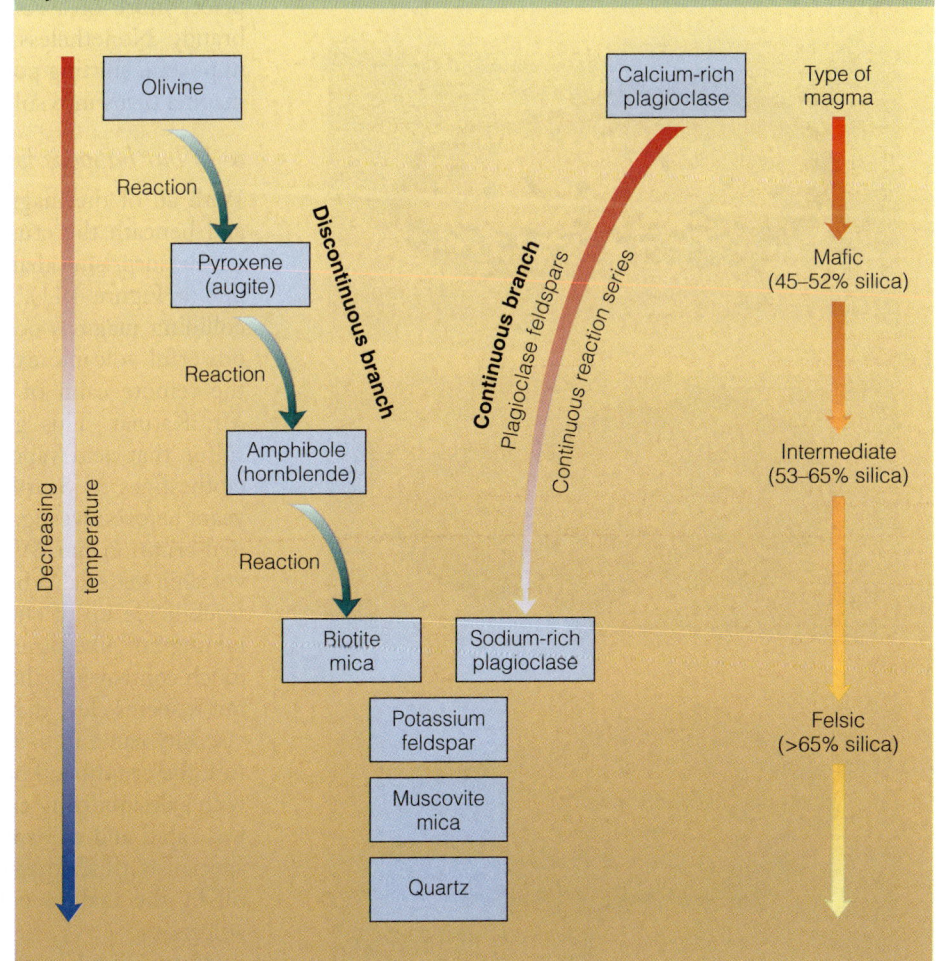

▶ **Figure 4.9 Bowen's Reaction Series** Bowen's reaction series consists of a discontinuous branch along which a succession of ferromagnesian silicates crystallize as the magma's temperature decreases, and a continuous branch along which plagioclase feldspars with increasing amounts of sodium crystallize. Notice also that the composition of the initial mafic magma changes as crystallization takes place along the two branches.

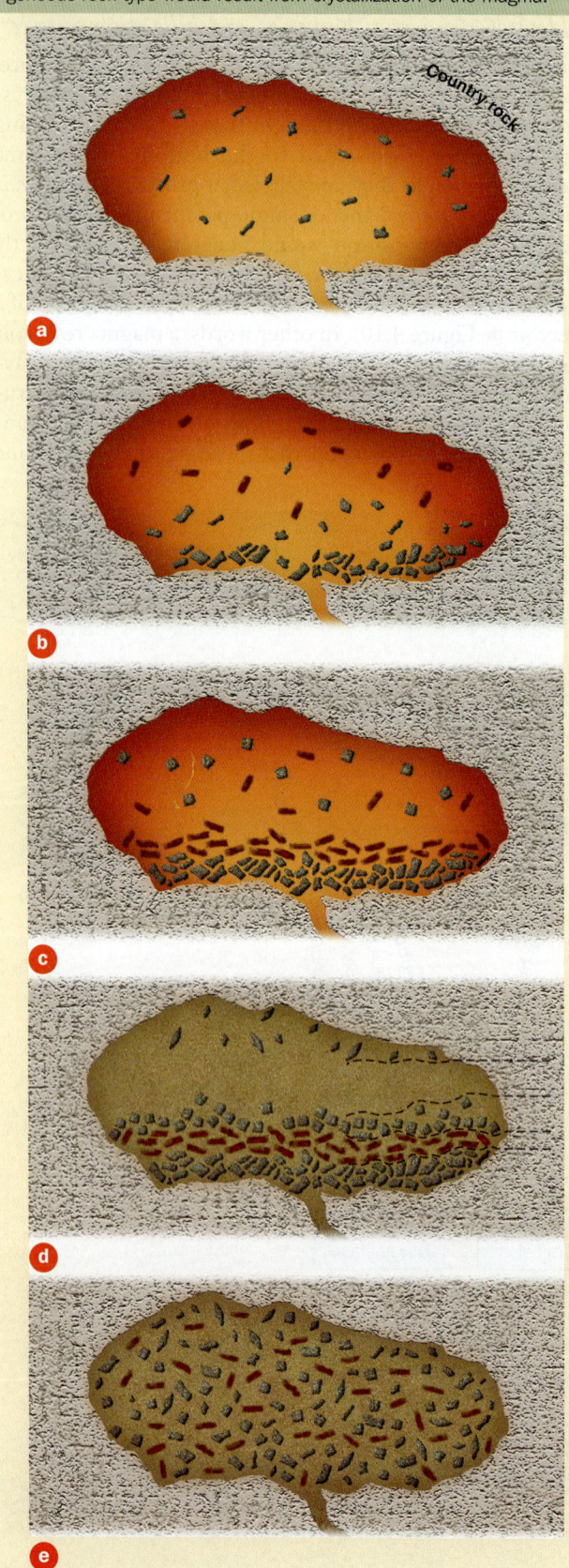

Geo-focus Figure 4.10 Gravitational differentiation of crystals settling in a melt (a), (b). Notice how rocks of five distinct compositions (**dashed lines in d**) result from a single parent magma that is crystallizing minerals at different temperatures. (**e**) Without gravitational differentiation, a single homogeneous rock type would result from crystallization of the magma.

magma in the world to generate the volume of granite that we observe via gravitational differentiation alone. The theory of plate tectonics did not exist in Bowen's day, and many geologists did not even believe that granite could form from cooling magma. Many imagined that it came from the fluid alteration of preexisting rocks, a much-debated process called *granitization*.

Bowen's *physical* vision of how a magma reservoir crystallizes was too simplistic. It is not merely the case that crystals form in the liquid at the top of a magma chamber where heat loss is greatest. Crystals also form close to the walls that extend deep down into the reservoir. Nor do they start settling the moment they begin crystallizing, if indeed they settle much at all. Convection currents in the deeper, hotter parts of the magma can keep crystals aloft near the roof, and viscosity constraints can be great. Many geologists think that a dense, crystal-rich mush zone develops across the top of a typical magma body. If the volumetric concentration of crystals exceeds 40–50% of the total magma at this level, the viscosity is too high either for gravitational differentiation or for an eruption to take place. In places, this mass of crystals can grow so heavy that the mush zone locally collapses into the melt below, shedding avalanches of crystals toward the bottom. But many crystals may end up reassimilated before they get there. So, in fact, an active magma reservoir is a very dynamic and complex place, more of a roiling stewpot than a still for making brandy. Nonetheless, Bowen's reaction series provides an important starting point for understanding how any silicate magma turns into solid rock.

▪ *What happens when two magmas come together?*

With all of the magma production going on both within and beneath the crust of many active tectonic belts, it is hardly surprising that rising magma bodies sometimes collide (▶ Figure 4.11). One of the most graphic examples of colliding magmas occurred in June 1912, when the most powerful volcanic explosion of the 20th century shattered the remote calm of the Alaska Peninsula at the foot of Mt. Katmai. This gigantic eruption produced so much sulfur that acid vapors ate holes in garments drying on clotheslines in Seattle, 3000 km downwind. It was four years before investigators were able to reach the uninhabited eruption site. What they found was a very interesting volcanic material: a bed of pyroclastic debris tens of meters thick made up of clumps of golden rhyolite pumice containing taffylike shreds of dark gray andesite. There is still much controversy about what exactly took place at Katmai, but it seems clear that one kind of magma of very different viscosity came in contact with another magma shortly before the eruption. The two did not mix but simply mingled with a destabilizing exchange of heat. The volatile response was rapid and an explosion ensued. Indeed, some volcano specialists think that many, if not most, eruptions are set off by the shallow collision of different magmas beneath volcanoes.

Many plutonic rocks also reveal evidence of magma mingling. For example, during the late Paleozoic, multiple

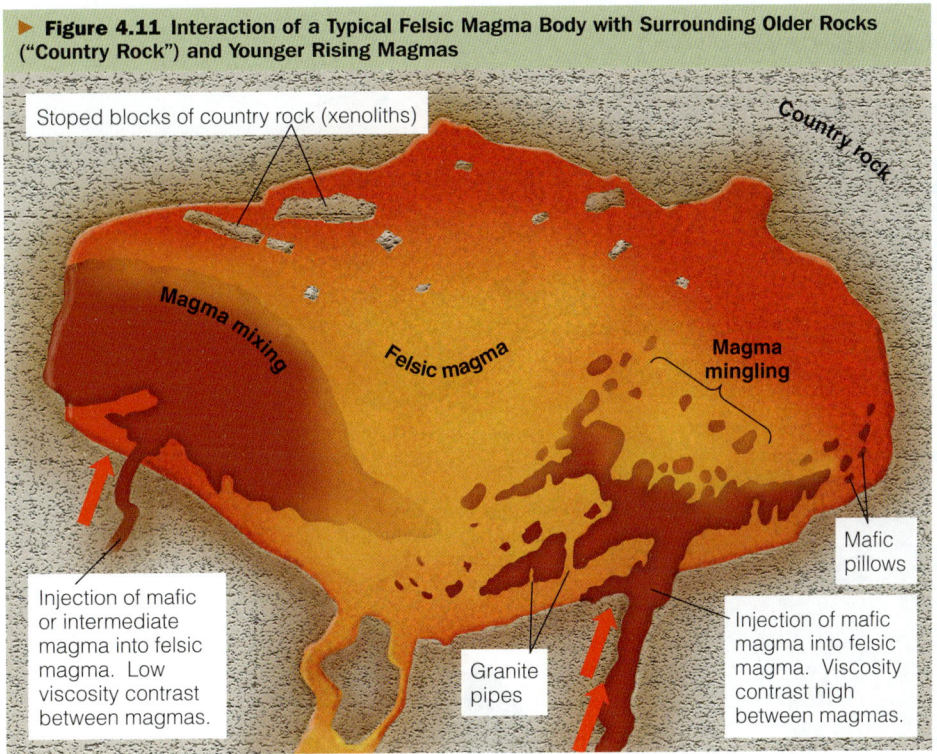

Figure 4.11 Interaction of a Typical Felsic Magma Body with Surrounding Older Rocks ("Country Rock") and Younger Rising Magmas

injections of gabbroic "guest" magma invaded the bottoms of felsic "host" magma chambers along what is now the central coast of Maine. In many parts of the host plutons, the gabbro cooled into sandbaglike piles of dark inclusions that resemble the pillow lava mounds of mid-oceanic rifts (▶ Figure 4.12). Instead of erupting into the much cooler seawater, however, the mafic melt "erupted" into somewhat cooler silica-rich magma. Locally, cooling gabbro appears to have squeezed trapped felsic melt beneath it, which squirted up into the gabbro like catsup out of a bottle via tubular intrusions called *granite pipes* (▶ Figure 4.13). Maine's coastal plutons tell a story of guest magma repeatedly invading host magma, only to be locally intruded by trapped host melt in the final stage of cooling and consolidation—a complicated magmatic interplay that created some extraordinary rocks.

■ *What are zoned plutons?*

Gravitational differentiation and multiple intrusions in a magma reservoir can lead to the development of a *zoned pluton* with concentric layers of different plutonic rocks. Typically, the boundaries between the zones are gradational. Some zoned plutons have mafic edges and felsic cores; others show the opposite pattern. In other cases, the pattern is more complex, such as in the plutons of coastal Maine.

Intermediate rocks in a zoned pluton commonly contain some crystals that match those found in felsic rocks of the same plutonic body, and they may have textures that grade into adjacent mafic rocks. The intermediate composition in such cases clearly arose from the blending together of two melts, a process called **magma mixing,** or *hybridization*. Melt viscosities must be fairly close to one another in order for this

Figure 4.12 Mafic Pillows in Granite, Vinalhaven Island, Maine

114 Chapter 4 Igneous Rocks and Intrusive Igneous Activity

Geo-focus Figure 4.13 Origins of Granite Pipes

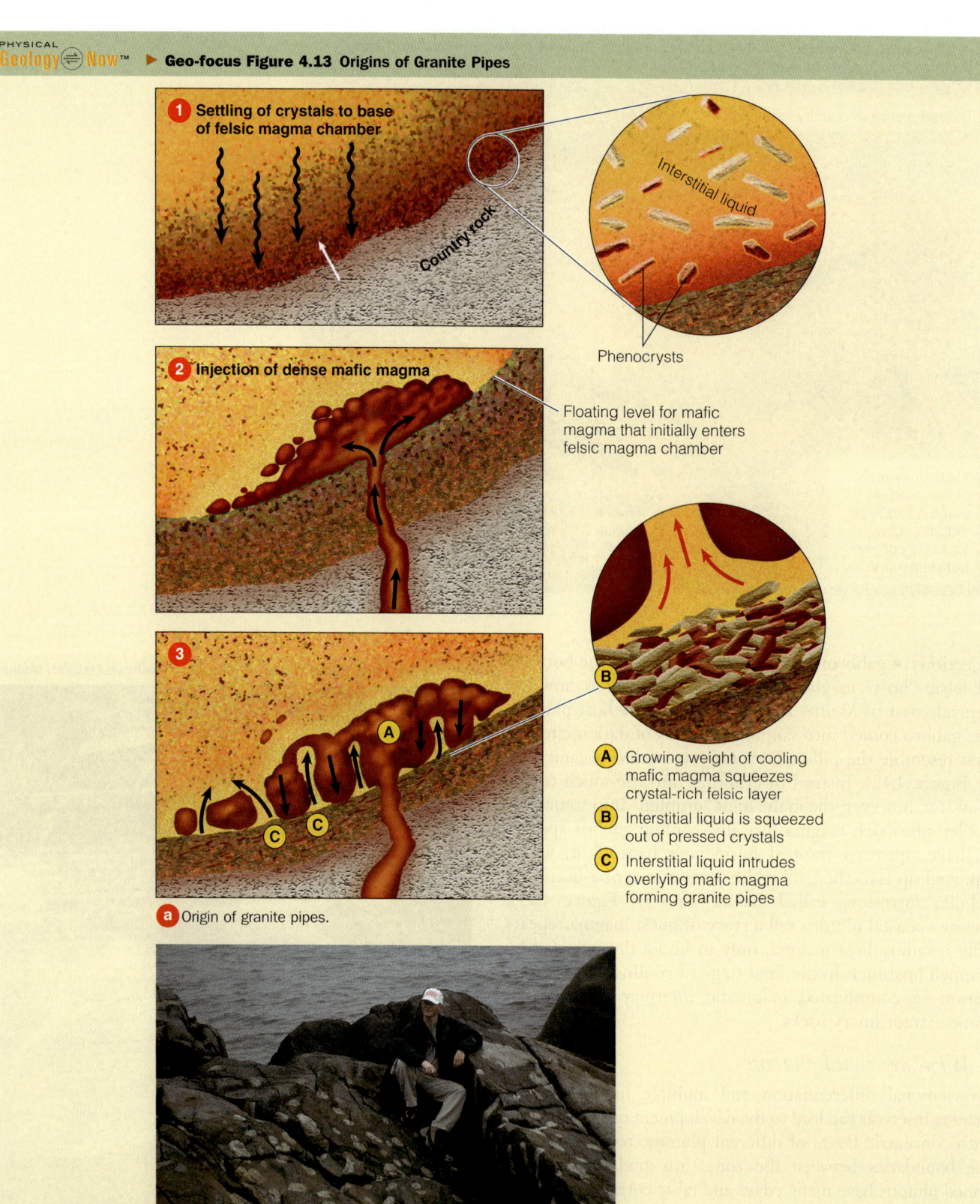

a Origin of granite pipes.

b Granite pipes on the coast of Maine.

to happen; otherwise, the result is **magma mingling**, the intermixing of two distinctive liquids that cannot blend together, like oil and water. Not all intermediate igneous rocks form by magma mixing, but evidence indicates that this is a more common process than geologists thought in Bowen's days.

> ### Section 4.2 Summary
>
> - Magma forms under restricted conditions of high temperature and low pressure inside Earth. Although rock is hot throughout the planet's interior, there are only a few zones where the pressure is low enough for the rock to melt. One is the outer core, and the other is the asthenosphere. Most of the molten rock produced at shallow levels in Earth originates where upwelling currents of hot mantle material rapidly rise into the lower-pressure environment of the asthenosphere. Magma transfers not only matter toward the surface but heat as well. Heat from mantle-derived magma contributes to melting continental sedimentary and metamorphic rocks at convergent plate margins to generate granitic magmas. The sedimentary and metamorphic rocks contain trapped water, which lowers the melting temperature of the rock in some cases by hundreds of degrees.
>
> - Magma may be defined chemically as mafic, with high magnesium and iron contents, or felsic, with high silica, aluminum, sodium, potassium and water contents. Intermediate magmas have compositions between those of mafic and felsic. The terms *mafic*, *felsic*, and *intermediate* also apply to igneous rocks.
>
> - Magma rises buoyantly because of the difference in density between the molten rock and the surrounding country rock. At the level of neutral buoyancy, where this density difference disappears, magma stagnates. Volatiles, including carbon dioxide, water vapor, sulfur gases, chlorine, fluorine, and other gases, build up in the magma reservoir and expand into bubbles whenever pressure drops. The force of expanding gases is the primary agent of volcanic eruptions. Viscosity, the resistance to flow, greatly influences the way that magmas erupt. High-viscosity molten rock oozes across the surface slowly. Low-viscosity lava flows can travel rapidly for tens of kilometers. Some volcanic eruptions are triggered by the collision of two magma bodies. Different magmas can mingle, or they can mix to produce hybridized rocks of intermediate composition.
>
> - Bowen's reaction series describes the crystallization of magma. It consists of two branches. The discontinuous reaction series describes the reaction of minerals with the magma at specific temperatures, leading to the formation of new minerals that are stable at lower temperatures. The continuous reaction series describes minerals that are continuously adjusting their compositions as the magma cools.

4.3 Igneous Rocks—Their Characteristics and Classification

So far, we have introduced two ways of classifying igneous rocks. They may be identified as volcanic (also called **extrusive**) or plutonic (**intrusive**), and they may be characterized chemically as mafic, felsic, or intermediate. But exactly how do we determine whether a rock is volcanic or plutonic, mafic, felsic, or intermediate? The answer is that two aspects of the rock must be examined: (1) its *texture* and (2) its *mineral composition*. Going one step further, we can assign specific technical names to igneous rocks, like "megacrystic granite" or "basalt porphyry," by knowing both the texture and the *specific* mineral proportions in each sample we examine. This is an important skill for new geologists to acquire, although it is sometimes difficult or even impossible to do in the real world because weathering and alteration can dramatically modify rocks. This does not stop determined geologists, however, who can take the additional step of slicing rocks very thinly to study them under a polarizing light microscope, or grinding them into powder and dissolving them in strong acids for high-powered chemical analysis in a laboratory. Microscope work often reveals features that the unaided eye cannot perceive, and chemical analyses can reveal things about a rock that would never be seen simply by looking at it, even with a microscope.

■ *What does the term* texture *mean with respect to rocks, and what are some examples of igneous textures?*

The term *texture* refers to the size, shape, and arrangement of mineral grains composing rocks (▶ Figure 4.14). Size is the most important because mineral crystal size is related to the cooling history of magma or lava and generally indicates whether an igneous rock is intrusive or extrusive. The atoms in magma and lava are in constant motion, but when cooling begins, some atoms bond to form small mineral nuclei. As other atoms in the liquid chemically bond to these nuclei, they do so in an orderly geometric arrangement and the nuclei grow into crystalline *mineral grains*, the individual particles that make up igneous rocks.

During rapid cooling, the rate at which mineral nuclei form exceeds the growth rate of crystals, and an aggregate of many small mineral grains is formed. The result is a fine-grained or **aphanitic texture,** in which individual minerals are too small to be seen without the aid of a microscope (▶ Figure 4.14a, b). With slow cooling, the rate of growth exceeds the rate of nuclei formation, and relatively large mineral grains form, thus yielding a coarse-grained or **phaneritic texture** in which minerals are clearly visible (▶ Figure 4.14c, d). Aphanitic textures generally indicate an extrusive origin, whereas rocks with phaneritic textures are usually intrusive. However, shallow plutons might have an aphanitic texture, and the rocks that form in the interiors of thick lava flows might be phaneritic.

Geo-focus Figure 4.14 The Various Textures of Igneous Rocks Texture is one criterion used to classify igneous rocks. **(a,b)** Rapid cooling as in lava flows results in many small minerals and an aphanitic (fine-grained) texture. **(c,d)** Slower cooling in plutons yields a phaneritic texture. **(e,f)** These porphyritic textures indicate a complex cooling history. **(g)** Obsidian has a glassy texture because magma cooled too quickly for mineral crystals to form. **(h)** Gases expand in lava and yield a vesicular texture. **(i)** Microscopic view of an igneous rock with a fragmental texture. The colorless, angular objects are pieces of volcanic glass measuring up to 2 mm.

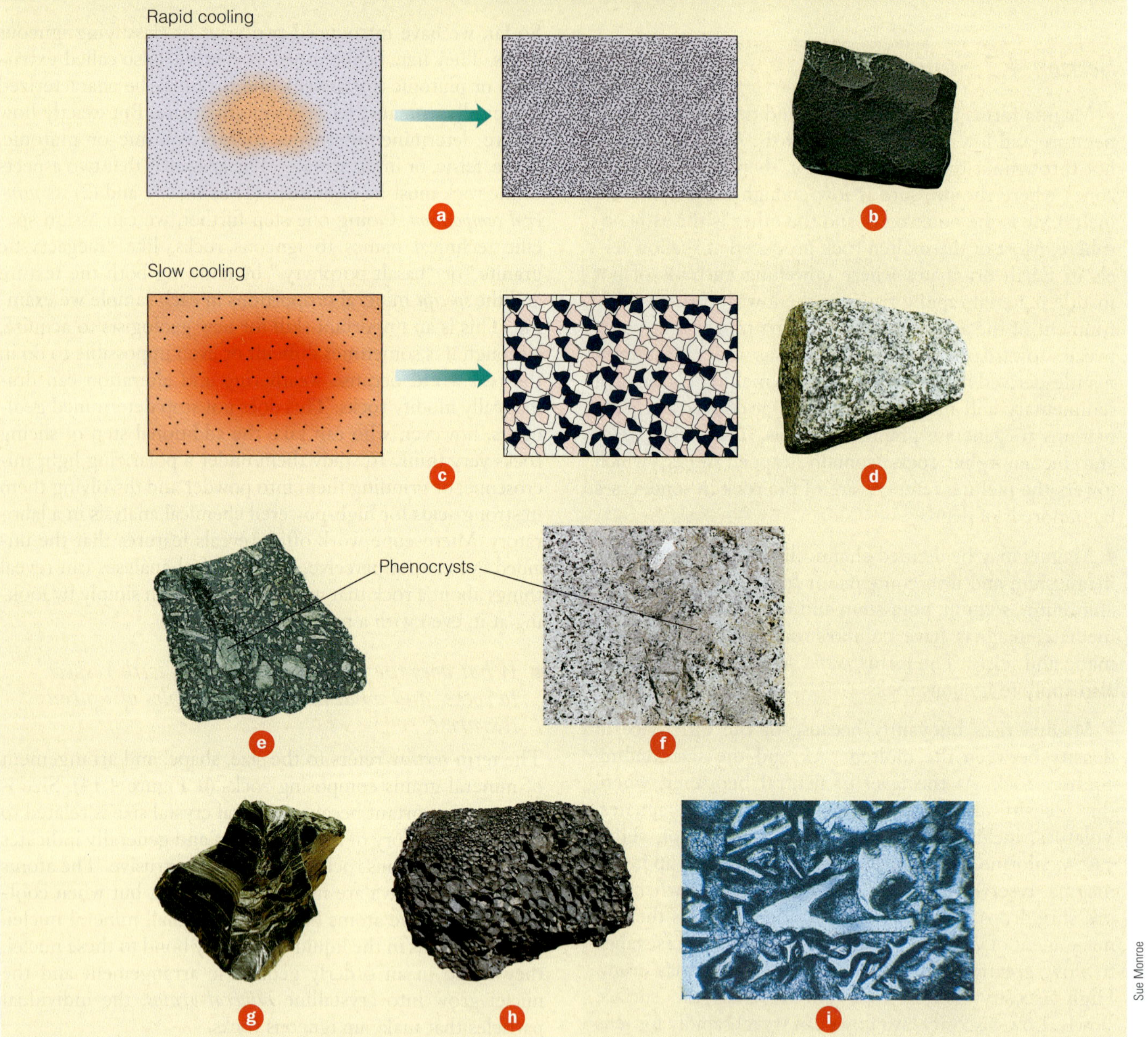

Another common texture in igneous rocks is called **porphyritic**, in which minerals of markedly different size are present in the same rock. The larger minerals are *phenocrysts* and the smaller ones collectively make up the *groundmass*, which is the grains between phenocrysts (▶ Figure 4.14e, f). The groundmass can be either aphanitic or phaneritic; the only requirement for a porphyritic texture is that the phenocrysts be considerably larger than the minerals in the groundmass. Igneous rocks with porphyritic textures are designated *porphyry* (e.g., basalt porphyry). These rocks have more complex cooling histories than rocks with aphanitic or phaneritic features. For example, magma partly cooling beneath the surface to form large crystals (future phenocrysts) can be followed by eruption and rapid cooling at the surface to create a fine-grained groundmass.

Granite porphyry is characterized by especially large phenocrysts of potassium feldspar, called *megacrysts*. Individual crystals may be 3–10 cm long. Because potassium feldspar is among the last minerals to crystallize in Bowen's reaction series, it is a mystery how such large, well-formed phenocrysts develop in granite porphyry. Some geologists think that they grow large under special conditions near the roof of the magma chamber and then settle down into the ordinary granitic magma below as it cools. Others pro-

▶ **Figure 4.15 Volcanic Breccia** A brown layer of volcanic breccia overlain by thin layers of white tuff, near Los Alamos, New Mexico.

David Buesch

pose chemical processes in which some crystals gobble up others to grow large, or suggest that megacrysts grow from hot fluids circulating through the granite after it solidifies. In any case, not all igneous textures have simple explanations or are well understood.

Lava may cool so rapidly that its constituent atoms do not have time to become arranged in the ordered, three-dimensional frameworks of minerals. As a consequence, *natural glass* such as *obsidian* forms (▶ Figure 4.14g). Even though obsidian with its glassy texture is not composed of minerals, geologists classify it as an igneous rock.

Some magmas contain large amounts of water vapor and other gases. These gases may be trapped in cooling lava where they form numerous small holes or cavities known as **vesicles;** rocks with many vesicles are called *vesicular*, as in "vesicular basalt" (▶ Figure 4.14h).

A **pyroclastic** or **fragmental texture** characterizes igneous rocks formed by explosive volcanic eruptions. The sizes and sorting of fragments are important to note in properly classifying pyroclastic deposits, which in a sense are as much sedimentary as igneous (volcanic) features. *Tuff* is a fine-grained pyroclastic rock that originated as a fall of powdery volcanic ash (see Chapter 5). *Volcanic breccia* consists of a poorly sorted mixture of fine grains and larger, angular fragments produced by especially violent eruptions, volcanic landslides, and mudflows near active volcanoes. Interlayers of tuff and breccia are common (▶ Figure 4.15). There are many other, more carefully defined types of pyroclastic rocks, all useful for deducing the behaviors of the eruptions that produced them.

■ *Why is mineral content important for classifying igneous rocks?*

The igneous rocks shown in ▶ Figure 4.16 are distinguished by not only texture but also mineral content. Reading across the chart from rhyolite to andesite to basalt, for example, we see that the relative proportions of nonferromagnesian

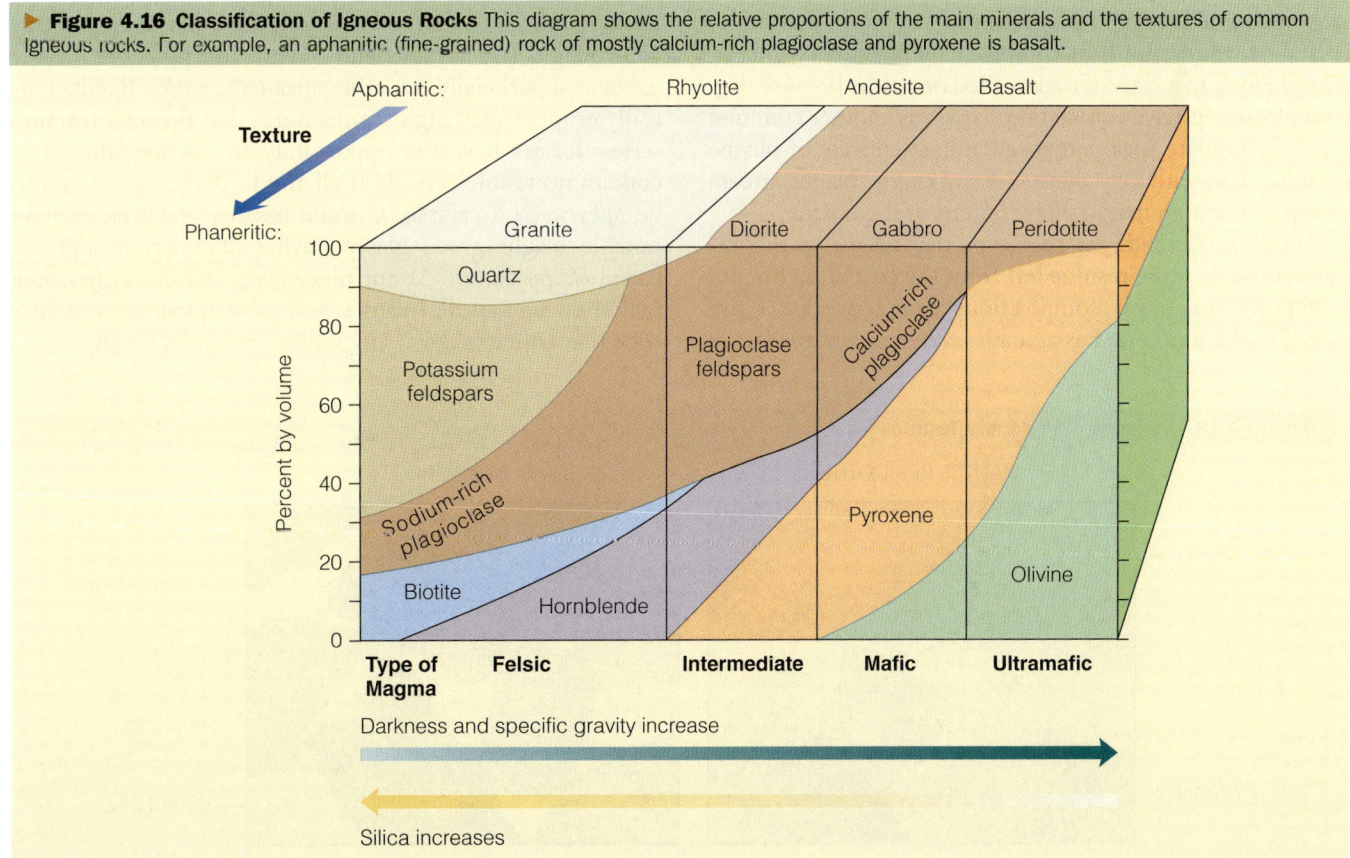

▶ **Figure 4.16 Classification of Igneous Rocks** This diagram shows the relative proportions of the main minerals and the textures of common igneous rocks. For example, an aphanitic (fine-grained) rock of mostly calcium-rich plagioclase and pyroxene is basalt.

▶ **Figure 4.17 Peridotite** This specimen of the ultramafic rock peridotite is made up mostly of olivine. Notice in Figure 4.16 that peridotite is the only phaneritic rock that does not have an aphanitic counterpart. Peridotite is rare at Earth's surface but is very likely the rock making up the mantle.

and ferromagnesian silicates differ. The differences in composition are gradual, however, so that a compositional continuum exists.

Ultramafic Rocks Ultramafic rocks (<45% silica) are composed largely of ferromagnesian silicates. The ultramafic rock *peridotite* contains mostly apple-green olivine, lesser amounts of dark green to black pyroxene, and usually a little white plagioclase feldspar (▶ Figures 4.16 and 4.17). Another ultramafic rock, *pyroxenite*, is composed predominately of pyroxene. Overall, ultramafic rocks are generally black or dark green, depending on the degree of weathering, and noticeably heavier than other igneous rocks.

Some peridotites and pyroxenites form by the settling of crystals to the bottoms of cooling mafic magma chambers. They belong to a class of rocks called *cumulates* because they result from crystal accumulation. *Dunite* is another common type of cumulate rock composed almost entirely of olivine crystals. Unweathered dunite is strikingly bright green, owing to the high magnesium content of the olivine.

Mantle peridotite is the rock type that makes up the upper mantle. It is the residue left from the partial melting of a hypothetical rock composition, called *pyrolite* ("fire stone"), which no one has actually ever seen but that must have been common at one time throughout the mantle to account for the development of Earth's crust. Mantle peridotite, then, is hardly a "conventional" igneous rock. Some geologists distinguish it by calling it a *restite*—a leftover from the melting of an originally much different looking parent composition. As you might expect, the texture of mantle peridotite is much different from that of cumulate peridotite. Grains of nearly equal size interlock tightly in mantle peridotite and do not appear to have piled up on top of one another by settling as phenocrysts through a liquid (▶ Figure 4.18).

The makeup and temperature of the upper mantle prior to 2.5 billion years ago must have been quite different from what they are today, because ultramafic lava flows called *komatiites*, with beautiful bladelike ferromagnesian phenocysts, were erupting worldwide at that time. Younger komatiites are rare; a few small flows of Cretaceous age are known from islands off the coast of Central America. The reason is that ultramafic lava must have a near-surface temperature of about 1600°C in order to erupt, but the surface temperatures of present-day mafic lavas are between 1000° and 1200°C, indicating that magmas today are cooler than they once were. During early Earth history, more radioactive decay heated the mantle perhaps as much as 300°C higher than at present. Today's Earth is plainly incapable of erupting the strange lavas that it once did.

Basalt-Gabbro *Gabbro* and *basalt* are the coarse-grained and fine-grained rocks, respectively, that crystallize from mafic magma (45–52% silica) (▶ Figure 4.19). Characteristically gabbros contain a large amount of lath-shaped calcium plagioclase crystals and olivine and pyroxene. Many gabbros superficially resemble ultramafic rocks. Basalts typically erupt at such high temperatures that Bowen's reaction series doesn't have the opportunity to operate; they may contain no visible crystals at all and in that case are said to be *aphyric* ("no crystals"). Some basalts, however, contain notable bright green glassy olivine phenocrysts and are called *olivine basalts*. Minor plagioclase and black pyroxene may show up as well. In any case, a gray, aphanitic, vesicular rock that contains few if any visible crystals is basalt.

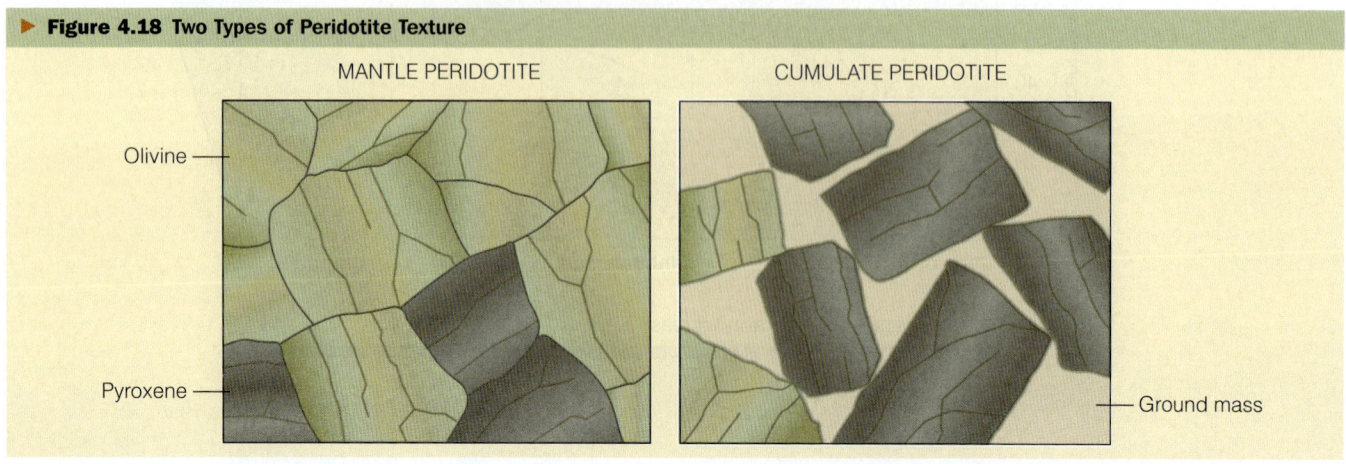

▶ **Figure 4.18 Two Types of Peridotite Texture**

Figure 4.19 Mafic Igneous Rocks

a Basalt is aphanitic.

b Gabbro is phaneritic. Notice the light reflected from crystal faces.

As mentioned earlier, basalts and gabbros erupt at mid-oceanic spreading ridges where new seafloor forms. They also erupt from oceanic hot-spot volcanoes such as in the Hawaiian Islands, and in continental areas that are undergoing tectonic extension (stretching) or rifting. They are the primary product, compositionally, of both the smallest and the largest types of volcanoes: cinder cones and shield volcanoes (see Chapter 5). Many basalt flows form beautiful columnar structures as they cool, leading early observers to gasp at their geometric symmetry and wonder whether they weren't the work of some god or devil.

Andesite-Diorite Magmas of intermediate composition (53–65% silica) crystallize to form *diorite* and *andesite*, coarse-grained and fine-grained igneous rocks, respectively, of equivalent mineral composition (▶ Figure 4.20). As mentioned earlier, diorite has a distinctive salt-and-peppery appearance. The white mineral in these rocks is plagioclase and the black ones are typically beautifully platy biotite, although some needlelike hornblende may be present too.

Andesites superficially resemble basalts, but generally there is a notably higher concentration of plagioclase phenocrysts, and olivine may be sparse or absent altogether. Pyroxene is a common phenocryst too, and it is usually black. Andesites rarely contain hornblende phenocrysts but, where found, the hornblende is large, black, needlelike, and hard to mistake for any other mineral. Typically, the plagioclase grains in andesite are elongate and randomly oriented, a texture some imaginative geologists call "turkey-track," like the pattern of birds' feet one sees in the sand on a beach.

Andesite is a common lava erupted at convergent plate boundaries where oceanic lithosphere subducts, heats up, and dehydrates beneath younger oceanic lithosphere or continental crust. The towering composite volcanoes of the Andes Mountains of South America and the Cascade Range in the western United States and Canada are composed largely of andesite, as are many of the arcuate island chains fringing the oceans. The origin of andesite and diorite is something of a mystery. Oceanic slab dewatering during subduction, mantle partial melting, and assimilation of the lower crust all appear to be involved. Partial melting of mafic rocks of the oceanic crust yields intermediate (53–65% silica) and felsic (>65% silica) magmas, both of which are richer in silica than the source rock. When some partial melting of the silica-rich continental sedimentary material carried downward with the subducted plate, and some low-melting-point, high-silica minerals from the lower continental crust are added in, diorite/andesite melt could result.

Some diorite plainly results from magma mixing, but it certainly also crystallizes directly from its own discrete

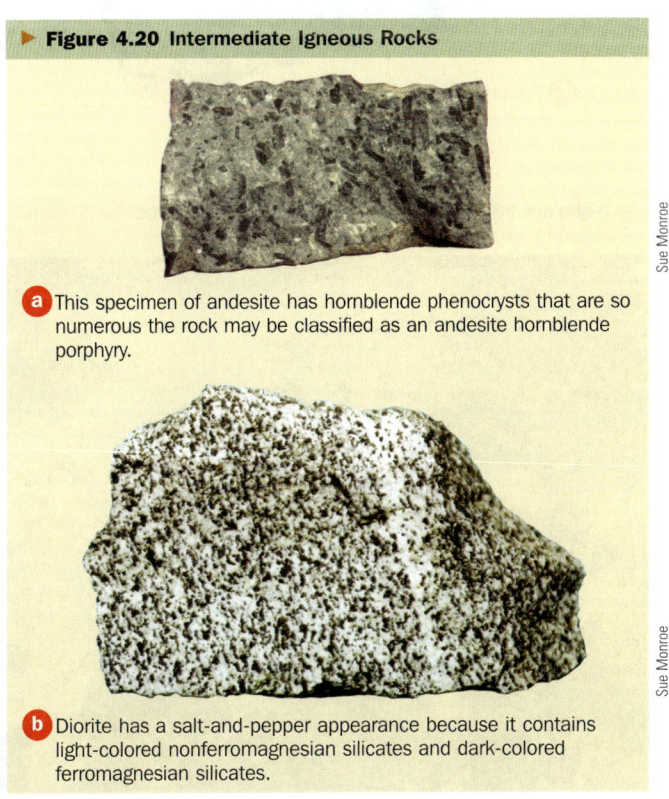

Figure 4.20 Intermediate Igneous Rocks

a This specimen of andesite has hornblende phenocrysts that are so numerous the rock may be classified as an andesite hornblende porphyry.

b Diorite has a salt-and-pepper appearance because it contains light-colored nonferromagnesian silicates and dark-colored ferromagnesian silicates.

▶ **Figure 4.21 Felsic Igneous Rocks** These rocks are typically light-colored because they contain mostly nonferromagnesian silicate minerals. The dark spots in the granite specimen are biotite mica. The white and pinkish minerals are feldspars, whereas the glassy-appearing minerals are quartz.

a Rhyolite.

b Granite.

▶ **Figure 4.22 Obsidian Boulders** A professor and students examine boulders of obsidian in the Medicine Lake Highland, California.

magma. Isolated rounded inclusions of diorite and mafic rock mingle with granite in many plutons. Such mingled inclusions are called *enclaves*.

Rhyolite-Granite *Granite* and *rhyolite* (>65% silica) crystallize from felsic magma and are therefore silica-rich (▶ Figure 4.21). Granite consists largely of potassium feldspar, sodium-rich plagioclase, and quartz, with perhaps some biotite and rarely amphibole (▶ Figure 4.21b). Some granites contain enormous (>3 cm long) crystals of potassium feldspar, called *megacrysts*, whose origin is unclear. Megacrystic granites are sometimes called *granite porphyries*.

In its unaltered state, nonglassy rhyolite is almost white like ordinary granite. But the iron in it often oxidizes during eruption, turning it a brick-red color. Rhyolite may contain some small irregular bluish phenocrysts of sanidine (a type of high-temperature potassium feldspar), minor fragments of quartz, biotite, and amphibole. Small plagioclase laths may be visible only in reflected light. Some rhyolite erupts as silvery gray pumice with as much as 75% vesicles, or as black to maroon-streaked aphyric obsidian, highly prized by Native Americans for making arrowheads and spear points (▶ Figure 4.22). One would hardly guess that such rocks are chemically identical to granite.

Extrusive material of rhyolitic composition is much less common than andesite and basalt flows.

It is more difficult for granitic magma to erupt than mafic melts because granitic magma has a lower temperature and a higher viscosity owing to silica enrichment. The abundance of volatiles, ironically, does not help granitic melt erupt either. Volatiles ordinarily inhibit the building of crystal bonds and keep magma molten to lower temperatures than if no volatiles were present. The escape of gases from magma can force the solidification of any melt left behind. The granitic melt that does manage to reach the surface usually does so explosively, given all the factors mentioned above. Hence, there is far more rhyolite *pyroclastic rock* than rhyolite *lava* on Earth's surface. What does flow out is a thick, pasty mass that may do no more than pile up in a fantastically spiny mound called a *lava dome*, as much as 100 m thick at the vent.

Pegmatite The term *pegmatite* has a dual meaning; it refers both to a kind of texture and to a particular igneous rock type that in most cases is similar to granite in composition. Granitic pegmatites contain mostly quartz, potassium feldspar, and sodium-rich plagioclase crystals, with some muscovite and biotite and in many cases lesser amounts of rare minerals, as considered in the following text. All of these minerals crystallize at low temperature in Bowen's reaction series and they are water rich; thus the minerals are easily dissolved and transported by hot solutions. The crystals are spectacularly large, ranging up to many centimeters long, even tens of centimeters in some cases. Typically pegmatites occur as dikes or veins in country rock and may be found as irregular pockets in ordinary granite, to which they are plainly related (▶ Figure 4.23). A few pegmatites are mafic or intermediate in composition and are appropriately called *gabbro* and *diorite pegmatites*.

Pegmatites develop during the end stages of magma crystallization. Heating and fracturing of surrounding country rock and the earlier-hardened portions of a pluton itself open up pathways for late-stage, gas-rich fluids to circulate. This water-rich melt often contains a number of elements that rarely enter into the common minerals that form granite, including lithium, beryllium, cesium, boron, and several others. Pegmatites that contain such minerals are called *complex pegmatites*. Some complex pegmatites contain as many as 300 different mineral species, a few of which are important economically. In addition, several gem minerals such as emerald and aquamarine, both of which are varieties of the silicate mineral beryl, and tourmaline (▶ Figure 4.23c) may occur in pegmatite. Many rare minerals of lesser value and well-formed crystals of common minerals, such as quartz, are also mined and sold to collectors and museums.

The formation and growth of mineral-crystal nuclei in pegmatites are similar to those processes in other magmas but with one critical difference: A high concentration of vapor, primarily steam, in the molten rock from which pegmatites crystallize inhibits the formation of crystal nuclei. Some nuclei do form,

▶ **Figure 4.23 Pegmatite**

a This pegmatite, the light-colored rock, is exposed in the Black Hills of South Dakota.

b Closeup view of a specimen from a pegmatite with minerals measuring 2 to 3 cm across.

c Tourmaline from the Dunton Pegmatite in Maine.

What Would You Do?

You are prospecting for emeralds and aquamarine in a steamy tropical jungle. You can approach outcrops on a distant mountain range only by following streambeds through the forest. You have a choice of two streams to follow. One contains cobbles of basalt, diorite, and gabbro. The other contains cobbles of granite, rhyolite, and pegmatite. Which stream will you follow, and why? If you encounter other stream junctions along the way, which stream branches will you ignore?

- Diorite is an intermediate rock containing plagioclase, hornblende, biotite, and minor quartz formed at oceanic–continental convergent plate boundaries. Andesite is the volcanic equivalent of diorite, and it contains abundant plagioclase crystals with minor olivine, pyroxene, and hornblende. Partial melting in the mantle above subducting slabs and interaction with the lower crust contribute to the formation of these rocks.

- Granite and rhyolite are felsic rocks formed by melting entirely within Earth's crust, mostly at convergent plate boundaries. Granites contain abundant quartz, potassium feldspar, plagioclase, hornblende, and biotite. Rhyolite may contain minor amounts of each of these minerals as phenocrysts. Obsidian is a glassy form of rhyolite.

---

however, and because the appropriate ions in the liquid can move easily through the water-rich magma and attach themselves to a growing crystal, individual minerals have the opportunity to grow very large. In those cases, where a vapor phase is not present, crystallization is controlled by rapid temperature loss and grain sizes are near-microscopic. Such "fine-grained" pegmatites are called *aplites*. They generally lack the interesting content of rare minerals that often occurs in granite pegmatites because many of the elements essential to making these minerals can be transported only by volatiles.

Many other kinds of igneous rocks exist in addition to the ones described above. However, these are the main igneous rocks that are important to the development of Earth's crust, and it is important for you to keep them in mind as you continue reading about geology.

Section 4.3 Summary

- The specific arrangement, sizes, and shapes of mineral grains in a rock define the rock's texture. Examples of igneous rock textures include aphanitic rocks, which are fine-grained, and phaneritic rocks, which are coarse-grained. Porphyritic rocks include large crystals set in a bed, or groundmass, of finer grains. Pyroclastic texture typifies explosively erupted igneous rocks.

- Mineral content and texture are both important features in classifying igneous rocks. Texture determines whether a rock is called plutonic or volcanic. Mineral content allows further classification.

- Ultramafic rocks are characterized by a high content of olivine and pyroxene. Most ultramafic rocks are plutonic; very few lavas are known in the young geologic record.

- Mafic rocks contain large amounts of plagioclase as well as olivine and pyroxene. Gabbro is the most abundant mafic plutonic rock; basalt is the lava equivalent. Most gabbro and basalt develop at mid-oceanic ridges.

4.4 Plutons—Their Characteristics and Origins

■ *What are the largest igneous intrusions, and what do their contacts look like?*

The structure of intrusive igneous rock bodies varies according to the size and depth of the intrusion and to a certain extent the composition. By far the largest intrusions are intermediate and felsic plutons, crystallized magma reservoirs that may individually contain hundreds or thousands of cubic kilometers of rock. **Batholiths** are exposed plutons that have at least 100 km^2 of surface area. Multiple injections of magma may have taken place to build up the batholith over time. The coastal batholith of Peru, for instance, was emplaced during a period of 60 to 70 million years and is made up of as many as 800 individual plutons. Exposures of plutonic rock with a surface area smaller than 100 km^2 are called **stocks,** but exposure area may be misleading; many stocks are simply parts of larger intrusive bodies that, once exposed by erosion, are batholithic in size (▶ Figure 4.24).

The contacts of most plutons with their enclosing country rock are **discordant,** meaning that the intrusions cut across preexisting rock layers. Near the roof and at shallow depths along the margins of some plutons, however, contact relations may be **concordant,** rock layering that parallels the contact with the intrusion. The overall shapes of plutons range from rounded to sharply angular. Some are vertically cylindrical. Many spread out as large, flattened, onion-shaped bodies in the shallow crust. Some seem to be nothing more than a complex maze of sheets and fingers enclosing huge slices of country rock. Many appear to be coherent and homogeneous igneous masses. Individual plutons may range up to 60 km in diameter and as much as 5–10 km thick. Plutons less than about 5–10 km deep can produce volcanic eruptions. Some plutons continue to ascend as the overlying volcanic cover accumulates, so that they may penetrate and intrude their own lavas and pyroclastic layers. Two adjacent plutons may be entirely different in structure, composition, and texture.

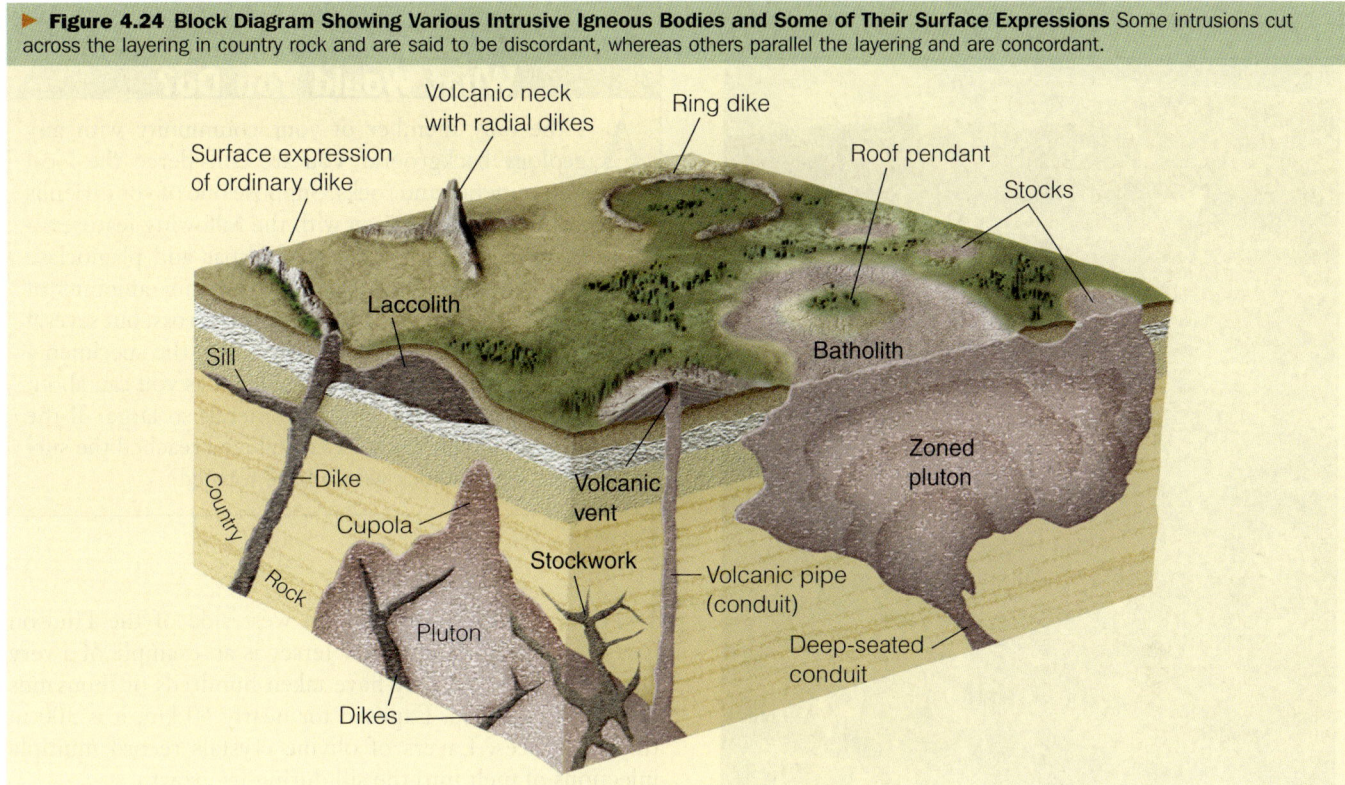

▶ **Figure 4.24 Block Diagram Showing Various Intrusive Igneous Bodies and Some of Their Surface Expressions** Some intrusions cut across the layering in country rock and are said to be discordant, whereas others parallel the layering and are concordant.

Erosion has exposed mafic plutons in many areas of ancient oceanic crust. Many geologists think that a typical magma chamber at a mid-oceanic ridge resembles a long, flattened tube stretching underneath the axial rift for as much as tens of kilometers. Transform faults and fracture zones are at the ends of the magma chamber, although molten rock may leak out along these boundaries for some distance. The chamber lies between about 1.5 and 10 km deep and is somewhat wider than it is thick. Multiple small batches of mafic melt rise through the mantle on either side of the axial chamber, some erupting to form seamounts and abyssal hills on the flanks of the mid-oceanic ridge (see Chapter 12).

The root zones of plutons are rarely exposed, and there is much uncertainty about what lies at their base. Networks of fine veins extending across a wide region of crust are thought to merge and feed molten rock to the base of a magma chamber. In other cases, a definite pipe links a chamber to even deeper chambers below. There is still much to be learned about the deeper architecture of the crust.

■ *What are the major types of sheetlike igneous intrusions?*

On a smaller scale, igneous intrusions form sheetlike bodies including dikes, sills, veins, laccoliths, and pipes. A **dike** is a discordant intrusion that appears as a band of igneous rock slicing across strata, whereas a **sill** is concordant and in most cases emplaced along the contact between two adjacent layers (▶ Figure 4.25). Dikes and sills commonly range in width from a few tens of centimeters to tens of meters. They may extend for hundreds of meters or even many kilometers. It is important to note that not all dikes are vertical sheets, nor are all sills horizontal, just as not all strata and layers in the Earth are horizontal. **Veins** are very thin (no more than a few centimeters), irregular, sheetlike intrusions that may be either concordant or discordant. Most dikes and sills are mafic. Veins are commonly felsic, some made up exclusively of quartz but others of aplite.

How are sheetlike intrusions that cut across an unlayered or "massive" rock body classified? There is no clear-cut nomenclature in this case, but it is customary for geologists to refer to all thick sheet intrusions (more than a few centimeters wide) as *dikes* and all thin ones (less than a few centimeters wide) as *veins*, irrespective of orientation. Obviously it helps to see the continuation of a particular intrusion into nearby country rock to ascertain its relationship to layered rock, but this is not always possible.

Dikes develop in crust that is stretching and splitting. The magma forces itself into fractures that may also act as faults while opening; that is, the crust can slide along the magma-filling crack during intrusion. Some dikes flatten out into sills, which propagate during intrusion by wedging apart rock layers and filling in between them. Imagine a rising sheet of magma that encounters a strong rock layer that won't break. Because there could be less resistance to feeding the magma sideways along the base of the layer rather than continuing to build up pressure to break through vertically, the magma starts intruding as a sill.

▶ Figure 4.25 Dikes and Sills

a Dike in the side of Herchenberg volcano, Eifel district, Germany.

b Dark, mafic sills in lighter-colored country rock, Santa Monica Mountains, California.

What Would You Do?

As the only member of your community with any geology background, you are considered the local expert on minerals and rocks. Suppose one of your friends brings you a rock specimen with the following features—composition: mostly potassium feldspar and plagioclase feldspar with about 10% quartz and minor amounts of biotite; texture: minerals average 3 mm across, but several potassium feldspars are up to 3 cm. Give the specimen a rock name and tell your friend as much as you can about the rock's history. Why are the minerals so large? If the magma from which the rock crystallized reached the surface, what kind of rock would have formed?

The Palisades Sill along the west side of the Hudson River in New York and New Jersey is an example of a very large intrusion that may have taken hundreds of thousands of years to develop. Exposed for nearly 60 km, it is 300 m thick in places. Layers of olivine crystals record multiple injections of melt into the sill during its growth.

Many shallow sills inflate into arching **laccoliths** as magma pressure increases within them and eventually lifts the overlying crust. Furthermore, resistance to further propagation along the length of the sill may play a role in causing a laccolith to grow. Laccoliths may be several hundred meters thick and several kilometers wide. The overlying surface rises into a domelike hill or even a mountain. The Henry Mountains in southern Utah are an outstanding example of large laccoliths.

Dikes, sills, and veins commonly extend from the roofs of plutons into the overlying country rock. Together with laccoliths, they characterize the internal structure of volcanoes. The intrusions stem from a central intrusive **volcanic pipe** or *conduit* that occupies the core of each volcano. As molten rock rises, it causes the shallow bedrock and the volcanic cone itself to swell and fracture, allowing dikes to penetrate the flanks in a radial manner. Deep erosion of an ancient volcano, such as Ship Rock in New Mexico, exposes the tough central pipe as a resistant pinnacle of veined intrusive rock, called a **volcanic neck.** Dikes stand out as eroded ribs radiating across the landscape from the central neck. Some ribs in turn feed sills and laccoliths at their termini, reflecting a change in stress with distance from the conduit during the time of intrusion.

The origin of volcanic pipes is unclear, but in many cases they lie at the intersections of faults or fractures where magma rises through the crust by exploiting a vertical path of weakness. Recent sideways drilling into Unzen, an intermediate-composition volcano in Japan, penetrated a 30-m-wide volcanic pipe formed in 1991–1995. This conduit is surrounded by a zone of older pipes and altered country rock several hundred meters wide.

Section 4.4 Summary

- The largest igneous intrusions are batholiths and stocks. These are the surface expressions of eroded plutons. If the area of exposure is larger than 100 km^2, then the body is a batholith. It is a stock otherwise. Plutons have diverse, irregular geometry inside the crust. Many are bulbous and kilometers wide. Sheetlike intrusions include dikes, sills, and veins. Dikes are discordant, cutting across preexisting layering. Sills are concordant, developing along the contacts between pre-existing layers. Veins are thin versions of dikes and sills and often are very irregular in geometry. Laccoliths are sill-like bodies that inflate into arching mounds, which may create hills and even mountains at Earth's surface. Sheetlike intrusions are common around volcanoes and in the roof crust overlying magma chambers. Volcanic sheet intrusions radiate from volcanic pipes, which erosion can expose as volcanic necks, such as Ship Rock in New Mexico.

Review Workbook

ESSENTIAL QUESTIONS SUMMARY

4.1 Introduction

■ *What are common plutonic and volcanic rocks?*
Gabbro/basalt, and granite/rhyolite are pairs of common rock types that represent two ends of the compositional range of most igneous rocks at Earth's surface. Gabbro and granite are plutonic rocks; basalt and rhyolite are volcanic rocks.

■ *How are plutonic and volcanic rocks related?*
Depending on whether magma erupts, it can produce plutonic (deep-seated) or volcanic (eruptive) rocks. It is best to learn the different kinds of igneous rocks as *pairs* of equivalent plutonic and volcanic compositions; that is, each plutonic rock has its volcanic compositional equivalent, and vica versa. For example, gabbro is chemically equivalent to basalt, and granite is chemically equivalent to rhyolite.

4.2 The Properties and Behavior of Magma and Lava

■ *What role does magma play in helping heat escape Earth?*
Magma transfers heat from Earth's deep interior as it rises. Magma carries heat up faster than the other main processes of heat release: conduction and solid-state convection. It is a very efficient way for heat to escape Earth.

■ *What special conditions are required to make magma?*
The temperature must be high enough and the pressure low enough for rock to melt. Also important is the convective upwelling of hot mantle rock beneath mid-oceanic ridges. The rock rises so quickly that it cannot release heat fast enough to keep from melting with the reduction of pressure during ascent. The most common lava in the world, basalt, results from this process.

■ *What is unique about the origin of granitic magma?*
Granitic magma forms in continental crust, and it is a secondary product of heat introduced by deeper-seated mantle magmas into the lower crust. Bonded mineral and pore water play an important role in generating granitic magma by enabling the melting of sedimentary and metamorphic rocks that might remain solid if they were "dry."

■ *Why and how does magma rise?*
Molten rock is less dense than solid rock, and so it is buoyant. Initially, the rate of magma ascent is related to the difference in density between the melt and the surrounding country rock. Later the magma may stall out and crystallize at a level of neutral buoyancy. A large body of magma can push aside the overlying crust and stope and thus assimilate blocks of crust on its way toward the surface.

■ *What are volatiles and why are they important?*
Volatiles are dissolved substances in magma that emerge as gases at low pressures. Increasing volatile pressure in a shallow magma chamber can help wedge open fractures in the chamber roof and drive the magma to the surface. In particular, expanding water vapor is a major agent in explosive volcanic eruptions.

■ *What is the chemical composition of magma?*
Most magma, like most minerals, consists of silicon and oxygen with lesser amounts of other elements, such as magnesium (Mg), iron (Fe), sodium (Na), potassium (K), and aluminum (Al). Mg-Fe–rich magma is called "mafic," and magma rich in Na, K, Al and H_2O is "felsic." Gabbros and basalts are products of mafic magmas; granites and rhyolites are products of felsic magmas.

■ *How hot is magma?*
No one knows for sure, but measured temperatures of lavas suggest that typical mafic magma is somewhat hotter than 1200°C, and felsic magma is somewhat hotter than 900°C.

■ *What is viscosity and what controls it?*
Viscosity is a measure of a substance's resistance to flow. The higher the viscosity, the stiffer and thicker the flow. Tar and glacial ice are good examples of highly viscous substances. Water and syrup have low viscosities and flow readily. The viscosity of magma is controlled by temperature, silica content, loss of volatiles, crystallinity, bubble content, and shear stress during movement. Characteristically, mafic magmas are less viscous than felsic magmas, primarily because they are hotter and have less silica.

■ *What is Bowen's reaction series?*
Bowen's reaction series describes the sequence of mineral crystallization in a cooling magma. There are two branches in the reaction series. Minerals that form by the reaction of other minerals with the magma at certain specific temperatures belong to the

discontinuous series, whereas minerals that react continuously with the magma to adjust their compositions during cooling belong to the discontinuous series. The reverse of Bowen's reaction series describes the melting of rock.

■ *How does magma change as crystallization occurs?*
During crystallization, the remaining melt becomes progressively more silica-enriched. This is shown by Bowen's reaction series, in which quartz (SiO_2) is the last mineral to crystallize from a cooling magma.

■ *What happens when two magmas come together?*
Two different magmas can blend together (mix) if their viscosities are very similar, or they can mingle without blending if their viscosities are very different. Mingling of magmas beneath a volcano can trigger a volcanic eruption.

■ *What are zoned plutons?*
Characteristically, zoned plutons have multiple igneous rock compositions, ranging from mafic to felsic. Many zoned plutons show evidence of partial blending of different magmas to produce hybrid igneous rocks. The different compositions in the pluton are not randomly distributed. For example, one rock type may predominate near the top and on the sides of a pluton, with another type in the interior and near the base.

4.3 Igneous Rocks—Their Characteristics and Classification

■ *What does the term* texture *mean with respect to rocks, and what are some examples of igneous textures?*
Texture refers to the size, shape, and arrangement of mineral grains composing a rock. Examples of igneous textures include aphanitic, meaning very fine-grained, phaneritic, meaning very coarse-grained, and porphyritic, meaning scattered large crystals set in a finer-grained groundmass. Layered volcanic ash and pumice deposits display pyroclastic texture.

■ *Why is mineral content important for classifying igneous rocks?*
After geologists identify a rock as plutonic or volcanic, the name it is given depends on the relative percentages of certain key minerals, such as quartz, orthoclase, plagioclase, olivine, and pyroxene. The most common igneous rocks are peridotite, gabbro, basalt, andesite, diorite, granite, and rhyolite. Pegmatite is a rock type closely related to granite that contains many minerals not ordinarily found in other igneous rocks.

4.4 Plutons—Their Characteristics and Origins

■ *What are the largest igneous intrusions and what do their contacts look like?*
Batholiths are plutons that have more than 100 km^2 in area of exposure. Stocks are somewhat smaller plutonic bodies. Large plutons tend to have discordant contacts; that is, they cut across pre-existing rock layers.

■ *What are the major types of sheetlike igneous intrusions?*
Dikes and sills are the most common sheetlike igneous intrusions. Dikes are discordant features (meaning they cut across layering in the country rock), whereas sills are concordant. Laccoliths are sill-like bodies with inflated cores. Volcanic pipes are magma-filled, cylindrical feeder channels beneath volcanoes. Pipes can become volcanic necks with deep erosion. Dikes, sills, and laccoliths radiate from many volcanic pipes and necks.

ESSENTIAL TERMS TO KNOW

aphanitic texture (p. 115)
assimilation (p. 107)
batholith (p. 122)
Bowen's reaction series (p. 110)
concordant (p. 122)
country rock (p. 107)
dike (p. 123)
discordant (p. 122)
extrusive (p. 115)
felsic (p. 109)
igneous rock (p. 102)
intermediate magma (p. 110)
intrusive (p. 115)

laccolith (p. 124)
lava (p. 102)
mafic (p. 109)
magma (p. 102)
magma chamber (reservoir) (p. 104)
magma mingling (p. 115)
magma mixing (p. 113)
phaneritic texture (p. 115)
pluton (p. 102)
plutonic (intrusive igneous) rock (p. 102)
porphyritic texture (p. 116)
pyroclastic materials (p. 102)

pyroclastic (fragmental) texture (p. 117)
sill (p. 123)
stock (p. 122)
stoping (p. 107)
vein (p. 123)
vesicle (p. 117)
viscosity (p. 110)
volcanic (p. 102)
volcanic neck (p. 124)
volcanic pipe (p. 124)
volcanic (extrusive igneous) rock (p. 102)

REVIEW QUESTIONS

1. Two aphanitic igneous rocks have the following compositions. Specimen 1: 15% biotite, 15% sodium-rich plagioclase, 60% potassium feldspar, and 10% quartz. Specimen 2: 10% olivine, 55% pyroxene, 5% hornblende, and 30% calcium-rich plagioclase. Use Figure 4.16 to classify these rocks. Which is the darkest and most dense?

2. How does a sill differ from a dike? Make a sketch showing the differences.

3. How do gravitational differentiation (crystal settling) and assimilation bring about compositional changes in magma? Also, give evidence that these processes actually take place.

4. How do laccoliths form? How might they be related to dikes and sills? Make a sketch showing a laccolith and its possible relationships to these other types of intrusions.

5. Describe a porphyritic texture and explain how it might originate.

6. Why are felsic lava flows so much more viscous than mafic ones?

7. How does pegmatite form, and why are the mineral crystals so large?

8. Compare the continuous and discontinuous branches of Bowen's reaction series. Why are potassium feldspar and quartz not part of either branch?

9. What evidence for magma mixing or mingling would you seek in the appearance of a rock?

10. Most rock inside Earth is hot enough to be molten under surface conditions, but it isn't molten inside the planet. Why not? What accounts for the formation of magma inside Earth?

11. Where does most granite form, and how does it originate?

12. Where does most basalt form, and how does it originate?

13. Why is it so much more difficult for felsic magma to erupt than mafic magma?

14. How and why do the textures of volcanic rocks differ from the textures of plutonic rocks?

FIELD QUESTION

1. The image below shows an imaginary landscape made up largely, though not entirely, of igneous rocks. Many features discussed in this chapter appear here. Identify as many as you can. Rock types include basalt (B), diorite (D), gabbro (Gb), granite (Gr), pegmatite (Pg), and red sandstone (SS). After making your identifications, reconstruct as much of the geologic history as you can. Do not be frustrated if you are left with questions hanging. Geologists face this dilemma all the time! (An answer is given in the answer section at the end of the book).

This imaginary landscape contains a number of igneous features discussed in this chapter.

Geo-focus Figure

GEOLOGY MATTERS

GEOLOGY IN YOUR LIFE

Of Mineral Water and Meteorites

Many surprising connections can be made between igneous processes and common occurrences in daily life. For example, let's consider two items—bottled mineral water and the façade of a church—and two rare kinds of rock. In the Eifel Range of western Germany, an 11,000-year-old volcano, Lacher See, slowly leaks carbon dioxide into the groundwater and atmosphere (▶ Figure 1). In the village of Wallenborn ("wallowing spring"), the CO_2 erupts at the top of a well once every 15 minutes to produce a geyser 1.5 m high. The water tastes awful and smells like rotten eggs. But visitors *have* to try it, and popular mineral waters bottled "for health" in the region are naturally carbonated, thanks to the cooling and degassing of shallow magma.

The large amount of CO_2 that exists or once existed beneath this part of Europe is reflected in the occurrence of *carbonatite*, a pale igneous rock that has the same composition as baking soda with a little silica thrown in. No carbonatite volcanoes are presently active in Germany, and in fact the only known active carbonatite volcano in the world is Oldoinyo Lengai in East Africa (▶ Figure 2). Temperatures of carbonatite melt are relatively low—only 600–700°C, too low to produce incandescence during eruption. Carbonatite lava looks like erupting mud, in fact. Because there is little or no silica present, however, the flows have very low viscosity. They may be only a few centimeters thick in contrast to the bulky rhyolite flows around lava domes.

Even more bizarre are the igneous products of meteorite impacts, including small aerodynamic spheres of glass called *tektites*, which can be scattered hundreds of kilometers from an impact site, and *suevites*, which resemble a class of pyroclastic rock because they contain vesicular fragments of obsidian and country rock mantled with obsidian, packed in ash. The ash comes from pulverized and melted rock. It is easy to see why, given that the impact speeds of meteorites are typically 15–25 km *per second*. Among the youngest substantial suevite deposits in the world are those near Nuremberg in southern Germany, where a meteorite struck 14.7 million years ago (▶ Figures 3 and 4). It instantaneously excavated a crater 18 km in diameter. Local residents dug up suevite blocks to make the huge, obsidian-speckled cathedral at Nordlingen, an interesting combination of astronomy and religion (▶ Figure 5)! To learn more about Oldoinyo Lengai in East Africa read *Geology in Unexpected Places* on p. 165.

Figure 1
Lacher See, Germany, may not look like a volcano, but the crater filled with this placid lake erupted violently 11,000 years ago, with 25 times the force of the 1980 Mount St. Helens eruption in Washington State.

Figure 2
Carbonatite lava erupting at Oldoinyo Lengai volcano, East Africa.

Figure 3
Suevite from the Nordlingen meteorite impact, Germany. The black fragments are obsidian. The white fragments are bits of country rock that did not melt.

Figure 4
Blocks of suevite breccia from the 14.7-million-year-old Nordlinger-Ries meteorite impact in Germany. The black inclusions are obsidian pieces from the impact-caused melting of granite. The white pieces are bits of unmelted granite and limestone.

Figure 5
The town of Nordlinger, in the center of the giant Nordlinger meteorite crater in Germany. Since medieval times, suevite has been used as a building stone in Nordlinger, most impressively for the town's huge cathedral. Unfortunately, the obsidian in the suevite weathers very easily in Germany's climate.

GEOLOGY MATTERS

GEOLOGY IN FOCUS Some Remarkable Volcanic Necks

As mentioned in the text, when an extinct volcano weathers and erodes, a remnant of the original mountain may persist as a volcanic neck. These isolated monoliths rising above otherwise rather flat land are scenic, awe inspiring, and the subject of legends. They are found in many areas of recently active volcanism. A rather small volcanic neck rising only 79 m above the surface in the town of Le Puy, France, is the scenic site of the 11th-century chapel of Saint Michel d'Aiguilhe (▶ Figure 1). It is so steep that the materials and tools used in its construction had to be hauled up in baskets.

Perhaps the most famous volcanic neck in the United States is Shiprock, New Mexico, which rises nearly 550 m above the surrounding plain and is visible from 160 km away. Radiating outward from this conical structure are three vertical dikes that stand like walls above the adjacent countryside (▶ Figure 2).

According to one legend, Shiprock, or *Tsae-bidahi*, meaning "winged rock," represents a giant bird that brought the Navajo people from the north. The same legend holds that the dikes are snakes that turned to stone. An absolute age determined for one of the dikes indicates that Shiprock is about 27 million years old. When the original volcano formed, apparently during explosive eruptions, rising magma penetrated various rocks including the Mancos Shale, the rock unit now exposed at the surface adjacent to Shiprock. The rock that makes up Shiprock itself is tuff-breccia, consisting of fragmented volcanic debris as well as pieces of metamorphic, sedimentary, and igneous rocks.

Geologists agree that Devil's Tower in northwestern Wyoming cooled from a small body of magma and that erosion has exposed it in its present form (▶ Figure 3). However, opinion is divided on whether it is a volcanic neck or an eroded laccolith. In either case, the rock that makes up Devil's Tower is 45 to 50 million years old, and President Theodore Roosevelt designated this impressive landform as our first national monument in 1906. At 260 m high, Devil's Tower is visible from 48 km away and served as a landmark for early travelers in this area. It achieved further distinction in 1977 when it was featured in the film "Close Encounters of The Third Kind."

Figure 2
Shiprock, a volcanic neck in northwestern New Mexico, rises nearly 550 m above the surrounding plain. One of the dikes radiating from Shiprock is in the foreground.

Figure 1
This volcanic neck in Le Puy, France, rises 79 m above the surface of the town. Workers on the Chapel of Saint Michel d'Aiguilhe had to haul building materials and tools up in baskets.

Figure 3
Devil's Tower in northeastern Wyoming rises about 260 m above its base. It may be a volcanic neck or an eroded laccolith. The vertical lines result from intersections of fractures called columnar joints. According to Cheyenne legend, though, a gigantic grizzly bear made the deep scratches.

GEOLOGY MATTERS

Geology
IN UNEXPECTED PLACES

Little Rock, Big Story

More than a million people visit Plymouth Rock at Pilgrim Memorial State Park in Plymouth, Massachusetts, each year (▶ Figure 1). Most are surprised to discover that it is a rather small boulder. Legend holds that it is the landing place where the Pilgrims first set foot in the New World in 1620. In fact, the Pilgrims first landed near Provincetown on Cape Cod and then went on to the Plymouth area, but even then probably landed north of Plymouth, not at Plymouth Rock.

Actually the boulder that is Plymouth Rock was much larger when the Pilgrims landed, but an attempt to move the symbolic stone to the town square in 1774 split it in two. The lower half of the rock remained near the seashore. In 1880 the two pieces of the rock were reunited, and in 1921 the boulder was moved once again, this time to a stone canopy erected over the original site.

Plymouth Rock has great symbolic value, but otherwise is a rather ordinary rock, although it is geologically interesting. The stone itself is from the 600-million-year-old Dedham Granodiorite, an intrusive igneous rock similar to granite, which was carried to its present site and deposited by a glacier during the Ice Age (1.8 million to 10,000 years ago). The Dedham Granodiorite is part of an association of rocks in New England that geologists think represent a chain of volcanic islands similar to the Aleutian Islands. These volcanic islands were incorporated into North America when plates collided and caused a period of mountain building called the Taconic orogeny about 450 million years ago.

Figure 1
Plymouth Rock is the first piece of land on which the Pilgrims supposedly set foot when they arrived in Massachusetts in 1620. The rock first became a patriotic icon during the Revolutionary War.

Mark E. Gibson/Corbis

GEOLOGY MATTERS

Geology AND CULTURAL CONNECTIONS

Hell Helped Early Earth Science

European naturalists looking at Earth materials in the 1600s and 1700s quickly came to think that granite and other similar rocks must have formed from molten material. No one had ever seen granite actually form in such a manner, but people knew about molten metals and much lower-temperature molten materials like liquid wax. They had seen these hot liquids cool and solidify. And everyone in northern Europe had lots of experience watching liquid water solidify, or freeze, always forming ice crystals as it did so.

Figure 1
Close-up of granite, showing its component minerals of quartz, feldspar, mica and hornblende. The large pink and white crystals are diffrent kinds of feldspar. The quartz is gray, and the mica and hornblende are the dark spots.

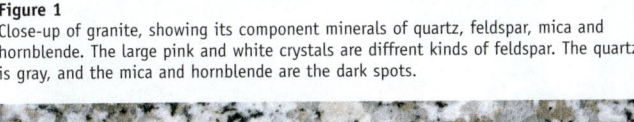

Figure 2
Hell, by Jan the Elder, a Flemish painter (1568–1625) better known for painting cut flowers.

The texture of granite was visible to the naked eye and clearly showed different types of interlocking crystals pointing in all directions (▶ Figure 1). From this it was reasonable to infer that the rock material had once been molten and that it had cooled to form solid granite, with crystals growing in different directions as it did so.

The idea of molten material deep in Earth made sense to early naturalists because it fit with broader cultural themes. The Western idea of hell was based on references in both the Old and the New Testaments of the Bible to "fire and brimstone" (sulfur) (▶ Figure 2). And both scripture and the traditions of the church linked hell to a place underground. Heaven was upward, hell was downward—a cultural assumption so basic that Westerners can hardly imagine reversing the two notions. In the Middle Ages, hell was a place frequently discussed by clerics and even laypeople. An enormously hot hell, full of suffering and located somewhere under the ground, was more real to many Europeans of the time than a place like Afghanistan is to North Americans today.

Early European naturalists could easily believe that very hot regions under Earth contained molten rock. The observation of certain rocks and inferences about them fit with cultural assumptions and religious beliefs. When observations go against the grain of a culture, they may be ignored for a long time, but what naturalists saw in igneous rocks fit easily into their basic cultural framework.

All educated Europeans also knew that volcanoes in places like Italy occasionally erupted lava (generally smelling of sulfur, as it happens), which solidified into solid rock. This emphatically confirmed the idea of the general mechanism required to form at least some rocks.

Thus, when the famous naturalist James Hutton wrote about the formation of granite in the 1700s, the notion of molten rock cooling to form interlocking mineral crystals was accepted by most of his contemporaries even though no one had actually seen the specific processes involved in the formation of granite (▶ Figure 3).

In Hutton's time, two general types of rocks were most clearly understood. The first were the igneous rocks (so named for an ancient term for "fire," which is also the root for English words like *ignite*). The second type were the sedimentary rocks, like sandstone, which clearly owed their origin to water. Metamorphic rocks were the most difficult for naturalists and early scientists to understand for two reasons. First, the conditions of their formation were outside direct observation, and second, the length of time required for their formation was well beyond ordinary experience.

Figure 3
James Hutton

Chapter 5

Volcanoes and Volcanism

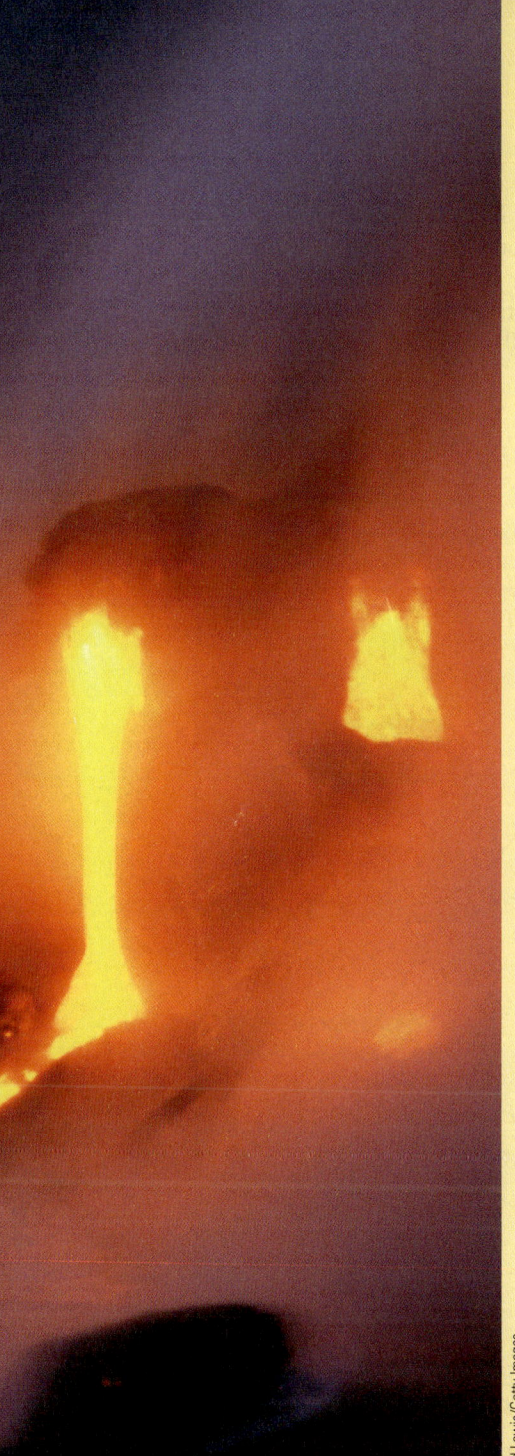

Kilauea Volcano, Hawaii
Radiant rivulets of lava delicately drip into the ocean like beams of amber and carnelian light. When lava streams into the ocean, the sea steams and boils. Brilliant bursts of molten lava and rock spray outward like explosions of fireworks. Kilauea is a dynamic volcano. If you visit, depending on the current volcanic activity you may see an active lava flow. To check Kilauea's activity go to http://hvo.wr.usgs.gov/.
—A. W.

ESSENTIAL QUESTIONS TO ASK

5.1 Introduction
- *How can volcanism be both constructive and destructive?*

5.2 Volcanism and Volcanoes
- *What gases do volcanoes commonly emit?*
- *How and why do aa and pahoehoe lava flows differ?*
- *What are pyroclastic materials, and how are they dangerous to air traffic?*

5.3 Types of Volcanoes
- *What are calderas and how do they form?*
- *What are cinder cones and what are they composed of?*
- *What are lava domes and why are they so dangerous?*

5.4 Other Volcanic Landforms
- *How do basalt plateaus form?*
- *What are pyroclastic sheet deposits?*

5.5 The Distribution of Volcanoes
- *Where are the three zones or belts with most of Earth's volcanoes?*

5.6 North America's Active Volcanoes
- *Where is the Cascade Range and what types of volcanoes are found there?*

5.7 Plate Tectonics, Volcanoes, and Plutons
- *What kinds of igneous rocks make up the oceanic crust?*
- *How do plate tectonics and volcanism account for the origin of the Hawaiian Islands?*

5.8 Volcanic Hazards, Volcano Monitoring, and Forecasting Eruptions
- *What are the most dangerous manifestations of volcanoes?*

GEOLOGY MATTERS

GEOLOGY IN FOCUS:
The Bronze Age Eruption of Santorini

GEOLOGY IN UNEXPECTED PLACES:
Oldoinyo Lengai Volcano

GEOLOGY IN YOUR LIFE:
Do Volcanic Gases Cause Ozone Depletion?

GEOLOGY AT WORK:
Volcano Observatories and Volcano Monitoring

 This icon, which appears throughout the book, indicates an opportunity to explore interactive tutorials, animations, or practice problems available on the Physical GeologyNow website at **http://now.brookscole.com/phygeo6.**

5.1 Introduction

No other geologic phenomenon has captured the public imagination more than erupting volcanoes, especially lava issuing forth in fiery streams or blasted into the atmosphere in sensational pyrotechnic displays. What better subject for a disaster movie? Several such movies of varying quality and scientific accuracy have appeared in recent years. One of the best was *Dante's Peak* in 1997. Certainly the writers and director exaggerated some aspects of volcanism, but the movie depicted rather accurately the phenomenal power of an explosive eruption. Incidentally, the volcano called Dante's Peak was a 10-m-high model built of wood and steel.

Figure 5.1 The Eruption of Mt. Vesuvius

a The Mount Vesuvius region on the shore of the Bay of Naples, Italy. Mount Vesuvius erupted in A.D. 79 and destroyed the cities of Pompeii, Herculaneum, and Stabiae.

b The excavated ruins of Pompeii are now a popular tourist attraction.

c Body casts of some of the volcano's victims in Pompeii.

Incandescent streams of molten rock are commonly portrayed in movies as posing a great danger to humans, and, in fact, on a few occasions lava flows have caused fatalities. Of course, lava flows may destroy homes, roadways, and croplands, but they are the least dangerous manifestation of volcanism. Explosive eruptions accompanied by little or no lava flow activity are quite dangerous, especially if they occur near populated areas. In this respect, *Dante's Peak* was accurate, although it would be most unusual for a volcano to both erupt explosively and produce fluid lava flows at the same time as depicted in the movie.

One of the best-known catastrophic eruptions was the A.D. 79 outburst of Mount Vesuvius, which destroyed the thriving Roman cities of Pompeii, Herculaneum, and Stabiae in what is now Italy (▶ Figure 5.1). Fortunately for us, Pliny the Younger recorded the event in great detail; his uncle, Pliny the Elder, died while trying to investigate the eruption. Pompeii, a city of about 20,000 people and only 9 km from Mount Vesuvius, was buried in nearly 3 m of pyroclastic materials that covered all but the tallest buildings. At least 2000 victims have been found in the city, but certainly far more were killed. Pyroclastic materials covered Pompeii rather gradually, but surges of incandescent volcanic materials in glowing avalanches swept though Herculaneum, quickly covering the city to a depth of 20 m. Since A.D. 79, Mount Vesuvius has erupted 80 times, most violently in 1631 and 1906; it last erupted in 1944. Volcanic eruptions and earthquakes in this area pose a continuing threat to the many cities and towns along the shores of the Bay of Naples (▶ Figure 5.1).

■ *How can volcanism be both constructive and destructive?*

The fact that lava flows and explosive eruptions cause property damage, injuries, and fatalities (Table 5.1) and at least short-term climate changes indicates that eruptions are destructive events, at least from the human perspective. Ironically, though, volcanism is actually a constructive process in the context of Earth history. Earth's atmosphere and surface waters most likely resulted from the emission of volcanic gases during the early history of the planet, and oceanic crust is continuously produced by volcanism at spreading ridges. Many oceanic islands such as Iceland, the Hawaiian Islands, the Azores, and the Galápagos Islands owe their existence to volcanic eruptions. In tropical areas, weathering converts lava, pyroclastic materials, and volcanic mudflows into fertile soils.

One reason to study volcanic activity is that volcanoes provide us with an excellent opportunity to see how Earth's systems interact. The emission of gases and pyroclastic materials has an immediate and profound impact on the atmosphere, hydrosphere, and biosphere, at least in the vicinity of an eruption. And in some cases, the effects are worldwide, as they were following the eruptions of Tambora in 1815, Krakatau in 1883, and Mount Pinatubo in 1991.

Section 5.1 Summary

- Interactions among systems are demonstrated by large volcanic eruptions because they have an impact on the hydrosphere, atmosphere, and biosphere.

TABLE 5.1

Some Notable Volcanic Eruptions

Date	Volcano	Deaths
Apr. 10, 1815	Tambora, Indonesia	92,000; includes deaths from eruption and famine and disease
Oct. 8, 1822	Galunggung, Java	Pyroclastic flows and mudflows killed 4011
Mar. 2, 1856	Awu, Indonesia	2806 died in pyroclastic flows
Aug. 27, 1883	Krakatau, Indonesia	More than 36,000 died; most killed by tsunami
June 7, 1892	Awu, Indonesia	1532 died in pyroclastic flows
May 8, 1902	Mount Pelée, Martinique	Nuée ardente engulfed St. Pierre and killed 28,000
Oct. 24, 1902	Santa Maria, Guatemala	5000 died during eruption
May 19, 1919	Kelut, Java	Mudflows devastated 104 villages and killed 5110
Jan. 21, 1951	Lamington, New Guinea	Pyroclastic flows killed 2942
Mar. 17, 1963	Agung, Indonesia	1148 perished during eruption
May 18, 1980	Mount St. Helens, Washington	63 killed; 600 km^2 of forest devastated
Mar. 28, 1982	El Chichón, Mexico	Pyroclastic flows killed 1877
Nov. 13, 1985	Nevado del Ruiz, Colombia	Minor eruption triggered mudflows that killed 23,000
Aug. 21, 1986	Oku volcanic field, Cameroon	Cloud of CO_2 released from Lake Nyos killed 1746
June 15, 1991	Mount Pinatubo, Philippines	~ 281 killed during eruption; 83 died in later mudflows; 358 died of illness
July 1999	Soufrière Hills, Montserrat	19 killed; 12,000 evacuated
Jan. 17, 2002	Nyiragongo, Zaire	Lava flow killed 147 in Goma

5.2 Volcanism and Volcanoes

What do we mean by the terms *volcanism* and *volcano*? The latter is a landform—that is, a feature on Earth's surface—whereas **volcanism** is the process in which magma rises through Earth's crust and issues forth at the surface as lava flows and/or pyroclastic materials and gases. We will discuss the origin and nature of volcanoes and other volcanic landforms in later sections, but here we point out that volcanism is also responsible for the origin of all extrusive igneous (volcanic) rocks, such as basalt, tuff, and obsidian (see Chapter 4).

Volcanism is a common phenomenon. About 550 volcanoes are *active*; that is, they are erupting or have erupted during historic time. Only about a dozen are erupting at any one time. Most of this activity is minor and goes unreported in the popular press unless an eruption, even a small one, takes place near a populated area or has tragic consequences. However, large eruptions that cause extensive property damage, injuries, and fatalities are not uncommon (Table 5.1). Indeed, a great amount of effort is devoted to better understanding and more effectively anticipating large eruptions.

In addition to active volcanoes, Earth has numerous *dormant* volcanoes that could erupt in the future. The distinction between *active* and *dormant* is not precise. Prior to its eruption in A.D. 79, Mount Vesuvius had not been active in human memory. The largest volcanic outburst since 1912 took place in 1991, when Mount Pinatubo in the Philippines erupted after lying dormant for 600 years. Some volcanoes have not erupted during historic time and show no evidence of erupting again; thousands of these *extinct* or *inactive* volcanoes are known.

All terrestrial planets and Earth's moon were volcanically active during their early histories, but now only Earth and a few other bodies in the solar system have active volcanoes. At least one active volcano is likely present on Venus, and Triton, a moon of Neptune, and Titan, a moon of Saturn, probably have active volcanoes. But Jupiter's moon Io is by far the most volcanically active body in the solar system. Many of its more than 100 volcanoes are erupting at any given time.

Volcanic Gases

- *What gases do volcanoes commonly emit?*

Samples from present-day volcanoes indicate that 50–80% of all volcanic gases are water vapor. Volcanoes also emit carbon dioxide, nitrogen, sulfur dioxide, hydrogen sulfide, and very small amounts of carbon monoxide, hydrogen, and chlorine. In many areas of recent and ongoing volcanism, such as Lassen Volcanic National Park in California, one cannot help but notice the rotten-egg odor of hydrogen sulfide gas (▶ Figure 5.2). In fact, this *hydrothermal (hot water) activity* is one potential source of energy (see Chapter 16).

Most volcanic gases quickly dissipate in the atmosphere and pose little danger to humans, but on occasion they have caused fatalities. In 1783 toxic gases, probably sulfur dioxide, from Laki fissure in Iceland had tragic effects. About 75% of the nation's livestock died, and the haze from the gases caused lower temperatures and crop failures. About

▶ **Figure 5.2 Fumeroles** Gases emitted from vents (fumeroles) at the Sulfur Works in Lassen Volcanic National Park in California. Hot, acidic gases and fluids have altered the original igneous rocks to clay. Several other vents are also present in this area, but the two shown here opened up only a few years ago.

24% of Iceland's population died from what was called the Blue Haze Famine. The eruption also produced a "dry fog" in the upper atmosphere that was likely responsible for dimming the intensity of sunlight and the severe winter of 1783–1784 in Europe and North America.

The 1815 eruption of Tambora in Indonesia, the largest and deadliest historic eruption, was probably responsible for the particularly cold spring and summer of 1816. The eruption of Mayon volcano in the Philippines during the previous year may have contributed to freezing temperatures, frost, and crop failures during the spring and summer in North America, or what residents at the time called the "year without a summer" or "1816 and froze to death."

In 1986, in the African nation of Cameroon, 1746 people died when a cloud of carbon dioxide engulfed them. The gas accumulated in the waters of Lake Nyos, which lies in a volcanic depression. Scientists are not sure why the gas burst forth suddenly, but once it did, it flowed downhill along the surface because it was denser than air. As it moved, it flattened trees and killed thousands of animals and many people, some as far as 23 km from the lake.

Residents of the island of Hawaii have coined the term *vog* for volcanic smog. Kilauea volcano has been erupting continuously since 1983, releasing small amounts of lava and copious quantities of carbon dioxide and sulfur dioxide every day. Car-

▶ **Figure 5.3 Lava Tubes** Lava tubes consisting of hollow spaces beneath the surfaces of lava flows are common in many areas.

a An active lava tube in Hawaii. Part of the tube's roof has collapsed, forming a skylight.

b A lava tube in Hawaii after the lava has drained out.

▶ **Figure 5.4 Pahoehoe and aa Lava Flows** Pahoehoe and aa were named for lava flows in Hawaii, but the same kinds of flows are found in many other areas.

a An excellent example of the taffylike appearance of pahoehoe.

b An aa lava flow advances over an older pahoehoe flow. Notice the rubbly nature of the aa flow.

bon dioxide has been no problem, but sulfur dioxide produces a haze and the unpleasant odor of sulfur. Vog probably poses no risk for tourists, but a long-term threat exists for people living on the west side of the island where vog is most common.

Lava Flows

Although lava flows are portrayed in movies and on television as a great danger to humans, they only rarely cause fatalities. The reason is that most lava flows do not move very fast, and because they are fluid, they follow low areas. Thus, once a lava flow erupts from a volcano, determining the path it will take is easy, and anyone in areas likely to be affected can be evacuated.

Even low-viscosity (fluid) lava flows usually do not move rapidly, but they flow much faster when their margins and upper surfaces cool and solidify to form a **lava tube**—that is, a tunnel-like structure insulated on all sides. Thus confined, lava may flow at speeds up to 50 km/hr; if part of a lava tube's roof collapses, forming a *skylight*, the active flow can be observed (▶ Figure 5.3a). When an eruption ceases, the tube drains, leaving an empty tube (▶ Figure 5.3b). In Hawaii, lava moves through lava tubes for many kilometers and some discharges into the sea.

■ *How and why do aa and pahoehoe lava flows differ?*
Geologists define two types of lava flows, both named for lava flows in Hawaii. The type of flow called **pahoehoe** (pronounced *pah-hoy-hoy*) has a smooth, ropy surface much like taffy (▶ Figure 5.4a). An **aa** (pronounced *ah-ah*) flow, in

contrast, is made up of jagged, angular blocks and fragments (▶ Figure 5.4b). Pahoehoe flows are less viscous than aa flows; indeed, aa flows are viscous enough to break up into blocks and move forward as a wall of rubble. A pahoehoe flow may change to aa along its length, but aa flows do not change to pahoehoe.

Many lava flows have a distinctive pattern of columns bounded by fractures, or what geologists call **columnar joints** (▶ Figure 5.5). Once a lava flow ceases moving, it contracts as it cools, thus producing forces that cause fractures called *joints* to open. On the surface of a lava flow, these joints intersect and outline polygons, which are commonly six-sided (▶ Figure 5.5b). The joints (fractures) extend down into the flow, thereby forming parallel columns with their long axes perpendicular to the principal cooling surface. Although found mostly in mafic lava flows, columnar joints are also present in some intrusive bodies.

We know that lava flows follow low areas such as stream valleys, and yet we see lava flows in many areas perched on the tops of ridges or hills. So why do they now stand high above the surrounding countryside? When a lava flow cools, it forms rock that is commonly harder and more resistant to erosion than rocks adjacent to it. Accordingly, as erosion proceeds, the rocks along a lava flow's margins erode more

▶ **Figure 5.5 Columnar Jointing** Columnar jointing is seen mostly in mafic lava flows and related intrusive rocks.

a As lava cools and contracts, three-pronged cracks form that grow and intersect to form four- to seven-sided columns, but most are six-sided.

b Columnar joints in a basalt lava flow at Devil's Postpile National Monument in California. The rubble in the foreground is collapsed columns.

c Surface view of the columns from (b). The straight lines and polish resulted from abrasion by a glacier that moved over this surface.

rapidly and what was a valley becomes a ridge or hilltop (▶ Figure 5.6). In short, what was a low area becomes a high area in what geologists call an inversion of topography.

Much of the upper oceanic crust is made up of bulbous masses of basalt that resemble pillows—hence the name **pillow lava** (▶ Figure 5.7). Geologists knew long ago that pillow lava forms when lava is rapidly chilled underwater, but its formation was not observed until 1971. Divers near Hawaii saw pillows form when a blob of lava broke through the crust of an underwater lava flow and cooled quickly, forming a pillow-shaped mass with a glassy exterior. Fluid lava then broke through the crust of the pillow just formed and formed another pillow, repeating the process and resulting in an interconnected accumulation of pillows (▶ Figure 5.7).

Pyroclastic Materials

As magma rises toward the surface, pressure decreases and the contained gases begin to expand. In highly viscous felsic magma, expansion is inhibited; gas pressure increases and may eventually cause an explosion and produce particulate matter known as **pyroclastic materials.** In contrast, low-viscosity mafic magma allows gases to expand and escape easily. Accordingly, mafic magma usually erupts rather quietly as fluid lava flows.

Ash is the name for pyroclastic materials that measure less than 2.0 mm (▶ Figure 5.8). In some eruptions, ash is ejected into the atmosphere and settles as an *ash fall*. In 1947 ash erupted from Mount Hekla in Iceland fell 3800 km away on Helsinki, Finland. In contrast to an ash fall, an *ash flow* is a cloud of ash and gas that flows along or close to the surface. Some ash flows move faster than 100 km/hr, and they may cover vast areas.

■ *What are pyroclastic materials, and how are they dangerous to air traffic?*

In populated areas adjacent to volcanoes, ash falls and ash flows pose serious problems, and volcanic ash in the atmosphere is a hazard to aviation. Since 1980, about 80 aircraft have been damaged when

▶ **Figure 5.6 Inversion of Topography** In some places we see lava flows on ridges or hilltops, yet when the flow occurred it must have followed a valley.

a Lava flows into a valley, where it cools and crystallizes, forming volcanic rock.

b The areas adjacent to the flow erode more easily than the flow, producing an inversion of topography.

c The basalt that caps this small hill near Orland, California, was originally a lava flow that followed a valley from its source far to the east.

▶ **Figure 5.7 Pillow Lava** Much of the upper part of the oceanic crust is made up of pillow lava that formed when lava erupted underwater.

a Pillow lava on the seafloor in the Pacific Ocean about 150 miles west of Oregon that formed about 5 years before the photo was taken.

b Ancient pillow lava now on land in Marin County, California. The largest pillow measures about 0.6 m across.

they encountered clouds of volcanic ash, some so diffuse that pilots cannot see them. The most serious incident took place in 1989, when ash from Redoubt Volcano in Alaska caused all four jet engines to fail on KLM Flight 867. The plane carrying 231 passengers nearly crashed when it fell more than 3 km before the crew could restart the engines. The plane landed safely in Anchorage, Alaska, but it required $80 million in repairs.

In addition to ash, volcanoes erupt *lapilli*, consisting of pyroclastic materials that measure 2–64 mm, and *blocks* and *bombs*, both of which are larger than 64 mm (▶ Figure 5.8). Bombs have a twisted, streamlined shape, indicating they were erupted as globs of magma that cooled and solidified during their flight through the air. Blocks are angular pieces of rock ripped from a volcanic conduit or pieces of a solidified crust of a lava flow. Because of their size, lapilli, bombs, and blocks are confined to the immediate area of an eruption.

Section 5.2 Summary

- A volcano is a landform, whereas volcanism is the process whereby magma and its contained gases rise to the surface.

- Water vapor is the most common volcanic gas, but several others, including carbon dioxide and sulfur gases, are also emitted.

- Aa lava flows are made up of jagged, angular blocks, whereas pahoehoe flows have a taffy-like texture. Lava tubes, pillow lava, and columnar joints are found in some lava flows.

- Pyroclastic materials are particulate matter ejected from volcanoes during explosive eruptions.

▶ **Figure 5.8 Pyroclastic Materials** Pyroclastic materials are all particles ejected from volcanoes, especially during explosive eruptions.

a The volcanic bomb is elongate because it was molten when it descended through the air. The lapilli was collected at a small volcano in Oregon, whereas the ash came from the 1980 eruption of Mount St. Helens in Washington.

b This volcanic block in Hawaii formed when partially solidified lava collapsed into the sea, resulting in a steam explosion.

5.3 Types of Volcanoes

Simply put, a **volcano** is a hill or mountain that forms around a vent where lava, pyroclastic materials, and gases erupt. Although volcanoes vary in size and shape, all have a conduit or conduits leading to a magma chamber beneath the surface. Vulcan, the Roman deity of fire, was the inspiration for calling these mountains *volcanoes*, and because of their danger and obvious connection to Earth's interior, they have been held in awe by many cultures.

In Hawaiian legends, the volcano goddess Pele resides in the crater of Kilauea on Hawaii. During one of her frequent rages, Pele causes earthquakes and lava flows, and she may hurl flaming boulders at those who offend her. Native Americans in the Pacific Northwest tell of a titanic battle between the volcano gods Skel and Llao to account for huge eruptions that took place about 7700 years ago in Oregon and California. Pliny the Elder (A.D. 23–79), mentioned in Section 5.1, believed that before eruptions "the air is extremely calm and the sea quiet, because the winds have already plunged into the earth and are preparing to reemerge."*

■ *What are calderas and how do they form?*

Most volcanoes have a circular depression known as a **crater** at their summit, or on their flanks, that forms by explosions or collapse. Most craters are less than 1 km across, whereas much larger rimmed depressions are called **calderas**. In fact, some volcanoes have a summit crater within a caldera. Calderas are huge structures that form following voluminous eruptions during which part of a magma chamber drains and the mountain's summit collapses into the vacated space below. An excellent example is misnamed Crater Lake in Oregon (▶ Figure 5.9). Crater Lake is actually a steep-rimmed caldera that formed 7700 years ago in the manner just described; it is more than 1200 m deep and measures 9.7 km long and 6.5 km wide. As impressive as Crater Lake is, it is not nearly as large as some other calderas, such as the Toba caldera in Sumatra, which is 100 km long and 30 km wide.

Geologists recognize several types of volcanoes, but one must realize that each volcano is unique in its history of eruptions and development. For instance, the frequency of eruptions varies considerably; the Hawaiian volcanoes and Mount Etna on Sicily have erupted repeatedly, whereas Pinatubo in the Philippines erupted in 1991 for the first time in 600 years. And some volcanoes are complex mountains that have characteristics of more than one type of volcano. Nevertheless, most volcanoes are conveniently classified as *shield volcanoes, cinder cones, composite volcanoes,* or *lava domes*.

Shield Volcanoes

A **shield volcano** looks like the outer surface of a shield lying on the ground with its convex side up (▶ Figure 5.10). Low-viscosity basalt lava flows issue from a shield volcano's crater or caldera and spread out as thin layers, forming gentle slopes that range from 2 to 10 degrees. Eruptions from shield volcanoes, commonly called *Hawaiian-type eruptions*, are nonexplosive because the fluid lava loses its gases easily and consequently poses little danger to humans. Lava fountains as high as 400 m form where gases escape and contribute pyroclastic materials to shield volcanoes, but otherwise shield volcanoes are made up mostly of basalt lava flows. About 99% of the Hawaiian volcanoes above sea level are composed of lava flows.

Although eruptions of shield volcanoes tend to be rather quiet, some of the Hawaiian volcanoes have, on occasion, produced sizable explosions when magma comes in contact with groundwater, causing it to vaporize instantly. In 1790 Chief Keoua led 250 warriors across the summit of Kilauea volcano to engage a rival chief in battle. About 80 of Keoua's warriors were killed by a cloud of hot volcanic gases.

The current activity of Kilauea is impressive because it has been erupting continuously since January 3, 1983, making it the longest recorded eruption. During these 22 years, more than 2.3 km^3 of molten rock has flowed out at the surface, much of it reaching the sea and forming 2.2 km^2 of new property on the island of Hawaii. Unfortunately, lava flows from Kilauea have also destroyed about 200 homes and caused some $61 million in damages.

Shield volcanoes are most common in the ocean basins, such as the Hawaiian Islands and Iceland, but some are also present on the continents—for example, in East Africa. The island of Hawaii consists of five huge shield volcanoes; two of them, Kilauea and Mauna Loa, are active much of the time. Mauna Loa, at nearly 100 km across its base and more than 9.5 km above the surrounding seafloor, is the largest vigorously active volcano in the world (▶ Figure 5.10). Its volume is estimated at about 50,000 km^3.

Cinder Cones

■ *What are cinder cones and what are they composed of?*

Small, steep-sided volcanoes made up of pyroclastic materials that resemble cinders are known as **cinder cones** (▶ Figure 5.11). Cinder cones are only rarely higher than 400 m, with slope angles up to 33 degrees, because they are made up of irregularly shaped particles. Many of these small volcanoes have large, bowl-shaped craters, and if they issue any lava flows at all, they usually break through the lower flanks rather than erupt from the crater (▶ Figure 5.12). A cinder cone may be a nearly perfect cone, but when some erupt, the prevailing winds cause the pyroclastic materials to build up

---
*Quoted from M. Krafft, *Volcanoes: Fire from the Earth* (New York: Harry N. Abrams, 1993), p. 40.

142   Chapter 5   Volcanoes and Volcanism

PHYSICAL Geology Now™ ► Geo-focus Figure 5.9 The Origin of Crater Lake, Oregon Events leading to the origin of Crater Lake, Oregon. Remember, Crater Lake is actually a caldera that formed by partial draining of a magma chamber.

a Eruption begins as huge quantities of ash are ejected from the volcano.

b The eruption continues as more ash and pumice are ejected into the air and pyroclastic flows move down the flanks of the mountain.

c The collapse of the summit into the partially drained magma chamber forms a huge caldera.

d Postcaldera eruptions partly cover the caldera floor, and the small cinder cone called Wizard Island forms.

e View from the rim of Crater Lake showing Wizard Island. The lake is 594 m deep, making it the second deepest in North America.

higher on the downwind side of the vent, resulting in a markedly asymmetric shape.

Many cinder cones form on the flanks or within the calderas of larger volcanoes. For instance, Newberry Volcano in Oregon has more than 400 cinder cones on its flanks, and Wizard Island is a small cinder cone that formed following the origin of the caldera we now call Crater Lake in Oregon (► Figure 5.9e). Hundreds of cinder cones are present in the southern Rocky Mountain states, particularly in New Mexico and Arizona, and many others are found in California, Oregon, and Washington.

Eruptions at cinder cones are rather short-lived. For instance, on February 20, 1943, a farmer in Mexico noticed fumes emanating from a crack in his cornfield, and a few minutes later ash and cinders were erupted. Within a month, a 300-m-high cinder cone had formed, later named Paríutin, from which lava flowed and covered two nearby towns (► Figure 5.11b). Activity ceased in 1952. In Iceland a new cinder cone rose to 100 m above the surrounding area in only two days after it began erupting on January 23, 1973. By February a massive aa lava flow 10–20 m thick at its leading edge was advancing toward the town of Vestmannaeyjar.

Figure 5.10 Shield Volcanoes

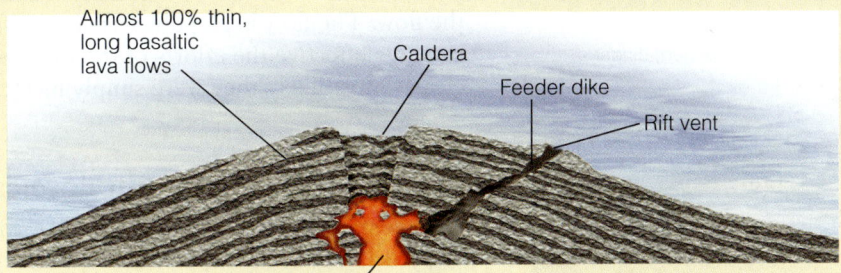

a Diagrammatic view of the internal structure of a shield volcano.

b Mauna Loa on Hawaii is an active shield volcano that has erupted 33 times since 1843; it erupted last in 1983. Its summit stands 17 km above its base on the seafloor.

c Crater Mountain in Lassen County, California, an extinct shield volcano, is about 10 km across and 460 m high. The depression at its summit is a 2-km-wide crater.

▶ **Figure 5.11 Cinder Cones** Cinder cones are small, steep-sided volcanoes made up of pyroclastic materials that resemble cinders.

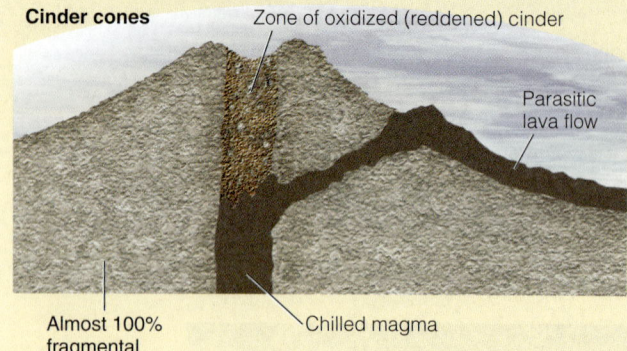

a Cross section of a cinder cone showing its internal structure. Although composed mostly of pyroclastic materials, lava flows may erupt through the base or flanks of cinder cones.

b This 400-m-high cinder cone named Paricutín formed in a short time in Mexico in 1943 when pyroclastic materials began to erupt in a farmer's field. Lava flows from the volcano covered two nearby villages, but all activity ceased by 1952.

Residents of the community sprayed the front of the flow with seawater in an effort to solidify it and divert the rest of the flow. The flow in fact diverted around most of the town, but how effective the efforts of the townspeople were is not clear—most likely they were simply lucky.

Composite Volcanoes (Stratovolcanoes)

Composite volcanoes, also called *stratovolcanoes*, are made up of pyroclastic layers and lava flows, although some, perhaps many, of the lava flows may actually be sills (▶ Figure 5.13a). Both the pyroclastic materials and lava have an intermediate composition, so the lava typically cools to form andesite. Another component of composite volcanoes is volcanic mudflows, or what geologists call **lahars.** A lahar may form when rain falls on unconsolidated pyroclastic materials and creates a mixture of particles and water that moves downslope (▶ Figure 5.14). A rather minor eruption of Nevado del Ruiz in Colombia on November 13, 1985, melted snow and ice on the mountain, causing lahars that killed about 23,000 people (Table 5.1).

Composite volcanoes differ from shield volcanoes and cinder cones in composition and shape. Remember that shield volcanoes have very low slopes, whereas cinder cones are small, steep-sided, conical mountains. In contrast, composite volcanoes are steep near their summits, with slope angles up to 30 degrees, but the slope decreases toward the base, where it may be no more than 5 degrees. Mayon volcano in the Philippines is one of the most nearly symmetrical composite volcanoes anywhere (▶ Figure 5.13b). It erupted in 1999 for the 13th time during the 1900s.

Composite volcanoes are what most people visualize when they think of volcanoes. And some of these mountains are indeed impressive. Mount Shasta in northern California is made up of an estimated 350 km³ of material and measures about 20 km across its base (▶ Figure 5.13c). It dominates the skyline when approached from any direction. Other familiar composite volcanoes are several others in the Cascade Range of the Pacific Northwest as well as Fujiyama in Japan and Mount Vesuvius in Italy.

Lava Domes

Volcanoes are complex landforms. Most are shield volcanoes, cinder cones, or composite volcanoes, but some, however, show features of more than one kind of volcano. For example, Mount Etna in Italy is partly a shield volcano and a composite volcano. And, of course, there are a few very unusual volcanoes.

■ *What are lava domes and why are they so dangerous?*

Some volcanic mountains are steep-sided, bulbous masses of viscous magma that geologists call **lava domes** or *volcanic domes*. Most are composed of felsic magma and occasionally intermediate magma that was forced upward under great

Figure 5.12 Cinder Cone, Lassen Volcanic National Park, California This aerial view of Cinder Cone in Lassen Volcanic National Park in California shows an area about 6.5 km long, with north toward the right.

a View of 230-m-high Cinder Cone from the ground.

b Large, bowl-shaped crater at the summit of Cinder Cone.

c View from the top of Cinder Cone showing the Painted Dunes, which is volcanic ash that fell on still-hot lava.

d The Fantastic Lava Beds are aa lava flows that broke through the base of Cinder Cone during the 1650s.

pressure but was too viscous to flow. Lava domes may stand as small, isolated volcanic mountains, or they may rise into the craters of composite volcanoes (▶ Figure 5.15). Unfortunately, lava domes are quite unstable and commonly collapse under their own weight, resulting in huge flows of debris. In June 1991, a lava dome in Japan's Unzen volcano collapsed and the hot debris and ash killed 43 people in a nearby town. During both the 1980 and 2004 eruptions of Mount St. Helens in Washington, lava domes formed and were subsequently destroyed (▶ Figure 5.15a).

Lava domes may be particularly dangerous. In 1902 viscous magma accumulated beneath the summit of Mount Pelée on the island of Martinique in the Caribbean Sea. The gas pressure increased until the side of the mountain blew out in a tremendous explosion, ejecting a mobile, dense cloud of pyroclastic materials and a cloud of gases and ash called a **nuée ardente,** a French term for "glowing cloud." The lower part of this mass, the pyroclastic flow, followed a valley to the sea, but the upper part, the nuée ardente, jumped a ridge and engulfed the city of St. Pierre (▶ Figure 5.16).

A tremendous blast hit St. Pierre and leveled buildings; hurled boulders, trees, and pieces of masonry down the streets; and moved a 3-ton statue 16 m. Accompanying the blast was a swirling cloud of incandescent ash and gases with an internal temperature of 700°C that incinerated everything in its path. The nuée ardente passed through St. Pierre in 2 or 3 minutes, only to be followed by a firestorm as combustible materials burned and casks of rum exploded. But by then most of the 28,000 residents of the city were already dead. In fact, in the area covered by the nuée ardente, only 2 survived!* One survivor was on the outer edge of the nuée ardente, but even there he was terribly burned and his family and neighbors were all killed. The other survivor, a stevedore incarcerated the night before for disorderly conduct, was in a windowless cell partly below ground level. He remained in his cell badly

---

*Although it is commonly reported that only 2 people survived the eruption, at least 69 and possibly as many as 111 people survived beyond the extreme margins of the nuée ardente and on ships in the harbor. Many, however, were badly injured.

146 Chapter 5 Volcanoes and Volcanism

▶ **Figure 5.13 Composite Volcanoes** Composite volcanoes, also known as stratovolcanoes, are the large picturesque volcanoes on the continents.

a Cross section of a composite volcano, which is made up largely of lava flows and pyroclastic materials, although the latter may be reworked as mudflows (lahars).

b Mayon volcano in the Philippines is almost perfectly symmetrical. It erupted 13 times during the 1900s, most recently in 1999.

c View of Mount Shasta in California from the north. The main peak is on the left; the one on the right is Shastina, a smaller cone on the flank of the volcano.

▶ **Figure 5.14 Volcanic Mudflows (Lahars)** Composite volcanoes are made up partly of volcanic mudflows or lahars, mixtures of pyroclastic materials and water that move rapidly downslide.

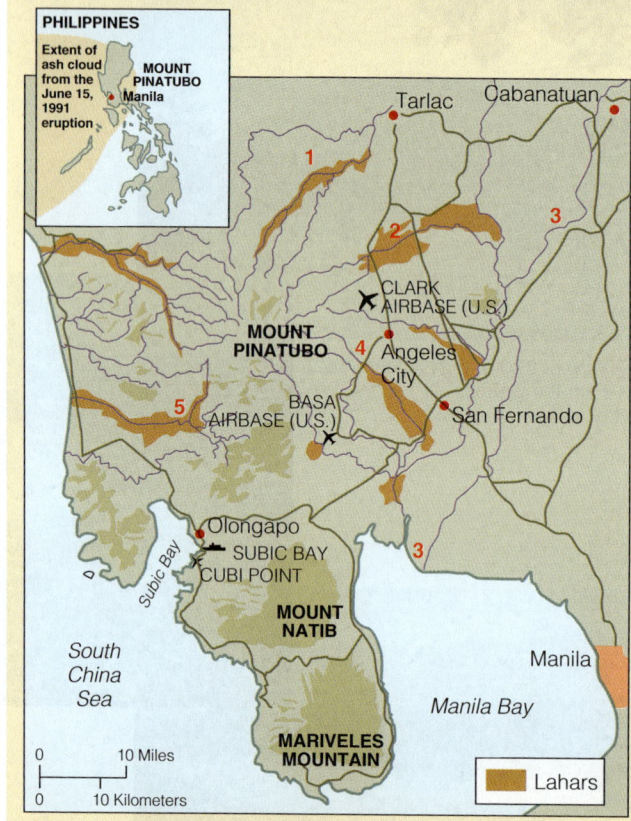

a Map showing the areas covered by lahars following the 1991 eruption of Mount Pinatubo in the Philippines. By 1993 lahars caused more damage in the lowlands around the volcano than the eruption did.

b Homes partly buried by a lahar from Mount Pinatubo only hours after the 1991 eruption.

burned for four days after the eruption until rescue workers heard his cries for help. He later became an attraction in the Barnum & Bailey Circus, where he was advertised as "the only living object that survived in the 'Silent City of Death' where 40,000 beings were suffocated, burned or buried by one belching blast of Mont Pelée's terrible volcanic eruption."†

---

†Quoted from A. Scarth, *Vulcan's Fury: Man Against the Volcano* (New Haven, CT: Yale University Press, 1999), p. 177.

Physical Geology Now™ Log into GeologyNow and select this chapter to work through a **Geology Interactive** activity on "Volcanic Landforms" (click Volcanism→Volcanic Landforms).

Eruptions and the Type of Volcanoes

The types of volcanoes we discussed in the preceding sections—shield, cinder cone, composite, and lava dome—differ in size, shape, and composition, largely because of the kinds

▶ **Figure 5.15 Lava Domes** Lava domes are bulbous masses of magma that are emplaced in craters of composite volcanoes or stand alone as irregularly shaped mountains flanked by debris shed from the dome.

a This image shows a lava dome in the crater of Mount St. Helens in 1984.

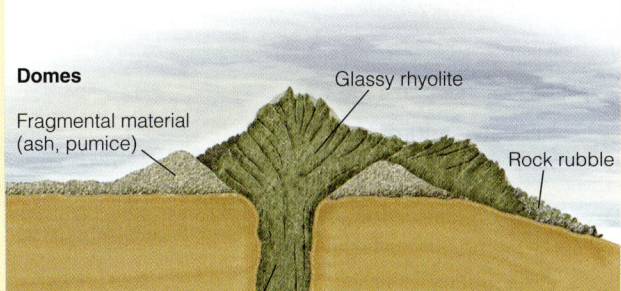

b Diagram of a mass of viscous magma forming a lava dome.

c Chaos Crags in the distance are made up of at least four lava domes that formed less than 1200 years ago in Lassen Volcanic National Park in California. The debris in the foreground is Chaos Jumbles, which resulted from the partial collapse of some of the domes.

▶ **Figure 5.16 Nuée Ardente**

a St. Pierre, Martinique, after it was destroyed by a nuée ardente from Mount Pelée in 1902. Only 2 of the city's 28,000 inhabitants survived.

b An April 1986 pyroclastic flow rushing down Augustine volcano in Alaska. This flow is similar to the one that wiped out St. Pierre.

▶ **Figure 5.17 Volcanic Eruptions** The corners of the triangle show the three kinds of materials ejected from volcanic vents. How a volcano erupts and the type of volcano that results depends on the relative mix of these ingredients.

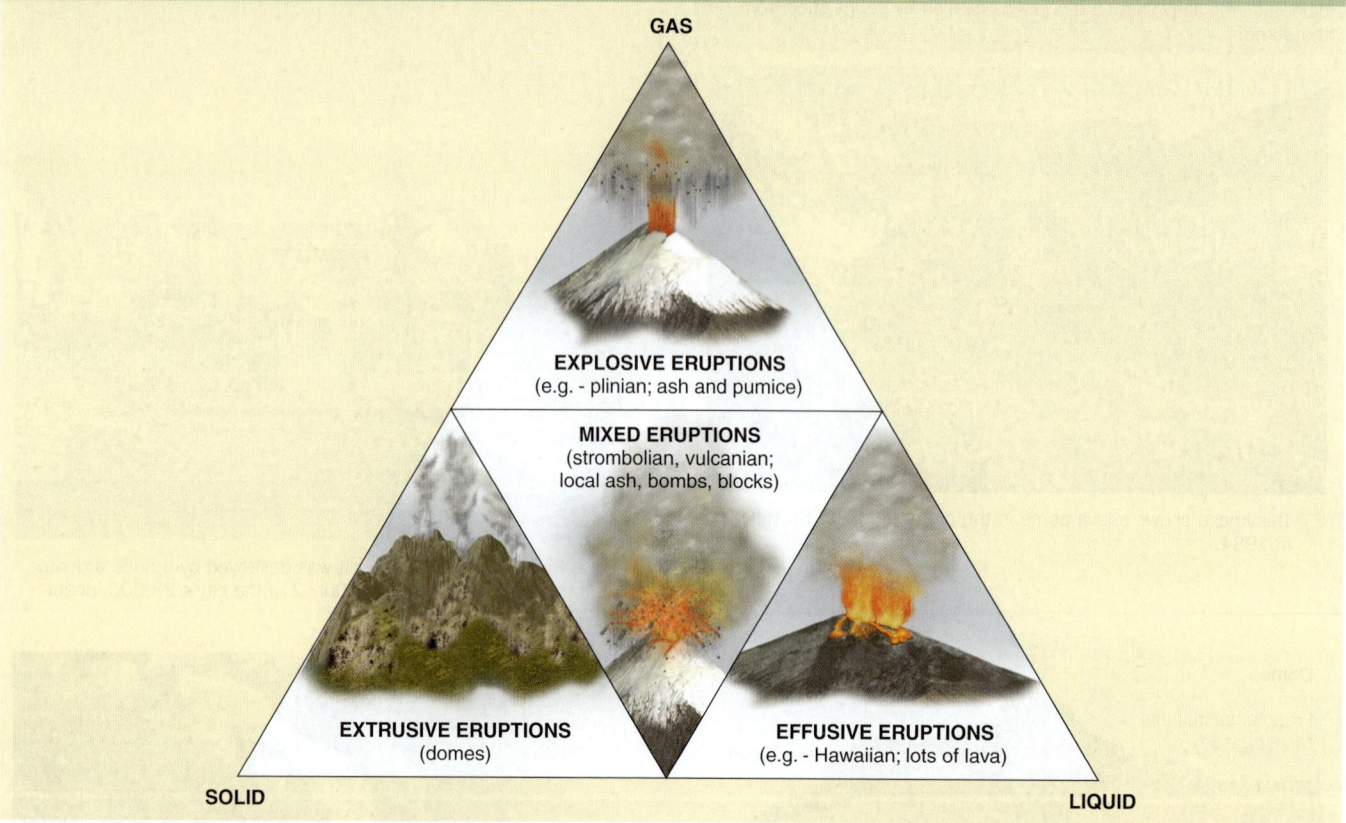

of materials erupted—gases, liquid (lava), and solids (we include very viscous lava here) (▶ Figure 5.17). How a volcano erupts and the type of volcano formed depends on the relative mix of these ingredients. Keep in mind, though, that a volcano's eruptive style may change through time. For example, Mount Etna in Italy is a shield volcano up to an elevation of about 2900 m, but the top 400 m is a composite volcano.

Notice in Figure 5.17 that geologists characterize eruptions as Hawaiian, Strombolian, Vulcanian, and Plinian depending on the mix of solids, liquids, and gases. These are simply names derived from active volcanoes that show the features within the diagram, except for Plinian, which comes from Pliny the Younger, who described the A.D. 79 eruption of Mount Vesuvius (see Section 5.1). For example, Hawaiian type eruptions involve mostly fluid lava, but not much gases or pyroclastic materials, whereas during Plinian eruptions large volumes of gases are discharged along with huge amounts of pumice and ash. We conclude this section with a summary diagram showing the four types of volcanoes we have discussed (▶ Figure 5.18).

What Would You Do?

You are a natural history enthusiast and would like to share your interests with your family. Accordingly, you plan a vacation that will take you to several of our national parks and monuments in Wyoming, Idaho, Washington, Oregon, and California. What specific areas might you visit, and what kinds of volcanic features would you see in these areas? Are there any other areas in the United States that you might visit in the future to see evidence of past volcanism or ongoing eruptions? If so, where would you go and what would you see?

Section 5.3 Summary

- All volcanoes, regardless of size or shape, form where lava and pyroclastic materials are erupted. Most have one or more craters or a caldera, a large oval to circular structure formed when a volcanic peak collapses into a partially drained magma chamber.

- Shield volcanoes have low, rounded profiles and are made mostly of fluid lava flows; cinder cones are small, steep-sided volcanoes composed of pyroclastic materials; and composite volcanoes consist of lava flows, pyroclastic layers, and lahars.

- Viscous bulbous masses of lava, generally of felsic composition, are lava domes, which are dangerous because they erupt explosively.

Geo-focus Figure 5.18 Types of Volcanoes A summary diagram showing the types of volcanoes. Keep in mind, though, that not all of these volcanoes would be found together in one area.

5.4 Other Volcanic Landforms

We mentioned that volcanoes are landforms—that is, features on Earth's surface that result from volcanism. What other kinds of volcanic landforms are there? The most notable are *basalt plateaus*, which form as a result of *fissure eruptions*, and *pyroclastic sheet deposits*, which, as their name implies, have a sheetlike shape.

Fissure Eruptions and Basalt Plateaus

■ *How do basalt plateaus form?*

Rather than erupting from central vents, the lava flows making up **basalt plateaus** issue from long cracks or fissures during **fissure eruptions**. The lava is so fluid (has such low viscosity) that it spreads out and covers vast areas. A good example is the Columbia River basalt in eastern Washington and parts of Oregon, and Idaho. This huge accumulation of 17- to 6-million-year-old overlapping lava flows covers about 164,000 km² (▶ Figure 5.19a,b). The Columbia River basalt has an aggregate thickness greater than 1000 m, and some individual flows are enormous; the 30-m-thick Roza flow advanced along a front about 100 km wide and covered 40,000 km². Geologists have identified some 300 huge flows here, one of which flowed 600 km from its source.

Similar accumulations of vast, overlapping lava flows are also found in the Snake River Plain in Idaho (▶ Figure 5.19a,c). However, these flows are 1.6 to 5.0 million years old, and they represent a style of eruption between flood basalts and those in Hawaii during which shield volcanoes form. In fact, there are small, low shields as well as fissure flows in the Snake River Plain.

Currently, fissure eruptions occur only in Iceland. Iceland has a number of volcanoes, but the bulk of the island is composed of basalt lava flows that issued from fissures. In fact, about half the lava erupted during historic time in Iceland came from two fissure eruptions, one in A.D. 930 and the other in 1783. The 1783 eruption from Laki fissure, which is more than 30 km long, accounted for lava that covered 560 km² and in one place filled a valley to a depth of about 200 m.

▶ **Figure 5.19 Basalt Plateaus** Basalt plateaus are vast areas of overlapping lava flows that issued from long fissures. Fissure eruptions take place today in Iceland, but in the past they formed basalt plateaus in several areas.

ⓐ Relief map of the northwestern United States showing the locations of the Columbia River basalt and the Snake River Plain.

ⓑ About 20 lava flows of the Columbia River basalt are exposed in the canyon of the Grand Ronde River in Washington.

ⓒ Basalt lava flows of the Snake River Plain near Twin Falls, Idaho.

Pyroclastic Sheet Deposits

■ *What are pyroclastic sheet deposits?*

Geologists think that vast areas covered by pyroclastic materials a few meters to hundreds of meters thick originated as **pyroclastic sheet deposits.** That is, deposits of pyroclastic materials with a sheetlike geometry. These deposits were known to geologists more than a century ago, and based on observations of present-day pyroclastic flows, such as the one erupted from Mount Pelée in 1902, led them to conclude that they formed in a similar manner. They cover much larger areas than any observed during historic time and evidently erupted from long fissures rather than from central vents. Remember that lithified ash is the volcanic rock known as *tuff* (see Chapter 4), but the ash in many of these flows was so hot that the particles fused to form *welded tuff*.

Some of these vast pyroclastic sheet deposits formed during the voluminous eruptions that were followed by the origin of calderas. For instance, pyroclastic flows were erupted during the formation of Crater Lake in Oregon (▶ Figure 5.9). Similarly, the Bishop Tuff of eastern California was erupted shortly before the origin of the Long Valley caldera. It is interesting that earthquakes in the Long Valley caldera and nearby areas beginning in 1978 and the escape of volcanic gases may indicate that magma is moving up beneath part of the caldera.

PHYSICAL
Geology≈Now™ Log into GeologyNow and select this chapter to work through a **Geology Interactive** activity on "Magma Chemistry and Explosivity" (click Volcanism→Magma Chemistry and Explosivity) and on "Volcano Watch, USA" (click Volcanism→Volcano Watch, USA).

Section 5.4 Summary

● Fluid mafic lava erupted from fissures spreads over large areas to form a basalt plateau.

● Pyroclastic sheet deposits result when huge eruptions of ash and other pyroclastic materials take place, especially when calderas form.

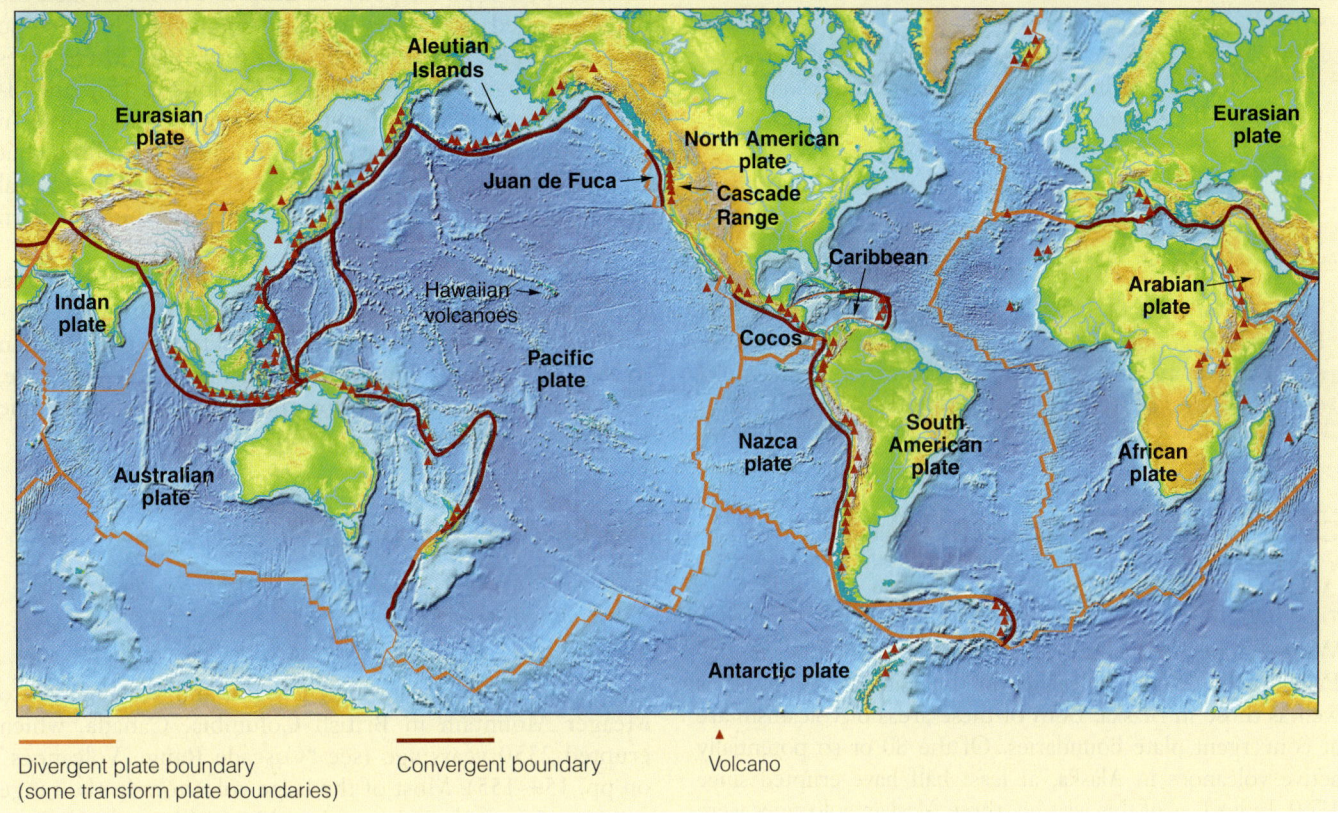

▶ **Figure 5.20 Volcanoes at Convergent and Divergent Plate Boundaries** Most volcanoes are at or near convergent and divergent plate boundaries in two main belts. The circum-Pacific belt has about 60% of all active volcanoes, and about 20% of all active volcanoes are in the Mediterranean belt. Most of the rest are near mid-oceanic ridges (divergent plate boundaries), but not all of them are shown on this map. For example, Axial volcano lies on the Juan de Fuca Ridge west of Oregon.

5.5 The Distribution of Volcanoes

■ *Where are the three zones or belts with most of Earth's volcanoes?*

Rather than being randomly distributed volcanoes are found mostly in three well-defined belts: the circum-Pacific belt, the Mediterranean belt, and along the mid-oceanic ridges. You have probably heard of the *Ring of Fire*, a phrase that alludes to the fact that a nearly continuous belt of volcanoes encircles the Pacific Ocean basin. Geologists refer to this as the **circum-Pacific belt,** where more than 60% of all active volcanoes are found. It includes the volcanoes in South and Central America, those in the Cascade Range of North America, and the volcanoes in Alaska, Japan, the Philippines, Indonesia, and New Zealand (▶ Figure 5.20). Also included in this belt are the southernmost active volcanoes at Mount Erebus in Antarctica, and a large caldera at Deception Island that last erupted in 1970.

The second major area of volcanism is the **Mediterranean belt,** with about 20% of all active volcanoes (▶ Figure 5.20). The famous Italian volcanoes include Mount Etna, which has issued lava flows on more than 200 occasions since 1500 B.C., when activity was first recorded. Mount Vesuvius, also in the Mediterranean belt, erupted violently in A.D. 79 and destroyed Pompeii, Herculaneum, and Stabiae (see Section 5.1). Remember that Mount Vesuvius has erupted 80 times since A.D. 79. Another important volcano in this belt is the Greek island of Santorini.

Most of the remaining 20% of active volcanoes are at or near mid-oceanic ridges or their extensions onto land (▶ Figure 5.20). The volcanoes at or near mid-oceanic ridges include those along the East Pacific Rise, the Mid-Atlantic Ridge, and the Indian Ridge, which account for submarine eruptions as well as the volcanic islands in the Pacific, Atlantic, and Indian Oceans. Iceland in the Atlantic Ocean, for instance, is found on the Mid-Atlantic Ridge. Branches of the Indian Ridge extend into the Red Sea and East Africa, where several volcanoes are found, including Kilamanjaro in Tanzania, Nyiragongo in the Democratic Republic of Congo, and Erta Ale in Ethiopia with its continuously active lava lake.

Anyone with a passing familiarity with volcanoes will have noticed that we have not mentioned the Hawaiian volcanoes. This is not an oversight; they are the notable exceptions to the distribution of active volcanoes in well-defined belts. We discuss their location and significance in Section 5.7, Plate Tectonics, Volcanoes, and Plutons.

PHYSICAL GeologyNow™ Log into GeologyNow and select this chapter to work through a **Geology Interactive** activity on "Distribution of Volcanism" (click Volcanism→ Distribution of Volcanism).

Section 5.5 Summary

- About 60% of all active volcanoes are in the circum-Pacific belt, another 20% are in the Mediterranean belt, and the remaining 20% are mostly at or near mid-oceanic ridges or their extensions onto land.

5.6 North America's Active Volcanoes

We mentioned in the previous section that part of the circum-Pacific belt includes volcanoes in the Pacific Northwest as well as those in Alaska. Both of these areas of volcanism are at convergent plate boundaries. Of the 80 or so potentially active volcanoes in Alaska, at least half have erupted since 1760. Indeed, as of this writing, three Alaskan volcanoes were erupting—Mount Spurr, Veniaminof volcano, and Shishaldin volcano.

The other active North American volcanoes are in the Cascade Range in the Pacific Northwest where the Juan de Fuca plate is subducted beneath North America. Many of these volcanoes have been historically active, although during the 1900s only Lassen Peak in California and Mount St. Helens in Washington erupted. And of course Mount St. Helens began erupting again during late September 2004.

Alaska's Volcanoes

Many of the volcanoes in mainland Alaska and in the Aleutian Islands are composite volcanoes, some with huge calderas (▶ Figure 5.20). Mount Spurr has erupted explosively at least 35 times during the last 5000 years, but its eruptions pale by comparison with that of Novarupta in 1912. Novarupta Volcano now lies in Katmai National Park and Preserve, which has a total of 15 active volcanoes. Its defining event was the June 1912 eruption, the largest in the world since the late 1800s. At least 15 km^3 and perhaps as much as 23 km^3 of volcanic materials, mostly pyroclastic materials, erupted during about 60 hours. "The expulsion of such a large volume of magma excavated a funnel-shaped vent 2 kilometers wide and triggered the collapse of Mount Katmai volcano 10 kilometers away."*

When the eruption was over, 120 km^2 of land was buried beneath pyroclastic deposits as deep as 213 m. In fact, the deposits filled the Valley of Ten Thousand Smokes—so named because of the hundreds of fumaroles where gases vented through the hot deposits for as long as 15 years following the eruption. Fortunately, the eruption took place in a remote area so there were no injuries or fatalities, but enough ash, gases, and pumice were ejected that for several days the sky was darkened over much of the Northern Hemisphere.

By the time you read this chapter, several more volcanoes in Alaska will have erupted as the Pacific Plate moves relentlessly northward only to be subducted at the Aleutian Trench. The Alaska Volcanoes Observatory in Anchorage, Alaska, continues to monitor these volcanoes and issue warnings about potential eruptions.

The Cascade Range

■ *Where is the Cascade Range and what types of volcanoes are found there?*

The **Cascade Range** stretches from Lassen Peak in northern California north through Oregon and Washington to Meager Mountain in British Columbia, Canada, which erupted 2350 years ago (see "Cascade Range Volcanoes" on pp. 154–155). Most of the large volcanoes in the range are composite volcanoes, such as Mount Shasta in California (▶ Figure 5.13c), Mount Hood in Oregon, and Mount St. Helens in Washington, but Lassen Peak in California is the world's largest lava dome. Actually it is a rather small volcano that developed 27,000 years ago on the flank of a much larger, deeply eroded composite volcano. It erupted from 1914 to 1917 but has since been quiet except for ongoing hydrothermal activity.

Two large shield volcanoes lie just to the east of the main Cascade Range volcanoes—Medicine Lake Volcano in California and Newberry Volcano in Oregon. Distinctive features at Newberry Volcano are a 1600-year-old obsidian flow and casts of trees that formed when lava flowed around them and solidified. Cinder cones are common throughout the range, such as Wizard Island in Crater Lake, Oregon (▶ Figure 5.9e), and Cinder Cone in Lassen Volcanic National Park, California (▶ Figure 5.12).

What was once a nearly symmetrical composite volcano changed markedly on May 6, 1980, when Mount St. Helens in Washington erupted explosively, killing 63 people and thousands of animals and leveling some 600 km^2 of forest (see "Cascade Range Volcanoes" on pp. 154–155). Geologists, citing Mount St. Helens's past explosive eruptions, warned that it was the most likely

---

*From Brantley, 1994. *Volcanoes in the United States.* USGS General Interest Publication, p. 30.

Cascade Range volcano to erupt violently. In fact, a huge lateral blast caused much of the damage and fatalities, but snow and ice on the volcano melted and pyroclastic materials displaced water in lakes and rivers, causing lahars and extensive flooding.

Mount St. Helens's renewed activity beginning in late September 2004 has resulted in dome growth and small steam and ash explosions. At the time of this writing (February 2005), scientists at the Cascades Volcano Observatory in Vancouver, Washington, have issued a low-level alert for an eruption, but they think that if one takes place it will be much less violent than the one in 1980.

Several of the Cascade Range volcanoes will almost certainly erupt again, but the most dangerous is probably Mount Rainier in Washington. Rather than lava flows or even a colossal explosion, the greatest danger from Mount Rainier is volcanic mudflows or huge debris flows. Of the 60 large flows that have occurred during the last 100,000 years, the largest, consisting of 4 km^3 of debris, covered an area now occupied by more than 120,000 people. Indeed, in August 2001 a sizable debris flow took place on the south side of the mountain, but it caused no injuries or fatalities. No one knows when the next flow will take place, but at least one community has taken the threat seriously enough to formulate an emergency evacuation plan. Unfortunately, the residents would have only 1 or 2 hours to carry out the plan.

Section 5.6 Summary

- Since 1760 more than 40 volcanoes have erupted in Alaska, including the Aleutian Islands, some of them many times. The largest volcanic outburst since the late 1800s took place at Novarupta in Alaska in 1912.

- The Cascade Range includes volcanoes in northern California, Oregon, Washington, and British Columbia, Canada. Only three eruptions have occurred since 1914, one at Lassen Peak in California (1914–1917) and two at Mount St. Helens in Washington (1980 and 2004–2006).

- The large volcanoes in Alaska and the Cascade Range are mostly composite volcanoes, although some shield volcanoes are present and both areas have many cinder cones.

5.7 Plate Tectonics, Volcanoes, and Plutons

In Chapter 2 we noted that plate tectonic theory is a unifying theory in geology that explains many seemingly unrelated geologic phenomena. So how do we relate the eruption of volcanoes and the emplacement of plutons to plate tectonics? You already know from Chapter 4 that (1) mafic magma is generated beneath spreading ridges and (2) intermediate magma and felsic magma form where an oceanic plate is subducted beneath another oceanic plate or where an oceanic plate is subducted beneath a continental plate. Accordingly, most of Earth's volcanism and emplacement of plutons take place at or near divergent and convergent plate boundaries.

Divergent Plate Boundaries and Igneous Activity

■ *What kinds of igneous rocks make up the oceanic crust?*

Much of the mafic magma that forms beneath spreading ridges is emplaced at depth as vertical dikes and gabbro plutons. But some rises to the surface, where it forms submarine lava flows and pillow lava (▶ Figure 5.7). Indeed, the oceanic crust is composed largely of gabbro and basalt. Much of this submarine volcanism goes undetected, but researchers in submersible craft have observed the results of these eruptions.

Pyroclastic materials are not common in this environment because mafic lava is very fluid, allowing gases to escape easily, and at great depth, water pressure prevents gases from expanding. Accordingly, the explosive eruptions that yield pyroclastic materials are not common. If an eruptive center along a ridge builds above sea level, however, pyroclastic materials may be erupted at lava fountains, but most of the magma issues forth as fluid lava flows that form shield volcanoes.

Excellent examples of divergent plate boundary volcanism are found along the Mid-Atlantic Ridge, particularly where it is above sea level as in Iceland (▶ Figure 5.20). In November 1963 a new volcanic island, later named Surtsey, rose from the sea just south of Iceland. The East Pacific Rise and the Indian Ridge are areas of similar volcanism. Not all divergent plate boundaries are beneath sea level as in the previous examples. For instance, divergence and igneous activity are taking place in Africa at the East African Rift system (▶ Figure 5.20).

Igneous Activity at Convergent Plate Boundaries

Nearly all of the large active volcanoes in both the circum-Pacific and Mediterranean belts are composite volcanoes near the leading edges of overriding plates at convergent plate boundaries (▶ Figure 5.20). The overriding plate, with its chain of volcanoes, may be oceanic as in the case of the Aleutian Islands, or it may be continental as is, for instance, the South American plate with its chain of volcanoes along its western edge.

As we have noted, these volcanoes at convergent plate boundaries consist largely of lava flows and pyroclastic materials of intermediate to felsic composition. Remember

Cascade Range Volcanoes

Several large volcanoes and hundreds of smaller volcanic vents are in the Cascade Range, which stretches from northern California into southern British Columbia, Canada. Medicine Lake Volcano and Newberry Volcano are shield volcanoes that lie just east of the main trend of the Cascade Range.

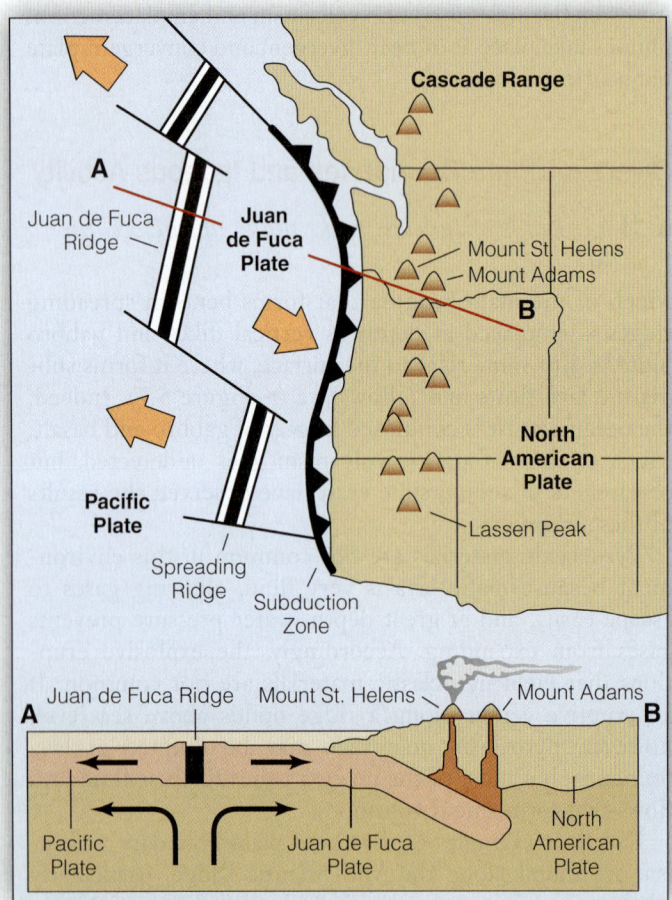

▲ **4.** A huge steam explosion called the Great Hot Blast leveled the area in the foreground in 1915. In the 89 years since the eruption, trees are becoming reestablished in this area, known as the Devastated Area.

▲ **1.** Plate tectonic setting for the Pacific Northwest. Subduction of the Juan de Fuca plate beneath North America accounts for the continuing volcanism in this region.

▶ **3.** Lassen Peak today. This most southerly peak in the Cascade Range is made up of 2 km³ of material, including the bulbous masses of rock visible in this image.

▶ **2.** Lassen Peak is a dacite lava dome that formed 27,000 years ago on the flank of an eroded composite volcano called Mount Tehama. Dacite has a composition between andesite and rhyolite.

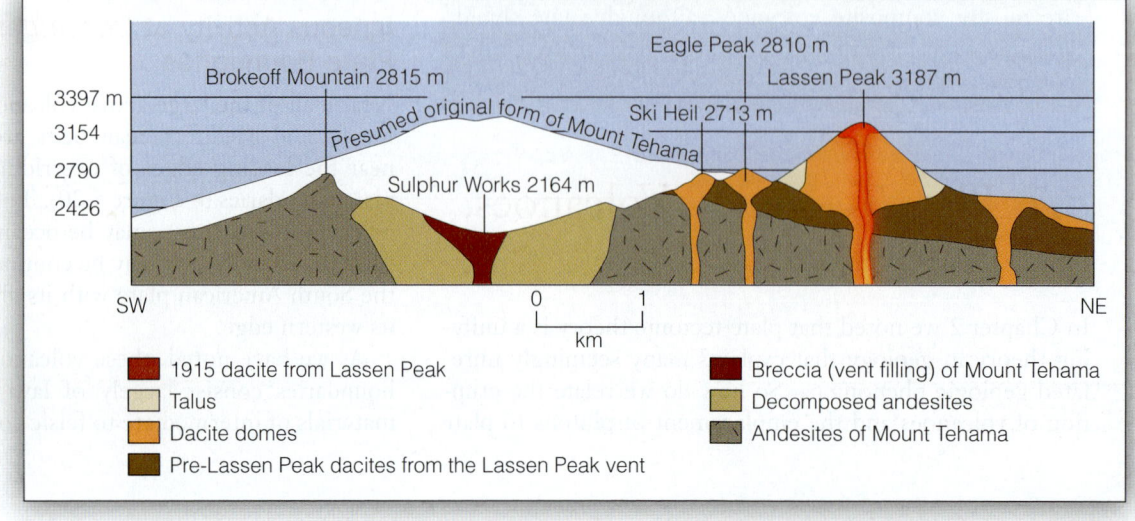

- 1915 dacite from Lassen Peak
- Talus
- Dacite domes
- Pre-Lassen Peak dacites from the Lassen Peak vent
- Breccia (vent filling) of Mount Tehama
- Decomposed andesites
- Andesites of Mount Tehama

◀ **5.** Lassen Peak erupted numerous times from 1914 to 1917. This eruption took place in 1915.

▶ **6.** Mount St. Helens, a composite volcano, as it appeared from the east in 1978.

D. R. Crandel/USGS

Courtesy of Keith Ronnholm

▲ **7.** Mount St. Helens. The lateral blast on May 18, 1980, took place when a bulge on the volcano's north face collapsed, reducing the pressure on gas-charged magma. The lateral blast killed 63 people and leveled 600 km^2 of forest.

◀ **9.** Shortly after the lateral blast, this 19-km-high ash and steam cloud erupted from Mount St. Helens.

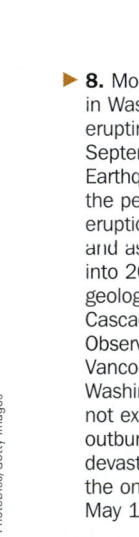

PhotoDisc/Getty Images

▶ **8.** Mount St. Helens in Washington began erupting during late September 2004. Earthquakes beneath the peak as well as eruptions of gases and ash continue into 2006, but geologists at the Cascade Volcano Observatory in Vancouver, Washington, do not expect an outburst as devastating as the one on May 18, 1980.

Steve Schilling/USGS

that when mafic oceanic crust partially melts, some of the magma generated is emplaced near plate boundaries as plutons and some is erupted to build up composite volcanoes. More viscous magmas, usually of felsic composition, are emplaced as lava domes, thus accounting for the explosive eruptions that typically occur at convergent plate boundaries.

In previous sections, we mentioned several eruptions at convergent plate boundaries. Good examples are the explosive eruptions of Mount Pinatubo and Mayon volcano in the Philippines, both of which are situated near a plate boundary beneath which an oceanic plate is subducted. Mount St. Helens in Washington is similarly situated, but it is on a continental rather than an oceanic plate (see "Cascade Range Volcanoes" on pp. 154–155). Mount Vesuvius in Italy, one of several active volcanoes in that region, lies on a plate that the northern margin of the African plate is subducted beneath.

Intraplate Volcanism

■ *How do plate tectonics and volcanism account for the origin of the Hawaiian Islands?*

Mauna Loa and Kilauea on the island of Hawaii and Loihi just 32 km to the south are within the interior of a rigid plate far from any divergent or convergent plate boundary (▶ Figure 5.20). The magma is derived from the upper mantle, as it is at spreading ridges, and accordingly is mafic so it builds up shield volcanoes. Loihi is particularly interesting because it represents an early stage in the origin of a new Hawaiian island. It is a submarine volcano that rises higher than 3000 m above the seafloor, but its summit is still about 940 m below sea level.

Even though the Hawaiian volcanoes are not at a spreading ridge near a subduction zone, their evolution is related to plate movements. Notice in Figure 2.22 that the ages of the rocks that make up the various Hawaiian Islands increase toward the northwest; Kauai formed 3.8 to 5.6 million years ago, whereas Hawaii began forming less than 1 million years ago, and Loihi began to form even more recently. Continuous movement of the Pacific plate over the hot spot, now beneath Hawaii and Loihi, has formed the islands in succession.

Section 5.7 Summary

- Most of the magma that is emplaced as plutons or that rises to the surface as lava at divergent plate boundaries is mafic, even where divergence takes place on land, as in East Africa.

- Igneous activity at convergent plate boundaries involves mostly intermediate and felsic magma and lava. Huge plutons such as batholiths as well as composite volcanoes are common in this geologic setting.

5.8 Volcanic Hazards, Volcano Monitoring, and Forecasting Eruptions

You no doubt suspect that living near an active volcano poses some risk, and of course this is a correct assessment. But what exactly are volcanic hazards, is there any way to anticipate eruptions, and what can we do to minimize the dangers from eruptions? We already mentioned that lava flows pose little danger to humans, but there are exceptions. In 1977 a lava lake in the crater of Nyiragongo volcano in the Democratic Republic of Congo (formerly Zaire) suddenly drained through a fracture and killed 70 people (300 in some reports) and a herd of elephants. More recently, on January 17, 2002, a lava flow from Nyiragongo sliced through the city of Goma, 19 km south, destroying everything in a 60-m-wide path (▶ Figure 5.21a). The lava ignited fires and huge explosions where it came into contact with gasoline storage tanks, killing 147 people. Nevertheless, lava flows are low on the list of direct dangers to humans, although they may destroy croplands, houses, roadways, and other structures.

■ *What are the most dangerous manifestations of volcanoes?*

Of much greater concern than lava flows are nuée ardentes, volcanic gases, lahars, and landslides (▶ Figure 5.21). Indeed, the latter two phenomena may take place even when no eruption has occurred for a long time. And remember that volcanic gases caused many fatalities in Cameroon and Iceland.

The areas most vulnerable to volcanic hazards in the United States are Alaska, Hawaii, California, Oregon, and Washington, but some other parts of the western states might also experience renewed volcanism. Canada's most recent eruption took place in northern British Columbia about 150 years ago, but there is the potential for future eruptions from volcanoes in the northern Cascade Range. The greatest threat may be from Mount Baker in Washington, which lies about 24 km south of the U.S.–Canadian border. It last erupted in 1870.

The Size and Duration of Eruptions

The most widely used indication of the size of a volcanic eruption is the **volcanic explosivity index (VEI)** (▶ Figure 5.22). The VEI has numerical values corresponding to eruptions characterized as gentle, explosive, and cataclysmic. It is based on several aspects of an eruption, particularly the volume of material explosively ejected and the height of the eruption cloud; the volume of lava, fatalities, and property damage are not considered. The 1985 eruption of Nevado del Ruiz in Colombia killed 23,000 people yet had a VEI value of only 3, whereas the huge eruption (VEI = 6) of Novarupta in Alaska in 1912 caused no fatalities or injuries. Since A.D. 1500, only the 1815 eruption of Tambora in Indonesia has had a VEI

▶ **Figure 5.21 Volcanic Hazards** A volcanic hazard is any manifestation of volcanism that poses a threat, including lava flows and, more importantly, volcanic gas, ash, and lahars.

ⓐ This 2002 lava flow in Goma, Democratic Republic of Congo, killed 147 people, mostly by causing gasoline storage tanks to explode.

ⓑ This sign at Mammoth Mountain volcano in California warns of the potential danger of CO_2 gas, which has killed 170 acres of trees.

ⓒ When Mount Pinatubo in the Philippines erupted on June 15, 1991, this huge cloud of ash and steam formed over the volcano.

value of 7; it was both large and deadly (Table 5.1). Geologists have assigned VEI values to nearly 5700 eruptions that took place during the last 10,000 years, but none has exceeded 7 and most (62%) have a value of 2.

The duration of eruptions varies considerably. Fully 42% of about 3300 historic eruptions lasted less than one month. About 33% erupted for one to six months, but some 16 volcanoes have been active more or less continuously for longer than 20 years. Stromboli and Mount Etna in Italy and Erta Ale in Ethiopia are good examples. In some explosive volcanoes, the time from the onset of their eruptions to the climactic event is weeks or months. A case in point is the explosive eruption of Mount St. Helens on May 18, 1980, which occurred two months after eruptive activity began. Unfortunately, many volcanoes give little or no warning of large-scale eruptions; of 252 explosive eruptions, 42% erupted most violently during their first day of activity. As one might imagine, predicting eruptions is complicated by those volcanoes that give so little warning of impending activity.

Forecasting Eruptions

Only a few of Earth's potentially dangerous volcanoes are monitored, including some in Japan, Italy, Russia, New Zealand, and the United States. Many of the methods now used to monitor volcanoes were developed at the Hawaiian Volcano Observatory.

Volcano monitoring involves recording and analyzing physical and chemical changes at volcanoes. Geologists may use various instruments to detect ground deformation—that is, changes in a volcano's slopes as it inflates when magma rises beneath it. Using some instruments requires that geologists actually visit a potentially dangerous volcano, but now

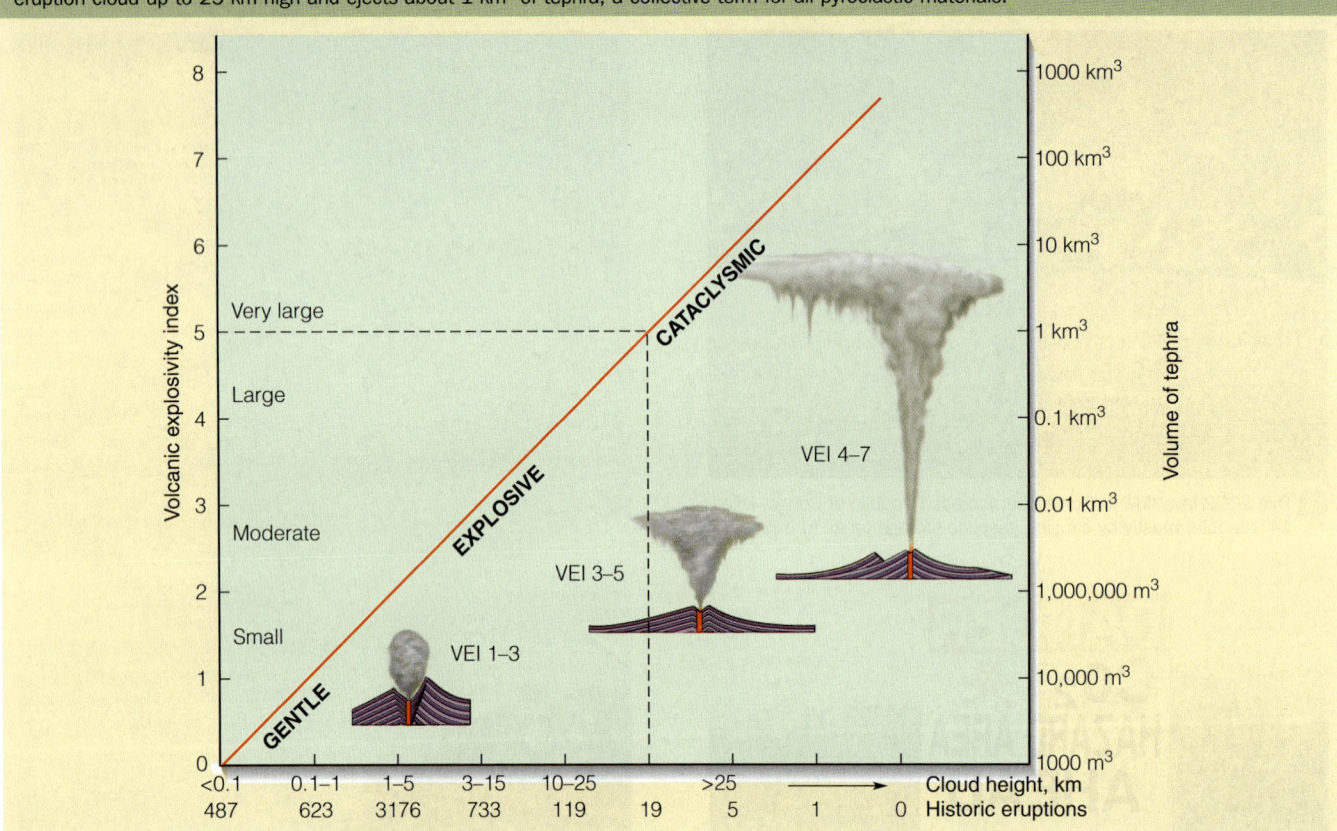

▶ **Figure 5.22 The Volcanic Explosivity Index** The volcanic explosivity index (VEI). In this example, an eruption with a VEI value of 5 has an eruption cloud up to 25 km high and ejects about 1 km³ of tephra, a collective term for all pyroclastic materials.

global positioning system (GPS) technology allows them to monitor a volcano from a safe distance (▶ Figure 5.23). During the renewed activity at Mount St. Helens beginning in September 2004, helicopters placed GPS instruments on the lava dome in the volcano's crater. Geologists also monitor gas emissions, changes in groundwater level and temperature, hot springs activity, and changes in local magnetic and electrical fields. Even the accumulating snow and ice, if any, are evaluated to anticipate hazards from floods should an eruption take place.

Of critical importance in volcano monitoring and warning of an imminent eruption is the detection of **volcanic tremor**, the continuous ground motion lasting for minutes to hours as opposed to the sudden, sharp jolts produced by most earthquakes. Volcanic tremor, also known as *harmonic tremor*, indicates that magma is moving beneath the surface.

Geologists study the record of past eruptions preserved in rocks to better anticipate the future activity of a volcano. Detailed studies before 1980 indicated that Mount St. Helens had erupted explosively 14 or 15 times during the last 4500 years, so geologists concluded that it was one of the most likely Cascade Range volcanoes to erupt again. In fact, the maps they prepared showing areas in which damage from an eruption could be expected were helpful in determining which areas should have restricted access and evacuations once an eruption did take place.

Geologists successfully gave timely warnings of impending eruptions of Mount St. Helens in Washington and Mount Pinatubo in the Philippines, but in both cases the climactic eruptions were preceded by eruptive activity of lesser intensity. In some cases, however, the warning signs are much more subtle and difficult to interpret. Numerous small earthquakes and other warning signs indicated to geologists of the U.S. Geological Survey (USGS) that magma was moving beneath the surface of the Long Valley caldera in eastern California, so in 1987 they issued a low-level warning and then nothing happened.

Volcanic activity in the Long Valley caldera occurred as recently as 250 years ago, and there is every reason to think it will occur again, but when it will take place is an unanswered question. Unfortunately, the local populace was largely unaware of the geologic history of the region, the USGS did a poor job in communicating its concerns, and premature news releases caused more concern than was justified. In any case, local residents where outraged because the warnings caused a decrease in tourism (Mammoth Mountain on the margins of the caldera is the second largest ski area in the country) and property values plummeted. Monitoring continues in the Long Valley caldera, and the signs of renewed volcanism, including earthquake swarms, trees being killed by carbon dioxide gas apparently emanating from magma (▶ Figure 5.21b), and hot spring activity, cannot be ignored.

▶ **Figure 5.23 Volcanic Monitoring** Some important techniques used to monitor volcanoes.

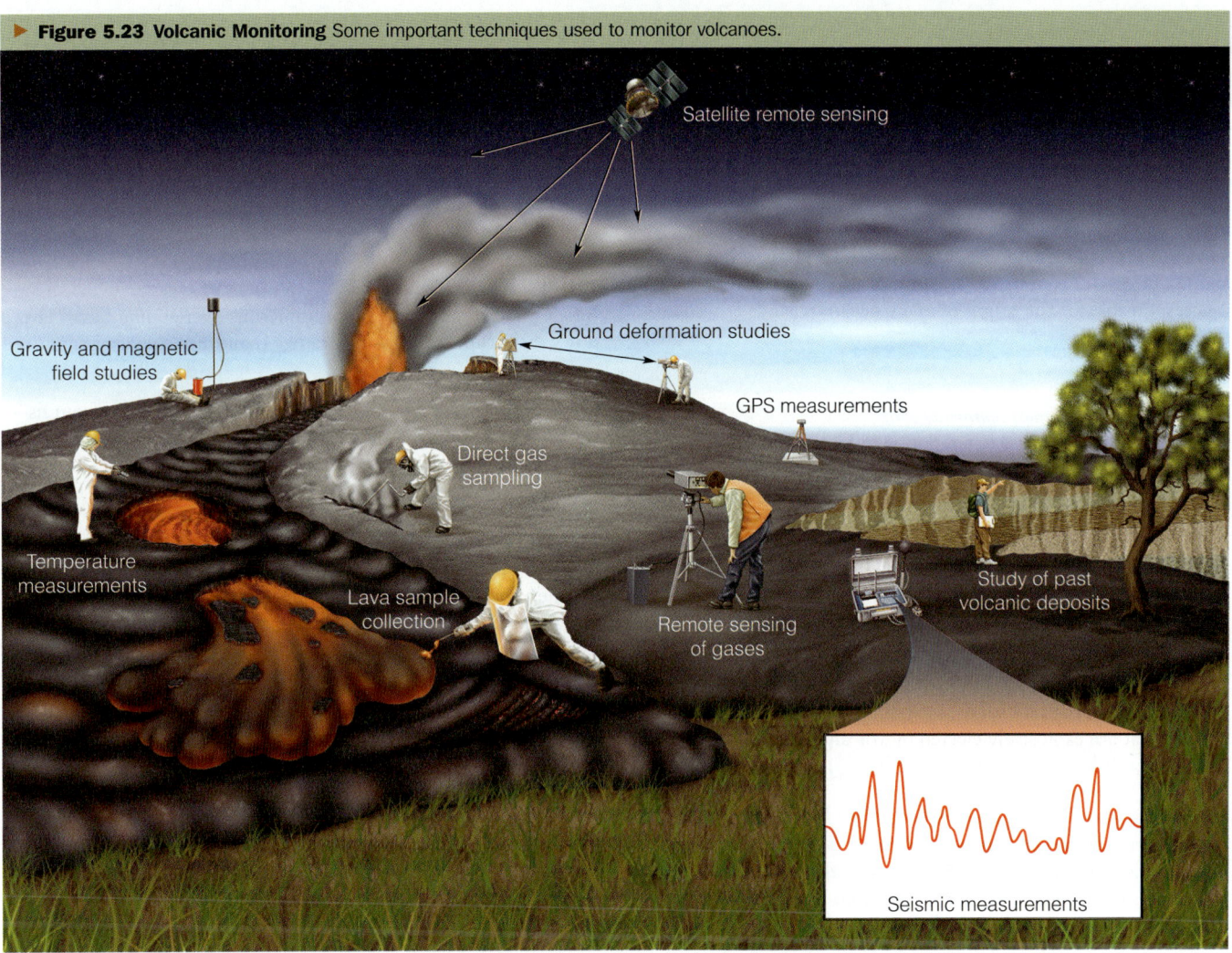

For the better-monitored volcanoes, it is now possible to make accurate short-term predictions of eruptions. But for many volcanoes, little or no information is available.

Section 5.8 Summary

- Geologists have devised a volcanic explosivity index (VEI) to indicate the size of an eruption. VEI values depend on the volume of material erupted and the height of an eruption plume; fatalities and property damage are not considered.

- Although lava flows and lava fountains are impressive, the most dangerous manifestations of volcanoes are eruptions of pyroclastic materials, especially nuée ardentes, as well as mudflows and debris flows, which may take place even when a volcano is not erupting.

- Volcano monitoring involves evaluating physical and chemical aspects of volcanoes. Especially important for anticipating eruptions is detecting volcanic tremor and determining the eruptive history of a volcano.

What Would You Do?

No one doubts that some of the Cascade Range volcanoes will erupt again (actually Mount St. Helens in Washington was erupting when this was written in February 2005). But no one knows when future eruptions will take place or how large they will be. A job transfer takes you to a community in Oregon with several nearby volcanoes, and you have some concerns about future eruptions. What kinds of information about specific threats would you seek out before buying a home in this area? Also, as a concerned citizen, could you make any suggestions about what members of your community should do in case of an eruption?

Review Workbook

ESSENTIAL QUESTIONS SUMMARY

5.1 Introduction
■ *How can volcanism be both constructive and destructive?*
Volcanism may destroy houses and farmland and cause injuries and fatalities so it is destructive, but it is also constructive because it is responsible for the origin of many oceanic islands as well as the oceanic crust.

5.2 Volcanism and Volcanoes
■ *What gases do volcanoes commonly emit?*
Most volcanic gases are water vapor, with lesser amounts of carbon dioxide, nitrogen, sulfur dioxide, and hydrogen sulfide, and very small amounts of carbon monoxide, hydrogen, and chlorine.

■ *How and why do aa and pahoehoe lava flows differ?*
Aa is made up of angular blocks and fragments, whereas pahoehoe has a smooth surface much like taffy. The flows differ mostly because aa is viscous enough to fragment.

■ *What are pyroclastic materials, and how are they dangerous to air traffic?*
Pyroclastic materials are solids, including ash, lapilli, blocks, and bombs, that are explosively ejected by volcanoes. Ash is dangerous to aircraft because it fouls jet engines.

5.3 Types of Volcanoes
■ *What are calderas and how do they form?*
A caldera is a large oval to circular volcanic depression that forms when the summit of a volcano collapses into its magma chamber following voluminous eruptions.

■ *What are cinder cones and what are they composed of?*
Cinder cones are small, steep-sided volcanoes made up of pyroclastic materials that resemble cinders.

■ *What are lava domes and why are they so dangerous?*
Lava domes are bulbous masses of viscous magma that commonly erupt explosively.

5.4 Other Volcanic Landforms
■ *How do basalt plateaus form?*
Basalt plateaus are made up of numerous overlapping basalt lava flows that erupt from fissures rather than from a central vent.

■ *What are pyroclastic sheet deposits?*
Huge eruptions of pyroclastic materials, especially ash, from fissures that form during the origin of calderas are responsible for pyroclastic sheet deposits.

5.5 The Distribution of Volcanoes
■ *Where are the three zones or belts with most of Earth's volcanoes?*
About 60% of all active volcanoes are in the circum-Pacific belt, another 20% are in the Mediterranean belt, and most of the remaining 20% are at or near mid-oceanic ridges or their extensions onto land.

5.6 North America's Active Volcanoes
■ *Where is the Cascade Range and what types of volcanoes are found there?*
The Cascade Range stretches from Lassen Peak in California north through Oregon and Washington to Meager Mountain in British Columbia, Canada. Most of the large volcanoes in the range are composite volcanoes, but there are also two huge shield volcanoes and numerous cinder cones.

5.7 Plate Tectonics, Volcanoes, and Plutons
■ *What kinds of igneous rocks make up the oceanic crust?*
The oceanic crust is made up of mafic igneous rocks. Gabbro is found in the lower part of the oceanic crust, whereas vertical dikes and pillow lava, both composed of basalt, make up the upper part.

■ *How do plate tectonics and volcanism account for the origin of the Hawaiian Islands?*

As the Pacific plate moved over a hot spot, a chain of volcanoes formed in succession, so the oldest in the chain is far to the northwest and active volcanism now occurs only on the island of Hawaii and at Loihi.

5.8 Volcanic Hazards, Volcano Monitoring, and Forecasting Eruptions

■ *What are the most dangerous manifestations of volcanoes?*

Lava dome eruptions during which huge amounts of pyroclastic materials and gases are ejected are the most dangerous volcanic eruptions. Lahars are also dangerous and they may take place long after an eruption.

ESSENTIAL TERMS TO KNOW

aa (p. 137)
ash (p. 139)
basalt plateau (p. 149)
caldera (p. 141)
Cascade Range (p. 152)
cinder cone (p. 141)
circum-Pacific belt (p. 151)
columnar joint (p. 138)
composite volcano (stratovolcano) (p. 144)

crater (p. 141)
fissure eruption (p. 149)
lahar (p. 144)
lava dome (p. 144)
lava tube (p. 137)
Mediterranean belt (p. 151)
nuée ardente (p. 145)
pahoehoe (p. 137)
pillow lava (p. 139)
pyroclastic materials (p. 139)

pyroclastic sheet deposits (p. 150)
shield volcano (p. 141)
volcanic explosivity index (VEI) (p. 156)
volcanic tremor (p. 158)
volcanism (p. 136)
volcano (p. 141)

REVIEW QUESTIONS

1. Suppose you find rocks on land that consist of layers of pillow lava overlain by deep-sea sedimentary rocks. Where and how did the pillow lava form, and what type of rock would you expect to find beneath the pillow lava?

2. How do columnar joints and lava tubes form? Where are good places to see each?

3. What geologic events would have to take place for a chain of composite volcanoes to form along the east coasts of the United States and Canada?

4. Explain why eruptions of mafic lava are nonexplosive but eruptions of felsic magma are commonly explosive.

5. What does volcanic tremor indicate, and how does it differ from the shaking caused by most earthquakes?

6. Describe a nuée ardente and explain why they are so dangerous.

7. Why do shield volcanoes have such gentle slopes whereas cinder cones have very steep slopes?

APPLY YOUR KNOWLEDGE

1. Considering what you know about the origin of magma and about igneous activity, why is the magma (lava) at divergent plate boundaries mostly mafic, whereas at convergent plate boundaries it is mostly intermediate and felsic?

2. What criteria are used to assign a volcanic explosivity index (VEI) value to an eruption? Why do you think the number of fatalities and property damage are not considered when assigning a value?

FIELD QUESTIONS

1. The Mona Schist at Marquette, Michigan (see the image), is basalt that has been changed slightly by metamorphism. How do you think the elliptical features in this ancient lava formed?

Geo-focus Figure

2. The accompanying image shows part of the Giant's Causeway in Northern Ireland. How did these vertical columns form?

3. Identify the types of volcanoes shown in the diagram. Which of these do you think would erupt mostly violently? Why?

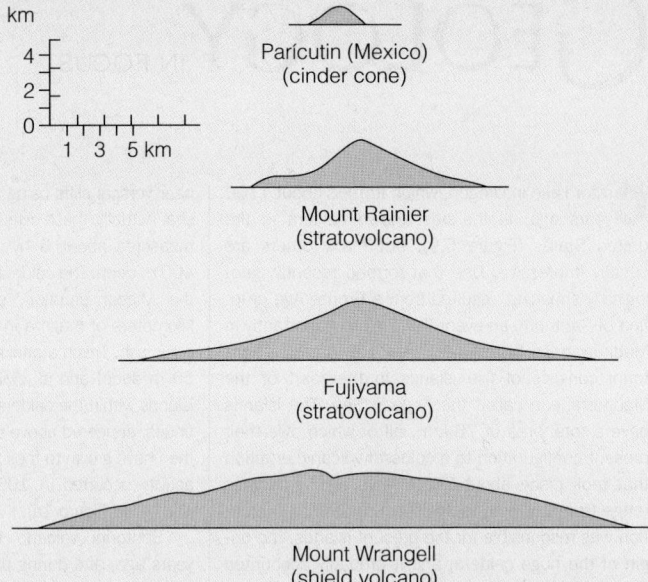

4. The accompanying image shows a basalt lava flow (black) overlying a layer of pyroclastic materials (light colored) in Colorado. Why is the upper part of the pyroclastic layer discolored?

Geo-focus Figure

GEOLOGY MATTERS

GEOLOGY IN FOCUS   The Bronze Age Eruption of Santorini

Crater Lake in Oregon, which formed about 7700 years ago, is the best-known caldera in the United States (Figure 5.9), but many others are equally impressive. One that formed recently, geologically speaking, resulted from a Bronze Age eruption of Santorini, an event that figured importantly in Mediterranean history (▶ Figure 1a). Actually, Santorini consists of five islands in that part of the Mediterranean called the Aegean Sea. The islands have a total area of 76 km^2, all of which owe their present configuration to a colossal volcanic eruption that took place about 3600 years ago (estimates range from 1596 BC to 1650 BC). Indeed, the eruption was responsible for the present islands, the origin of the huge caldera, and it probably accounted for or at least contributed to the demise of the Minoan culture on Crete. Furthermore, some authorities think that the disappearance of much of the original island during this eruption was the basis for Plato's story about Atlantis (see Chapter 12).

As you approach Santorini from the sea, the first impression is snow-covered cliffs in the distance. On closer inspection, though, the "snow" is actually closely spaced white buildings that cover much of the higher parts of the largest island (▶ Figure 1b). Perhaps the most impressive features of Santorini are the near vertical cliffs rising as much as 350 m from the sea. Actually, these cliffs are the walls of a caldera that measures about 6 by 12 km and is as much as 400 m deep. The caldera-forming eruption, "known as the 'Minoan eruption,' ejected into the air 30 cubic kilometers of magma in the form of pumice and volcanic ash. This material buried the island [as much as 50 m deep] and its civilization. . . ."[1] The two small islands within the caldera, where volcanic activity continues, appeared above sea level in 197 BC, and since then have grown to their present size. The most recent activity occurred in 1950 on the larger of the two islands (▶ Figure 1a).

Santorini volcano began forming two million years ago, and during the last 400,000 years it has erupted at least 100 times, each eruption adding new layers to the island, making it larger. Today, about 8000 people live on the islands, and we know from archaeological evidence that several tens of thousand of people resided there before the Minoan eruption when Santorini was larger. However, a year or so before the catastrophic eruption, a devastating earthquake occurred and many people left the island then; perhaps there were signs of an impending eruption by this time.

The fact that the island's residents escaped is indicated by the lack of human and animal skeletons in the ruins of the civilization, the only exception being one pig skeleton. In fact, archaeological excavations, many still in progress, show that the people had time to collect their valuables and tools before evacuating the island. Their destination, however, remains a mystery.

[1]Vougioukalakis, G., *Santorini: The Volcano* (Institute for the Study and Monitoring of the Santorini Volcano, 1995), p. 7.

Figure 1
The Minoan eruption of Santorini that took place between 1596 and 1650 B.C. formed a large caldera. It may have contributed to the demise of the Minoan culture on Crete and may be the basis for Plato's account of the sinking of Atlantis. **(a)** Map showing Santorini and nearby areas in the Aegean Sea, which is part of the northwestern Mediterranean Sea. **(b)** These 350-m-high cliffs are part of the caldera wall just west of Fira.

(a)

(b)

GEOLOGY MATTERS

Geology in Unexpected Places

Oldoinyo Lengai Volcano

The East African Rift is not an unexpected place for volcanoes, but the one known as Oldoinyo Lengai (or Ol Doinyo Lengai) in Tanzania is certainly among Earth's most peculiar volcanoes. Oldoinyo Lengai, which means "Mountain of God" in the Masai language, is an active composite volcano standing about 2890 m high (▶ Figure 1a). It is one of 20 or so volcanoes in an east-west belt near the southern part of the East African Rift, a divergent plate boundary (see Figure 2.16a). Oldoinyo Lengai is peculiar because it is the world's only volcano that erupts carbonatite lava rather than lava with abundant silica. *Carbonatite* refers to an igneous rock with at least 50% carbonatite minerals, that is, minerals with the $(CO_3)^{-2}$ radical. In fact, carbonatite looks much like the metamorphic rock marble (see Chapter 8).

Why is the carbonatite lava at Oldoinyo Lengai unique? Remember that based on silica content most magma ranges from mafic to felsic, or even ultramafic; only rarely does magma have the elements necessary to form carbonatite-rich rocks. Carbonatite is found at about 330 localities worldwide, but most of it is in small plutons and thus intrusive. So the carbonatite lava that erupts at Oldoinyo Lengai makes it unusual because it is extrusive. Another feature of the Oldoinyo Lengai carbonatite is that it contains significant amounts of sodium and potassium, whereas most carbonatites are made up of calcite ($CaCO_3$) and dolomite [$CaMg(CO_3)_2$].

The carbonatite lava at Oldoinyo Lengai has some surprising characteristics. For one thing it is fluid at temperatures of only 540°C and 595°C, and because it has so little silica it has very low viscosity and flows quickly. Indeed, the temperature is so low that the lava is not incandescent, so we do not see the red glow we expect in lava flows. In fact, it looks more like black mud. And because the minerals in carbonatite are unstable, they react with water in the atmosphere and their color quickly changes to pale gray (▶ Figure 1b).

Until the 1960s, there was disagreement among geologists about the origin of carbonatite. Some thought that the carbonatite plutons noted above were made up of marble that was altered from limestone by heat and as a result looked like igneous rock. Observations at Oldoinyo Lengai, however, provided compelling evidence that carbonatite is found in lava flows as well as in plutons.

We now know that carbonatite is found in small plutons and erupts as lava, but geologists continue to debate the origin of carbonatite magma. Some think that it forms by partial melting of carbonatite rocks in Earth's crust, but others think it is not a primary magma at all but rather forms by alteration of sodium-rich carbonatite from which hydrothermal fluids removed the sodium.

Although the rocks at Oldoinyo Lengai have no economic importance, carbonatites elsewhere are mined because they have higher percentages of rare earth elements[1] that any other type of igneous rock. They are also sources of niobium, phosphates, and fluorite.

Rare earth elements are 17 chemical elements that have similar atomic structures, atomic radii, and 3+ or 4+ electrical charges.

Figure 1
Of the hundreds of volcanoes in the world, Oldoinyo Lengai in Tanzania is one of the most unusual because it erupts lava that cools to form igneous rock rich in carbonate minerals. **(a)** Lava spattering from a small cone in the crater of Oldoinyo Lengai in 1990. **(b)** In 1994 this black lava from Oldoinyo Lengai flowed over a lava flow only a few months old that had already turned gray.

(a) (b)

GEOLOGY MATTERS

GEOLOGY IN YOUR LIFE

Do Volcanic Gases Cause Ozone Depletion?

Earth supports life because of its distance from the Sun, and the fact that it has abundant liquid water and an oxygen-rich atmosphere. An ozone layer (O_3) in the stratosphere (10 to 48 km above the surface) protects Earth because it blocks out most of the harmful ultraviolet radiation that bombards our planet. During the early 1980s, scientists discovered an *ozone hole* over Antarctica that has continued to grow. In fact, depletion of the ozone layer is now also recognized over the Arctic region and elsewhere. Any depletion in ozone levels is viewed with alarm because it would allow more dangerous radiation to reach the surface, increasing the risk of skin cancer, among other effects.

This discovery unleashed a public debate about the primary cause of ozone depletion, and how best to combat the problem. Scientists proposed that one cause of ozone depletion is chlorofluorocarbons (CFCs), which are used in various consumer products; for instance, in aerosol cans. According to this theory, CFCs rise into the upper atmosphere where reactions with ultraviolet radiation liberate chlorine, which in turn reacts with and depletes ozone (▶ Figure 1). As a result of this view, an international agreement called the Montreal Protocol was reached in 1983, limiting the production of CFCs, along with other ozone-depleting substances.

However, during the 1990s this view was challenged by some radio talk show hosts as well as a few government officials. They proposed an alternative idea that ozone depletion was because of natural causes rather than commercial products such as CFCs. They pointed out that volcanoes release copious quantities of hydrogen chloride (HCl) gas that rises into the stratosphere and which could be responsible for ozone depletion. Furthermore, they claimed that because CFCs are heavier than air they would not rise into the stratosphere.

It is true that volcanoes release HCl gas as well as several other gases, some of which are quite dangerous—recall the cloud of CO_2 gas in Cameroon that killed 1746 people and the Blue Haze Famine in Iceland. However, most eruptions are too weak to inject gases of any kind high into the stratosphere. Even when it is released, HCl gas from volcanoes is very soluble and quickly removed from the atmosphere by rain and even by steam (water vapor) from the same eruption that released HCl gas in the first place. Measurements of chlorine concentrations in the stratosphere show that only temporary increases occur following huge eruptions. For example, the largest volcanic outburst since 1912, the eruption of Mount Pinatubo in 1991, caused little increase in upper atmosphere chlorine. The impact of volcanic eruptions is certainly not enough to cause the average rate of ozone depletion taking place each year.

Although it is true that CFCs are heavier than air, this does not mean that they cannot rise into the stratosphere. Earth's surface heats differentially, meaning that more heat may be absorbed in one area than in an adjacent one. The heated air above a warmer area becomes less dense, rises by convection, and carries with it CFCs and other substances that are actually denser than air. Once in the stratosphere, ultraviolet radiation, which is usually absorbed by ozone, breaks up CFC molecules and releases chlorine that reacts with ozone. Indeed, a single chlorine atom can destroy 100,000 ozone molecules (▶ Figure 1). In contrast to the HCl gas produced by volcanoes, CFCs are absolutely insoluble; it is the fact that they are inert that made them so desirable for various uses. Because a CFC molecule can last for decades, any increase in CFCs is a long-term threat to the ozone layer.

Another indication that CFCs are responsible for the Antarctic ozone hole is that the rate of ozone depletion has slowed since the implementation of the Montreal Protocol. A sound understanding of the science behind these atmospheric processes helped world leaders act quickly to address this issue. Now scientists hope that with continued compliance with the Protocol, the ozone layer will recover by the middle of this century.

Figure 1
Ozone is destroyed by chlorofluorocarbons (CFCs). Chlorine atoms are continuously regenerated, so one chlorine atom can destroy many ozone molecules.

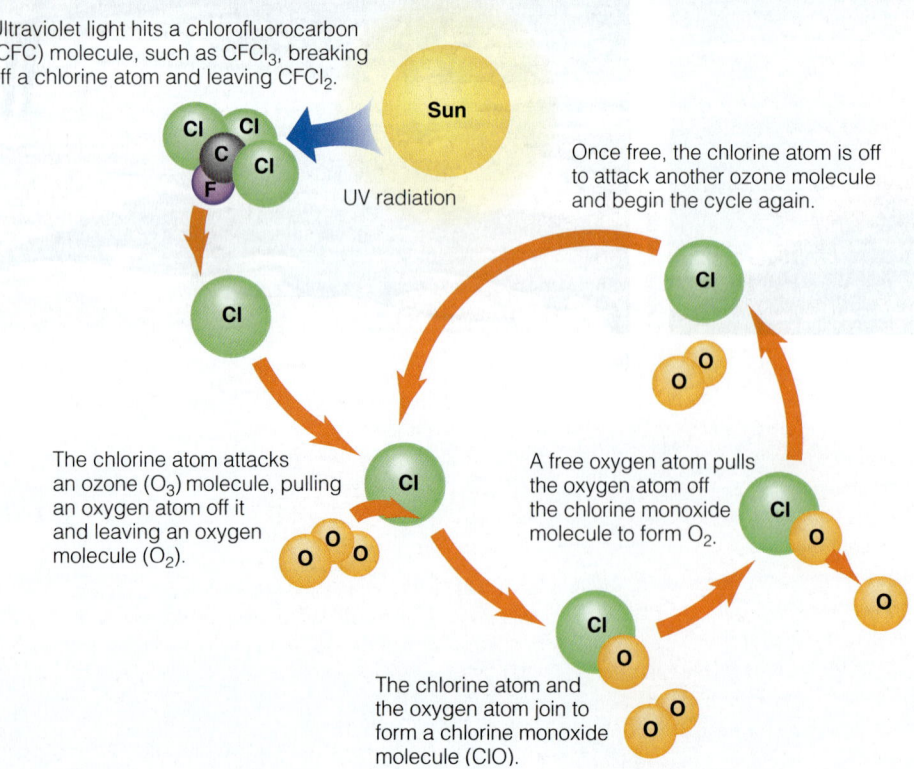

Ultraviolet light hits a chlorofluorocarbon (CFC) molecule, such as $CFCl_3$, breaking off a chlorine atom and leaving $CFCl_2$.

Once free, the chlorine atom is off to attack another ozone molecule and begin the cycle again.

The chlorine atom attacks an ozone (O_3) molecule, pulling an oxygen atom off it and leaving an oxygen molecule (O_2).

A free oxygen atom pulls the oxygen atom off the chlorine monoxide molecule to form O_2.

The chlorine atom and the oxygen atom join to form a chlorine monoxide molecule (ClO).

GEOLOGY MATTERS

Geology AT WORK

Volcano Observatories and Volcano Monitoring

The United States Geological Survey maintains several volcano observatories around the country, each with the goal of monitoring local volcanic activity to add to our scientific understanding of volcanoes as well as to ensure public safety. Geologists at the Hawaii Volcano Observatory developed many of the methods now used to monitor volcanoes elsewhere, although other observatories have also contributed. The Alaska Volcano Observatory in Fairbanks monitors active volcanoes in that state, and issues warnings of imminent eruptions, particularly to aircraft. The David A. Johnston Cascades Volcano Observatory in Vancouver, Washington, was established following the 1980 eruption of Mount St. Helens and named for a geologist who was killed in the eruption. And the Long Valley Observatory at Menlo Park, California, monitors activity at Long Valley Caldera in eastern California. Personnel at the US Geological Survey operate the Yellowstone Volcano Observatory in cooperation with the University of Utah and the National Park Service. Their function is to monitor and assess possible hazards, and to conduct research at Yellowstone National Park in Wyoming, Montana, and Idaho.

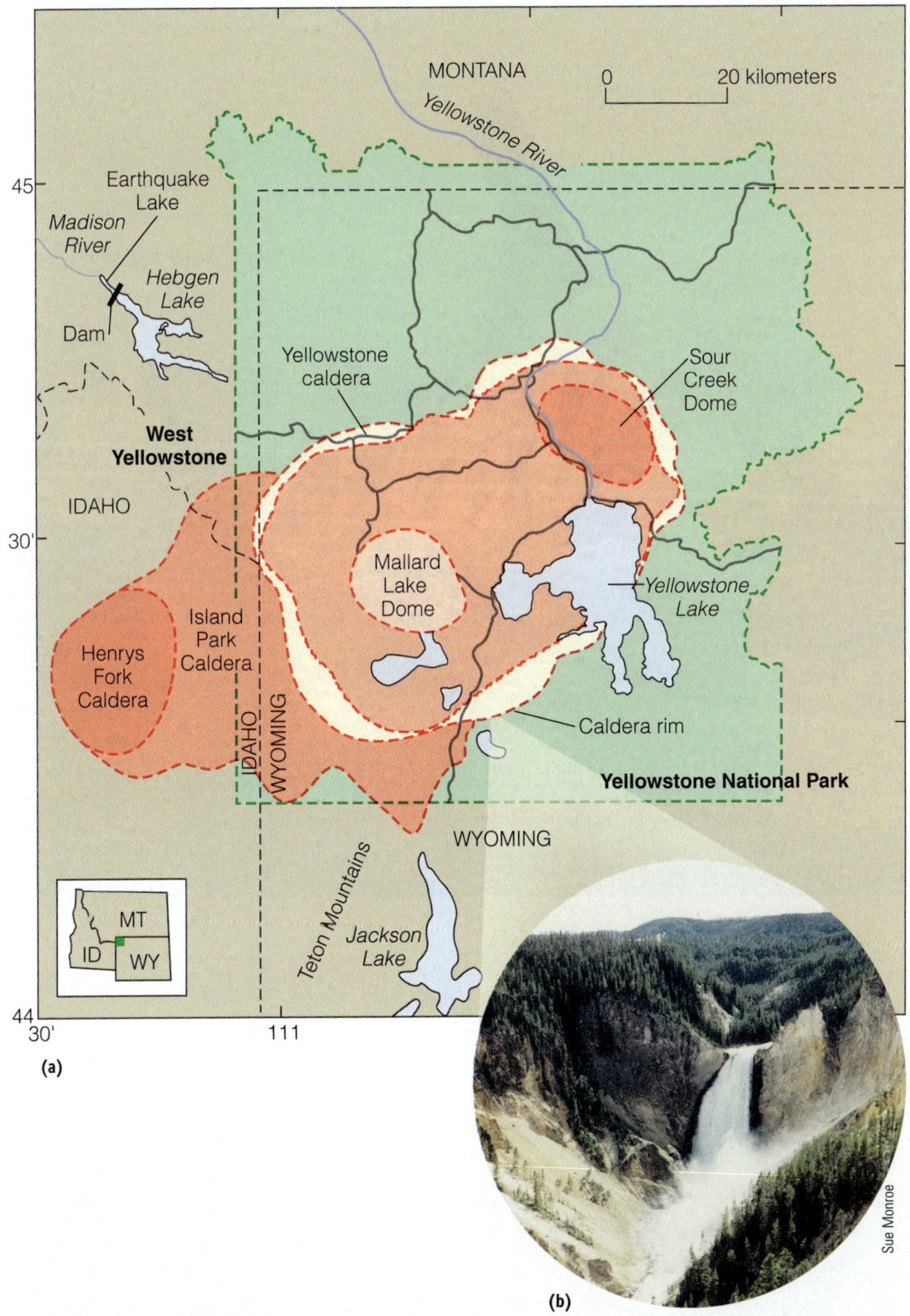

The Yellowstone region has had a history of huge volcanic eruptions, although the most recent one was about 70,000 years ago. Why monitor an area so closely given that no volcanic outbursts have taken place for thousands of years? Continuing earthquake activity, 80 cm of uplift of the ground surface since 1923, and the hydrothermal activity (see Chapter 16) for which the park is famous, indicate that future eruptions are possible. And judging from the geologic record, renewed eruptions may be catastrophic.

Geologist have no formal definition for *supervolcano*, but we can take it to mean a volcano that erupts explosively, ejecting hundreds of cubic kilometers of pyroclastic materials followed by the origin of a vast caldera. No supervolcano eruption has occurred in recorded history, but geologists know of several that occurred during the past two million years—Long Valley in eastern California, Toba in Indonesia, and Taupo in New Zealand, for example. The origin of the Yellowstone caldera in Yellowstone National Park in Wyoming is a good example of a caldera formed by a supervolcano eruption.

On three separate occasions, supervolcano eruptions followed the accumulation of rhyolitic magma beneath the Yellowstone region, each yielding a widespread blanket of volcanic ash and pumice. In each instance, collapse of the magma chamber led to the formation of a huge caldera (▶ Figure 1a). We can summarize Yellowstone's volcanic history by noting that supervolcano eruptions took place 2 million years ago when 2500 km³ of pyroclastic materials were ejected, 1.3 million years ago (280 km³), and 600,000 years ago (1000 km³). It was during this last eruption that the present-day Yellowstone caldera formed, which is actually part of a larger composite caldera that resulted from the three cataclysmic eruptions. Between 150,000 and 75,000 years ago, an additional 1000 km³ of pyroclastic materials were erupted within the caldera (▶ Figure 1b).

What caused such voluminous eruptions in the Yellowstone area? Many geologists are convinced that a mantle plume, a cylindrical mass of magma rising from the mantle, underlies the area. As this rising mass of magma nears the surface, it triggers volcanic eruptions, and because the magma is rhyolitic, and thus viscous, the eruptions are particularly explosive. Thanks to the personnel at Yellowstone Volcano Observatory, who continue to collect data and assess the possibility of renewed eruptions, we should have ample warning before another supervolcano eruption.

Figure 1
(a) The Yellowstone caldera (shown in yellow) has been partly filled with younger volcanic rocks. (b) The walls of the Grand Canyon of the Yellowstone River are made up of the hydrothermally altered Yellowstone Tuff that partly fills the Yellowstone caldera.

Chapter 6

Weathering, Erosion, and Soil

Arches National Park, Utah
A window in the rock delicately carved by Earth's sculptors: Wind, Water, and Time. Look out and see a turret arch still standing against the horizon like the ancient ruins of a castle's watchtower.

—A. W.

ESSENTIAL QUESTIONS TO ASK

6.1 Introduction
- How does weathering differ from erosion?

6.2 Alteration of Minerals and Rocks
- What accounts for differential weathering and erosion, and what are their effects?

6.3 Mechanical Weathering—Disaggregation of Earth Materials
- How do freezing and thawing contribute to weathering?
- What are sheet joints and how do they form?
- How do organisms contribute to mechanical and chemical weathering?

6.4 Chemical Weathering—Decomposition of Earth Materials
- What happens to rocks composed of carbonate minerals when they are exposed to acidic solutions?
- What takes place when potassium feldspars undergo hydrolysis?
- How does mechanical weathering contribute to the rate of chemical weathering?

6.5 Soil and Its Origin
- What factors determine the rate of soil formation?
- What is laterite, and where and how does it form?

6.6 Expansive Soils and Soil Degradation
- Why do some soils expand and contract, and does expansion and contraction cause any problems?
- What is soil degradation, and how does it take place?

6.7 Weathering and Natural Resources
- How does the ore of aluminum form?

GEOLOGY MATTERS

GEOLOGY IN FOCUS:
The Dust Bowl

GEOLOGY IN UNEXPECTED PLACES:
Gravestones and Geology

GEOLOGY IN YOUR LIFE:
Industrialization and Acid Rain

GEOLOGY AND CULTURAL CONNECTIONS:
The Agricultural Revolution

This icon, which appears throughout the book, indicates an opportunity to explore interactive tutorials, animations, or practice problems available on the Physical GeologyNow website at **http://now.brookscole.com/phygeo6**.

6.1 Introduction

■ *How does weathering differ from erosion?*

Weathering includes several physical and chemical processes that alter surface and near-surface rocks and minerals, whereas **erosion** involves the removal of weathered materials, by running water, for example, from the area where weathering takes place. Actually, weathering alters rocks and minerals so that they are more nearly in equilibrium with a new set of environmental conditions. For instance, many igneous and metamorphic rocks form within Earth's crust where pressure and temperature are high and little or no water or oxygen is present. These same rocks at the surface, though, are exposed to low pressure and temperature, water, the atmosphere, and the activities of organisms. Thus, interactions of Earth materials with the hydrosphere, atmosphere, and biosphere bring about changes as they break down physically (*disintegrate*) and change chemically (*decompose*).

Weathering is such a pervasive phenomenon that it is easy to overlook. Nevertheless, it takes place continuously at variable rates on all surface and near-surface rocks and minerals (▶ Figure 6.1), including rocklike substances used in construction. Roadways and runways, bricks, and concrete in sidewalks, foundations, and bridges change with time as they are relentlessly attacked by the elements. Obviously, weathering of sidewalks, foundations, and other structures requires costly repairs or replacement.

From the academic point of view, weathering is an important group of processes that once again illustrates the interactions among Earth's systems. But there are other good reasons to study weathering. One reason is so that we can develop construction materials that are more resistant to physical and chemical changes, or develop more effective methods to protect them from the elements. Another reason is that weathering is an essential part of the *rock cycle* (see Figure 1.13). When **parent material**—that is, rocks and minerals exposed to weathering—breaks down into smaller pieces or perhaps dissolves, this weathered material may be eroded from the weathering site and transported elsewhere, by running water or wind, and deposited as *sediment*, the raw materials for sedimentary rocks (see Chapter 7).

In addition to providing the raw materials for sedimentary rocks, weathering is responsible for the origin of *soils*. Needless to say, we depend on soils directly or indirectly for our existence. In some areas, however, erosion takes place faster than soil-forming processes operate, thereby decreasing the amount of productive soil. Of course, erosion is an ongoing natural process, so some soil losses are expected as a normal consequence of evolution of the land. Unfortunately, some human activities have compounded the problem, leading to less fertile soil and reduced agricultural production.

Another reason to study weathering is that it accounts for the origin or concentration of some natural resources, the ore of aluminum, for instance. Weathering is also responsible for many clay deposits that are used in ceramics and the manufacture of paper, and weathering coupled with erosion and deposition yields deposits of tin, gold, and diamonds. Finally, weathering and erosion are important topics in several of the chapters on surface processes such as running water, glaciers, wind, and shoreline processes.

▶ Figure 6.1 **Weathering of Granite**

ⓐ This exposure of granite has been so thoroughly weathered that only a few spherical masses of the original rock are visible.

ⓑ Closeup view of the weathered material. Mechanical weathering has predominated, so the particles are mostly small pieces of granite and minerals such as quartz and feldspars.

Section 6.1 Summary

- Weathering is a pervasive phenomenon that alters rocks and minerals so that they are more nearly in equilibrium with a new set of environmental conditions.

- Erosion removes weathered materials which are transported elsewhere and deposited as sediment that may be converted to sedimentary rock.

6.2 Alteration of Minerals and Rocks

Purely physical processes bring about *mechanical weathering*, whereas chemical alteration is responsible for *chemical weathering*. Both types of weathering proceed simultaneously on all surface and near-surface minerals and rocks at their source areas as well as the same materials during transport and deposition elsewhere. Depending on such variables as climate and the types of minerals or rocks, one kind of weathering may predominate, but both take place constantly. We discuss mechanical weathering and chemical weathering separately in the following sections only for convenience.

■ *What accounts for differential weathering and erosion, and what are their effects?*

Although weathering is a near-surface phenomenon, the rocks it acts on are not structurally and compositionally homogeneous throughout, which accounts for **differential weathering.** That is, weathering takes place at different rates even on the same body of rock, which commonly yields uneven surfaces. The intensity of weathering is not likely to vary over a few meters' distance, but rock properties might. For example, weathering and erosion act on fractured rather than nonfractured parts of the same rock body. Also, some minerals in rocks are more susceptible to alteration than others. Differential weathering coupled with *differential erosion*—that is, erosion at variable rates—yields some unusual and even bizarre features, such as hoodoos, spires, arches, and pedestals (▶ Figure 6.2). Indeed, these processes have produced a remarkable landscape at Arches National Park in Utah (see "Arches National Park" on pp. 176–177).

Section 6.2 Summary

- Earth materials are not structurally and compositionally homogeneous, so they are attacked by weathering processes at different rates, even in the same area.

6.3 Mechanical Weathering—Disaggregation of Earth Materials

What do we mean by mechanical weathering? Simply put, **mechanical weathering** takes place when physical forces break minerals and rocks into smaller pieces that retain the composition of the parent material. For example, mechanical weathering of granite yields small pieces of granite (rock fragments) plus individual minerals such as quartz, potassium feldspars, plagioclase feldspars, and other minerals in lesser quantities (▶ Figure 6.1). The physical processes responsible for mechanical weathering include frost action, pressure release, thermal expansion and contraction, salt crystal growth, and the activities of organisms.

Frost Action

■ *How do freezing and thawing contribute to weathering?*

Frost action involving repeated freezing and thawing of water in cracks and pores in rocks is a particularly effective physical weathering process. When water freezes, it expands by about 9% and exerts great force on the walls of cracks thereby widening and extending them by

▶ **Figure 6.2 Differential Weathering** In this example of differential weathering, the rock layers have been tilted so that the planes separating adjacent layers are vertical. Weathering and erosion along these surfaces have yielded the pillars, spires, and isolated knobs at this rock exposure in Montana.

frost wedging. As a result of repeated freezing and thawing, pieces of rock eventually detach from the parent material (▶ Figure 6.3a). Freezing and thawing are most effective in areas where temperatures commonly fluctuate above and below freezing, as in the high mountains of the western United States and Canada. It is of little or no importance in the tropics or where water remains permanently frozen.

The debris produced by frost wedging and other weathering processes in mountains commonly accumulates as large cones of **talus** lying at the bases of slopes (▶ Figure 6.3b). The debris forming talus is simply angular pieces of rock from a larger body that has been mechanically and to a lesser degree chemically weathered. Most rocks have a system of fractures called *joints* along which frost action is particularly effective. Water seeps along the joint surfaces and eventually wedges pieces of rock loose, which then tumble downslope to accumulate with other loosened rocks.

In the phenomenon known as *frost heaving*, a mass of sediment or soil undergoes freezing, expansion, and actual lifting, followed by thawing, contraction, and lowering of the mass. Frost heaving is particularly evident where water freezes beneath roadways and sidewalks.

Pressure Release

■ *What are sheet joints and how do they form?*

Some rocks form at depth and are stable under tremendous pressure. Granite, for instance, crystallizes far below the surface, so when it is uplifted and the overlying material is eroded, its contained energy is released by outward expansion, a phenomenon called **pressure release.** Outward expansion results in the origin of fractures called **sheet joints** that more or less parallel the exposed rock surface. Sheet-joint–bounded slabs of rock slip or slide off the parent rock—a process known as **exfoliation**—leaving large, rounded masses of rock called **exfoliation domes.** Excellent examples of exfoliation domes are found in many areas—at Yosemite National Park in California and at Stone Mountain in Georgia, for example (▶ Figure 6.4).

That solid rock expands and produces fractures is counterintuitive but is nevertheless a well-known phenomenon. In deep mines, masses of rock suddenly detach from the sides of the excavation, often with explosive violence. Spectacular examples of these *rock bursts* take place in deep

▶ **Figure 6.3 Frost Wedging**

a Frost wedging takes place when water seeps into cracks, expands as it freezes, and pries loose angular pieces of rock.

b The parent material at the top of this slope is fractured and susceptible to frost wedging. Notice the angular pieces of rock that have accumulated at the base of the slope.

c These huge cones of talus are in the Rocky Mountains in Canada.

mines, where they and the related but less violent phenomenon called *popping* pose a danger to mine workers. In South Africa about 20 miners are killed by rock bursts every year.

In some quarrying operations,* the removal of surface materials to a depth of only 7 or 8 m has led to the formation of sheet joints in the underlying rock (▶ Figure 6.5). At quarries in Vermont and Tennessee, the excavation of marble exposed rocks that were formerly buried and under great pressure. When the overlying rock was removed, the marble expanded and sheet joints formed. Some slabs of rock bounded by sheet joints burst so violently that quarrying machines weighing more than a ton were thrown from their tracks, and some quarries had to be abandoned because fracturing rendered the stone useless.

Thermal Expansion and Contraction

During **thermal expansion and contraction,** the volume of rocks changes in response to heating and cooling. In a desert, where the temperature may vary as much as 30°C in one day, rocks expand when heated and contract as they cool. Rock is a poor conductor of heat, so its outside heats up more than its inside; the surface expands more than the interior, producing stresses that may cause fracturing. Furthermore, dark minerals absorb heat faster than light-colored ones, so differential expansion occurs even between the minerals in some rocks.

Repeated thermal expansion and contraction is a common phenomenon, but are the forces generated sufficient to overcome the internal strength of a rock? Experiments in which rocks were heated and cooled many times to simulate years of such activity indicate that thermal expansion and contraction are not an important agent of mechanical weathering.† Despite these experimental results, some rocks in deserts do indeed appear to show the effects of this process.

Daily temperature variation is the most common cause of alternating expansion and contraction, but these changes take place over periods of hours. In contrast, fire causes very rapid expansion. During a forest fire, rocks may heat very rapidly, especially near the surface, because they conduct heat so poorly. The heated surface layer expands more rapidly than the interior, and thin sheets paralleling the rock surface become detached.

Growth of Salt Crystals

Under some circumstances, salt crystals that form from solution cause disaggregation of rocks. Growing crystals exert enough force to widen cracks and crevices or dislodge particles in porous, granular rocks such as sandstone. Even in crystalline rocks such as granite, **salt crystal growth** may pry loose individual minerals. To the extent that salt crystal growth produces forces that expand openings in rocks, it is similar to frost wedging. Most salt crystal growth occurs in hot, arid areas, although it probably affects rocks in some coastal regions as well.

▶ **Figure 6.4 Exfoliation Domes**

(a) Slabs of granitic rock bounded by sheet joints in the Sierra Nevada of California. The slabs are inclined downward toward the roadway visible at the lower left.

(b) Stone Mountain in Georgia is a large exfoliation dome.

---

*A *quarry* is a surface excavation, generally for the extraction of building stone.
†Thermal expansion and contraction may be more significant on the Moon, where extreme temperature changes occur quickly.

▶ **Figure 6.5 Expansion** Sheet joint formed by expansion in the Mount Airy Granite in North Carolina. The hammer is about 30 cm long.

Organisms

■ *How do organisms contribute to mechanical and chemical weathering?*

Animals, plants, and bacteria all participate in the mechanical and chemical alteration of rocks. Burrowing animals, such as worms, termites, reptiles, rodents, and many others, constantly mix soil and sediment particles and bring material from depth to the surface where further weathering occurs. Even materials ingested by worms are further reduced in size, and animal burrows give gases and water easier access to greater depths. The roots of plants, especially large bushes and trees, wedge themselves into cracks in rocks and further widen them (▶ Figure 6.6).

▶ **Figure 6.6 Organisms and Weathering**

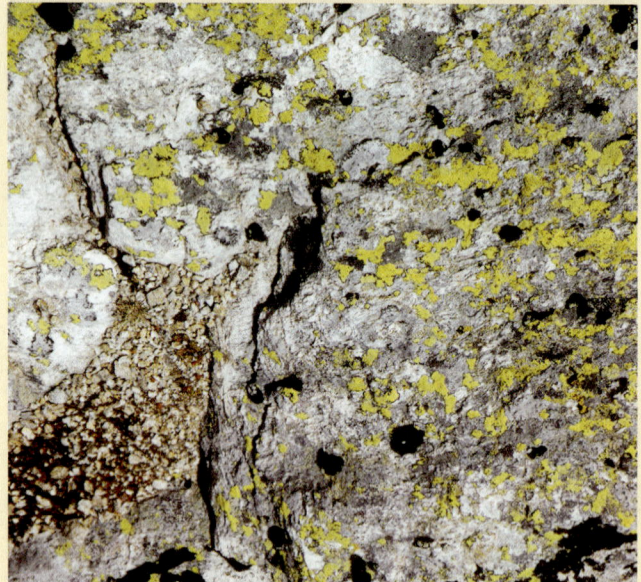

a The yellow and black masses on this rock in Switzerland are lichens, composite organisms made up of fungi and algae. Lichens derive their nutrients from the rock and thus contribute to chemical weathering.

b These trees and low-growing vegetation in Lassen Volcanic National Park in California help break the parent material into smaller pieces and thus contribute to mechanical weathering.

Tree roots growing under or through sidewalks and foundations do considerable damage.

> ### Section 6.3 Summary
>
> - Mechanical weathering processes break parent material into smaller pieces but do not change its composition.
>
> - Frost action, pressure release, thermal expansion and contraction, salt crystal growth, and the activities of organisms account for mechanical weathering.

6.4 Chemical Weathering—Decomposition of Earth Materials

How does chemical weathering differ from mechanical weathering? Remember that mechanical weathering yields smaller pieces of parent material with no other change, but **chemical weathering** includes those processes that bring about changes in the structure, composition, or both of parent material. Several clay minerals (sheet silicates) form by the structural and chemical alteration of feldspar minerals, all of which are framework silicates. In fact, some minerals may be completely dissolved during chemical weathering and their ions taken into solution, but some of the more chemically stable minerals such as quartz may simply be liberated from the parent material.

Organisms also play an important role in chemical weathering. Plants remove ions from soil water and reduce the chemical stability of soil minerals, and plant roots release organic acids. Mosses and lichens (composite organisms made up of fungi and algae) also contribute to chemical weathering (▶ Figure 6.6). Neither mosses nor lichens have roots, but both have hairlike filaments (rhizoids) that penetrate between minerals or grains and anchor them to rock surfaces. In addition, they secrete weak acids, and mosses especially are effective in retaining moisture. Other chemical weathering processes include solution, oxidation, and hydrolysis.

Solution

Solution is a chemical weathering process during which the ions in minerals separate in a liquid and the solid substance dissolves. Water is a remarkable solvent because its molecules of oxygen and hydrogen are arranged so that the angle between the two hydrogen atoms is about 104 degrees (▶ Figure 6.7a). As a result of this asymmetry, the oxygen end of the

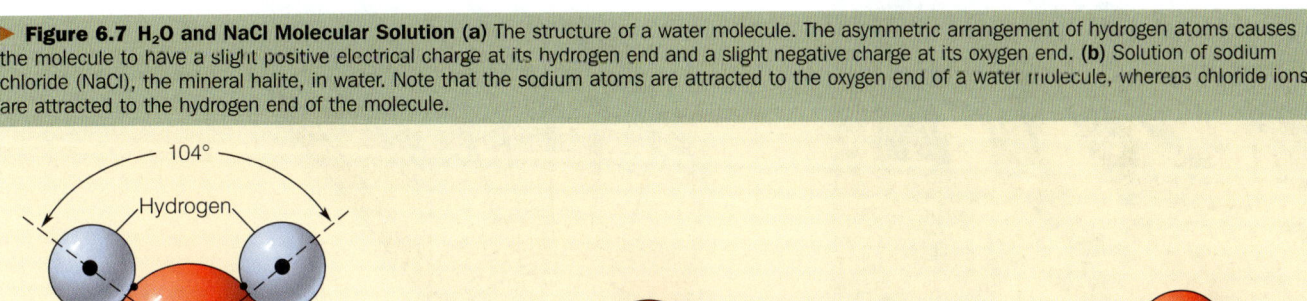

▶ **Figure 6.7 H$_2$O and NaCl Molecular Solution** (a) The structure of a water molecule. The asymmetric arrangement of hydrogen atoms causes the molecule to have a slight positive electrical charge at its hydrogen end and a slight negative charge at its oxygen end. (b) Solution of sodium chloride (NaCl), the mineral halite, in water. Note that the sodium atoms are attracted to the oxygen end of a water molecule, whereas chloride ions are attracted to the hydrogen end of the molecule.

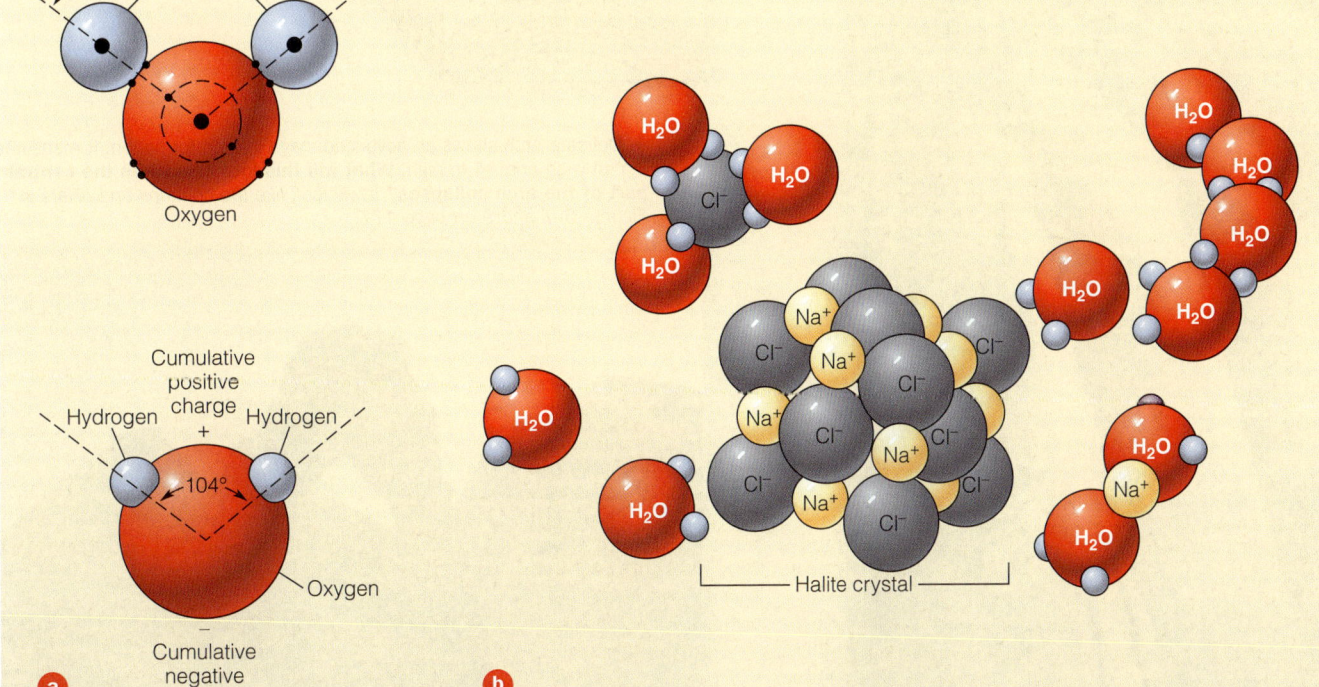

Arches National Park, Utah

The sandstone arches at Arches National Park near Moab, Utah, are in the Jurassic-aged Entrada Sandstone. The park is noted for its arches, spires, balanced rocks, pinnicales, and other features that resulted from differential weathering and erosion.

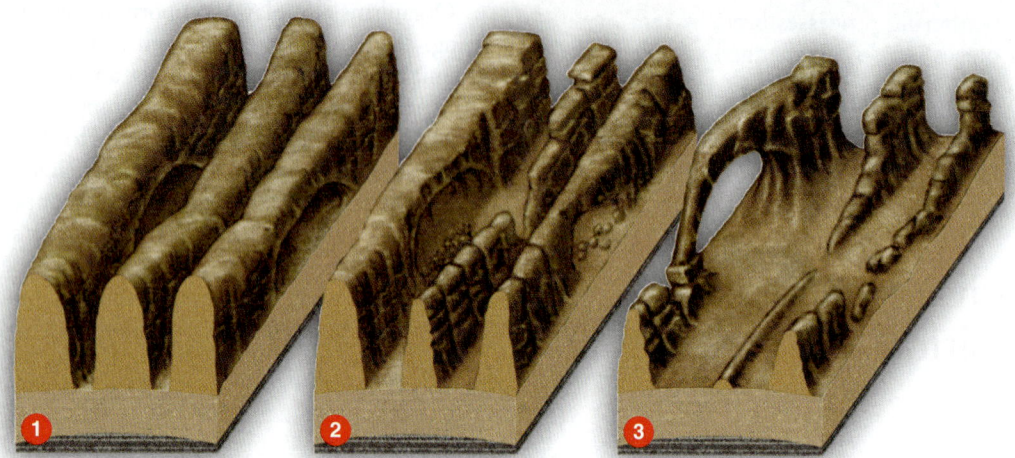

▶ **1–3.** When the area was uplifted (below), cracks (joints) developed. Differential weathering and erosion enlarged the joints, leaving fins (1 above and left). Continued weathering and erosion of the fins (2 above) yielded arches, many of which have collapsed (3 above).

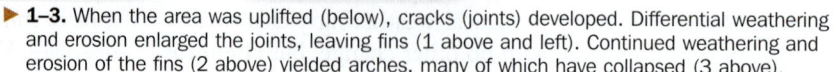

▼ **4.** Delicate Arch shows an advanced stage in arch formation. It measures 9.7 m wide and 14 m high. **What will this look like when the central part of the arch collapses?** *It will look like the columns and pillars in Figures 5–7 and 9 on page 177.*

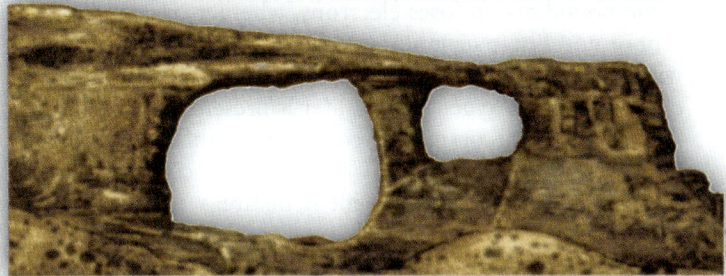

▶ **5–7.** Probable evolution of Sheep Rock and the fin with Baby Arch or Hole-in-the-Wall (left and above). Close-up view of Baby Arch (below). It measures 7.6 m wide and 4.5 m high.

▼ **8.** You may recall that the opening scene in the movie *Indiana Jones and the Last Crusade* showed a young Indiana Jones, played by River Phoenix, at Double Arch.

◀ **9.** Balanced Rock rises 39 m above its base. Isolated columns and pillars like this one are called *hoodoos,* many of which are in the park.

molecule retains a slight negative electrical charge, but the hydrogen end has a slight positive charge. So when in water, the positively charged sodium ions in the mineral halite (NaCl) are attracted to the negative end of a water molecule, and the negatively charged chlorine ions are attracted to the positive end of the water molecule (▶ Figure 6.7b). Thus ions are liberated from the mineral and the solid substance dissolves; that is, it goes into solution.

■ *What happens to rocks composed of carbonate minerals when they are exposed to acidic solutions?*

Most minerals are not very soluble in pure water because the attractive forces of water molecules are not strong enough to overcome the forces bonding ions together in minerals. For example, the carbonate mineral calcite ($CaCO_3$), the main component of the sedimentary rock limestone and the metamorphic rock marble, is nearly insoluble in pure water, but it rapidly dissolves if the water is slightly acidic. An easy way for water to become acidic is by liberating the hydrogen ions in carbonic acid.

$$\underset{\text{WATER}}{H_2O} + \underset{\substack{\text{CARBON} \\ \text{DIOXIDE}}}{CO_2} \rightleftharpoons \underset{\substack{\text{CARBONIC} \\ \text{ACID}}}{H_2CO_3} \rightleftharpoons \underset{\substack{\text{HYDROGEN} \\ \text{ION}}}{H^+} + \underset{\substack{\text{BICARBONATE} \\ \text{ION}}}{HCO_3^-}$$

According to this chemical equation, water and carbon dioxide combine to form *carbonic acid*, a small amount of which dissociates (breaks down into other substances) to yield hydrogen and bicarbonate ions. A solution's acidity depends on the concentration of hydrogen ions; the more hydrogen ions present, the stronger the acid.

The carbon dioxide that reacts with water to form carbonic acid comes from several sources. Rainwater is slightly acidic because about 0.3% of the atmosphere is made up of carbon dioxide and some comes from volcanic gases. Human activities have added gases to the atmosphere that contribute to acid rain, acid snow, and acid fog. But the main source of carbon dioxide for chemical weathering is decaying organic matter and the respiration of organisms, especially those that live in soil. Accordingly, in humid regions where vegetation is abundant, groundwater tends to be slightly acidic. In arid regions, though, there is little vegetation and few soil organisms, so the supply of carbon dioxide is limited and groundwater tends to be alkaline—that is, has a low concentration of hydrogen ions.

Regardless of the source of carbon dioxide, once it is present, carbonic acid forms and calcite rapidly goes into solution according to the following reaction:

$$\underset{\substack{\text{MINERAL} \\ \text{CALCITE}}}{CaCO_3} + \underset{\substack{\text{CARBONIC} \\ \text{ACID}}}{H_2CO_3} \rightleftharpoons \underset{\substack{\text{IONS IN SOLUTION} \\ \text{CALCIUM} \\ \text{ION}}}{Ca^{+2}} + \underset{\substack{\text{BICARBONATE} \\ \text{ION}}}{2HCO_3^-}$$

The solution of the calcite in limestone and marble has had dramatic effects, ranging from small cavities to enormous caverns such as Mammoth Caves in Kentucky and Carlsbad Caverns in New Mexico (see Chapter 16). Limestone and related carbonate rocks such as dolostone are certainly soluble in acidic solutions, but their solubility varies depending on climate. In the semiarid to arid West where groundwater is alkaline, carbonate rocks are not soluble and they form bold cliffs, but subdued exposures of the same rocks are typical of humid regions where groundwater is acidic.

Oxidation

To a chemist, **oxidation** is any chemical reaction in which a compound or ion loses electrons, whether oxygen is present or not. But in chemical weathering it refers to reactions with oxygen to form oxides (one or more metallic elements combined with oxygen) or, if water is present, hydroxides (a metallic element or radical combined with the hydroxyl, OH ion). For example, iron rusts when it combines with oxygen and forms the iron oxide hematite:

$$\underset{\text{IRON}}{4Fe} + \underset{\text{OXYGEN}}{3O_2} \rightarrow \underset{\substack{\text{IRON OXIDE} \\ \text{(HEMATITE)}}}{2Fe_2O_3}$$

Oxygen is, of course, abundant in the atmosphere, but oxidation is a slow process unless water is present, so most oxidation takes place in water with dissolved oxygen.

Oxidation of iron in the ferromagnesian silicates such as olivine, pyroxenes, amphiboles, and biotite forms the red iron oxide hematite (Fe_2O_3) or the yellow or brown hydroxide limonite [$FeO(OH) \cdot nH_2O$]. In fact, small amounts of these two minerals account for the yellow, brown, and red colors of many sedimentary rocks (see Chapter 7). Some iron-bearing minerals such as pyrite (FeS_2) form and are stable in the absence of oxygen but are rapidly oxidized in water. Of particular concern is pyrite in debris from coal mining, where it oxidizes to form sulfuric acid (H_2SO_4) and iron oxide. A serious environmental problem is acid soils and runoff that form in this manner (▶ Figure 6.8).

Hydrolysis

■ *What takes place when potassium feldspars undergo hydrolysis?*

When **hydrolysis** takes place, hydrogen ions (H^+) or the hydrogen in (OH^-) of water reacts with and replaces positive ions in minerals, thereby changing their composition and liberating soluble compounds and iron that may then be oxidized. Potassium feldspars such as orthoclase ($KAlSi_3O_8$) and plagioclase feldspars (which vary from $CaAl_2Si_2O_8$ to $NaAlSi_3O_8$) are framework silicates, but when altered by hydrolysis they yield materials in solution and clay minerals, which are sheet silicates.

The alteration of potassium feldspar by hydrolysis looks complex, but the main points of the reaction are fairly straightforward.

$$\underset{\substack{\text{MINERAL} \\ \text{ORTHOCLASE}}}{KAlSi_3O_8} + \underset{\text{CARBONIC ACID}}{2H^+ + 2HCO_3^-} + \underset{\text{WATER}}{H_2O} \rightarrow$$

$$\underset{\substack{\text{MINERAL} \\ \text{CLAY (KAOLINITE)}}}{Al_2Si_2O_5(OH)_4} + \underset{\text{IONS AND SILICA IN SOLUTION}}{K^+ + HCO_3^- + 4SiO_2}$$

▶ **Figure 6.8 Acid Runoff** The oxidation of pyrite in mine tailings forms acid water, as in this small stream. More than 11,000 km of U.S. streams, mostly in the Appalachian region, are contaminated by abandoned coal mines that leak sulfuric acid.

▶ **Figure 6.9 Weathering Along Fractures** These granitic rocks in Joshua Tree National Park in California have been chemically weathered more intensely along fractures than in unfractured parts of the same rock outcrop.

In this reaction, hydrogen ions attack the ions in orthoclase, and some liberated ions are incorporated into a developing clay mineral, whereas others go into solution. The dissolved silica on the right side of the equation, which would not fit into the crystal structure of the clay mineral, is one source of cement that binds sedimentary particles together to form sedimentary rocks (see Chapter 7).

The hydrolysis reaction for plagioclase feldspars is the same as the one above, except it yields soluble calcium and sodium compounds. In fact, the presence of these compounds, especially calcium compounds, is what makes hard water hard. Calcium in water inhibits the reaction of detergents with dirt and precipitates as scaly minerals in water pipes and water heaters (see Chapter 16).

The Rate of Chemical Weathering

Chemical weathering operates on the surface of particles, so rocks and minerals alter from the outside inward. In fact, rocks commonly have a rind of weathered material near the surface but are completely unaltered inside. The rate at which chemical weathering proceeds depends on several factors. One is simply the presence or absence of fractures, because fluids seep along fractures, accounting for more intense chemical weathering along these surfaces (▶ Figure 6.9). Thus, given the same rock type under similar conditions, the more fractures, the more rapid the chemical weathering. Of course, other factors also control the rate of chemical weathering, including particle size, climate, and parent material.

Particle Size Because chemical weathering affects particle surfaces, the greater the surface area, the more effective the weathering. It is important to realize that small particles have larger surface areas compared to their volume than do large particles. Notice in ▶ Figure 6.10 that a block measuring 1 m on a side has a total surface area of 6 m^2, but when the block is broken into particles measuring 0.5 m on a side, the total surface area increases to 12 m^2. And if these particles are all reduced to 0.25 m on a side, the total surface area increases to 24 m^2 but the total volume remains the same at 1 m^3.

■ *How does mechanical weathering contribute to the rate of chemical weathering?*

We can make two important statements regarding the block in Figure 6.10. First, as it is divided into a number of smaller blocks, its total surface area increases. Second, the smaller any single block is, the more surface area it has compared to its volume. We can conclude that mechanical weathering, which reduces the size of particles, contributes to chemical weathering by exposing more surface area.

Your own experiences with particle size verify our contention regarding surface area and volume. Because of its very small particle size, powdered sugar gives an intense

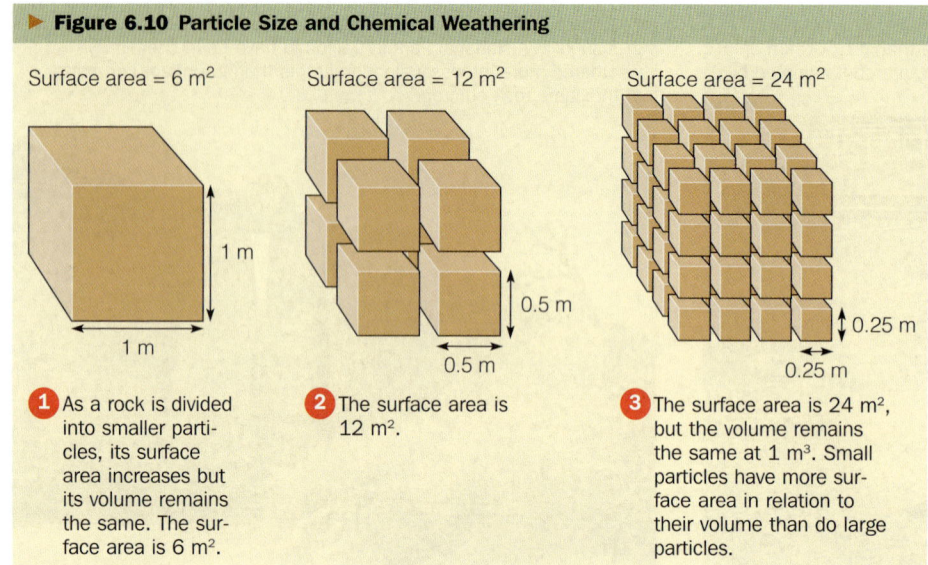

▶ **Figure 6.10 Particle Size and Chemical Weathering**

① As a rock is divided into smaller particles, its surface area increases but its volume remains the same. The surface area is 6 m².

② The surface area is 12 m².

③ The surface area is 24 m², but the volume remains the same at 1 m³. Small particles have more surface area in relation to their volume than do large particles.

burst of sweetness as the tiny pieces dissolve rapidly, but otherwise it is the same as the granular sugar we use on our cereal or in our coffee. As an experiment, see how rapidly crushed ice and an equal volume of block ice melt, or determine the time it takes to boil an entire potato as opposed to one cut into small pieces.

Climate and Chemical Weathering Chemical processes proceed more rapidly at high temperatures and in the presence of liquids, so chemical weathering is more effective in the tropics than in arid and arctic regions because temperatures and rainfall are high and evaporation rates are low (▶ Figure 6.11). In addition, vegetation and animal life are much more abundant in the tropics. Consequently, the effects of weathering extend to depths of several tens of meters, but commonly extend only centimeters to a few meters deep in arid and arctic regions. You should realize, though, that chemical weathering goes on everywhere, except perhaps where earth materials are permanently frozen.

Parent Material Some rocks are more resistant to chemical alteration than others. The metamorphic rock quartzite, composed of quartz (SiO_2), is an extremely stable substance that alters very slowly compared with most other rock types. In contrast, basalt, with its large amount of calcium-rich plagioclase and pyroxene minerals, decomposes rapidly because these minerals are chemically unstable. In fact, the stability of common minerals is just the opposite of their order of crystallization in Bowen's reaction series (Table 6.1). The minerals that form last in this series are chemically stable, whereas those that form early are more easily altered by chemical processes.

One manifestation of chemical weathering is **spheroidal weathering** (▶ Figure 6.12). In spheroidal weathering, a stone, even a rectangular one, weathers into a more spherical shape because that is the most stable shape it can assume. The reason? On a rectangular stone, the corners are attacked by weathering processes from three sides, and the edges are attacked from two sides, but the flat surfaces weather more or less uniformly (▶ Figure 6.13). Consequently, the corners and edges alter more rapidly, the material sloughs off them, and a more spherical shape develops. Once a spherical shape is present, all surfaces weather at the same rate.

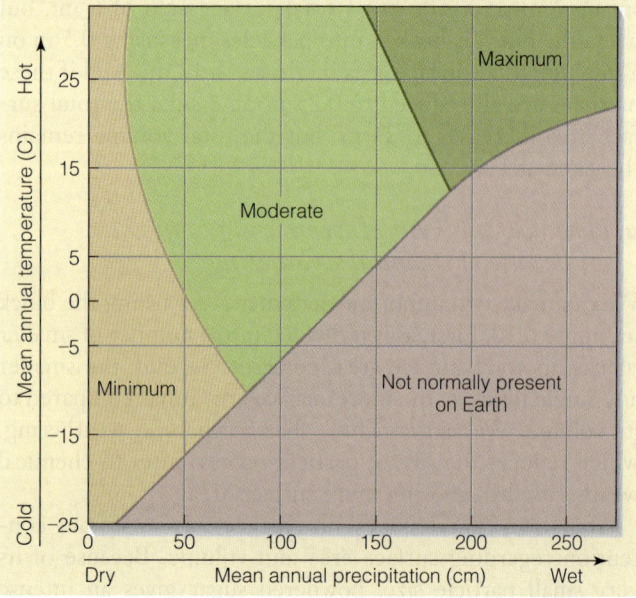

▶ **Figure 6.11 Chemical Weathering and Climate** Chemical weathering is most intense where temperature and rainfall are high, and at a minimum in arid environments whether hot or cold.

TABLE 6.1
Stability of Silicate Minerals

Ferromagnesian Silicates	Nonferromagnesian Silicates
Olivine	Calcium plagioclase
Pyroxene	
Amphibole	Sodium plagioclase
Biotite	Potassium feldspar
	Muscovite
	Quartz

Increasing Stability ↓

▶ **Figure 6.12 Spheroidal Weathering** These illustrations show four stages in the chemical weathering phenomenon called spheroidal weathering, which reduces rocks with intersecting fractures to more spherical shapes.

The effects of spheroidal weathering are obvious in many rock bodies, particularly those that have a rectangular pattern of fractures (joints) similar to those in ▶ Figure 6.9. Fluids seep along the joint surfaces, resulting in more intense weathering at the edges and corners of the rectangular blocks, thus yielding more nearly spherical objects. Fractured granitic rocks are especially susceptible to spheroidal weathering, but good examples can be found in all rock types (▶ Figure 6.13).

Section 6.4 Summary

• Chemical weathering processes include solution, oxidation, hydrolysis, and the activities of organisms. They result in a change in the structure, composition, or both of parent material.

• Particle size, climate, and type of parent material determine how quickly chemical weathering takes place.

What Would You Do?

Acid rain is one consequence of industrialization, but it is a problem that can be solved. In your community, a local copper smelter is obviously pouring out gases, thereby contributing to the problem both locally and regionally. As chairperson of a committee of concerned citizens, you make some recommendations to the smelter owners to clean up their emissions. What specific recommendations would you favor? Suppose further that the smelter owners say that the remedies are too expensive and if they are forced to implement them, they will close the smelter. Would you still press forward even though your efforts might mean economic disaster for your community?

6.5 Soil and Its Origin

Most of Earth's land surface is covered by an unconsolidated layer of **regolith,** a collective term for sediment, regardless of how it was deposited, as well as layers of pyroclastic materials and the residue formed by weathering. Only some of this regolith is **soil,** which by definition is made up of

▶ Figure 6.13 Spheroidal Weathering

a Spheroidal weathering in granite at Aswan, Egypt.

b This basalt lava flow near Susanville, California, is riddled with fractures where weathering took place more rapidly than in other parts of the same rock, thereby yielding spherical blocks.

weathered rock, air, water, and organic matter and supports vegetation. Soil is an essential link between the parent material below and life above. Indeed, nearly all land-dwelling organisms depend on soil for their existence. Plants grow in soil from which they derive their nutrients and most of their water, and animals depend directly or indirectly on plants for sustenance.

Only about 45% of a good soil for farming or gardening is composed of weathered material, mostly sand, silt, and clay. Much of the remaining 55% is simply void spaces filled with either air or water and a small but important amount of organic matter, mostly humus (▶ Figure 6.14a). **Humus** consists of carbon derived by bacterial decay of organic matter and is highly resistant to further decay. Even a fertile soil might contain as little as 5% humus, but it is nevertheless an important source of plant nutrients and it enhances moisture retention. Furthermore, it gives the upper layers of many soils their dark color; soils with little humus are lighter colored and not as productive.

Some weathered materials in soils are simply sand- and silt-sized mineral grains, especially quartz, but other weathered materials may be present as well. These solid particles hold soil particles apart, allowing oxygen and water to circulate more freely. Clay minerals are also important constituents of soils and aid in the retention of water as well as suppling nutrients to plants. Soils with excess clay minerals, however, drain poorly and are sticky when wet and hard when dry.

The Soil Profile

Observed in vertical cross section, a soil has distinct layers, or **soil horizons,** that differ from one another in texture, structure, composition, and color (▶ Figure 6.14b). Starting from the top, the horizons are designated O, A, B, and C, but the boundaries between horizons are transitional rather than sharp. Because soil-forming processes begin at the surface and work downward, the upper soil layer is more altered from the parent material than are the layers below.

Horizon O, which is only a few centimeters thick, consists of organic matter. The remains of plants are clearly recognizable in the upper part of horizon O, but its lower part consists of humus.

Horizon A, called *topsoil*, contains more organic matter than horizons B and C, and it is also characterized by intense biological activity because plant roots, bacteria, fungi, and animals such as worms are abundant. Threadlike soil bacteria give freshly plowed soil its earthy aroma. In soils developed over a long period of time, horizon A consists mostly of clays and chemically stable minerals such as quartz. Water percolating down through horizon A dissolves soluble minerals and caries them away or down to lower levels in the soil by a process called *leaching*, so horizon A is also known as the **zone of leaching** (▶ Figure 6.14).

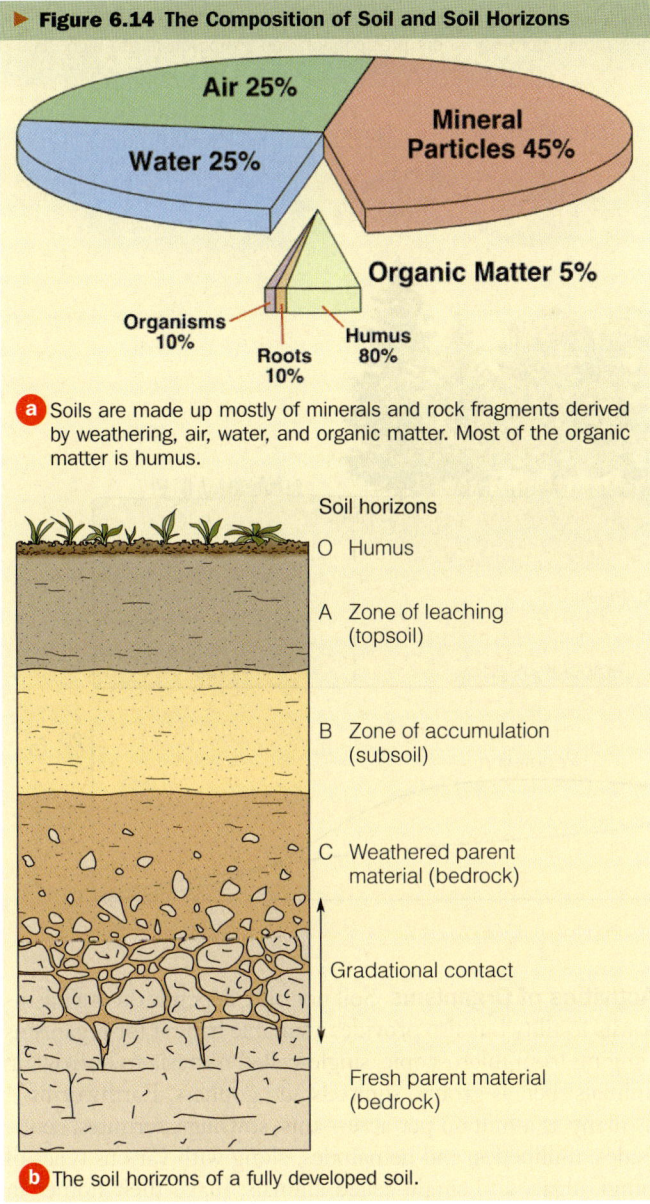

▶ Figure 6.14 The Composition of Soil and Soil Horizons

a. Soils are made up mostly of minerals and rock fragments derived by weathering, air, water, and organic matter. Most of the organic matter is humus.

b. The soil horizons of a fully developed soil.

Factors in Soil Formation

■ *What factors determine the rate of soil formation?*

All soils form by mechanical and chemical weathering, but they differ in texture, color, thickness, and fertility. Accordingly, we are interested in the factors that control the attributes and locations of various soils as well as how rapidly soil-forming processes operate. Climate, parent material, organic activity, relief and slope, and time are the critical factors in soil formation. Complex interactions among these determine soil type, thickness, and fertility.

Climate and Soil Soil scientists know that climate is the single most important factor influencing soil type and depth. Intense chemical weathering in the tropics yields deep soils from which most of the soluble minerals have been removed by leaching. In arctic and desert climates, soils tend to be thin, contain significant quantities of soluble minerals, and be composed mostly of materials derived by mechanical weathering (▶ Figure 6.15).

A very general classification recognizes three major soil types characteristic of different climate settings. Soils that develop in humid regions, such as the eastern United States and much of Canada, are **pedalfers,** a name derived from the Greek word *pedon*, meaning "soil," and from the chemical symbols for aluminum (Al) and iron (Fe) (▶ Figure 6.16a). Because pedalfers form where moisture is abundant, most of the soluble minerals have been leached from horizon A. Although it may be gray, horizon A is commonly dark because of abundant organic matter, and aluminum-rich clays and iron oxides tend to accumulate in horizon B.

Soils found in much of the arid and semiarid western United States, especially the Southwest, are **pedocals.** Pedocal derives its name in part from the first three letters of *calcite* (▶ Figure 6.16b). These soils contain less organic matter than pedalfers, so horizon A is lighter colored and contains more unstable minerals because of less intense chemical weathering. As soil water evaporates, calcium carbonate leached from above precipitates in horizon B, where it forms irregular masses of *caliche* (▶ Figure 6.17a). Precipitation of sodium salts in some desert areas where soil water evaporation is intense yields *alkali soils* that are so alkaline that they support few plants (▶ Figure 6.17b).

■ *What is laterite, and where and how does it form?*

Laterite is a soil formed in the tropics where chemical weathering is intense and leaching of soluble minerals is complete. These soils are red, commonly extend to depths of several tens of meters, and are composed largely of aluminum hydroxides, iron oxides, and clay minerals; even quartz, a chemically stable mineral, is leached out (▶ Figure 6.18a).

Although laterites support lush vegetation, as in tropical rain forests, they are not very fertile. The native vegetation

Horizon B, or *subsoil*, contains fewer organisms and less organic matter than horizon A. Horizon B is known as the **zone of accumulation** because soluble minerals leached from horizon A accumulate as irregular masses. If horizon A is stripped away by erosion, leaving horizon B exposed, plants do not grow as well; and if horizon B is clayey, it is harder when dry and stickier when wet than other soil horizons.

Horizon C, the lowest soil layer, consists of partially altered parent material grading down into unaltered parent material (▶ Figure 6.14). In horizons A and B, the composition and texture of the parent material have been so thoroughly altered that it is no longer recognizable. In contrast, rock fragments and mineral grains of the parent material retain their identity in horizon C. Horizon C contains little organic matter.

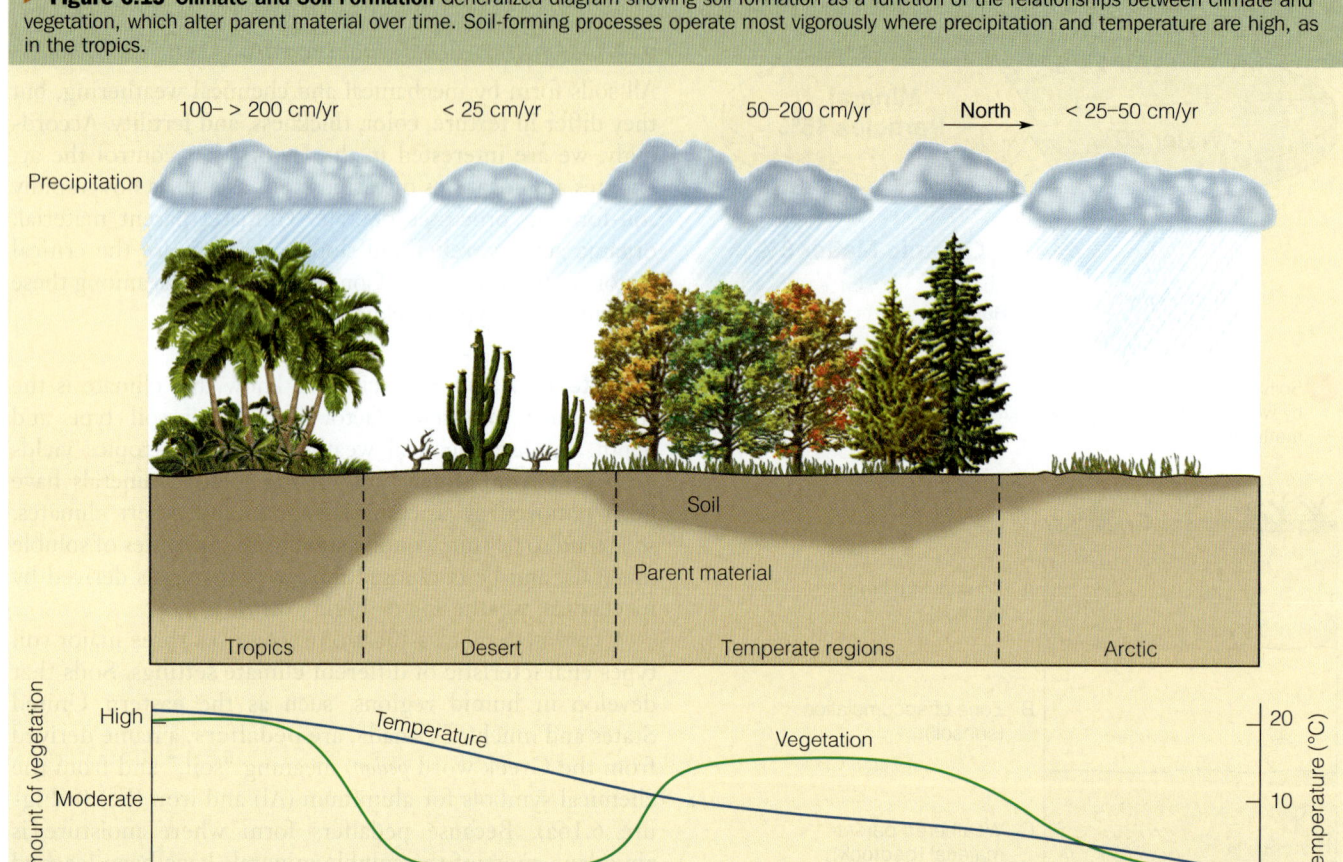

▶ **Figure 6.15 Climate and Soil Formation** Generalized diagram showing soil formation as a function of the relationships between climate and vegetation, which alter parent material over time. Soil-forming processes operate most vigorously where precipitation and temperature are high, as in the tropics.

is sustained by nutrients derived mostly from the surface layer of organic matter. When these soils are cleared of their native vegetation, the surface accumulation of organic matter rapidly oxidizes, and there is little to replace it. Consequently, when societies practicing slash-and-burn agriculture clear these soils, they can raise crops for only a few years at best (▶ Figure 6.19). Then the soil is depleted of plant nutrients, the clay-rich–laterite bakes brick hard in the tropical sun, and the farmers move on to another area, where the process is repeated.

Parent Material The same rock type can yield different soils in different climatic regimes, and in the same climatic regime the same soils can develop on different rock types. Thus it seems that climate is more important than parent material in determining the type of soil. Nevertheless, rock type does exert some control. For example, the metamorphic rock quartzite will have a thin soil over it because it is chemically stable, whereas an adjacent body of granite will have a much deeper soil (▶ Figure 6.20a).

Soil that develops on basalt will be rich in iron oxides because basalt contains abundant ferromagnesian silicates, but rocks lacking these minerals will not yield an iron oxide-rich soil no matter how thoroughly they are weathered. Also, weathering of a pure quartz sandstone yields no clay, whereas weathering of clay yields no sand.

Activities of Organisms Soil not only depends on organisms for its fertility but also provides a suitable habitat for organisms ranging from microscopic, single-celled bacteria to burrowing animals such as ground squirrels and gophers. Earthworms—as many as a million per acre—ants, sowbugs, termites, centipedes, millipedes, and nematodes, along with various types of fungi, algae, and single-celled animals, make their homes in soil. All contribute to the formation of soils and provide humus when they die and are decomposed by bacterial action.

Much of the humus in soils is provided by grasses or leaf litter that microorganisms decompose to obtain food. In so doing, they break down organic compounds within plants and release nutrients back into the soil. In addition, organic acids produced by decaying soil organisms are important in further weathering of parent materials and soil particles.

Burrowing animals constantly churn and mix soils, and their burrows provide avenues for gases and water. Soil organisms, especially some types of bacteria, are extremely important in changing atmospheric nitrogen into a form of soil nitrogen suitable for use by plants.

The Lay of the Land—Relief and Slope *Relief* is the difference in elevation between high and low points in a region. In some mountainous areas relief is measured in hundreds of meters, whereas it rarely exceeds a few meters in, for instance, the Great Plains of the United States and

PHYSICAL Geology Now™ ▶ **Geo-focus Figure 6.16 Pedalfer, Pedocal, and Laterite (a)** Pedocal, **(b)** pedalfer, and **(c)** laterite form under different climatic conditions.

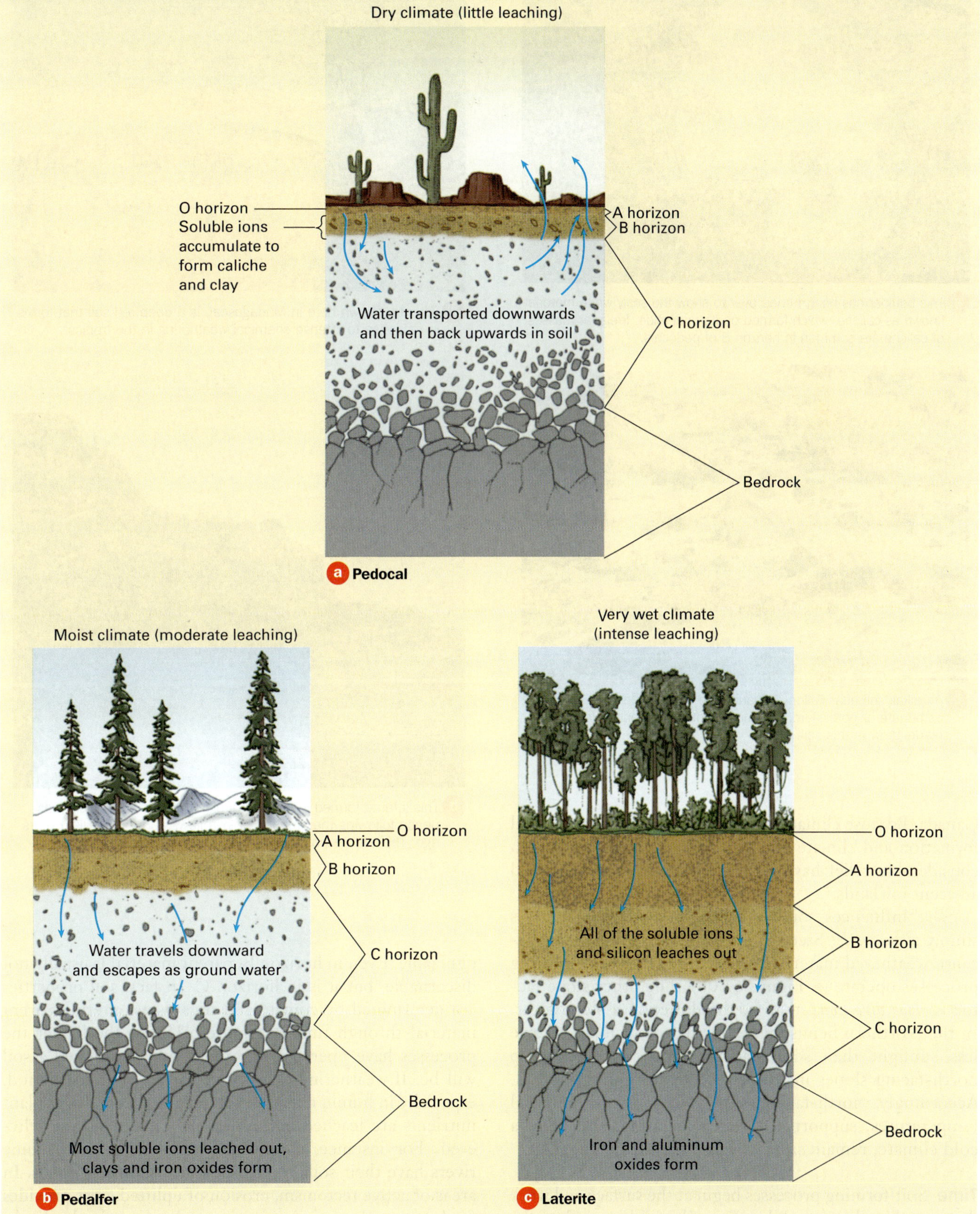
Dry climate (little leaching)
O horizon
Soluble ions accumulate to form caliche and clay
A horizon
B horizon
Water transported downwards and then back upwards in soil
C horizon
Bedrock
a Pedocal
Moist climate (moderate leaching)
O horizon
A horizon
B horizon
Water travels downward and escapes as ground water
C horizon
Bedrock
Most soluble ions leached out, clays and iron oxides form
b Pedalfer
Very wet climate (intense leaching)
O horizon
A horizon
All of the soluble ions and silicon leaches out
B horizon
C horizon
Bedrock
Iron and aluminum oxides form
c Laterite

Figure 6.17 Features of Soils in Arid Regions

a This boulder has been turned over to show the scaly white material known as *caliche*, which formed on its underside. Irregular masses of caliche are common in horizon B of pedocals.

b An alkali soil near Fallon, Nevada. The white material is sodium carbonate or potassium carbonate. Notice that only a few hardy plants grow in this soil.

Figure 6.18 Laterites

a Laterite, shown here in Madagascar, is a deep red soil that forms in response to intense chemical weathering in the tropics.

b This laterite formed in Northern Ireland during a 2-million-year period following the eruption of lava flows about 60 million years ago. The laterite was then buried beneath more lava flows.

Canada. Because climate is such an important factor in soil formation and climate changes with elevation, areas with considerable relief have different soils in mountains and adjacent lowlands.

Slope influences soil formation in two ways. One is simply *slope angle:* Steep slopes have little or no soil because weathered materials erode faster than soil-forming processes operate (▶ Figure 6.20b). The other slope factor is *slope direction*—that is, the direction a slope faces. In the Northern hemisphere, north-facing slopes receive less sunlight than south-facing slopes. In fact, steep north-facing slopes may not receive any sunlight at all. Accordingly, north-facing slopes have cooler internal temperatures, support different vegetation, and, if in a cold climate, remain snow covered or frozen longer.

Time Soil-forming processes begin at the surface and work downward, so horizon A has been altered longer than the other horizons, and thus parent material is no longer recognizable. Even in horizon B, parent material is usually not discernable, but it is in horizon C. In fact, soil properties are determined by climate and organisms, altering parent material through time (▶ Figure 6.15), so the longer the processes have operated, the more fully developed a soil will be. If weathering takes place for an extended period, especially in humid climates, soil fertility decreases as plant nutrients are leached out, unless new materials are delivered. For instance, agricultural lands adjacent to some rivers have their soils replenished yearly during floods. In areas of active tectonism, erosion of uplifted areas provides fresh materials that are transported to nearby lowlands, where they contribute to soils.

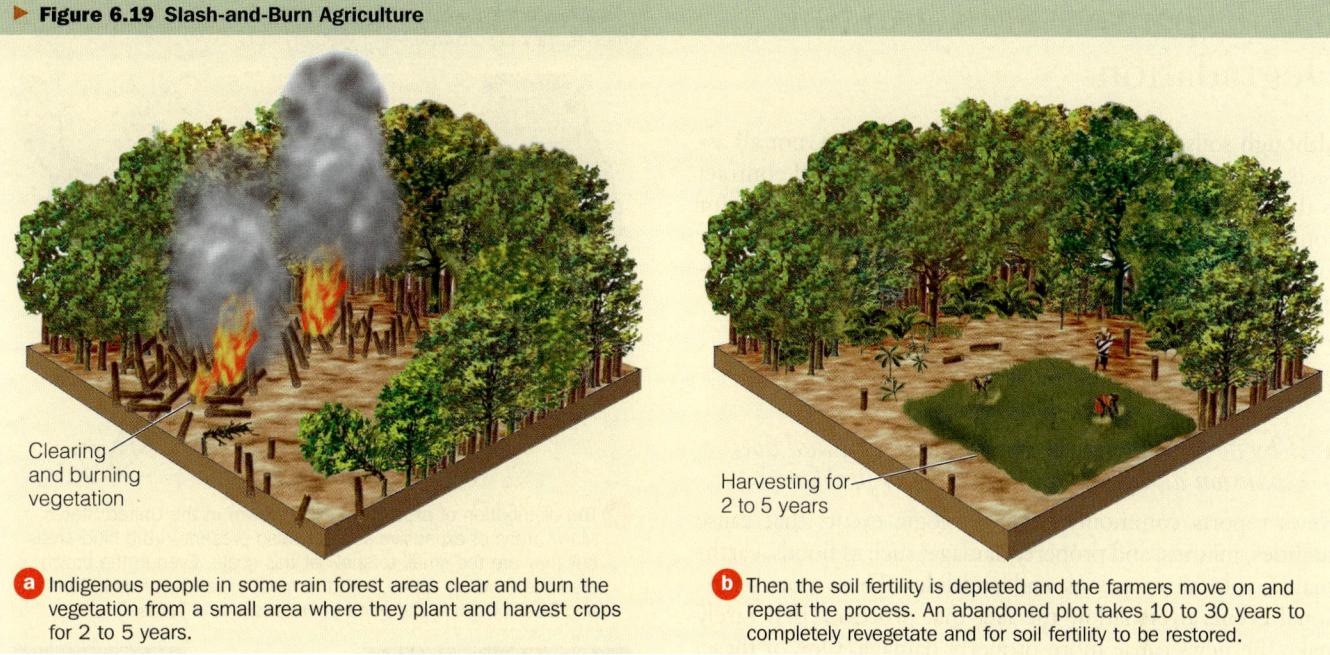

▶ **Figure 6.19 Slash-and-Burn Agriculture**

a Indigenous people in some rain forest areas clear and burn the vegetation from a small area where they plant and harvest crops for 2 to 5 years.

b Then the soil fertility is depleted and the farmers move on and repeat the process. An abandoned plot takes 10 to 30 years to completely revegetate and for soil fertility to be restored.

How much time is needed to develop a centimeter of soil or a fully developed soil a meter or so deep? We can give no definitive answer because weathering proceeds at vastly different rates depending on climate and parent material, but an overall average might be about 2.5 cm per century. However, a lava flow a few centuries old in Hawaii may have a well-developed soil on it, whereas a flow the same age in Iceland has considerably less soil. Given the same climatic conditions, soil develops faster on unconsolidated sediment than it does on bedrock.*

Under optimum conditions, soil-forming processes operate rapidly in the context of geologic time. From the human perspective, though, soil formation is a slow process; consequently, soil is a nonrenewable resource.

Section 6.5 Summary

• Soil is made up of weathered materials, air, water, and organic matter and can support vegetation.

• Soil profiles for pedalfer, pedocal, and laterite show horizons designated O, A, B, and C that differ in structure, composition, and color.

• Climate, parent material, organisms, angle of slope, the direction a slope faces, and time are the factors that control the rate of soil formation.

Bedrock is a general term for the rock underlying soil or unconsolidated sediment.

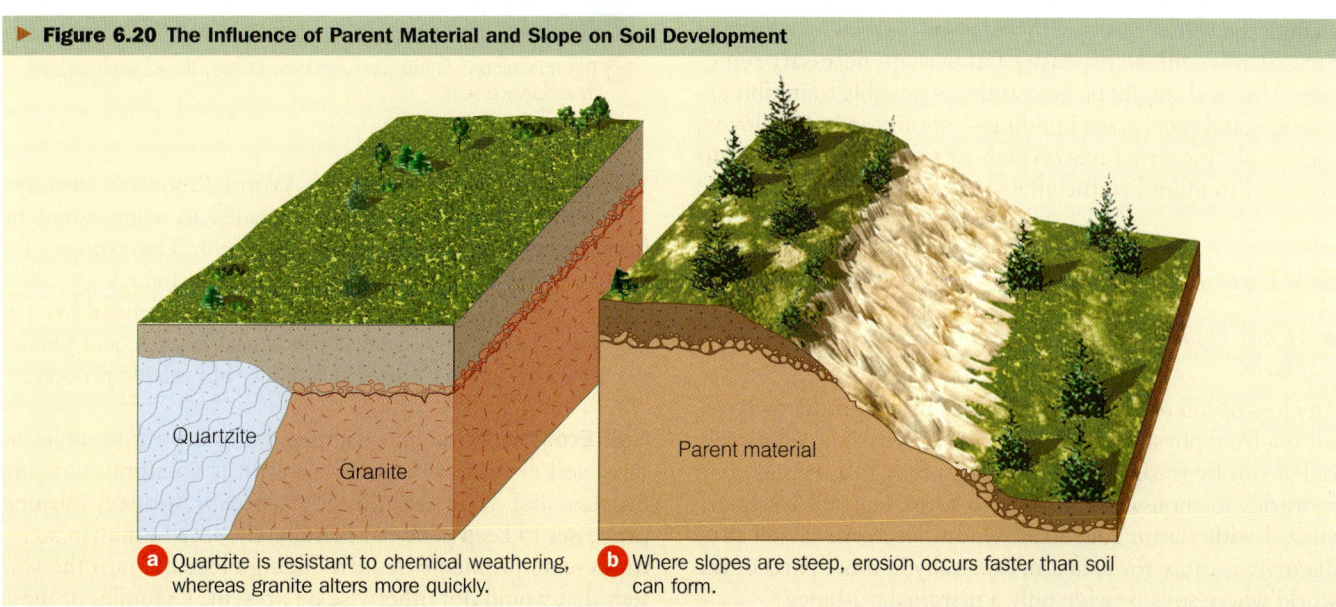

▶ **Figure 6.20 The Influence of Parent Material and Slope on Soil Development**

a Quartzite is resistant to chemical weathering, whereas granite alters more quickly.

b Where slopes are steep, erosion occurs faster than soil can form.

6.6 Expansive Soils and Soil Degradation

Although soils are essential for food production, not all aspects of soils are beneficial. Some soils expand and contract as they absorb water and then dry out, posing problems for homeowners, developers, and engineers. And given that our survival depends on soils, soil degradation, which is any decrease in soil fertility or soil losses, is viewed with alarm.

Expansive Soils

- *Why do some soils expand and contract, and does expansion and contraction cause any problems?*

News reports commonly cover geologic events that cause fatalities, injuries, and property damage, such as floods, earthquakes, volcanic eruptions, and landslides. These more sensational events overshadow the fact that processes that rarely make the news cause more property damage. One of these, soil creep, is considered in Chapter 14, but here we are concerned with **expansive soils,** soils containing clay minerals that increase in volume when wet and shrink when they dry out. Soils with clays that expand 6% are considered highly expansive, and some are even more expansive. About $6 billion in damage to foundations, roadways, sidewalks, and other structures takes place each year in the United States, mostly in the Rocky Mountain states, the Southwest, and some of the states along the Gulf of Mexico (▶ Figure 6.21a).

When soil expands and contracts, overlying structures are first uplifted and then subside, thus experiencing forces that usually are not equally applied. Part of a house or sidewalk might be uplifted more than an adjacent part of the same structure (▶ Figure 6.21b). Avoiding areas of expansive soils is the best way to prevent damage, but what can be done to minimize the damage to existing structures? In some cases the soil is removed, or mixed with chemicals that change the way it reacts with water, or covered by a layer of nonexpansive fill, all expensive but perhaps necessary remedies. Also, soil should be kept as dry as possible to inhibit expansion, and specialized building methods can be employed, such as placing structures on piers or reinforced foundations designed to minimize the effects of expansion.

Soil Degradation

- *What is soil degradation, and how does it take place?*

Any loss of soil to erosion or decrease in soil productivity resulting from physical and chemical phenomena is collectively called **soil degradation.** Given that a fully mature soil takes centuries to thousands of years to form, any soil losses are viewed with alarm. And likewise, any decrease in soil productivity is cause for concern, especially in those parts of the world where soils provide only a marginal existence.

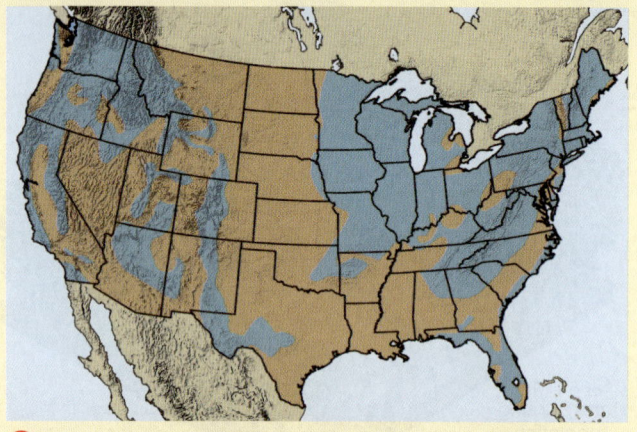

▶ **Figure 6.21 Expansive Soils** Expansive soils expand and contract as they absorb water and then dry out.

a The distribution of expansive soils (brown) in the United States. Many areas of expansive soils are also present in the blue areas, but they are too small to show at this scale. Even in the brown area the problem varies, with the most expansive soils in North and South Dakota and parts of Nebraska and Texas.

b The undulations in this sidewalk near Dallas, Texas, were caused by expansive soils.

According to studies by the World Resources Institute, 17% of the world's soils were degraded to some extent by human activities between 1945 and 1990. The estimate for North America is only 5.3%, whereas the figures for other parts of the world are much higher. The three general types of soil degradation are *erosion*, *chemical deterioration*, and *physical deterioration*, each with several separate but related processes.

Soil Erosion Wind and running water are responsible for most soil erosion. Of course, erosion is a natural, ongoing process, but it is usually slow enough for soil-forming processes to keep pace. Unfortunately, several human activities create problems by introducing elements into the system that would not otherwise be present. Examples of these

▶ **Figure 6.22 Soil Degradation Resulting from Erosion**

a Rill erosion in a field in Michigan during a rainstorm. The rill was later plowed over.

b A large gully in the upper basin of the Rio Reventado in Costa Rica.

c Accelerated soil erosion on a bare surface in Madagascar that was once covered by lush forest.

human activities are the removal of natural vegetation by agricultural practices such as plowing and overgrazing, and overexploitation for firewood and deforestation by logging operations. Exposed soil is simply more easily eroded than vegetation-covered soil.

Just how easily soil pulverized by plowing can be eroded by wind was clearly demonstrated during the 1930s, when a large area of the western United States lost millions of tons of topsoil. Several states in the Great Plains were particularly hard hit by drought, and wind eroded the exposed soils.

Wind is certainly effective in some areas, especially on exposed soils, but running water is capable of much greater erosion. Some soil is removed by **sheet erosion,** which erodes thin layers of soil over an extensive, gently sloping area by water not confined to channels. In contrast, **rill erosion** takes place when running water scours small, trough-like channels. If these channels are shallow enough to be eliminated by plowing, they are called *rills*, but if deeper than about 30 cm, they cannot be plowed over and are called *gullies* (▶ Figure 6.22a, b). Where *gullying* is extensive, croplands can no longer be tilled and must be abandoned.

Increased soil erosion invariably follows clearing of tropical rain forests for agriculture or logging operations (▶ Figure 6.22c). In either case, more rainwater runs off at the surface, gullying becomes more common, and flooding is more frequent because trees with their huge water-holding capacity are no longer present. Clearing woodlands in less humid regions such as much of the eastern United States also results in accelerated erosion, at least initially. Studies of lake deposits in several areas indicate that years of rapid erosion follow clearing and then fall off markedly when the land is covered by crops. Nevertheless, erosion rates may be 10 times greater than they were before clearing took place.

If soil losses to erosion are minimal, soil-forming processes keep pace and the soil remains productive, unless, of course, the soil is chemically or physically degraded, too. If the loss rate exceeds the rate of renewal, though, the most productive soil layer, horizon A, is removed, exposing horizon B, which is much less productive. High rates of soil erosion are obviously problems, but there are additional consequences. For one thing, eroded soil is transported elsewhere, perhaps onto roadways or into streams and rivers. Also, sediment

Table 6.2

Soil Conservation Practices

Terracing	Creating flat areas on sloping ground. One of the oldest and most effective ways of preserving soil and water.
Strip-cropping	Growing of different crops on alternate, parallel strips of ground to minimize wind and water erosion. Alternating strips of corn and alfalfa, for instance.
Crop Rotation	Yearly alternation of crops on the same land. Significant reduction in soil erosion when soil-depleting crops are alternated with soil-enriching crops.
Contour Plowing	Plowing along a slope's contours so that furrows and ridges are perpendicular to the slope.
No-till Planting	Planting seeds through the residue of a previously harvested crop.
Windbreaks	Planting trees or large shrubs along the margins of a field. Especially effective in reducing wind erosion.

accumulates in canals and irrigation ditches, and fertilizers and pesticides are carried into waterways and lakes.

Soil erosion problems experienced during the past, especially during the 1930s, motivated federal and state agencies to develop practices to minimize soil erosion on agricultural lands (Table 6.2). Crop rotation, contour plowing and strip-cropping (▶ Figure 6.23), and terracing have all proved effective and are now used routinely. No-till planting, in which the residue from a harvested crop is left on the ground to inhibit erosion, has also been effective.

Chemical and Physical Soil Degradation Soil undergoes chemical deterioration when its nutrients are depleted and its productivity decreases. Loss of soil nutrients is most notable in many of the populous developing nations where soils are overused to maintain high levels of agricultural productivity. Chemical deterioration is also caused by improper use of fertilizers and by the removal or reduction of certain kinds of vegetation. Examples of chemical deterioration are found everywhere, but it is most prevalent in South America, where it accounts for 30% of all soil degradation.

Other types of chemical deterioration are pollution and *salinization*, which occurs when the concentration of salts increases in soil, making it unfit for agriculture. Pollution can be caused by improper disposal of domestic, industrial, and mining wastes, oil and chemical spills, and the concentration of insecticides and pesticides in soils. Soil pollution is a particularly severe problem in eastern Europe.

Physical deterioration of soils results when soil particles are compacted under the weight of heavy machinery and livestock, especially cattle. Compacted soils are more costly to plow, and plants have a more difficult time emerging from them. Furthermore, water does not readily infiltrate, so more runoff occurs, which in turn accelerates the rate of water erosion.

Soil degradation is a serious problem in many parts of the world. In North America, with local exceptions, it is moderate but nevertheless of some concern. The rich prairie soils of the Midwestern United States and the Great Plains of the United States and Canada remain productive, although their overall productivity has decreased somewhat over the last several decades. In any case, lessons learned during the past have convinced farmers, government agencies, and the public in general that they can no longer take soil for granted or regard it as a resource that needs no nurturing.

▶ **Figure 6.23 Soil Conservation Practices** Contour plowing and strip-cropping are two soil conservation practices used on this farm. Contour plowing involves plowing parallel to the contours of the land to inhibit runoff and soil erosion. In strip-cropping row crops such as corn alternate with other crops such as alfalfa.

What Would You Do?

You have inherited a piece of property ideally located for everything you consider important. Unfortunately, as you prepare to have a house built, your contractor tells you that the soil is rich in clay that expands when wet and contracts when it dries. You nevertheless go ahead with construction but must now decide what measures to take to prevent damage to the structure. Make several proposals that might solve this problem. Which one or ones do you think would be the most cost effective?

▶ **Figure 6.24 Bauxite** Bauxite, the ore of aluminum, is made up of aluminum oxides and hydroxides as well as silica, silt, iron oxides, and clay minerals. It forms by intense chemical weathering of aluminum-rich rocks in the tropics.

Section 6.6 Summary

- Expansive soils contain clay minerals that expand when wet and shrink when they dry.

- Soil degradation is any loss of soil or decrease in soil fertility resulting from erosion, compaction, and contamination by pollutants.

6.7 Weathering and Natural Resources

We have already discussed various aspects of soils, which are certainly one of our most precious natural resources. Indeed, if it were not for soils, food production on Earth would be vastly different and capable of supporting far fewer people. In addition, other aspects of soils are important economically. We discussed the origin of laterites in response to intense chemical weathering in the tropics, and noted further that they are not very productive.

■ *How does the ore of aluminum form?*

Although soils in the tropics are not very productive, if the parent material is rich in aluminum, the ore of aluminum called *bauxite* forms because aluminum is nearly insoluble and accumulates in horizon B (▶ Figure 6.24). Some bauxite is found in several states, but at present it is cheaper to import it rather than mine these deposits, so both the United States and Canada depend on foreign sources of aluminum ore, mostly from Australia.

Bauxite and other accumulations of valuable minerals formed by the selective removal of soluble substances during chemical weathering are known *as residual concentrations*. Certainly bauxite is the best-known example of a residual concentration, but other deposits that formed in a similar fashion include ores of iron, manganese, clays, nickel, phosphate, tin, diamonds, and gold. Some of the sedimentary iron deposits in the Lake Superior region of the United States and Canada were enriched by chemical weathering when soluble parts of the deposits were carried away.

Most of the economic deposits of various clay minerals formed by sedimentary processes or by hydrothermal alteration of granitic rocks, but some are residual concentrations. A number of kaolinite deposits in the southern United States formed when chemical weathering altered feldspars in pegmatites or as residual concentrations of clay-rich limestones and dolostones. Kaolinite is a type of clay mineral used in the manufacture of paper and ceramics.

Chemical weathering is also responsible for gossans and ore deposits that lie beneath them. A *gossan* is a yellow to red deposit made up mostly of hydrated iron oxides that formed by oxidation and leaching of sulfide minerals such as pyrite (FeS_2). The dissolution of pyrite and other sulfides forms sulfuric acid, which causes other metallic minerals to dissolve, and these tend to be carried down toward the groundwater table, where the descending solutions form minerals containing copper, lead, and zinc. And below the water table the concentration of sulfide minerals forms the bulk of the ore minerals. Gossans have been mined for iron, but they are far more important as indicators of underlying ore deposits.

Section 6.7 Summary

- Weathering is responsible for the origin of some natural resources and for the concentration of others, such as bauxite and other residual concentrations. And of course soils are one product of weathering.

Review Workbook

ESSENTIAL QUESTIONS SUMMARY

6.1 Introduction
■ *How does weathering differ from erosion?*
By definition, weathering is the mechanical and chemical alteration of Earth materials at or near the surface, whereas erosion involves removing weathered materials from their place of origin—by running water or wind, for example.

6.2 Alteration of Minerals and Rocks
■ *What accounts for differential weathering and erosion, and what are their effects?*
Weathering and erosion take place at different rates even on the same body of rock because rocks are not compositionally and structurally homogeneous throughout, thereby producing uneven surfaces.

6.3 Mechanical Weathering—Disaggregation of Earth Materials
■ *How do freezing and thawing contribute to weathering?*
When water freezes in cracks in rocks it expands and then it contracts when it thaws, thus exerting pressure and opening the cracks wider. Repeated freezing and thawing diaggregates rocks into angular pieces that may tumble downslope and accumulate as talus.

■ *What are sheet joints and how do they form?*
Sheet joints are fractures that more or less parallel exposed rock surfaces, especially rocks now at the surface that formed under great pressure at depth. These joints form in response to pressure release; that is, when the rocks formed, they contained energy that is released by outward expansion.

■ *How do organisms contribute to mechanical and chemical weathering?*
Any organic activity such as tree roots growing in cracks contributes to mechanical weathering, whereas organic acids and the tendrils of mosses and lichens aid in the chemical alteration of parent material.

6.4 Chemical Weathering—Decomposition of Earth Materials
■ *What happens to rocks composed of carbonate minerals when they are exposed to acidic solutions?*
Rocks such as limestone ($CaCO_3$) are nearly insoluble in neutral or alkaline solutions, but they rapidly dissolve in acidic solutions because the atoms making up the minerals disassociate, that is, they separate and the rock dissolves.

■ *What takes place when potassium feldspars undergo hydrolysis?*
During hydrolysis hydrogen ions react with and replace positive ions in potassium feldspar, yielding clay minerals and substances in solution such as potassium and silica.

■ *How does mechanical weathering contribute to the rate of chemical weathering?*
The main effect of mechanical weathering is to reduce Earth materials into smaller pieces that have more surface area compared to their volume. Because chemical weathering is a surface process, the more surface exposed, the faster the weathering.

6.5 Soil and Its Origin
■ *What factors determine the rate of soil formation?*
Certainly climate is the most important factor because chemical processes operate faster where it is warm and wet. The type of parent material also plays a role because some rocks such as quartzite weather more slowly than basalt, for example.

■ *What is laterite, and where and how does it form?*
Laterite is a deep red soil typical of the tropics where chemical weathering is intense. In fact, laterites are made up of clays and the most insoluble compounds that were present in the parent material.

6.6 Expansive Soils and Soil Degradation
■ *Why do some soils expand and contract, and does expansion and contraction cause any problems?*
Soils with some types of clay minerals expand when wet because the clays absorb water, and then they contract as they dry out. Such volume changes damage structures such as foundations, roadways, and sidewalks that are sited on expansive soils.

■ *What is soil degradation and how does it take place?*
Any soil losses, physical changes, or chemical alteration is called soil degradation, and all lead to reduced soil productivity. Causes include erosion, compaction, and any kind of chemical pollution that inhibits plant growth.

6.7 Weathering and Natural Resources
■ *How does the ore of aluminum form?*
The ore of aluminum, called bauxite, forms in the tropics where chemical weathering is so intense that only the most insoluble compounds accumulate in the soil, which includes iron and aluminum compounds.

ESSENTIAL TERMS TO KNOW

chemical weathering (p. 175)
differential weathering (p. 171)
erosion (p. 170)
exfoliation (p. 172)
exfoliation dome (p. 172)
expansive soil (p. 188)
frost action (p. 171)
frost wedging (p. 172)
humus (p. 182)
hydrolysis (p. 178)
laterite (p. 183)

mechanical weathering (p. 171)
oxidation (p. 178)
parent material (p. 170)
pedalfer (p. 183)
pedocal (p. 183)
pressure release (p. 172)
regolith (p. 181)
rill erosion (p. 189)
salt crystal growth (p. 173)
sheet erosion (p. 189)
sheet joint (p. 172)

soil (p. 181)
soil degradation (p. 188)
soil horizon (p. 182)
solution (p. 175)
spheroidal weathering (p. 180)
talus (p. 172)
thermal expansion and contraction (p. 173)
weathering (p. 170)
zone of accumulation (p. 183)
zone of leaching (p. 182)

REVIEW QUESTIONS

1. What are residual concentrations and how do they form? Are any of them of economic value?
2. Why is groundwater in arid regions alkaline, whereas in humid regions it is acidic?
3. How do exfoliation domes form? Where would you go to see examples?
4. Draw a soil profile for a humid region and an arid area, and list the characteristics of each.
5. Why do people in tropical areas practice slash-and-burn agriculture?
6. Discuss three soil conservation practices that counteract losses to soil degradation.
7. Describe the result of spheroidal weathering and explain how it takes place.
8. Why is thermal expansion and contraction *not* a particularly effective type of mechanical weathering on Earth?
9. What are the similarities between the mechanical weathering processes called frost wedging and salt crystal growth?
10. Do gossans have any economic significance?

APPLY YOUR KNOWLEDGE

1. How do you account for the fact that some of the rock layers in Figure 7.13 have steep slopes whereas others have gentle slopes? Also, given that limestone is soluble in the presence of acidic groundwater, why does it have such a bold exposure at this location?

FIELD QUESTION

1. The two images below show exposures of granite. Explain what weathering phenomena account for their present surface expression. (An answer is given in the answer section in the back of the book.)

► Geo-focus Figure

GEOLOGY MATTERS

GEOLOGY IN FOCUS — The Dust Bowl

Prior to the 1930s, farmers had enjoyed a degree of success unparalleled in U.S. history. During World War I (1914–1918), the price of wheat soared, and after the war when Europe was recovering, the government subsidized wheat prices. High prices and mechanized farming resulted in more and more land being tilled. Even the weather cooperated, and land in the western United States that would otherwise have been marginally productive was plowed. Deep-rooted prairie grasses that held the soil in place were replaced by shallow-rooted wheat.

Beginning in about 1930, drought conditions prevailed throughout the country, but its consequences were particularly severe in the southern Great Plains. Some rain fell, but not enough to maintain agricultural production. And since the land, even marginal land, had been tilled, the native vegetation was no longer available to keep the topsoil from blowing away. And blow away it did—in huge quantities.

A large region in the southern Great Plains that was particularly hard hit by the drought, dust storms, and soil erosion came to be known as the Dust Bowl. Although its boundaries were not well defined, it included parts of Kansas, Colorado, and New Mexico, as well as the panhandles of Oklahoma and Texas.

Dust storms were common during the 1930s, and some reached phenomenal sizes (▶ Figure 1a). One of the largest occurred in 1934 and covered more than 3.5 million km². It lifted dust nearly 5 km into the air, obscured the sky over large parts of six states, and blew hundreds of millions of tons of soil eastward. The Soil Conservation Service reported dust storms of regional extent on 140 occasions during 1936 and 1937. Dust was everywhere. It seeped into houses, suffocated wild animals and livestock, and adversely affected human health.

The dust was, of course, the topsoil derived from the tilled lands. Blowing dust was not the only problem; sand piled up along fences, drifted against houses and farm machinery, and covered what otherwise might have been productive soils. Agricultural production fell precipitously, farmers could not meet their mortgage payments, and by 1935 tens of thousands were homeless, on relief, or leaving (▶ Figure 1b). Many of these people went west to California and became the migrant farm workers immortalized in John Steinbeck's novel *The Grapes of Wrath*.

The Dust Bowl was an economic disaster of great magnitude. Droughts had stricken the southern Great Plains before, and have done so since, but the drought of the 1930s was especially severe. Political and economic factors also contributed to the disaster. Due in part to the artificially inflated wheat prices, many farmers were deeply in debt; they had purchased farm machinery in order to produce more and benefit from the high prices. Feeling economic pressure because of their huge debts, they tilled marginal land and employed few, if any, soil conservation measures.

If the Dust Bowl has a bright side, it is that the government, farmers, and the public no longer take soil for granted or regard it as a substance that needs no nurturing. A number of soil conservation methods developed then have now become standard practices.

Figure 1
The Dust Bowl of the 1930s was a time of drought, dust storms resulting from wind erosion, and economic hardship. **(a)** This huge dust storm was photographed at Lamar, Colorado, in 1934. **(b)** By the mid- to late-1930s, thousands of people were on relief, homeless, or had left the Dust Bowl. Dorothea Lange took this photograph in 1939 of a homeless family of seven in Pittsburgh County, Oklahoma.

GEOLOGY MATTERS

GEOLOGY IN UNEXPECTED PLACES

Gravestones and Geology

Cemeteries are good places to study geology for at least two reasons. First, a variety of different rock types have been used for gravestones, monuments, and memorials. Two of the most popular rocks used are granite (and closely related rocks such as diorite) and marble, but any readily available rock may be used, including sandstone, slate, limestone, and gneiss. Second, most gravestones carry inscriptions with dates, so you can see how various rocks have weathered over time. For example, ▶ Figure 1a shows a gravestone in the cemetery in Deerfield, Massachusetts, that dates from 1733. Although you can still read the inscription, it is becoming faint. Furthermore, notice that the lower part of the gravestone shows more alteration than the top. Why? Because moisture tends to concentrate there.

Notice in ▶ Figure 1b that this bench, also in the Deerfield, Massachusetts, cemetery, is badly weathered. In addition to support at its four corners, it has a central pillar beneath it. The reason is that rock slabs supported only at their ends are weak (see Chapter 13). You cannot see any detail on the gravestones in the background, but the inscriptions are easily read on those made of slate (the black ones), some of which date from the 1660s. The gravestones made of sandstone that date from the mid- to late-1700s are barely legible. Why? Slate does not weather as rapidly as sandstone.

Another observation from Figure 1b is that some of the gravestones in the background are at odd angles; that is, they are tilted from their original vertical position. Assuming no vandalism, how can you account for this observation? (*Hint:* Remember that frost action is an agent of weathering.)

Certainly rock type plays a role in how rapidly gravestones weather, but so does climate. Lettering on gravestones in arid regions is legible for much longer than is lettering on the same type of stone in tropical areas. ▶ Figure 2 shows a burial vault placed on the ground because of the high water table on the tropical island of Rarotonga, which is part of the Cook Islands in the South Pacific. The vault is only 90 years old, and yet it is badly decomposed and the cross shows obvious signs of alteration. Notice too the vegetation growing between the pedestal for the cross and the burial vault.

In addition to the weathering expected from natural processes, gravestones and other stone monuments in industrial areas commonly show the effects of acid precipitation and staining by soot and particulate matter. Indeed, any stone used for any purpose changes with age; it simply does so more rapidly in some areas than in others. Even the pyramids in Egypt, though in an arid environment where weathering is less intense, show the signs of age.

Figure 1
(a) A gravestone in the cemetery at Deerfield, Massachusetts, that dates from 1733. **(b)** Bench and gravestones made of slate (black) and sandstone (gray).

(b)

Figure 2
This burial vault on the island of Rarotonga in the Cook Islands is on the ground surface because of the high water table. The vault is only 90 years old but shows obvious signs of decomposition, especially near the ground.

GEOLOGY MATTERS

GEOLOGY IN YOUR LIFE

Industrialization and Acid Rain

One result of industrialization is atmospheric pollution, which causes smog, possible disruption of the ozone layer, global warming, and acid rain. Acidity, a measure of hydrogen ion concentration, is measured on the pH scale (▶ Figure 1a). A pH value of 7 is neutral, whereas acidic conditions correspond to values less than 7, and values greater than 7 denote alkaline, or basic, conditions. Normal rain has a pH value of about 5.6, making it slightly acidic, but acid rain has a pH of less than 5.0. In addition, some areas experience acid snow and even acid fog with a pH as low as 1.7.

Several natural processes, including soil bacteria metabolism and volcanism, release gases into the atmosphere that contribute to acid rain. Human activities also produce added atmospheric stress, especially burning fossil fuels that release carbon dioxide and nitrogen oxide from internal combustion engines. Both of these gases add to acid rain, but the greatest culprit is sulfur dioxide released mostly by burning coal that contains sulfur that oxidizes to form sulfur dioxide (SO_2). As sulfur dioxide rises into the atmosphere, it reacts with oxygen and water droplets to form sulfuric acid (H_2SO_4), the main component of acid rain.

Robert Angus Smith first recognized acid rain in England in 1872, but not until 1961 did it become an environmental concern when scientists realized that acid rain is corrosive and irritating, kills vegetation, and has a detrimental effect on surface waters. Since then, the effects of acid rain are apparent in Europe (especially in eastern Europe) and the eastern part of North America, where the problem has been getting worse for the last three decades (▶ Figure 1b).

The areas affected by acid rain invariably lie downwind from plants that emit sulfur gases, but the effects of acid rain in these areas may be modified by the local geology. For instance, if the area is underlain by limestone or alkaline soils, acid rain tends to be neutralized, but granite has little or no modifying effect. Small lakes lose their ability to neutralize acid rain and become more and more acidic until various types of organisms disappear, and in some cases all life-forms eventually die.

Acid rain also causes increased chemical weathering of limestone and marble and, to a lesser degree, sandstone. The effects are especially evident on buildings, monuments, and tombstones as in Gettysburg National Military Park in Pennsylvania.

The devastation caused by sulfur gases on vegetation near coal-burning plants is apparent, and many forests in the eastern United States show signs of stress than cannot be attributed to other causes.

Millions of tons of sulfur dioxide are released yearly into the atmosphere in the United States. Power plants built before 1975 have no emission controls, but the problems they pose must be addressed if emissions are to be reduced to an acceptable level. The most effective way to reduce emissions from these older plants is with flue-gas desulfurization, a process that removes up to 90% of the sulfur dioxide from exhaust gases.

Flue-gas desulfurization has some drawbacks. One is that some plants are simply too old to be profitably upgraded. Other problems include disposal of sulfur wastes, the lack of control on nitrogen gas emissions, and reduced efficiency of the power plant, which must burn more coal to make up the difference.

Other ways to control emissions are burning low-sulfur coal, fluidized bed combustion, and conservation of electricity. Natural gas contains practically no sulfur, but converting to this alternative energy source would require the installation of expensive new furnaces in existing plants.

Acid rain, like global warming, is a worldwide problem that knows no national boundaries. Wind may blow pollutants from the source in one country to another where the effects are felt. For instance, much of the acid rain in eastern Canada actually comes from sources in the United States.

Figure 1
(a) Values less than 7 on the pH scale indicate acidic conditions, whereas those greater than 7 are alkaline. The pH scale is a logarithmic scale, so a decrease of one unit is a 10-fold increase in acidity. (b) Areas where acid rain is now a problem, and areas where the problem may develop.

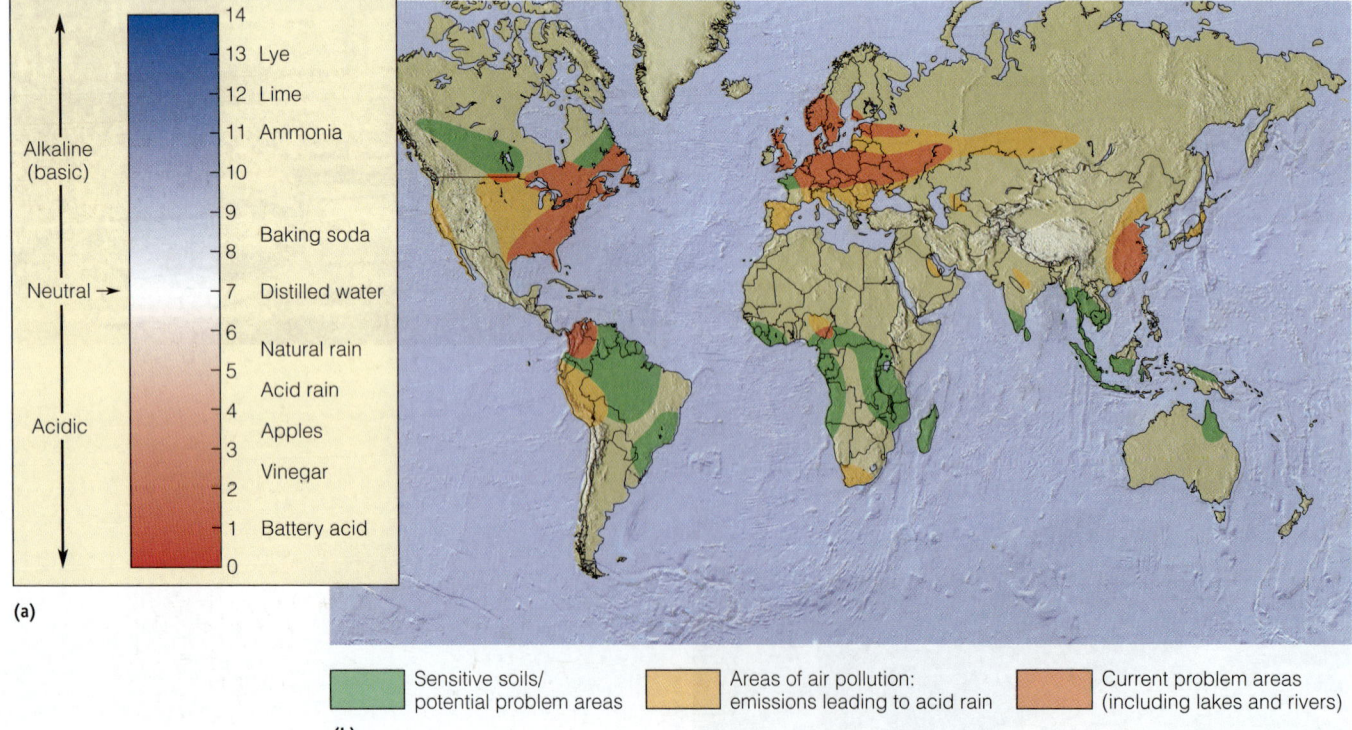

GEOLOGY MATTERS

Geology AND CULTURAL CONNECTIONS

The Agricultural Revolution

World civilization depends on the soils that produce the food that sustains more than 6.4 billion human beings on our planet. In that sense, soil is the foundation of our lives and all world culture.

Throughout most of our history, though, humans did not know how to farm. For millions of years we and our ancestors were hunter-gatherers, not farmers (▶ Figure 1). Fully modern *Homo sapiens,* people just like us, spent their lives gathering wild foodstuffs rather than farming. Hunter-gatherers are nomads, often on the move. People in such groups all do similar work. Life in such cultures is generally brief and difficult.

For reasons that we don't understand, we humans changed our way of living at several places around the globe some 8000 to 14,000 years ago. Different groups, separated by vast oceans, came to understand the advantages of agriculture at roughly the same time, near the end of the most recent cold spell in the Pleistocene Epoch (the Ice Age). The climate change does not explain this shift, however, because there had been earlier periods in the Pleistocene when cold spells had ended and people didn't alter their way of life.

The adoption of farming as a way of life was a momentous change for humans (▶ Figure 2). Farming ties people to a particular place, thus ending the nomadic lifestyle. If the crops grow well, farming can produce more food than any other system that came before it, allowing people to have at least small amounts of "free time." With more time, people learned to throw pots (needed to store grain and liquids like olive oil), make simple bricks (ideal for building simple homes), and weave sturdy baskets (useful at harvest). It wasn't long after these specialists appeared in village society that systems of writing down words and numbers appeared in human history. Thus we see that the division of labor made possible by farming is the foundation of modern society.

Once people were committed to farming, they began to domesticate large animals and taught them to pull plows. This dramatically increased the productivity of human farming efforts. For the first time in our history it wasn't only human muscle power that could be put to use for our benefit. Animal waste was also a natural fertilizer for farm fields.

Although farming brought astounding benefits, agriculture can be highly destructive to the very foundation of farming: soil. Today we have more areas of the planet under the plow than ever before. Erosion and soil degradation make some modern farm fields infertile, while many others are kept in production only through the use of artificial fertilizers. Individual farmers have a strong incentive to do whatever they can to increase their harvest, rather than doing what could protect their soils.

Sometimes governments can step in to promote practices that conserve soils. In the United States and Canada, farmers have been subsidized for planting different crops side by side in the same field, a practice that reduces erosion because the whole field isn't disturbed at the same time. In some parts of Central America, in contrast, there has been no regulation of agriculture and some peasant farmers burn down forests and jungles (▶ Figure 3), farm the land until the soil is exhausted, and then repeat the process in a new area, to the detriment of the soils (Figure 6.19).

Recently the technology of planting cereal grains has changed in a manner that allows farmers in the industrialized world to avoid plowing or disking their fields. The practice is called "no-till" farming and depends on a drill-like device that plants seeds in soil under the stubble of the harvested grain. So far it looks like no-till farming is a success for cereal crops in several areas around the globe, greatly reducing soil erosion because the farmer need not turn the soil over each year.

Our future as a species is clearly linked to the wise management of our soil resources. Like other conservation efforts, using soils sustainably depends on a combination of scientific understanding and political will. Your generation will likely see new pressures on world soil resources and may potentially contribute new solutions to perennial soil problems.

Figure 1
Humans were hunters and gatherers during most of their existence. Prehistoric artists, from 'L'Homme Primitif' by Louis Figuier, published Hachette, 1870 (engraving) (b/w photo) by French School, (19th century).

Figure 3
Slash and burn agriculture is practiced in some areas, particularly in Central America.

Figure 2
Farming tied people to a particular location and yielded more food than hunting and gathering. This image shows farmers in southern Egypt irrigating their field with water from the Nile River.

Chapter 7

Sediment and Sedimentary Rocks

ESSENTIAL QUESTIONS TO ASK

7.1 Introduction
- What are sediment and sedimentary rocks?

7.2 Sediment Sources, Transport, and Deposition
- Why are rounding and sorting important in sediments and sedimentary rocks?
- What is a depositional environment?

7.3 Lithification: Converting Sediment into Sedimentary Rock
- What processes bring about lithification of sediments?
- What is cementation and what are the common cements in sedimentary rocks?

7.4 The Types of Sedimentary Rocks
- How do conglomerate and sedimentary breccia differ?
- Why is quartz the most common mineral in sandstone?
- What are carbonate rocks?

7.5 Sedimentary Facies
- What criteria do geologists use to define sedimentary facies?
- What vertical sequence of facies develops during a marine transgression?

7.6 Read the Story Told by Sedimentary Rocks
- How is cross-bedding used to determine ancient current directions?
- What are fossils and how are they preserved?
- How do we know that the Navajo Sandstone formed as a desert dune deposit?

7.7 Important Resources in Sediments and Sedimentary Rocks
- What are stratigraphic and structural traps?
- What is oil shale?
- Why is banded iron formation such an important sedimentary rock?

Monument Valley, Utah
Branchlets of lightning like meandering rivers in the sky flood the darkness and illuminate the reddened edges of sedimentary rock below. This image was photographed in Monument Valley, Utah. Monument Valley crosses Southern Utah and Northern Arizona, and is protected as a Navajo Tribal Park. Whereas you can glimpse several buttes and spires on your own, many areas of the valley can only be seen on special tours led by official Navajo guides.
—A. W.

GEOLOGY MATTERS

GEOLOGY IN FOCUS:
Fossils and Fossilization

GEOLOGY IN UNEXPECTED PLACES:
Sandstone Lion

GEOLOGY AT WORK:
Exploring for Oil and Natural Gas—Susan M. Landon

GEOLOGY AND CULTURAL CONNECTIONS:
Deep Time

 This icon, which appears throughout the book, indicates an opportunity to explore interactive tutorials, animations, or practice problems available on the Physical GeologyNow website at http://now.brookscole.com/phygeo6.

7.1 Introduction

Suppose you are the only person in your community with any training in geology and you need to explain how the processes of weathering, erosion, deposition, and mineral crystallization are essential to the origin of sedimentary rocks. How would you explain all of this as well as convince your audience that there are practical reasons for understanding the origin of sedimentary rocks? To find your answers, you must combine what you already know with what you are about to learn in this chapter.

■ *What are sediment and sedimentary rock?*

Sediment, all of which comes from preexisting rocks, includes (1) all solid particles derived by weathering, (2) minerals that come from solutions such as seawater that contain chemical elements, and (3) minerals extracted from water by organisms to build their shells. Sediment, whatever its origin, is a loose aggregate of solids such as sand on a beach, gravel in a river channel, or mud on the seafloor. If these loose solids are bound together, they become **sedimentary rocks,** which by definition are made up of sediment; sand becomes sandstone, for example.

Earth's crust is made up mostly of *crystalline rocks*, a term that refers loosely to metamorphic rocks and igneous rocks, with the exception of igneous rocks composed of pyroclastic materials. Nevertheless, sediment and sedimentary rocks, making up perhaps only 5% of the crust, are the most commonly encountered Earth materials. Indeed, they cover about two-thirds of the continents and most of the seafloor except spreading ridges. Given what you know about plate tectonics, why are sediments lacking on the seafloor at and very near spreading ridges but found elsewhere on the seafloor?

All rocks are critical to our inquiries into Earth history, but sedimentary rocks are particularly significant because they preserve evidence of the surface processes that were responsible for them. For instance, a variety of rocks are present in John Day Fossil Beds National Monument in Oregon, including lava flows, layers of tuff, and sedimentary rocks composed of sand, gravel, and mud. Geologists have thoroughly studied the rocks and now know much about the igneous and sedimentary history of the area for a time spanning from

▶ **Figure 7.1a Sheep Rock (John Day Fossil Beds National Monument, Oregon)** The rocks and fossils at John Day Fossil Beds National Monument in Oregon provide a good record of the physical and biological events that took place there between 6 and 54 million years ago. At Sheep Rock the rocks are mostly sedimentary, but a small remnant of a lava flow is present on the hilltop. Also notice the small fault (fracture) that cuts through the rocks.

▶ **Figure 7.1b Mammal and vegetation restoration (John Day Fossil Beds National Monument)** A painting of the vegetation and mammals that lived here between 37 and 55 million years ago. From left to right: a carnivore, titanotheres, rhinoceroses, tapirs, and ancient horses. The climate was subtropical, but now it is semiarid.

54 million to 6 million years ago. Numerous fossils of land-dwelling animals and plants are found in some of the rock layers in the monument (▶ Figure 7.1), so we know something about the ancient ecosystem and climate of this area.

In addition to their importance in deciphering Earth and life history, some sediments such as sand and gravel, and sedimentary rocks, including coal and phosphate-rich rocks, are resources themselves. Or they are the host rocks for resources such as gold, petroleum, and natural gas. Most of the world's iron ore is derived from sedimentary rocks, and some kinds of sedimentary rocks may be alternative sources of petroleum in the future.

Section 7.1 Summary

- An important part of the rock cycle is weathering because it yields the raw materials for sedimentary rocks that are transported from their source and deposited elsewhere.

- Sediment and sedimentary rocks make up only a small part of Earth's crust, but they are the most commonly encountered Earth materials.

7.2 Sediment Sources, Transport, and Deposition

We noted in Chapter 6 that mechanical and chemical weathering yield solid particles and ions and compounds in solution that were derived from preexisting rocks. These materials, which are the raw materials for sedimentary rocks, may be eroded from their site of origin and transported some distance, but eventually transport ceases and the sediment is deposited in some geographic area called a *depositional environment* (▶ Figure 7.2). Our task here is to understand sediment transport and the processes operating in depositional environments.

Sediment Transport

Sediment is transported by running water, wind, glaciers, waves, and currents near the seashore. Sediment may be transported from a mountain range into an adjacent valley only a few kilometers away, or it may move from the most distant sources of the Mississippi River to the Gulf of Mexico. In all cases, though, the solid particles of sediment are modified during transport, so the sizes and shapes of particles in sedimentary rocks give us some idea about how they were transported in the first place.

One important criterion for classifying sedimentary particles is their size, particularly for *detrital sediment*—that is, the solid particles yielded by weathering. For instance, granite may break down into individual minerals (quartz, feldspars, and biotite) or small pieces of granite (rock fragments) (see Figure 6.1). Any detrital particle larger than 2 mm, regardless of composition, is gravel, whereas sand measures $\frac{1}{16}$–2 mm. We refer to mixtures of silt ($\frac{1}{256}$–$\frac{1}{16}$ mm) and clay ($>\frac{1}{256}$ mm) as mud.

Chemical sediment, in contrast, is composed of solids that were derived by inorganic chemical processes, such as the evaporation of seawater, as well as the activities of organisms. Clams, oysters, corals, and several other animals and some plants construct their skeletons of minerals they extract from seawater or more rarely lake water, especially aragonite or calcite ($CaCO_3$) and silicon dioxide (SiO_2).

■ *Why are rounding and sorting important in sediments and sedimentary rocks?*

During the transport of detrital sediment, *abrasion* reduces the size of particles, and the sharp corners and edges of gravel and sand are worn smooth as particles collide, a

▶ **Geo-focus Figure 7.2 Origin and Transport of Sediment** Whether derived from preexisting rocks by mechanical or chemical weathering, solid particles and ions and compounds in solution are transported and deposited elsewhere. If they are lithified, they become detrital and chemical sedimentary rocks.

SOURCE OF SEDIMENTARY MATERIALS

MECHANICAL WEATHERING
(gravel, sand, silt, clay–sized particles)

TRANSPORT

CHEMICAL WEATHERING
(clay minerals and ions, compounds in solution)

TO SITES OF DEPOSITION

processes called **rounding. Sorting,** which refers to the size distribution of particles in sediment or sedimentary rocks, also takes place during transport. Sediment and sedimentary rocks are well sorted if their particles are all about the same size, but if there is a wide range of sizes, they are poorly sorted

Rounding and sorting may seem like rather trivial rock properties, but both have important implications for other aspects of sediments and sedimentary rocks, such as how groundwater, liquid petroleum, and natural gas as well as various pollutants move through them. They are also useful for determining transport and depositional history, which may be important economically, a topic we cover more fully in a later section.

Deposition of Sediment

■ *What is a depositional environment?*

Regardless of how or how far sediment is transported, it is eventually deposited. That is, it accumulates in a specific area called its **depositional environment,** where physical, chemical, and biological processes impart distinctive features to the sediment. There is no completely satisfactory classification of depositional environments, but geologists recognize three broad depositional settings: continental, transitional, and marine, each with several specific depositional environments (▶ Figure 7.3). The continental depositional setting includes all areas on land where sediment accumulates, as in stream channels and their floodplains, lakes, deserts, and areas covered by and adjacent to glaciers. Deltas, beaches, tidal flats, and barrier islands are transitional environments because processes operating on land as well as those in the marine realm affect the deposits. A stream or river flowing into the sea deposits a delta, but waves and tides modify the deposit. Seaward of the transitional environments are the marine environments, where only marine processes are important as on the continental shelf and beyond (▶ Figure 7.3).

Section 7.2 Summary

● Sedimentary particles are classified in order of decreasing size as gravel, sand, silt, and clay (a mixture of silt and clay is mud).

● During transport, sediment may be sorted and rounded; both processes are important in how sediment and sedimentary rocks transmit fluids.

● Sediment may be transported and deposited by running water, wind, glaciers, or waves. It accumulates in depositional environments on the continents, along the seashore, or in the oceans.

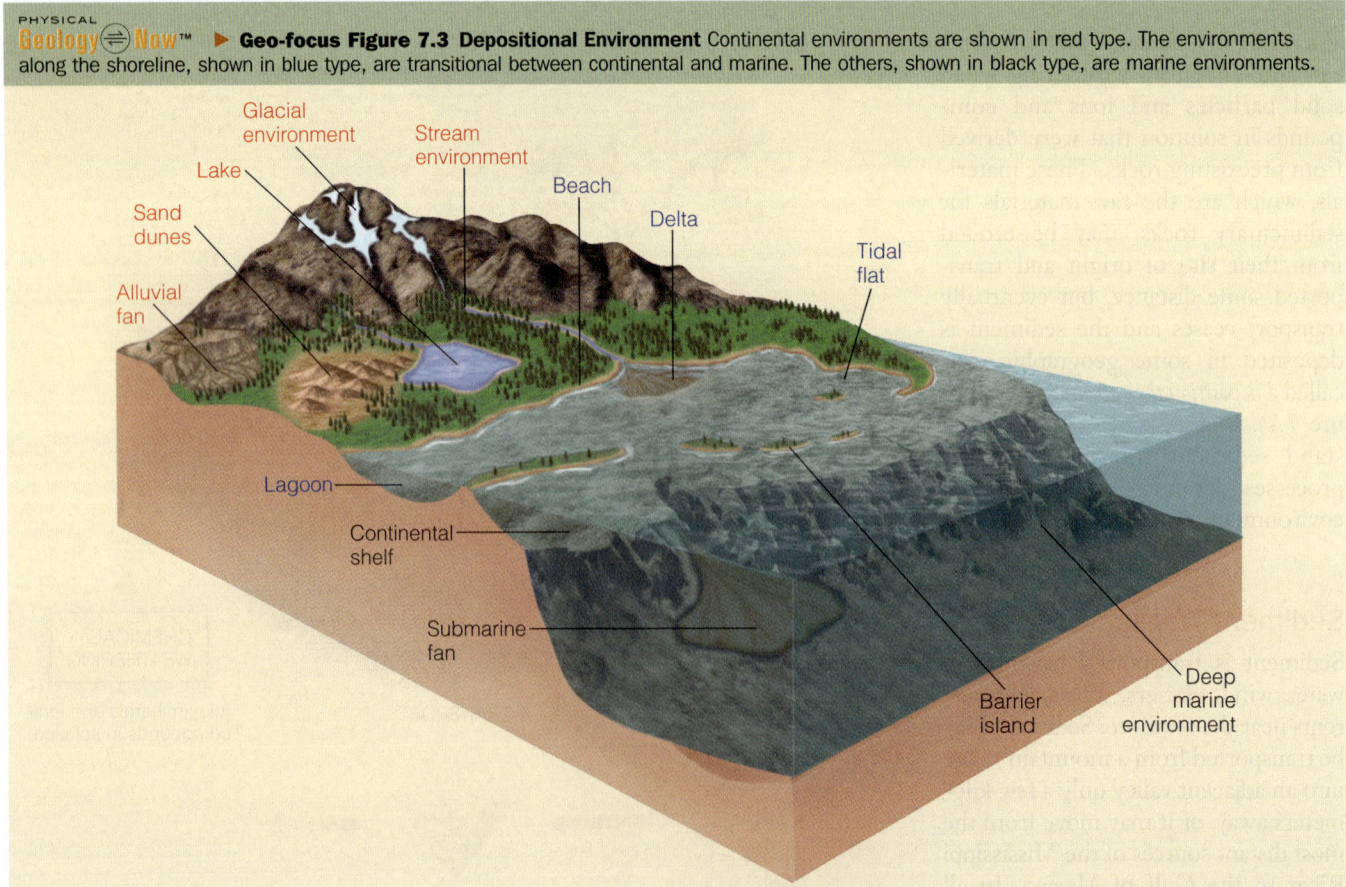

▶ **Geo-focus Figure 7.3 Depositional Environment** Continental environments are shown in red type. The environments along the shoreline, shown in blue type, are transitional between continental and marine. The others, shown in black type, are marine environments.

7.3 Lithification: Converting Sediment into Sedimentary Rock

We have presented examples of detrital sediment such as gravel in a stream channel, sand on a beach, and mud on the seafloor, all of which are loose aggregates of solids. The process of converting these deposits into sedimentary rocks is **lithification,** which involves compaction, cementation, or both (▶ Figure 7.4).

Compaction

■ *What processes bring about lithification of sediments?*

Both compaction and cementation are involved in lithification, but the importance of one or the other varies depending on the type of sediment. Consider two detrital deposits: one made up of mud (a silt and clay mixture) and the other of sand. In both cases, the sediment consists of solid particles and *pore spaces*, the voids between particles. **Compaction** reduces the volume of a deposit by decreasing the amount of pore space as particles fit more closely together because the

Geo-focus Figure 7.4 Lithification and Classification of Detrital and Chemical Sedimentary Rocks Detrital sediment (solid load) and chemical sediment (chemical load) are transported to a depositional environment, where they accumulate. If lithified—that is, compacted and cemented—the sediment is converted to sedimentary rock.

deposit settles under its own weight and the weight of any sediment deposited on top of it. Our hypothetical mud deposit may have 80% water-filled pore space, but after compaction its volume may be reduced by as much as 40%. The sand deposit with as much as 50% pore space is also compacted, but far less than the mud deposit.

Compaction is easily understood when you consider pouring a granular substance such as sugar or salt into a container. If the container is full and you have a small amount left over, what do you do? You shake the container, thereby causing compaction as the grains fit more closely together, thus creating more space for the remaining material.

Cementation

■ *What is cementation and what are the common chemical cements in sedimentary rocks?*

Cementation takes place when minerals crystallize in pore spaces and effectively bind the sedimentary particles to one another. The most common cements in sedimentary rocks are calcium carbonate ($CaCO_3$), silicon dioxide (SiO_2), and iron oxides and hydroxides, such as hematite (Fe_2O_3) and limonite [$FeO(OH) \cdot nH_2O$], respectively. In Chapter 6 we noted that calcium carbonate readily dissolves in water that contains a small amount of carbonic acid, and that chemical weathering of feldspars and other silicate minerals yields silica in solution. Minerals crystallize in pore spaces from circulating groundwater with these compounds in solution, thereby cementing sedimentary particles together. Compaction alone is enough for lithification of mud, but for sand and gravel, cementation is also necessary.

Iron oxide and hydroxide cements are much less common than calcium carbonate and silicon dioxide, but they account for the red, yellow, and brown sedimentary rocks found in many areas. Much of the iron cement in these rocks comes from the oxidation of ferromagnesian silicates in the original deposits, but circulating groundwater carries some of it in.

So far we have discussed the lithification of detrital sediments, but what about chemical sediments? Remember that both detrital and chemical sediments accumulate as aggregates of solids, but there is one example in which the unconsolidated sediment stage is skipped. Wave-resistant structures known as *reefs* are made up of skeletons of corals and mollusks as well as encrusting algae. Reefs are solids when they form, and they constitute a type of *limestone*. However, if waves rip apart a reef, the pieces so derived are sedimentary particles and these too are characterized as gravel-, sand-, silt-, and clay-sized.

By far the most common accumulations of chemical sediment consist of calcium carbonate ($CaCO_3$) mud, sand, and gravel. Compaction and cementation also take place in these deposits, converting them into various types of limestone, but compaction is generally less effective than in detrital sediment because cementation takes place soon after deposition. In any case, the cement is calcium carbonate derived by partial solution of some of the particles in the deposit.

PHYSICAL **Geology Now**™ Log into GeologyNow and select this chapter to work through a **Geology Interactive** activity on "The Rock Cycle" (click Rocks and the Rock Cycle→Rock Cycle).

Section 7.3 Summary

- Lithification takes place when sediment is compacted, cemented, or both, thereby converting it into sedimentary rock.

- Compaction involves a decrease in the volume of a deposit as it settles under its own weight and the weight of any overlying deposit. Cementation involves the crystallization of minerals in the pore spaces of sediment.

7.4 The Types of Sedimentary Rocks

Geologists recognize two broad categories of sedimentary rocks, *detrital* and *chemical*, and with the exception of coal all are derived from preexisting rocks. If you live in the northern part of Ontario or in northern Michigan, you see mostly metamorphic rocks, but if your home is in the Cascade Range in western North America, most of the nearby rocks are volcanic. Nevertheless, the continents as a whole have far more sediments and sedimentary rocks simply because they blanket much of the land surface. Our discussion so far has focused on the origin of sediment and its transport, deposition, and lithification. Now we consider the types of sedimentary rocks and their classification.

Detrital Sedimentary Rocks

All **detrital sedimentary rocks** are made up of solid particles (gravel, sand, silt, and clay) derived from preexisting rocks by mechanical and chemical weathering. They have a *clastic texture*, which means that they are composed of particles or fragments known as *clasts*. Although composition is used to modify some detrital rock names, they are classified primarily by the size of their constituent particles (▶ Figure 7.4).

■ *How do conglomerate and sedimentary breccia differ?*

Conglomerate and Sedimentary Breccia *Conglomerate* and *sedimentary breccia* are made up of gravel-sized particles—that is, detrital particles larger than 2 mm (▶ Figures 7.4 and 7.5). The only difference between the two is that the gravel in conglomerate is rounded, whereas sedimentary breccia has angular gravel called *rubble*. The composition of the gravel is irrelevant; it may be rock fragments or large pieces of individual minerals.

Conglomerate is fairly common, but sedimentary breccia is rare because gravel-sized particles become rounded quickly during sediment transport or by processes in the environments where gravel is deposited. For instance, gravel transported only a few kilometers in a stream or moved to and fro on a beach is invariably rounded because the particles grind against one another, thereby smoothing off corners and edges. Thus, if you encounter sedimentary breccia, you can be sure that its angular gravel was transported a very short distance.

Considerable energy is necessary to transport gravel, so it is deposited in only high-energy environments such as on beaches, in channels of swiftly flowing streams, and where glaciers move over the surface. In contrast, it is most unusual to find gravel far offshore in a lake or where wind transport and deposition took place, as in desert dunes.

■ *Why is quartz the most common mineral in sandstone?*

Sandstone Sandstone is rock made up of sand—that is, detrital particles that measure $\frac{1}{16}$ to 2 mm—regardless of composition. To be sandstone a rock must meet this criterion, but geologists use composition to define different types of sandstone (▶ Figures 7.4 and 7.6). *Quartz sandstone* is the most common variety and, as its name implies, is composed mostly of sand-sized particles of quartz. Another variety of sandstone known as *arkose* has at least 25% feldspar minerals.

Although there are local exceptions, quartz is by far the most common mineral in sandstone. But why should this be? After all, Earth's crust is made up of an estimated 51% feldspar minerals, 24% ferromagnesian silicates, 12% quartz, and 13% other minerals. Feldspars and quartz are both common, but quartz is hard, lacks cleavage, and is chemically stable, whereas feldspars are not quite as hard, possess two cleavage planes, and are much more susceptible to chemical weathering. Likewise, the ferromagnesian silicates have cleavage planes and are chemically unstable (see Table 6.1). There are in addition to quartz other chemically stable, hard minerals such as zircon and tourmaline, but they are rare in source rocks so they are found in very small quantities in sandstone.

Sandstone forms in several depositional environments, such as stream channels, desert dunes, beaches and barrier islands, deltas, and the continental shelf. Sandstone is a common sedimentary rock that is easily recognized by its gritty appearance and feel.

Mudrocks *Mudrock* is a general term that encompasses all detrital sedimentary rocks composed of silt- and clay-sized particles (▶ Figures 7.4 and 7.7). Among the mudrocks, *siltstone* is composed mostly of silt-sized particles, *mudstone* is a mixture of silt and clay, and *claystone* has mostly clay-sized particles. Some mudstones and claystones are designated as *shale* if they are fissile, meaning that they break along closely spaced parallel planes (▶ Figure 7.7a).

About 40% of all sedimentary rocks are mudrocks, making them more abundant than conglomerate and sandstone. Because silt- and clay-sized particles are so small, they are transported by weak currents and kept suspended in water by minor turbulence. As a result, deposition takes place only where currents and turbulence are at a minimum, as in the quiet offshore waters of lakes, in lagoons, and on river floodplains.

Chemical Sedimentary Rocks

Remember from Chapter 6 that during chemical weathering various compounds and ions are taken into solution and transported elsewhere. These are the raw materials for **chemical sedimentary rocks,** so named because they result from inorganic chemical processes or from the

▶ **Figure 7.5 Conglomerate and Sedimentary Breccia**

ⓐ A layer of conglomerate overlying sandstone. The largest clast measures about 30 cm across.

ⓑ Sedimentary breccia is made up of angular gravel.

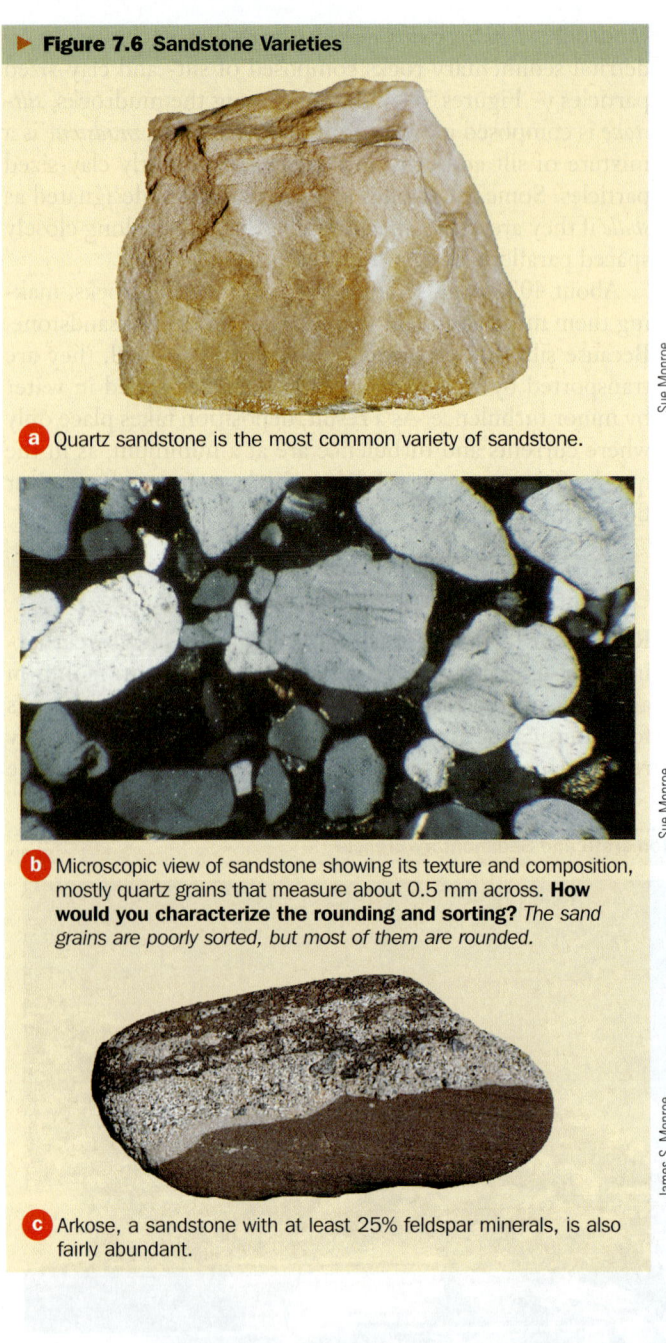

▶ Figure 7.6 Sandstone Varieties

a Quartz sandstone is the most common variety of sandstone.

b Microscopic view of sandstone showing its texture and composition, mostly quartz grains that measure about 0.5 mm across. **How would you characterize the rounding and sorting?** *The sand grains are poorly sorted, but most of them are rounded.*

c Arkose, a sandstone with at least 25% feldspar minerals, is also fairly abundant.

chemical activities of organisms (▶ Figure 7.4). Because organisms are so important in the origin of some of these rocks, the rocks are assigned to a subcategory called **biochemical sedimentary rocks** (Table 7.1).

Some chemical sedimentary rocks have a *crystalline texture*, meaning that they are made up of an interlocking mosaic of mineral crystals just as in metamorphic and most igneous rocks. Many others, however, have a clastic texture much like detrital sedimentary rocks; some limestones, for example, are made up of fragments of shells.

■ *What are carbonate rocks?*

Limestone and Dolostone *Limestone*, composed of calcite ($CaCO_3$), and *dolostone*, composed of dolomite [$CaMg(CO_3)_2$], are **carbonate rocks** because both contain the carbonate radical $(CO_3)^{-2}$ (see Figure 3.12). In Chapter 6 we noted that calcite dissolves rapidly in acidic water, but the chemical reaction leading to dissolution is reversible, so under some circumstances calcite and thus limestone form by inorganic chemical precipitation. A finely crystalline type of limestone called *travertine* that forms around hot springs is one example. And some limestones contain small spherical grains called *ooids* (▶ Figure 7.8a), which form as layers of calcite are chemically precipitated around a nucleus such as a sand grain or shell fragment. Lithified deposits of ooids are *oolitic limestones*.

Much of the world's limestone has a large component of calcite that was extracted from seawater by the activities of organisms and is thus biochemical sedimentary rock (Table 7.1). Many marine organisms, including corals, clams, and oysters, construct their shells of aragonite, an unstable form of calcium carbonate that alters to calcite, the main constituent of limestone. The limestone known as *coquina* is made up entirely of shell fragments cemented by calcium carbonate, *chalk* is a type of limestone consisting of microscopic shells, and fragmented or entire shells are quite common in many other limestones (▶ Figure 7.8b, c, d).

Dolostone resembles limestone, but nearly all of it is formed secondarily by the alteration of limestone. Geologists agree that dolostone forms when magnesium replaces some of the calcium in calcite, thereby converting calcite ($CaCO_3$) to the mineral dolomite [$CaMg(CO_3)_2$]. One way this might happen is in a lagoon where seawater evaporates, and much of the calcium in the water is used in calcite ($CaCO_3$) and gypsum ($CaSO_4 \cdot 2H_2O$). Under these conditions, magnesium (Mg) becomes concentrated in the water, which then becomes denser, and permeates any preexisting limestone, converting it to dolostone.

Evaporites Seawater and some lake waters are salty, that is, they contain ions and radicals such as sodium (Na), chlorine (Cl), calcium (Ca), magnesium (Mg), potassium (K), carbonate (CO_3), sulfate (SO_4), and many others. If these waters begin to evaporate, the concentration of dissolved materials increases relative to the volume of water, and eventually the saturation point is reached and minerals begin to crystallize and settle to the bottom. Minerals and their corresponding sedimentary rocks that form by this process are **evaporites.** Many evaporites are known, but only two are particularly common, *rock gypsum*, composed of the mineral gypsum ($CaSO_4 \cdot 2H_2O$),* and *rock salt*, which is made up of halite (NaCl) (Table 7.1 and ▶ Figure 7.9). Both have a crystalline texture.

Evaporites are deposited in bodies of water like lagoons in arid regions, such as the Persian Gulf, and in some arid region lakes, including the Great Salt Lake in Utah and the Dead Sea on the border between Israel and Jordan. In all cases, the first thing to form when evaporation begins is a small amount of calcium carbonate (calcite) because it is the

---

*When deeply buried, gypsum loses its water and is converted to anhydrite ($CaSO_4$).

▶ **Figure 7.7** Mudrock

a Exposure of shale in Tennessee. Notice that the rock breaks along closely spaced planes, so it is fissile.

b This sedimentary rock exposure in Montana consists of mudstone.

least soluble, followed by rock gypsum and rock salt, and then a variety of others in very small quantities.

Evaporites are not very common compared with the abundance of mudrocks, sandstone, and limestone/dolostone. Nevertheless, vast deposits of rock salt and rock gypsum are found in parts of Michigan, Ohio, and New York; the Louann Salt of the Gulf Coast underlies a large area; and similar deposits are present in Texas, Canada, Germany, and elsewhere. Rock salt, rock gypsum, and several other evaporites are important resources (discussed in a later section).

TABLE 7.1
Classification of Chemical and Biochemical Sedimentary Rocks

Chemical Sedimentary Rocks

Texture	Composition	Rock Name	
Varies	Calcite ($CaCO_3$)	Limestone	Carbonate rocks
Varies	Dolomite [$CaMg(CO_3)_2$]	Dolostone	
Crystalline	Gypsum ($CaSO_4 \cdot 2H_2O$)	Rock gypsum	Evaporites
Crystalline	Halite (NaCl)	Rock salt	

Biochemical Sedimentary Rocks

Texture	Composition	Rock Name
Clastic	Calcite ($CaCO_3$) shells	Limestone (various types such as chalk and coquina)
Usually crystalline	Altered microscopic shells of SiO_2	Chert (various color varieties)
—	Carbon from altered land plants	Coal (lignite, bituminous, anthracite)

Figure 7.8 Limestone Varieties

a This limestone is made up partly of ooids (see inset), which are rather spherical grains of calcium carbonate.

b Coquina is limestone composed of broken shells.

c The rock in these sea cliffs in Denmark is chalk, a type of limestone consisting of microscopic shells.

d Limestone with numerous fossil shells.

Chert *Chert*, composed of microscopic crystals of quartz (SiO_2), is hard and breaks with a conchoidal (smoothly curved) fracture like that seen in quartz and obsidian (Table 7.1). Color varieties include *flint*, which is black because it contains organic matter, and *jasper*, which is red, brown, or yellow due to iron oxides. Because chert is so hard and lacks cleavage, it can be shaped into sharp cutting edges and points for knives, spear points, and arrowheads.

Chert is found as irregular masses or *nodules* within other rocks, especially limestones, and as distinct layers of *bedded chert* (▶ Figure 7.10). Chert nodules are clearly secondary; that is, they have replaced some of the host limestone by chemical precipitation. Some bedded chert may also form by chemical precipitation, but because so little silica is dissolved in seawater, a biochemical origin is more likely. It probably forms as deposits of shells of silica-secreting organisms such as radiolarians and diatoms accumulate on the seafloor. Unfortunately, these shells are easily altered, so the biochemical origin of bedded chert is obscured.

Coal Compressed, altered remains of land plants make up *coal*, so it is a biochemical sedimentary rock. Coal forms in oxygen-deficient swamps and bogs where there is too little oxygen to oxidize the wastes of bacteria. Accordingly, the bacteria die, decay ceases, and the vegetation does not completely decompose but rather becomes an organic muck, which when buried and compressed becomes *peat* (▶ Figure 7.11a). Where peat is abundant, as in Ireland and Scotland, it is used for fuel.

If peat, which represents the first step in forming coal, is further altered, it is converted to dull black or brown coal called *lignite*, in which plant remains are clearly visible (▶ Figure 7.11b). During the change from peat to lignite, the easily vaporized or volatile elements of the vegetation, such as oxygen, nitrogen, and hydrogen, are driven off, enriching the residue in carbon; peat has about 50% carbon,

Figure 7.9 Evaporites

a This cylindrical core of rock salt was taken from an oil well in Michigan.

b Rock gypsum. When deeply buried, gypsum ($CaSO_4 \cdot 2H_2O$) loses its water and is converted to anhydrite ($CaSO_4$).

Figure 7.10 Chert Nodules and Bedded Chert

a These chert nodules are in limestone at the Valley of the Kings in Egypt. Part of the lower nodule has broken away, but notice that it is not a continuous layer.

b Bedded chert exposed in Marin County, California. Most of the layers are about 5 cm thick.

whereas lignite has about 70%. Further changes may take place that change lignite to *bituminous coal*, which has about 80% carbon. It is black and so thoroughly altered that plant remains are rarely seen (▶ Figure 7.11c). The highest-grade coal—that is, coal with the most carbon, about 98%, and thus the best fuel—is *anthracite*. It is essentially a metamorphic type of coal (see Chapter 8).

PHYSICAL GeologyNow™ Log into GeologyNow and select this chapter to work through a **Geology Interactive** activity on "Rock Laboratory" (click Rocks and the Rock Cycle→Rock Laboratory).

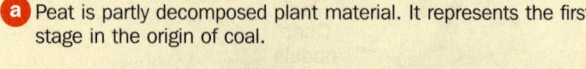

Figure 7.11 The Origin of Coal

a Peat is partly decomposed plant material. It represents the first stage in the origin of coal.

b Lignite is a dull variety of coal in which plant remains are still visible.

c Bituminous coal is shinier and darker than lignite, and only rarely are plant remains visible.

Section 7.4 Summary

• Sedimentary rocks are classified as *detrital* or *chemical*, although a subcategory of the latter called *biochemical* recognizes the importance of organisms in the origin of some rocks.

• Detrital sedimentary rocks include conglomerate and sedimentary breccia, sandstone, and several varieties of mudrocks, some of which are designated as shale if they are fissile.

• Carbonate rocks include *limestone* and *dolostone*, both of which contain the carbonate radical $(CO_3)^{-2}$. Dolostone forms when magnesium replaces some of the calcium in the calcium carbonate of limestones.

• Evaporites include rock salt and rock gypsum, both of which form when water with ions and radicals evaporates. Chert forms when silicon dioxide replaces other rocks or shells accumulate, and coal results from the altered remains of land plants.

7.5 Sedimentary Facies

■ *What criteria do geologists use to define sedimentary facies?*

Deposition of sediment under specific conditions yields **sedimentary facies,** which are bodies of sediment with distinctive physical, chemical, and biological attributes. Geologists use any aspect of sediments or sedimentary rocks that makes them recognizably different from adjacent rocks of the same or approximately the same age to define a sedimentary facies. ▶ Figure 7.12, for instance, shows a sandstone facies, a shale facies, and a limestone facies. The reason the sediments in Figure 7.12 vary laterally is that different processes operated simultaneously in adjacent depositional environments. Sand was deposited in the high-energy nearshore environment, whereas mud and carbonate sediments accumulated at the same time in the adjacent low-energy environments.

Marine Transgressions and Regressions

■ *What vertical sequence of facies develops during a marine transgression?*

In the lower part of the Grand Canyon of Arizona three facies are found in a vertical succession consisting of sandstone followed upward by shale and finally limestone (▶ Figure 7.13). All of these rocks contain evidence, including fossils, showing that they were deposited in transitional and marine environments. The question is: What accounts for the presence of marine facies far from the sea, and how were these facies deposited in the order observed? These and similar deposits elsewhere formed during a time when sea level rose with respect to the land, giving rise to a **marine transgression.** During a marine transgression, the shoreline migrates landward, as do the environments and facies that parallel the shoreline thereby covering more and more of a continent (▶ Fig-

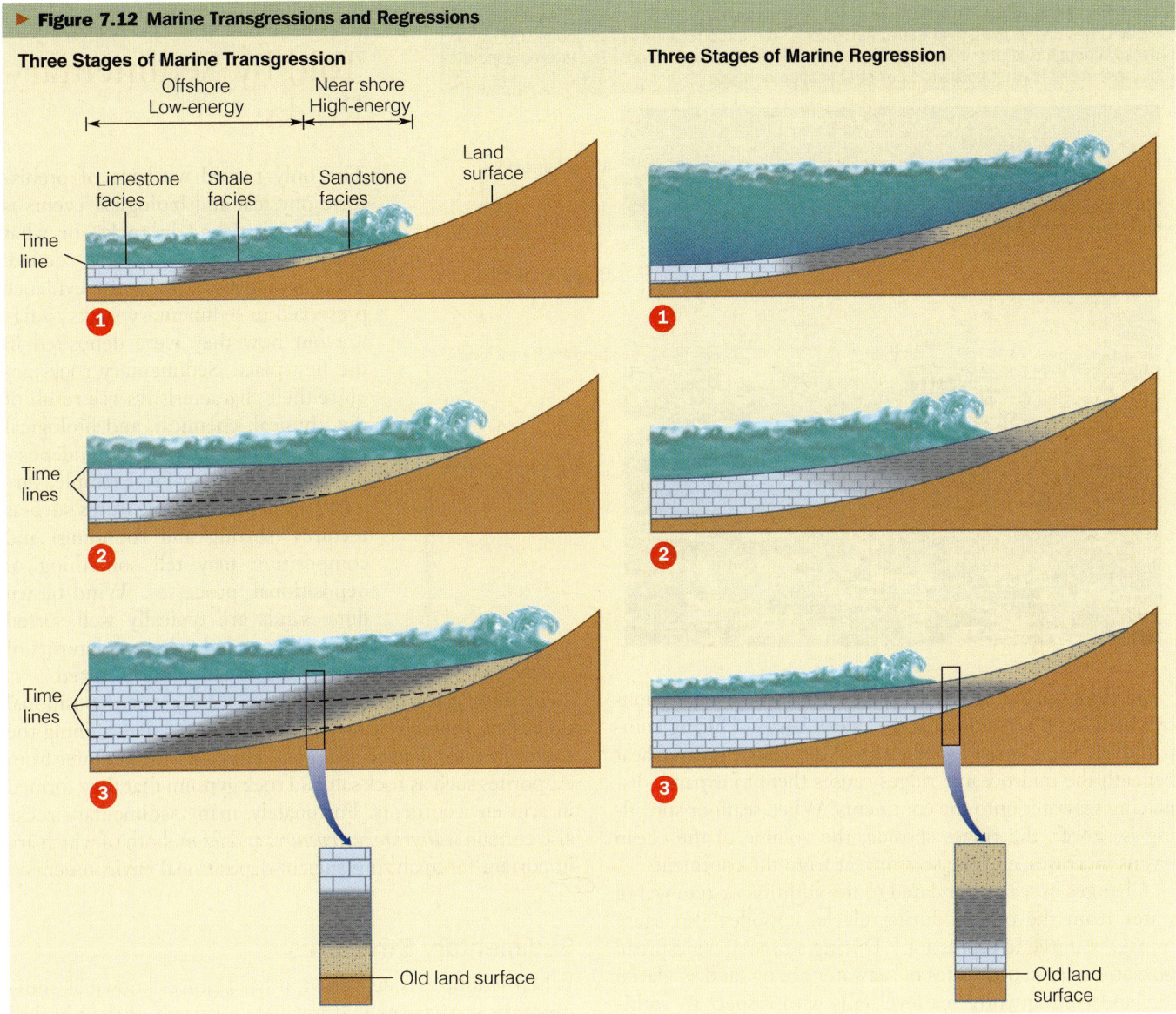

▶ Figure 7.12 Marine Transgressions and Regressions

ure 7.12a). Accordingly, the offshore facies are deposited on the facies of the nearshore environment.

Another important aspect of marine transgressions is that individual facies are deposited over vast geographic areas. It is true that the nearshore environment is long and narrow at any particular time, but deposition takes place continuously as the nearshore environment migrates landward. The sand deposited in this environment may be tens to hundreds of meters thick, but its length and width are measured in hundreds of kilometers. Indeed, the Tapeats Sandstone in Figure 7.13, although it goes by different names elsewhere, is found over a large area from southern California to South Dakota.

Obviously the sea is no longer present in Arizona, so the transgression responsible for the vertical facies sequence in Figure 7.13 must have ended. In fact it did, during a **marine regression,** when sea level fell with respect to the continent and the environments that paralleled the shoreline migrated seaward. In other words, a marine regression is the opposite of a marine transgression, and it yields a vertical sequence with nearshore facies overlying offshore facies (▶ Figure 7.12b). Marine regressions also account for the deposition of facies over huge geographic areas.

Causes of Marine Transgressions and Regressions

Uplift or subsidence of the continents, rates of seafloor spreading, and the amount of seawater frozen in glaciers are sufficient to cause marine transgressions and regressions. If a continent is uplifted, it rises with respect to sea level, the shoreline moves seaward, and a regression ensues. The opposite type of movement, subsidence, results in a transgression.

▶ **Figure 7.13 Grand Canyon** View of the Tapeats Sandstone, Bright Angel Shale, and Muav Limestone in the Grand Canyon in Arizona. The three rock layers are about 300 m thick, although part of the Tapeats Sandstone is buried beneath debris. The layered aspect of these rocks is what geologists call stratification or bedding.

Muav Limestone

Bright Angel Shale

Tapeats Sandstone

Alan Mayo/GeoPhoto Publishing

Seafloor spreading causes transgressions and regressions by changing the volumes of the ocean basins. During comparatively rapid episodes of seafloor spreading, greater heat beneath the mid-oceanic ridges causes them to expand, displacing seawater onto the continents. When seafloor spreading is slower, the ridges subside, the volume of the ocean basins increases, and the seas retreat from the continents.

Changes in sea level related to the addition or removal of water from the oceans during glacial episodes also cause transgressions and regressions. During a time of widespread glaciation, large quantities of seawater are on land as glacial ice, and consequently sea level falls with respect to continents. When glaciers melt, the water returns to the seas, and sea level rises (see Chapter 17).

Section 7.5 Summary

- Sedimentary facies are bodies of sediment or sedimentary rocks that are recognizably different from adjacent sediments or rocks of the same or about the same age. They result from simultaneous deposition in adjacent but different environments.

- Vertical sequences of rocks with offshore facies overlying nearshore facies form when sea level rises, causing a marine transgression. A rise in the land relative to sea level causes a marine regression, which results in nearshore facies overlying offshore facies.

- Marine transgressions and regressions are caused by uplift or subsidence of the continents, rates of seafloor spreading, and the amount of seawater frozen in glaciers on land.

7.6 Read the Story Told by Sedimentary Rocks

The only record we have of prehistoric physical and biological events is the one preserved in rocks, or what geologists call the **geologic record.** Thus geologists evaluate the evidence preserved in sedimentary rocks to figure out how they were deposited in the first place. Sedimentary rocks acquire their characteristics as a result of the physical, chemical, and biological processes that operated in the depositional environment. For example, various sedimentary rock features such as textures (sorting and rounding) and composition may tell something of depositional processes. Wind-blown dune sands are typically well sorted and well rounded, whereas deposits of glaciers are usually poorly sorted.

In most cases composition tells little about depositional processes, although it may be important for determining the source area for detrital sediment. However, we can infer from evaporites such as rock salt and rock gypsum that they formed in arid environments. Fortunately, many sedimentary rocks also contain *sedimentary structures* and *fossils*, both of which are important for analyzing ancient depositional environments.

Sedimentary Structures

When sediment is deposited, it has features known as **sedimentary structures** that formed as a result of the physical and biological processes that operated in the depositional environment. For instance, most sedimentary rocks show some kind of layering or, more precisely, **strata** or **beds,** which vary from less than a millimeter to many meters thick

What Would You Do?

You live in the continental interior where sedimentary rocks are well exposed. A local resident tells you of a nearby location where sandstone and mudstone with dinosaur fossils are overlain first by seashell-bearing sandstone, followed upward by shale, and finally by limestone with fossil corals and sea lilies. How would you explain the presence of marine fossils so far from the sea? How would you summarize the events responsible for the deposition of all of the rock layers?

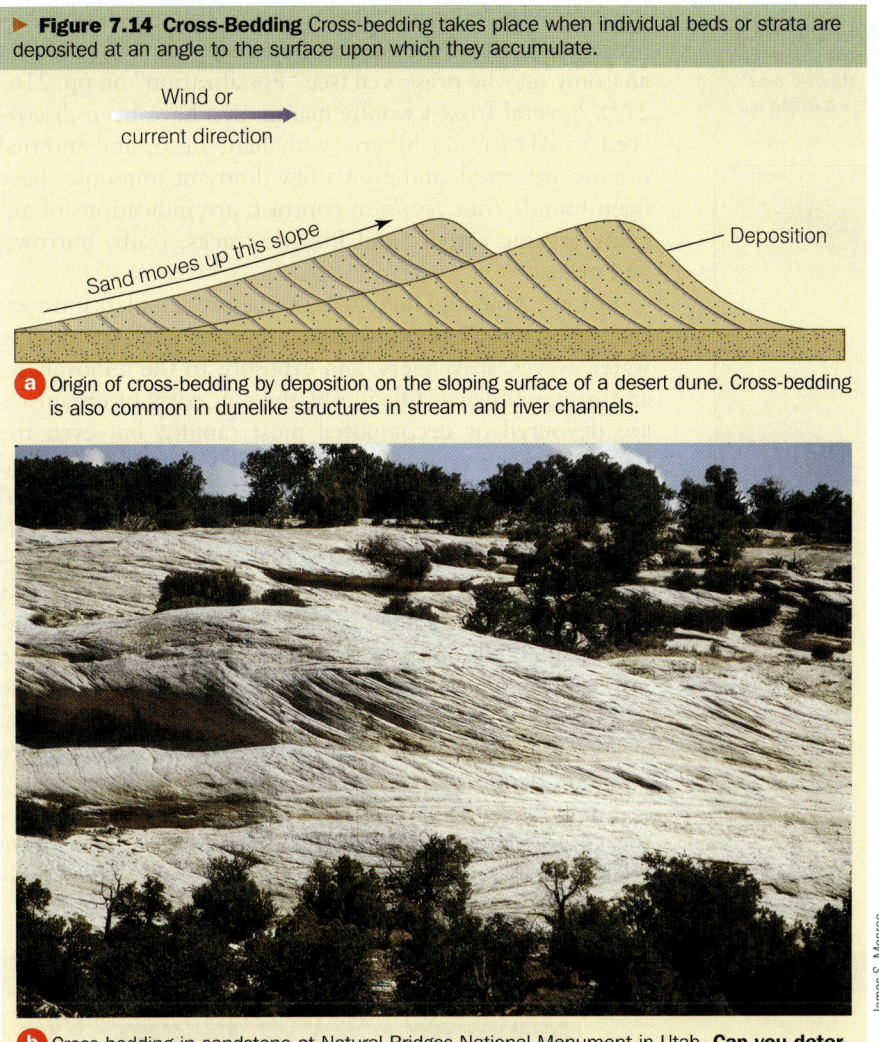

▶ **Figure 7.14 Cross-Bedding** Cross-bedding takes place when individual beds or strata are deposited at an angle to the surface upon which they accumulate.

a Origin of cross-bedding by deposition on the sloping surface of a desert dune. Cross-bedding is also common in dunelike structures in stream and river channels.

b Cross-bedding in sandstone at Natural Bridges National Monument in Utah. **Can you determine the current direction from the cross-beds?** *The current moved from left to right.*

(▶ Figure 7.13). These layers may be separated from one another by a distinct surface called a *bedding plane*, above and below which the rocks differ in color, composition, texture, or a combination of features. Abrupt changes like this indicate either rapid changes in the type of sediment deposited or a period of deposition followed by nondeposition and then renewed deposition. In contrast, a layer may grade upward from one rock type into another, indicating a gradual change in the type of deposits.

■ *How is cross-bedding used to determine ancient current directions?*

Many sedimentary rocks possess **cross-bedding,** in which layers are arranged at an angle to the surface upon which they accumulate (▶ Figure 7.14a). Cross-bedding is common in desert dunes that have an asymmetrical profile with a steep downwind slope and a gentle upwind side. Wind transports sand up the gentle slope, and then the sand cascades down the steep slope to form a cross-bed (▶ Figure 7.14b). Water currents are also responsible for cross-bedding in shallow marine environments and in stream channels. Because cross-beds form on the downwind or downcurrent side of dunelike structures, they are good indications of ancient current directions, or *paleocurrents*.

In **graded bedding,** grain size decreases upward within a single layer. Although some graded bedding forms in stream channels when the flow velocity gradually diminishes, most of it develops quickly from turbidity currents, which are underwater flows of sediment–water mixtures that are denser than sediment-free water. Turbidity currents move downslope along the seafloor or a lake floor until they reach a comparatively level area, where they slow down and deposit sediment, with the largest particles followed in succession by progressively smaller ones (▶ Figure 7.15).

Small (less than 2 cm high) alternating ridges and troughs called **ripple marks** are common on the surfaces of sand beds. Some ripple marks are asymmetrical in cross section, with a steep down-current slope and a gentle up-current slope. These are known as *current ripple marks* and result from currents that flow in one direction, as in stream channels (▶ Figure 7.16a, b). Wind blowing over sand also creates current ripple marks, but these have much straighter crests than those generated by running water (▶ Figure 7.16c). In contrast, the to and fro motion of waves produces ripple marks that tend to be symmetrical in cross section and have ridges that nearly parallel one another. Known as *wave-formed ripple marks*, they form mostly in shallow, nearshore waters of oceans and lakes (▶ Figure 7.17).

Intersecting fractures known as **mud cracks** form when clay-rich sediments dry and shrink (▶ Figure 7.18). Mud cracks in ancient sedimentary rocks indicate that deposition took place where periodic drying was possible, as on tidal flats, floodplains, and near lakeshores. Mud cracks, some ripple marks, graded bedding, and cross-bedding are useful in areas where sedimentary rocks have been deformed—that is, crumpled and fractured—to determine which layers are youngest and oldest.

Fossils

■ *What are fossils and how are they preserved?*

Fossils are the remains or traces of ancient organisms preserved in rocks. The actual remains of organisms, such as bones, shells, and teeth, are known as *body fossils*. Although

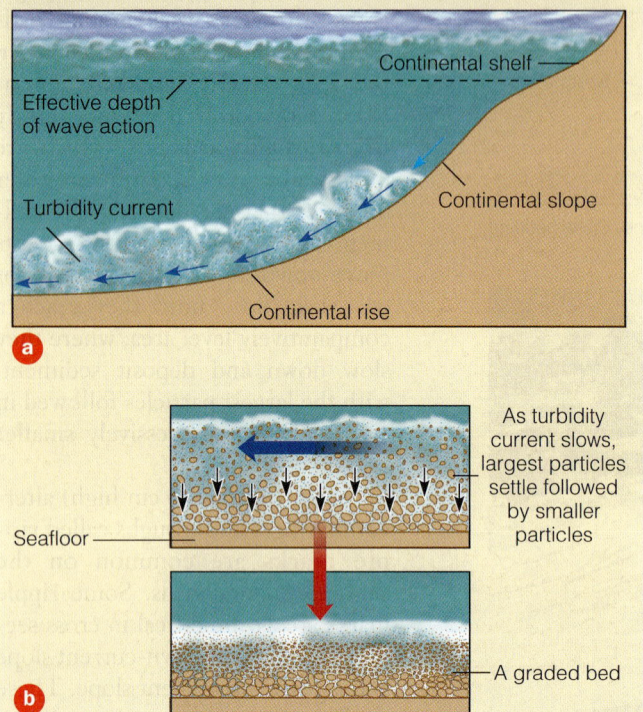

▶ **Figure 7.15 Origin of Graded Bedding in Which Particle Size Decreases Upward Within a Single Layer** (a) A turbidity current flows downslope along the seafloor or a lake bottom. (b) The flow slows and deposits progressively smaller particles, thus forming a graded bed.

the hard skeletal parts of organisms are the most easily preserved as fossils, under some conditions even the soft anatomy may be preserved (see "Fossilization" on pp. 216–217). Several frozen woolly mammoths have been discovered in Alaska and Siberia with hair, flesh, and internal organs preserved, and even a few dinosaur mummies have been found. *Trace fossils*, in contrast, are indications of ancient organic activity and include tracks, trails, burrows, and nests.

For any potential body fossil to be preserved, it must escape the ravages of destructive processes such as running water, waves, scavengers, and exposure to the atmosphere, and bacterial decay. Obviously, the soft parts of organisms are devoured or decomposed most rapidly, but even the hard skeletal parts are destroyed unless buried and protected in mud, sand, or volcanic ash. Even if buried, bones and shells may be dissolved by groundwater or destroyed by alteration of the host rock during metamorphism. Nevertheless, fossils are common. The remains of microscopic plants and animals are the most common, but these require specialized methods of recovery, preparation, and study and are not sought out by casual fossil collectors. Shells of marine animals are also common and easily collected in many areas, and even the bones and teeth of dinosaurs are much more common than most people realize.

Some fossils retain their original composition and structure and are preserved as unaltered remains, but many have been altered in some way. Dissolved minerals may precipi-

▶ **Figure 7.16 Current Ripple Marks** These are small (< 2 cm high) sedimentary structures that have an asymmetric profile.

a Current ripple marks form where water or wind flows in one direction over sand. The enlargement shows the internal structure of one ripple mark.

b Current ripple marks in a stream.

c Ripple marks formed by wind.

tate in the pores of bones, teeth, and shells or fill the spaces within cells of wood. Wood may be preserved by silica replacing the woody tissues; it then is referred to as *petrified*, a term that means "to become stone." Under some conditions silicon dioxide (SiO_2) or iron sulfide (FeS_2) completely replaces the calcium carbonate ($CaCO_3$) shells of marine animals. Insects and leaves, stems, and roots of plants are commonly preserved as thin carbon films that show the details of the original organism. Shells and bones in sediment may be dissolved, leaving a cavity called a *mold* shaped like the shell or bone. If a mold is filled in, it becomes a *cast*. (See "Fossilization" on pages 216–217)

If it were not for fossils, our only record of ancient plants and animals, we would have no knowledge of trilobites, dinosaurs, and other extinct organisms. But fossils are not simply curiosities. In many geologic studies, it is necessary to correlate or determine the age equivalence of sedimentary rocks in different areas. Such correlations are most commonly demonstrated with fossils (see Chapter 9).

Figuring Out the Environment of Deposition

Suppose you encounter a layer of sandstone. What kinds of evidence would allow you to figure out how it was deposited in the first place? Certainly you would consider texture—that is, sorting and rounding—and also note the kinds of sedimentary structures and fossils, if any. In addition, you might compare the features in the sandstone with those seen in sand deposits forming today. But are you justified in using present-day processes and deposits to make inferences about what happened when no human observers were present?

▶ **Figure 7.17 Wave-Formed Ripple Marks Tend to Be Symmetrical in Profile** Just like current ripple marks, these are small-scale sedimentary structures found on sand beds.

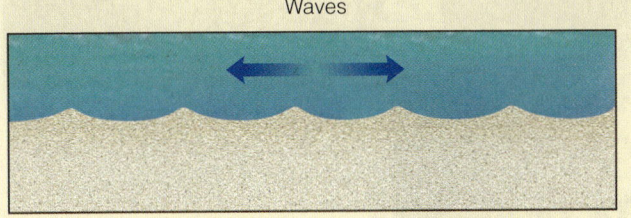

a Wave-formed ripple marks form where waves move to and fro.

b Wave-formed ripple marks in shallow seawater.

c These are also wave-formed ripple marks, but after one set of ripples formed, the waves approached from a different direction, thereby superimposing another set of ripples on the original ones.

▶ **Figure 7.18 Mud Cracks** They form in clay-rich sediments when they dry and contract.

a Mud cracks in a present-day environment.

b Ancient mud cracks in Glacier National Park in Montana. Notice that the cracks have been filled in by sediment.

Fossilization

Although the chance that any one organism will be preserved in the fossil record is slight, fossils are nevertheless common because so many billions of organisms lived during the past several hundred million years. Hard skeletal parts of organisms living where burial was likely are most common.

▶ **1.** The bones of this dinosaur (below) and the shells of these marine animals called ammonites (right) have had minerals added to their pores, making them more durable.

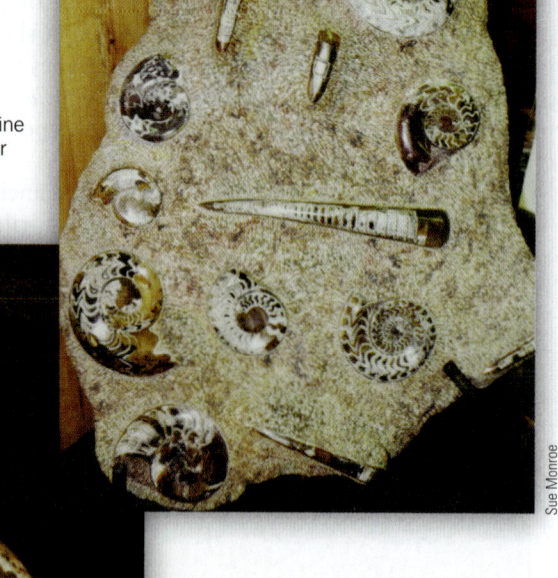

▶ **2.** Trace fossils do not include actual remains—only tracks, burrows, nests, and droppings—such as this tiny amphibian track.

▲ **4.** A coprolite (fossilized feces) from a carnivorous mammal. The coprolite is about 5.5 cm long.

▼ **5.** Fossil insect replaced by silicon dioxide (SiO_2).

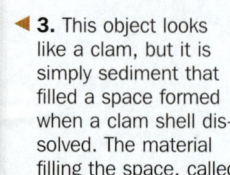

◀ **3.** This object looks like a clam, but it is simply sediment that filled a space formed when a clam shell dissolved. The material filling the space, called a mold, is a cast.

◀ **6.** Fossilized tree stump at Florissant Fossil Beds National Monument, Colorado. The woody tissue has been replaced by silicon dioxide. Mudflows 3 to 6 m deep buried the lower parts of many trees at this site.

▶ **7.** A palm frond and an insect preserved by carbonization.

▲ **9.** Insects preserved in amber—hardened resin secreted by coniferous trees.

▲ **8.** This 6- or 7-month-old frozen baby mammoth was found in Siberia in 1977. It measures 1.15 m long and 1 m high. Most of its hair has fallen out except around the feet. The carcass is about 40,000 years old.

▶ **10.** This insect from the La Brea Tar Pits in Los Angeles is just one of hundreds of species of animals found in the asphalt-like substance in the pits.

Actually you have probably interpreted other events that you did not witness. Skid marks on a street, broken glass, and a damaged power pole almost certainly indicate that a vehicle hit the pole. If you saw a shattered, badly burned tree in the forest you could conclude that a bomb damaged it, but in the absence of bomb fragments or residue, you would most likely decide that it was hit by lightning. Geologists use exactly the same reasoning when they evaluate the evidence preserved in sedimentary rocks. That is, they use their knowledge of natural processes and the effects of those processes preserved in rocks. When the rocks formed is irrelevant. Perhaps some examples will clarify this point.

A sandstone layer has symmetrical ripple marks just like those forming today by the to-and-fro motion of waves (▶ Figure 7.17). Accordingly, you conclude that the sandstone was deposited near a lakeshore or seashore. Suppose further that the sandstone contains fossil corals and clams, thus ruling out the possibility of deposition in a lake. However, if this sandstone had contained wave-formed ripple marks and dinosaur fossils, you would no doubt opt for the lake interpretation.

Various other features of sedimentary rocks are useful for environmental analyses. Particles such as ooids (▶ Figure 7.8a) form today in shallow marine environments where currents are vigorous. Glacial deposits are typically poorly sorted, show little or no stratification, and are found with other features indicating glacial transport and deposition (see Chapter 17). Dinosaur fossils may be washed into transitional or marine environments, but most of them are found in deposits that accumulated on land. Large-scale cross-bedding is characteristic of—but not restricted to—desert dunes, whereas a succession of numerous graded beds was likely deposited on the deep seafloor (see Chapter 12).

■ *How do we know that the Navajo Sandstone formed as a desert dune deposit?*

The Navajo Sandstone in the southwestern United States is an ancient desert dune deposit. But what evidence justifies this conclusion? This 300-m-thick sandstone is made up of well-sorted, well-rounded sand grains measuring 0.2–0.5 mm in diameter. Furthermore, it has cross-beds as high as 30 m (▶ Figure 7.19) and current ripple marks, both typical of desert dunes. Some of the sand layers have preserved dinosaur tracks and footprints of other land-dwelling animals, ruling out the possibility of deposition in the ocean. In short, the Navajo Sandstone has features that indicate a desert dune depositional environment.

Finally, the cross-beds are inclined downward toward the southwest, indicating that the prevailing winds were from the northeast.

Determining how sedimentary rocks were deposited satisfies our curiosity about Earth history, but such studies are not only of academic interest. Quite the contrary. Knowing how sedimentary rocks were deposited is useful in locating petroleum and natural gas, finding and recovering adequate supplies of groundwater, and evaluating an area's potential for other natural resources. Indeed, the professional journals of such organizations as the American Association of Petroleum Geologists often have articles about the depositional environments of resource-bearing sedimentary rocks.

Section 7.6 Summary

- Sedimentary structures such as ripple marks, cross-bedding, and mud cracks form in sediments during or shortly after deposition.

- Fossils, which are the remains or traces of ancient organisms common in some sedimentary rocks, provide our only record of prehistoric life. They are useful in correlation and environmental analyses.

- Geologists determine the depositional environments of ancient sedimentary rocks by evaluating sedimentary textures and structures, examining fossils, and making comparisons with present-day sediments deposited by known processes.

▶ **Figure 7.19 Jurassic-Aged Navajo Sandstone, Zion National Park, Utah** Navajo Sandstone is a wind-blown dune deposit. Vertical fractures intersect cross-beds, giving this cliff its checkerboard appearance—hence, its name Checkerboard Mesa. It stands about 300 m high.

What Would You Do?

No one was present millions of years ago to record data about the climate, the fauna and flora, the geography, and geologic processes. So how is it possible to decipher unobserved past events? In other words, what features in sedimentary rocks would you look for to determine what happened during the distant past? Can you think of any economic reasons to interpret Earth history from the record preserved in rocks?

7.7 Important Resources in Sediments and Sedimentary Rocks

Sediments and sedimentary rocks or the materials they contain have many uses. Sand and gravel are essential to the construction industry, particularly as aggregate for concrete. Silica sand, composed mostly of quartz, is used to manufacture glass, refractory bricks for blast furnaces, and molds for casting iron, aluminum, and copper alloys. Clay minerals are used for ceramics, and limestone is used to manufacture cement and in blast furnaces where iron ore is refined to make steel. Evaporites are the source of table salt as well as several chemical compounds, and rock gypsum is used to manufacture wallboard. Even some of the less common evaporites are important resources—sylvite (KCl) is used in fertilizers, dyes, and soaps; trona, a sodium carbonate, is a source of sodium; and borax, which contains barium, is used in glass, ceramics, and pharmaceuticals.

Placer deposits are surface accumulations resulting from the separation and concentration of materials of greater density from those of less density in streams and on beaches. Much of the gold recovered during the California gold rush (1848-1853) was mined from placer deposits, and placers of other minerals such as diamonds and tin are important.

The tiny island nation of Nauru, with one of the highest per capita incomes in the world, has an economy based almost entirely on mining and exporting phosphate-bearing sedimentary rock used in fertilizers. More than half of Florida's mineral value comes from mining phosphate rock (see Chapter 12). Dolostones in Missouri are the host rocks for ores of lead and zinc. Diatomite, a lightweight, porous sedimentary rock composed of the microscopic silica (SiO_2) shells of single-celled plants, is used in gas purification and to filter fluids such as molasses, fruit juices, water, and sewage.

Petroleum and Natural Gas

Petroleum and natural gas are *hydrocarbons*, meaning they are composed solely of hydrogen and carbon. Hydrocarbons form when the remains of microorganisms settle to the seafloor or lake floor where little oxygen is available to decompose them. As they are buried under layers of sediment, they are heated and transformed into petroleum and natural gas.

The rock in which hydrocarbons formed is the *source rock*, but for them to accumulate in economic quantities, they must migrate into and become trapped in some kind of *reservoir rock* beneath cap rock. Otherwise, they would migrate upward, and eventually seep out at the surface (▶ Figure 7.20). Effective reservoir rocks contain considerable pore space where appreciable quantities of hydrocarbons accumulate. Furthermore, reservoir rocks must possess high *permeability*, or the capacity to transmit fluids; otherwise, hydrocarbons cannot be extracted in reasonable quantities.

■ *What are stratigraphic and structural traps?*

Many oil and gas traps are called *stratigraphic traps* (▶ Figure 7.20a), because they owe their existence to variations in the rock layers or strata. Ancient coral reefs are good stratigraphic traps; some of the oil in the Persian Gulf region is trapped in ancient reefs. *Structural traps* result when rocks are deformed by folding, fracturing, or both (▶ Figure 7.20b). In sedimentary rocks deformed into a series of folds, hydrocarbons migrate to and accumulate in the high parts of these structures. Displacement of rocks along faults (fractures along which movement has occurred) also yields situations conducive to trapping hydrocarbons.

In the Gulf Coast region, hydrocarbons are commonly found in folds and faults adjacent to salt domes. A vast layer of ancient rock salt is present, which is a low-density sedimentary rock. When deeply buried beneath more dense sediments such as sand and mud, it rises toward the surface in pillars known as *salt domes*. As the rock salt rises, it penetrates and deforms the overlying rock layers, forming structures along its margins that may trap petroleum and gas (▶ Figure 7.20c).

Although large concentrations of petroleum are present in many areas of the world, more than 50% of all proven reserves are in the Persian Gulf region! Many nations are heavily dependent on imports of Persian Gulf oil. The United States is the third largest oil producer and yet must import about 13 million barrels daily, nearly 60% of what it uses, much of it from the Middle East, although it also buys large quantities from Venezuela and Mexico. Most geologists think that all the truly gigantic oil fields have already been found but concede that some significant discoveries are yet to be made. One must view these potential discoveries in the proper perspective, however. For example, the discovery of an oil field comparable to that of the North Slope of Alaska (about 10 billion barrels) constitutes about a 2-year supply for the United States at the current consumption rate.

Oil Shale and Tar Sands

■ *What is oil shale?*

Another source of petroleum is *oil shale*, a fine-grained, thinly layered sedimentary rock containing an organic substance called *kerogen* from which liquid oil and combustible

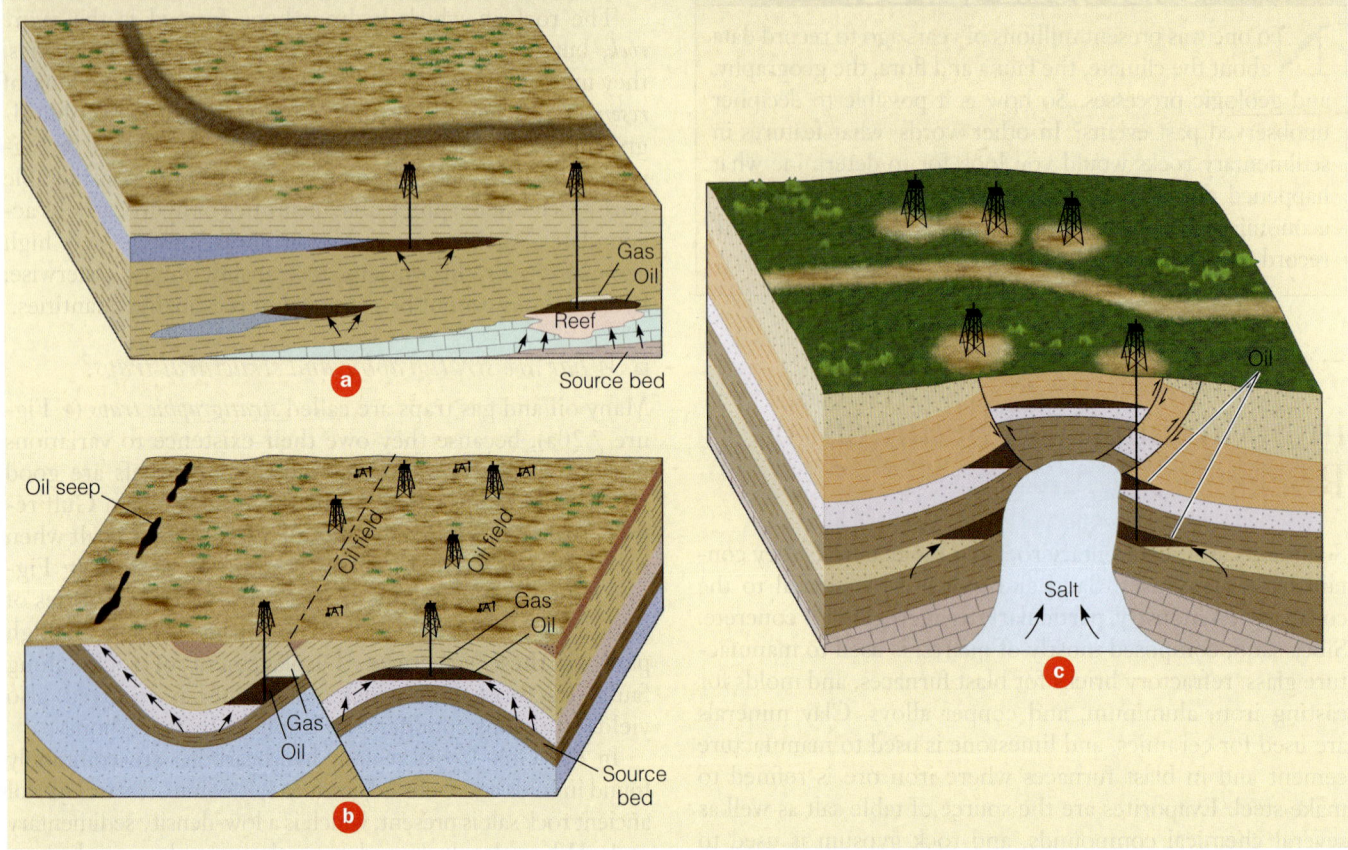

▶ **Figure 7.20 Oil and Natural Gas Traps** The arrows in the diagrams indicate the migration direction of hydrocarbons. **(a)** Two examples of stratigraphic traps: one in a sand body within mudrock and the other a buried reef. **(b)** Two examples of structural traps: one formed by folding and the other by faulting. **(c)** An example of structures adjacent to a salt dome where oil and natural gas may be trapped.

gases can be extracted (▶ Figure 7.21). During the Middle Ages (from about A.D. 600 to 1350), Europeans used oil shale as a solid fuel for heating. Small oil shale industries existed in the eastern United States during the 1850s but were discontinued when drilling and pumping liquid oil began in 1858.

About two-thirds of all known oil shale is found in the United States, most of it in the Green River Formation of Wyoming, Colorado, and Utah, but large deposits are also present in South America. No liquid oil or combustible gases are currently produced from oil shale because extracting them is more expensive than conventional drilling and pumping. Nevertheless, oil shale represents a huge untapped resource; according to one estimate, 80 billion barrels of oil could be recovered from the Green River Formation with present technology.

If the Green River Formation were mined for its oil shale, large-scale excavations would be necessary as would huge quantities of water, in a region where water is already in short supply. Overall, a venture of this magnitude would have a profound impact on the environment. And given the current and expected future rates of consumption, even if the full potential of oil shale could be realized, it would still not solve all our energy needs.

In addition to oil shale, liquid petroleum can be derived from *tar sand*, a type of sandstone with its pore spaces filled with viscous, asphaltlike hydrocarbons, the sticky residue of once-liquid petroleum from which the volatile elements have been lost. The Athabaska tar sands in Alberta, Canada, is one of the largest deposits of this type, but the United States has few tar sands and cannot look to this source as a significant future energy resource. Large-scale extraction of oil is currently taking place from the Athabaska tar sands, which contain an estimated 300 billion barrels of recoverable petroleum.

Coal

Historically, most coal mined in the United States was bituminous coal from Kentucky, Ohio, West Virginia, and Pennsylvania that formed in coastal swamps during the Pennsylvanian Period between 299 and 318 million years ago. Now, though, huge lignite and subbituminous coal deposits in the western United States are becoming increasingly important. During 2002, more than a billion metric tons of coal were mined in the United States, with nearly 60% from mines in Wyoming, West Virginia, and Kentucky.

▶ **Figure 7.21 Oil Extraction from Shale** The sedimentary rock oil shale (left) and the oil extracted from it. There are vast oil shale deposits in the United States, but no oil shale is currently processed for its oil because conventional methods of drilling and pumping are cheaper.

▶ **Figure 7.22 Banded Iron Formation is an Important Sedimentary Rock Because it is the Main Source of Iron Ore.**

a This banded iron formation at Ishpeming, Michigan, consists of alternating layers of red chert and silver-colored iron minerals.

b Much of the iron ore in the Great Lakes region is processed and shaped into pellets about 1 cm across, and then either shipped directly to steel mills or stored for later shipment.

Anthracite coal is an especially desirable resource because it burns hot with a smokeless flame. Unfortunately, it is the least common type of coal, so most coal used for heating buildings and for generating electrical energy is bituminous (▶ Figure 7.11c). *Coke* is a hard, gray substance consisting of the fused ash of bituminous coal. It is prepared by heating coal and driving off the volatile matter and is used to fire blast furnaces for steel production. Synthetic oil and gas and a number of other products are also made from bituminous coal and lignite.

Uranium

Most uranium used in nuclear reactors in North America comes from the complex potassium-, uranium-, vanadium-bearing mineral *carnotite* found in some sedimentary rocks. Some uranium is also derived from *uraninite* (UO_2), a uranium oxide found in granitic rocks and hydrothermal veins. Uraninite is easily oxidized and dissolved in groundwater, transported elsewhere, and chemically reduced and precipitated in the presence of organic matter.

The richest uranium ores in the United States are in the Colorado Plateau area of Colorado and adjoining parts of Wyoming, Utah, Arizona, and New Mexico. These ores, consisting of fairly pure masses and encrustations of carnotite, are associated with plant remains, including petrified trees in sandstones that formed in ancient stream channels.

Large reserves of low-grade uranium ore also are found in the Chattanooga Shale. The uranium is finely disseminated in this black, organic-rich mudrock that underlies large parts of several states, including Illinois, Indiana, Ohio, Kentucky, and Tennessee. Canada is the world's largest producer and exporter of uranium.

Banded Iron Formation

■ *Why is banded iron formation such an important sedimentary rock?*

Banded iron formation (BIF), consisting of alternating thin layers of chert and iron minerals, mostly the iron oxides hematite (Fe_2O_3) and magnetite (Fe_3O_4), is a chemical sedimentary rock of great economic importance. (▶ Figure 7.22). Banded iron formations are present on all the continents and account for most of the iron ore mined in the world today. Vast BIF are present in the Lake Superior region of the United States and Canada and in the Labrador trough of eastern Canada.

The origin of BIF is not fully understood, and none are currently forming. Fully 92% of all BIF were deposited in shallow seas between 2.5 and 2.0 billion years ago during the Proterozoic Eon. A highly reactive element, iron in the presence of oxygen combines to form rustlike oxides that are not readily soluble in water. During early Earth history, little oxygen was present in the atmosphere, so little was dissolved in seawater. However, soluble reduced iron (Fe^{+2}) and silica were present in seawater.

Geologic evidence indicates that abundant photosynthesizing organisms were present about 2.5 billion years ago. These organisms, such as bacteria, release oxygen as a by-product of respiration; thus they released oxygen into seawater and caused large-scale precipitation of iron oxides and silica as banded iron formations.

Section 7.7 Summary

- Many sediments and sedimentary rocks, including sand, gravel, evaporites, and coal, are important natural resources. Petroleum and natural gas as well as placer deposits of gold and diamonds are found in sedimentary rocks.

- Most of the world's iron ore comes from a chemical sedimentary rock known as banded iron formation that formed long ago during the Proterozoic Eon.

Review Workbook

ESSENTIAL QUESTIONS SUMMARY

7.1 Introduction

■ *What are sediment and sedimentary rocks?*
Sediment is a loose aggregate of solids derived from the mechanical breakdown of rocks, or solids derived from solution by inorganic chemical processes or the activities of organisms. Sedimentary rock is a solid aggregate of sediment.

7.2 Sediment Sources, Transport, and Deposition

■ *Why are rounding and sorting important in sediments and sedimentary rocks?*
Both are important in determining how fluids move through sediments and sedimentary rocks, and they are useful for determining how sedimentary rocks were deposited.

■ *What is a depositional environment?*
Any geographic area where sediment is deposited is a depositional environment. Physical, chemical, and biological processes operating in an environment impart distinctive features, such as ripple marks, mud cracks, and fossils, to sediment that may be preserved in sedimentary rocks.

7.3 Lithification: Converting Sediment into Sedimentary Rock

■ *What processes bring about lithification of sediments?*
Compaction results in the reduction in the volume of a deposit as the weight of overlying sediment causes a reduction in pore space as particles fit more closely together. Compaction alone is sufficient for lithification of mud, but for sand and gravel cementation is necessary too.

■ *What is cementation and what are the common cements in sedimentary rocks?*
Cementation takes places when minerals crystallize in the pore spaces of sediment and bind it together. By far the most common cements are calcium carbonate ($CaCO_3$) and silicon dioxide (SiO_2), but iron oxides and iron hydroxides are found in some rocks.

7.4 The Types of Sedimentary Rocks

■ *How do conglomerate and sedimentary breccia differ?*
Both are composed of gravel-sized detrital particles, but the gravel in conglomerate is rounded whereas in sedimentary breccia it is angular.

■ *Why is quartz the most common mineral in sandstone?*
Quartz is a common mineral in many source rocks, and in addition it is very stable chemically and quite durable mechanically.

■ *What are carbonate rocks?*
Any rock that contains the carbonate radical $(CO_3)^{-2}$ is a carbonate, but only limestone and dolostone are very common.

7.5 Sedimentary Facies

■ *What criteria do geologists use to define sedimentary facies?*
Any aspect of sedimentary rocks that makes them recognizably different from adjacent rocks of about the same age is used to define a sedimentary facies. One example is a cross-bedding sandstone facies versus a shale facies.

■ *What vertical sequence of facies develops during a marine transgression?*
A marine transgression takes place when the shoreline moves landward and the depositional environments parallel with the shoreline do likewise. As a result, the different sediments that accumulate in these environments become superposed with offshore facies overlying nearshore facies.

7.6 Read the Story Told by Sedimentary Rocks

■ *How is cross-bedding used to determine ancient current directions?*
Not all sedimentary structures can be used in this endeavor, but certainly current ripple marks and cross-bedding can. Current ripple marks are asymmetrical, with their steep slope inclined in the same direction as flow, and cross-beds dip or are inclined downward in the flow direction.

■ *What are fossils and how are they preserved?*
Fossils are the remains or traces of prehistoric organisms preserved in rocks. Remains (body fossils)—that is, bones, shells, and teeth—may be preserved unaltered (having the original composition and structure) or, more commonly, as altered remains—for example, when minerals are added to their pores or they are replaced by substances with a different composition.

■ *How do we know that the Navajo Sandstone formed as a desert dune deposit?*
Although no one witnessed the deposition of the Navajo Sandstone, it possesses features that point to this interpretation. For example, the tracks of land-dwelling animals rule out a marine origin, and other features such as current ripple marks and large-scale cross-bedding as well as sorting of the sand grains are much like the same features seen in desert dune deposits today.

7.7 Important Resources in Sediments and Sedimentary Rocks

■ *What are stratigraphic and structural traps?*
Both are areas where petroleum, natural gas, or both accumulate in economic quantities, but stratigraphic traps form because of variations in the rock layers (strata), whereas structural traps form as the result of folding or fracturing (faulting) of rocks.

■ *What is oil shale?*
Oil shale is a fine-grained sedimentary rock that contains kerogen from which liquid oil and combustible gases can be derived. None is mined at present in the United States because oil and gas from conventional sources is cheaper.

■ *Why is banded iron formation such an important sedimentary rock?*
Banded iron formation consists of alternating layers of chert and iron minerals, mostly iron oxides. Nearly all of Earth's iron ore is mined from ancient banded iron formations.

ESSENTIAL TERMS TO KNOW

bed (bedding) (p. 212)
biochemical sedimentary rock (p. 206)
carbonate rock (p. 206)
cementation (p. 204)
chemical sedimentary rock (p. 205)
compaction (p. 203)
cross-bedding (p. 213)
depositional environment (p. 202)
detrital sedimentary rock (p. 204)

evaporite (p. 206)
fossil (p. 213)
geologic record (p. 212)
graded bedding (p. 213)
lithification (p. 203)
marine regression (p. 211)
marine transgression (p. 210)
mud crack (p. 213)
ripple mark (p. 213)

rounding (p. 202)
sediment (p. 200)
sedimentary facies (p. 210)
sedimentary rock (p. 200)
sedimentary structure (p. 212)
sorting (p. 202)
strata (stratification) (p. 212)

REVIEW QUESTIONS

1. Why is knowledge of mineral crystallization important for understanding how cementation takes place?
2. Explain how limestone may be converted to dolostone.
3. Explain how deposits of sand and mud are lithified.
4. Under what circumstances does coal form, and what varieties of coal do geologists recognize?
5. How are all three of the major rock families related?
6. Why are fossils of marine-dwelling animals with shells so common even though very few were actually preserved as fossils?
7. What is meant by the terms clastic texture and crystalline texture? Name a sedimentary rock with each texture.
8. How is it possible to determine how sedimentary rocks were deposited given that no one was present to witness their deposition?
9. What are structural and stratigraphic traps and why are they important?
10. How do the following pairs of detrital sedimentary rocks compare? Conglomerate and sedimentary breccia; quartz sandstone and arkose.

APPLY YOUR KNOWLEDGE

1. The United States uses about 860 million metric tons of coal per year from its reserve of 243 billion metric tons. Assuming that all of this coal can be mined, how long will it last at the current rate of consumption. Why is it improbable that all of this reserve can be mined?

FIELD QUESTIONS

1. The cross section below shows layers of sedimentary rocks that are no longer in their original horizontal position. Refer to Geology In Unexpected Places on p. 227 and see if you can determine which layer is youngest (place a large X on it). Also, indicate on the cross section what the *original* current direction was for the currents that generated the ripple marks. (An answer is provided at the end of the book.)

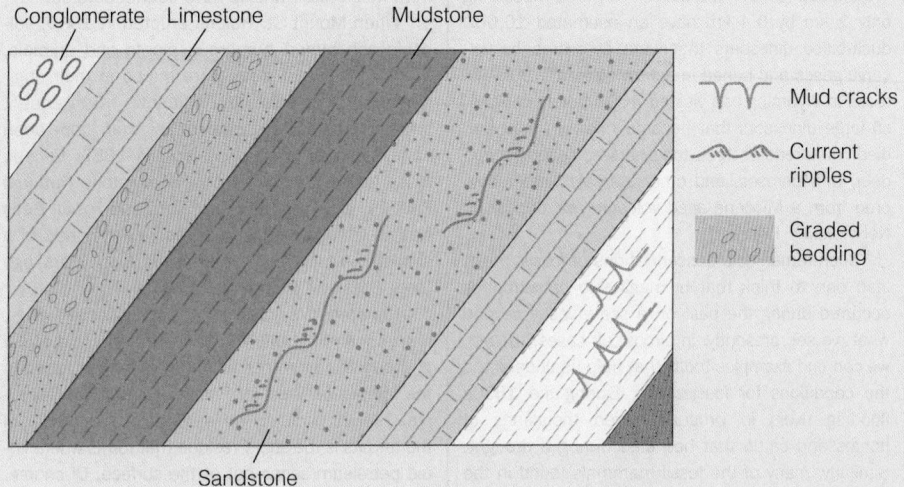

2. Suppose you encounter the sedimentary rocks shown in the cross section below. Interpret the geologic history of this area as thoroughly as possible. (An answer is provided at the end of the book.)

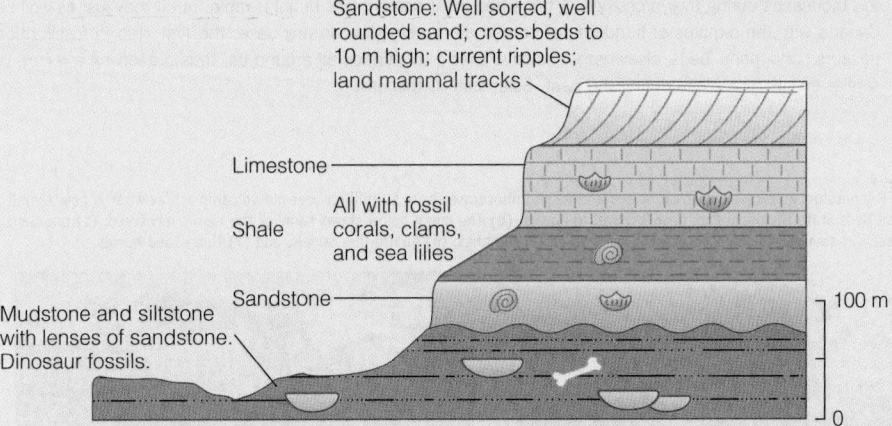

GEOLOGY MATTERS

GEOLOGY IN FOCUS Fossils and Fossilization

Even though very few of the organisms that ever lived were actually fossilized, fossils are far more common than most people realize. If even a tiny percentage of the hundreds of billions of organisms that have lived during hundreds of millions of years were preserved, the number of fossils would still be truly phenomenal. And although it is true that some organisms have a very poor or even nonexistent fossil record, the record for others is good.

Certainly marine-dwelling organisms with shells, such as corals and clams, are the most easily preserved and most easily collected fossils, even by casual collectors. But even the remains of vertebrate animals, those with a segmented vertebral column such as fish, amphibians, reptiles, birds, and mammals, are found in large numbers in some areas. Cretaceous rocks in Montana in an area measuring only 2 km by 0.4 km have an estimated 10,000 duck-billed dinosaurs that were overcome by volcanic gases and buried in ash. A deposit in Jurassic rocks in Wyoming has yielded 4000 bones of 20 or so large dinosaurs that became mired in mud and died. Hundreds of three-toed horses, saber-toothed deer, rhinoceroses, and other animals were recovered from a Miocene-aged volcanic ash deposit in Nebraska (▶ Figure 1).

Such remarkable concentrations of fossils might lead one to think that unimaginable catastrophes occurred during the past on a scale vastly beyond what we see presently. In almost all cases, though, we can find examples today that help us understand the conditions for fossilization. During the 1830s flooding rivers in Uruguay buried thousands of horses and cattle that had died during a drought. Similarly, many of the fossil mammals found in the White River Badlands in South Dakota were buried in floodplain deposits.

African elephants today become mired in mud and perish just as Jurassic dinosaurs did in Wyoming, although it happens rarely. In 1984 thousands of caribou died in the Caniapiscau River of Quebec, Canada, and in Africa wildebeests die by the thousands during river crossings. A bone bed in Canada with the remains of hundreds of horned dinosaurs, and bone beds elsewhere, indicate that similar events took place during the past. Scientists in a submersible in the Gulf of California saw the remains of a whale that were slowly being covered by sediment, like the sediment in southern California from which fossil whales have been recovered.

When Mount St. Helens erupted in 1980, volcanic ash buried numerous plants and animals, and thousands of trees were buried in their original growth position by mudflows. Subsequent stream erosion exposed trees that had been buried in their original position in 1885. Eocene fossil forests in what is now Yellowstone National Park in Wyoming and at Florissant Fossil Beds National Monument in Colorado were buried in a similar manner more than 33 million years ago (see "Fossilization" on pp. 216–217).

Hundreds of fossils of saber-toothed cats, horses, camels, vultures, armored mammals called glyptodonts, and other animals have been found in the famous La Brea Tar Pits in Los Angeles, California. The tar (actually an asphaltlike substance) in the tar pits is the sticky residue that forms where liquid petroleum seeps out at the surface. Of course, the area no longer supports saber-toothed cats, camels, and glyptodonts, but small mammals, birds, and insects are still trapped in the tar exactly as ancient animals were.

These examples show that the first step in fossilization, the burial of organic remains, requires only natural processes with which we are familiar. In some cases burial is rapid, but it may just as well be very slow. In any case, the first step in fossilization continues all around us, thus preserving a record of this era.

Figure 1
(a) Paleontologists excavating fossil horses and fossil rhinoceroses from 10-million-year-old volcanic ash at what is now Ashfall Fossil Beds State Historical Park near Orchard, Nebraska. **(b)** The mural below shows some of the fossils recovered: (1) one-toed horses, (2) small camels, (3) turtles, (4) rhinoceroses, (5) cranes, (6) giraffe-like camels, and (7) three-toed horses.

GEOLOGY MATTERS

GEOLOGY IN UNEXPECTED PLACES

Sandstone Lion

A 9-m-long Lion Monument in Lucerne, Switzerland, was chiseled into sandstone in 1821 as a memorial to about 850 Swiss soldiers who died during the French Revolution of 1792 in Paris (▶ Figure 1a). Lukas Ahorn chiseled the monument into the sandstone wall of a quarry; the inscription above the lion pays honor to the "loyalty and courage of the Swiss." An officer on leave from the army at the time of the battle in Paris took the first steps to set up the monument.

Notice in ▶ Figure 1a that the sandstone layers are inclined downward, or dip, to the left at about 50 degrees. We could postulate that (1) the original layers were horizontal and then tilted 50 degrees into this position, or (2) perhaps they were rotated 140 degrees from their original position so that the layers are now upside down, or overturned in geologic parlance. To resolve this problem, you must determine which of the layers shown was at the top of the original sequence of beds and thus the youngest. In ▶ Figure 1b, observe that the cross-beds have a sharp angular contact with the younger rocks above them, whereas they are nearly parallel with the older rocks below, as in Figure 7.14b. Accordingly, we conclude that the youngest rock layer is the one toward the upper left and the rock layers have not been overturned.

Having determined which layer is oldest and which is youngest, we now know that any rocks exposed to the right of the image are older than the ones shown and, of course, any to the left are younger. However, it is important to note that we have determined relative ages only—that is, which layers are older versus younger. Nothing in this image tells us the absolute age in number of years before the present. We will consider relative and absolute dating more fully in Chapter 9.

Figure 1
(a) Lion Monument in Lucerne, Switzerland. (b) Cross-bedding shows sharp, angular contact with younger rocks above and nearly parallel contact with older rocks below.

GEOLOGY MATTERS
Geology AT WORK

Exploring for Oil and Natural Gas
Susan M. Landon

Susan M. Landon began her career in 1974 with Amoco Production Company and in 1989 opened her own consulting office in Denver, Colorado. She is currently a partner in the exploration group Thomasson Partner Associates. In 1990 she was elected president of the American Institute of Professional Geologists and in 1992 treasurer of the American Association of Petroleum Geologists. Landon shares her experiences in the following narrative.

I am an independent petroleum geologist. I specialize in applying geologic principles to oil and gas exploration in frontier areas—places where little or no exploration has occurred and few or no hydrocarbons have been discovered. It is very much like solving a mystery. The Earth provides a variety of clues—rock type, organic content, stratigraphic relationships, structure, and the like—that geologists must piece together to determine the potential for the presence of hydrocarbons.

For example, as part of my work for Amoco, I explored the Precambrian Midcontinent Rift frontier in the north central portion of the United States (▶ Figure 1). Some rifts, like the Gulf of Suez and the North Sea, are characterized by significant hydrocarbon reserves, and the presence of an unexplored rift basin in the center of North America is intriguing. A copper mine in the Upper Peninsula of Michigan, the White Pine Mine, has historically been plagued by oil bleeding out of fractures in the shale. For many years, this had been documented as academically interesting because the rocks are much older than those that typically have been associated with hydrocarbon production.

Field and laboratory work documented that the copper-bearing shale at the White Pine Mine contained adequate organic material to be the source of the oil. The thermal history of the basin was modeled to determine the timing of hydrocarbon generation. If hydrocarbons had been generated prior to deposition of an effective seal and formation of a trap, the hydrocarbons would have leaked out naturally into the atmosphere.

Further work identified sandstones with enough porosity to serve as reservoirs for hydrocarbons. Seismic data were acquired and interpreted to identify specific traps. We then had to convince Amoco management that this prospect had high enough potential to contain hydrocarbon reserves to offset the significant risks and costs. In this case, management agreed that the risk was offset by the potential for a very large accumulation of hydrocarbons, and a well was authorized. Amoco drilled a 5441-m well in Iowa to test the prospect, at a cost of nearly $5 million. The well was dry (economically unsuccessful), but the geologic information obtained as a result of drilling the well will be used to continue to define prospective drilling sites in the Midcontinent Rift.

My career in the petroleum industry began with Amoco Production Company, and after 15 years, I made the decision to leave the company to work independently. For several years, I consulted for a variety of companies, assisting them in exploration projects. I am now involved in a partnership with several other exploration geologists and geophysicists, and we are actively exploring for oil and natural gas in the United States. We develop ideas, defining areas that we believe may be prospective, and attract other companies as partners to assist us in further exploration and drilling. Currently, I am working on projects located in Wyoming, southern Illinois, southern Michigan, Iowa, and Minnesota. I was Manager of Exploration Training when I left Amoco, and as a result, I currently teach a few courses (such as Petroleum Geology for Engineers), which allows me to travel to places like Cairo, Egypt; Quito, Ecuador; and Houston, Texas.

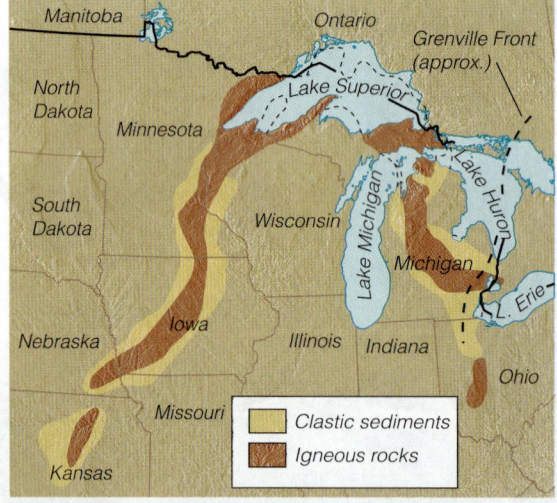

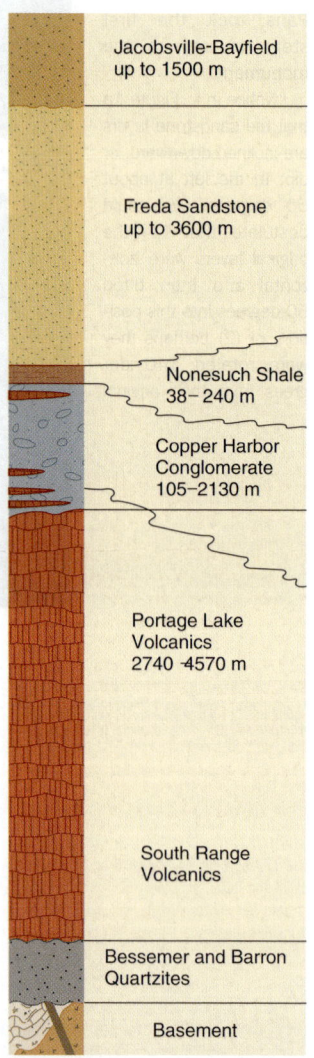

Figure 1
Location of the Midcontinent Rift
(a) The Midcontinent rift and the rocks filling it are well exposed in the Lake Superior region, but they are deeply buried elsewhere. **(b)** Vertical relationships among the rocks in the rift near Lake Superior. **(c)** The Copper Harbor Conglomerate and **(d)** the Portage Lake Volcanics.

GEOLOGY MATTERS

Geology AND CULTURAL CONNECTIONS

Deep Time

Sedimentary rocks were well known to early geologists. Then, as now, it was relatively easy to look at a bed of sandstone and understand in general terms how it formed.

In the 1700s, James Hutton looked at many of the sedimentary rocks in Britain. He quickly came to realize two things. First, he understood that the quartz sand he saw on the beach was linked to quartz sandstone. In other words, he made the connection between loose sediment and sedimentary rock. Second, he reasoned that the grains of quartz sand he found on beaches and in rivers came from the weathering of rocks on land. One of those rocks was granite, and Hutton realized that as granite weathers, its mafic minerals and much of its feldspar decompose but its quartz crystals are moved downstream. As they slowly travel down rivers, they are rounded and sorted, ending up either on the seafloor or on beaches.

Hutton was one of the first naturalists to think of Earth history in terms of gradual processing like weathering, erosion, and transport. He thought that bit by bit, over great expanses of time, mountains could be worn away and sediments carried to an area of deposition. Gradually they were buried, compressed, and cemented together.

The emphasis on gradual processes and gradual change led Hutton to think that the age of the Earth was great indeed. He is famous as the first geologist to realize that the Earth is so old that our most ancient civilization seems in comparison like something that happened just a moment ago. Geologists still print bumper stickers and T-shirts with a phrase from Hutton's writings emblazoned on them. Earth history, the T-shirts proclaim, "has no vestige of a beginning, no prospect of an end."

Although Hutton lived and worked in the British Isles, you can make observations similar to his in North America. If you stand at the top of the Grand Canyon, you can see a mile-thick stack of sedimentary rock (▶ Figure 1). If you seriously consider that the sediments in question weathered slowly out of earlier rocks and then were transported and deposited, you will grasp at a gut level that Earth's age is not to be understood on a human time scale.

A single rock can convince you of this truth as well. Imagine that you pick up a piece of conglomerate or are handed one in your geology class. The smaller pieces of rock imbedded in the conglomerate might be rounded pieces of granite and small chunks of sedimentary rock like chert. When you look at the granite, you are seeing evidence of a plutonic rock cooling slowly underground and then being uplifted to the surface of the Earth where it could be weathered, eroded, transported, and deposited. Clearly a great deal of time passed while these processes were at work. The chert pieces in the conglomerate are evidence of the weathering of preexisting rocks, the transport of silica to an area of deposition, and then the formation of the chert. The chert then also had to be exposed to weathering and erosion so that it could end up being transported to the place where the conglomerate formed. In short, from one sedimentary rock you can get a feel for what geologists call "deep time."

The idea of deep time has direct cultural implications. Once we understand the scope of geologic time, it puts our civilization in a new perspective. It's not that we are unimportant; it's just that we are only one tiny part of the long, long story of Earth history.

From your study of sedimentary rocks in this chapter you can infer that Earth processes take place over vast reaches of time. Sedimentary rocks can make clear how very recent our culture is and how many interesting things happened on Earth long before our oldest civilization emerged.

Figure 1
The rocks from the depths of the Grand Canyon to its rim represent about 1 billion years of Earth history. Time of this magnitude is difficult to comprehend, but geologists routinely deal with time measured in millions or even billions of years.

Chapter 8

Metamorphism and Metamorphic Rocks

Roman Sculpture Busts
Faces of white marble look out against a canvas of cobalt sky and sea. Who were they? What thoughts lay frozen behind their stony gaze?
These Roman sculpture busts line the Belvedere of Infinity balcony at the Hotel Villa Cimbrone in Ravello, Italy. In the background are the Amalfi coast, and the Tyrrhenian Sea, which has mountain chains and active volcanoes in its depths. White marble is the result of metamorphism of very pure limestone. Since classical times it has been preferred by sculptors of the human body, because of its softness, relative resistance to shattering, and waxy, lifelike glow.
—A. W.

ESSENTIAL QUESTIONS TO ASK

8.1 Introduction
- What is metamorphism?
- Why is metamorphism worth studying?

8.2 Equilibrium and the Causes of Metamorphism
- What are the principal agents of metamorphism?
- Why do rocks metamorphose?
- What are metamorphic fluids and why are they important?

8.3 The Main Types of Metamorphism
- What are the conditions of contact metamorphism?
- What are lithostatic and differential pressures, and why are they important?
- Where does dynamic metamorphism occur?
- What is shock metamorphism?
- What is the most common type of metamorphism?

8.4 Classification of Metamorphic Rocks
- What is foliated texture, and what are some examples of foliated rocks?
- What is a common nonfoliated metamorphic texture, and what are some examples of nonfoliated metamorphic rocks?

8.5 Metamorphic Zones and Facies
- What are isograds and metamorphic zones?
- What is a metamorphic facies?
- How do metamorphic zones and metamorphic facies differ?

8.6 Plate Tectonics and Metamorphism
- How does metamorphism relate to plate boundaries?
- How is an orogenic metamorphic belt exposed?

8.7 Metamorphism and Global Climate Change
- What does metamorphism have to do with climate?

8.8 Some Economic Uses of Metamorphic Materials
- What are some economically valuable metamorphic rocks?
- What are some economically valuable metamorphic minerals?

GEOLOGY MATTERS

GEOLOGY IN UNEXPECTED PLACES:
Starting Off with a Clean Slate

GEOLOGY AT WORK:
What Do Metamorphic Petrologists Do?

GEOLOGY IN FOCUS:
Asbestos: Good or Bad?

This icon, which appears throughout the book, indicates an opportunity to explore interactive tutorials, animations, or practice problems available on the Physical GeologyNow website at **http://now.brookscole.com/phygeo6.**

8.1 Introduction

Its homogeneity, softness, and various textures have made marble, a metamorphic rock formed from limestone or dolostone, a favorite rock of sculptors throughout history. As the value of authentic marble sculptures has increased over the years, the number of forgeries has also increased. With the price of some marble statues in the millions of dollars, private collectors and museums need some means of ensuring the authenticity of the work they are buying. Aside from the monetary considerations, it is important that forgeries not become part of the historical and artistic legacy of human endeavor.

Experts have traditionally relied on artistic style and weathering characteristics to determine whether a marble sculpture is authentic or a forgery. Because marble is not very resistant to weathering, however, forgers have been able to produce the weathered appearance of an authentic work.

Using newly developed techniques, geologists can now distinguish a naturally weathered marble surface from one that has been artificially altered. Yet, there are examples in which expert opinion is still divided on whether or not a sculpture is authentic. One of the best examples is the Greek *kouros* (a sculpted figure of a Greek youth) at the J. Paul Getty Museum in Malibu, California, purchased for a reputed price of $7 million in 1984 (see "Marble, Plaster, and Art" on pages 234–235). Because some of its stylistic features caused some experts to question its authenticity, the museum had a variety of geochemical and mineralogical tests performed in an effort to determine the authenticity of the sculpture.

Although numerous scientific tests have not unequivocally proven authenticity, they have shown that the weathered surface layer of the kouros bears more similarities to naturally occurring weathered surfaces of dolomitic marble than to known artificially produced surfaces. Furthermore, no evidence indicates that the surface alteration of the sculpture is of modern origin.

Unfortunately, despite intensive study by scientists, archaeologists, and art historians, opinion is still divided on the authenticity of the Getty kouros. Most scientists accept that it was carved sometime around 530 B.C. Pointing to inconsistencies in its style of sculpture for that period, other art historians think it is a modern forgery.

Regardless of whether the kouros is proven to be authentic or a forgery, geologic testing to authenticate marble sculptures is now an important part of many museums' curatorial functions. To help geologists in the authentication of sculptures, a large body of data about the characteristics and origin of marble is being amassed as more sculptures and their quarries are analyzed. Interpreting these data properly belongs to the field of metamorphic petrology, a major branch of geology.

■ What is metamorphism?

It is appropriate that we turn to metamorphic rocks after considering igneous and sedimentary rocks because metamorphic rocks result from the transformation of these other rocks by metamorphic processes that are widespread deep within Earth's crust and mantle (see Figure 1.13). The term *metamorphism*, in fact, derives from the ancient Greek *meta*, "change," and *morpho*, "shape." During **metamorphism**, the crystals within a preexisting rock begin to grow again, a process called **recrystallization.** Minerals can also react with one another to become new kinds of minerals that look very different from the earlier ones. The whole appearance of the rock may change, so that ultimately it bears no resemblance to its predecessor. Consider, for example, the contrast in appearance between limestone, which looks a lot like cement, and the marble that forms from it, which has large, sparkly crystals. A useful analogy for metamorphism is baking a cake. Just like a metamorphic rock, the resulting cake depends on the ingredients, their proportions, how they are mixed together, how much water or milk is added, and the temperature and length of time used for baking the cake.

■ Why is metamorphism worth studying?

Except for marble and slate, most people are not familiar with metamorphic rocks, and therefore they seem to be one of the more abstract topics of scientific study. Some metamorphic minerals are of great value economically, however, and metamorphic rocks provide important information about the physical conditions of plate tectonics. Together with igneous rocks, they make up most of Earth's continental crust. In fact, metamorphism is vital for the growth of continents. It may even play a significant role in making Earth's climate warm and habitable. So, clearly metamorphism is a relevant subject for exploration.

Section 8.1 Summary

- Metamorphism is the transformation through recrystallization of preexisting rocks inside Earth and is closely linked to plate tectonics.

- Many metamorphic rocks and minerals are economically valuable, including marble, which has long been used for sculpting.

What Would You Do?

As the director of a major museum, you have the opportunity to purchase, for a considerable sum of money, a newly discovered marble bust by a famous ancient sculptor. You want to be sure it is not a forgery. What would you do to ensure that the bust is authentic and not a clever forgery? After all, you are spending a large sum of the museum's money. As a nonscientist, how would you go about making sure the proper tests are being performed to authenticate the bust?

8.2 Equilibrium and the Causes of Metamorphism

■ *What are the principal agents of metamorphism?*

The three principal agents of change that cause metamorphism are *heat*, *pressure*, and *fluid activity*. *Time* also plays a role because the agents that trigger metamorphism must be applied long enough for chemical reactions to take place in a rock.

Every sedimentary and igneous rock forms under its own particular conditions of temperature, pressure, and fluid activity. In the case of sedimentary rocks, these conditions are fairly close to what you and I experience in our daily lives because sedimentary rocks form at or near Earth's surface. When it comes to igneous rocks, we have to think more abstractly because these rocks crystallize at much higher temperatures than we experience. Physically and chemically speaking, each rock represents as close an adjustment to its environment of formation as possible. Salt deposits, for example, are what one would expect to form from evaporating saltwater, not limestone or granite.

■ *Why do rocks metamorphose?*

Metamorphism takes place because of a change in a rock's physical and chemical environment. To use a more technical term, the rock is no longer in **equilibrium** with its environment; that is, other combinations of elements and mineral textures within the rock may be more *stable* than the existing ones given the new circumstances. Metamorphic reactions take place as the rock attempts to achieve equilibrium under these new conditions.

For instance, consider the metamorphism of an impure limestone consisting of calcite ($CaCO_3$) and quartz (SiO_2). As pressure and temperature increase, the ionic bonds holding these minerals together deform. If the bonds rupture entirely, the rock will melt. This doesn't happen, however, because a mineral exists that can substitute for calcite and quartz at high temperature with less strain in its bonds: wollastonite ($CaSiO_3$). So, the following metamorphic reaction spontaneously occurs: $CaCO_3 + SiO_2 \rightarrow CaSiO_3 + CO_2$. (The CO_2 escapes the rock to become carbonated "mineral" water.) This reaction illustrates the nature of change toward equilibrium during metamorphic reactions; it is a minimization of the strain energy that builds up in mineral bonds. Nature tries to hold matter together with the least amount of effort.

The preceding example illustrates a kind of metamorphic reaction in which the mineral composition of the rock changes: a **heterogeneous metamorphic reaction.** Metamorphic transformation can also take place without creating new kinds of minerals. Consider the metamorphism of quartz sandstone. If the rock is hot for a long enough time, individual ions will diffuse across the crystal lattices of the constituent quartz grains and along grain boundaries into positions that reduce the overall strain as temperature and pressure increase. Smaller quartz grains become incorporated by larger ones that continue to grow. Crystals throughout the rock develop planar faces and straight edges (▶ Figure 8.1). From a strictly geometric point of view, the total area of grain-to-grain contact in the rock decreases as metamorphism progresses, strengthening the rock. The orientations of crystal lattices also change, so that grains are less likely to rupture under pressure. This type of metamorphism requires only recrystallization, and no change in the *types* of minerals present.

In many cases, both heterogeneous reactions and recrystallization leading to increasing grain size occur during metamorphism. Thus mudstones and shales composed of

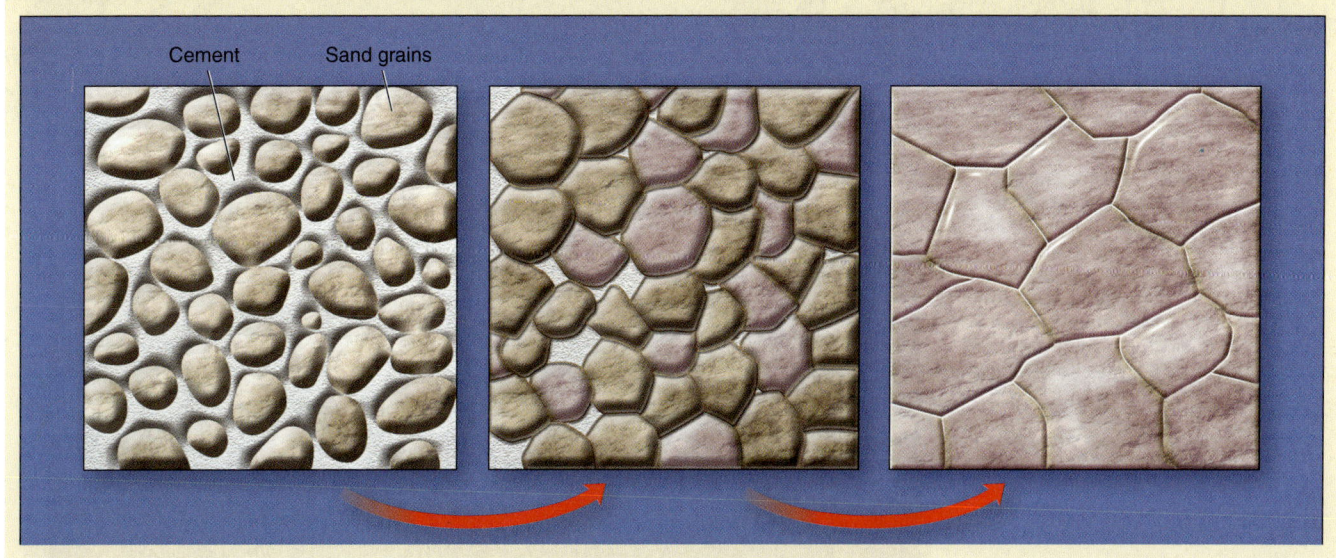

▶ **Figure 8.1 How the Shapes of Crystals in a Quartz Sandstone Change as Metamorphism Progresses** Notice that the total length of crystal borders decreases during metamorphism.

Marble, Plaster, and Art

Marble is a remarkable stone that has a variety of uses. Formed from limestone or dolostone by the metamorphic processes of heat and pressure, marble comes in a variety of colors and textures.

Marble has been used by sculptors and architects for many centuries in statuary, monuments, as a facing and main stone in buildings and structures, as well as for floor tiling and other ornamental and structural uses. Ground marble can also be found in toothpaste and is a source of lime in agricultural fertilizers.

▶ 1. *Aphrodite of Melos*, also known as *Venus de Milo*, is one of the most recognizable works of art in the world. Dated around 150 B.C., *Venus de Milo* was created by an unknown artist during the Hellenistic period and carved from the world-famous Parian marble from Paros in the Cyclades. Today *Venus de Milo* attracts thousands of visitors a year to the Louvre Museum in Paris, where she can be viewed and appreciated.

◀ 2. Marble has been used extensively as a building stone through the ages and throughout the world. For example, the Greek Parthenon was constructed of white Pentelic marble from Mt. Pentelicus in Attica.

▶ 3. The Taj Mahal in India is largely constructed of Makrana marble quarried from hills just southwest of Jaipur in Rajasthan. In addition to its main use as a building material, marble was used throughout the structure in art works and intricately carved marble flowers (right). All in all, it took more than 20,000 workers 17 years to build the Taj Mahal from A.D. 1631 to 1648.

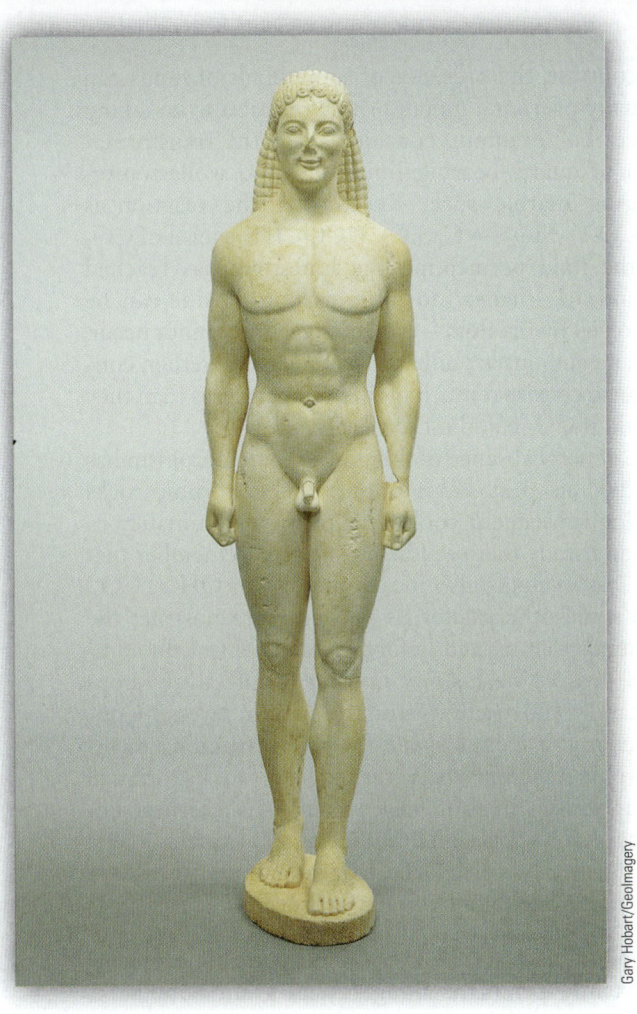

◀ **4.** This Greek kouros, which stands 206 cm tall, has been the object of an intensive authentication study by the Getty Museum. Using a variety of geologic tests, scientists have determined that the kouros was carved from dolomitic marble and probably came from the Cape Vathy quarries on the island of Thasos.

▶ **5.** The Peace Monument at Pennsylvania Avenue on the west side of the Capitol is constructed from white marble from Carrara, Italy a locality famous for its marble.

▼ **6.** A marble quarry in northcentral Vermont. Vermont is known for producing some of the finest marble in the United States.

▲ **7.** A public bath in the ancient Roman town of Herculaneum, preserved following the A.D. 79 eruption of Mt. Vesuvius in southern Italy. Marble heads ornament the upper wall. A thin layer of *plaster* still covers other parts of the wall, and includes a decorative scene made of blue mosaic tile near the floor.

"Plaster" is an artificial metamorphic material made from the baking of marble and limestone to drive off volatiles and convert calcite to *lime,* or pure calcium oxide. The calcium oxide is mixed with grains of sand, commonly including non-reactive quartz and feldspar. We make cement, which is a plaster-like substance, in much the same way.

microscopically fine grains of clay, quartz, and plagioclase can be transformed during mountain building into coarse-grained schist containing quartz, plagioclase, chlorite, muscovite, orthoclase, sillimanite, cordierite, and pyroxene. Nevertheless, some of the same minerals that existed in the original mudstones and shales may survive in the schist. They simply grow larger and straighter-edged as metamorphism progresses.

Some minerals respond to increasing temperature and pressure by simply repacking their atoms to build more stable structures. For example, at a depth of 400–600 km, the common mantle mineral olivine alters its crystal lattice to match that of another mantle mineral, spinel. In so changing, the olivine becomes about 12% more compact per unit volume; hence more of it can exist stably in one space as pressure increases. This type of metamorphic reaction is called **polymorphic transformation.** It is distinguished from simple recrystallization by the fact that the olivine grains change not only their shapes and sizes but also the basic arrangement of their atomic structure.

■ *What are metamorphic fluids and why are they important?*

The term *fluid* has a more specific meaning when applied to metamorphism than it has in everyday language. **Metamorphic fluids** are not true liquids but are highly mobile substances that combine the properties of liquids and gases under the high temperature and pressure conditions of Earth's deep interior. A typical metamorphic fluid consists of H_2O plus carbon dioxide plus dissolved mineral matter such as silica. As the fluid approaches the surface, the carbon dioxide separates out as ordinary gas, the silica precipitates as a solid, and the H_2O becomes groundwater or it seeps out of the ground.

Rocks undergoing metamorphism have little pore space and they lack permeability. A highly porous sedimentary rock may consist of 30% pore space, whereas a typical metamorphic rock contains less than 1%. The fluids that probably exist in a metamorphic rock probably form as only a thin film a few molecules thick along the boundaries between individual mineral grains. Nevertheless, the presence or absence of fluids in metamorphism is vital for three reasons:

1. Fluids are essential for some kinds of metamorphic reactions because they supply OH^{-2} (the hydroxyl ion from the breakdown of H_2O) for the creation of certain key minerals. For example, seawater can enter fractured basalt and peridotite at mid-oceanic ridges, making them "wet." As heat and pressure increase, minerals in the ocean floor rocks undergo metamorphic reactions and form chlorite, actinolite, and other minerals that require OH^{-2} from the trapped seawater. One especially important reaction is $2Mg_2SiO_4$ (olivine) $+ 2H_2O \rightarrow Mg_3Si_2O_5(OH)_4$ (serpentine) $+ MgO$ (available to make brucite and spinel). Among other things, this reaction can create the serpentine form of asbestos. Nearby "dry" rocks may not show any metamorphic change at all.

2. In contrast, the presence of some kinds of fluids can actually prevent a metamorphic reaction from taking place. For example, consider again the transformation of quartz-bearing limestone into wollastonite-bearing marble rock. Recall that the reaction is $CaCO_3 + SiO_2 \rightarrow CaSiO_3 + CO_2$. If the level of CO_2 in the fluid permeating the limestone has reached saturation—that is, no more carbon dioxide can be dissolved by the fluid—then the reaction cannot occur. Calcite and quartz, unhappy as they are together, continue to coexist at much higher temperatures than they would if CO_2 could escape easily.

3. The general absence of fluids after metamorphism has reached its peak helps preserve metamorphic rocks during subsequent cooling, allowing us ultimately to see and study them at Earth's surface. Remember that the loss of fluids also means the loss of OH^{-2}, CO_2, and some other materials needed to reconstruct the minerals that existed before metamorphism. In addition, loss of heat slows down and effectively stops metamorphic reactions from running in reverse. Consequently, it is impossible for reverse-running metamorphic reactions to return a rock to its original composition. In those few instances where heat decreases slowly enough and fluid is still present, a partial form of reverse or **retrograde metamorphism** can take place, but it can never proceed very far.

How does fluid escape from rock during metamorphism? It can certainly percolate along crystal grain boundaries away from areas of high pressure and temperature; however, this very slow process is likely to be effective over only short distances. Fluids exit far more easily through fractures and

What Would You Do?

The problem of removing asbestos from public buildings is an important national health and political issue. The current policy of the Environmental Protection Agency (EPA) mandates that all forms of asbestos are treated as identical hazards. Yet studies indicate that only one form of asbestos is a known health hazard. Because the cost of asbestos removal has been estimated to be as high as $100 billion, many people are questioning whether it is cost effective to remove asbestos from all public buildings where it has been installed.

As a leading researcher on the health hazards of asbestos, you have been asked to testify before a congressional committee on whether it is worthwhile to spend so much money for asbestos removal. How would you address this issue in terms of formulating a policy that balances the risks and benefits of removing asbestos from public buildings? What role would geologists play in formulating this policy?

along joints in the rock. These natural pathways often seal up with mineral deposits, especially silica, as the fluid cools down and stagnates at shallow levels. Some geologists think that new fractures open when the pressure of trapped fluids increases in the metamorphosing rocks farther down, so that fluids tend to be released in pulses.

Fluids can also escape along **dissolution seams,** which form when certain types of rock dissolve at high temperatures and pressures. The dissolution does not occur everywhere but tends to become concentrated along parallel, near-planar surfaces at a right angle to the direction of strongest pressure acting on the rock. For instance, fluid-bearing limestone will partly dissolve as it is squeezed and transformed into marble.

The carbonate-rich solution exits the limestone along seams called *stylolites* that have a characteristic sawtooth shape in cross section. Stylolites are popular features in building facades, countertops, and monuments (▶ Figure 8.2). They show up because certain dark or brightly colored trace minerals in the marble, such as clays and iron oxides, do not dissolve and end up concentrated along the seams.

Section 8.2 Summary

- Heat, pressure, fluid activity, and time all play a determining role in the metamorphism of rocks.
- Rocks undergo metamorphism because the physical conditions to which they are exposed change inside Earth.
- Minerals seek to adjust to changing conditions in ways that maximize their stability. These may involve chemical reactions that change the mineral composition of a rock.
- Metamorphic fluids, which are very active in Earth's crust, greatly influence the ease with which certain mineral reactions take place.
- The absence of fluids and the loss of heat prevent retrograde metamorphic reactions from undoing the effects of metamorphism when erosion exposes these rocks to surface conditions.

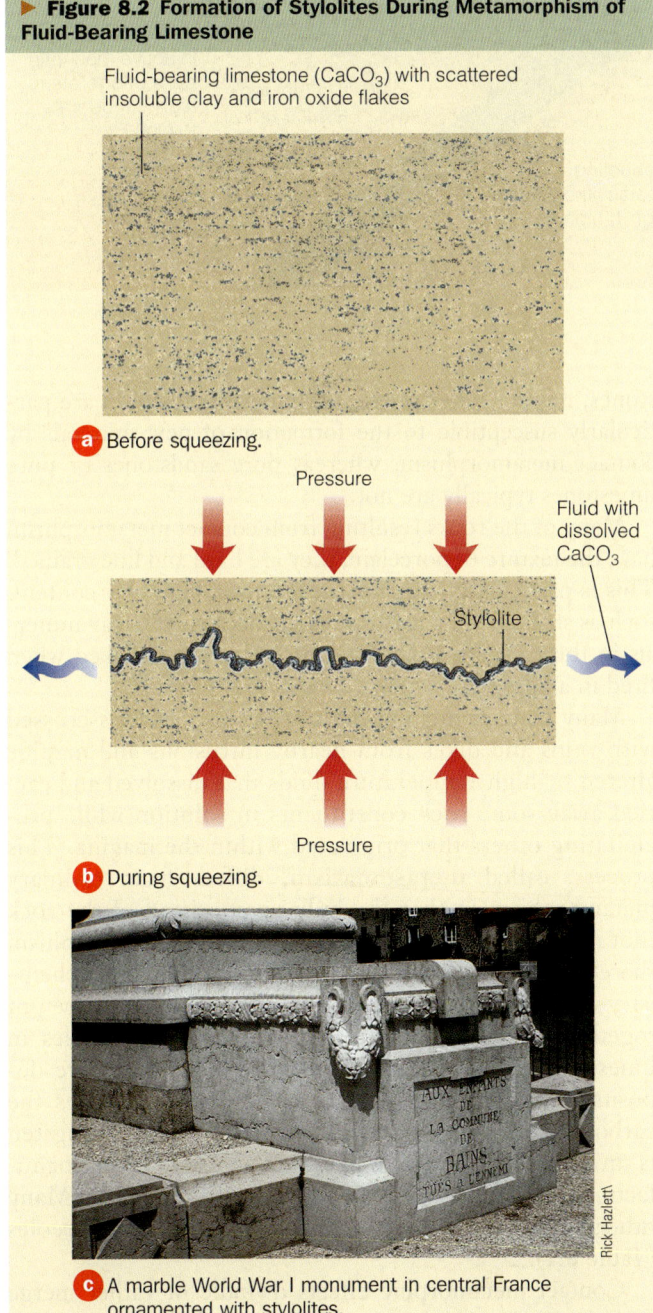

▶ **Figure 8.2** Formation of Stylolites During Metamorphism of Fluid-Bearing Limestone

a Before squeezing.

b During squeezing.

c A marble World War I monument in central France ornamented with stylolites.

8.3 The Main Types of Metamorphism

Metamorphism occurs in a wide range of environments worldwide, most of which are closely associated with tectonic processes (▶ Figure 8.3). Geologists recognize three major types of metamorphism: contact metamorphism, in which magmatic heat and fluids act to produce change; dynamic metamorphism, which is principally the result of high differential pressures associated with intense deformation; and regional metamorphism, which occurs within a large area and is associated with major mountain-building episodes. Even though we discuss each type of metamorphism separately, the boundary between them is not always distinct and depends largely on which of the three metamorphic agents is dominant.

■ *What are the conditions of contact metamorphism?*

Contact (thermal) metamorphism takes place when a body of magma alters the surrounding country rock (▶ Figure 8.4). At shallow depths, intruding magma raises the temperature of the surrounding rock, causing thermal alteration. Furthermore, the release of hot fluids into the country rock during intrusion can aid in the formation of new minerals.

Important factors in contact metamorphism are the initial temperature, the size of the intrusion, and the fluid content of the magma and the country rock. The initial temperature

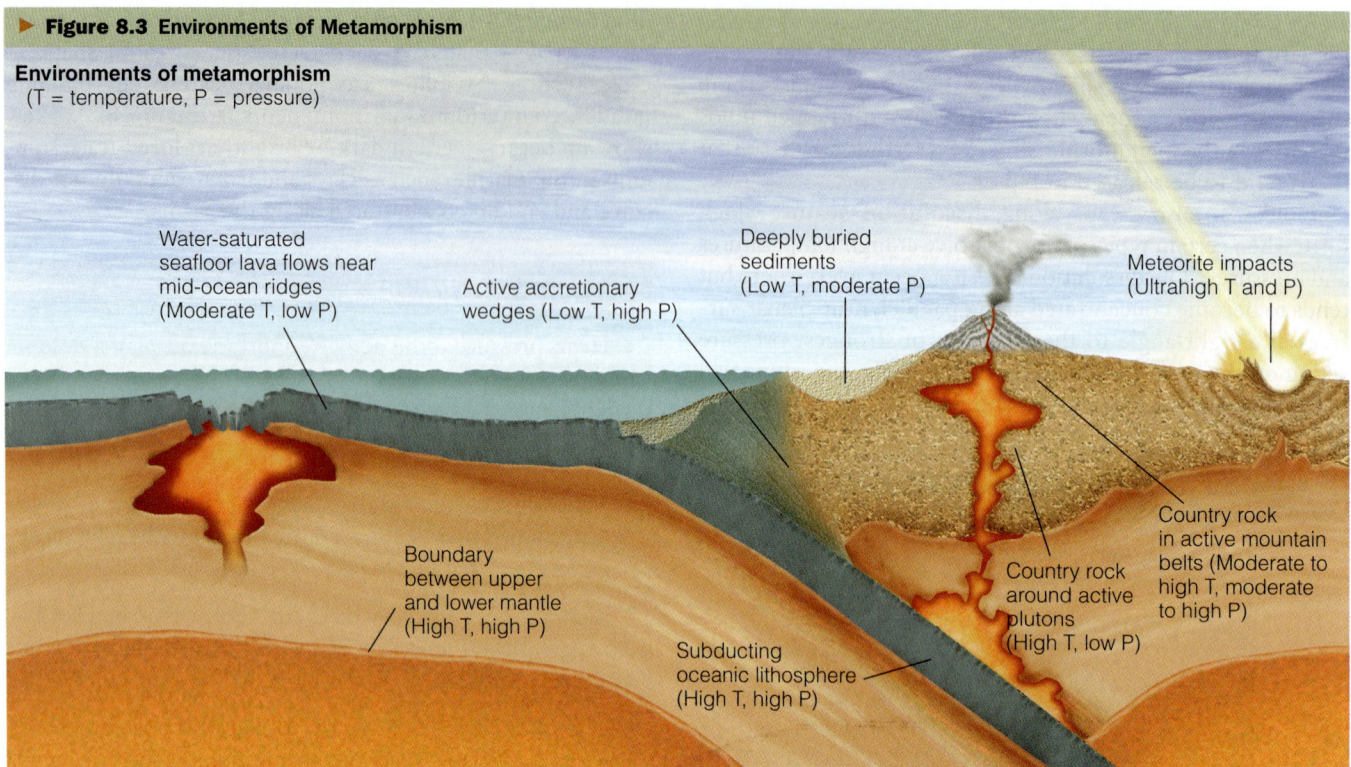

Figure 8.3 Environments of Metamorphism

of an intrusion depends, in part, on its composition; mafic magmas are hotter than felsic magmas (see Chapter 4) and hence have a greater thermal effect on the rocks surrounding them. The size of the intrusion is also important. In the case of small intrusions such as dikes and sills, usually only those rocks in immediate contact with the intrusion are affected. Because large intrusions such as batholiths take a long time to cool, the increased temperature in the surrounding rock may last long enough for a larger area to be affected.

An **aureole** is the area of metamorphism surrounding an intrusion. Metamorphic aureoles vary in width depending on the size, temperature, and composition of the intruding magma as well as the mineralogy of the surrounding country rock. Aureoles range from a few centimeters wide bordering small dikes and sills, to several hundred meters, or even several kilometers wide around large plutons. The degree of metamorphic change within an aureole typically diminishes outward from its contact with an intrusion reflecting decrease in temperature with distance from the original heat source (▶ Figure 8.5). The portion of the aureole closest to the intrusion—in other words that part subject to the highest temperatures—commonly contains metamorphic minerals that crystallized in equilibrium with a higher-temperature environment, such as sillimanite. The more distal outer part may be characterized by lower-temperature metamorphic minerals such as chlorite, talc, and epidote.

The formation of new minerals by contact metamorphism depends not only on proximity to the intrusion but also on the composition of the country rock. Shales, mudstones, impure limestones, and impure dolostones are particularly susceptible to the formation of new minerals by contact metamorphism, whereas pure sandstones or pure limestones typically are not.

Many of the rocks resulting from contact metamorphism have the texture of porcelain; they are hard and fine grained. This is particularly true for rocks with a high clay content, such as shale. Such a texture results because the clay minerals in the rock are baked, just as a clay pot is baked when fired in a kiln.

Many contact metamorphic rocks also are crisscrossed with veins and dikes from nearby intrusions and may be altered by high-temperature fluids that dissolved and carried away some rock constituents in solution while precipitating others that originated within the magma. This process, called **metasomatism,** differs from ordinary metamorphism in that the bulk composition of the rock (not merely the fluid content) changes. In metamorphism there is little if any change in the overall elemental chemistry—merely a change in the way those elements are put together. Metasomatic alteration of contact aureoles in limestone has produced some valuable tungsten ore deposits in which the tungstate ion (WO_4^{-2}) replaces the carbonate ion (CO_3^{-2}) in calcite (▶ Figure 8.6). Tungsten is an important alloy in steel and is used in the manufacture of filaments for incandescent light bulbs. Many other kinds of ore deposits also form in contact aureoles (Table 8.1).

Contact metamorphic effects dwindle or rather merge with those of regional metamorphism around an igneous in-

▶ **Figure 8.4 Contact (Thermal) Metamorphism** A sharp and clearly defined boundary (red line) occurs between the intruding light-colored igneous rock on the left and the dark metamorphosed country rock on the right. The intrusion is part of the Peninsular Ranges Batholith east of San Diego, California.

trusion, becoming indistinguishable from them below around 3–5 km. This reflects the increasing role of pressure with depth in the metamorphic process.

- *What are lithostatic and differential pressures, and why are they important?*

With deeper levels of burial, greater pressure is exerted on rocks, just as you feel greater pressure the deeper you dive in a swimming pool. The pressure on your submerged body is called *hydrostatic*, whereas in the case of rocks it is called **lithostatic pressure.** It doesn't matter how you turn your body in the pool, provided that you keep it positioned at the same general depth; the stress (force per unit area) feels no different coming from the water above than it does from the water to one side or the other, or below. If a rock could speak and feel, most often it would say the same thing about where it sits in the crust as it is deeply buried (▶ Figure 8.7).

During tectonic activity, however, this uniform stress field no longer applies. Stresses coming from some directions are definitely stronger than others. For example, when two tectonic plates collide, the horizontal stress from collision can greatly exceed the vertical stress from burial. This condition is called **differential pressure.**

The structures and mineral textures that develop within metamorphic rocks depend upon whether pressure was lithostatic or differential during metamorphism. Differential pressure plays an especially important role in dynamic and regional metamorphism.

- *Where does dynamic metamorphism occur?*

Significant **dynamic metamorphism,** or pressure-dominated recrystallization, is associated with fault zones in the shallow crust where rocks are subjected to concentrated high levels of differential pressure. The metamorphic rocks that result from pure dynamic metamorphism are called *mylonites*, and typically they are restricted to narrow zones adjacent to faults. Mylonites are irregular, banded, black and white rocks with clots of light-colored mineral matter that are smeared out as though caught in a giant millstone (as

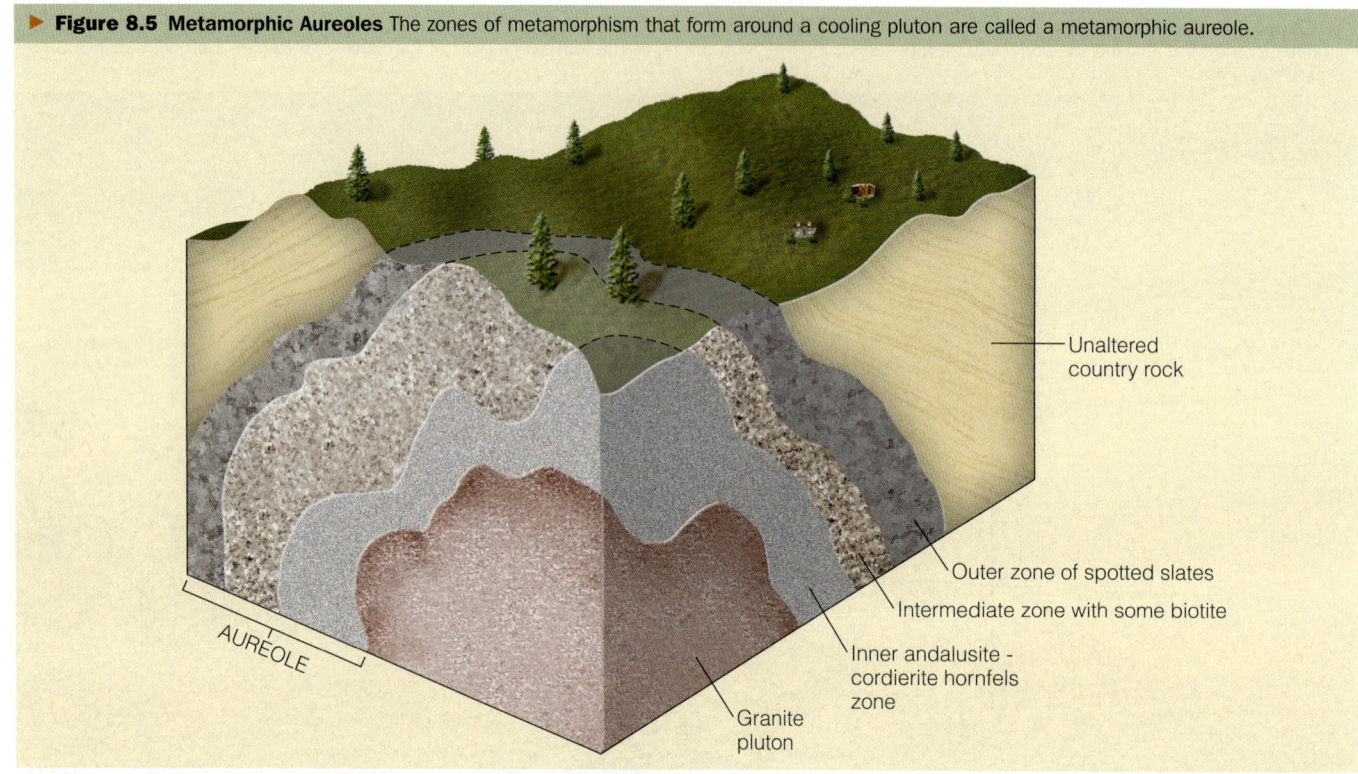

▶ **Figure 8.5 Metamorphic Aureoles** The zones of metamorphism that form around a cooling pluton are called a metamorphic aureole.

indeed they are!) (▶ Figure 8.8). Locally, heat and pressure generated along an active fault can actually cause the rock to melt, forming *pseudotachylite*, a black, glassy material that resembles obsidian. Where milling is less intense, the crust may simply be shattered into chaotic chunks of rock separated by veins of pulverized rock debris—a structure called *cataclastic*. The effects of dynamic metamorphism can be seen even at the surface in active fault zones, where mylonites and cataclastites grade upward into a narrow (centimeters to tens of meters wide) zone of *fault gouge*, a gray powdery material that looks a lot like volcanic ash if derived from granitic rock.

On a broader scale, dynamic metamorphism may be spread over a large region of crust subjected to a rapid buildup in pressure at low temperatures. A case in point is accretionary wedges, the sedimentary and volcanic prows of continental plates that override oceanic plates at convergent plate boundaries. One of the salient rock types produced by this kind of metamorphism is *blueschist*, which owes its color to the bluish amphibole glaucophane. Blueschist bodies tend to be cut up and dismembered by numerous faults, essentially forming a large-scale cataclastic structure.

■ *What is shock metamorphism?*

Meteorite impacts cause an even more extreme type of dynamic metamorphism called **shock metamorphism.** Shock waves from a large impacting meteorite radiate out through the crust with pressures ranging from 20 to 500 kilobars (kb). To put this in perspective, 1 kb is equivalent to 1000 atmospheres of pressure, and the lithostatic pressure of rock buried 3 km deep is only 1 kb. Obviously, these are extreme forces even for metamorphic rocks. They are also short-lived, on the order of seconds—just long enough to cause quartz grains to undergo polymorphic transformation into stishovite, an ultra-dense form of silica. The crust may also show evidence of shock wave strain by developing telltale conical bundles of fractures, the tips of which all point in the direction of the impact (▶ Figure 8.9). The presence of these *shatter cones* and of stishovite has helped geologists identify two of the world's largest ore-producing

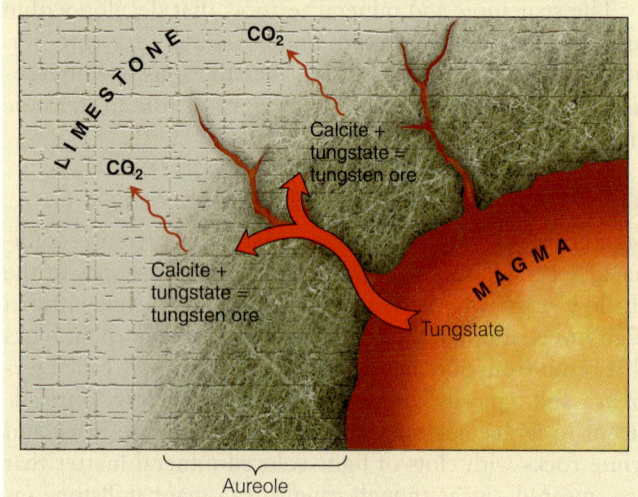

▶ **Figure 8.6 An Example of Metasomatism, Which Is Not Quite the Same Thing as Metamorphism**

TABLE 8.1
The Main Ore Deposits Resulting from Contact Metamorphism

Ore Deposit	Major Mineral	Formula	Use
Copper	Bornite Chalcopyrite	Cu_5FeS_4 $CuFeS_2$	Important sources of copper, which is used in various aspects of manufacturing, transportation, communications, and construction
Iron	Hematite Magnetite	Fe_2O_3 Fe_3O_4	Major sources of iron for manufacture of steel, which is used in nearly every form of construction, manufacturing, transporation, and communications
Lead	Galena	PbS	Chief source of lead, which is used in batteries, pipes, solder, and elsewhere where resistance to corrosion is required
Tin	Cassiterite	SnO_2	Principal source of tin, which is used for tin plating, solder, alloys, and chemicals
Tungsten	Scheelite Wolframite	$CaWO_4$ $(Fe, Mn)WO_4$	Chief sources of tungsten, which is used in hardening metals and manufacturing carbides
Zinc	Sphalerite	$(Zn, Fe)S$	Major source of zinc, which is used in batteries and in galvanizing iron and making brass

mafic plutons as the products of major meteorite strikes: the Bushveld Complex in South Africa and the Sudbury pluton in Ontario, Canada.

■ *What is the most common type of metamorphism?*

Most metamorphic rocks result from **regional**, or *dynamothermal*, **metamorphism,** which occurs over a large area and is usually caused by tremendous temperatures, pressures, and deformation all occurring together within the deeper portions of the crust. Regional metamorphism is most obvious along convergent plate boundaries where rocks are intensely deformed and recrystallized during convergence and subduction. These metamorphic rocks usually reveal a gradation of metamorphic intensity from areas that were subjected to the most intense pressures and/or highest temperatures to areas of lower pressures and temperatures. Such a gradation in metamorphism can be recognized by the metamorphic minerals that are present. Regional metamorphism is not confined to only convergent margins. It also occurs in areas where plates diverge, although usually at much shallower depths because of the high geothermal gradient associated with these areas.

The term **metamorphic grade** generally characterizes the degree to which a rock has undergone metamorphic change. There are no hard and fast boundaries between low-, medium-, and high-grade metamorphic conditions, but the distinction is nonetheless useful for communicating in a general way about metamorphism. For example, when a clay-rich rock such as shale undergoes regional metamorphism, the mineral chlorite first begins to crystallize under relatively low temperatures of about 200°C and its presence in these rocks indicates "low-grade" metamorphism. If temperatures and pressures continue to increase, however, new minerals crystallize and replace chlorite because they are more stable under the changing conditions. Garnet and biotite are good indicators of "medium-grade" metamorphism, and the presence of sillimanite, which forms at around 550°C, indicates "high-grade" conditions.

▶ **Figure 8.7 Lithostatic Pressure**

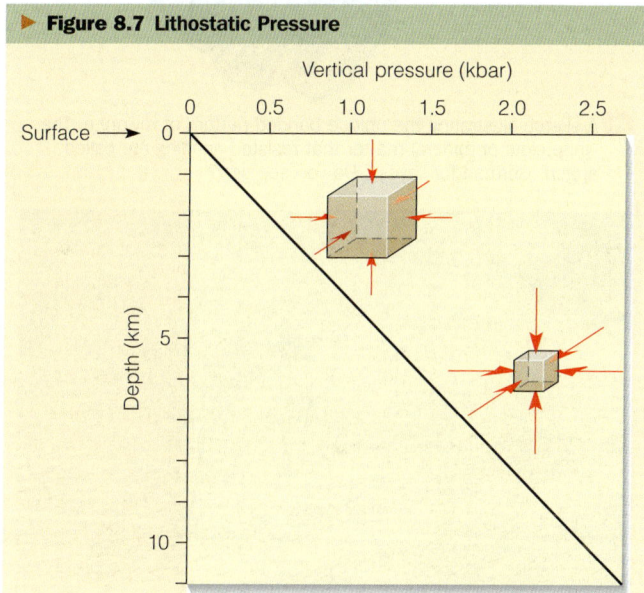

1 kilobar (kbar) = 1000 bars
Atmospheric pressure at sea level = 1 bar

a Lithostatic pressure is applied equally in all directions in Earth's crust due to the weight of overlying rocks. Thus pressure increases with depth, as indicated by the sloping black line.

b A similar situation occurs when 200-ml Styrofoam cups are lowered to ocean depths of approximately 750 m and 1500 m. Increased water pressure is exerted equally in all directions on the cups, and they consequently decrease in volume while maintaining their general shape. Source: (a): From C. Gillen, *Metamorphic Geology*, Figure 4.4, p. 73. Copyright © 1982. Reprinted with the kind permission of Kluwer Academic Publishers and C. Gillen.

▶ **Figure 8.8 Mylonite**

Augen

a A sketch illustrating the unique banded pattern of mylonite. The white clots of mineral matter that resisted grinding are called *augen*, German for "eyes." Do you see why?

b Mylonite from the Adirondack Highlands, New York.

▶ **Figure 8.9 Shatter Cones Shown by Fine Radiating Lines in Broken Rock Faces** The cones form when powerful shock waves travel through the crust.

Section 8.3 Summary

- The three main kinds of metamorphism are contact, dynamic, and regional.

- Heat is the dominant agent driving contact metamorphism, which is restricted to the vicinity of igneous intrusions at shallow levels in the crust, where lithostatic pressure is the dominant type of pressure acting on the rock.

- Differential pressure, a nonuniform pressure characteristic of rock undergoing deformation, is the principal driving force in dynamic metamorphism, which is common around fault zones and in accretionary wedges.

- Shock metamorphism is a special kind of high-intensity dynamic metamorphism caused by meteorite impacts.

- Regional metamorphism, the most widespread type of metamorphism, involves heat and pressure and is associated with mountain building at convergent plate boundaries.

Far more than contact or dynamic metamorphism, regional metamorphism provides a great amount of information about the conditions of temperature and pressure in the crust associated with tectonic processes. It isn't merely that this type of metamorphism is widespread, as the name *regional* implies, but the combination of heat and pressure applied to clay-rich rocks (so-called **pelitic rocks**) triggers an especially large number of metamorphic reactions whose temperatures and pressures can be constrained in laboratory experiments. Many geologists also regard metamorphosed pelites as the most beautiful rocks of all.

8.4 Classification of Metamorphic Rocks

For purposes of classification, metamorphic rocks are commonly divided into two groups: those exhibiting a foliated (from the Latin *folium*, "leaf") texture and those with a nonfoliated texture (Table 8.2).

■ *What is foliated texture, and what are some examples of foliated rocks?*

Rocks subjected to heat and differential pressure during metamorphism typically have minerals arranged in a parallel, layered fashion, giving them a **foliated texture** (▶ Figure 8.10). Low-grade metamorphic rocks have a finely foliated texture in which mineral grains are so tiny that they cannot be distinguished without the aid of a microscope. High-grade foliated rocks, as you might expect, are coarse grained, with individual grains easily seen with the unaided eye.

Slate is the lowest-grade foliated metamorphic rock. It is also well known because it is used to make chalkboards and roofing tiles. (Next time you write on an old-fashioned chalkboard, reflect that you are using a fossiliferous sedimentary material—the chalk—to write on a metamorphosed sedimentary surface—the board.) Pelitic slate, by far the most common variety, is typically dark green, gray, or brown and uniform in color. The finely laminated foliation, or *slaty cleavage*, results in platy foliation, which allows slate to be split easily into thin sheets (▶ Figure 8.11).

With a slight increase in grain size, a pelitic rock that otherwise looks like a slate acquires a shiny luster on its foliated faces. These rocks are called *phyllites*. Phyllites have a distinctive iridescent sheen due to the reflection of light from many tiny crystal faces. However, it is not possible to discern individual grains in phyllites without the aid of a microscope.

Further coarsening in grain size produces *schist*—a medium-grade metamorphic rock. Like phyllites, schist is a shiny rock with individually discernable grains no larger than a few millimeters across. Metamorphosed mafic rocks produce *greenschist*, which is rich in chlorite and plagioclase. Pelitic rocks produce a variety of schists ranging from silvery gray owing to the abundance of muscovite, to black from the concentration of biotite, to green, also due to chlorite. These micas define the foliation fabric of the rock, which is called *schistose foliation* or *schistosity*. Unlike the platy breakage of slates, schistosity tends to produce wavy surfaces when the rock is split (▶ Figure 8.12). **Porphyroblasts** are larger crystals that nucleate and grow in a solid state and push aside or partly incorporate the foliation planes as they grow. They may be as large as several centimeters. Examples of porphyroclasts commonly found in pelitic schists include garnet, andalusite, and staurolite.

Further coarsening of grains accompanies the development of segregated lenses and layers of minerals within the

TABLE 8.2 Classification of Common Metamorphic Rocks

Texture	Metamorphic Rock	Typical Minerals	Metamorphic Grade	Characteristics of Rocks	Parent Rock
Foliated	Slate	Clays, micas, chlorite	Low	Fine-grained, splits easily into flat pieces	Mudrocks, volcanic ash
	Phyllite	Fine-grained quartz, micas, chlorite	Low to medium	Fine-grained, glossy or lustrous sheen	Mudrocks
	Schist	Micas, chlorite, quartz, talc, hornblende, garnet, staurolite, graphite	Low to high	Distinct foliation, minerals visible	Mudrocks, carbonates, mafic igneous rocks
	Gneiss	Quartz, feldspars, hornblende, micas	High	Segregated light and dark bands visible	Mudrocks, sandstones, felsic igneous rocks
	Amphibolite	Hornblende, plagioclase	Medium to high	Dark, weakly foliated	Mafic igneous rocks
	Migmatite	Quartz, feldspars, hornblende, micas	High	Streaks or lenses of granite intermixed with gneiss	Felsic igneous rocks mixed with sedimentary rocks
Nonfoliated	Marble	Calcite, dolomite	Low to high	Interlocking grains of calcite or dolomite, reacts with HCl	Limestone or dolostone
	Quartzite	Quartz	Medium to high	Interlocking quartz grains, hard, dense	Quartz sandstone
	Greenstone	Chlorite, epidote, hornblende	Low to high	Fine-grained, green	Mafic igneous rocks
	Hornfels	Micas, garnets, andalusite, cordierite, quartz	Low to medium	Fine-grained, equidimensional grains, hard, dense	Mudrocks
	Anthracite	Carbon	High	Black, lustrous, subconcoidal fracture	Coal

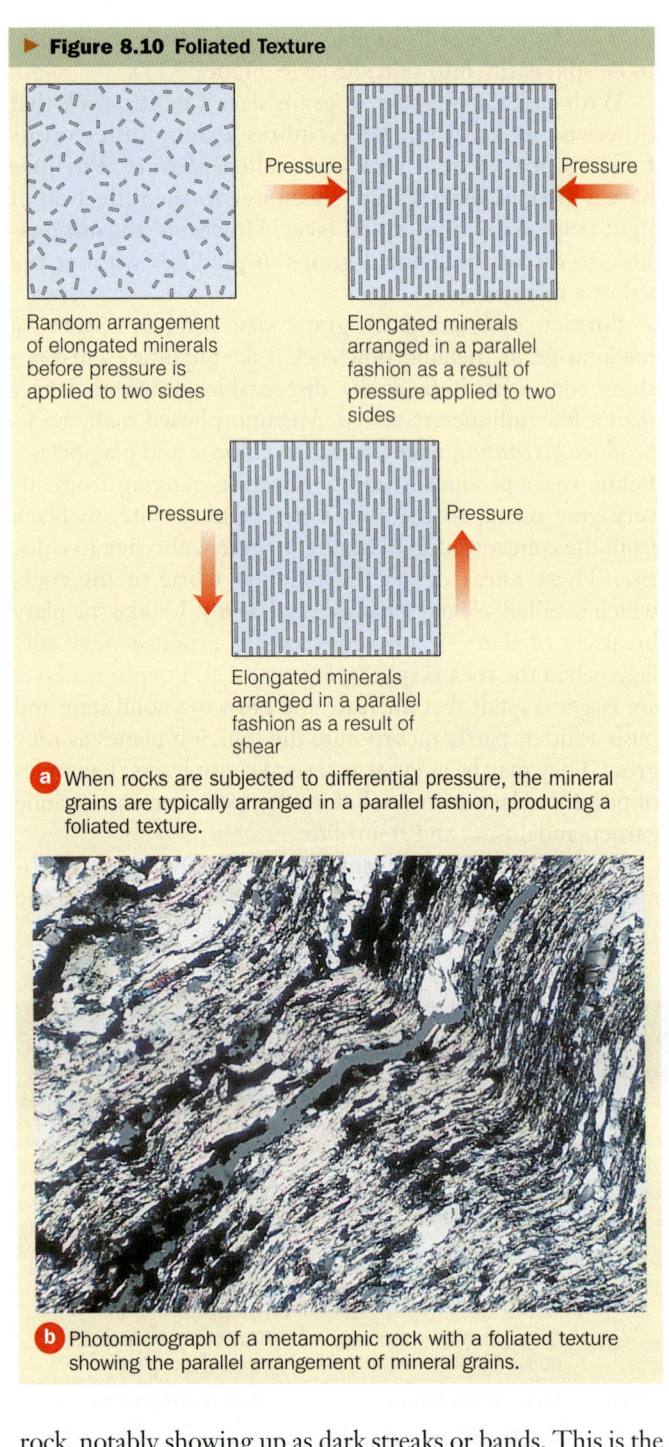

▶ **Figure 8.10 Foliated Texture**

a When rocks are subjected to differential pressure, the mineral grains are typically arranged in a parallel fashion, producing a foliated texture.

Random arrangement of elongated minerals before pressure is applied to two sides

Elongated minerals arranged in a parallel fashion as a result of pressure applied to two sides

Elongated minerals arranged in a parallel fashion as a result of shear

b Photomicrograph of a metamorphic rock with a foliated texture showing the parallel arrangement of mineral grains.

▶ **Figure 8.11 Slate**

a Hand specimen of slate.

b Slaty cleavage in a streambed near Lexington, Virginia.

c Slate roof of Chalet Enzian, Switzerland.

rock, notably showing up as dark streaks or bands. This is the defining feature of *gneiss*, a high-grade metamorphic rock that is typically pelitic in origin but also can result from metamorphism of intermediate and felsic igneous rocks. Gneisses consist of mostly granular minerals such as quartz, feldspar, or both, with lesser percentages of platy or elongated minerals such as micas or amphiboles (▶ Figure 8.13). Quartz and feldspar characteristically make up the light-colored bands, whereas biotite and hornblende compose the dark bands. Typically, gneiss breaks in an irregular manner, much like coarsely crystalline nonfoliated rocks.

Amphibolite is another fairly common foliated metamorphic rock. A dark rock, it is composed of mainly hornblende and plagioclase. The alignment of the hornblende crystals produces a slightly foliated texture. Many amphibolites result from medium- to high-grade metamorphism of basalt and ferromagnesian-rich mafic rocks.

Regional metamorphism of conglomerates produces spectacular *stretched-pebble metaconglomerates* (▶ Figure 8.14).

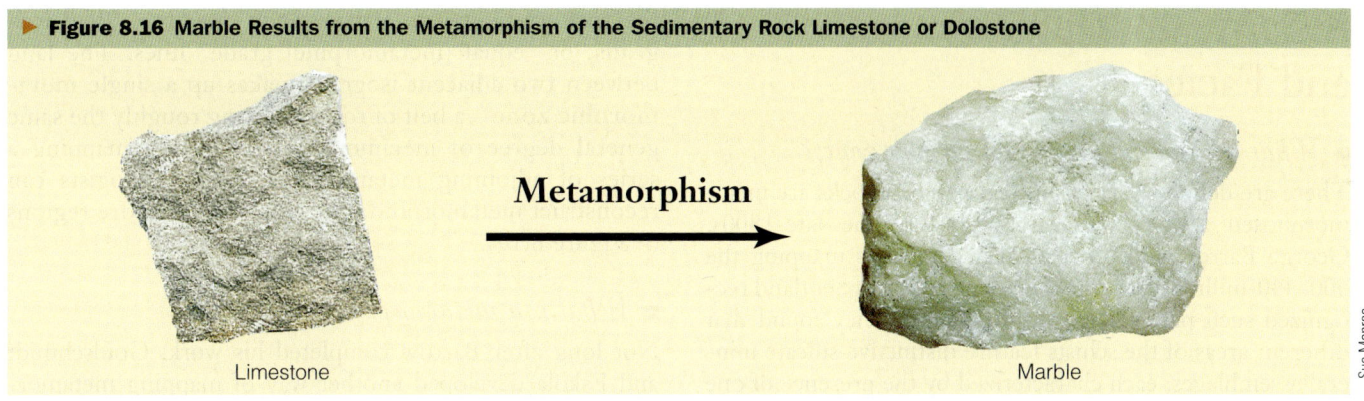

Figure 8.16 Marble Results from the Metamorphism of the Sedimentary Rock Limestone or Dolostone

Limestone → Metamorphism → Marble

low- to high-grade metamorphic conditions. The green color results from the presence of chlorite, epidote, and hornblende.

Hornfels is a common, fine-grained, nonfoliated metamorphic rock resulting from contact metamorphism consisting of various equidimensional mineral grains. The composition of hornfels depends directly on the composition of the original rock, and many compositional varieties are known. The majority of hornfelses, however, are apparently derived from contact metamorphism of pelitic rocks or impure dolostones.

Anthracite is a black, lustrous, hard coal that contains a high percentage of fixed carbon and a low percentage of volatile matter. It is highly valued by people who burn coal for heating and power. Anthracite usually forms from the metamorphism of lower-grade coals by heat and pressure, and many geologists consider it to be a metamorphic rock.

Coal is not the only formerly living material that is caught up in metamorphism. Some marble contains *graphite*, a mineral of pure carbon that we mine to make high-temperature laboratory furnaces and the lead in pencils. The graphite originates as living tissues, but under the severe conditions imposed by metamorphism, the organic bonds break down; heat drives off the oxygen, hydrogen, nitrogen, sulfur, and other elements in the dead residue; and a pure carbon residue remains behind.

Section 8.4 Summary

- Metamorphic rocks can be divided texturally into foliated and nonfoliated types.

- Foliated rocks show a parallel arrangement of platy minerals that produces a crudely layered structure. Most foliated rocks are pelitic (clay-rich) or mafic in origin.

- Foliated metamorphic rocks show an increase in mean grain size with slates and phyllites being fine-grained, schists medium-grained, and gneisses coarse-grained as well as displaying banding of light and dark minerals.

- Nonfoliated rocks display an equigranular mineral texture and include quartzites, marbles, and hornfelses.

- Migmitites show evidence of partial melting, whereas metaconglomerates show a stretched-pebble structure.

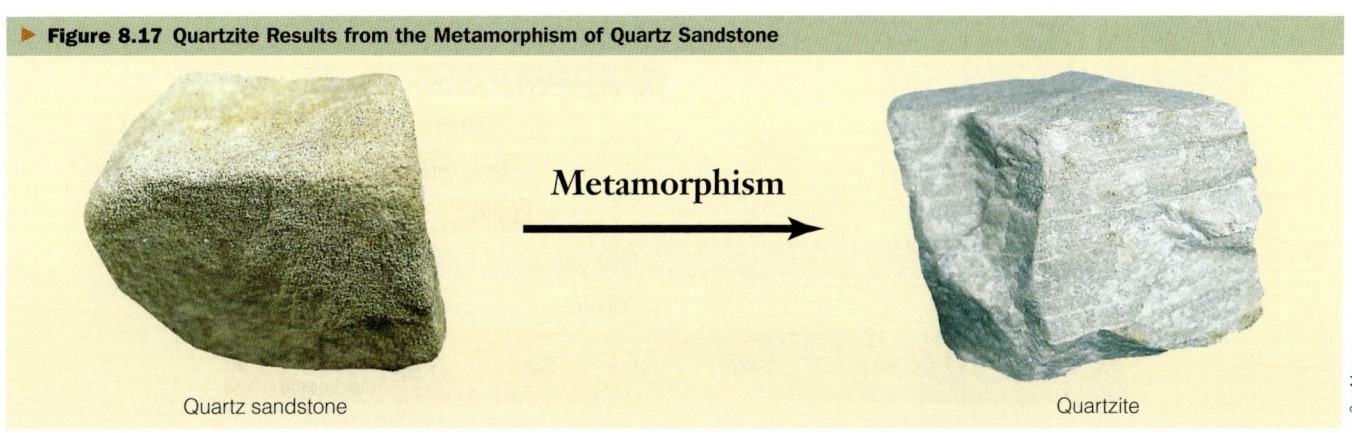

Figure 8.17 Quartzite Results from the Metamorphism of Quartz Sandstone

Quartz sandstone → Metamorphism → Quartzite

8.5 Metamorphic Zones and Facies

■ *What are isograds and metamorphic zones?*

There are definite patterns in the way that rocks are metamorphosed across broad areas; and in the late-1800s, George Barrow and other British geologists mapping the 400–440-million-year-old Dalradian schists of Scotland recognized such patterns for the first time. They found that different areas of the schists feature distinctive silicate mineral assemblages, each characterized by the presence of one or more telltale *index minerals*, indicating different degrees of metamorphism. The index minerals Barrow selected to represent progressively increasing metamorphic intensity were chlorite, biotite, garnet, staurolite, kyanite, and sillimanite (▶ Figure 8.18), which we now know all result from the recrystallization of pelitic rocks.

Later work in Norway and Finland by geologists V.M. Goldschmidt and Pentii Eskola revealed that other mineral assemblages and index minerals are produced from rocks having different original bulk compositions. If the Dalradian countryside had been made of mafic rocks, for example, the index minerals identified by Barrow and his co-workers would likely have been chlorite and hornblende.

To map index minerals, Barrow worked from lower grade to higher grade metamorphic terrain, plotting all the positions where a particular mineral first appeared in the rocks as a line on a map. Such lines are called **isograds**, or "equal [metamorphic] grade" lines. The land between two adjacent isograds makes up a single **metamorphic zone**—a belt of rocks showing roughly the same general degree of metamorphism. Through mapping a series of adjoining metamorphic zones, geologists can reconstruct metamorphic conditions across entire regions (▶ Figure 8.19).

■ *What is a metamorphic facies?*

Not long after Barrow completed his work, Goldschmidt and Eskola developed another way of mapping metamorphism that proved even more useful than the metamorphic zone approach, given the great variety of rock types they encountered while studying contact aureoles in Scandanavia. Eskola defined **metamorphic facies** as a group of metamorphic rocks, with each rock type characterized by its own distinctive mineral assemblage formed under broadly similar temperature and pressure conditions (▶ Figure 8.20). He named each facies he encountered after its most characteristic rock or mineral. For example, the green metamorphic mineral chlorite, which forms under relatively low temperatures and pressures, yields rocks belonging to the *greenschist facies*. Under increasingly higher temperatures and pressures, mineral assemblages indicative of the *amphibolite* and *granulite facies* develop. Ultimately, geologists identified ten metamorphic facies, spanning the full range of metamorphic conditions possible within Earth.

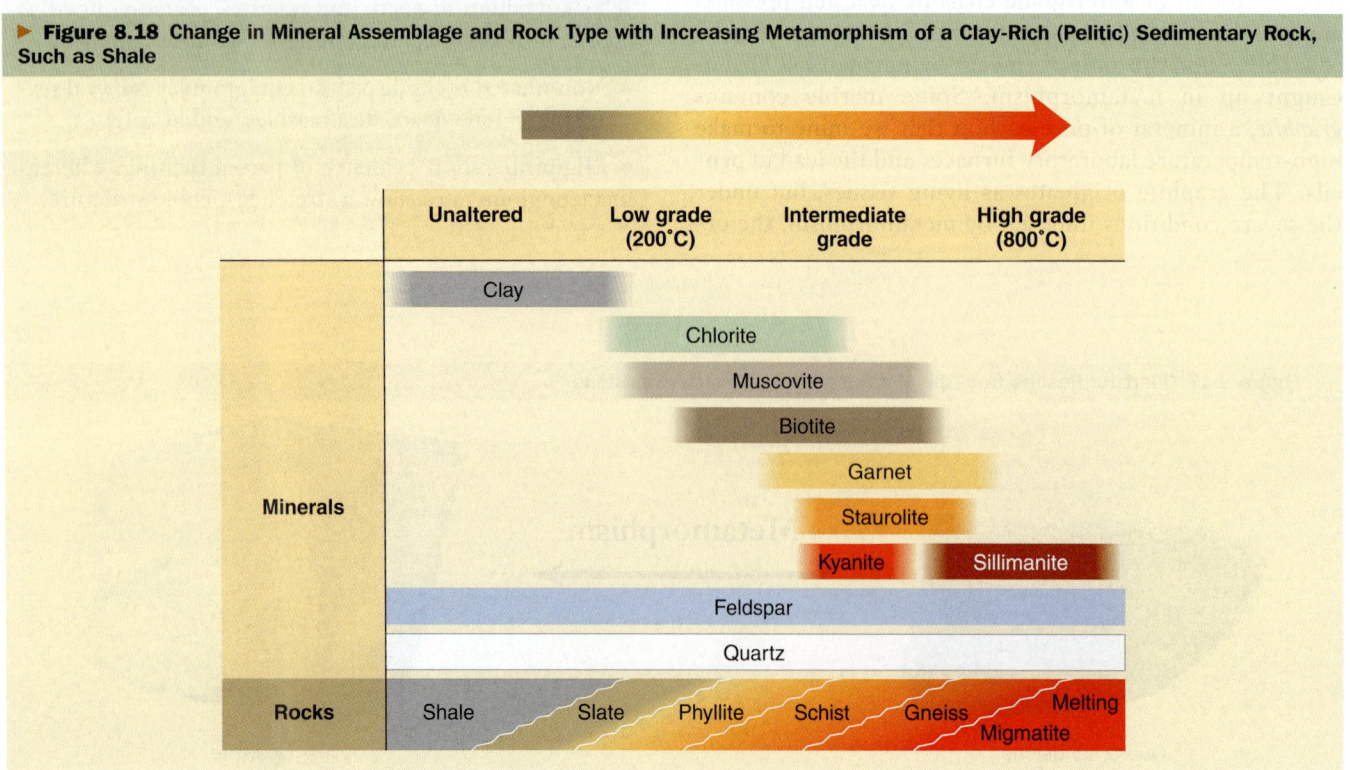

▶ **Figure 8.18** Change in Mineral Assemblage and Rock Type with Increasing Metamorphism of a Clay-Rich (Pelitic) Sedimentary Rock, Such as Shale

■ *How do metamorphic zones and metamorphic facies differ?*

The concepts of metamorphic facies and metamorphic zones are easily confused by students new to the study of metamorphic rocks. Bear in mind that metamorphic zones are identified by single index minerals that appear in a predictable sequence within rocks of the same general composition throughout an area. In contrast, rocks of greatly different composition within an area can belong to the same metamorphic facies. Each rock type has its own characteristic assemblage of minerals whose presence indicates metamorphism within the range of pressures and temperatures unique to that facies. Whether a geologist approaches the study of metamorphism by mapping isograds and zones or facies boundaries depends on the particular conditions encountered in the field.

Under certain circumstances, neither metamorphic zones nor facies can be identified. All one can say for sure is that "metamorphism happened!" This is particularly the case in areas where the original rocks were pure quartz sandstones or pure limestones or dolostones. Such rocks would yield only quartzites and marbles, regardless of imposed temperature and pressure conditions.

Section 8.5 Summary

● Geologists map metamorphic rocks based on occurrences of characteristic minerals or mineral assemblages.

● Isograd maps delineate metamorphic zones, whereas distinctive sets of minerals in particular rock types indicate metamorphic facies.

● Pelitic rocks are the most informative rock type for recording the effects of metamorphism. Some rocks, however, such as pure quartz sandstone or pure limestone, provide no clue as to the intensity of metamorphism.

▶ **Figure 8.19 Metamorphic Zones in the Upper Peninsula of Michigan** The zones in this region are based on the presence of distinctive silicate mineral assemblages resulting from the metamorphism of sedimentary rocks during an interval of mountain building and minor granitic intrusion during the Proterozoic Eon, about 1.5 billion years ago. The lines separating the different metamorphic zones are isograds. Source: From H. L. James, *Geological Society of America Bulletin*, vol. 66, plate 1, page 1454, with permission of the publisher, the Geological Society of America, Boulder, Colorado, USA. Copyright © 1955 Geological Society of America.

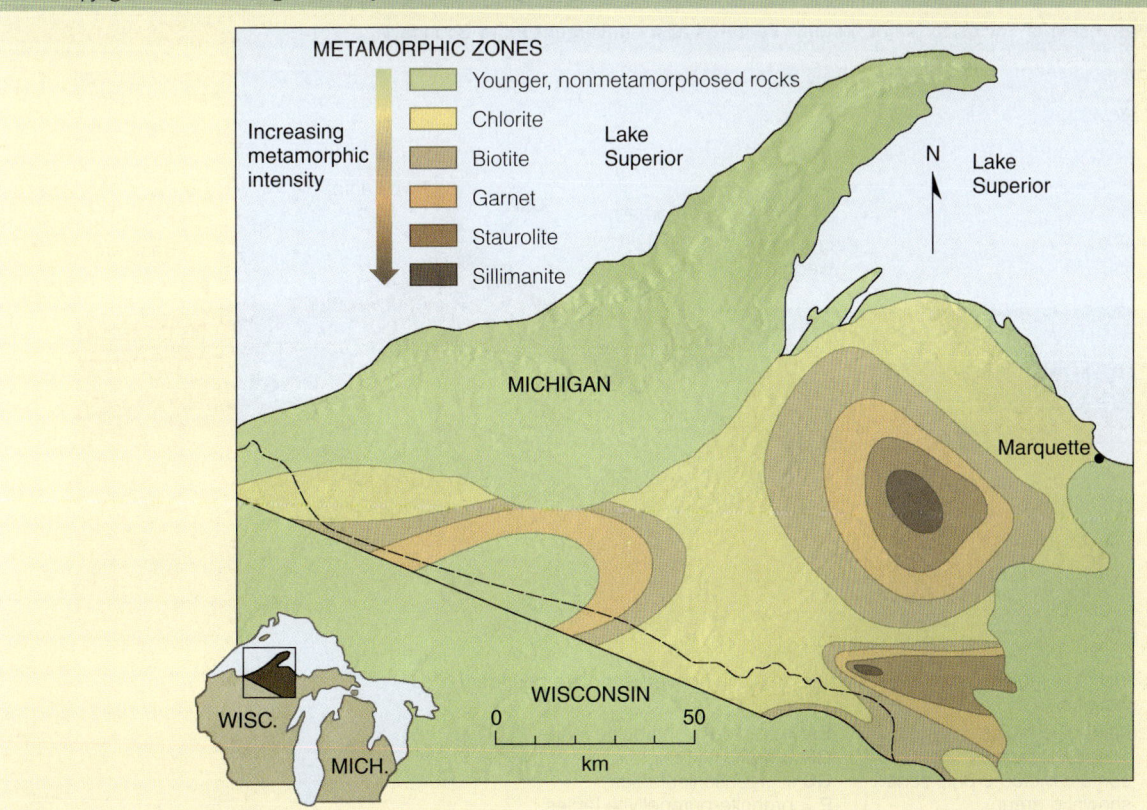

8.6 Plate Tectonics and Metamorphism

■ *How does metamorphism relate to plate boundaries?*

Metamorphism is associated with all three types of plate boundaries. At mid-oceanic ridges, infiltration of seawater into hot crust and mantle leads to patchy low- to medium-grade metamorphism. Along transform fault boundaries, dynamic metamorphism creates mylonites and cataclastites, especially in continental settings. At oceanic–continental convergent plate boundaries, temperatures and pressures create **paired metamorphic belts.** A **metamorphic belt** is a zone of metamorphic rocks sharing the same general conditions and age of formation. A typical paired metamorphic belt consists of a zone of dynamically metamorphosed rocks within the accretionary wedge and forearc basin adjoining a zone of regional and contact metamorphic rocks within the arc. ▶ Figure 8.21 illustrates the various metamorphic facies conditions present at a typical oceanic–continental convergent plate boundary. Where two continents converge, regional metamorphism becomes even more widespread, overprinting the dynamic metamorphism of forearcs and accretionary wedges.

One of the most important consequences of metamorphism at convergent plate boundaries is the production of mafic magma. Recall that subducting oceanic lithosphere

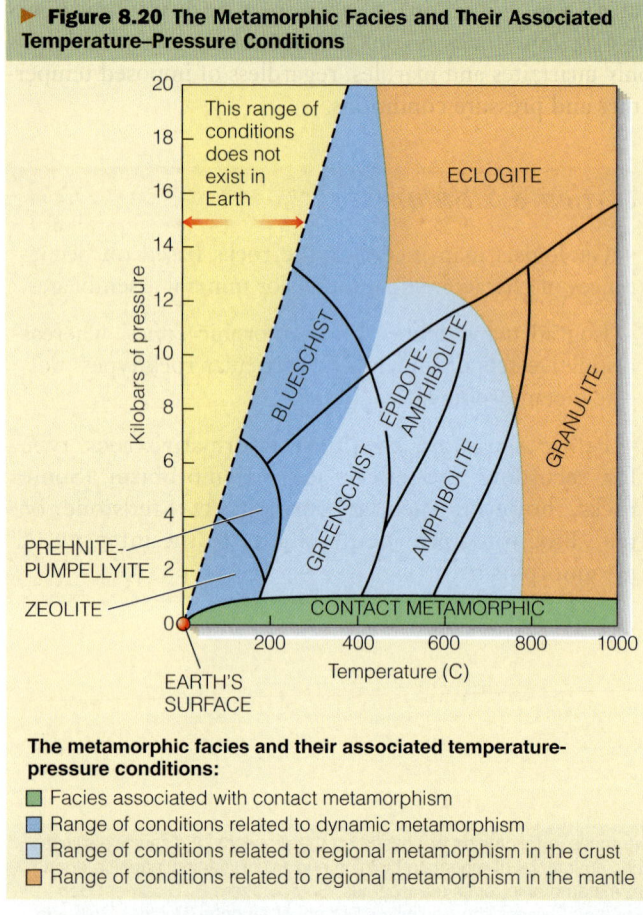

▶ Figure 8.20 The Metamorphic Facies and Their Associated Temperature–Pressure Conditions

The metamorphic facies and their associated temperature-pressure conditions:
- Facies associated with contact metamorphism
- Range of conditions related to dynamic metamorphism
- Range of conditions related to regional metamorphism in the crust
- Range of conditions related to regional metamorphism in the mantle

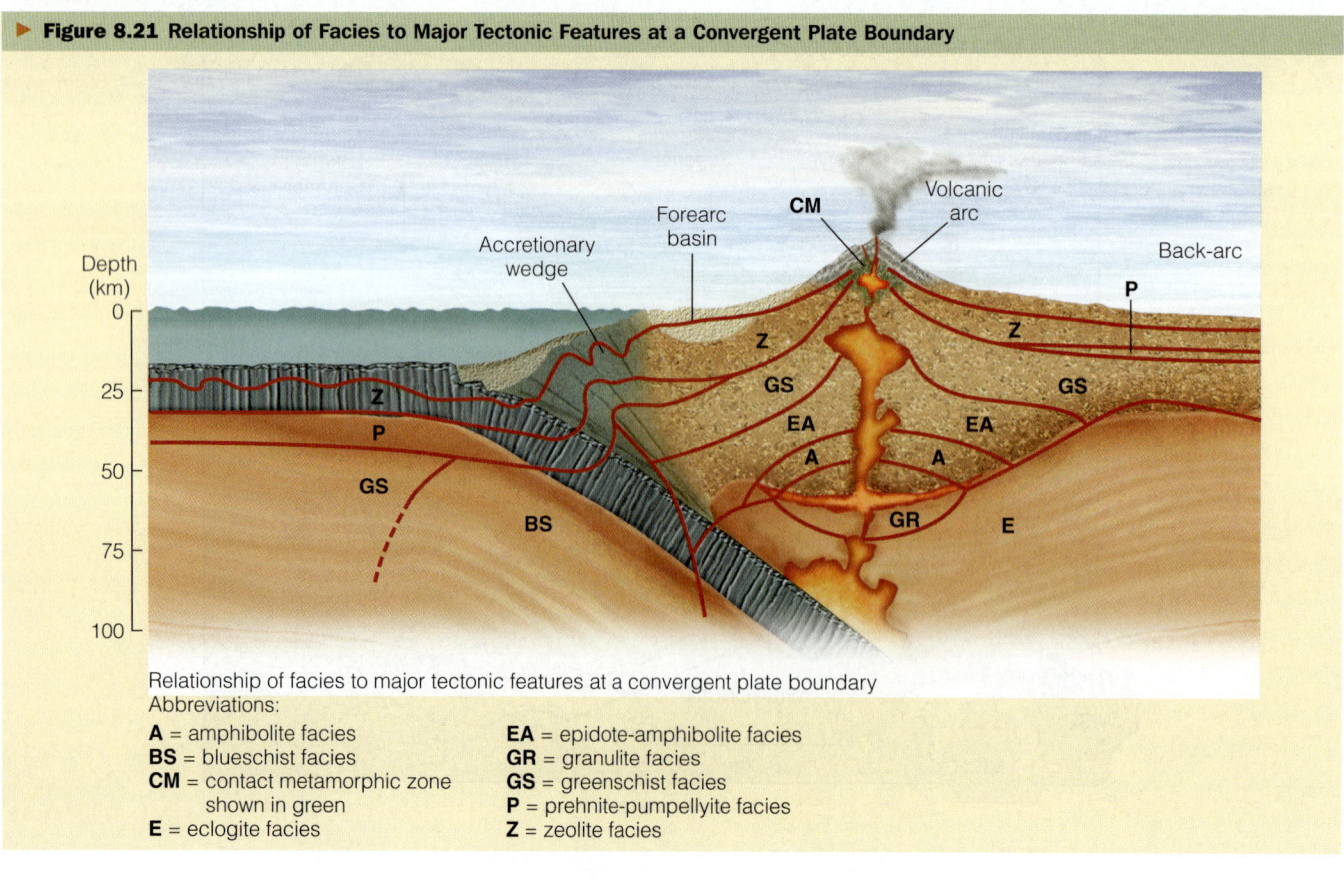

▶ Figure 8.21 Relationship of Facies to Major Tectonic Features at a Convergent Plate Boundary

Relationship of facies to major tectonic features at a convergent plate boundary
Abbreviations:
A = amphibolite facies
BS = blueschist facies
CM = contact metamorphic zone shown in green
E = eclogite facies
EA = epidote-amphibolite facies
GR = granulite facies
GS = greenschist facies
P = prehnite-pumpellyite facies
Z = zeolite facies

carries with it large amounts of trapped seawater. Much of this is stored as H_2O or OH^{-2} bonded in the lattices of metamorphic minerals like chlorite, actinolite, and glaucophane. Once the slab reaches a depth of 75–100 km, heat and pressure combine to drive off the H_2O and replace these minerals with newer ones that lack H_2O, such as the "dry" minerals garnet and pyroxene. The fluid escapes from the subducting slab like bubbles from a sinking ship, and this promotes partial melting in the mantle wedge between the slab and continental crust. What happens next, you'll remember, is described in detail in Chapter 4 on igneous rocks.

Some geologists think that metamorphism plays an important role in *causing* subduction. A deeply sinking slab transforms into metamorphic rock of the eclogite facies as it subsides past the point of fluid release. Eclogite is an especially dense rock, made all the heavier by the fact that it is somewhat cooler than the enclosing mantle. The eclogitic slabs may act like anchors, pulling the oceanic lithosphere hundreds of kilometers beneath the surface.

During regional metamorphism and mountain building, huge thrust faults called *nappes* may form, allowing the highly compressed central crust of a young mountain range to spread out horizontally within a few kilometers of the surface, somewhat like putty being squeezed from a tube. The presence of numerous garnet porphyroblasts in certain pelitic rocks shows that intensive metamorphism takes place concurrently with nappe thrusting. The shear stress generated by thrusting causes the garnets to rotate as they grow. The **rotated garnets** incorporate inclusions of other smaller mineral grains, giving them striking S-shaped or spiral inclusion patterns (▶ Figure 8.22). Careful radiometric dating of rotated garnets can enable geologists indirectly to ascertain the rate at which thrust faulting related to mountain building takes place.

■ *How is an orogenic metamorphic belt exposed?*

As plate convergence abates or ceases, erosion combined with isostatic uplift discussed in Chapter 11 gradually unroofs the metamorphic heart of the mountain range. It may

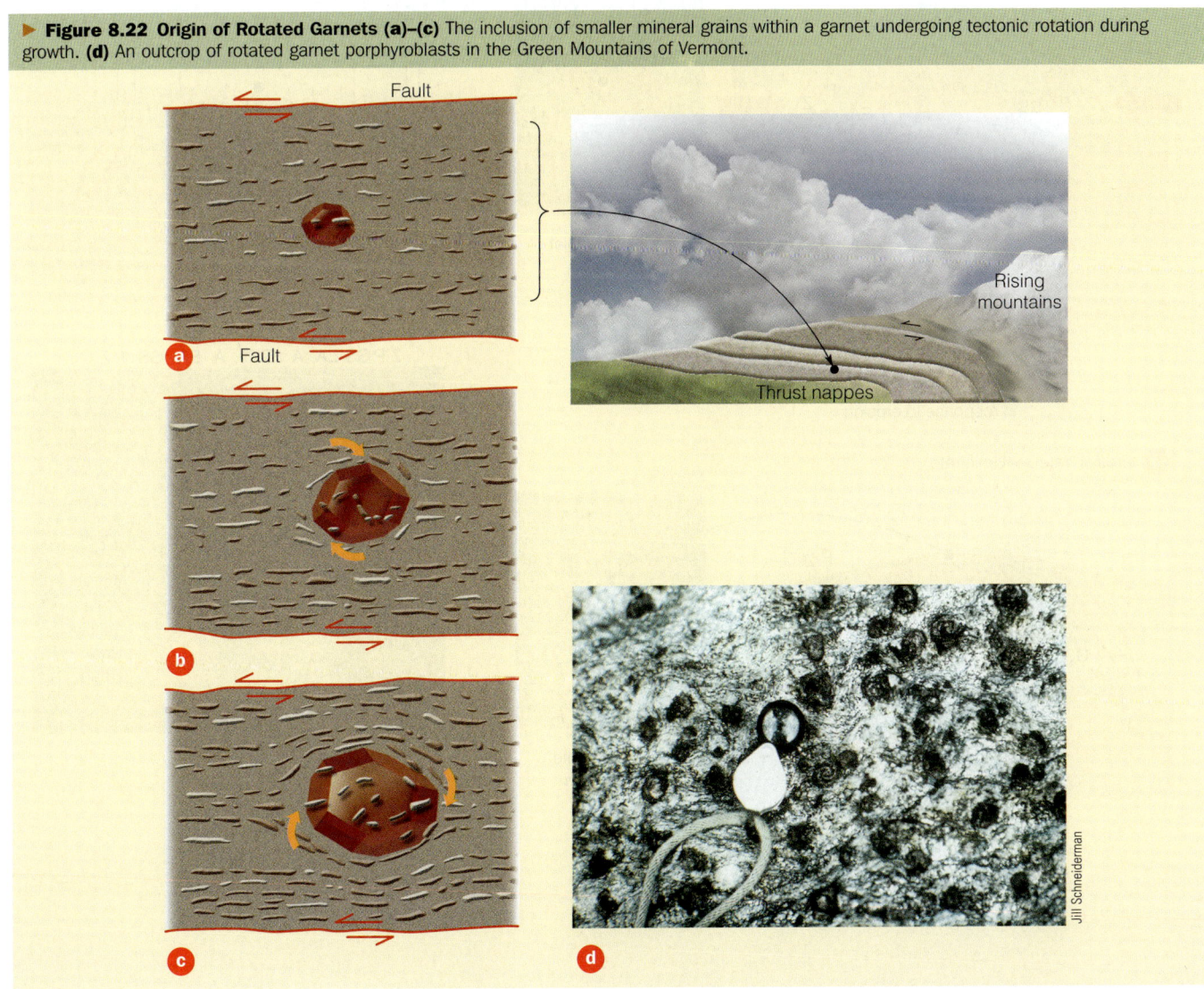

▶ **Figure 8.22 Origin of Rotated Garnets (a)–(c)** The inclusion of smaller mineral grains within a garnet undergoing tectonic rotation during growth. **(d)** An outcrop of rotated garnet porphyroblasts in the Green Mountains of Vermont.

252 Chapter 8 Metamorphism and Metamorphic Rocks

▶ Figure 8.23 Development of a Metamorphic Belt During and Following Mountain Building

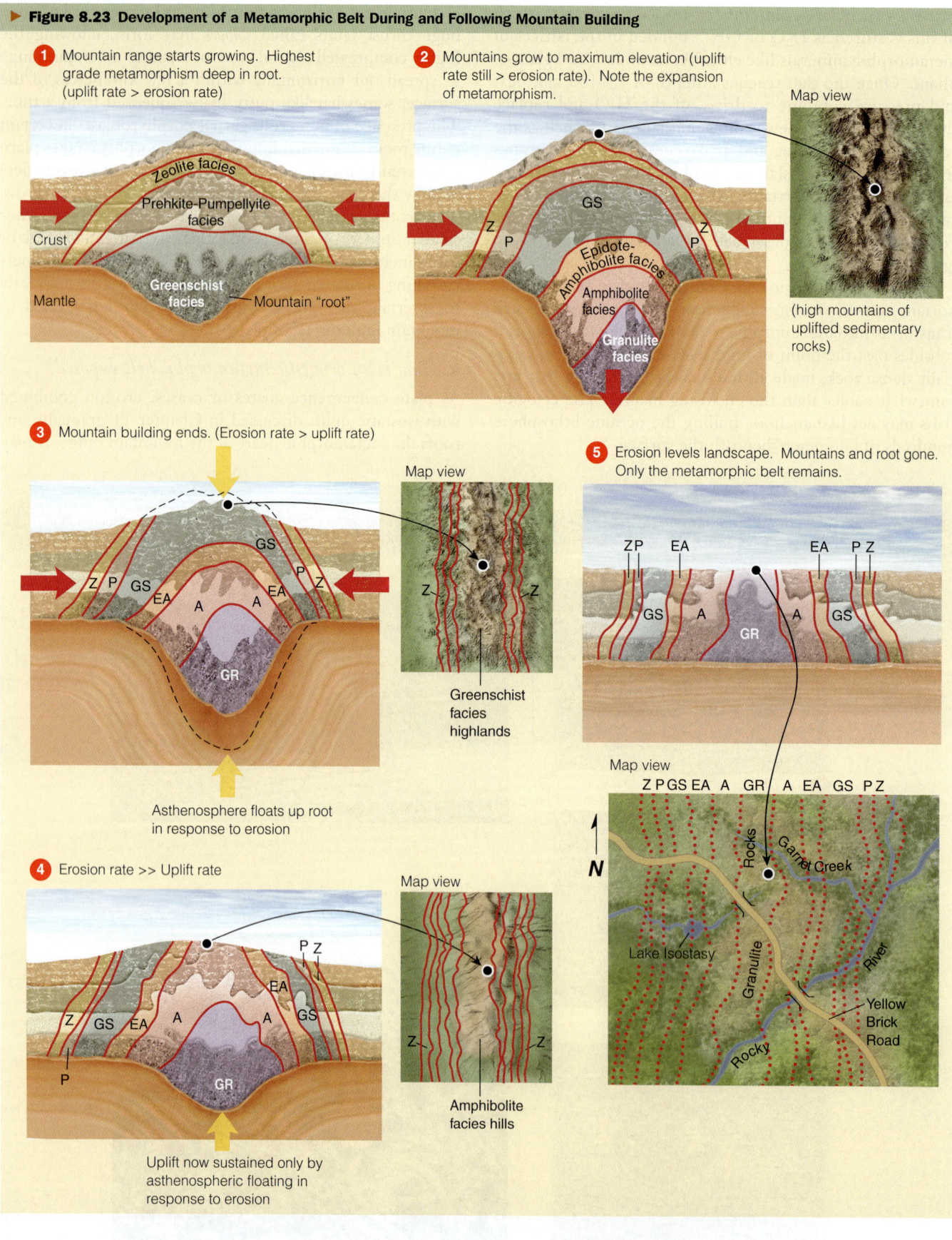

seem paradoxical, but the most likely place to find the highest-grade, deepest-formed metamorphic rocks is in the highest mountains because this is where the most uplift and erosion have occurred during and after plate collision. A facies map of a deeply eroding mountain range shows that the lowest grade of metamorphism commonly occurs on the flanks of the mountains, with the highest grade at the center (▶ Figure 8.23). The whole pattern defines an *orogenic metamorphic belt* (see Chapter 13, where orogenic activity is discussed). Detailed studies of metamorphic belts reveal that some parts of mountain ranges rise faster and erode more deeply than others, in response to factors that in most cases are poorly understood. The southern part of California's Sierra Nevada Range, for instance, features metamorphic rocks formed at depths of 20–25 km, whereas little or no metamorphic rock crops out at the much less deeply eroded northern end. In some older parts of continents, all topographic expression of mountain ranges may have worn away and the land has become flat. The only indication that mountains may once have existed in a region is the presence of a metamorphic belt exposed in scattered outcrops or revealed in drill cores and mine shafts.

Section 8.6 Summary

- Metamorphic rocks develop at all three types of plate boundaries.

- They form paired metamorphic belts at convergent margins, with conditions ranging from low-temperature, high-pressure metamorphism in accretionary wedges to high-pressure, high-temperature metamorphism beneath volcanic arcs and growing mountain ranges.

- Certain minerals such as rotated garnets record the strains and rates of mountain building.

- An orogenic metamorphic belt is a widely exposed zone of metamorphism produced by uplift and erosion of a mountain belt during and after plate convergence.

8.7 Metamorphism and Global Climate Change

■ *What does metamorphism have to do with climate?*

Medium-grade metamorphism of the sort that produces greenschist facies rocks can release large amounts of carbon dioxide and water as a result of heterogeneous metamorphic reactions. Some geologists think that these products may build up in the atmosphere and contribute significantly to global warming as they gradually find their way to the surface. But because this process takes place over a period of millions of years, evidence for it is largely circumstantial.

During Eocene times (56–34 million years ago) Earth's atmosphere was much warmer than it is today. Tropical ecosystems spread over twice or three times as much area as at present, in some places extending as far from the equator as England and southern Alaska. Atmospheric carbon dioxide levels were probably in the range of 700–2000 parts per million (ppm), compared with the modern level of 380 ppm. There is no clear correlation of this anomalous warming with any astronomical event or the release of a large amount of gas from the deep mantle. But during the Eocene the collision of India with Asia began building the Himalayan Mountains, accompanied by a large amount of metamorphism involving carbonate, organic, and OH^{-2}-rich rocks. Other metamorphic events simultaneously took place in the Mediterranean and around the Pacific Rim, so that *metamorphic outgassing* took place virtually worldwide. Thus the warm global Eocene climate and an unusually large amount of metamorphism may not be coincidental.

Currently, metamorphism is still active above active subduction zones in South and Central America, Asia, and New Zealand, but it appears to have tapered off in the collision zones marked by the Alps of Europe and elsewhere. A smaller amount of metamorphism in the post-Eocene world is certainly consistent with a cooler post-Eocene world, but other factors certainly have been more influential in causing the ice ages of the geologically recent past.

Section 8.7 Summary

- Metamorphic events may influence global climate by releasing gases into the atmosphere as a result of metamorphism along convergent plate boundaries.

- The warm Eocene climate may have resulted, in part, due to the early growth of the Himalayan Mountains as India collided with southern Asia.

8.8 Some Economic Uses of Metamorphic Materials

■ *What are some economically valuable metamorphic rocks?*

As mentioned in Section 8.1, marble has been a favorite sculpting and building stone since ancient times, including countless monuments from the ancient pyramids to Michelangelo's *David* and the Washington Monument and Lincoln Memorial (▶ Figure 8.24). People have used slate not only to make blackboards, writing tablets, and roofing tiles, but also to pave pathways and floors. In Germany, greenschist tombstones are common in older graveyards.

▶ **Figure 8.24 The Washington Monument** The marble-faced Washington Monument rises 169 m above the Mall in Washington, DC. The monument is built from three different kinds of marble. The first 46 m, built between 1848 and 1854, is faced with marble from the Texas, Maryland, quarry. Following 25 years of virtual inactivity, construction resumed with four rows of white marble from Lee, Massachusetts, added above the Texas marble. But this new stone proved to be too costly, so the monument was finished with slightly darker marble from the Cockeysville, Maryland, quarry. The three different kinds of marbles can be distinguished by subtle differences in color.

■ *What are some economically valuable metamorphic minerals?*

Metamorphic minerals of great economic importance include garnets, talc, kyanite, and asbestos. Garnets, for example, are used as gemstones or abrasives; talc is used in cosmetics and lubricants and in manufacturing paint; and kyanite is used in producing heat-resistant porcelain in sparkplugs. Asbestos is widely used for insulation and fireproofing and is widespread in buildings and building materials. It is also widely regarded as an environmental hazard because a concentration of asbestos fibers in your lungs could have fatal consequences. There are different forms of asbestos, however, and they do not all pose the same health hazards. Recognizing the different forms of asbestos would have been useful during the debates over the dangers posed to the public's health by this controversial mineral, and the proper use of this information could have saved taxpayers billions of dollars in unnecessary asbestos removal.

Section 8.8 Summary

- Metamorphic rocks and minerals yield products of value in ordinary construction, art, metallurgy, jewelry, and cultural expression.

- Some metamorphic minerals, most notably asbestos, have controversial aspects.

Review Workbook

ESSENTIAL QUESTIONS SUMMARY

8.1 Introduction

■ *What is metamorphism?*
Metamorphism is the physical transformation of rocks through recrystallization. It takes place beneath Earth's surface and it affects all rock types, including sedimentary and igneous. Some metamorphic rocks bear little resemblance to their earlier forms.

■ *Why is metamorphism worth studying?*
Metamorphism is an important process that is closely related to plate tectonics, the growth of continents, and even climate change. It is also responsible for producing a number of economically valuable materials. Though largely hidden from view, metamorphism is part of the world around you. The world would be a much different and less interesting place without it.

8.2 Equilibrium and the Causes of Metamorphism

■ *What are the principal agents of metamorphism?*
Temperature, pressure, fluid activity, and time all contribute in important ways to transforming rocks deep inside Earth.

■ *Why do rocks metamorphose?*
The environment in which rocks exist changes as they are buried, subjected to intrusion, and deformed by tectonic forces. Minerals are no longer stable (no longer "in equilibrium") as conditions change. Minerals can react with one another to create more stable kinds of minerals (heterogeneous metamorphic reactions) or

undergo rearrangements in their atoms to reduce the strain that builds up within them (crystal lattice reorientation and polymorphic transformations).

■ *What are metamorphic fluids and why are they important?*
Metamorphic fluids are high-pressure, high-temperature solutions containing water, carbon dioxide, and other dissolved substances. They result from certain heterogeneous metamorphic reactions. They can trigger some reactions by adding OH^{-2} to mineral bonds, and they can suppress others by making it impossible for particular reaction products to escape. Fluids flow through the crust away from areas of metamorphism, following cracks, joints, and dissolution seams to escape. Their absence as metamorphism wanes prevents metamorphic rocks from returning to their original mineral contents and structures through retrograde metamorphism.

8.3 The Main Types of Metamorphism
■ *What are the conditions of contact metamorphism?*
Contact metamorphic rocks form under conditions of high temperature and low pressure. They are arrayed in aureoles, or metamorphosed zones, around plutons and other intrusive igneous bodies.

■ *What are lithostatic and differential pressures, and why are they important?*
Lithostatic pressure is a uniform field of pressure experienced by most rocks beneath Earth's surface. Like the hydrostatic pressure experienced by divers underwater, the pressure acting on a rock embedded in the crust "feels" the same from all directions. Differential pressure is a nonuniform field of pressure; the pressure acting on a rock in some directions is stronger than it is in others. Many metamorphic rocks form under conditions of differential pressure, which influences the development of metamorphic structures and textures in significant ways.

■ *Where does dynamic metamorphism occur?*
Dynamic metamorphism is associated with faults and areas where lots of pressure builds up in the crust, but the temperature is not very great, such as in the accretionary wedges at convergent plate boundaries.

■ *What is shock metamorphism?*
Shock metamorphism is a type of dynamic (pressure-dominated) metamorphism associated with meteorite impacts. The pressure is extremely high relative to other natural processes that operate on Earth.

■ *What is the most common type of metamorphism?*
Regional metamorphism is the most common type of metamorphism. As the name implies, regional metamorphism has a broad range. Temperature and pressure both act as driving forces for metamorphic reactions in regional metamorphism.

8.4 Classification of Metamorphic Rocks
■ *What is foliated texture, and what are some examples of foliated rocks?*
Foliation is the parallel alignment of platy crystals, especially micas and clay minerals. The overall appearance of a foliated rock is roughly layered. Some foliations are planar; others are wavy. Examples of foliated rocks include slates, phyllites, schists, and gneisses.

■ *What is a common nonfoliated metamorphic texture, and what are some examples of nonfoliated metamorphic rocks?*
Equigranular texture is a common nonfoliated texture. It consists of crystals of nearly equal size that come in contact with one another along straight boundaries. Examples of nonfoliated rocks are marbles and quartzites. Hornfels is the most common type of nonfoliated metamorphic rock.

8.5 Metamorphic Zones and Facies
■ *What are isograds and metamorphic zones?*
In some regions geologists can look for particular key minerals, called index minerals, to determine the degree of metamorphism recorded in metamorphic rocks. The locations of index minerals can be mapped as isograds ("equal grade" of metamorphism), which enclose metamorphic zones, or areas of rock that all have similar grades of metamorphism.

■ *What is a metamorphic facies?*
A facies is a group of metamorphic rocks whose minerals, grouped together, all indicate formation under a particular range of temperatures and pressures. In other words, they share a common "environment of formation." Facies are named for a characteristic rock type or mineral, such as amphibolite or greenschist.

■ *How do metamorphic zones and metamorphic facies differ?*
Metamorphic zones show the gradational metamorphic change within a single rock composition—for example, pelitic. Metamorphic facies are groups of many different rock compositions whose mineral contents all indicate common temperature and pressure conditions during metamorphism.

8.6 Plate Tectonics and Metamorphism
■ *How does metamorphism relate to plate boundaries?*
Plate interactions along all three types of plate boundaries cause metamorphism. Low- to medium-grade regional metamorphism characterizes divergent boundaries. Dynamic metamorphism takes place along transform boundaries. Dynamic, contact, and regional metamorphism occur where plates collide, with regional metamorphism being the dominant metamorphic process, especially during collisions of continents.

■ *How is an orogenic metamorphic belt exposed?*
Erosion accompanied by uplift of mountain roots gradually exposes metamorphic rocks that may have originated many kilometers beneath the surface. Commonly, the highest grades of metamorphism are localized where the highest mountains occur—or where the highest mountains once existed because this corresponds to the area of greatest uplift and erosion.

8.7 Metamorphism and Global Climate
■ *What does metamorphism have to do with climate?*
Metamorphic fluids can release greenhouse gases into the atmosphere. The correlation of Eocene global warming and the rise of the Himalayan Mountains may be a good example of this phenomenon.

8.8 Some Economic Uses of Metamorphic Materials
■ *What are some economically valuable metamorphic rocks?*
Marble and slate are outstanding examples of economically important metamorphic rocks. Other examples include quartzite and coal (e.g., anthracite), although coal is not often thought of as being metamorphic.

■ *What are some economically valuable metamorphic minerals?*
Examples include many metallic ores and serpentine asbestos. Metals derived from metamorphic rocks include copper, tin, tungsten, lead, iron, and zinc.

ESSENTIAL TERMS TO KNOW

aureole (p. 238)
contact (thermal) metamorphism (p. 237)
differential pressure (p. 239)
dissolution seam (p. 237)
dynamic metamorphism (p. 239)
equilibrium (p. 233)
foliated texture (p. 243)
heterogeneous metamorphic reaction (p. 233)
isograd (p. 248)
lithostatic pressure (p. 239)
metamorphic belt (p. 250)
metamorphic facies (p. 248)
metamorphic fluid (p. 236)
metamorphic grade (p. 241)
metamorphic zone (p. 248)
metamorphism (p. 232)
metasomatism (p. 238)
nonfoliated texture (p. 246)
paired metamorphic belt (p. 250)
pelitic rock (p. 242)
polymorphic transformation (p. 236)
porphyroblast (p. 243)
recrystallization (p. 232)
regional metamorphism (p. 241)
retrograde metamorphism (p. 236)
rotated garnet (p. 251)
shock metamorphism (p. 240)

REVIEW QUESTIONS

1. Discuss the role each of the major agents of metamorphism plays in transforming any rock into a metamorphic rock.

2. How do metamorphic rocks record the influence of differential pressure in their structures and mineral textures?

3. Where does metasomatism occur? How does it differ from ordinary metamorphism?

4. How do nonfoliated metamorphic rocks record metamorphism in their textures?

5. Name several economically valuable metamorphic minerals or rocks, and discuss why they are valuable.

6. What specific features about foliated metamorphic rocks would make them unsuitable as foundations for dams? Are there any metamorphic rocks that would make good foundations? Explain your answer.

7. If plate tectonic movement did not exist, could there be metamorphism? Do you think metamorphic rocks exist on other planets in our solar system? Why?

8. Why is metamorphism important to understanding environmental issues and global environmental conditions?

APPLY YOUR KNOWLEDGE

1. What metamorphic facies is formed at conditions of 450°C and 6 kb pressure? If the pressure is raised to 10 kb, what facies is represented by the new conditions? What change in depth of burial is required to effect the pressure change of 6 to 10 kb?

FIELD QUESTION

1. Try to get as much information as you can from the geology shown in this scene. What geologic history can you reconstruct? (An answer is given at the end of the book.)

GEOLOGY MATTERS

GEOLOGY IN UNEXPECTED PLACES

Starting Off with a Clean Slate

Slate is a common metamorphic rock that has many uses. Two familiar uses are in the playing surface of billiard tables and roofing shingles.

Although slate is abundant throughout the world, most of it is unsuitable for billiard tables. For billiard tables, the slate must have a very fine grain so it can be honed to a smooth surface, somewhat elastic so it will expand and contract with the table's wood frame, and essentially nonabsorbent. Presently Brazil, China, India, and Italy are the major exporters of billiard table–quality slate, with the best coming from the Liguarian region of northern Italy. Most quality tables use at least 1-inch-thick slate that is split into three pieces. Although using three slabs requires extra work to ensure a tight fit and smooth surface, a table with three pieces is preferred over a single piece because it is less likely to fracture. Furthermore, the slate is usually slightly larger than the playing surface so that it extends below the rails of the table, thus giving additional strength to the rails and stability to the table. In addition, a quality table will have a wood backing glued to the underside of the slate so the felt cloth that is stretched tightly over the slate's surface can be stapled to the wood to provide a smooth playing surface.

Slate has been used as a roofing material for centuries. When properly installed and maintained, slate normally lasts for 60 to 125 years; many slate roofs have been around for more than 200 years. In the United States, slate roofing tiles typically come in shades of gray, green, purple, black, and red (▶ Figure 1). There are 36 common sizes of tiles, ranging from 12 to 24 inches long with the width about half the length. The typical slate tile is usually ¼ inch thick. Thicker tiles may be used, but they are harder to work with and greatly increase the weight of the roof.

The years between 1897 and 1914 witnessed the height of the U.S. roofing slate industry in both quantity and value of output. By the end of the 19th century, more than 200 slate quarries were operating in 13 states. With the introduction of asphalt shingles, which can be mass produced, easily transported, and installed at a much lower cost than slate shingles, the slate shingle industry in the United States began to decline around 1915. The renewed popularity of historic preservation and the recognition of slate's durability, however, have brought about a resurgence in the slate roofing industry. It's not that unusual these days for geology to be overhead as well as underfoot.

Figure 1
Different colored slates make up the roof of this elementary school in Mount Pleasant, Michigan.

GEOLOGY MATTERS

Geology AT WORK

What Do Metamorphic Petrologists Do?

Metamorphic petrologists are geologists who study metamorphic rocks. Most metamorphic petrologists work for universities and colleges, but some are employed by the U.S. Geological Survey and various state or provincial geologic agencies. Metamorphic petrologists are highly skilled at both fieldwork and laboratory analyses. Sophisticated lab work is critical for determining and interpreting the complex compositions of metamorphic rocks. In fact, a good metamorphic petrologist will have solid training in physical and analytical chemistry as well as geology, especially in *thermodynamics*, the study of how heat energy relates to changes in nature.

Some metamorphic petrologists map the distribution of rock bodies in metamorphosed landscapes in order to provide basic information about what is there for future studies or for economic purposes. Other collect samples to analyze the detailed makeup of porphyroblasts and other minerals with various lab instruments, such as electron microprobes. This information enables petrologists to pinpoint the peak intensities of past metamorphic events and to precisely determine how metamorphic conditions changed throughout the histories of individual rocks. A complete reconstruction of a rock's metamorphism is called a *P–T path,* or a "Pressure–Temperature" path. If the minerals in a rock are also radiometrically dated, then a *P–T–t* path ("Pressure–Temperature–time" path) can be unraveled. In an exciting new development in metamorphic petrology, geologists are beginning to correlate certain P–T paths with particular kinds of tectonic behavior, such as thrust faulting and uplift, and with the intrusion of magmas, which can have important economic implications.

Figure 1
A sample *P-T* path, showing the subduction, metamorphism, and uplift of a rock.

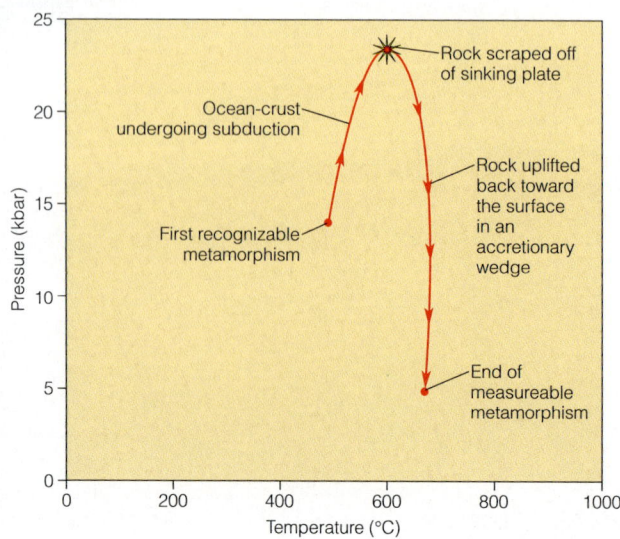

GEOLOGY IN FOCUS — Asbestos: Good or Bad?

Asbestos (from the Latin, meaning "unquenchable") is a general term applied to any silicate mineral that easily separates into flexible fibers. The combination of such features as fire resistance and flexibility makes asbestos an important industrial material of considerable value. In fact, asbestos has more than 3000 known uses, including brake linings, fireproof fabrics, and heat insulators.

Asbestos is divided into two broad groups: *serpentine asbestos* and *amphibole asbestos*. Chrysotile is the fibrous form of serpentine asbestos (▶ Figure 1); it is the most valuable type and constitutes the bulk of all commercial asbestos. Its strong, silky fibers are easily spun and can withstand temperatures as high as 2750°C.

The vast majority of chrysotile asbestos is in serpentine, a type of rock formed by the alteration of ultramafic igneous rocks such as peridotite under low- and medium-grade metamorphic conditions. Other chrysotile results when the metamorphism of magnesium limestone or dolostone produces discontinuous serpentine bands within the carbonate beds.

Among the varieties of amphibole asbestos, *crocidolite* is the most common. Also known as blue asbestos, crocidolite is a long, coarse, spinning fiber that is stronger but more brittle than chrysotile and also less resistant to heat. Crocidolite is found in such metamorphic rocks as slates and schists and is thought to form by the solid-state alteration of other minerals as a result of deep burial.

Despite the widespread use of asbestos, the U.S. Environmental Protection Agency (EPA) instituted a gradual ban on all new asbestos products. The ban was imposed because some forms of asbestos can cause lung cancer and scarring of the lungs if fibers are inhaled.

Because the EPA apparently paid little attention to the issue of risks versus benefits when it enacted this rule, the U.S. Fifth Circuit Court of Appeals overturned the EPA ban on asbestos in 1991.

The threat of lung cancer has also resulted in legislation mandating the removal of asbestos already in place in all public buildings, including all public and private schools. However, important questions have been raised concerning the threat posed by asbestos and the additional potential hazards that may arise from its improper removal.

Current EPA policy mandates that all forms of asbestos are to be treated as identical hazards. Yet studies indicate that only the amphibole forms constitute a known health hazard. Chrysotile, whose fibers tend to be curly, does not become lodged in the lungs. Furthermore, its fibers are generally soluble and disappear in tissue. In contrast, crocidolite has long, straight, thin fibers that penetrate the lungs and stay there. These fibers irritate the lung tissue and over a long period of time can lead to lung cancer (▶ Figure 2). Thus crocidolite, and not chrysotile, is overwhelmingly responsible for asbestos-related lung cancer. Because about 95% of the asbestos in place in the United States is chrysotile, many people question whether the dangers from asbestos are exaggerated.

Removing asbestos from buildings where it has been installed could cost as much as $100 billion. Unless the material containing the asbestos is disturbed, asbestos does not shed fibers and thus does not contribute to airborne asbestos that can be inhaled. Furthermore, improper removal of asbestos can lead to contamination. In most cases of improper removal, the concentration of airborne asbestos fibers is far higher than if the asbestos had been left in place.

The problem of asbestos contamination is a good example of how geology affects our lives and why we should have a basic knowledge of science before making decisions that could have broad economic and societal impacts.

Figure 1
Specimen of chrysotile from Thetford, Quebec, Canada. Chrysotile is the fibrous form of serpentine asbestos.

Figure 2
Lung cancer. Colored tomography (CT) scan of an axial section through the chest of a patient with a mesothelioma cancer (light red). It is surrounding and constricting the lung at right (pink). The other lung (dark blue) has a healthy pleura (dark red). The spine (lower center, light blue), the descending aorta (green) and the heart (dark green, between lungs) are also seen. Mesothelioma is a malignant cancer of the pleura, the membrane lining the chest cavity and lungs. It is usually caused by asbestos exposure. It often reaches a large size, as here, before diagnosis, and the prognosis is then poor.

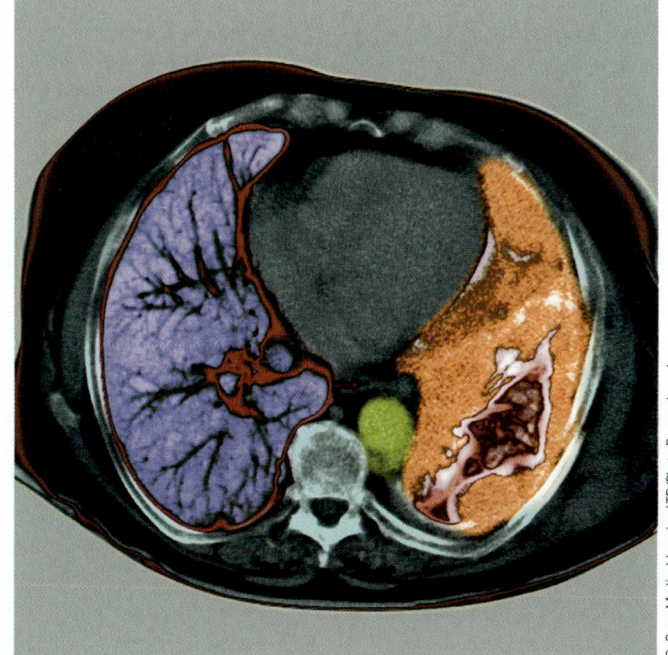

Chapter 9

Geologic Time: Concepts and Principles

Vermillion Cliffs Wilderness, Arizona
Swirls of stone sweep across Earth like a rage of sea waves suspended in time.
"The Wave," eroded sandstone, stretches across the slopes of Coyote Buttes in Paria Canyon—Vermillion Cliffs Wilderness Area. The brilliant bands of color in the sandstone mirror a sunset: shades of red, yellow, orange, and pink. These layers of color are caused by the weathering of minerals like iron, which turns the stone vermillion-red.
—A. W.

ESSENTIAL QUESTIONS TO ASK

9.1 Introduction
- How is geologic time measured?
- Why is the study of geologic time important?

9.2 Early Concepts of Geologic Time and the Age of Earth
- How has our concept of geologic time and Earth's age changed throughout human history?

9.3 James Hutton and the Recognition of Geologic Time
- Who is James Hutton and why is he important?
- How did Lord Kelvin almost overturn the uniformitarian foundation of geology?

9.4 Relative Dating Methods
- What is relative dating and why are the principles of relative dating important?
- What are the six fundamental principles of relative dating?
- What are unconformities?
- What are the three specific types of unconformities?
- How are the principles of relative dating applied to interpret the geologic history of an area?

9.5 Correlating Rock Units
- What is correlation?
- How are subsurface units correlated?

9.6 Absolute Dating Methods
- Why was the discovery of radioactivity important to geology?
- What are atoms, elements, and isotopes?
- What is radioactive decay?
- What are half-lives?
- What are some of the sources of uncertainty in radiometric dating?
- What are the common long-lived radioactive isotope pairs?
- What is fission-track dating?
- What is the carbon-14 dating technique?
- What is tree-ring dating?

9.7 Development of the Geologic Time Scale
- What is the geologic time scale?
- How did the geologic time scale develop?

9.8 Geologic Time and Climate Change
- How are climate change and geologic time linked?

GEOLOGY MATTERS

GEOLOGY IN FOCUS:
Denver's Weather—280 Million Years Ago!

GEOLOGY IN UNEXPECTED PLACES:
Time Marches On—The Great Wall of China

GEOLOGY AT WORK:
Geology in Hollywood

Physical Geology Now™ This icon, which appears throughout the book, indicates an opportunity to explore interactive tutorials, animations, or practice problems available on the Physical GeologyNow website at **http://now.brookscole.com/phygeo6**.

9.1 Introduction

In 1869 Major John Wesley Powell, a Civil War veteran who lost his right arm in the battle of Shiloh, led a group of hardy explorers down the uncharted Colorado River through the Grand Canyon. With no maps or other information, Powell and his group ran the many rapids of the Colorado River in fragile wooden boats, hastily recording what they saw. Powell wrote in his diary that "all about me are interesting geologic records. The book is open and I read as I run."

From this initial reconnaissance, Powell led a second expedition down the Colorado River in 1871. This second trip included a photographer, a surveyor, and three topographers. Members of the expedition made detailed topographic and geologic maps of the Grand Canyon area as well as the first photographic record of the region.

Probably no one has contributed as much to the understanding of the Grand Canyon as Major Powell. In recognition of his contributions, the Powell Memorial was erected on the South Rim of the Grand Canyon in 1969 to commemorate the 100th anniversary of this history-making first expedition.

Most tourists today, like Powell and his fellow explorers in 1869, are astonished by the seemingly limitless time represented by the rocks exposed in the walls of the Grand Canyon. For most visitors, viewing a 1.5 km deep cut into Earth's crust is the only encounter they'll ever have with the enormity of geologic time. When standing on the rim and looking down into the Grand Canyon, we are really looking far back in time, all the way back to the early history of our planet. In fact, more than 1 billion years of history are preserved in the rocks of the Grand Canyon.

Vast periods of time set geology apart from most of the other sciences, and an appreciation of the immensity of geologic time is fundamental to understanding the physical and biological history of our planet. In fact, understanding and accepting the magnitude of geologic time are major contributions geology has made to the sciences.

■ How is geologic time measured?

In some respects, time is defined by the methods used to measure it. Geologists use two different frames of reference when discussing geologic time. **Relative dating** is placing geologic events in a sequential order as determined from their positions in the geologic record. Relative dating will not tell us how long ago a particular event took place, only that one event preceded another. A useful analogy for relative dating is a television guide that does not list the times programs are shown. You cannot tell what time a particular program will be shown, but by watching a few shows and checking the guide, you can determine whether you have missed the show or how many shows are scheduled before the one you want to see.

The various principles used to determine relative dating were discovered hundreds of years ago, and since then they have been used to construct the *relative geologic time scale* (▶ Figure 9.1). Furthermore, these principles are still widely used by geologists today.

Absolute dating provides specific dates for rock units or events expressed in years before the present. In our analogy of the television guide, the times when the programs are actually shown would be the absolute dates. In this way, you not only can determine whether you have missed a show (relative dating), but also know how long it will be until a show you want to see will be shown (absolute dating).

Radiometric dating is the most common method of obtaining absolute ages. Dates are calculated from the natural decay rates of various radioactive elements present in trace amounts in some rocks. It was not until the discovery of radioactivity near the end of the 19th century that absolute ages could be accurately assigned to the relative geologic time scale. Today the geologic time scale is really a dual scale: a relative scale based on rock sequences with radiometric dates expressed as years before the present (▶ Figure 9.1).

■ Why is the study of geologic time important?

One of the most important lessons to be learned in this chapter is how to reason and apply the fundamental geologic principles to solve geologic problems. The logic used in applying the principles of relative dating to interpret the geologic history of an area involves basic reasoning skills that can be transferred to and used in almost any profession or discipline.

Advances and refinements in absolute dating techniques during the 20th century have changed the way we view Earth in terms of when events occurred in the past and the rates of geologic change through time. The ability to accurately determine past climatic changes and their causes has important implications for the current debate on global warming and its effects on humans.

Section 9.1 Summary

● Geologic time can be measured in two ways. Relative dating places events in a sequential order as determined from their positions in the geologic record. Absolute dating provides specific dates for rock units or events expressed in years before the present. Radiometric dating is the most common method of obtaining absolute dates.

● The logic used in applying the principles of relative dating to interpret the geologic history of an area involves basic reasoning skills that are useful in almost any profession or discipline a student pursues.

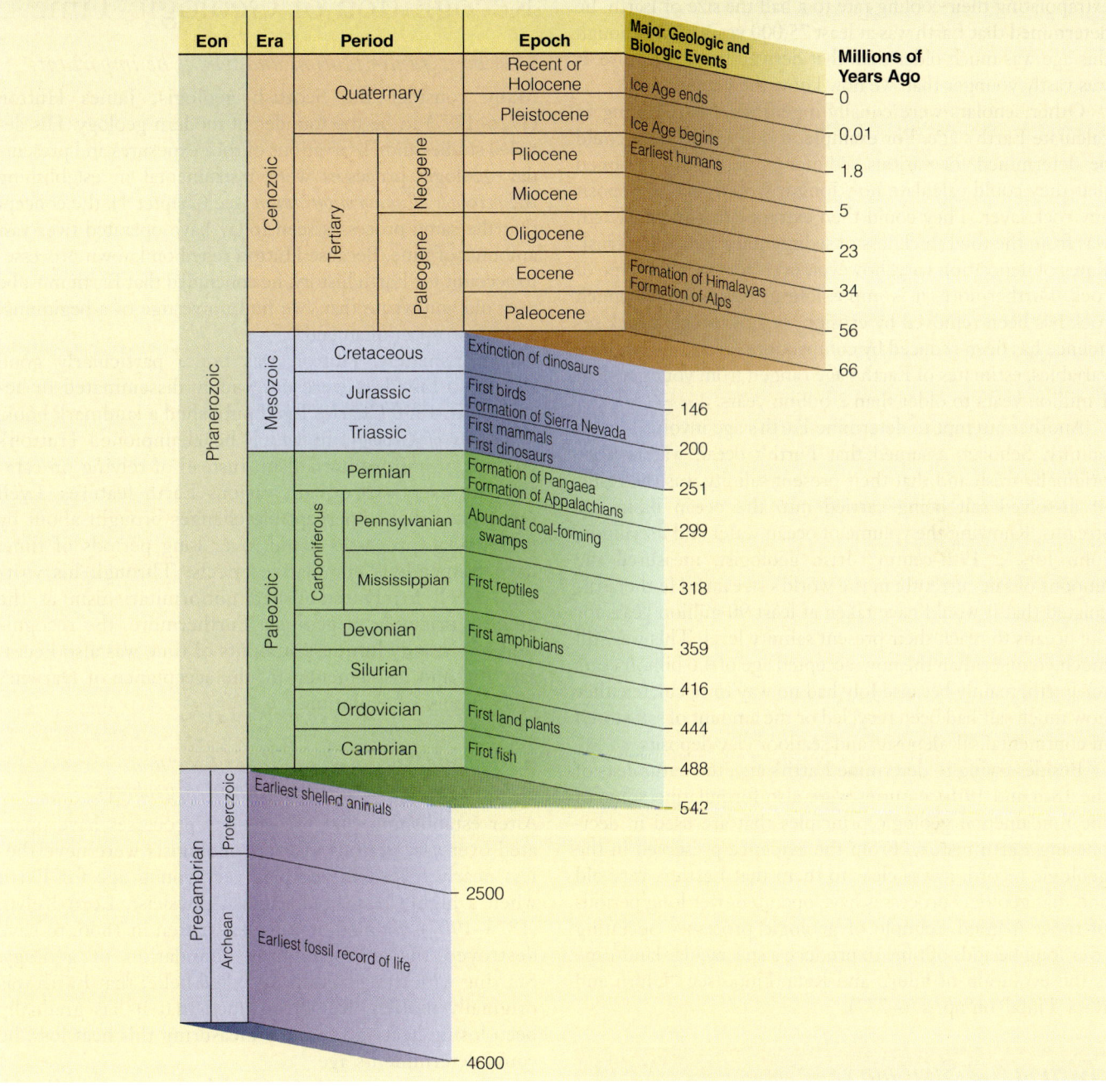

Figure 9.1 The Geologic Time Scale Some of the major geologic and biologic events are indicated along the right-hand margin. Dates are from Gradstein, F., Ogg, J. and Smith, A., *A Geologic Timescale 2004* (Cambridge, UK: Cambridge University Press, 2005), Figure 1.2.

9.2 Early Concepts of Geologic Time and the Age of Earth

- *How has our concept of geologic time and Earth's age changed throughout human history?*

The concept of geologic time and its measurement have changed throughout human history. Many early Christian scholars and clerics tried to establish the date of creation by analyzing historical records and the genealogies found in Scripture. One of the most influential scholars was James Ussher (1581–1656), Archbishop of Armagh, Ireland, who based upon Old Testament genealogy asserted that God created Earth on Sunday October 23, 4004 B.C. In 1701, an authorized version of the Bible made this date accepted Church doctrine. The idea of a very young Earth provided the basis for most Western chronologies of Earth history prior to the 18th century.

During the 18th and 19th centuries, several attempts were made to determine Earth's age on the basis of scientific evidence rather than revelation. The French zoologist Georges Louis de Buffon (1707–1788) assumed Earth gradually cooled to its present condition from a molten beginning. To simulate this history, he melted iron balls of various diameters and allowed them to cool to the surrounding temperature. By extrapolating their cooling rate to a ball the size of Earth, he determined that Earth was at least 75,000 years old. Although this age was much older than that derived from Scripture, it was vastly younger than we now know the planet to be.

Other scholars were equally ingenious in attempting to calculate Earth's age. For example, if deposition rates could be determined for various sediments, geologists reasoned that they could calculate how long it would take to deposit any rock layer. They could then extrapolate how old Earth was from the total thickness of sedimentary rock in its crust. Rates of deposition vary, however, even for the same type of rock. Furthermore, it is impossible to estimate how much rock has been removed by erosion, or how much a rock sequence has been reduced by compaction. As a result of these variables, estimates of Earth's age ranged from younger than 1 million years to older than 2 billion years.

Another attempt to determine Earth's age involved ocean salinity. Scholars assumed that Earth's ocean waters were originally fresh and that their present salinity was the result of dissolved salt being carried into the ocean basins by streams. Knowing the volume of ocean water and its salinity, John Joly, a 19th-century Irish geologist, measured the amount of salt currently in the world's streams. He then calculated that it would have taken at least 90 million years for the oceans to reach their present salinity level. This was still much younger than the now-accepted age of 4.6 billion years for Earth, mainly because Joly had no way to calculate either how much salt had been recycled or the amount of salt stored in continental salt deposits and seafloor clay deposits.

Besides trying to determine Earth's age, the naturalists of the 18th and 19th centuries were also formulating some of the fundamental geologic principles that are used in deciphering Earth history. From the evidence preserved in the geologic record, it was clear to them that Earth is very old and that geologic processes have operated over long periods of time. A good example of geologic processes operating over long periods of time to produce a spectacular landscape is the evolution of Uluru and Kata Tjuta (see "Uluru and Kata Tjuta" on pp. 276–277).

Section 9.2 Summary

- The concept of geologic time and the methods used to measure it have changed throughout human history. Many Christian scholars and clerics used historical records and the genealogies from Scripture to determine when Earth was created. Two other attempts involved extrapolating the cooling rate of iron balls to a ball the size of Earth and calculating how long it would take to reach the present-day salinity of the world's oceans, assuming the early oceans contained freshwater and received their dissolved salts from streams draining into the oceans. All of these attempts, though ingenious, were flawed and yielded very young ages for Earth.

9.3 James Hutton and the Recognition of Geologic Time

■ *Who is James Hutton and why is he important?*

Many consider the Scottish geologist James Hutton (1726–1797) to be the founder of modern geology. His detailed studies and observations of rock exposures and present-day geologic processes were instrumental in establishing the *principle of uniformitarianism* (see Chapter 1), the concept that the same processes seen today have operated over vast amounts of time. Because Hutton relied on known processes to account for Earth history, he concluded that Earth must be very old and wrote that "we find no vestige of a beginning, and no prospect of an end."

Unfortunately, Hutton was not a particularly good writer, so his ideas were not widely disseminated or accepted. In 1830 Charles Lyell published a landmark book, *Principles of Geology*, in which he championed Hutton's concept of uniformitarianism. Instead of relying on catastrophic events to explain various Earth features, Lyell recognized that imperceptible changes brought about by present-day processes could, over long periods of time, have tremendous cumulative effects. Through his writings, Lyell firmly established uniformitarianism as the guiding principle of geology. Furthermore, the recognition of virtually limitless amounts of time was also necessary for, and instrumental in, the acceptance of Darwin's 1859 theory of evolution.

■ *How did Lord Kelvin almost overturn the uniformitarian foundation of geology?*

After establishing that present-day processes have operated over vast periods of time, geologists were nevertheless nearly forced to accept a very young age for Earth when a highly respected English physicist, Lord Kelvin (1824–1907), claimed, in a paper written in 1866, to have destroyed the uniformitarian foundation of geology. Starting with the generally accepted belief that Earth was originally molten, Kelvin assumed that it has gradually been losing heat and that, by measuring this heat loss, he could determine its age.

Kelvin knew from deep mines in Europe that Earth's temperature increases with depth, and he reasoned that Earth is losing heat from its interior. By knowing the size of Earth, the melting temperatures of rocks, and the rate of heat loss, Kelvin calculated the age at which Earth was entirely molten. From these calculations, he concluded that Earth could be neither older than 400 million years nor younger than 20 million years. This wide discrepancy in age reflected uncertainties in average temperature increases with depth and the various melting points of Earth's constituent materials.

After establishing that Earth was very old and that present-day processes operating over long periods of time account for geologic features, geologists were in a quandary. If they accepted Kelvin's dates, they would have to abandon the concept of seemingly limitless time that was the underpinning of uniformitarian geology and one of the foundations of Darwinian evolution and squeeze events into a shorter time frame.

Kelvin's reasoning and calculations were sound, but his basic premises were false, thereby invalidating his conclusions. Kelvin was unaware that Earth has an internal heat source, radioactivity, that has allowed it to maintain a fairly constant temperature through time.* His 40-year campaign for a young Earth ended with the discovery of radioactivity near the end of the 19th century and the insight that natural radioactive decay can be used to date the age of formation of rocks in 1905. As geologists had long thought, Earth indeed is very old.

Although the discovery of radioactivity destroyed Kelvin's arguments, it provided geologists with a clock that could measure Earth's age and validate what geologists had been saying all along—namely, that Earth was indeed very old!

Section 9.3 Summary

- James Hutton is considered by most geology historians to be the father of modern geology. His observations of present-day processes and detailed studies of rock exposures were instrumental in establishing the principle of uniformitarianism and that Earth had a very long history.

- Lord Kelvin, an English physicist, nearly overturned Hutton's concept of an ancient Earth by calculating that, based on heat loss from its interior, Earth could be no younger than 20 million years and no older than 400 million years. Although Kelvin's mathematical calculations were sound, the discovery of radioactivity, which allowed for Earth to maintain a fairly constant temperature through time, invalidated his conclusions.

9.4 Relative Dating Methods

■ *What is relative dating and why are the principles of relative dating important?*

Before the development of radiometric dating techniques, geologists had no reliable means of absolute dating and therefore depended solely on relative dating methods. Relative dating places events in sequential order but does not tell us how long ago an event took place. Although the principles of relative dating may now seem self-evident, their discovery was an important scientific achievement because they provided geologists with a means to interpret geologic history and develop a relative geologic time scale.

---

*Actually Earth's temperature has decreased through time because the original amount of radioactive materials has been decreasing and thus is not supplying as much heat. However, the temperature is decreasing at a rate considerably slower than would be required to lend any credence to Kelvin's calculations.

■ *What are the six fundamental principles of relative dating?*

The six fundamental geologic principles used in relative dating are superposition, original horizontality, lateral continuity, cross-cutting relationships, inclusions, and fossil succession.

The 17th century was an important time in the development of geology as a science because of the widely circulated writings of the Danish anatomist Nicolas Steno (1638–1686). Steno observed that when streams flood, they spread out across their floodplains and deposit layers of sediment that bury organisms dwelling on the floodplain. Subsequent floods produce new layers of sediments that are deposited or superposed over previous deposits. When lithified, these layers of sediment become sedimentary rock. Thus, in an undisturbed succession of sedimentary rock layers, the oldest layer is at the bottom and the youngest layer is at the top. This **principle of superposition** is the basis for relative-age determinations of strata and their contained fossils (▶ Figure 9.2).

Steno also observed that, because sedimentary particles settle from water under the influence of gravity, sediment is deposited in essentially horizontal layers, thus illustrating the **principle of original horizontality** (▶ Figure 9.2). Therefore a sequence of sedimentary rock layers that is steeply inclined from the horizontal must have been tilted after deposition and lithification.

Steno's third principle, the **principle of lateral continuity,** states that a layer of sediment extends laterally in all directions until it thins and pinches out or terminates against the edge of the depositional basin (▶ Figure 9.2).

James Hutton is credited with discovering the **principle of cross-cutting relationships.** Based on his detailed studies and observations of rock exposures in Scotland, Hutton recognized that an igneous intrusion or fault must be younger than the rocks it intrudes or displaces (▶ Figure 9.3).

Although this principle illustrates that an intrusive igneous structure is younger than the rocks it intrudes, the association of sedimentary and igneous rocks may cause problems in relative dating. Buried lava flows and sills look very similar in a sequence of strata (▶ Figure 9.4). A buried lava flow, however, is older than the rocks above it (principle of superposition), whereas a sill, resulting from later igneous intrusion, is younger than all the beds below it and younger than the immediately overlying bed as well.

To resolve such relative-age problems as these, geologists look to see whether the sedimentary rocks in contact with the igneous rocks show signs of baking or alteration by heat (see the section on contact metamorphism in Chapter 8, p. 237). A sedimentary rock that shows such effects must be older than the igneous rock with which it is in contact. In Figure 9.4, for example, a sill produces a zone of baking immediately above and below it because it intruded into previously existing sedimentary rocks. A lava flow, in contrast, bakes only those rocks below it.

Another way to determine relative ages is by using the **principle of inclusions.** This principle holds that inclusions, or fragments of one rock contained within a layer of another,

▶ **Figure 9.2 Grand Canyon, Arizona** The Grand Canyon of Arizona illustrates three of the six fundamental principles of relative dating. The sedimentary rocks of the Grand Canyon were originally deposited horizontally in a variety of marine and continental environments (principle of original horizontality). The oldest rocks are at the bottom of the canyon, and the youngest rocks are at the top, forming the rim (principle of superposition). The exposed rock layers extend laterally for some distance (principle of lateral continuity).

are older than the rock layer itself. The batholith shown in ▶ Figure 9.5a contains sandstone inclusions, and the sandstone unit shows the effects of baking. Accordingly, we conclude that the sandstone is older than the batholith. In Figure 9.5b, however, the sandstone contains granite rock fragments, indicating that the batholith was the source rock for the inclusions and is therefore older than the sandstone.

Fossils have been known for centuries (see Chapter 7), yet their utility in relative dating and geologic mapping was not fully appreciated until the early 19th century. William Smith (1769–1839), an English civil engineer involved in surveying and building canals in southern England, independently recognized the principle of superposition by reasoning that the fossils at the bottom of a sequence of strata are older than those at the top of the sequence. This recognition served as the basis for the **principle of fossil succession,** or the *principle of faunal and floral succession,* as it is sometimes called (▶ Figure 9.6).

According to this principle, fossil assemblages succeed one another through time in a regular and predictable order.

The validity and successful use of this principle depend on three points: (1) Life has varied through time, (2) fossil assemblages are recognizably different from one another, and (3) the relative ages of the fossil assemblages can be determined. Observations of fossils in older versus younger strata clearly demonstrate that life-forms have changed. Because this is true, fossil assemblages (point 2) are recognizably different. Furthermore, superposition can be used to demonstrate the relative ages of the fossil assemblages.

■ *What are unconformities?*

Our discussion so far has been concerned with vertical relationships among conformable strata—that is, sequences of rocks in which deposition was more or less continuous. A bedding plane between strata may represent a depositional break of anywhere from minutes to tens of years, but it is inconsequential in the context of geologic time. However, in some sequences of strata, surfaces known as **unconformities** may be present, representing times of nondeposition, erosion, or both. Unconformities encompass long periods of geologic time, perhaps millions or tens of millions of years. Accordingly, the geologic record is incomplete wherever an unconformity is present, just as a book with missing pages is incomplete, and the interval of geologic time not represented by strata is called a *hiatus* (▶ Figure 9.7).

■ *What are the three specific types of unconformities?*

The general term *unconformity* encompasses three specific types of surfaces. First, a **disconformity** is a surface of erosion or nondeposition separating younger from older rocks, both of which are parallel with one another (▶ Figure 9.8). Unless the erosional surface separating the older from the younger parallel beds is well defined or distinct, the disconformity frequently resembles an ordinary bedding plane. Hence, many disconformities are difficult to recognize and must be identified on the basis of fossil assemblages.

Second, an **angular unconformity** is an erosional surface on tilted or folded strata over which younger rocks were deposited (▶ Figure 9.9). The strata below the unconformable surface generally dip more steeply than those above, producing an angular relationship.

The angular unconformity illustrated in Figure 9.9b is probably the most famous in the world. It was here at Siccar Point, Scotland, that James Hutton realized that severe upheavals had tilted the lower rocks and formed mountains that were then worn away and covered by younger, flat-lying rocks. The erosional surface between the older tilted rocks and the younger flat-lying strata meant that a significant gap existed in the geologic record. Although Hutton did not use the term *unconformity*, he was the first to understand and explain the significance of such discontinuities in the geologic record.

A **nonconformity** is the third type of unconformity. Here an erosion surface cut into metamorphic or igneous rocks is covered by sedimentary rocks (▶ Figure 9.10). This type of unconformity closely resembles an intrusive igneous contact with sedimentary rocks. The principle of inclusions is helpful in determining whether the relationship between the underlying igneous rocks and the overlying sedimentary rocks is the result of an intrusion or erosion (▶ Figure 9.5). A nonconformity is also marked in many places by an ancient zone of weathering, or even a reddened, brick-like soil horizon, or paleosol. In the case of an intrusion, the igneous rocks are younger, whereas in the case of erosion, the sedimentary rocks are younger. Being able to distinguish between a nonconformity and an intrusive contact is very important because they represent different sequences of events.

■ *How are the principles of relative dating applied to interpret the geologic history of an area?*

We can decipher the geologic history of the area represented by the block diagram in ▶ Figure 9.11 by applying the various relative dating principles just discussed. The methods and logic used in this example are the same as those applied by 19th-century geologists in constructing the geologic time scale.

According to the principles of superposition and original horizontality, beds A–G were deposited horizontally;

▶ **Figure 9.3** The Principle of Cross-Cutting Relationships

a A dark gabbro dike cuts across granite in Acadia National Park, Maine. The dike is younger than the granite it intrudes.

b A small fault (arrows show direction of movement) cuts across, and thus displaces, tilted sedimentary beds along the Templin Highway in Castaic, California. The fault is therefore younger than the youngest beds that are displaced.

PHYSICAL Geology Now™ ▶ Geo-focus Figure 9.4 Differentiating between a Buried Lava Flow and a Sill In frames **(a)** and **(b)** below, the ages of sedimentary strata are shown by numbering from 1–6, with (1) being the oldest sedimentary rock layer in each frame, and (6) the youngest. Remember that the lava flow took place *during* deposition of an ordinary sedimentary sequence, whereas the sill intruded the sedimentary strata *after* they had accumulated. The notes in both frames highlight the physical features one would look for to distinguish between a lava flow and a sill, which can look quite similar in an outcrop.

a A buried lava flow has baked underlying bed 2 when it flowed over it. Clasts of the lava were deposited along with other sediments during deposition of bed 3. The lava flow is younger than bed 2 and older than beds 3, 4, and 5.

Strands and shreds of underlying sedimentary material may be in the base of the flow

Baking zone at bottom of flow

Clasts of lava may be present in the overlying, younger layer

Rubble zones may be present at the top and bottom of the flow

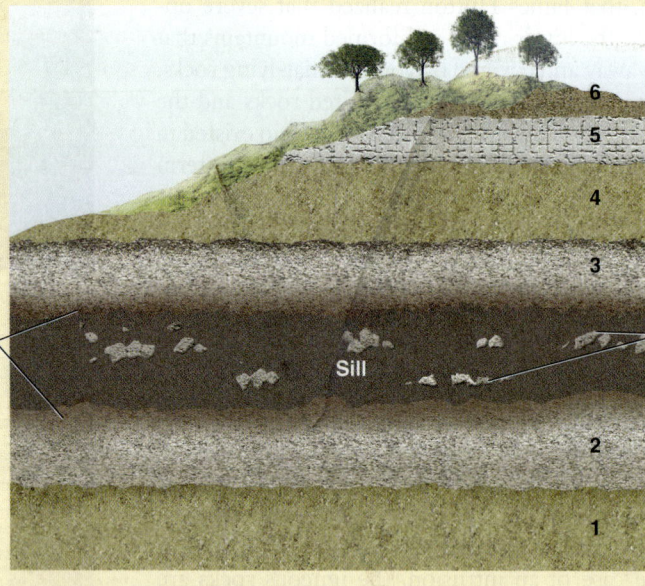

b The rock units above and below the sill have been baked, indicating that the sill is younger than beds 2 and 3, but its age relative to beds 4–6 cannot be determined. **Why is this?** *Because beds 2 and 3 had to have been there in order for the sill to have intruded between them. However, the sill could have been intruded at any time after the lithification of bed 3—that is, before beds 4, 5, and 6 were deposited, after bed 4 was deposited but before beds 5 and 6 were deposited, after bed 5 was deposited but before bed 6 was deposited, or after beds 4, 5, and 6 were deposited. All we know for sure is that the sill intruded after the lithification of bed 3.*

Baking zones on both sides of sill

Inclusions of rock from both layers above (3) and below (2) may exist in the sill

c This buried lava flow in Yellowstone National Park, Wyoming, displays columnar jointing. A baked zone is present below, and not above, the igneous structure, indicating it is a buried lava flow and not a sill.

Figure 9.5 The Principle of Inclusions

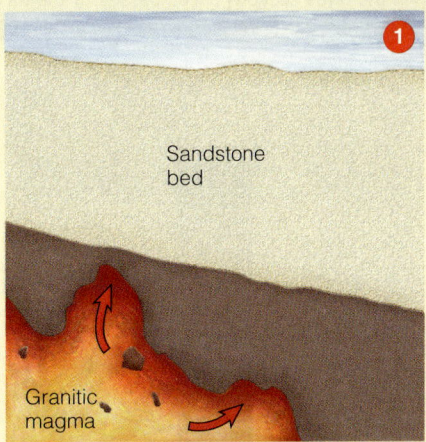
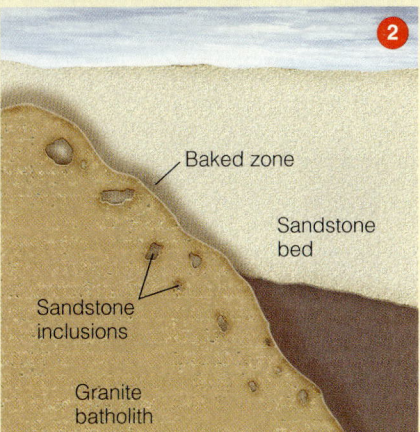

a The sandstone is older than the granite batholith because there are inclusions of sandstone inside the granite. The sandstone also shows evidence of having been baked along its contact with the granite batholith when the granitic magma intruded the overlying sedimentary beds.

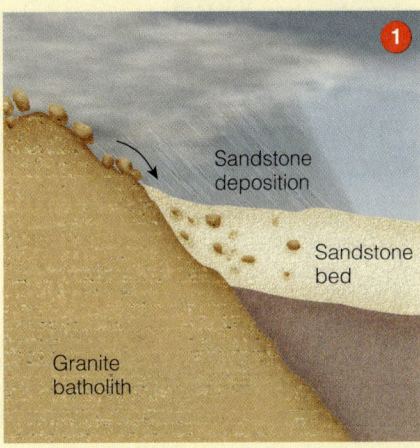
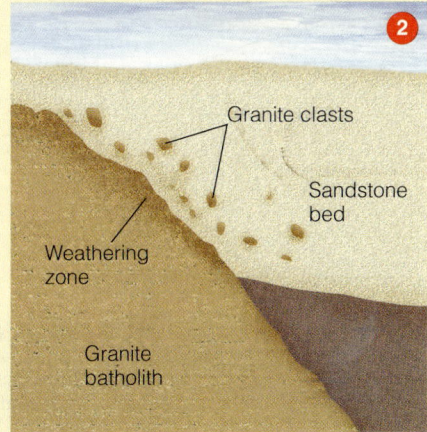

b The sandstone is younger than the granite batholith because it contains pieces (clasts) of granite. The granite is also weathered along the contact with the sandstone, indicating it was the source of the granite clasts, and must therefore be older than the sandstone.

c Outcrop in northern Wisconsin showing basalt inclusions (dark gray) in granite (white). Accordingly, the basalt inclusions are older than the granite.

▶ **Figure 9.6 The Principle of Fossil Succession** This generalized diagram shows how geologists use the principle of fossil succession to identify strata of the same age in different areas. The rocks in the three sections encompassed by the dashed lines contain similar fossils and are therefore the same age. Note that the youngest rocks in this region are in section B, whereas the oldest rocks are in section C.

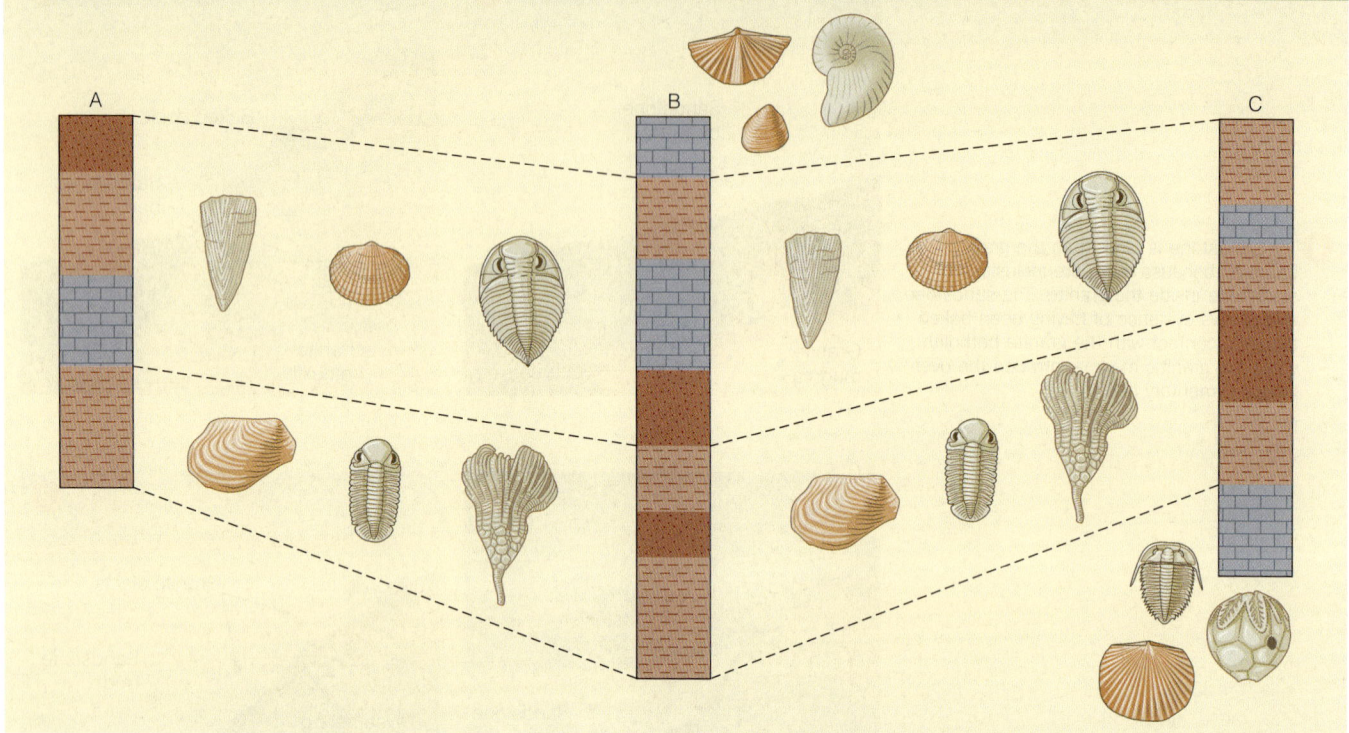

▶ **Figure 9.7 The Development of a Hiatus and an Unconformity** (a) Deposition began 12 million years ago (MYA) and continued more or less uninterrupted until 4 MYA. (b) Between 3 and 4 MYA, an episode of erosion occurred. During that time, some of the strata deposited earlier was eroded. (c) A hiatus of 3 million years thus exists between the older strata and the strata that formed during a renewed episode of deposition that began 3 MYA. (d) The actual stratigraphic record as seen in an outcrop today. The unconformity is the surface separating the strata and represents a major break in our record of geologic time.

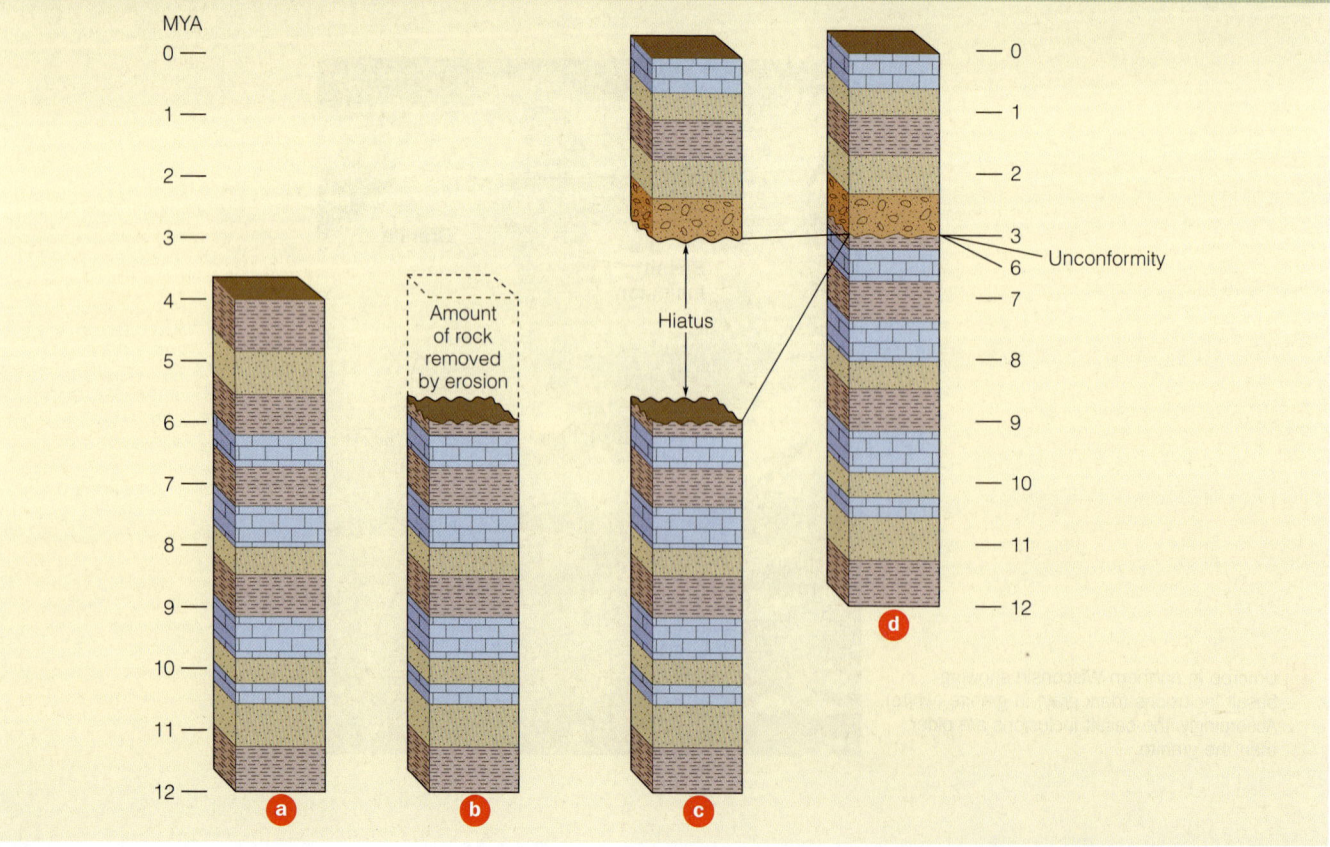

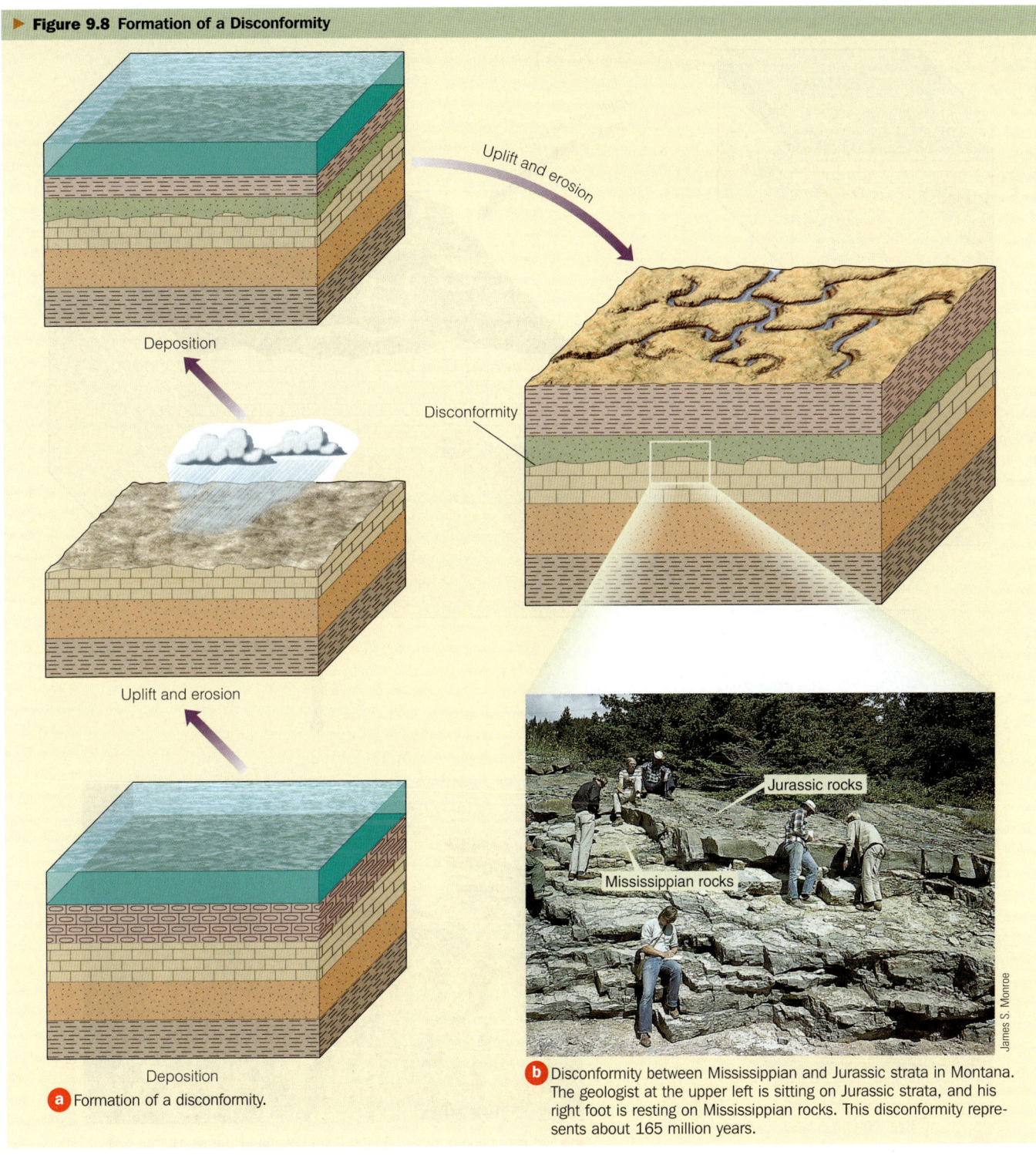

▶ Figure 9.8 Formation of a Disconformity

a Formation of a disconformity.

b Disconformity between Mississippian and Jurassic strata in Montana. The geologist at the upper left is sitting on Jurassic strata, and his right foot is resting on Mississippian rocks. This disconformity represents about 165 million years.

then either they were tilted, faulted (H), and eroded, or after deposition, they were faulted (H), tilted, and then eroded (▶ Figure 9.12a–c). Because the fault cuts beds A–G, it must be younger than the beds according to the principle of cross-cutting relationships.

Beds J–L were then deposited horizontally over this erosional surface, producing an angular unconformity (I) (▶ Figure 9.12d). Following deposition of these three beds, the entire sequence was intruded by a dike (M), which, according to the principle of cross-cutting relationships, must be younger than all the rocks it intrudes (▶ Figure 9.12e).

The entire area was then uplifted and eroded; next beds P and Q were deposited, producing a disconformity (N) between beds L and P and a nonconformity (O) between the igneous intrusion M and the sedimentary bed P (▶ Figure 9.12f,g). We know that the relationship between igneous intrusion M and the overlying sedimentary bed P is a nonconformity because of the presence of inclusions of M in P (principle of inclusions).

Figure 9.9 Formation of an Angular Unconformity

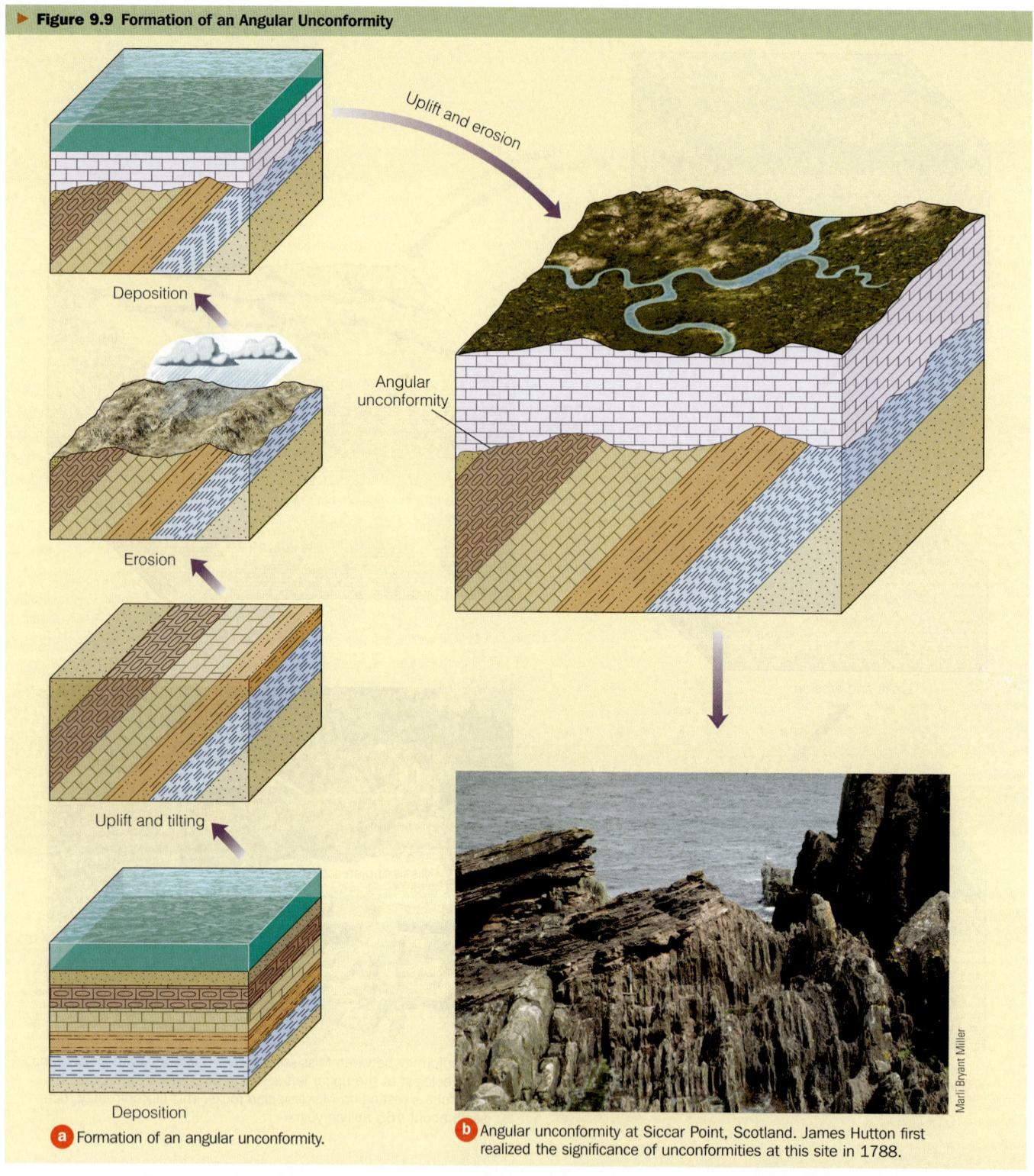

a Formation of an angular unconformity.

b Angular unconformity at Siccar Point, Scotland. James Hutton first realized the significance of unconformities at this site in 1788.

At this point, there are several possibilities for reconstructing the geologic history of this area. According to the principle of cross-cutting relationships, dike R must be younger than bed Q because it intrudes into it. It could have intruded anytime *after* bed Q was deposited; however, we cannot determine whether R was formed right after Q, right after S, or after T was formed. For purposes of this history, we will say that it intruded after the deposition of bed Q (▶ Figure 9.12g,h).

Following the intrusion of dike R, lava S flowed over bed Q, followed by the deposition of bed T (▶ Figure 9.12i,j). Although the lava flow S is not a sedimentary unit, the principle of superposition still applies because it flowed onto the surface, just as sediments are deposited on Earth's surface.

9.4 Relative Dating Methods 273

▶ **Figure 9.10 Formation of a Nonconformity**

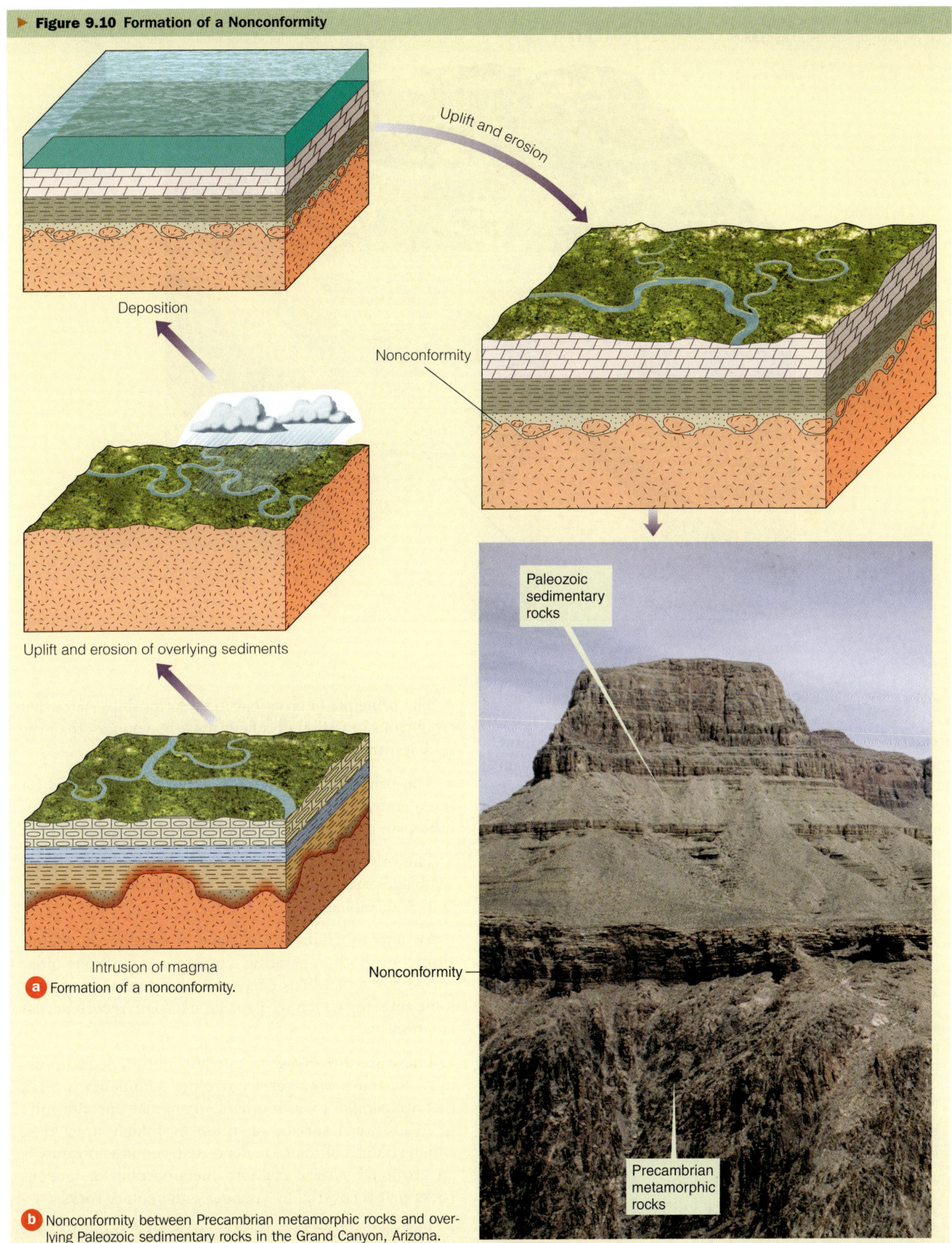

a Formation of a nonconformity.

b Nonconformity between Precambrian metamorphic rocks and overlying Paleozoic sedimentary rocks in the Grand Canyon, Arizona.

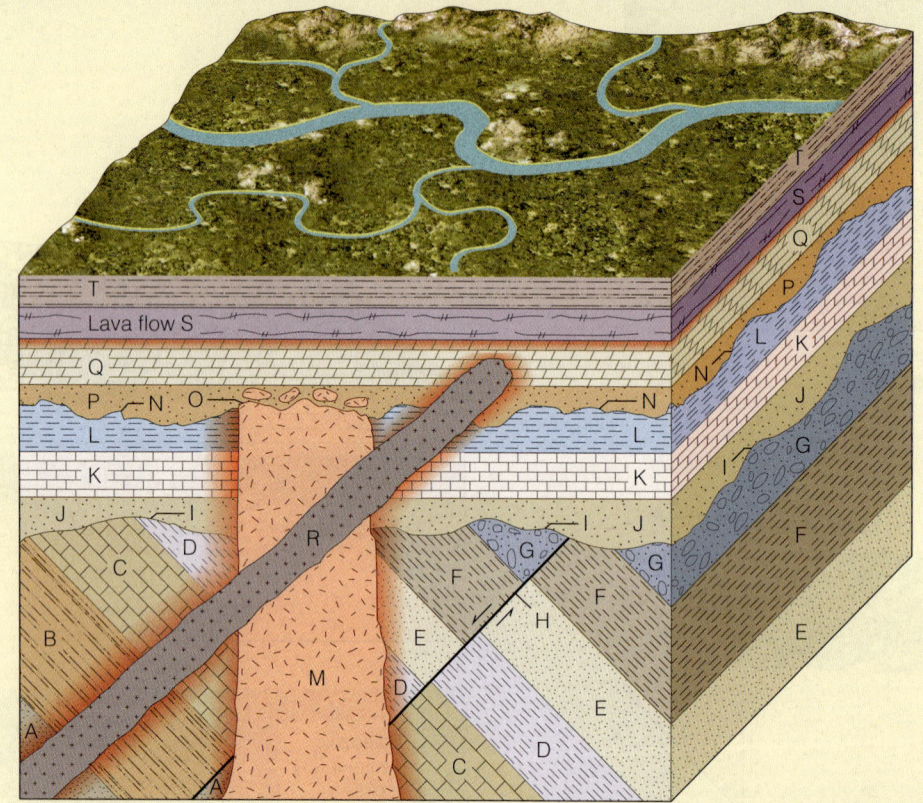

▶ **Figure 9.11 Block Diagram of a Hypothetical Area** A block diagram of a hypothetical area in which the various relative dating principles can be applied to determine its geologic history. See Figure 9.12 to see how the geologic history was determined using relative dating principles.

We have established a relative chronology for the rocks and events of this area by using the principles of relative dating. Remember, however, that we have no way of knowing how many years ago these events occurred unless we can obtain radiometric dates for the igneous rocks. With these dates, we can establish the range of absolute ages between which the different sedimentary units were deposited and also determine how much time is represented by the unconformities.

Section 9.4 Summary

- Relative dating places events in sequential order without regard to how long ago an event took place. There are six fundamental principles of relative dating.

- The principle of superposition states that in an undisturbed succession of sedimentary rocks, the oldest layer is on the bottom and the youngest layer is at the top.

- The principle of original horizontality states that sediments are originally deposited horizontally under the influence of gravity.

- The principle of lateral continuity states that a layer of sediment extends laterally in all directions until it thins and pinches out or terminates against the edge of the depositional basin.

- The principle of cross-cutting relationships states that an igneous intrusion or a fault must be younger than the rock it intrudes or displaces.

- The principle of inclusions states that inclusions, or fragments of one rock contained within a layer of another, are older than the rock layer itself.

- The principle of fossil succession states that fossil assemblages succeed one another through time in a regular and predictable order.

- An unconformity is a surface of erosion, nondeposition, or both separating younger rocks from older rocks. These surfaces encompass long periods of geologic time for which we have no geologic record at that location.

- There are three types of unconformities. A disconformity separates younger from older sedimentary rocks that are parallel to each other. An angular unconformity is an erosional surface on tilted or folded strata over which younger strata were deposited. A nonconformity is an erosional surface cut into metamorphic or igneous rocks that is covered by younger sedimentary rocks.

- The principles of relative dating can be applied in deciphering the geologic history of an area.

▶ **Figure 9.12 Using Relative Dating Principles to Interpret the Geologic History of a Hypothetical Area** (a) Beds A–G are deposited. (b) The preceding beds are tilted and faulted. (c) Erosion. (d) Beds J–L are deposited, producing an angular unconformity I. (e) The entire sequence is intruded by a dike. (f) The entire sequence is uplifted and eroded. (g) Beds P and Q are deposited, producing a disconformity (N) and a nonconformity (O). (h) Dike R intrudes. (i) Lava S flows over bed Q, baking it.

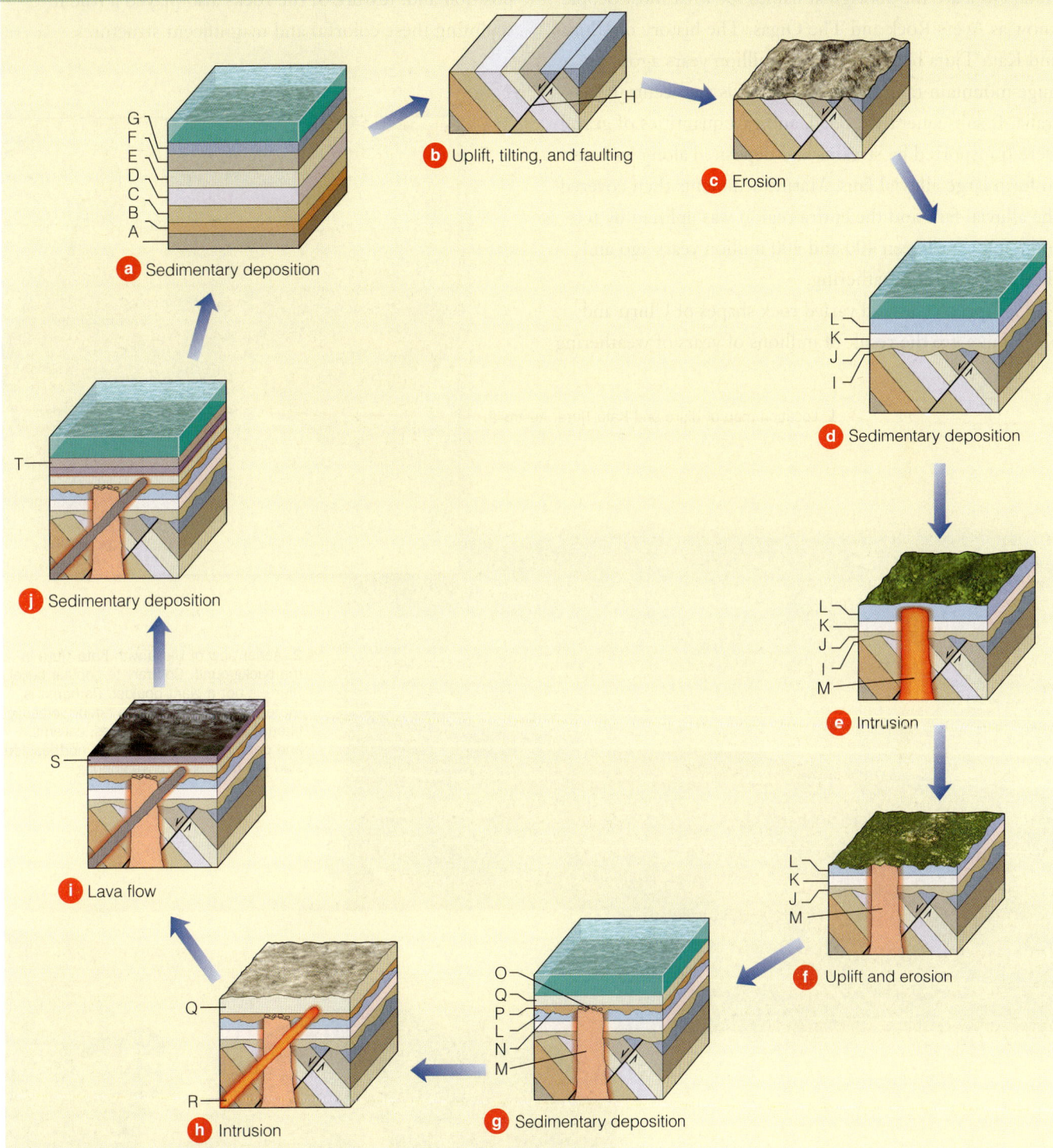

Uluru and Kata Tjuta

Rising majestically above the surrounding flat desert of central Australia are Uluru and Kata Tjuta. Uluru and Kata Tjuta are the aboriginal names for what most people know as Ayers Rock and The Olgas. The history of Uluru and Kata Tjuta began about 550 million years ago when a huge mountain range formed in what is now central Australia. It subsequently eroded, and vast quantities of gravel were transported by streams and deposited along its base to form large alluvial fans. Marine sediments then covered the alluvial fans and the entire region was uplifted by tectonic forces between 400 and 300 million years ago and then subjected to weathering.

The spectacular and varied rock shapes of Uluru and Kata Tjuta are the result of millions of years of weathering and erosion by water and, to a lesser extent, wind acting on the fractures formed during uplift. Differences in the composition and texture of the rocks also played a role in sculpting these colorful and magnificent structures.

▶ **1.** Location map of Uluru and Kata Tjuta, Australia.

◀ **2.** Aerial view of Uluru with Kata Tjuta in the background. Contrary to popular belief, Uluru is not a giant boulder. Rather, it is the exposed portion of the nearly vertically tilted Uluru Arkose. The caves, caverns, and depressions visible on the northeastern side are the result of weathering.

▶ **3.** A close-up view of the brain- and honeycomb-like small caves seen on the northeastern side of Uluru.

4. Uluru at sunset. The near-vertical tilting of the sedimentary beds of the Uluru Arksoe that make up Uluru can be seen clearly. Differential weathering of the sedimentary layers has produced the distinct parallel ridges and other features characteristic of Uluru.

5. Aerial view of Kata Tjuta with Uluru in the background. Kata Tjuta is composed of the Mount Currie Conglomerate, a coarse-grained and poorly sorted conglomerate. The sediments that were lithified into the Mount Currie Conglomerate were deposited, like the Uluru Arkose, as an alluvial fan beginning approximately 550 million years ago.

6. A view of the rounded domes of Kata Tjuta and typical vegetation as seen from within one of its canyons. The red color of the rocks is the result of oxidation of iron in the sediments.

7. An aerial close-up view of Kata Tjuta. The distinctive dome shape of the rocks is the result of weathering and erosion of the Mount Currie Conglomerate. In addition to weathering, the release of pressure on the buried rocks as they were exposed at the surface by tectonic forces contributed to the characteristic rounded shapes of Kata Tjuta.

9.5 Correlating Rock Units

■ *What is correlation?*

To decipher Earth history, geologists must demonstrate the time equivalency of rock units in different areas. This process is known as **correlation.**

If surface exposures are adequate, units may simply be traced laterally (principle of lateral continuity), even if occasional gaps exist (▶ Figure 9.13). Other criteria used to correlate units are similarity of rock type, position in a sequence, and key beds. *Key beds* are units, such as coal beds or volcanic ash layers, that are sufficiently distinctive to allow identification of the same unit in different areas (▶ Figure 9.13).

Generally, no single location in a region has a geologic record of all events that occurred during its history; therefore geologists must correlate from one area to another in order to determine the complete geologic history of the region. An excellent example is the history of the Colorado Plateau (▶ Figure 9.14). This region provides a record of events occurring over approximately 2 billion years. Because of the forces of erosion, the entire record is not preserved at any single location. Within the walls of the Grand Canyon are rocks of the Precambrian and Paleozoic eras, whereas Paleozoic and Mesozoic Era rocks are found in Zion National Park, and Mesozoic and Cenozoic Era rocks are exposed in Bryce Canyon National Park (▶ Figure 9.14). By correlating the uppermost rocks at one location with the lowermost equivalent rocks of another area, geologists can decipher the history of the entire region.

Although geologists can match up rocks on the basis of similar rock type and superposition, correlation of this type can be done only in a limited area where beds can be traced from one site to another. To correlate rock units over a large area or to correlate age-equivalent units of different composition, fossils and the principle of fossil succession must be used.

Fossils are useful as relative time indicators because they are the remains of organisms that lived for a certain length of time during the geologic past. Fossils that are easily identified, are geographically widespread, and existed for a rather short interval of geologic time are particularly useful. Such fossils are **guide fossils** or *index fossils* (▶ Figure 9.15).

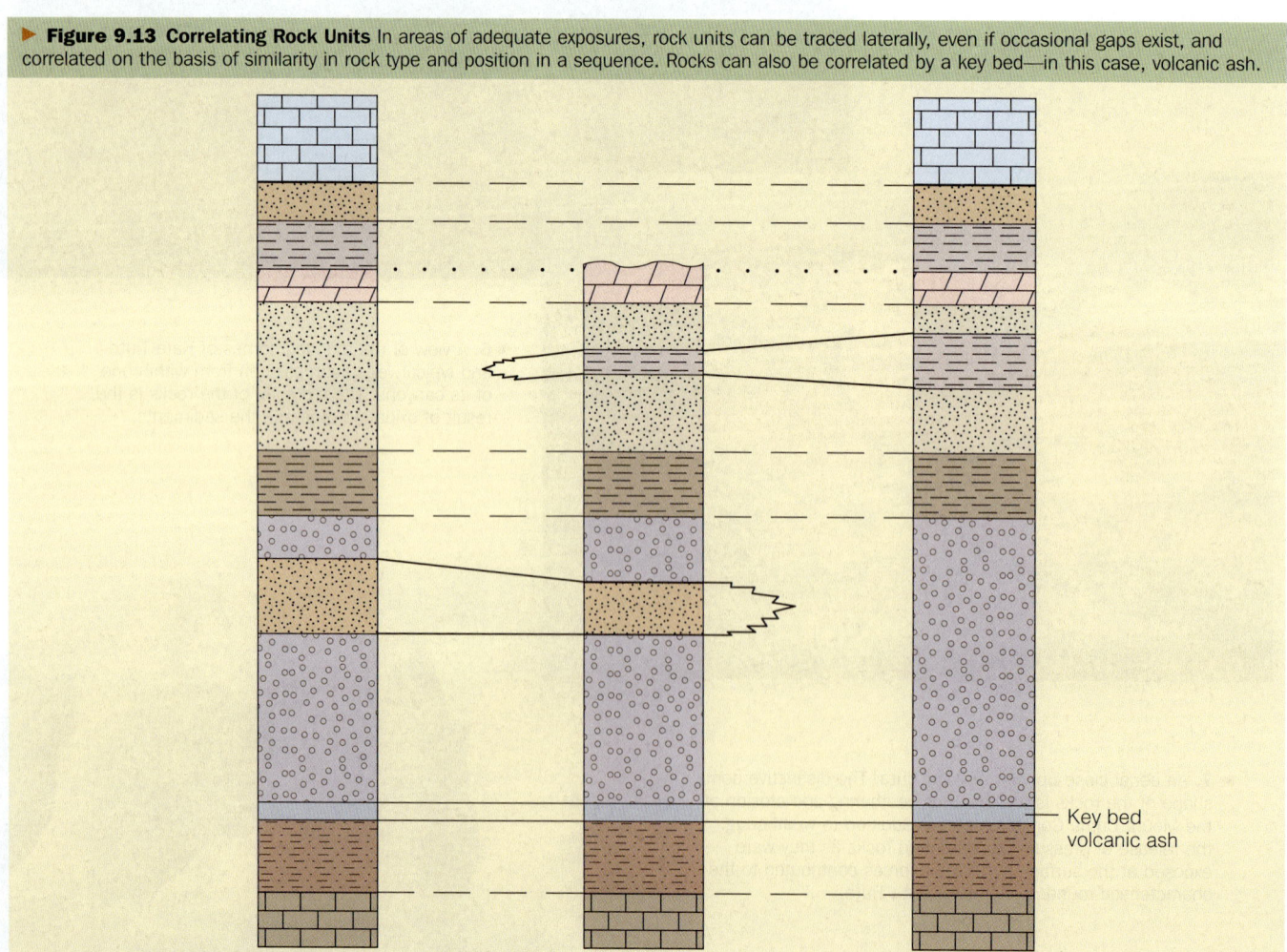

▶ **Figure 9.13 Correlating Rock Units** In areas of adequate exposures, rock units can be traced laterally, even if occasional gaps exist, and correlated on the basis of similarity in rock type and position in a sequence. Rocks can also be correlated by a key bed—in this case, volcanic ash.

▶ **Figure 9.14 Correlation of Rock Units within the Colorado Plateau** At each location, only a portion of the geologic record of the Colorado Plateau is exposed. By correlating the youngest rocks at one exposure with the oldest rocks at another exposure, geologists can determine the entire geologic history of the region. For example, the rocks forming the rim of the Grand Canyon, Arizona, are the Kaibab Limestone and Moenkopi Formation and are the youngest rocks exposed in the Grand Canyon. The Kaibab Limestone and Moenkopi Formation are the oldest rocks exposed in Zion National Park, Utah, and the youngest rocks are the Navajo Sandstone and Carmel Formation. The Navajo Sandstone and Carmel Formation are the oldest rocks exposed in Bryce Canyon National Park, Utah. By correlating the Kaibab Limestone and Moenkiopi Formation between the Grand Canyon and Zion National Park, geologists have extended the geologic history from the Precambrian to the Jurassic. And by correlating the Navajo Sandstone and Carmel Formation between Zion and Bryce Canyon National Parks, geologists can extend the geologic history through the Tertiary. Thus, by correlating the rock exposures between these areas and applying the principle of superposition, scientists can reconstruct the geologic history of the region.

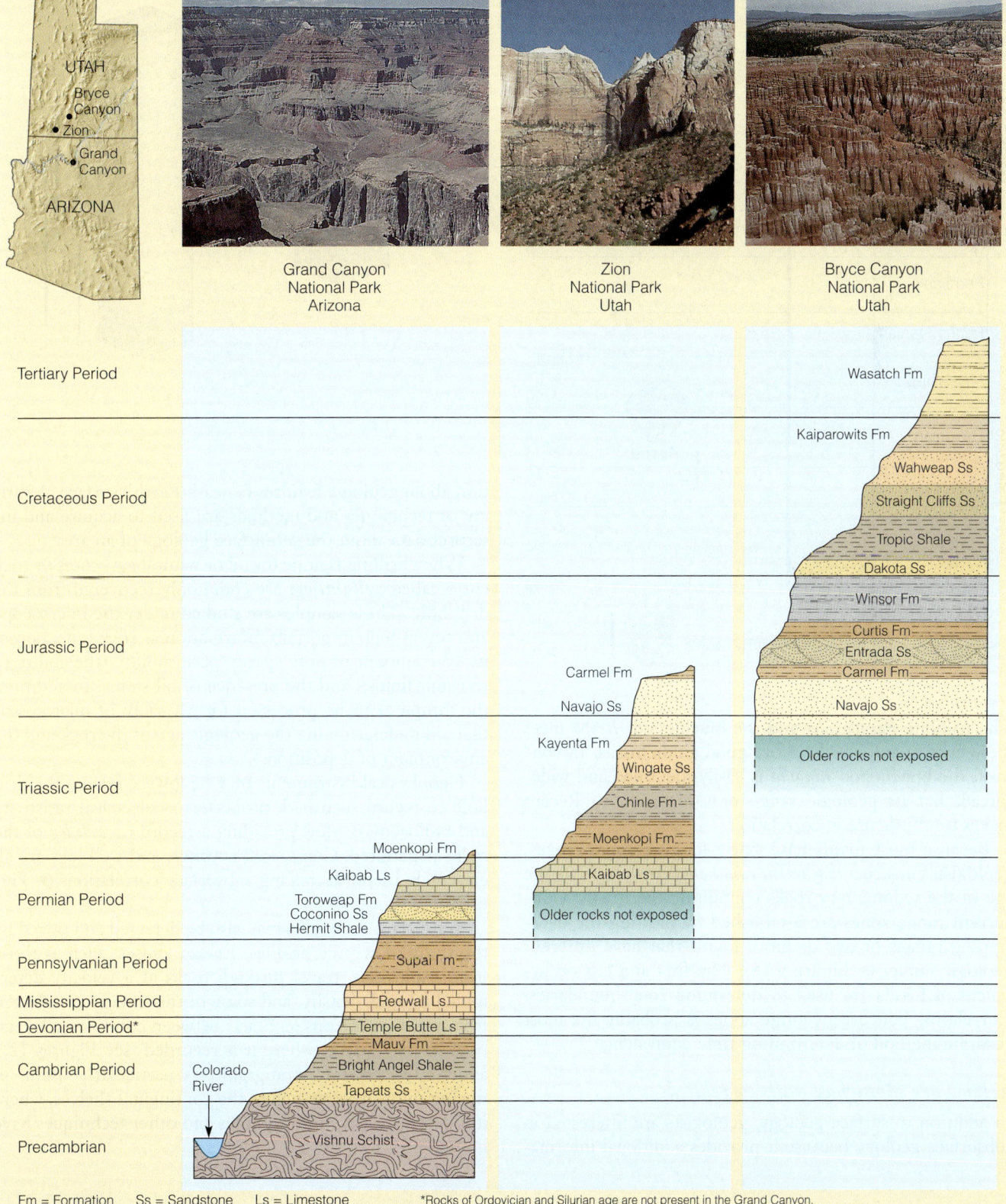

Fm = Formation   Ss = Sandstone   Ls = Limestone   *Rocks of Ordovician and Silurian age are not present in the Grand Canyon.

Figure 9.15 Guide Fossils Comparison of the geologic ranges (heavy vertical lines) of three marine invertebrate animals. *Lingula* is of little use in correlation because it has such a long range. But *Atrypa* and *Paradoxides* are good guide fossils because both are widespread, easily identified, and have short geologic ranges. Thus both can be used to correlate rock units that are widely separated and to establish the relative age of a rock that contains them.

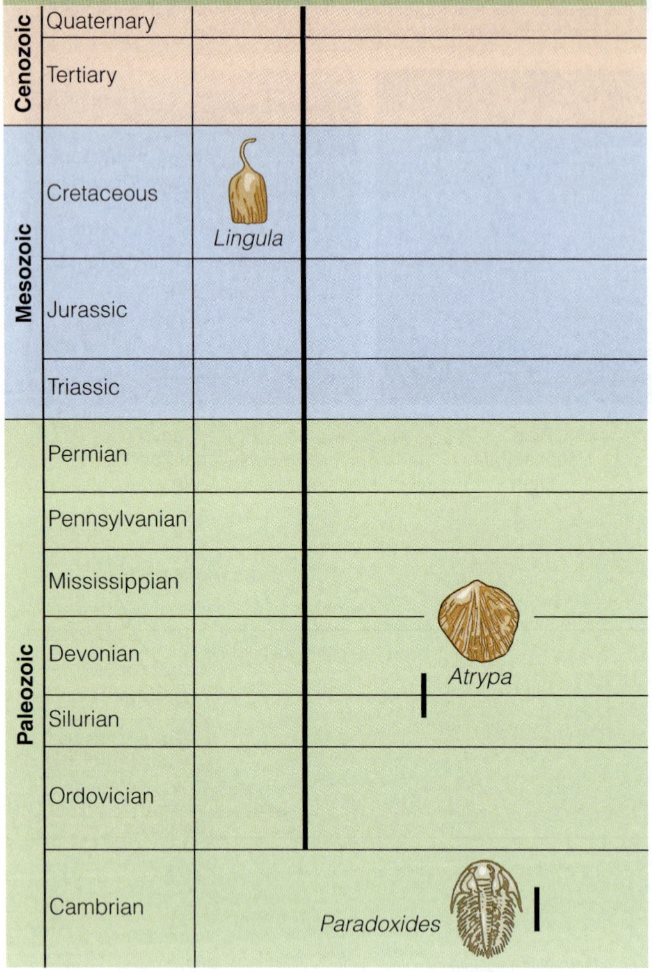

Cenozoic	Quaternary	
	Tertiary	
Mesozoic	Cretaceous	*Lingula*
	Jurassic	
	Triassic	
Paleozoic	Permian	
	Pennsylvanian	
	Mississippian	
	Devonian	*Atrypa*
	Silurian	
	Ordovician	
	Cambrian	*Paradoxides*

Figure 9.16 Correlation of Two Sections Using Concurrent Range Zones This concurrent range zone was established by the overlapping geologic ranges of fossils symbolized here by the letters A through E. The concurrent range zone is of shorter duration than any of the individual fossil geologic ranges. Correlating by concurrent range zones is probably the most accurate method of determining time equivalence.

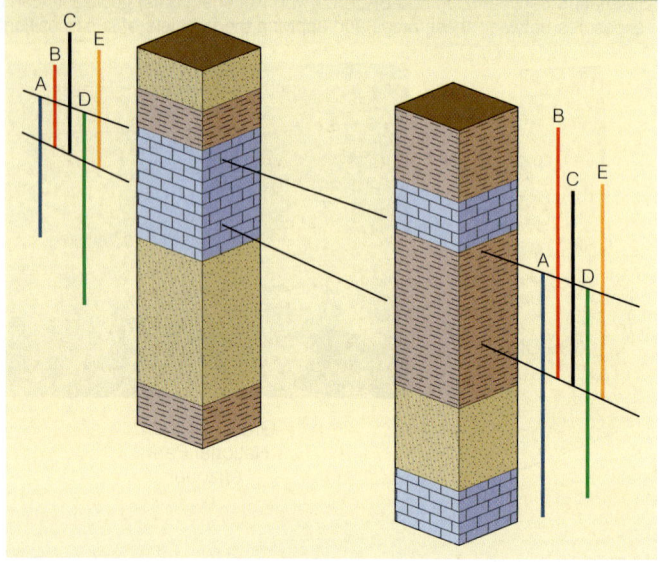

The trilobite *Paradoxides* and the brachiopod *Atrypa* meet these criteria and are therefore good guide fossils. In contrast, the brachiopod *Lingula* is easily identified and widespread, but its geologic range of Ordovician to Recent makes it of little use in correlation.

Because most fossils have fairly long geologic ranges, geologists construct *concurrent range zones* to determine the age of the sedimentary rocks containing the fossils. Concurrent range zones are established by plotting the overlapping ranges of two or more fossils that have different geologic ranges (▶ Figure 9.16). The first and last occurrences of fossils are used to determine zone boundaries. Correlating concurrent range zones is probably the most accurate method of determining time equivalence.

■ *How are subsurface units correlated?*

In addition to surface geology, geologists are interested in subsurface geology because it provides additional information about geologic features beneath Earth's surface. A variety of techniques and methods are used to acquire and interpret data about the subsurface geology of an area.

When drilling is done for oil or natural gas, cores or rock chips called *well cuttings* are commonly recovered from the drill hole. These samples are studied under the microscope and reveal such important information as rock type, porosity (the amount of pore space), permeability (the ability to transmit fluids), and the presence of oil stains. In addition, the samples can be processed for a variety of microfossils that aid in determining the geologic age of the rock and the environment of deposition.

Geophysical instruments may be lowered down the drill hole to record such rock properties as electrical resistivity and radioactivity, thus providing a record or *well log* of the rocks penetrated. Cores, well cuttings, and well logs are all extremely useful in making subsurface correlations (▶ Figure 9.17).

Subsurface rock units may also be detected and traced by the study of seismic profiles. Energy pulses, such as those from explosions, travel through rocks at a velocity determined by rock density, and some of this energy is reflected from various horizons (contacts between contrasting layers) back to the surface, where it is recorded (see Figures 12.3 and 12.13). Seismic stratigraphy is particularly useful in tracing units in areas such as the continental shelves, where it is very expensive to drill holes and other techniques have limited use.

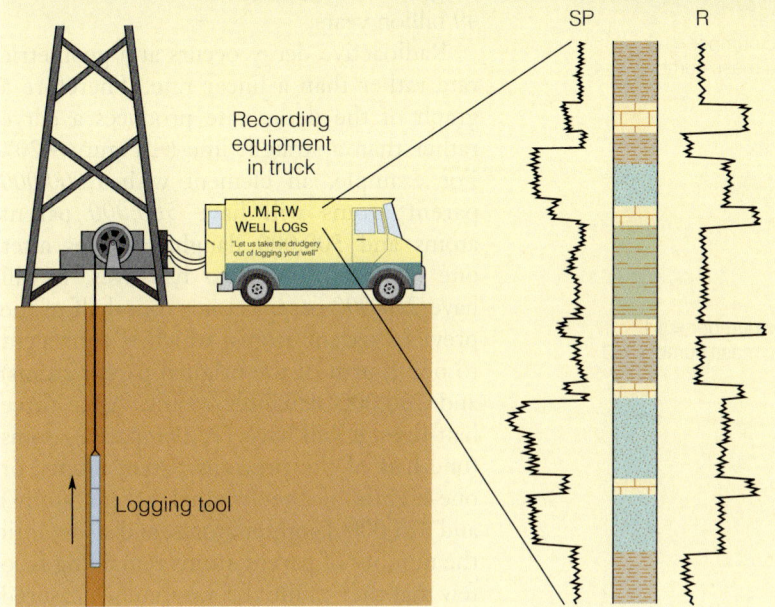

Figure 9.17 Well Logs A schematic diagram showing how well logs are made. As the logging tool is withdrawn from the drill hole, data are transmitted to the surface, where they are recorded and printed as a well log. The curve labeled SP in this diagrammatic electric log is a plot of self-potential (electrical potential caused by different conductors in a solution that conducts electricity) with depth. The curve labeled R is a plot of electrical resistivity with depth. Electric logs yield information about the rock type and fluid content of subsurface formations. Electric logs are also used to correlate from well to well.

Section 9.5 Summary

- Correlation is the process by which geologists match up rock units in different areas by showing time equivalency. Surface exposures can be correlated by using similarity of rock type, position in a sequence, key beds, and fossil assemblages as well as tracing units laterally. Subsurface techniques of correlation are the same as those used for surface correlation, with the addition of such geophysical methods as electrical resistivity, radioactivity, and seismic profiles.

9.6 Absolute Dating Methods

■ *Why was the discovery of radioactivity important to geology?*

Although most of the isotopes of the 92 naturally occurring elements are stable, some are radioactive and spontaneously decay to other more stable isotopes of elements, releasing energy in the process. The discovery, in 1903 by Pierre and Marie Curie, that radioactive decay produces heat meant that geologists finally had a mechanism for explaining Earth's internal heat that did not rely on residual cooling from a molten origin. Furthermore, geologists now had a powerful tool to date geologic events accurately and to verify the long time periods postulated by Hutton, Lyell, and Darwin.

■ *What are atoms, elements, and isotopes?*

As we discussed in Chapter 3, all matter is made up of chemical elements, each composed of extremely small particles called *atoms*. The nucleus of an atom is composed of *protons* (positively charged particles) and *neutrons* (neutral particles) with *electrons* (negatively charged particles) encircling it (see Figure 3.3). The number of protons is an element's *atomic number* and helps determine its properties and characteristics. The combined number of protons and neutrons in an atom is its *atomic mass number*. However, not all atoms of the same element have the same number of neutrons in their nuclei. These variable forms of the same element are called *isotopes* (see Figure 3.5). Most isotopes are stable, but some are unstable and spontaneously decay to a more stable form. It is the decay rate of unstable isotopes that geologists measure to determine the absolute age of rocks.

■ *What is radioactive decay?*

Radioactive decay is the process by which an unstable atomic nucleus is spontaneously transformed into an atomic nucleus of a different element. Scientists recognize three types of radioactive decay, all of which result in a change of atomic structure (▶ Figure 9.18). In *alpha decay*, 2 protons and 2 neutrons are emitted from the nucleus, resulting in the loss of 2 atomic numbers and 4 atomic mass numbers. In *beta decay*, a fast-moving electron is emitted from a neutron in the nucleus, changing that neutron to a proton and consequently increasing the atomic number by 1, with no resultant atomic mass number change. *Electron capture* is when a proton captures an electron from an electron shell and thereby converts to a neutron, resulting in the loss of 1 atomic number but not changing the atomic mass number.

Some elements undergo only 1 decay step in the conversion from an unstable form to a stable form. For example, rubidium 87 decays to strontium 87 by a single beta emission, and potassium 40 decays to argon 40 by a single electron capture. Other radioactive elements undergo several decay steps. Uranium 235 decays to lead 207 by 7 alpha and 6 beta steps, whereas uranium 238 decays to lead 206 by 8 alpha and 6 beta steps (▶ Figure 9.19).

■ *What are half-lives?*

When we discuss decay rates, it is convenient to refer to them in terms of half-lives. The **half-life** of a radioactive element

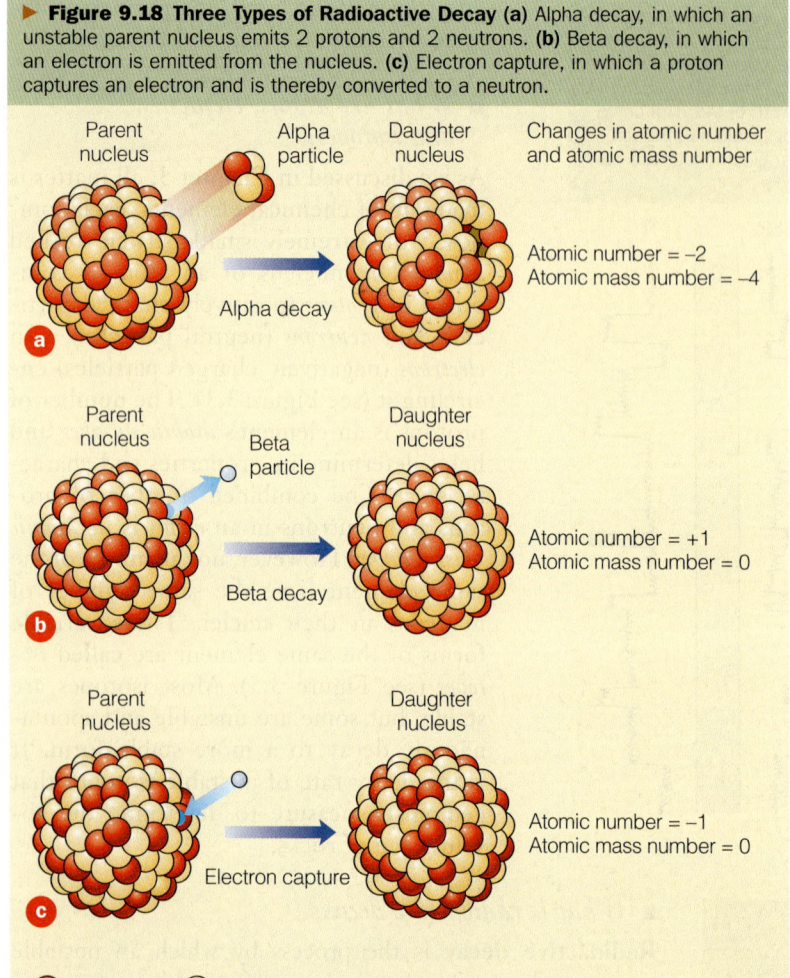

▶ **Figure 9.18 Three Types of Radioactive Decay** (a) Alpha decay, in which an unstable parent nucleus emits 2 protons and 2 neutrons. (b) Beta decay, in which an electron is emitted from the nucleus. (c) Electron capture, in which a proton captures an electron and is thereby converted to a neutron.

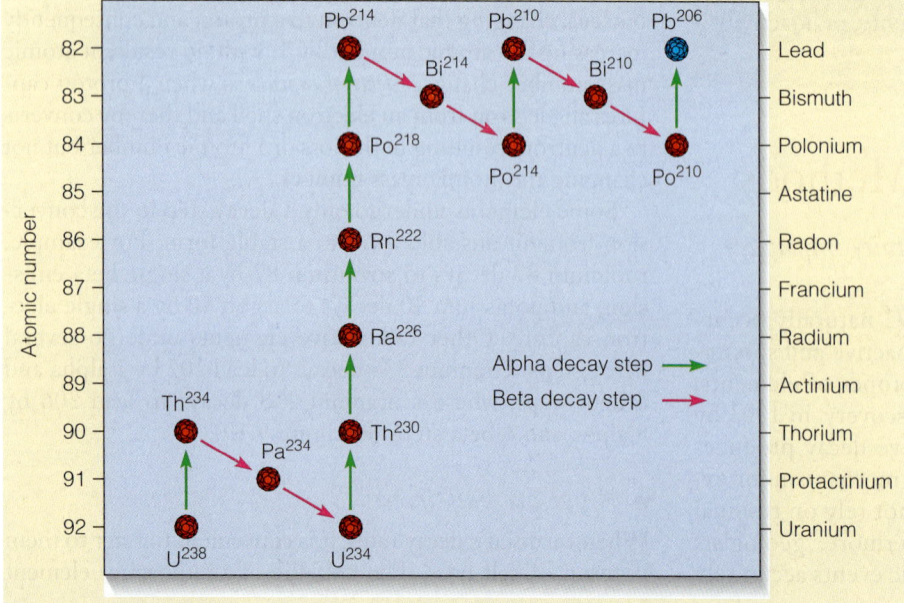

▶ **Figure 9.19 Radioactive Decay Series for Uranium 238 and Lead 206** Radioactive uranium 238 decays to its stable daughter product, lead 206, by 8 alpha and 6 beta decay steps. A number of different isotopes are produced as intermediate steps in the decay series.

is the time it takes for half of the atoms of the original unstable *parent element* to decay to atoms of a new, more stable *daughter element*. The half-life of a given radioactive element is constant and can be precisely measured. Half-lives of various radioactive elements range from less than a billionth of a second to 49 billion years.

Radioactive decay occurs at a geometric rate rather than a linear rate. Therefore a graph of the decay rate produces a curve rather than a straight line (▶ Figure 9.20). For example, an element with *1,000,000* parent atoms will have *500,000* parent atoms and 500,000 daughter atoms after one half-life. After two half-lives, it will have *250,000* parent atoms (one-half of the previous parent atoms, which is equivalent to one-fourth of the original parent atoms) and 750,000 daughter atoms. After three half-lives, it will have *125,000* parent atoms (one-half of the previous parent atoms, or one-eighth of the original parent atoms) and 875,000 daughter atoms, and so on until the number of parent atoms remaining is so few that they cannot be accurately measured by present-day instruments.

By measuring the parent–daughter ratio and knowing the half-life of the parent (which has been determined in the laboratory), geologists can calculate the age of a sample that contains the radioactive element. The parent–daughter ratio is usually determined by a *mass spectrometer*, an instrument that measures the proportions of atoms of different masses.

- **What are some of the sources of uncertainty in radiometric dating?**

The most accurate radiometric dates are obtained from igneous rocks. As magma cools and begins to crystallize, radioactive parent atoms are separated from previously formed daughter atoms. Because they are the right size, some radioactive parent atoms are incorporated into the crystal structure of certain minerals. The stable daughter atoms, though, are a different size from the radioactive parent atoms and consequently cannot fit into the crystal structure of the same mineral as the parent

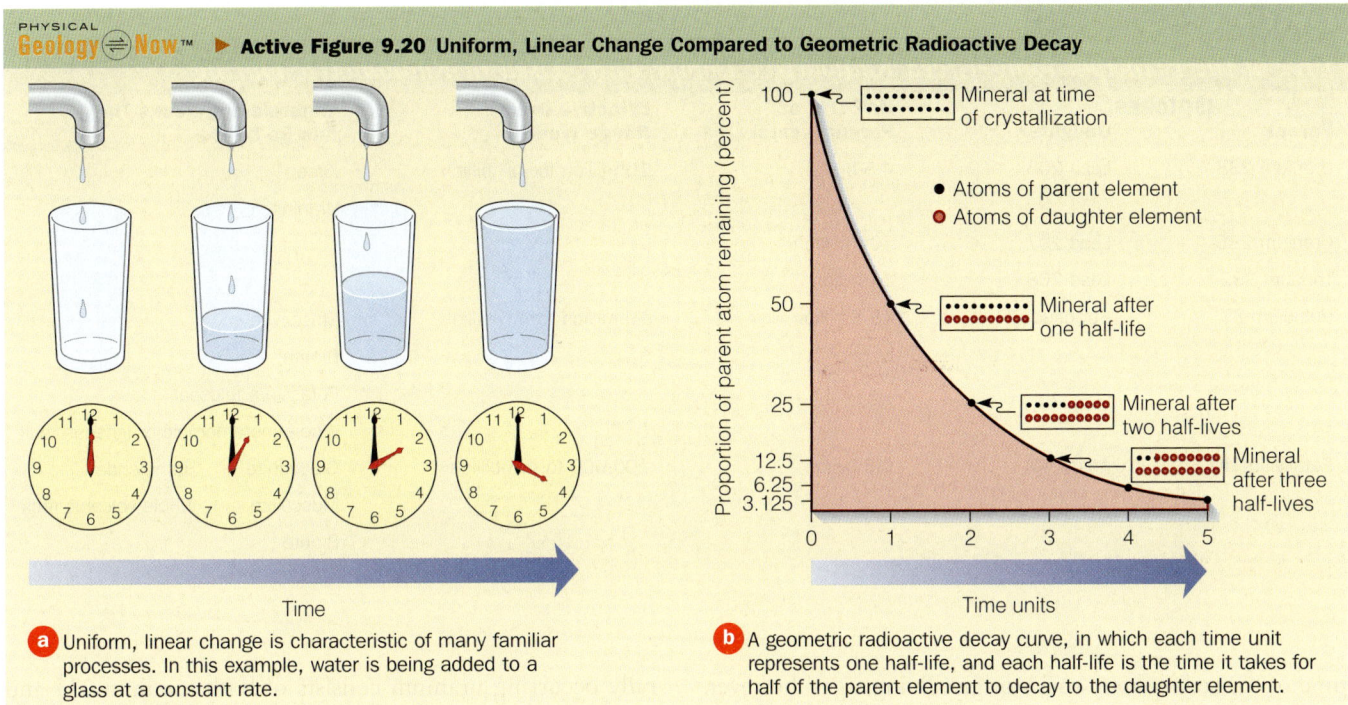

Active Figure 9.20 Uniform, Linear Change Compared to Geometric Radioactive Decay

(a) Uniform, linear change is characteristic of many familiar processes. In this example, water is being added to a glass at a constant rate.

(b) A geometric radioactive decay curve, in which each time unit represents one half-life, and each half-life is the time it takes for half of the parent element to decay to the daughter element.

atoms. Therefore a mineral that crystallizes in a cooling magma will contain radioactive parent atoms but no stable daughter atoms (▶ Figure 9.21). Thus the time that is being measured is the time of crystallization of the mineral containing the radioactive atoms, and not the time of formation of the radioactive atoms.

Except in unusual circumstances, sedimentary rocks cannot be radiometrically dated because one would be measuring the age of a particular mineral rather than the time that it was deposited as a sedimentary particle. One of the few instances in which radiometric dates can be obtained on sedimentary rocks is when the mineral glauconite is present. Glauconite is a greenish mineral containing radioactive potassium 40, which decays to argon 40 (Table 9.1). It forms in certain marine environments as a result of chemical reactions with clay minerals during the conversion of sediments to sedimentary rock. Thus glauconite forms when the sedimentary rock forms, and a radiometric date indicates the

▶ **Figure 9.21 Crystallization of Magma Containing Radioactive Parent and Stable Daughter Atoms** (a) Magma contains both radioactive parent atoms and stable daughter atoms. The radioactive parent atoms are larger than the stable daughter atoms. (b) As magma cools and begins to crystallize, some of the radioactive parent atoms are incorporated into certain minerals because they are the right size and can fit into the crystal structure. In this example, only the larger radioactive parent atoms fit into the crystal structure. Therefore, at the time of crystallization, minerals in which the radioactive parent atoms can fit into the crystal structure will contain 100% radioactive parent atoms and 0% stable daughter atoms. (c) After one half-life, 50% of the radioactive parent atoms will have decayed to stable daughter atoms, such that those minerals that had radioactive parent atoms in their crystal structure will now have 50% radioactive parent atoms and 50% stable daughter atoms.

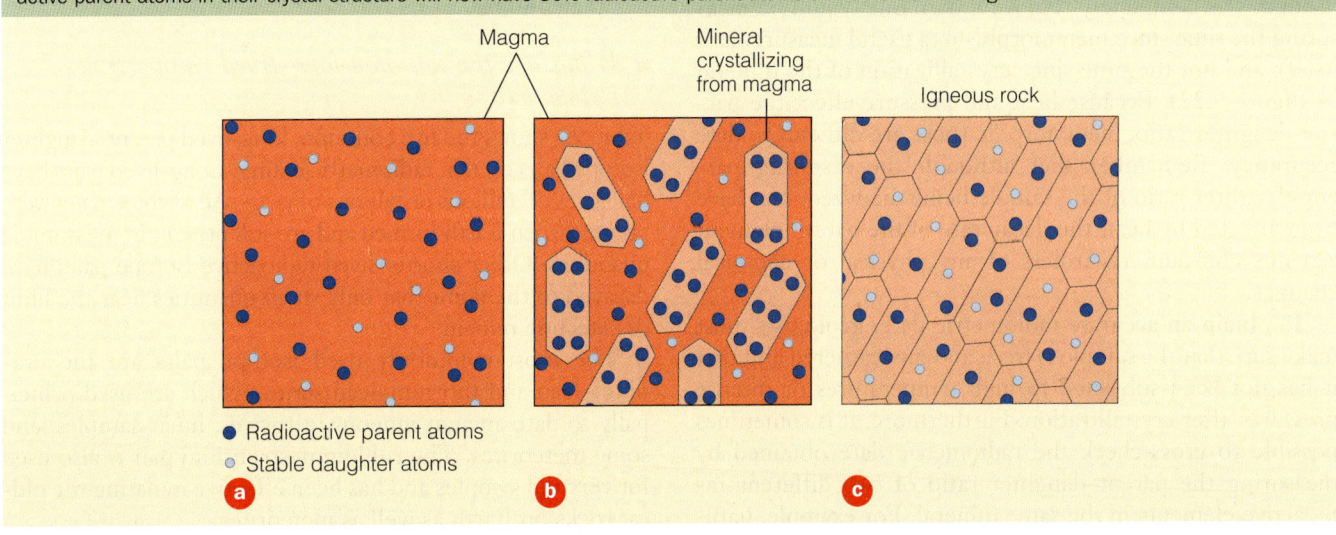

TABLE 9.1
Five of the Principal Long-Lived Radioactive Isotope Pairs Used in Radiometric Dating

ISOTOPES		Half-Life of	Effective Dating	Minerals and Rocks That
Parent	Daughter	Parent (years)	Range (years)	Can Be Dated
Uranium 238	Lead 206	4.5 billion	10 million to 4.6 billion	Zircon
				Uraninite
Uranium 235	Lead 207	704 million		
Thorium 232	Lead 208	14 billion		
Rubidium 87	Strontium 87	48.8 billion	10 million to 4.6 billion	Muscovite
				Biotite
				Potassium feldspar
				Whole metamorphic or igneous rock
Potassium 40	Argon 40	1.3 billion	100,000 to 4.6 billion	Glauconite, Hornblende
				Muscovite, Whole volcanic rock
				Biotite

time of the sedimentary rock's origin. Being a gas, however, the daughter product argon can easily escape from a mineral. Therefore any date obtained from glauconite, or any other mineral containing the potassium 40 and argon 40 pair, must be considered a minimum age.

To obtain accurate radiometric dates, geologists must be sure that they are dealing with a *closed system*, meaning that neither parent nor daughter atoms have been added or removed from the system since crystallization and that the ratio between them results from only radioactive decay. Otherwise, an inaccurate date will result. If daughter atoms have leaked out of the mineral being analyzed, the calculated age will be too young; if parent atoms have been removed, the calculated age will be too old.

Leakage may take place if the rock is heated or subjected to intense pressure, as can sometimes occur during metamorphism. If this happens, some of the parent or daughter atoms may be driven from the mineral being analyzed, resulting in an inaccurate age determination. If the daughter product was completely removed, then one would be measuring the time since metamorphism (a useful measurement itself), and not the time since crystallization of the mineral (▶ Figure 9.22). Because heat and pressure affect the parent–daughter ratio, metamorphic rocks are difficult to date accurately. Remember that although the resulting parent–daughter ratio of the sample being analyzed may have been affected by heat, the decay rate of the parent element remains constant, regardless of any physical or chemical changes.

To obtain an accurate radiometric date, geologists must make sure that the sample is fresh and unweathered and that it has not been subjected to high temperatures or intense pressures after crystallization. Furthermore, it is sometimes possible to cross-check the radiometric date obtained by measuring the parent–daughter ratio of two different radioactive elements in the same mineral. For example, naturally occurring uranium consists of both uranium 235 and uranium 238 isotopes. Through various decay steps, uranium 235 decays to lead 207, whereas uranium 238 decays to lead 206 (▶ Figure 9.19). If the minerals that contain both uranium isotopes have remained closed systems, the ages obtained from each parent–daughter ratio should agree closely and therefore should indicate the time of crystallization of the magma. If the ages do not closely agree, then other samples must be taken and ratios measured to see which, if either, date is correct.

Recent advances and the development of new techniques and instruments for measuring various isotope ratios have enabled geologists to analyze not only increasingly smaller samples, but with a greater precision than ever before. Presently the measurement error for many radiometric dates is typically less than 0.5% of the age, and in some cases it is even better than 0.1%. Thus, for a rock 540 million years old (near the beginning of the Cambrian Period), the possible error could range from nearly 2.7 million years to less than 540,000 years.

■ *What are the common long-lived radioactive isotope pairs?*

Table 9.1 shows the five common, long-lived parent–daughter isotope pairs used in radiometric dating. Long-lived pairs have half-lives of millions or billions of years. All of these pairs were present when Earth formed and are still present in measurable quantities. Other shorter-lived radioactive isotope pairs have decayed to the point that only small quantities near the limit of detection remain.

The most commonly used isotope pairs are the uranium–lead and thorium–lead series, which are used principally to date ancient igneous intrusives, lunar samples, and some meteorites. The rubidium–strontium pair is also used for very old samples and has been effective in dating the oldest rocks on Earth as well as meteorites.

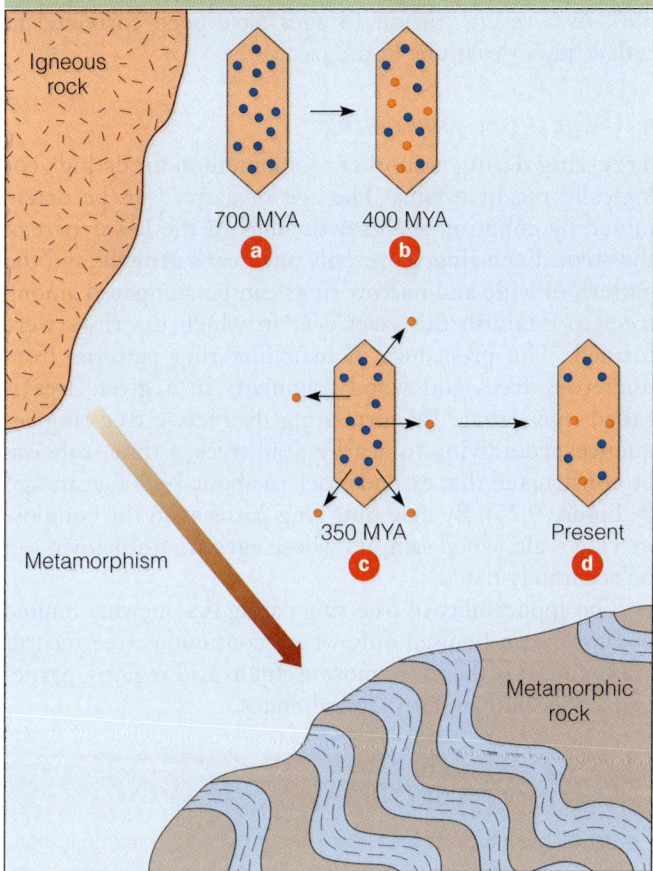

Figure 9.22 Effects of Metamorphism on Radiometric Dating The effect of metamorphism in driving out daughter atoms from a mineral that crystallized 700 million years ago (MYA). The mineral is shown immediately after crystallization (a), then at 400 MYA (b), when some of the parent atoms had decayed to daughter atoms. Metamorphism at 350 MYA (c) drives the daughter atoms out of the mineral into the surrounding rock. (d) If the rock has remained a closed chemical system throughout its history, dating the mineral today yields the time of metamorphism, whereas dating the whole rock provides the time of its crystallization, 700 MYA.

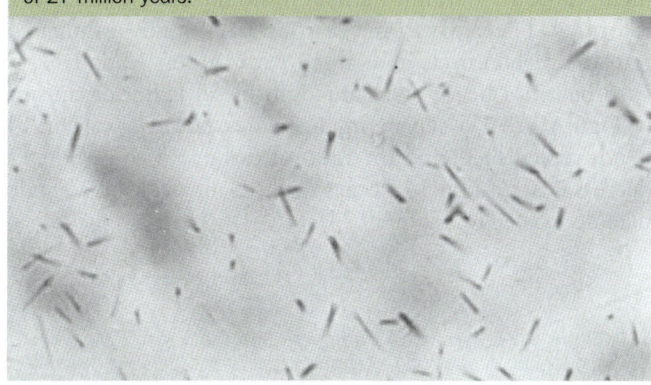

Figure 9.23 Fission-Track Dating Each fission track (about 16 microns long) in this apatite crystal is the result of the radioactive decay of a uranium atom. The apatite crystal, which has been etched with hydrofluoric acid to make the fission tracks visible, comes from one of the dikes at Shiprock, New Mexico, and has a calculated age of 27 million years.

The potassium–argon pair is typically used for dating fine-grained volcanic rocks from which individual crystals cannot be separated; hence the whole rock is analyzed. Because argon is a gas, great care must be taken to ensure that the sample has not been subjected to heat, which would allow argon to escape; such a sample would yield an age that is too young. Other long-lived radioactive isotope pairs exist, but they are rather rare and are used in only special situations.

■ *What is fission-track dating?*

The emission of atomic particles resulting from the spontaneous decay of uranium within a mineral damages its crystal structure. The damage appears as microscopic linear tracks that are visible only after the mineral is etched with hydrofluoric acid, an acid so powerful that its vapors can destroy one's sense of smell without careful handling. The age of the sample is determined from the number of fission tracks present and the amount of uranium the sample contains: the older the sample, the greater the number of tracks (▶ Figure 9.23).

Fission-track dating is of particular interest to archaeologists and geologists because the technique can be used to date samples ranging from only a few hundred to hundreds of millions of years old. It is most useful for dating samples from the time between about 40,000 and 1.5 million years ago, a period for which other dating techniques are not always particularly suitable. One of the problems in fission-track dating occurs when the rocks have later been subjected to high temperatures. If this happens, the damaged crystal structures are repaired by annealing, and consequently the tracks disappear. In such instances, the calculated age will be younger than the actual age.

■ *What is the carbon-14 dating technique?*

Carbon is an important element in nature and one of the basic elements found in all forms of life. It has three isotopes; two of these, carbon 12 and 13, are stable, whereas carbon 14 is radioactive (see Figure 3.5). Carbon 14 has a half-life of 5730 years plus or minus 30 years. The **carbon-14 dating technique** is based on the ratio of carbon 14 to carbon 12 and is generally used to date once-living material.

The short half-life of carbon 14 makes this dating technique practical only for specimens younger than about 70,000 years. Consequently, the carbon-14 dating method is especially useful in archaeology and has greatly helped unravel the events of the latter portion of the Pleistocene Epoch. For example, carbon-14 dates of maize from the Tehuacan Valley of Mexico have forced archeologists to rethink their ideas of where the first center for maize domestication in Mesoamerica arose. Carbon-14 dating is also helping to answer the question of when humans began populating North America.

Carbon 14 is constantly formed in the upper atmosphere when cosmic rays, which are high-energy particles (mostly protons), strike the atoms of upper-atmospheric gases, splitting their nuclei into protons and neutrons. When a neutron strikes the nucleus of a nitrogen atom (atomic number 7, atomic mass number 14), it may be absorbed into the nucleus and a proton emitted. Thus the atomic number of the atom decreases by 1, whereas the atomic mass number stays the same. Because the atomic number has changed, a new element, carbon 14 (atomic number 6, atomic mass number 14), is formed. The newly formed carbon 14 is rapidly assimilated into the carbon cycle and, along with carbon 12 and 13, is absorbed in a nearly constant ratio by all living organisms (▶ Figure 9.24). When an organism dies, however, carbon 14 is not replenished, and the ratio of carbon 14 to carbon 12 decreases as carbon 14 decays back to nitrogen by a single beta decay step (▶ Figure 9.24).

▶ **Figure 9.24 Carbon 14 Dating Method** The carbon cycle showing the formation of carbon 14 in the upper atmosphere, its dispersal and incorporation into the tissue of all living organisms, and its decay back to nitrogen 14 by beta decay.

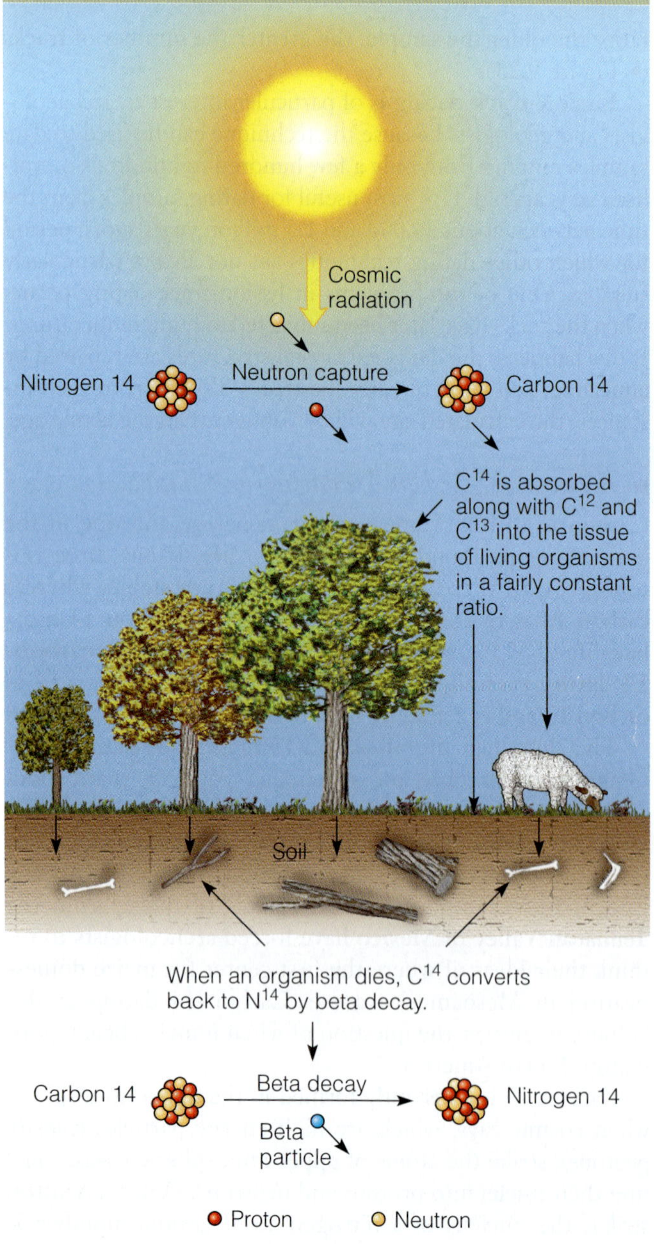

Currently the ratio of carbon 14 to carbon 12 is remarkably constant in both the atmosphere and living organisms. There is good evidence, however, that the production of carbon 14, and thus the ratio of carbon 14 to carbon 12, has varied somewhat during the past several thousand years. This was determined by comparing ages established by carbon-14 dating of wood samples with ages established by counting annual tree rings in the same samples. As a result, carbon-14 ages have been corrected to reflect such variations in the past.

■ *What is tree-ring dating?*

Tree-ring dating is another useful method for dating geologically recent events. The age of a tree can be determined by counting the growth rings in the lower part of the stem. Each ring represents one year's growth, and the pattern of wide and narrow rings can be compared among trees to establish the exact year in which the rings were formed. The procedure of matching ring patterns from numerous trees and wood fragments in a given area is called *cross-dating*. By correlating distinctive tree-ring sequences from living to nearby dead trees, a time scale can be constructed that extends back to about 14,000 years ago (▶ Figure 9.25). By matching ring patterns to the composite ring scale, wood samples whose ages are not known can be accurately dated.

The applicability of tree-ring dating is somewhat limited because it can be used only where continuous tree records are found. It is therefore most useful in arid regions, particularly the southwestern United States.

Section 9.6 Summary

- Radioactive decay is the process by which an unstable atomic nucleus is spontaneously transformed into an atomic nucleus of a different element. Three types of radioactive decay are recognized: alpha decay, beta decay, and electron capture.

- A half-life is the time it takes for half of the atoms of the original unstable parent atom to decay to a new, more stable daughter element. By measuring the parent–daughter ratio and knowing the half-life of the parent, geologists can calculate the age of a sample containing that radioactive element.

- To obtain accurate radiometric dates, fresh, unweathered samples must be used, in which neither parent nor daughter atoms have migrated into or out of the sample being analyzed.

- Five long-lived parent–daughter isotope pairs (two uranium–lead pairs and one thorium–lead, rubidium–strontium, and potassium–argon pair) are commonly used in radiometric dating. These isotope pairs are used to date ancient igneous and metamorphic rocks as well as lunar samples and some meteorites.

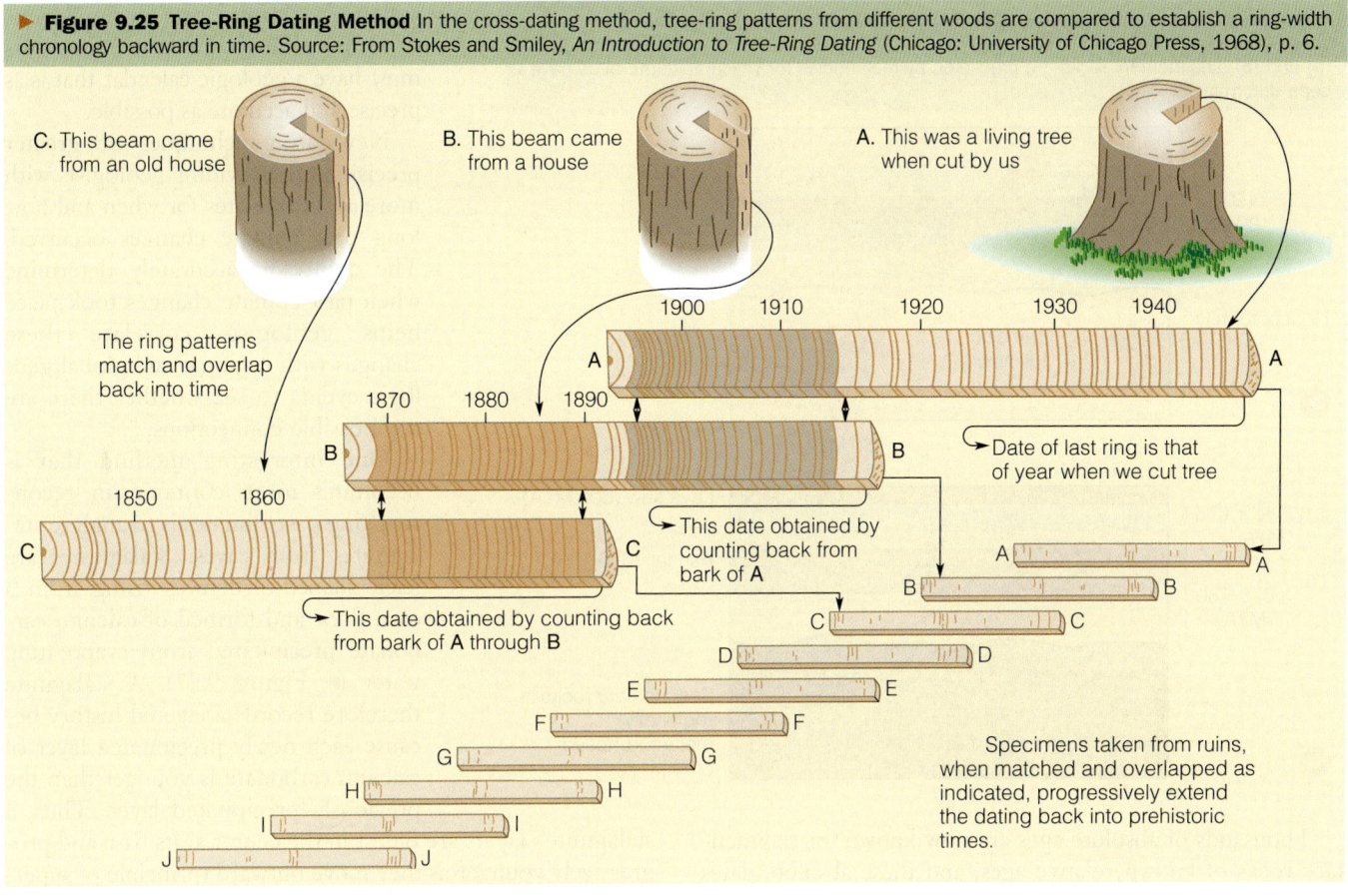

▶ **Figure 9.25 Tree-Ring Dating Method** In the cross-dating method, tree-ring patterns from different woods are compared to establish a ring-width chronology backward in time. Source: From Stokes and Smiley, *An Introduction to Tree-Ring Dating* (Chicago: University of Chicago Press, 1968), p. 6.

- Fission-track dating, which measures the number of microscopic linear tracks left by the decay of uranium, is useful for dating samples from about 40,000 to 1.5 million years old, a period of time for which other dating techniques are not always suitable.

- The carbon-14 dating technique is based on the ratio of carbon 14 to carbon 12 and is generally used to date once-living material back to approximately 70,000 years.

- Tree-ring dating is used for dating geologically recent events and involves counting the annual growth rings in the lower part of a tree's stem. It is most useful in arid regions.

9.7 Development of the Geologic Time Scale

■ *What is the geologic time scale?*

The geologic time scale is a hierarchical scale in which the 4.6-billion-year history of Earth is divided into time units of varying duration (▶ Figure 9.1). It did not result from the work of any one individual but rather evolved, primarily during the 19th century, through the efforts of many people.

■ *How did the geologic time scale develop?*

By applying relative-dating methods to rock outcrops, geologists in England and Western Europe defined the major geologic time units without the benefit of radiometric dating techniques. Using the principles of superposition and fossil succession, they correlated various rock exposures and pieced together a composite geologic section. This composite section is, in effect, a relative time scale because the rocks are arranged in their correct sequential order.

By the beginning of the 20th century, geologists had developed a relative geologic time scale but did not yet have any absolute dates for the various time-unit boundaries. Following the discovery of radioactivity near the end of the 19th century, radiometric dates were added to the relative geologic time scale (▶ Figure 9.1).

Because sedimentary rocks, with rare exceptions, cannot be radiometrically dated, geologists have had to rely on interbedded volcanic rocks and igneous intrusions to apply absolute dates to the boundaries of the various subdivisions of the geologic time scale (▶ Figure 9.26). An ashfall or lava flow provides an excellent marker bed that is a time-equivalent surface, supplying a minimum age for the sedimentary rocks below and a maximum age for the rocks above. Ashfalls are particularly useful because they may fall over both marine and nonmarine sedimentary environments and can provide a connection between these different environments.

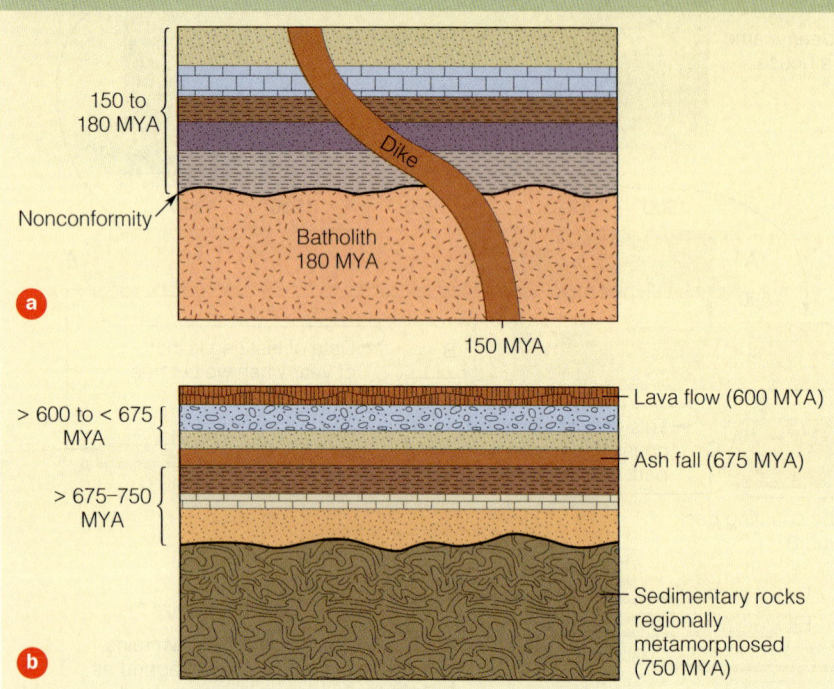

▶ **Figure 9.26 Determining Absolute Dates for Sedimentary Rocks** The absolute ages of sedimentary rocks can be determined by dating associated igneous rocks. In **(a)** and **(b)**, sedimentary rocks are bracketed by rock bodies for which absolute ages have been determined.

Thousands of absolute ages are now known for sedimentary rocks of known relative ages, and these absolute dates have been added to the relative time scale. In this way, geologists have been able to determine the absolute ages of the various geologic periods and their duration (▶ Figure 9.1). In fact, the dates for the era, period, and epoch boundaries of the geologic time scale are still being refined as more accurate dating methods are developed and new exposures dated. The ages shown in Figures 1.16 and 9.1 are the most recently published ages as of 2004.

Section 9.7 Summary

- The geologic time scale is a hierarchical scale in which Earth's 4.6-billion-year history is divided into time units of varying duration. It has evolved from being a relative time scale to one in which absolute age dates could be assigned for the various time-unit boundaries. The dates for the various time-unit boundaries are still being refined as more accurate dating methods are developed and new exposures dated.

9.8 Geologic Time and Climate Change

■ *How are climate change and geologic time linked?*

Given the debate concerning global warming and its possible implications, it is extremely important to be able to reconstruct past climatic regimes as accurately as possible. To model how Earth's climate system has responded to changes in the past and use that information for simulations of future climate scenarios, geologists must have a geologic calendar that is as precise and accurate as possible.

New dating techniques with greater precision are providing geologists with more accurate dates for when and how long past climate changes occurred. The ability to accurately determine when past climate changes took place helps geologists correlate these changes with regional and global geologic events to see whether there are any possible connections.

One interesting method that is becoming more common in reconstructing past climates is to analyze stalagmites from caves. Stalagmites are icicle-shaped structures rising from a cave floor and formed of calcium carbonate precipitated from evaporating water (▶ Figure 9.27). A stalagmite therefore records a layered history because each newly precipitated layer of calcium carbonate is younger than the previously precipitated layer. Thus, a stalagmite's layers are oldest in the center at its base and progressively younger as they move outward (principle of superposition). Using techniques based on ratios of uranium 234 to thorium 230, geologists can achieve very precise radiometric dates on individual layers of a stalagmite. This technique enables geologists to determine the age of materials much older than they can date by the carbon-14 method, and it is reliable back to about 500,000 years.

A study of stalagmites from Crevice Cave in Missouri revealed a history of climatic and vegetation change in the midcontinent region of the United States during the interval between 75,000 and 25,000 years ago. Dates obtained from the Crevice Cave stalagmites were correlated with major changes in vegetation and average temperature fluctuations, obtained from carbon 13 and oxygen 18 isotope profiles, to reconstruct a detailed picture of climate changes during this time period.

What Would You Do?

You have been chosen to be part of the first astronaut crew to land on Mars. You were selected because you are a geologist, and therefore your primary responsibility is to map the geology of the landing site area. An important goal of the mission is to work out the geologic history of the area. How will you go about this? Will you be able to use the principles of relative dating? How will you correlate the various rock units? Will you be able to determine absolute ages? How will you do this?

Geo-focus Figure 9.27 Stalagmites and Climate Change (a) Stalagmites are icicle-shaped structures rising from the floor of a cave, and formed by the precipitation of calcium carbonate from evaporating water. A stalagmite is thus layered, with the oldest layer in the center and the youngest layers on the outside. Uranium 234 frequently substitutes for the calcium ion in the calcium carbonate of the stalagmite. Uranium 234 decays to thorium 230 at a predictable and measurable rate. Therefore, the age of each layer of the stalagmite can be dated by measuring its ratio of U^{234} to Th^{230}. **(b)** There are two isotopes of oxygen, a light one O^{16} and a heavy one O^{18}. Because the oxygen 16 isotope is lighter than the oxygen 18 isotope, it vaporizes more readily then the oxygen 18 isotope when water evaporates. Therefore, as the climate becomes warmer, evaporation increases, and the O^{18}/O^{16} ratio becomes higher in the remaining water. Water in the form of rain or snow percolates into the ground, and becomes trapped in the pores between the calcite forming the stalagmites.

(continues)

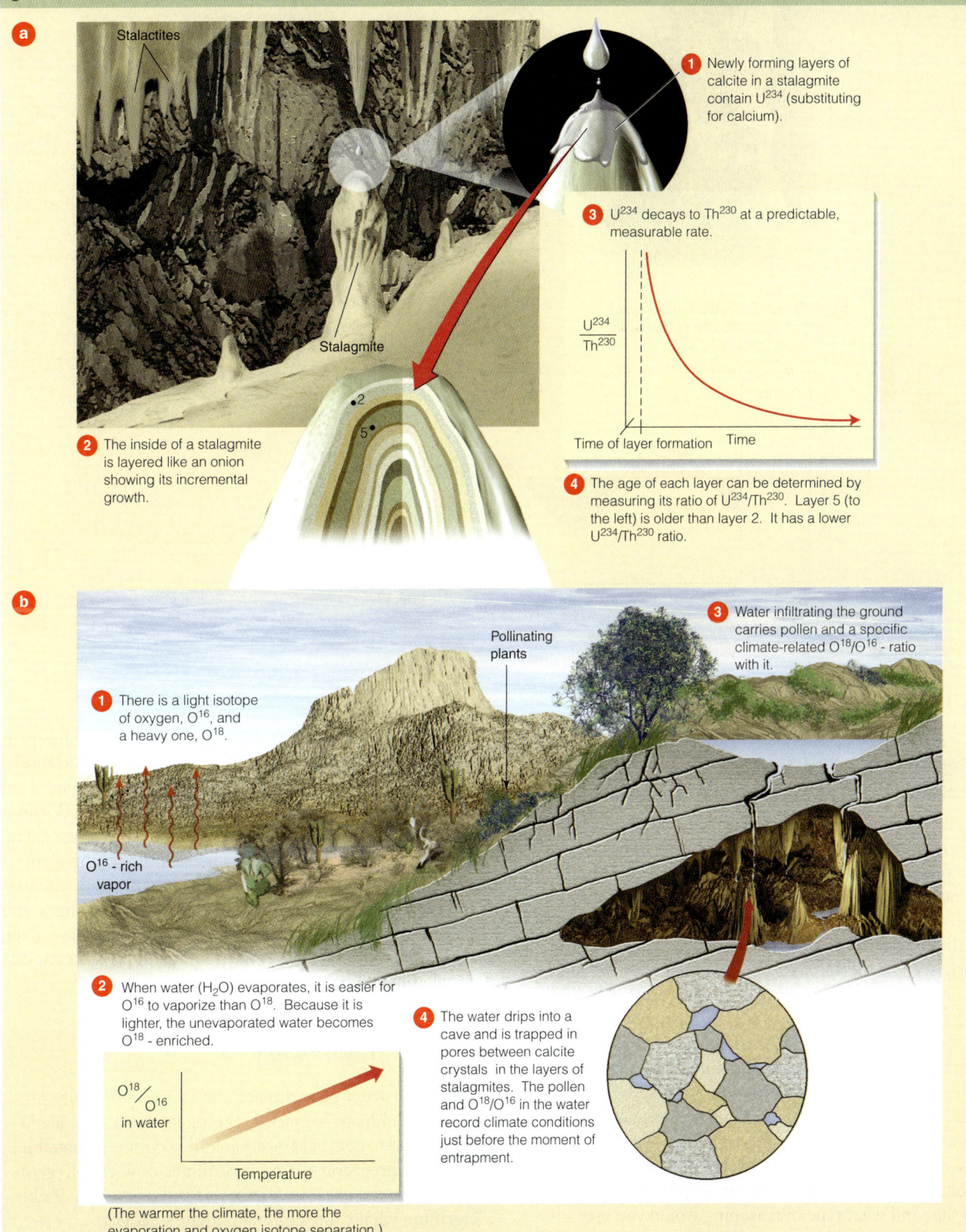

(continued)

▶ **Figure 9.27 Stalagmites and Climate Change (continued) (c)** The layers of a stalagmite can be dated by measuring the U^{234}/Th^{230} ratio, and the O^{18}/O^{16} ratio determined for the pore water trapped in each layer. Thus, a detailed record of climatic change for the area can be determined by correlating the climate of the area as determined by the O^{18}/O^{16} ratio to the time period determined by the U^{234}/Th^{230} ratio.

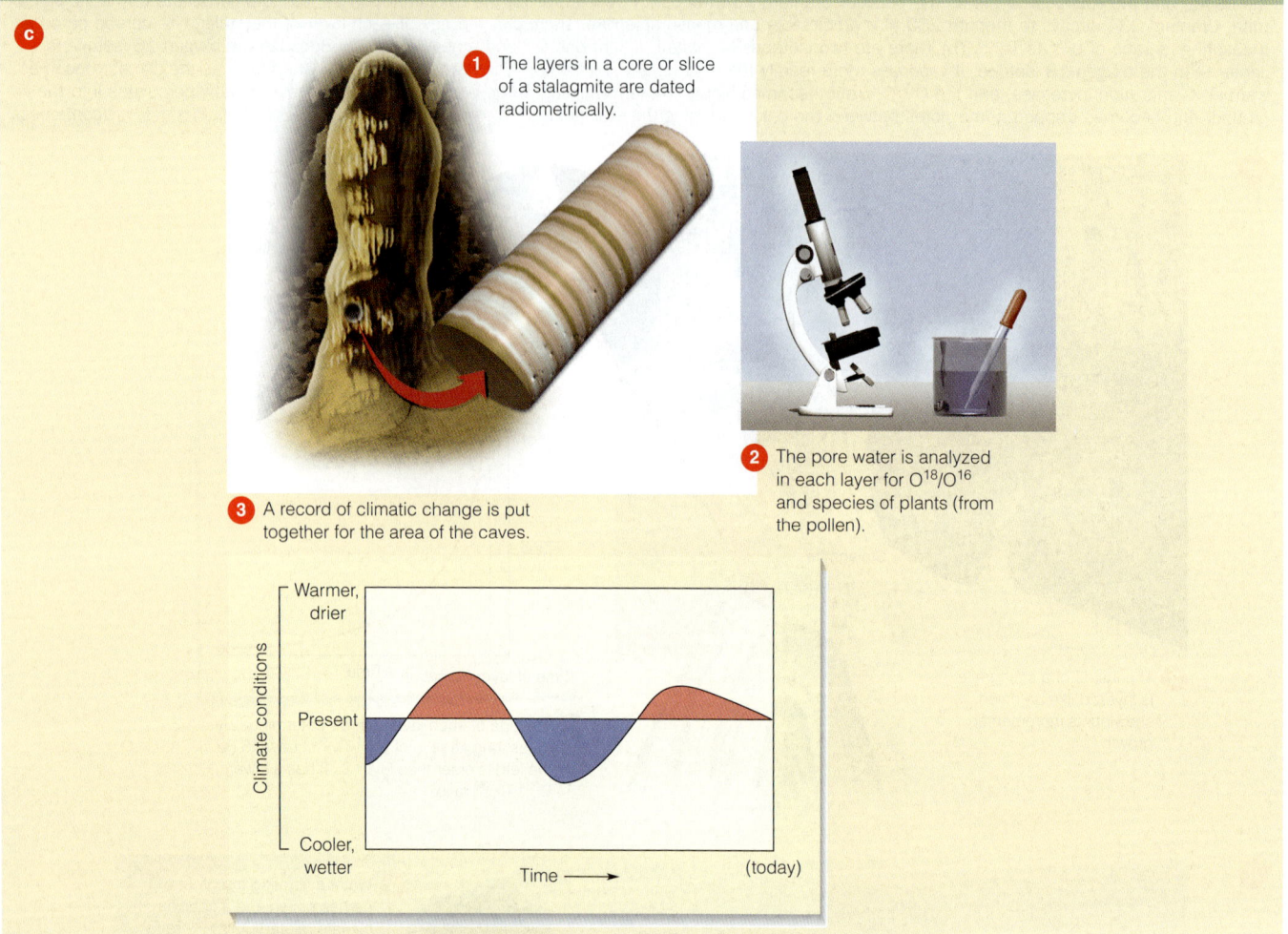

Thus precise dating techniques in stalagmite studies provide an accurate chronology that allows geologists to model climate systems of the past and perhaps to determine what causes global climate changes and their duration. Without these sophisticated dating techniques and others like them, geologists would not be able to make precise correlations and accurately reconstruct past environments and climates. By analyzing past environmental and climate changes and their duration, geologists hope they can use these data, sometime in the near future, to predict and possibly modify regional climate changes.

What Would You Do?

You are a member of a regional planning commission that is considering a plan for constructing what is said to be a much-needed river dam that will create a recreational lake. Opponents of the dam project have come to you with a geologic report and map showing that a fault underlies the area of the proposed dam, and the fault trace can be clearly seen at the surface. The opponents say the fault may be active and thus some day it will move suddenly, bursting the dam and sending a wall of water downstream. You seek the advice of a local geologist who has worked in the dam area. She tells you that she found a lava flow covering the fault less than a mile from the proposed dam project site. Can you use this information along with a radiometric date from the lava flow to help convince the opponents that the fault has not moved in any direction (vertically or laterally) anytime in the recent past? How would you do this, and what type of reasoning would you use?

Section 9.8 Summary

• To reconstruct past climate changes and possibly predict future climate scenarios, geologists must have a geologic calendar that is as precise and accurate as possible. New high-precision dating techniques now enable geologists to model climate systems of the past and possibly determine what causes global climate changes.

Review Workbook

ESSENTIAL QUESTIONS SUMMARY

9.1 Introduction

■ *How is geologic time measured?*

Time is defined by the methods used to measure it. Relative dating places geologic events in sequential order as determined from their positions in the geologic record. Absolute dating provides specific dates for geologic rock units or events expressed in years before the present. Radiometric dating is the most common method used in obtaining absolute dates.

■ *Why is the study of geologic time important?*

The logic used in applying the principles of relative dating to interpret the geologic history of an area involves basic reasoning skills that are useful in almost any profession or discipline. Furthermore, an accurate and precise geologic calendar is critical in determining the onset, duration, and possible causes of such past events as global climate changes and their potential effects on humans.

9.2 Early Concepts of Geologic Time and the Age of Earth

■ *How has our concept of geologic time and Earth's age changed throughout human history?*

The concept of geologic time, the methods used to measure it, and calculations of Earth's age have changed throughout human history. Most early attempts, though ingenious, were flawed and yielded very young ages for Earth. It was not until the discovery of radioactivity in the late 19th century that an accurate age for Earth and past geologic events could be determined.

9.3 James Hutton and the Recognition of Geologic Time

■ *Who is James Hutton and why is he important?*

James Hutton is considered to be the father of modern geology. His observations of present-day processes and his detailed studies of the geology in his native Scotland were instrumental in establishing the principle of uniformitarianism and the fact that Earth was much older than earlier scientists had thought.

■ *How did Lord Kelvin almost overturn the uniformitarian foundation of geology?*

By knowing the size of Earth, the melting temperature of rocks, and the present-day rate of heat loss from Earth's interior, the English physicist Lord Kelvin calculated the age at which Earth was entirely molten. By assuming a molten origin for Earth, Kelvin calculated that Earth was between 400 million and 20 million years old. These ages were much younger than geologists had thought, but based on the prevailing hypothesis that Earth had a molten origin, Kelvin's mathematics was irrefutable. With the discovery of radioactivity in the late 19th century, Kelvin's basic premise that Earth has been cooling since its origin was shown to be false, thus invalidating his conclusions.

9.4 Relative Dating Methods

■ *What is relative dating and why are the principles of relative dating important?*

Relative dating places events in sequential order but does not tell us how long ago an event took place. The principles of relative dating provide geologists with a means to interpret geologic history and develop a relative geologic time scale.

■ *What are the six fundamental principles of relative dating?*

The six fundamental principles of relative dating and their definition are as follows. Superposition: In an undisturbed sequence of sedimentary rocks, the oldest rocks are on the bottom and the youngest rocks are on the top. Original horizontality: Sediments are originally deposited horizontally under the influence of gravity. Lateral continuity: A layer of sediment extends laterally in all directions until it thins and pinches out or terminates against the edge of the depositional basin. Cross-cutting relationships: An igneous intrusion or a fault must be younger than the rock it intrudes into or displaces. Inclusions: Inclusions, or the fragments of one rock contained within a layer of another, are older than the rock layer itself. Fossil succession: Fossil assemblages succeed one another through time in a regular and predictable order.

■ *What are unconformities?*

An unconformity is a surface of erosion, nondeposition, or both separating younger strata from older strata. These surfaces encompass long periods of geologic time for which there is no geologic record at that location.

■ *What are the three specific types of unconformities?*

A disconformity separates younger from older sedimentary strata that are parallel to each other. An angular unconformity is an erosional surface on tilted or folded rocks, over which younger rocks were deposited. A nonconformity is an erosional surface cut into igneous or metamorphic rocks and overlain by younger sedimentary rocks.

■ *How are the principles of relative dating applied to interpret the geologic history of an area?*

The principles of relative dating can be used to reconstruct the geologic history of an area. Although no specific dates can be applied, the relative sequence of events can be determined by using the principles of relative dating.

9.5 Correlating Rock Units

■ *What is correlation?*

Correlation is the demonstration of time equivalency of rock units in different areas. Similarity of rock type, position within a rock sequence, key beds, and fossil assemblages can all be used to correlate rock units.

■ *How are subsurface units correlated?*

In addition to the methods mentioned above, such rock properties as electrical resistivity, radioactivity, and seismic profiles can be used to correlate subsurface units.

9.6 Absolute Dating Methods

■ *Why was the discovery of radioactivity important to geology?*

The discovery that radioactivity produces heat meant that geologists had a mechanism to explain Earth's internal heat that did not rely on residual cooling from a molten origin. Furthermore, geologists now had a tool to accurately date geologic events, and show that Earth was indeed very old.

■ *What are atoms, elements, and isotopes?*

Atoms are the smallest units of matter that retain the characteristics of an element. An element is a substance composed of atoms that all have the same properties. Isotopes of an element behave the same chemically but have different atomic mass numbers. Some isotopes are radioactive and are useful for radiometric dating.

■ *What is radioactive decay?*
Radioactive decay is the process in which an unstable atomic nucleus is spontaneously transformed into an atomic nucleus of a different element.

■ *What are half-lives?*
The half-life of a radioactive element is the time it takes for one-half of the original unstable parent element to decay into a new, more stable daughter element. The half-lives of various radioactive elements range from less than a billionth of a second to 49 billion years. The half-life of a radioactive element is constant and can be precisely measured.

■ *What are some of the sources of uncertainty in radiometric dating?*
The most accurate dates are obtained from igneous rocks because during the cooling of magma, radioactive parent atoms are separated from previously formed daughter atoms and incorporated into the crystal structure of a mineral. In radiometric dating, it is important that no parent or daughter atoms have been added or removed from the sample being tested. Furthermore, the sample must be fresh and unweathered and it must not have been subjected to high temperatures or intense pressures after crystallization. Although heat and pressure do not affect the rate of radioactive decay, they can cause the migration of parent and daughter atoms after crystallization, thus affecting the calculated age.

■ *What are the common long-lived radioactive isotope pairs?*
The five common long-lived radioactive isotope pairs are the uranium–lead and thorium–lead series, which are used primarily to date ancient igneous intrusions, lunar samples, and some meteorites; the rubidium–strontium pair, which is typically used on very old rocks, including the oldest known rocks on Earth as well as some meteorites; and potassium–argon, which is typically used to date fine-grained volcanic rocks from which individual crystals cannot be separated.

■ *What is fission-track dating?*
Fission-track dating involves counting the number of microscopic linear tracks revealed after a mineral is etched in hydrofluoric acid. These tracks result from crystal damage caused by the spontaneous decay of uranium within a mineral. The age of a sample is determined by the amount of uranium the sample contains and the number of tracks present; the more tracks, the older the sample.

■ *What is the carbon-14 dating technique?*
Carbon, one of the basic elements found in all forms of life, has three isotopes: carbon 12, 13, and 14. Carbon 12 and 13 are stable, whereas carbon 14 is radioactive with a half-life of 5730 years plus or minus 30 years. Carbon 14 is produced in the upper atmosphere as a result of the bombardment of cosmic rays and is rapidly assimilated into the carbon cycle, where, along with carbon 12 and 13, it is absorbed in a nearly constant ratio by all living organisms. Carbon 14 decays back to nitrogen. Thus, as long as an organism is alive, the ratio of carbon 12 to carbon 14 remains constant, with new carbon 14 replacing that lost to radioactive decay. When an organism dies, however, carbon 14 is not replenished, and the ratio of carbon 14 to carbon 12 decreases. Therefore, the older a fossil is, the lower the ratio of carbon 14 to carbon 12.

■ *What is tree-ring dating?*
Tree-ring dating involves counting the annular rings in a tree's stem to determine how old the tree is. The applicability of tree-ring dating is somewhat limited because it can be used for dating only geologically recent events where continuous tree records are found. It is most useful in arid regions.

9.7 Development of the Geologic Time Scale
■ *What is the geologic time scale?*
The geologic time scale is a hierarchical scale in which the Earth's 4.6-billion-year history is divided into time units of varying duration.

■ *How did the geologic time scale develop?*
The geologic time scale was developed primarily during the 19th century through the efforts of many people. It was originally a relative time scale, but with the discovery of radioactivity and the development of radiometric dating methods, absolute age dates were added at the beginning of the 20th century. Since then, refinement of the time-unit boundary dates has continued.

9.8 Geologic Time and Climate Change
■ *How are climate change and geologic time linked?*
To reconstruct past climate changes and link them to possible causes, geologists must have a geologic calendar that is as precise and accurate as possible. Thus they must be able to date geologic events and the onset and duration of climate changes as precisely as possible, so it is critical to develop new dating methods and refine the dates of the geologic time scale.

ESSENTIAL TERMS TO KNOW

absolute dating (p. 262)
angular unconformity (p. 267)
carbon-14 dating technique (p. 285)
correlation (p. 278)
disconformity (p. 267)
fission-track dating (p. 285)
guide fossil (p. 278)

half-life (p. 281)
nonconformity (p. 267)
principle of cross-cutting relationships (p. 265)
principle of fossil succession (p. 266)
principle of inclusions (p. 265)
principle of lateral continuity (p. 265)

principle of original horizontality (p. 265)
principle of superposition (p. 265)
radioactive decay (p. 281)
relative dating (p. 262)
tree-ring dating (p. 286)
unconformity (p. 266)

REVIEW QUESTIONS

1. What is the difference between relative dating and absolute dating?

2. In addition to the methods mentioned in this chapter, can you think of some other nonradiometric dating methods that could be used to determine the age of Earth? What would be the drawbacks to the methods you propose?

3. Why were Lord Kelvin's arguments and calculations so compelling, and what was the basic flaw in his assumption? What do you think the course of geology would have been if radioactivity had not been discovered?

4. Can the various principles of relative dating be used to reconstruct the geologic history of Mars? Which principles might not apply to interpreting the geologic history of another planet?

5. When geologists reconstruct the geologic history of an area, why is it important for them to differentiate between a sill and a lava flow? How could you tell the difference between a sill and a lava flow at an outcrop if both structures consisted of basalt? What features would you look for in an outcrop to positively identify the structure as either a sill or a lava flow?

6. In some places where disconformities are particularly difficult to discern from a physical point of view, how could you use the principle of fossil succession to recognize a disconformity? How could the principle of inclusions be used to recognize a nonconformity?

7. Why do igneous rocks yield the most accurate radiometric dates? Why can't sedimentary rocks be dated radiometrically? What problems are encountered in dating metamorphic rocks?

8. An igneous rock was radiometrically dated using the uranium 235 to lead 207 and potassium 40 to argon 40 isotope pairs. The isotope pairs yielded distinctly different ages. What possible explanation could be offered for this result? How can one rock have two different ages? What would you do to rectify the discrepancy in ages?

9. What is the major difference between the carbon-14 dating technique and the techniques used for the five common, long-lived radioactive isotope pairs?

10. If you wanted to calculate the absolute age of an intrusive body, what information would you need?

APPLY YOUR KNOWLEDGE

1. If a radioactive element has a half-life of 16 million years, what fraction of the original amount of parent material will remain after 96 million years? What percentage will that be?

2. How many half-lives are required to yield a mineral with 625,000,000 atoms of thorium^{232} and 19,375,000,000 atoms of lead^{208}?

3. Given the current debate over global warming and the many short-term consequences for humans, can you visualize how the world might look in 100,000 years or even 10 million years? Use what you've learned about plate tectonics and the direction and rate of movement of plates, as well as how plate movement and global warming will affect ocean currents, weather patterns, weathering rates, and other factors, to make your prediction. Do you think such short-term changes can be extrapolated to long-term trends in trying to predict what Earth will be like using a geologic time perspective?

FIELD QUESTIONS

1. From the following diagram, provide a geologic history of the area. Name the two types of unconformities (indicated by wavy lines), and give as many absolute ages as possible for the sedimentary rocks. The batholith has been dated as 400 million years, the dike as 100 million years, and the sill as 25 million years. All of the other units shown are sedimentary rocks. (An answer appears at the end of the book.)

2. What type of unconformity is shown here? Provide a geologic history of this area based on what is shown. (An answer appears at the end of the book.)

3. What principle(s) of relative dating are shown here? Provide a geologic history of this area based on what is shown. (An answer appears at the end of the book.)

GEOLOGY IN FOCUS

Denver's Weather—280 Million Years Ago!

With all the concern about global climate change, it might be worthwhile to step back a bit and look at climate change from a geologic perspective! We're all aware that some years are hotter than others and some years we have more rain, but generally things tend to average out over time. We know it will be hot in the summer in Arizona and it will be very cold in Minnesota in the winter. We also know that scientists, politicians, and concerned people everywhere are debating whether humans are partly responsible for the global warming Earth seems to be experiencing.

What about long-term climate change? We know that Earth has experienced periods of glaciation in the past—for instance, during the Precambrian, the end of the Ordovician Period, and most recently the Pleistocene Epoch. Earth has also undergone widespread periods of aridity, such as during the end of the Permian and the beginning of the Triassic periods. Such long-term climatic changes are probably the result of slow geographic changes related to plate tectonic activity. Not only are continents carried into higher and lower latitudes, but their movement affects ocean circulation and atmospheric circulation patterns, which in turn affect climate and result in climate changes.

Even though we can't physically travel back in time, geologists can reconstruct what the climate was like in the past. The distribution of plants and animals is controlled, in part, by climate. Plants are particularly sensitive to climate change, and many can live in only particular environments. The fossils of plants and animals can tell us something about the environment and climate at the time these organisms were living. Furthermore, climate-sensitive sedimentary rocks can be used to interpret past climatic conditions. Desert dunes are typically well sorted and exhibit large-scale cross-bedding. Coals form in freshwater swamps where climatic conditions promote abundant plant growth. Evaporites such as rock salt result when evaporation exceeds precipitation, such as in desert regions or along hot, dry shorelines. Tillites (glacial sediments) result from glacial activity and indicate cold, wet environments. So by combining all relevant geologic and paleontologic information, geologists can reconstruct what the climate was like in the past and how it has changed over time at a given locality.

In his recently published book, *Ancient Denvers: scenes from the past 300 million years of the Colorado Front Range*, Kirk R. Johnson depicts what Denver, Colorado, looked like at 13 different time periods in the past. The time slices begin during the Pennsylvanian Period, 300 million years ago, and end with a view of the Front Range amid a spreading wave of houses on the southern edge of metropolitan Denver.

The information for piecing together Denver's geologic past was derived mainly from a 688-m-deep well drilled by the Denver Museum of Nature and Science beneath Kiowa, Colorado, in 1999. Using the information gleaned from the rocks recovered from the well, plus additional geologic evidence from other parts of the area, the museum scientists and artists were able to reconstruct Denver's geologic past.

Beginning 300 million years ago (Pennsylvanian Period), the Denver area had coastlines on its eastern and western borders and a mountain range (not the Rocky Mountains of today). The climate was mostly temperate with lots of seedless vascular plants, like ferns, as well as very tall trees related to the modern horsetail rush. Huge insects such as millipedes, cockroaches, and dragonflies shared this region with relatively small fin-backed reptiles and a variety of amphibians.

By 280 million years ago, the area was covered by huge sand seas, much like the Sahara today (▶ Figure 1). This change in climate and landscape was the result of the formation of Pangaea. As the continents collided, arid and semiarid conditions prevailed over much of the supercontinent, and the Denver area was no exception.

During the Jurassic (150 million years ago), herds of plant-eating dinosaurs like *Apatosaurus* roamed throughout the Denver area, feasting on the succulent and abundant vegetation. Grasses and flowering plants had not yet evolved, so the dinosaurs ate the ferns and gymnosperms that were abundant at this time.

As a result of rising sea level, Denver was covered by a warm, shallow sea 70 million years ago (late Cretaceous). Marine reptiles such as plesiosaurs and mosasaurs ruled these seas, while overhead, pterosaurs soared through the skies looking for food (▶ Figure 2).

Beginning around 66 million years ago, the Rocky Mountains began to form as tectonic forces started a mountain-building episode known as the *Laramide orogeny* that resulted in the present-day Rocky Mountains. Dinosaurs still roamed the land around Denver, and flowering plants began their evolutionary history.

By 55 million years ago (Eocene), the world was in the grip of an intense phase of global warming. A subtropical rainforest with many trees that would be recognizable today filled the landscape. Primitive mammals were becoming more abundant, and many warm-climate-loving animals could be found living north of the Arctic Circle.

Although ice caps still covered portions of North America, mammoths and other mammals wandered among the plains of Denver 16,000 years ago (▶ Figure 3). Mastodons, horses, bison, lions, and giant ground sloths, to name a few, all lived in this region, and their fossils can be found in the sedimentary rocks from this area.

What was once a rainforest, desert, warm shallow sea, and mountainous region is now home to thousands of people. What the Denver region will be like in the next several million years is anyone's guess. Whereas humans can effect change, what change we will cause is open to debate. Certainly the same forces that have shaped the Denver area in the past will continue to determine its future. With the rise of humans and technology, we, as a species, will also influence what future Denvers will be like. Let us hope the choices we make are good ones.

Figure 1
Denver as it appeared 280 million years ago. As a result of the collision of continents and the formation of Pangaea, the world's climate was generally arid, and Denver was no exception. Denver was probably covered by great seas of sand, much as the Sahara is today.

Figure 2
Pterosaurs (flying reptiles) soar over Denver 70 million years ago. At this time, Denver was below a warm, shallow sea that covered much of western North America. Marine reptiles such as plesiosaurs and mosasaurs swam in these seas in search of schools of fish.

Figure 3
Mammoths and camels wander on the prairie during a summer day 16,000 years ago. Whereas much of northern North America was covered by an ice sheet, Denver had pine trees and prairie grass. This area was home to a large variety of mammals, including mammoths, camels, horses, bison, and giant ground sloths. Humans settled in this area about 11,000 years ago, hunting the plentiful game that was available.

GEOLOGY MATTERS

GEOLOGY IN UNEXPECTED PLACES

Time Marches On—The Great Wall of China

The Great Wall of China, built over many centuries as a military fortification against invasion by enemies from the north, has largely succumbed to the ravages of nature and human activity. Originally begun as a series of short walls during the Zhou Dynasty (770–476 B.C.), the wall grew as successive dynasties connected different parts, with final improvements made during the Ming Dynasty (A.D. 1368–1644). Even though the Great Wall is not continuous, it stretches more than 5000 km across northern China from the east coast to the central part of the country (▶ Figure 1).

Contrary to popular belief, the Great Wall of China is not the only human-made structure visible from space. From low Earth orbit, many artificial objects are visible, such as highways, cities, and railroads. In the view from a distance of a few thousand kilometers, no human-made objects are visible and the Great Wall can barely be seen with the naked eye from the shuttle according to NASA. In fact, China's first astronaut, Yang Liwei, told state television on his return from space, "I did not see the Great Wall from space" (▶ Figure 2).

So with that short history and debunking of an urban legend, what is the Great Wall made of? Basically, the Great Wall was constructed with whatever material was available in the area. This included sedimentary, metamorphic, and igneous rocks, bricks, sand, gravel, and even dirt and straw. Regardless of the material, the wall was built by hand by thousands of Chinese over many centuries.

In the Badaling area of Beijing, which is the part of the Great Wall most tourists visit and has been restored, the wall was built using the igneous and sedimentary rocks from the mountains around Badaling. The sedimentary rocks are mudstones, sandstones, and limestones, whereas the igneous rocks are granite. The sides of the walls in this area are constructed of rectangular slabs of granite, and the top or roof of the Great Wall here is paved with large gray bricks (▶ Figure 3).

The average height of the Great Wall is 8.5 m, and it is 6.5 m wide along its base. The wall along the top averages 5.7 m and is wide enough for 5 horses or 10 warriors to walk side by side.

Most visitors to the Great Wall are so impressed by its size and history that they don't even notice what it is made from. Now, however, you know, and should you visit this impressive structure you can tell your fellow travelers all about it.

Figure 1
The Great Wall of China winding across the top of the hills at Badaling, just outside Beijing.

Figure 2
China's first astronaut, Yang Liwei, waves as his space capsule door was opened after returning from space. Yang Liwei landed on the Inner Mongolian grassland of northern China on October 16, 2003, after spending 21 hours in orbit. He reported that he never did see the Great Wall of China from space.

Figure 3
The top of the Great Wall at Badaling. Note the original rocks in the lower portion of the side of the wall and the paving bricks in the foreground comprising the top of the wall.

Geology AT WORK

Geology in Hollywood
John R. Horner

John "Jack" Horner attended the University of Montana, where he majored in geology and zoology. In 1975 he was hired as a research assistant in the Museum of Natural History at Princeton University, where he worked until 1982. From 1982 until the present he has been Curator of Paleontology at the Museum of the Rockies in Bozeman, Montana. Jack's research teams discovered the first dinosaur egg clutches in the Western Hemisphere and six new species of dinosaurs.

Not very many scientists are likely to get a phone call from Steven Spielberg, but that's what happened to me back in the early 1990s. Steven called to ask if I would consider being a technical advisor for a movie he was working on, based on the book by Michael Crichton called *Jurassic Park*. I told Mr. Spielberg that I'd be happy to work with him on his movie, so I went to Universal Studios in Los Angeles and began working on one of the coolest movies ever made. My job was to make sure that the dinosaurs looked as accurate as possible, based on the available scientific understanding, and to help the actors like Sam Neil understand paleontology. One of my first duties was to help Stan Winston, the person whose studio made the animatronics, with the *T. rex* and velociraptors. All the animatronics in the Jurassic Park movies are life-size. They are first sculpted out of clay, then scanned in 3-D, and finally built as life-size puppet robots. When the animatronics were built and the sets finished, I was called back on several occasions to assist Steven while he shot the movie. I was there to answer questions and make sure that the dinosaurs didn't do something they weren't supposed to, or to be sure the actors didn't say something scientifically wrong. One of my favorite scenes was in the kitchen where the children were being pursued by the raptors. It's hard to imagine, but I and a couple hundred other people were in that kitchen with piles of electronics and hundreds of feet of wire and cable. Someone did a great job of editing the film and removing all of our reflections from those shiny appliances. It gave me a great appreciation for how a camera captures what the director wants us to see.

On the day they were getting ready to shoot the sequence where the raptors first come into the kitchen, Steven said that the raptors would come in with their forked tongues waving in the air like a lizard, tasting the air for their prey. I jumped in and explained that we know dinosaurs didn't have forked tongues, and that the forked tongues would suggest that dinosaurs were cold-blooded rather than warm-blooded. Steven changed that scene so that just before they come into the kitchen, one dinosaur snorts at the window and fogs it up, revealing their warm-bloodedness. It was a small change, but one that made a big difference in how we perceive the dinosaurs.

I suggested other changes to Steven as we continued to shoot the movie. Some of the suggestions he took, and others he didn't (▶ Figure 1). It often depended on whether or not the suggestion was based on scientific data or was just my guess. If it was an accepted scientific idea, Steven was inclined to listen, but if it was an idea still in the process of being researched, he'd go with whatever he thought would be more exciting for the movie. Personally, I think that is what makes Steven a great director: He wasn't making a documentary, but rather a fictional movie. He wasn't tied to facts like documentaries, but instead he could exaggerate for the sake of a good movie. For example, the velociraptors in the movie are larger than any known to have lived, but there are larger dinosaurs very closely related to velociraptor, so it's not much of an exaggeration to make them big in the movie.

One of the funny things that happened was that Steven had actually based the Alan Grant character on me, but I didn't discover this until the movie was finished. Movies are made of lots of parts, and only the director knows what it will look like in the end. It's kind of like finding a dinosaur skeleton and having to wait until it's all excavated, prepared, and mounted before everyone else knows what you found.

In the second movie, *The Lost World*, I didn't have much to do because most of the dinosaurs were the same as the ones we'd used in *Jurassic Park*. I did help with the *T. rex* nest, a lot of the footprints, and even some of the *T. rex* walking scenes.

In *Jurassic Park III* I did a great deal of work because there wasn't a book. Steven, the executive producer, Joe Johnston, the director, a group of writers, and I all got together and created the story line. I suggested bringing in the *Spinosaurus* to kill the *T. rex*, and I helped with virtually all aspects of the movie except the finishing touches. I even got to help in the editing trailer and worked very closely with the computer graphics people at Industrial Light and Magic.

For me it has been a great experience to work on these movies, but I wouldn't trade my job for any of theirs—not the director, the movie stars, or anyone else. Being a scientist and discovering new dinosaur skeletons and information that helps us unravel the geological history of our planet is the best job a person could ever have.

Figure 1
Jack Horner (second from right) on the *Jurassic Park* set with Steven Spielberg (right) and crew.

Chapter 10

Earthquakes

ESSENTIAL QUESTIONS TO ASK

10.1 Introduction
- What is an earthquake?
- Why should we study earthquakes?

10.2 Elastic Rebound Theory
- What is the elastic rebound theory, and what does it explain about earthquakes?

10.3 Seismology
- What is seismology?
- What are the focus and epicenter of an earthquake?

10.4 The Frequency and Distribution of Earthquakes
- Where do most earthquakes occur?
- How many earthquakes occur per year?

10.5 Seismic Waves
- What are seismic waves?
- What are the two types of body waves?
- What are the two major types of surface waves?

10.6 Locating an Earthquake
- How is the location of an earthquake determined?

10.7 Measuring the Strength of an Earthquake
- How is an earthquake's strength measured?
- What is intensity and how is it determined?
- What factors determine an earthquake's intensity?
- What is the Richter Magnitude Scale, and what does it measure?
- How is an earthquake's magnitude determined?

10.8 The Destructive Effects of Earthquakes
- What factors determine an earthquake's destructiveness?
- What are the destructive effects of earthquakes?

10.9 Earthquake Prediction
- Can earthquakes be predicted?
- What are earthquake precursors?
- Which nations have earthquake prediction programs, and what do they involve?

10.10 Earthquake Control
- Can earthquakes be controlled?

Big Fault
As seen from above, the San Andreas Fault looks like the skeletal imprint of a spine in Earth where it crosses the Carrizo Plain in central California.
The fault, which is famous for generating large and destructive earthquakes, stretches approximately 1300 km across California. Of particular interest to seismologists is the area of the fault that runs through Parkfield, California, where earthquakes occur almost every 22 years. This is also the site of current scientific research by Earthscopes's San Andreas Fault Observatory at Depth (SAFOD), a deep borehole observatory designed to assess the physical conditions under which plate boundary earthquakes occur.
—A. W.

GEOLOGY MATTERS

GEOLOGY IN FOCUS:
Paleoseismology

GEOLOGY IN FOCUS:
Designing Earthquake-Resistant Structures

GEOLOGY IN UNEXPECTED PLACES:
It's Not My Fault

GEOLOGY AT WORK:
Pacific Tsunami Warning Center

 This icon, which appears throughout the book, indicates an opportunity to explore interactive tutorials, animations, or practice problems available on the Physical GeologyNow website at http://now.brookscole.com/phygeo6.

10.1 Introduction

At 3:02 A.M. on August 17, 1999, violent shaking from an earthquake awakened millions of people in Turkey. When the earthquake was over, an estimated 17,000 people were dead, at least 50,000 were injured, and tens of thousands of survivors were left homeless.

The amount of destruction this earthquake caused is staggering. More than 150,000 buildings were moderately to heavily damaged, and another 90,000 suffered slight damage. Collapsed buildings were everywhere, streets were strewn with rubble, and all communications were knocked out. And if this weren't enough, the same area was struck again only three months later on November 12, 1999, by an aftershock nearly as large as the original earthquake, which killed an additional 374 people and injured about 3000 more.

All in all, this was a disaster of epic proportions. Yet it was not the first, nor will it be the last major devastating earthquake in this region or other parts of the world (Table 10.1).

Earthquakes, along with volcanic eruptions, are manifestations of Earth's dynamic and active makeup. As one of nature's most frightening and destructive phenomena, earthquakes have always aroused feelings of fear and have been the subject of numerous myths and legends. What makes an earthquake so frightening is that when it begins, there is no way to tell how long it will last or how violent it will be. About 13 million people have died in earthquakes during the past 4000 years, with about 2.7 million of these deaths occurring during the past century alone (Table 10.1). This increase in fatalities shows that the rapid rise in numbers of humans living in hazardous conditions has trumped our improved understanding of how to build and live safely in earthquake-prone areas.

■ What is an earthquake?

Geologists define an **earthquake** as the shaking or trembling of the ground caused by the sudden release of energy, usually as a result of faulting, which involves displacement of rocks along fractures (we will discuss the different types of faults in Chapter 13). After an earthquake, continuing adjustments along a fault may generate a series of earthquakes known as *aftershocks*. Most aftershocks are smaller than the main shock, but they can still cause considerable damage to already weakened structures, as happened in the 2003 Iran earthquake.

Although the geologic definition of an earthquake is accurate, it is not nearly as imaginative or colorful as the explanations many people held in the past. Many cultures attributed the cause of earthquakes to movements of some kind of animal on which Earth rested. In Japan it was a giant catfish, in Mongolia a giant frog, in China an ox, in South America a whale, and to the Algonquin of North America an immense tortoise. And a legend from Mexico holds that earthquakes occur when the devil, El Diablo, rips open the crust so that he and his friends can reach the surface.

If earthquakes are not the result of animal movement or the devil ripping open the crust, what does cause earthquakes? Geologists know that most earthquakes result from energy released along plate boundaries, and as such, earthquakes are a manifestation of Earth's dynamic nature and the fact Earth is an internally active planet.

TABLE 10.1
Some Significant Earthquakes

Year	Location	Magnitude (estimated before 1935)	Deaths (estimated)
1556	China (Shanxi Province)	8.0	1,000,000
1755	Portugal (Lisbon)	8.6	70,000
1906	USA (San Francisco, California)	8.3	3000
1923	Japan (Tokyo)	8.3	143,000
1960	Chile	9.5	5700
1976	China (Tangshan)	8.0	242,000
1985	Mexico (Mexico City)	8.1	9500
1988	Armenia	6.9	25,000
1990	Iran	7.3	50,000
1993	India	6.4	30,000
1995	Japan (Kobe)	7.2	6000+
1998	Afghanistan	6.9	5000+
1999	Turkey	7.4	17,000
2001	India	7.9	14,000+
2003	Iran	6.6	43,000
2004	Indonesia	9.0	>220,000
2005	Pakistan	7.6	>86,000

■ *Why should we study earthquakes?*

The obvious reason to study earthquakes is that they are destructive and cause many deaths and injuries to the people living in earthquake-prone areas. Earthquakes also affect the economies of many countries in terms of cleanup costs, lost jobs, and lost business revenues. From a purely personal standpoint, you someday may be caught in an earthquake. Even if you don't plan to live in an area subject to earthquakes, you probably will sooner or later travel where there is the threat of earthquakes, and you should know what to do if you experience one. Such knowledge may help you avoid serious injury or even death!

Section 10.1 Summary

• An earthquake is the shaking of the ground caused by the sudden release of energy, usually as a result of faulting.

• Because earthquakes are very destructive and kill and injure thousands of people every year, it is important to know what to do before, during, and after an earthquake. Even though you might not presently live in an earthquake-prone area, you might move to one or travel to an area subject to earthquakes.

10.2 Elastic Rebound Theory

■ *What is the elastic rebound theory, and what does it explain about earthquakes?*

Based on studies conducted after the 1906 San Francisco earthquake, H. F. Reid of The Johns Hopkins University proposed the **elastic rebound theory** to explain how energy is released during earthquakes. Reid studied three sets of measurements taken across a portion of the San Andreas fault that had broken during the 1906 earthquake. The measurements revealed that points on opposite sides of the fault had moved 3.2 m during the 50-year period prior to breakage in 1906, with the west side moving northward (▶ Figure 10.1).

According to Reid, rocks on opposite sides of the San Andreas fault had been storing energy and bending slightly for at least 50 years before the 1906 earthquake. Any straight line such as a fence or road that crossed the San Andreas fault was gradually bent because rocks on one side of the fault moved relative to rocks on the other side (▶ Figure 10.1). Eventually, the internal strength of the rocks was exceeded, the rocks on opposite sides of the fault rebounded or "snapped back" to their former undeformed shape, and the energy stored in them was released as earthquake waves radiating out from the break.

▶ **Active Figure 10.1** The Elastic Rebound Theory

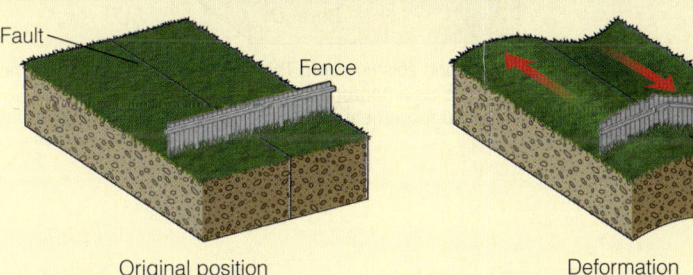
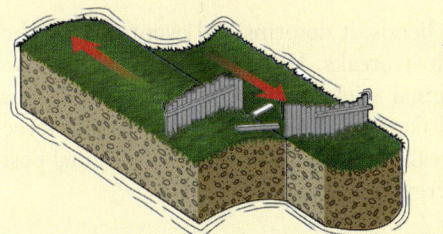
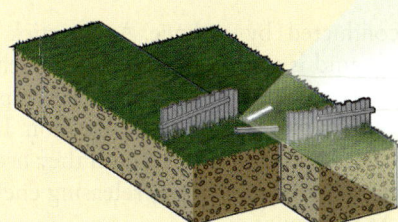

a According to the elastic rebound theory, rocks experiencing deformation store energy and bend. When the internal strength of the rocks is exceeded, they rupture, releasing their accumulated energy, and "snap back" or rebound to their former undeformed shape. This sudden release of energy is what causes an earthquake.

b During the 1906 San Francisco earthquake, this fence in Marin County was displaced by 2.5 m. Whereas many people would see a broken fence, a geologist sees that the fence has moved or been displaced and would look for evidence of a fault. A geologist would also notice that the ground has been displaced toward the right side, relative to his or her view. Regardless of which side of the fence you stand on, you must look to the right to see the other part of the fence. Try it!

▶ **Figure 10.2 Seismographs**

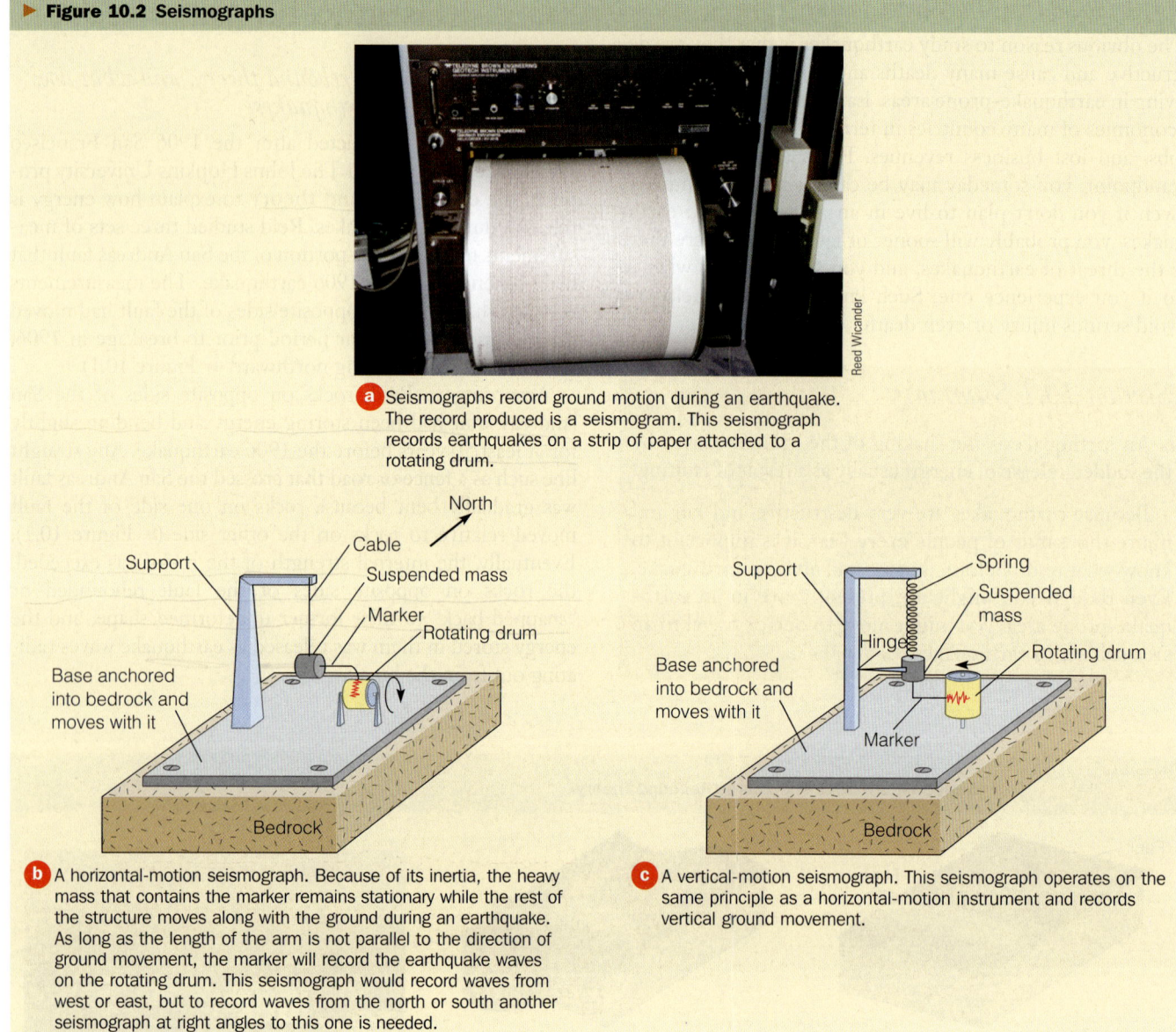

a Seismographs record ground motion during an earthquake. The record produced is a seismogram. This seismograph records earthquakes on a strip of paper attached to a rotating drum.

b A horizontal-motion seismograph. Because of its inertia, the heavy mass that contains the marker remains stationary while the rest of the structure moves along with the ground during an earthquake. As long as the length of the arm is not parallel to the direction of ground movement, the marker will record the earthquake waves on the rotating drum. This seismograph would record waves from west or east, but to record waves from the north or south another seismograph at right angles to this one is needed.

c A vertical-motion seismograph. This seismograph operates on the same principle as a horizontal-motion instrument and records vertical ground movement.

Additional field and laboratory studies conducted by Reid and others have confirmed that elastic rebound is the mechanism by which energy is released during earthquakes. In laboratory studies, rocks subjected to forces equivalent to those occurring in the crust initially change their shape. As more force is applied, however, they resist further deformation until their internal strength is exceeded. At that point, they break and snap back to their original undeformed shape, releasing internally stored energy.

The energy stored in rocks undergoing deformation is analogous to the energy stored in a tightly wound watch spring. The tighter the spring is wound, the more energy is stored, thus making more energy available for release. If the spring is wound so tightly that it breaks, then the stored energy is released as the spring rapidly unwinds and partially regains its original shape. Perhaps an even more meaningful analogy is simply bending a long, straight stick over your knee. As the stick bends, it deforms and eventually reaches the point at which it breaks. When this happens, the two pieces of the original stick snap back into their original straight position. Likewise, rocks subjected to intense forces bend until they break and then return to their original position, releasing energy in the process.

Section 10.2 Summary

- The elastic rebound theory is an explanation for how energy is released during an earthquake. As force is applied to rocks on opposite sides of a fault, energy accumulates and slowly deforms the rocks. When the internal strength of the rocks is exceeded, a sudden movement occurs along the fault, the stored energy is released, and the rocks snap back to their original undeformed shape.

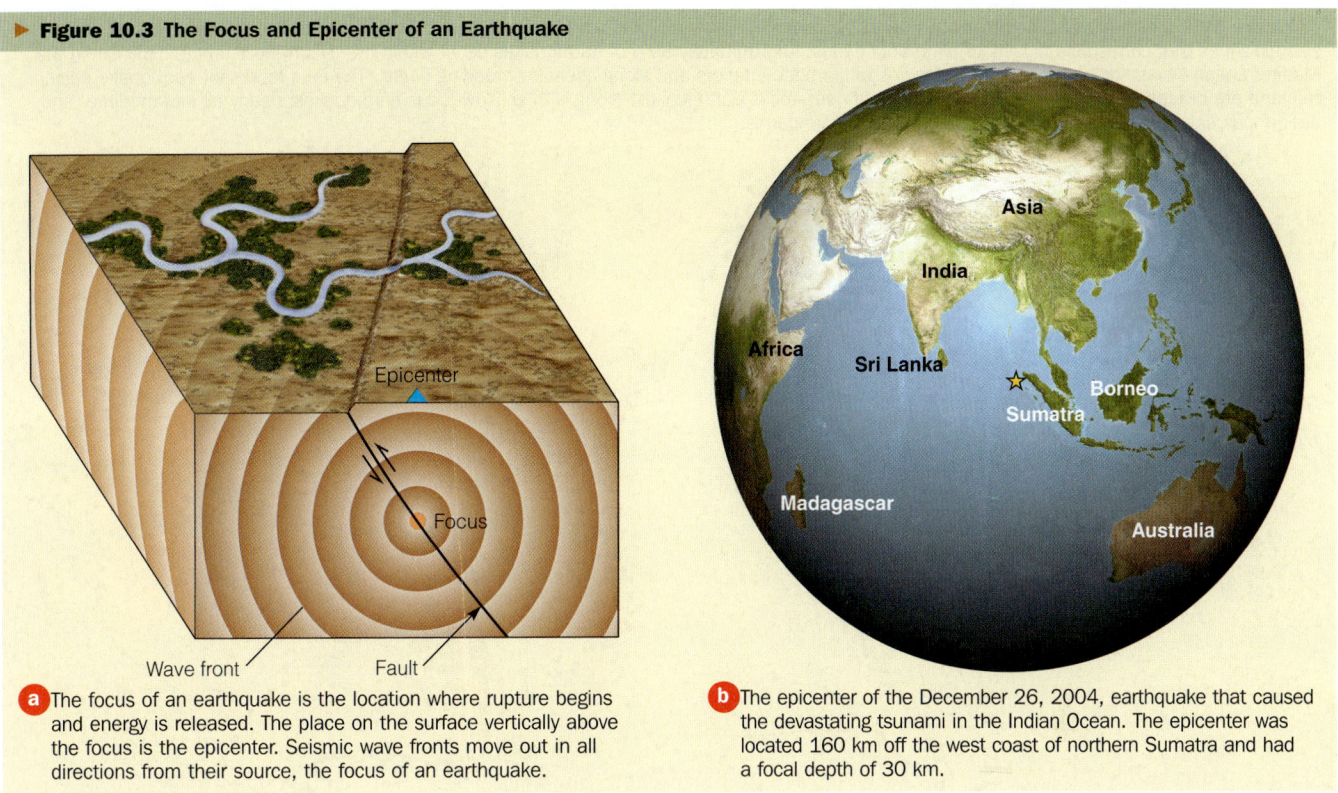

▶ **Figure 10.3** The Focus and Epicenter of an Earthquake

a The focus of an earthquake is the location where rupture begins and energy is released. The place on the surface vertically above the focus is the epicenter. Seismic wave fronts move out in all directions from their source, the focus of an earthquake.

b The epicenter of the December 26, 2004, earthquake that caused the devastating tsunami in the Indian Ocean. The epicenter was located 160 km off the west coast of northern Sumatra and had a focal depth of 30 km.

10.3 Seismology

■ What is seismology?

Seismology, the study of earthquakes, emerged as a true science during the 1880s with the development of **seismographs,** instruments that detect, record, and measure the vibrations produced by an earthquake (▶ Figure 10.2). The record made by a seismograph is called a *seismogram*. Modern seismographs have electronic sensors and record moments precisely using computers rather than simply relying on the drum strip-charts commonly used in older seismographs.

When an earthquake occurs, energy in the form of *seismic waves* radiates out from the point of release (▶ Figure 10.3). These waves are somewhat analogous to the ripples that move out concentrically from the point where a stone is thrown into a pond. Unlike waves on a pond, however, seismic waves move outward in all directions from their source.

Earthquakes take place because rocks are capable of storing energy but their strength is limited, so if enough force is present, they rupture and thus release their stored energy. In other words, most earthquakes result when movement occurs along fractures (faults), most of which are related, at least indirectly, to plate movements. Once a fracture begins, it moves along the fault at several kilometers per second for as long as conditions for failure exist. The longer the fracture along which movement occurs, the more time it takes for the stored energy to be released, and therefore the longer the ground will shake. During some very large earthquakes, the ground might shake for 3 minutes, a seemingly brief time but interminable if you are experiencing the earthquake firsthand!

■ What are the focus and epicenter of an earthquake?

The location within Earth's lithosphere where fracturing begins—that is, the point at which energy is first released—is an earthquake's **focus,** or *hypocenter*. What we usually hear in news reports, however, is the location of the **epicenter,** the point on Earth's surface directly above the focus (▶ Figure 10.3a). For example, according to the U.S. Geological Survey, the December 26, 2004, earthquake that triggered the devastating tsunami in the Indian Ocean had an epicenter 160 km off the west coast of northern Sumatra (3°18′ N and 95°52′ E) and a focal depth of 30 km (▶ Figure 10.3b).

Seismologists recognize three categories of earthquakes based on focal depth. *Shallow-focus* earthquakes have focal depths of less than 70 km from the surface, whereas those with foci between 70 and 300 km are *intermediate focus*, and the foci of those characterized as *deep focus* are more than 300 km deep. Earthquakes, however, are not evenly distributed among these three categories. Approximately 90% of all earthquake foci are at depths of less than 100 km, whereas only about 3% of all earthquakes are deep. Shallow-focus earthquakes are, with few exceptions, the most destructive because the energy they release has little time to dissipate before reaching the surface.

A definite relationship exists between earthquake foci and plate boundaries. Earthquakes generated along divergent or transform plate boundaries are invariably shallow focus, whereas many shallow- and nearly all intermediate- and

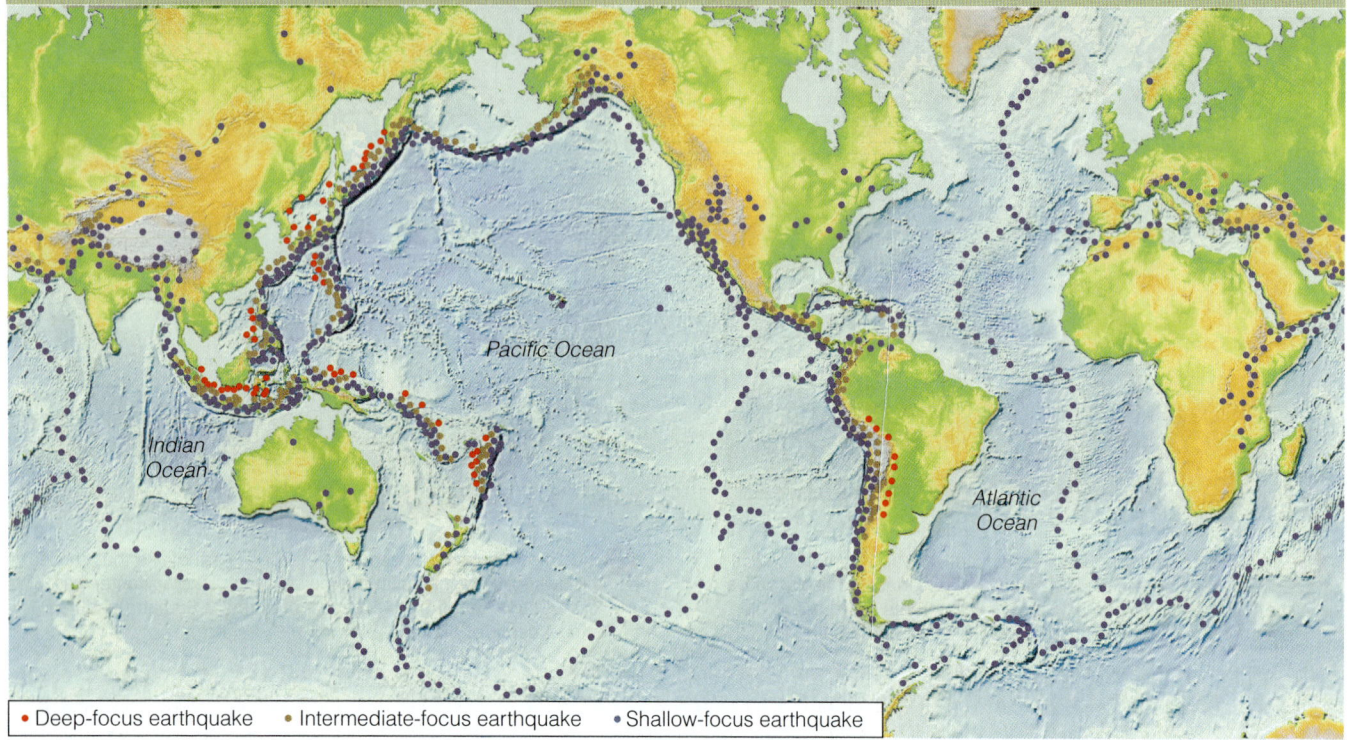

► **Figure 10.4 Earthquake Epicenters and Plate Boundaries** This map of earthquake epicenters shows that most earthquakes occur within seismic zones that correspond closely to plate boundaries. Approximately 80% of earthquakes occur within the circum-Pacific belt, 15% within the Mediterranean–Asiatic belt, and the remaining 5% within plate interiors and along oceanic spreading ridges. The dots represent earthquake epicenters and are divided into shallow-, intermediate-, and deep-focus earthquakes. Along with shallow-focus earthquakes, nearly all intermediate- and deep-focus earthquakes occur along convergent plate boundaries.

• Deep-focus earthquake • Intermediate-focus earthquake • Shallow-focus earthquake

deep-focus earthquakes occur along convergent margins (► Figure 10.4). Furthermore, a pattern emerges when the focal depths of earthquakes near island arcs and their adjacent ocean trenches are plotted. Notice in ► Figure 10.5 that the focal depth increases beneath the Tonga Trench in a narrow, well-defined zone that dips approximately 45 degrees. Dipping seismic zones, called *Benioff* or *Benioff-Wadati zones*, are common along convergent plate boundaries where one plate is subducted beneath another. Such dipping seismic zones indicate the angle of plate descent along a convergent plate boundary.

Section 10.3 Summary

- Seismology is the study of earthquakes. Earthquakes are recorded on seismographs, and the record of an earthquake is a seismogram.

- The location where rupture within Earth's lithosphere occurs and energy is released is an earthquake's focus. The point on Earth's surface directly above the focus is the epicenter.

- Seismologists recognize three categories of earthquakes based on focal depth: shallow-focus earthquakes with focal depths of less than 70 km, intermediate-focus earthquakes with focal depths between 70 and 300 km, and deep-focus earthquakes with focal depths greater than 300 km.

10.4 The Frequency and Distribution of Earthquakes

■ *Where do most earthquakes occur?*

No place on Earth is immune to earthquakes, but almost 95% take place in seismic belts corresponding to plate boundaries where plates converge, diverge, and slide past each other. Earthquake activity distant from plate margins is minimal but can be devastating when it occurs. The relationship between plate margins and the distribution of earthquakes is readily apparent when the locations of earthquake epicenters are superimposed on a map showing the boundaries of Earth's plates (► Figure 10.4).

The majority of all earthquakes (approximately 80%) occur in the **circum-Pacific belt,** a zone of seismic activity nearly encircling the Pacific Ocean basin. Most of these earthquakes result from convergence along plate margins, as in the case of the 1995 Kobe, Japan, earthquake (► Figure 10.6a). The earthquakes along the North American Pacific Coast, especially in California, are also in this belt, but here plates slide past one another rather than converge. The October 17, 1989, Loma Prieta earthquake in the San Francisco area (► Figure 10.6b) and the January 17, 1994, Northridge earthquake (► Figure 10.6c) happened along this plate boundary.

▶ **Figure 10.5 Benioff Zones** Focal depth increases in a well-defined zone that dips approximately 45 degrees beneath the Tonga volcanic arc in the South Pacific. Dipping seismic zones are called *Benioff* or *Benioff–Wadati* zones.

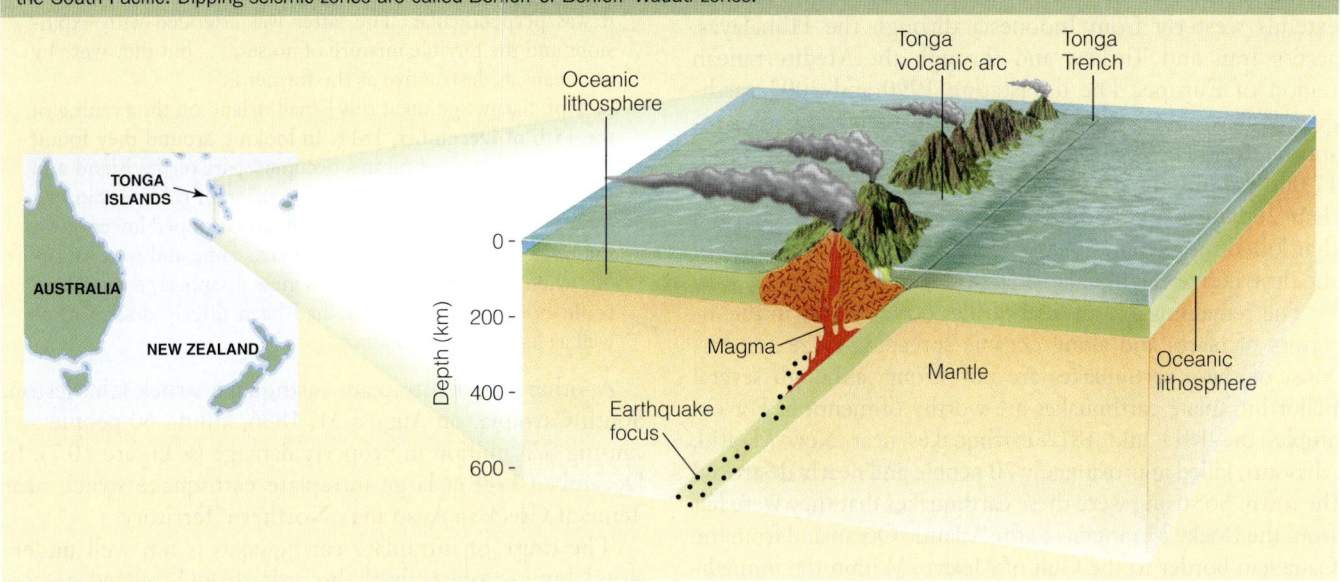

▶ **Figure 10.6 Earthquake Damage in the Circum-Pacific Belt**

a Some of the damage in Kobe, Japan, caused by the January 1995 earthquake in which more than 5000 people died.

b Damage in Oakland, California, resulting from the October 1989 Loma Prieta earthquake. The columns supporting the upper deck of Interstate 880 failed, causing the upper deck to collapse onto the lower one.

c View of the severe exterior damage to the Northridge Meadows apartments in which 16 people were killed as a result of the January 1994 Northridge, California, earthquake.

The second major seismic belt, accounting for 15% of all earthquakes, is the **Mediterranean–Asiatic belt.** This belt extends westerly from Indonesia through the Himalayas, across Iran and Turkey, and through the Mediterranean region of Europe. The devastating 1990 and 2003 earthquakes in Iran that killed 40,000 and 43,000 people, respectively, the 1999 Turkey earthquake that killed about 17,000 people, the 2001 India earthquake that killed more than 20,000 people, and the 2005 earthquake in Pakistan that killed more than 86,000 people are recent examples of the destructive earthquakes that strike this region.

The remaining 5% of earthquakes occur mostly in the interiors of plates and along oceanic spreading-ridge systems. Most of these earthquakes are not strong, although several major intraplate earthquakes are worthy of mention. For example, the 1811 and 1812 earthquakes near New Madrid, Missouri, killed approximately 20 people and nearly destroyed the town. So strong were these earthquakes that they were felt from the Rocky Mountains to the Atlantic Ocean and from the Canadian border to the Gulf of Mexico. Within the immediate area, numerous buildings were destroyed and forests were flattened. The land sank several meters in some areas, causing flooding; and reportedly the Mississippi River reversed its flow during the shaking and changed its course slightly. Eyewitnesses described the scene at New Madrid as follows:

> The earth was observed to roll in waves a few feet high with visible depressions between. By and by these swells burst throwing up large volumes of water, sand, and coal.
>
> ... Undulations of the earth resembling waves, increasing in elevations as they advanced, and when they attained a certain fearful height the earth would burst.
>
> The shocks were clearly distinguishable into two classes, those in which the motion was horizontal and those in which it was perpendicular. The latter was attended with explosions and the terrible mixture of noises, ... but they were by no means as destructive as the former.
>
> Cpt. Sarpy tied up at this [small] island on the evening of the 15th of December, 1811. In looking around they found that a party of river pirates occupied part of the island and were expecting Sarpy with the intention of robbing him. As soon as Sarpy found that out he quietly dropped lower down the river. In the night the earthquake came and next morning when the accompanying haziness disappeared the island could no longer be seen. It had been utterly destroyed as well as its pirate inhabitants.*

Another major intraplate earthquake struck Charleston, South Carolina, on August 31, 1886, killing 60 people and causing $23 million in property damage (▶ Figure 10.7). In December 1988 a large intraplate earthquake struck near Tennant Creek in Australia's Northern Territory.

The cause of intraplate earthquakes is not well understood, but geologists think they arise from localized stresses caused by the compression that most plates experience along their margins. A useful analogy is moving a house. Regardless of how careful the movers are, moving something so large without its internal parts shifting slightly is impossible. Similarly, plates are not likely to move without some internal stresses that occasionally cause earthquakes. It is interesting that many intraplate earthquakes are asso-

---

*C. Officer and J. Page, *Tales of the Earth* (New York: Oxford University Press, 1993), pp. 49–50.

▶ **Figure 10.7 Damage Resulting from an Intraplate Earthquake** Damage done to Charleston, South Carolina, by the intraplate earthquake of August 31, 1886. This earthquake is the largest reported in the eastern United States.

ciated with ancient and presumed inactive faults that are reactivated at various intervals.

■ *How many earthquakes occur per year?*

More than 900,000 earthquakes are recorded annually by the worldwide network of seismograph stations. Many of these, however, are too small to be felt but are nonetheless recorded. These small earthquakes result from the energy released as continuous adjustments take place between the various plates. However, more than 31,000 earthquakes, on average per year, are strong enough to be felt, and can cause various amounts of damage, depending on how strong they are and where they occur.

Section 10.4 Summary

- Approximately 80% of all earthquakes occur in the circum-Pacific belt, 15% within the Mediterranean–Asiatic belt, and the remaining 5% mostly in the interiors of plates and along oceanic spreading ridges.

- More than 900,000 earthquakes occur each year, with more than 31,000 of those strong enough to be felt.

10.5 Seismic Waves

■ *What are seismic waves?*

Many people have experienced an earthquake but are probably unaware that the shaking they feel and the damage to structures are caused by the arrival of various *seismic waves*, a general term encompassing all waves generated by an earthquake. When movement on a fault takes place, energy is released in the form of two kinds of seismic waves that radiate outward in all directions from an earthquake's focus. **Body waves,** so called because they travel through the solid body of Earth, are somewhat like sound waves, and **surface waves,** which travel along the ground surface, are analogous to undulations or waves on water surfaces.

■ *What are the two types of body waves?*

An earthquake generates two types of body waves: P-waves and S-waves (▶ Figure 10.8). **P-waves,** or *primary waves*, are the fastest seismic waves and can travel through solids, liquids, and gases. P-waves are compressional, or push–pull, waves and are similar to sound waves in that they move material forward and backward along a line in the same direction that the waves themselves are moving

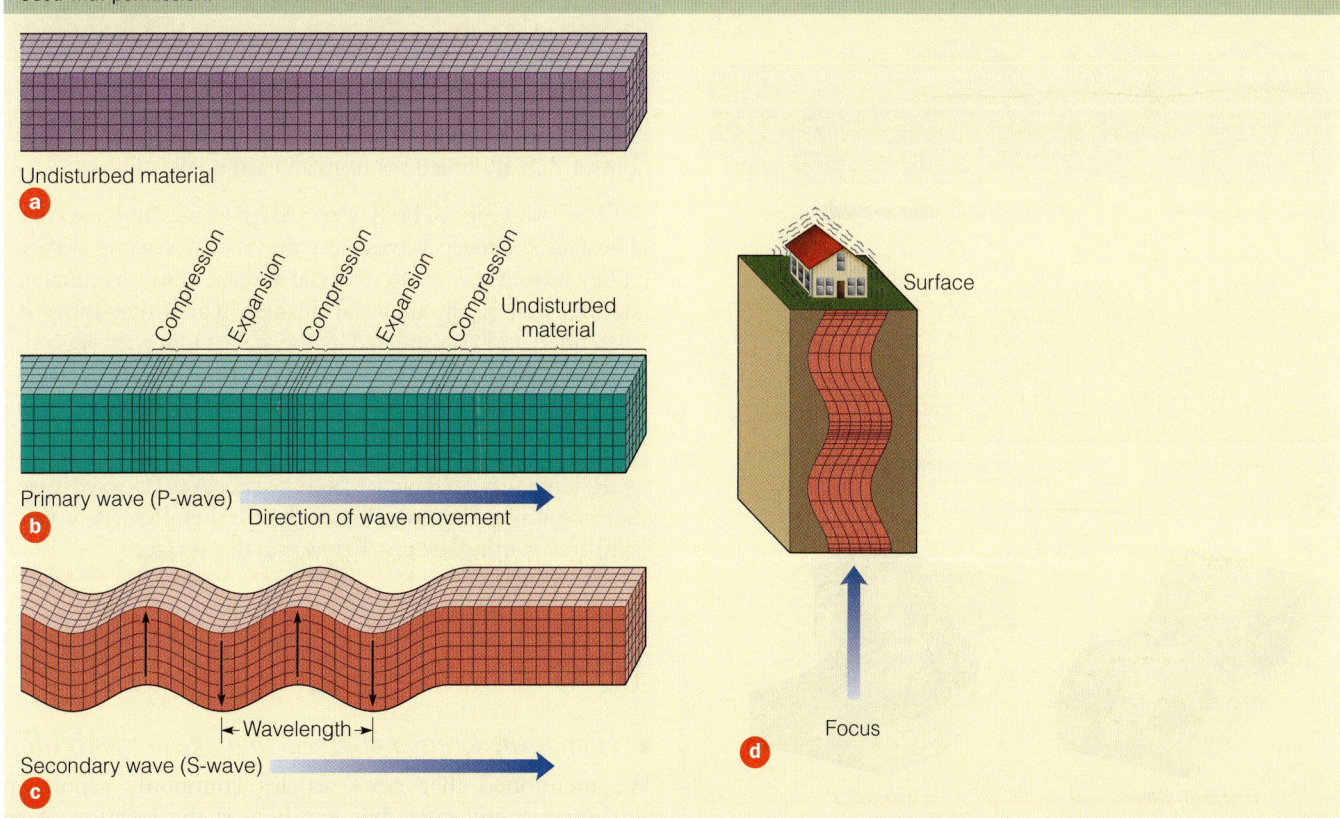

Active Figure 10.8 Primary and Secondary Seismic Body Waves (a) Undisturbed material for reference. **(b)** and **(c)** show how body waves travel through Earth. Primary waves (P-waves) compress and expand material in the same direction they travel. **(c)** Secondary waves (S-waves) move material perpendicular to the direction of wave movement. **(d)** P- and S-waves and their effect on surface structures. Source: (a, b, c) From *Nuclear Explosions and Earthquakes: The Parted Veil*, by Bruce A Bolt. Copyright © 1976 by W. H. Freeman and Company. Used with permission.

(▶ Figure 10.8b). Thus the material through which P-waves travel is expanded and compressed as the waves move through it and returns to its original size and shape after the waves pass by. In fact, some P-waves emerging from Earth are transmitted into the atmosphere as sound waves, which at certain frequencies can be heard by humans and animals.

S-waves, or *secondary waves,* are somewhat slower than P-waves and can travel only through solids. S-waves are *shear waves* because they move the material perpendicular to the direction of travel, thereby producing shear stresses in the material they move through (▶ Figure 10.8c). Because liquids (as well as gases) are not rigid, they have no shear strength, and S-waves cannot be transmitted through them.

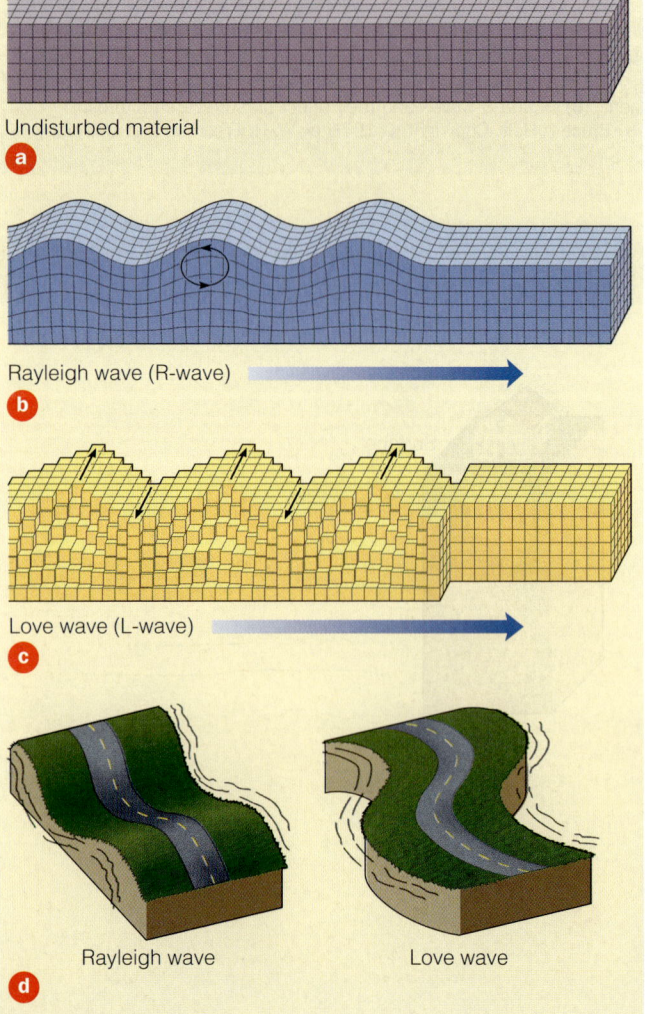

▶ **Active Figure 10.9 Rayleigh and Love Seismic Surface Waves (a)** Undisturbed material for reference. **(b)** and **(c)** show how surface waves travel along Earth's surface or just below it. Rayleigh waves (R-waves) move material in an elliptical path in a plane oriented parallel to the direction of wave movement. **(c)** Love waves (L-waves) move material back and forth in a horizontal plane perpendicular to the direction of wave movement. **(d)** The arrival of R- and L-waves causes the surface to undulate and shake from side to side. (Source: a, b, c: From *Nuclear Explosions and Earthquakes: The Parted Veil,* by Bruce A. Bolt. Copyright © 1976 by W. H. Freeman and Company. Used with permission.)

The velocities of P- and S-waves are determined by the density and elasticity of the materials through which they travel. For example, seismic waves travel more slowly through rocks of greater density but more rapidly through rocks with greater elasticity. *Elasticity* is a property of solids, such as rocks, and means that once they have been deformed by an applied force, they return to their original shape when the force is no longer present. Because P-wave velocity is higher than S-wave velocity in all materials, P-waves always arrive at seismic stations first.

■ *What are the two major types of surface waves?*

Surface waves travel along the surface of the ground, or just below it, and are slower than body waves. Unlike the sharp jolting and shaking that body waves cause, surface waves generally produce a rolling or swaying motion, much like the experience of being in a boat.

Several types of surface waves are recognized. The two most important are Rayleigh waves and Love waves, named after the British scientists who discovered them, Lord Rayleigh and A. E. H. Love. **Rayleigh waves (R-waves)** are generally the slower of the two and behave like water waves in that they move forward while the individual particles of material move in an elliptical path within a vertical plane oriented in the direction of wave movement (▶ Figure 10.9b).

The motion of a **Love wave (L-wave)** is similar to that of an S-wave, but the individual particles of the material move only back and forth in a horizontal plane perpendicular to the direction of wave travel (▶ Figure 10.9c). This type of lateral motion can be particularly damaging to building foundations.

Section 10.5 Summary

• Seismic waves are produced by the energy released when an earthquake occurs. Body waves and surface waves move outward in all directions from an earthquake's focus.

• The two types of body waves are P-waves and S-waves. P-waves (primary waves) are the fastest seismic waves. They move through all material and are compressional in nature; that is, the material through which they move is expanded and compressed as the waves move through it. S-waves (shear waves) are slower than P-waves, travel only through solids, and move material perpendicular to their direction of travel.

• Raleigh waves (R-waves) and Love waves (L-waves) are surface waves that are slower than either P- or S-waves and travel along, or just below, Earth's surface.

10.6 Locating an Earthquake

■ *How is the location of an earthquake determined?*

We mentioned that news articles commonly report an earthquake's epicenter, but just how is the location of an epicenter determined? Once again, geologists rely on the study of seismic waves. We know that P-waves travel faster than S-waves, nearly twice as fast in all substances, so P-waves arrive at a seismograph station first, followed some time later by S-waves. Both P- and S-waves travel directly from the focus to the seismograph station through Earth's interior, but L- and R-waves arrive last because they are the slowest, and they also travel the longest route along the surface (▶ Figure 10.10a). L- and R-waves cause much of the damage during earthquakes, but only P- and S-waves need concern us here because they are the ones important in determining an epicenter.

Seismologists, geologists who study seismology, have accumulated a tremendous amount of data over the years and now know the average speeds of P- and S-waves for any specific distance from their source. These P- and S-wave travel times are published in *time–distance graphs* that illustrate the difference between the arrival times of the two waves as a function of the distance between a seismograph and an earthquake's focus (▶ Figure 10.10b). That is, the farther the waves travel, the greater the *P–S time interval* or simply the time difference between the arrivals of P- and S-waves (▶ Figure 10.10a,b).

If the P–S time intervals are known from at least three seismograph stations, then the epicenter of any earthquake can be determined (▶ Figure 10.11). Here is how it works. Subtracting the arrival time of the first P-wave from the arrival time of the first S-wave gives the P–S time interval for each seismic station. Each of these time intervals is then plotted on

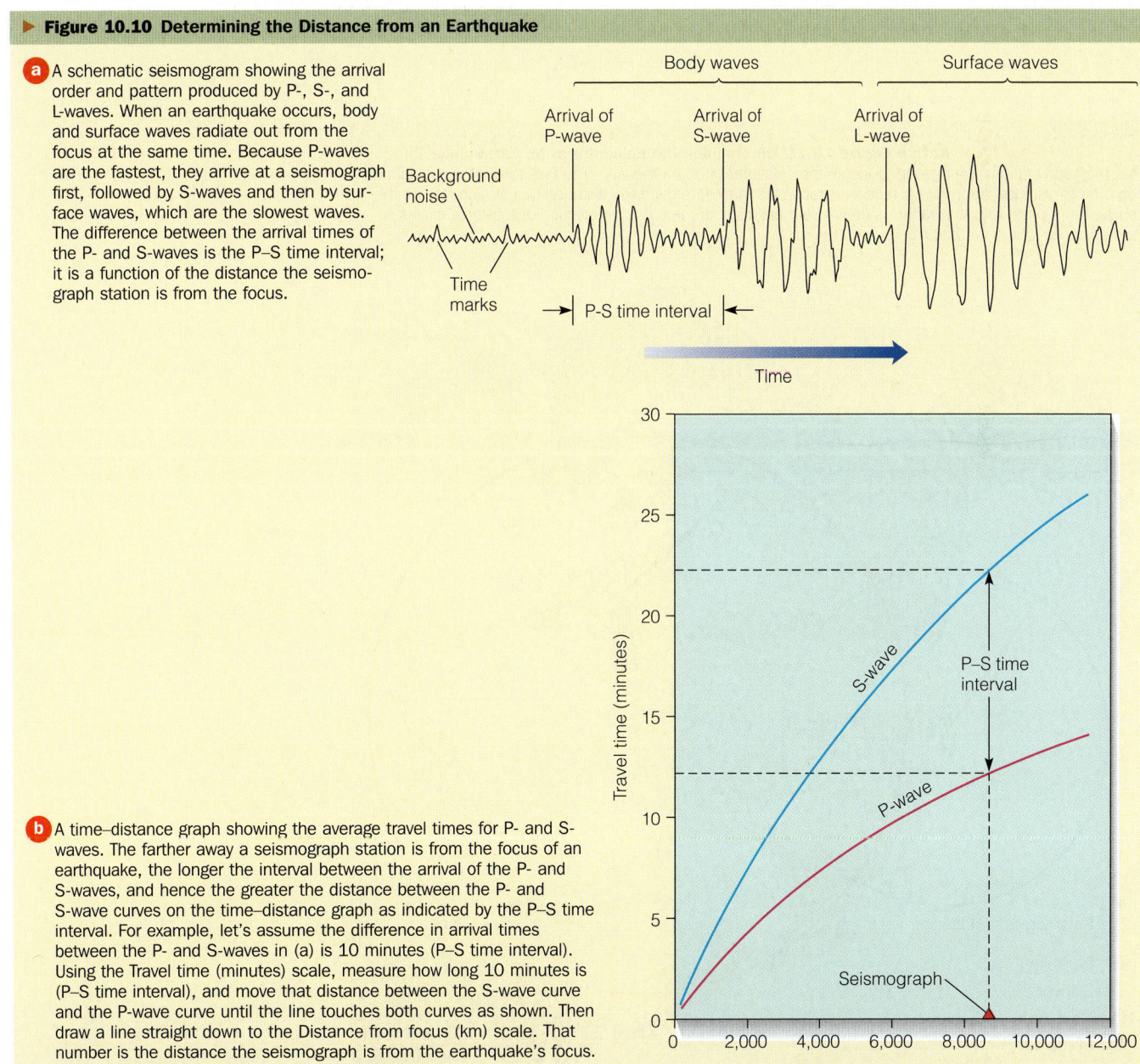

▶ **Figure 10.10 Determining the Distance from an Earthquake**

a A schematic seismogram showing the arrival order and pattern produced by P-, S-, and L-waves. When an earthquake occurs, body and surface waves radiate out from the focus at the same time. Because P-waves are the fastest, they arrive at a seismograph first, followed by S-waves and then by surface waves, which are the slowest waves. The difference between the arrival times of the P- and S-waves is the P–S time interval; it is a function of the distance the seismograph station is from the focus.

b A time–distance graph showing the average travel times for P- and S-waves. The farther away a seismograph station is from the focus of an earthquake, the longer the interval between the arrival of the P- and S-waves, and hence the greater the distance between the P- and S-wave curves on the time–distance graph as indicated by the P–S time interval. For example, let's assume the difference in arrival times between the P- and S-waves in (a) is 10 minutes (P–S time interval). Using the Travel time (minutes) scale, measure how long 10 minutes is (P–S time interval), and move that distance between the S-wave curve and the P-wave curve until the line touches both curves as shown. Then draw a line straight down to the Distance from focus (km) scale. That number is the distance the seismograph is from the earthquake's focus. In this example, the distance is almost 9000 km.

a time–distance graph, and a line is drawn straight down to the distance axis of the graph, thus giving the distance from the focus to each seismic station (▶ Figure 10.10b). Next, a circle whose radius equals the distance shown on the time–distance graph from each of the seismic stations is drawn on a map (▶ Figure 10.11). The intersection of the three circles is the location of the earthquake's epicenter. It should be obvious from ▶ Figure 10.11 that P–S time intervals from at least three seismic stations are needed. If only one were used, the epicenter could be at any location on the circle drawn around that station, and using two stations would give two possible locations for the epicenter.

Determining the focal depth of an earthquake is much more difficult and considerably less precise than finding its epicenter. The focal depth is usually found by making computations based on several assumptions, comparing the results with those obtained at other seismic stations, and then recalculating and approximating the depth as closely as possible. Even so, the results are not highly accurate, but they do tell us that most earthquakes, probably about 75%, have foci no deeper than 10 to 15 km and that a few are as deep as 680 km.

Section 10.6 Summary

- An earthquake's epicenter is determined by using a time–distance graph and the arrival times of the P- and S-waves at three seismic stations. The difference in the arrival times of the P- and S-waves is a function of how far away the seismic station is from the earthquake. The greater the difference in arrival times, the farther away the seismic station is from the earthquake. By determining the distance each seismic station is from an earthquake using a time–distance graph, seismologists draw a circle around each seismic station whose radius equals the distance the seismic station is from the earthquake. The point where all three circles intersect is the epicenter.

Active Figure 10.11 Determining the Epicenter of an Earthquake Three seismograph stations are needed to locate the epicenter of an earthquake. The P–S time interval is plotted on a time–distance graph for each seismograph station to determine the distance that station is from the epicenter. A circle with that radius is drawn from each station, and the intersection of the three circles is the epicenter of the earthquake.

10.7 Measuring the Strength of an Earthquake

■ *How is an earthquake's strength measured?*

Following any earthquake that causes extensive damage, fatalities, and injuries, graphic reports of the quake's violence and human suffering are common. Headlines tell us that thousands died, many more were injured or homeless, and property damage is in the millions and possibly billions of dollars. Few other natural processes have such tragic consequences. Although descriptions of fatalities and damage give some indication of the size of an earthquake, geologists are interested in more reliable methods of determining an earthquake's size.

Two measures of an earthquake's strength are commonly used. One is *intensity*, a qualitative assessment of the kinds of damage done by an earthquake. The other, *magnitude*, is a quantitative measure of the amount of energy released by an earthquake. Each method provides important information that can be used to prepare for future earthquakes.

■ *What is intensity and how is it determined?*

Intensity is a subjective or qualitative measure of the kind of damage done by an earthquake, as well as people's reaction to it. Since the mid-19th century, geologists have used intensity as a rough approximation of the size and strength of an earthquake. The most common intensity scale used in the United States is the **Modified Mercalli Intensity Scale,** which has values ranging from I to XII (Table 10.2).

Intensity maps can be constructed for regions hit by earthquakes by dividing the affected region into various intensity zones. The intensity value given for each zone is the maximum intensity that the earthquake produced for that zone. Even though intensity maps are not precise because of the subjective nature of the measurements, they do provide geologists with a rough approximation of the location of the earthquake, the kind and extent of the damage done, and the effects of local geology on different types of building construction (▶ Figure 10.12). Because intensity is a measure of the kind of damage done by an earthquake, insurance companies still classify earthquakes on the basis of intensity.

■ *What factors determine an earthquake's intensity?*

Generally, a large earthquake will produce higher intensity values than a small earthquake, but many other factors besides the amount of energy released by an earthquake also affect its intensity. These include distance from the epicenter, focal depth of the earthquake, population density and geology of the area, type of building construction employed, and duration of shaking.

A comparison of the intensity map for the 1906 San Francisco earthquake and a geologic map of the area shows a strong correlation between the amount of damage done and the underlying rock and soil conditions (▶ Figure 10.12). Damage was greatest in those areas underlain by poorly consolidated material or artificial fill because the effects of shaking are amplified in these materials, whereas damage was less in areas of solid bedrock. The correlation between the geology and the amount of damage done by an earthquake was further reinforced by the 1989 Loma Prieta earthquake, when many of the same areas that were extensively damaged in the 1906 earthquake were once again heavily damaged.

■ *What is the Richter Magnitude Scale, and what does it measure?*

If earthquakes are to be compared quantitatively, we must use a scale that measures the amount of energy released and is independent of intensity. Charles F. Richter, a seismologist at

TABLE 10.2

Modified Mercalli Intensity Scale

I	Not felt except by a very few under especially favorable circumstances.
II	Felt by only a few people at rest, especially on upper floors of buildings.
III	Felt quite noticeably indoors, especially on upper floors of buildings, but many people do not recognize it as an earthquake. Standing automobiles may rock slightly.
IV	During the day felt indoors by many, outdoors by few. At night some awakened. Sensation like heavy truck striking building, standing automobiles rocked noticeably.
V	Felt by nearly everyone, many awakened. Some dishes, windows, etc. broken, a few instances of cracked plaster. Disturbance of trees, poles, and other tall objects sometimes noticed.
VI	Felt by all, many frightened and run outdoors. Some heavy furniture moved, a few instances of fallen plaster or damaged chimneys. Damage slight.
VII	Everybody runs outdoors. Damage negligible in buildings of good design and construction; slight to moderate in well-built ordinary structures; considerable in poorly built or badly designed structures; some chimneys broken. Noticed by people driving automobiles.
VIII	Damage slight in specially designed structures; considerable in normally constructed buildings with possible partial collapse; great in poorly built structures. Fall of chimneys, monuments, walls. Heavy furniture overturned. Sand and mud ejected in small amounts.
IX	Damage considerable in specially designed structures. Buildings shifted off foundations. Ground noticeably cracked. Underground pipes broken.
X	Some well-built wooden structures destroyed; most masonry and frame structures with foundations destroyed; ground badly cracked. Rails bent. Landslides considerable from river banks and steep slopes. Water splashed over river banks.
XI	Few, if any (masonry) structures remain standing. Bridges destroyed. Broad fissures in ground. Underground pipelines completely out of service.
XII	Damage total. Waves seen on ground surfaces. Objecs thrown upward into the air.

Source: U.S. Geological Survey.

▶ **Figure 10.12 Relationship between Intensity and Geology for the 1906 San Francisco Earthquake** A close correlation exists between the geology **(a)** of the San Francisco area and **(b)** intensity during the 1906 earthquake. Areas underlain by bedrock correspond to the lowest intensity values, followed by areas underlain by thin alluvium (sediment) and thick alluvium. Bay mud, artificial fill, or both lie beneath the areas shaken most violently.

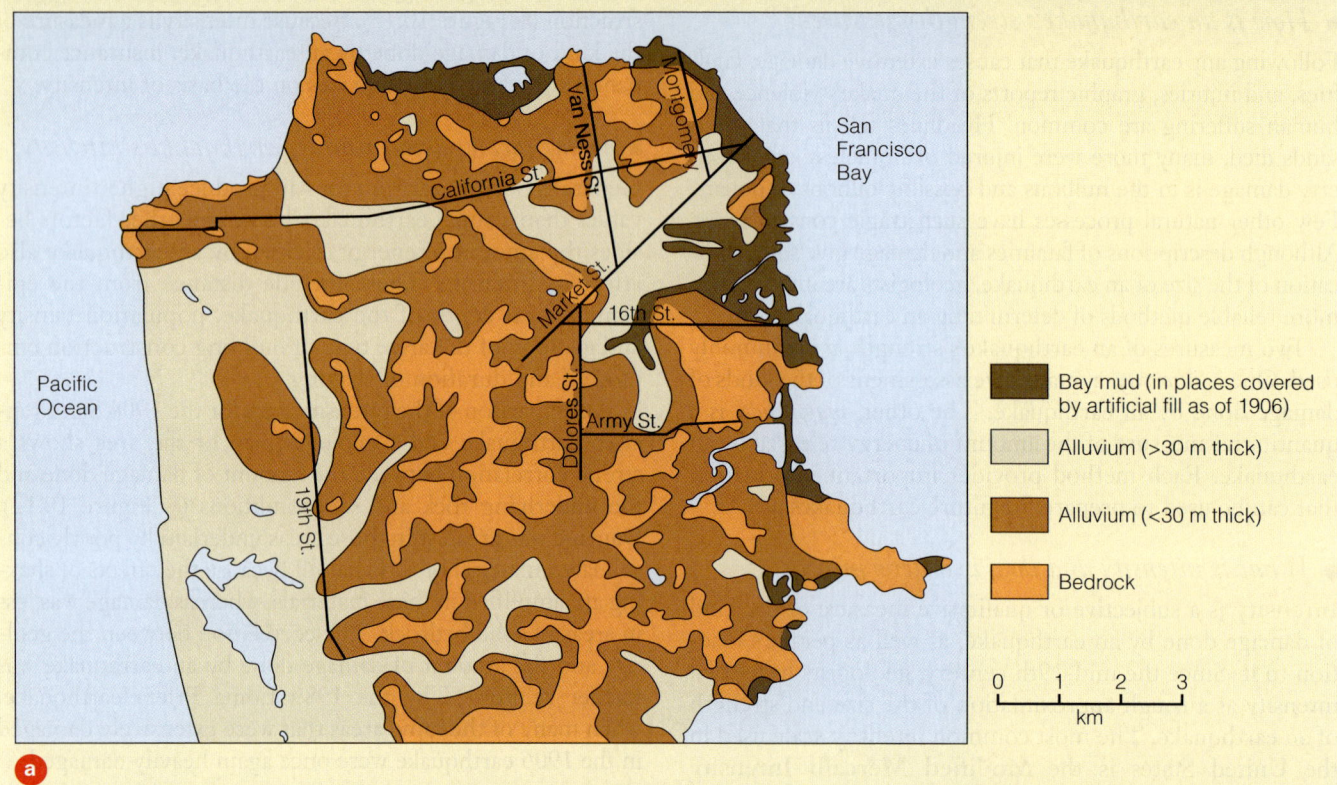

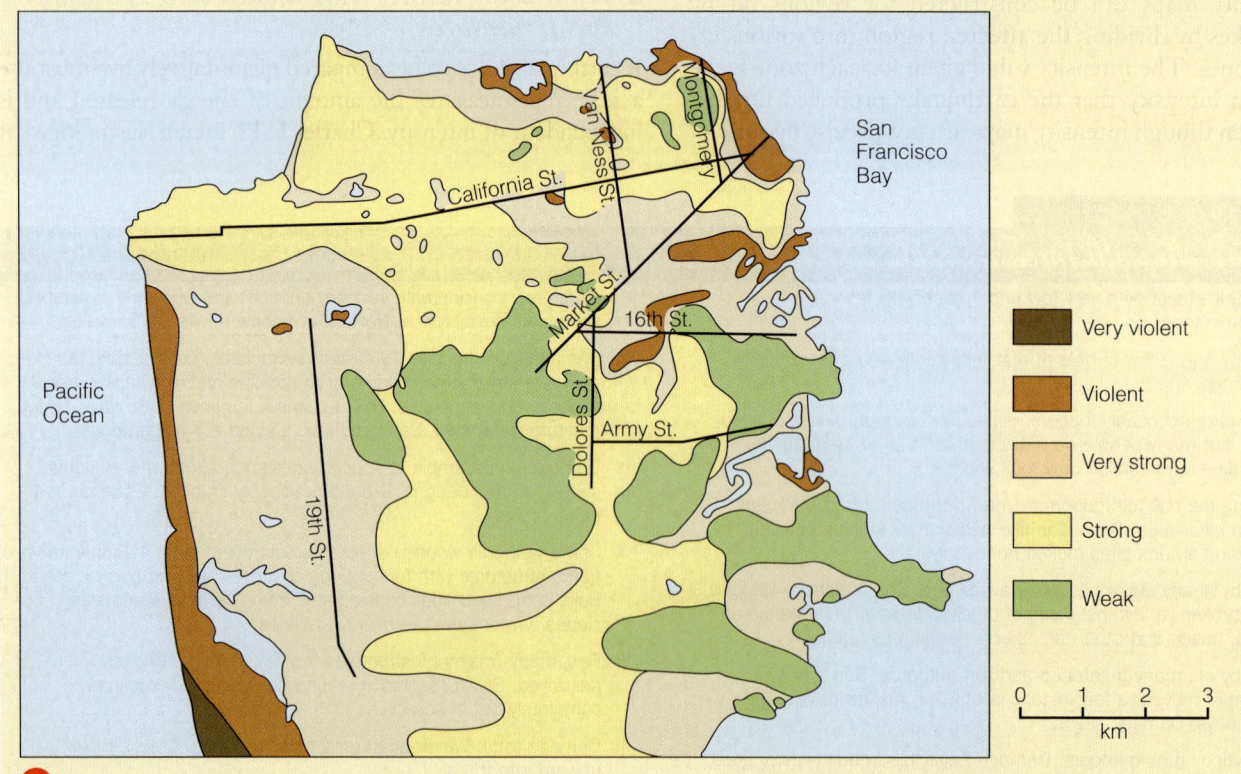

the California Institute of Technology, developed such a scale in 1935. The **Richter Magnitude Scale** measures earthquake **magnitude,** which is the total amount of energy released by an earthquake at its source. It is an open-ended scale with values beginning at zero. The largest magnitude recorded was a magnitude-9.5 earthquake in Chile on May 22, 1960.

■ *How is an earthquake's magnitude determined?*

Seismologists determine the magnitude of an earthquake by measuring the amplitude of the largest seismic wave as recorded on a seismogram (▶ Figure 10.13). To avoid large numbers, Richter used a conventional base-10 logarithmic scale to convert the amplitude of the largest recorded seismic wave to a numeric magnitude value (▶ Figure 10.13). Therefore each whole-number increase in magnitude represents a 10-fold increase in wave amplitude. For example, the amplitude of the largest seismic wave for an earthquake of magnitude 6 is 10 times that produced by an earthquake of magnitude 5, 100 times as large as a magnitude-4 earthquake, and 1000 times that of an earthquake of magnitude 3 ($10 \times 10 \times 10 = 1000$).

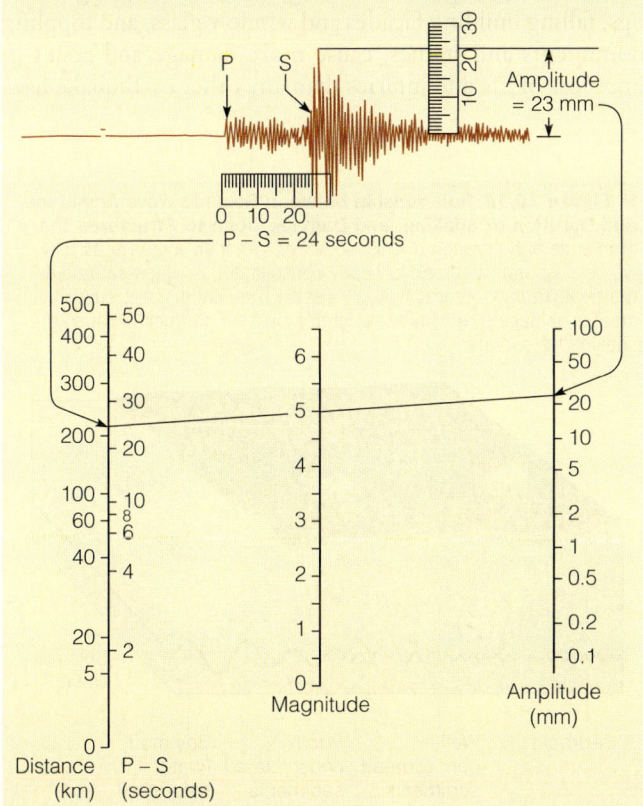

▶ **Figure 10.13 Richter Magnitude Scale** The Richter Magnitude Scale measures the total amount of energy released by an earthquake at its source. The magnitude is determined by measuring the maximum amplitude of the largest seismic wave and marking it on the right-hand scale. The difference between the arrival times of the P- and S-waves (recorded in seconds) is marked on the left-hand scale. When a line is drawn between the two points, the magnitude of the earthquake is the point at which the line crosses the center scale. (Source: From *Earthquakes*, by Bruce A Bolt. Copyright © 1988 by W. H. Freeman and Company. Used with permission.)

A common misconception about the size of earthquakes is that an increase of one unit on the Richter Magnitude Scale—a 7 versus a 6, for instance—means a 10-fold increase in size. It is true that each whole-number increase in magnitude represents a 10-fold increase in the wave amplitude, but each magnitude increase of one unit corresponds to a roughly 30-fold increase in the amount of energy released (actually it is 31.5, but 30 is close enough for our purposes). This means that it would take about 30 earthquakes of magnitude 6 to equal the energy released in one earthquake of magnitude 7.

The 1964 Alaska earthquake with a magnitude of 8.6 released almost 900 times more energy than the 1994 Northridge, California, earthquake with a magnitude of 6.7! And the Alaska earthquake released more than 27,000 times as much energy as an earthquake with a magnitude of 5.6 would have.

We mentioned that more than 900,000 earthquakes are recorded around the world each year. This figure can be placed in better perspective by reference to Table 10.3, which shows that the vast majority of earthquakes have a Richter magnitude of less than 2.5 and that great earthquakes (those with a magnitude greater than 8.0) occur, on average, only once every five years.

The Richter Magnitude Scale was devised to measure earthquake waves on a particular seismograph and at a specific distance from an earthquake. One of its limitations is that it underestimates the energy of very large earthquakes because it measures the highest peak on a seismogram, which represents only an instant during an earthquake. For large earthquakes, though, the energy might be released over several minutes and along hundreds of kilometers of a fault. For instance, during the 1857 Fort Tejon, California, earthquake, the ground shook for longer than 2 minutes and energy was released for 360 km along the fault. Despite their shortcomings, however, Richter magnitudes still usually appear in news releases.

Seismologists now commonly use a somewhat different scale to measure magnitude. Known as the *seismic-moment magnitude scale*, this scale takes into account the strength of

TABLE 10.3

Average Number of Earthquakes of Various Magnitudes per Year Worldwide

Magnitude	Effects	Average Number per Year
<2.5	Typically not felt but recorded	900,000
2.5–6.0	Usually felt; minor to moderate damage to structures	31,000
6.1–6.9	Potentially destructive, especially in populated areas	100
7.0–7.9	Major earthquakes; serious damage results	20
>8.0	Great earthquakes; usually result in total destruction	1 every 5 years

Source: Modified from *Earthquake Information Bulletin*, and B. Gutenberg and C. F. Richter, *Seismicity of the Earth and Associated Phenomena* (Princeton: Princeton University Press, 1949).

the rocks, the area of a fault along which rupture occurs, and the amount of movement of rocks adjacent to the fault. Because larger earthquakes rupture more rocks than smaller earthquakes and rupture usually occurs along a longer segment of a fault and therefore for a longer duration, these very large earthquakes release more energy. For example, the December 26, 2004, Sumatra earthquake that generated the devastating tsunami created the longest fault rupture and had the longest duration ever recorded.

Thus magnitude is now frequently given in terms of both Richter magnitude and seismic-moment magnitude. For example, the 1964 Alaska earthquake is given a Richter magnitude of 8.6 and a seismic-moment magnitude of 9.2. Because the Richter Magnitude Scale is most commonly used in the news, we will use that scale here.

Section 10.7 Summary

- An earthquake's strength can be measured qualitatively or quantitatively. Intensity is a qualitative, or subjective, measure of the damage done by an earthquake. It is expressed in values from I to XII in the Modified Mercalli Intensity Scale, which indicates the degree of damage done by an earthquake. Factors that determine an earthquake's intensity include distance from the epicenter, focal depth of the earthquake, population density and local geology of the area, type of building construction employed, and the duration of ground shaking.

- Magnitude is a quantitative measure of the total amount of energy released by an earthquake at its source. It is expressed by the Richter Magnitude Scale, which is an open-ended scale with values beginning at 1. Each unit increase on the Richter Magnitude Scale corresponds to an approximately 30-fold increase in energy released.

- Seismologists now commonly use the seismic-moment magnitude scale because it more effectively measures the amount of energy released by very large earthquakes.

10.8 The Destructive Effects of Earthquakes

■ *What factors determine an earthquake's destructiveness?*

Certainly earthquakes are one of nature's most destructive phenomena. Little or no warning precedes earthquakes, and once they begin, essentially nothing can be done to minimize their destructive effects, although planning before an earthquake can help. The number of deaths and injuries as well as the amount of property damage depend on several factors. Generally speaking, earthquakes that occur during working hours and school hours in densely populated urban areas are the most destructive and cause the most fatalities and injuries. However, magnitude, duration of shaking, dis-

tance from the epicenter, geology of the affected region, and type of structures are also important considerations. Given these variables, it should not be surprising that a comparatively small earthquake can have disastrous effects, whereas a much larger one might go largely unnoticed, except perhaps by seismologists.

■ *What are the destructive effects of earthquakes?*

The destructive effects of earthquakes include ground shaking, fire, seismic sea waves, and landslides, as well as panic, disruption of vital services, and psychological shock. In some cases, rescue attempts are hampered by inadequate resources or planning, conditions of civil unrest, or simply the magnitude of the disaster.

Ground Shaking

Ground shaking, the most obvious and immediate effect of an earthquake, varies depending on the earthquake's magnitude, distance from the epicenter, and the type of underlying materials in the area—unconsolidated sediment or fill versus bedrock, for instance. Certainly ground shaking is terrifying, and it might be violent enough for fissures to open in the ground. Nevertheless, contrary to popular myth, fissures do not swallow up people and buildings and then close on them. And although California will no doubt have big earthquakes in the future, rocks cannot store enough energy to displace a landmass as large as California into the Pacific Ocean, as the tabloids sometimes suggest will happen.

The effects of ground shaking, such as collapsing buildings, falling building facades and window glass, and toppling monuments and statues, cause more damage and result in more loss of life and injuries than any other earthquake haz-

▶ **Figure 10.14 Relationship between Seismic Wave Amplitude and Duration of Shaking, and Damage Done to Structures** The amplitude and duration of seismic waves generally increase as the waves pass from bedrock to poorly consolidated or water-saturated material. Thus structures built on weaker material typically suffer greater damage than similar structures built on bedrock because the shaking lasts longer.

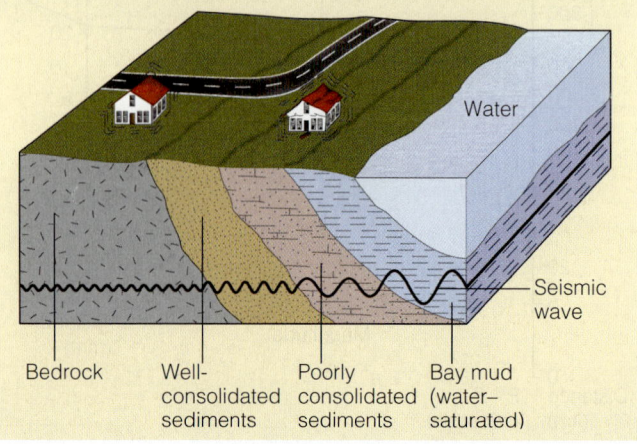

ard. Structures built on solid bedrock generally suffer less damage than those built on poorly consolidated material such as water-saturated sediments or artificial fill.

Structures built on poorly consolidated or water-saturated material are subjected to ground shaking of longer duration and higher S-wave amplitude than those on bedrock (▶ Figure 10.14). In addition, fill and water-saturated sediments tend to liquefy, or behave like a fluid, a process known as *liquefaction*. When shaken, the individual grains lose cohesion and the ground flows. Two dramatic examples of damage resulting from liquefaction are Niigata, Japan, where large apartment buildings were tipped on their sides after the water-saturated soil of the hillside collapsed (▶ Figure 10.15), and Turnagain Heights, Alaska, where many homes were destroyed when the Bootlegger Cove Clay lost all of its strength when it was shaken by the 1964 earthquake (see Figure 14.19).

Besides the magnitude of an earthquake and the underlying geology, the material used and the type of construction also affect the amount of damage done. Adobe and mud-walled structures are the weakest and almost always collapse during an earthquake. Unreinforced brick structures and poorly built concrete structures are also particularly susceptible to collapse, such as was the case in the 1999 Turkey earthquake in which an estimated 17,000 people died (▶ Figure 10.16). The 1976 earthquake in Tangshan, China, completely leveled the city because hardly any structures were built to resist seismic forces. In fact, most had unreinforced brick walls, which have no flexibility, and consequently they collapsed during the shaking.

The magnitude-6.4 earthquake that struck India in 1993 killed about 30,000 people, whereas the magnitude-6.7 Northridge, California, earthquake one year later resulted in only 61 deaths. What is the reason for such a difference in the death toll? Both earthquakes occurred in densely populated regions, but in India the brick and stone buildings could not withstand ground shaking; most collapsed, entombing their occupants.

Fire

In many earthquakes, particularly in urban areas, fire is a major hazard. Almost 90% of the damage done in the 1906 San Francisco earthquake was caused by fire. The shaking severed many of the electrical and gas lines, which touched off flames and started fires all over the city. Because the earthquake ruptured water mains, there was no effective way to fight the fires that raged out of control for three days, destroying much of the city.

Eighty-three years later, during the 1989 Loma Prieta earthquake, a fire broke out in the Marina district of San Francisco (▶ Figure 10.17). This time, however, the fire was contained within a small area because San Francisco had a system of valves throughout its water and gas pipeline system so that lines could be isolated from breaks (see "The San Andreas Fault" on pp. 316–317).

During the September 1, 1923, earthquake in Japan, fires destroyed 71% of the houses in Tokyo and practically all the houses in Yokohama. In all, 576,262 houses were destroyed by fire, and 143,000 people died, many as a result of the fire. A horrible example occurred in Tokyo where thousands of people gathered along the banks of the Sumida River to escape the raging fires. Suddenly, a firestorm swept over the area, killing more than 38,000 people. The fires from this earthquake were particularly devastating because most of the buildings were constructed of wood; many fires were started by chemicals and fanned by 20 km/hr winds.

Tsunami: Killer Waves

On December 26, 2004, a magnitude-9.0 earthquake struck 160 km off the west coast of northern Sumatra, Indonesia, generating the deadliest tsunami in history. Within hours, walls of water as high as 10.5 m pounded the coasts of Indonesia, Sri Lanka, India, Thailand, Somalia, Myanmar, Malaysia, and the Maldives, killing more than 220,000 people and causing billions of dollars in damage (▶ Figure 10.18).

This earthquake generated what is popularly called a "tidal wave" but more correctly termed a *seismic sea wave* or **tsunami,** a Japanese term meaning "harbor wave." The term *tidal wave* nevertheless persists in popular literature and some news

▶ **Figure 10.15 Liquefaction** The effects of ground shaking on water-saturated soil are dramatically illustrated by the collapse of these buildings in Niigata, Japan, during a 1964 earthquake. The buildings were designed to be earthquake resistant and fell over on their sides intact when the ground below them underwent liquefaction.

The San Andreas Fault

The circum-Pacific belt is well-known for its volcanic activitiy and earthquakes. Indeed, about 60% of all volcanic eruptions and 80% of all earthquakes take place in this belt, which nearly encircles the Pacific Ocean basin (see Figure 10.4).

One well-known and well-studied segment of the circum-Pacific belt is the 1300-km-long San Andreas fault, extending from the Gulf of California north through coastal California until it terminates at the Mendocino fracture zone off California's north coast. In plate tectonic terminology, it marks a transform plate boundary between the North American and Pacific plates (see Chapter 2).

Earthquakes along the San Andreas and related faults will continue to occur. But other segments of the circum-Pacific belt as well as the Mediterranean–Asiatic belt are also quite active and will continue to experience earthquakes.

▲ **1.** Aerial view of the San Andreas fault.

▲ **2.** View across the San Andreas fault at Tomales Bay, north of San Francisco. The low area occupied by the bay is underlain by shattered rocks of the San Andreas fault zone. Rocks underlying the hills in the distance are on the North American plate, whereas those at the point where this photograph was taken are on the Pacific plate.

▶ **3.** This shop in Olema, California, is rather whimsically called The Epicenter, alluding to the fact that it is in the San Andreas fault zone.

San Francisco following the 1906 earthquake. This view along Sacramento Street shows damaged buildings and the approaching fire.

▶ **4.** Rocks on opposite sides of the San Andreas fault periodically lurch past one another, generating large earthquakes. The most famous one destroyed San Francisco on April 18, 1906. It resulted when 465 km of the fault ruptured, causing about 6 m of horizontal displacement in some areas (see Figure 10.1b). It is estimated that 3000 people died. The shaking lasted nearly 1 minute and caused property damage estimated at $400 million in 1906 dollars! About 28,000 buildings were destroyed, many of them by the three-day fire that raged out of control and devastated about 12 km^2 of the city.

▶ **5.** Since 1906 the San Andreas fault and its subsidiary faults have spawned many more earthquakes; one of the most tragic was centered at Northridge, California, a small community north of Los Angeles. During the early morning hours of January 17, 1994, Northridge and surrounding areas were shaken for 40 seconds. When it was over, 61 people were dead and thousands injured; an oil main and at least 250 gas lines had ruptured, igniting numerous fires; nine freeways were destroyed; and thousands of homes and other buildings were damaged or destroyed by ground shaking. The nearly total destruction to this apartment complex resulted in 16 deaths.

▲ **6.** In Santa Clarita, California, a portion of Interstate 5 (Golden State Freeway) collapsed onto Interstate 14 during the 1994 Northridge earthquake. Only the supporting structure for this portion of Interstate 5 remains standing.

▶ **7.** A spectacular fire on Balboa Boulevard, in Northridge, was caused by a gas-main explosion during the earthquake.

▶ **Figure 10.16 Ground Shaking** Most of the buildings collapsed or were severely damaged as a result of ground shaking during the August 17, 1999, Turkey earthquake, which killed more than 17,000 people.

accounts, but these waves are not caused by or related to tides. Indeed, tsunami are destructive sea waves generated when the sea floor undergoes sudden, vertical movements. Many result from submarine earthquakes, such as the 2004 Indonesian earthquake, but volcanoes at sea or submarine landslides can also cause them. For example, the 1883 eruption of Krakatau between Java and Sumatra generated a large sea wave that killed 36,000 people on nearby islands.

Once a tsunami is generated, it can travel across an entire ocean and cause devastation far from its source. In the open sea, tsunami travel at several hundred kilometers per hour and commonly go unnoticed as they pass beneath ships because they are usually less than 1 m high and the distance between wave crests is typically hundreds of kilometers. When they enter shallow water, however, the wave slows down and water piles up to heights anywhere from a meter or two to many meters high (▶ Figure 10.19). The 1946 tsunami that struck Hilo, Hawaii, was 16.5 m high! In any case, the tremendous energy possessed by a tsunami is concentrated on a shoreline when it hits either as a large breaking wave or, in some cases, as what appears to be a very rapidly rising tide.

A common popular belief is that a tsunami is a single large wave that crashes onto a shoreline. Any tsunami consists of a series of waves that pour onshore for as long as 30 minutes followed by an equal time during which water rushes back to sea. Furthermore, after the first wave hits, more waves follow at 20- to 60-minute intervals. About 80 minutes after the 1755 Lisbon, Portugal, earthquake, the first of three tsunami, the largest more than 12 m high, destroyed the waterfront area and killed thousands of people. Following the arrival of a 2-m-high tsunami in Crescent City, California, in 1964, curious people went to the waterfront to inspect the damage. Unfortunately, 10 were killed by a following 4-m-high wave!

One of nature's warning signs of an approaching tsunami is a sudden withdrawal of the sea from a coastal region. In fact, the sea might withdraw so far that it cannot be seen and the seafloor is laid bare over a huge area. On more than one occasion, people have rushed out to inspect exposed reefs or to collect fish and shells, only to be swept away when the tsunami arrived. During the December 2004 tsunami, however, a 10-year-old British girl saved numerous lives because she recognized the warning signs she had learned in a school lesson on tsunami only two weeks

▶ **Figure 10.17 Fire** San Francisco Marina district fire caused by broken gas lines during the 1989 Loma Prieta earthquake.

before! While vacationing with her mother on the island of Phuket, Thailand, a popular resort area, the girl noticed the water quickly receding from the beach. She immediately told her mother that she thought a tsunami was coming, and her mother, along with the resort staff, quickly warned everyone standing around watching the water recede to clear the beach area. Their quick action resulted in many lives being saved.

Following the tragic 1946 tsunami that hit Hilo, Hawaii, the U.S. Coast and Geodetic Survey established a Pacific Tsunami Warning System in Ewa Beach, Hawaii (▶ Figure 10.20). This system combines seismographs and instruments that detect earthquake-generated waves. Whenever a strong earthquake takes place anywhere within the Pacific Ocean basin, its location is determined and instruments are checked to see whether a tsunami has been generated. If it has, a warning is sent out to evacuate people from low-lying areas that may be affected. Nevertheless, tsunami remain a threat to people in coastal areas, especially around the Pacific Ocean (Table 10.4). Unfortunately, no such warning system exists for the Indian Ocean. If one had been in place, it is possible that the death toll from the December 26, 2004, tsunami might not have been as high.

Ground Failure

Earthquake-triggered landslides are particularly dangerous in mountainous regions and have been responsible for tremendous amounts of damage and many deaths. The 1959 earthquake in Madison Canyon, Montana, for example, caused a huge rock slide (▶ Figure 10.21), and the 1970 Peru earthquake caused an avalanche that destroyed the town of Yungay and killed an estimated 66,000 people. Most of the 100,000 deaths from the 1920 earthquake in Gansu, China, resulted when cliffs composed of loess (wind-deposited silt) collapsed. More than 20,000 people were killed when two-thirds of the town of Port Royal, Jamaica, slid into the sea following an earthquake on June 7, 1692.

▶ **Figure 10.18 2004 Indian Ocean Tsunami** The magnitude-9.0 earthquake off the coast of northwest Sumatra in December 2004 generated a devastating tsunami throughout the Indian Ocean.

a Map showing the epicenter of the earthquake and its aftershocks equal to or greater than magnitude 4.

b Satellite image of the north shore of Banda Aceh, the capital city of Aceh Province, Sumatra, Indonesia, taken on June 23, 2004.

c A similar satellite image of the same area on December 28, 2004, two days after the tsunami struck. Notice the total destruction of all the buildings and one bridge.

d View of shoreline south of Aceh a few days after the tsunami. Tsunami flood waters continue to drain the land, but some flooding is also due to tectonic subsidence caused by the earthquake itself.

PHYSICAL Geology⇌Now™ ▶ Geo-focus/Active Figure 10.19 Anatomy of a Tsunami Displacement of the seafloor along a fault produces a tsunami that radiates outward in all directions. Tsunami in deep water can travel at high speeds. As the waves move into shallower water, they are slowed, producing the large walls of water that are so destructive when they reach the shoreline and flood coastal areas.

PHYSICAL Geology⇌Now™ ▶ Active Figure 10.20 The Pacific Tsunami Warning System The Pacific Tsunami Warning System consists of seismographs and various instruments that record large earthquakes in the Pacific Ocean basin that might generate a tsunami. If a tsunami has been generated, a warning is sent out to those regions likely to be affected by it. Shown here are the reporting stations and tsunami travel times to Honolulu, Hawaii.

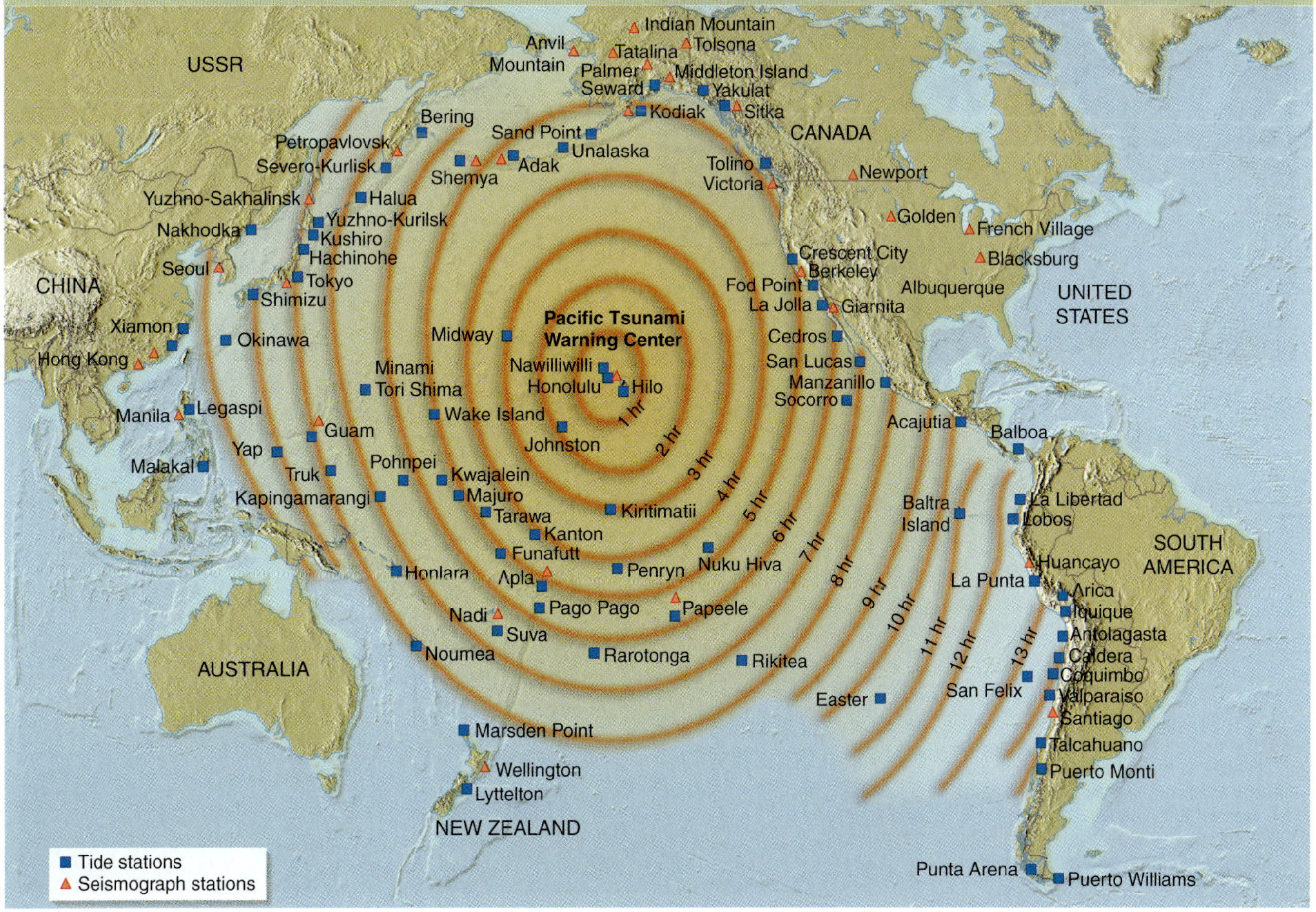

TABLE 10.4
Tsunami Fatalities Since 1990

Date	Location	Maximum Wave Height	Fatalities
September 2, 1992	Nicaragua	10 m	170
December 12, 1992	Flores Island	26 m	>1000
July 12, 1993	Okushiri, Japan	31 m	239
June 2, 1994	East Java	14 m	238
November 14, 1994	Mindoro Island	7 m	49
October 9, 1995	Jalisco, Mexico	11 m	1
January 1, 1996	Sulawesi Island	3.4 m	9
February 17, 1996	Irian Jaya	7.7 m	161
February 21, 1996	North coast of Peru	5 m	12
July 17, 1998	Papua New Guinea	15 m	>2200
December 26, 2004	Sumatra, indonesia	10.5 m	>220,000

Source: F. I. Gonzales, Tsunami! *Scientific American* 280, no. 5 (1999): 59, and United States Geological Survey.

▶ **Figure 10.21 Ground Failure** On August 17, 1959, an earthquake with a Richter magnitude of 7.3 shook southwestern Montana and a large area in adjacent states.

a The fault scarp in this image was produced when the block in the background moved up several meters relative to the one in the foreground.

b The earthquake triggered a landslide (visible in the distance) that blocked the Madison River in Montana and created Earthquake Lake (foreground). The slide entombed about 26 people in a campground at the valley bottom.

What Would You Do?

Your city has experienced moderate to large earthquakes in the past and as a result, the local planning committee, of which you are a member, has been charged with making recommendations as to how your city can best reduce damage as well as potential injuries and fatalities resulting from future earthquakes. You are told to consider zoning regulations; building codes for private dwellings, hospitals, public buildings, and high-rise structures; and emergency contingency plans. What kinds of recommendations would you make? What and whom would you ask for professional guidance?

Section 10.8 Summary

- Many factors determine an earthquake's destructiveness, including the time an earthquake strikes, population density, duration of the earthquake, the earthquake's magnitude, geology of the area, and the type of building construction.

- The destructive effects of earthquakes include ground shaking, fire, tsunami, and landslides. Ground shaking is the most destructive of all earthquake hazards. The loss of life, injuries, and amount of damage done depend on the earthquake's magnitude, distance from the epicenter, underlying geology, and type of building construction. Fire is a major hazard in urban areas and can cause more destruction than ground shaking in some instances.

- Tsunami are seismic sea waves produced by earthquakes, submarine landslides, and eruptions of volcanoes at sea. They are particularly destructive to coastal areas, even thousands of kilometers from the epicenter. Landslides are especially dangerous in mountainous areas.

10.9 Earthquake Prediction

■ *Can earthquakes be predicted?*

A successful prediction must include a time frame for the occurrence of an earthquake, its location, and its strength. Despite the tremendous amount of information geologists have gathered about the cause of earthquakes, successful predictions are still rare. Nevertheless, if reliable predictions can be made, they can greatly reduce the number of deaths and injuries.

From an analysis of historic records and the distribution of known faults, geologists construct **seismic risk maps** that indicate the likelihood and potential severity of future earthquakes based on the intensity of past earthquakes. An international effort by scientists from several countries resulted in the publication of the first Global Seismic Hazard Assessment Map in December 1999 (▶ Figure 10.22). Although such maps cannot be used to predict when an earthquake will

▶ **Figure 10.22 Global Seismic Hazard Assessment Map** The Global Seismic Hazard Assessment Program published this seismic hazard map showing peak ground accelerations. The values are based on 90% probability that the indicated horizontal ground acceleration during an earthquake is not likely to be exceeded in 50 years. The higher the number, the greater the hazard. As expected, the greatest seismic risks are in the circum-Pacific belt and the Mediterranean–Asiatic belt. **Why is this?** *Because approximately 80% of all earthquakes occur in the circum-Pacific belt and 15% occur in the Mediterranean–Asiatic belt.*

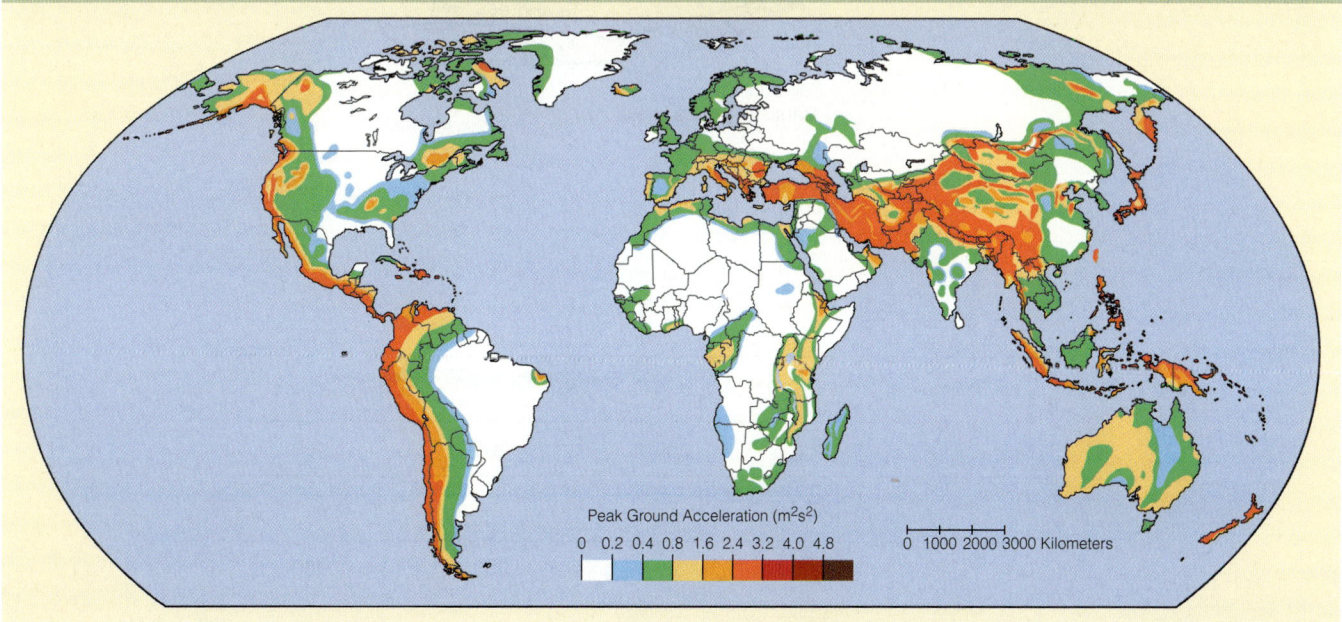

take place in any particular area, they are useful in anticipating future earthquakes and helping people plan and prepare for them.

■ *What are earthquake precursors?*

Studies conducted during the past several decades indicate that most earthquakes are preceded by both short-term and long-term changes within Earth. Such changes are called *precursors*.

One long-range prediction technique used in seismically active areas involves plotting the location of major earthquakes and their aftershocks to detect areas that have had major earthquakes in the past but are currently inactive. Such regions are locked and not releasing energy. Nevertheless, pressure is continuing to accumulate in these regions due to plate motions, making these *seismic gaps* prime locations for future earthquakes. Several seismic gaps along the San Andreas fault have the potential for future major earthquakes (▶ Figure 10.23). A major earthquake that damaged Mexico City in 1985 occurred along a seismic gap in the convergence zone along the west coast of Mexico.

Changes in elevation and tilting of the land surface have frequently preceded earthquakes and may be warnings of impending quakes. Extremely slight changes in the angle of the ground surface can be measured by tiltmeters. Tiltmeters have been placed on both sides of the San Andreas fault to measure tilting of the ground surface that is thought to result from increasing pressure in the rocks. Data from measurements in central California indicate significant tilting immediately preceding small earthquakes. Furthermore, extensive tiltmeter work performed in Japan prior to the 1964 Niigata earthquake clearly showed a relationship between increased tilting and the main shock. Although more research is needed, such changes appear to be useful in making short-term earthquake predictions.

Other earthquake precursors include fluctuations in the water level of wells and changes in Earth's magnetic field and the electrical resistance of the ground. These fluctuations are thought to result from changes in the amount of pore space in rocks because of increasing pressure.

The Chinese used the precursors just mentioned, except seismic gaps, to successfully predict a large earthquake in Haicheng on February 4, 1975. The earthquake had a magnitude of 7.5 and destroyed hundreds of buildings but claimed very few lives because most people had been evacuated from the buildings and were outdoors when it occurred.

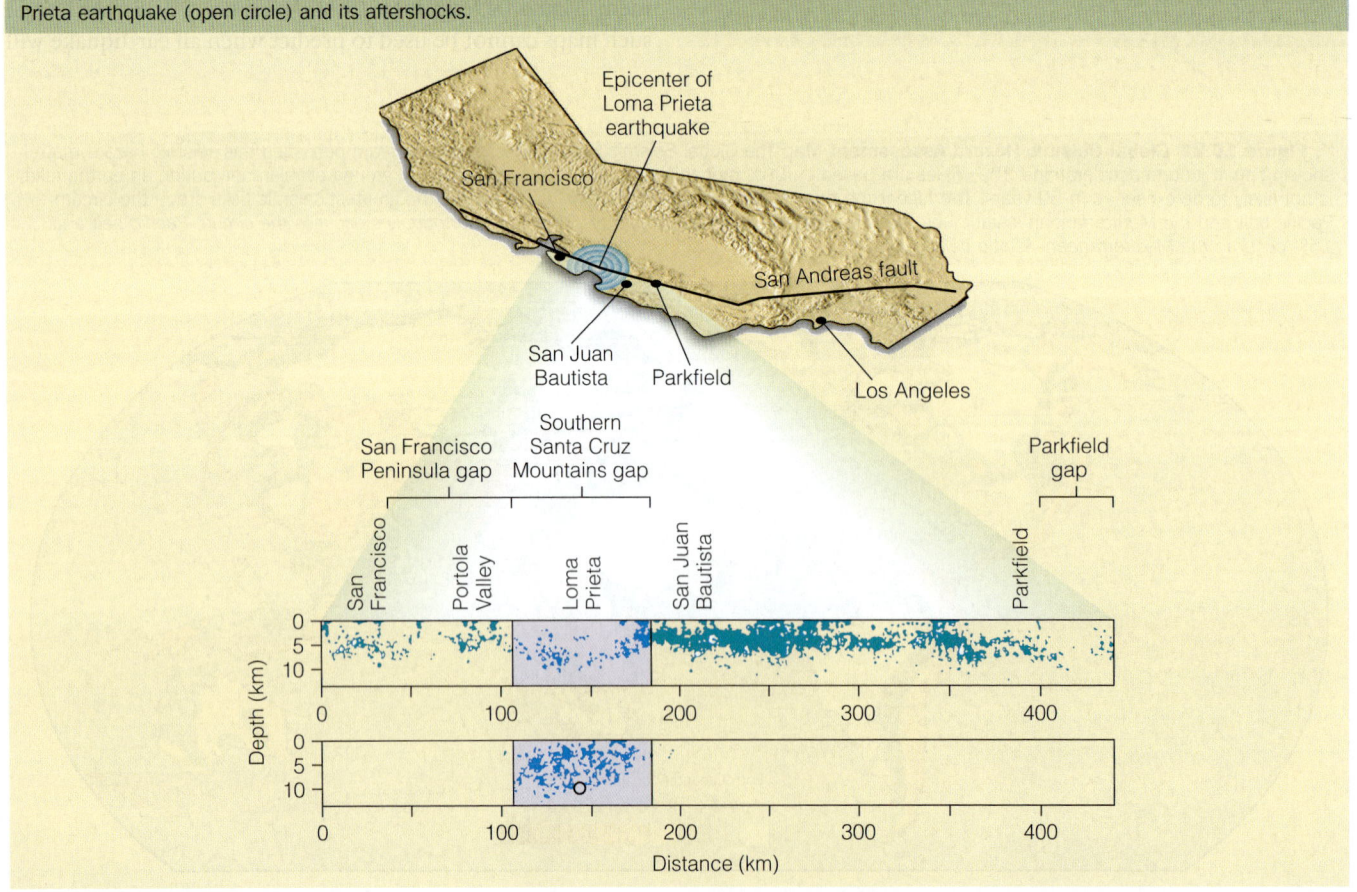

▶ **Figure 10.23 Earthquake Precursors** Seismic gaps are one type of earthquake precursor that can indicate a potential earthquake in the future. Seismic gaps are regions along a fault that are locked; that is, they are not moving and releasing energy. Three seismic gaps are evident in this cross section along the San Andreas fault from north of San Francisco to south of Parkfield. The first is between San Francisco and Portola Valley, the second near Loma Prieta Mountain, and the third southeast of Parkfield. The top section shows the epicenters of earthquakes between January 1969 and July 1989. The bottom section shows the southern Santa Cruz Mountains gap after it was filled by the October 17, 1989, Loma Prieta earthquake (open circle) and its aftershocks.

Unfortunately, visits by U.S. scientists revealed that the prediction resulted from unique circumstances that really could not be applied to earthquake prediction elsewhere. And for that matter, the Chinese failed to predict a 1976 earthquake that killed 242,000 people.

Many of the precursors just discussed can be related to the *dilatancy model*, which is based on changes occurring in rocks subjected to very high pressures, such as occurs along major faults. Laboratory experiments have shown that rocks undergo an increase in volume, known as *dilatancy*, just before rupturing. As pressure builds in rocks along faults, numerous small cracks are produced that alter the physical properties of the rocks. Water enters the cracks and increases the fluid pressure; this further increases the volume of the rocks and decreases their inherent strength until failure eventually occurs, producing an earthquake.

The dilatancy model is consistent with many earthquake precursors. Although additional research is needed, this model has the potential for predicting earthquakes under certain circumstances.

■ *Which nations have earthquake prediction programs, and what do they involve?*

Currently only four nations—the United States, Japan, Russia, and China—have government-sponsored earthquake prediction programs. These programs conduct laboratory and field studies of rock behavior before, during, and after large earthquakes, as well as monitoring activity along major active faults. Most earthquake prediction work in the United States is done by the U.S. Geological Survey (USGS) and involves research into all aspects of earthquake-related phenomena.

The Chinese have perhaps one of the most ambitious earthquake prediction programs in the world, which is understandable considering their long history of destructive earthquakes. Their earthquake prediction program was initiated soon after two large earthquakes occurred at Xingtai (300 km southwest of Beijing) in 1966. This program includes extensive study and monitoring of all possible earthquake precursors. In addition, the Chinese emphasize changes in phenomena that can be observed and heard without the use of sophisticated instruments. They successfully predicted the 1975 Haicheng earthquake, as we noted, but failed to predict the devastating 1976 Tangshan earthquake that killed at least 242,000 people.

Progress is being made toward dependable, accurate earthquake predictions, and studies are under way to assess public reactions to long-, medium-, and short-term earthquake warnings. However, unless short-term warnings are actually followed by an earthquake, most people will probably ignore the warnings, as they frequently do now for hurricanes, tornadoes, and tsunami. Perhaps the best we can hope for is that people in seismically active areas will take measures to minimize the risk of injury during and after the next major earthquake (Table 10.5).

TABLE 10.5

What You Can Do to Prepare for an Earthquake

Anyone who lives in an area that is subject to earthquakes or who will be visiting or moving to such an area can take certain precautions to reduce the risks and losses resulting from an earthquake.

Before an earthquake:

1. Become familiar with the geologic hazards of the area where you live and work.
2. Make sure your house is securely attached to the foundation by anchor bolts and that the walls, floors, and roof are all firmly connected together.
3. Heavy furniture such as bookcases should be bolted to the walls; semiflexible natural gas lines should be used so that they can give without breaking; water heaters and furnaces should be strapped and the straps bolted to wall studs to prevent gas-line rupture and fire. Brick chimneys should have a bracket or brace that can be anchored to the roof.
4. Maintain a several-day supply of freshwater and canned foods, and keep a fresh supply of flashlight and radio batteries as well as a fire extinguisher.
5. Maintain a basic first-aid kit and have a working knowledge of first-aid procedures.
6. Learn how to turn off the various utilities at your house.
7. Above all, have a planned course of action for when an earthquake strikes.

During an earthquake:

1. Remain calm and avoid panic.
2. If you are indoors, get under a desk or table if possible, or stand in an interior doorway or room corner as these are the structurally strongest parts of a room; avoid windows and falling debris.
3. In a tall building, do not rush for the stairwells or elevators.
4. In an unreinforced or other hazardous buliding, it may be better to get out of the building rather than to stay in it. Be on the alert for fallen power lines and the possibility of falling debris.
5. If you are outside, get to an open area away from buildings if possible.
6. If you are in an automobile, stay in the car, and avoid tall buildings, overpasses, and bridges if possible.

After an earthquake:

1. If you are uninjured, remain calm and assess the situation.
2. Help anyone who is injured.
3. Make sure there are no fires or fire hazards.
4. Check for damage to utilities and turn off gas valves if you smell gas.
5. Use your telephone only for emergencies.
6. Do not go sightseeing or move around the streets unnecessarily.
7. Avoid landslide and beach areas.
8. Be prepared for aftershocks.

Section 10.9 Summary

- Seismic risk maps help geologists determine the likelihood and potential severity of future earthquakes based on the intensity of past earthquakes. However, generally speaking, successful earthquake predictions are still rare because so many variables affect when and where an earthquake may occur.

- Earthquake precursors are changes that precede an earthquake. They include seismic gaps, which are locations along a fault that are locked and thus not releasing energy. Changes in surface elevations may also indicate a buildup of energy in rocks preceding a major earthquake.

- The dilatancy model is based on changes occurring in rocks subject to very high pressures, such as occurs along major faults. Rocks undergo an increase in volume (dilatancy) just before rupturing, and this may presage an earthquake.

- Various earthquake research programs are under way in the United States, Japan, Russia, and China. They involve laboratory and field studies of rock behavior before, during, and after large earthquakes, as well as monitoring major active faults.

10.10 Earthquake Control

■ *Can earthquakes be controlled?*

Reliable earthquake prediction is still in the future, but can anything be done to control or at least partly control earthquakes? Because of the tremendous energy involved, it seems unlikely that humans will ever be able to prevent earthquakes. However, it may be possible to gradually release the energy stored in rocks, thus decreasing the probability of a large earthquake and extensive damage.

During the early to mid-1960s, Denver, Colorado, experienced numerous small earthquakes. This was surprising because Denver had not been prone to earthquakes in the past. In 1962 geologist David M. Evans suggested that Denver's earthquakes were directly related to the injection of contaminated wastewater into a disposal well 3674 m deep at the Rocky Mountain Arsenal, northeast of Denver (▶ Figure 10.24a). The U.S. Army initially denied that a connection existed, but a USGS study concluded that the pumping of waste fluids into fractured rocks beneath the disposal well decreased the friction on opposite sides of fractures and, in effect, lubricated them so that movement occurred, causing the earthquakes that Denver experienced.

Figure 10.24b shows the relationship between the average number of earthquakes in Denver per month and the average amount of contaminated fluids injected into the disposal well per month. Obviously, a high degree of correlation between the two exists, and the correlation is particularly convincing considering that during the time when no waste fluids were injected, earthquake activity decreased dramatically.

Experiments conducted in 1969 at an abandoned oil field near Rangely, Colorado, confirmed the arsenal hypothesis. Water was pumped into and out of abandoned oil wells, the pore-water pressure in these wells was measured, and seismographs were installed in the area to measure any seismic activity. Monitoring showed that small earthquakes were occurring in the area when fluid was injected and that earthquake activity declined when the fluids were pumped out. What the geologists were doing was starting and stopping earthquakes at will, and the relationship between pore-water pressures and earthquakes was established.

Based on these results, some geologists have proposed that fluids be pumped into the locked segments or seismic gaps of active faults to cause small- to moderate-sized earthquakes. They think this would relieve the pressure on the fault and prevent a major earthquake from occurring. Although this plan is intriguing, it also has many potential problems. For instance, there is no guarantee that only a small earthquake might result. Instead, a major earthquake might occur, causing tremendous property damage and loss of life. Who would be responsible? Certainly, a great deal more research is needed before such an experiment is performed, even in an area of low population density.

As mentioned earlier, it appears that until such time as earthquakes can be accurately predicted or controlled, the best means of defense is careful planning and preparation (Table 10.5).

Section 10.10 Summary

- Because of the tremendous energy involved, it seems unlikely that humans will ever be able to prevent earthquakes. However, it might be possible to gradually release small amounts of the energy stored in rocks along faults and thus decrease the probability of a large earthquake.

- One promising possibility is to inject fluids into the locked portions of faults to cause small- to moderate-sized earthquakes, thus relieving the buildup of pressure that has the potential to cause a major earthquake. However, one potential problem with such a scenario is triggering a major earthquake and causing tremendous property damage and loss of life.

What Would You Do?

Some geologists think that by pumping liquids into locked segments of active faults, they can generate small- to moderate-sized earthquakes. These earthquakes would relieve the buildup of pressure along a fault and thus prevent very large earthquakes from taking place. What do you think of this proposal? What kind of social, political, and economic consequences would there be? Do you think such an effort will ever actually reduce the threat of earthquakes?

▶ **Figure 10.24 Controlling Earthquakes**

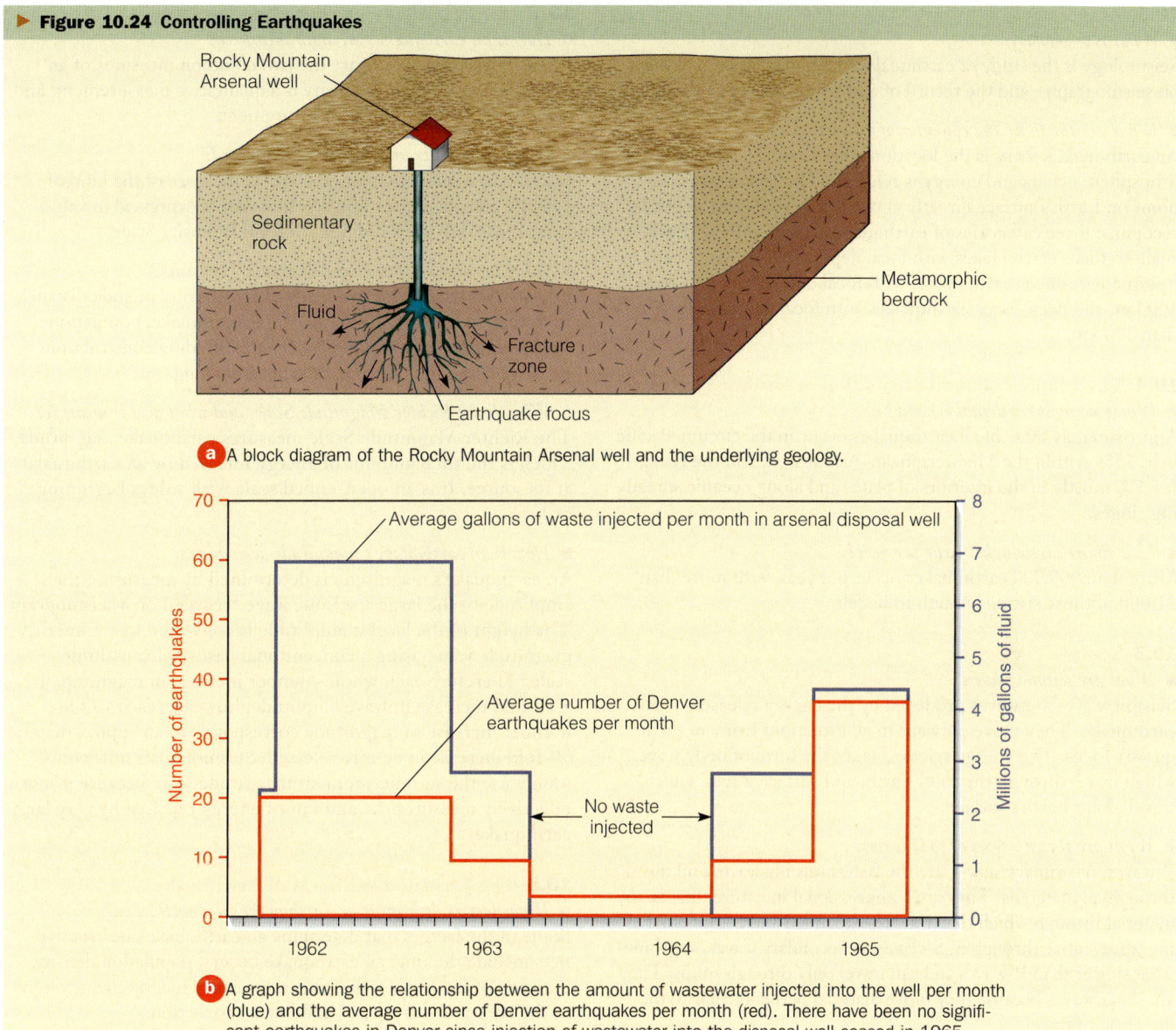

a A block diagram of the Rocky Mountain Arsenal well and the underlying geology.

b A graph showing the relationship between the amount of wastewater injected into the well per month (blue) and the average number of Denver earthquakes per month (red). There have been no significant earthquakes in Denver since injection of wastewater into the disposal well ceased in 1965.

Review Workbook

ESSENTIAL QUESTIONS SUMMARY

10.1 Introduction

■ *What is an earthquake?*
An earthquake is the shaking of the ground caused by the sudden release of energy, usually as a result of faulting.

■ *Why should we study earthquakes?*
Earthquakes are very destructive and cause many deaths and injuries every year. Even if you don't live in an earthquake-prone area, you might someday travel where there is the threat of an earthquake. Knowing what to do before, during, and after an earthquake could save your life or prevent serious injury.

10.2 Elastic Rebound Theory

■ *What is the elastic rebound theory, and what does it explain about earthquakes?*
The elastic rebound theory is an explanation for how energy is released during earthquakes. As rocks on opposite sides of a fault are subjected to force, they accumulate energy and slowly deform until their internal strength is exceeded. At that time, a sudden movement occurs along the fault, releasing the accumulated energy, and the rocks snap back to their original undeformed shape.

10.3 Seismology
■ *What is seismology?*
Seismology is the study of earthquakes. Earthquakes are recorded on seismographs, and the record of an earthquake is a seismogram.

■ *What are the focus and epicenter of an earthquake?*
An earthquake's focus is the location where rupture within Earth's lithosphere occurs and energy is released. The epicenter is the point on Earth's surface directly above the focus. Seismologists recognize three categories of earthquakes based on focal depth: shallow-focus earthquakes with focal depths of less than 70 km, intermediate-focus earthquakes with focal depths between 70 and 300 km, and deep-focus earthquakes with focal depths greater than 300 km.

10.4 The Frequency and Distribution of Earthquakes
■ *Where do most earthquakes occur?*
Approximately 80% of all earthquakes occur in the circum-Pacific belt, 15% within the Mediterranean–Asiatic belt, and the remaining 5% mostly in the interiors of plates and along oceanic spreading ridges.

■ *How many earthquakes occur per year?*
More than 900,000 earthquakes occur per year, with more than 31,000 of those strong enough to be felt.

10.5 Seismic Waves
■ *What are seismic waves?*
Seismic waves are waves produced by the energy released by an earthquake. They move outward in all directions from an earthquake's focus. The energy released takes the form of body waves, which travel through the solid Earth, and surface waves, which travel along Earth's surface.

■ *What are the two types of body waves?*
P-waves, or primary waves, are the fastest seismic waves and travel through all materials. They are compressional in nature; that is, the material through which they travel is expanded and compressed as the waves move through it. S-waves, or secondary waves, are somewhat slower than P-waves and can travel only through solids. They are shear waves because they move material perpendicular to the direction of travel.

■ *What are the two major types of surface waves?*
Rayleigh waves (R-waves) and Love waves (L-waves) move along or just below Earth's surface. R-waves are the slower of the two and behave like water waves in that the individual particles of material move in an elliptical path within a vertical plane oriented in the direction of wave movement. L-wave motion is similar to S-wave motion, but the individual particles of material move back and forth in a horizontal plane perpendicular to the direction of wave travel.

10.6 Locating an Earthquake
■ *How is the location of an earthquake determined?*
An earthquake's epicenter is determined by using a time–distance graph of the P- and S-waves to calculate how far away a seismic station is from an earthquake. The greater the difference in arrival times between the two waves, the farther away the seismic station is from the earthquake. By drawing a circle whose radius equals that distance for each of three different seismic stations, seismologists can determine the epicenter of the earthquake. The point where all three circles intersect marks the location of the epicenter.

10.7 Measuring the Strength of an Earthquake
■ *How is an earthquake's strength measured?*
Intensity and magnitude are the two common measures of an earthquake's strength. Intensity is a qualitative measurement, and magnitude is a quantitative measurement.

■ *What is intensity and how is it determined?*
Intensity is a subjective, or qualitative, measure of the kind of damage done by an earthquake. Intensity is expressed in values from I to XII in the Modified Mercalli Intensity Scale.

■ *What factors determine an earthquake's intensity?*
Factors that determine an earthquake's intensity include distance from the epicenter, focal depth of the earthquake, population density and geology of the area, type of building construction employed, and the duration of ground shaking.

■ *What is the Richter Magnitude Scale, and what does it measure?*
The Richter Magnitude Scale measures earthquake magnitude, which is the total amount of energy released by an earthquake at its source. It is an open-ended scale with values beginning at 1.

■ *How is an earthquake's magnitude determined?*
An earthquake's magnitude is determined by measuring the amplitude of the largest seismic wave recorded on a seismogram. The height of the largest amplitude is converted to a numeric magnitude value using a conventional base-10 logarithmic scale. Therefore each whole-number increase in magnitude is a 10-fold increase in wave amplitude; however, each whole-number increase in magnitude corresponds to an approximately 30-fold increase in energy released. Seismologists now commonly use the seismic-moment magnitude scale because it more effectively measures the amount of energy released by very large earthquakes.

10.8 The Destructive Effects of Earthquakes
■ *What factors determine an earthquake's destructiveness?*
Some of the factors that determine an earthquake's destructiveness include the time an earthquake occurs, population density, duration of the earthquake, the earthquake's magnitude, geology of the area, and the type of building construction.

■ *What are the destructive effects of earthquakes?*
These include ground shaking, fire, tsunami, landslides, and disruption of vital services.

10.9 Earthquake Prediction
■ *Can earthquakes be predicted?*
Seismic risk maps help geologists in determining the likelihood and potential severity of future earthquakes based on the intensity of past earthquakes. Generally speaking, however, successful earthquake predictions are still rare because so many variables affect when and where an earthquake may occur.

■ *What are earthquake precursors?*
Precursors are changes preceding an earthquake and include seismic gaps, changes in surface elevation, and fluctuations of water levels in wells. Such monitored changes can be useful in earthquake prediction. The dilatancy model is based on changes occurring in rocks subject to very high pressures, such as along major faults. Rocks undergo an increase in volume (dilatancy) just before rupturing, which may presage an earthquake.

- Which nations have earthquake prediction programs, and what do they involve?

A variety of earthquake research programs are under way in the United States, Japan, Russia, and China. They involve laboratory and field studies of rock behavior before, during, and after large earthquakes, as well as monitoring major active faults.

10.10 Earthquake Control
- Can earthquakes be controlled?

Because of the tremendous energy involved, it seems unlikely that humans will ever be able to prevent earthquakes. However, it might be possible to release small amounts of the energy stored in rocks and thus avoid a large earthquake and the extensive damage that typically results.

ESSENTIAL TERMS TO KNOW

body wave (p. 307)
circum-Pacific belt (p. 304)
earthquake (p. 300)
elastic rebound theory (p. 301)
epicenter (p. 303)
focus (p. 303)
intensity (p. 311)

Love wave (L-wave) (p. 308)
magnitude (p. 313)
Mediterranean–Asiatic belt (p. 306)
Modified Mercalli Intensity Scale (p. 311)
P-wave (p. 307)
Rayleigh wave (R-wave) (p. 308)

Richter Magnitude Scale (p. 313)
seismic risk map (p. 323)
seismograph (p. 303)
seismology (p. 303)
surface wave (p. 307)
S-wave (p. 308)
tsunami (p. 317)

REVIEW QUESTIONS

1. How does the elastic rebound theory account for the energy released during an earthquake?
2. What are the differences between intensity and magnitude? How are the two measured?
3. Describe how a tsunami is generated, how it travels, and what impact it has on shorelines.
4. What is the difference between an earthquake's focus and its epicenter? Why is an earthquake's epicenter the location that is usually reported in the news?
5. How are plate boundaries and earthquake foci related?
6. What are precursors, and how can they be used to predict earthquakes?
7. Why are structures built on bedrock usually not as severely damaged during an earthquake as those sited on unconsolidated material?
8. Describe the various ways earthquakes are destructive.
9. How does the seismic-moment magnitude scale differ from the Richter Magnitude Scale?
10. What are the two types of seismic body waves? How do they affect material as they pass through it? Why are the differences between these two waves important in determining the location of an earthquake?

APPLY YOUR KNOWLEDGE

1. From the arrival times of P- and S-waves shown in the table below and from the graph in Figure 10.10, calculate how far away from each seismograph station the earthquake occurred. How would you determine the epicenter of this earthquake?

	Arrival Time of P-Wave	Arrival Time of S-Wave
Station A	2:59:03 P.M.	3:04:03 P.M.
Station B	2:51:16 P.M.	3:01:16 P.M.
Station C	2:48:25 P.M.	2:55:55 P.M.

2. Refer to the graph in Figure 10.13. A seismograph in Berkeley, California, records the arrival time of an earthquake's P-waves at 6:59:54 P.M. and the S-waves at 7:00:02 P.M. The maximum amplitude of the S-waves as recorded on the seismogram was 75 mm. What was the magnitude of the earthquake, and how far away from Berkeley did it occur?

3. Approximately how much more energy was released by the 1960 Chile earthquake with a magnitude of 9.5 than by the 2003 Iran earthquake with a magnitude of 6.6? Even though the 1960 Chile earthquake was much more powerful in terms of energy released than the 2003 Iran earthquake, only 5700 people died in the Chile earthquake compared to 43,000 people in the Iran earthquake. Can you think of some possible reasons for the more powerful earthquake resulting in fewer deaths?

GEOLOGY MATTERS

GEOLOGY IN FOCUS   Paleoseismology

Paleoseismology is the study of prehistoric earthquakes. As more people move into seismically active areas, it is important to know how frequently earthquakes in the area have occurred in the past, and how strong those earthquakes were. In this way, prudent decisions can be made about what precautions need to be taken in developing an area and how stringent the building codes for a region need to be.

A typical technique in paleoseismology is to excavate trenches across active faults in an area to be studied and date the sediments disturbed by prehistoric earthquakes (▶ Figure 1). By exposing the upper few meters of material along an active fault, geologists can find evidence of previous earthquakes in the ancient soil layers. Furthermore, by dating the paleosoils by carbon 14 or other dating techniques, geologists can determine the frequency of past earthquakes and when the last earthquake occurred, and thus have a basis for estimating the probability of future earthquakes.

Paleoseismic studies are currently under way in many areas of North America, particularly along the San Andreas fault in California and in the coastal regions of Washington. An interesting case in point concerns an ancient earthquake in what is now Seattle, Washington.

Data from a variety of sources have convinced many geologists that a shallow-focus earthquake of at least magnitude 7 occurred beneath Seattle, Washington, less than 1100 years ago. In a point not lost on officials, they noted the catastrophic effects a similar-sized earthquake would have if it occurred in the same area today.

The first link in the chain of evidence for a paleoearthquake came from the discovery of a marine terrace that had been uplifted some 7 m at Restoration Point, 5 km west of Seattle. Carbon-14 analysis of peat within sediments of the terrace indicates that uplift occurred between 500 and 1700 years ago. Carbon-14 dating of other sites to the north, south, and east also indicates a sudden uplift in the area within the same time period. The amount of uplift suggests to geologists a magnitude-7 or greater earthquake.

Because many earthquakes in and around the Pacific Ocean basin can cause tsunami, geologists looked for evidence that a tsunami occurred, and they found it in the form of unusual sand layers in nearby tidal marsh deposits. Carbon-14 dating of organic matter associated with the sands yielded an age of 850 to 1250 years ago, well within the time period during which the terraces were uplifted.

Geologists also found evidence of rock avalanches in the Olympic Mountains that dammed streams, thereby forming lakes. Drowned trees in the lakes were dated as having died between 1000 and 1300 years ago, again fitting in nicely with the date of the ancient earthquake.

One of the final pieces of evidence is the deposits found on the bottom of Lake Washington. The earthquake apparently caused the bottom sediment of the lake to be resuspended and to move downslope as a turbidity current. Dating of these sediments indicates they were deposited between 940 and 1280 years ago, consistent with their being caused by an earthquake.

All evidence points to a large (magnitude 7 or greater) shallow-focus earthquake occurring in the Seattle area about 1000 years ago. If history and events of the geologic past are any guide, it is very likely that another large earthquake will hit the Seattle area in the future. When this will occur can't yet be predicted, but it would be wise to plan for such an eventuality. After all, metropolitan Seattle has a population of more than 2.5 million people, and its entire port area is built on fill that would probably be hard hit by an earthquake. Furthermore, most of Seattle's schools, hospitals, utilities, and fire and police stations are not built to withstand a strong earthquake! What the future holds for Seattle has yet to be determined. Geologists have provided a window on what has happened seismically in the past, and it is up to today's government and its various agencies to decide how they want to use this information.

Figure 1
Geologists examine a trench across an active fault in California to determine possible seismic hazards. Excavating trenches is a common method used by geologists to gather information about ancient earthquakes in a region and to help assess the potential for future earthquakes and the damage they might cause.

GEOLOGY IN FOCUS

Designing Earthquake-Resistant Structures

One way to reduce property damage, injuries, and loss of life is to design and build structures that are as earthquake resistant as possible. Many things can be done to improve the safety of current structures and of new buildings.

To design earthquake-resistant structures, engineers must understand the dynamics and mechanics of earthquakes, including the type and duration of the ground motion and how rapidly the ground accelerates during an earthquake. An understanding of the area's geology is also important because certain ground materials such as water-saturated sediments or landfill can lose their strength and cohesiveness during an earthquake. Finally, engineers must be aware of how different structures behave under different earthquake conditions.

With the level of technology currently available, a well-designed, properly constructed building should be able to withstand small, short-duration earthquakes of less than 5.5 magnitude with little or no damage. In moderate earthquakes (5.5–7.0 magnitude), the damage suffered should not be serious and should be repairable. In a major earthquake of greater than 7.0 magnitude, the building should not collapse, although it may later have to be demolished.

Many factors enter into the design of an earthquake-resistant structure, but the most important is that the building be tied together; that is, the foundation, walls, floors, and roof should all be joined together to create a structure that can withstand both horizontal and vertical shaking (▶ Figure 1). Almost all the structural failures resulting from earthquake ground movement occur at weak connections, where the various parts of a structure are not securely tied together. Buildings with open or unsupported first stories are particularly susceptible to damage. Some reinforcement must be done, or collapse is a distinct possibility (▶ Figure 1).

Tall buildings, such as skyscrapers, must be designed so that a certain amount of swaying or flexing can occur, but not so much that they touch neighboring buildings during swaying. If a building is brittle and does not give, it will crack and fail. Besides designed flexibility, engineers must ensure that a building does not vibrate at the same frequency as the ground does during an earthquake. When that happens, the force applied by the seismic waves at ground level is multiplied several times by the time they reach the top of the building.

Damage to high-rise structures can also be minimized or prevented by using diagonal steel beams to help prevent swaying. In addition, tall buildings in earthquake-prone areas are now commonly placed on layered steel and rubber structures and devices similar to shock absorbers that help decrease the amount of sway.

What about structures built many years ago? Just as in new buildings, the most important thing that can be done to increase the stability and safety of older structures is to tie together the different components of each building. This can be done by adding a steel frame to unreinforced parts of a building such as a garage, bolting the walls to the foundation, adding reinforced beams to the exterior, and using beam and joist connectors whenever possible. Although such modifications are expensive, they are usually cheaper than having to replace a building that was destroyed by an earthquake.

Figure 1
This illustration shows some of the things a homeowner can do to reduce damage to a building because of ground shaking during an earthquake. Notice that the structure must be solidly attached to its foundation, and bracing the walls helps prevent damage from horizontal motion.

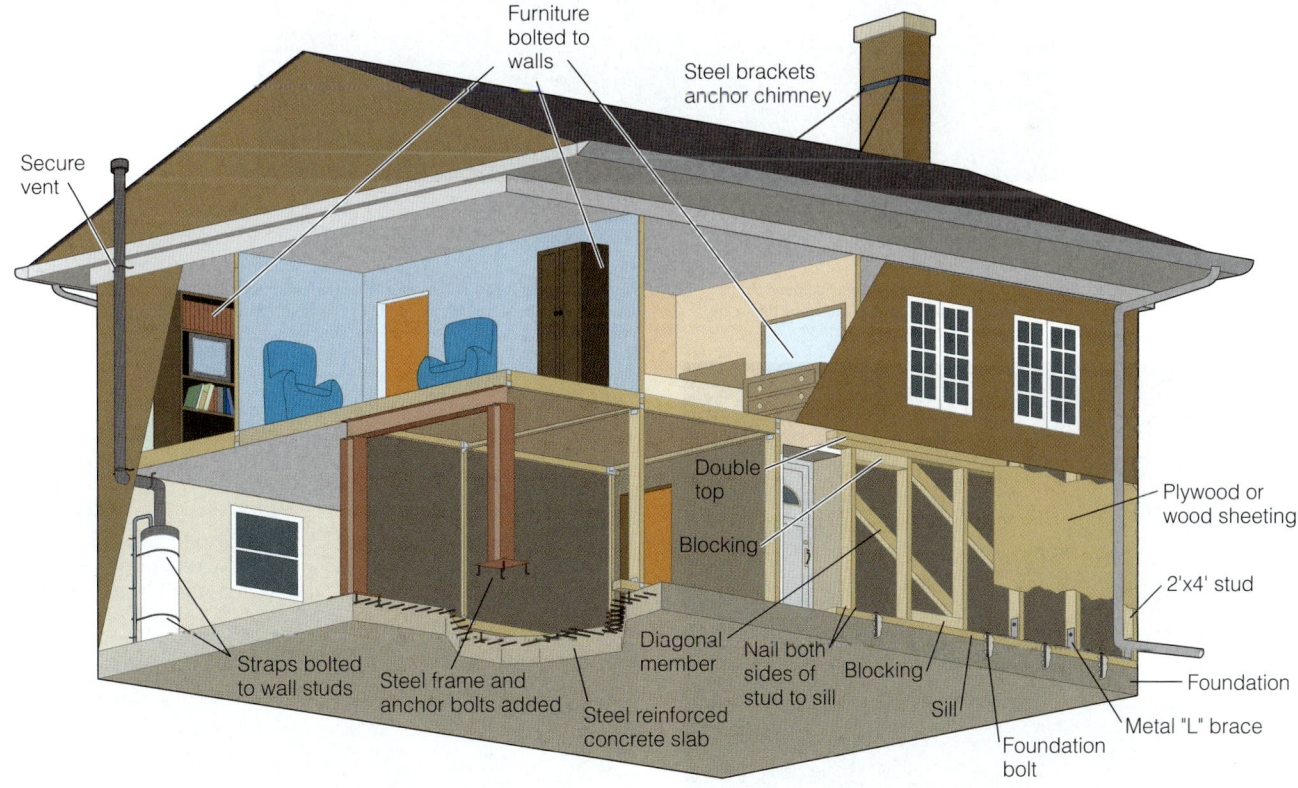

GEOLOGY MATTERS

GEOLOGY IN UNEXPECTED PLACES

It's Not My Fault

Are you the type of person who finds fault with others, or are you a geologist who just finds faults? Faults can be seen in some rather unexpected places. And what does finding a fault have to do with earthquakes? Remember that most earthquakes are caused by movement along a fault. This movement releases energy in the form of seismic waves, which are what we feel during an earthquake. So when we find faults or evidence of fault movement, the energy released along that fault may have generated an earthquake.

The most direct evidence of a fault is a fault scarp. Imagine my surprise when I was visiting a tranquil lake in Nanjing, China, many years ago and, while walking along a trail, noticed the afternoon sun reflecting off an ancient fault scarp, which is a bluff or cliff formed by the vertical movement of rocks along a fault (▶ Figure 1). That discovery certainly ranks as one of my all-time-great "geology in unexpected places" experiences!

Another time, while hiking in the Gallatin National Forest in Montana, I came across the approximately 6-m-high Red Canyon Fault scarp (▶ Figure 2). Vertical movement along this fault on August 17, 1959, produced a magnitude-7.3 earthquake that triggered a landslide that blocked the Madison River in Montana and created Earthquake Lake (see Figure 10.21). Because of the magnitude of the earthquake that occurred, I was not surprised to find a fault scarp in the area, but the height of the scarp and its preservation 18 years later were unexpected.

Sometimes we can fault others for not recognizing a potential problem when it presents itself. Such is the case in Dana Point, California, where houses have been built next to and over an ancient fault (▶ Figure 3). Although it is unlikely that the fault is active and movement will occur along it, erosion is proceeding along the fault at a faster rate than it is elsewhere in the area, causing the homeowners to take remedial action to slow the rate of erosion.

Figure 1
Light reflecting off a fault scarp along a lake trail in Nanjing, China. Note the highly polished scarp surface, indicating movement along this surface as rocks moved past each other.

Figure 2
An approximately 6-m-high fault scarp is exposed at Cabin Creek Campground in the Gallatin National Forest, Montana. Movement along the fault that produced this scarp, generated a magnitude-7.3 earthquake in the region on August 17, 1959.

Figure 3
This inactive fault can be recognized by the pink rocks (left) juxtaposed against white rocks (right). Because movement along this fault has pulverized the rocks, erosion has proceeded more rapidly along the fault, forming a narrow valley. Note the retaining wall in the upper right of the photo and the house behind it. Both the retaining wall and the house are built directly over this fault.

GEOLOGY MATTERS

Geology AT WORK

Pacific Tsunami Warning System

How would you like to work on a tropical island in a job that potentially may save tens of thousands of lives? Sounds good, doesn't it? And where do such jobs exist? At the Pacific Tsunami Warning Center in Ewa Beach, Hawaii. Established in 1949, the Pacific Tsunami Warning Center is the operational center of the Pacific Tsunami Warning System. Composed of 26 member states, the Pacific Tsunami Warning System has as its objectives to detect and locate major earthquakes within the Pacific Ocean basin that might generate a tsunami and to deliver timely warnings to those areas potentially affected by a tsunami, so as to minimize injuries and loss of life.

As we witnessed in the Indian Ocean at the end of 2004, powerful earthquakes can generate tsunami that cause tremendous destruction and loss of life along coastal areas up to thousands of kilometers from the epicenter of the earthquake. In an effort to reduce the loss of life and property damage in the Pacific Ocean basin, the National Oceanic and Atmospheric Administration's (NOAA) National Weather Service operates two tsunami warning centers. The Alaska Tsunami Warning Center in Palmer, Alaska, serves as the regional warning system for Alaska, British Columbia, Washington, Oregon, and California. The Pacific Tsunami Warning Center not only is responsible for issuing tsunami warnings for Hawaii but also acts as the international warning center for tsunami that might occur throughout the Pacific Ocean (▶ Figure 1).

How does the Tsunami Warning System work? Whenever an earthquake of approximately 6.5 magnitude or greater is detected anywhere in the Pacific Ocean region by participating seismic stations, an alarm is activated, and information on the earthquake is sent to the Pacific Tsunami Warning Center. After analyzing the data, center personnel issue a Tsunami Information Bulletin to participating warning system members.

A Tsunami Information Bulletin advises participants that a major earthquake in the Pacific region has occurred, resulting in one of three possibilities: (1) The center declares that a Pacific-wide tsunami was not generated and no further bulletins will be issued; (2) additional data are gathered and analyzed to determine whether a Pacific-wide tsunami was generated and additional bulletins will be issued hourly or sooner as more information becomes available; (3) or some areas might experience minor sea-level changes, but a destructive Pacific-wide tsunami threat does not exist and no further bulletins will be issued unless additional information warrants a Pacific-wide tsunami threat.

If an earthquake in the Pacific Ocean region is of 7.5 magnitude or greater (greater than 7.0 in the Aleutian Island region), then a Regional Tsunami Warning/Watch Bulletin is issued. This alerts all participants that a tsunami is possible and that a tsunami investigation is taking place. Areas within a 3-hour tsunami travel-time receive a tsunami warning, and those areas within a 3- to 6-hour tsunami travel-time are placed under tsunami watch status. Hourly updates are issued until either the warning/watch is cancelled or it is upgraded to a Pacific-wide tsunami warning.

A Pacific-wide tsunami warning is issued after the Pacific Tsunami Warning Center has determined that a tsunami has been generated that is capable of causing destruction beyond the local area and poses a threat to the coastal regions of the Pacific Ocean basin. Additional updates are issued hourly until the warning is cancelled.

How does the Pacific Tsunami Warning Center get the information needed to alert its Pacific Ocean basin member states? Earthquake information is relayed to the center by seismic stations operated by the Pacific Tsunami Warning Center, the Alaska Tsunami Warning Center, the U.S. Geological Survey's National Earthquake Information Center, as well as other international seismic stations. These raw data are fed into computers and analyzed to determine the appropriate level of response in terms of issuing a warning.

In addition to seismic stations, the Pacific Tsunami Warning System makes use of six deep-ocean buoys deployed throughout the Pacific Ocean (▶ Figure 2). These buoys consist of water pressure sensors anchored to the seafloor. A pressure sensor measures the weight of the overlying water. When a tsunami passes over the sensor, the water pressure increases, thus triggering the buoy to send a signal via a satellite transmitter to alert the center to tsunami activity. Although only six buoys are currently in operation, the George W. Bush administration has committed funds for the deployment of 32 new advanced-technology buoys within the Pacific, Caribbean, and Atlantic. These will provide an enhanced tsunami warning system by 2007 that will allow a response within minutes after an earthquake occurs.

Does the Pacific Tsunami Warning System work? One of the biggest problems for any type of natural disaster warning system is getting the word out to the public in a timely fashion, and making sure people know what to do when they are notified of an impending event. For example, when an alert is issued in Hawaii, beachside sirens go off and urgent messages are shown on television and broadcast over the radio. In addition, evacuation maps on telephone book covers show people how to get to higher ground. As Paul Whitmore of the West Coast and Alaska Tsunami Warning Center told the BBC News website, "Although I couldn't put a number on it, a lot of lives have been saved by the Pacific early warning system."

However, remote areas or islands that don't have modern communication systems will still suffer high causality rates. The biggest problem is establishing an effective communication infrastructure that can reach everyone. Even something as simple as hand-operated sirens along the beach or disaster personnel dispersed throughout low-lying areas blowing whistles to alert the populace can greatly reduce the loss of life. Without education of the population about tsunami and how to respond when warned, as well as the cooperation of governments, tens and hundreds of thousands of people will continue to lose their lives, in spite of the best efforts of scientists and engineers to develop a technologically sophisticated warning system.

Figure 1
The Pacific Tsunami Warning Center in Ewa Beach, Hawaii, is home to the Pacific Tsunami Warning System. From this unassuming building are issued tsunami bulletins, watches, and warnings to the 26 member states that participate in the warning system.

Figure 2
Part of the Pacific Tsunami Warning System, these deep-ocean buoys measure changes in water pressure resulting from the passage of a tsunami. When such changes are detected, the information is relayed via a satellite transmitter to the Pacific Tsunami Warning Center and the Alaska Tsunami Warning Center for analysis.

Chapter 11

Earth's Interior

On Hokusai's "Red Fugi"
 Mountain full of fire
Burning russet red against
 Indigo blue sky

Mt. Fuji, an example of a composite volcano, is the highest mountain in Japan. It is an active volcano with a low risk of an eruption. Like many of Earth's geological activities, volcanic eruptions are set into motion by processes within Earth's interior. Mt. Fuji's last documented eruption was in 1707. Throughout the ages it has been the focus of poems and paintings like the one pictured above by Japanese artist Hokusai. Each year thousands of people spend the night climbing the mountain to see nature's painting in the sky—sunrise.

—A. W.

ESSENTIAL QUESTIONS TO ASK

11.1 Introduction
- Why is it important to comprehend Earth's internal processes to more fully understand phenomena taking place at the surface?

11.2 Earth's Size, Density, and Internal Structure
- Why is it unlikely that vast caverns exist far below Earth's surface?
- What is Earth's overall density, how was it determined, and why must it increase with depth?
- How do geologists study Earth's interior?

11.3 Earth's Crust—Its Outermost Part
- How does continental crust differ from oceanic crust?
- What is the crust–mantle boundary called, and how was it discovered?

11.4 Earth's Mantle—The Layer Below the Crust
- What is the low-velocity zone in the mantle?
- How did geologists come to the conclusion that the mantle is probably made up of peridotite?

11.5 The Core
- What are the P- and S-wave shadow zones, and what do they indicate about Earth's interior?
- What are the density and composition of the inner and outer cores?

11.6 Earth's Internal Heat
- What is the geothermal gradient, and can it be used to determine the temperature of Earth's core?

11.7 Gravity and How Its Force Is Determined
- What are positive and negative gravity anomalies?

11.8 Floating Continents—The Principle of Isostasy
- How is it possible for a solid (Earth's crust) to "float" in another solid (the mantle)?

11.9 Earth's Magnetic Field
- Where and how is Earth's magnetic field generated?

GEOLOGY MATTERS

GEOLOGY IN FOCUS:
Planetary Alignments, Gravity, and Catastrophes

GEOLOGY IN UNEXPECTED PLACES:
Diamonds and Earth's Interior

This icon, which appears throughout the book, indicates an opportunity to explore interactive tutorials, animations, or practice problems available on the Physical GeologyNow website at http://now.brookscole.com/phygeo6.

335

11.1 Introduction

Although you may never have experienced a volcanic eruption, an earthquake, or a flood, you have certainly seen reports or TV specials on these topics or at least read about them. In contrast, most people have not given much thought to Earth's interior, and as a result it is a difficult topic for beginning students. You might appreciate the stunning beauty of the aurora borealis (northern lights) and be completely unaware that Earth's magnetic field is partly responsible for them. Given that Earth's interior is completely inaccessible, most people have little idea of its internal structure and composition or how they were determined.

■ *Why is it important to comprehend Earth's internal processes to more fully understand phenomena taking place at the surface?*

Even though Earth's interior may seem a rather esoteric topic, there are nevertheless important reasons to study it. Volcanic eruptions, earthquakes, moving tectonic plates, and mountain building are seen or experienced at the surface, but they are generated by dynamic processes operating within our planet—particularly Earth's internal heat. Indeed, the slow release of heat from the interior is one important factor accounting for the fact that Earth is such a dynamic planet.

Scientists now have a reasonably good idea of Earth's overall density, composition, and internal structure (▶ Figure 11.1). During most of historic time, though, people perceived of Earth's interior as an underground world of vast caverns, heat, and sulfur gases populated by demons or the souls of those waiting to be judged. Many cultures have myths to explain volcanoes and earthquakes as manifestations of the activities of deities that dwell far below the surface. The Romans believed that Vulcan, the god of fire, had underground workshops where his beating on anvils caused the ground to rumble and volcanoes to erupt. And in Mexico earthquakes were attributed to El Diablo (the devil) as he tried to rip open Earth's crust.

From our perspective these ancient myths may seem naïve, but some rather bizarre ideas about Earth's interior persist even now (see "Earth's Place in the Cosmos" on pp. 338–339). For instance, in 1869 Cyrus Reed Teed claimed that Earth is hollow and that humans live on the inside. And in 1913 Marshall B. Gardener held that Earth is a large, hollow sphere with a 1300-km-thick outer shell surrounding a central Sun, and access can be gained through holes at the poles. Even today, books and articles promote the hollow Earth concept, and several websites are devoted to this thesis. According to some, Earth's interior has been visited by groups of explorers in recent times, and the government is aware of the openings to Earth's interior but keeps this information from us. Other books, articles, and websites hold that Earth is flat, at the center of the universe, or both.

Jules Verne made no claim to present a reliable picture of Earth's interior in his 1864 novel *Journey to the Center of the Earth*, in which he described the adventures of Professor Hardwigg, his nephew, and an Icelandic guide. The trio descended through a volcano in Iceland and fol-

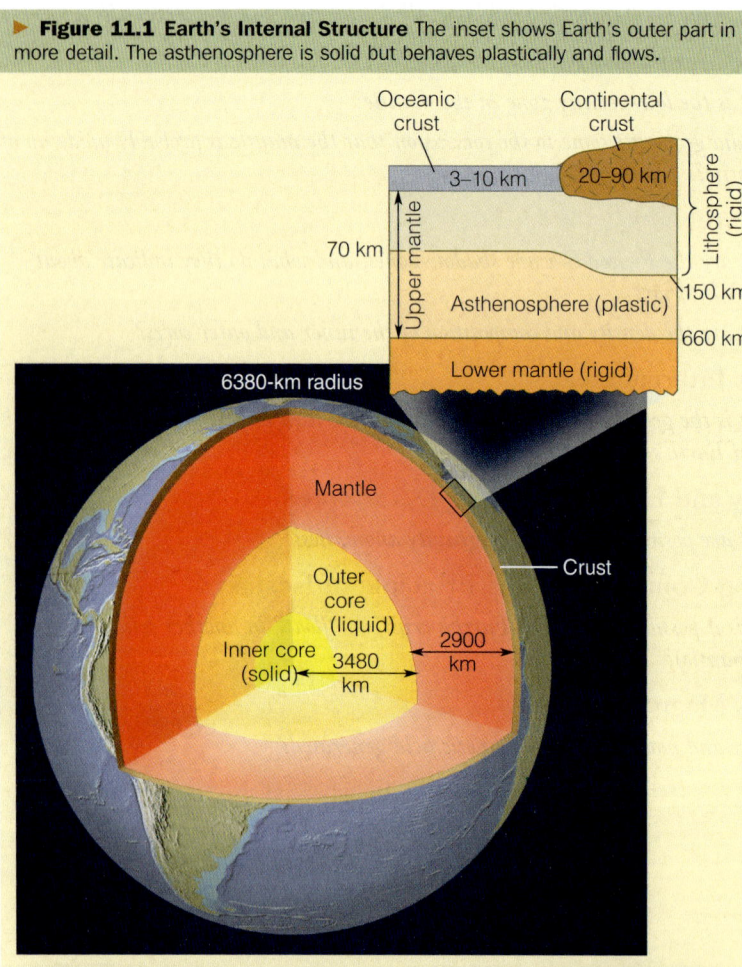

▶ **Figure 11.1 Earth's Internal Structure** The inset shows Earth's outer part in more detail. The asthenosphere is solid but behaves plastically and flows.

Earth's Composition and Density

	Composition	Density (g/cm³)
Continental crust	Average composition of granodiorite	~2.7
Oceanic crust	Upper part basalt, lower part gabbro	~3.0
Mantle	Peridotite (made up of ferromagnesian silicates)	3.3–5.7
Outer core	Iron with perhaps 12% sulfur silicon, oxygen, nickel, and potassium	9.9–12.2
Inner core	Iron with 10–20% nickel	12.6–13.0

lowed a labyrinth of passageways until they arrived 140 km below the surface. (In the 1959 movie of the same name, three men and one woman reached the center of the planet.) Professor Hardwigg and his companions encountered a vast cavern illuminated by some electrical phenomenon, and they saw herds of mastodons (extinct relatives of elephants) and a gigantic human shepherd, Mesozoic-aged reptiles, and huge turtles in what Verne call the "central sea." Their adventure ended when they were carried back to the surface on a rising plume of water (lava in the movie).

Section 11.1 Summary

- Earth's internal heat is responsible for several phenomena that we experience at the surface, such as volcanism, earthquakes, and moving plates.

11.2 Earth's Size, Density, and Internal Structure

Scientists can make no direct observations of Earth's interior, so how is it possible to determine Earth's overall density and what it looks like inside? Its size is less problematic. Scientists can calculate the density from the gravitational attraction exerted by Earth or other objects, such as the Moon, and its size. Additionally, they study seismic waves, meteorites, and inclusions in igneous rocks, and then evaluate the results of laboratory experiments to figure out the internal structure.

Even when Jules Verne wrote his novel in 1864 scientists knew that Earth's overall density is about 5.5 g/cm^3 and that pressure and temperature increase with depth. Little else was known, although humans had probed beneath the surface with mines and wells for centuries. Even the deepest mines extend down to only about 3 km, however, whereas the deepest drill hole, about 12 km, penetrates only 0.18% of the distance to the center of the planet (▶ Figure 11.2).

■ *Why is it unlikely that vast caverns exist far below Earth's surface?*

No vast caverns exist far below Earth's surface, even at the rather modest depth of 140 km as in Jules Verne's story, where rock is so hot, yet solid, and under such tremendous pressure that it yields by flowage to any applied force. Accordingly, even if an open space somehow existed, rock flowage would quickly eliminate it. In fact, rocks encountered in very deep mines and drill holes behave very differently than they do at the surface. Remember from Chapter 6 that, because of pressure release, rock bursts and popping are constant problems in mines and that rocks that form under great pressure expand outward and fracture when exposed at the surface (see Figure 6.5).

Among the terrestrial planets, Earth, with a diameter of 12,760 km at the equator, is slightly larger than Venus and much larger than Mercury, Mars, and Earth's Moon. Its equatorial diameter is slightly greater than its polar diameter, so it is an oblate spheroid rather than a true sphere. Scientists have known for more than 200 years that Earth is not homogeneous throughout. Sir Isaac Newton (1642–1727) noted that Earth's average density—that is, its mass per unit of volume—is 5.0 to 6.0 g/cm^3, and in 1797 Henry Cavendish calculated a density value very close to the 5.52 g/cm^3 now accepted.

■ *What is Earth's overall density, how was it determined, and why must it increase with depth?*

To accurately determine Earth's density, first its size had to be known. In the third century B.C. the Greek librarian Eratosthenes (c.276–c.195 B.C.) determined that Earth's circumference is about 40,000 km—a figure only slightly different from the one accepted now (see "Earth's Place in the Cosmos" on pp. 338–339). From this, one can calculate Earth's diameter and radius, and then determine the volume of our nearly spherical planet.

▶ **Figure 11.2 Drill Rig That Penetrated 12 km into Earth's Crust** This 30-story structure on the Kola Peninsula in northwestern Russia houses a drill rig that penetrated about 12 km into Earth's crust. A 12-km-deep drill hole is impressive, but put into perspective it would be about 0.5 mm deep if the vertical dimension of this page represented Earth's radius.

Earth's Place in the Cosmos

Various views of Earth's size, shape, and its place in the cosmos. During much of historic time, people viewed Earth as flat, hollow, inhabited by demons, and the center of the cosmos. We now know that Earth's internal heat is responsible for plate movements, earthquakes, and volcanism, and electrical currents in Earth's core generate the magnetic field.

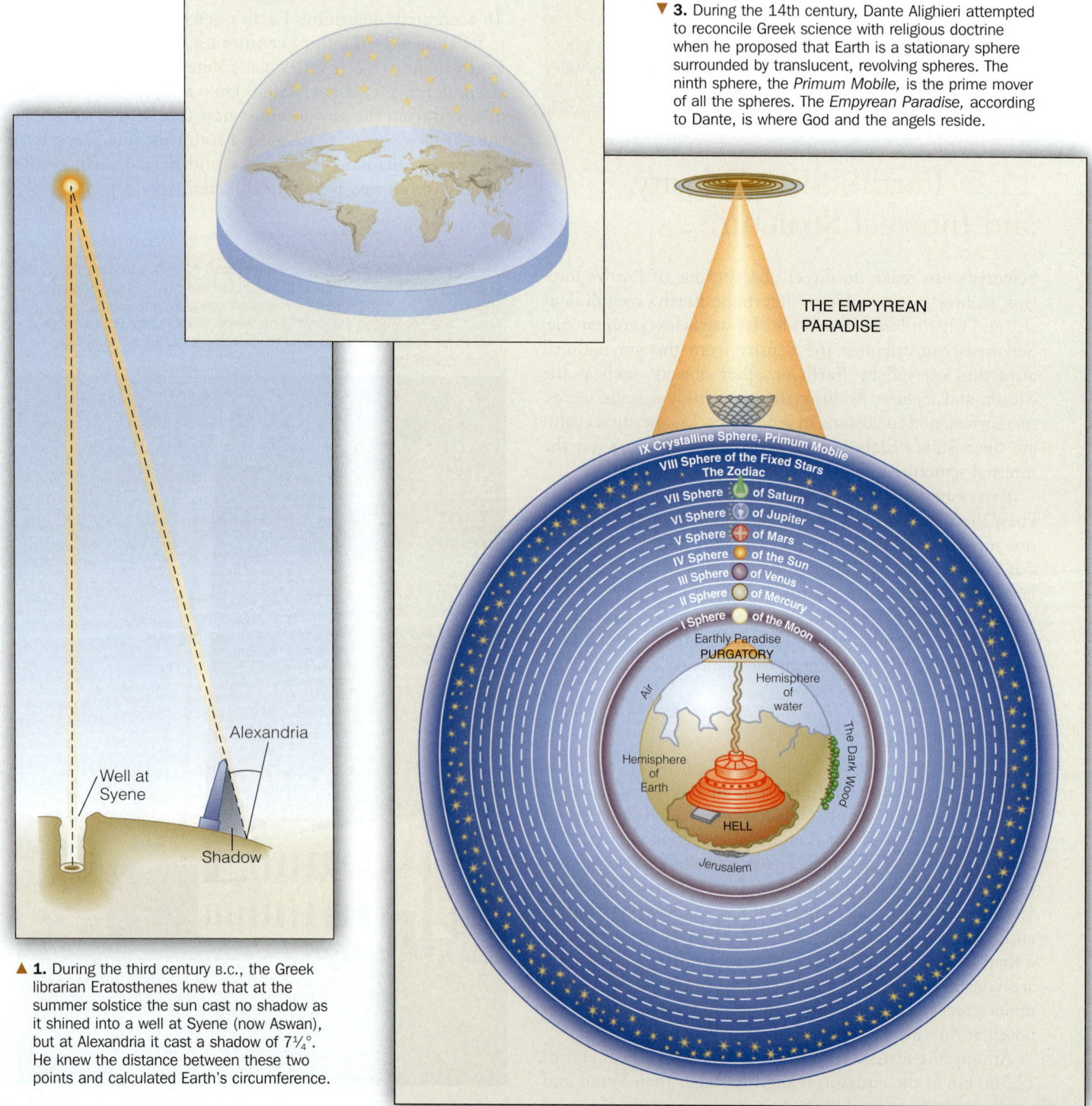

▼ 2. Even though the ancient Greeks accepted that Earth is spherical, the idea of a flat Earth persisted for centuries. In this depiction, the flat Earth has a crystal half-sphere over it that keeps the oceans from spilling off the edges.

▼ 3. During the 14th century, Dante Alighieri attempted to reconcile Greek science with religious doctrine when he proposed that Earth is a stationary sphere surrounded by translucent, revolving spheres. The ninth sphere, the *Primum Mobile,* is the prime mover of all the spheres. The *Empyrean Paradise,* according to Dante, is where God and the angels reside.

▲ 1. During the third century B.C., the Greek librarian Eratosthenes knew that at the summer solstice the sun cast no shadow as it shined into a well at Syene (now Aswan), but at Alexandria it cast a shadow of 7¼°. He knew the distance between these two points and calculated Earth's circumference.

▼ 5. Jules Verne made no claims of scientific accuracy in his book *A Journey to the Center of the Earth,* but the ideas of passages leading to great depths and a hollow Earth persist even now. The figure on the left is from an 1871 translation of the book; the Central Sea with its inhabitants is shown.

▲ 6. In this view from the 1959 movie *Journey to the Center of the Earth,* actors James Mason and Arlene Dahl explore the remains of a lost civilization as members of an expedition to Earth's center.

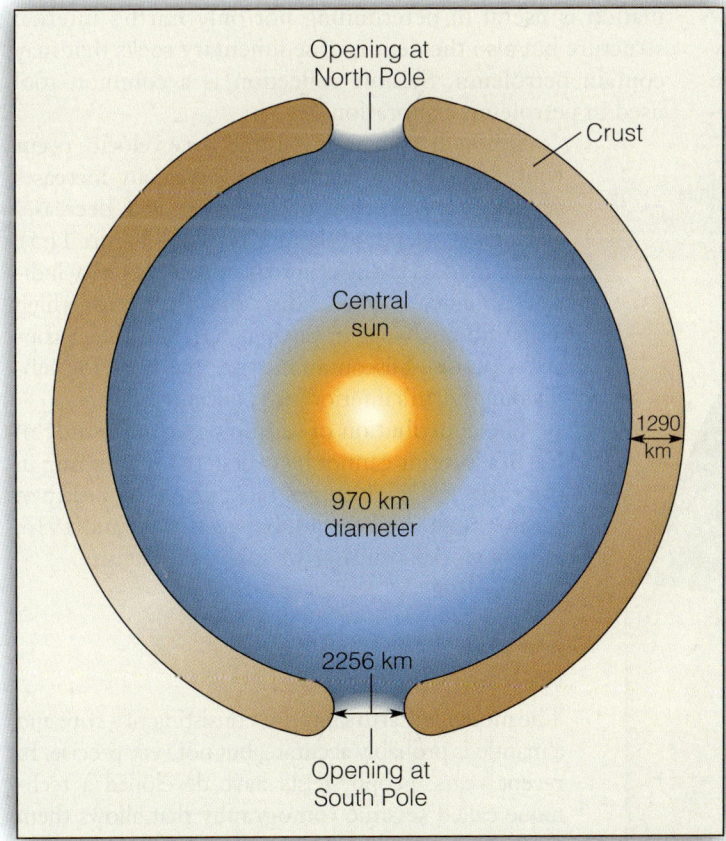

▲ 4. A variety of ideas persist on the concept of a hollow Earth. In this view, Earth consists of a 1290-km-thick crust with vast openings at the poles, and it has a central sun about 970 km in diameter. People live on Earth's outer surface, some live in middle Earth, that is, within the crust, and some live on the crust's inner surface.

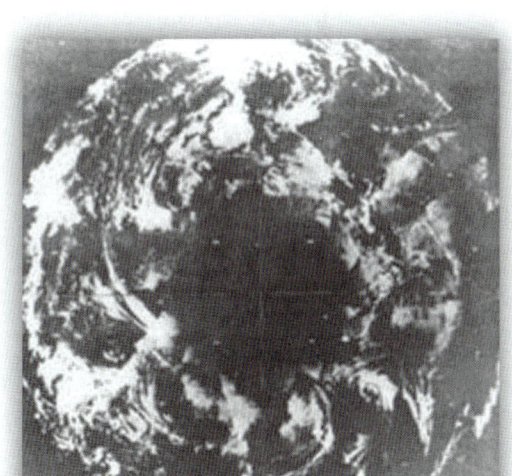

▲ 7. This 1967 satellite image shows what hollow-Earth advocates claim is a 2256-km-diameter hole at the North Pole. Notice that the entire area beyond the "hole" is in sunlight, an impossible situation unless Earth's North Pole points toward the Sun. The image is a composite of many images taken during January, a time when the North Polar region receives no sunlight. Also, if Earth has an internal sun, why isn't the hole illuminated?

Density, however, is another matter; it can be found only indirectly by, for example, comparing the gravitational attraction between Earth and its Moon and between metal spheres of known mass. Scientists carried out these calculations and determined an average density for the planet of 5.52 g/cm^3. But we know that most of Earth's surface and near-surface rocks have densities of only 2.5 to 3.0 g/cm^3. Accordingly, the materials that make up Earth's interior must be much denser than those near the surface, otherwise, the overall density would be less than 5.52 g/cm^3.

Seismic Waves and What They Tell Us about Earth's Interior

■ *How do geologists study Earth's interior?*

The behavior and travel times of P- and S-waves provide geologists with information about Earth's internal structure. Seismic waves travel outward as wave fronts from their source areas, although it is most convenient to depict them as *wave rays*, which are lines showing the direction of movement of small parts of wave fronts (▶ Figure 11.3). Any disturbance such as a passing train or construction equipment causes seismic waves, but only those generated by large earthquakes, explosive volcanism, asteroid impacts, and nuclear explosions travel completely through Earth.

As we noted in Chapter 10, P- and S-wave velocity is determined by the density and elasticity of the materials they travel through, both of which increase with depth. Wave velocity is slowed by increasing density but increases in materials with greater elasticity. Because elasticity increases with depth faster than density, a general increase in seismic wave velocity takes place as the waves penetrate to greater depths. P-waves travel faster than S-waves under all circumstances, but unlike P-waves, S-waves are not transmitted through liquids because liquids have no shear strength (rigidity); liquids simply flow in response to shear stress.

If Earth were a homogeneous body, P- and S-waves would travel in straight paths as shown in ▶ Figure 11.4a. But as a seismic wave travels from one material into another of different density and elasticity, its velocity and direction of travel change. That is, the wave is bent, a phenomenon known as **refraction,** in much the same way light waves are refracted as they pass from air into more dense water. Because seismic waves pass through materials of differing density and elasticity, they are continuously refracted so that their paths are curved; wave rays travel in a straight line only when their direction of travel is perpendicular to a boundary (▶ Figure 11.4b, c).

In addition to refraction, seismic waves are **reflected,** much as light is reflected from a mirror. When seismic waves encounter a boundary separating materials of different density or elasticity, some of a wave's energy is *reflected* back to the surface (▶ Figure 11.4c). If we know the wave velocity and the time required for the wave to travel from its source to the boundary and back to the surface, we can calculate the depth of the reflecting boundary. Such information is useful in determining not only Earth's internal structure but also the depths of sedimentary rocks that may contain petroleum. Seismic reflection is a common tool used in petroleum exploration.

Although changes in seismic wave velocity occur continuously with depth, P-wave velocity increases suddenly at the base of the crust and decreases abruptly at a depth of about 2900 km (▶ Figure 11.5). These marked changes in seismic wave velocity indicate a boundary called a **discontinuity** across which a significant change in earth materials or their properties occurs. Discontinuities are the basis for subdividing Earth's interior into concentric layers.

The contribution of seismology to the study of Earth's interior cannot be overstated. Beginning in the early 1900s, scientists recognized the utility of seismic wave studies and, between 1906 and 1936, largely worked out Earth's internal structure.

Seismic Tomography and Earth's Interior

The model of Earth's interior consisting of a core and a mantle is probably accurate, but not very precise. In recent years, geophysicists have developed a technique called **seismic tomography** that allows them to develop more accurate models of Earth's interior. In seismic tomography, numerous crossing seismic waves are analyzed much as CAT (computerized axial tomography) scans are analyzed. In CAT scans, X rays penetrate the body, and a two-dimensional image of its interior is formed. Repeated CAT scans

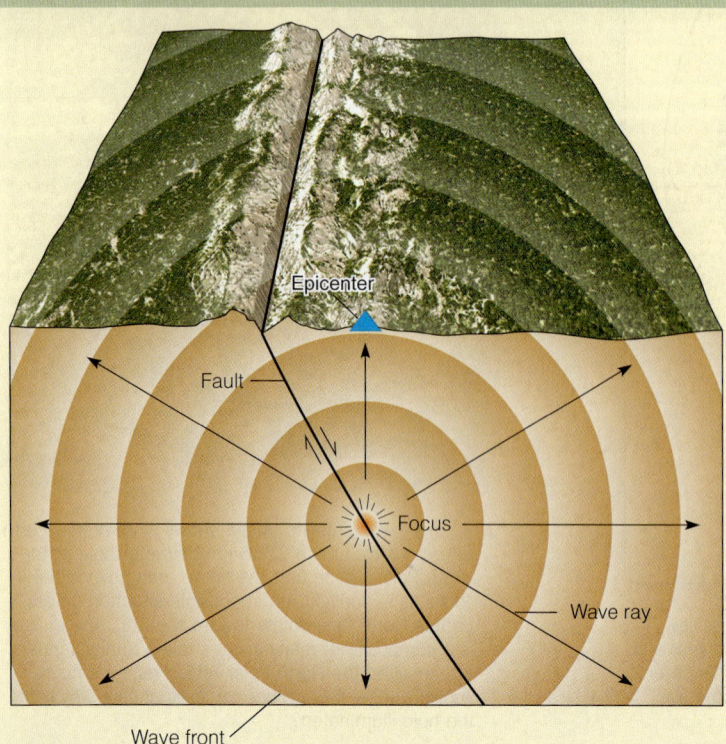

▶ **Figure 11.3 Seismic Wave Fronts** Seismic wave fronts move out in all directions from their source, the focus of an earthquake in this example. Wave rays are lines drawn perpendicular to wave fronts.

▶ **Figure 11.4** What if Earth Were Homogeneous?

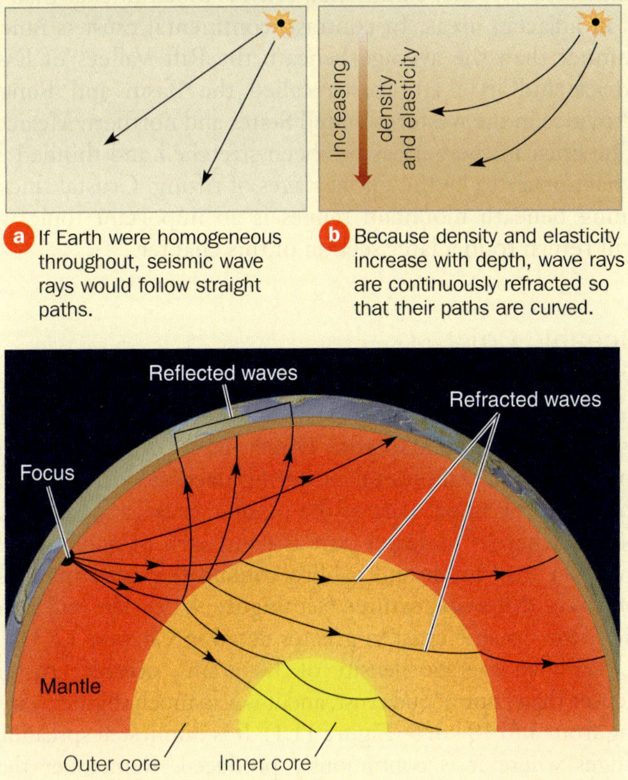

a If Earth were homogeneous throughout, seismic wave rays would follow straight paths.

b Because density and elasticity increase with depth, wave rays are continuously refracted so that their paths are curved.

c Refraction and reflection of P-waves as they encounter boundaries separating materials of different density or elasticity. Notice that the only wave ray not refracted is the one perpendicular to boundaries.

from slightly different angles are stacked to produce a three-dimensional image.

In a similar manner, geophysicists use seismic waves to probe Earth's interior. In seismic tomography, the average velocities of numerous crossing seismic waves are analyzed so that "slow" and "fast" areas of wave travel are detected. Remember that seismic wave velocity depends partly on elasticity; cold rocks have greater elasticity and thus transmit seismic waves faster than hotter rocks.

As a result of studies in seismic tomography, a much clearer picture of Earth's interior is emerging. It has already given us a better understanding of complex convection within the mantle and a clearer picture of the nature of the mantle–core boundary.

What Would You Do?

Of course, novels such as *Journey to the Center of the Earth* are fiction, but it is surprising how many people think that vast caverns and cavities exist deep within the planet. How would you explain that even though we have no direct observations at great depth, we can still be sure that these proposed openings do not exist?

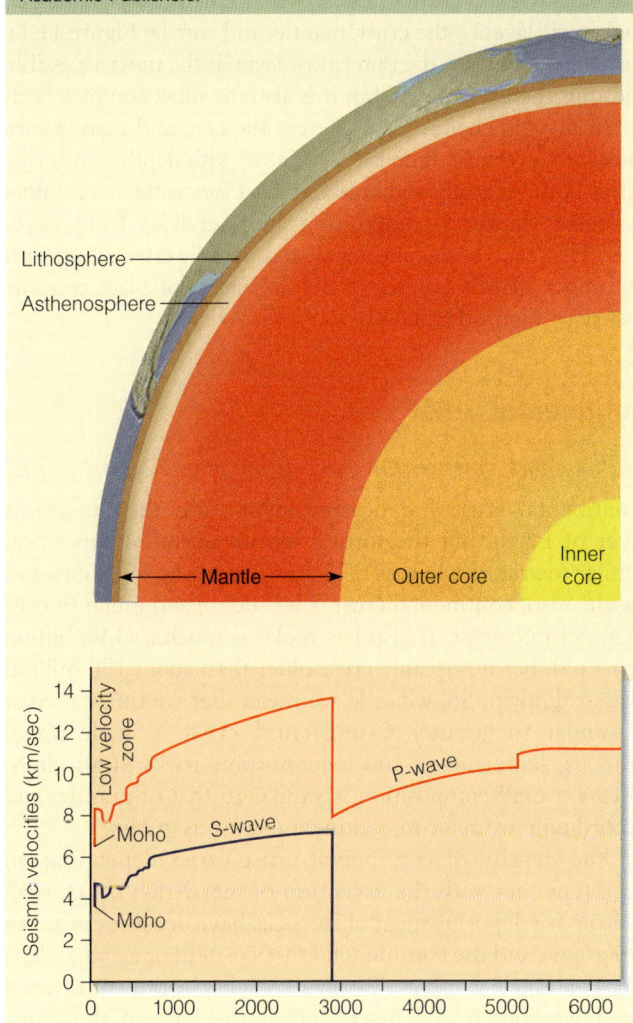

▶ **Figure 11.5** Seismic Wave Velocities Profiles showing seismic wave velocities versus depth. Several discontinuities are shown across which seismic wave velocities change rapidly. Source: From G. C. Brown and A. E. Musset, *The Inaccessible Earth* (Kluwer Academic Publishers, 1981), Figure 12.7a. Reprinted with the kind permission of Kluwer Academic Publishers.

Section 11.2 Summary

- Earth's overall density is 5.52 g/cm³, but the density of near-surface rocks is mostly between 2.5 to 3.0 g/cm³, so Earth's density must increase with depth.

- As seismic waves travel to greater depths, they are refracted and reflected, which provides some evidence for Earth's internal composition and structure.

- Much remains unknown about Earth's interior, but now scientists have a good idea of its internal structure.

Log into GeologyNow and select this chapter to work through a **Geology Interactive** activity on "Reflection and Refraction" (click Earth's Layers→Reflection and Refraction).

11.3 Earth's Crust—Its Outermost Part

You already know from Chapter 1 that Earth consists of three concentric layers—the crust, mantle, and core (▶ Figure 11.1). Certainly the **crust,** the outermost layer, is the most accessible and thus the best known, but it is also the most complex both physically and chemically. Whereas the core and mantle vary mostly in a vertical dimension—that is, with depth—the crust varies both vertically and horizontally. *Continental crust* differs in density, thickness, composition, and overall age from *oceanic crust*. However, both continental and oceanic crust along with the upper mantle constitute the *lithosphere* of plate tectonic theory (▶ Figure 11.1).

Continental Crust

■ *How does continental crust differ from oceanic crust?*

Continental crust and oceanic crust form the outermost layer of Earth, but the former has an overall composition similar to granite whereas the latter is made up of gabbro and basalt. Also, continental crust is less dense and much thicker than oceanic crust. It also has rocks as much as 3.96 billion years old, but no oceanic crust older than about 180 million years is known. So, what do we mean that continental crust is similar to granite? **Continental crust** is made up of igneous, sedimentary, and metamorphic rocks of all kinds, but its overall composition is similar to that of granite, and accordingly we refer to continental crust as *granitic.*

The density of continental crust varies depending on rock type, but with the exception of metal-rich rocks, such as iron ore deposits, most of its rocks have densities of 2.5 to 3.0 g/cm³, and the continental crust's overall or average density is about 2.7 g/cm³. P-wave velocity in the continental crust is about 6.75 km/sec, but at the base of the crust, P-wave velocity abruptly increases to about 8 km/sec. Continental crust averages about 35 km thick, but its thickness varies from 20 to 90 km. Beneath mountain ranges such as the Rocky Mountains, the Alps in Europe, and the Himalayas in Asia, continental crust is much thicker than it is in adjacent areas. In contrast, continental crust is much thinner than the average beneath the Rift Valleys of East Africa and in a large area called the Basin and Range Province in the western United States and northern Mexico. The crust in these areas has been stretched and thinned in what appear to be the initial stages of rifting. Crustal thickening beneath mountain ranges is an important topic we will return to in a later section of this chapter.

Oceanic Crust

Variations are also found in **oceanic crust**—that is, Earth's crust beneath the ocean basins—but they are not as notable as those in continental crust. Direct observations from submersibles and samples taken during deep-sea drilling confirm that the upper part of the oceanic crust is made up of basalt, whereas its lower part is composed of gabbro. Remember that basalt and gabbro contain the same minerals but have different textures (see Figure 4.19). We will consider the oceanic crust in greater detail in Chapter 12.

With an average density of 3.0 g/cm³, oceanic crust is denser than continental crust, and it is also much thinner, varying from 5 to 10 km (▶ Figure 11.1). It is thinnest at spreading ridges where it is continuously produced. Remember that oceanic crust is consumed at subduction zones, thereby accounting for the fact that none of it is more than 180 million years old. P-waves travel through oceanic crust at about 7 km/sec, but just as at the base of the continental crust, their velocity increases when they enter the underlying mantle.

The Crust–Mantle Boundary

■ *What is the crust–mantle boundary called, and how was it discovered?*

In 1909 Yugoslavian seismologist Andrija Mohorovičić made a remarkable discovery—he detected a seismic discontinuity at a depth of about 30 km. He was studying seis-

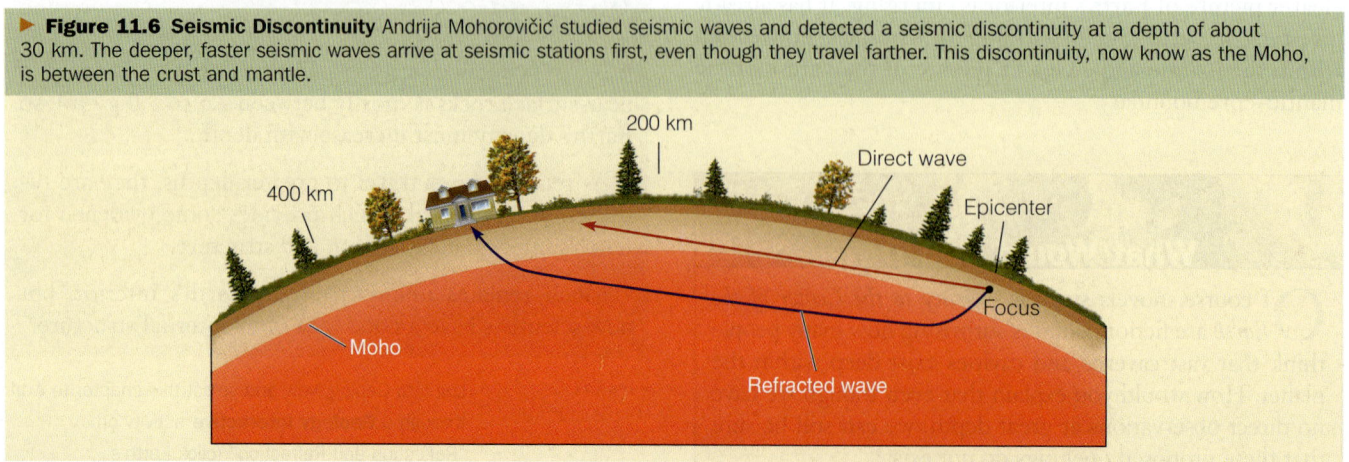

▶ **Figure 11.6 Seismic Discontinuity** Andrija Mohorovičić studied seismic waves and detected a seismic discontinuity at a depth of about 30 km. The deeper, faster seismic waves arrive at seismic stations first, even though they travel farther. This discontinuity, now know as the Moho, is between the crust and mantle.

mic waves from earthquakes in the Balkan region (part of southeastern Europe) and noticed that seismic stations a few hundred kilometers from an earthquake's epicenter recorded two distinct sets of P- and S-waves.

Mohorovičić concluded that a sharp boundary, a discontinuity, separating rocks with different properties is present at about 30 km below the surface. This discontinuity separates the crust from the mantle and is now called the **Mohorovičić discontinuity,** or simply the **Moho.** To account for the two sets of P- and S-waves, Mohorovičić postulated that below this boundary P-waves travel at 8 km/sec and S-waves travel at 6.75 km/sec. When an earthquake occurs, some waves travel directly from the focus to a seismic station, but others travel through the deeper layer and some of their energy is refracted back to the surface (▶ Figure 11.6). The waves traveling through the deeper layer (the mantle) travel farther to a seismic station, but they do so more rapidly and arrive before those that travel more slowly in the shallower layer.

The Moho is present everywhere except beneath spreading ridges. However, its depth varies: Beneath continents, it ranges from 20 to 90 km with an average of 35 km; beneath the seafloor, it is only 5 to 10 km deep.

Section 11.3 Summary

- Continental crust is granitic, meaning that it has an overall composition similar to granite, it has an average density of 2.7 g/cm³, and it varies from 20 to 90 km thick.

- Oceanic crust is made up of basalt and gabbro, its density is 3.0 g/cm³, and it is only 5 to 10 km thick.

- The Moho, the boundary between Earth's crust and mantle, is present everywhere except beneath spreading ridges.

11.4 Earth's Mantle— The Layer Below the Crust

All of Earth's interior from the base of the crust to the surface of the core at 2900 km is the **mantle.** It makes up 83% of the volume of the planet, but because it is less dense than the core, it is only 67.8% of Earth's mass. Remember from the discussion of the crust–mantle boundary that the Moho marks the base of the crust, or the top of the mantle, and that the Moho varies from 5 to 90 km deep. Beginning in 1958, project *Mohole* was funded to drill through the crust into the mantle, and it initially received considerable support. But in 1966 it was abandoned when funding dried up because of technological problems. As a consequence, the only direct observations of the mantle have been made from submersibles that descended to the walls of seafloor fractures (see Chapter 12).

■ *What is the low-velocity zone in the mantle?*

Seismic wave velocity in the mantle increases with depth, but several discontinuities exist. Between depths of 100 and 250 km, both P- and S-wave velocities decrease markedly (▶ Figure 11.7). This 100- to 250-km-deep layer is the **low-velocity zone,** which corresponds closely to the **asthenosphere,** a layer in which the rocks are close to their melting point and are less elastic, accounting for the observed decrease in seismic wave velocity. The asthenosphere is an important zone because it is where most magma is generated, especially under the ocean basins. Furthermore, it lacks strength, flows plastically, and is thought to be the layer over which the plates of the outer, rigid **lithosphere** move.

Even though the low-velocity zone and the asthenosphere closely correspond, they are still distinct. The asthenosphere appears to be present worldwide, but the low-velocity zone is not. In fact, the low-velocity zone appears to be poorly defined or even absent beneath the ancient shields of continents.

Other discontinuities are also present at deeper levels within the mantle. But unlike those between the crust and mantle or between the mantle and core, these probably represent structural changes in minerals rather than compositional changes. In other words, the mantle is composed of the same material throughout, but the structural states of minerals such as olivine change with depth. At a depth of 410 km, seismic wave velocity increases slightly as a consequence of changes in mineral structure (▶ Figure 11.7). Another velocity increase occurs

▶ **Figure 11.7 Variation in P-wave Velocity** Variations in P-wave velocity in the upper mantle and transition zone. Source: From G. C. Brown and A. E. Musset, *The Inaccessible Earth* (Kluwer Academic Publishers, 1981), Figure 7.11. Reprinted with the kind permission of Kluwer Academic Publishers.

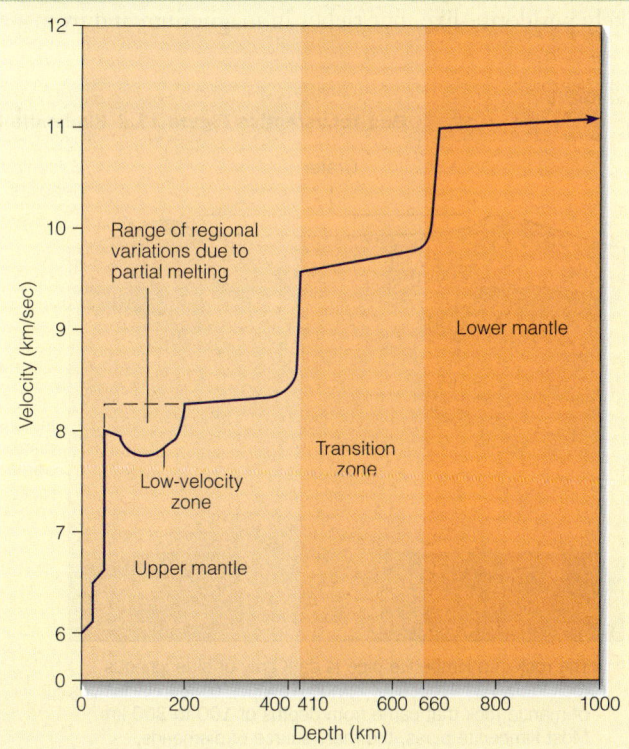

at about 660 km, where the minerals break down into metal oxides, such as FeO (iron oxide) and MgO (magnesium oxide), and silicon dioxide (SiO_2). These two discontinuities define the top and base of a *transition zone* separating the upper mantle from the lower mantle (▶ Figure 11.7).

A decrease in seismic wave velocity in a zone extending 200 to 300 km above the core–mantle boundary is recognized as the D″ layer. Although commonly included within the lower mantle, the D″ layer might be considerably different in composition. Experiments indicate that silicates in the mantle react with liquid from the outer core, thus forming a vertically and laterally heterogeneous layer. Some geologists think that mantle plumes originate in the D″ layer.

■ *How did geologists come to the conclusion that the mantle is probably made up of peridotite?*

Although the mantle's density, which varies from 3.3 to 5.7 g/cm³, can be inferred rather accurately from seismic waves, its composition is less certain. The igneous rock *peridotite* is considered the most likely component of the upper mantle. Peridotite contains mostly ferromagnesian silicates (60% olivine and 30% pyroxene) and about 10% feldspars (see Figures 4.17 and 4.18). Peridotite is considered the most likely candidate for three reasons. First, laboratory experiments indicate that it possesses physical properties that account for the mantle's density and observed rates of seismic wave transmissions. Second, peridotite forms the lower parts of igneous rock sequences thought to be fragments of the oceanic crust and upper mantle, called *ophiolites*, emplaced on land (see Chapter 12). And third, peridotite is found as inclusions in volcanic rock bodies such as *kimberlite pipes* that are known to have come from the mantle at depths of 100 to 300 km (▶ Figure 11.8). The deeper mantle is likely peridotite-like, but richer in magnesium and iron.

Section 11.4 Summary

- Earth's mantle lies beneath the crust to a depth of about 2900 km. It makes up more than 80% of Earth's volume and about two-thirds of its mass.

- The mantle probably consists of peridotite, and it is made up of three parts—the upper mantle, which includes the asthenosphere, a transition zone, and the lower mantle.

- At depths of 100 to 250 km, seismic wave velocity decreases in the low-velocity zone. It corresponds closely with the asthenosphere but is distinct in that the asthenosphere is present everywhere but the low-velocity zone is not.

11.5 The Core

Once again seismic waves, experiments, meteorites, and Earth's overall density indicate that Earth's deep interior must be made up of materials that are much more dense than those at the surface or in the mantle. So although geologists are confident that Earth does have a core, several aspects of this remote region are not understood. Based on the best available information, the **core** lies about 2900 km below the surface (▶ Figure 11.1). Its diameter of 6960 km makes it about the same size as Mars, and it constitutes 16.4% of Earth's volume and nearly one-third of its mass.

Seismic waves have indicated much of what we know about the core, but seismic tomography reveals that the core–mantle boundary is not as smooth as implied in Fig-

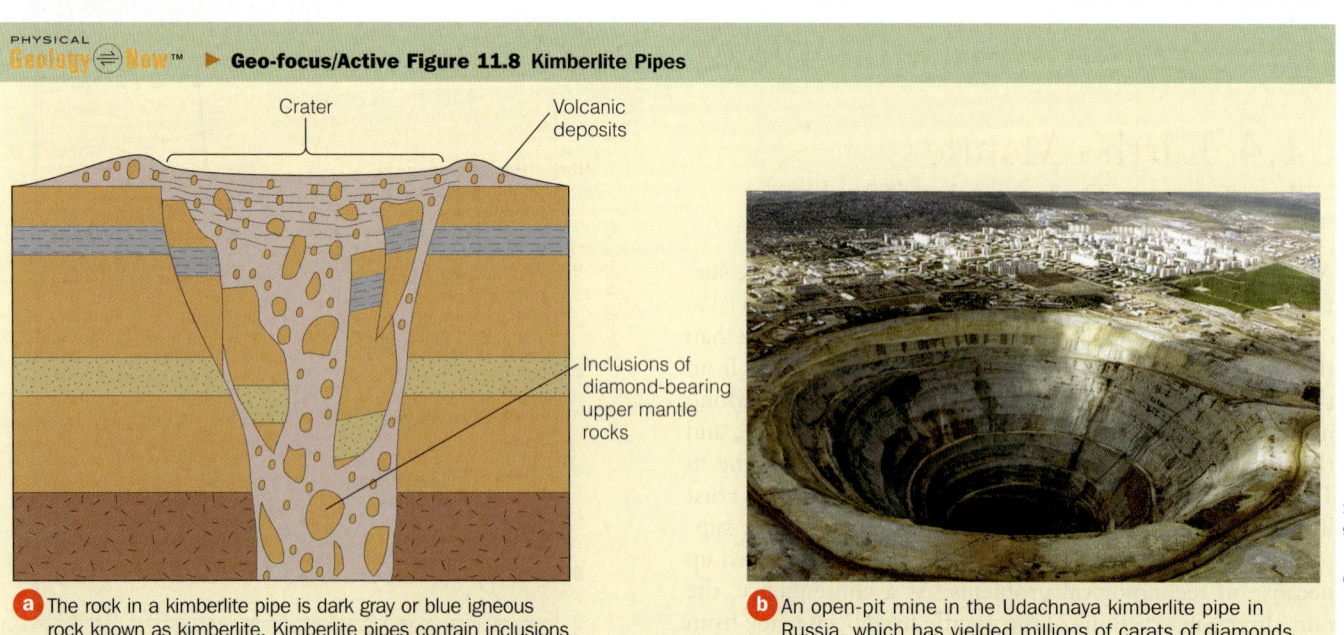

▶ Geo-focus/Active Figure 11.8 Kimberlite Pipes

a The rock in a kimberlite pipe is dark gray or blue igneous rock known as kimberlite. Kimberlite pipes contain inclusions of mantle rock that came from depths of 100 to 300 km. Most kimberlite pipes, the main source of diamonds, measure less than 500 m across at the surface.

b An open-pit mine in the Udachnaya kimberlite pipe in Russia, which has yielded millions of carats of diamonds. The benches in the excavation are about 20 m high.

ure 11.1. In fact, the surface of the core has broad depressions and rises extending several kilometers into the mantle. Of course, the mantle has the same features in reverse, and it now seems that the core's outer surface is continuously deformed by sinking and rising masses of mantle material.

Discovery of Earth's Core

In 1906 R. D. Oldham of the Geological Survey of India realized that seismic waves arrived later than expected at seismic stations more than 130 degrees from an earthquake focus. He postulated that Earth has a core that transmits seismic waves more slowly than shallower Earth materials. We now know that P-wave velocity decreases markedly at a depth of 2900 km, which indicates an important discontinuity recognized as the core–mantle boundary (▶ Figure 11.5).

■ *What are the P- and S-wave shadow zones, and what do they indicate about Earth's interior?*

Because of the sudden decrease in P-wave velocity at the core–mantle boundary, P-waves are refracted in the core so that little P-wave energy reaches the surface between 103 and 143 degrees from an earthquake focus (▶ Figure 11.9b). This **P-wave shadow zone**, as it is called, was discovered in 1914 by the German seismologist Beno Gutenberg. However, the P-wave shadow zone is a not a perfect shadow zone because some weak P-wave energy is recorded within it. Scientists proposed several hypotheses to account for this observation, but all were rejected by the Danish seismologist Inge Lehmann (▶ Figure 11.9a), who in 1936 postulated that the core is not entirely liquid as previously thought. She proposed that seismic wave reflection from a solid inner core accounts for the arrival of weak P-wave energy in the P-wave shadow zone (▶ Figure 11.9c), a proposal that was quickly accepted by seismologists.

In 1926 the British physicist Harold Jeffreys realized that S-waves were not simply slowed by the core but were completely blocked by it. So, besides a P-wave shadow zone, a much larger and more complete **S-wave shadow zone** also exists (▶ Figure 11.10). At locations greater than 103 degrees from an earthquake focus, no S-waves are recorded, which indicates that S-waves cannot be transmitted through the core. S-waves will not pass through a liquid, so it seems that the outer core must be liquid or behave as a liquid. The inner core, however, is thought to be solid because P-wave velocity increases at the base of the outer core.

Density and Composition of the Core

■ *What are the density and composition of the inner and outer cores?*

We can estimate the core's density and composition by using seismic evidence and laboratory experiments. For instance, geologists use a diamond-anvil pressure cell in which small

PHYSICAL Geology Now™ ▶ **Active Figure 11.9 P-wave Shadow**

a In 1936 Inge Lehmann, a Danish seismologist, postulated that Earth has a solid inner core.

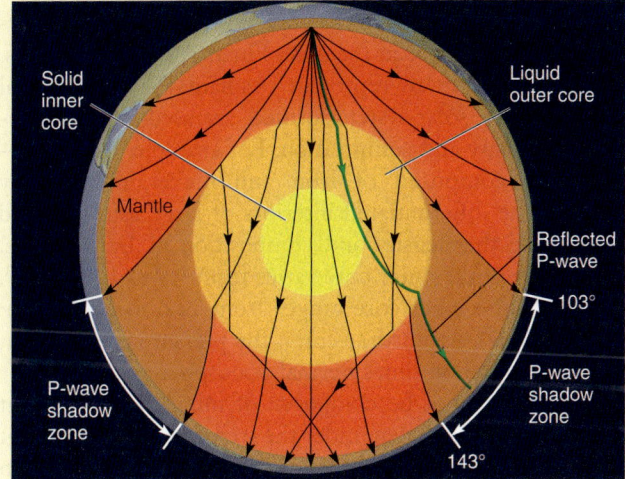

b According to Lehmann, reflection from an inner core explains the arrival of weak P-wave energy in the P-wave shadow zone.

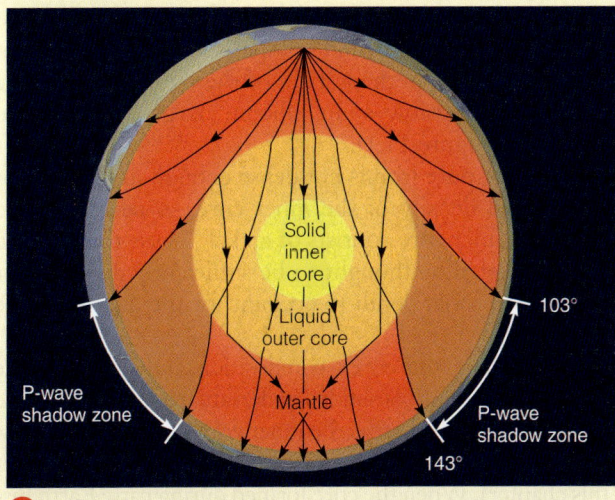

c P-waves are refracted so that no direct P-wave energy reaches the surface in the P-wave shadow zone.

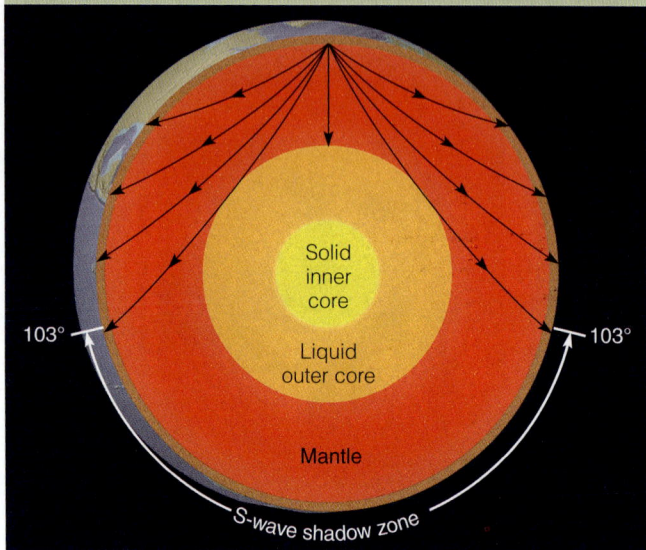

▶ **Figure 11.10 S-wave Shadow** The presence of an S-wave shadow zone indicates that S-waves are being blocked within Earth.

samples are studied while subjected to pressures and temperatures similar to those in the core. Furthermore, meteorites, which are thought to represent remnants of the material from which the solar system formed, are used to make estimates of density and composition. For example, the irons—meteorites composed of iron and nickel alloys—may represent the differentiated interiors of large asteroids and approximate the density and composition of Earth's core. The density of the outer core varies from 9.9 to 12.2 g/cm^3, and that of the inner core ranges from 12.6 to 13.0 g/cm^3 (▶ Figure 11.1). At Earth's center, the pressure is equivalent to about 3.5 million times normal atmospheric pressure.

The core cannot be composed of minerals common at the surface because, even under the tremendous pressures at great depth, they would still not be dense enough to yield an average density of 5.52 g/cm^3 for Earth. Both the outer and inner cores are thought to be composed largely of iron, but pure iron is too dense to be the sole constituent of the outer core. It must be "diluted" with elements of lesser density. Laboratory experiments and comparisons with iron meteorites indicate that perhaps 12% of the outer core consists of sulfur and possibly some silicon, oxygen, nickel, and potassium (▶ Figure 11.1).

In contrast, pure iron is not dense enough to account for the estimated density of the inner core, so perhaps 10 to 20% of the inner core consists of nickel. These metals form an iron–nickel alloy thought to be sufficiently dense under the pressure at that depth to account for the density of the inner core.

Any model of the core's composition and physical state must explain not only variations in density but also (1) why the outer core is liquid whereas the inner core is solid and (2) how the magnetic field is generated within the core (discussed later in this chapter). When the core formed during early Earth history, it was probably entirely molten and has since cooled so that its interior has crystallized. Indeed, the inner core continues to grow as Earth slowly cools, and liquid of the outer core crystallizes as iron. Recent evidence also indicates that at present the inner core rotates faster than the outer core, moving about 20 km/yr relative to the outer core.

The temperature at the core–mantle boundary is estimated at 2500° to 5000°C, yet the high pressure within the inner core prevents melting. In contrast, the outer core is under less pressure, but more important than the differences in pressure are compositional differences between the inner and outer cores. The sulfur content of the outer core helps depress its melting temperature. An iron–sulfur mixture melts at a lower temperature than does pure iron, or an iron–nickel alloy, so despite the high pressure, the outer core is molten.

Section 11.5 Summary

- Earth's deep interior is made up of an inner (solid) and an outer (liquid) core, both composed largely of iron and nickel with several other elements as well. Its density varies from 9.9 to 13 g/cm^3.

- Much of what scientists know about the core is based on studies of the P- and S-wave shadow zones, experimental evidence, and comparisons with meteorites.

PHYSICAL Geology Now™ Log into GeologyNow and select this chapter to work through a **Geology Interactive** activity on "Core Studies" (click Earth's Layers→Core Studies).

11.6 Earth's Internal Heat

How hot is it inside our planet, and where does the heat come from? The answer to the first part of this question depends on depth because Earth's temperature increases with depth, but not in a linear fashion. The source of heat is twofold: First, some is residual heat from Earth's origin, and second, much of Earth's internal heat is generated by radioactive decay. In any case, internal heat is responsible for Earth's ongoing seismicity, volcanism, moving plates, and mountain building, or, as previously stated, it is one feature that makes Earth such a dynamic planet.

■ *What is the geothermal gradient, and can it be used to determine the temperature of Earth's core?*

During the 19th century, scientists realized that the temperature in deep mines increases with depth. More recently, the same trend has been observed in deep drill holes. This temperature increase with depth, or **geothermal gradient**, near the surface is about 25°C/km. In areas of active or recently active volcanism, the geothermal gradient is greater than in adjacent nonvolcanic areas, and temperature rises faster beneath spreading ridges than elsewhere beneath the seafloor (▶ Figure 11.11).

Much of Earth's internal heat is generated by radioactive decay, especially the decay of isotopes of uranium and thorium and to a lesser degree of potassium 40. When these isotopes

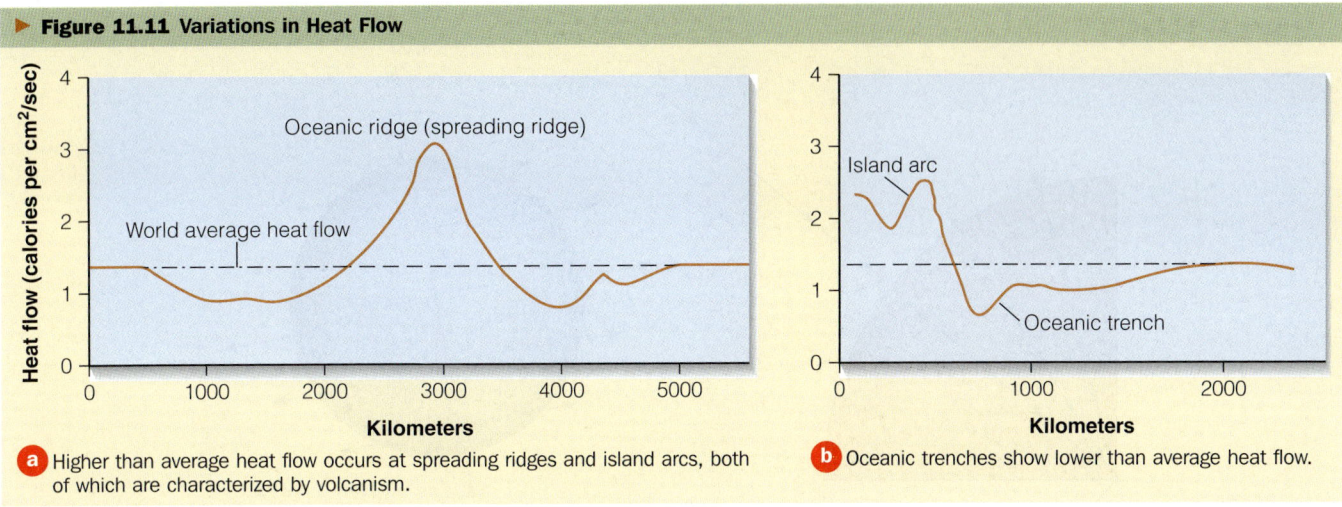

Figure 11.11 Variations in Heat Flow

a Higher than average heat flow occurs at spreading ridges and island arcs, both of which are characterized by volcanism.

b Oceanic trenches show lower than average heat flow.

decay, they emit energetic particles and gamma rays that heat surrounding rocks. Because rock is such a poor conductor of heat, it takes little radioactive decay to build up considerable heat, given enough time.

Unfortunately, the geothermal gradient is not useful for estimating temperatures at great depth. If we were simply to extrapolate from the surface downward, the temperature at 100 km would be so high that, despite the great pressure, all known rocks would melt. Yet except for pockets of magma, it appears that the mantle is solid rather than liquid because it transmits S-waves. Accordingly, the geothermal gradient must decrease markedly.

Current estimates of the temperature at the base of the crust are 800° to 1200°C. The latter figure seems to be an upper limit: If it were any higher, melting would be expected. Furthermore, fragments of mantle rock in kimberlite pipes that came from depths of 100 to 300 km appear to have reached equilibrium at these depths and at about 1200°C. At the core–mantle boundary, the temperature is probably between 2500° and 5000°C; the wide range of values indicates the uncertainties of such estimates. If these figures are reasonably accurate, the geothermal gradient in the mantle is only about 1°C/km.

Because the core is so remote and its composition uncertain, only very general estimates of its temperature are possible. Based on various experiments, the maximum temperature at the center of the core is estimated to be 6500°C, very close to the estimated temperature for the surface of the Sun!

Even though rocks are poor conductors of heat, detectable amounts of heat from Earth's interior escape at the surface by **heat flow.** The amount of heat lost is small and is detected only by sensitive instruments. Heavy, cylindrical probes are dropped into soft seafloor sediments, and temperatures are measured at various depths along the cylinder. On the continents, temperature measurements are made at various depths in drill holes and mines.

More than 70% of the total heat lost by Earth is lost through the seafloor, but heat-flow values for both oceanic basins and continents decrease with increasing age. This result is expected in the ocean basins because high heat-flow values occur at spreading ridges where oceanic crust is continuously formed by igneous activity. Spreading ridges are also the sites of hydrothermal vents, where heat is transported upward by hydrothermal convection. Heat flow through the continental crust is not as well understood, but it too shows lower values for older crust.

It should be apparent that if heat is escaping from within Earth, the interior should be cooling unless a mechanism exists to replenish it. Radioactive decay generates heat continuously, but the quantity of radioactive isotopes (except carbon 14) decreases with time as they decay to stable daughter products. Accordingly, during its early history Earth possessed more internal heat and has been cooling continuously since then.

Section 11.6 Summary

- Earth's temperature increases with depth, its geothermal gradient averages about 25°C/m, but it is higher in areas of active or recently active volcanism.

- Heat escapes at Earth's surface, a phenomenon known as heat flow.

11.7 Gravity and How Its Force Is Determined

Sir Isaac Newton (1642–1727) devised the *law of universal gravitation*, in which the force of gravitational attraction (F) between two masses (m_1 and m_2) is directly proportional to the products of the two masses and inversely proportional to the distance (D) between their centers of mass:

$$F = \frac{G\, m_1 \times m_2}{D^2}$$

G in this equation is the universal gravitational constant. Accordingly, **gravity** is the attractive force that exists between

Figure 11.12 Earth's Gravity

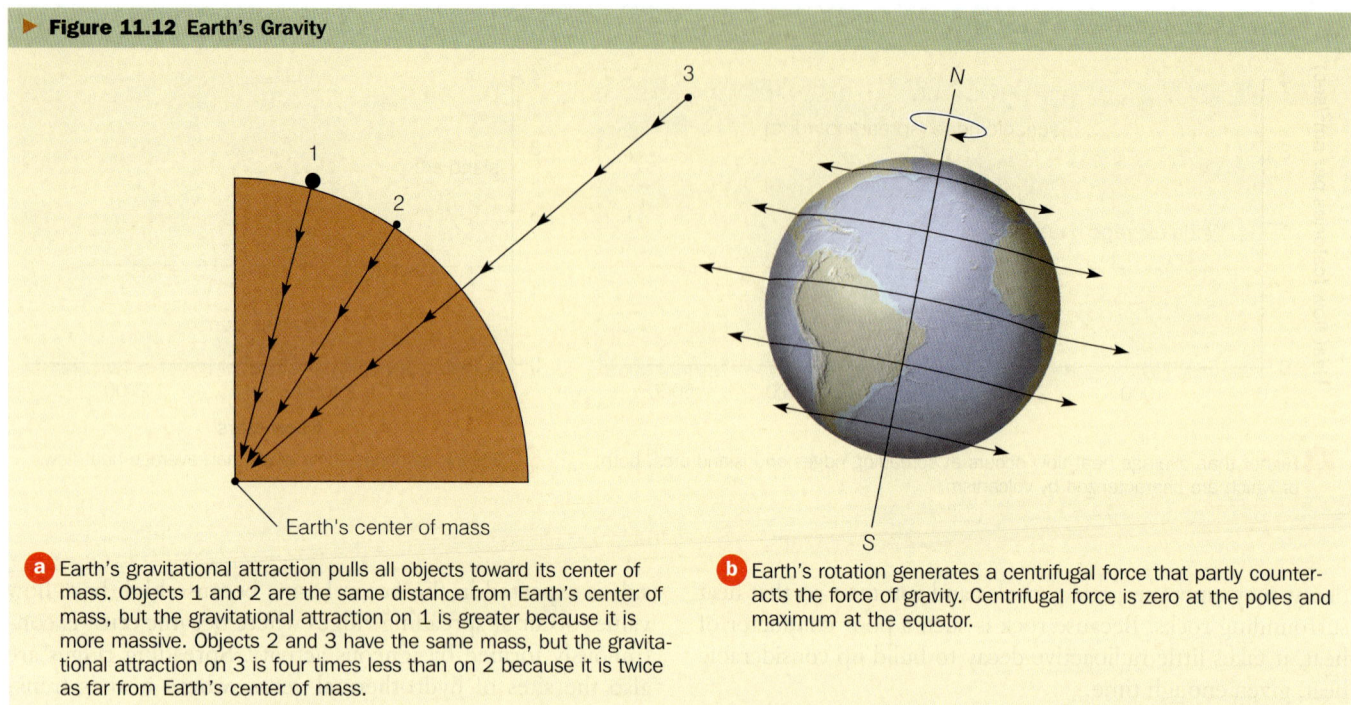

a Earth's gravitational attraction pulls all objects toward its center of mass. Objects 1 and 2 are the same distance from Earth's center of mass, but the gravitational attraction on 1 is greater because it is more massive. Objects 2 and 3 have the same mass, but the gravitational attraction on 3 is four times less than on 2 because it is twice as far from Earth's center of mass.

b Earth's rotation generates a centrifugal force that partly counteracts the force of gravity. Centrifugal force is zero at the poles and maximum at the equator.

any two bodies. Our main concern here is with the attraction between any two bodies on Earth, but the gravitational attraction on Earth's surface waters by the Moon and the Sun, which is responsible for tides, is also important (see Chapter 19).

The equation tells us that the gravitational force (F) is greater between two massive bodies—the Moon and Earth, for instance—if the distance between the centers of mass remain the same (▶ Figure 11.12a). But because F is inversely proportional to the square of distance, it decreases by a factor of 4 when the distance is doubled. We usually refer to the gravitational force between an object and Earth as its *weight*.

Figure 11.13 Gravity Anomalies

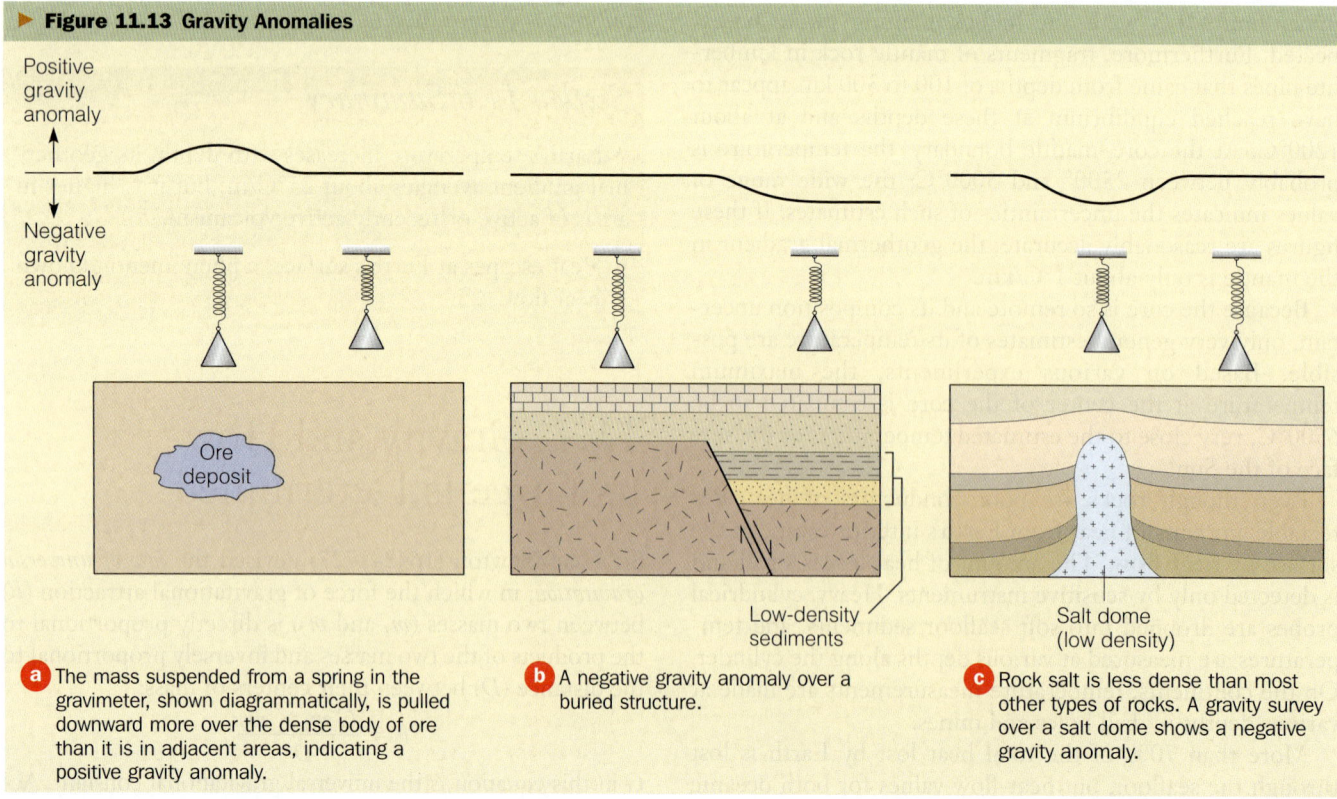

a The mass suspended from a spring in the gravimeter, shown diagrammatically, is pulled downward more over the dense body of ore than it is in adjacent areas, indicating a positive gravity anomaly.

b A negative gravity anomaly over a buried structure.

c Rock salt is less dense than most other types of rocks. A gravity survey over a salt dome shows a negative gravity anomaly.

Gravity

Gravitational attraction would be the same everywhere on the surface if Earth were perfectly spherical, homogeneous throughout, and not rotating. As a consequence of rotation, however, a centrifugal force is generated that partly counteracts the force of gravity (▶ Figure 11.12b), so an object at the equator weighs slightly less than the same object would at the poles. The force of gravity also varies with distance between the centers of masses, so an object would weigh slightly less above the surface than if it were at sea level (▶ Figure 11.12a).

Gravity Anomalies

■ *What are positive and negative gravity anomalies?*

Geologists use a sensitive instrument called a *gravimeter* to measure variations in the force of gravity. A gravimeter is simple in principle; it contains a weight suspended on a spring that responds to variations in gravity (▶ Figure 11.13). Long ago, geologists realized that anomalous gravity values should exist over buried bodies of ore minerals and salt domes and that geologic structures such as faulted strata could be located by surface gravity surveys (▶ Figure 11.13).

Gravity measurements are higher over an iron ore deposit than over unconsolidated sediment because of the ore's greater density (▶ Figure 11.13a). Such departures from the expected force of gravity are **gravity anomalies.** In other words, the measurement over the body of iron ore indicates an excess of dense material, or simply a *mass excess*, between the surface and the center of Earth and is considered a **positive gravity anomaly.** A **negative gravity anomaly** indicating a *mass deficiency* exists over low-density sediments because the force of gravity is less than expected (▶ Figure 11.13b). Large negative gravity anomalies also exist over salt domes (▶ Figure 11.13c) and at subduction zones, indicating that the crust is not in equilibrium.

Departures from Earth's expected gravitational attraction (gravity anomalies) certainly exist, but what of the tourist sites around the United States that claim gravity has somehow gone awry? All kinds of mysterious gravity-defying effects are claimed to occur in these areas, including objects rolling uphill, unsupported objects clinging to walls, and the famous plank illusion in which the heights of people on a level plank change when they switch positions. All are actually clever optical illusions that can be duplicated by anyone with the interest in doing so.

Section 11.7 Summary

- Gravitational attraction between any two bodies depends on the products of their masses and the distance between their centers of mass.

- Positive and negative gravity anomalies are found where mass excesses and deficiencies occur. Gravity surveys are useful in exploration for minerals and hydrocarbons.

11.8 Floating Continents—The Principle of Isostasy

■ *How is it possible for a solid (Earth's crust) to "float" in another solid (the mantle)?*

We previously said that Earth's crust and mantle are solids except for pockets of magma, so how is it possible for a solid (continental crust) to float in another solid (the mantle)? Floating brings to mind a ship at sea or a block of wood in water, but continents certainly do not behave in this fashion. Or do they? Actually, they do float, in a manner of speaking, in the denser mantle below, but a complete answer requires much more discussion.

More than 150 years ago, British surveyors in India detected a discrepancy of 177 m when they compared the results of two measurements between points 600 km apart. Even though this discrepancy was small, only about 0.03%, it was an unacceptably large error. The surveyors realized that the gravitational attraction of the nearby Himalaya Mountains probably deflected the plumb line (a cord with a suspended weight) of their surveying instruments from the vertical, thus accounting for the error. Calculations revealed, however, that if the Himalayas were simply thicker crust piled on denser material, the error should have been greater than that observed (▶ Figure 11.14a).

In 1865 George Airy proposed that, in addition to projecting high above sea level, the Himalayas—and other mountains as well—project far below the surface and thus have a low-density root (▶ Figure 11.14b). In effect, he was saying that mountains float on denser rock at depth. Their excess mass above sea level is compensated for by a mass deficiency at depth, which would account for the observed deflection of the plumb line during the British survey (▶ Figure 11.14).

Another explanation was proposed by J. H. Pratt, who thought that the Himalayas were high because they were composed of rocks of lower density that those in adjacent regions. Although Airy was correct with respect to the Himalayas, and mountains in general, Pratt was correct in that there are indeed places where the crust's elevation is related to its density. For example, (1) continental crust is thicker and less dense than oceanic crust and thus stands high, and (2) the mid-oceanic ridges stand higher than adjacent areas because the crust there is hot and less dense than cooler oceanic crust elsewhere.

Gravity and seismic studies have revealed that mountains do indeed have a low-density "root" projecting deep into the mantle. If it were not for this low-density root, a gravity survey across a mountainous area would reveal a huge positive gravity anomaly. The fact that no such anomaly exists indicates that a mass excess is not present, so some of the dense mantle at depth must be displaced by lighter crustal rocks, as shown in ▶ Figure 11.14b.

Both Airy and Pratt agreed that Earth's crust is in floating equilibrium with the more dense mantle below, and

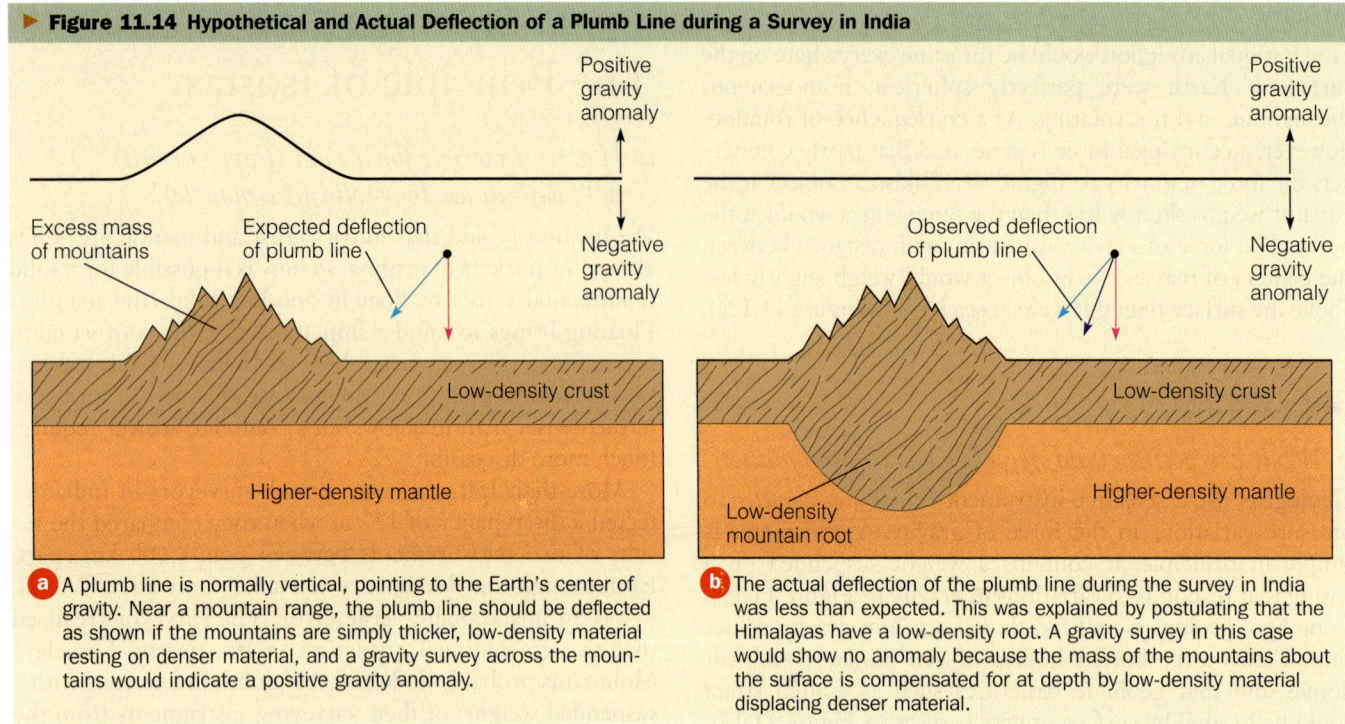

Figure 11.14 Hypothetical and Actual Deflection of a Plumb Line during a Survey in India

a A plumb line is normally vertical, pointing to the Earth's center of gravity. Near a mountain range, the plumb line should be deflected as shown if the mountains are simply thicker, low-density material resting on denser material, and a gravity survey across the mountains would indicate a positive gravity anomaly.

b The actual deflection of the plumb line during the survey in India was less than expected. This was explained by postulating that the Himalayas have a low-density root. A gravity survey in this case would show no anomaly because the mass of the mountains about the surface is compensated for at depth by low-density material displacing denser material.

now their proposal is known as the **principle of isostasy.** This phenomenon is easy to understand by analogy to a ship or an iceberg (▶ Figure 11.15). Ice is slightly less dense than water, and thus it floats. According to Archimedes's principle of buoyancy, an iceberg sinks in water until it displaces a volume of water whose weight is equal to that of the ice. When the iceberg has sunk to an equilibrium position, only about 10% of its volume is above water level. If some of the ice above water level should melt, the iceberg rises in order to maintain the same proportion of ice above and below water (▶ Figure 11.15b).

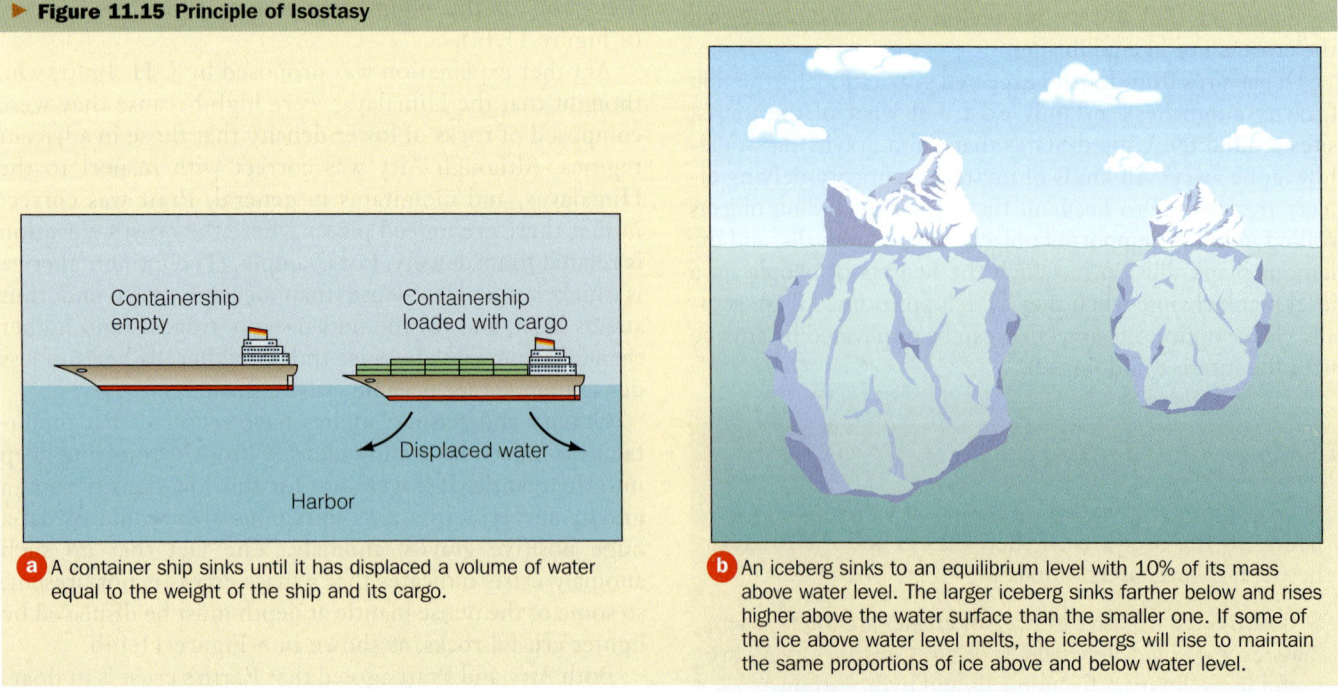

Figure 11.15 Principle of Isostasy

a A container ship sinks until it has displaced a volume of water equal to the weight of the ship and its cargo.

b An iceberg sinks to an equilibrium level with 10% of its mass above water level. The larger iceberg sinks farther below and rises higher above the water surface than the smaller one. If some of the ice above water level melts, the icebergs will rise to maintain the same proportions of ice above and below water level.

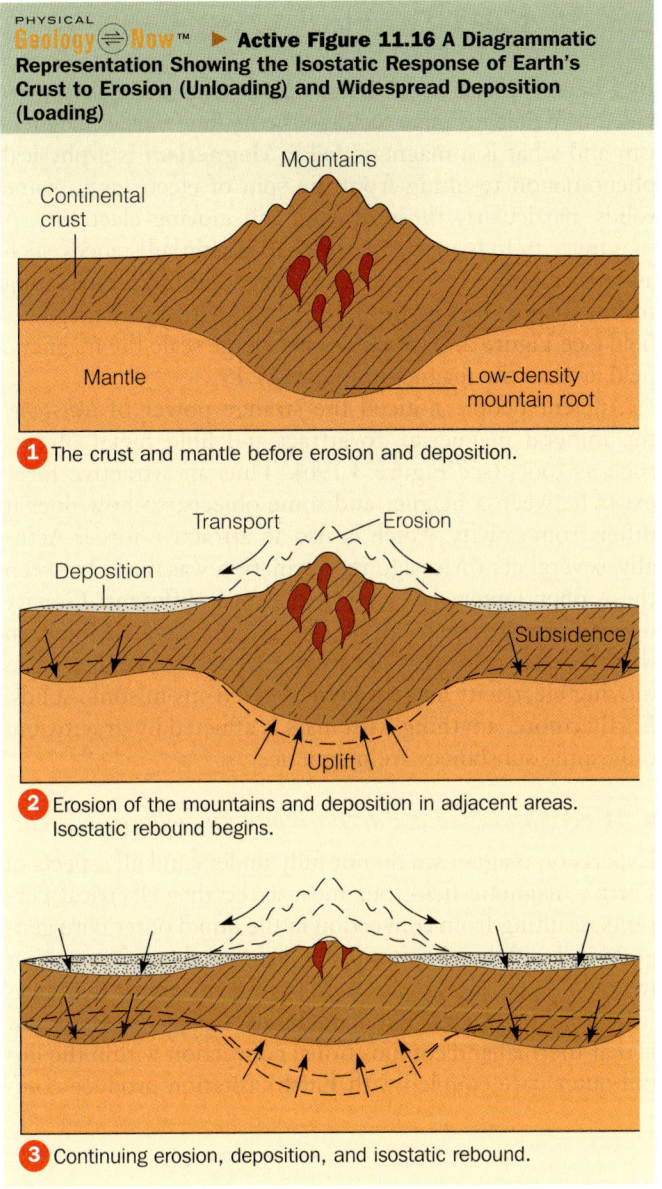

Active Figure 11.16 A Diagrammatic Representation Showing the Isostatic Response of Earth's Crust to Erosion (Unloading) and Widespread Deposition (Loading)

1. The crust and mantle before erosion and deposition.
2. Erosion of the mountains and deposition in adjacent areas. Isostatic rebound begins.
3. Continuing erosion, deposition, and isostatic rebound.

Figure 11.17 A Diagrammatic Representation of the Response of Earth's Crust to the Added Weight of Glacial Ice

1. The crust and mantle before glaciation.
2. The weight of glacial ice depresses the crust into the mantle.
3. When the glacier melts, isostatic rebound begins and the crust rises to its former position.
4. Isostatic rebound is complete.

Earth's crust is similar to the iceberg, or a ship, in that it sinks into the mantle to its equilibrium level. Where the crust is thickest, as beneath mountain ranges, it sinks farther down into the mantle but also rises higher above the equilibrium surface (▶ Figure 11.14b). Continental crust, being thicker and less dense than oceanic crust, stands higher than the ocean basins. Earth's crust responds isostatically to widespread erosion and sediment deposition (▶ Figure 11.16). It also responds to loading when vast glaciers form and depress the crust into the mantle to maintain equilibrium (▶ Figure 11.17). In Greenland and Antarctica, the crust has been depressed below sea level by the weight of glacial ice.

Unloading of the crust causes it to respond by rising upward until equilibrium is again attained. This phenomenon, known as **isostatic rebound,** takes place in areas that are deeply eroded and that were formerly covered by vast glaciers. Scandinavia, which was covered by an ice sheet until about 10,000 years ago, is still rebounding at a rate of up to 1 m per century (▶ Figure 11.18). Coastal cities in Scandinavia have been uplifted rapidly enough that docks constructed several centuries ago are now far from shore. Isostatic rebound has also occurred in eastern Canada, where the land has risen as much as 100 m during the last 6000 years.

If the principle of isostasy is correct, it implies that the mantle behaves like a liquid. In preceding discussions, though, we said that the mantle must be solid because it transmits S-waves, which will not move through a liquid. How can this apparent paradox be resolved? When considered in terms of the short time necessary for S-waves to pass through it, the mantle is indeed solid. But when subjected to stress over long periods, it yields by flowage and at these timescales is a viscous liquid. Silly Putty, a familiar substance that has the properties of a solid or a liquid depending on how rapidly deforming forces are applied, will flow under its own weight if given enough time, but shatters as a brittle solid if struck a sharp blow.

▶ **Figure 11.18 Isostatic Rebound in Scandinavia** The lines show rates of uplift in centimeters per century. Source: From Beno Gutenburg, *Physics of the Earth's Interior* (Orlando, Florida: Academic Press, 1959), 194, Figure 9.1.

Section 11.8 Summary

- According to the principle of isostasy, Earth's less dense crust "floats" in the denser mantle below.

- If the crust is loaded where sediment or vast glaciers accumulate, it responds by subsiding, but it rises (isostatic rebound) when deep erosion takes place or glaciers melt.

- Continental crust stands higher than oceanic crust because it is thicker and less dense.

What Would You Do?

While teaching a high school science class, you mention that Earth's crust behaves as if floating in the denser mantle below. It's obvious that your students are having difficulties grasping the concept. After all, how can a solid float in a solid? Can you think of any analogies that might help them understand? Also, what kinds of experiments might you devise to demonstrate that both crustal composition and thickness play a role in isostasy?

11.9 Earth's Magnetic Field

In Chapter 2 you learned that Earth's magnetic field provided evidence for plate tectonic theory, but what is magnetism and what is a magnetic field? **Magnetism** is a physical phenomenon resulting from the spin of electrons in some solids, particularly those of iron, and moving electricity. A **magnetic field** is an area in which magnetic substances, such as iron, are affected by lines of magnetic force emanating from a magnet (▶ Figure 11.19). Note that Earth's magnetic field (see Figure 2.7) resembles on a large scale the magnetic field around the magnet in Figure 11.19.

Ancient people noticed the strange power of *lodestone*, the mineral magnetite, to attract and hold metal objects such as tools (see Figure 3.19b). Thus an attractive force exists between a magnet and some objects, so how does it differ from gravity, which is also an attractive force? Actually, several centuries ago no distinction was made between these phenomena, but they are in fact different. Gravity depends on the masses of objects and the distance between their centers of mass, whereas magnetism results from moving electricity and the spin of electrons in some solids. Furthermore, anything with mass is affected by gravity, but only some substances are magnetic.

■ *Where and how is Earth's magnetic field generated?*

Experts on magnetism do not fully understand all aspects of Earth's magnetic field, but most agree that electrical currents resulting from convection in the liquid outer core generate it. Furthermore, it must be generated continuously or it would decay and Earth would have no magnetic field in as little as 20,000 years. The model most widely accepted now is that thermal and compositional convection within the liquid outer core coupled with Earth's rotation produce com-

▶ **Figure 11.19 Magnetic Field** Iron filings align along the lines of magnetic force emanating from a bar magnet. Earth's magnetic field also has similar lines of force but obviously on a much larger scale (see Figure 2.7).

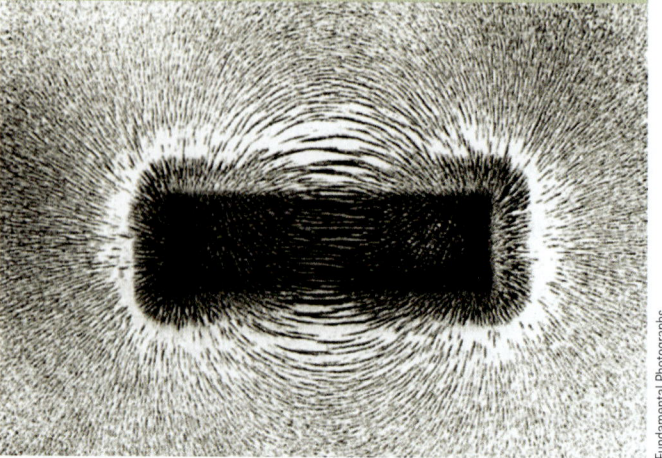

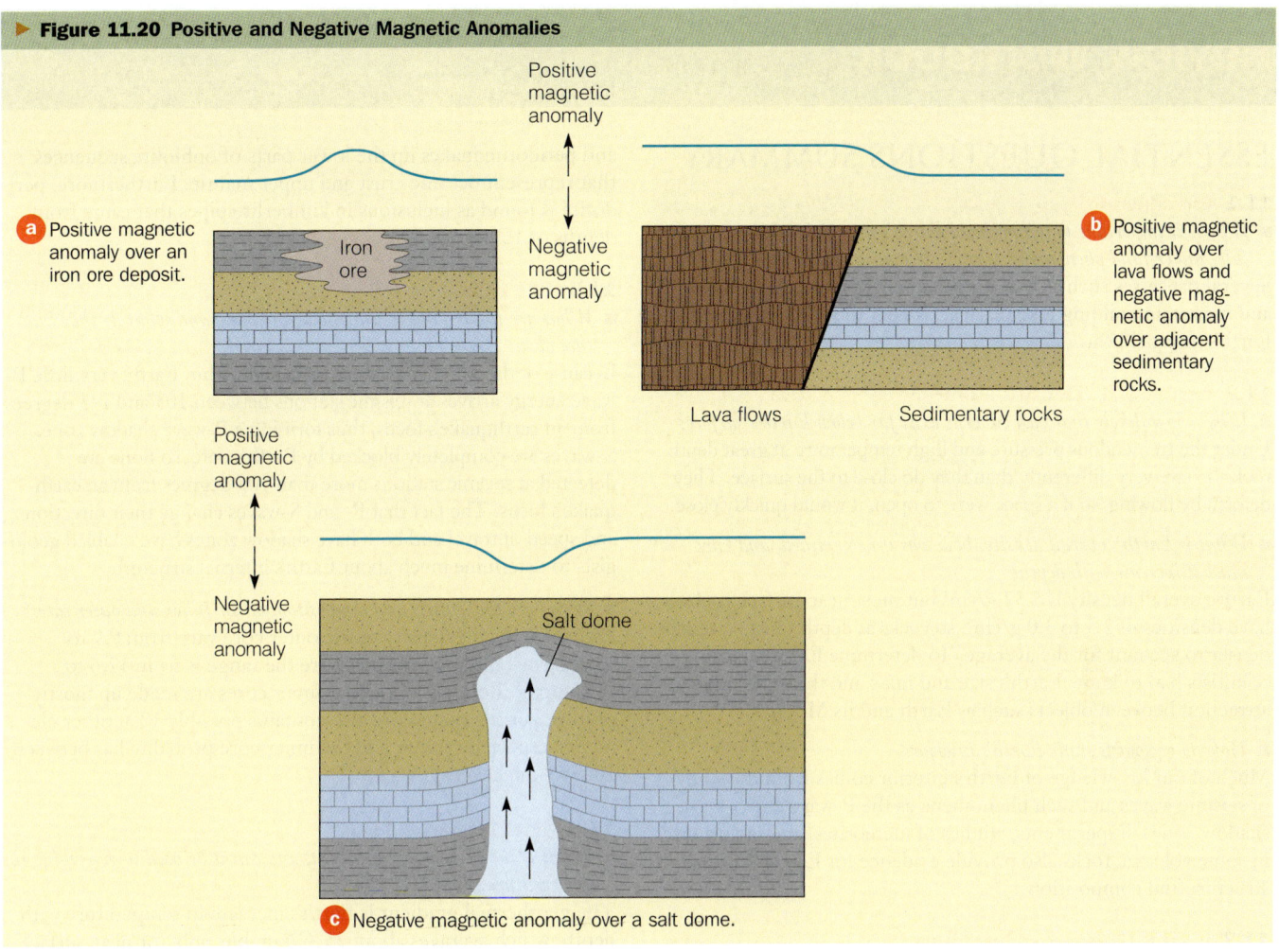

▶ Figure 11.20 Positive and Negative Magnetic Anomalies

(a) Positive magnetic anomaly over an iron ore deposit.
(b) Positive magnetic anomaly over lava flows and negative magnetic anomaly over adjacent sedimentary rocks.
(c) Negative magnetic anomaly over a salt dome.

plex electrical currents or a *self-exciting dynamo* that in turn generates the magnetic field.

In Chapter 2 we mentioned *magnetic anomalies*, deviations from the average strength of Earth's magnetic field, when we discussed seafloor spreading (see Figure 2.11). Anomalies may be either positive or negative, and anomalies occur on both regional and local scales. Regional anomalies are most likely related to complexities of convection in the outer core, but local anomalies result from vertical and lateral variations of rock types within the crust.

In the Great Lakes region of the United States and Canada, huge iron ore deposits add their magnetism to Earth's magnetic field, resulting in a positive gravity anomaly (▶ Figure 11.20a). Areas underlain by thick basalt lava flows also show positive anomalies, but an adjacent area of sedimentary rocks shows a negative anomaly (▶ Figure 11.20b). Geologists use an instrument called a *magnetometer* to detect slight variations in the strength of the magnetic field. In fact, magnetometers have been used for decades to locate deeply buried iron ore deposits and to find buried structures, such as salt domes that show negative magnetic anomalies (▶ Figure 11.20c).

Section 11.9 Summary

- Scientists think that thermal and compositional convection in the outer core generates Earth's magnetic field.

Review Workbook

ESSENTIAL QUESTIONS SUMMARY

11.1 Introduction
- *Why is it important to comprehend Earth's internal processes to more fully understand phenomena taking place at the surface?*

Several processes such as volcanism, moving plates, earthquakes, and mountain building take place at or near the surface because of Earth's internal heat.

11.2 Earth's Size, Density, and Internal Structure
- *Why is it unlikely that vast caverns exist far below Earth's surface?*

Under the tremendous pressure and high temperature at great depth, rocks behave very differently than they do close to the surface. They deform by flowing, so if a space were to open, it would quickly close.

- *What is Earth's overall density, how was it determined, and why must it increase with depth?*

Earth's overall density is 5.52 g/cm³ but most near-surface rocks have densities of 2.5 to 3.0 g/cm³, so rocks at depth must be much denser to account for the average. To determine Earth's density, scientists had to know Earth's size and mass and the gravitational attraction between objects such as Earth and its Moon.

- *How do geologists study Earth's interior?*

Much of our knowledge of Earth's interior comes from the study of seismic waves and such phenomena as the P-wave and S-wave shadow zones. Experiments, studies of meteorites, and inclusions in some volcanic rocks also provide evidence for Earth's internal structure and composition.

11.3 Earth's Crust—Its Outermost Part
- *How does continental crust differ from oceanic crust?*

Continental crust is composed of all rock types but has an overall composition similar to granite, and its rocks vary from recent to as old as 3.96 billion years. In contrast, oceanic crust is made up of basalt and gabbro and none is known to be older than about 180 million years. Finally, continental crust is thicker and less dense than oceanic crust.

- *What is the crust–mantle boundary called, and how was it discovered?*

The boundary between the crust and mantle is the Mohorovičić discontinuity, or simply the Moho, which is present everywhere except beneath spreading ridges. It is a boundary across which seismic waves travel more rapidly. It was discovered when Mohorovičić noticed that seismic stations received two sets of P- and S-waves. He reasoned that one set traveled deeper but more quickly (below this discontinuity) than the shallower waves.

11.4 Earth's Mantle—The Layer Below the Crust
- *What is the low-velocity zone in the mantle?*

The speeds of P- and S-waves decrease markedly in this 100- to 250-km-deep layer, which corresponds to the asthenosphere. The asthenosphere is present everywhere, but the low-velocity zone is poorly defined or absent beneath the ancient shields of continents. Decreased elasticity accounts for decreased seismic wave velocity in the low-velocity zone.

- *How did geologists come to the conclusion that the mantle is probably made up of peridotite?*

Experiments indicate that peridotite has the physical properties and density to account for seismic wave velocity in the mantle, and peridotite makes up the lower parts of ophiolite sequences that represent oceanic crust and upper mantle. Furthermore, peridotite is found as inclusions in kimberlite pipes that came from depths of 100 to 300 km.

11.5 The Core
- *What are the P- and S-wave shadow zones, and what do they indicate about Earth's interior?*

Because of the way P-waves are refracted within Earth, very little P-wave energy arrives at seismic stations between 103 and 143 degrees from an earthquake's focus, thus forming a P-wave shadow zone. S-waves are completely blocked by Earth's core, so none are detected at seismic stations more than 103 degrees from an earthquake's focus. The fact that P- and S-waves change their direction and speed of travel and both have shadow zones have enabled geologists to determine much about Earth's internal structure.

- *What are the density and composition of the inner and outer cores?*

Estimates for the density of the outer core vary from 9.9 to 12.2 g/cm³, and for the inner core the range is from 12.6 to 13.0 g/cm³. Both the outer and inner cores are made up mostly of iron, but the outer core also contains possibly 12% other elements, especially sulfur, and the inner core probably has between 10–20% nickel.

11.6 Earth's Internal Heat
- *What is the geothermal gradient, and can it be used to determine the temperature of Earth's core?*

The geothermal gradient is Earth's increase in temperature with depth, which averages about 25°C/km, but only for near-surface materials. If the temperature increased at the same rate to the center of the planet, all known materials would be molten. Accordingly, the geothermal gradient for the mantle and core is probably no more than 1°C/km.

11.7 Gravity and How Its Force Is Determined
- *What are positive and negative gravity anomalies?*

Any departure from the expected force of gravity is a gravity anomaly. An anomaly is positive if an excess of mass is present or negative if there is a mass deficiency.

11.8 Floating Continents—The Principle of Isostasy
- *How is it possible for a solid (Earth's crust) to "float" in another solid (the mantle)?*

Although both the crust and mantle are solid, the crust is less dense and it is buoyed up by the mantle, which because it is under great pressure and high temperature, behaves much like a fluid.

11.9 Earth's Magnetic Field
- *How does magnetism differ from gravity?*

Both are attractive forces, but gravity depends on the mass of any two objects and the distance between their centers of mass whereas moving electricity generates magnetism. Also, anything with mass is affected by gravity, but only some substances are magnetic.

- *Where and how is Earth's magnetic field generated?*

Scientists think that thermal and compositional convection in the liquid outer core generates electrical currents, which in turn create the magnetic field.

ESSENTIAL TERMS TO KNOW

asthenosphere (p. 343)
continental crust (p. 342)
core (p. 344)
crust (p. 342)
discontinuity (p. 340)
geothermal gradient (p. 346)
gravity (p. 347)
gravity anomalies (p. 349)
heat flow (p. 347)
isostatic rebound (p. 351)
lithosphere (p. 343)
low-velocity zone (p. 343)
magnetic field (p. 352)
magnetism (p. 352)
mantle (p. 343)
Mohorovičić discontinuity (Moho) (p. 343)
negative gravity anomaly (p. 349)
oceanic crust (p. 342)
positive gravity anomaly (p. 349)
P-wave shadow zone (p. 345)
principle of isostasy (p. 350)
reflection (p. 340)
refraction (p. 340)
S-wave shadow zone (p. 345)
seismic tomography (p. 340)

REVIEW QUESTIONS

1. Why do continents stand higher than ocean basins?
2. Why do scientists think that the inner core is solid and the outer core is liquid? What comprises the core?
3. If Earth were completely solid and homogeneous throughout, how would P- and S- waves behave as they traveled through the planet? How do they actually behave?
4. What accounts for the two seismic discontinuities in the mantle?
5. How did scientists determine Earth's overall density and its internal composition?
6. What is the likely explanation for changes with depth in the mantle?
7. What is the Moho and how was it discovered?
8. How does the lithosphere differ from the asthenosphere?
9. What are kimberlite pipes, and how do they provide evidence for the composition of the mantle?
10. How was Earth's solid inner core discovered?

APPLY YOUR KNOWLEDGE

1. Earth's crust is deeply eroded in a vast area, but widespread thick sedimentary layers are deposited elsewhere. How do you think the crust will respond to the decreased load in the first area and the increased load in the other?

FIELD QUESTION

1. Label the diagram at the right and show what each area is likely composed of.

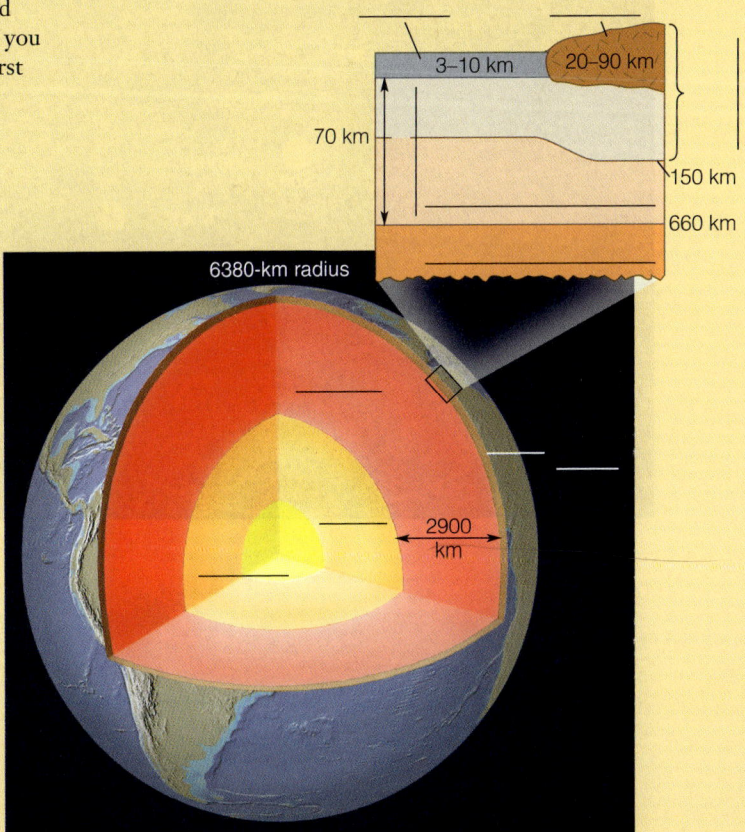

GEOLOGY MATTERS

GEOLOGY IN FOCUS Planetary Alignments, Gravity, and Catastrophes

Do you remember the grim predictions for May 5, 2000, that appeared in the tabloids, sensational books, at least one television show, and several radio talk shows? On or about this date, the five visible planets—Mercury, Venus, Mars, Jupiter, and Saturn—as well as the Sun and Moon lined up on one side of Earth in a Grand Alignment, as it was called by some. One author claimed that these celestial bodies would be in nearly a straight line for the first time in 6000 years.

Actually, this was not the first alignment in 6000 years; 71 alignments at least as notable have taken place during the last 8200 years. In addition, in this most recent Grand Alignment, the planets, Sun, and Moon were not nearly in a straight line. They were only roughly aligned, spread over about 25 degrees of space (▶ Figure 1). Nevertheless, according to several predictions, the greater gravitational attraction on Earth during this alignment would lead to several catastrophes—more earthquakes, some of huge magnitude, a sea-level rise of up to 90 m, huge shifts in Earth's crust, and winds up to 3200 km/hr. Of course, May 5, 2000, passed without any of these predicted events taking place.

The predictions for May 5, 2000, were not the first made for times of planetary alignments. In 1974 two astronomers published *The Jupiter Effect*, in which they predicted that increased gravitational attraction during a planetary alignment in 1982 would trigger catastrophes. In their view, tidal forces would cause solar flares, and streams of solar particles entering Earth's atmosphere would bring about weather changes, slow Earth's rate of rotation, and cause earthquakes. Actually, *The Jupiter Effect* concentrated on California's San Andreas fault but did not mention other earthquake-prone areas, probably because "in California better than any other place on Earth, one finds a fear of earthquakes combined with a proven market for sensational books."*

During the 1982 planetary alignment, the planets were not precisely aligned but rather spread over about 60 degrees of space. There were indeed earthquakes in 1982 and some areas experienced unusual weather, but earthquakes and unusual weather occur somewhere on Earth every year. In short, there were no noticeable changes in weather patterns, frequency of earthquakes, or Earth's rate of rotation during 1982 or any time thereafter. Actually, there were fewer large earthquakes (Richter magnitude of 7.0 or higher) during 1982 than the long-term average.

Intuitively, it seems that planetary alignments would cause increased gravitational attraction and perhaps cause more earthquakes. So why don't they? The answer is related to the equation for the universal law of gravitation. Remember the denominator in the equation tells us that the gravitational attraction between any two bodies decreases by a factor of 4 when the distance between them is doubled. (see Figure 11.12a). So even when planets, the Moon, and Sun roughly align on one side of Earth, their combined effect is so small that they have a minimal impact on our planet. Indeed, even Jupiter, a giant among planets, when in an alignment has only about 1/500,000 of the tidal force exerted by the Moon. In short, scientists have never found any relationship between planetary alignments and changes in Earth's surface or internal processes.

When will the next planetary alignment take place? The six inner planets align every 50 to 100 years, but a Grand Alignment similar to the one of May 5, 2000, will not occur until 2438. Still, predictions of catastrophes are popular and will no doubt be made again prior to the next planetary alignment, be it a minor or Grand Alignment.

*J. Mosley, "Planetary Alignments in 2000," http://www.griffithobs.org/Sky/Alignments.html.

Figure 1
The Grand Alignment of May 2000 as seen from deep space far above the plane of the solar system. Earth's Moon is not shown, but it too was part of this alignment. The alignment was a remarkable event, but the planets, Moon, and Sun were not in a straight line as some people claimed.

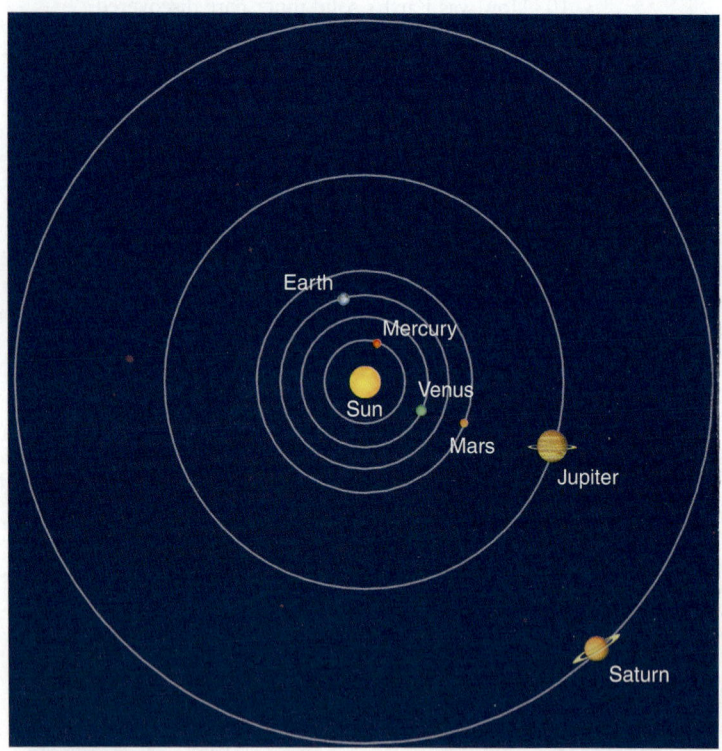

GEOLOGY MATTERS

GEOLOGY IN UNEXPECTED PLACES

Diamonds and Earth's Interior

Diamond—a gemstone, a shape, a tool. Chances are you or someone you know owns a diamond ring, necklace, or bracelet, because diamond, which symbolizes strength and purity, is the most popular and sought-after gemstone. The value of a gem-quality diamond depends on the four Cs: color (colorless ones are most desirable), clarity (lack of flaws), carat (its size; 1 carat = 200 milligrams), and cut. A diamond's cut refers to the way it is cleaved, cut, and polished to yield small plane surfaces known as *facets*, which enhance the quality of reflected light (▶ Figure 1).

Many of the world's diamonds come from stream and beach placer deposits, but their ultimate source is cone-shaped igneous bodies known as kimberlite pipes composed of dark gray or blue rock that originated at great depth (see Figure 11.8). In fact, diamonds, composed of carbon, form at pressures found at least 100 km deep—that is, in Earth's mantle. Diamond establishes the minimum depth for the magma that cooled to form kimberlite, but a variety of silica also found in kimberlite indicates the magma originated between 100 and 300 km below the surface. In addition, kimberlite contains inclusions of peridotite, the rock that geologists think makes up the mantle (see Figure 4.17). Kimberlite pipes are emplaced explosively in Earth's crust in perhaps as little as a few hours to days because if the magma were to rise slowly, the diamond would convert to a lower temperature form of carbon such as graphite.

The greatest concentration of kimberlite pipes is in southern Africa and Siberia, but they are known in many other areas as well. In North America they are found in the Canadian Arctic and several U.S. states. One kimberlite pipe at Murfreesboro, Arkansas, was briefly worked for diamonds; it is now Crater of Diamonds State Park, where visitors can prospect for, and keep any diamonds they find (▶ Figure 2). Now the only operating diamond mine in the United States is on the Colorado–Wyoming border, but Canada produces diamonds at several locations.

Fully 80% of all diamonds found are used for industrial purposes. Because diamond is so hard (see Table 3.2) it is used for abrasives and for tools that cut other hard substances, including other gemstones, ceramics, metals, concrete, eyeglasses, and computer chips. In road construction, diamonds are used to grind down old pavement before a new layer of blacktop is poured. Even petroleum companies use diamond-studded drill bits for drilling wells.

Traditionally, South Africa was the world's leader in diamond production, particularly of gem-quality diamonds, and it still is an important industry there. Today, however, Australia followed by Botswana, Russia, and Zaire are the leading producers.

Figure 1
The 3.5-cm-long pendant in this necklace is the 67.9-carat Victoria Transvaal diamond. It was cut from a 240-carat diamond. The 106 smaller diamonds in the necklace total about 45 carats.

Figure 2
This 37-acre plowed field at Crater of Diamonds State Park in Arkansas is the eroded surface of an ancient kimberlite pipe. Diamonds were discovered here in 1906, and since then more than 75,000 diamonds have been found, including the 40.23-carat Uncle Sam diamond, the largest ever found in the United States.

Chapter 12

The Seafloor

ESSENTIAL QUESTIONS TO ASK

12.1 Introduction
- Why is it important to know about the seafloor?

12.2 Methods Used to Study the Seafloor
- What is echo sounding?

12.3 Oceanic Crust—Its Structure and Composition
- How have scientists determined the structure and composition of the oceanic crust?

12.4 The Continental Margins
- What is the continental shelf?
- What is the continental rise and how does it form?
- How do active and passive continental margins compare?

12.5 Features Found in the Deep-Ocean Basins
- Why are abyssal plains found mostly near passive continental margins?
- What are submarine hydrothermal vents?
- How do seamounts and guyots form?

12.6 Sediments on the Deep Seafloor
- Why are there no deposits of sand and gravel in the deep oceans?
- What are the two main types of sediment found on the seafloor?

12.7 Reefs—Rocks Made by Organisms
- What events lead to the origin of an atoll?

12.8 Resources from the Oceans
- What is the Exclusive Economic Zone?
- What is methane hydrate, and why might it be an energy resource in the future?

Heart Reef, Australia
Beneath warm blue sea a heart of coral pulses with life.
Like cloud shapes in the sky, the Heart Reef naturally formed into the shape of a heart. It is located in the Whitsunday region of Australia and is part of the Great Barrier Reef. The Great Barrier Reef, a system of approximately 900 islands and 3000 coral reefs is the largest coral reef in the world.
—A. W.

GEOLOGY MATTERS

GEOLOGY IN FOCUS:
Oceanic Circulation and Resources from the Sea

GEOLOGY IN UNEXPECTED PLACES:
Ancient Seafloor in San Francisco

GEOLOGY AT WORK:
The Future Beneath the Sea

This icon, which appears throughout the book, indicates an opportunity to explore interactive tutorials, animations, or practice problems available on the Physical GeologyNow website at http://now.brookscole.com/phygeo6.

12.1 Introduction

In about 350 B.C. the Greek philosopher Plato wrote two dialogues in which he described a huge continent called Atlantis that, according to him, existed in the Atlantic Ocean west of the Pillars of Hercules, or what we now call the Strait of Gibraltar (▶ Figure 12.1). According to Plato's account, Atlantis controlled a large area that extended through the Mediterranean region to Egypt. Yet despite its vast wealth, advanced technology, and large army and navy, Atlantis was defeated in war by Athens. Plato wrote that, following the conquest of Atlantis, "there were violent earthquakes and floods and one terrible day and night came when . . . Atlantis . . . disappeared beneath the sea. And for that reason even now the sea there has become unnavigable and unsearchable, blocked as it is by the mud shallows, which the island produced, as it sank."*

There are no "mud shallows" in the Atlantic as Plato asserted, but present-day proponents of Atlantis have claimed that the Azores, Bermuda, the Bahamas, and the Mid-Atlantic Ridge are remnants of the lost continent. Indeed, in the centuries since Plato lived, supporters of Atlantis as an actual continent have reported it in nearly every imaginable location on the planet. If a continent actually had sunk in the ocean, however, it could easily be detected by a gravity survey. Furthermore, the oceanic crust has been drilled in hundreds of places and no evidence has ever come to light to support the idea of a lost continent.

Why then has the legend of Atlantis persisted for so long? One reason is that sensational stories of lost continents with advanced civilizations are popular, but another is that until only a few decades ago no one had much knowledge of what lies beneath the oceans. Much like Earth's interior, the seafloor is largely a hidden domain, so myths and legends about what lies below the ocean's surface were widely accepted. Indeed, the most basic observation we can make about Earth is that it has vast water-covered areas hidden from direct view and continents that at first glance may seem to be nothing more than parts of the planet not underwater.

Nevertheless, there are considerable differences between continents and ocean basins. Ocean basins are obviously lower than continents, but why should this be so? Remember from Chapter 11 that continental crust is thicker and less dense than oceanic crust, so according to the principle of isostasy, continental crust should stand higher. Furthermore, oceanic crust is made up of basalt and gabbro, whereas continental crust is composed of all rock types, but its overall composition compares closely to granite. Oceanic crust is produced continuously at spreading ridges and consumed at subduction zones, so none is more than 180 million years old, quite young by geologic standards. In contrast, rocks on continents vary from recent to 3.96 billion years old.

---

*From the *Timaeus*, quoted in E. W. Ramage, Ed., *Atlantis: Fact or Fiction?* (Bloomington: Indiana University Press, 1978), p. 13.

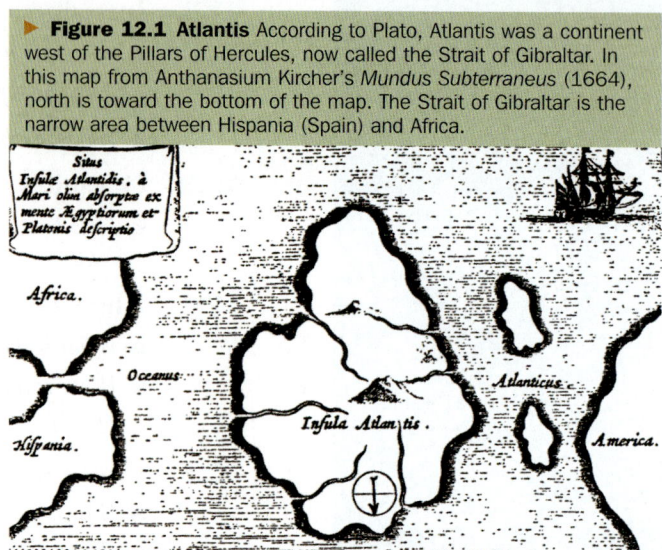

▶ **Figure 12.1 Atlantis** According to Plato, Atlantis was a continent west of the Pillars of Hercules, now called the Strait of Gibraltar. In this map from Anthanasium Kircher's *Mundus Subterraneus* (1664), north is toward the bottom of the map. The Strait of Gibraltar is the narrow area between Hispania (Spain) and Africa.

■ *Why is it important to know about the seafloor?*

One important reason to study the seafloor is that it constitutes the largest part of Earth's surface, and despite the commonly held misconception that it is flat and featureless, it has topography as varied as that on the continents (▶ Figure 12.2). In addition, many features found on the seafloor as well as the nature of the oceanic crust provide important evidence for plate tectonic theory (see Chapter 2). And finally, several natural resources are present on the shallowly submerged continental margins, in seawater, and on the seafloor.

Our investigation of the seafloor will focus on (1) the physical nature and composition of the oceanic crust, (2) the origin and evolution of the continental margins, (3) seafloor topography, and (4) the composition and distribution of seafloor sediments. We should point out that whereas the oceans and their marginal seas (▶ Figure 12.2) are underlain by oceanic crust, the same is not true of such bodies of water as the Dead Sea, Salton Sea, and Caspian Sea; these are actually saline lakes on the continents.

Section 12.1 Summary

- The seafloor makes up the largest part of Earth's surface; it has varied topography, provides evidence for plate tectonic theory, and contains natural resources.

12.2 Methods Used to Study the Seafloor

Little was known about the seafloor until a few decades ago when instruments were developed that aided in exploration. Deep-diving craft (submersibles) came into use,

▶ **Figure 12.2 The Seafloor Constitutes the Largest Part of the Earth's Surface** Map showing the four major oceans and many of the world's seas, which are marginal parts of oceans.

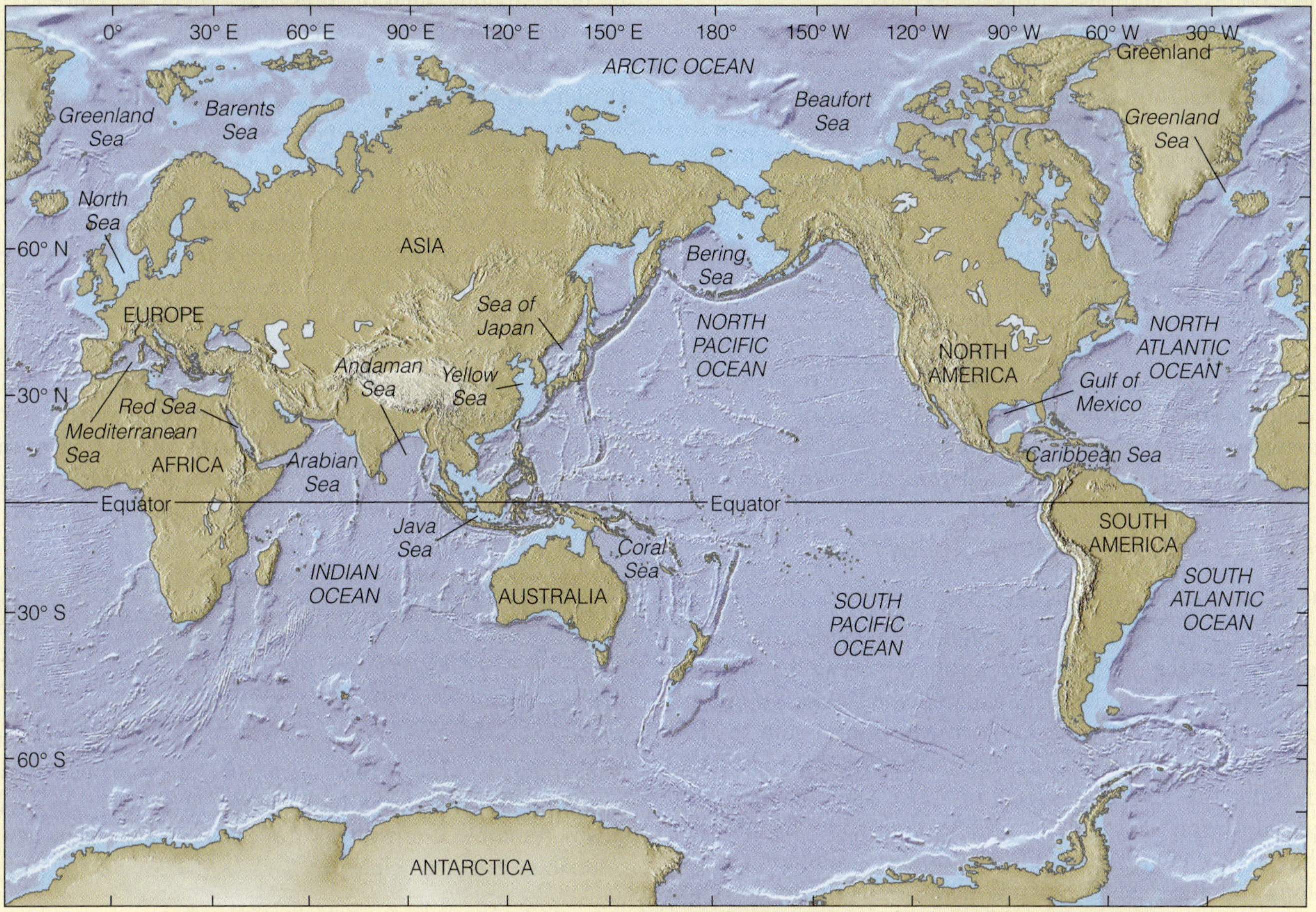

Numeric Data for the Oceans

Ocean*	Surface Area (Million KM²)	Water Volume (Million KM³)	Average Depth (FM)	Maximum Depth (KM)D
Pacific	180	700	4.0	11.0
Atlantic	93	335	3.6	9.2
Indian	77	285	3.7	7.5
Arctic	15	17	1.1	5.2

Source: P. R. Pinet, 1992. Oceanography. (St. Paul: West.)
*Excludes adjacent seas, such as the Caribbean Sea and Sea of Japan, which are marginal parts of oceans.

and methods were developed to drill and sample the seafloor. As a result scientists now have a good idea of what the oceanic crust is composed of and what kinds of features are found on the seafloor. During most of historic time, though, people knew little about the seafloor and perceived of it as a flat, featureless plain. Indeed, the seafloor, in one sense, was more remote than the Moon's surface because it could not be observed except in the shallowest ocean waters.

Early Exploration

The ancient Greeks had determined Earth's size and shape rather accurately, but western Europeans were not aware of Earth's size or the vastness of the oceans until the 1400s and 1500s, when explorers sought trade routes to the Indies. Even when Christopher Columbus set sail on August 3, 1492, in an effort to find a route to the Indies, he greatly underestimated the width of the Atlantic Ocean. Contrary to popular belief, he was not attempting to demonstrate Earth's spherical shape;

its shape was well accepted by then. The controversy was over Earth's circumference and the shortest route to China; on these points, Columbus's critics were correct.

These voyages added considerably to our knowledge of the oceans, but truly scientific investigations did not begin until the late 1700s. Great Britain was the dominant maritime power, and to maintain that dominance, the British sought to increase their knowledge of the oceans. The earliest British scientific voyages were led by Captain James Cook in 1768, 1772, and 1777. In 1872 the converted British warship HMS *Challenger* began a 4-year voyage to sample seawater, collect and analyze seafloor sediment and rocks, and name and classify thousands of species of marine organisms.

Continuing exploration of the oceans revealed that the seafloor is not flat and featureless as formerly thought. Indeed, scientists discovered that the seafloor possesses varied topography including oceanic trenches, submarine ridges, deep canyons, broad plateaus, hills, and vast plains.

How Are Oceans Explored Today?

■ *What is echo sounding?*

Measuring the length of a weighed line lowered to the seafloor was the method for determining ocean depths, but now scientists use an instrument called an **echo sounder,** which detects sound waves that travel from a ship to the seafloor and back (▶ Figure 12.3). Depth is calculated by knowing the speed of sound in water and the time required for the waves to reach the seafloor and return to the ship, thus yielding a continuous profile of seafloor depths along the ship's route. **Seismic profiling** is similar to echo sounding but even more useful. Strong waves from an energy source reflect from the seafloor, and some of the waves penetrate seafloor layers and reflect from various horizons back to the surface (▶ Figure 12.3). Seismic profiling is particularly useful for mapping the structure of the oceanic crust where it is buried beneath seafloor sediments.

The Deep Sea Drilling Project, an international program sponsored by several oceanographic institutions, began in 1968. Its first research vessel, the *Glomar Challenger*, could drill in water more than 6000 m deep and recover long cores of seafloor sediment and oceanic crust. The *Glomar Challenger* drilled more than 1000 holes in the seafloor during the 15 years of the program. The Deep Sea Drilling Project ended in 1983, but beginning in 1985 the Ocean Drilling Program with its research vessel the JOIDES* *Resolution* continued to explore the seafloor (▶ Figure 12.4a). Research vessels also use *clamshell samplers* that retrieve sediment from the seafloor and *piston corers* that are designed to penetrate as much as 25 m of seafloor sediment and bring up long plugs or cores of sediment (▶ Figure 12.5).

In addition to surface vessels, submersibles are now important vehicles for seafloor exploration. Some, such as the *Argo*, are remotely controlled and towed by a surface vessel. In 1985 the *Argo*, equipped with sonar and television systems, provided the first views of the British ocean liner HMS *Titanic* since it sank in 1912. The U.S. Geological Survey uses a towed device with sonar to produce seafloor images resembling aerial photographs. Scientists aboard submersibles such as *Alvin* (▶ Figure 12.4b) have descended to the seafloor in many areas to make observations and collect samples.

Scientific investigations have yielded important information about the oceans for more than 200 years, but much of our current knowledge has been acquired since World War II (1939–1945). This is particularly true of the seafloor because only in recent decades has instrumentation been available to study this largely hidden domain.

> ## Section 12.2 Summary
>
> • Myths about the seafloor persisted until fairly recently because the technology to explore this largely hidden domain was developed only during the last few decades.
>
> • Geologists and others study the seafloor by echo sounding, seismic profiling, sampling, and collecting data from remotely controlled submersibles and from submersibles carrying scientists.

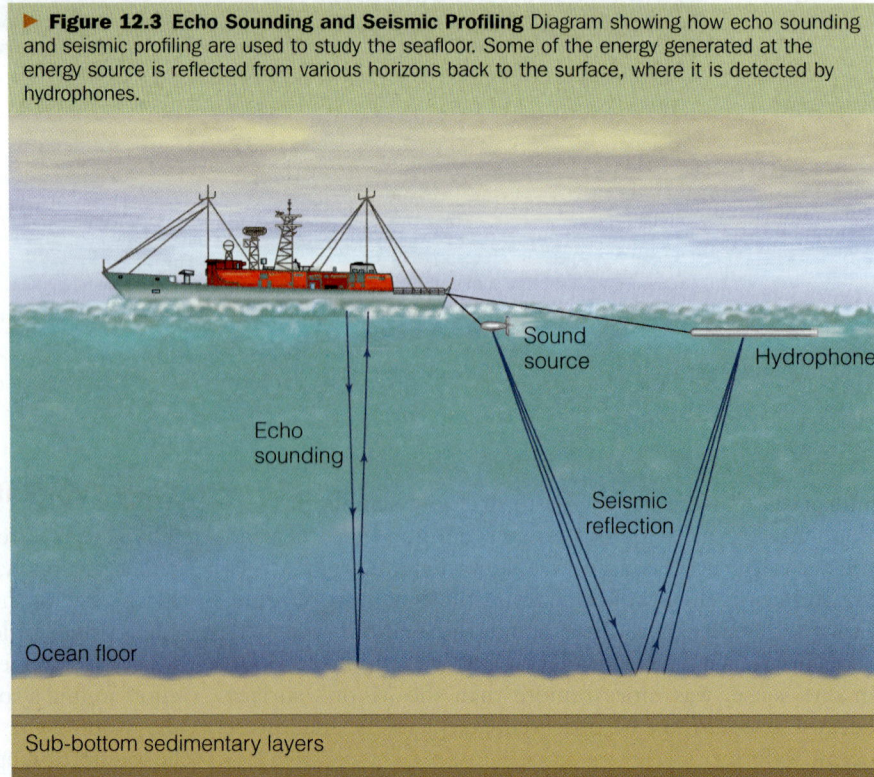

▶ **Figure 12.3 Echo Sounding and Seismic Profiling** Diagram showing how echo sounding and seismic profiling are used to study the seafloor. Some of the energy generated at the energy source is reflected from various horizons back to the surface, where it is detected by hydrophones.

*JOIDES is an acronym for Joint Oceanographic Institutions for Deep Earth Sampling.

▶ Figure 12.4 Oceanographic Research Vessels

a The JOIDES *Resolution* is capable of drilling the deep seafloor.

b The submersible *Alvin* is used for observing and sampling the deep seafloor.

▶ Figure 12.5 Seafloor Sampling

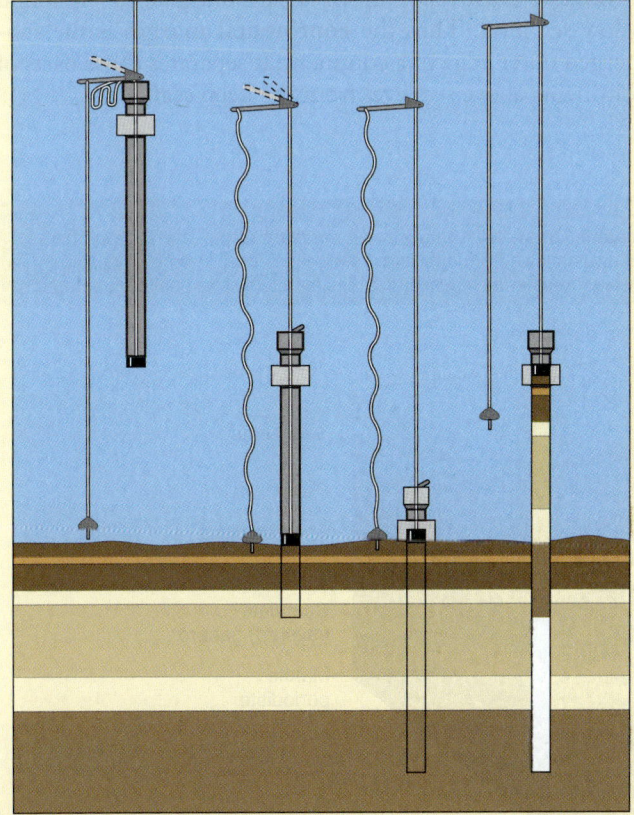

a A piston corer falls to the seafloor, penetrates the sediment, and then is retrieved.

 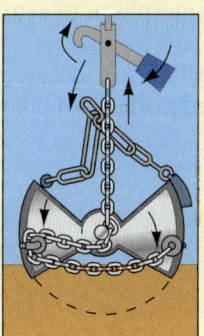 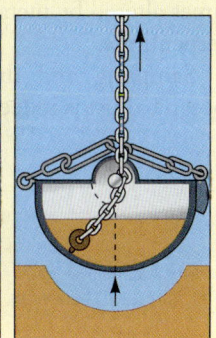

b A clamshell sampler taking a seafloor sample.

c A core from the Gulf of California shows alternating thin layers of seafloor sediment. Empty places in the core show where material was removed for study.

12.3 Oceanic Crust— Its Structure and Composition

- *How have scientists determined the structure and composition of the oceanic crust?*

Drilling into the oceanic crust has provided some information on its structure and composition, but the crust has never been completely penetrated and sampled. So how do we know what it is composed of and how it varies with depth? Of course, we already know that its upper part is made up of basalt, and even before it was sampled and observed, geologists had a good idea of its overall structure and composition because fragments of oceanic crust along with part of the underlying mantle are found on land.

Remember that oceanic crust is thinner and denser than continental crust, and it is consumed at subduction zones and thus most of it is recycled. A small amount is found in mountain ranges on land where it was emplaced by moving along large fractures called thrust faults (faults are discussed more fully in Chapter 13). These preserved slivers of oceanic crust along with part of the underlying upper mantle are known at **ophiolites.** Detailed studies reveal that an ideal ophiolite consists of rocks of the upper oceanic crust, especially pillow lava and sheet lava flows, underlain by a sheeted dike complex, consisting of vertical basaltic dikes, and massive gabbro and layered gabbro that probably formed in the upper part of a magma chamber (▶ Figure 12.6). And finally, the lowermost unit is peridotite, representing the upper mantle; this is sometimes altered by metamorphism to a greenish rock known as serpentinite.

Sampling and drilling at oceanic ridges reveal that oceanic crust is indeed made up of pillow lava and lava flows underlain by a sheeted dike complex, just as predicted from studies of ophiolites. But it was not until 1989 that a submersible carrying scientists descended to the walls of a seafloor fracture in the North Atlantic and verified what lay below the sheeted dike complex. Just as expected, the lower oceanic crust consists of gabbro and the upper mantle is made up of peridotite.

Section 12.3 Summary

- Oceanic crust is thinner and denser than continental crust, and it is made up of basalt and gabbro.

- Ophiolites are fragments of upper mantle and oceanic crust emplaced on land during deformation. From bottom to top they are made up of peridotite (sometimes altered to serpentinite), massive gabbro, a sheeted dike complex, and pillow lava.

12.4 The Continental Margins

Many people perceive of continents as land areas outlined by sea level or, put another way, that part of Earth's crust that is not submerged by the oceans. The true margin of a continent—that is, where granitic continental crust changes to oceanic crust made up of basalt and gabbro—is found below sea level. Thus the **continental margin** is the submerged outer edge of a continent; it separates those parts of continents above sea level from the deep seafloor.

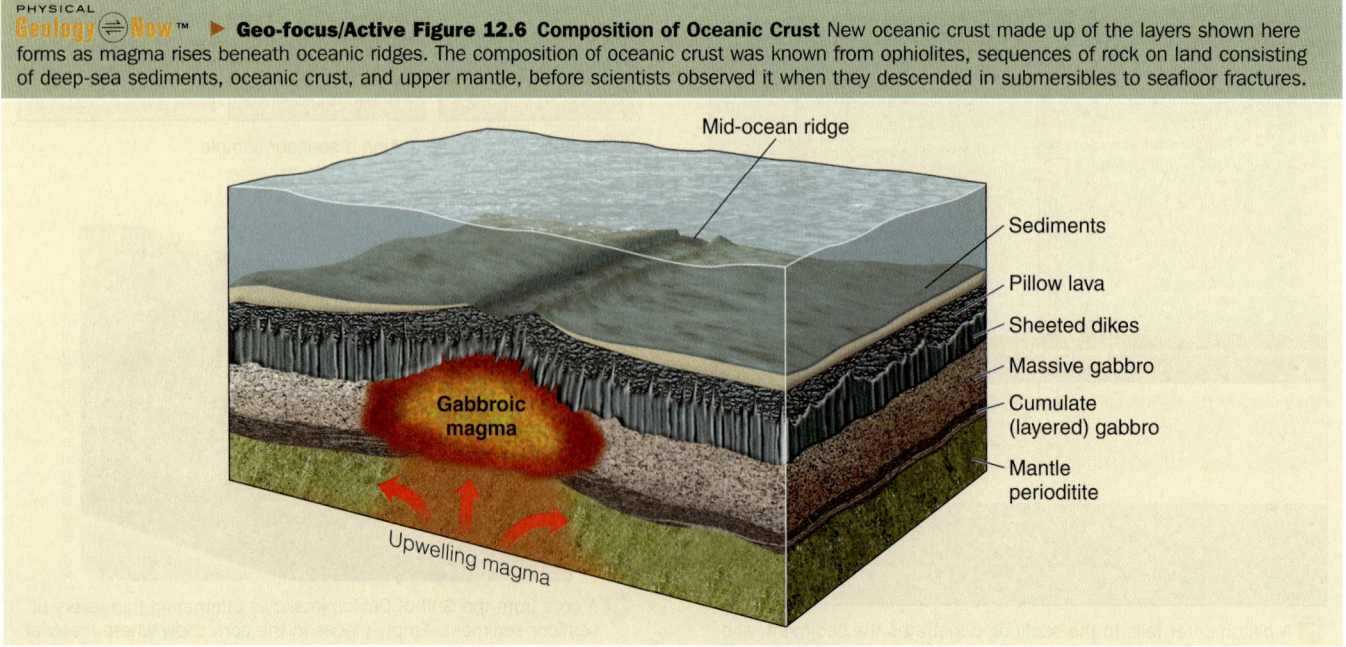

▶ **Geo-focus/Active Figure 12.6 Composition of Oceanic Crust** New oceanic crust made up of the layers shown here forms as magma rises beneath oceanic ridges. The composition of oceanic crust was known from ophiolites, sequences of rock on land consisting of deep-sea sediments, oceanic crust, and upper mantle, before scientists observed it when they descended in submersibles to seafloor fractures.

A continental margin is made up of a gently sloping continental shelf, a more steeply inclined continental slope, and in some cases, a deeper, gently sloping continental rise (▶ Figure 12.7). Seaward of the continental margin lies the deep-ocean basin. Thus, the continental margins extend to increasingly greater depths until they merge with the deep seafloor. Continental crust changes to oceanic crust somewhere beneath the continental rise, so part of the continental slope and the continental rise actually rest on oceanic crust.

The Continental Shelf

■ *What is the continental shelf?*

As one proceeds seaward from the shoreline across the continental margin, the first area encountered is a gently sloping **continental shelf** lying between the shore and the more steeply dipping continental slope (▶ Figure 12.7). The width of the continental shelf varies considerably, ranging from a few tens of meters to more than 1000 km; the shelf terminates where the inclination of the seafloor increases abruptly from 1 degree or less to several degrees. The outer margin of the continental shelf, or simply the *shelf–slope break*, is at an average depth of 135 m, so by oceanic standards the continental shelves are covered by shallow water.

At times during the Pleistocene Epoch (1.8 million to 10,000 years ago), sea level was as much as 130 m lower than it is now. As a result, the continental shelves were above sea level and were areas of stream channel and floodplain deposition. In addition, in many parts of northern Europe and North America, glaciers extended well out onto the continental shelves and deposited gravel, sand, and mud. At the end of the Pleistocene, sea level rose, submerging these deposits, which are now being reworked by marine processes. Evidence that these sediments were in fact deposited on land includes remains of human settlements and fossils of a variety of land-dwelling animals.

The Continental Slope and Rise

■ *What is the continental rise and how does it form?*

The seaward margin of the continental shelf is marked by the *shelf–slope break* (at an average depth of 135 m) where the more steeply inclined **continental slope** begins (▶ Figure 12.7). In most areas around the margins of the Atlantic, the continental slope merges with a more gently sloping **continental rise.** This rise is absent around the margins of the Pacific where continental slopes descend directly into oceanic trenches (▶ Figure 12.7).

An interesting feature found on the continental slope adjacent to the Yucatán Peninsula of Mexico is layers of asphalt that were erupted much like lava flows beneath water 3000 m deep. In fact, the flows look much like aa and pahoehoe. The solid asphalt covers about 1 km^2 in an area known as the Campeche Knolls, which are salt domes that have risen through sedimentary rocks (see Figure 7.20c). Apparently, hydrocarbons seep up through cracks in the rocks adjacent to the salt domes.

The shelf-slope break is an important feature in terms of sediment transport and deposition. Landward of the break—that is, on the shelf—sediments are affected by waves and tidal currents, but these processes have no effect on sediments seaward of the break, where gravity is responsible for their transport and deposition on the slope and rise. In fact, much of the land-derived sediment crosses the shelves and is eventually deposited on the continental slopes and rises, where more than 70% of all sediments in the oceans are found. Much of this sediment is transported through submarine canyons by turbidity currents.

▶ **Figure 12.7 Features of a Continental Margin** A generalized profile of the seafloor showing features of the continental margins. The vertical dimensions of the features in this profile are greatly exaggerated because the vertical and horizontal scales differ.

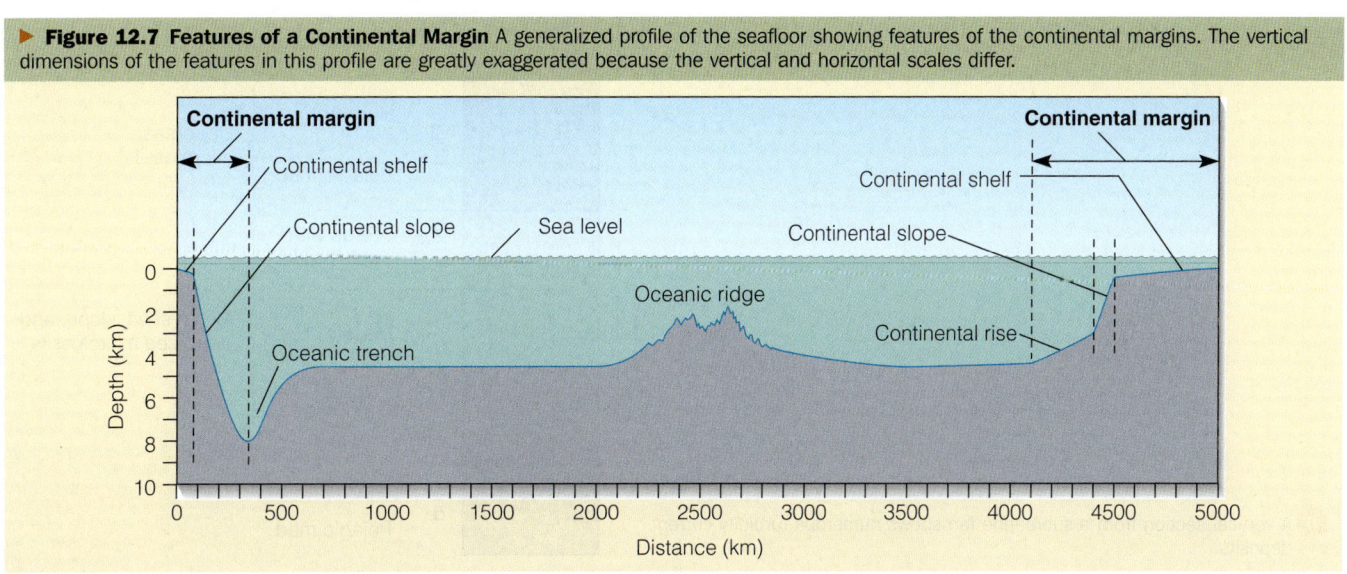

Turbidity Currents, Submarine Fans, and Submarine Canyons

In Chapter 7 we discussed the origin of graded bedding, most of which results from **turbidity currents,** underwater flows of sediment–water mixtures with densities greater than that of sediment-free water. When a turbidity current flows onto the relatively flat seafloor, it slows down and begins depositing its largest particles followed successively by smaller ones, thus forming a graded bed (see Figure. 7.15). Repeated deposition by turbidity currents yields a series of overlapping **submarine fans,** which constitute a large part of the continental rise (▶ Figure 12.8). Submarine fans are distinctive, but their outer margins are difficult to discern because they merge with deep-sea sediments.

Until recently no one had ever seen a natural turbidity current in progress in the oceans (note photo in ▶ Figure 12.9 shows a turbidity current), so for many years some doubted their existence. Perhaps the most compelling evidence for turbidity currents is the pattern of submarine cable breaks that occurred in the North Atlantic near Newfoundland in 1929. Scientists assumed initially that an earthquake had severed several telegraph and telephone cables. However, the breaks on the continental shelf near the epicenter occurred when the earthquake struck, the cables farther seaward were broken later and in succession. The last cable to break was 720 km from the earthquake's epicenter and it did not snap until 13 hours after the first break.

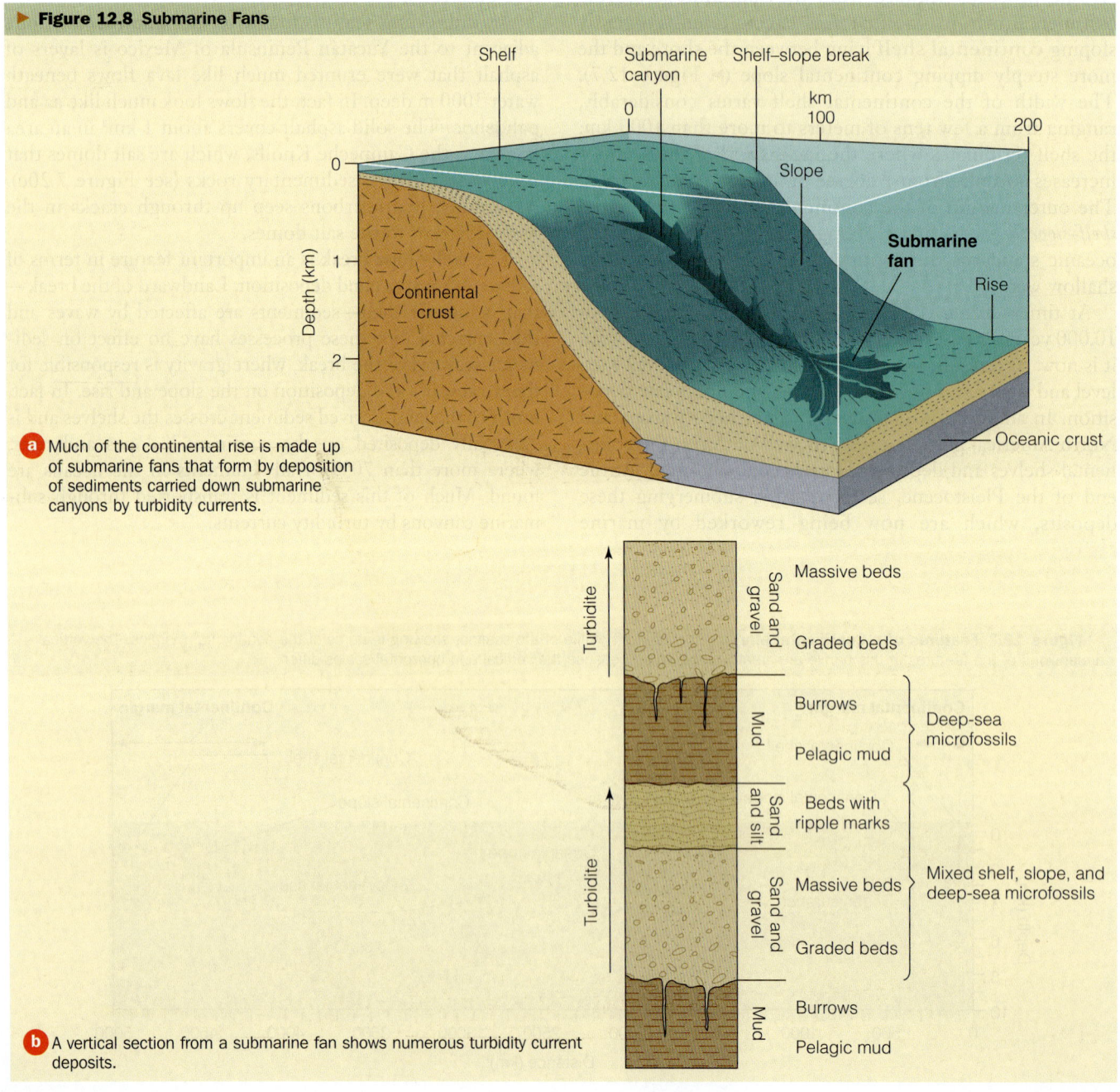

▶ Figure 12.8 Submarine Fans

a) Much of the continental rise is made up of submarine fans that form by deposition of sediments carried down submarine canyons by turbidity currents.

b) A vertical section from a submarine fan shows numerous turbidity current deposits.

In 1949 geologists realized that an earthquake-generated turbidity current had moved downslope, breaking the cables in succession (▶ Figure 12.9). The precise time at which each cable broke was known as were the distances between cables, so calculating the speed of the turbidity current was simple. It moved at about 80 km/hr on the continental slope and slowed to about 27 km/hr when it reached the continental rise.

In more recent years evidence for turbidity currents has been found in seafloor samples showing a succession of layers with graded bedding and the remains of shallow-water organisms that were displaced into deeper water. Also research along the continental margins of North America has revealed displacement of scientific instruments placed on the seafloor, and repeated sampling in the same areas indicate that turbidity current deposits have accumulated.

Steep-sided **submarine canyons** that resemble stream- and river-cut canyons on land are present on the continental shelves, but they are best developed on continental slopes (▶ Figures 12.8 and 12.10). Some of these canyons extend across the continental shelf to rivers on land and apparently were eroded by rivers when sea level was lower during the Pleistocene. But many have no association with rivers on land and they extend to depths greater than can be accounted for by river erosion during times of lower sea level. Geologists are now convinced that scouring and erosion generated by turbidity currents, in many instances triggered by collapse of shallow-water sediment accumulating off the mouths of rivers, is the principal reason that these canyons form.

Types of Continental Margins

■ *How do active and passive continental margins compare?*

Continental margins are *active* or *passive*, depending on their relationship to plate boundaries. An **active continental margin** develops at the leading edge of a continental plate where oceanic lithosphere is subducted (▶ Figure 12.11). The western margin of South America is a good example. Here, an oceanic plate is subducted beneath the continent, resulting in seismic activity, a geologically young mountain

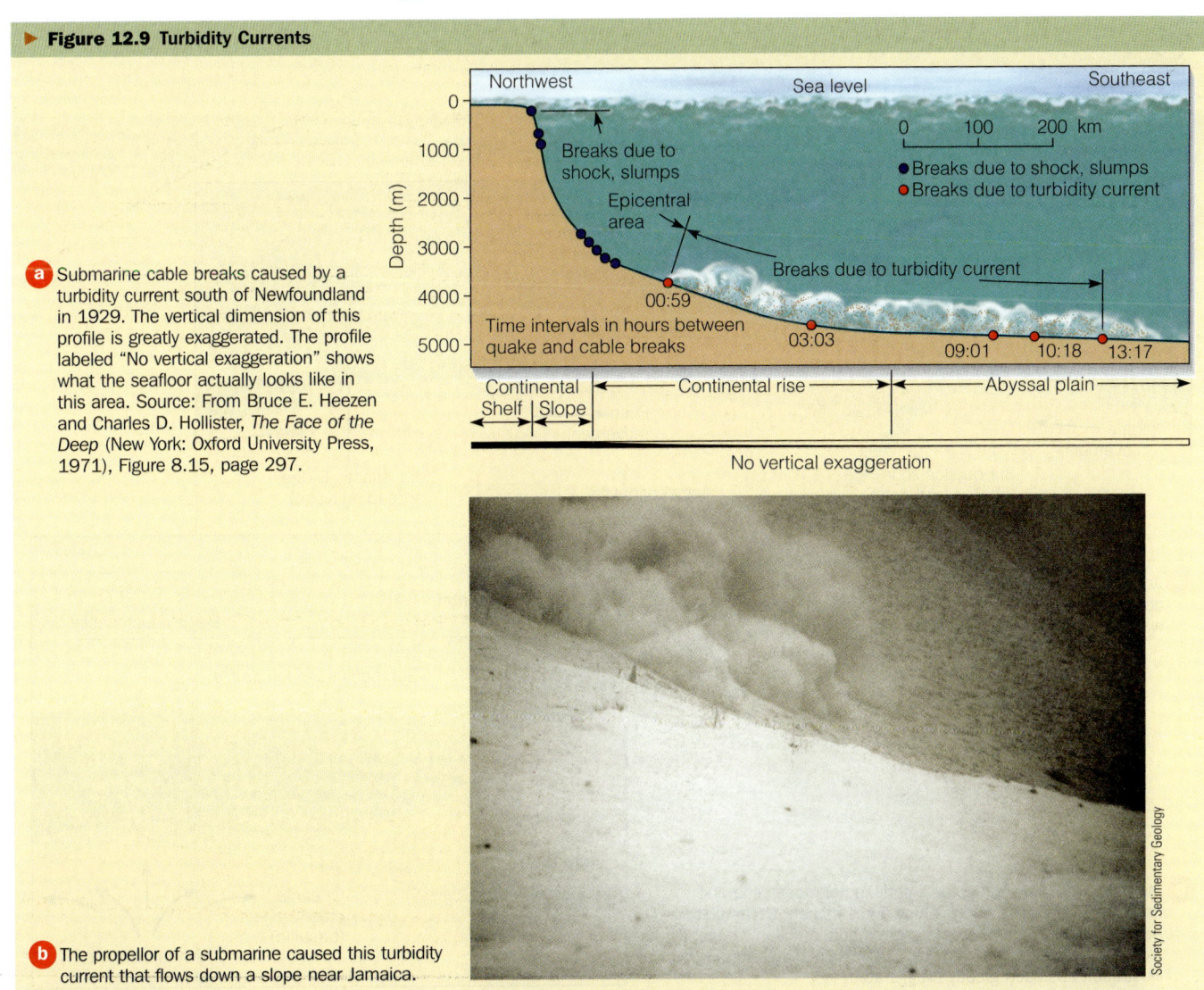

▶ **Figure 12.9 Turbidity Currents**

a Submarine cable breaks caused by a turbidity current south of Newfoundland in 1929. The vertical dimension of this profile is greatly exaggerated. The profile labeled "No vertical exaggeration" shows what the seafloor actually looks like in this area. Source: From Bruce E. Heezen and Charles D. Hollister, *The Face of the Deep* (New York: Oxford University Press, 1971), Figure 8.15, page 297.

b The propellor of a submarine caused this turbidity current that flows down a slope near Jamaica.

▶ **Figure 12.10 Submarine Canyons** Sediment cascading down San Lucas submarine canyon off the coast of Mexico.

▶ **Active Figure 12.11 Passive and Active Continental Margins**

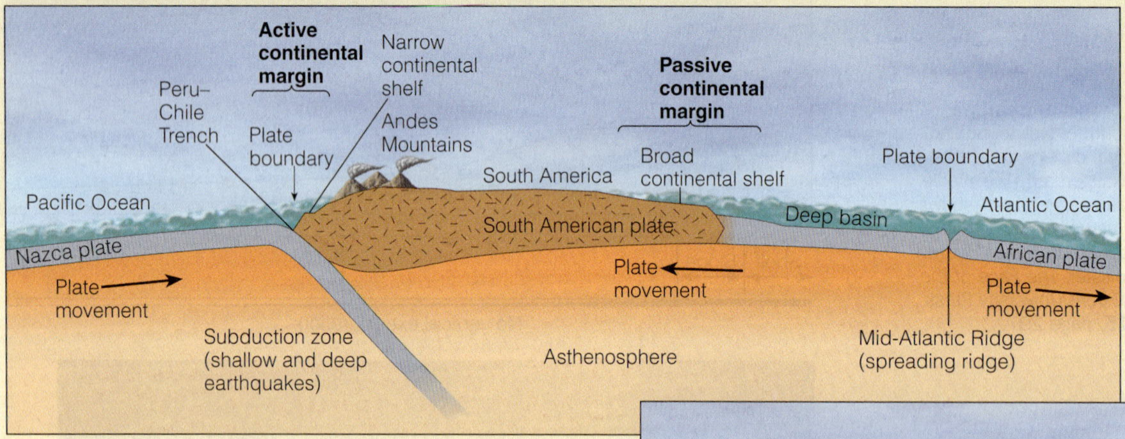

a Active and passive continental margins along the west and east coasts of South America. Notice that passive margins are much wider than active margins. Seafloor sediment is not shown.

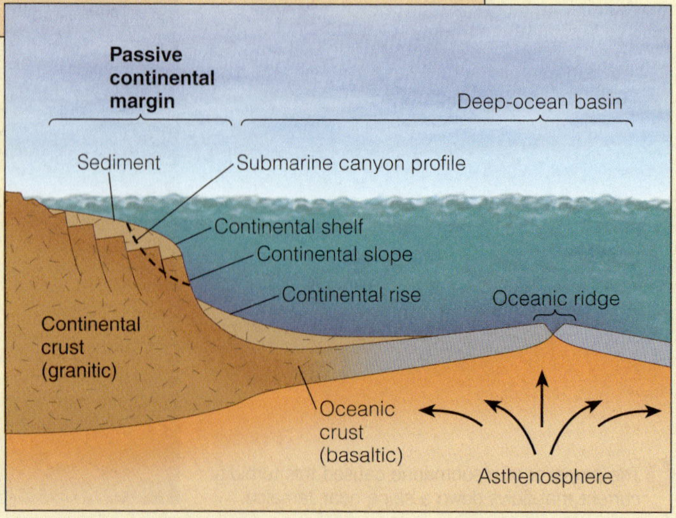

b More detailed view of a passive continental margin showing fractures that formed during rifting and sediment accumulation. Sediment is also found on the shelves and in oceanic trenches along active continental margins. The vertical dimension in this illustration has been greatly exaggerated.

range, and active volcanism. In addition, the continental shelf is narrow, and the continental slope descends directly into the Peru–Chile Trench, so sediment is dumped into the trench and a continental rise does not develop. The western margin of North America is also considered an active continental margin, although much of it is now bounded by transform faults rather than a subduction zone. However, plate convergence and subduction still take place in the Pacific Northwest along the continental margins of northern California, Oregon, and Washington.

The continental margins of eastern North America and South America differ considerably from their western margins. For one thing, they possess broad continental shelves as well as a continental slope and rise; also, vast, flat areas known as *abyssal plains* are present adjacent to the rises (▶ Figure 12.11). Furthermore, these **passive continental margins** are within a plate rather than at a plate boundary. The continental margin for eastern North America, for example, is far from the boundary of the North American plate, which is at the Mid-Atlantic Ridge. In addition, passive continental margins have no active volcanoes and seismicity is minimal, although large earthquakes do occur occasionally—the 1886 Charleston, South Carolina, earthquake, for example (see Figure 10.7).

Active and passive continental margins are notably different in the widths of their continental shelves, and active margins have an oceanic trench but no continental rise. Why the differences? At both types of continental margins, turbidity currents transport sediment into deeper water. At passive margins, the sediment forms a series of overlapping submarine fans and thus develops a continental rise, whereas at an active margin, sediment is simply dumped into the trench, where much of it is eventually subducted, and no rise forms. The proximity of a trench to a continent also explains why the continental shelves of active margins are so narrow. In contrast, land-derived sedimentary deposits at passive margins have built a broad platform extending far out into the ocean.

Section 12.4 Summary

- Continental margins are made up of a broad, gently sloping continental shelf and a more steeply inclined continental slope that in some areas lies adjacent to a continental rise that merges with the deep seafloor. In other areas the slope descends directly into an oceanic trench.

- Much of the continental rise is made up of overlapping submarine fans that form as turbidity currents carry sediment through submarine canyons cut into the continental shelf and slope.

- Seismicity, volcanism, an oceanic trench, a narrow shelf, and a young mountain range characterize active continental margins, whereas passive continental margins have a much wider shelf, no volcanism, and minimal seismicity.

12.5 Features Found in the Deep-Ocean Basins

Investigations during the 1800s revealed that the seafloor has broad plateaus and ridges, but the prevailing view was that it was mostly a flat, featureless plain. In fact it does have flat, featureless plains, but we now know that it also has submarine mountains, plateaus, canyons, volcanoes, trenches, and huge fractures. In short, the seafloor has varied topography much like the continents do. Of course, we know less about the seafloor than the continents because it is more difficult to observe.

The oceans average more than 3.8 km deep, so most of the seafloor lies far below the depth of sunlight penetration, which is rarely more than 100 m. Accordingly, most of the seafloor is completely dark, the temperature is just above 0°C, and the pressure varies from 200 to more than 1000 atmospheres depending on depth, enough to squash a full-grown adult down to the size of a grapefruit. Scientists in submersibles have descended to the greatest oceanic depths, the oceanic ridges, and elsewhere, so some of the seafloor has been observed. Nevertheless, much of the deep-ocean basin has been studied only by echo sounding, seismic profiling, sediment and oceanic crust sampling, and remote devices that have descended in excess of 11,000 m.

Abyssal Plains

■ *Why are abyssal plains found mostly near passive continental margins?*

Beyond the continental rises of passive continental margins are **abyssal plains,** flat surfaces covering vast areas of the seafloor. In some places, they are interrupted by peaks rising more than 1 km, but they are nevertheless the flattest, most featureless areas on Earth (▶ Figure 12.12). The flat topography is a result of sediment deposition; where sediment accumulates in sufficient quantities it buries the rugged seafloor (▶ Figure 12.13).

Seismic profiles and seafloor samples reveal that abyssal plains are covered with fine-grained sediment derived mostly from the continents and deposited by turbidity currents. Abyssal plains are invariably found adjacent to the continental rises, which are composed mostly of overlapping submarine fans that owe their origin to deposition by turbidity currents (▶ Figures 12.8 and 12.9). Along active continental margins, sediments derived from the shelf and slope are trapped in an oceanic trench, and abyssal plains fail to develop. Accordingly, abyssal plains are common in the Atlantic Ocean basin but rare in the Pacific Ocean basin (▶ Figure 12.12).

Oceanic Trenches

Oceanic trenches constitute a small percentage of the seafloor, probably less than 2%, but they are very important, for it is here that lithospheric plates are consumed by

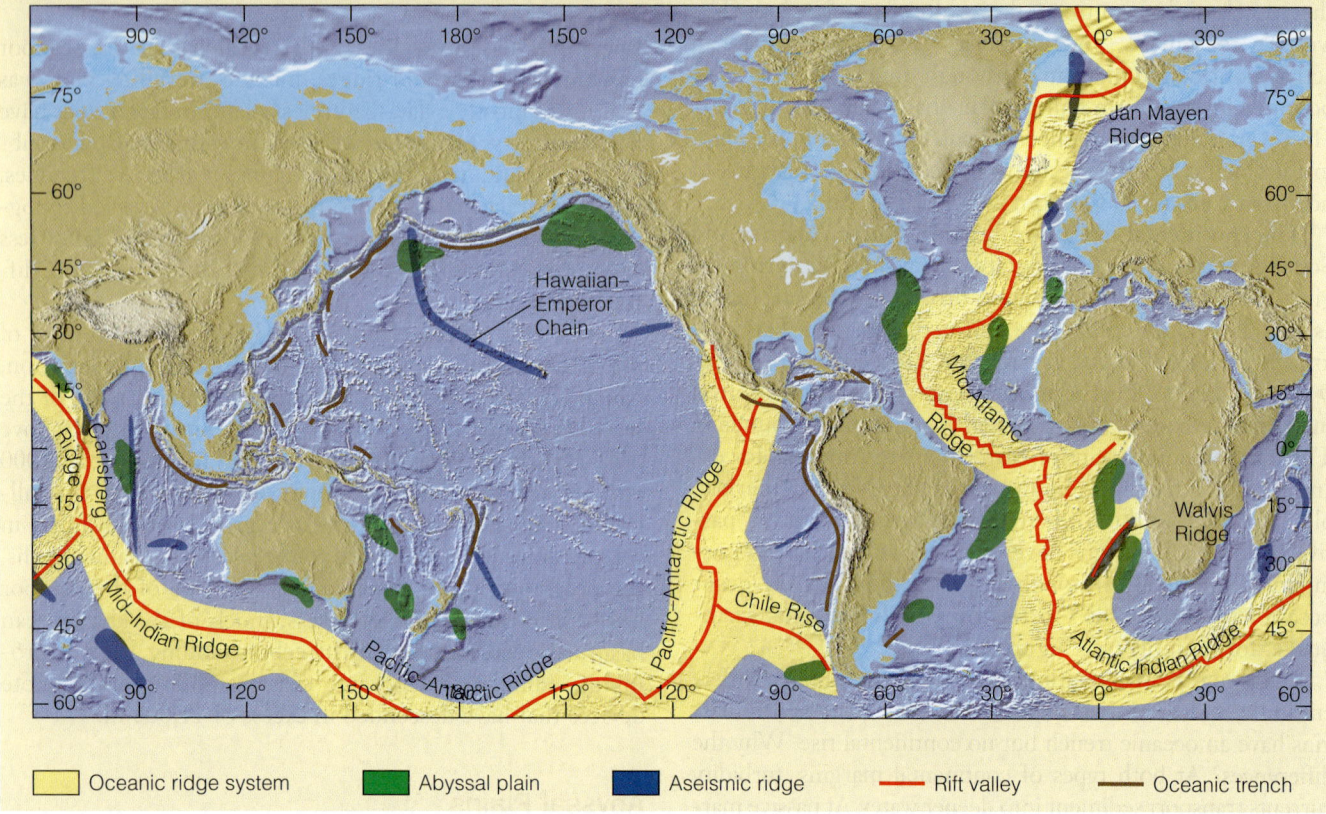

▶ **Figure 12.12 Deep Seafloor Features** Features found on the deep seafloor include oceanic trenches (brown), abyssal plains (green), the oceanic ridge system (yellow), rift valleys (red), and some aseismic ridges (blue). Other features such as seamounts and guyots are shown in Figure 12.17. Source: From Alyn and Alison Duxbury, *An Introduction to the World's Oceans*. Copyright 1984 McGraw-Hill.

subduction (see Chapter 2). Oceanic trenches are long, narrow features restricted to active continental margins, so they are common around the margins of the Pacific Ocean basin (▶ Figure 12.12). For instance, the 8000-m-deep Peru–Chile Trench west of South America is 5900 km long but only 100 km wide. On the landward side of oceanic trenches, the continental slope descends at angles of up to 25 degrees. Oceanic trenches are also the sites of the greatest oceanic depths; a depth of more than 11,000 m has been recorded in the Challenger Deep of the Marianas Trench in the Pacific Ocean.

Oceanic trenches show anomalously low heat flow compared with other areas of oceanic crust, indicating that the crust at trenches is cooler and slightly more dense than elsewhere. Gravity surveys across trenches reveal huge negative gravity anomalies because the crust is not in isostatic equilibrium. In fact, oceanic crust at trenches is subducted, giving rise to *Benioff-Wadati* zones in which earthquake foci become progressively deeper in the direction the subducted plate descends (see Figure 10.5). These inclined seismic zones account for most intermediate- and deep-focus earthquakes—such as the June 1994 magnitude–8.2 earthquake in Bolivia, which had a focal depth of 640 km. Finally, subduction at oceanic trenches also results in volcanism, either as an arcuate chain of volcanic islands on oceanic crust or as a chain of volcanoes along the margin of a continent, as in western South America.

Oceanic Ridges

When the first submarine cable was laid between North America and Europe during the late 1800s, a feature called the Telegraph Plateau was discovered in the North Atlantic. Using data from the 1925–1927 voyage of the German research vessel *Meteor*, scientists proposed that the plateau was actually a continuous ridge extending the length of the Atlantic Ocean basin. Subsequent investigations revealed that this proposal was correct, and we now call this feature the Mid-Atlantic Ridge (▶ Figure 12.12).

The Mid-Atlantic Ridge is more than 2000 km wide and rises 2 to 2.5 km above the adjacent seafloor. Furthermore, it is part of a much larger **oceanic ridge** system of mostly submarine mountainous topography. It runs from the Arctic Ocean through the middle of the Atlantic, curves around South Africa where the Indian Ridge continues into the Indian Ocean, then the Pacific–Antarctic Ridge extends eastward and a branch of this, the East Pacific Rise, trends northeast until it reaches the Gulf of California (▶ Figure 12.12). The entire system is at least 65,000 km long, far longer than any mountain range on land. Oceanic ridges are composed almost entirely of basalt and gabbro and possess features produced by tensional forces, such as long, deep fractures. Mountain ranges on land, in contrast, consist of igneous, metamorphic, and sedimentary rocks, and formed when rocks were folded and fractured by compressive forces (see Chapter 13).

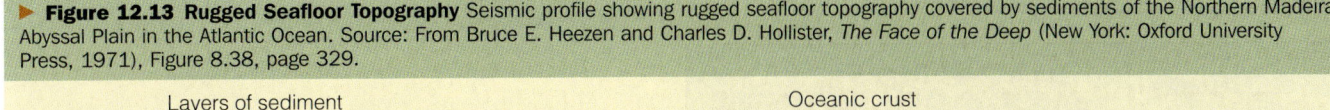

▶ **Figure 12.13 Rugged Seafloor Topography** Seismic profile showing rugged seafloor topography covered by sediments of the Northern Madeira Abyssal Plain in the Atlantic Ocean. Source: From Bruce E. Heezen and Charles D. Hollister, *The Face of the Deep* (New York: Oxford University Press, 1971), Figure 8.38, page 329.

Oceanic ridges are mostly below sea level, but they rise above the sea in Iceland, the Azores, Easter Island, and several other places. Of course, oceanic ridges are the sites where new oceanic crust is generated and plates diverge (see Chapter 2). The rate of plate divergence is important because it determines the cross-section profile of a ridge. For example, the Mid-Atlantic Ridge has a comparatively steep profile because divergence is slow, allowing the new oceanic crust to cool, shrink, and subside closer to the ridge crest than it does in areas of faster divergence such as at the East Pacific Rise. A ridge may also have a rift along its crest that opens in response to tension (▶ Figure 12.14). A rift is particularly obvious along the Mid-Atlantic Ridge but is absent along parts of the East Pacific Rise. Rifts are commonly 1 to 2 km deep and several kilometers wide. They open as seafloor spreading takes place and are characterized by shallow-focus earthquakes, basaltic volcanism, and high heat flow.

Even though most oceanographic research is still done by echo sounding, seismic profiling, and seafloor sampling, scientists have been making direct observations of oceanic ridges and their rifts since 1974. As part of project FAMOUS (French-American Mid-Ocean Undersea Study), submersibles have descended to the ridges and into their rifts in several areas. Researchers have not seen any active volcanism, but they have seen pillow lavas (see Figure 5.7a), lava tubes, and sheet lava flows, some of which formed very recently. In fact, on return visits to a site they have seen the effects of volcanism that occurred since their last visit. And on January 25, 1998, submarine Axial Volcano began erupting along the Juan de Fuca Ridge west of Oregon.

Submarine Hydrothermal Vents

■ *What are submarine hydrothermal vents?*

Near spreading ridges, cold seawater seeps down through cracks in the oceanic crust where hot rocks heat it to far above the boiling temperature. The superheated water then rises and is discharged into seawater at **submarine hydrothermal**

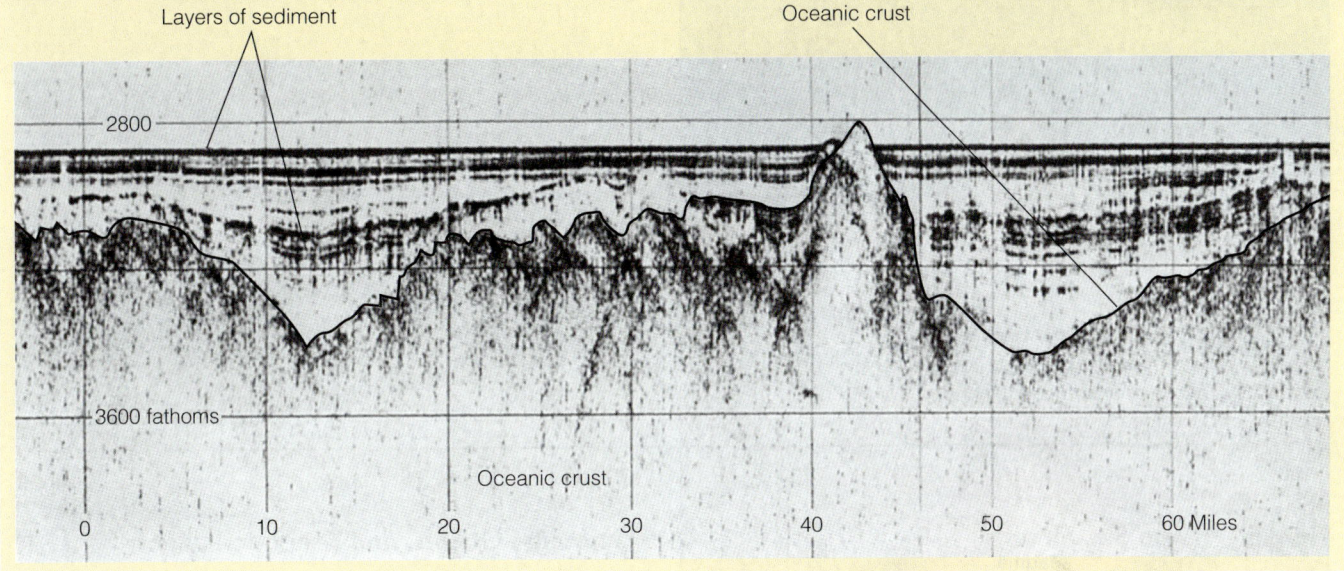

▶ **Figure 12.14 A Central Rift** Cross section of the Mid-Atlantic Ridge showing its central rift where recent moundlike accumulations of volcanic rocks, mostly pillow lava, are found.

Figure 12.15 Submarine Hydrothermal Vents

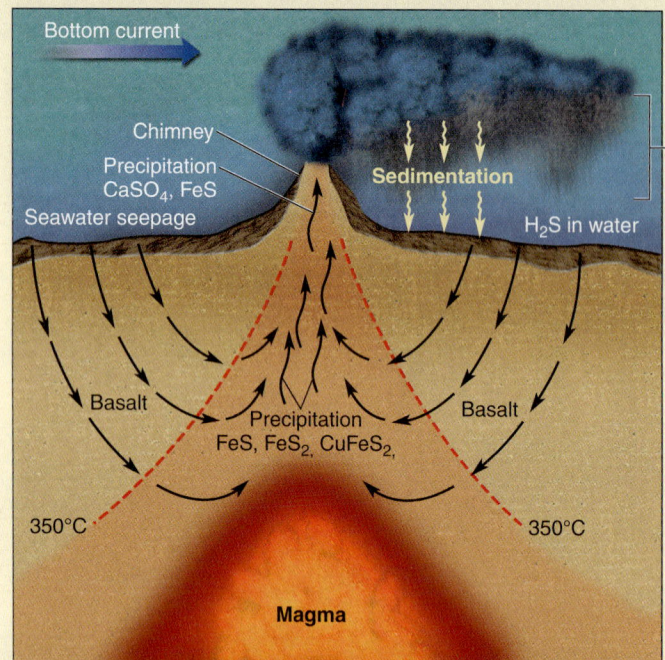

a Cross section showing the origin of a submarine hydrothermal vent called a black smoker.

b This black smoker on the East Pacific Rise is at a depth of 2800 m. The plume of "black smoke" is heated seawater with dissolved minerals.

c Several types of organisms including these tube worms live near black smokers. Tube worms may be as long as 3 m.

vents. Many of these plumes of hot water, which are up to 400°C, are black because dissolved minerals give them the appearance of black smoke—hence the name **black smoker** (▶ Figure 12.15a, b). Despite the high temperature, the water does not boil because of the tremendous pressure at great ocean depths.

Since the 1970s, when black smokers were first observed, they have been found at many other locations in the Pacific, Atlantic, and Indian Oceans. When the submersible *Alvin* descended 2500 m into the Galápagos Rift in 1979, scientists saw a diverse biological community made up of many organisms that no one had ever seen before, including tube worms (▶ Figure 12.15c), albino crabs, mussels, and starfish. No sunlight is available at such great depths so bacteria that derive their nutrients by oxidizing sulfur in the hot vent fluids lie at the base of the food chain.

Black smokers are also interesting as a potential source of natural resources, including several minerals and ores of iron, copper, and zinc. As the descending seawater is heated, it reacts with the oceanic crust and becomes a metal-rich solution, but it cools as it is discharged, accounting for the accumulation of a chimney made up of minerals. The chimneys so formed eventually collapse to form a mound of sediments rich in the minerals and metals mentioned above.

The chimneys that build around submarine hydrothermal vents grow rapidly. A 10-m-high chimney accidentally knocked over by a submersible in 1991 grew 6 m in only three months. Also in 1991 scientists aboard *Alvin* saw the results of a submarine lava flow they missed by less then two weeks; fresh lava covered the area as well as the remains of tube worms. And in a nearby area, a new fissure opened in the seafloor and in less than two years a new hydrothermal vent community had become well established.

Black smokers are at or very near spreading ridges, but in 2000 scientists announced the discovery of another kind of seafloor vent in the North Atlantic. They called it the Lost City hydrothermal vent field because of its numerous massive spires and pillars made up of light-colored carbonate minerals derived from the oceanic crust. Some of these structures are as high as 60 m, but they are 14 or 15 km from spreading ridges.

Seafloor Fractures

Oceanic ridges do not wind without interruption around the globe. They abruptly terminate where they are offset along fractures oriented more or less at right angles to ridge axes (▶ Figure 12.16). These fractures run for hundreds of kilometers, although they are difficult to trace where buried beneath seafloor sediments. Many geologists are convinced that some geologic features on continents can best be accounted for by the extension of these fractures into continents.

Where oceanic ridges are offset, they are characterized by shallow seismic activity only in the area between the displaced ridge segments (▶ Figure 12.16). Furthermore, because ridges are higher than the seafloor adjacent to them, the offset segments yield vertical relief on the seafloor. Nearly vertical escarpments 2 or 3 km high develop, as illustrated in Figure 12.16. The reason oceanic ridges show so many fractures is that plate divergence takes place irregularly on a sphere, resulting in stresses that cause fracturing.

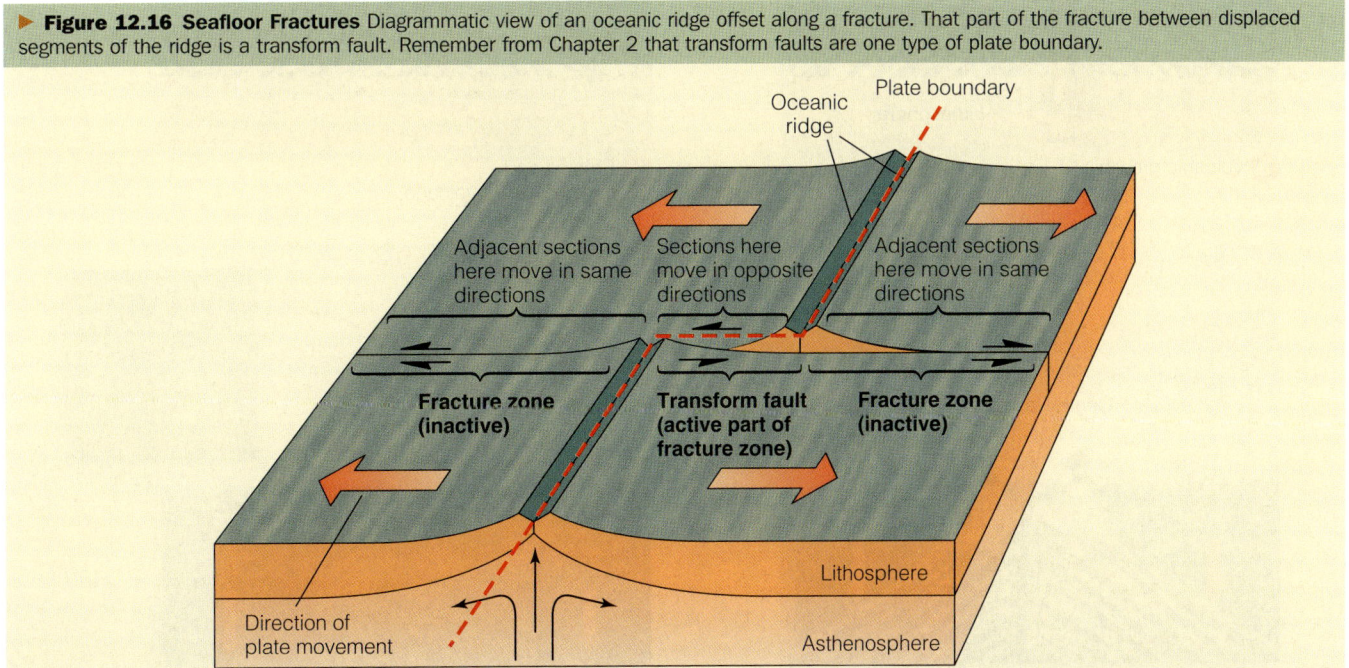

▶ **Figure 12.16 Seafloor Fractures** Diagrammatic view of an oceanic ridge offset along a fracture. That part of the fracture between displaced segments of the ridge is a transform fault. Remember from Chapter 2 that transform faults are one type of plate boundary.

Seamounts, Guyots, and Aseismic Ridges

■ *How do seamounts and guyots form?*

As noted, the seafloor is not a flat, featureless plain except for the abyssal plains, and even these are underlain by rugged topography (▶ Figure 12.13). In fact, a large number of volcanic hills, seamounts, and guyots rise above the seafloor in all ocean basins, but they are particularly abundant in the Pacific. All are volcanic and differ mostly in size. **Seamounts** rise more than 1 km above the seafloor; if they are flat topped, they are called **guyots** rather than seamounts (▶ Figure 12.17). Guyots are volcanoes that originally extended above sea level. But as the plate they rested on continued to move, they were carried away from a spreading ridge, and the oceanic crust cooled and descended to greater oceanic depths. So what was an island was eroded by waves as it slowly sank beneath the sea, making it flat-topped.

Many other volcanic features smaller than seamounts are also present on the seafloor, but they probably originated in the same way. These so-called *abyssal hills*, averaging only about 250 m high, are common on the seafloor and underlie thick sediments on the abyssal plains.

Other features in the ocean basins are long, narrow ridges and broad plateaulike features rising as much as 2 to 3 km above the surrounding seafloor. Known as **aseismic ridges,** they lack seismic activity. A few of these ridges are probably small fragments separated from continents during rifting. These are referred to as *microcontinents* and are represented by such features as the Jan Mayen Ridge in the North Atlantic (▶ Figure 12.12).

Most aseismic ridges form as a linear succession of hotspot volcanoes. These may develop at or near an oceanic ridge, but each volcano so formed is carried laterally with the plate on which it originated. The net result is a sequence of seamounts/guyots extending from an oceanic ridge; the Walvis Ridge in the South Atlantic is a good example (▶ Figure 12.12). Aseismic ridges also form over hot spots unrelated to ridges. The Emperor Seamount–Hawaiian Island chain in the Pacific formed in such a manner (▶ Figure 12.12).

PHYSICAL Geology Now™ ▶ **Geo-focus/Active Figure 12.17 The Origin of Seamounts and Guyots** A seamount may start out as a volcano that extends above sea level. However, as the plate on which the volcano rests moves, it carries the volcano to greater depths and it subsides below sea level. If the volcano is eroded and becomes flat-topped, it is called a guyot.

a Volcanic growth

b Erosion and subsidence of volcano

c Atoll development

d Atoll subsidence, guyot stage

What Would You Do?

As the only person in your community with any geologic training, you are often called on to identify rocks and fossils. Several school children on a natural history field trip picked up some rocks you recognize as peridotite. When you visit the site where the rocks were collected, you notice that overlying the peridotite are rocks you identify as gabbro, basalt dikes, pillow lava, and thinly bedded chert. How would you explain this association of rocks? How did they come to be in their present location on land?

Section 12.5 Summary

- The deep seafloor is not flat and featureless as once thought; it has topography as varied as that on the continents. In addition to abyssal plains, which are flat, the deep seafloor has oceanic trenches, ridges, submarine hydrothermal vents, seafloor fractures, seamounts, guyots, and aseismic ridges.

12.6 Sediments on the Deep Seafloor

■ *Why are there no deposits of sand and gravel in the deep ocean?*

Most seafloor sediments, except those very near landmasses, are fine-grained, consisting of silt- and clay-sized particles, because there are few mechanisms to transport larger particles of sand and gravel very far from land. Certainly icebergs transport sand and gravel from land, and broad bands of glacial-marine sediments are found adjacent to Antarctica and Greenland (▶ Figure 12.18).

In Chapter 2 you learned that the seafloor, except for spreading ridges, is covered with sediment. Remember though that the thickness of seafloor sediment increases with increasing distance from spreading ridges (see Figure 2.13). Where does the sediment come from? After all, much of the seafloor is far from any potential source, at least for detrital sediment.

Most of the fine-grained sediment in the deep sea is wind-blown dust and volcanic ash from the continents and oceanic islands and the shells of microscopic organisms that live in the near-surface waters. Other sediment sources include cosmic dust and deposits resulting from chemical reactions in seawater. The manganese nodules that are fairly common in all the ocean basins are a good example of the latter (▶ Figure 12.19). They are composed mostly of manganese and iron oxides but also contain copper, nickel, and cobalt. Nodules may be an important source of some metals in the future; the United States, which imports most of its manganese and cobalt, is particularly interested in this potential resource.

The contribution of cosmic dust to deep-sea sediment is negligible. Even though some researchers estimate that as much as 40,000 metric tons of cosmic dust falls to Earth each year, this is a trivial quantity compared to the volume of sediments derived from other sources.

■ *What are the two main types of sediment found on the deep seafloor?*

Pelagic clay and ooze are the most common sediments on the deep seafloor. Most deep-sea sediments are *pelagic*, meaning that they settled from suspension far from land. Pelagic sediment is further characterized as pelagic clay or ooze (▶ Figure 12.20). **Pelagic clay** is made up of clay-sized

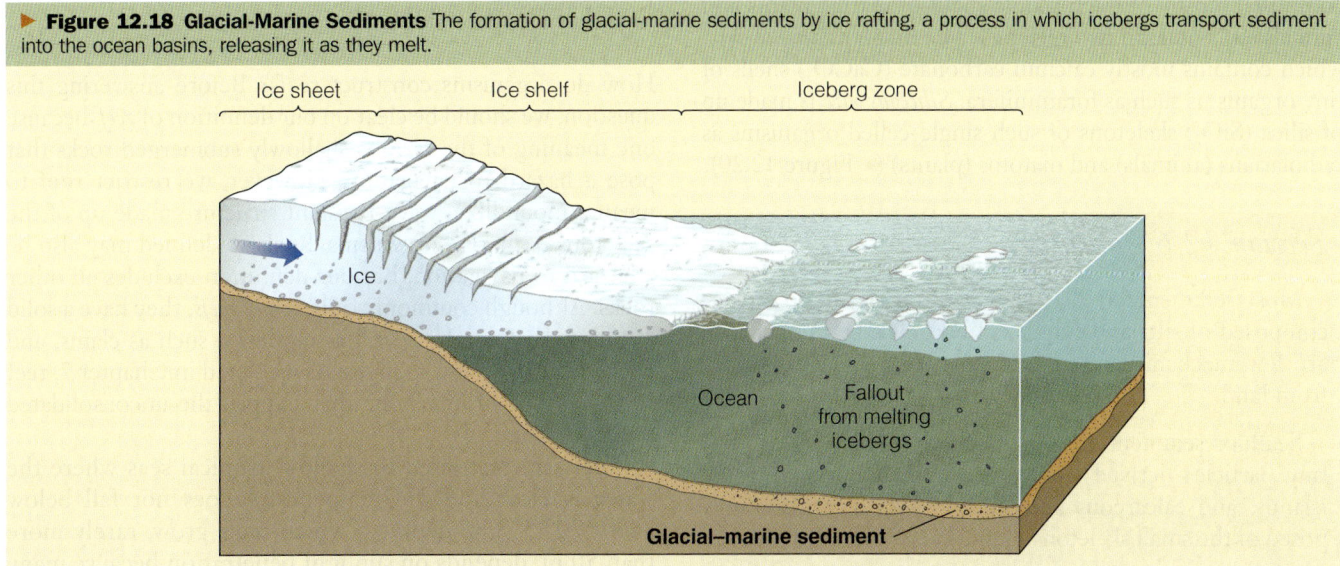

▶ **Figure 12.18 Glacial-Marine Sediments** The formation of glacial-marine sediments by ice rafting, a process in which icebergs transport sediment into the ocean basins, releasing it as they melt.

Figure 12.19 Manganese Modules

a Manganese nodules on the seafloor.

b A sectioned manganese nodule showing its internal structure consisting of concentric layers. This nodule measures about 6 cm across.

particles derived from the continents and oceanic islands and covers most of the deeper parts of the ocean basins. **Ooze** is composed mostly of microscopic shells of marine plants and animals and comes in two varieties. The first is *calcareous ooze*, which contains mostly calcium carbonate ($CaCO_3$) shells of tiny organisms such as foraminifera. *Siliceous ooze* is made up of silica (SiO_2) skeletons of such single-celled organisms as radiolarians (animals) and diatoms (plants) (▶ Figure 12.20).

Section 12.6 Summary

- Most sediment on the deep seafloor is fine-grained, composed of silt- and clay-sized particles, because there are few mechanisms that transport sand and gravel far from land.
- Seafloor sediment consists of pelagic clay made up of tiny particles derived from the continents and oceanic islands, and calcareous and siliceous oozes, both composed of the small skeletons of marine plants and animals.

12.7 Reefs—Rocks Made by Organisms

How do organisms construct reefs? Before answering this question, we should be clear on our definition of *reef* because one meaning of the term is shallowly submerged rocks that pose a hazard to navigation. However, we restrict **reef** to mean a moundlike, wave-resistant structure made up of the skeletons of marine organisms. Reefs so defined may also be a hazard to navigation, but our definition excludes all other rocks. Although commonly called *coral reefs*, they have a solid framework of corals as well as mollusks, such as clams, and encrusting algae and sponges. As we noted in Chapter 7, reef rock is a type of limestone that skipped the unconsolidated sediment stage.

Reefs are restricted to shallow, tropical seas where the water is clear and the temperature does not fall below about 20°C. The depth to which reefs grow, rarely more than 50 m, depends on sunlight penetration because many

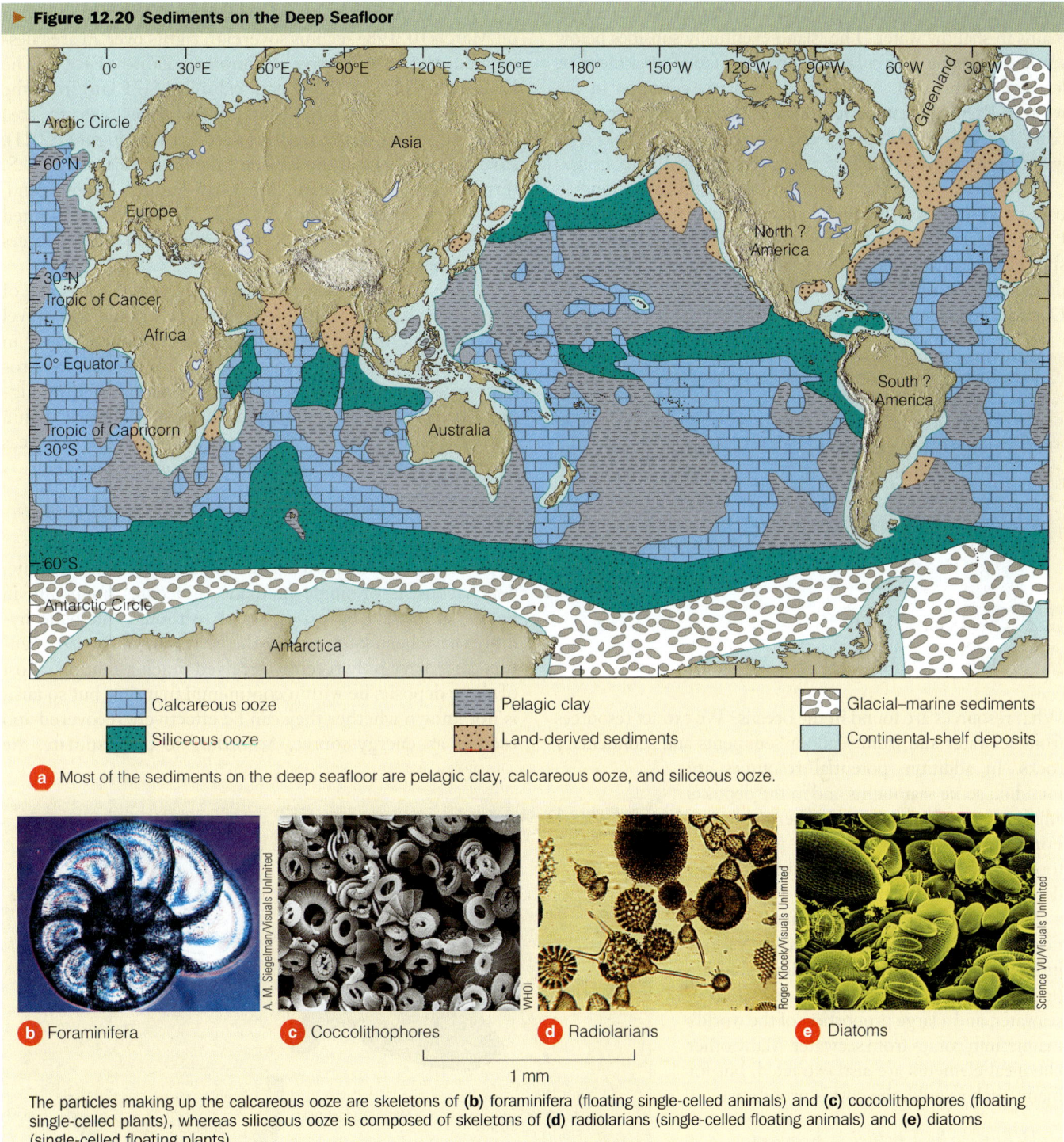

▶ **Figure 12.20 Sediments on the Deep Seafloor**

a Most of the sediments on the deep seafloor are pelagic clay, calcareous ooze, and siliceous ooze.

b Foraminifera   **c** Coccolithophores   **d** Radiolarians   **e** Diatoms

The particles making up the calcareous ooze are skeletons of **(b)** foraminifera (floating single-celled animals) and **(c)** coccolithophores (floating single-celled plants), whereas siliceous ooze is composed of skeletons of **(d)** radiolarians (single-celled floating animals) and **(e)** diatoms (single-celled floating plants).

of the corals rely on symbiotic algae that depend on sunlight for energy.

Reefs are found in a variety of shapes and sizes, but most are one of three basic types: fringing, barrier, and atoll (see "Reefs: Rocks Made by Organisms" on pages 380–381). *Fringing reefs* are solidly attached to the margins of an island or continent. They have a rough, tablelike surface, are as much as 1 km wide, and, on their seaward side, slope steeply down to the seafloor. *Barrier reefs* are similar to fringing reefs, except that a lagoon separates them from the mainland. Probably the best-known barrier reef in the world is the 2000-km-long Great Barrier Reef of Australia.

■ *What events lead to the origin of an atoll?*

An *atoll* is a circular to oval reef surrounding a lagoon. Atolls form around volcanic islands that subside below sea level as the plate they rest on is carried progressively farther from an oceanic ridge. As subsidence proceeds, the reef organisms construct the reef upward so that the living part of the reef remains in shallow water. The island eventually subsides below sea level, leaving a circular lagoon surrounded by a more or less continuous reef. Atolls are particularly common in the western Pacific Ocean basin. Many of these began as fringing reefs, but as the plate they were on was carried into deeper water, they evolved first to barrier reefs and finally to atolls.

This scenario for the evolution of reefs from fringing to barrier to atoll was proposed more than 150 years ago by Charles Darwin while he was serving as a naturalist on the ship HMS *Beagle*. Drilling into atolls has revealed that they do indeed rest on a basement of volcanic rocks, confirming Darwin's hypothesis.

Section 12.7 Summary

- Reefs are wave-resistant structures composed of animal skeletons, especially corals and mollusks, as well as encrusting algae and sponges. Reefs vary considerably in size and shape but most are conveniently classified as fringing reefs, barrier reefs, and atolls.

12.8 Resources from the Oceans

What resources are found in the oceans? We extract resources from seawater and from seafloor sediments and sedimentary rocks. In addition, potential resources are found on some seamounts and in the deposits adjacent to submarine hydrothermal vents. For some resources, extraction is currently too expensive or the technology has not been developed to exploit them.

Seawater contains many elements in solution, some of which are extracted for various industrial and domestic uses. Sodium chloride (table salt) is produced by the evaporation of seawater, and a large proportion of the world's magnesium comes from seawater. Many other chemical elements are also extracted, but for many, such as gold, the cost is prohibitive.

■ *What is the Exclusive Economic Zone?*

In addition to substances in seawater, deposits on the seafloor or within seafloor sediments are becoming increasingly important. Many of these potential resources lie well beyond continental margins, so their ownership is a political and legal problem that has not yet been resolved. Most nations bordering an ocean claim those resources within their adjacent continental margin.

The United States, by a presidential proclamation issued on March 10, 1983, claims sovereign rights over an area designated as the **Exclusive Economic Zone (EEZ)**. The EEZ extends seaward 200 nautical miles (371 km) from the coast, giving the United States jurisdiction over an area about 1.7 times larger than its land area (▶ Figure 12.21). Also included within the EEZ are the areas adjacent to U.S. territories, such as Guam, American Samoa, Wake Island, and Puerto Rico (▶ Figure 12.21). In short, the United States claims a huge area of the seafloor and any resources on or beneath it.

Numerous resources are found within the EEZ, some of which have been exploited for many years. Sand and gravel for construction are mined from the continental shelf in several areas. About 17% of U.S. oil and natural gas production comes from wells on the continental shelf (▶ Figure 12.22). Some 30 sedimentary basins are known within the EEZ, several of which contain hydrocarbons, whereas others are areas of potential hydrocarbon production.

■ *What is methane hydrate, and why might it be an energy resource in the future?*

A potential resource within the EEZ is methane hydrate, which consists of single methane molecules bound up in networks formed by frozen water. Although methane hydrates have been known since the early part of the 19th century, they have only recently received much attention. Most of these deposits lie within continental margins, but so far it is not known whether they can be effectively recovered and used as an energy source. According to one estimate, the

▶ **Figure 12.21 Exclusive Economic Zone (EEZ)** The EEZ, shown in dark blue, includes a vast area adjacent to the United States and its possessions.

amount of carbon in methane hydrates is double that in all coal, oil, and conventional natural gas reserves.

Methane hydrates are stable at water depths exceeding 500 m and at near-freezing temperatures. Along the Atlantic continental margin of the United States, numerous submarine landslide scars are present where the seafloor should be stable. Geologists at the U.S. Geological Survey think that methane hydrates contribute to these slides, during which methane is released into the atmosphere, where along with carbon dioxide it contributes to global warming. Its contribution to global warming must be assessed because a volume of methane 3000 times greater than that in the atmosphere is present in seafloor sediments, and it is 10 times more effective than carbon dioxide as a factor in global warming.

Other seafloor resources of interest include massive sulfide deposits that form by submarine hydrothermal activity at spreading ridges. These deposits containing iron, copper, zinc, and other metals have been identified within the EEZ at the Gorda Ridge off the coasts of California and Oregon; similar deposits occur at the Juan de Fuca Ridge within the Canadian EEZ.

Other potential resources include the manganese nodules, discussed previously (▶ Figure 12.19), and metalliferous oxide crusts found on seamounts. Manganese nodules contain manganese, cobalt, nickel, and copper; the United States is heavily dependent on imports of the first three of these elements. Within the EEZ, manganese nodules are found near Johnston Island in the Pacific Ocean and on the Blake Plateau off the coast of South Carolina and Georgia. In addition, seamounts and seamount chains within the EEZ in the Pacific are known to have metalliferous oxide crusts several centimeters thick from which cobalt and manganese could be mined.

Another important resource forming in shallow seawater is phosphate-rich sedimentary rock known as *phosphorite*.

▶ **Figure 12.22 Oil and Natural Gas Wells on the Continental Shelf** This is Shell Oil Company's tension-leg platform *Ursa* in the Gulf of Mexico about 265 km south of New Orleans, Louisiana. It is anchored in about 1160 m of water by steel cables that hold the partly submerged platform in place. It is designed to withstand waves up to 22 m high and hurricane-force winds of 225 km/hr. Another tension-leg platform deployed in January 2004 in the Gulf of Mexico is in water more than 1300 m deep.

Section 12.8 Summary

• An area extending 371 km from the shoreline of the United States and its possessions is the Exclusive Economic Zone in which the United States claims the rights to all resources.

• Resources from the seafloor and seafloor rocks include sand and gravel from the continental shelf, petroleum and natural gas, and various metal-rich minerals. A potential future energy resource may be methane hydrate, which is found within the continental margins.

What Would You Do?

Hydrothermal vents on the seafloor are known sites of several metals of great importance to industrialized societies. Furthermore, it appears that these metals are being deposited even now, so if we mine one area, more of the same resources form elsewhere. Given these conditions, it would seem that our problems of diminishing resources are solved. So why not simply mine the seafloor? Also, many chemical elements are present in seawater. The technology exists to extract elements such as gold, uranium, and others, so why not do so?

Reefs: Rocks Made by Organisms

Reefs are wave-resistant structures composed of the skeletons of corals, mollusks, sponges, and encrusting algae. Most reefs are classified as fringing, barrier, or atolls, all of which actively grow in shallow, warm seawater where there is little influx of detrital sediment, especially mud. Reef rock is a type of limestone that forms as a solid rather than from sediment that is later lithified. Ancient reefs are important reservoirs for hydrocarbons in some areas.

▶ **1.** Stages in the evolution of a reef. A fringing reef forms around a volcanic island, but as the island is carried into deeper water on a moving plate, the reef is separated from the island by a lagoon and becomes a barrier reef. With continued plate movement, the island disappears below the sea and the reef grows upward, forming an atoll.

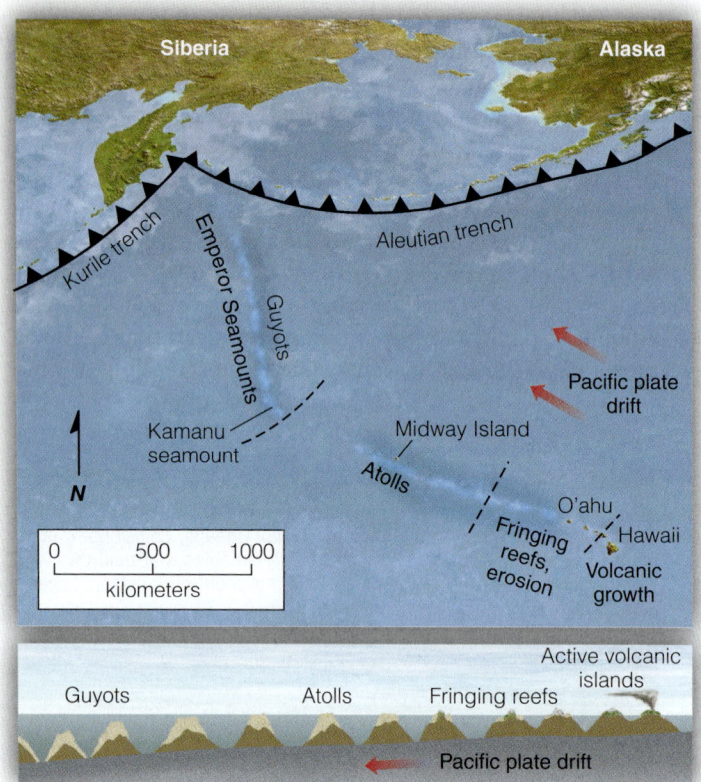

◀ **2.** ▶ **3.** Underwater views of reefs in the Red Sea (2) and in Hawaii (3).

▼ **4.** The white line of breaking waves marks the site of a barrier reef around Rarotonga in the Cook Islands in the Pacific Ocean. The island is only about 12 km long.

▲ **5.** This oval reef with a central lagoon in the Pacific Ocean is an atoll. How does it differ from the reef shown at the bottom of the previous page?

▲ **6.** An ancient reef in Australia. You can see the reef talus on the left side of the image sloping away from the reef core with no layering, and on the right side, the back reef deposits, which show horizontal bedding. It looks much like ▼**7.** below.

▲ **8.** Fossils provide the information necessary for this restoration of a Middle Devonia reef from the Great Lakes area. Reefs have existed for hundreds of millions of years, but the reef organisms have changed through time.

Review Workbook

ESSENTIAL QUESTIONS SUMMARY

12.1 Introduction
■ *Why is it important to know about the seafloor?*
The seafloor constitutes the largest part of Earth's surface. Studies of the seafloor provide some of the evidence for plate tectonic theory, and various resources are found in seawater and on or in the seafloor.

12.2 Methods Used to Study the Seafloor
■ *What is echo sounding?*
In echo sounding, ocean depth is calculated from the speed of sound in water and the time required for sound waves to travel from their source at the surface to the seafloor and back to a detector at the surface.

12.3 Oceanic Crust—Its Structure and Composition
■ *How have scientists determined the structure and composition of the oceanic crust?*
Although the oceanic crust has never been completely penetrated by drilling, drill core samples, direct observations from submersibles, and fragments of oceanic crust now on land have enabled scientists to determine its structure and composition.

12.4 The Continental Margins
■ *What is the continental shelf?*
The continental shelf is the gently sloping part of a continental margin that lies between the shore and the continental slope. It varies from a few tens of meters to more than 1000 km wide, but rarely is it inclined more than 1 degree.

■ *What is the continental rise and how does it form?*
A continental margin consists of a gently sloping shelf, more steeply dipping slope, and, in some cases, a gently sloping continental rise made up largely of overlapping submarine fans. Much of the sediment in the fans was transported from the continents and carried by turbidity currents through submarine canyons incised into the slope and perhaps the shelf.

■ *How do active and passive continental margins compare?*
Active continental margins are found at the leading edge of a tectonic plate where an oceanic plate is subducted beneath a continental plate. A deep-sea trench, volcanism, seismicity, and folded mountains are found at these margins, which are narrow. Passive continental margins are within a plate and are much wider. Although they may show some seismicity, no volcanism takes place.

12.5 Features Found in the Deep-Ocean Basins
■ *Why are abyssal plains found mostly near passive continental margins?*
Because there is no oceanic trench offshore, land-derived sediment is carried far seaward where it forms submarine fans that grade into abyssal plains.

■ *What are submarine hydrothermal vents?*
These vents on the seafloor at or very near spreading ridges are where circulating water is heated and discharged into seawater. As the hot water circulates through the oceanic crust it becomes a metal-rich brine from which deposits of various minerals and metals are deposited.

■ *How do seamounts and guyots form?*
Both are submarine mountains of volcanic origin that form at or near a spreading ridge. However, as the plate they are on moves away from the ridge, those that extend to sea level are eroded and become flat-topped (guyots), whereas others that did not rise as high retain their original shape.

12.6 Sediments on the Deep Seafloor
■ *Why are there no deposits of sand and gravel in the deep oceans?*
Sand and gravel deposits are found adjacent to continents and oceanic islands, but there is no mechanism to transport large particles in quantity far out to sea. Of course icebergs may carry sand and gravel out to sea, but their contribution to the quantity of sediment deposited in the deep oceans is very small.

■ *What are the two main types of sediment found on the deep seafloor?*
Pelagic clay is made up of tiny particles derived from the continents and from oceanic islands. Ooze, in contrast, is composed of the tiny shells of marine plants and animals. It may be calcareous ooze or siliceous ooze depending on the composition of the shells it contains.

12.7 Reefs—Rocks Made by Organisms
■ *What events lead to the origin of an atoll?*
An atoll is an oval to circular reef enclosing a lagoon that started out as a fringing reef—that is, a reef attached to an island. As the island is carried into deeper water by plate movement, the reef becomes a barrier reef lying offshore, and finally as the island sinks below sea level, the reef becomes an atoll.

12.8 Resources from the Oceans
■ *What is the Exclusive Economic Zone?*
The EEZ is an area extending 200 nautical (371 km) miles from the coast of the United States and its possessions in which the United States claims the rights to all resources. Other nations bordering the oceans make similar claims.

■ *What is methane hydrate, and why might it be an energy resource in the future?*
Single methane molecules bound up in ice crystals are found in huge quantities in sediments of the continental margins. The amount of carbon in methane hydrate exceeds that in all coal, oil, and natural gas, so if a technology is developed to recover it economically, it may become an energy resource.

ESSENTIAL TERMS TO KNOW

abyssal plain (p. 369)
active continental margin (p. 367)
aseismic ridge (p. 374)
black smoker (p. 373)
continental margin (p. 364)
continental rise (p. 365)
continental shelf (p. 365)
continental slope (p. 365)
echo sounder (p. 362)

Exclusive Economic Zone (EEZ) (p. 378)
guyot (p. 374)
oceanic ridge (p. 370)
oceanic trench (p. 369)
ooze (p. 376)
ophiolite (p. 364)
passive continental margin (p. 369)
pelagic clay (p. 375)

reef (p. 376)
seamount (p. 374)
seismic profiling (p. 362)
submarine canyon (p. 367)
submarine fan (p. 366)
submarine hydrothermal vent (p. 371)
turbidity current (p. 366)

REVIEW QUESTIONS

1. Identify the types of continental margins shown in Figure 12.7. What are the characteristics of each?

2. The most distant part of a 30-million-year-old aseismic ridge is 1000 km from an oceanic ridge. How fast, on average, did the plate with this ridge move in centimeters per year?

3. How did geologists figure out the nature of the upper mantle and oceanic crust even before they observed mantle and crust rocks in the ocean basins?

4. During the Pleistocene Epoch (Ice Age), sea level was 130 m lower than it is now. What effect did this lower sea level have on rivers on land? Is there any evidence from the continental shelves that might bear on this question? If so, what?

5. How do the submarine hydrothermal vents called black smokers differ from the vents at the Lost City?

6. Although seafloor fractures may extend for hundreds of kilometers, earthquakes occur on only those parts of the fractures between offset ridge segments. Why?

7. What kind(s) of evidence would you need to demonstrate that an atoll evolved from a fringing reef to a barrier reef and finally to an atoll?

8. How do the mountains of mid-oceanic ridges differ from the mountain ranges on land?

APPLY YOUR KNOWLEDGE

1. Suppose you are a member of an oceanographic research team. As you expected, you find pelagic clay and ooze in the deeper parts of the North Atlantic Ocean. However, once in a while you find a small amount of sand and a few pebbles. How did they get so far from land?

FIELD QUESTION

1. The two images below show alternating layers of sandstone and mudrocks that were deposited by turbidity currents (see Figure 12.9). Notice in the upper left image there are more and thicker sandstone layers, whereas the lower shows thinner sandstone layers and much more mudrock. Considering how turbidity currents deposit on submarine fans, how can you account for these differences?

GEOLOGY MATTERS

GEOLOGY IN FOCUS Oceanic Circulation and Resources from the Sea

Earth's oceans are in constant motion. The Gulf Stream and South Equatorial Current carry great quantities of water toward the poles and have an important modifying effect on climate. In addition to surface and deep currents that carry water horizontally, vertical circulation takes place when *upwelling* slowly transfers cold water from depth to the surface and *downwelling* transfers warm surface water to depth.

Upwelling not only transfers water from depth to the surface but also carries nutrients, especially nitrates and phosphate, into the zone of sunlight penetration where they sustain high concentrations of floating organisms, which in turn support other organisms. In fact, other than the continental shelves and areas adjacent to hydrothermal vents on the seafloor, areas of upwelling are the only parts of the oceans where biological productivity is very high. Less than 1% of the ocean surfaces are areas of upwelling, yet they support more than 50% (by weight) of all fishes.

Scientists recognize three types of upwelling, but only one need concern us here—*coastal upwelling*. Most coastal upwelling takes place along the west coasts of Africa, North America, and South America, although one notable exception is present in the Indian Ocean. Coastal upwelling involves movement of water offshore, which is replaced by water rising from depth (▶ Figure 1). For instance, the winds blowing south along the coast of Peru, coupled with the Coriolis effect, transport surface water seaward, and cold, nutrient-rich water from depth rises to replace it. This area, too, is a major fishery, and changes in the surface-water circulation every 3 to 7 years adjacent to South America are associated with the onset of El Niño, a weather phenomenon with far-reaching consequences.

Our interest here is that among the nutrients in upwelling oceanic waters is considerable phosphorus, an essential element for animal and plant nutrition. Although present in minute quantities in many sedimentary rocks, most commercial phosphorus is derived from *phosphorite*, a sedimentary rock containing such phosphate-rich minerals as fluorapatite $[Ca_5(PO_4)_3F]$. Areas of upwelling along the outer margins of continental shelves are the sites of deposition of most of these so-called bedded phosphorites, which are interlayered with carbonate rocks, chert, and detrital rocks such as mudrocks and sandstone.

Upwelling accounts for most of Earth's phosphate-rich sedimentary rocks, but some form by other processes. In *phosphatization*, carbonate grains such as animal skeletons and ooids are replaced by phosphate, and *guano* is made up of calcium phosphate from bird or bat excrement. Another type of phosphate deposit is essentially a placer deposit in which the skeletons of vertebrate animals are found in large numbers. Vertebrate skeletons are made up mostly of hydroxyapatite $[Ca_5(PO_4)_3OH]$. The 3- to 15-million-year-old Bone Valley Formation of Florida is a good example (▶ Figure 2).

The United States is the world leader in production and consumption of phosphate rock, most of it coming from deposits in Florida and North Carolina, but some is also mined in Idaho and Utah. More than 90% of all phosphate rock mined in this country is used to make chemical fertilizers and animal-feed supplements. It also has other uses in metallurgy, preserved foods, ceramics, and matches.

Figure 1
Wind from the north along the west coast of a continent coupled with the Coriolis effect causes surface water to move offshore, resulting in upwelling of cold, nutrient-rich, deep water.

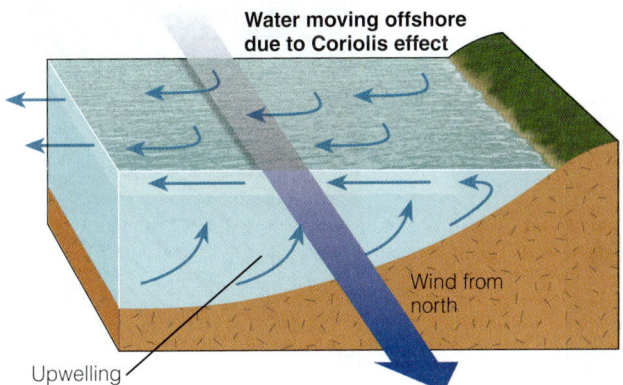

Figure 2
Miocene-aged phosphate-rich rocks in Florida's Bone Valley Member of the Peace River Formation in the IMC Four Corners Mine, Polk County, Florida.

GEOLOGY MATTERS

GEOLOGY IN UNEXPECTED PLACES

Ancient Seafloor in San Francisco

San Francisco is certainly one of America's most beautiful and interesting cities, but it also has a fascinating geologic history and ongoing geologic processes of some concern. Just offshore to the west is the famous San Andreas fault, which spawns many small earthquakes and has the potential to generate large ones. Indeed, at 5:13 A.M. on April 18, 1906, a huge earthquake shook the area, killing thousands and causing extensive property damage (see Chapter 10). Before there was a San Andreas fault, though, the West Coast of North America was a convergent plate boundary where an oceanic plate was subducted beneath the continent.

The rocks that underlie San Francisco are ancient seafloor sediments and volcanic rocks that were once part of the oceanic crust. In fact, the city's bedrock consists of a succession of rocks scraped off against the continental margin as an oceanic plate was subducted during the Jurassic and Cretaceous periods (▶ Figure 1). Some of the rocks are called *mélange*, a geologic term for a chaotic mixture of rock fragments set in a matrix of clay that formed during intense shearing as subduction took place.

The rocks designated mélange are not well exposed because they weather and erode so rapidly, but excellent exposures of sedimentary and igneous rocks are present in several locations. A type of sandstone known as greywacke, which has abundant rock fragments and clay, is present on Alcatraz Island (▶ Figure 1c). The island served as a maximum-security prison from 1934 to 1963, holding such infamous criminals as Al Capone. These rocks and similar ones on the west side of the city probably formed in an oceanic trench much like the present-day Peru–Chile Trench along the West Coast of South America.

Other rocks of interest in and around San Francisco include pillow lava (see Figure 5.7b) that formed at or near a spreading ridge and became part of the oceanic crust. As this oceanic crust moved east, it was subducted, but parts of it were deformed, uplifted, and incorporated into the western margin of North America. You can appreciate the intensity of deformation by observing the thinly bedded, complexly folded, bedded chert that also formed as sedimentary rock on the seafloor (▶ Figure 1b).

Following the origin of the rocks discussed above, the nature of the tectonic activity along the West Coast changed, and the convergent plate boundary changed to a transform plate boundary. Since then, the area has continued to evolve as more sedimentary rocks formed and additional uplift took place. During the Ice Age, sea level was much lower and the shoreline was about 30 km west of where it is now. Only during the last few thousands of years did the sea reach its current level and the San Francisco area took on its present appearance.

Figure 1
(a) Map showing the distribution of rocks underlying San Francisco and nearby areas. (b) Intensely deformed bedded chert of the Marin Headlands terrane. (c) Rocks of the Alcatraz terrane, seen here on Alcatraz Island, are mostly a type of sandstone called graywacke.

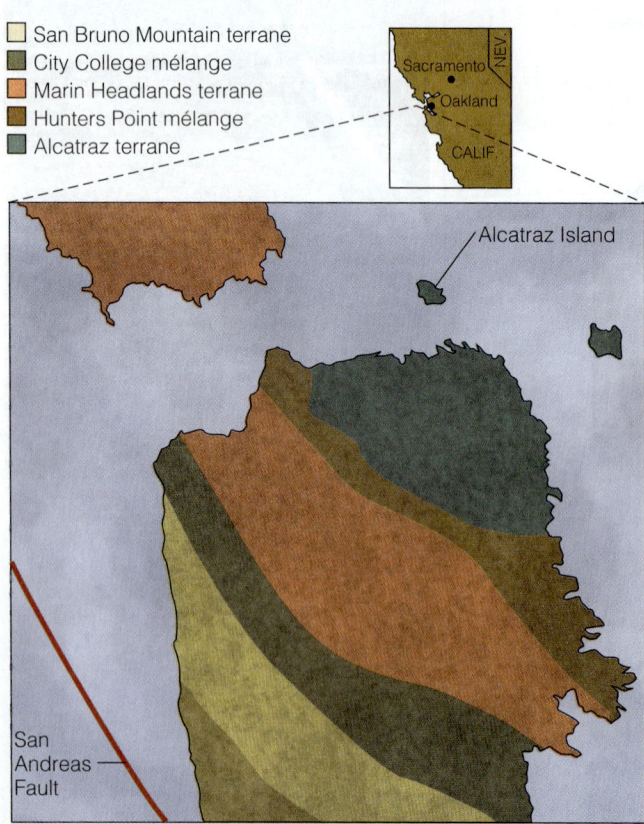

Geology AT WORK

GEOLOGY MATTERS

The Future beneath the Sea

Robert D. Ballard is president of the Institute for Exploration in Mystic, Connecticut, and Scientist Emeritus in the Department of Applied Physics and Engineering at Woods Hole Oceanographic Institution. He is also the founder and chairman of the Jason Foundation for Education. Ballard earned a B.S. in physical science at the University of California, Santa Barbara, and a Ph.D. in marine geology and geophysics at the University of Rhode Island, Graduate School in Oceanography.

After 30 years as an undersea explorer first at the Woods Hole Oceanographic Institution and now at the Institute for Exploration in Mystic, Connecticut—I am still fascinated by the sea. Given our exploding population, our diminished interest in the promises of the space program, and the continued development of advanced technology, I truly believe the 21st century will usher in an explosion in human activity in the sea. I am convinced the next generation will explore more of Earth—that is, the 71% that lies underwater—than all previous generations combined.

Just as Lewis and Clark's explorations led to the settling of the West, the exploration of the sea will lead to its colonization. The gathering and hunting of the living resources of the sea, an activity characteristic of primitive societies on land, will be replaced at sea by farming and herding. High-tech barbwire in the form of acoustic, thermal, or other barrier techniques will emerge to control and manage the living resources of the sea.

Oil and gas exploration and exploitation will continue moving into deeper and deeper depths. We have already discovered and mapped oil and gas reserves down to 12,000 ft, which represents the average depth of the ocean, and each year the oil industry brings production wells on line in waters deeper than the previous year.

In recent years, we have discovered major mineral deposits in the deep sea similar to those mined for centuries on Cyprus. They contain high concentrations of copper, lead, and sulfur, as well as silver and gold. Their formation continues today in the vast hydrothermal vent systems of the Mid-Ocean Ridge. These mineral deposits will be processed using the very geothermal energy that drives the crustal processes that lead to their formation. Some of these magnificent vent areas will also become the Yellowstone Parks of the deep sea, leading to future arguments over their commercial versus tourist value.

The unique chemosynthetic life-forms that process the toxic material associated with the vent communities will hopefully be bioengineered to convert a portion of our waste products into less harmful or even commercially valuable by-products.

These exotic creatures will also help us understand the early origin of life on our planet as well as the potential for life on other planets we once ruled out for their lack of a friendly nearby Sun.

Whether all this occurs during the next generation's time on Earth, only time will tell. But the seeds can be found in programs already under way. With this exploration will come better understanding of the ocean and the surface beneath it, which is critical to our understanding of the planet as a whole.

Ballard and his crew embarked on a mission that was to make headlines around the world. Ballard had resolved to find the sunken hulk of HMS Titanic, the supposedly "unsinkable" ocean liner which had sunk, with massive loss of life, after she struck an iceberg on her maiden voyage in 1912 (▶ Figure 1). Drawing on all of Ballard's accumulated expertise in undersea exploration, he and his crew located the wreck, more than two miles beneath the waves of the North Atlantic, on September 1, 1985 (▶ Figure 2). Ballard was then forced to wait an excruciating year for weather conditions favorable to a manned mission to view the wreck at close range.

The next year, he and a two-man crew, in the Alvin submersible, made the two-and-a-half hour descent to the ocean floor to view the wreck at firsthand. Over the next few days, they descended again and again and, using the Jason Jr. remote camera, recorded eerie scenes of the ruined interior of the luxury liner.

http://www.achievement.org/autodoc/page/bal0bio-1

Figure 1
An illustration depicting HMS *Titanic* as she set off on her maiden voyage in 1912.

Figure 2
Bow of HMS *Titanic* as it appeared in 1986 in more than 3600 m of water east of Newfoundland. The supposedly unsinkable ship sank on its maiden voyage in 1912 after hitting an iceberg.

Chapter 13

Deformation, Mountain Building, and the Evolution of Continents

The Andes, Northern Chile
Like a painting that negates the bareness of a wall, the foothills of the Andes seemingly challenge the desolate canvas of The Atacama Desert with a brilliant collage of volcano, salt pan, and gorge.
The Atacama Desert is the driest desert on Earth. It is blocked from moisture by the Andes and other coastal mountains. The Andean mountain range which is the highest range outside of Asia formed by deformation along a convergent plate boundary.
—A. W.

ESSENTIAL QUESTIONS TO ASK

13.1 Introduction
- What do geologists mean by the term deformation?
- What kinds of geologic activities indicate that Earth is an internally active planet?

13.2 Rock Deformation
- How can you explain stress and strain by using an example of an ice-covered pond?
- How do compression and tension differ?
- What do the terms ductile and brittle indicate about rocks?

13.3 Folded Rock Layers
- What is a monocline and how might one form?
- How is an eroded anticline different from an eroded syncline?

13.4 Joints and Faults—Deformation by Fracturing
- How do joints differ from faults?
- What kinds of dip-slip faults do geologists recognize, and what features distinguish each type?

13.5 Deformation and the Origin of Mountains
- At what kinds of plate boundaries do most of Earth's large mountain systems form?
- What sequence of events takes place during an orogeny at an oceanic–continental plate boundary?

13.6 The Formation and Evolution of Continents
- What are shields, platforms, and cratons?
- What happened during the Precambrian evolution of North America?

GEOLOGY MATTERS

GEOLOGY IN UNEXPECTED PLACES:
Ancient Ruins and Geology

GEOLOGY IN FOCUS:
Geology Maps—Their Construction and Uses

GEOLOGY AT WORK:
Engineering and Geology

 This icon, which appears throughout the book, indicates an opportunity to explore interactive tutorials, animations, or practice problems available on the Physical GeologyNow website at **http://now.brookscole.com/phygeo6**.

13.1 Introduction

■ *What do geologists mean by the term* deformation?

"Solid as a rock" implies durability and permanence, but you know from earlier chapters that physical and chemical processes disaggregate and decompose rocks, and rocks behave differently at great depth than they do at or near Earth's surface. Indeed, under the tremendous pressure and high temperature at several kilometers below the surface, rock layers actually crumple or fold and yet do not break, and at shallower depths, they yield by fracturing or a combination of folding and fracturing. In either case, dynamic forces within Earth cause **deformation,** a general term encompassing all changes in the shape or volume (or both) of rocks (▶ Figure 13.1).

■ *What kinds of geologic activities indicate that Earth is an internally active planet?*

That there are dynamic forces within Earth is obvious from ongoing seismic activity, volcanism, plate movements, and the continuing evolution of vast mountain systems in South America, Asia, and elsewhere. In short, Earth is an active planet in which *internal heat* accounts for plate movements. Most of Earth's seismic activity, volcanism, and deformation take place at divergent, convergent, and transform plate boundaries.

Plate divergence and igneous activity account for the mountainous topography of the oceanic ridges (see Chapter 12). However, the origin of the large mountain systems on land involves tremendous deformation, volcanism, metamorphism, and the emplacement of plutons at convergent plate boundaries. Thus mountains on land differ markedly from those in the ocean basins. The Appalachians in North America as well as the Alps in Europe, the Himalayas in Asia, and the Andes in South America all owe their existence to geologic processes at convergent plate boundaries. In fact, the evolution of mountains is ongoing in several areas. Thus, deformation and mountain building are closely related topics and accordingly we consider both in this chapter.

The past and continuing evolution of continents involves not only deformation at their margins but also additions of new material to existing continents, a phenomenon known as *continental accretion*. North America, for example, has not always had its present shape and size. Indeed, it began evolving during the Archean Eon (4.6–2.5 billion years ago) as new material was added to the continent at deformation belts along its margins.

Much of this chapter is devoted to *geologic structures*—that is, folded and fractured rock layers, their descriptive terminology, and the forces responsible for them. Even so, there are several practical reasons to study deformation and mountain building. For one thing, crumpled and fractured rocks provide a record of the kinds and intensities of forces that operated during the past. Studies of these structures enable us to satisfy our curiosity about Earth history and are also essential to engineering endeavors such as choosing sites for dams, bridges, and nuclear power plants, especially if they are in areas of continuing deformation. Also, many aspects of mining and exploration for petro-

▶ **Figure 13.1 Deformation** Many rocks show the effect of deformation. These rocks have been deformed by folding and fracturing. Notice the light pole for scale. The nearly vertical fracture where light-colored rocks were displaced is a fault, a fracture along which rocks on opposite sides of the fracture have moved parallel with the fracture surface.

leum and natural gas rely on correctly identifying geologic structures.

> ### Section 13.1 Summary
>
> - Deformation involves any change in shape or volume or both of rocks. Correctly identifying the types of folds and fractures in rock layers is important in several engineering applications and is essential for finding and recovering natural resources.
>
> - Volcanism, seismicity, plate movements, and ongoing mountain building indicate that Earth is an internally active planet.

13.2 Rock Deformation

We defined *deformation* as a general term that refers to any change in the shape or volume or both of rocks. Rocks may be crumpled into folds or fractured as a result of **stress,** which is the force applied to a given area of rock. If the intensity of the stress is greater than the rock's internal strength, the rock undergoes **strain,** which is the deformation caused by stress. The terminology is a little confusing at first, but keep in mind that *deformation* and *strain* are synonyms, and *stress* is the force that causes deformation or strain. The following discussion and reference to ▶ Figure 13.2 will help clarify the meaning of stress and the distinction between stress and strain.

Stress and Strain

■ *How can you explain stress and strain by using an example of an ice-covered pond?*

The stress or force exerted by a person walking on an ice-covered pond is a function of the person's weight and the area beneath her or his feet. The ice's internal strength resists the stress unless the stress is too great, in which case the ice may bend or crack as it is strained (deformed) (▶ Figure 13.2). If the ice shows signs of deformation, the person may lie down; this does not reduce the weight of the person on the ice, but it does distribute it over a larger area, thus reducing the force per unit area.

■ *How do compression and tension differ?*

Stress is force per unit and area, it comes in three varieties—*compression*, *tension*, and *shear*—depending on the direction of the applied force. In **compression,** rocks, or any other object, are squeezed or compressed by forces directed toward one another along the same line, as when you hold a rubber ball between a thumb and a finger and squeeze. Rock layers in compression tend to be shortened in the direction of stress by either folding or fracturing (▶ Figure 13.3a). **Tension** results from forces acting along the same line but in opposite directions, and it tends to lengthen rock layers or pull them apart (▶ Figure 13.3b). Incidentally, rocks are much stronger in compression than they are in tension. In **shear stress** forces act parallel to one another but in opposite directions, resulting in deformation by displacement along closely spaced planes (▶ Figure 13.3c).

▶ Figure 13.2 Stress and Strain Exerted on an Ice-Covered Pond

① The woman weighs 65 kg (6500 g). Her weight is imparted to the ice through her feet, which have a contact area of 120 cm². The stress she exerts on the ice (6500 g/120 cm²) is 54 g/cm². This is sufficient stress to cause the ice to crack.

② To avoid plunging into the freezing water, the woman lays out flat, thereby decreasing the stress she exerts on the ice. Her weight remains the same, but her contact area with the ice is 3150 cm², so the stress is only about 2 g/cm² (6500 g/3150 cm²), which is well below the threshold needed to crack the ice.

▶ **Figure 13.3 Stress and Possible Types of Resulting Deformation**

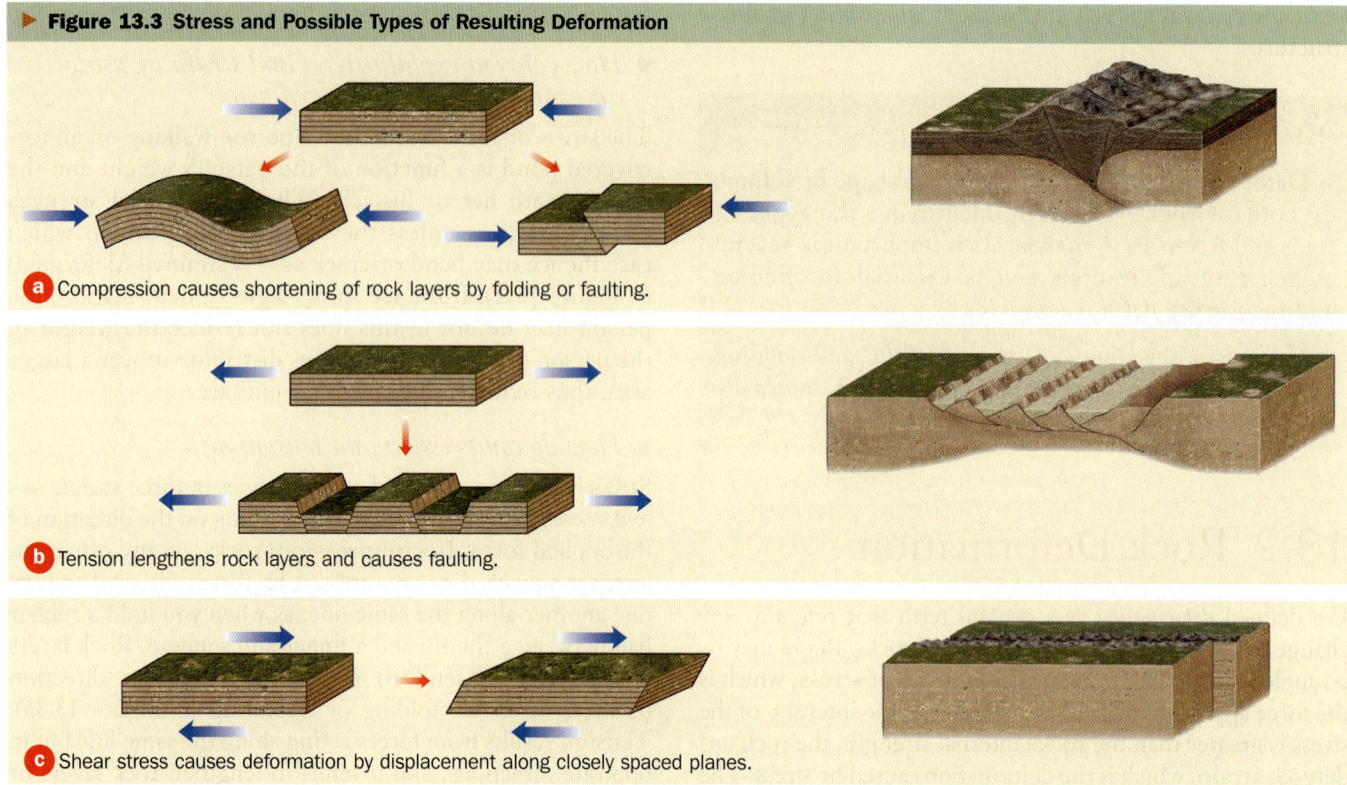

a Compression causes shortening of rock layers by folding or faulting.

b Tension lengthens rock layers and causes faulting.

c Shear stress causes deformation by displacement along closely spaced planes.

Types of Strain

Geologists characterize strain as **elastic strain** if deformed rocks return to their original shape when the deforming forces are relaxed. In Figure 13.2, the ice on the pond may bend under a person's weight but return to its original shape once the person leaves. As you might expect, rocks are not very elastic, but Earth's crust behaves elastically when loaded by glacial ice and is depressed into the mantle and when it rebounds after the load of glacial ice is no longer present (see Figure 11.17).

As stress is applied, rocks respond first by elastic strain, but when strained beyond their elastic limit, they undergo **plastic strain,** as when they yield by folding, or they behave like brittle solids and **fracture** (▶ Figure 13.4). In either folding or fracturing, the stain is permanent, that is, the rocks do not recover their original shape or volume even if the stress is removed.

■ *What do the terms* ductile *and* brittle *indicate about rocks?*

Whether strain is elastic, plastic, or brittle depends on the kind of stress applied, pressure and temperature, rock type, and time. A small stress applied over a long period of time as on a mantelpiece supported only at its ends, causes the rock to sag—that is, to deform plastically (▶ Figure 13.4). In contrast, a large stress applied quickly to the same object, as when struck by a hammer, results in fracture. Rock type is also important because not all rocks have the same yield strength and thus respond to stress differently. Some rocks are *ductile* whereas others are *brittle* under the same conditions. Brittle rocks show little or no plastic strain before they fracture, but ductile rocks exhibit a great deal (▶ Figure 13.4).

Many rocks show the effects of plastic deformation that must have taken place deep within the crust. At or near the

▶ **Figure 13.4 Rock Response to Stress** Rocks initially respond to stress by elastic deformation and then return to their original shape when the stress is released. If the elastic limit is exceeded, as in curve A, rocks deform plastically, which is permanent deformation. The amount of plastic deformation rocks exhibit before fracturing depends on their ductility. If they are ductile, they show considerable plastic deformation (curve A), but if they are brittle, they show little or no plastic deformation before failing by fracture (curve B).

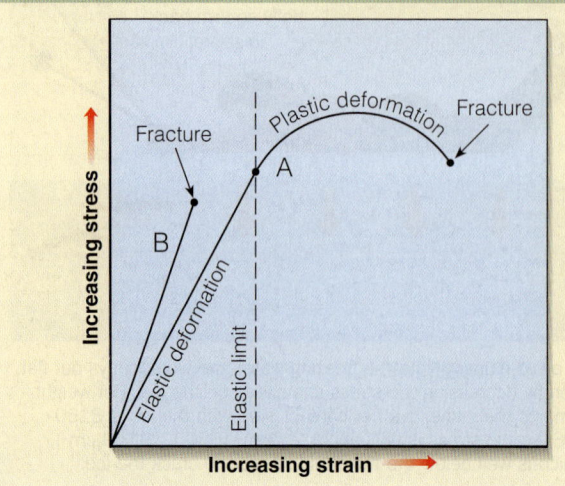

surface, rocks commonly behave like brittle solids and fracture, but at depth they more often yield by plastic deformation; they become more ductile with increasing pressure and temperature. Most earthquake foci are at depths of less than 30 km, indicating that deformation by fracturing becomes increasingly difficult with depth, and no fracturing is known to occur deeper than about 700 km, within subducted slabs of lithosphere.

During deformation, rock layers may crumple into folds, much like the folds produced in a tablecloth when you lay your hands on it and move them together. Unlike the tablecloth, though, folds in rocks are permanent; that is, they cannot be unfolded. On the other hand, rock layers may fracture in response to stress; that is, they behave like brittle solids. In either case the features resulting from deformation are called **geologic structures.** Geologic structures of one kind or another are present almost everywhere that rocks are exposed, and many are detected far below the surface by drilling or geophysical techniques.

Strike and Dip of Deformed Rocks

During the 1660s, Nicolas Steno, a Danish anatomist, proposed the *principle of original horizontality*, which holds that sediments accumulate in horizontal or nearly horizontal layers. Thus, if we see steeply inclined sedimentary rock layers, we are justified in concluding that they were originally deposited horizontally, lithified, and then tilted into their present position (▶ Figure 13.5a). To describe the orientation of deformed rocks, geologists use the concepts of *strike* and *dip*.

By definition, **strike** is the direction of a line formed by the intersection of a horizontal plane with an inclined plane. For instance, the surfaces of the rock layers in Figure 13.5b are inclined planes, and the water surface is a horizontal plane. The direction of the line formed at the intersection of these two planes, the shoreline in other words, is the strike; its orientation is determined by using a compass to measure its angle with respect to north. **Dip** is an angular measurement of an inclined plane's deviation from horizontal, so it must be measured at right angles to the strike direction.

Geologic structures are depicted on geologic maps that use a special symbol to indicate strike and dip. A long line oriented in the appropriate direction shows strike, and a short line perpendicular to the strike line shows the direction of dip (▶ Figure 13.5b). The number adjacent to the strike and dip symbol is the dip angle. The usefulness of strike and dip will become apparent in the following sections on folds and faults.

▶ **Figure 13.5 Strike and Dip of Deformed Rocks**

a We can infer that these sandstone beds in Colorado were deposited horizontally, lithified, and then deformed. To describe their orientation geologists use the terms *strike* and *dip*.

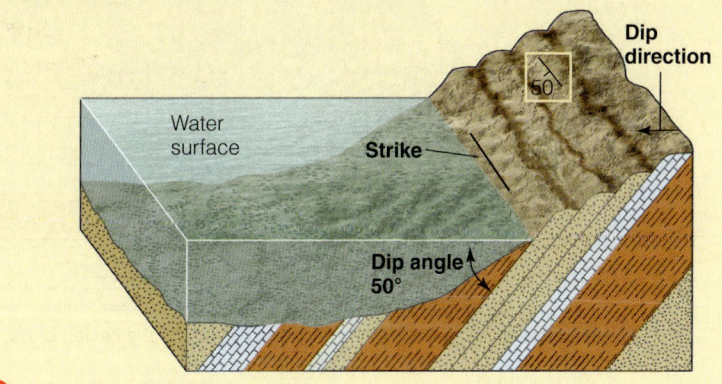

b Strike is the intersection of a horizontal plane (the water surface) with an inclined plane (the surface of the rock layer). Dip is the maximum angular deviation of the inclined layer from horizontal. Notice the strike and dip symbol.

Section 13.2 Summary

- Stress is the force applied per unit area of rock and may be compression, tension, or shear stress. Strain is deformation caused by stress. Strain may be elastic, which is recoverable, or plastic, or fracture which are permanent.

- Rocks characterized as ductile exhibit a great deal of plastic strain before they fracture, whereas brittle rocks show little plastic strain before they fracture.

What Would You Do?

The types of stresses as well as elastic versus plastic strain might seem rather esoteric, but perhaps understanding these concepts has some practical applications. What relevance do you think knowing about stress and strain has to some professions other than geology, and what professions might these be? Can you think of stresses and strain that we contend with in our daily lives? For example, what happens when you step on a soda can?

13.3 Folded Rock Layers

Geologic structures known as **folds,** in which rock layers have been crumpled and bent, are quite common. Most folds form in response to compression, as in the example above of moving your hands together on a tablecloth, but of course folds in rocks are permanent. Most folding probably takes place deep in the crust where rocks are more ductile than they are at shallow depths or at the surface. The configuration of folds and the intensity of folding vary considerably, but there are only three basic types of folds: *monoclines*, *anticlines*, and *synclines*.

▶ Figure 13.6 Monoclines

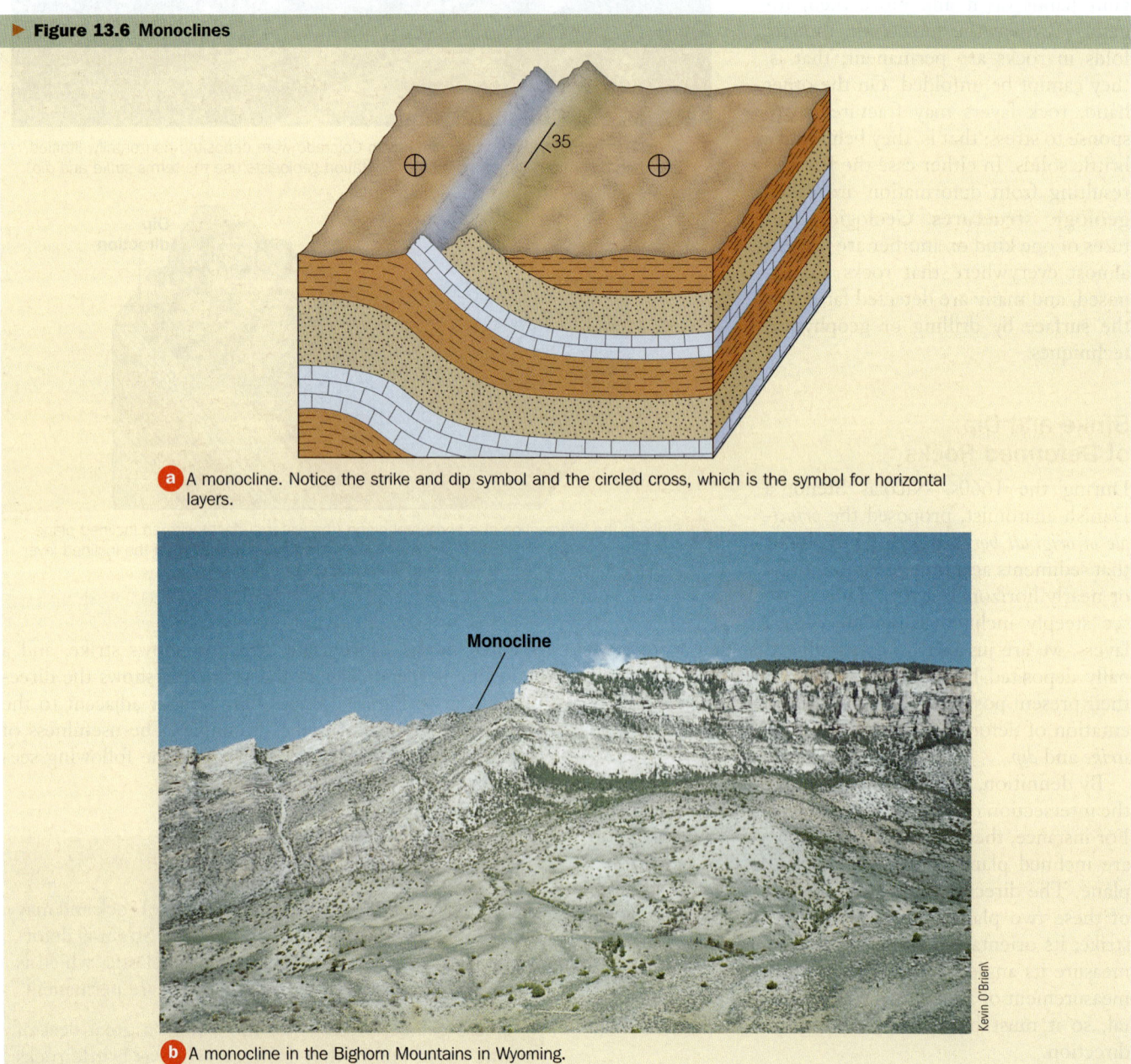

a A monocline. Notice the strike and dip symbol and the circled cross, which is the symbol for horizontal layers.

b A monocline in the Bighorn Mountains in Wyoming.

Monoclines, Anticlines, and Synclines

■ *What is a monocline and how might one form?*

A **monocline** is a bend or flexure in otherwise horizontal or uniformly dipping strata (▶ Figure 13.6). The large monocline in Figure 13.6b formed when the Bighorn Mountains of Wyoming were uplifted along a fracture that did not penetrate to the surface. As uplift proceeded, the near-surface rock layers were bent and now appear to be draped over the margin of the uplifted block.

Monoclines are not rare, but they are not nearly as common as anticlines and synclines. An **anticline** is an up-arched or convex upward fold with the oldest rock layers in its core, whereas a **syncline** is a down-arched or concave upward fold in which the youngest rock layers are in its core (▶ Figure 13.7).

Anticlines and synclines have an axial plane connecting the points of maximum curvature in each folded layer (▶ Figure 13.8); the axial plane divides folds into halves, each half being a *limb*. Because folds most often are found as a series of anticlines alternating with synclines, an anticline and adjacent syncline share a limb. It is important to remember that anticlines and synclines are simply folded rock layers and do not necessarily correspond to high and low areas at Earth's surface, except in young, uneroded landscapes (▶ Figure 13.9).

■ *How is an eroded anticline different from an eroded syncline?*

Anticlines and synclines are commonly exposed to view in areas of deep erosion, but even if eroded, they are easily distinguished by their strike and dip and the relative ages of the folded layers. Notice in ▶ Figure 13.10 that in the surface view of an anticline, each limb dips outward or away from the center of the fold where the oldest exposed rocks are found. In a syncline, however, each limb dips inward toward the center of the fold where the youngest exposed rocks are found.

Thus far, we have described upright folds in which the axial plane is vertical and each fold limb dips at the same angle (▶ Figure 13.10). If the axial plane is inclined, however, the limbs dip at different angles, and the fold is *inclined* (▶ Figure 13.11a). In an *overturned fold*, both limbs dip in the same direction because one fold limb has been rotated more than 90 degrees from its original position so that it is now upside down (▶ Figure 13.11b). Folds in which the axial plane is horizontal are called *recumbent*

▶ **Figure 13.7 Synclines and Anticlines** Folded rocks in the Calico Mountains of southeastern California. Three folds are visible from left to right: a syncline, an anticline, and another syncline. We can infer that compression was responsible for these folds.

(▶ Figure 13.11c). Overturned and recumbent folds are particularly common in mountain ranges that formed by compression at convergent plate boundaries (discussed later in this chapter).

Our definitions of anticlines and synclines as up- and down-arched folds are not sufficient for folds tipped on their sides or upside down. In Figure 13.11c, for instance, you cannot determine which fold is an anticline or a syncline with the information available. Even strike and dip will not help in this case, but the relative ages of the folded rock layers will resolve the problem. Recall that an anticline has the oldest rocks in its core, so the fold nearest the surface is an anticline and the lower fold is a syncline.

Plunging Folds

Geologists further characterize folds as *nonplunging* or *plunging*. In the former, the fold *axis*, a line formed by the intersection of the axial plane with the folded beds, is horizontal (▶ Figure 13.8). But it is much more common for the axis to be inclined so that it appears to plunge beneath the surrounding rocks. Folds possessing an inclined axis are **plunging folds** (▶ Figure 13.12a). To differentiate plunging anticlines from plunging synclines, geologists use exactly the same criteria used for nonplunging folds: All rocks dip away from the fold axis in plunging anticlines; in plunging synclines, all rocks dip inward toward the axis. The oldest exposed rocks are in the center of an eroded plunging anticline, whereas the youngest exposed rocks are in the center of an eroded plunging syncline (▶ Figure 13.12b).

In Chapter 7 we noted that anticlines form one type of structural trap for petroleum and natural gas (see Figure 7.20b). As a matter of fact, most of the world's petroleum

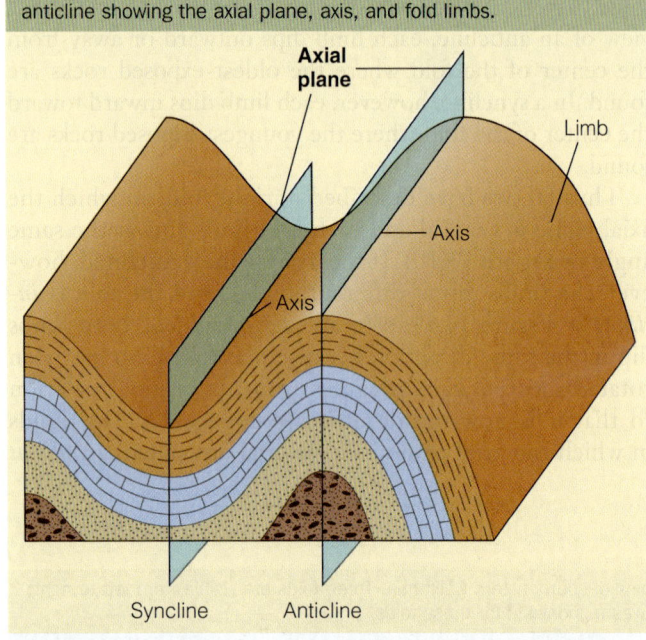

▶ **Figure 13.8 Syncline and Anticline Axial Planes** Syncline and anticline showing the axial plane, axis, and fold limbs.

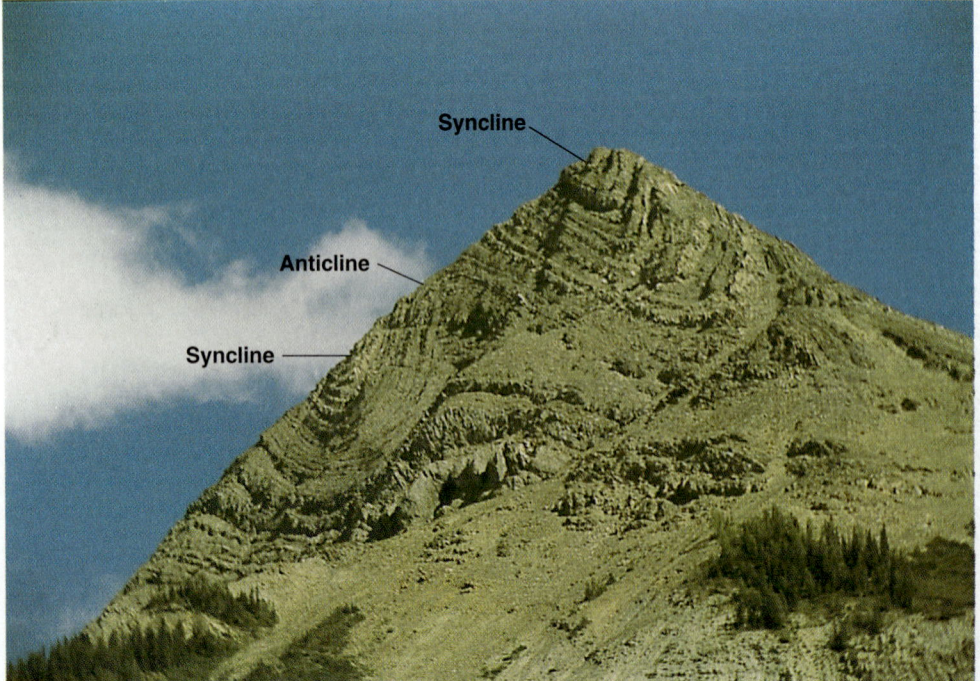

▶ **Figure 13.9 Folds and Their Relationship to Topography** Anticlines and synclines do not necessarily correspond to high and low areas of the surface. A syncline is at the peak of this mountain in Kootenay National Park, British Columbia, Canada. Lower on the left flank of the mountain, an anticline and another syncline are also visible.

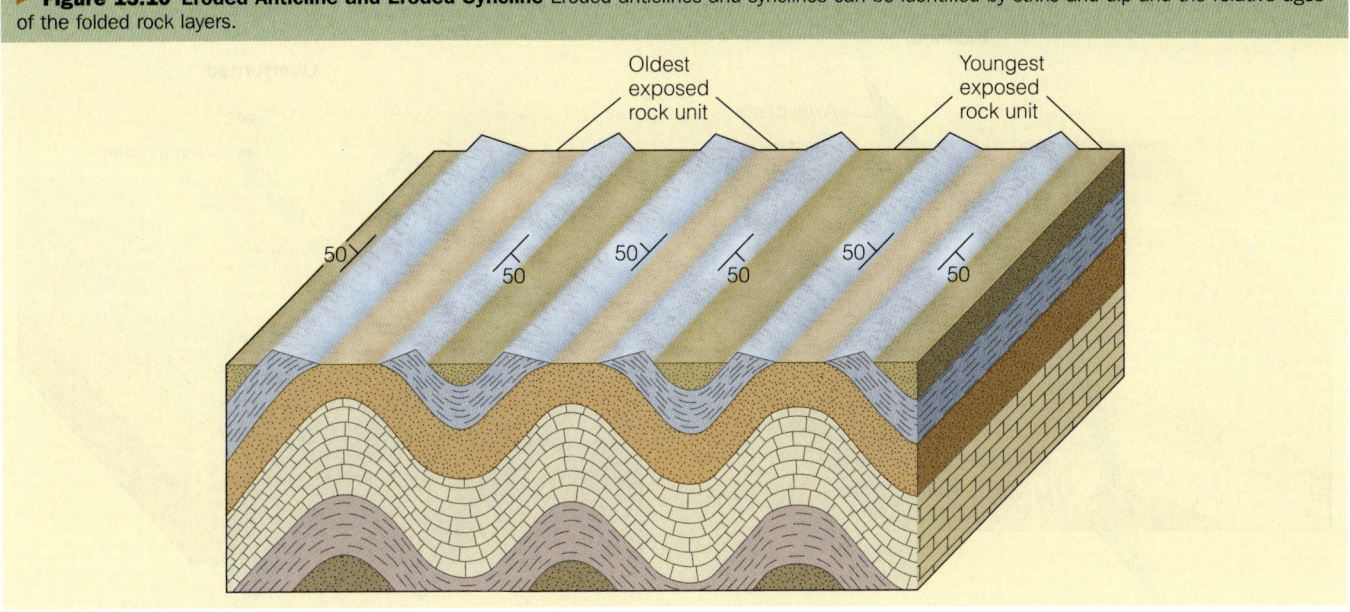

▶ **Figure 13.10 Eroded Anticline and Eroded Syncline** Eroded anticlines and synclines can be identified by strike and dip and the relative ages of the folded rock layers.

production comes from anticlinal traps, although several other types of traps are also important. Accordingly, geologists are particularly interested in correctly identifying the geologic structures in areas of potential petroleum and natural gas production.

Domes and Basins

Anticlines and synclines are elongate structures; that is, their length exceeds their width. **Domes** and **basins** are the oval and circular equivalents of anticlines and synclines, respectively (▶ Figure 13.13). Geologists use exactly the same criteria to identify domes and basins that they use for anticlines and synclines. For example, in an eroded dome the oldest exposed rocks are at the center, whereas in a basin the opposite is true. All strata in a dome dip away from a central point (as opposed to dipping away from a fold axis, which is a line). In contrast, all strata in a basin dip inward toward a central point (▶ Figure 13.13).

The terms *dome* and *basin* are also used to designate high and low areas on Earth's surface, but domes and basins resulting from deformation do not necessarily correspond to mountains and valleys. In several instances in the following discussions and chapters, we will mention basins in other contexts, and we have already used the word *dome*, as in "exfoliation dome," so here we will modify these two terms with *structural*. Accordingly, a structural dome or structural basin is an area of deformation corresponding to our previous definition, with no reference to surface elevations.

Some structural domes and basins are small structures that are easily recognized by their surface exposure patterns, but many are so large that they can be visualized only on geologic maps or aerial photographs. Many of these large-scale structures formed in the continental interior, not by compression but as a result of vertical uplift of parts of the crust with little additional folding and faulting.

The Black Hills of South Dakota are a large eroded structural dome with a core of ancient rocks surrounded by progressively younger rocks (▶ Figure 13.13c). The rocks dip outward rather uniformly from the Black Hills, which have been uplifted so that they now stand higher than 2000 m above the adjacent plains. One of the best-known structural basins in the United States is the Michigan Basin. Most of the Michigan Basin is buried beneath younger rocks, so it cannot be observed at the surface. Nevertheless, strike and dip of exposed rocks near the basin margin and thousands of drill holes for oil and gas clearly show that the rock layers beneath the surface are deformed into a large structural basin.

Section 13.3 Summary

- Most folds form in response to compression, but domes and basins probably result mostly from vertical uplift.

- The strike and dip of folded strata as well as their relative ages help us distinguish anticlines from synclines and domes from basins.

- Many anticlines and synclines plunge, thus making their surface expression complex, but geologists use the same criteria to identify them that they use for nonplunging folds.

▶ **Figure 13.11** Types of Folds

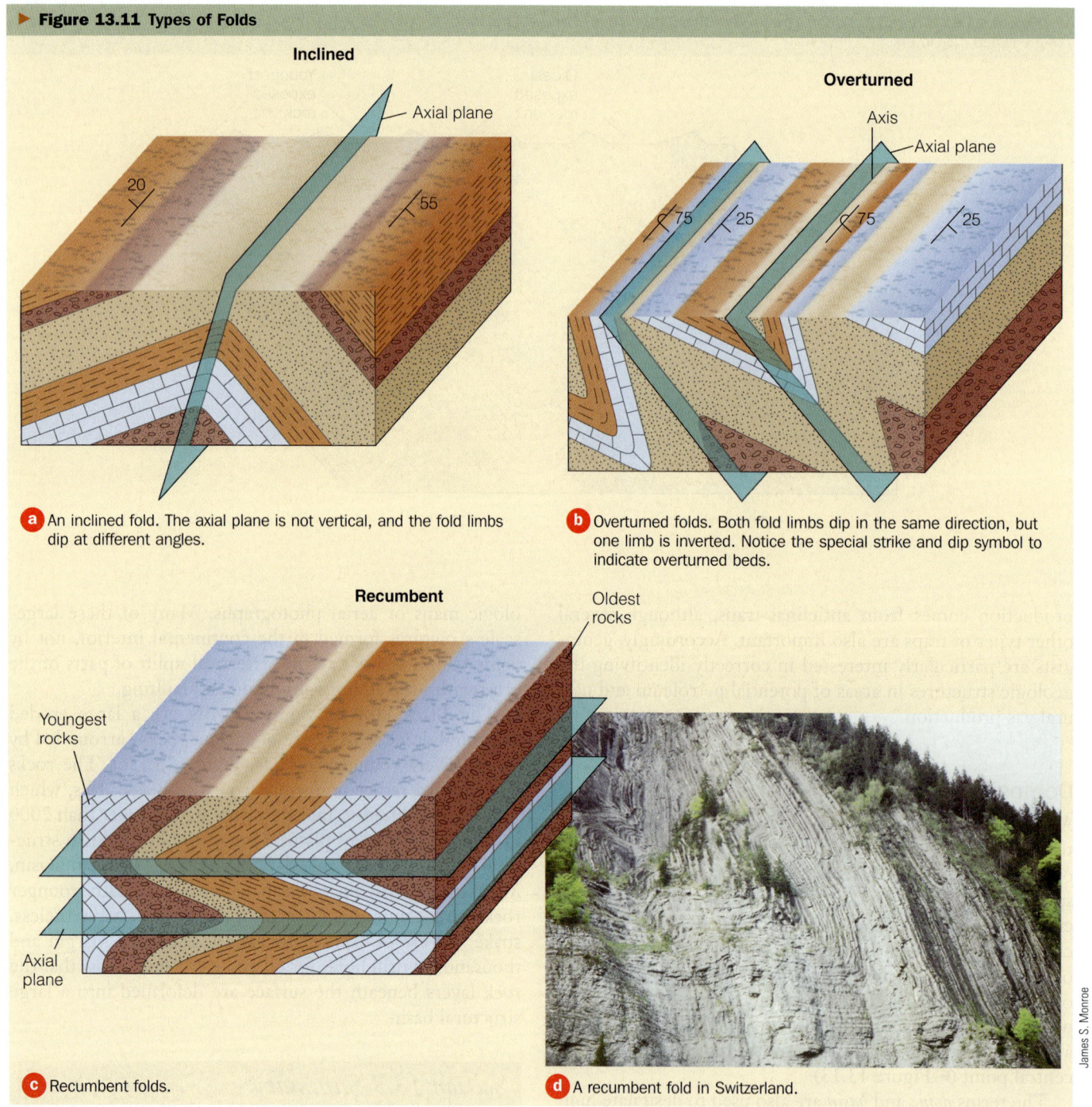

a) An inclined fold. The axial plane is not vertical, and the fold limbs dip at different angles.

b) Overturned folds. Both fold limbs dip in the same direction, but one limb is inverted. Notice the special strike and dip symbol to indicate overturned beds.

c) Recumbent folds.

d) A recumbent fold in Switzerland.

13.4 Joints and Faults— Deformation by Fracturing

In the preceding section we discussed monoclines, anticlines, synclines, domes, and basins, all of which formed by plastic deformation; that is, the rock layers have been crumpled, mostly by compression. Now we turn to *joints* and *faults*, both of which are fractures resulting from compression, tension, or shear stresses.

■ *How do joints differ from faults?*

Joints

In addition to folding, fracturing is another way in which rocks are permanently deformed; that is, they behave like brittle solids and break in response to stress. Some breaks in rocks are called **joints**, which are fractures along which no movement has taken place parallel with the fracture surface (▶ Figure 13.14), meaning that the rocks on opposite sides of the fracture surface have not slid up, down, or sideways along the fracture surface. Coal miners used the term *joint*

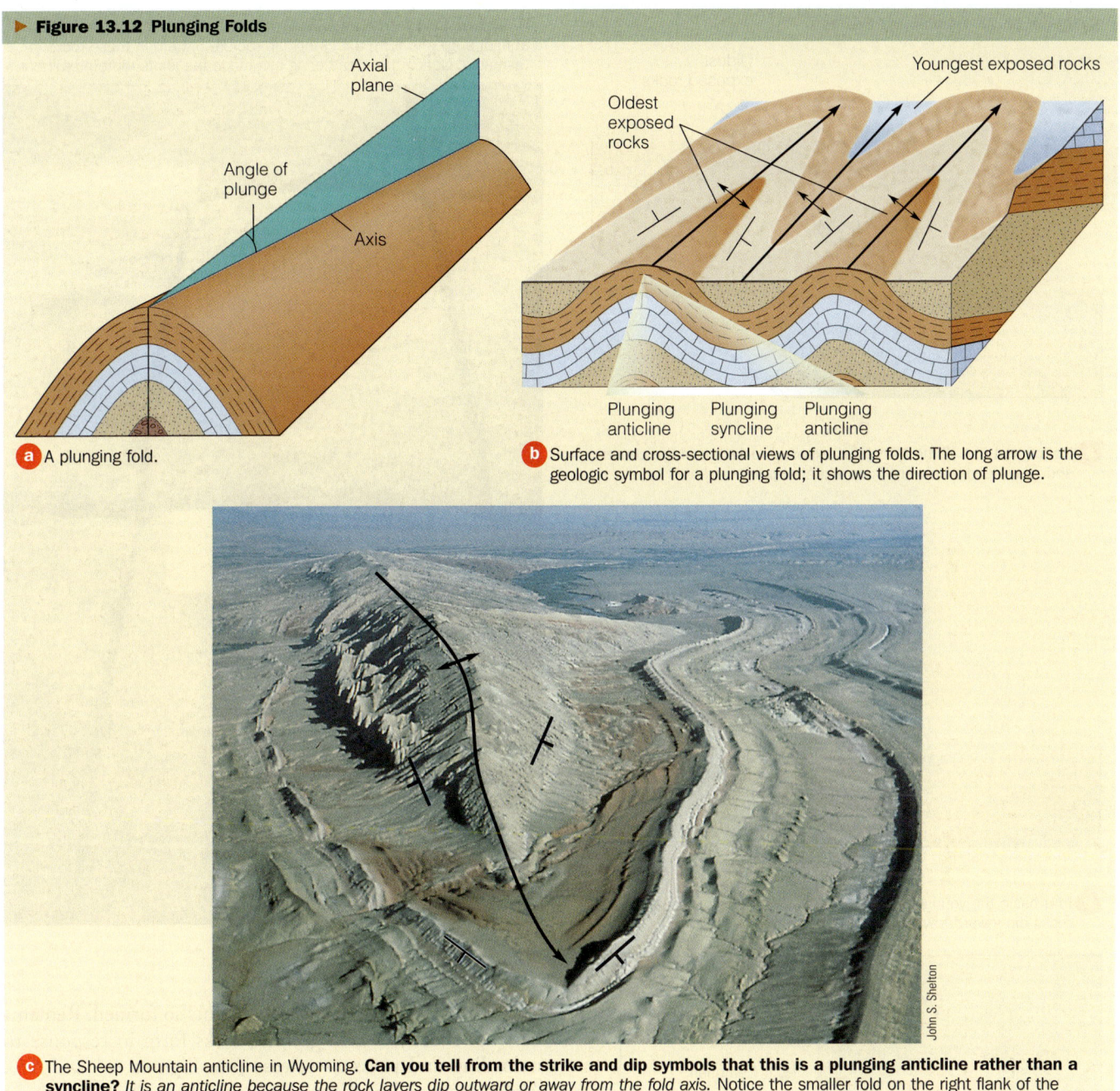

Figure 13.12 Plunging Folds

a A plunging fold.

b Surface and cross-sectional views of plunging folds. The long arrow is the geologic symbol for a plunging fold; it shows the direction of plunge.

c The Sheep Mountain anticline in Wyoming. **Can you tell from the strike and dip symbols that this is a plunging anticline rather than a syncline?** *It is an anticline because the rock layers dip outward or away from the fold axis.* Notice the smaller fold on the right flank of the anticline. **What kind of fold is it?** *Syncline.*

long ago for cracks that they thought were surfaces along which adjacent blocks of rock were "joined" together (see Figures 6.10 and 7.19).

Remember that near-surface rocks are brittle and tend to fail by fracturing when subjected to stress. In fact, nearly all near-surface rocks have joints that form in response to any type of stress—compression, tension, or shearing. Joints vary from minute cracks to huge fractures extending for many kilometers, and joints are commonly arranged in two or perhaps three predominant sets. Regional mapping reveals that sets of joints are usually related to other geologic structures such as folds and faults.

It may seem odd that joints, which indicate brittle behavior, are so common in folded rocks that have undergone plastic deformation. Consider the crest of an anticline and the stress generated as folding proceeds. An anticline results from compression, but the rock layers in the fold are arched upward so that tension takes place perpendicular to the fold's crest. Accordingly, tension joints commonly form parallel to the long axis of the fold (see "Arches National Park" on pp. 176–177).

All joints result from stress, and most of them are formed when rocks are deformed by externally applied stresses. Some, however, form by internal stresses or during cooling.

▶ **Figure 13.13 Domes and Basins**

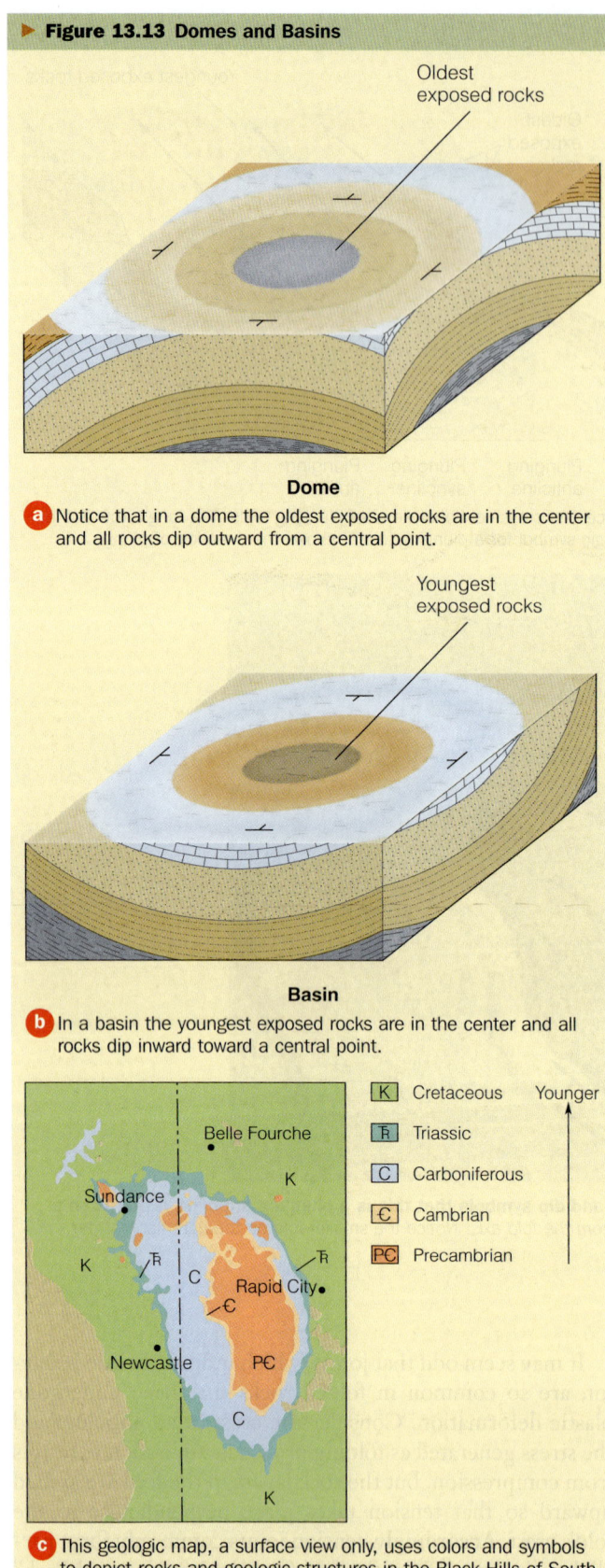

Dome

a Notice that in a dome the oldest exposed rocks are in the center and all rocks dip outward from a central point.

Basin

b In a basin the youngest exposed rocks are in the center and all rocks dip inward toward a central point.

c This geologic map, a surface view only, uses colors and symbols to depict rocks and geologic structures in the Black Hills of South Dakota. **From the information provided, can you determine whether this map depicts a dome or a basin?** *It is a dome because the oldest rock layers are exposed at its center.*

▶ **Figure 13.14 Rectangular Fracture** Two sets of joints intersect at right angles to form a rectangular fracture pattern in these rocks in Tasmania. Notice that weathering has made the joints more prominent.

In fact, we have already discussed joints so formed. Remember that sheet joints in granitic rocks form in response to unloading and pressure release (see Figures 6.4 and 6.5). And columnar joints form in some lava flows and some plutons as the lava or magma cools and contracts and polygonal fractures develop (see Figure 5.5).

Faults

We have defined a *joint* as a fracture along which no movement takes places, but when the blocks on opposite sides of a fracture move up, down, or sideways, the fracture is called a **fault.** In other words, on a fault the rock on one side of the fracture surface, the **fault plane,** is displaced relative to the rock on the opposite side of the fracture (▶ Figure 13.15). Not all faults penetrate to the surface, but those that do might show a *fault scarp,* a bluff or cliff formed by vertical movement (▶ Figure 13.16a). When movement

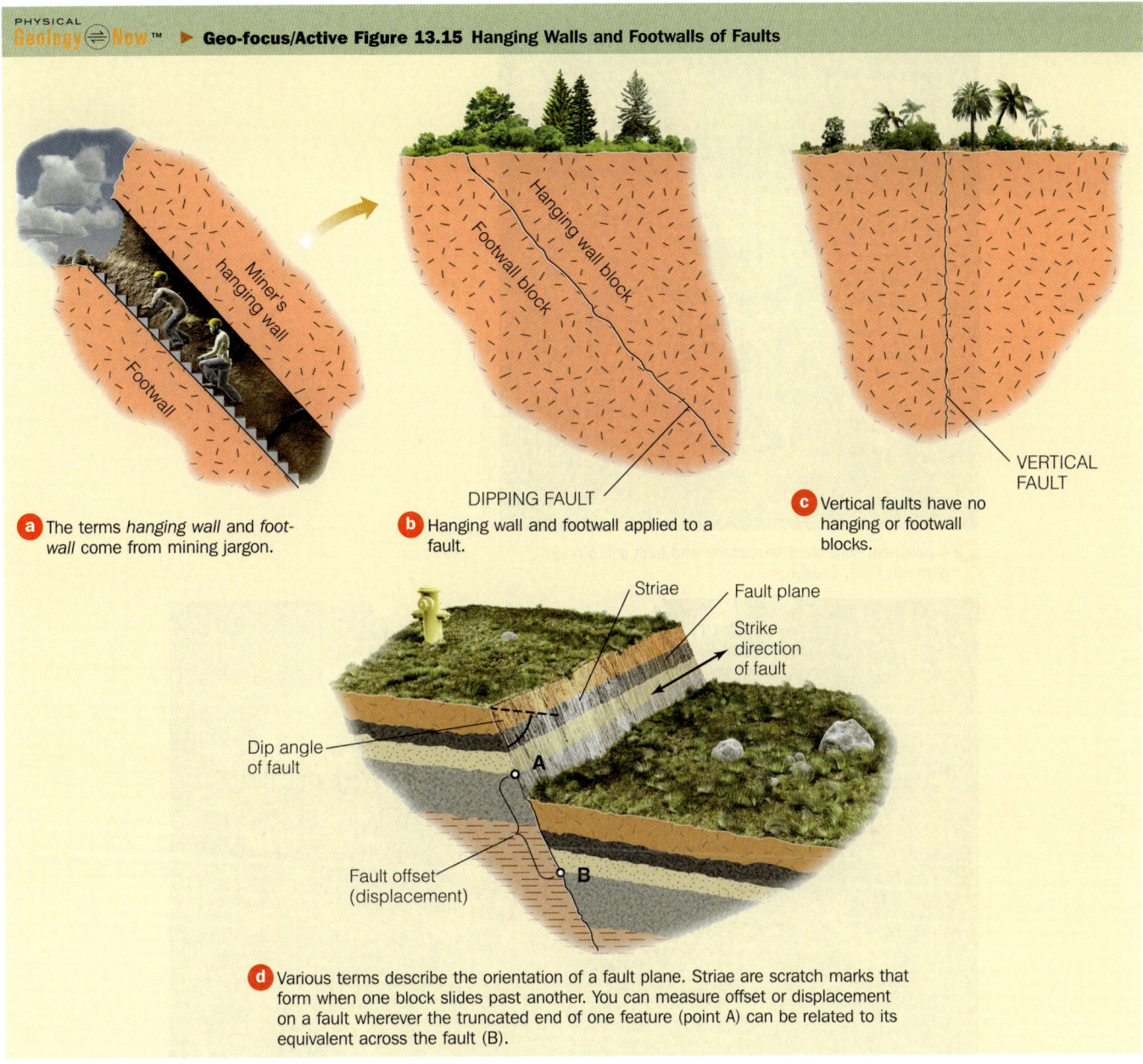

Geo-focus/Active Figure 13.15 Hanging Walls and Footwalls of Faults

a The terms *hanging wall* and *footwall* come from mining jargon.

b Hanging wall and footwall applied to a fault.

c Vertical faults have no hanging or footwall blocks.

d Various terms describe the orientation of a fault plane. Striae are scratch marks that form when one block slides past another. You can measure offset or displacement on a fault wherever the truncated end of one feature (point A) can be related to its equivalent across the fault (B).

takes place along a fault plane, the rocks on opposite sides may be scratched and polished (▶ Figure 13.15d), or they may be crushed and shattered into angular blocks forming *fault breccia* (▶ Figure 13.16b).

Notice the distinction between *hanging wall block* and *footwall block* in Figure 13.15. The **hanging wall block** consists of the rock overlying or above an inclined fault plane, whereas the **footwall block** is the rock underlying or beneath an inclined fault plane. We emphasize the term *inclined* in the preceding sentence because no such distinction can be made on a vertical fault—that is, one that dips at 90 degrees (▶ Figure 13.16c). The hanging wall block–footwall block distinction is important because to identify some specific types of faults, you must not only correctly identify these two blocks but also determine which one moved relatively up or down.

We use the term *relative movement* when discussing faults because you cannot usually determine which block actually moved or if both moved. For example, no matter how closely you examine Figure 13.15d, you cannot tell whether the footwall block moved up or the hanging wall block moved down, or both could have moved simultaneously. Nevertheless, the footwall block appears to have moved up relative to the hanging wall block, or put the other way around, the hanging wall block appears to have moved down relative to the footwall block.

Remember the definitions of strike and dip that geologists use to define the orientation of inclined planes such as the surfaces of rock layers (▶ Figure 13.5b). Likewise, fault planes are inclined planes and they too are characterized by strike and dip (▶ Figure 13.15d). Strike and dip are important

▶ **Figure 13.16 Faults**

a A polished, scratched fault plane and fault scarp near Klamath Falls, Oregon.

b Fault breccia, the zone of rubble along a fault in the Bighorn Mountains, Wyoming. **If arrows were not shown, could you determine which side of the fault moved relatively up?**

considerations because the two basic varieties of faults are defined by whether the blocks on opposite sides of the fault plane moved parallel to the direction of dip (dip-slip faults) or along the direction of strike (strike-slip faults). (See "Types of Faults" on pp. 410–411.)

- *What kinds of dip-slip faults do geologists recognize, and what features distinguish each type?*

Dip-Slip Faults As their name implies, all movement (slip) on **dip-slip faults** is parallel to the dip of the fault; that is, movement is vertical, either up or down the fault plane. Notice in ▶ Figure 13.17a-d that all are dip-slip faults, but in ▶ Figure 13.17a-b the hanging wall block has moved down relative to the footwall block, giving rise to a **normal fault**. In ▶ Figure 13.17c-d, however, the hanging wall block has moved up relative to the footwall block, so it is a **reverse fault**. The fault in ▶ Figure 13.17d looks like a reverse fault too but the fault plane dips at less than 45 degrees: it is a special variety of reverse fault called a **thrust fault**.

Notice in Figure 13.3b that normal faults form in response to tension. Numerous normal faults are present along one or both sides of mountain ranges in the Basin and Range Province of the western United States where the crust has been stretched and thinned. Huge normal faults are present at the western margin of the Basin and Range Province along which the Sierra Nevada has risen more than 3000 m relative to the lowlands to the east. An active normal fault along the eastern margin of the Teton Range in Wyoming accounts for the 2100-m elevation difference between the valley floor and the highest peaks in the mountains.

Normal faults result from tension, but the opposite stress, compression, is responsible for reverse and thrust faults (▶ Figure 13.3a). Large-scale examples of both reverse and thrust faults are found in mountains that formed at convergent plate boundaries (discussed later in this chapter). A well-known example of a huge thrust fault is the Lewis overthrust in Montana. (An overthrust is a low-angle thrust fault with movement measured in kilometers.) A huge slab of Precambrian-aged rocks moved at least 75 km eastward on this fault and rests on much younger Cretaceous-aged rocks.

Strike-Slip Faults Unlike the vertical movements on dip-slip faults, all movement on **strike-slip faults** is along the strike of the fault plane. In other words, horizontal movement takes place as one or both blocks on opposite sides of the fault plane move relative to one another (▶ Figure 13.17e-f). Many large strike-slip faults are well known, but certainly the most famous one is the San Andreas fault that cuts through coastal California. Remember from Chapter 2 that the San Andreas fault is also called a transform fault in plate tectonic terminology because it separates two tectonic plates. Most strike-slip faults are not transform faults, but all transform faults are strike-slip.

We need not distinguish between hanging wall and footwall blocks for strike-slip faults, but we still must determine the direction of relative movement of the rocks on opposite sides of the fault plane. In ▶ Figure 13.17e-f, for example, one or the other of the blocks could have moved in the directions indicated by the arrows or both blocks could have moved. Suppose you were standing on the block with the "slow" sign in Figure 13.17e. Notice that the block on the opposite side of the fault appears to have moved to the right, so this is a *right-lateral strike-slip fault*. Even if you were on the block oppposite the sign and looked at the block across the fault, the movement would still appear to be to the right. On the other hand, Figure 13.17f shows a *left-lateral strike-slip fault*. Can you determine whether the San Andreas fault on page 411 is a left-lateral or right-lateral strike-slip fault?

Oblique-Slip Faults The only possible movements on a fault are up or down the fault plane (dip-slip faults), along the strike of a fault plane (strike-slip faults), or a combination of the two. In other words, a strike-slip fault may show right-lateral movement, and the hanging wall block may also move down relative to the footwall block. Such complex movements give rise to an **oblique-slip fault** (▶ Figure 13.17g). Any combination of dip-slip and strike-slip is possible; the one in Figure 13.17g shows both normal fault displacement and left-lateral strike-slip movement.

Folds, faults, and joints along with colors and symbols for different types and ages of rocks are portrayed on *geologic maps*. As you would expect, geologists construct and use geologic maps, but people in several other professions also rely on information on geologic maps.

Section 13.4 Summary

- Joints and faults are both fractures. Movement has taken place parallel to the fracture surface along faults but not on joints.

- Dip-slip faults result from tension or compression and include normal, reverse, and thrust faults. In all cases, all movement is up or down the dip of the fault plane. Only horizontal movement along a fault plane's strike takes place on strike-slip faults, which are further characterized as left-lateral or right-lateral. Oblique-slip faults exhibit a combination of dip-slip and strike-slip movements.

What Would You Do?

Refer to Figure 13.10 and consider the following. Which layer(s) of sedimentary rock would make good source rock, cap rock, and hydrocarbon reservoirs? Explain. Suppose you are trying to line up investors to finance an oil well in this area. How would you convince your prospective investors where you should drill, and why you chose this location(s)?

What Would You Do?

As a member of a planning commission, you are charged with developing zoning regulations and building codes for an area with known active faults, steep hills, and deep soils. A number of contractors as well as developers and citizens in your community are demanding action because they want to begin several badly needed housing developments. How might geologic maps and an appreciation of geologic structures influence you in this endeavor? Do you think, considering the probable economic gains for your community, that the regulations you draft should be rather lenient or very strict? If the latter, how would you explain why you favored regulations that would involve additional cost for houses?

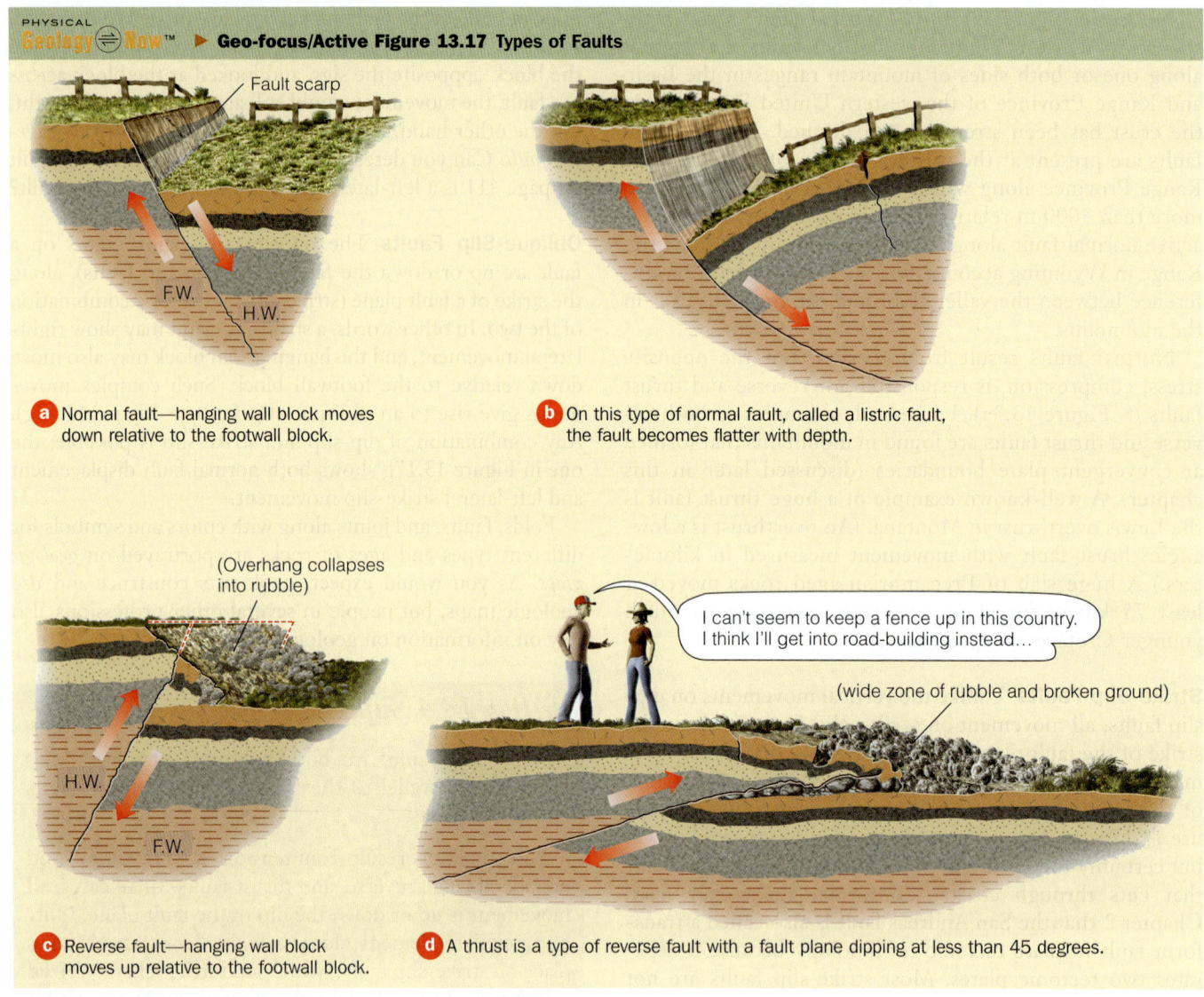

Geo-focus/Active Figure 13.17 Types of Faults

a Normal fault—hanging wall block moves down relative to the footwall block.

b On this type of normal fault, called a listric fault, the fault becomes flatter with depth.

c Reverse fault—hanging wall block moves up relative to the footwall block.

d A thrust is a type of reverse fault with a fault plane dipping at less than 45 degrees.

13.5 Deformation and the Origin of Mountains

Mountain is the designation for any area of land that stands significantly higher, at least 300 m, than the surrounding country and has a restricted summit area. Some mountains are single, isolated peaks, but more commonly they are part of a linear association of peaks and ridges known as *mountain ranges* that are related in age and origin. A *mountain system* is made up of many mountain ranges. The Appalachian Mountains in the eastern United States and Canada include, among others, the Great Smokey Mountains of North Carolina and Tennessee, the Adirondack Mountains of New York, the Green Mountains of Vermont, and the Blue Mountains in Maine. Likewise, the Rocky Mountains are made up of many ranges, such as the Teton Range and the Wind River Range in Wyoming, the Mission Mountains in Montana, and the Wasatch Range in Utah.

Mountain Building

■ *At what kinds of plate boundaries do most of Earth's large mountain systems form?*

Mountains form in several ways, some of them involving little or no deformation, but the large mountain ranges on land form by compression and intense deformation at convergent plate boundaries. Differential weathering and erosion in the southwestern United States have created high areas with adjacent lowlands, but these are flat-topped or pinnacle-shaped mesas and buttes (see Chapter 18). Also, the oceanic ridge system is made up of mountains, but these mountains are composed entirely of volcanic rocks and show features produced by tension at divergent plate boundaries.

Block faulting is yet another way for mountains to form; it involves movement on normal faults so that one or more blocks are elevated relative to adjacent areas (▶ Figure 13.18). A good example of large-scale block faulting is in the Basin

▶ **Figure 13.17 (continued)**

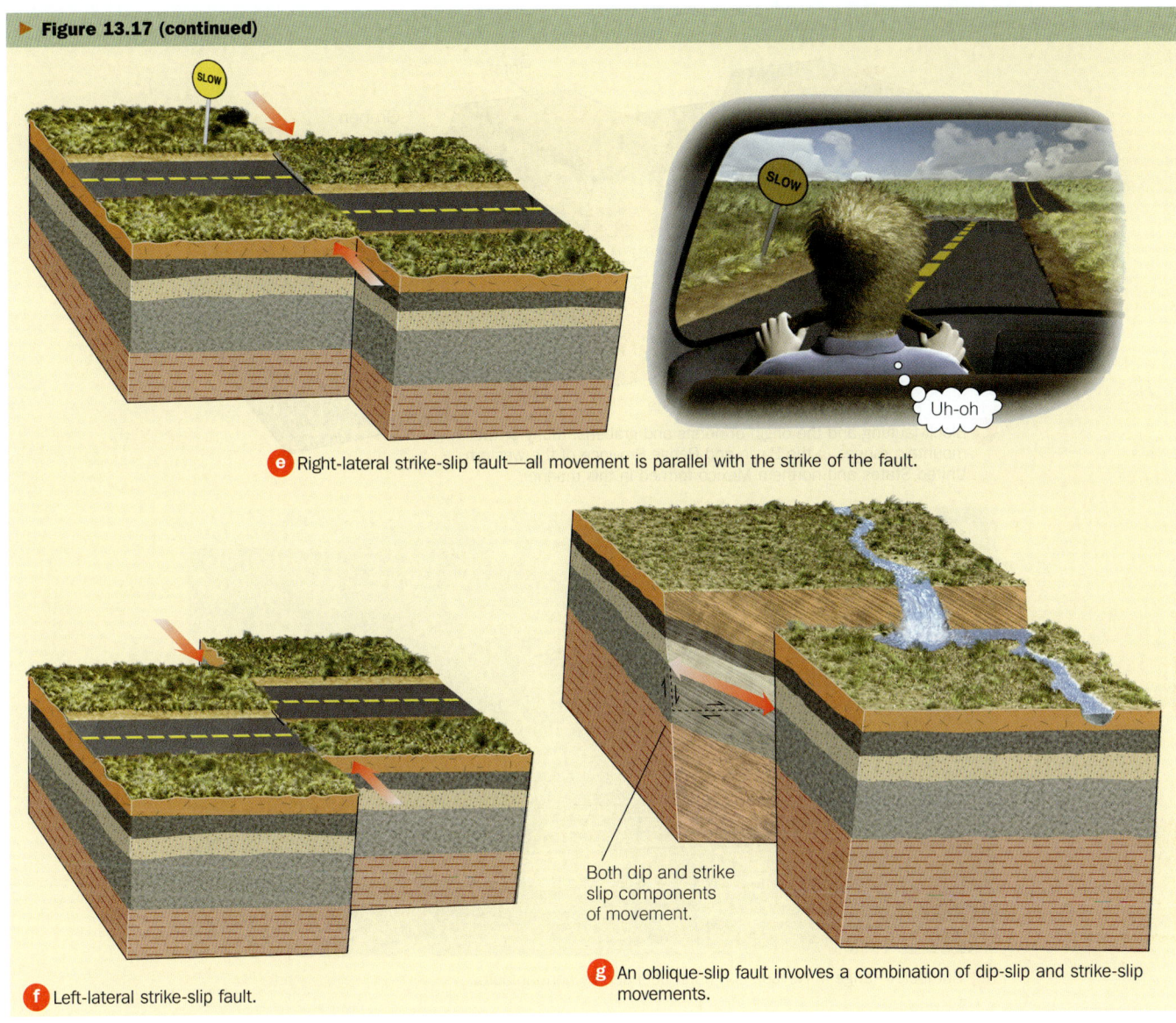

e Right-lateral strike-slip fault—all movement is parallel with the strike of the fault.

f Left-lateral strike-slip fault.

Both dip and strike slip components of movement.

g An oblique-slip fault involves a combination of dip-slip and strike-slip movements.

and Range Province of the western United States, a large area centered on Nevada but extending into several adjacent states and northern Mexico. Here Earth's crust is being stretched in an east–west direction and tensional stresses produce north–south faults on one or both sides of uplifted blocks called *horsts* and downdropped blocks known as *grabens* (▶ Figure 13.18). In other words, the horsts are the ranges and the grabens are the basins. Remember that we used the term *structural basin* earlier for rather circular folds, but here *basin* refers to a relatively low area.

Plate Tectonics and Mountain Building

Now we turn our attention to the truly large mountain systems on the continents—the Appalachian Mountains of North America, the Andes in South America, the Himalayas in Asia, and the Alps of Europe. All these mountain systems formed at convergent plate boundaries and, although each system is unique, they share several characteristics. In fact,

geologists call an episode of mountain building an **orogeny** during which intense deformation takes place, usually in response to compression, accompanied by metamorphism, the emplacement of plutons, and local thickening of Earth's crust.

Any theory that accounts for orogeny must adequately explain the characteristics of mountain systems, such as their geometry and location. Most mountain ranges tend to be long and narrow and with few exceptions are at or near continental margins. Mountain systems also show intense deformation, especially compression-induced overturned and recumbent folds as well as reverse and thrust faults. Furthermore, granitic plutons and regional metamorphism are found in the interiors or cores of mountain systems. Finally, many have sedimentary rocks now far above sea level that were clearly deposited in marine environments.

Although not all aspects of orogenies are fully understood, geologists are convinced that plate tectonic theory provides an adequate explanation for many features of

▶ **Figure 13.18 Origins of Horsts and Grabens**

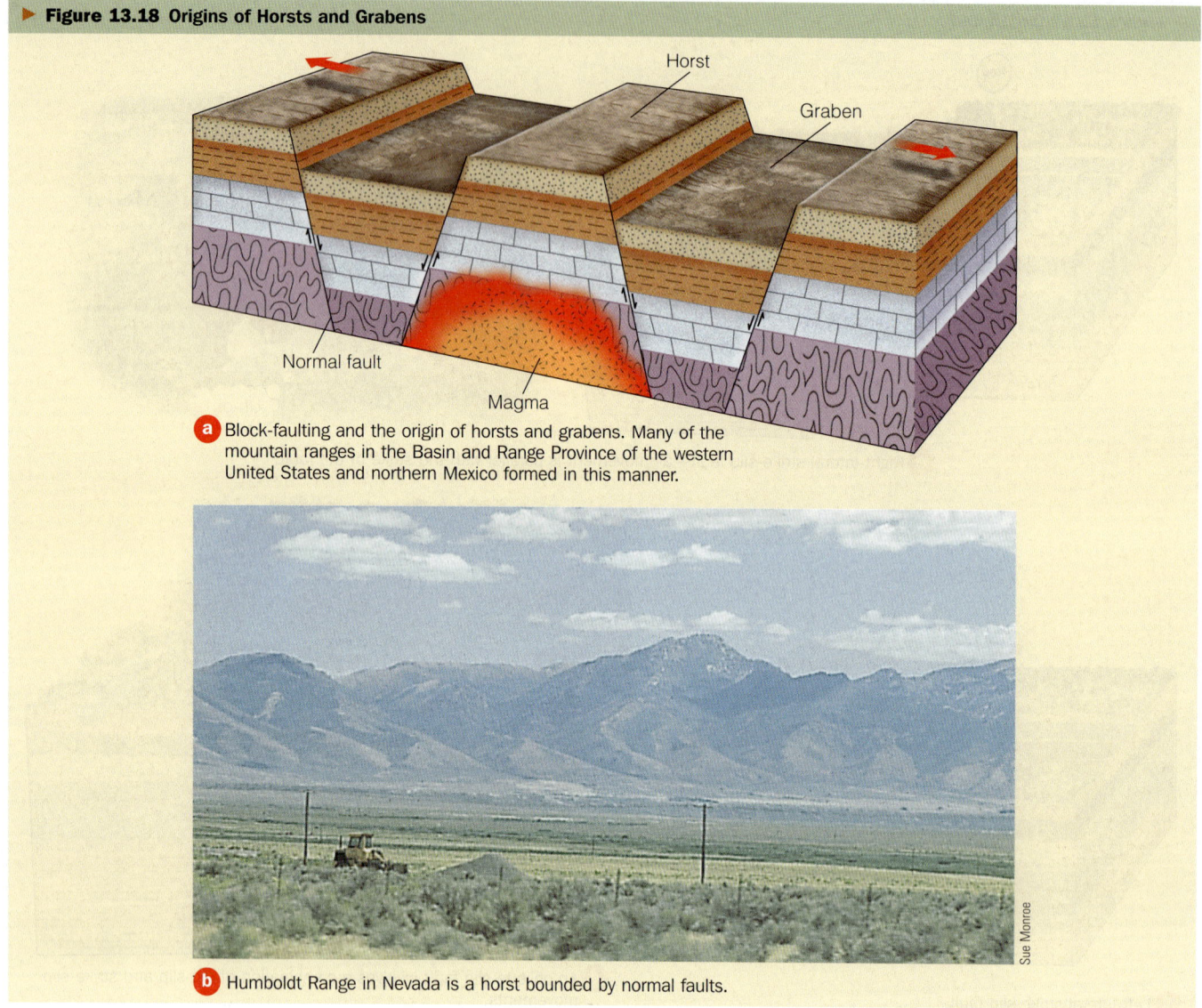

a Block-faulting and the origin of horsts and grabens. Many of the mountain ranges in the Basin and Range Province of the western United States and northern Mexico formed in this manner.

b Humboldt Range in Nevada is a horst bounded by normal faults.

mountain systems. Convergence of lithospheric plates accounts for the geometry and location of mountain systems as well as the complex geologic structures, plutons, metamorphism, and crustal thickening. Yet the present-day topographic expression of mountains is also related to erosion by surface processes such as mass wasting (gravity-driven processes including landslides), running water, and glaciers.

Most of Earth's geologically recent and ongoing orogenies are found in two major zones or belts: the *Alpine–Himalayan orogenic belt* and the *circum-Pacific orogenic belt* (▶ Figure 13.19). Each belt consists of a number of smaller segments known as *orogens*, each of which is an area of deformation. In fact, most of Earth's past and present orogenies take place at convergent plate boundaries, but recall from Chapter 2 that convergent boundaries may be oceanic–oceanic, oceanic–continental, or continental–continental.

Orogenies at Oceanic–Oceanic Plate Boundaries Deformation, igneous activity, and the origin of a volcanic island arc characterize orogenies that take place where oceanic lithosphere is subducted beneath oceanic lithosphere. Sediments derived from the evolving island arc are deposited in an adjacent oceanic trench, and then deformed and scraped off against the landward side of the trench (▶ Figure 13.20). These deformed sediments are part of a subduction complex, or *accretionary wedge*, of intricately folded rocks cut by numerous thrust faults. In addition, orogenies in this setting show the low-temperature, high-pressure metamorphism of the blueschist facies (see Figure 8.20).

Emplacement of plutons also causes deformation in the island arc system, and many of the rocks show the effects of high-temperature, low-pressure metamorphism. The overall effect of an island arc orogeny is the origin of two more or less parallel belts consisting of a landward volcanic island

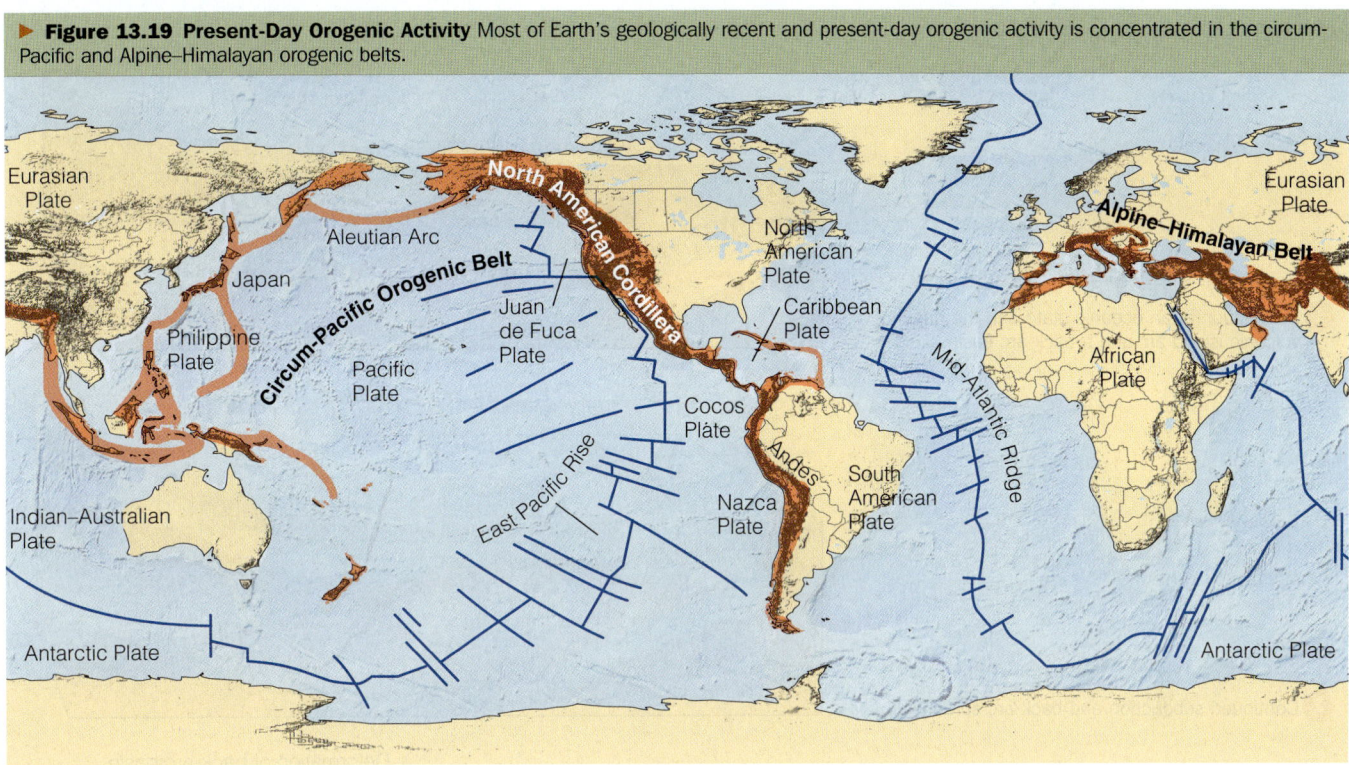

Figure 13.19 Present-Day Orogenic Activity Most of Earth's geologically recent and present-day orogenic activity is concentrated in the circum-Pacific and Alpine–Himalayan orogenic belts.

arc intruded by batholiths and a seaward belt of deformed trench rocks (▶ Figure 13.20). The Japanese islands are a good example of this type of orogeny.

In addition to deformation in the island arc, volcanic rocks and sediments deposited in the *back-arc basin*, the area between the island arc and its nearby continent, are also deformed, especially along low-angle thrust faults, as plate convergence continues. Eventually the entire island arc complex is fused to the edge of the continent, and the back-arc basin sediments are thrust onto the continent (▶ Figure 13.20).

■ *What sequence of events takes place during an orogeny at an oceanic–continental plate boundary?*

Orogenies at Oceanic–Continental Plate Boundaries
The Andes Mountains in South America are the best example of a continuing orogeny at an oceanic–continental plate boundary (▶ Figure 13.21). Among the ranges of the Andes are many active volcanoes and the highest peaks in the world except for the Himalayas in Asia. In addition, the west coast of South America is a very active part of the circum-Pacific earthquake belt, and one of Earth's great oceanic trenches, the Peru–Chile Trench, lies just off the west coast.

Prior to 200 million years ago, the western edge of South America was a passive continental margin much like its eastern margin is now. When Pangaea started to separate along the Mid-Atlantic Ridge, however, the South American plate began moving west and oceanic lithosphere began subducting beneath the continent along its western margin (▶ Figure 13.21). What had been a passive continental margin was now an active one and partial melting of the subducted plate yielded magma that rose to form huge batholiths and a chain of andesitic volcanoes.

As a result of the events described above, the Andes consist of a central core of granitic rocks capped by andesitic volcanoes. To the west of the central core lie deformed rocks of an accretionary wedge, and to the east are folded sedimentary rocks that were thrust eastward onto the continent (▶ Figure 13.21). Continuing volcanism and seismic activity along South America's western margin indicate that the Andes remain an evolving mountain system.

Orogenies at Continental–Continental Plate Boundaries
Several areas of colliding continents and the origin of mountain systems are well documented, but the best example taking place now is the Himalayas in Asia. Before this vast mountain system began to form, India lay far to the south and was separated from Asia by an ocean basin (▶ Figure 13.22a). At this stage of development, there was a subduction zone along Asia's southern edge, partial melting generated magma that rose to form a volcanic arc, and large granitic plutons were emplaced in what is now Tibet. In short, it was much like the present situation along the west coast of South America (▶ Figure 13.21, part 3).

As the ocean separating India from Asia continued to close, India eventually collided with Asia and what had been two continental plates were sutured together (▶ Figure 13.22b). Thus the Himalayas are within a continent, far from a continental margin. This collision probably began 40 to 50 million

Figure 13.20 Orogeny and the Origin of a Volcanic Island Arc at an Oceanic–Oceanic Plate Boundary

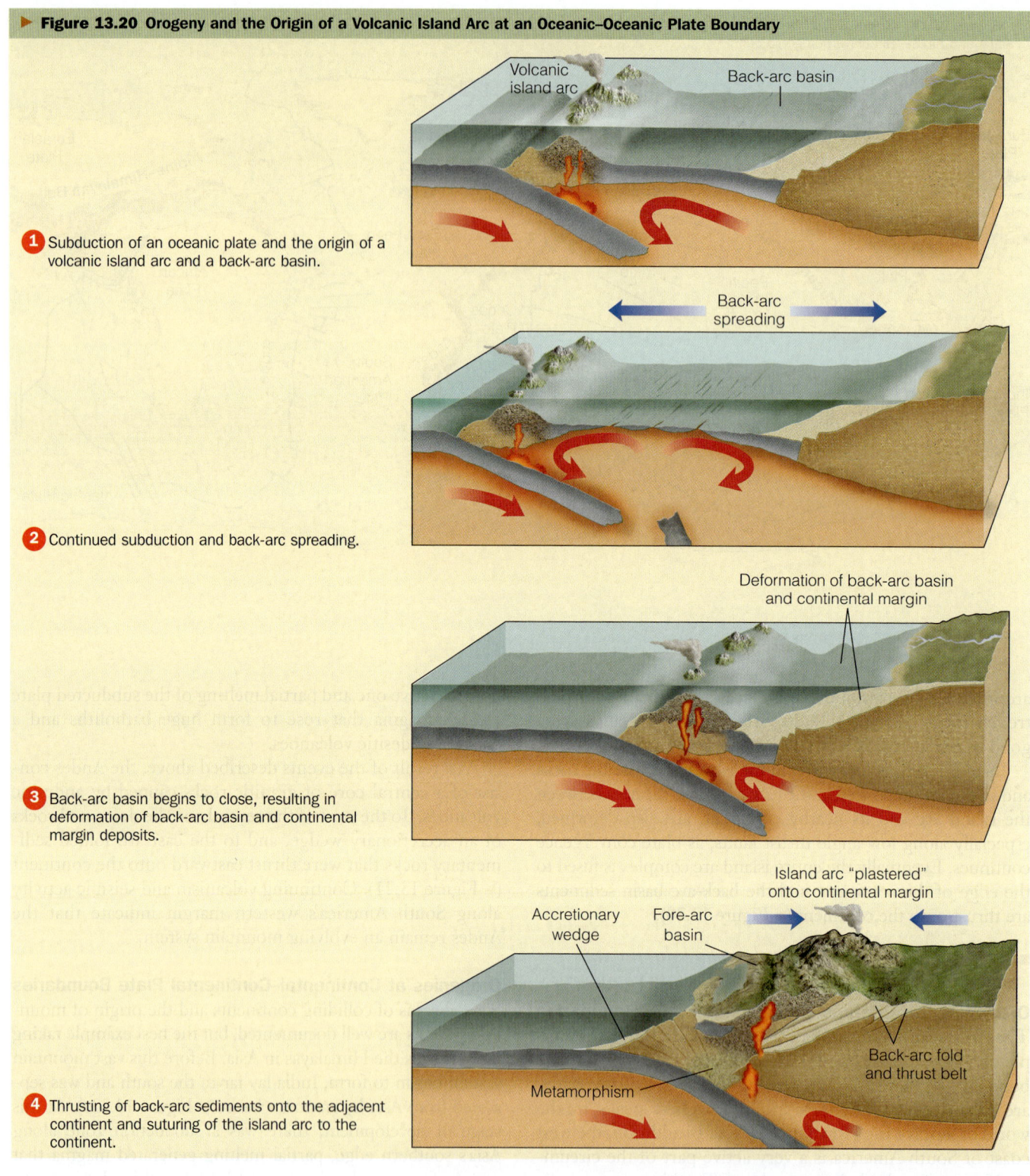

1. Subduction of an oceanic plate and the origin of a volcanic island arc and a back-arc basin.

2. Continued subduction and back-arc spreading.

3. Back-arc basin begins to close, resulting in deformation of back-arc basin and continental margin deposits.

4. Thrusting of back-arc sediments onto the adjacent continent and suturing of the island arc to the continent.

years ago because at that time India's northward rate of movement decreased abruptly from 15–20 cm per year to about 5 cm per year. Continental lithosphere is less dense than oceanic lithosphere, so this decrease marks the time of collision and India's resistance to subduction.

As a result of this continental–continental collision, the leading margin of India has been thrust 2000 km beneath Asia and continues moving north at about 5 cm per year. Sedimentary rocks that were deposited in the sea south of Asia were thrust northward, and two large thrust faults carried rocks of Asian origin onto the Indian plate (▶ Figure 13.22b). In short, tremendous deformation occurred and Earth's crust was thickened and uplifted far above sea level (remember the principle of isostasy). In fact, sedi-

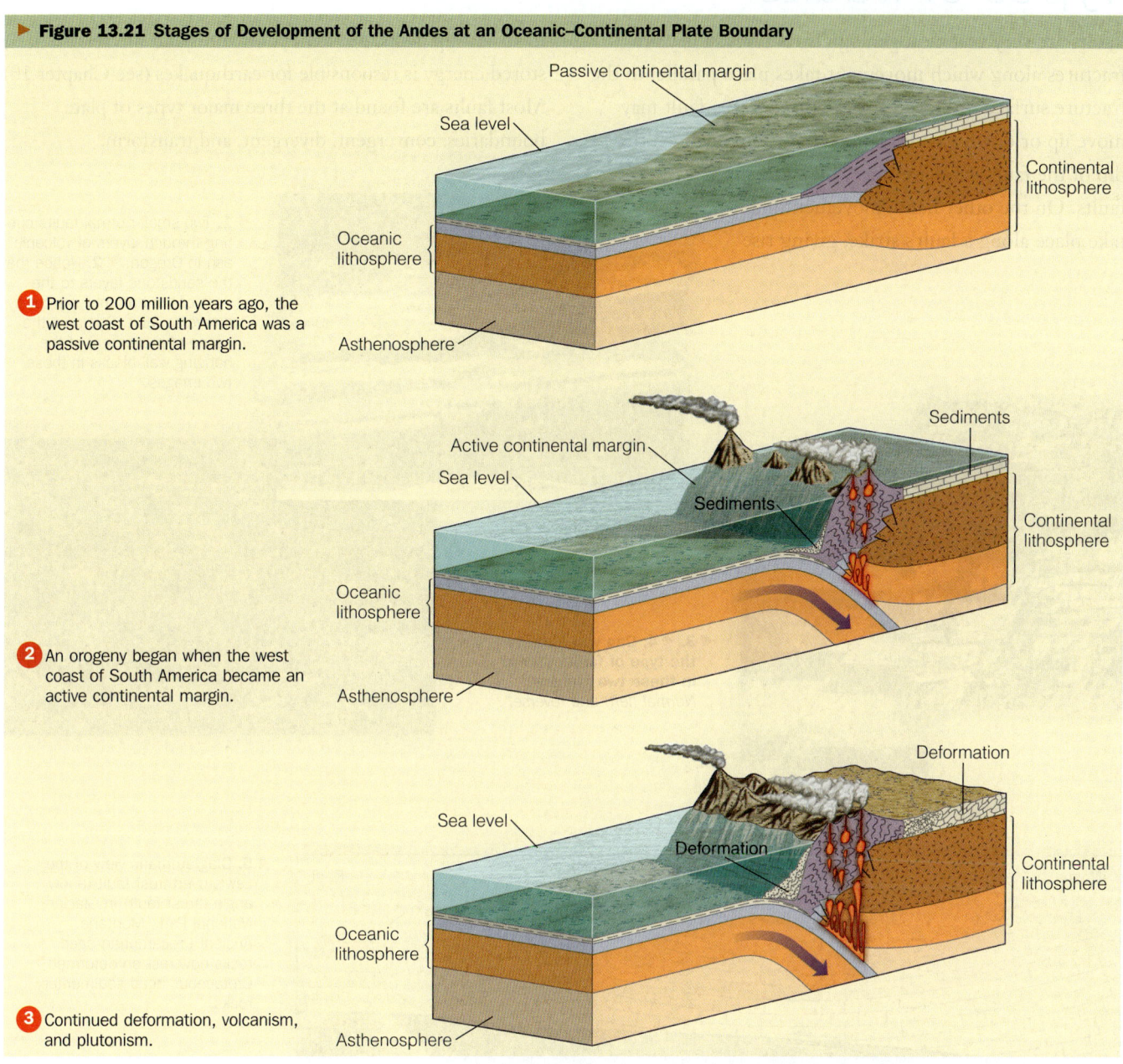

▶ **Figure 13.21** Stages of Development of the Andes at an Oceanic–Continental Plate Boundary

❶ Prior to 200 million years ago, the west coast of South America was a passive continental margin.

❷ An orogeny began when the west coast of South America became an active continental margin.

❸ Continued deformation, volcanism, and plutonism.

mentary rocks that formed in shallow seas now make up the higher part of the Himalayas.

Other mountain systems with a similar history of continental–continental collision are the Urals in Russia and the ongoing collision of the Arabian plate with Asia along the Zagros Mountains of Iran. Even the Appalachian mountain system formed in a similar manner during the assembly of Pangaea, although subsequent rifting has separated this landmass.

Terranes and the Origin of Mountains

In the preceding section, we discussed orogenies along convergent plate boundaries that result in the addition of material to the margin of a continent, a process called **continental accretion.** Much of the material added to continental margins is eroded older continental crust, but some plutonic and volcanic rocks are new additions. During the 1970s and 1980s, however, geologists discovered that parts of many mountain systems are also made up of small, accreted lithospheric blocks that clearly originated elsewhere. These **terranes,*** as they are called, are fragments of seamounts, island arcs, and small pieces of continents that were carried on oceanic plates that collided with continental plates, thus adding them to the continental margins (▶ Figure 13.23).

---

*Some geologists prefer the terms *suspect terrane*, *exotic terrane*, and *displaced terrane*. Notice the spelling of *terrane* as opposed to the more familiar *terrain*; the latter is a geographic term referring to a particular area of land.

Types of Faults

Faults are very common geologic structures. They are fractures along which movement takes place parallel to the fracture surface. A block of rock adjacent to a fault may move up or down a fault plane—that is, up or down the dip of the fault—and are thus called dip-slip faults. On the other hand, movement may take place along a fault's strike, giving rise to strike-slip faults. Movement on faults and the release of stored energy is responsible for earthquakes (see Chapter 10). Most faults are found at the three major types of plate boundaries: convergent, divergent, and transform.

◀ **1.** Two small normal faults cutting through layers of volcanic ash in Oregon. ▼ **2.** Notice that the sandstone layers to the right of the hammer are cut by a reverse fault. Compare the sense of movement of the hanging wall blocks in these two images.

◀ **3.** ◀ **4.** Can you identify the type of faults shown in these two images? *Normal (left) and reverse.*

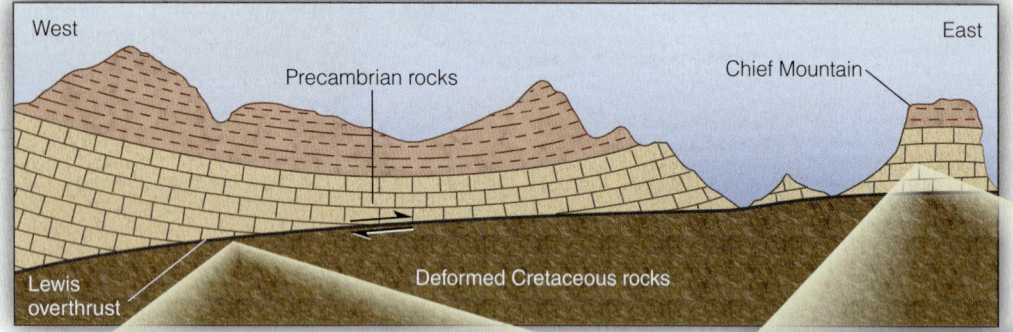

◀ **5.** Diagrammatic view of the Lewis overthrust fault (a low angle thrust fault) in Glacier National Park, Montana. Ancient Precambrian-aged rocks now rest on deformed Cretaceous-aged sedimentary rocks.

▲ **6.** View from Marias Pass reveals the fault as a light-colored line on the mountainside.

▲ **7.** Erosion has isolated Chief Mountain from the rest of the slab of overthrust rock.

Sue Monroe

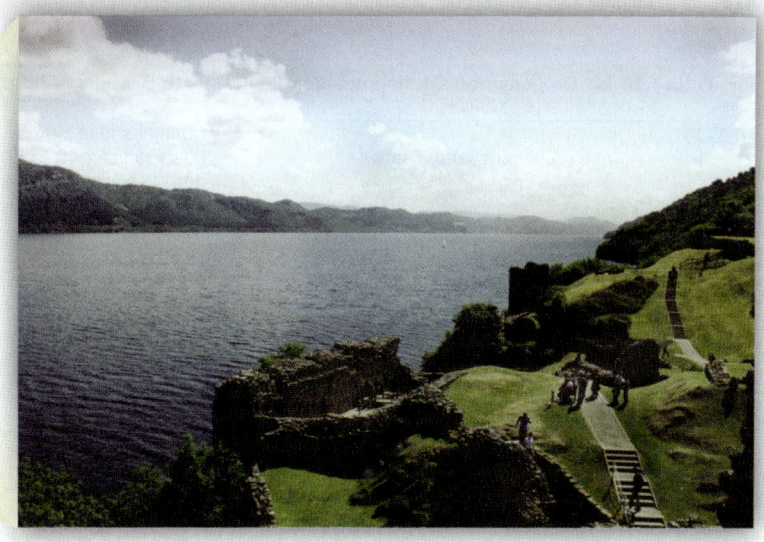

▲ **9.** View from the southwest along Loch Ness, Scotland, lying in the Great Glen fault zone, which at this point is more than 1.5 km wide.

◀ **8.** Map showing the location of the Great Glen fault, a left-lateral strike-slip fault that cuts across Scotland.

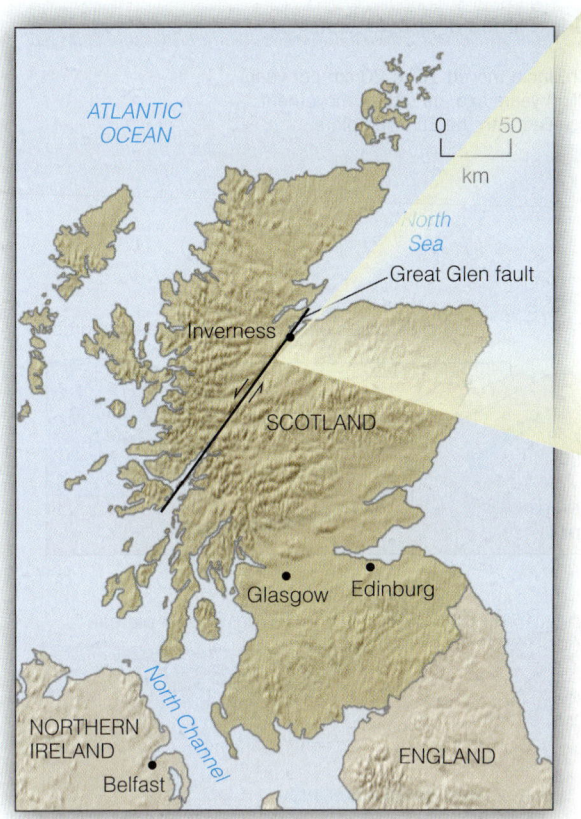

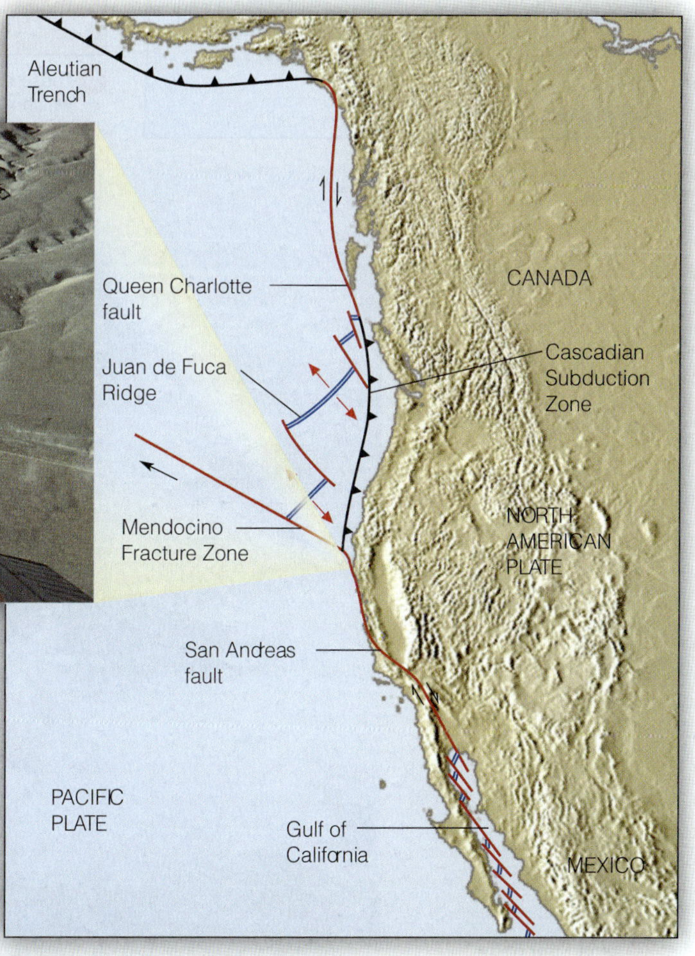

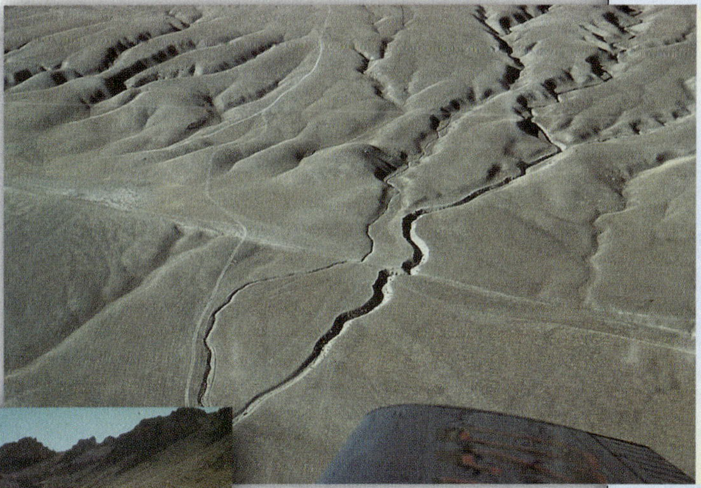

John S. Shelton

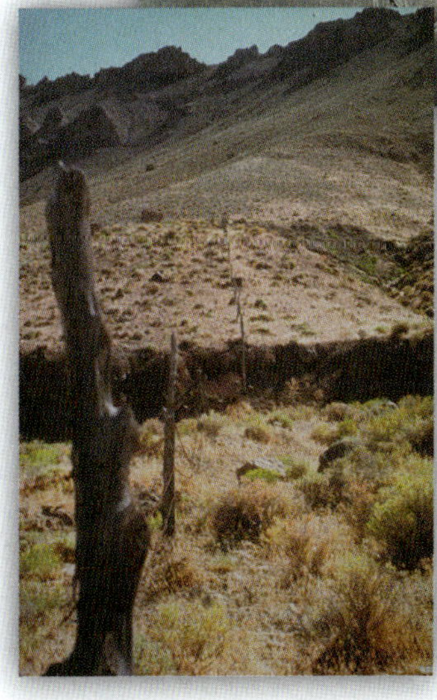

Lindie Brewer/USGS

▲ **11.** Right-lateral offset of a gully by the San Andreas fault in southern California. The gully is offset about 21 m.

◀ **12.** Oblique-slip took place on this fault in north-central Nevada during a 1915 earthquake. Notice the fence that shows right-lateral displacement and dip-slip displacement.

▲ **10.** Plate tectonic setting for the San Andreas fault, a strike-slip fault. Remember that in plate tectonics terminology this is a transform fault.

▶ **Figure 13.22 Orogeny at a Continental–Continental Plate Boundary and the Origin of the Himalayas of Asia**

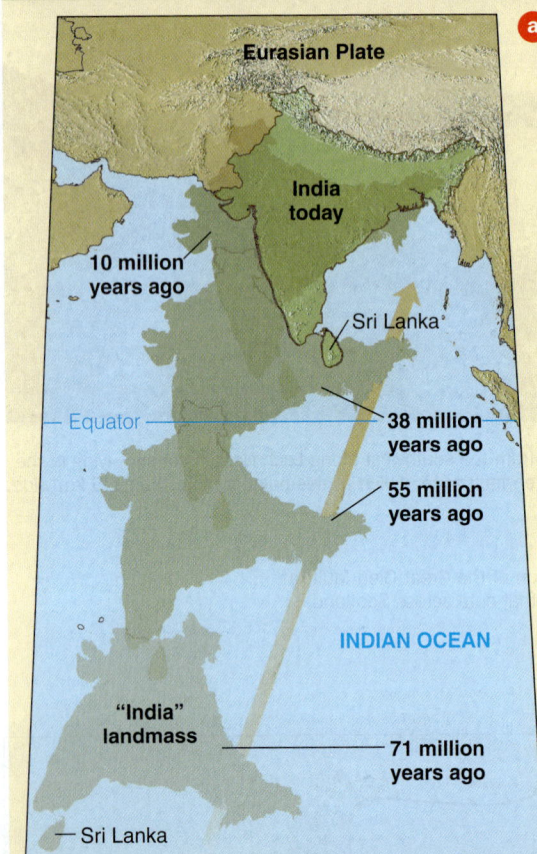

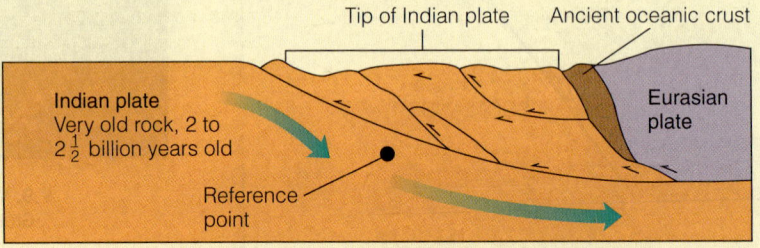

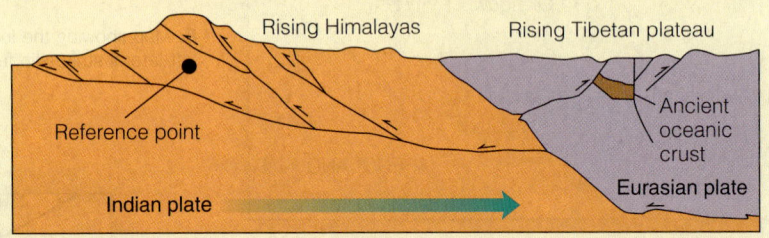

a During its long journey north, India moved 15 to 20 cm per year, but beginning 40 to 50 million years ago, its rate of movement decreased markedly as it collided with the Eurasian plate.

b These cross sections show the Indian and Eurasian plates during their collision, which resulted in uplift of the Himalayas and the Tibetan Plateau. The top illustration shows an earlier time than the lower one.

c Aerial view of the Himalaya Mountains in Tibet.

Geologic evidence indicates that much of the Pacific margin of North America from Alaska to Baja California is made up of accreted terranes or igneous intrusions. According to one estimate, more than 100 terranes have been added along the West Coast during the last 200 million years

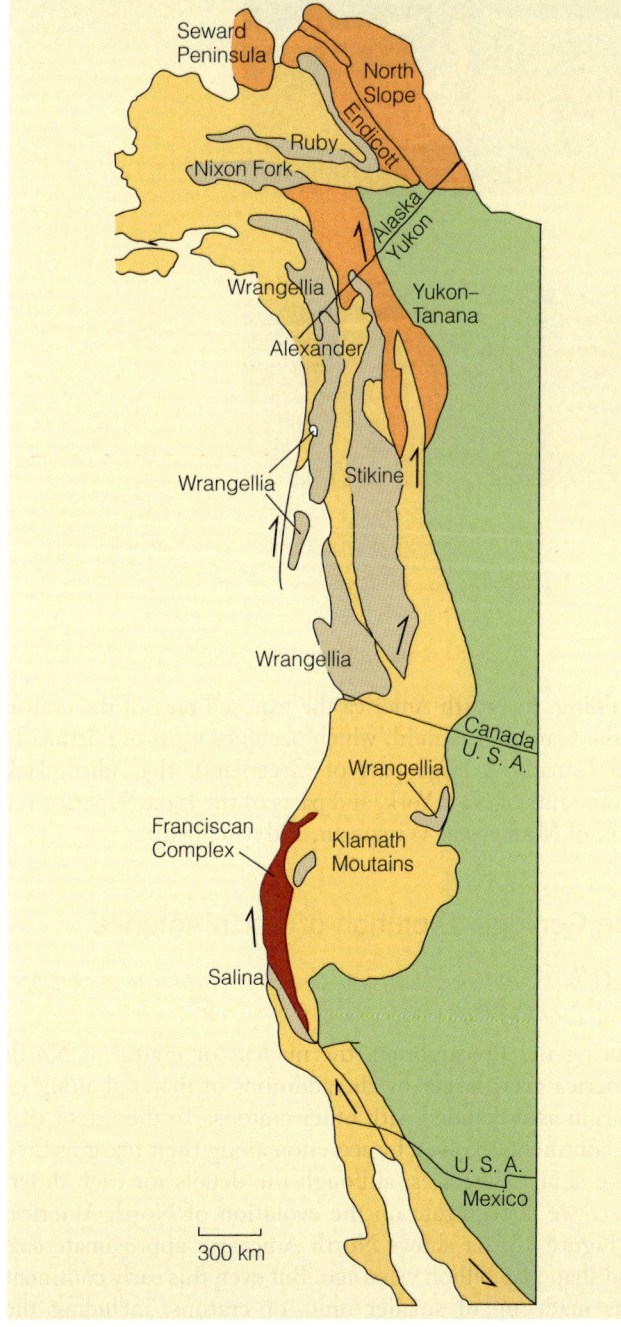

▶ **Figure 13.23 The Origin of Terranes** Some of the accreted lithospheric blocks called *terranes* that form the western margin of North America. The light brown blocks probably originated as parts of continents other than North America. The reddish brown blocks are possibly displaced parts of North America. The Franciscan Complex consists of a variety of seafloor rocks. Source: From Zvi Ben-Avraham, "The Movement of Continents," *American Scientist*, 69:291–299, Figure 9, p. 298, Journal of Sigma Xi, The Scientific Research Group.

(▶ Figure 13.23). Numerous potential terranes such as seamounts and aseismic ridges are present in the ocean basins today (see Figure 12.12).

Most terranes identified so far are in the mountains of western North America, but a number of others are probably present in other mountain systems. About a dozen terranes have been identified in the Appalachian Mountains, but they are more difficult to identify in older mountain systems.

The basics of orogenies along convergent plate boundaries discussed in the preceding sections remain unchanged, but in view of terrane accretion some details are certainly different. For one thing, accreted terranes are mostly new additions to continents rather than reworked older continental material.

Section 13.5 Summary

● The large mountain systems on the continents formed by compression at convergent plate boundaries.

● Two more or less parallel belts of deformation are found where mountains form at oceanic–oceanic plate boundaries. An orogeny at an oceanic–continental plate boundary yields a mountain system on the continental plate consisting of metamorphic rocks, plutons, deformed sedimentary rocks, and andesitic volcanoes. A collision between two continental plates yields a mountain system at the juncture of the two colliding plates, so the mountains are now located far from a plate boundary.

13.6 The Formation and Evolution of Continents

If we could somehow go back and visit Earth shortly after it formed 4.6 billion years ago, we would see a hot, barren, waterless planet bombarded by comets and meteorites, no continents, intense cosmic radiation, and ubiquitous volcanism. Judging from the oldest known rocks, the 3.96-billion-year-old Acasta Gneiss in Canada and ancient rocks in Montana, some continental crust existed by that time. And sedimentary rocks in Australia contain detrital zircons ($ZrSiO_4$) dated at 4.2 billion years, so source rocks at least this old existed.

According to one model for the origin of continental crust, the earliest crust was thin, unstable, and made up of ultramafic rocks. Upwelling basaltic magma disrupted this ultramafic crust and formed island arcs, and the ultramafic crust was destroyed at subduction zones (▶ Figure 13.24a). Collisions between island arcs formed a continental core (▶ Figure 13.24b–c), which increased in volume to form a proto-continent (▶ Figure 13.24d). The larger proto-continents also collided with one another thereby building a full-scale craton (▶ Figure 13.24e-f). By 3.96 billion years ago, several continental nuclei or cratons existed.

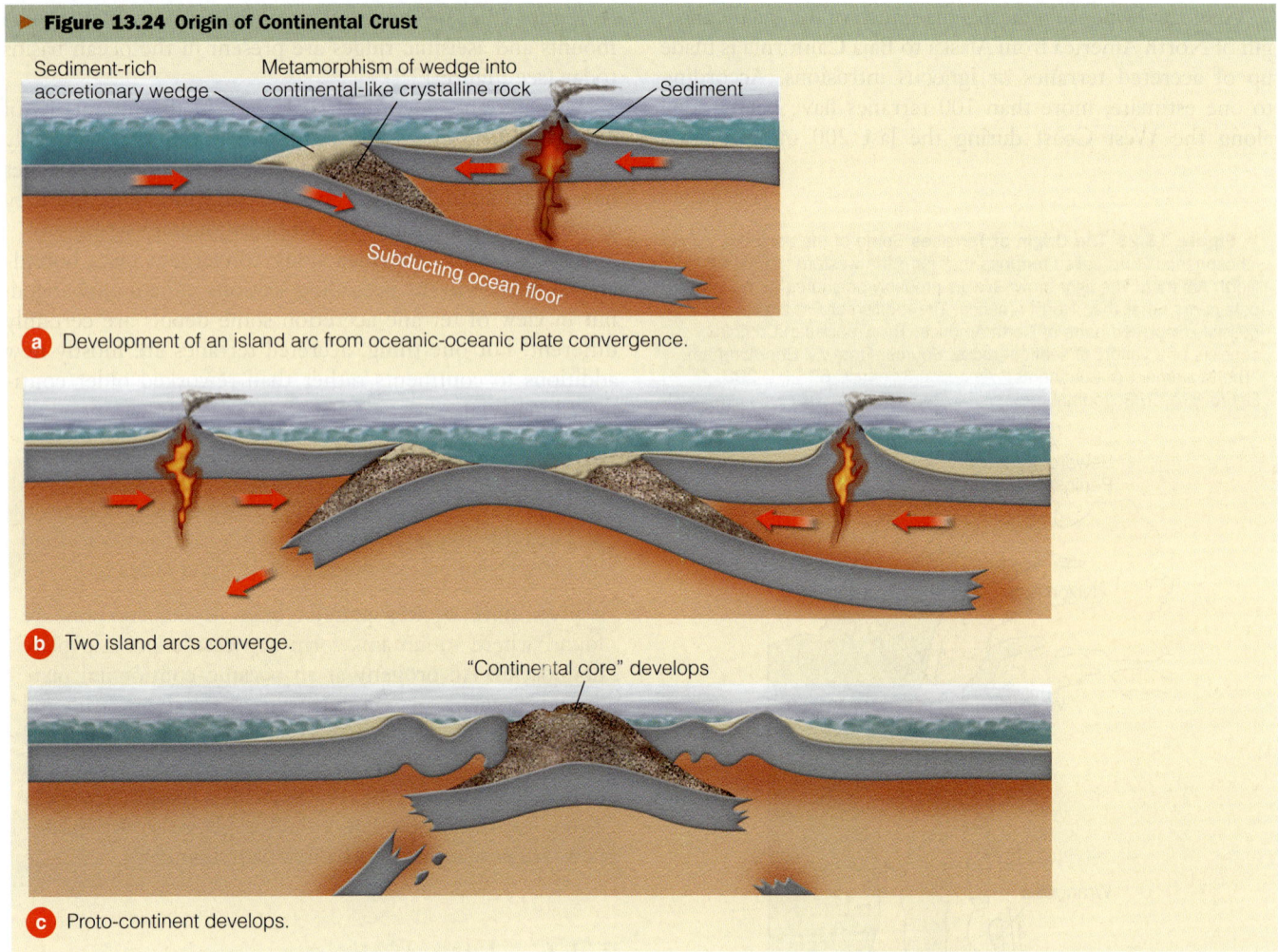

▶ **Figure 13.24 Origin of Continental Crust**

a Development of an island arc from oceanic-oceanic plate convergence.

b Two island arcs converge.

c Proto-continent develops.

Shields, Platforms, and Cratons

■ *What are shields, platforms, and cratons?*

Continents are more than simply land areas above sea level. You already know that they have an overall composition similar to granite and that continental crust is thicker and less dense than oceanic crust, which is made up of basalt and gabbro. In addition, a **shield** consisting of a vast area or areas of exposed ancient (Precambrian) rocks is found on all continents. Continuing outward from shields are broad **platforms** of buried ancient rocks, merely extensions of the shields, that underlie much of each continent. Collectively, a shield and platform made up a **craton,** which we can think of as a continent's ancient nucleus.

The cratons are the foundations of the continents, and along their margins more continental crust was added as they evolved to their present size and shapes, a phenomenon called *continental accretion.* Many of the rocks along the margins of the cratons show evidence of deformation accompanied by metamorphism, igneous activity, and mountain building. In North America the exposed part of the craton is the **Canadian shield,** which occupies most of northeastern Canada, a large part of Greenland, the Adirondack Mountains of New York, and parts of the Lake Superior region of Minnesota, Wisconsin, and Michigan.

The Geologic Evolution of North America

■ *What happened during the Precambrian evolution of North America?*

During the Precambrian, the nucleus or craton of North America grew larger by the additions of material along its margin as it collided with other cratons. To the extent that all continents evolved by accretion along their margins they have similar histories, although the details for each differ. Here we concentrate on the evolution of North America. ▶ Figure 13.25a shows North America's approximate size and shape 1.8 billion years ago. But even this early continent was made up of smaller units or cratons, including the

▶ Figure 13.24 (continued)

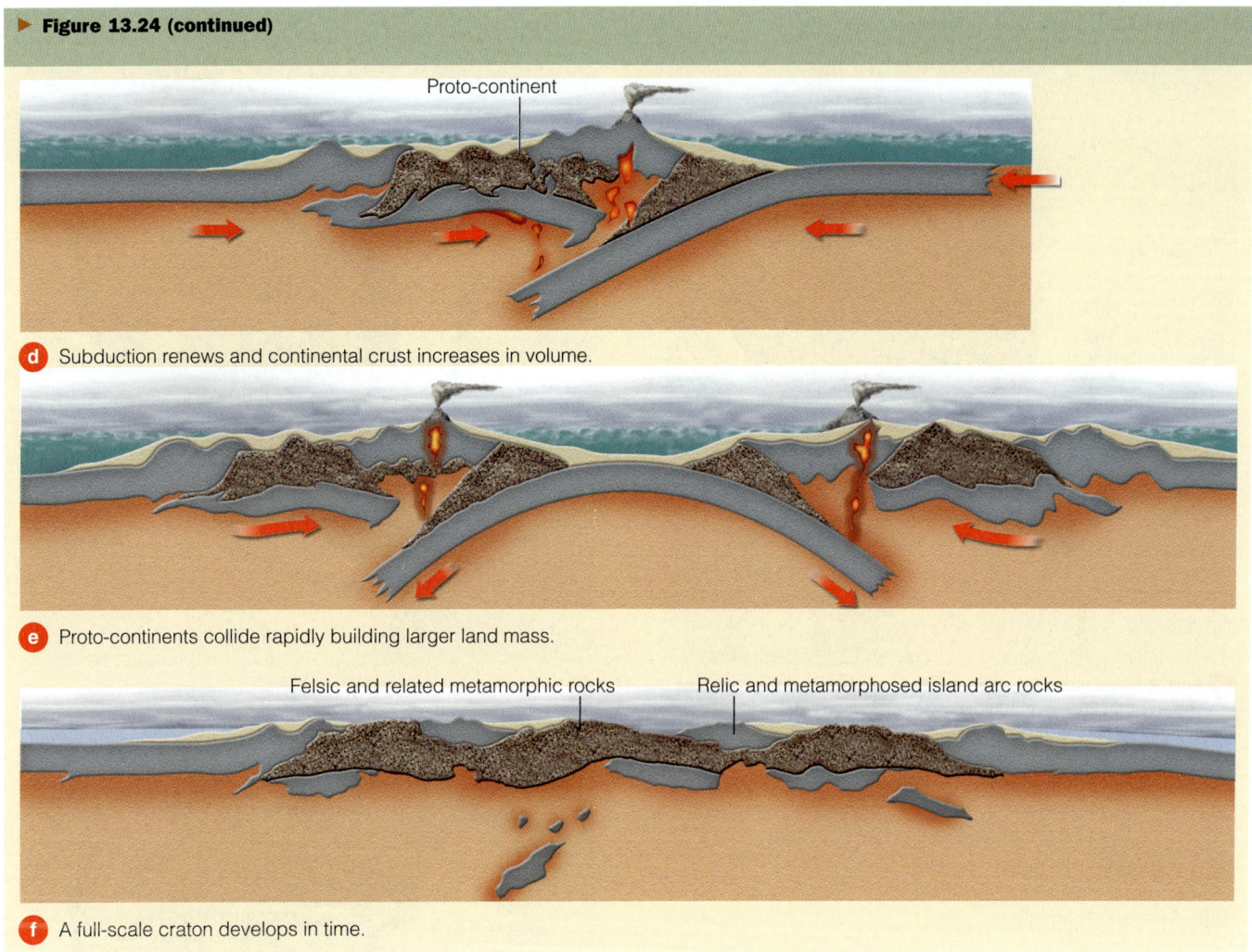

d Subduction renews and continental crust increases in volume.

e Proto-continents collide rapidly building larger land mass.

f A full-scale craton develops in time.

Superior and Wyoming cratons that had collided along deformation belts such as the Trans-Hudson orogen (remember an orogeny is an episode of mountain building). By 1.6 billion years ago, more accretion had taken place along the southern margin of the continent (▶ Figure 13.25b), and by about 900 million years ago, additional accretion along the continent's eastern margin gave rise to a landmass approximating the size and shape of that in Figure 13.25c.

The sequence of events shown in Figure 13.25 took place during that part of geologic time that geologists call the Precambrian, which ended 542 million years ago (see Figure 1.16). Since the Precambrian—that is, during the Paleozoic, Mesozoic, and Cenozoic eras—continental accretion has taken place, mostly along the eastern, southern, and western margins of the growing continent, giving rise to the present configuration of North America. During this time the craton itself has been remarkably stable, but it has been invaded six times by the seas during marine transgressions followed by regressions. It has been only mildly deformed into a number of large structural domes and basins.

The Continuing Evolution of North America

We have briefly discussed the phenomenon of continental accretion and how it accounts for the present size and configuration of North America. We ignored many important geologic events that are more appropriately covered in courses in historical geology, but we should make it clear that North America continues to evolve. For example, plate collisions are occurring along the continental margin in the Pacific Northwest from northern California to southern British Columbia, Canada, and where the Pacific plate is subducted along the Aleutian Trench of Alaska. In fact, the entire western margin of the continent was an area of plate collision until the origin of the San Andreas fault beginning about 30 million years ago (▶ Figure 13.26). Ongoing movements on the San Andreas fault and dozens of subsidiary faults account for continuing deformation, especially the origin of mountains and depressions called basins in southern California. Rifting in the Gulf of California accounts for the continuing separation of Baja California from mainland Mexico (▶ Figure 13.26).

▶ **Figure 13.25 Origin of San Andreas Fault**

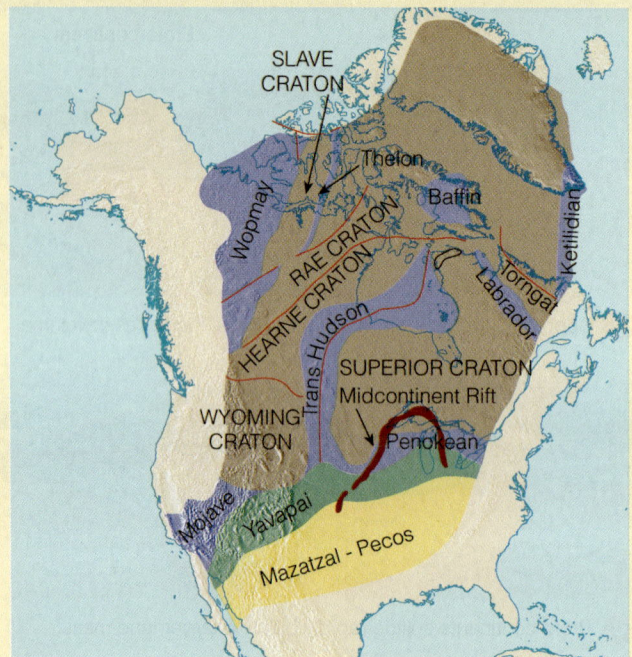

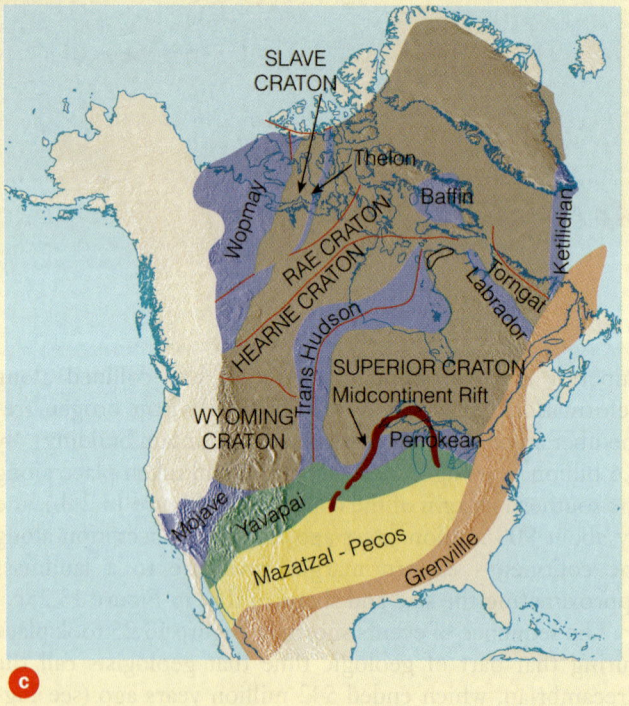

a By 1.8 billion years ago, the continent consisted of the elements shown here. The several orogens formed when older cratons collided to form a larger craton.

Years ago:
- 900 million–1.2 billion
- 1.6 billion–1.75 billion
- 1.75 billion–1.8 billion
- 1.8 billion–2.0 billion
- 2.5 billion–3.0 billion

b c Continental accretion took place along the southern and eastern margins of North America. By about 1.0 billion years ago, North America had the approximate size and shape shown in (c). The rest of North America evolved during the Phanerozoic Eon—that is, during the last 542 million years. Source: Reprinted with permission from Kent C. Condie, *Plate Tectonics and Crustal Evolution*, 4d, p. 65 (Fig. 2.26), © 1997, Butterworth-Heinemann.

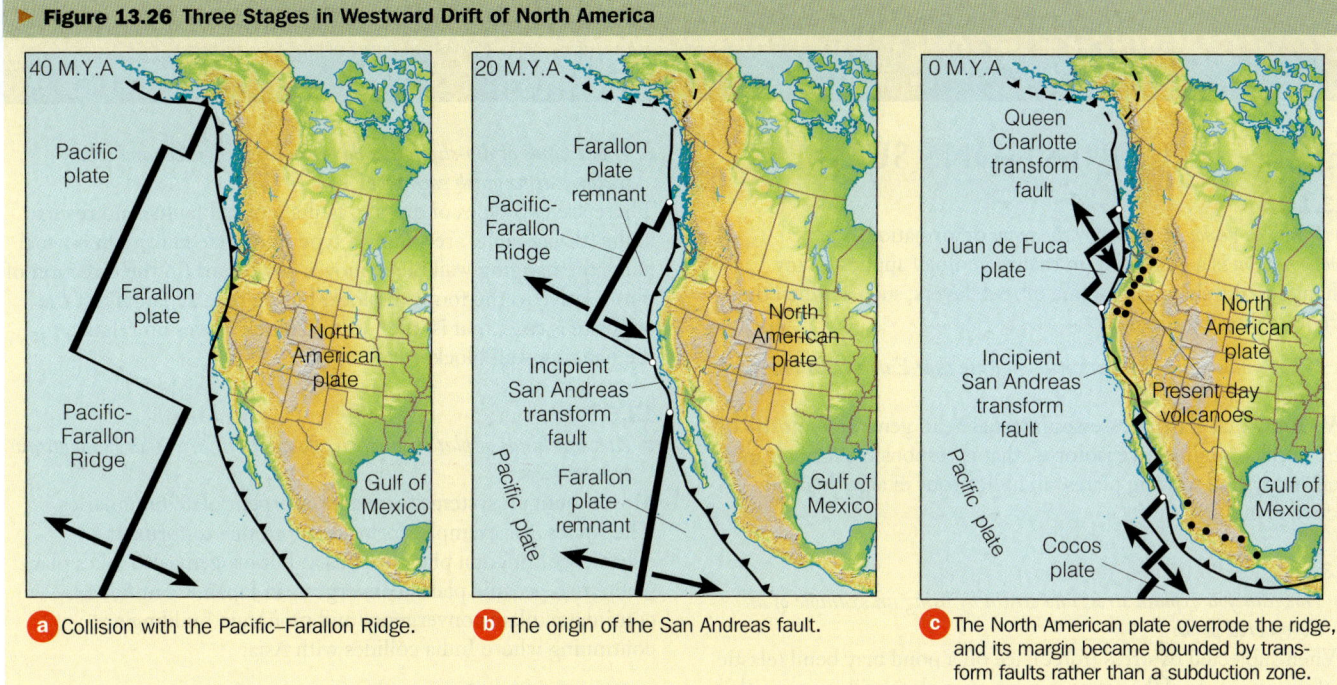

▶ **Figure 13.26** Three Stages in Westward Drift of North America

a Collision with the Pacific–Farallon Ridge.

b The origin of the San Andreas fault.

c The North American plate overrode the ridge, and its margin became bounded by transform faults rather than a subduction zone.

The continuing evolution of eastern North America and the Gulf Coast is not at all like what is going on in the west. Since Pangaea began breaking apart during the Triassic Period, eastern North America and the Gulf region have been bounded by passive continental margins. These areas are still evolving geologically. For example, erosion of the continental interior continues, and rivers and streams carry huge quantities of sediment to the continental margin where it is deposited on, for example, the Mississippi River delta. And the continental shelf, slope, and rise continue to receive sediment and grow seaward. Some of these deposits, especially on the continental shelf, are areas of prolific hydrocarbon production. In addition, the Florida section of the Gulf Coast is an area of deposition of sediments that when lithified will be limestone.

In the broader view, it is important to note that Earth remains a dynamic planet because of internal heat, the driving mechanism for plate tectonics, and processes such as running water, wind, glaciers, and waves, which continue to modify its surface. In fact, the remaining chapters in this book are concerned with the dynamic interactions among these surface processes and how they account for present-day landscapes.

Section 13.6 Summary

• Some continental crust existed by about 3.96 billion years ago as island arcs collided and formed cratons. Each continent has a craton made up of an exposed shield and a more extensive buried platform.

• The cratons of all continents grew during the Precambrian by additions of new material along their margins, a process that continues even now.

Review Workbook

ESSENTIAL QUESTIONS SUMMARY

13.1 Introduction
■ *What do geologists mean by the term* deformation?
Deformation is a general term that in geology applies to any change in the shape or volume of rock layers, such as when they are folded or fractured.

■ *What kinds of geologic activities indicate that Earth is an internally active planet?*
We know that Earth possesses internal heat, generated mostly by the decay of radioactive isotopes, that is responsible for volcanism, seismicity, and moving plates, all indications of a geologically active planet.

13.2 Rock Deformation
■ *How can you explain stress and strain by using an example of an ice-covered pond?*
When subjected to stress (force), ice on a pond may bend (elastic deformation), or if the stress is great enough, it will fracture, that is, the ice is strained or deformed in response to stress.

■ *How do compression and tension differ?*
Both are types of stress (force) applied to an area of rock, but they differ in the directions of the deforming stress. In compression the forces are directed toward one another along the same line, whereas in tension the forces along the same line act in opposite directions.

■ *What do the terms* ductile *and* brittle *indicate about rocks?*
Ductile rocks show a great amount of plastic strain before they fracture, but brittle rocks fracture after only a small amount of plastic strain.

13.3 Folded Rock Layers
■ *What is a monocline and how might one form?*
A monocline is a flexure in otherwise horizontal or uniformly dipping rock layers. Uplift of rock layers above a fault that does not penetrate to the surface is one way in which a monocline may form.

■ *How is an eroded anticline different from an eroded syncline?*
Both show similar patterns as seen from above or on a map, but in an anticline all rock layers dip outward from the center where the oldest exposed rocks are found, whereas in a syncline all rock layers dip inward toward the center where the youngest exposed rocks are located.

13.4 Joints and Faults—Deformation by Fracturing
■ *How do joints differ from faults?*
Both are fractures, but joints show no movement up, down, or sideways of the rocks on opposite sides of the fracture surface. Faults are fractures along which blocks of rock on opposite sides of the fracture have moved relative to one another.

■ *What kinds of dip-slip faults do geologists recognize, and what features distinguish each type?*
There are two types of dip-slip faults, normal faults, and reverse faults (thrust faults are simply a type of reverse fault). On normal faults the hanging wall block moves downward (in the direction of dip) relative to the footwall block, but on reverse faults just the opposite is true, that is, the footwall block moves up relative to the hanging wall block.

13.5 Deformation and the Origin of Mountains
■ *At what kinds of plate boundaries do most of Earth's large mountain systems form?*
Most mountain systems form at convergent plate boundaries. The Andes, for example, form, and continue to form, at an oceanic–continental plate boundary, but orogeny also takes place where two oceanic plates converge as in Japan. Continental–continental plate convergence and mountain building are continuing where India collides with Asia.

■ *What sequence of events takes place during an orogeny at an oceanic–continental plate boundary?*
Where an oceanic plate is subducted beneath a continental plate, as along the west coast of South America, Earth's crust is thickened as batholiths are emplaced and rocks along the continental margin are deformed by compression. As a result, the mountains have a central core of granitic rocks flanked on the seaward side by an accretionary wedge and on the other side by deformed sedimentary rocks.

13.6 The Formation and Evolution of Continents
■ *What are shields, platforms, and cratons?*
A craton, defined as the ancient stable nucleus of a continent, is made up of a shield, which is the exposed part of a craton, and a platform which is the extension of a shield that is buried beneath younger rocks.

■ *What happened during the Precambrian evolution of North America?*
During the Precambrian new material was added to North America along its margins by continental accretion. But even at the end of the Precambrian, about 542 million years ago, the continent had a different size and shape than it has now.

ESSENTIAL TERMS TO KNOW

anticline (p. 395)
basin (p. 397)
Canadian shield (p. 414)
compression (p. 391)
continental accretion (p. 409)
craton (p. 414)
deformation (p. 390)
dip (p. 393)
dip-slip fault (p. 402)
dome (p. 397)
elastic strain (p. 392)
fault (p. 400)
fault plane (p. 400)

fold (p. 394)
footwall block (p. 401)
fracture (p. 392)
geologic structure (p. 393)
hanging wall block (p. 401)
joint (p. 398)
monocline (p. 395)
normal fault (p. 402)
oblique-slip fault (p. 403)
orogeny (p. 405)
plastic strain (p. 392)
platform (p. 414)
plunging fold (p. 396)

reverse fault (p. 402)
shear stress (p. 391)
shield (p. 414)
strain (p. 391)
stress (p. 391)
strike (p. 393)
strike-slip fault (p. 403)
syncline (p. 395)
tension (p. 391)
terrane (p. 409)
thrust fault (p. 402)

REVIEW QUESTIONS

1. What information would you need to resolve whether the fold in Figure 13.11d is a recumbent anticline or a recumbent syncline?

2. Draw a surface view of a plunging anticline and an adjacent plunging syncline. Show strike and dip symbols and indicate on your diagram which rock layer is oldest and which is youngest.

3. What is meant by the term *terrane?* How are terranes incorporated into continents?

4. Describe how time, rock type, pressure, and temperature influence rock deformation.

5. Suppose rocks were displaced 200 km along a strike-slip fault during a period of 5 million years. What was the average rate of movement per year? Is the average likely to represent the actual rate of displacement on this fault? Explain.

6. What kinds of evidence would you look for in the Canadian shield, where there are no mountains, to verify that mountain building took place there long ago?

7. How does a structural basin differ from a syncline?

8. Suppose that an area along a continent's margin was subjected to compression for millions of years. What kinds of geologic structures would you expect to find in this area?

APPLY YOUR KNOWLEDGE

1. Most of Earth's internal heat is generated by decay of radioactive isotopes, especially isotopes of potassium and uranium. Certainly the amount of internally generated heat will diminish in the far, far distant future and activities such as volcanism and moving plates will cease. Why will this occur? (*Hint:* Refer to the discussion of radioactive decay in Chapter 9.)

2. Is there any connection between the principle of isostasy and mountain building? If so, what?

FIELD QUESTIONS

1. The geologic cross section below shows a view of several rock layers and a fault. Your task is to decipher the history of this area. What event took place first, second, and so on? What kind of fold and fault are present? To fully answer these questions, refer to the discussions of sedimentary structures in Chapter 7 and superposition in Chapter 9.

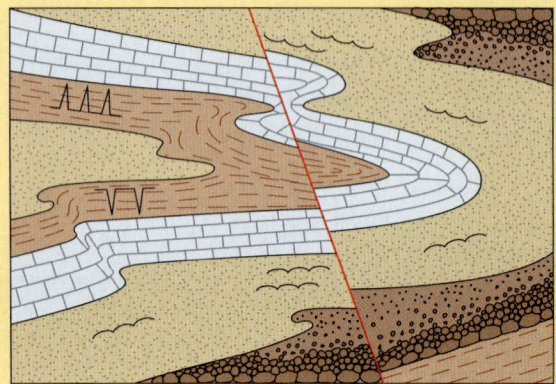
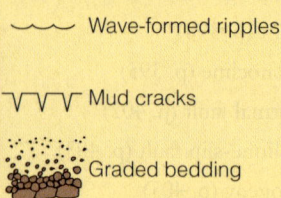

2. Look at the image of the fault to the right. If you examine it closely, you should be able to determine which side of the fault moved relatively down and whether it is a normal or reverse fault. Reference to Figure 13.16b should help you solve this problem.

GEOLOGY MATTERS

GEOLOGY IN UNEXPECTED PLACES

Ancient Ruins and Geology

You cannot help but be impressed by the colossal stone structures erected by ancient Egyptians, Greeks, and Romans. Although stone is very heavy as a building material, it is readily available in many areas and it is strong, but its strength depends on how it is used. For instance, stone chimneys, fences, walls, and pyramids are built so that all stresses act vertically, with stones below supporting the weight of those above; that is, they are in compression.

In ancient Egyptian, Greek, and Roman temples and other buildings, the builders used horizontal stone beams supported by closely spaced columns (▶ Figure 1). Wouldn't it have been more efficient to use longer beams and eliminate every other column? Of course longer beams would weigh more and be more difficult to place in position, but the builders could have made longer beams more slender. This wouldn't work either. The problem is that stone is not very strong when it is in tension. If the beams spanned a great distance, they would collapse under their own weight because they would be compressed on their upper sides but subjected to tension on their undersides. The stress on the top of a beam is about the same as it is on the bottom, but in this configuration it adds little to the beam's overall strength. So, if the beams were used in this way, tension fractures would form on the bottom and travel upward to the top, causing failure.

The same consideration of stone's strength in tension holds true for rocklike substances such as concrete. Concrete beams and floors in buildings as well as slabs of concrete in elevated highways must be reinforced with steel rods or wire mesh placed on their sides in tension. If the rods or wire mesh were placed in the part of a concrete slab in compression, they would add little to its strength.

Stone as a building material also has limitations in long structures such as aqueducts and bridges. The Romans solved this problem by developing the *Roman arch,* in which all structural elements are in compression. In this arrangement, stress is directed downward and outward toward the adjacent arch, and the outermost arches are anchored to bedrock or some kind of supporting structure (▶ Figure 2). Unfortunately for the Romans, though, the arches could not be very large because they had to be supported by some kind of framework until the last stone, or keystone, was in place.

Even the magnificent Gothic cathedrals in Europe depend on a modification of the Roman arch. To build the huge cathedrals with large, open interiors, the high, steep *Gothic arch* was developed, in which all of the elements are in compression. Just as in the Roman arch, stress is directed downward and outward, but cathedrals cannot be anchored to bedrock, so *buttresses* were built to oppose the outward stress from the arches (▶ Figure 3).

Because of their tremendous weight and huge compressive stresses, buildings of stone or brick are limited to only a few stories. In contrast, structural steel is comparatively lightweight and equally strong in compression and tension. With its development, builders could make truly large structures such as skyscrapers and long suspension bridges. Stone, however, remains popular for use as decorative facing stones on buildings and for various other uses, such as mantelpieces, tabletops, and paving stones for walkways. Of course stone is used extensively for gravestones, mausoleums, and monuments.

Figure 1
The Temple of Amun at Luxor, Egypt. Notice that the horizontal beams are supported by closely spaced columns.

Figure 2
Roman arches were constructed so that each arch was in compression and the outermost arches were anchored to bedrock. Arches could not be very large, though, because they were not self-supporting until a keystone was in place.

Figure 3
Part of Notre Dame Cathedral in Paris, France. Built between 1163 and 1330, its huge arches are supported by buttresses.

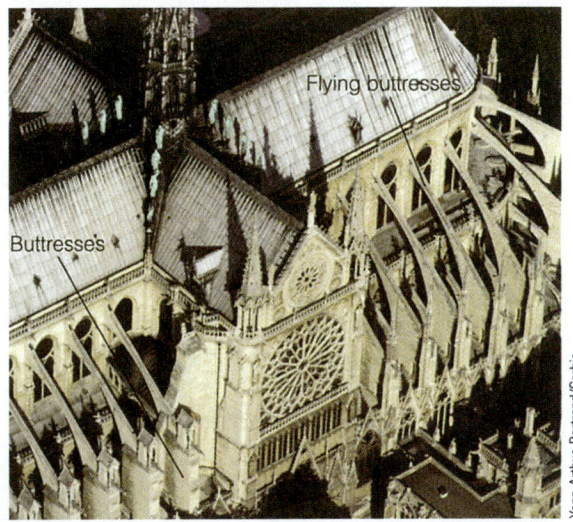

GEOLOGY MATTERS

GEOLOGY IN FOCUS   Geologic Maps—Their Construction and Uses

Let us suppose you own property near another property on which some valuable resources were discovered (▶ Figure 1). The rock with the resources is not exposed on your property but it may lie beneath the surface, so drilling or excavating to the rock layer, if present, may be very lucrative. How will you decide whether to embark on this venture? Figure 1 is a geologic map—that is, a map using lines, symbols for strike and dip, and colors to depict the surface distribution of rocks, show age relationships among rocks, and delineate geologic structures such as folds and faults. Given the information in the map, can you determine whether the resource-bearing rock layer is beneath your property?

The resource-bearing rock is indeed below your property, but you still must decide whether it is worth the time, effort, and expense to dig down to it. In other words, how far below the surface does it lie? To answer this question let us examine in more detail how geologic maps are constructed and used.

Most geologic maps are printed on some kind of base map—that is, a map showing geographic locations and perhaps elevations. Like all maps they have a scale and a legend to explain the colors and symbols used. The area depicted on a geologic map depends on the purpose of the original study. Some maps show entire countries or continents, which are useful for showing the regional distribution of rocks and large-scale geologic structures, but geologic maps of much smaller areas are of more immediate use (▶ Figure 2).

Constructing a geologic map is simple in principle but not always easy in practice. Geologists visit surface rock exposures (outcrops) where they record strike and dip, rock composition and internal structure, fossils, if present, and the relative ages of the rocks in the area. Notice in Figure 2 that outcrops are discontinuous, with most of the area covered by vegetation and soil. Nevertheless, the outcrops allow us to infer what lies beneath the surface between outcrops. Thus the completed geologic map shows the area as if there were no soil cover (▶ Figure 2).

Having completed a geologic map, geologists determine the geologic history of the area shown. Because strike and dip are recorded at many locations, geologists construct cross sections, thereby illustrating three-dimensional relationships among the various rocks. As you might expect, cross sections are useful in evaluating the potential for recovering resources.

Besides geologists, people in several other professions use geologic maps. You, for example, may one day be a member of a safety planning board of a city with known active faults. Geologic maps would be useful in developing zoning regulations and construction codes. For example, suppose that an active fault is present in your area. Approving a housing development in the fault area would not be prudent, but the area might be perfectly satisfactory for agriculture or some other land use. Perhaps you will become a land-use planner or member of a county commission charged with selecting a site for a sanitary landfill or securing an adequate supply of groundwater for your community. Geologic maps would almost certainly be used in these endeavors.

Geologic maps are used extensively in planning and constructing dams, choosing the best route for a highway through a mountainous region, and selecting the sites for nuclear reactors. It would certainly be unwise to build a dam in a valley with a known active fault or with rocks too weak to support such a large structure. And in areas of continuing volcanic activity, geologic maps provide essential information about the kinds of volcanic processes that might occur in the future, such as lava flows, ashfalls, or mudflows. Recall from Chapter 5 that one way in which volcanic hazards are assessed is by mapping and determining the geologic history of an area.

Figure 1
This geologic map is a surface view of an area in which valuable natural resources were discovered.

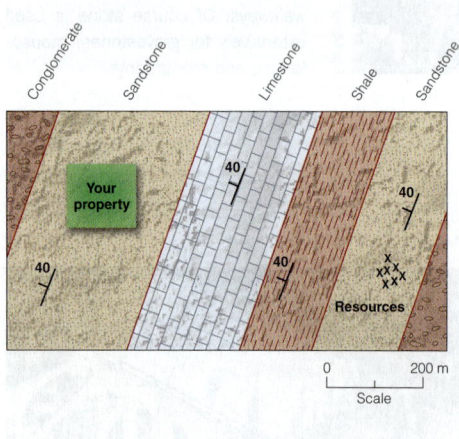

Figure 2
Construction of a geologic map and cross section from surface rock exposures (outcrops). **(a)** Valley with outcrops. In much of the area, the rocks are covered by soil. **(b)** Data from the outcrops are used to infer what is present in the covered areas between outcrops. The lines shown represent boundaries between different types of rock. **(c)** A geologic map (top) showing the areas as if the soil had been removed. Strike and dip would be recorded at many places, but only two are shown here. The orientation of the rocks as recorded by strike and dip is used to construct the cross section (bottom).

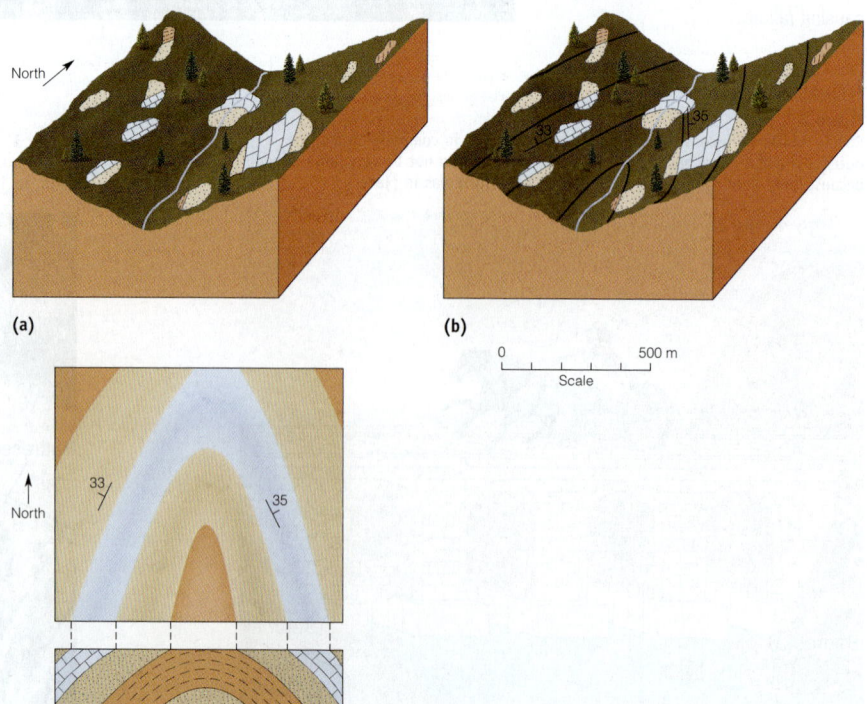

Geology AT WORK

GEOLOGY MATTERS

Engineering and Geology

As you might expect, engineering geologists apply the principles and concepts of geology to engineering practices. Geologists with this specialized training may be involved in slope stability studies and in studies of acceptable areas for power plants, for highways in mountainous regions, for tunnels and canals, and for structures to protect riverbank and seashore communities.

A good example is the concern prior to building the Mackinac (pronounced "mack-in-aw") Bridge, a huge suspension bridge that connects the Upper and Lower Peninsulas of Michigan (▶ Figure 1). Geologists and engineers were aware that some of the rock in the area, called the Mackinac Breccia, was a collapse breccia, or rubble that formed when caverns collapsed. The concern was whether the breccia or any uncollapsed caverns beneath the area would support the weight of the huge piers and abutments for the bridge. Obviously the project was completed successfully, but detailed studies were done before construction began.

Geologic engineers are invariably involved in planning for large-scale structures such as bridges, dams, power plants, and highways, especially in tectonically active regions. For example, the bridge in ▶ Figure 2 is only a short distance from the San Andreas fault, so engineers had to take into account the near certainty that it would be badly shaken during an earthquake. Many other structures on or near the San Andreas fault were constructed when codes were much less stringent, and now they are being retrofitted to make them safer during earthquakes.

Being aware of a problem and taking remedial action sometimes come too late. For instance, engineers were aware that the Santa Monica Freeway in the Los Angeles area would likely be damaged during an earthquake and retrofitting was scheduled for February 1994. Unfortunately, the Northridge earthquake struck on January 17, 1994, and part of the freeway collapsed. Of course, this and other similar events provide important information that can be incorporated into engineering practice to make freeways, buildings, and bridges safer during earthquakes.

Natural hazards include wildfires, swarms of insects, and severe weather phenomena as well as various geologic hazards such as flooding, landslides, volcanic eruptions, earthquakes, tsunami, land subsidence, soil creep, and radon gas. Geologic hazards account for thousands of fatalities and billions of dollars in property damages every year, and whereas the incidence of hazards has not increased, fatalities and damages have grown because more and more people live in disaster-prone areas.

We cannot eliminate geologic hazards, but we can better understand these events, design structures to protect shoreline communities and withstand shaking during earthquakes, enact zoning and land-use regulations, and at the very least decrease the amount of damage and human suffering. Unfortunately, geologic information that is readily available is often ignored or overlooked. A case in point is the Turnagain Heights subdivision in Anchorage, Alaska, that was so heavily damaged when the soil liquefied during the 1964 earthquake (see Figure 14.19). Not only were reports on soil stability ignored or overlooked before homes were built there, but since 1964 new homes have been built on part of the same site!

In many mountainous or even hilly areas, highways are notoriously unstable and slump or slide from hillsides. When this happens, engineering geologists are consulted for their recommendations for stabilizing slopes. They may suggest building retaining walls or drainage systems to keep the slopes dry, planting vegetation, or in some cases simply rerouting the highway if it is too costly to maintain in its present position.

As you read the following chapters on surface processes, keep in mind that engineering geologists are involved in many aspects of stabilizing slopes, designing and constructing dams and flood-control projects, and building structures to protect seaside communities.

Figure 1
Before the Mackinac Bridge that connects the Lower and Upper Peninsulas of Michigan could be built, geologists and engineers had to determine whether the Mackinac Breccia could support such a large structure.

Figure 2
This freeway crosses a small valley a short distance from the San Andreas fault in California. Engineers had to take into account the near certainty that this structure would be badly shaken during an earthquake.

Chapter 14

Mass Wasting

Fallingwater
Is the house part of Earth, or is Earth part of the house? Both.
"Fallingwater" was Frank Lloyd Wright's architectural recipe that successfully stirred together two disparate ingredients: a house and a waterfall. He created a living environment in which the lines between people and Earth were blurred. Suspended over the waterfall, the house mirrors the rhythm of the sandstone shelves it is fastened to. Anchoring "Fallingwater" to the rock ledges helps prevent it from falling into the spill of water below.
—A. W.

Fallingwater, Courtesy of the Western Pennsylvania Conservancy

ESSENTIAL QUESTIONS TO ASK

14.1 Introduction
- What is mass wasting?

14.2 Factors That Influence Mass Wasting
- What is a slope's shear strength and how is it related to the force of gravity?
- What are the major factors that cause mass wasting?
- How does the relationship between the topography and geology of an area affect slope stability?
- What are the triggering mechanisms of mass wasting?

14.3 Types of Mass Wasting
- How are mass movements classified?
- What are rockfalls?
- What are slides and how are they recognized?
- What are the different types of flows?
- What are complex movements?

14.4 Recognizing and Minimizing the Effects of Mass Movements
- What can be done to eliminate or minimize the effects of mass wasting?

GEOLOGY MATTERS

GEOLOGY IN FOCUS:
The Vaiont Dam Disaster

GEOLOGY IN UNEXPECTED PLACES:
New Hampshire Says Good-Bye to the "Old Man"

GEOLOGY IN YOUR LIFE:
Southern California Landslides

This icon, which appears throughout the book, indicates an opportunity to explore interactive tutorials, animations, or practice problems available on the Physical GeologyNow website at **http://now.brookscole.com/phygeo6**.

425

14.1 Introduction

Triggered by relentless torrential rains that began in December 1999, the floods and mudslides that devastated Venezuela were some of the worst ever to strike that country. Although a reliable death toll is impossible to determine, it is estimated that at least 19,000 people were killed, as many as 150,000 were left homeless, 35,000 to 40,000 homes were destroyed or buried by mudslides, and between $10 billion and $20 billion in damage was done before the rains and slides abated. It is easy to cite the numbers of dead and homeless, but the human side of the disaster was most vividly brought home by a mother who described standing helplessly by and watching her four small children buried alive in the family car as a raging mudslide carried it away.

Mudslides engulfed and buried not only homes, buildings, and roads, but also entire communities. Some areas were covered with as much as 7 m of mud. In addition, flooding and the accompanying mudslides swept away large parts of many of Venezuela's northern coastal communities, leaving huge areas uninhabitable.

This terrible tragedy illustrates how geology affects all of our lives. The underlying causes of the mudslides in Venezuela can be found anywhere in the world. In fact, *landslides* (a general term for mass movements) cause, on average, between 25 and 50 deaths and more than $2 billion in damage annually in the United States. By being able to recognize and understand how landslides occur and what the result may be, we can find ways to reduce hazards and minimize damage in terms of both human suffering and misery as well as property damage.

■ What is mass wasting?

Mass wasting (also called *mass movement*) is defined as the downslope movement of material under the direct influence of gravity. Most types of mass wasting are aided by weathering and usually involve surficial material. The material moves at rates ranging from almost imperceptible, as in the case of creep, to extremely fast, as in a rockfall or slide. Although water can play an important role, the relentless pull of gravity is the major force behind mass wasting.

Mass wasting is an important geologic process that can occur at any time and almost any place. It is thus important to study this phenomenon because it affects all of us, no matter where we live (Table 14.1). Although all major landslides have natural causes, many smaller ones are the result of human activity and could have been prevented or their damage minimized.

Section 14.1 Summary

● Mass wasting, also known as mass movement, is the downslope movement of material under the direct influence of gravity. Mass wasting is an important geologic process that can occur almost anywhere and at any time and is responsible for the loss of many lives and extensive property damage annually.

Table 14.1

Selected Landslides, Their Cause, and the Number of People Killed

Date	Location	Type	Deaths
218 B.C.	Alps (Europe)	Avalanche—destroyed Hannibal's army	18,000
1556	China (Hsian)	Landslides—earthquake triggered	1,000,000
1806	Switzerland (Goldau)	Rock slide	457
1903	Canada (Frank, Alberta)	Rock slide	70
1920	China (Kansu)	Landslides—earthquake triggered	~200,000
1941	Peru (Huaraz)	Avalanche and mudflow	7000
1962	Peru (Mt. Hauscarán)	Ice avalanche and mudflow	~4000
1963	Italy (Vaiont Dam)	Landslide—subsequent flood	~3000
1966	United Kingdom (Aberfan, South Wales)	Debris flow—collapse of mining-waste tip	144
1970	Peru (Mt. Hauscarán)	Rockfall and debris avalanche—earthquake triggered	25,000
1981	Indonesia (West Irian)	Landslide—earthquake triggered	261
1987	El Salvador (San Salvador)	Landslide	1000
1989	Tadzhikistan	Mudflow—earthquake triggered	274
1994	Colombia (Paez River Valley)	Avalanche—earthquake triggered	>300
1999	Venezuela	Mudflow	>19,000
2005	U.S.A. (Southern California)	Mudflow	10

Source: Data from J. Whittow, *Disasters: The Anatomy of Environmental Hazards* (Athens: University of Georgia Press, 1979); Geotimes; Earth; and U.S.G.S.

14.2 Factors That Influence Mass Wasting

■ *What is a slope's shear strength and how is it related to the force of gravity?*

When the gravitational force acting on a slope exceeds its resisting forces, slope failure (mass wasting) occurs. The resisting forces that help to maintain slope stability include the slope material's strength and cohesion, the amount of internal friction between grains, and any external support of the slope (▶ Figure 14.1). These factors collectively define a slope's **shear strength.**

Opposing a slope's shear strength is the force of gravity. Gravity operates vertically but has a component acting parallel to the slope, thereby causing instability (▶ Figure 14.1). The steeper a slope's angle, the greater the component of force acting parallel to the slope, and the greater the chance for mass wasting. The steepest angle that a slope can maintain without collapsing is its *angle of repose*. At this angle, the shear strength of the slope's material exactly counterbalances the force of gravity. For unconsolidated material, the angle of repose normally ranges from 25 to 40 degrees. Slopes steeper than 40 degrees usually consist of unweathered solid rock.

All slopes are in a state of *dynamic equilibrium*, which means that they are constantly adjusting to new conditions. Although we tend to view mass wasting as a disruptive and usually destructive event, it is one of the ways that a slope adjusts to new conditions. Whenever a building or road is constructed on a hillside, the equilibrium of that slope is affected. The slope must then adjust, perhaps by mass wasting, to this new set of conditions.

■ *What are the major factors that cause mass wasting?*

Many factors can cause mass wasting: a change in slope angle, weakening of material by weathering, increased water content, changes in the vegetation cover, and overloading. Although most of these are interrelated, we will examine them separately for ease of discussion, but we will also show how they individually and collectively affect a slope's equilibrium.

Slope Angle

Slope angle is probably the major cause of mass wasting. Generally speaking, the steeper the slope, the less stable it is. Therefore steep slopes are more likely to experience mass wasting than gentle ones.

A number of processes can oversteepen a slope. One of the most common is undercutting by stream or wave action (▶ Figure 14.2). This process removes the slope's base, increases the slope angle, and thereby increases the gravitational force acting parallel to the slope. Wave action, especially during storms, often results in mass movements along the shores of oceans or large lakes (▶ Figure 14.3).

Excavations for road cuts and hillside building sites are another major cause of slope failure (▶ Figure 14.4). Grading the slope too steeply or cutting into its side increases the stress in the rock or soil until it is no longer strong enough to remain at the steeper angle and mass movement ensues. Such action is analogous to undercutting by streams or waves and has the same result, thus explaining why so many mountain roads are plagued by frequent mass movements.

Weathering and Climate

Mass wasting is more likely to occur in loose or poorly consolidated slope material than in bedrock. As soon as rock is exposed at Earth's surface, weathering begins to disintegrate and decompose it, reducing its shear strength and increasing its susceptibility to mass wasting. The deeper the weathering zone extends, the greater the likelihood of some type of mass movement.

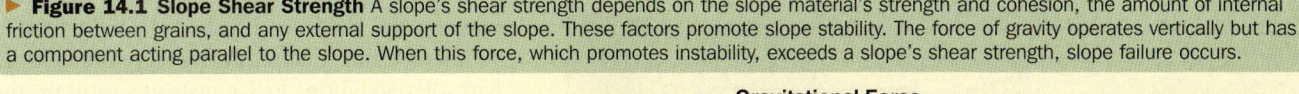

▶ **Figure 14.1 Slope Shear Strength** A slope's shear strength depends on the slope material's strength and cohesion, the amount of internal friction between grains, and any external support of the slope. These factors promote slope stability. The force of gravity operates vertically but has a component acting parallel to the slope. When this force, which promotes instability, exceeds a slope's shear strength, slope failure occurs.

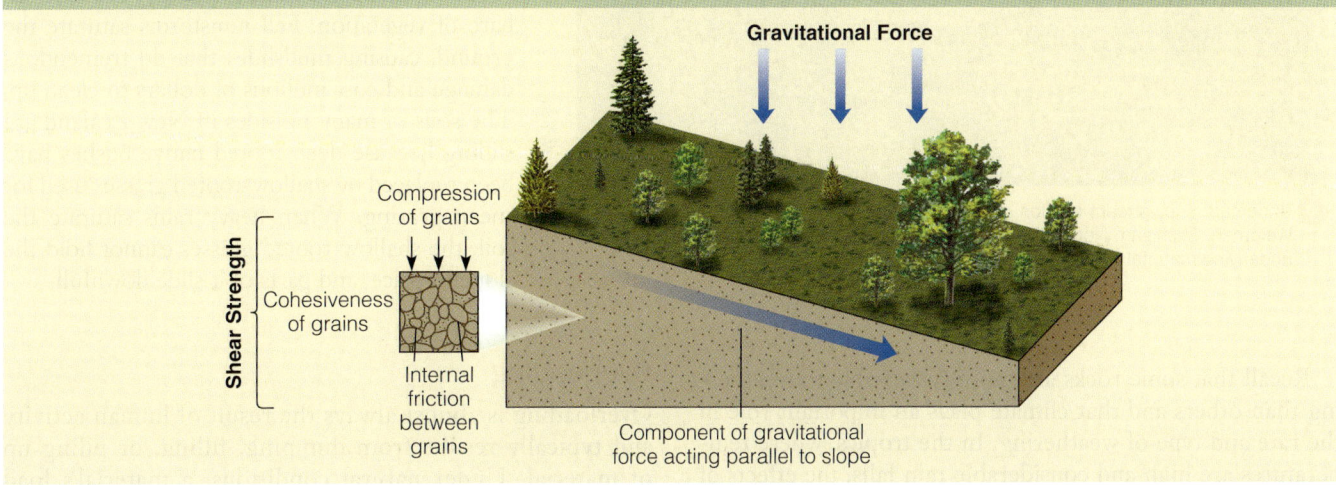

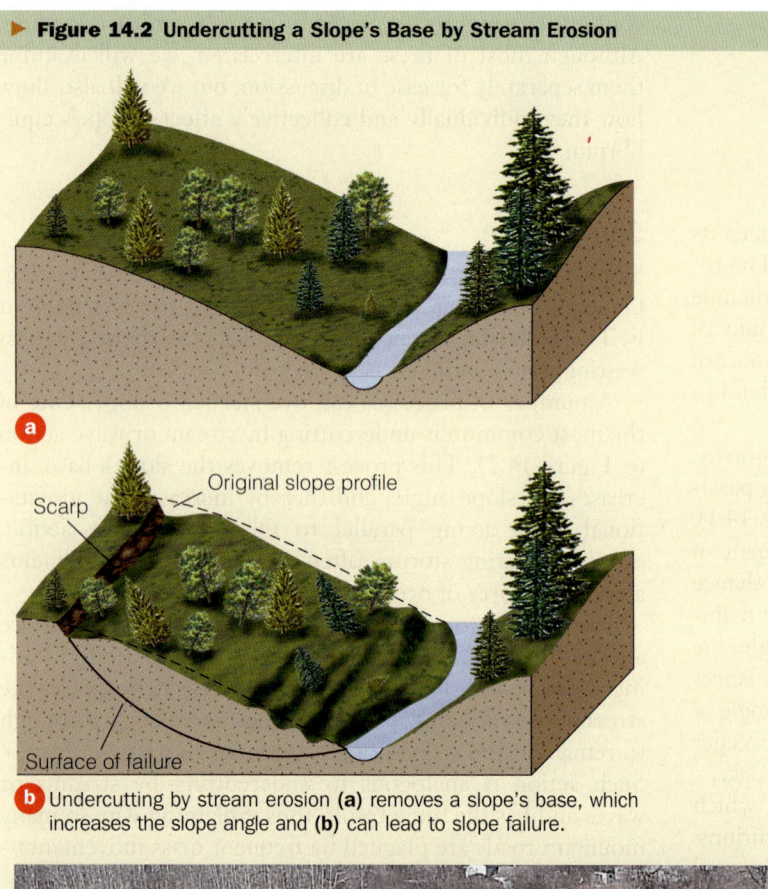

▶ **Figure 14.2 Undercutting a Slope's Base by Stream Erosion**

(a)

(b) Undercutting by stream erosion **(a)** removes a slope's base, which increases the slope angle and **(b)** can lead to slope failure.

(c) Undercutting by stream erosion caused slumping along this stream near Weidman, Michigan. Notice the scarp, which is the exposed surface of the underlying material following slumping.

mass movements most commonly occur in the deep weathering zone. In arid and semiarid regions, the weathering zone is usually considerably shallower. Nevertheless, intense, localized cloudbursts can drop large quantities of water on an area in a short time. With little vegetation to absorb this water, runoff is rapid and frequently results in mudflows.

Water Content

The amount of water in rock or soil influences slope stability. Large quantities of water from melting snow or heavy rainfall greatly increase the likelihood of slope failure. The additional weight that water adds to a slope can be enough to cause mass movement. Furthermore, water percolating through a slope's material helps decrease friction between grains, contributing to a loss of cohesion. For example, slopes composed of dry clay are usually quite stable, but when wetted they quickly lose cohesiveness and internal friction and become an unstable slurry. This occurs because clay, which can hold large quantities of water, consists of platy particles that easily slide over each other when wet. For this reason, clay beds are frequently the slippery layer along which overlying rock units slide downslope.

Vegetation

Vegetation affects slope stability in several ways. By absorbing the water from a rainstorm, vegetation decreases water saturation of a slope's material that would otherwise lead to a loss of shear strength. Vegetation's root system also helps stabilize a slope by binding soil particles together and holding the soil to bedrock.

The removal of vegetation by either natural or human activity is a major cause of many mass movements. Summer brush and forest fires in southern California frequently leave the hillsides bare of vegetation. Fall rainstorms saturate the ground, causing mudslides that do tremendous damage and cost millions of dollars to clean up. The soils of many hillsides in New Zealand are sliding because deep-rooted native bushes have been replaced by shallow-rooted grasses used for sheep grazing. When heavy rains saturate the soil, the shallow-rooted grasses cannot hold the slope in place, and parts of it slide downhill.

Recall that some rocks are more susceptible to weathering than others and that climate plays an important role in the rate and type of weathering. In the tropics, where temperatures are high and considerable rain falls, the effects of weathering extend to depths of several tens of meters, and

Overloading

Overloading is almost always the result of human activity and typically results from dumping, filling, or piling up of material. Under natural conditions, a material's load is carried by its grain-to-grain contacts, with the friction

▶ **Figure 14.3 Undercutting a Slope's Base by Wave Action** This sea cliff north of Bodega Bay, California, was undercut by waves during the winter of 1997–1998. As a result, part of the land slid into the ocean, damaging several houses.

James S. Monroe

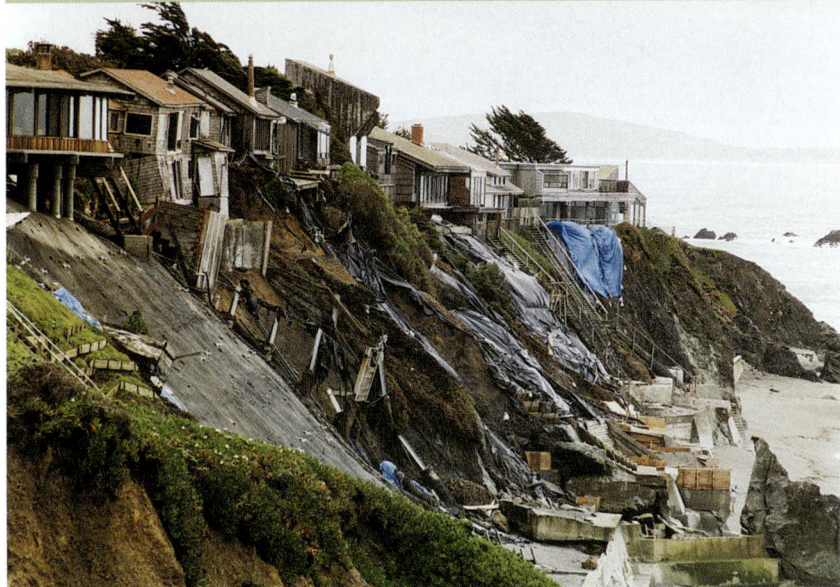

▶ **Figure 14.4 Highway Excavation (a)** Highway excavations disturb the equilibrium of a slope by **(b)** removing a portion of its support as well as oversteepening it at the point of excavation, which can result in **(c)** landslides along the highway

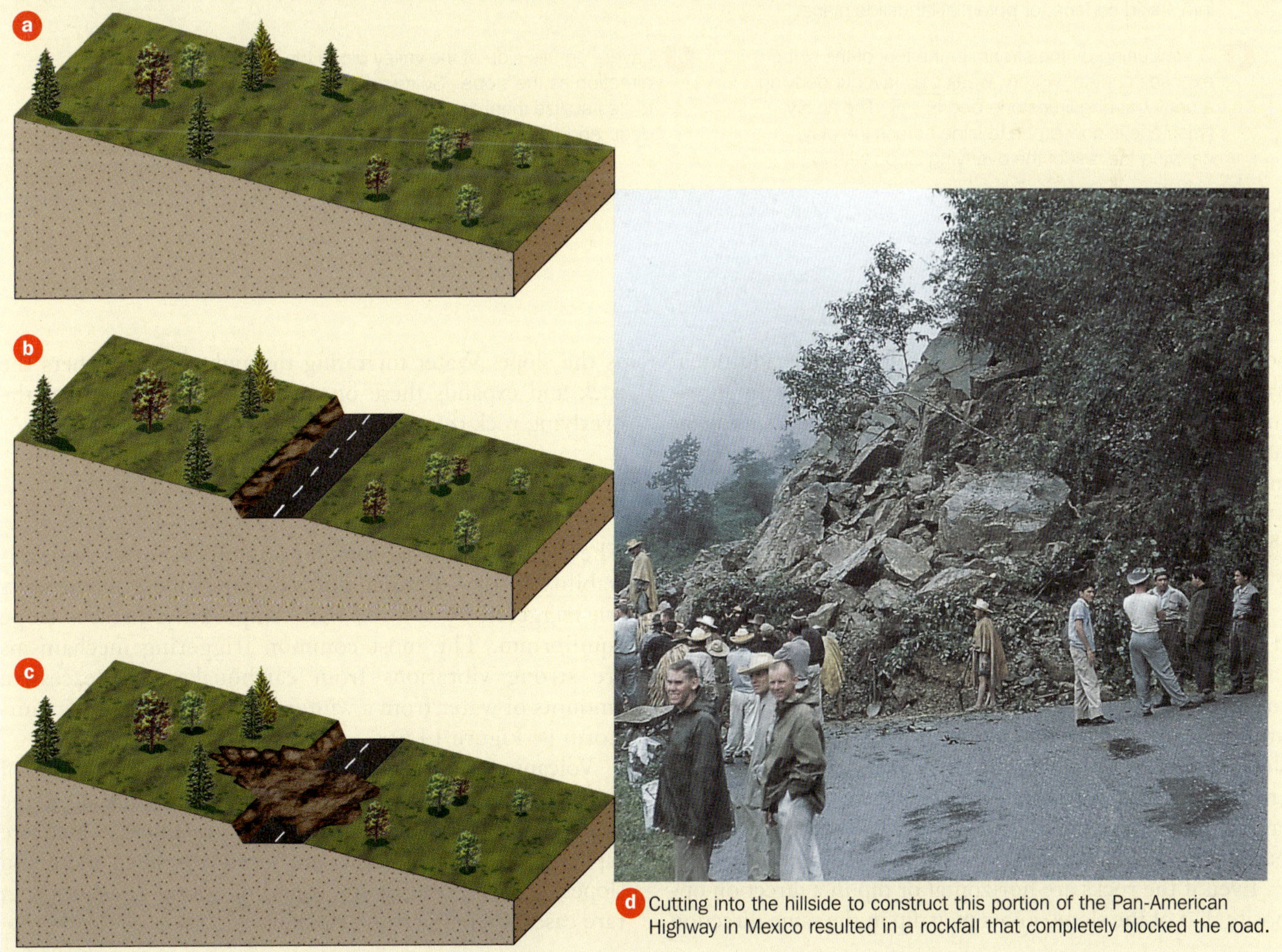

d Cutting into the hillside to construct this portion of the Pan-American Highway in Mexico resulted in a rockfall that completely blocked the road.

Courtesy of R. V. Dietrich

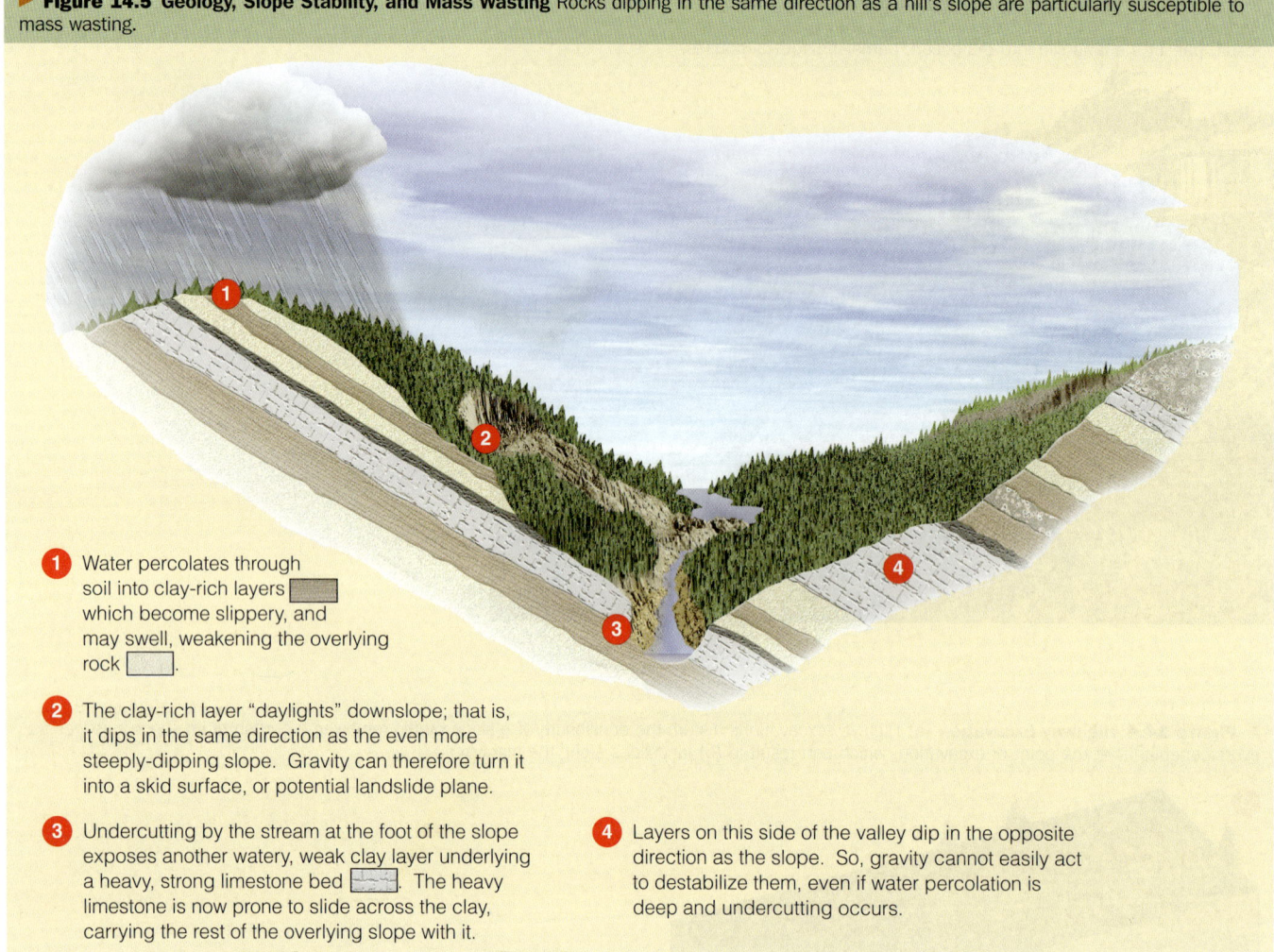

▶ **Figure 14.5 Geology, Slope Stability, and Mass Wasting** Rocks dipping in the same direction as a hill's slope are particularly susceptible to mass wasting.

1. Water percolates through soil into clay-rich layers ▭ which become slippery, and may swell, weakening the overlying rock ▭.

2. The clay-rich layer "daylights" downslope; that is, it dips in the same direction as the even more steeply-dipping slope. Gravity can therefore turn it into a skid surface, or potential landslide plane.

3. Undercutting by the stream at the foot of the slope exposes another watery, weak clay layer underlying a heavy, strong limestone bed ▭. The heavy limestone is now prone to slide across the clay, carrying the rest of the overlying slope with it.

4. Layers on this side of the valley dip in the opposite direction as the slope. So, gravity cannot easily act to destabilize them, even if water percolation is deep and undercutting occurs.

between the grains maintaining a slope. The additional weight created by overloading increases the water pressure within the material, which in turn decreases its shear strength, thereby weakening the slope material. If enough material is added, the slope will eventually fail, sometimes with tragic consequences.

■ *How does the relationship between the topography and geology of an area affect slope stability?*

If the rocks underlying a slope dip in the same direction as the slope, mass wasting is more likely to occur than if the rocks are horizontal or dip in the opposite direction (▶ Figure 14.5). When the rocks dip in the same direction as the slope, water can percolate along the various planes and decrease the cohesiveness and friction between adjacent rock layers. This is particularly true when clay layers are present because clay becomes slippery when wet.

Even if the rocks are horizontal or dip in a direction opposite that of the slope, joints may dip in the same direction as the slope. Water migrating through them weathers the rock and expands these openings until the weight of the overlying rock causes collapse and mass wasting to occur.

■ *What are the triggering mechanisms of mass wasting?*

The factors discussed thus far all contribute to slope instability. Most—though not all—rapid mass movements are triggered by a force that temporarily disturbs slope equilibrium. The most common triggering mechanisms are strong vibrations from earthquakes and excessive amounts of water from a winter snowmelt or a heavy rainstorm (▶ Figure 14.6).

Volcanic eruptions, explosions, and even loud claps of thunder may also be enough to trigger a landslide if the slope is sufficiently unstable. Many *avalanches*, which are rapid movements of snow and ice down steep mountain slopes, are triggered by the sound of a loud gunshot or, in rare cases, even a person's shout.

▶ **Figure 14.6 Landslide Triggered by Heavy Rains, La Conchita, California** Heavy winter rains caused this 200,000-m³ landslide in March 1995 at La Conchita, California, 120 km northeast of Los Angeles. Although no casualties occurred, nine homes were destroyed or badly damaged. Another landslide 10 years later occurred in the same area and under similar conditions, resulting in 10 deaths.

Section 14.2 Summary

• A slope's shear strength is those factors that help to maintain slope stability, such as the slope material's strength and cohesion, the amount of internal friction between grains, and any external support. Opposing a slope's shear strength is the force of gravity, which promotes instability.

• Various factors combine to cause mass wasting. These include a change in slope angle, weakening of material by weathering, increased water content, changes in the vegetation cover, and overloading. In addition, the relationship between topography and the geology of an area plays an important role in determining slope stability.

• The most common triggering mechanisms of mass wasting are earthquakes and excessive amounts of water, although anything that disturbs the slope's equilibrium will result in mass wasting.

14.3 Types of Mass Wasting

■ *How are mass movements classified?*

Mass movements are generally classified on the basis of three major criteria (Table 14.2): (1) rate of movement (rapid or slow); (2) type of movement (primarily falling, sliding, or flowing); and (3) type of material involved (rock, soil, or debris). Even though many slope failures are combinations of different materials and movements, the resulting mass movements are typically classified according to their dominant behavior.

Rapid mass movements involve a visible movement of material. Such movements usually occur quite suddenly, and the material moves very quickly downslope. Rapid mass movements are potentially dangerous and frequently result in loss of life and property damage. Most rapid mass movements occur on relatively steep slopes and can involve rock, soil, or debris.

Slow mass movements advance at an imperceptible rate and are usually detectable only by the effects of their movement, such as tilted trees and power poles or cracked foundations. Although rapid mass movements are more dramatic, slow mass movements are responsible for the downslope transport of a much greater volume of weathered material.

■ *What are rockfalls?*

Rockfalls are a common type of extremely rapid mass movement in which rocks of any size fall through the air (▶ Figure 14.7). Rockfalls occur along steep canyons, cliffs, and road cuts and build up accumulations of loose rocks and rock fragments, called *talus*, at their base (see Figure 6.3c).

Rockfalls result from failure along joints or bedding planes in the bedrock and are commonly triggered by natural or human undercutting of slopes, or by earthquakes. Many rockfalls in cold climates are the result of frost wedging. Chemical weathering caused by water percolating through the fissures in carbonate rocks (limestone, dolostone, and marble) is also responsible for many rockfalls.

Rockfalls range in size from small rocks falling from a cliff to massive falls involving millions of cubic meters of debris that destroy buildings, bury towns, and block highways (▶ Figure 14.8). Rockfalls are a particularly common hazard in mountainous areas where roads have been built by blasting and grading through steep hillsides of bedrock. Anyone who has ever driven through the Appalachians, the Rocky Mountains, or the Sierra Nevada is familiar with the "Watch for Falling Rocks" signs posted to warn drivers of the danger. Slopes that are particularly susceptible to rockfalls are sometimes covered with wire mesh in an effort to prevent dislodged rocks from falling to the road below (▶ Figure 14.9a). Another tactic is to put up wire mesh fences along the base of the slope to catch or slow down bouncing or rolling rocks (▶ Figure 14.9b).

Table 14.2

Classification of Mass Movements and Their Characteristics

Type of Movement	Subdivision	Characteristics	Rate of Movement
Falls	Rockfall	Rocks of any size fall through the air from steep cliffs, canyons, and road cuts	Extremely rapid
Slides	Slump	Movement occurs along a curved surface of rupture; most commonly involves unconsolidated or weakly consolidated material	Extremely slow to moderate
	Rock slide	Movement occurs along a generally planar surface	Rapid to very rapid
Flows	Mudflow	Consists of at least 50% silt- and clay-sized particles and up to 30% water	Very rapid
	Debris flow	Contains larger-sized particles and less water than mudflows	Rapid to very rapid
	Earthflow	Thick, viscous, tongue-shaped mass of wet regolith	Slow to moderate
	Quick clays	Composed of fine silt and clay particles saturated with water; when disturbed by a sudden shock, lose their cohesiveness and flow like a liquid	Rapid to very rapid
	Solifluction	Water-saturated surface sediment	Slow
	Creep	Downslope movement of soil and rock	Extremely slow
Complex movements		Combination of different movement types	Slow to extremely rapid

■ *What are slides and how are they recognized?*

A **slide** involves movement of material along one or more surfaces of failure. The type of material may be soil, rock, or a combination of the two, and it may break apart during movement or remain intact. A slide's rate of movement can vary from extremely slow to very rapid (Table 14.2).

Two types of slides are generally recognized: (1) slumps or rotational slides, in which movement occurs along a curved surface, and (2) rock or block slides, which move along a more or less planar surface.

A **slump** involves the downward movement of material along a curved surface of rupture and is characterized by the backward rotation of the slump block (▶ Figure 14.10). Slumps usually occur in unconsolidated or weakly consolidated material and range in size from small individual sets, such as occur along stream banks, to

▶ **Figure 14.7 Rockfall** Rockfalls result from failure along cracks, fractures, or bedding planes in the bedrock and are common features in areas of steep cliffs.

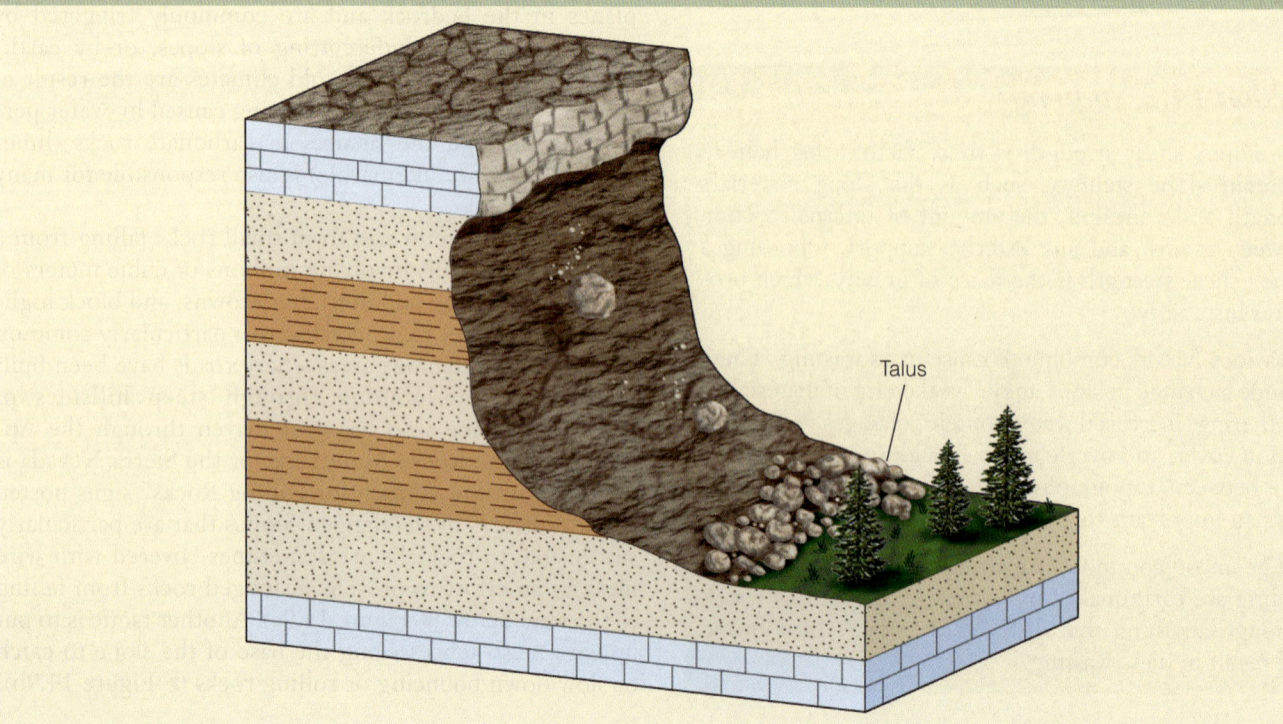

▶ **Figure 14.8 Rockfall along Highway 70, California** A huge rockfall closed both lanes of traffic on Highway 70 near Rogers Flat, California, on July 25, 2003. Rocks the size of large dump trucks had to be blasted into smaller pieces to clear the highway. Despite the pavement cracking caused by the falling boulders, geologists determined that the roadbase was undamaged and the road would be safe following cleanup operations.

▶ **Figure 14.9 Minimizing Damage from Rock Falls**

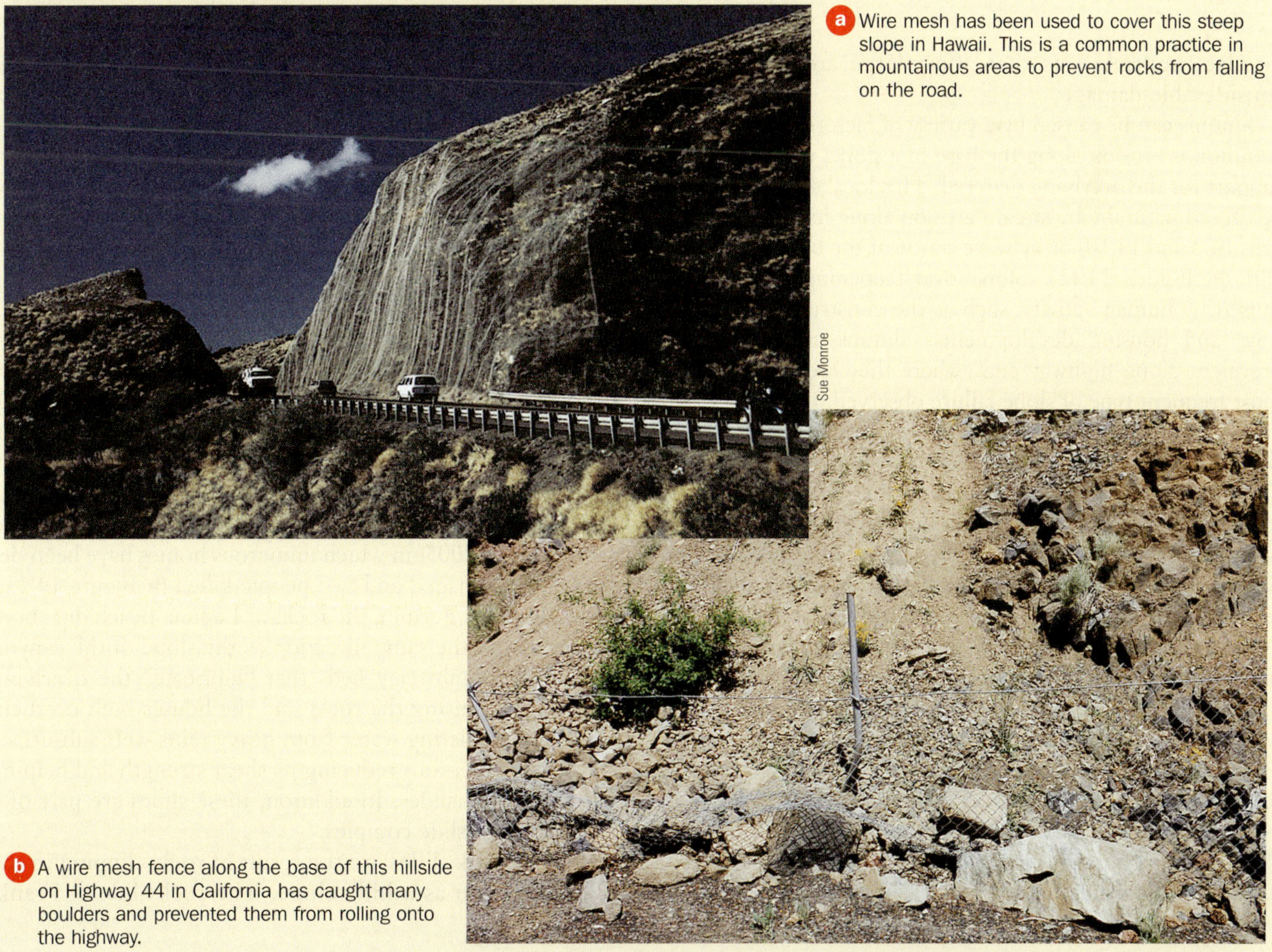

a Wire mesh has been used to cover this steep slope in Hawaii. This is a common practice in mountainous areas to prevent rocks from falling on the road.

b A wire mesh fence along the base of this hillside on Highway 44 in California has caught many boulders and prevented them from rolling onto the highway.

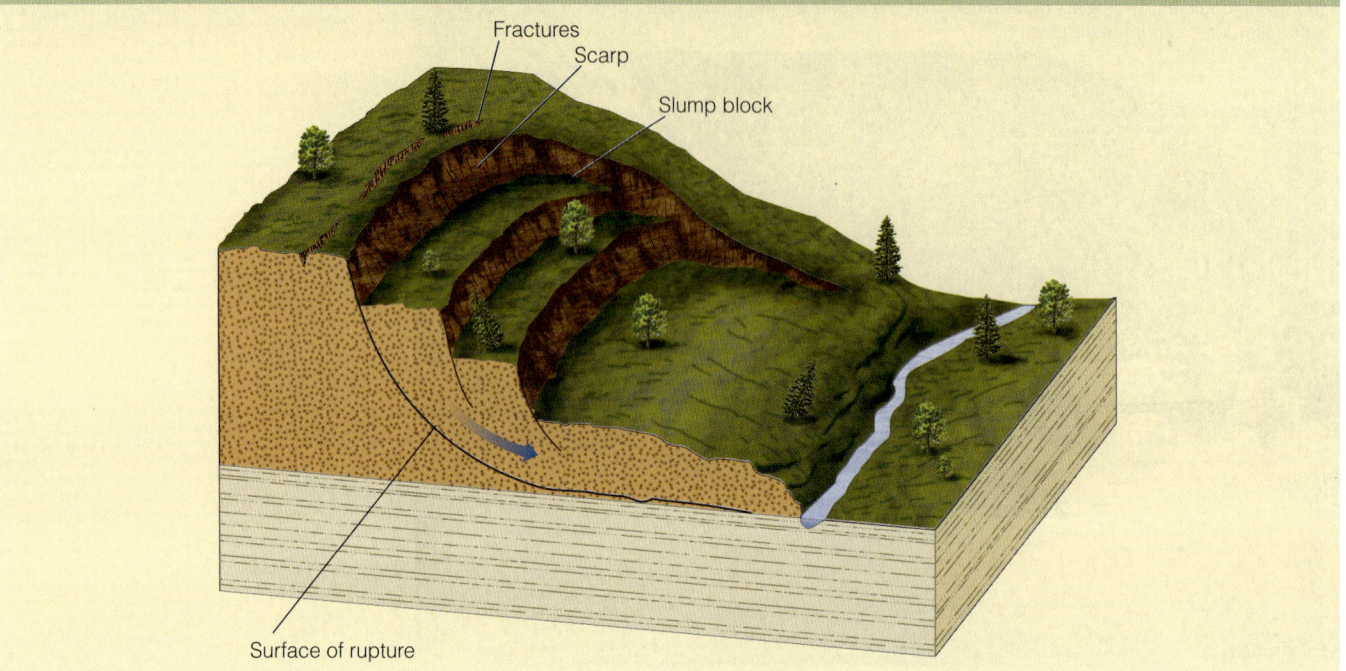

▶ **Figure 14.10 Slumping** In a slump, material moves downward along the curved surface of a rupture, causing the slump block to rotate backward. Most slumps involve unconsolidated or weakly consolidated material and are typically caused by erosion along the slope's base, such as a stream in this example.

massive, multiple sets that affect large areas and cause considerable damage.

Slumps can be caused by a variety of factors, but the most common is erosion along the base of a slope, which removes support for the overlying material. This local steepening may be caused naturally by stream erosion along its banks (▶ Figures 14.2 and 14.10) or by wave action at the base of a coastal cliff (▶ Figure 14.11). Slope oversteepening can also be caused by human activity, such as the construction of highways and housing developments. Slumps are particularly prevalent along highway cuts, where they are generally the most frequent type of slope failure observed.

Although many slumps are merely a nuisance, large-scale slumps involving populated areas and highways can cause extensive damage. Such is the case in coastal southern California, where slumping and sliding have been a constant problem. Many areas along the coast are underlain by poorly to weakly consolidated silts, sands, and gravels interbedded with clay layers, some of which are weathered ashfalls. In addition, southern California is tectonically active so that many of these deposits are cut by faults and joints, which allow the infrequent rains to percolate downward rapidly, wetting and lubricating the clay layers.

Southern California lies in a semiarid climate and is dry most of the year. When it does rain, typically between November and March, large amounts of rain can fall in a short time. Thus the ground quickly becomes saturated, leading to landslides along steep canyon walls as well as along coastal cliffs (▶ Figure 14.11). Most of the slope failures along the southern California coast are the result of slumping. These slumps have destroyed many expensive homes and forced many roads to be closed and relocated.

A **rock** or *block* **slide** occurs when rocks move downslope along a more or less planar surface. Most rock slides take place because the local slopes and rock layers dip in the same direction (▶ Figure 14.12), although they can also occur along fractures parallel to a slope. In addition to slumping, rock slides are common occurrences along the southern California coast. At Point Fermin, seaward-dipping rocks with interbedded slippery clay layers are undercut by waves, causing numerous slides (see "Point Fermin—Slip Sliding Away" on pp. 436–437).

Farther south, in the town of Laguna Beach, residents have been hit by rock slides and mudslides in 1978, 1998, and as recently as 2005, in which numerous homes have been destroyed or damaged and two people killed (▶ Figure 14.13). Just as at Point Fermin, the rocks at Laguna Beach dip about 25 degrees in the same direction as the slope of the canyon walls and contain clay beds that "lubricate" the overlying rock layers, causing the rocks and the houses built on them to slide. Percolating water from heavy rains wets subsurface clayey siltstone, thus reducing its shear strength and helping to activate the slides. In addition, these slides are part of a larger ancient slide complex.

Not all rock slides are the result of rocks dipping in the same direction as a hill's slope. The rock slide at Frank,

▶ **Figure 14.11 Slumping in the Pacific Palisades, Southern California** Undercutting of steep sea cliffs by wave action resulted in massive slumping in the Pacific Palisades area of southern California on March 31 and April 3, 1958. Highway 1 was completely blocked. Note the heavy earth-moving equipment for scale. **Can anything be done to prevent or minimize future slumping here?** Probably very little can be done to prevent the massive slumping shown here because of the underlying geology and periodic heavy winter rains. However, lessening the slope angle where possible, providing support along the base of the cliffs, ensuring draining of the sediments, and planting vegetation can all help to reduce the potential for large-scale slumping.

Alberta, Canada, on April 29, 1903, illustrates how nature and human activity can combine to create a situation with tragic results (▶ Figure 14.14).

It would appear at first glance that the coal-mining town of Frank, lying at the base of Turtle Mountain, was in no danger from a landslide (▶ Figure 14.14). After all, many of the rocks dipped away from the mining valley, unlike the situations at Point Fermin and Laguna Beach. The joints in the massive limestone composing Turtle Mountain, however, dip steeply toward the valley and are essentially parallel with the slope of the mountain itself. Furthermore, Turtle Mountain is supported by weak limestones, shales, and coal layers that underwent slow plastic deformation from the weight of the overlying massive limestone. Coal mining along the base of the valley also contributed to the stress on the rocks by removing some of the underlying support. All these factors, as well as frost action and chemical weathering that widened the joints, finally resulted in a massive rock slide. Almost 40 million m³ of rock slid down Turtle Mountain along joint planes, killing 70 people and partially burying the town of Frank.

■ *What are the different types of flows?*

Mass movements in which material flows as a viscous fluid or displays plastic movement are called *flows*. Their rate of movement ranges from extremely slow to extremely rapid (Table 14.2). In many cases, mass movements begin as falls, slumps, or slides and change into flows farther downslope.

Of the major mass movement types, **mudflows** are the most fluid and move most rapidly (at speeds of up to

Point Fermin—Slip Sliding Away

Dubbed the "sunken city" by residents of the area, Point Fermin in southern California is famous for its numerous examples of mass wasting. The area is underlain by fine-grained sedimentary rocks interbedded with diatomite* layers and volcanic ash. When these layers get wet, they get slippery and tend to slide easily. The rocks also dip slightly toward the ocean and form steep coastal bluffs that are being undercut by constant wave action at their base. This wave action results in oversteeping of the cliffs, which causes slumping.

Mass wasting began in 1929 with minor slumping in the area. In the early 1940s, water mains in the region were broken and several individual blocks began slumping. Movement largely ceased following this main phase of slumping, but it has continued intermittently until the present, and residents have paid the price for living in an unstable coastal area.

*A sedimentary rock composed of the shells of diatoms, which are microscopic siliceous algae.

▲ 1. A map of southern California, showing the location of Point Fermin and an aerial view at low tide of the sliding that has taken place. Note the numerous slump blocks and oversteepened cliffs. The continuous pounding of waves and surf along the base of the cliffs further erodes and undercuts them, leading to even more slumping and sliding.

▼ 2. A view of one portion of the Point Fermin slide area, showing the fine-grained sedimentary rocks dipping slightly toward the ocean and the oversteepened cliffs resulting from slumping and sliding in the foreground.

◀ **3.** A view of one slump block shows remnants of a former road and a palm tree still growing as if nothing has happened.

▶ **4.** An abandoned house on the edge of an oversteepened coastal bluff. Note the concrete blocks placed along the beach to absorb the erosive energy of the incoming waves and slow down the erosion to the cliffs.

▼ **5.** Creep and minor slumping are evident in this photo. Note the two small slump scarps. The smaller one in the background is mostly grass-covered, whereas the one in the foreground has bare spots and is apparently moving at a slightly faster rate. Notice the effect of creep on the right-hand wall of the house. The bottom part of the wall is moving toward the right of the photo as a result of creep, producing a bend of the wall that can be seen clearly near its base.

▲ **6.** Slumping along the oversteepened cliffs at Point Fermin. The abandoned house in the photo above right is just to the right of this view at the top of the cliff.

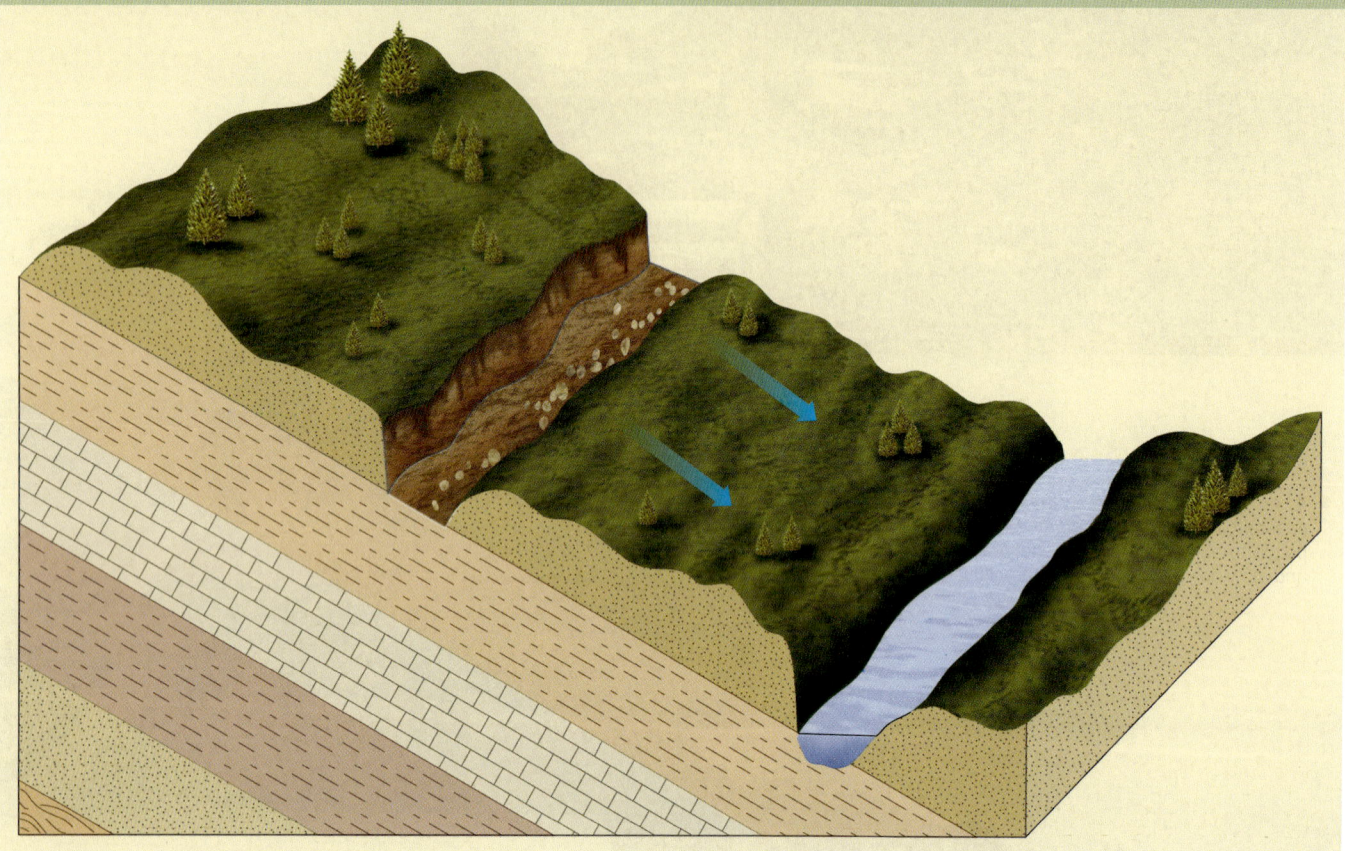

▶ **Figure 14.12 Rock Slide** Rock slides occur when material moves downslope along a generally planar surface. Most rock slides result when the underlying rocks dip in the same general angle as the slope of the land. Undercutting along the base of the slope and clay layers beneath porous rock or soil layers increase the chance of rock slides.

80 km/hr). They consist of at least 50% silt- and clay-sized material combined with a significant amount of water (up to 30%). Mudflows are common in arid and semiarid environments where they are triggered by heavy rainstorms that quickly saturate the regolith, turning it into a raging flow of mud that engulfs everything in its path. Mudflows can also occur in mountain regions (▶ Figure 14.15) and in areas covered by volcanic ash, where they can be particularly destructive (see Chapter 5). Because mudflows are so fluid, they generally follow preexisting channels until the slope decreases or the channel widens, at which point they fan out.

As urban areas in arid and semiarid climates continue to expand, mudflows and the damage they create are becoming problems. Mudflows are common, for example, in the steep hillsides around Los Angeles, where they have damaged or destroyed many homes.

Debris flows are composed of larger particles than mudflows and do not contain as much water. Consequently, they are usually more viscous than mudflows, typically do not move as rapidly, and rarely are confined to preexisting channels. Debris flows can be just as damaging, though, because they can transport large objects (▶ Figure 14.16).

Earthflows move more slowly than either mudflows or debris flows. An earthflow slumps from the upper part of a hillside, leaving a scarp, and flows slowly downslope as a thick, viscous, tongue-shaped mass of wet regolith (▶ Figure 14.17). Like mudflows and debris flows, earthflows can be of any size and are frequently destructive. They occur most commonly in humid climates on grassy, soil-covered slopes, following heavy rains.

Some clays spontaneously liquefy and flow like water when they are disturbed. Such **quick clays** have caused serious damage and loss of lives in Sweden, Norway, eastern Canada (▶ Figure 14.18), and Alaska (Table 14.1). Quick clays are composed of fine silt and clay particles made by the grinding action of glaciers. Geologists think these fine sediments were originally deposited in a marine environment where their pore space was filled with saltwater. The ions in saltwater helped establish strong bonds between the clay particles, thus stabilizing and strengthening the clay. When the clays were subsequently uplifted above sea level, the saltwater was flushed out by fresh groundwater, reducing the effectiveness of the ionic bonds between the clay particles and thereby reducing the overall strength and cohesiveness of the clay. Consequently, when the clay is dis-

▶ **Figure 14.13 Rock Slide, Laguna Beach, California** A combination of interbedded clay beds that become slippery when wet, rocks dipping in the same direction as the slope of the sea cliffs, and undercutting of the sea cliffs by wave action activated a rock slide at Laguna Beach, California, that destroyed numerous homes and cars on October 2, 1978. This same area was hit by another rock slide in 2005. **What lesson should be learned from this?** *If you are going to build in an area that has experienced landslides in the past and the factors causing those landslides are still present, don't be surprised when mass wasting once again occurs.*

turbed by a sudden shock or shaking, it essentially turns to a liquid and flows.

An example of the damage that can be done by quick clays occurred in the Turnagain Heights area of Anchorage, Alaska, in 1964 (▶ Figure 14.19). Underlying most of the Anchorage area is the Bootlegger Cove Clay, a massive clay unit of poor permeability (▶ Figure 14.19a). Because the Bootlegger Cove Clay forms a barrier that prevents groundwater from flowing through the adjacent glacial deposits to the sea, considerable hydraulic pressure builds up behind the clay. Some of this water has flushed out the saltwater in the clay and has saturated the lenses of sand and silt associated with the clay beds. When the magnitude-8.6 Good Friday earthquake struck on March 27, 1964, the shaking turned parts of the Bootlegger Cove Clay into a quick clay and precipitated a series of massive slides in the coastal bluffs that destroyed most of the homes in the Turnagain Heights subdivision (▶ Figure 14.19b).

Solifluction is the slow downslope movement of water-saturated surface sediment. Solifluction can occur in any climate where the ground becomes saturated with water, but is most common in areas of permafrost.

Permafrost, ground that remains permanently frozen, covers nearly 20% of the world's land surface (▶ Figure 14.20a). During the warmer season when the upper portion of the

▶ **Figure 14.14 Rock Slide, Turtle Mountain, Canada**

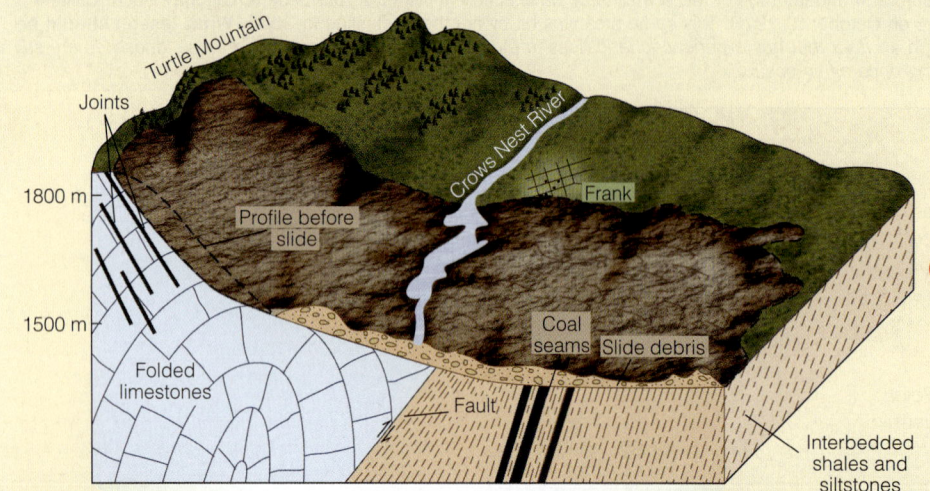

a The tragic Turtle Mountain rock slide that killed 70 people and partially buried the town of Frank, Alberta, Canada, on April 29, 1903, was caused by a combination of factors. These included joints that dipped in the same direction as the slope of Turtle Mountain, a fault partway down the mountain, weak shale and siltstone beds underlying the base of the mountain, and mined-out coal seams.

b Results of the 1903 rock slide at Frank.

B. Bradley and the University of Colorado's Geology Department, National Geophysical Data Center

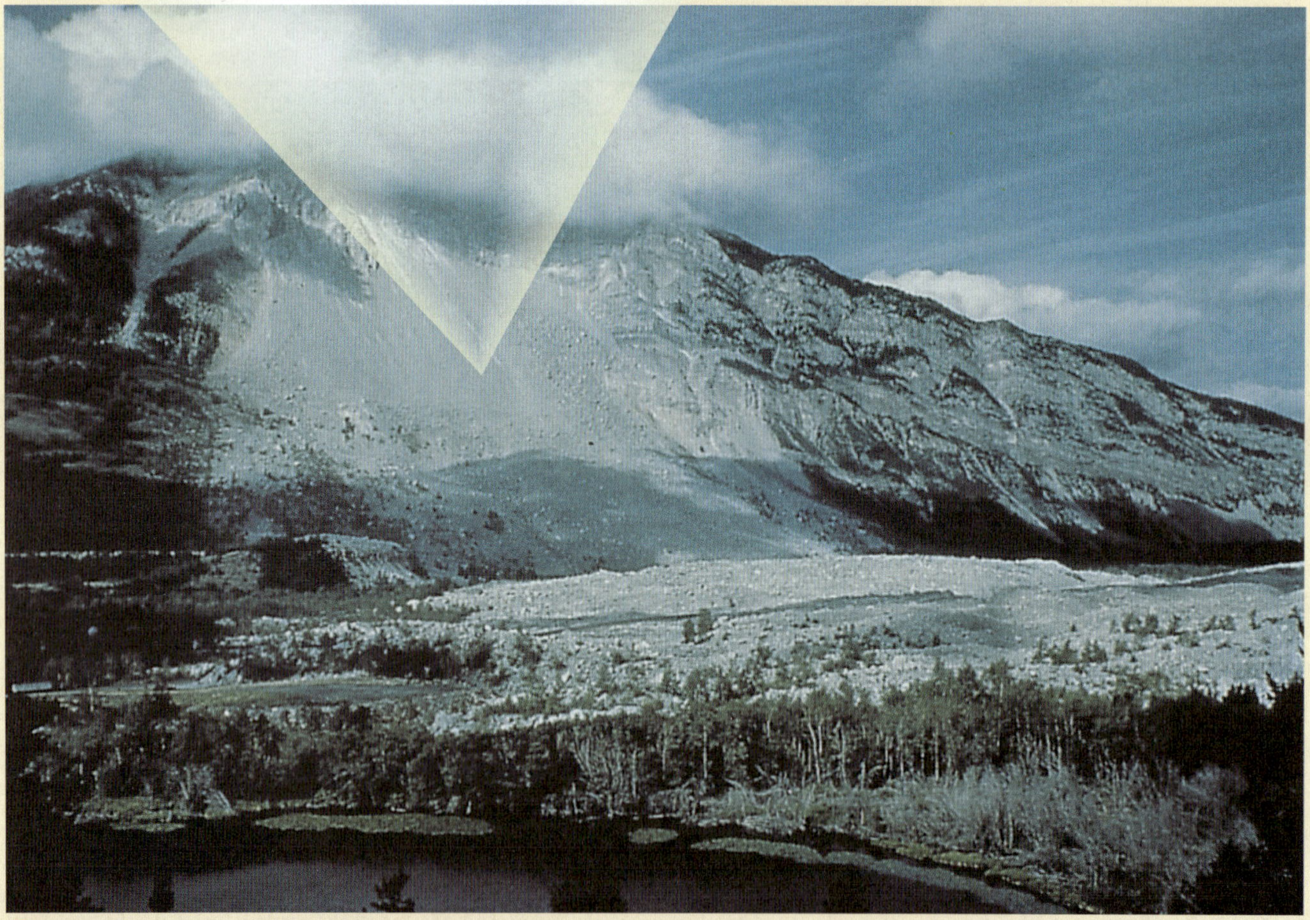

▶ **Figure 14.15 Mudflow, Estes Park, Colorado** Mudflows move swiftly downslope, engulfing everything in their path. Note how this mudflow has fanned out at the base of the hill.

James S. Monroe

permafrost thaws, water and surface sediment form a soggy mass that flows by solifluction and produces a characteristic lobate topography (▶ Figure 14.20b).

As might be expected, many problems are associated with construction in a permafrost environment. A good example is what happens when an uninsulated building is constructed directly on permafrost. Heat escapes through the floor, thaws the ground below, and turns it into a soggy, unstable mush. Because the ground is no longer solid, the building settles unevenly into the ground, and numerous structural problems result (▶ Figure 14.21).

Construction of the Alaska pipeline from the oil fields in Prudhoe Bay to the ice-free port of Valdez raised numerous concerns about the effect the pipeline might have on the permafrost and the potential for solifluction. Some thought that oil flowing through the pipeline would be warm enough to melt the permafrost, causing the pipeline to sink farther into the ground and possibly rupture. After numerous studies were conducted, scientists concluded that the pipeline, completed in 1977, could safely be buried for more than half of its 1280-km length; where melting of the permafrost might cause structural problems to the pipe, it was insulated and installed above ground.

Creep, the slowest type of flow, is the most widespread and significant mass-wasting process in terms of the total amount of material moved downslope and the monetary damage it does annually. Creep involves extremely slow downhill movement of soil or rock. Although it can occur anywhere and in any climate, it is most effective and significant as a geologic agent in humid regions. In fact, it is the most common form of mass wasting in the

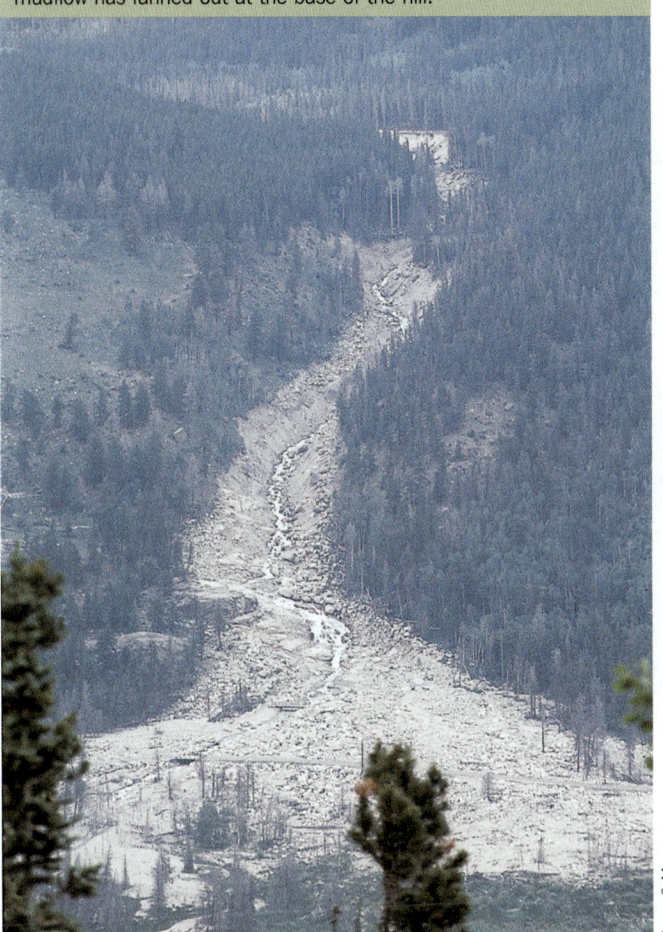

▶ **Figure 14.16 Debris Flow, Ophir Creek, Nevada** A debris flow and damaged house in lower Ophir Creek, western Nevada. Note the many large boulders that are part of the debris flow. Debris flows do not contain as much water as mudflows and typically are composed of larger particles.

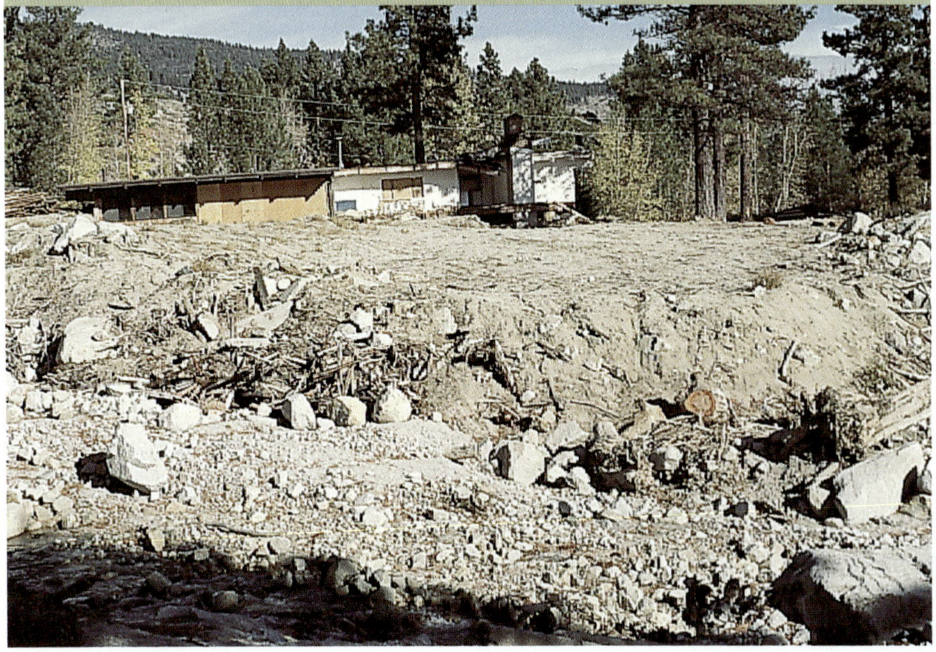

B. Pipkin, University of Southern California

▶ **Figure 14.17 Earthflow**

a Earthflows form tongue-shaped masses of wet regolith that move slowly downslope. They occur most commonly in humid climates on grassy, soil-covered slopes.

b An earthflow near Baraga, Michigan. **Can you see the outline of the tongue-shaped mass?** *It forms a lobate structure just below the scarp.*

▶ **Figure 14.18 Quick-Clay Slide, Nicolet, Quebec, Canada** The house on the slide (to the right of the bridge and circled in red) traveled several hundred feet with relatively little damage.

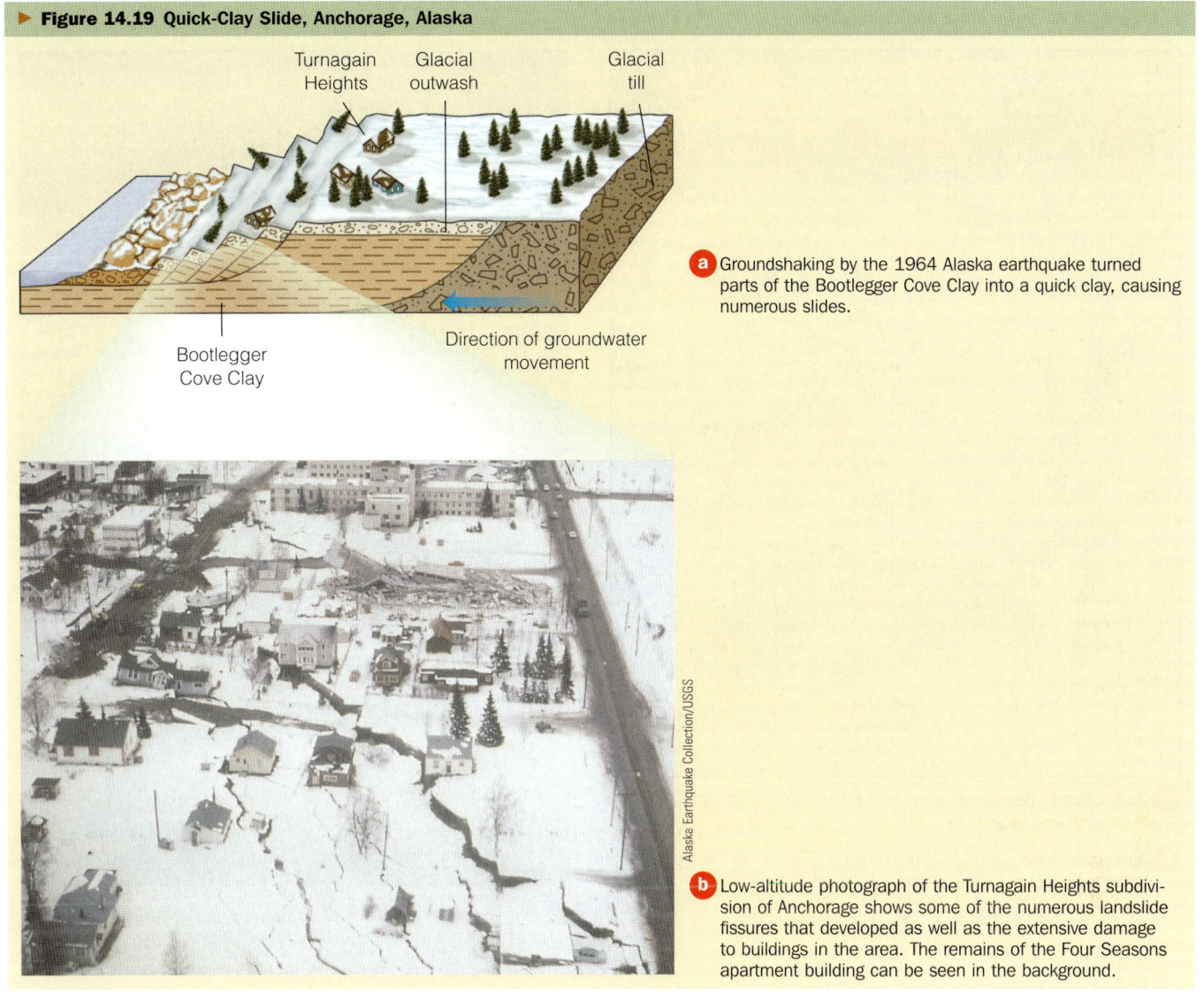

Figure 14.19 Quick-Clay Slide, Anchorage, Alaska

a Groundshaking by the 1964 Alaska earthquake turned parts of the Bootlegger Cove Clay into a quick clay, causing numerous slides.

b Low-altitude photograph of the Turnagain Heights subdivision of Anchorage shows some of the numerous landslide fissures that developed as well as the extensive damage to buildings in the area. The remains of the Four Seasons apartment building can be seen in the background.

southeastern United States and the southern Appalachian Mountains.

Because the rate of movement is essentially imperceptible, we are frequently unaware of creep's existence until we notice its effects: tilted trees and power poles, broken streets and sidewalks, or cracked retaining walls or foundations (▶ Figure 14.22). Creep usually involves the whole hillside and probably occurs, to some extent, on any weathered or soil-covered, sloping surface.

Creep is difficult not only to recognize but also to control. Although engineers can sometimes slow or stabilize creep, many times the only course of action is to simply avoid the area if at all possible or, if the zone of creep is relatively thin, design structures that can be anchored into the solid bedrock.

■ *What are complex movements?*

Recall that many mass movements are combinations of different movement types. When one type is dominant, the movement can be classified as one of those described thus far. If several types are more or less equally involved, it is called a **complex movement.**

The most common type of complex movement is the slide-flow, in which there is sliding at the head and then some type of flowage farther along its course. Most slide-flow landslides involve well-defined slumping at the head, followed by a debris flow or earthflow (▶ Figure 14.23). Any combination of different mass movement types can, however, be classified as a complex movement.

A *debris avalanche* is a complex movement that often occurs in very steep mountain ranges. Debris avalanches typically start out as rockfalls when large quantities of rock, ice, and snow are dislodged from a mountain-side, frequently as a result of an earthquake. The material then slides or flows down the mountainside, picking up additional surface material and increasing in speed. The 1970 Peru earthquake (Table 14.1) set in motion the debris avalanche that destroyed the towns of Yungay and Ranrahirca, Peru, and killed more than 25,000 people (▶ Figure 14.24).

▶ **Figure 14.20 Permafrost and Solifluction**

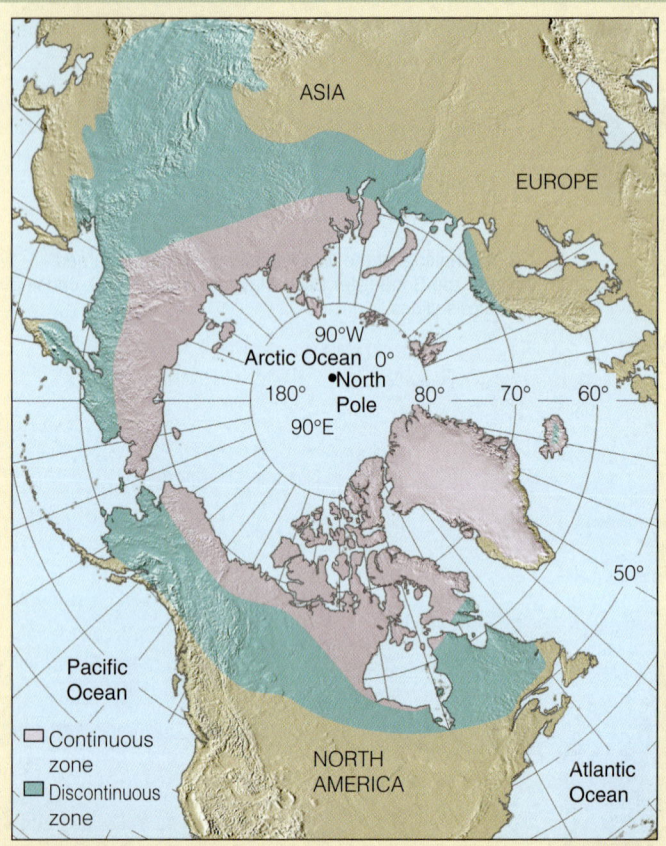

ⓐ Distribution of permafrost areas in the Northern Hemisphere.

ⓑ Solifluction flows near Suslositna Creek, Alaska, show the typical lobate topography that is characteristic of solifluction conditions.

▶ **Figure 14.21 Permafrost Damage** This house, south of Fairbanks, Alaska, has settled unevenly because the underlying permafrost in fine-grained silts and sands has thawed.

▶ **Figure 14.22 Creep**

a Some evidence of creep: (A) curved tree trunks; (B) displaced monuments; (C) tilted power poles; (D) displaced and tilted fences; (E) roadways moved out of alignment; (F) hummocky surface.

b Trees bent by creep, Wyoming.

c Creep has bent these sandstone and shale beds of the Haymond Formation near Marathon, Texas.

d Stone wall tilted due to creep, Champion, Michigan.

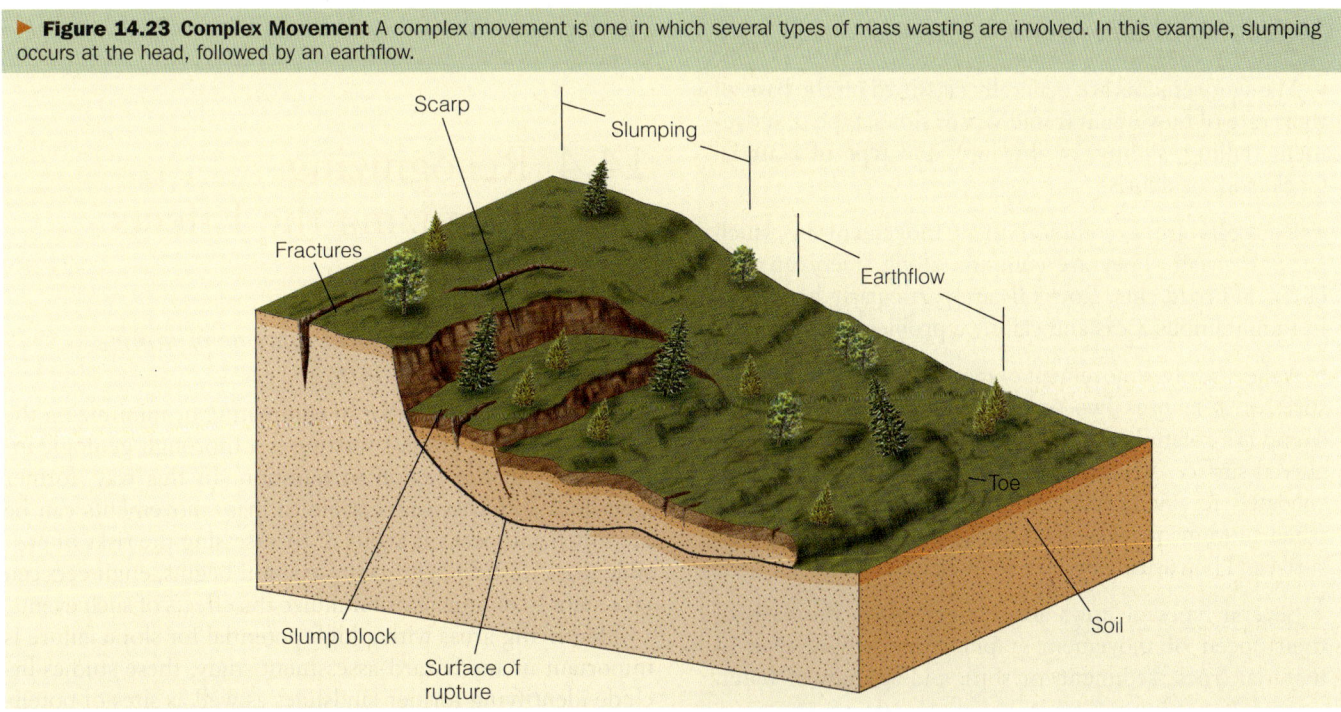

▶ **Figure 14.23 Complex Movement** A complex movement is one in which several types of mass wasting are involved. In this example, slumping occurs at the head, followed by an earthflow.

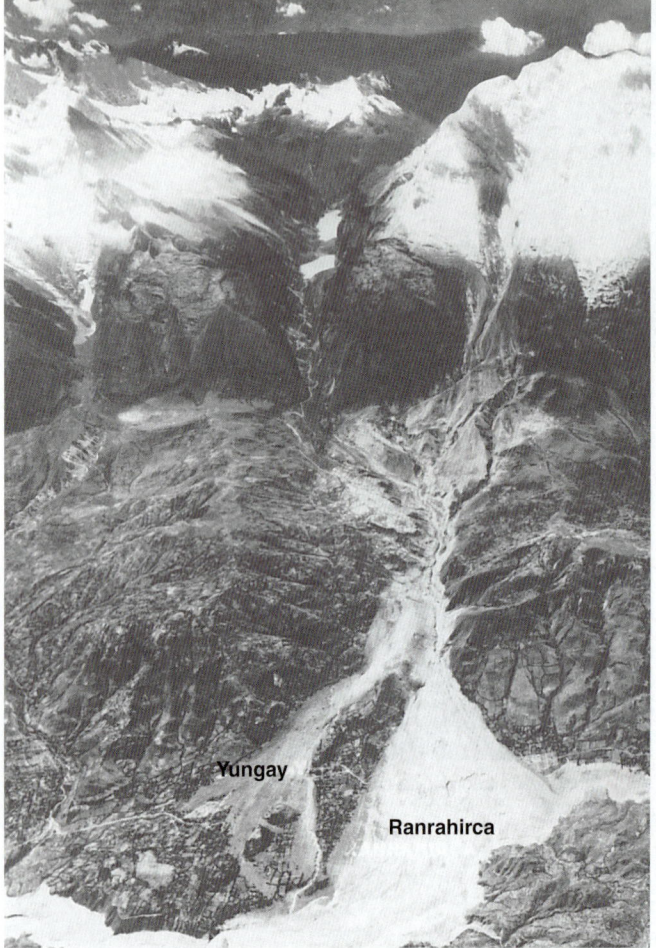

▶ **Figure 14.24 Debris Avalanche** An earthquake 65 km away triggered this debris avalanche on Nevado Huascarán, Peru, that destroyed the towns of Yungay and Ranrahirca and killed more than 25,000 people.

These include mudflows, debris flows, earthflows, quick clays, and solifluction.

- Creep is the slowest type of flow and is the imperceptible downslope movement of rock or soil. Most of the time, the evidence of creep comes from its effects, such as tilted trees and power poles, broken streets and sidewalks, and cracked retaining walls or foundations. Creep is the most widespread and significant mass-wasting process in terms of monetary damage done annually.

- Complex movements are combinations of different types of mass movements in which no single type is dominant. The most common type of complex movement is the slide-flow, in which slumping usually occurs at the head of the landslide, followed by one of the types of flow such as a mudflow or debris flow.

What Would You Do?

You are a member of a planning board for your seaside community. A developer wants to rezone some coastal property to build 20 condominiums. This would be a boon to the local economy because it would provide jobs and increase the tax base. However, because the area is somewhat hilly and fronts the ocean, you are concerned about how safe the buildings would be in this area. What types of studies should be done before any rezoning can take place? Is it possible to build safe structures along a hilly coastline? What specifically would you ask the environmental consulting firm the planning board has hired to look for in terms of actual or potential geologic hazards, if the condominiums are built?

Section 14.3 Summary

- Mass movements are generally classified on the basis of their rate of movement (rapid versus slow), type of movement (falling, sliding, or flowing), and type of material (rock, soil, or debris).

- Rockfalls are a common mass movement in which rocks free-fall. They are common along steep canyons, cliffs, and road cuts. Rockfalls are particularly hazardous in mountainous areas and can be a problem along roads.

- A slide involves movement of material along one or more surfaces of failure. Two types of slides are recognized. A slump is a rotational slide that involves movement along a curved surface. Slumps are most common in poorly consolidated to unconsolidated material. Rock slides occur when movement takes place along a more or less planar surface. They usually involve solid pieces of rock.

- Several types of flows are recognized on the basis of their speed of movement (rapid versus slow), type of material (rock, sediment, or soil), and amount of water.

14.4 Recognizing and Minimizing the Effects of Mass Wasting

■ *What can be done to eliminate or minimize the effects of mass wasting?*

The most important factor in eliminating or minimizing the damaging effects of mass wasting is a thorough geologic investigation of the region in question. In this way, former landslides and areas susceptible to mass movements can be identified and perhaps avoided. By assessing the risks of possible mass wasting before construction begins, engineers can take steps to eliminate or minimize the effects of such events.

Identifying areas with a high potential for slope failure is important in any hazard-assessment study; these studies include identifying former landslides as well as sites of poten-

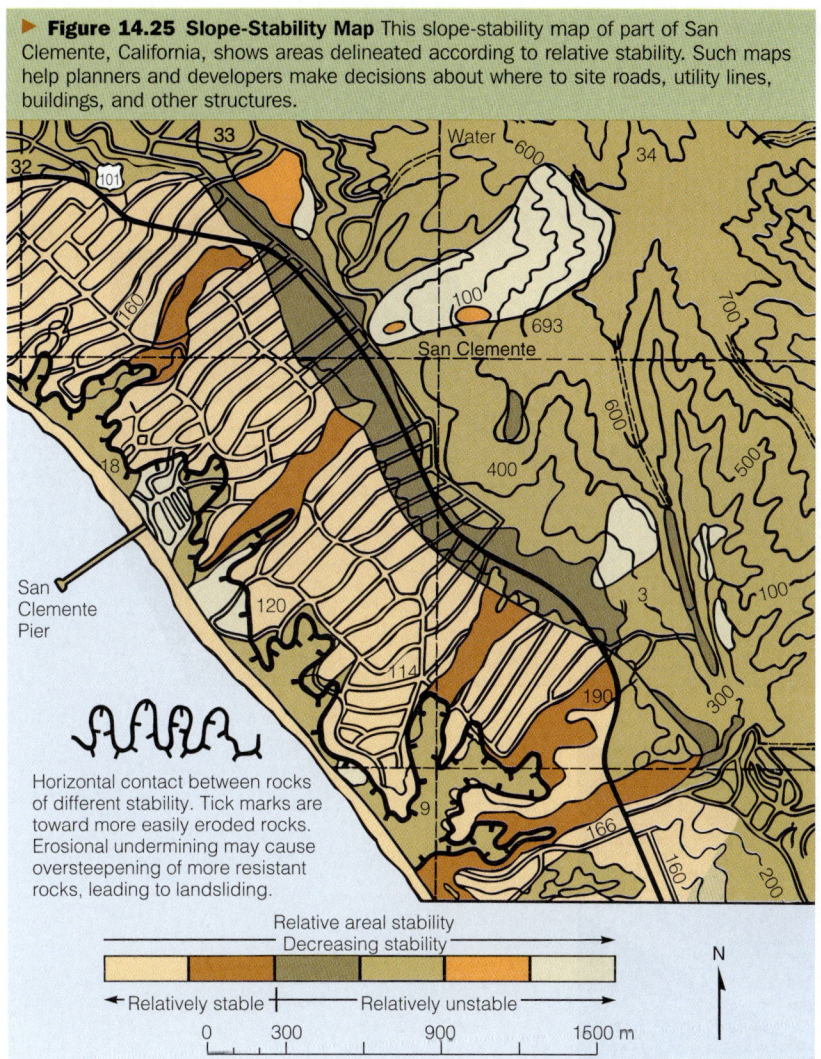

▶ **Figure 14.25 Slope-Stability Map** This slope-stability map of part of San Clemente, California, shows areas delineated according to relative stability. Such maps help planners and developers make decisions about where to site roads, utility lines, buildings, and other structures.

tial mass movement. Scarps, open fissures, displaced or tilted objects, a hummocky surface, and sudden changes in vegetation are some of the features that indicate former landslides or an area susceptible to slope failure. The effects of weathering, erosion, and vegetation may, however, obscure the evidence of previous mass wasting.

Soil and bedrock samples are also studied, in both the field and laboratory, to assess such characteristics as composition, susceptibility to weathering, cohesiveness, and ability to transmit fluids. These studies help geologists and engineers predict slope stability under a variety of conditions.

The information derived from a hazard-assessment study can be used to produce *slope-stability maps* of the area (▶ Figure 14.25). These maps allow planners and developers to make decisions about where to site roads, utility lines, and housing or industrial developments based on the relative stability or instability of a particular location. In addition, the maps indicate the extent of an area's landslide problem and the type of mass movement that may occur. This information is important for grading slopes or building structures, to prevent or minimize slope-failure damage.

Although most large mass movements usually cannot be prevented, geologists and engineers can employ various methods to minimize the danger and damage resulting from them. Because water plays such an important role in many landslides, one of the most effective and inexpensive ways to reduce the potential for slope failure or to increase existing slope stability is through surface and subsurface drainage of a hillside. Drainage serves two purposes. It reduces the weight of the material likely to slide and increases the shear strength of the slope material by lowering pore pressure.

Surface waters can be drained and diverted by ditches, gutters, or culverts designed to direct water away from slopes. Drainpipes perforated along one surface and driven into a hillside can help remove subsurface water (▶ Figure 14.26). Finally, planting vegetation on hillsides helps stabilize slopes by holding the soil together and reducing the amount of water in the soil.

Another way to help stabilize a hillside is to reduce its slope. Recall that overloading and oversteepening by grading are common causes of slope failure. Reducing the angle

Figure 14.26 Using Drainpipes to Remove Subsurface Water

a Driving drainpipes that are perforated on one side into a hillside, with the perforated side up, can remove some subsurface water and help stabilize a hillside.

b A drainpipe driven into the hillside at Point Fermin, California, helps to reduce the amount of subsurface water in these porous beds. **What features in this photograph indicate this is an area particularly susceptible to mass wasting?** *The area has very steep slopes and rock layers that dip in the same direction as the slopes and the entire area is being undercut by wave action.*

of a hillside decreases the potential for slope failure. Two methods are usually employed to reduce a slope's angle. In the *cut-and-fill* method, material is removed from the upper part of the slope and used as fill at the base, thus providing a flat surface for construction and reducing the slope (▶ Figure 14.27). The second method, which is called *benching*, involves cutting a series of benches or steps into a hillside (▶ Figure 14.28). This process reduces the overall average slope, and the benches serve as collecting sites for small landslides or rockfalls that might occur. Benching is most commonly used on steep hillsides in conjunction with a system of surface drains to divert runoff.

In some situations, retaining walls are constructed to provide support for the base of the slope (▶ Figure 14.29). The walls are usually anchored well into bedrock, backfilled with crushed rock, and provided with drain holes to prevent the buildup of water pressure in the hillside.

Rock bolts, similar to those employed in tunneling and mining, can sometimes be used to fasten potentially unstable rock masses into the underlying stable bedrock (▶ Figure 14.30). This technique has been used successfully on the hillsides of Rio de Janeiro, Brazil, and to help secure the slopes at the Glen Canyon Dam on the Colorado River.

Recognition, prevention, and control of landslide-prone areas are expensive, but not nearly as expensive as the damage can be when such warning signs are ignored or not recognized. Unfortunately, there are numerous examples of landfill and dam collapses that serve as tragic reminders of the price paid in loss of lives and property damage when the warning signs of impending disaster are ignored.

▶ **Figure 14.27 Stabilizing a Hillside by the Cut-and-Fill Method** One common method used to help stabilize a hillside and reduce its slope is the cut-and-fill method. Material from the steeper upper part of the hillside is removed, thereby decreasing the slope angle, and is used to fill in the base. This provides some additional support at the base of the slope.

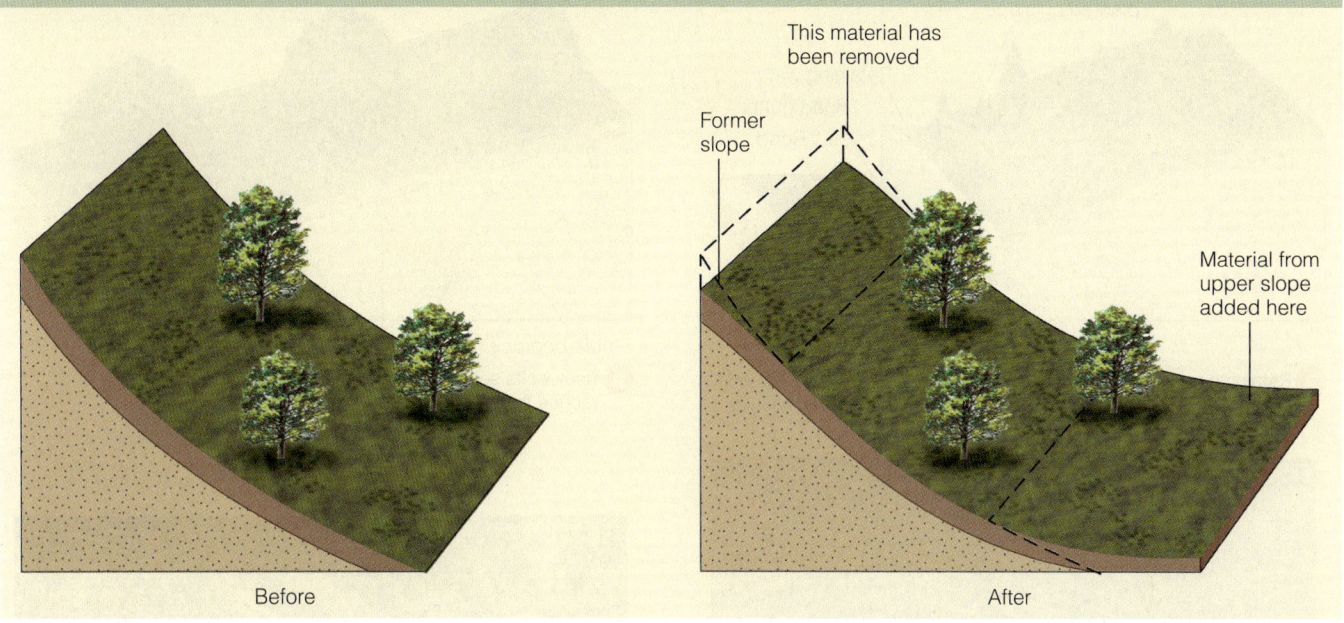

▶ **Figure 14.28 Stabilizing a Hillside by Benching**

a Another common method used to stabilize a hillside and reduce its slope is benching. This process involves making several cuts along a hillside to reduce the overall slope. Furthermore, individual slope failures are now limited in size, and the material collects on the benches.

b Benching is used in many road cuts and can be clearly seen in this photograph. **What else could be done to minimize mass wasting along the highway?** Placing fencing along the benches will stop larger rocks from falling on the highway, and planting vegetation will reduce erosion of the slope and decrease the amount of subsurface water.

▶ **Figure 14.29** Retaining Walls Help Reduce Landslides

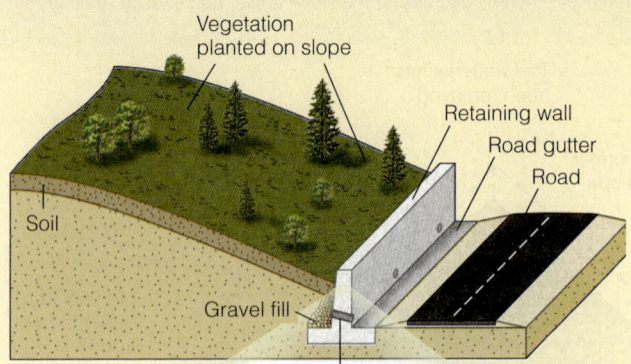

a Retaining walls anchored into bedrock, backfilled with gravel, and provided with drainpipes can support a slope's base and reduce landslides.

b A steel retaining wall built to stabilize the slope and keep falling and sliding rocks off the highway.

▶ **Figure 14.30** Rock Bolts and Wire Mesh Help Reduce Landslides

a Rock bolts secured in bedrock can help stabilize a slope and reduce landslides.

b Rock bolts and wire mesh are used to secure rock on a steep hillside in Brisbane, Australia.

Section 14.4 Summary

- A thorough geologic investigation of an area is the most important thing that can be done to eliminate or minimize the damaging effects of mass wasting. Such an investigation includes mapping and recognizing former landslide areas, evidence of past mass wasting, and present conditions of potential mass wasting, as well as soil and bedrock analysis. The information derived from such a study is used to produce slope-stability maps of the area in question, which can be used by citizens, private, and government agencies to assess the risks of mass wasting in the area.

- Although most large mass movements cannot be prevented, areas susceptible to mass wasting can be partially or wholly stabilized by building retaining walls, draining excess water, regrading slopes, and planting vegetation.

What Would You Do?

You've found your dream parcel of land in the hills of northern Baja California, where you plan to retire someday. Because you want to make sure the area is safe to build a house, you decide to do your own geologic investigation of the area to make sure there aren't any obvious geologic hazards. What specific things would you look for that might indicate mass wasting in the past? Even if there is no obvious evidence of rapid mass wasting, what features would you look for that might indicate a problem with slow types of mass wasting such as creep?

Review Workbook

ESSENTIAL QUESTIONS SUMMARY

14.1 Introduction
- *What is mass wasting?*

Mass wasting is the downslope movement of material under the direct influence of gravity. It frequently results in loss of life as well as millions of dollars in damage annually.

14.2 Factors That Influence Mass Wasting
- *What is a slope's shear strength and how is it related to the force of gravity?*

The forces that help maintain slope stability, including the material's strength and cohesion, internal friction between grains, and any external support of the slope, determine a slope's shear strength. Opposing a slope's shear strength is the force of gravity. Whenever the gravitational force acting on a slope exceeds the slope's shear strength, slope failure and thus mass movement occur.

- *What are the major factors that cause mass wasting?*

The major factors causing mass wasting include slope angle, weathering and climate, water content, vegetation, and overloading. Slope angle is probably the major cause of slope failure. Generally the steeper the slope, the less stable it is and the more susceptible it is to failure. Mass wasting is more likely to occur in loose or poorly consolidated slope material than in bedrock. Large quantities of water from melting snow or heavy storms increase the likelihood of slope instability. Vegetation can help absorb water from rains, and its root network can help stabilize a slope. Overloading, almost always from human activity such as dumping, filling, or piling up of material, can increase water pressure within the slope material, which decreases its shear strength and thus weakens the slope material.

- *How does the relationship between the topography and geology of an area affect slope stability?*

Generally, rocks that are horizontal or dipping in the opposite direction of a hillside's slope are more stable than rocks that dip in the same direction as the slope.

- *What are the triggering mechanisms of mass wasting?*

The most common triggering mechanisms are earthquakes and excessive amounts of water, although anything that disturbs the slope's equilibrium will result in mass wasting.

14.3 Types of Mass Wasting
- *How are mass movements classified?*

Mass movements are generally classified on the basis of their rate of movement (rapid versus slow), type of movement (falling, sliding, or flowing), and type of material (rock, soil, or debris).

- *What are rockfalls?*

Rockfalls are a common mass movement in which rocks free-fall. They are common along steep canyons, cliffs, and road cuts.

- *What are slides and how are they recognized?*

Slides involve movement of material along one or more surfaces of failure. Two types of slides are generally recognized. Slumps are rotational slides that involve movement along a curved surface and are most common in poorly consolidated or unconsolidated material. Rock slides occur when movement takes place along a more or less planar surface, and they usually involve solid pieces of rock.

- *What are the different types of flows?*

Several types of flows are recognized on the basis of their rate of movement (rapid versus slow), type of material (rock, sediment, or soil), and amount of water. Mudflows consist of mostly clay- and silt-sized particles and contain up to 30% water. They are most common in semiarid and arid environments. Debris flows are composed of larger particles and contain less water than mudflows. Earthflows move more slowly than either debris flows or mudflows. Quick clays are clays that spontaneously liquefy and flow like water when they are disturbed. Solifluction is the slow downslope movement of water-saturated surface material and is most common in areas of permafrost. Creep is imperceptible downslope movement of soil or rock and is the most widespread of all types of mass wasting.

- *What are complex movements?*

Complex movements are combinations of different types of mass movements in which no single type is dominant. Most complex movements involve sliding and flowing.

14.4 Recognizing and Minimizing the Effects of Mass Movement
- *What can be done to eliminate or minimize the effects of mass wasting?*

The most important factor in reducing or eliminating the damaging effects of mass wasting is a thorough geologic investigation to outline areas susceptible to mass movements. Although mass movement cannot be eliminated, its effects can be minimized by building retaining walls, draining excess water, regrading slopes, and planting vegetation.

ESSENTIAL TERMS TO KNOW

- complex movement (p. 443)
- creep (p. 441)
- debris flow (p. 438)
- earthflow (p. 438)
- mass wasting (p. 426)
- mudflow (p. 435)
- permafrost (p. 439)
- quick clay (p. 438)
- rapid mass movement (p. 431)
- rockfall (p. 431)
- rock slide (p. 434)
- shear strength (p. 427)
- slide (p. 432)
- slow mass movement (p. 431)
- slump (p. 432)
- solifluction (p. 439)

REVIEW QUESTIONS

1. Discuss some of the ways slope stability can be maintained so as to reduce the likelihood of mass movements.
2. Discuss how topography and the underlying geology contribute to slope failure.
3. Why are slumps such a problem along highways and railroad tracks in areas with topographic relief?
4. Discuss how the different factors that influence mass wasting are interconnected.
5. How could mass wasting be recognized on other planets or moons? What would that tell us about the geology and perhaps the atmosphere of the planet or moon on which it occurred?
6. If an area has a documented history of mass wasting that has endangered or taken human life, how should people and governments prevent such events from happening again? Are most large mass-wasting events preventable or predictable?
7. Why is it important to know about the different types of mass wasting?
8. How would removing preexisting vegetation tend to affect most slopes in humid regions with respect to mass wasting?
9. Why is creep so prevalent? Why does it do so much damage? What are some of the ways that creep might be controlled?
10. Where are rockfalls most common? What are some ways to reduce damage from them?

APPLY YOUR KNOWLEDGE

1. What potential value would a slope-stability map have for a person seeking to purchase property for a new home? Using the slope-stability map in Figure 14.25, locate the line that shows a horizontal contact between rocks of different stability. What is the potential for mass wasting along this line and why?
2. Using Figure 14.12 as an example, explain how the geologic planes of weakness in this slope plus water from rainfall influenced development of the depicted slide.
3. What features would you look for to determine whether the site on which you want to build your dream house has not been or is not likely to be subject to mass wasting?

FIELD QUESTION

1. What features of slope stabilization do you see in this photograph of a new housing development in Concord, California? You should be able to recognize at least three features.

GEOLOGY MATTERS

GEOLOGY IN FOCUS   The Vaiont Dam Disaster

On October 9, 1963, more than 240 million m³ of rock and soil slid into the Vaiont Reservoir in Italy, triggering a destructive flood that killed nearly 3000 people (▶ Figure 1). To fully appreciate the immensity of this catastrophe, consider the following: Within a period of 15 to 30 seconds, the slide filled the reservoir with a mass of debris 2 km long and as high as 175 m above the reservoir level. The impact of the debris created a wave of water that overflowed the dam by 100 m and was still more than 70 m high 1.6 km downstream. The slide also set off a blast of wind that shook houses, broke windows, and even lifted the roof off one house in the town of Casso, which is 260 m above the reservoir on the opposite side of the valley; it also set off shock waves that were recorded by seismographs throughout Europe. Given the forces generated by the slide, it is a tribute to the designer and construction engineer that the dam itself survived the disaster.

The dam was built in a glacial valley underlain by thick layers of folded and faulted limestones and clay layers that were further weakened by jointing (▶ Figure 2). Signs of previous slides in the area were obvious, and the few boreholes in the valley slopes revealed clay layers and small-scale slide planes. Despite the geologic evidence of previous mass wasting and objections to the site by some of the early investigators, construction of the Vaiont Dam began. Completed in 1960, the 264-m-high, thin-arch concrete dam was designed to provide both flood control and hydroelectric power for the region.

A combination of adverse geologic features and conditions resulting from the dam's construction contributed to the massive landslide. Among the geologic causes were the rocks themselves, which were weak to begin with and dipped in the same direction as the valley walls of the reservoir. Fractured limestones make up the bulk of the rocks and are interbedded with numerous clay beds that are particularly susceptible to slippage. Active solution of the limestones by slightly acid groundwater further weakened them by developing and expanding an extensive network of cracks, joints, fissures, and other openings.

During the 2 weeks before the slide occurred, heavy rains saturated the ground, adding extra weight and reducing the shear strength of the rocks. In addition to water from the rains, water from the reservoir infiltrated the rocks of the lower valley walls, further reducing their strength.

Soon after the dam was completed, a small slide of 1 million m³ of material occurred on the south side of the reservoir. Following this slide, it was decided to limit the amount of water in the reservoir and monitor the slope upstream from the dam for creep and small landslides. During the next 3 years, the eventual slide area moved an average of 1 cm per week. By September 1963, the rate had increased to 25 cm per day, and by October 8, it had increased to 100 cm per day. By now, the engineers realized that these were not individual blocks moving, but the entire slide area. They quickly began lowering the reservoir level, but it was too late. At 10:41 P.M. on October 9, 1963, during yet another rainstorm, the south bank of the Vaiont valley slid into the reservoir, triggering the Vaiont Dam disaster.

In addition to the lives lost, the loss of the dam and reservoir cost $490 million. Other property damage downstream cost a similar amount, and the lawsuits for personal injury and loss of life cost even more.

What can be learned from this tragedy? First of all, before construction on any dam begins, a complete and systematic appraisal of the area must be conducted. Such a study should examine the geology of the area, identify past mass movements, assess their potential for recurrence, and evaluate the effects that the project will have on the rocks, including how it will alter their shear strength over time. Without these precautions, similar disasters will occur and lives will needlessly be lost.

The Vaiont Dam disaster remains a classic example and constant reminder of the consequences of inadequate geologic and engineering studies before, during, and after a dam is built. Whether the obvious evidence of previous mass wasting coupled with unsuitable geologic conditions was ignored or simply underestimated, the combined effects of all of these factors resulted in a catastrophe of epic proportions.

Figure 1
The location of the Vaiont Dam disaster and features associated with the landslide.

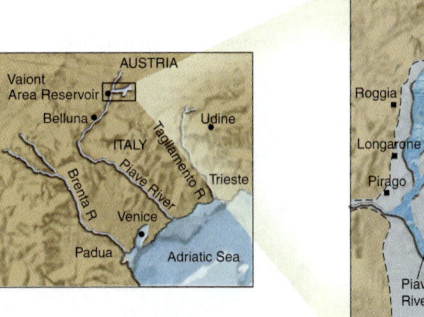
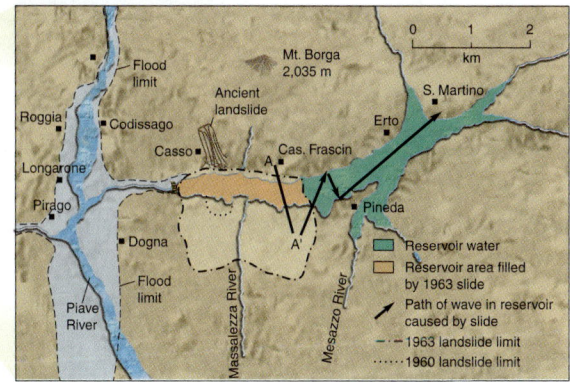

Figure 2
A generalized geologic cross section through the slide area of the Vaiont Reservoir area. The line of the section is shown in Figure 1.

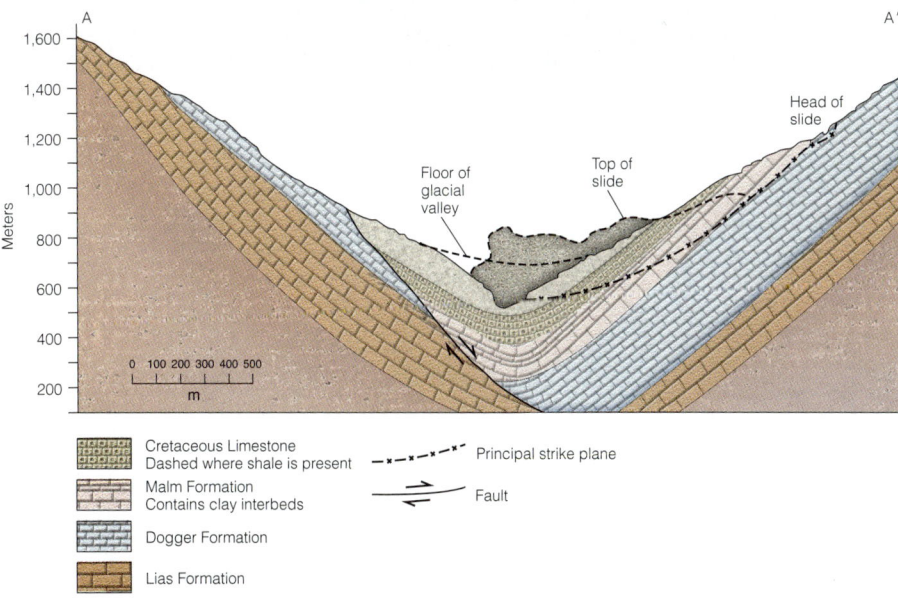

GEOLOGY MATTERS
GEOLOGY IN UNEXPECTED PLACES

New Hampshire Says Good-Bye to the "Old Man"

Imagine waking up one morning and discovering that your state symbol had suddenly disappeared during the night. That's exactly what happened to the residents of New Hampshire when they discovered that "The Old Man of the Mountain," a landmark symbolizing their state's independence and stubbornness, had collapsed from natural causes sometime during the night of May 5, 2003.

The Old Man of the Mountain, located in Franconia Notch State Park, was composed of a series of five horizontal granite ledges that had weathered over millions of years to form a man's profile (▶ Figure 1). Overlooking Profile Lake 400 m below, the Old Man measured about 13 m from chin to forehead and jutted out about 8 m from the main part of the mountain.

The natural forces that shaped the Old Man also brought about its demise. Freezing temperatures, high winds, and heavy rains all finally overcame the forces that held the rock together. Despite nearly 100 years of effort to protect the landmark from destruction, the forces of nature eventually won (▶ Figure 2). All that's left of the Old Man now is a pile of rubble at the mountain's base and some stabilizing cables and epoxy where the Old Man once looked out on the state it symbolized for so long.

Figure 1
The Old Man of the Mountain before it collapsed. The horizontal ledges that weathered to give the outcrop its distinctive shape can be clearly seen to the right of the Old Man.

Figure 2
Viewed from the opposite side of the Old Man of the Mountain, a volunteer crew in July 1997 works to protect New Hampshire's landmark symbol from the ravages of nature. Shown are workers applying epoxy to cracks in the granite. Annual maintenance also includes tightening the cables supporting the ledges and freshening and replacing brittle epoxy.

GEOLOGY MATTERS

GEOLOGY IN YOUR LIFE

Southern California Landslides

Southern California is no stranger to landslides. La Conchita, Point Fermin, Pacific Palisades, and Laguna Beach are all locations in Southern California that have suffered damaging mass movements during the past 50 years. Two regions in particular, La Conchita and Laguna Beach, have been in the news because of the landslides that destroyed numerous homes.

La Conchita is a small community along the coast at the base of a 100-m-high terrace 120 km northeast of Los Angeles, California. On March 4, 1995, following a period of heavy rains, some residents of this beach community noticed that the steep slope above their homes was slowly moving and that cracks were appearing in the walls of their houses, indicating the homes were also moving. Shortly thereafter, a 200,000-m³ slide destroyed or damaged nine homes in its path (▶ Figure 1).

Almost 10 years later, following a week of heavy rainfall in southern California, in which the hillside and previous landslide deposits were saturated with water, another landslide occurred in the same area. This time 10 people were killed and 15 homes buried under 10 m and 400,000 tons of mud (▶ Figure 2).

What went wrong and why was this situation repeated? The rocks that make up the steep sloped terrace behind La Conchita consist of soft, weak, and porous sediments that are not well lithified, and thus are easily weathered and susceptible to mass wasting. In addition, an irrigated avocado orchard sits on top of the hill, contributing to the water percolating through the porous sediments and rocks and contributing to the instability of the hillside. Add in heavy rainfall over an extended period, and you have all the ingredients for a landslide in the making. An ancient landslide area to begin with, a steep slope that has been undercut at its base by a road, well-saturated sediments decreasing the cohesion of the sediments that holds the hillside together, and continuing rains all contribute to the making of a landslide. And the potential is still there for another landslide, with no guarantee it won't happen again in the next 10 years!

Farther south in Laguna Beach, another landslide, in this case a rock slide, destroyed 18 expensive hillside homes and severely damaged about 20 others on June 1, 2005 (▶ Figure 3). Just as happened in 1978, the main triggering mechanism was probably unusually heavy winter rains, in this case the second-rainiest season on record. In this area of Southern California, the rocks dip in the same direction as the slope and contain numerous clay beds interbedded with porous sandstones. Such conditions, when combined with heavy rainfall, are ideal for rock slides. It should come as no surprise that the area where both the 1978 and 2005 rock slides occurred is also part of an ancient slide complex.

Can anything be done to prevent future landslides? The short answer is probably no. Decreasing the slope, benching the hillside, and making sure there is sufficient drainage and a good cover of vegetation are all steps that can minimize future mass wasting. But the sad fact is the geologic conditions are such that future landslides are inevitable as the landscape seeks equilibrium conditions by adjusting its slope. Add in the fact that the coastal terraces of Laguna Beach offer some of the most breathtaking views of the Pacific, and people are willing to pay a premium to live here, and you have the formula for future landslides, loss of life, and property damage.

Figure 1
La Conchita, California, is located at the base of a steep-sloped terrace. Heavy rains and irrigation of an avocado orchard (visible at the top of the terrace) contributed to the landslide that destroyed nine homes in 1995.

Figure 2
La Conchita, California, 10 years later (2005). Similar factors caused another massive landslide in the same area. The landslide can clearly be seen in the center of this photograph, and the scarp and the remains of the 1995 landslide are still visible on the right side, as is the avocado orchard in the foreground.

Figure 3
A rock slide on June 1, 2005, destroyed 18 expensive homes and damaged at least 20 others in Laguna Beach, California. Heavy rains combined with unstable underlying geology contributed to this most recent landslide in this area.

Chapter 15

Running Water

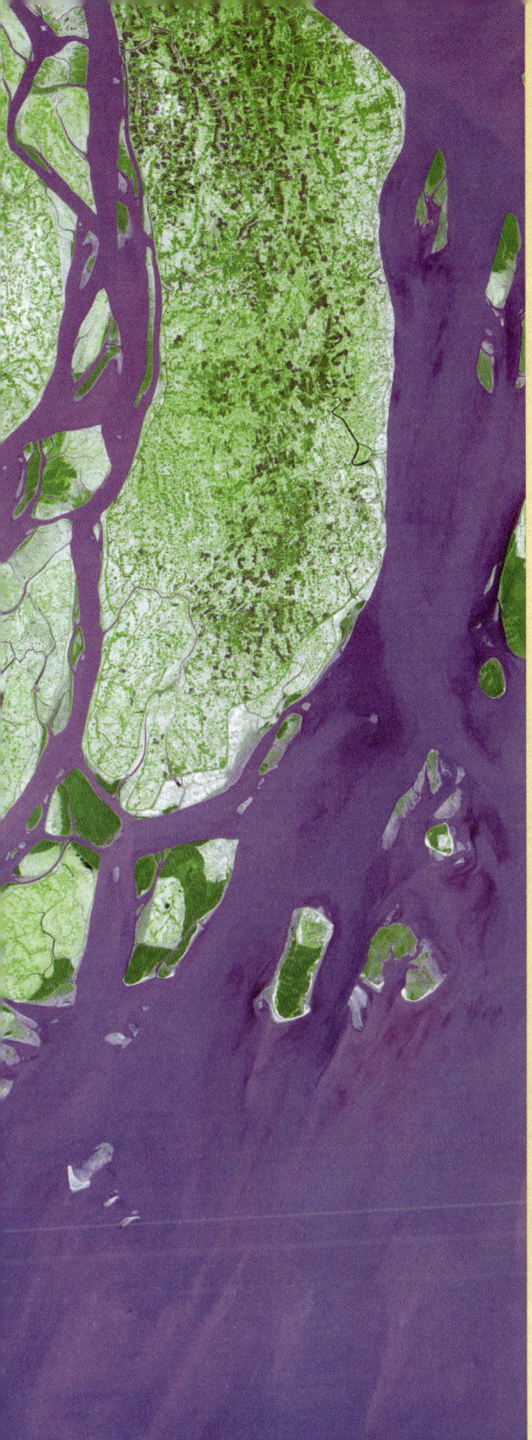

ESSENTIAL QUESTIONS TO ASK

15.1 Introduction
- Why is Earth the only terrestrial planet with abundant surface water?

15.2 Water on Earth
- What is the hydrologic cycle and why is it important?
- How does laminar flow differ from turbulent flow, and which is more important in rivers and streams?

15.3 Running Water
- Why on the average does stream velocity increase downstream even though gradient decreases in the same direction?
- How is daily discharge calculated?

15.4 Running Water, Erosion, and Sediment Transport
- How does running water erode by abrasion and hydraulic action?
- What are the dissolved, suspended, and bed loads of a stream?

15.5 Deposition by Running Water
- How and why do braided streams differ from meandering streams?
- How and where do point bars form?
- What are the similarities between deltas and alluvial fans?

15.6 Predicting and Controlling Floods
- What do scientists mean when they refer to a 20-year flood?
- Why are predictions of 100-year floods unreliable?

15.7 Drainage Systems
- What are a drainage basin and a divide?
- What is a graded stream, and what conditions might upset the graded condition?

15.8 The Evolution of Valleys
- How can headward erosion result in stream piracy?
- How does a superposed stream originate?

The Ganges River Delta
Like a spill of ink across the page: violet ribbons spreading vine-like into green.
The Ganges River forms a vast delta where it flows into the Bay of Bengal. The area surrounding the delta is called The Sundarbans, or "beautiful forest." It is one of the major habitats of the Royal Bengal Tiger, which is the national animal of both India and Bangladesh.
—A. W.

GEOLOGY MATTERS

GEOLOGY IN FOCUS:
The River Nile and the History of Egypt

GEOLOGY AT WORK:
Dams, Reservoirs, and Hydroelectric Power Plants

GEOLOGY IN UNEXPECTED PLACES:
Floating Burial Chambers and the Mississippi River Delta

 This icon, which appears throughout the book, indicates an opportunity to explore interactive tutorials, animations, or practice problems available on the Physical GeologyNow website at http://now.brookscole.com/phygeo6.

15.1 Introduction

■ *Why is Earth the only terrestrial planet with abundant surface water?*

Earth is the only terrestrial planet with abundant surface water because of its gravitational attraction and its surface temperatures. Mercury, Venus, Earth, and Mars share a similar early history of accretion, differentiation, and volcanism, but Mercury is too small and hot and the runaway greenhouse effect on Venus precludes the possibility of liquid water on these planets. Mars is too small and too cold to retain liquid water, but it does have some frozen water and trace amounts of water vapor in its atmosphere. However, satellite images reveal areas of winding valleys and deep canyons that were probably carved by running water during the planet's early history. In contrast, oceans and seas cover 71% of Earth's surface.

When we consider the interactions among Earth's systems, certainly the hydrosphere has a tremendous impact on the land surface. Of course, the hydrosphere consists mostly of water in the oceans (see Figure 12.2), but it also includes water vapor in the atmosphere and water frozen in glaciers (see Chapter 17). And finally, there is a small but important amount of water on land as groundwater (see Chapter 16), water in lakes, swamps, and bogs and, our main concern here, water confined to channels, that is, running water.

Many of us experience the power of running water when we swim, tube, or canoe in a rapidly flowing stream or river, but to really appreciate the power of moving water, we need only read the vivid accounts of floods. For example, at 4:07 P.M. on May 31, 1889, residents of Johnstown, Pennsylvania, heard a "roar like thunder" and within 10 minutes the town was destroyed. An 18-m-high wall of water roared through the town at more than 60 km/hr, sweeping up houses, debris, and hundreds of people. Entire families disappeared, and according to one account, "Thousands of people desperately tried to escape the wave. Those caught by the wave found themselves swept up in a torrent of oily, muddy water, surrounded by tons of grinding debris. . . . Many became hopelessly entangled in miles of barbed wire from the destroyed wire works."* When the flood was over, at least 2200 people were dead.

The Johnstown flood, the most deadly river flood in U.S. history, resulted from heavy rainfall and the failure of a dam. Now a catastrophe that is continuing to unfold along the U.S. Gulf Coast may surpass the Johnstown flood in terms of fatalities and certainly in the level of damage. When Hurricane Katrina roared ashore on August 29, 2005, high winds, a huge storm surge, and coastal flooding destroyed nearly everything in an area of 230,000 km². We will discuss the nature of storm surges more fully in Chapter 19, so here we will simply note that storm surges and intense rainfall cause coastal flooding during hurricanes as water is driven onshore.

A storm surge generated by hurricane Katrinia leveled much of Gulfport and Biloxi, Mississippi, and it had an enormous effect on New Orleans, Louisiana, too. Even when New Orleans was founded in 1718, engineers warned that building a community on a swampy, subsiding parcel of land between Lake Pontchartrain to the north and the Mississippi River to the south was risky; now most of the city lies below sea level (▶ Figure 15.1). In any case, New Orleans was nearly surrounded by levees (earthen embankments) and seawalls (structures of concrete and steel) to protect it from the lake and river.

When Hurricane Katrina came ashore, the levees initially held, but on the next day they were breached and about

▶ **Figure 15.1 Hurricane Katrina, 2005**

a On August 29, 2005, Hurricane Katrina made landfall along the Gulf Coast. In this image the eye of the storm is passing just to the east of New Orleans, Louisiana.

b Although many areas were hard hit by Katrina, New Orleans was extensively flooded when the levees failed that were built to protect the city from Lake Pontchartrain and the Mississippi River.

---

*National Park Service—U.S. Department of Interior, Johnstown Information Service Online.

80% of the city was flooded (▶ Figure 15.1). And because New Orleans is mostly below sea level the floodwaters cannot drain out naturally. In fact, the city has 22 pumping stations to remove water from normal rainstorms, but as the city flooded the pumps were overwhelmed, and when the power failed the pumps were useless.

Scientists, engineers, and some politicians had warned of just such a catastrophe for many years. They were aware that during a hurricane the size of Katrina that the levees would likely fail. In fact, the political leaders in Louisiana had pleaded for years for funds to strengthen the levees, but for one reason or another the funds were never allocated. "What do we do now?" is a question that will be debated for many years. Perhaps some of you will be involved in the decision-making regarding what to do to protect the Gulf Coast from future storm surges and coastal flooding.

Section 15.1 Summary

- Earth is the only terrestrial planet with a hydrosphere made up of oceanic water as well as water on land in lakes, swamps, bogs, and stream and river channels.

15.2 Water on Earth

According to one estimate, there are 1.36 billion km³ of water on Earth, nearly all of it (97.2%) in the oceans. Most of the rest (2.15%) is frozen in glaciers on land, especially in Antarctica and Greenland, and 0.625% of all water is in the atmosphere, lakes, swamps, and bogs, leaving about 0.0001% in streams and rivers at any one time. Nevertheless, as this tiny quantity of water moves downslope through channels, it modifies Earth's land surface more than any other surface process, except in areas covered by glaciers. Some parts of deserts are little modified by running water, but even in most desert areas where channels are dry most of the time, the effects of running water are conspicuous (▶ Figure 15.2).

Much of our discussion of running water is descriptive, but always be aware that streams and rivers are dynamic systems that continuously adjust to change. For instance, paving in urban areas increases surface runoff to waterways, and other human activities such as building dams and impounding reservoirs also alter the dynamics of stream and river systems. Of course, natural changes, too, affect the complex interacting parts of these systems.

The Hydrologic Cycle

■ *What is the hydrologic cycle and why is it important?*

The hydrologic cycle involves the continuous recycling of water from the oceans, through the atmosphere, to the continents, and back to the oceans in the **hydrologic cycle.** It is our only source of fresh (nonsaline) water. The connection between clouds and precipitation is obvious, but where does the moisture for rain and snow come from in the first place? We noted that 97.2% of all water on Earth is in the oceans, so you might think that the oceans are the ultimate source of precipitation, which is correct. The hydrologic cycle is powered by solar radiation and is possible because water changes easily from liquid to gas (water vapor) under Earth's surface conditions (▶ Figure 15.3). About 85% of all water that enters the atmosphere each year comes from a layer about 1 m thick that evaporates from the oceans. The remaining 15% comes from water on land, but this water originally came from the oceans as well.

Regardless of its source, water vapor rises into the atmosphere, where the complex processes of cloud formation and condensation take place. About 80% of the world's precipitation falls directly back into the oceans, so in this case the hydrologic cycle is a three-step process of evaporation, condensation, and precipitation. But for the 20% of all precipitation that falls on land, the hydrologic cycle is more complex, involving evaporation, condensation, movement of water vapor from the oceans to land, and precipitation that eventually makes it back to the oceans, although its return trip may be convoluted. Some precipitation evaporates as it falls and reenters the cycle, but about 36,000 km³ falls on land per year and returns to the oceans by **runoff**—that is, the surface flow in streams and rivers.

Of course not all precipitation is carried directly back to the oceans. Some is temporarily stored in lakes and swamps, snowfields and glaciers, or it seeps below the surface where it enters the groundwater system (see Chapter 16).

▶ **Figure 15.2 Golden Canyon after a Flash Flood** This boulder-littered channel is Golden Canyon in Death Valley National Park, California. Water rarely flows through the canyon, but when it does, it erodes vigorously and transports large quantities of sediment. In 1976 nearly 6 cm of rain in 4 days caused a flash flood that destroyed a paved road.

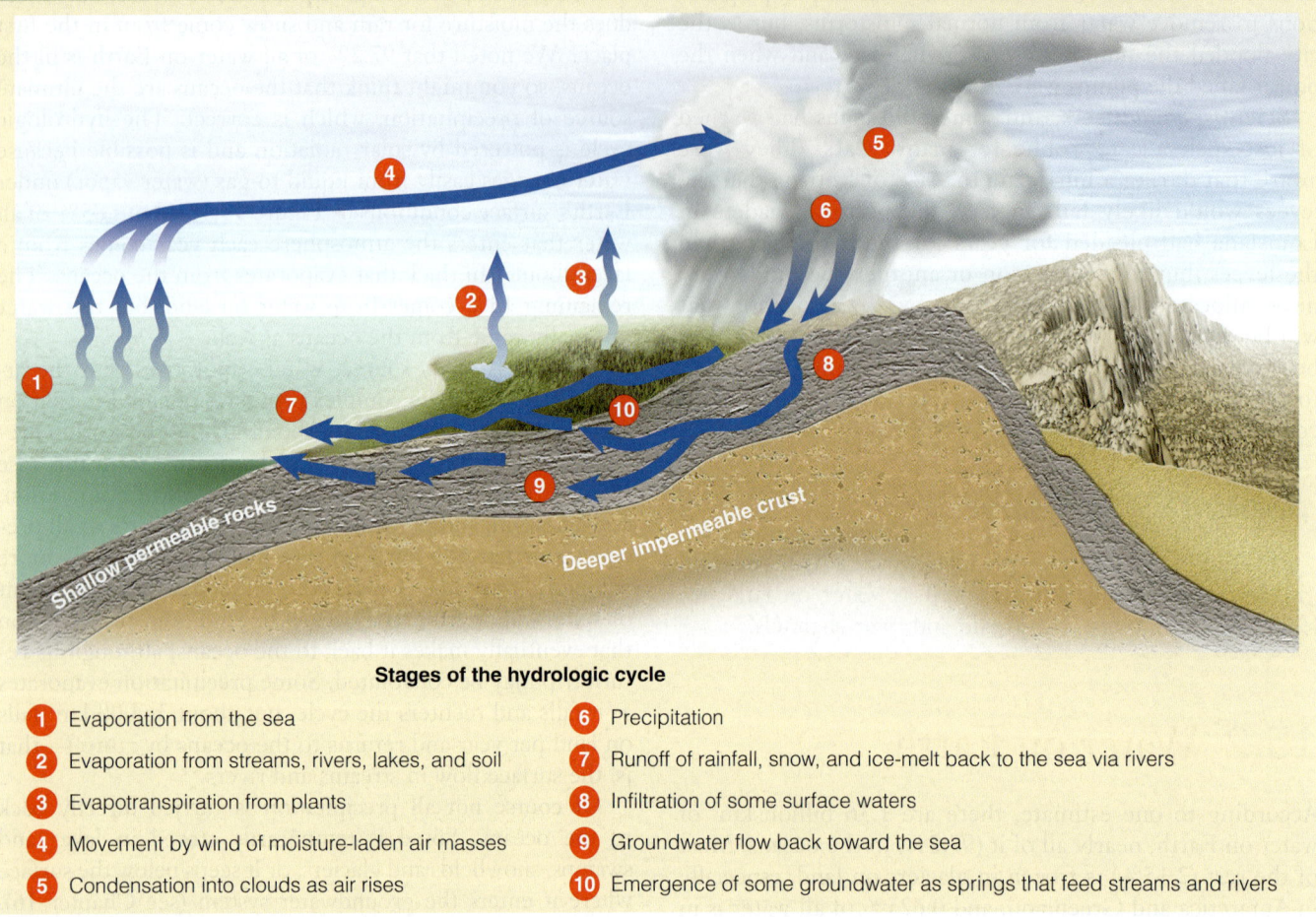

Geo-focus Figure 15.3 Stages of the Hydrologic Cycle Water is recycled from the oceans to land and back to the oceans. Note the term *permeable* in the illustration. It refers to how readily fluids move through soil, sediment, and rocks.

Stages of the hydrologic cycle

1. Evaporation from the sea
2. Evaporation from streams, rivers, lakes, and soil
3. Evapotranspiration from plants
4. Movement by wind of moisture-laden air masses
5. Condensation into clouds as air rises
6. Precipitation
7. Runoff of rainfall, snow, and ice-melt back to the sea via rivers
8. Infiltration of some surface waters
9. Groundwater flow back toward the sea
10. Emergence of some groundwater as springs that feed streams and rivers

Water may be effectively stored in these reservoirs in the hydrologic cycle for thousands of years, but eventually glaciers melt, lakes and groundwater feed streams and rivers, so this water also returns to the oceans. Even the water used by plants evaporates, a process known as *evapotransporation*, and returns to the atmosphere. So, in the long run all precipitation that falls on land makes it back to the oceans (▶ Figure 15.3).

Fluid Flow

■ *How does laminar flow differ from turbulent flow, and which is more important in rivers and streams?*

In laminar flow, all flow is in one direction with little or no mixing, whereas in turbulent flow, which is most common in streams and rivers, complex mixing takes place in the moving fluid. Any solid is a rigid substance that retains its shape unless deformed by a force, but fluids—that is, liquids and gases—have no strength so they flow in response to any force no matter how slight. Liquid water, our concern here, flows downslope in response to gravity, but its flow may be either *laminar* or *turbulent*. In laminar flow, the paths taken by individual water molecules, called *streamlines*, parallel one another with little or no mixing between adjacent layers in the fluid (▶ Figure 15.4a). All flow is in one direction only, and it remains unchanged through time. In turbulent flow, however, streamlines intertwine and cause complex mixing within the moving fluid (▶ Figure 15.4b). If we could trace the path of a single water molecule in turbulent flow, it may move in any direction at a particular time, although its overall movement would be in the direction of flow.

You can easily observe laminar flow in viscous fluids such as cold motor oil or syrup, both of which flow slowly and with difficulty. When heated, though, they flow more readily. Viscosity is also a consideration in running water, but water's viscosity is so low that it most often moves by turbulent flow. Temperature also influences viscosity, but the primary controls are velocity and the roughness of the surface over which water flows.

Some mudflows and lava flows certainly move by laminar flow, but examples of laminar flow in running water are difficult to find. It takes place when groundwater moves slowly through the tiny pores in soil, sediment, and rocks (see Chapter 16), but velocity and roughness in almost all surface flow ensure that it is fully turbulent. Even if laminar flow

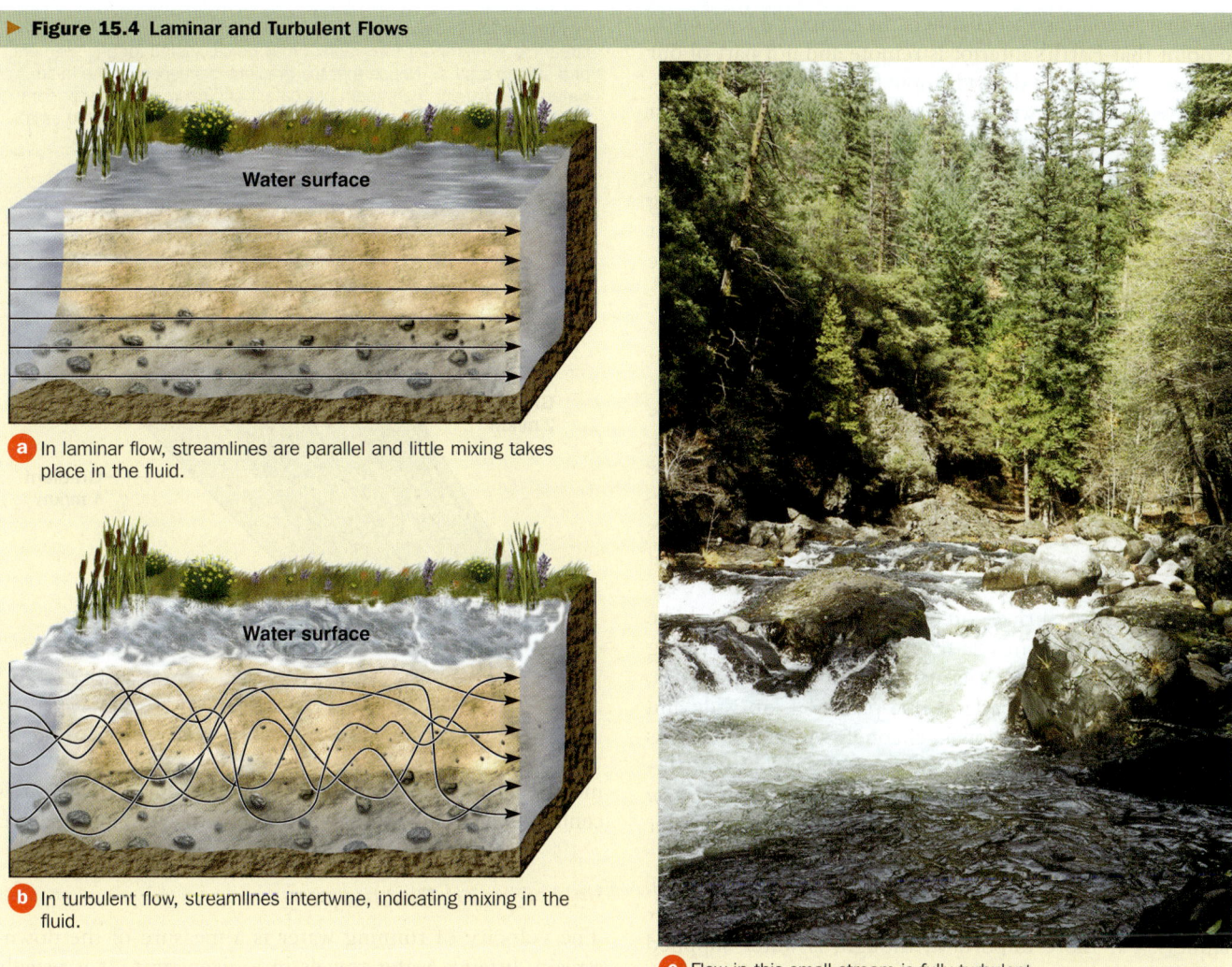

Figure 15.4 Laminar and Turbulent Flows

a In laminar flow, streamlines are parallel and little mixing takes place in the fluid.

b In turbulent flow, streamlines intertwine, indicating mixing in the fluid.

c Flow in this small stream is fully turbulent.

should take place at the surface, it is too slow and too shallow to cause any erosion. Turbulent flow, in contrast, is much more energetic and therefore capable of causing erosion and sediment transport.

Infiltration Capacity

Surface runoff during a rainstorm depends on **infiltration capacity,** the rate at which the ground soaks up water. Only when infiltration capacity is exceeded does water accumulate on the surface and runoff begins. Several factors influence how surface materials absorb water. The intensity and duration of rainfall are the most obvious; if rain is absorbed as fast as it falls, no surface runoff takes place. The harder and longer it rains, the more likely infiltration capacity will be exceeded. Loosely packed, dry soil absorbs water faster than tightly packed, wet soil, and accordingly more rain must fall on the former before runoff begins. But regardless of the initial condition of surface materials, once they are saturated, excess water collects on the surface and, if on a slope, it moves downhill.

Section 15.2 Summary

- Of the estimated 1.36 billion km³ of water on Earth more than 97% is in the oceans, but water continuously evaporates from the oceans, condenses, falls as precipitation on land, and returns to the oceans in the hydrologic cycle.

- Running water may move by either laminar or turbulent flow, but the latter is far more common in surface runoff and much more important for erosion and sediment transport. For surface runoff to take place, the infiltration capacity of surface materials must be exceeded.

15.3 Running Water

What do we mean by *running water*? Simply put, it is any surface water that moves from higher to lower areas as more or less continuous sheets or, much more commonly, the water that is confined to channels. In Chapter 11 we made the point that Earth's interior is remote and not part of our experience and thus a difficult topic for beginning students. Not so with running water—we are all familiar with at least some aspects of this phenomenon, although the terminology used by geologists may be new to you. Nevertheless, everyone has some appreciation for the power of running water and erosion caused by runoff. We have made the point that, other than glacier-covered areas and parts of some deserts, the effects of running water are ubiquitous.

Sheet Flow and Channel Flow

Even on steep slopes, flow is initially slow and hence causes little or no erosion. As water moves downslope, though, it accelerates and may move by *sheet flow*, a more or less continuous film of water flowing over the surface. Sheet flow is not confined to depressions, and it accounts for *sheet erosion*, a particular problem on some agricultural lands (see Chapter 6).

In *channel flow*, surface runoff is confined to troughlike depressions that vary in size from tiny rills with a trickling stream of water to the Amazon River in South America, which is 6450 km long and at one place 2.4 km wide and 90 m deep. We describe flow in channels with terms such as *rill, brook, creek, stream,* and *river*, most of which are distinguished by size and volume. Here we use the terms *stream* and *river* more or less interchangeably, although the latter usually refers to a larger body of running water.

Streams and rivers receive water from several sources, including sheet flow and rain that falls directly into their channels. Far more important, though, is water supplied by soil moisture and groundwater, both of which flow downslope and discharge into waterways. In areas where groundwater is plentiful, streams and rivers maintain a fairly stable flow year-round because their water supply is continuous. In contrast, the amount of water in streams and rivers in arid and semiarid regions fluctuates widely because they depend more on infrequent rainstorms and surface runoff for their water.

Stream Gradient

Water in any channel flows downhill over a slope known as its **gradient.** Suppose a river has its headwaters (source) 1000 m above sea level and it flows 500 km to the sea, so it drops vertically 1000 m over a horizontal distance of 500 km. Its gradient is found by dividing the vertical drop by the horizontal distance, which in this example is 1000 m/500 km = 2 m/km (▶ Figure 15.5). On the average this river drops vertically 2 m for every kilometer along its course.

In the preceding example we calculated the average gradient for a hypothetical river. Gradients vary not only among channels but even along the course of a single channel. Rivers and streams are steeper in their upper reaches (near their headwaters) where they may have gradients of several tens of

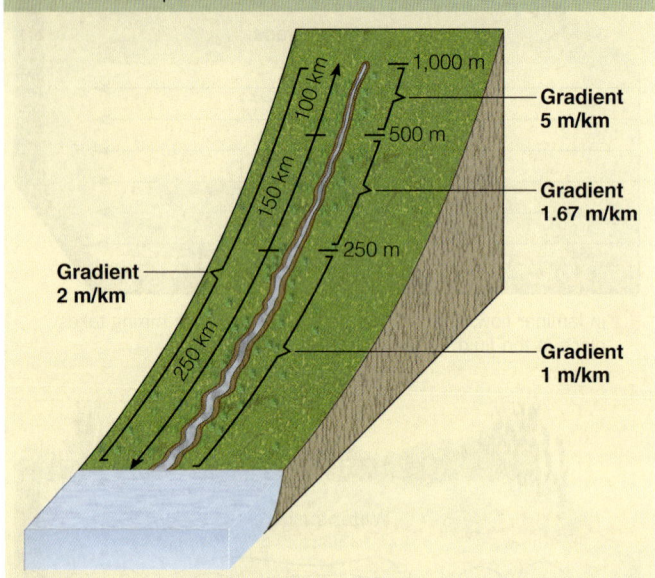

▶ **Figure 15.5 Calculating Gradient** The average gradient for this stream is 2 m/km, but the gradient may be calculated for any segment of a stream system. Notice that the gradient is steepest in the headwaters area (where the stream originates) and decreases downstream. In the lower reaches of some large rivers, the gradient is as little as a few centimeters per kilometer.

meters per kilometer, but they have gradients of only a few centimeters per kilometer where they discharge into the sea.

Velocity and Discharge

The **velocity** of running water is a measure of the downstream distance water travels in a given time. It is usually expressed in meters per second (m/sec) or feet per second (ft/sec), and it varies across a channel's width as well as along its length. Water moves more slowly and with greater turbulence near a channel's bed and banks because friction is greater there than it is some distance from these boundaries (▶ Figure 15.6a). Channel shape and roughness also influence flow velocity. Broad, shallow channels and narrow, deep channels have proportionately more water in contact with their perimeters than channels with semicircular cross sections (▶ Figure 15.6b). So if other variables are the same, water flows faster in a semicircular channel because there is less frictional resistance. As one would expect, rough channels, such as those strewn with boulders, offer more frictional resistance to flow than do channels with a bed and banks composed of sand or mud.

- *Why on the average does stream velocity increase downstream even though gradient decreases in the same direction?*

Intuitively you might suspect that gradient is the most important control on velocity—the steeper the gradient, the higher the velocity. In fact, a river's average velocity actually increases downstream even though its gradient decreases! Keep in mind that we are talking about average velocity for a long segment

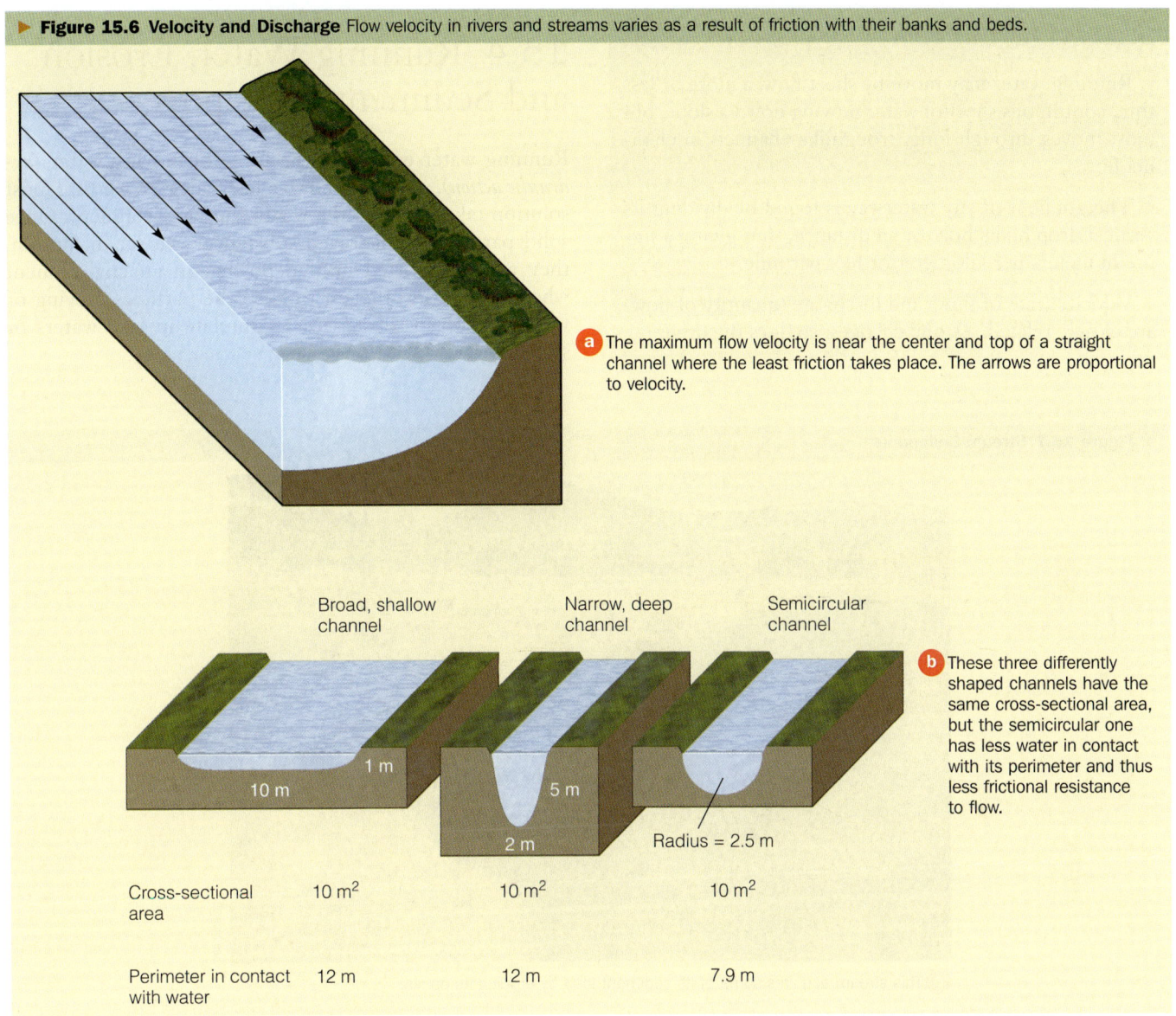

▶ **Figure 15.6 Velocity and Discharge** Flow velocity in rivers and streams varies as a result of friction with their banks and beds.

a The maximum flow velocity is near the center and top of a straight channel where the least friction takes place. The arrows are proportional to velocity.

b These three differently shaped channels have the same cross-sectional area, but the semicircular one has less water in contact with its perimeter and thus less frictional resistance to flow.

of a channel, not velocity at a single point. Three factors account for this downstream increase in velocity. First, velocity increases even with decreasing gradient in response to the acceleration of gravity. Second, the upstream parts of channels tend to be boulder-strewn, broad, and shallow, so frictional resistance to flow is high, whereas downstream segments of the same channels are more semicircular and have banks composed of finer materials. And finally, the number of smaller tributaries joining a larger channel increases downstream. Thus the total volume of water (discharge) increases, and increasing discharge results in higher velocity.

■ *How is daily discharge calculated?*

We mentioned discharge in the preceding paragraph but noted only that it refers to the volume of water. More specifically, **discharge** is the volume of water that passes a particular point in a given period of time. It is found by knowing the dimensions of a water-filled channel—that is, its cross-sectional area (A) and its flow velocity (V). Discharge (Q) is then calculated with the formula $Q = VA$ and is expressed in cubic meters per second (m^3/sec) or cubic feet per second (ft^3/sec). So, consider a stream with a channel 148 m wide and 2.6 m deep, with a flow velocity of 0.3 m/sec. The discharge for a stream with this flow velocity and these dimensions is 115.44 m^3/sec and 9,974,016 m^3/day. The Mississippi River has an average discharge of 18,000 m^3/sec, and the average discharge for the Amazon River in South America is 200,000 m^3/sec.

In most rivers and streams, discharge increases downstream as more and more water enters a channel. However, there are a few exceptions. Because of high evaporation rates and infiltration, the flow in some desert waterways actually decreases downstream until they disappear. And even in perennial rivers and streams, discharge is obviously greatest during times of heavy rainfall and at a minimum during the dry season.

Section 15.3 Summary

- Running water may move by sheet flow, a more or less thin, continuous sheet of water moving down a slope, but most moves through long, troughlike channels in channel flow.

- The gradient of any waterway is found by dividing its vertical drop by its horizontal distance, thus giving a figure in meters per kilometer or feet per mile.

- Velocity (rate of flow) and discharge (quantity of flow) are closely related. As one changes, so does the other.

15.4 Running Water, Erosion, and Sediment Transport

Running water erodes by the force of moving water *(hydraulic action)*, *abrasion*, and to a lesser degree *solution*; most solution takes place in the weathering environment. Once solid particles or dissolved substances are set in motion, they are transported until deposited in an environment where flow is insufficient to keep the particles moving or where dissolved substances accumulate in lake waters or seawater.

▶ Figure 15.7 Stream Sediments

a This stream acquires some of its sediment load by eroding its banks.

b Some of the sediment in the Snake River of Idaho comes from these talus cones that accumulated as a result of mass wasting.

Streams and rivers possess two kinds of energy: potential and kinetic. *Potential energy* is the energy of position, such as the energy of water at high elevation or on the upstream side of a dam. During stream flow, potential energy is converted to *kinetic energy*, the energy of motion. Much of this kinetic energy is dissipated within streams by fluid turbulence, but some of it is available for erosion and transport. You already know that erosion involves the removal of minerals, rock fragments, and dissolved materials from their source area, so the sediment in a stream or river consists of a *dissolved load* and solid particles.

Because the dissolved load of a stream is invisible, it is commonly overlooked, but it is an important part of the total sediment load. Some of it is acquired from the streambed and banks where soluble rocks such as limestone and dolostone are exposed, but much of it is carried into streams by sheet flow and by groundwater.

■ *How does running water erode by abrasion and hydraulic action?*

The solid sediment ranges from clay-sized particles to large boulders, much of it supplied by mass wasting, but some is derived directly from the streambed and banks (▶ Figure 15.7). The power of running water, called **hydraulic action,** is sufficient to set particles in motion. Everyone has seen the results of hydraulic action, although perhaps not in streams. For example, if the flow from a garden hose is directed onto loose soil, hydraulic action soon gouges out a hole.

Another process of erosion in streams is **abrasion,** in which exposed rock is worn and scraped by the impact of solid particles. If running water contains no sediment, no abrasion of rock surfaces results, but if the water is transporting sand and gravel, the impact of these particles abrades exposed rock surfaces (▶ Figure 15.8a). *Potholes* in the beds of streams are

▶ **Figure 15.8 Abrasion** Running water that carries sand and gravel effectively erodes by abrasion.

a The rocks just above water level have been abraded, which accounts for their polish. Notice that the rocks higher above the water have not been polished.

b These potholes in the bed of the Chippewa River in Ontario, Canada, are about 1 m across. Two potholes have merged to form a larger composite pothole.

c These stones from a pothole measure 7–9 cm across. They are spherical and smooth because of abrasion.

another manifestation of abrasion (▶ Figure 15.8b). These circular to oval holes form where eddying currents containing sand and gravel swirl around and erode depressions into rock.

- **What are the dissolved, suspended, and bed loads of a stream?**

A stream's dissolved load is made up of materials in solution, whereas its suspended load consists of solid particles kept up in the water by turbulence, and its bed load is sand- and gravel-sized particles that are too large to be lifted above the stream's bed. Running water transports a **dissolved load** consisting of materials taken into solution during chemical weathering, and it also transports sedimentary particles. The smallest of these particles, mostly silt and clay, are kept suspended by fluid turbulence and transported as a **suspended load** (▶ Figure 15.9a). The Mississippi River transports nearly 200 million metric tons of suspended load past Vicksburg, Mississippi, each year, and the Yellow River in China caries almost four times as much. Particles transported in suspension are deposited only where turbulence is minimal, as in lakes and lagoons.

Streams and rivers also transport a **bed load,** consisting of larger particles of sand and gravel (▶ Figure 15.9a). Because fluid turbulence is insufficient to keep sand and gravel suspended, they move along the streambed. However, part of the bed load can be suspended temporarily, as when an eddying current swirls across a streambed and lifts sand grains into the water. These particles move forward at approximately the flow velocity, but at the same time they settle toward the streambed where they come to rest, to be moved again later by the same process. This process of intermittent bouncing and skipping along the streambed is *saltation* (▶ Figure 15.9a).

Particles too large to be suspended even temporarily are transported by rolling or sliding (▶ Figure 15.9a). Obviously, greater flow velocity is required to move particles of these sizes. The maximum-sized particles that a stream can move define its *competence*, a factor related to flow velocity. Figure 15.9b shows the velocities required to erode, transport, and deposit particles of various sizes. As expected, high velocity is necessary to erode and transport gravel-sized particles, whereas sand is eroded and transported at lower velocities. Notice, though, that high velocity is also needed to erode clay because clay deposits are very cohesive; the tiny clay particles adhere to one another and only energetic flow can disrupt them. Once eroded, however, very little energy is needed to keep clay particles in motion.

Capacity is a measure of the total load a stream can carry. It varies as a function of discharge; with greater discharge, more sediment can be carried. Capacity and competence may seem similar, but they are actually related to different aspects of stream transport. For instance, a small, swiftly flowing stream may have the competence to move gravel-sized particles but not to transport a large volume of sediment, so it has a low capacity. A large, slow-flowing stream, on the other hand, has a low competence but may have a very large suspended load and hence a large capacity.

Section 15.4 Summary

- Running water erodes mostly by hydraulic action (the direct impact of moving water) and by abrasion (the scraping and wearing of rock by sediment transported in running water). It also erodes by solution, but most solution takes place in the weathering environment.

- Streams and rivers transport sediment as dissolved load or materials in solution, as suspended load, consisting of silt and clay, and as a bed load of larger particles, mostly sand and gravel.

▶ **Figure 15.9 Sediment Transport and Deposition**

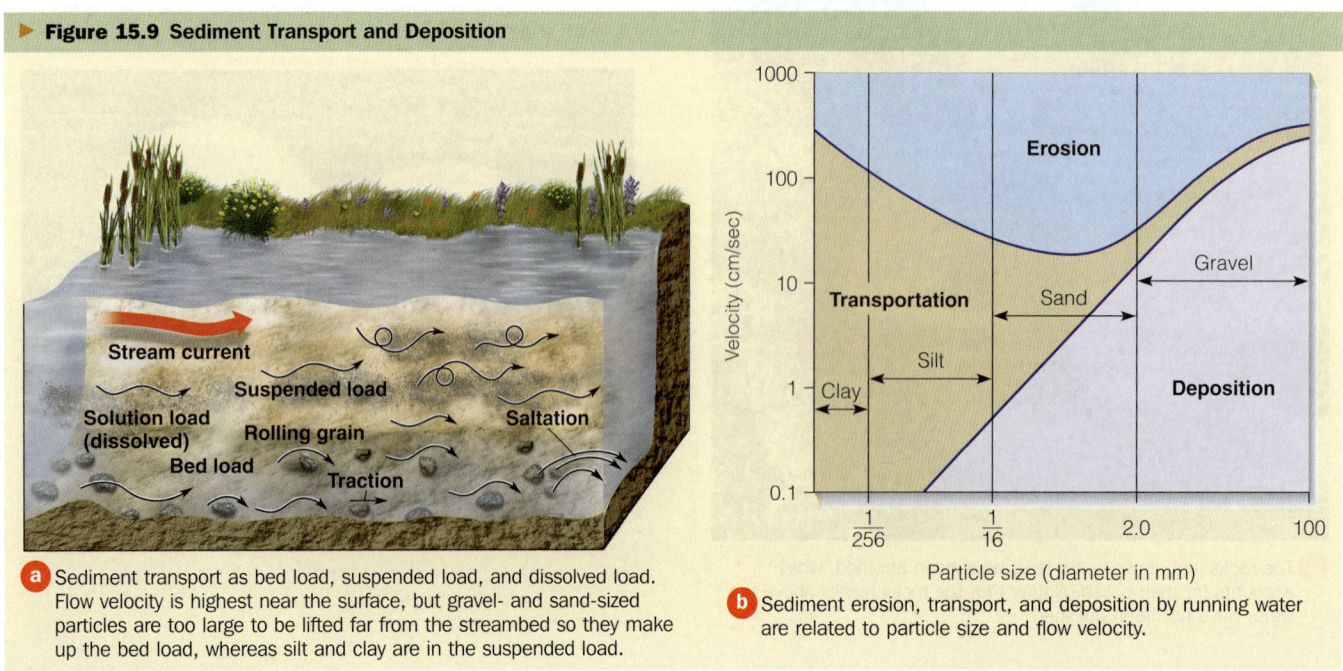

a Sediment transport as bed load, suspended load, and dissolved load. Flow velocity is highest near the surface, but gravel- and sand-sized particles are too large to be lifted far from the streambed so they make up the bed load, whereas silt and clay are in the suspended load.

b Sediment erosion, transport, and deposition by running water are related to particle size and flow velocity.

15.5 Deposition by Running Water

Sediment transport might be lengthy, but deposition eventually takes place. Some of the sediment now being deposited in the Gulf of Mexico by the Mississippi River came from such distant sources as Pennsylvania, Minnesota, and Alberta, Canada. Deposits may accumulate along the way in channels, on adjacent floodplains, or where rivers and streams discharge from mountains onto nearby lowlands or where they flow into lakes or seas.

Rivers and streams constantly erode, transport, and deposit sediment, but they do most of their geologic work when they flood. Consequently, their deposits, collectively called **alluvium,** do not represent the day-to-day activities of running water, but rather the periodic sedimentation that takes place during floods. Remember from Chapter 7 that sediments accumulate in *depositional environments* characterized as continental, transitional, and marine. Deposits of rivers and streams are found in the first two of these settings; however, much of the detrital sediment found on continental margins is derived from the land and transported to the oceans by running water.

■ *How and why do braided streams differ from meandering streams?*

The Deposits of Braided Streams

Braided streams have too much sediment and deposit sand and gravel bars in their channels, thus yielding a complex of interconnected channels (▶ Figure 15.10). Meandering streams, on the other hand, have a single sinuous channel. Seen from above, braided channels resemble the complex strands of a braid. Braided channels develop when sediment exceeds transport capacity, resulting in the deposition of sand and gravel bars. During high-water stages, the bars are submerged, but when the water is low, they are exposed and divide a single channel into multiple channels. Braided streams have broad, shallow channels and are characterized as bed-load transport streams because they transport and deposit mostly sand and gravel (▶ Figure 15.10).

Braided channels are common in arid and semiarid regions that have little vegetation and high erosion rates. Streams with easily eroded banks are also likely to be braided. In fact, a stream that is braided where its banks are easily eroded may have a single sinuous or meandering channel when it flows into an area of more resistant materials. Streams fed by melting glaciers are also commonly braided because melting glacial ice yields so much sediment (see Chapter 17).

Meandering Streams and Their Deposits

Meandering streams have a single, sinuous channel with broadly looping curves known as *meanders* (▶ Figure 15.11). The channels are semicircular in cross section along straight reaches, but at meanders they are markedly asymmetric,

▶ **Figure 15.10** Braided Streams and Their Deposits

a A braided stream in Denali National Park, Alaska.

b This small braided stream is near Grindelwald, Switzerland. The deposits of both streams are mostly sand and gravel.

being deepest near the outer bank, which commonly descends vertically into the channel (▶ Figure 15.11). The outer bank is called the *cut bank* because greater velocity and turbulence on that side of the channel erode it. Despite the known dynamics of erosion along the outer banks of meanders, people commonly build houses and other structures atop cut banks. The life expectancy of these structures is short because during a single flood, several meters of erosion might take place along a cut bank. In contrast, flow velocity is at a minimum near the inner bank, which slopes gently into the channel (▶ Figure 15.11).

■ *How and where do point bars form?*

As a result of the unequal distribution of flow velocity across meanders, the cut bank erodes and deposition of a point bar takes place along the opposite side of the channel. The net effect is that a meander migrates laterally, and the channel maintains a more or less constant width because erosion on the cut bank is offset by an equal amount of deposition on the opposite side of the channel. The deposit formed in this manner is a **point bar** consisting of cross-bedded sand or, in some cases, gravel (▶ Figure 15.12). Although point bars are distinctive, meandering stream deposits are mostly mud deposited on broad floodplains (see the next section) and channel deposits (point bars) are subordinate.

It is not uncommon for meanders to become so sinuous that the thin neck of land separating adjacent meanders is eventually cut off during a flood. The valley floors of meandering streams are marked by crescent-shaped **oxbow lakes,** which are actually cutoff meanders (▶ Figures 15.11 and 15.13). Oxbow lakes may persist as lakes for some time, but they eventually fill with organic matter and fine-grained sediment carried by floods. Even after they are filled, they remain visible on floodplains.

One immediate effect of meander cutoff is an increase in flow velocity; following cutoff, the stream abandons part of its old course and flows a shorter distance, thereby increasing its gradient. Numerous cutoffs would, of course, significantly shorten a meandering stream, but streams usually establish new meanders elsewhere when old ones are cut off.

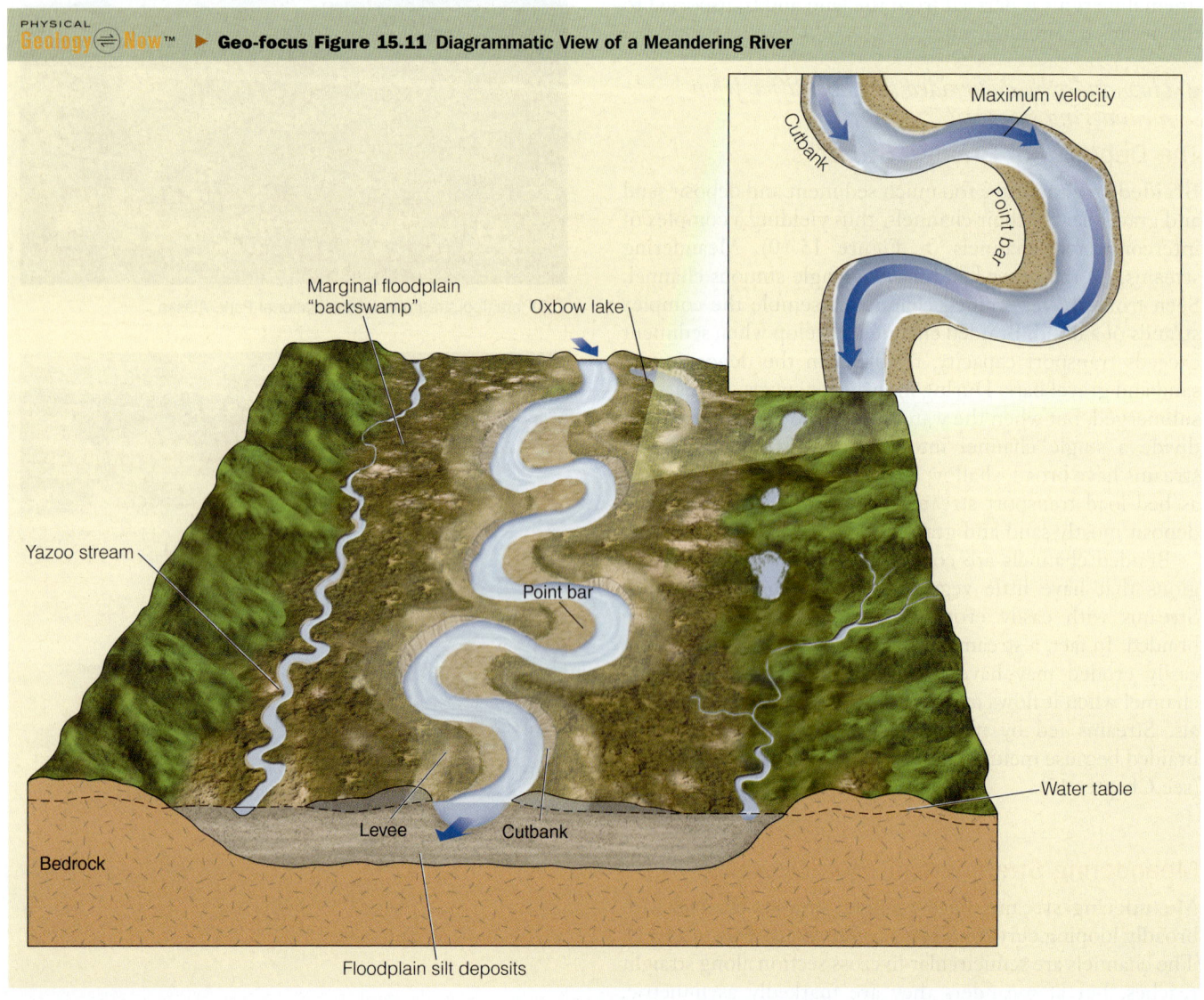

Geo-focus Figure 15.11 Diagrammatic View of a Meandering River

Floodplain Deposits

Channels periodically receive more water than they can carry, so the excess water spreads across low-lying, relatively flat **floodplains** adjacent to the channels (▶ Figure 15.11). Even small streams commonly have a floodplain, but floodplains are usually proportional to the size of the stream; thus small streams have narrow floodplains, whereas the lower Mississippi and other large rivers have floodplains many kilometers wide. Streams restricted to deep, narrow valleys have little or no floodplain.

When a meandering stream erodes its cut bank and deposits on the opposite bank, it migrates laterally across its floodplain. As it migrates laterally, a succession of point bars develops by *lateral accretion* (▶ Figure 15.14a). That is, the deposits build laterally as a result of repeated episodes of sedimentation on the inner banks of meanders, and the floodplains are composed mostly of sand and gravel.

When a stream overflows its banks and floods, the velocity of the water spilling onto the floodplain diminishes rapidly because of greater frictional resistance to flow as the

▶ **Figure 15.12 Point Bars** Point bars form on the side of a meandering channel where the flow velocity is lowest (Figure 15.11).

a) Two small point bars composed of sand in Otter Creek in Yellowstone National Park, Wyoming. Notice that the point bars are inclined into the deeper part of the channel.

b) Point bar in the Middle Fork of the San Joaquin River at Devils Postpile National Monument, California. This point bar is made up of gravel.

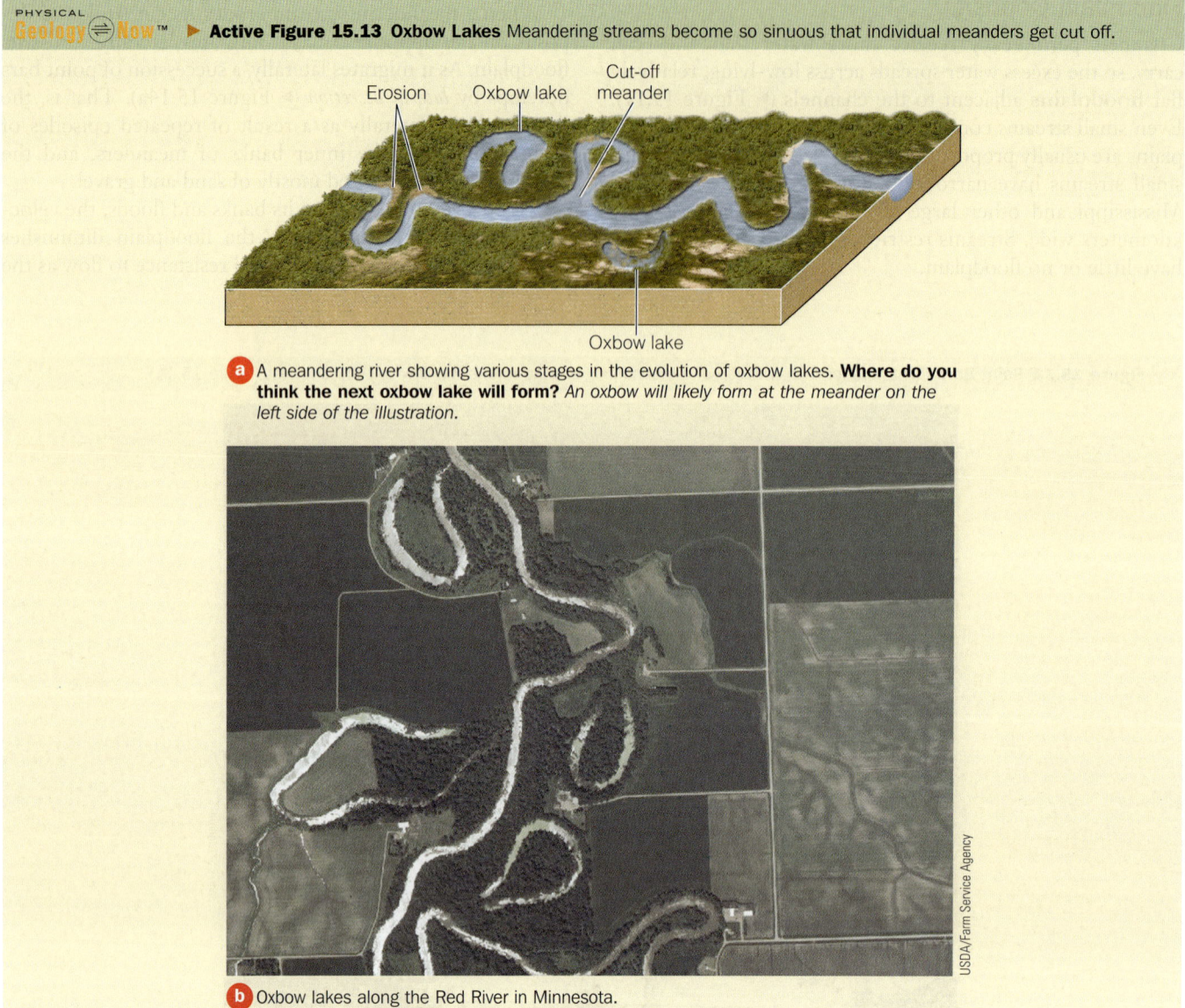

Active Figure 15.13 Oxbow Lakes Meandering streams become so sinuous that individual meanders get cut off.

a A meandering river showing various stages in the evolution of oxbow lakes. **Where do you think the next oxbow lake will form?** *An oxbow will likely form at the meander on the left side of the illustration.*

b Oxbow lakes along the Red River in Minnesota.

water spreads out as a broad, shallow sheet. In response to the diminished velocity, ridges of sandy alluvium known as **natural levees** are deposited along the margins of the channel (▶ Figure 15.14b,c). Natural levees build up by repeated deposition of sediment during numerous floods. These natural levees separate most of the floodplain from the stream channel, so floodplains are commonly poorly drained and swampy. In fact, certain tributary (or "yazoo") streams may parallel the main stream for many kilometers until they find a way through the natural levee system (▶ Figure 15.11).

Floodwaters spilling from a main channel carry large quantities of silt- and clay-sized sediment beyond the natural levees and onto the floodplain. During the waning stages of a flood, floodwaters may flow very slowly or not at all, and the suspended silt and clay eventually settle as layers of mud that build upward by deposition during successive floods, a process known as *vertical accretion* (▶ Figure 15.14b).

■ *What are the similarities between deltas and alluvial fans?*

Deltas

The process of delta formation is rather simple: Where running water flows into another body of water, its flow velocity decreases rapidly and it deposits sediment. As a result, a **delta** forms, causing the local shoreline to build out, or *prograde* (▶ Figure 15.15).

The simplest prograding deltas exhibit a characteristic vertical sequence in which *bottomset beds* are successively overlain by *foreset beds* and *topset beds* (▶ Figure 15.15). This sequence develops when a stream enters another body of water and the finest sediments are carried some distance beyond the stream's mouth, where they settle from suspension and form bottomset beds. Nearer the stream's mouth, foreset beds form as sand and silt are deposited in gently inclined layers. The topset beds

▶ **Figure 15.14 Floodplain Deposits**

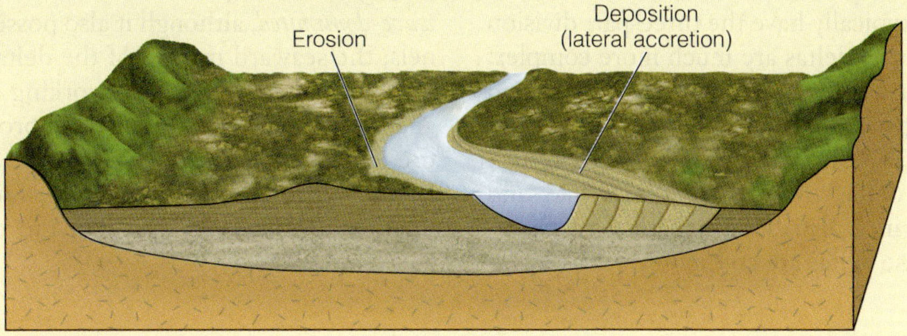

a Floodplain deposits formed by lateral accretion of point bars are made up mostly of sand.

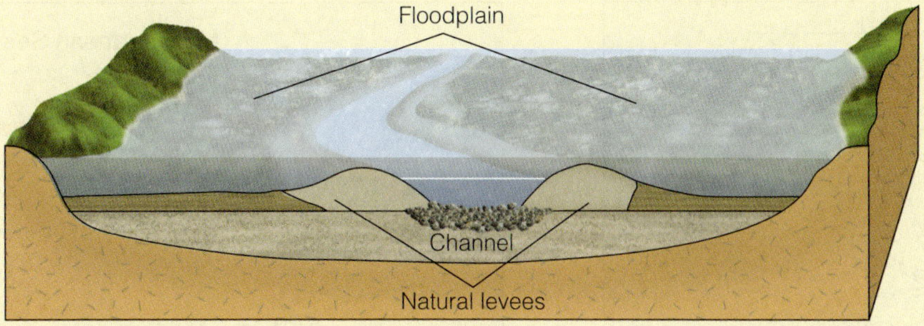

b The origin of vertical accretion deposits. During floods, streams deposit natural levees, and silt and mud settle from suspension on the floodplain.

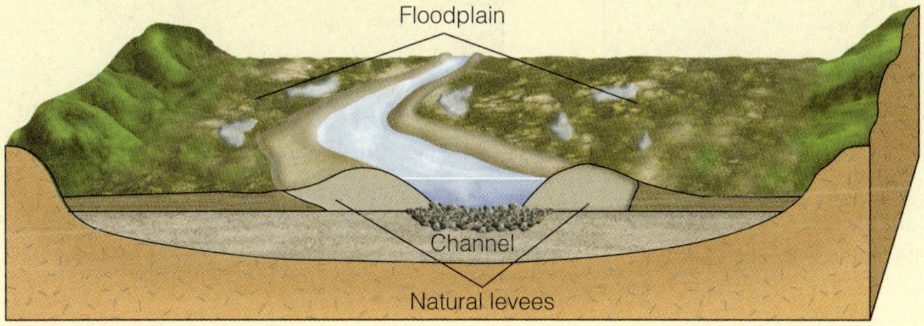

c After flooding.

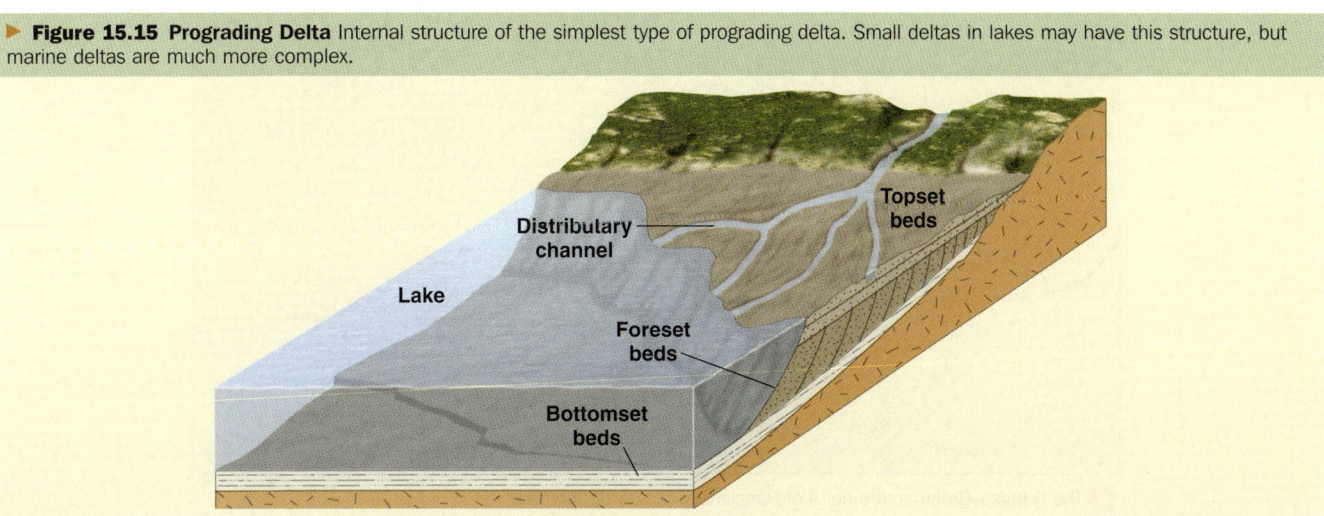

▶ **Figure 15.15 Prograding Delta** Internal structure of the simplest type of prograding delta. Small deltas in lakes may have this structure, but marine deltas are much more complex.

consist of coarse-grained sediments deposited in a network of *distributary channels* traversing the top of the delta.

Small deltas in lakes typically have the three-part division described, but large marine deltas are much more complex. Depending on the relative importance of stream, wave, and tidal processes, geologists identify three major types of marine deltas (▶ Figure 15.16). *Stream-dominated deltas*, such as the Mississippi River delta, consist of long fingerlike sand bodies, each deposited in a distributary channel that progrades far seaward. These deltas are commonly called *bird's-foot deltas* because the projections resemble the toes of a bird (▶ Figure 15.16a). In contrast, the Nile delta of Egypt is *wave-dominated*, although it also possesses distributary channels; the seaward margin of the delta consists of a series of barrier islands formed by reworking of sediments by waves, and the entire margin of the delta progrades seaward (▶ Figure 15.16b). *Tide-dominated deltas*, such as the Ganges–Brahmaputra of Bangladesh, are continually modified into tidal sand bodies that parallel the direction of tidal flow (▶ Figure 15.16c).

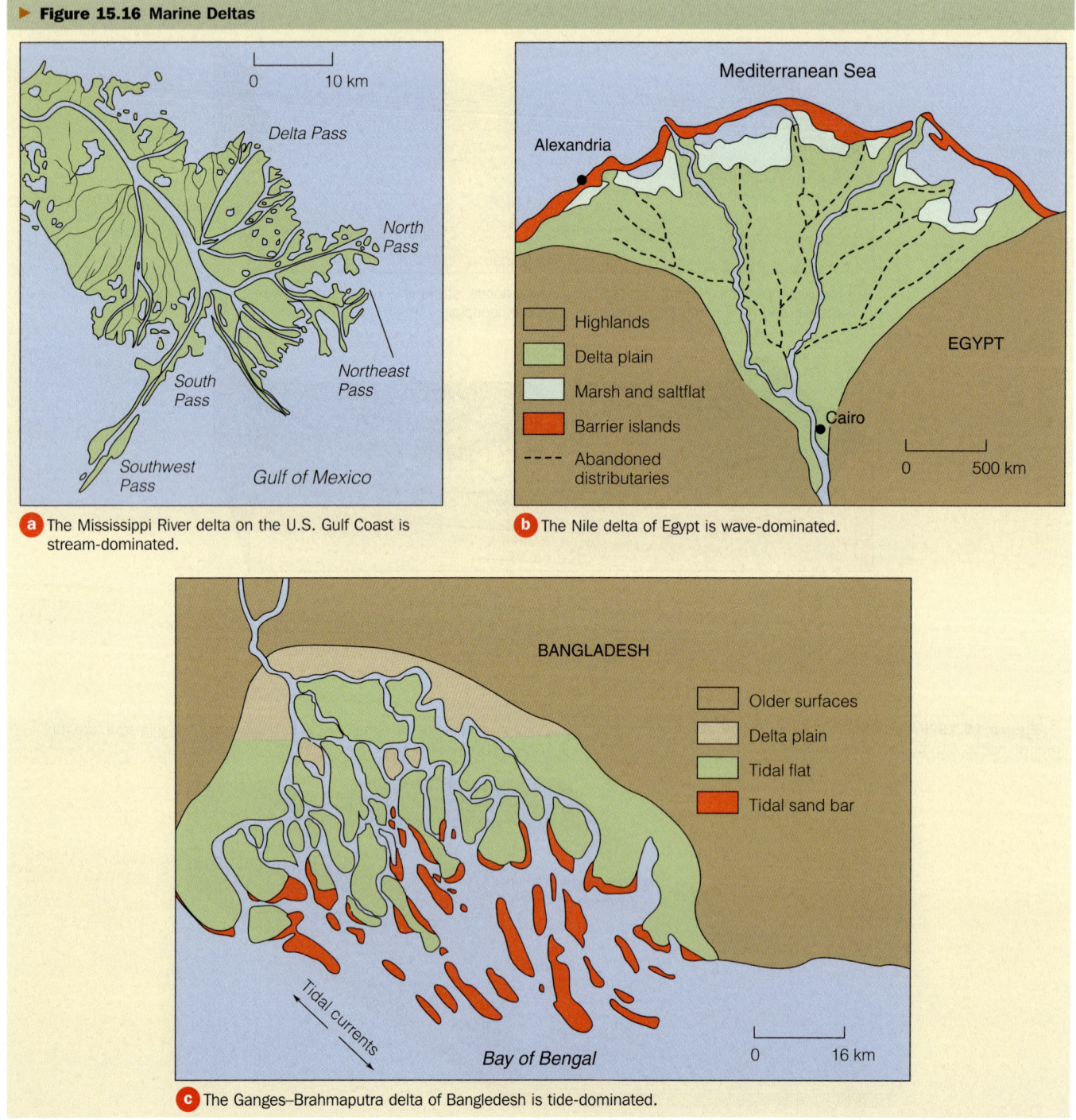

▶ **Figure 15.16 Marine Deltas**

a The Mississippi River delta on the U.S. Gulf Coast is stream-dominated.

b The Nile delta of Egypt is wave-dominated.

c The Ganges–Brahmaputra delta of Bangledesh is tide-dominated.

Delta progradation is one way that potential reservoirs for oil and gas form. Because the sediments of distributary sand bodies are porous and in proximity to organic-rich marine sediments, they commonly contain economic quantities of oil and gas. Much of the oil and gas production of the Gulf Coast of Texas comes from buried delta deposits. Some of the older deposits of the Niger River delta of Africa and the Mississippi River delta also contain vast reserves of oil and gas.

Alluvial Fans

Lobate deposits of alluvium on land known as **alluvial fans** form best on lowlands with adjacent highlands in arid and semiarid regions where little vegetation exists to stabilize surface materials (▶ Figure 15.17). During periodic rainstorms, surface materials are quickly saturated and surface runoff is funneled into a mountain canyon leading to adjacent lowlands. In the mountain canyon, the runoff is confined so it cannot spread laterally, but when it discharges onto the lowlands, it quickly spreads out, its velocity diminishes, and deposition ensues. Repeated episodes of sedimentation result in the accumulation of a fan-shaped body of alluvium.

Deposition by running water in the manner just described is responsible for many alluvial fans. In this case they are composed mostly of sand and gravel, both of which contain a variety of sedimentary structures. In some cases, though, the water flowing through a canyon picks up so much sediment that is becomes a viscous debris flow. Consequently, some alluvial fans consist mostly of debris flow deposits that show little or no layering. Of course, the dominant type of deposition can change through time, so a particular fan might have both types of deposits.

What Would You Do?

As the science teacher in your small community, you are asked to investigate some sedimentary rock exposures of interest to local boys and girls clubs and render an opinion on how the rocks were deposited. You note that at one location there is a thick (50 m) sequence of sand and gravel beds with current ripple marks and cross-bedding, whereas the other rock exposures are made up of mud with subordinate bodies of cross-bedded sandstone. You are convinced that both were deposited by streams or rivers, but how would you make a convincing case for braided versus meandering stream deposition?

Section 15.5 Summary

- Braided streams have a network of dividing and rejoining channels and their deposits are mostly sheets of sand and gravel, whereas meandering streams have a single, sinuous channel and their deposits are dominated by mud with subordinate channel deposits, mostly point bars.

- Floodplains are broad, fairly flat areas adjacent to stream channels. They are made up of fine-grained deposits, mostly mud, carried in during floods.

- Deltas form where a stream or river discharges into a standing body of water, flow velocity diminishes, and deposition takes places. Deltas in lakes commonly have a three-part division of bottomset, foreset, and topset beds, but marine deltas are much larger and more complex.

- An alluvial fan forms where a mountain stream discharges onto an adjacent lowland where it spreads out and deposition occurs. Alluvial fans develop best in arid and semiarid regions.

▶ **Figure 15.17 Alluvial Fans and Their Deposits**

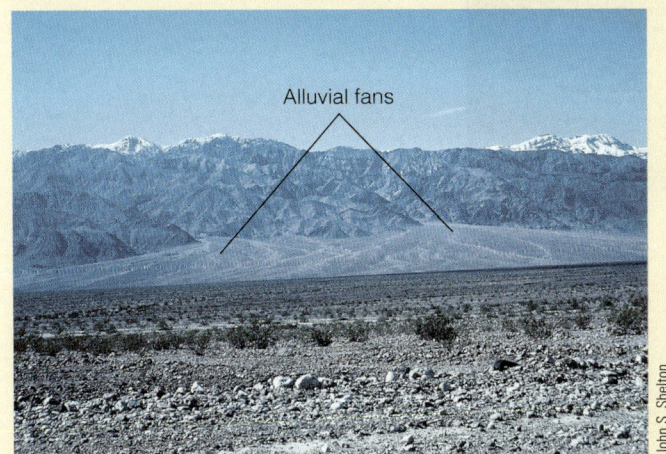

a These alluvial fans at the base of the Panamint Range in Death Valley, California, were deposited where streams discharged from mountain canyons onto adjacent lowlands.

b Most alluvial fan deposits are made up of sand and gravel and many show little or no stratification.

15.6 Predicting and Controlling Floods

Can floods be predicted and controlled? Before answering, we must define the types of floods one might encounter (Table 15.1). Our main interest is stream/river floods, which can be predicted but only in the broadest sense of stream/river behavior, and they can be controlled but only within the design limits of flood-control projects. Flash floods, on the other hand, commonly occur without warning when channels fill to overflowing in a very short time, perhaps minutes (▶ Figure 15.18).

Understanding why streams and rivers flood is simple. When they receive more water than their channels can handle, they spill over their banks and occupy some or all of their floodplains—in short, too much water in a short time. Floods are so common that unless they cause fatalities or extensive damage they rate little more than a passing reference in the news. Widespread flooding takes place every year in many locations around the world, but in the United States the last truly vast flooding occurred in 1993, especially in Missouri, Illinois, Iowa, and Minnesota (see "The Flood of '93" on pp. 476–477).

More and more flood-control projects are completed in North America every year, yet the damages from flooding are not decreasing. In the United States property damage exceeds $5 billion per year, not including coastal flooding from hurricanes. The combination of fertile soils, level surfaces, and proximity to water for agriculture, industry, and domestic use makes floodplains popular sites for development. Unfortunately, urbanization increases surface runoff because surface materials are compacted or covered by asphalt or concrete, reducing their infiltration capacity. Storm drains in urban areas quickly carry water to nearby channels, many of which flood more frequently than they did in the past.

TABLE 15.1
Types of Floods

Type of Flood	Comments
Stream/River Flood	Streams and rivers overflow their banks when they receive too much water in a short time.
Coastal Flood	Caused by wind-generated waves and storm surge (see Chapter 19) and tsunami (see Chapter 10).
Urban Flood	Former field and woodlands are paved over reducing their infiltration capacity leading to increased flooding. Backed up storm drains also contribute.
Flash Flood	A normally dry gully quickly fills with fast-moving water. Flooding may take place far from where precipitation fell.
Ice Jam Flood	Floating ice in cold climates may accumulate and form an ice dam, causing flooding upstream from the obstruction.

▶ **Figure 15.18 Flash Floods** This flash flood hit McDowell County, West Virginia, on July 8, 2001. On May 2, 2002, the area was hit again by flash floods. Many of the valleys in this area are narrow and deep, so following heavy rainfall the water rushes down the valleys, causing widespread destruction.

River and Stream Monitoring

To monitor stream behavior, the U.S. Geological Survey maintains more than 11,000 stream-gauging stations, and various state and county agencies record and analyze stream and river velocity and discharge and variations in these parameters through time. Data collected may be used to construct a *hydrograph*, which shows how a stream's discharge varies and is useful in planning irrigation and water-supply projects (▶ Figure 15.19). Hydrographs also give planners a better idea of what to expect during floods.

■ What do scientists mean when they refer to a 20-year flood?

A 20-year flood is one of a specified discharge or greater that is likely to take place within a 20-year period. Data from stream/river-gauging stations are used to construct *flood-frequency curves* (▶ Figure 15.20). To make a flood-frequency curve, the peak discharges for a river are first arranged in order of volume; the flood with the greatest discharge has a magnitude rank of 1, the second largest is 2, and so on (▶ Figure 15.20). The *recurrence interval*—that is, the time period during which a flood of a given magnitude or larger can be expected over an average of many years—is determined by the equation shown in Figure 15.20. For this river, floods with magnitude ranks of 1 and 23 have recurrence intervals of 77.00 and 3.35 years, respectively. Once the recurrence interval has been calculated, it is plotted on a graph against discharge, and a line is drawn through the data points.

According to the data in Figure 15.20, a 10-year flood for the Rio Grande near Lobatos, Colorado, has a discharge of 245 m^3/sec. This means that on average we can expect one flood of this size or larger within a 10-year period. However, we cannot predict that a flood of this size will take place in any particular year, only that it has a probability of 1 in 10 (1/10) of occurring in any particular year. Furthermore, 10-year floods are not necessarily separated by 10 years—two 10-year floods could occur in the same year or in successive years, but over a long period, perhaps centuries, their average occurrence would be once every 10 years.

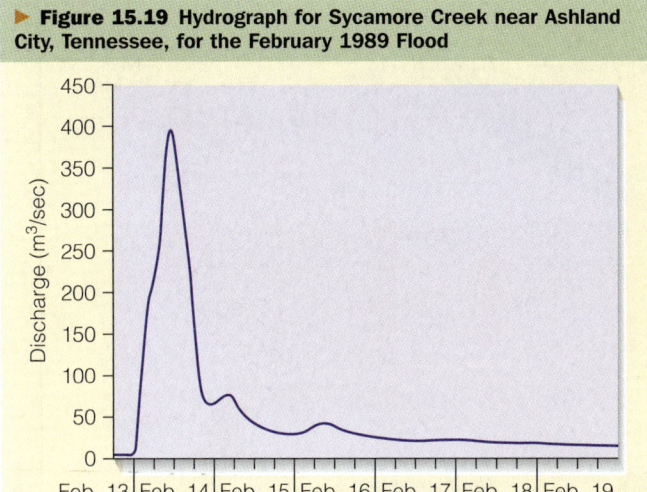

▶ **Figure 15.19** Hydrograph for Sycamore Creek near Ashland City, Tennessee, for the February 1989 Flood

▶ **Figure 15.20** Rio Grande Flood Data

Year	Discharge (m³/sec.)	Rank	Recurrence Interval
1900	133	23	3.35
1901	103	35	2.20
1902	16	69	1.12
1903	362	2	38.50
1904	22	66	1.17
1905	371	1	77.00
1906	234	10	7.70
1907	249	7	11.00
1908	61	45	1.71
1909	211	13	5.92
1974	22	64	1.20
1975	68	43	1.79

The greatest yearly discharge is given a magnitude rank (*m*) ranging from 1 to *N* (*N* = 76 in this example), and the recurrence interval (*R*) is calculated by the equation $R = (N + 1)/m$.

Source: U.S. Geological Survey Open-File Report 79–681.

a Selected data and recurrence intervals for the Rio Grande near Lobatos, Colorado.

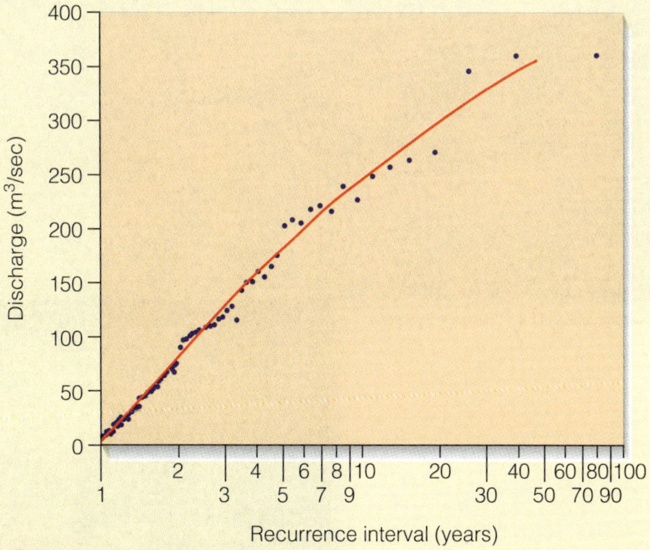

b The data in **(a)** were used to construct this flood-frequency curve for the Rio Grande near Lobatos, Colorado.

The Flood of '93

Although several floods take place each year that cause damage, injuries, and fatalities, the last truly vast flooding in North America occurred during June and July of 1993. Now called the Flood of '93, it was responsible for 50 deaths and 70,000 people left homeless. Extensive property damage occurred in several states, but particularly hard hit were Missouri and Iowa (see Figure 7). Unusual behavior of the jet stream and the convergence of air masses over the Midwest were responsible for numerous thunderstorms that caused the flooding.

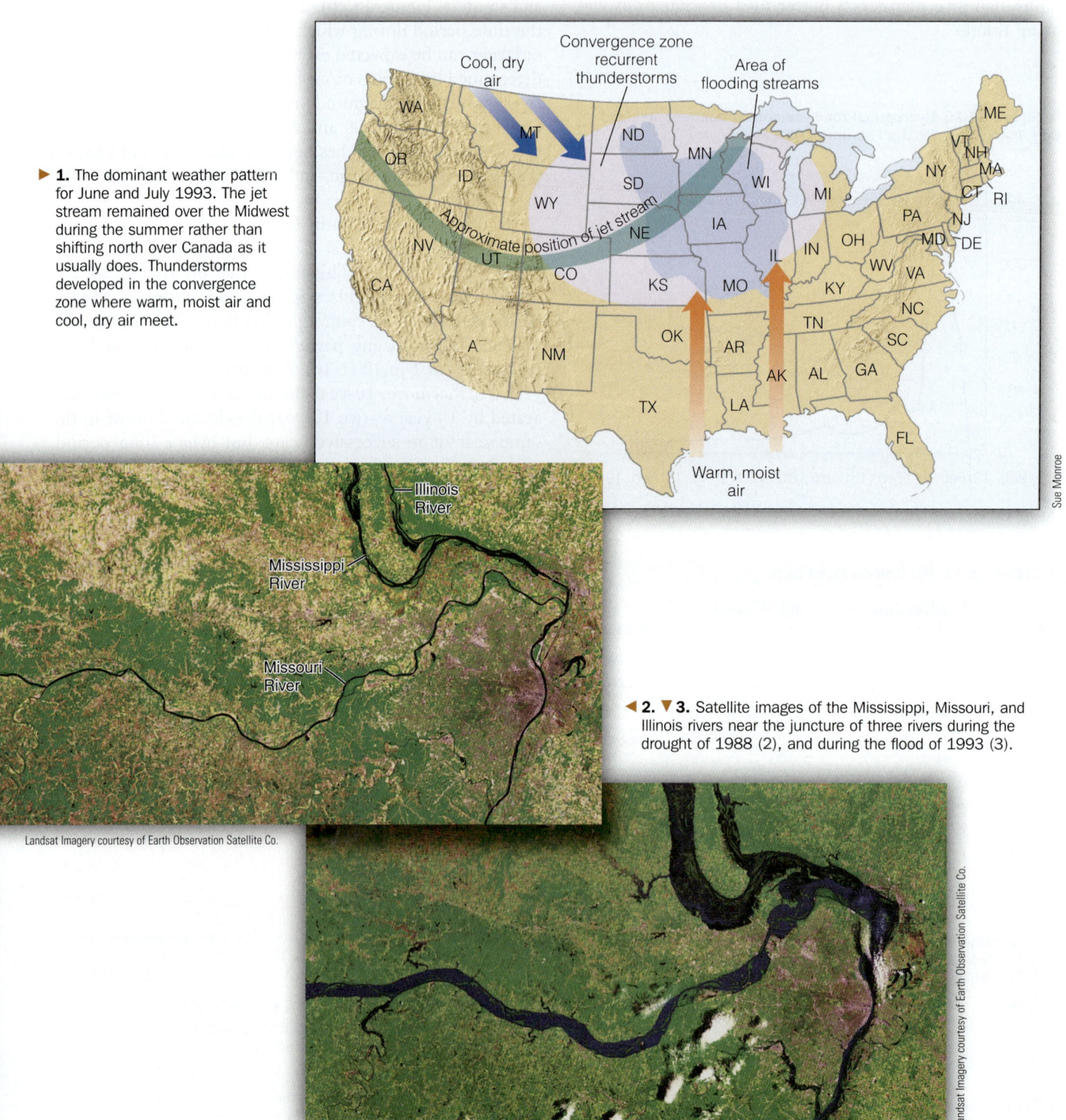

▶ **1.** The dominant weather pattern for June and July 1993. The jet stream remained over the Midwest during the summer rather than shifting north over Canada as it usually does. Thunderstorms developed in the convergence zone where warm, moist air and cool, dry air meet.

◀ **2.** ▼ **3.** Satellite images of the Mississippi, Missouri, and Illinois rivers near the juncture of three rivers during the drought of 1988 (2), and during the flood of 1993 (3).

▲ **4.** Portage des Sioux, St. Charles County, Missouri, on July 16, 1993. The channel of the Mississippi River is on the far right.

▲ **5.** Floodwaters in Portage des Sioux covered the 5.5-m-high pedestal of this statue on the bank of the Mississippi River.

◀ **6.** Breached levee on the Mississippi River near Davenport, Iowa. This is one of 800 levees that failed or was overtopped during the flood.

▶ **7.** Damage caused by the Flood of '93. Compiled by the U.S. Army Corps of Engineers, figures are rounded to the nearest $1000.

State	Residential	Agricultural	Other	Total
Illinois	$176,833,000	$166,502,000	$409,020,000	$752,355,000
Iowa	57,827,000	1,030,030,000	334,835,000	1,422,692,000
Kansas	35,829,000	855,849,000	176,162,000	1,067,840,000
Minnesota	16,940,000	694,041,000	286,540,000	997,521,000
Missouri	405,175,000	540,666,000	1,255,191,000	2,201,032,000
Nebraska	18,584,000	120,521,000	67,899,000	207,004,000
North Dakota	10,138,000	35,039,000	25,806,000	70,983,000
South Dakota	21,919,000	276,218,000	195,408,000	493,545,000
Wisconsin	17,747,000	133,835,000	135,917,000	287,499,000
Total—All States	760,992,000	3,852,701,000	2,886,778,000	7,500,471,000

■ *Why are predictions of 100-year floods unreliable?*

Predictions are unreliable for 100-year floods because stream gauge data in North America have been available for only a few decades and rarely for more than a century. Accordingly, we have a good idea of stream behavior over short periods—the 2- and 5-year floods, for example—but our knowledge of long-term behavior is limited by the short recordkeeping period. Accordingly, predictions of 50- or 100-year floods from Figure 15.20 are unreliable. In fact, the largest magnitude flood shown in Figure 15.20 may have been a unique event for this stream that will never be repeated. On the other hand, it may actually turn out to be a magnitude-2 or magnitude-3 flood when data for a longer time are available.

Although flood-frequency curves have limited applicability, they are nevertheless helpful in making flood-control decisions. Careful mapping of floodplains identifies areas at risk for floods of a given magnitude. For a particular river or stream, planners must decide what magnitude of flood to protect against because the cost goes up faster than the increasing sizes of floods would indicate.

Federal, state, and local agencies and land-use planners use flood-frequency analyses to develop recommendations and regulations concerning construction on and use of floodplains. Geologists and engineers are interested in these analyses for planning appropriate flood-control projects. They must decide where dams, levees, and floodways should be constructed to handle the excess water of floods.

Flood Control

Flood control has been practiced for thousands of years by constructing dams that impound reservoirs (▶ Figure 15.21a) and building levees along stream banks. Levees raise the banks of a stream, thereby restricting flow during floods (▶ Figure 15.21b). Unfortunately, deposition within

▶ **Figure 15.21 Flood Control** Dams and reservoirs, levees, floodways, and floodwalls are some of the structures used to control floods.

a Oroville Dam in California, at 235 m high, is the highest dam in the United States. It helps control floods, provides water for irrigation, and produces electricity at its power plant.

b This levee, an artificial embankment along a waterway, helps protect nearby areas from floods. A university campus lies out of view just to the right of the levee.

c This floodway carries excess water from a river (not visible) around a small community.

d This floodwall in Parkersburg, West Virginia, helps protect the city from floods.

the channel results in raising the streambed, making the levees useless unless they too are raised. Levees along the banks of the Huang Ho in China caused the streambed to rise more than 20 m above its surrounding floodplain in 4000 years. When the Huang Ho breached its levees in 1887, more than 1 million people died. In Section 15.1, we noted that the levees built to protect New Orleans from Lake Pontchartrain and the Mississippi River failed when Hurricane Katrina came ashore, and about 80% of the city flooded (▶ Figure 15.1).

Dams and levees alone are insufficient to control large floods, so in many areas floodways (also called bypasses) are also used (▶ Figure 15.21c). These usually consist of a channel constructed to divert part of the excess water in a steam around populated areas or areas of economic importance. Reforestation of cleared land also helps reduce the potential for flooding because vegetated soil helps prevent runoff by absorbing more water.

Hannibal, Missouri, and Parkersburg, West Virginia, and many other communities depend partly on *floodwalls* for protection. Parkersburg has had more than 40 damaging floods since 1832, so in 1946 the city built a high wall of concrete that extends for more than 3 km and a system of levees (▶ Figure 15.21d). Floodwalls have gates that permit access to the waterway and can be closed when the water rises. When flood-control projects are well planned and constructed, they are functional. What many people fail to realize is that these projects are designed to contain floods of a given size; if larger floods occur, flood waters spill onto floodplains anyway. Furthermore, dams occasionally collapse, and reservoirs eventually fill with sediment unless dredged. In short, flood-control projects are not only initially expensive but also require constant, costly maintenance. For example, some of the levees protecting Sacramento, California, one of the most flood-prone cities in the United States, are 150 years old and in poor condition. Unfortunately, the cost of repairing them has risen to as much as $250,000 per 100 m! Such costs must be weighed against the cost of damage if no control projects or maintenance were undertaken.

Section 15.6 Summary

- Stream/river monitoring and analyses of data enable scientists to predict floods of a given size within a specified range of years. These analyses are useful for planning flood-control projects.

- Unfortunately, our period of recordkeeping is limited, so determining the long-term behavior of waterways is more difficult than it is for short time periods.

- Dams, levees, floodways (bypasses), and floodwalls are used to contain or divert the excess water in streams and rivers during floods.

15.7 Drainage Systems

■ *What are a drainage basin and a divide?*

A **drainage basin** is an area from which a stream or river and its tributaries carry all surface runoff. High areas called divides separate drainage basins from one another (▶ Figure 15.22).Thousands of waterways, which are parts of larger drainage systems, flow directly or indirectly into the oceans. The only exceptions are some rivers and streams that flow into desert basins surrounded by higher areas. But even these basins of internal drainage are parts of larger systems consisting of a main channel with all its tributaries— that is, streams that contribute water to another stream. The Mississippi River and its tributaries, such as the Ohio, Missouri, Arkansas, and Red Rivers and thousands of smaller ones, or any other drainage system for that matter, carry runoff from an area known as a drainage basin. Topographically high areas called **divides** separate drainage basins from adjoining ones (▶ Figure 15.22). The continental divide along the crest of the Rocky Mountains in North America, for instance, separates drainage in opposite directions; drainage to the west goes to the Pacific, whereas drainage to the east eventually reaches the Gulf of Mexico.

Various arrangements of channels within an area are characterized as different types of **drainage patterns.** *Dendritic drainage*, consisting of a network of channels resembling tree branching, is the most common (▶ Figure 15.23a). It develops on gently sloping surfaces composed of materials that respond more or less homogeneously to erosion, such as areas underlain by nearly horizontal sedimentary rocks.

In dendritic drainage, tributaries join larger channels at various angles, but *rectangular drainage* is characterized by right-angle bends and tributaries joining larger channels at right angles (▶ Figure 15.23b). Such regularity in channels is strongly controlled by geologic structures, particularly regional joint systems that intersect at right angles.

Trellis drainage, consisting of a network of nearly parallel main streams with tributaries joining them at right angles, is common in some parts of the eastern United States. In Virginia and Pennsylvania, for example, erosion of folded sedimentary rocks developed a landscape of alternating ridges on resistant rock and valleys underlain by easily eroded rocks. Main waterways follow the valleys, and short tributaries flowing from the nearby ridges join the main channels at nearly right angles (▶ Figure 15.23c).

In *radial drainage*, streams flow outward in all directions from a central high point such as a large volcano (▶ Figure 15.23d). Many of the volcanoes in the Cascade Range of western North America have radial drainage patterns.

In all the types of drainage mentioned so far, some kind of pattern is easily recognized. *Deranged drainage*, in contrast, is characterized by irregularity, with streams flowing into and out of swamps and lakes, streams with only a few short tributaries, and vast swampy areas between channels (▶ Figure 15.23e). This kind of drainage developed recently

▶ **Figure 15.22 Drainage Basins**

a Small drainage basins are separated by divides (dashed lines), which are along the crests of ridges between channels (solid lines).

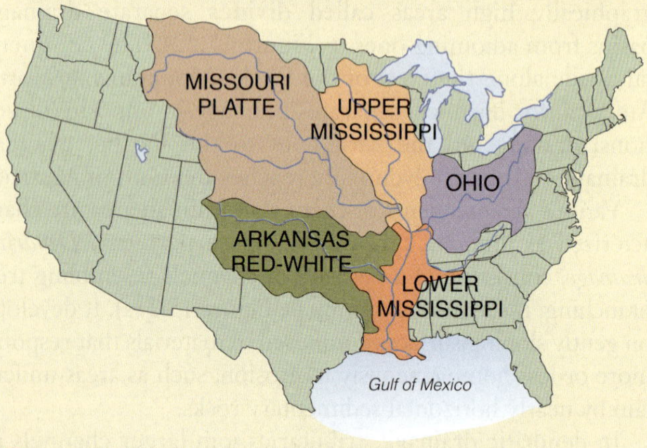

b The drainage basin of the Mississippi River and its main tributaries.

c A detailed view of the Wabash River's drainage basin, a tributary of the Ohio River. All tributary streams within the drainage basin, such as the Vermilion River, have their own smaller drainage basins. Divides are shown by red lines.

and has not yet formed a fully organized drainage system. In parts of Minnesota, Wisconsin, and Michigan, where glaciers obliterated the previous drainage, only 10,000 years have elapsed since the glaciers melted. As a result, drainage systems have not fully developed and large areas remain undrained.

The Significance of Base Level

Just how deeply can a stream or river erode? The Grand Canyon in Arizona is more than 1700 m deep, but the bottom of the canyon is still far above sea level. Obviously, streams and rivers must have some gradient to maintain the flow, so they can erode no lower than the body of water into which they flow. Most streams and rivers flow directly or indirectly into the oceans, so they can erode no lower than sea level, or what geologists call **base level.** Sea level is *ultimate base level*, and theoretically a channel could erode deeply enough so that its gradient rises ever so gently upstream (▶ Figure 15.24). Ultimate base level applies to an entire stream or river system that flows to the ocean, but channels may also have *local* or *temporary base levels* such as waterfalls and lakes (▶ Figure 15.24).

Ultimate base level is sea level, but suppose that sea level should drop or rise with respect to the land, or suppose that the land should subside or rise. In these cases base level changes and brings about changes in the entire system. For example, at times during the Pleistocene Epoch (Ice Age), sea level was about 130 m lower than it is now. Streams adjusted by eroding deeper valleys (they had steeper gradients) and extended well out onto the present-day continental shelves. Rising sea level at the end of the Ice Age accounted for rising base levels, decreased stream gradients, and deposition within channels.

Natural changes, such as fluctuations in sea level during the Pleistocene, alter the dynamics of rivers and streams, but so does human intervention. Geologists and engineers are well aware that building a dam to impound a reservoir creates a local base level (▶ Figure 15.25a). A stream entering a reservoir slows down and deposits sediment, so unless

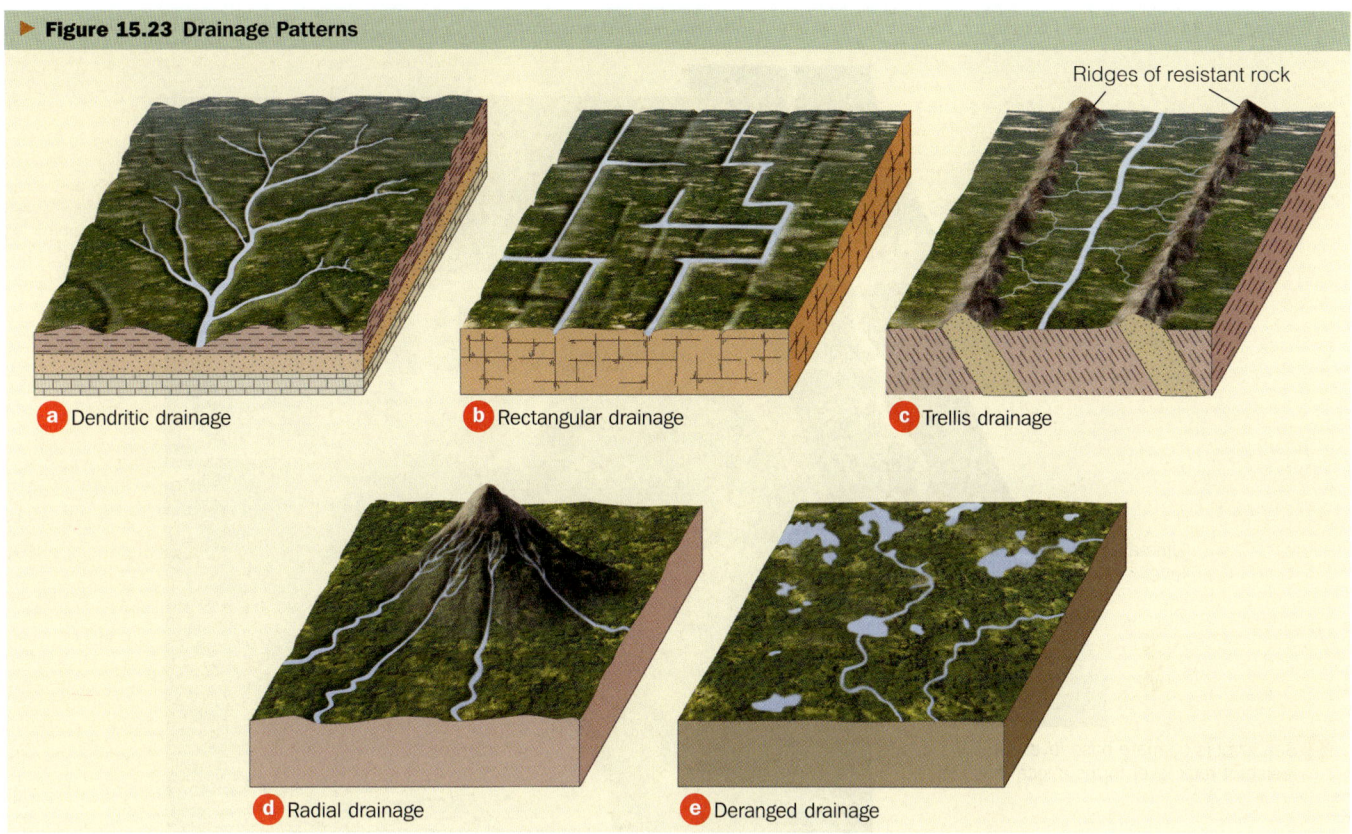

▶ Figure 15.23 Drainage Patterns

dredged, reservoirs eventually fill. In addition, the water discharged at a dam is largely sediment-free but still possesses energy to carry a sediment load. As a result it is not uncommon for streams to erode vigorously downstream from a dam to acquire a sediment load.

Draining a lake may seem like a small change and well worth the time and expense to expose dry land for agriculture or commercial development. But draining a lake eliminates a local base level, and a stream that originally flowed into the lake responds by rapidly eroding a deeper valley as it adjusts to a new base level (▶ Figure 15.25b).

Graded Streams

■ *What is a graded stream, and what conditions might upset the graded condition?*

A **graded stream** is one that has a balance among gradient, discharge, velocity, channel shape, and sediment load so that neither significant deposition nor erosion takes place. A change in any of these parameters upsets the graded condition. The *longitudinal profile* of any waterway shows the elevations of a channel along its length as viewed in cross section (▶ Figure 15.26). For some rivers and streams, the longitudinal profile is smooth, but others show a number of irregularities such as lakes and waterfalls, all of which are local base levels (▶ Figure 15.24). Over time these irregularities tend to be eliminated because deposition takes place where the gradient is insufficient to maintain sediment transport, and erosion decreases the gradient where it is steep. So, given enough time, rivers and streams develop a smooth, concave longitudinal profile of equilibrium, meaning that all parts of the system dynamically adjust to one another.

Even though the concept of a graded stream is an ideal, we can anticipate the response of a graded stream to changes that alter its equilibrium. For instance, a change in base level would cause a stream to adjust as previously discussed. Increased rainfall in a stream's drainage basin would

What Would You Do?

You own a farm several kilometers upstream from a small lake that developers drained to put up apartment buildings. Within a few years you notice that several small rills on your farm are becoming noticeably deeper and wider at an alarming rate. In fact, even now several are too deep to be plowed over. Why, after decades of stability, is the land eroding so rapidly? Is there anything you can do to stop or at least minimize the effects of erosion? Do you have any legal options?

▶ **Figure 15.24 Base Levels** Ultimate base level and local or temporary base level limit the depth to which streams and rivers can erode.

a Sea level is ultimate base level, whereas a resistant rock layer forms a local base level.

b Cumberland Falls on the Big South Fork River in Cumberland Falls State Resort Park, Kentucky, plunges 18 m over a local base level.

c Ultimate base level and a local base level where a stream flows into a lake.

d The Au Train River in Michigan flows into Lake Superior, its local base level.

result in greater discharge and flow velocity. In short, the stream would now possess greater energy—energy that must be dissipated within the stream system by, for example, a change from a semicircular to a broad, shallow channel that would dissipate more energy by friction. On the other hand, the stream may respond by eroding a deeper valley and effectively reduce its gradient until it is once again graded.

Vegetation inhibits erosion by stabilizing soil and other loose surface materials. So a decrease in vegetation in a drainage basin might lead to higher erosion rates, causing more sediment to be washed into a stream than it can effectively carry. Accordingly, the stream may respond by deposition within its channel, which increases its gradient until it is sufficiently steep to transport the greater sediment load.

▶ **Figure 15.25 Local Base Levels**

a A stream deposits much of its sediment load where it flows into a reservoir.

b Local base levels disappear as streams erode their beds into graded profiles matching the ultimate base level.

▶ **Figure 15.26 Longitudinal Profiles of Streams and the Concept of the Graded Stream**

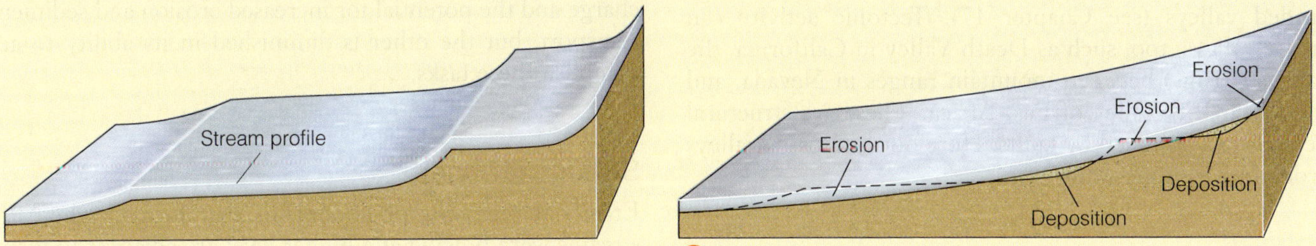

a An ungraded stream has irregularities in its longitudinal profile.

b Erosion and deposition along the course of a stream eliminate irregularities and yield the smooth, concave profile of a graded stream.

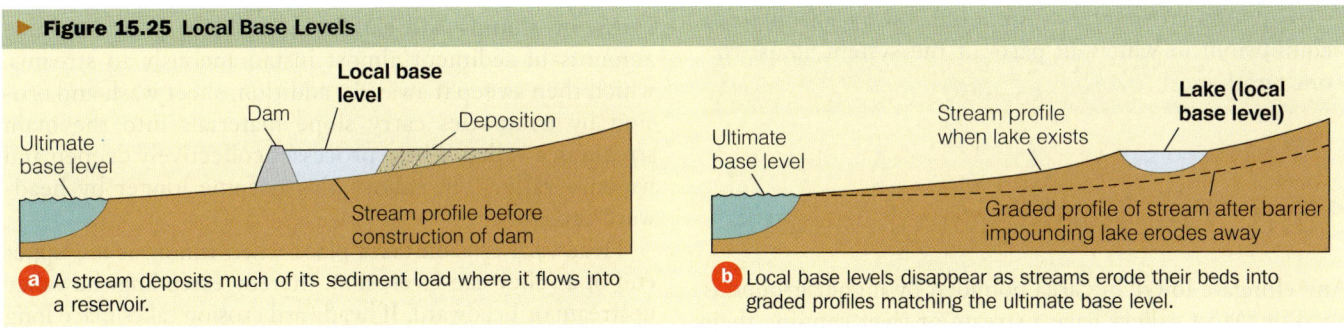

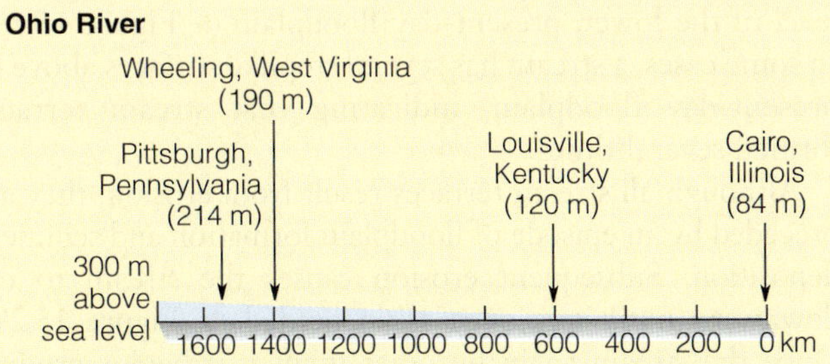

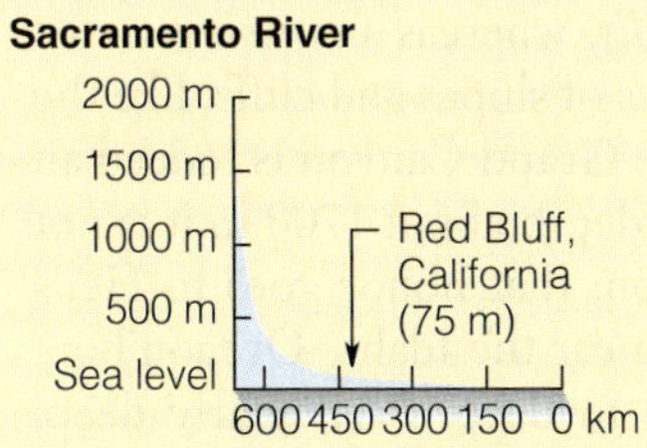

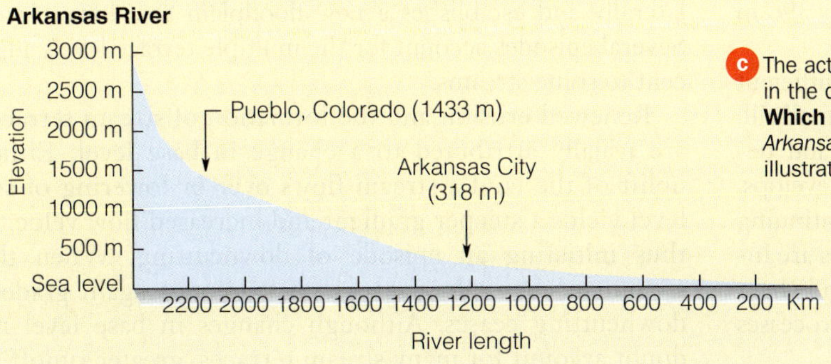

c The actual profiles of three rivers. All profiles are concave but differ in the degree of concavity because each has a different gradient. **Which one compares most closely with the stream in (a)?** *The Arkansas River.* The vertical scale is exaggerated 275 times in this illustration.

> **Section 15.7 Summary**
>
> - A stream or river and its tributaries carry water from its drainage basin, which is separated from other drainage basins by divides. Geologists recognize several drainage patterns, such as dendritic, radial, and deranged depending on the arrangement of channels in a drainage basin.
>
> - Ultimate base level, the lowest level to which streams and rivers can erode, is sea level, but waterfalls and lakes may form temporary or local base levels.
>
> - A graded stream has a smooth, concave profile of equilibrium in which all parts of the system adjust to one another.

15.8 The Evolution of Valleys

Any elongate low-lying area bounded by higher ground is a **valley.** Most valleys have a stream or river running their length, with tributaries draining the adjacent highlands. Valleys are very common landforms, and with few exceptions they form and evolve in response to running water, although other processes, especially mass wasting, contribute. In addition, glaciers may modify stream/river-eroded valleys (see Chapter 17). Tectonic activity can create valleys, too, such as Death Valley in California, the valleys (basins) between mountain ranges in Nevada, and the Great Rift Valley of East Africa. These are structural depressions bounded by faults. Here our interest is valleys eroded by running water.

Erosion and the Origin of Valleys

Valleys come in various sizes and shapes. For instance, a small, steep-sided valley is a *gully,* whereas a *canyon* is a vast valley with steep walls or a series of slopes and cliffs, like the Grand Canyon in Arizona. The Grand Canyon is indeed an awe-inspiring sight, but with a depth about 1700 m it is not North America's deepest canyon; that honor goes to Hells Canyon along the Snake River near the Idaho–Oregon border, which is more than 2400 m deep. Particularly deep, narrow valleys with near-vertical walls are known as *gorges*—the Black Canyon of the Gunnison River in Colorado is a good example; it is as much as 533 m deep but only 396 m wide (▶ Figure 15.27b).

Erosion of a valley may start where runoff has sufficient energy to dislodge surface materials and excavate a small rill. Once formed, a rill collects more surface runoff and becomes deeper and wider until a full-fledged valley develops. Running water is responsible for the origin and continuing development of valleys, but several distinct processes are involved, particularly downcutting, lateral erosion, sheet wash, and headward erosion. Several mass wasting processes are also important.

■ *How can headward erosion result in stream piracy?*

Headward erosion takes place at the upper end of a stream thereby lengthening it. In some cases, headward erosion breaches a divide and diverts some or all of the drainage of another stream system by stream piracy. Downcutting takes place when a stream or river has enough energy to erode its valley deeper through sediment abrasion. If downcutting were the only process involved, valleys would be narrow and steep-sided, but in most cases valley walls are undercut as a stream or river erodes laterally, creating unstable slopes that fail by mass wasting, particularly by slumps and slides. Landsliding provides huge amounts of sediment almost instantaneously to streams, which then sweep it away. In addition, sheet wash and erosion by tributaries carry slope materials into the main stream in a valley. These processes collectively deepen and widen a valley, but valleys also become longer by headward erosion.

Headward erosion takes place when runoff at the upper end of a valley causes erosion, thereby extending the valley upstream or headward. If headward erosion takes place long enough, it commonly results in *stream piracy*, which involves the breaching of a drainage divide and diversion of part of the drainage of another stream system (▶ Figure 15.28). Once stream piracy takes place, both drainage systems adjust to the new conditions; one system now has greater discharge and the potential for increased erosion and sediment transport, but the other is diminished in its ability to accomplish these tasks.

Stream Terraces

Erosional remnants of floodplains that formed when the streams were flowing at a higher level are adjacent to many channels. These surfaces, called **stream terraces,** consist of a fairly flat upper surface and a steep slope descending to the level of the lower, present-day floodplain (▶ Figure 15.29). In some cases, a stream has several steplike surfaces above its present-day floodplain, indicating that stream terraces formed several times.

Although all stream terraces result from erosion, they are preceded by an episode of floodplain formation and sediment deposition. Subsequent erosion causes the stream to cut downward until it is once again graded (▶ Figure 15.29). Once the stream again becomes graded, it begins eroding laterally and establishes a new floodplain at a lower level. Several episodes account for the multiple terrace levels adjacent to some streams.

Renewed erosion and the formation of stream terraces are usually attributed to a change in base level. Either uplift of the land a stream flows over or lowering of sea level yields a steeper gradient and increased flow velocity, thus initiating an episode of downcutting. When the stream reaches a level at which it is once again graded, downcutting ceases. Although changes in base level no doubt account for many stream terraces, greater runoff in

Figure 15.27 Valleys of Various Shapes and Sizes Form Mostly by Running Water and Mass Wasting

a This small valley in Alaska has steep walls that descend to a narrow valley bottom.

b The Black Canyon of the Gunnison River in Colorado. It has nearly vertical walls 533 m high and is thus a gorge.

a stream's drainage basin can also result in the formation of terraces.

Incised Meanders

Some streams are restricted to deep, meandering canyons cut into bedrock, where they form features called **incised meanders** (▶ Figure 15.30). For example, the Colorado River in Utah occupies a meandering canyon more than 600 m deep. Streams restricted by rock walls are generally ineffective in eroding laterally; thus they lack a floodplain and occupy the entire width of the canyon floor.

It is not difficult to understand how a stream cuts downward into solid rock, but how a stream forms a meandering pattern in bedrock is another matter. Because lateral erosion is inhibited once downcutting begins, one must infer that the meandering course was established when the stream flowed across an area covered by alluvium. For example, suppose that a stream near base level has established a meandering pattern. If the land the stream flows over is uplifted, then erosion begins and the meanders become incised into the underlying bedrock.

Superposed Streams

Water runs downhill in response to gravity, so the direction of flow in streams and rivers is determined by topography. Yet a number of waterways seem, at first glance, to have defied this fundamental control. For example, the Delaware, Potomac, and Susquehanna Rivers in the eastern United States flow in valleys that cut directly through ridges that lie in their paths.

Figure 15.28 Two Stages in the Evolution of a Valley

a The stream widens its valley by lateral erosion and mass wasting while simultaneously extending its valley by headward erosion.

b As the larger stream continues to erode headward, stream piracy takes place when it captures some of the drainage of the smaller stream. Notice also that the valley is wider in (b) than it was in (a).

Figure 15.29 Origin of Stream Terraces

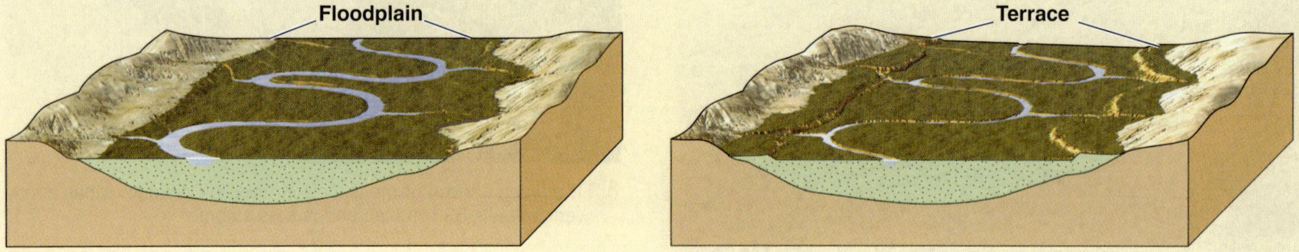

a A stream has a broad floodplain.

b The stream erodes downward and establishes a new floodplain at a lower level. Remnants of its old, higher floodplain are stream terraces.

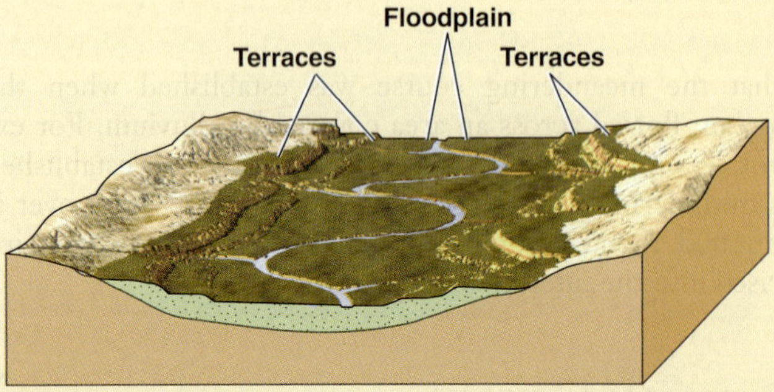

c Another level of stream terraces forms as the stream erodes downward again.

d Stream terraces along the Madison River in Montana.

J. R. Stacey/USGS

▶ **Figure 15.30** Incised Meanders

(a) When an area is uplifted, a meandering river flowing across a gently sloping surface becomes incised (b) as it erodes downward, thus occupying a meandering canyon. (c) In some cases the thin wall of rock between adjacent incised meanders is cut off, forming a natural bridge (e).

d The Colorado River at Dead Horse State Park, Utah, is incised to a depth of 600 m.

e Sipapu Bridge in Natural Bridges National Monument, Utah, formed as shown in (c). It stands 67 m above the canyon floor and has a span of 81.5 m.

■ *How does a superposed stream originate?*

Some streams cut directly across ridges in their paths because at one time they flowed on a higher surface, but subsequently eroded down and cut through resistant rocks. All of the examples noted previously are **superposed streams,** meaning that they once flowed on a surface at a higher level but as they eroded downward they eroded into resistant rocks and cut narrow canyons or what geologists call *water gaps* (▶ Figure 15.31a, b). During the Mesozoic Era, the Appalachian region was eroded to a sediment-covered plain across which numerous streams and rivers flowed generally eastward. Regional uplift began during the Cenozoic Era, and as a result the Delaware, Potomac, and Susquehanna Rivers were superposed directly upon evolving ridges, and instead of changing their courses, they cut narrow, steep-walled valleys. A similar sequence of events took place when the Jefferson River in Montana was superposed (▶ Figure 15.31c).

A water gap has a stream flowing though it, but if the stream is diverted elsewhere, perhaps by stream piracy, the abandoned gap is then called a *wind gap*. The Cumberland Gap in Kentucky is a good example; it was the avenue through which settlers migrated from Virginia to Kentucky from 1790 until well into the 1800s. Furthermore, several water gaps and wind gaps through the ridges of the Appalachian Mountains played important strategic roles during the Civil War.

Valley Development—A Summary

According to one concept, stream erosion of an area uplifted above sea level yields a distinctive series of landscapes. When erosion begins, streams erode downward; their valleys are deep, narrow, and V-shaped in cross section; and their profiles show a number of irregularities (▶ Figure 15.32a). As

▶ **Figure 15.31 Origin of a Superposed Stream**

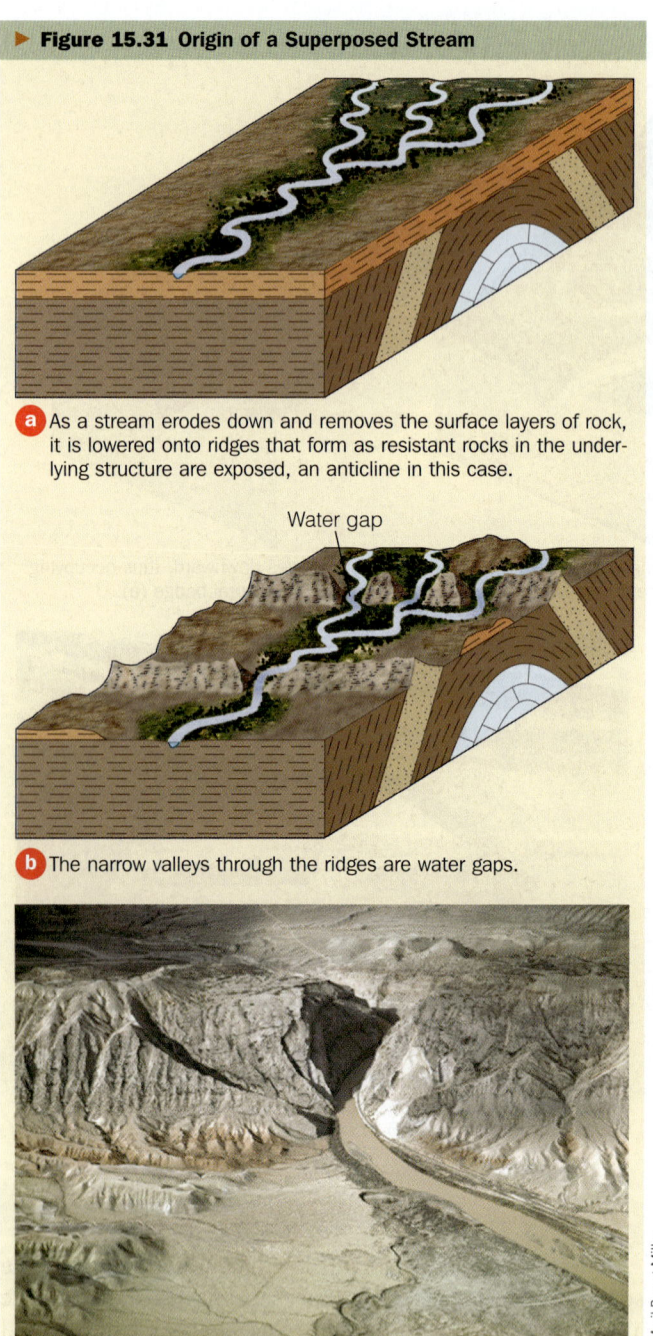

a As a stream erodes down and removes the surface layers of rock, it is lowered onto ridges that form as resistant rocks in the underlying structure are exposed, an anticline in this case.

b The narrow valleys through the ridges are water gaps.

Maril Bryant Miller

c View of a water gap cut by the Jefferson River in Montana.

▶ **Figure 15.32 Idealized Stages in the Development of a Stream and Its Associated Landform**

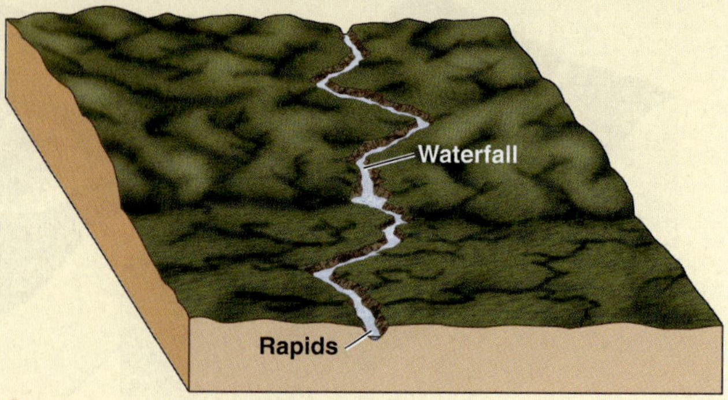

a Initial stage

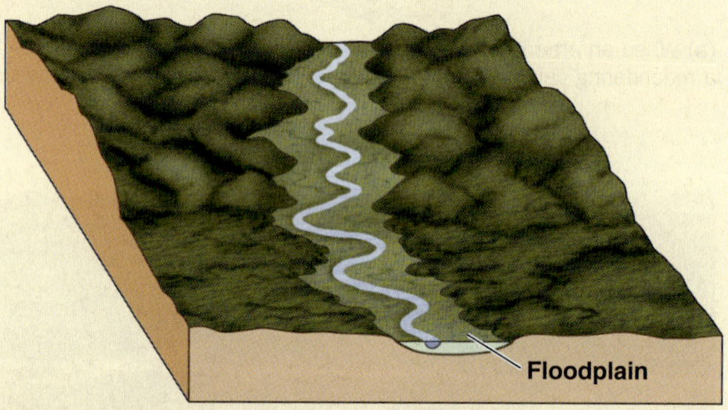

b Intermediate stage

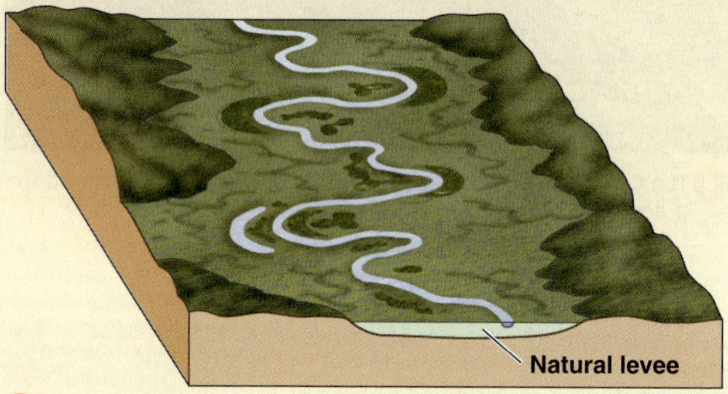

c Advanced stage

streams cease eroding downward, they start eroding laterally, establishing a meandering pattern and a broad floodplain (▶ Figure 15.32b). Finally, with continued erosion, a vast, rather featureless plain develops (▶ Figure 15.32c).

Many streams do indeed show an association of features typical of these stages. For instance, the Colorado River flows through the Grand Canyon and closely matches the features in the initial stage shown in Figure 15.32a. Streams in many areas approximate the second stage of development, and certainly the lower Mississippi closely resembles the last stage. Nevertheless, the idea of a sequential development of stream-eroded landscapes has been largely abandoned because there is no reason to think that streams necessarily follow this idealized cycle. Indeed, a stream on a gently sloping surface near sea level could develop features of the last stage very early in its history. In addition, as long as the rate of uplift exceeds the rate of downcutting, a stream will continue to erode downward and be confined to a narrow canyon.

Although it is doubtful that streams go through the three-stage sequence just outlined, they do indeed evolve in response to changing conditions in the variables that influence them. Of course, human intervention can change a stream's capacity for erosion and sediment transport, and increase the frequency of flooding. However, natural processes such as uplift or subsidence, climatic change, cutting a deeper outlet and draining a lake, or encountering a particularly resistant layer of rock can also influence stream activity. In short, they are dynamic systems that continuously adjust to changes.

Section 15.8 Summary

• Running water is responsible for the erosion of most valleys, but other processes, especially mass wasting, are important, too. Erosion not only deepens and widens valleys but also takes place at the upper end of the valley, thereby lengthening it.

• Many stream and river systems have stream terraces adjacent to them, which indicates that they had a period of deposition and floodplain formation followed by a time of renewed erosion.

• Incised meanders and superposed streams form when a stream erodes downward and cuts a meandering pattern in bedrock or when it is lowered onto resistant rock and cuts a narrow valley through a developing ridge.

• Stream/river-eroded landscapes match the three stages in an idealized cycle, but it is unlikely that streams actually go through these stages in succession. More likely they simply adjust to the prevailing conditions and retain their characteristics as long as those conditions last.

What Would You Do?

Despite your vigorous protests, your naïve brother-in-law insisted on building his new home on the cut bank of a meandering stream. He reasoned that the view is good and it's close to the river where he can go fishing. Now, after only two years, the riverbank that was 20 m from his house is only 12 m away. What is going on? At this point, is there anything he can do to save his house? After all, in only about three years it will be gone if the river continues to erode at its present rate.

Review Workbook

ESSENTIAL QUESTIONS SUMMARY

15.1 Introduction
■ *Why is Earth the only terrestrial planet with abundant surface water?*
Mercury is too small and too hot, Venus has a runaway greenhouse effect so it is too hot, and Mars is too small and too cold, although it probably had surface water during its early history.

15.2 Water on Earth
■ *What is the hydrologic cycle and why is it important?*
Water vapor from the oceans rises into the atmosphere where clouds form, some of which move over land and release their moisture as precipitation that eventually gets back to the oceans. This cycle is important because it is the source of all freshwater on land.

■ *How does laminar flow differ from turbulent flow, and which is more important in rivers and streams?*
Almost all surface flow is fully turbulent, which is more capable of erosion and sediment transport. It differs from laminar flow in that lines of flow are complexly intertwined, whereas in laminar flow little mixing takes place between adjacent flow lines within the fluid.

15.3 Running Water
■ *Why on the average does stream velocity increase downstream even though gradient decreases in the same direction?*
Gradient is one determinant of velocity, but so are the acceleration of gravity, changes in the roughness of the bed and banks, and changes in the channel shape. Channels tend to be relatively wide and shallow in a stream's upper reaches but more semicircular downstream. Also, more and more tributaries join a stream along its length, accounting for greater discharge and higher velocity.

■ *How is daily discharge calculated?*
For a stream 148 m wide and 2.6 m deep with a flow velocity of 0.3 m/sec, the discharge is 115.44 m^3/sec, which must be multiplied by 60 to calculate its discharge per minute (6926.4 m^3/min), then multiplied by 60 to get its discharge per hour (415,584 m^3/hr), and finally multiplied by 24 for its daily discharge (9,974,016 m^3/day).

15.4 Running Water, Erosion, and Sediment Transport
■ *How does running water erode by abrasion and hydraulic action?*
Abrasion takes place because running water carries sand and gravel that wear away any rock surface they move over. Hydraulic action is simply the impact of water itself, which is quite effective in dislodging loose soil and sediment.

■ *What are the dissolved, suspended, and bed loads of a stream?*
Each of these loads in a stream is defined by how it is transported. The dissolved load is made up of materials in solution and cannot be seen. Both the suspended load and bed load are made up of solid particles, but the former consists of silt and clay kept up in the water by turbulence, whereas the bed load of sand and gravel is transported along a stream's bed by saltation and rolling and sliding.

15.5 Deposition by Running Water
■ *How and why do braided streams differ from meandering streams?*
Braided streams have too much sand and gravel to effectively transport, so the sand and gravel are deposited as bars that divide the main channel into many channels. Meandering streams carry much more suspended load and have a single, sinuous channel. Their deposits are mostly mud with subordinate channel deposits.

■ *How and where do point bars form?*
In meandering channels, flow velocity varies across the channel, being fastest on the outside of a meander where erosion occurs and slowest on the opposite side. Because velocity on the inner side of the meander is insufficient to maintain sediment transport, deposition takes place. The point bar itself is commonly a body of cross-bedded sand, but in some streams it may be made up of gravel.

■ *What are the similarities between deltas and alluvial fans?*
Both are deposits of alluvium that form where streams and rivers lose their transport capacity and sediment accumulates. However, deltas are deposited along lakeshores and seashores, whereas alluvial fans form where a waterway discharges from a mountain canyon onto adjacent lowlands. Furthermore, alluvial fans are made up mostly of sand and gravel, but deltas have sand and a large component of mud.

15.6 Predicting and Controlling Floods
■ *What do scientists mean when they refer to a 20-year flood?*
A 20-year flood is a flood with at least a specified discharge that on the average can be expected within a period of 20 years, but not necessarily in any particular year.

■ *Why are predictions of 100-year floods unreliable?*
Our period of recordkeeping is too short to effectively evaluate the long-term behavior of streams and rivers. For instance, we may have only an 80-year record for a particular stream, so extrapolating the data to predict a 100-year flood is risky. These data, however, are adequate for anticipating 5- and 10-year floods, for instance.

15.7 Drainage Systems
■ *What are a drainage basin and a divide?*
A drainage system—that is, a main stream or river and all its tributaries—carries all the water from a particular geographic area, a drainage basin, that is separated from adjacent drainage basins by topographically higher areas called divides.

■ *What is a graded stream, and what conditions might upset the graded condition?*
A graded stream has a smooth, concave profile of equilibrium in which all elements are adjusted to one another—gradient, velocity, discharge, channel shape, and sediment load—so that neither significant erosion nor deposition takes place in its channel. The graded condition is an ideal approached by streams, but it is easily upset by changes in base level or climate or increased runoff because of fire.

15.8 The Evolution of Valleys

■ *How can headward erosion result in stream piracy?*

Headward erosion takes place when entering runoff at the upper end of a stream or river valley causes erosion and upstream extension of the valley. If this persists, erosion may breach a drainage divide and capture the headwaters of a stream in another drainage basin.

■ *How does a superposed stream originate?*

Superposition takes place when a stream erodes a land surface and encounters resistant rocks below that, when exposed, form ridges. The stream is lowered onto the evolving ridge and cuts a narrow canyon directly through it, which is called a water gap. If a water gap is later abandoned by a stream, it is called a wind gap.

ESSENTIAL TERMS TO KNOW

abrasion (p. 465)
alluvial fan (p. 473)
alluvium (p. 467)
base level (p. 480)
bed load (p. 466)
braided stream (p. 467)
delta (p. 470)
discharge (p. 463)
dissolved load (p. 466)
divide (p. 479)

drainage basin (p. 479)
drainage pattern (p. 479)
floodplain (p. 469)
graded stream (p. 481)
gradient (p. 462)
hydraulic action (p. 465)
hydrologic cycle (p. 459)
incised meander (p. 485)
infiltration capacity (p. 461)
meandering stream (p. 467)

natural levee (p. 470)
oxbow lake (p. 468)
point bar (p. 468)
runoff (p. 459)
stream terrace (p. 484)
superposed stream (p. 487)
suspended load (p. 466)
valley (p. 484)
velocity (p. 462)

REVIEW QUESTIONS

1. About 10.75 km^3 of sediment erodes from the continents each year, and the volume of the continents above sea level is about 93,000,000 km^3. Thus the continents should erode to sea level in just over 8,600,000 years. The figures given are reasonably accurate, but something is seriously wrong with the conclusion that the continents will be leveled in so short a time. Explain.

2. A river with its headwaters 2000 m above sea level flows 1500 km to the sea. What is its gradient? Do you think your calculated gradient is valid for all segments of this river? Explain.

3. The steeper a river or stream's gradient, the higher the flow velocity. Right? Explain.

4. Why do you think some valleys are narrow and deep, whereas others are deep but have gently sloping sides, and some have a series of cliffs and slopes like the ones in Figure 7.13?

5. Explain how stream terraces form and what they indicate about the dynamics of a stream or river system.

6. The discharge in most streams and rivers increases downstream, but in a few discharge actually decreases and they finally disappear. Why? Give an example.

7. What data are used to plan flood-control projects? Explain why these projects, even if well designed and constructed, still have limited effectiveness.

8. How do the deposits of meandering streams compare with those of braided streams?

9. On the average, how often does a flood with a discharge of 200 m^3/sec take place on the Colorado River near Lobatos, Colorado (see Figure 15.20b)? Explain why you would or would not expect this flood to recur at a regular interval.

APPLY YOUR KNOWLEDGE

1. A depositional basin measures 200 km by 100 km and averages 2 km deep. Streams and rivers flow into this basin from all sides and it is filled to the top—that is with 40,000 km³ of sediment, mostly sand and mud. However, the basin will actually hold far more sediment than its original dimensions indicate. Why? (*Hint:* Remember the concept of isostasy we discussed in Chapter 11 and the discussion of lithification in Chapter 7.)

2. Explain how a river with several local base levels may eventually develop into a graded stream.

FIELD QUESTION

1. The image below shows an aerial view of the Sacramento River in California. What is the large deposit indicated by the line, and how did it form? Where do you think flow velocity would be highest in this river?

GEOLOGY MATTERS

GEOLOGY IN FOCUS

The River Nile and the History of Egypt

Egypt is well known for its remarkable antiquities, such as the pyramids and sphinx at Giza, the tombs in the Valley of the Kings, and numerous temples and monuments. However, Egypt also has some fascinating geologic features, particularly the Nile River. In fact, in no other country has development and prosperity been so closely tied to a single geologic feature (▶ Figure 1a). Although the Nile at 6825 km long is the world's longest river, its discharge of 1584 m^3/sec makes it a modest-sized river when compared with the great rivers of the world. Nevertheless it is the lifeblood of an entire country.

The Nile River rises as the Blue Nile in Ethiopia and the White Nile in Sudan, which join and flow north into Egypt and eventually into the Mediterranean Sea. If it were not for the abundant precipitation in the Ethiopian highlands, Egypt, which is nearly all desert, would have no river at all. Indeed, the Atbara River in Sudan is its only other tributary.

Once the Nile enters Egypt it has no tributaries, although a number of gullies called *wadis* do carry water to the river following infrequent rainstorms. The Nile is flanked on the west and east by desert, and the only agricultural land is on the Nile's floodplain (▶ Figure 1b) and delta, at the Fayum Depression, and at a few scattered oases.

Although people have occupied what is now Egypt for hundreds of thousands of years, Egyptian civilization took root there about 5000 years ago and persisted for more than 3000 years. Even when ancient Greece flourished, Egyptian civilization was already ancient. But it was a civilization totally dependent on the annual flooding of the Nile, during which fertile deposits accumulated on the floodplain. Indeed, in ancient Egypt taxes were even determined by the extent of the annual flooding. And of course the Nile was also the main avenue of commerce throughout the region.

The Nile is unique in that it has so few tributaries and flows such a great distance through a desert, but in most other respects it is like any other river. It responds to short- and long-term climatic changes, interruptions in the system where dams are built, and changes in sea level. During the Late Miocene Epoch, about 6 or 7 million years ago, base level for the Nile was much lower than it is now because the Mediterranean Sea was closed off from the Atlantic at the Strait of Gibraltar. As a result the sea dried up, leaving a vast plain more than 3000 m below sea level. The Nile and other rivers around the Mediterranean basin responded by cutting deep canyons as they adjusted to this new lower base level. By the Pliocene Epoch (about 5 million years ago), however, the barrier at the Strait of Gibraltar was breached and the Mediterranean began to refill, thus raising base level.*

During the time of deep erosion, the Nile cut a canyon about 10–20 km wide and 2500 m deep near Cairo. Even 465 km upstream at Aswan, the canyon was cut 200 m below present-day sea level. As sea level rose, the ancient river canyon became an arm of the sea, but by about 3.3 million years ago the river began depositing gravel, sand, and mud in its canyon and it took on an appearance much like it has today.

Since the completion of the Aswan High Dam in 1971, flooding on the Nile has been drastically reduced, so now it does not deposit mud on its floodplain each year. As a result Egyptian farmers have had to apply chemical fertilizers to maintain agricultural productivity.

*Actually, the isolation of the Mediterranean from the Atlantic and its refilling probably occurred several times.

Figure 1
(a) Satellite image of the Nile River as it flows north through Egypt. The only arable land is on the river floodplain and its delta and at a few scattered oases. The floodplain measures about 20 km across at its widest point. (b) At this location between Luxor and Aswan, the floodplain is only a few hundred meters wide. The view is toward the Eastern Desert, which begins a short distance from the river.

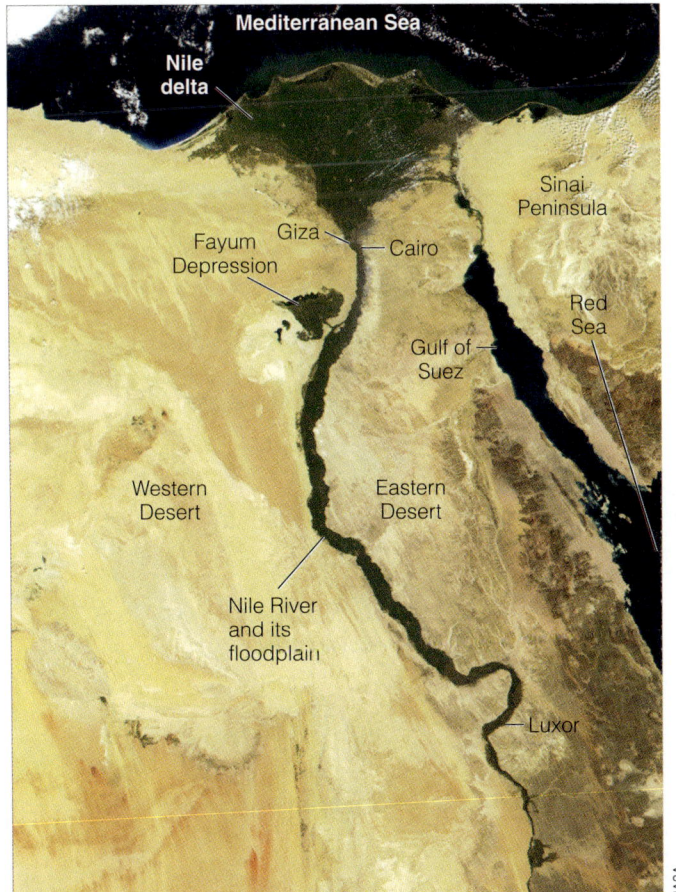

(a)

(b)

GEOLOGY MATTERS
Geology AT WORK

Dams, Reservoirs, and Hydroelectric Power Plants

Flip a switch and we illuminate our homes and workplaces; turn a dial and we heat our homes or perhaps cook our food or wash our clothes; and some of our public transportation relies on an unseen but important energy source—electricity. In fact, about 40% of the total energy used in industrialized nations is converted to electricity. Most electricity is generated at power plants that burn fossil fuels (oil, natural gas, and especially coal), but nuclear power plants are common in some areas. Geothermal energy (see Chapter 16), wind, and tidal power (see Chapter 19) account for only a small percentage of all electricity.

Electricity generated at *hydroelectric power plants*—that is, power plants that use moving water—accounts for only 8% of the total electricity generated in North America, but its importance varies from area to area. Less than 2% of the electricity generated in Florida, Ohio, and Texas comes from hydroelectric plants, whereas Washington, Oregon, and Idaho derive more than 80% of their electricity from this source.

At all power plants a spinning turbine connected to a generator containing a coil of wire produces electricity. In fossil-fuel-burning plants, coal, oil, or natural gas is burned to heat water and generate steam, which in turn serves as the energy source to turn turbines.

In hydroelectric power plants, moving water is used to turn turbines. To provide the necessary water, a dam is built impounding a reservoir where the water is higher than the power plant. Water moves from the reservoir through a large pipe called a penstock and spins a turbine (▶ Figure 1). In other words, the potential energy of the water in the reservoir is converted to electrical energy at the power plant.

Falling water was first used to generate electricity in 1892 at Appleton, Wisconsin, and since then hydroelectric power plants have become common features on many of the world's waterways. Currently the largest is Three Gorges Dam on the Yangtze River in the People's Republic of China.

One might be curious about why agencies and governments do not simply increase their hydroelectric output. After all, hydroelectric power generation has several appealing aspects, not the least of which is that it is a renewable resource. However, not all areas have this potential; suitable sites for dams and reservoirs might not be present, for instance. In addition, dams are very expensive to build, reservoirs fill with sediment, and during droughts enough water might not be available to keep reservoirs sufficiently full. And, of course, people must be relocated from areas where reservoirs are impounded and from the discharge areas downstream from dams. So although hydroelectric dams remain important, we cannot realistically look forward to very much additional use of this energy source.

Figure 1
Electricity is produced at a generator when water rushes through a penstock, where it spins a water wheel connected by a shaft to an electromagnet within a coil of wire. Each penstock shown here measures about 4 m in diameter and carries water at 16–22 km/hr. The dam from which the water enters the penstocks is visible in the background.

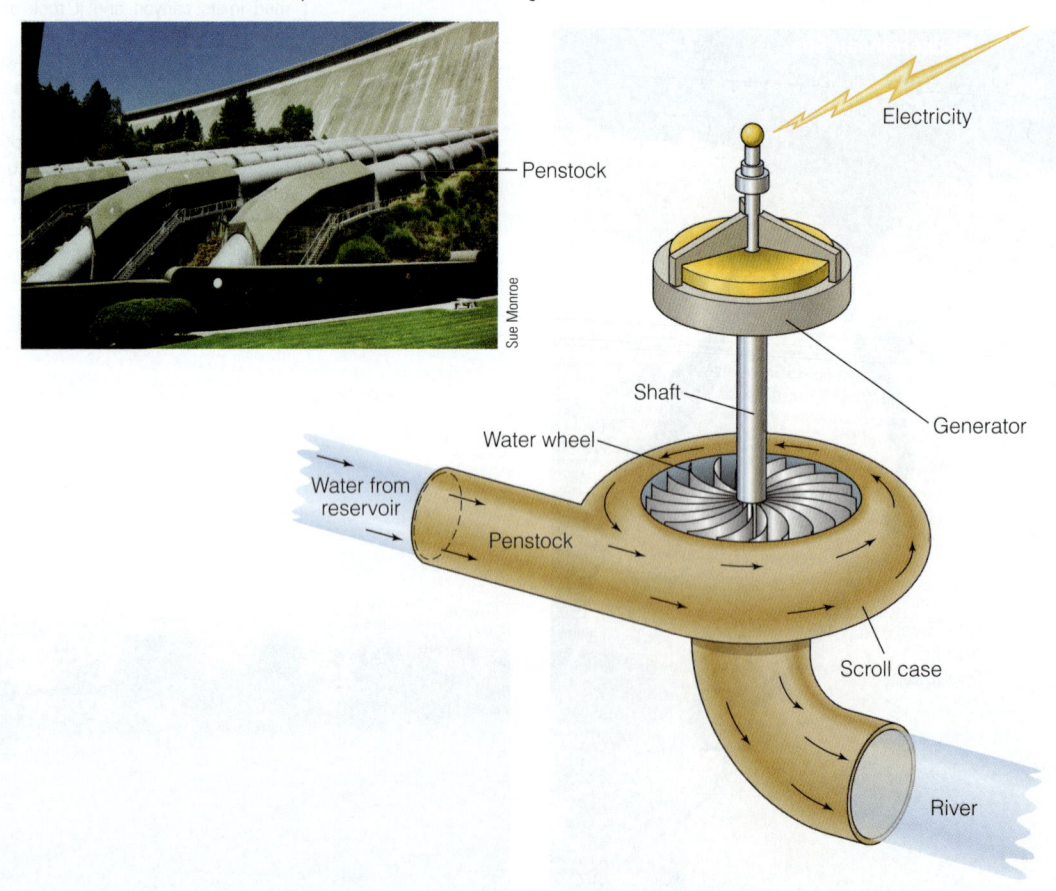

GEOLOGY MATTERS

GEOLOGY IN UNEXPECTED PLACES

Floating Burial Chambers and the Mississippi River Delta

In Geology in Unexpected Places: Gravestones and Geology (see Chapter 6), we noted that cemeteries are good places to study some aspects of geology. Once again we begin a brief investigation by visiting a cemetery, but this one is in Leeville, Louisiana. You are probably not aware that architects design cemeteries, and in doing so they must consider the geology of the site. So what has happened at Leeville, Louisiana, that sparks the interest of geologists? Leeville is located on one of the older parts of the Mississippi River delta, and because it has an unusually high water table, burial chambers are placed on the surface rather than buried.

So what's the problem? Burial chambers are placed on the ground surface wherever the water table is very near the surface. The problem is that many of the burial chambers at Leeville are floating or nearly afloat because the land is subsiding, with a concurrent rise in sea level. And because Leeville is near the seashore, the water table is rising, too. In fact, the cemeteries at Leeville are flooded at high tide because of subsidence—all in all, not a good situation for a cemetery.

This subsidence–rising water table problem is not confined to Leeville but affects much of southern Louisiana, especially the Mississippi River delta (▶ Figure 1). Flood-control projects on the Mississippi River prevent most sediment from reaching the delta, which contributes to loss of land due to wave erosion. Even more important, though, is the natural propensity for delta deposits, especially mud, to compact under their own weight and thus lower the surface.* Geologists estimate that during the last 4400 years the region has subsided at the rate of about 12 cm per century. The problem of delta subsidence goes far beyond the cemetery at Leeville. Indeed, as we noted in Section 15.1, much of New Orleans is below sea level because of subsidence, and its levees failed when Hurricane Katrina came ashore (Figure 15.1).

Each year about 65 km^2 of Louisiana's coastline is lost to subsidence and erosion. Ironically, human efforts to prevent floods have only worsened the situation. And what about global warming, which causes thermal expansion of the ocean's surface waters and melting of glaciers? This will certainly contribute to the area's problems as sea level rises and much of the delta continues to subside, especially because it is starved for sediment by flood-control projects.

Detailed geologic investigations reveal that the Mississippi River has followed seven different paths to the Gulf of Mexico during the last 6000 years, so the delta is actually a composite of several deltas deposited at different times (▶ Figure 1). During the last century or so engineers have built levees along the course of the river to prevent it from abandoning its present channel as it has in the past. In fact, if it were not for these efforts, the lower 500 km of the river would very likely switch to the Atchafalaya River and discharge into the Gulf of Mexico nearly 100 km to the west.

A final note on Leeville: When this was written a few days after Hurricane Katrina devastated more than 230,000 km^2 of the Gulf Coast, the fate of Leeville was not known. Certainly, we can assume that the burial chambers are now truly afloat, and given Leeville's location and its slight elevation above sea level, it was heavily damaged.

---
*Pumping oil and natural gas from the delta sediments also contributes to subsistence.

Figure 1
During the last 6000 years the Mississippi River has followed the seven routes to the Gulf of Mexico indicated in this diagram; 7 is the youngest, 6 next oldest, and so on. Notice the location of Leeville, Louisiana.

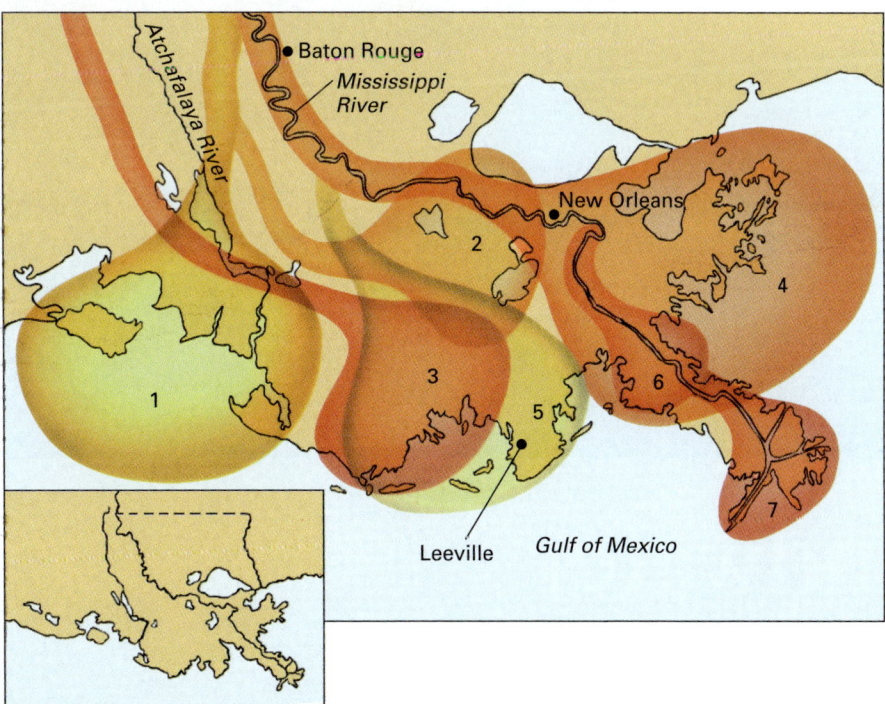

Chapter 16

Groundwater

Hidden Beauty
Light falls into the cave like ropes being lowered to help you out. Stalactites drip from the ceiling like quills of ice. Pools of milky blue groundwater bathe the rock covered floor.
Caves like this one near Valladoid, Mexico, are like secret rooms beneath Earth filled with a deluge of natural treasures that will dazzle you.
—A. W.

ESSENTIAL QUESTIONS TO ASK

16.1 Introduction
- Why is it important to study groundwater?

16.2 Groundwater and the Hydrologic Cycle
- What is groundwater?

16.3 Porosity and Permeability
- What is the difference between porosity and permeability?

16.4 The Water Table
- What is the water table and what does it look like?

16.5 Groundwater Movement
- How does groundwater move?

16.6 Springs, Water Wells, and Artesian Systems
- What are springs?
- What are water wells?
- What is a cone of depression?
- What is an artesian system?

16.7 Groundwater Erosion and Deposition
- What are sinkholes?
- What is karst topography?
- How do caves form and what are the common cave deposits?

16.8 Modifications of the Groundwater System and Their Effects
- What are some of the consequences of modification of the groundwater system?
- What determines groundwater quality?

16.9 Hydrothermal Activity
- What are the different types of hydrothermal activity?
- What is geothermal energy and why is it important?

GEOLOGY MATTERS

GEOLOGY IN FOCUS:
Arsenic and Old Lace

GEOLOGY IN UNEXPECTED PLACES:
Water-Treatment Plants

GEOLOGY AND CULTURAL CONNECTIONS:
Dowsing

PHYSICAL Geology Now™ This icon, which appears throughout the book, indicates an opportunity to explore interactive tutorials, animations, or practice problems available on the Physical GeologyNow website at http://now.brookscole.com/phygeo6.

16.1 Introduction

Within the limestone region of western Kentucky lies the largest cave system in the world. In 1941 approximately 51,000 acres were set aside and designated as Mammoth Cave National Park. In 1981 it became a World Heritage Site.

From ground level, the topography of the area is unimposing with gently rolling hills. Beneath the surface, however, are more than 540 km of interconnected passageways whose spectacular geologic features have been enjoyed by millions of cave explorers and tourists.

During the War of 1812, approximately 180 metric tons of saltpeter, used in the manufacture of gunpowder, was mined from Mammoth Cave. At the end of the war, the saltpeter market collapsed and Mammoth Cave was developed into a tourist attraction, easily overshadowing other caves in the area. During the next 150 years, the discovery of new passageways and links to other caverns helped establish Mammoth Cave as the world's premier cave and the standard against which all others were measured.

The formation of the caves themselves began about 3 million years ago when groundwater began dissolving the region's underlying St. Genevieve Limestone to produce a complex network of openings, passageways, and huge chambers that constitute present-day Mammoth Cave. Flowing through the various caverns is the Echo River, a system of streams that eventually joins the Green River at the surface.

The colorful cave deposits are the primary reason millions of tourists have visited Mammoth Cave over the years. Hanging down from the ceiling and growing up from the floor are spectacular icicle-like structures as well as columns and curtains in a variety of colors. Moreover, intricate passageways connect various-sized rooms. The cave is also home to more than 200 species of insects and other animals, including about 45 blind species.

■ *Why is it important to study groundwater?*

In addition to the beautiful caves, caverns, and cave deposits produced by groundwater movement, groundwater is an important source of freshwater for agriculture, industry, and domestic users. More than 65% of the groundwater used in the United States each year goes for irrigation, with industrial use second, followed by domestic needs. These demands have severely depleted the groundwater supply in many areas and led to such problems as ground subsidence and saltwater contamination. In other areas, pollution from landfills, toxic waste, and agriculture has rendered the groundwater supply unsafe.

As the world's population and industrial development expand, the demand for water, particularly groundwater, will increase. Not only must new groundwater sources be located, but once found, these sources must be protected from pollution and managed properly to ensure that users do not withdraw more water than can be replenished. It is therefore important that people become aware of what a valuable resource groundwater is, so that they can assure future generations of a clean and adequate supply of this water source.

Section 16.1 Summary

- Groundwater is an important source of freshwater for agriculture, industry, and domestic use. As the world's population and industrial needs increase, the demand for water, particularly groundwater, will grow. New sources will need to be found and protected from pollution and overuse. Groundwater is a valuable natural resource and must be treated as such.

16.2 Groundwater and the Hydrologic Cycle

■ *What is groundwater?*

Groundwater, water that fills open spaces in rocks, sediment, and soil beneath Earth's surface, is one reservoir in the hydrologic cycle, representing approximately 22% (8.4 million km^3) of the world's supply of freshwater (see Figure 15.3). Like all other water in the hydrologic cycle, the ultimate source of groundwater is the oceans, but its more immediate source is the precipitation that infiltrates the ground and seeps down through the voids in soil, sediment, and rocks. Groundwater may also come from water infiltrating from streams, lakes, swamps, artificial recharge ponds, and water treatment systems.

Regardless of its source, groundwater moving through the tiny openings between soil and sediment particles and the spaces in rocks filters out many impurities such as disease-causing microorganisms and various pollutants. Not all soils and rocks are good filters, though, and sometimes so much undesirable material may be present that it contaminates the groundwater. Groundwater movement and its recovery at wells depend on two critical aspects of the materials it moves through: *porosity* and *permeability*.

Section 16.2 Summary

- Groundwater is the water that fills open spaces in rocks, sediments, and soil beneath the surface. It is one reservoir in the hydrologic cycle and accounts for approximately 22% of the world's freshwater supply.

16.3 Porosity and Permeability

■ *What is the difference between porosity and permeability?*

Porosity and permeability are important physical properties of Earth materials and are largely responsible for the amount, availability, and movement of groundwater. Water soaks into the ground because soil, sediment, and rock have open spaces or pores. **Porosity** is the percentage of a material's total volume that is pore space. Porosity most often consists of the spaces between particles in soil, sediments, and sedimentary rocks, but other types of porosity include cracks, fractures, faults, and vesicles in volcanic rocks (▶ Figure 16.1).

Porosity varies among different rock types and depends on the size, shape, and arrangement of the material composing the rock (Table 16.1). Most igneous and metamorphic rocks, as well as many limestones and dolostones, have very low porosity because they consist of tightly interlocking crystals. Their porosity can be increased, however, if they have been fractured or weathered by groundwater. This is particularly true for massive limestone and dolostone, whose fractures can be enlarged by acidic groundwater.

By contrast, detrital sedimentary rocks composed of well-sorted and well-rounded grains can have high porosity because any two grains touch at only a single point, leaving relatively large open spaces between the grains (▶ Figure 16.1a). Poorly sorted sedimentary rocks, on the other hand, typically have low porosity because smaller grains fill in the spaces between the larger grains, further reducing porosity (▶ Figure 16.1b). In addition, the amount of cement between grains can decrease porosity.

Porosity determines the amount of groundwater Earth materials can hold, but it does not guarantee that the water can be easily extracted. So, in addition to being porous, Earth materials must have the capacity to transmit fluids, a property known as **permeability.** Thus both porosity and permeability play important roles in groundwater movement and recovery.

Permeability is dependent not only on porosity but also on the size of the pores or fractures and their interconnections. For example, deposits of silt or clay are typically more porous than sand or gravel, but they have low permeability because the pores between the particles are very small, and molecular attraction between the particles and water is great, thereby preventing movement of the water. In contrast, the pore spaces between grains in sandstone and conglomerate are much larger, and molecular attraction on the water is therefore low. Chemical and biochemical sedimentary rocks, such as limestone and dolostone, and many igneous and metamorphic rocks that are highly fractured can also be very permeable provided the fractures are interconnected.

The contrasting porosity and permeability of familiar substances are well demonstrated by sand versus clay. Pour some water on sand and it rapidly sinks in, whereas water poured on clay simply remains on the surface. Furthermore, wet sand dries quickly, but once clay absorbs water, it may take days to dry out because of its low permeability. Neither sand nor clay makes a good substance in which to grow crops or gardens, but a mixture of the two plus some organic matter in the form of humus makes an excellent soil for farming and gardening (see Chapter 6).

A permeable layer transporting groundwater is an *aquifer*, from the Latin *aqua*, "water." The most effective aquifers are deposits of well-sorted and well-rounded sand and gravel. Limestones in which fractures and bedding planes have been enlarged by solution are also good aquifers. Shales and many igneous and metamorphic rocks make poor aquifers because they are typically impermeable

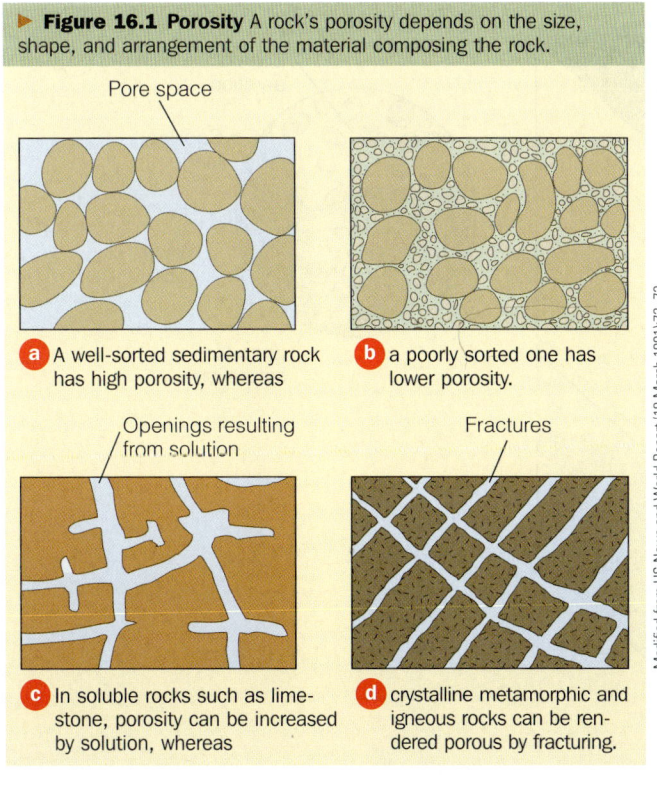

▶ **Figure 16.1 Porosity** A rock's porosity depends on the size, shape, and arrangement of the material composing the rock.

a. A well-sorted sedimentary rock has high porosity, whereas
b. a poorly sorted one has lower porosity.
c. In soluble rocks such as limestone, porosity can be increased by solution, whereas
d. crystalline metamorphic and igneous rocks can be rendered porous by fracturing.

Table 16.1	
Porosity Values for Different Materials	
Material	**Percentage Porosity**
Unconsolidated sediment	
Soil	55
Gravel	20–40
Sand	25–50
Silt	35–50
Clay	50–70
Rocks	
Sandstone	5–30
Shale	0–10
Solution activity in limestone, dolostone	10–30
Fractured basalt	5–40
Fractured granite	10

Source: U.S. Geological Survey, Water Supply Paper 2220 (1983) and others.

unless fractured. Rocks such as these and any other materials that prevent the movement of groundwater are *aquicludes*.

> ### Section 16.3 Summary
>
> - Porosity is the percentage of a material's total volume that is pore space. It most often consists of the spaces between particles in soil, sediments, and sedimentary rocks, but it can also include cracks, fractures, joints, and faults.
>
> - Permeability is a material's ability to transmit fluids. Permeability is dependent not only on porosity but also on the size of the pores or fractures and their interconnections.
>
> - Permeable layers that transport groundwater are called aquifers, and materials that prevent the flow of groundwater are called aquicludes.

16.4 The Water Table

■ *What is the water table and what does it look like?*

Some of the precipitation on land evaporates, and some enters streams and returns to the oceans by surface runoff; the remainder seeps into the ground. As this water moves down from the surface, a small amount adheres to the material it moves through and halts its downward progress. With the exception of this *suspended water*, however, the rest seeps farther downward and collects until it fills all the available pore spaces. Thus two zones are defined by whether their pore spaces contain mostly air, the **zone of aeration**, or mostly water, the underlying **zone of saturation**. The surface separating these two zones is the **water table** (▶ Figure 16.2).

The base of the zone of saturation varies from place to place but usually extends to a depth where an impermeable layer is encountered or to a depth where confining pressure

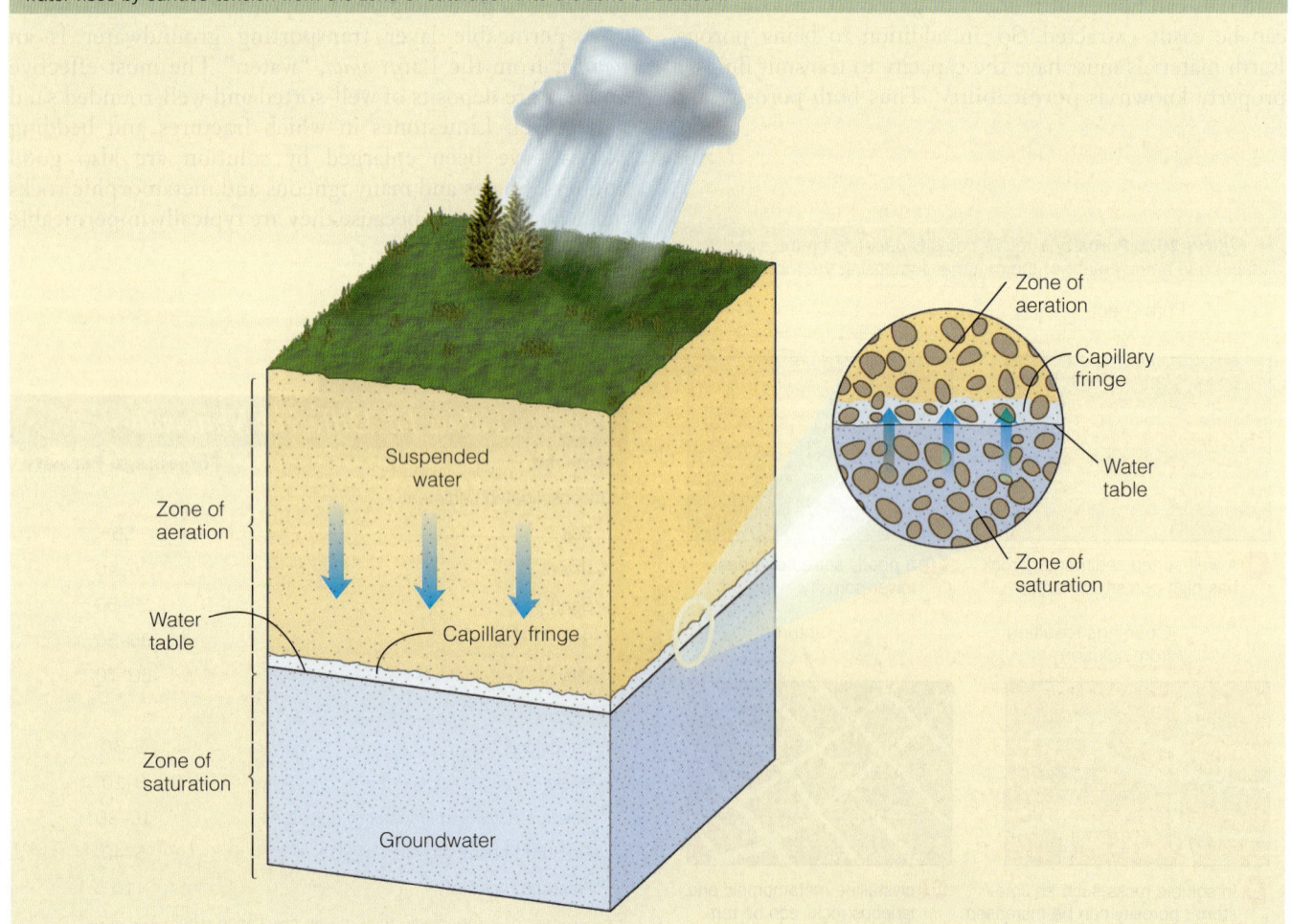

▶ **Figure 16.2 Water Table** The zone of aeration contains both air and water within its pore spaces, whereas all pore spaces in the zone of saturation are filled with groundwater. The water table is the surface separating the zones of aeration and saturation. Within the capillary fringe, water rises by surface tension from the zone of saturation into the zone of aeration.

closes all open space. Extending irregularly upward a few centimeters to several meters from the zone of saturation is the *capillary fringe*. Water moves upward in this region because of surface tension, much as water moves upward through a paper towel.

In general, the configuration of the water table is a subdued replica of the overlying land surface; that is, it rises beneath hills and has its lowest elevations beneath valleys. Several factors contribute to the surface configuration of a region's water table, including regional differences in amount of rainfall, permeability, and rate of groundwater movement. During periods of high rainfall, groundwater tends to rise beneath hills because it cannot flow fast enough into adjacent valleys to maintain a level surface. During droughts, the water table falls and tends to flatten out because it is not being replenished. In arid and semiarid regions, the water table is usually quite flat regardless of the overlying land surface.

Section 16.4 Summary

- The water table is a surface separating the zone of aeration, which contains both air and water in its open spaces, and the underlying zone of saturation, in which all the pore space is filled with groundwater. The capillary fringe is a region where water moves upward from the base of the zone of saturation due to surface tension.

- The configuration of the water table is generally a subdued replica of the overlying surface topography, except in arid and semiarid regions where it is usually quite flat regardless of the overlying land surface.

16.5 Groundwater Movement

■ *How does groundwater move?*

Gravity provides the energy for the downward movement of groundwater. Water entering the ground moves through the zone of aeration to the zone of saturation (▶ Figure 16.3). When water reaches the water table, it continues to move through the zone of saturation from areas where the water table is high toward areas where it is lower, such as streams, lakes, or swamps. Only some of the water follows the direct route along the slope of the water table. Most of it takes longer curving paths down and then enters a stream, lake, or swamp from below, because it moves from areas of high pressure toward areas of lower pressure within the saturated zone.

Groundwater velocity varies greatly and depends on many factors. Velocities range from 250 m per day in some extremely permeable material to less than a few centimeters per year in nearly impermeable material. In most ordinary aquifers, the average velocity of groundwater is a few centimeters per day.

Section 16.5 Summary

- Groundwater moves downward under the influence of gravity. It moves through the zone of aeration to the zone of saturation. Some of it then moves along the slope of the water table, while the rest moves through the zone of saturation from areas of high pressure to areas of low pressure.

- Groundwater velocity varies and depends on various factors. In general, it averages a few centimeters per day.

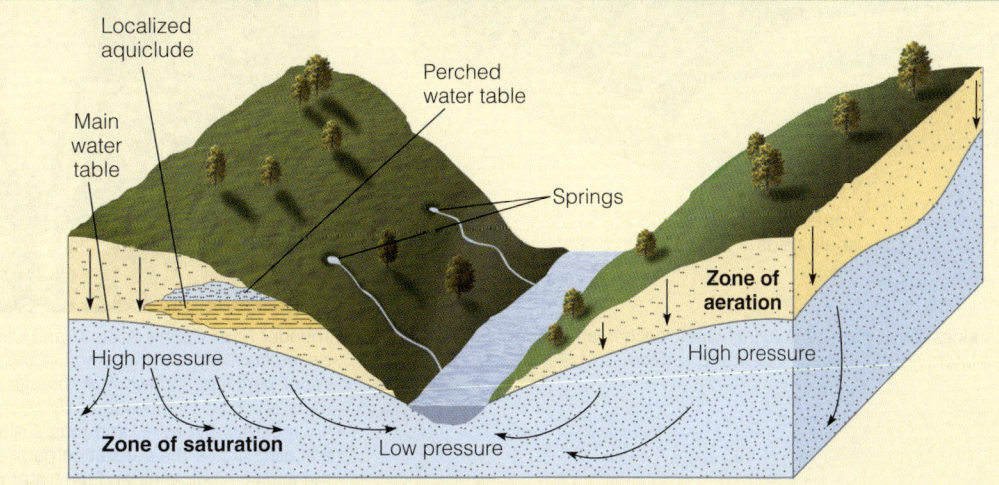

▶ **Figure 16.3 Groundwater Movement** Groundwater moves down through the zone of aeration to the zone of saturation. Then some of it moves along the slope of the water table, and the rest moves through the zone of saturation from areas of high pressure toward areas of low pressure. Some water might collect over a local aquiclude, such as a shale layer, thus forming a perched water table.

16.6 Springs, Water Wells, and Artesian Systems

You can think of the water in the zone of saturation much like a reservoir whose surface rises or falls depending on additions as opposed to natural and artificial withdrawals. *Recharge*—that is, additions to the zone of saturation—comes from rainfall or melting snow, or water might be added artificially at wastewater-treatment plants or recharge ponds constructed for just this purpose. But if groundwater is discharged naturally or withdrawn at wells without sufficient recharge, the water table drops, just as a savings account diminishes if withdrawals exceed deposits. Withdrawals from the groundwater system take place where groundwater flows laterally into streams, lakes, or swamps, where it discharges at the surface as springs, and where it is withdrawn from the system at water wells.

■ *What are springs?*

Places where groundwater flows or seeps out of the ground as **springs** have always fascinated people. The water flows out of the ground for no apparent reason and from no readily identifiable source. So it is not surprising that springs have long been regarded with superstition and revered for their supposed medicinal value and healing powers. Nevertheless, there is nothing mystical or mysterious about springs.

Although springs can occur under a wide variety of geologic conditions, they all form in basically the same way (▶ Figure 16.4a). When percolating water reaches the water table or an impermeable layer, it flows laterally, and if this flow intersects the surface, the water discharges as a spring (▶ Figure 16.4b). The Mammoth Cave area in Kentucky is underlain by fractured limestones whose fractures have been enlarged into caves by solution activity. In this geologic environment, springs occur where the fractures and caves intersect the ground surface, allowing groundwater to exit onto the surface. Most springs are along valley walls where streams have cut valleys below the regional water table.

Springs can also develop wherever a perched water table intersects the surface (▶ Figure 16.3). A *perched water table* may occur wherever a local aquiclude is present within a larger aquifer, such as a lens of shale within sandstone. As water migrates through the zone of aeration, it is stopped by the local aquiclude, and a localized zone of saturation

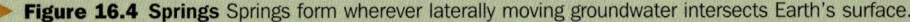

▶ **Figure 16.4 Springs** Springs form wherever laterally moving groundwater intersects Earth's surface.

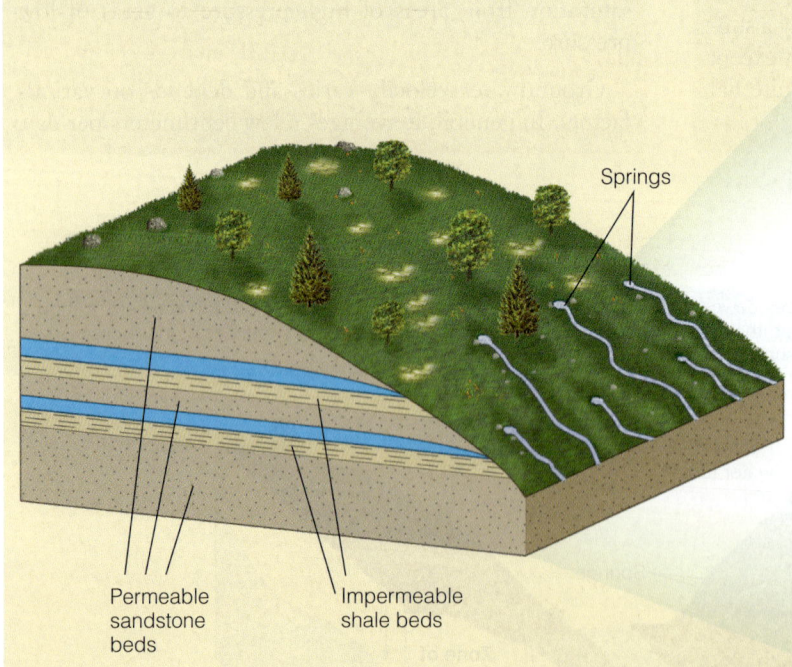

a Most commonly, springs form when percolating water reaches an impermeable layer and migrates laterally until it seeps out at the surface.

b Thunder River Spring in the Grand Canyon, Arizona, issues from rocks along a wall of the Grand Canyon. Water percolating downward through permeable rocks is forced to move laterally when it encounters an impermeable zone, and thus gushes out along this cliff. Notice the vegetation parallel to and below the springs, indicating enough water flows from springs along the cliff wall to support the vegetation.

"perched" above the main water table forms. Water moving laterally along the perched water table may intersect the surface to produce a spring.

■ *What are water wells?*

Water wells are openings made by drilling or digging down into the zone of saturation. Once the zone of saturation has been penetrated, water percolates into the well, filling it to the level of the water table. A few wells are free flowing (see the next section), but for most the water must be brought to the surface by pumping.

In some parts of the world, water is raised to the surface with nothing more than a bucket on a rope or a hand-operated pump. In many parts of the United States and Canada, one can see windmills from times past that employed wind power to pump water. Most of these are no longer in use, having been replaced by more efficient electric pumps.

■ *What is a cone of depression?*

When groundwater is pumped from a well, the water table in the area around the well is lowered, forming a **cone of depression.** A cone of depression forms because the rate of water withdrawal from the well exceeds the rate of water inflow to the well, thus lowering the water table around the well. This lowering normally does not pose a problem for the average domestic well, provided the well is drilled deep enough into the zone of saturation.

The tremendous amounts of water used by industry and for irrigation, however, may create a large cone of depression that lowers the water table sufficiently to cause shallow wells in the immediate area to go dry (▶ Figure 16.5). This situation is not uncommon and frequently results in lawsuits by the owners of the shallow dry wells. Furthermore, lowering of the regional water table is becoming a serious problem in many areas, particularly in the southwestern United States where rapid growth has placed tremendous demands on the groundwater system. Unrestricted withdrawal of groundwater cannot continue indefinitely, and the rising costs and decreasing supply of groundwater should soon limit the growth of this region of the United States.

■ *What is an artesian system?*

The word *artesian* comes from the French town and province of Artois (called Artesium during Roman times) near Calais, where the first European artesian well was drilled in 1126 and is still flowing today. The term **artesian system** can be applied to any system in which groundwater is confined and builds up high hydrostatic (fluid) pressure. Water in such a system is able to rise above the level of the aquifer if a well is drilled through the confining layer, thereby reducing the pressure and forcing the water upward. For an artesian system to develop, three geologic conditions must be present (▶ Figure 16.6): (1) The aquifer must be confined above and below by aquicludes to prevent water from escaping; (2) the rock sequence is usually tilted and exposed at the surface, enabling the aquifer to be recharged; and (3) there is sufficient precipitation in the recharge area to keep the aquifer filled.

The elevation of the water table in the recharge area and the distance of the well from the recharge area determine the height to which artesian water rises in a well. The surface defined by the water table in the recharge area, called the *artesian-pressure surface*, is indicated by the sloping dashed line in Figure 16.6. If there were no friction in the aquifer, well water from an artesian aquifer would rise exactly to the elevation of the artesian-pressure surface. Friction, however, slightly reduces the pressure of the aquifer water and consequently the level to which artesian water rises. This is why the pressure surface slopes.

An artesian well will flow freely at the ground surface only if the wellhead is at an elevation below the artesian-pressure surface. In this situation, the water flows out of the well because it rises toward the artesian-pressure surface, which is at a higher elevation than the wellhead. In a nonflowing artesian well, the wellhead is above the artesian-pressure surface,

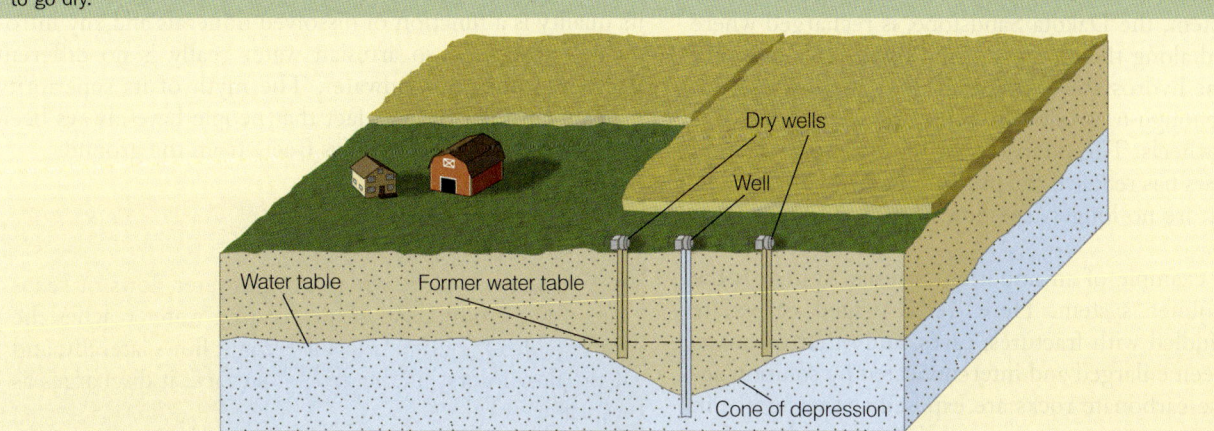

▶ **Figure 16.5 Cone of Depression** A cone of depression forms whenever water is withdrawn from a well. If water is withdrawn faster than it can be replenished, the cone of depression will grow in depth and circumference, lowering the water table in the area and causing nearby shallow wells to go dry.

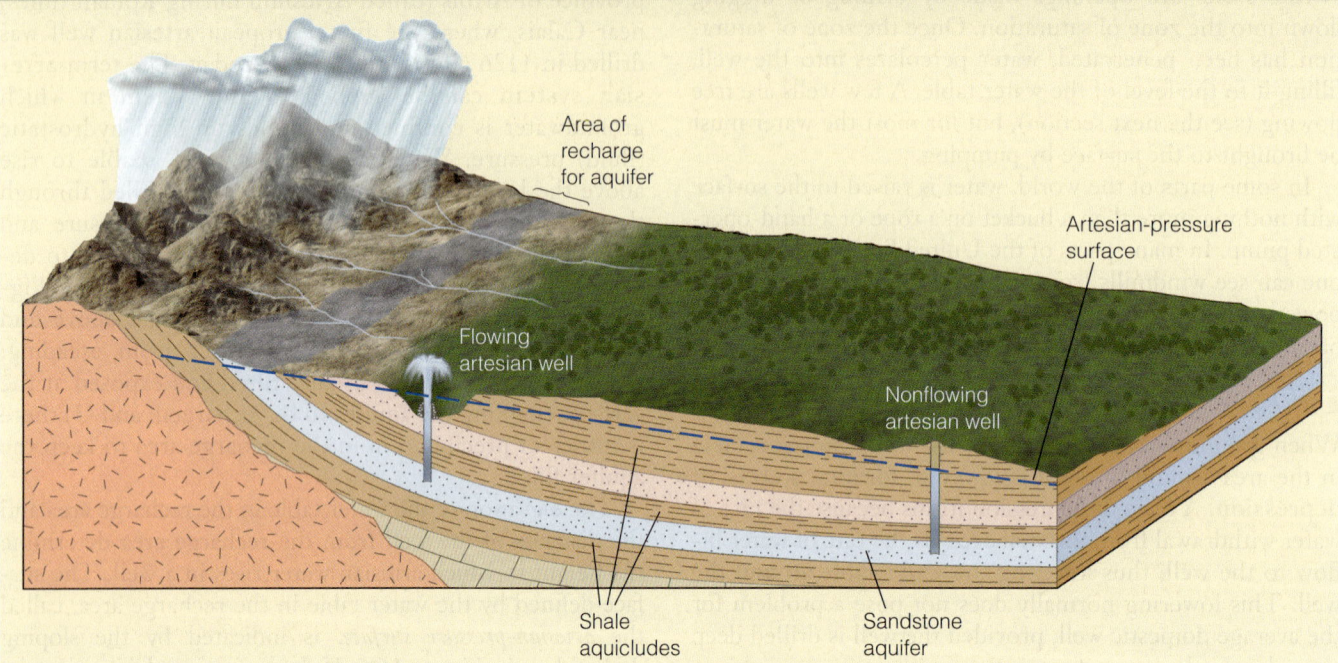

Active Figure 16.6 Artesian System An artesian system must have an aquifer confined above and below by aquicludes, the aquifer must be exposed at the surface, and there must be sufficient precipitation in the recharge area to keep the aquifer filled. The elevation of the water table in the recharge area, which is indicated by a sloping dashed line (the artesian-pressure surface), defines the highest level to which well water can rise. If the elevation of a wellhead is below the elevation of the artesian-pressure surface, the well will be free flowing because the water will rise toward the artesian-pressure surface. If the elevation of a wellhead is at or above that of the artesian-pressure surface, the well will be nonflowing. **Why is this?** *Because water in a well can rise only to the artesian-pressure surface.*

and the water will rise in the well only as high as the artesian-pressure surface.

In addition to artesian wells, many artesian springs exist. Artesian springs form if a fault or fracture intersects the confined aquifer, allowing water to rise above the aquifer. Oases in deserts are commonly artesian springs.

Because the geologic conditions necessary for artesian water can occur in a variety of ways, artesian systems are common in many areas of the world underlain by sedimentary rocks. One of the best-known artesian systems in the United States underlies South Dakota and extends southward to central Texas. The majority of the artesian water from this system is used for irrigation. The aquifer of this artesian system, the Dakota Sandstone, is recharged where it is exposed along the margins of the Black Hills of South Dakota. The hydrostatic pressure in this system was originally high enough to produce free-flowing wells and to operate waterwheels. The extensive use of water for irrigation over the years has reduced the pressure in many of the wells so that they are no longer free flowing and the water must be pumped.

Another example of an important artesian system is the Floridan aquifer system. Here Tertiary-aged carbonate rocks are riddled with fractures, caves, and other openings that have been enlarged and interconnected by solution activity. These carbonate rocks are exposed at the surface in the northwestern and central parts of the state where they are recharged, and they dip toward both the Atlantic and Gulf coasts where they are covered by younger sediments. The carbonates are interbedded with shales, forming a series of confined aquifers and aquicludes. This artesian system is tapped in the southern part of the state, where it is an important source of freshwater and one that is being rapidly depleted.

One final comment on artesian wells/springs and the water derived from them: It is not unusual for advertisers to tout the quality of artesian water as somehow being superior to other groundwater. Some artesian water might in fact be of excellent quality, but its quality is not dependent on the fact that water rises above the surface of an aquifer. Rather, its quality is a function of dissolved minerals and any introduced substances, so artesian water really is no different from any other groundwater. The myth of its superiority probably arises from the fact that people have always been fascinated by water that flows freely from the ground.

Section 16.6 Summary

- A spring is a place where groundwater flows or seeps out of the ground. When percolating water reaches the water table or an impermeable layer, it flows laterally, and if the flow intersects the ground surface, it discharges as a spring.

- A water well is an opening made by digging or drilling into the zone of saturation. Water percolates into the well and rises to the level of the water table.

- A cone of depression forms around a well when the rate of water withdrawn from the well exceeds the rate of water inflow to the well, thus lowering the water table around the well.

- In an artesian system, groundwater is confined and builds up high hydrostatic pressure. Three geologic conditions must be present for an artesian system to develop: The aquifer must be confined above and below by aquicludes; the rock sequence is usually tilted and exposed at the surface so that the aquifer can be recharged; and precipitation in the recharge area is sufficient to keep the aquifer filled.

16.7 Groundwater Erosion and Deposition

When rainwater begins to seep into the ground, it immediately starts to react with the minerals it contacts, weathering them chemically. In an area underlain by soluble rock, groundwater is the principal agent of erosion and is responsible for the formation of many major features of the landscape.

Limestone, a common sedimentary rock composed primarily of the mineral calcite ($CaCO_3$), underlies large areas of Earth's surface (▶ Figure 16.7). Although limestone is practically insoluble in pure water, it readily dissolves if a small amount of acid is present. Carbonic acid (H_2CO_3) is a weak acid that forms when carbon dioxide combines with water ($H_2O + CO_2 \rightarrow H_2CO_3$) (see Chapter 6). Because the atmosphere contains a small amount of carbon dioxide (0.03%) and carbon dioxide is also produced in soil by the decay of organic matter, most groundwater is slightly acidic. When groundwater percolates through the various openings in limestone, the slightly acidic water readily reacts with the calcite to dissolve the rock by forming soluble calcium bicarbonate, which is carried away in solution (see Chapter 6).

■ *What are sinkholes?*

In regions underlain by soluble rock, the ground surface may be pitted with numerous depressions that vary in size and shape. These depressions, called **sinkholes** or merely *sinks*, mark areas with underlying soluble rock (▶ Figure 16.8). Most sinkholes form in one of two ways. The first is when soluble rock below the soil is dissolved by seeping water, and openings in the rock are enlarged and filled in by the overlying soil. As the groundwater continues to dissolve the rock, the soil is eventually removed, leaving shallow depressions with gently sloping sides. When adjacent sinkholes merge, they form a network of larger, irregular, closed depressions called *solution valleys*.

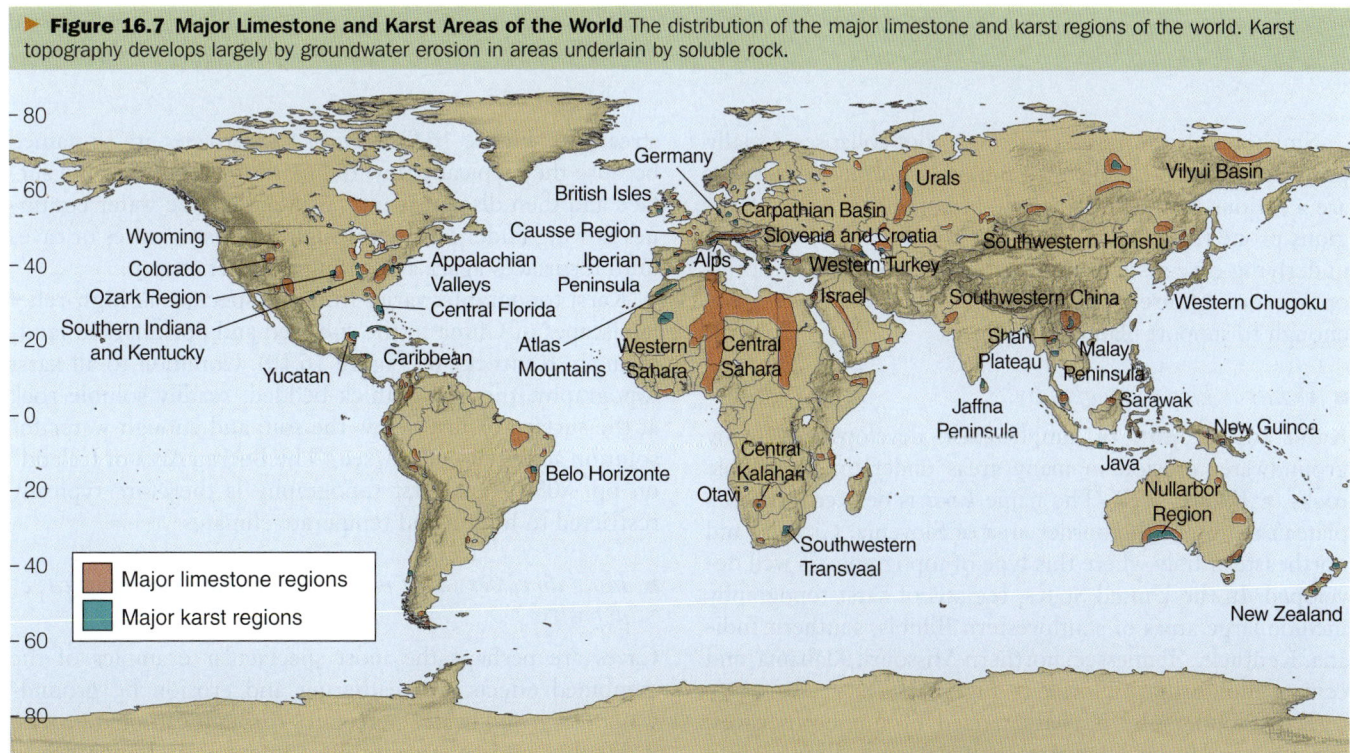

▶ **Figure 16.7 Major Limestone and Karst Areas of the World** The distribution of the major limestone and karst regions of the world. Karst topography develops largely by groundwater erosion in areas underlain by soluble rock.

Figure 16.8 Sinkholes

a This sinkhole formed on May 8 and 9, 1981, in Winter Park, Florida. It formed in previously dissolved limestone following a drop in the water table. The 100-m-wide, 35-m-deep sinkhole destroyed a house, numerous cars, and a municipal swimming pool.

b A small sinkhole in Montana now occupied by a lake. The water enters the lake from a hot spring so it remains warm year round. In fact, tropical fish have been introduced into the lake and thus live in an otherwise inhospitable climate.

Sinkholes also form when a cave's roof collapses, usually producing a steep-sided crater. Sinkholes formed in this way are a serious hazard, particularly in populated areas. In regions prone to sinkhole formation, the depth and extent of underlying cave systems must be mapped before any development to ensure that the underlying rocks are thick enough to support planned structures.

■ *What is karst topography?*

Karst topography, or simply *karst*, develops largely by groundwater erosion in many areas underlain by soluble rocks (▶ Figure 16.9). The name *karst* is derived from the plateau region of the border area of Slovenia, Croatia, and northeastern Italy where this type of topography is well developed. In the United States, regions of karst topography include large areas of southwestern Illinois, southern Indiana, Kentucky, Tennessee, northern Missouri, Alabama, and central and northern Florida (▶ Figure 16.7).

Karst topography is characterized by numerous caves, springs, sinkholes, solution valleys, and disappearing streams (▶ Figure 16.9). *Disappearing streams* are so named because they typically flow only a short distance at the surface and then disappear into a sinkhole. The water continues flowing underground through various fractures or caves until it surfaces again at a spring or other stream.

Karst topography varies from the spectacular high-relief landscapes of China to the subdued and pockmarked landforms in Kentucky (▶ Figure 16.10). Common to all karst topography, though, is thick-bedded, readily soluble rock at the surface or just below the soil, and enough water for solution activity to occur (see "The Burren Area of Ireland" on pp. 508–509). Karst topography is therefore typically restricted to humid and temperate climates.

■ *How do caves form and what are the common cave deposits?*

Caves are perhaps the most spectacular examples of the combined effects of weathering and erosion by groundwater. As groundwater percolates through carbonate rocks, it dissolves and enlarges original fractures and openings to

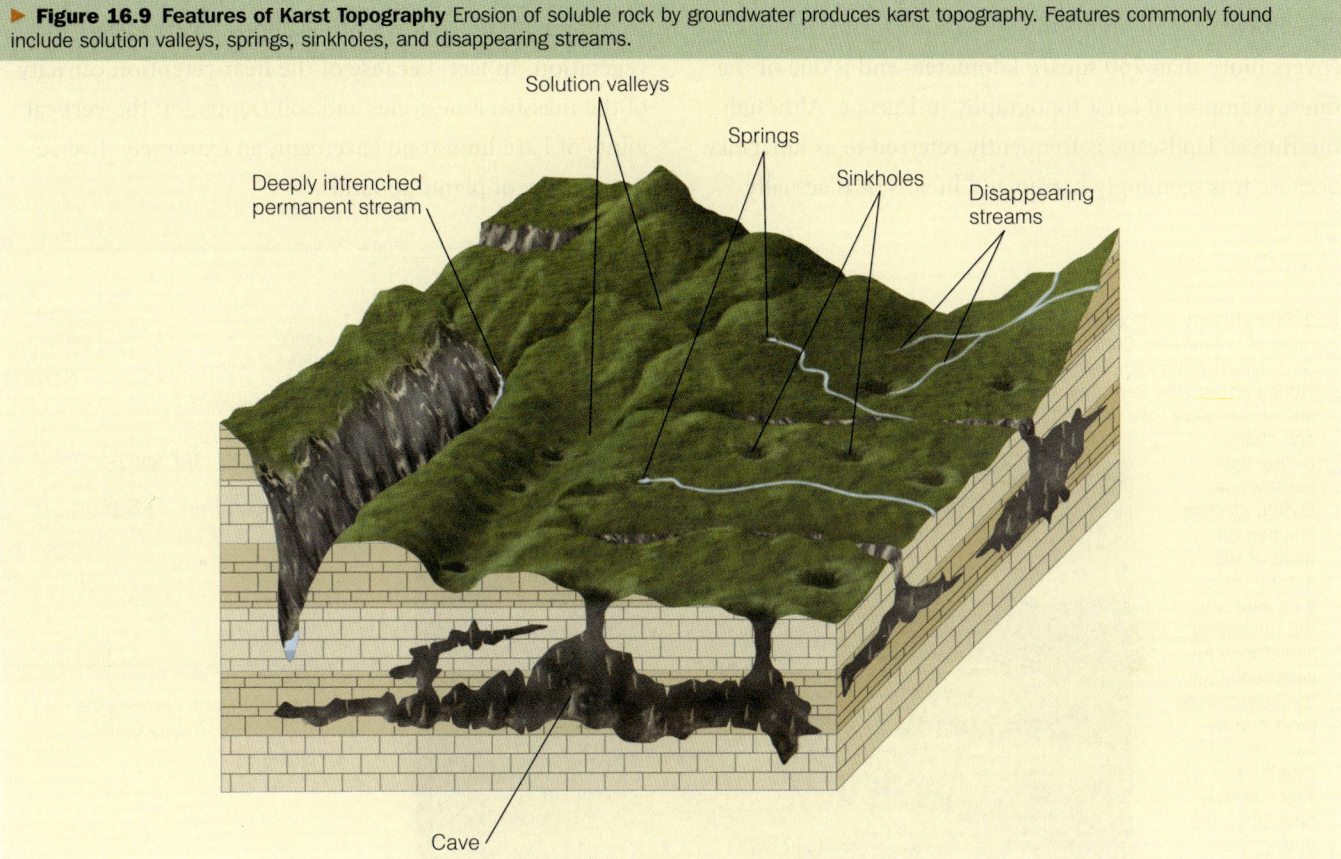

▶ **Figure 16.9 Features of Karst Topography** Erosion of soluble rock by groundwater produces karst topography. Features commonly found include solution valleys, springs, sinkholes, and disappearing streams.

form a complex interconnecting system of crevices, caves, caverns, and underground streams. A **cave** is usually defined as a naturally formed subsurface opening that is generally connected to the surface and is large enough for a person to enter. A *cavern* is a very large cave or a system of interconnected caves.

More than 17,000 caves are known in the United States. Most of them are small, but some are large and spectacular. Some of the more famous ones are Mammoth Cave, Kentucky; Carlsbad Caverns, New Mexico; Lewis and Clark Caverns, Montana; Wind Cave and Jewel Cave, South Dakota; Lehman Cave, Nevada; and Meramec Caverns, Missouri, which Jesse James and his outlaw band used as a hideout. The United States has many famous caves, but so has Canada, including 536-m-deep Arctomys Cave in Mount Robson Provincial Park, British Columbia, the deepest known cave in North America.

Caves and caverns form as a result of the dissolution of carbonate rocks by weakly acidic groundwater (▶ Figure 16.11). Groundwater percolating through the zone of aeration slowly dissolves the carbonate rock and enlarges its fractures and bedding planes. On reaching the water table, groundwater migrates toward the region's surface streams. As the groundwater moves through the zone of saturation, it continues to dissolve the rock and gradually forms a system of horizontal passageways through which the dissolved rock is carried to the streams. As the surface streams erode deeper valleys, the water table drops in response to the lower elevation of the streams. The water that flowed through the system of horizontal passageways now percolates to the lower water table where a new system of passageways begins to form. The abandoned channelways form an interconnecting system of caves and caverns. Caves eventually become unstable and collapse, littering the floor with fallen debris.

When most people think of caves, they think of the seemingly endless variety of colorful and bizarre-shaped deposits found in them. Although a great many different types of cave deposits exist, most form in essentially the same manner and are collectively known as *dripstone*. As water seeps into a cave, some of the dissolved carbon dioxide in the water escapes, and a small amount of calcite is precipitated. In this manner, the various dripstone deposits are formed.

Stalactites are icicle-shaped structures hanging from cave ceilings that form as a result of precipitation from dripping water (▶ Figure 16.12). With each drop of water, a thin layer of calcite is deposited over the previous layer, forming a cone-shaped projection that grows down from the ceiling. The water that drips from a cave's ceiling also precipitates a small amount of calcite when it hits the floor. As additional calcite is deposited, an upward-growing projection called a *stalagmite* forms (▶ Figure 16.12). If a stalactite and stalagmite meet,

The Burren Area of Ireland

The Burren region in northwest County Clare, Ireland, covers more than 260 square kilometers and is one of the finest examples of karst topography in Europe. Although the Burren landscape is frequently referred to as lunar-like because it is seemingly barren and lifeless, it is actually teeming with life and is world-famous for its variety of vegetation. In fact, because of the heat-retention capacity of the massive limestones and soil trapped in the vertical joints of bare limestone pavement, an extremely diverse community of plants abounds.

▶ 2. The present landscape is best described as glaciated karst. Like most of Ireland, the Burren was covered by a warm, shallow sea some 340 million years ago. As much as 780 m of interbedded marine limestones and shales were deposited at this time. These marine limestones and shales were then covered by nearly 330 m of sandstones, siltstones, and shales. During the Pleistocene Epoch, glaciers stripped off most of the detrital rocks, thus exposing the underlying limestones to weathering and the humic acids produced by localized vegetation. Together they have been the driving force producing the distinctive karst topography we find today.

▲ 1. A map of Ireland showing the Burren region of County Clare.

▼ 3. Bare limestone pavement with a small wedge tomb from the Neolithic period—approximately 6000 years ago—in the background.

▲ 4. Bare limestone pavements typically display a blocky appearance. The network of vertical cracks is the result of weathering of the joint pattern produced in the limestone during uplift of the region.

◀ 5. A close-up shows the characteristic karren weathering pattern produced by solution of the limestone. *Karren* is used to describe the various microsolutional features of limestone pavement.

6. Despite a lack of significant soil cover, a profusion of plants live in the soil trapped in the cracks, joints, and solution cavities of the limestone beds in the Burren.

7. Perhaps the most famous Irish Neolithic dolmen is the Poulnabrone portal tomb, which dates from about 5800 years ago. The name Poulnabrone literally means "the hole of the sorrows." The captsone of the Poulnabrone dolmen rests on two 1.8-m-high portal stones to create a chamber in a low circular cairn, 9 m in diameter.

9. A typical small wedge tomb. The Burren is famous for its monuments, characteristic of every period from the Neolithic to the present. The legacy of early settlers is the many Neolithic tombs as well as stone structures and walls.

8. Excavation of the Poulnabrone dolmen revealed that the tomb contained the remains of up to 22 people, buried over a period of six centuries, as well as a polished stone axe, two stone disc beads, flint and chert arrowheads and scrapers, and more than 60 shards of pottery.

10. A Burren limestone wall. Burren limestone has been used for centuries in the construction of tombs, forts, houses, and walls. Today it is in demand for limestone-finished houses, walls, and garden features.

▶ **Figure 16.10** Karst Landscape in Kunming, China and Bowling Green, Kentucky

ⓐ The Stone Forest, 125 km southeast of Kunming, China, is a high-relief karst landscape formed by the dissolution of carbonate rocks.

ⓑ Solution valleys, sinkholes, and sinkhole lakes dominate the subdued karst topography east of Bowling Green, Kentucky.

Geo-focus/Active Figure 16.11 Cave Formation

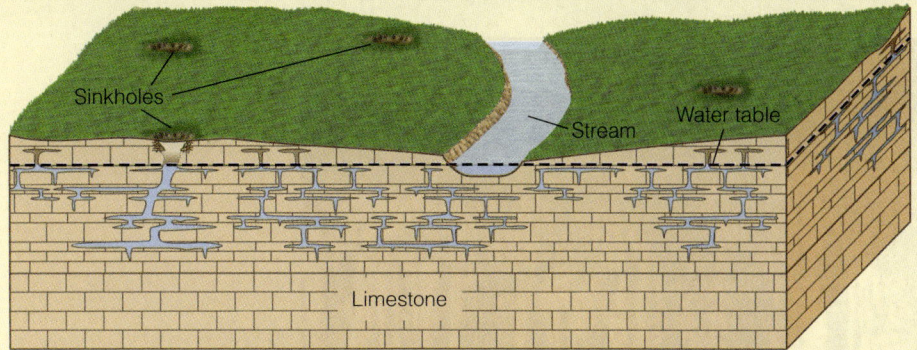

a As groundwater percolates through the zone of aeration and flows through the zone of saturation, it dissolves the carbonate rocks and gradually forms a system of passageways.

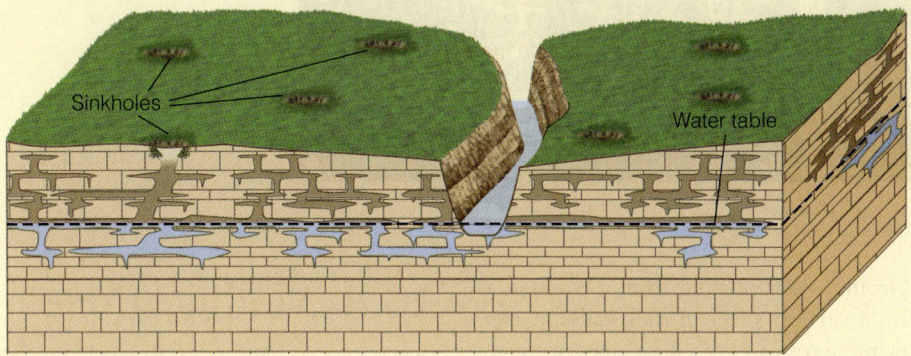

b Groundwater moves along the surface of the water table, forming a system of horizontal passageways through which dissolved rock is carried to the surface streams, thus enlarging the passageways.

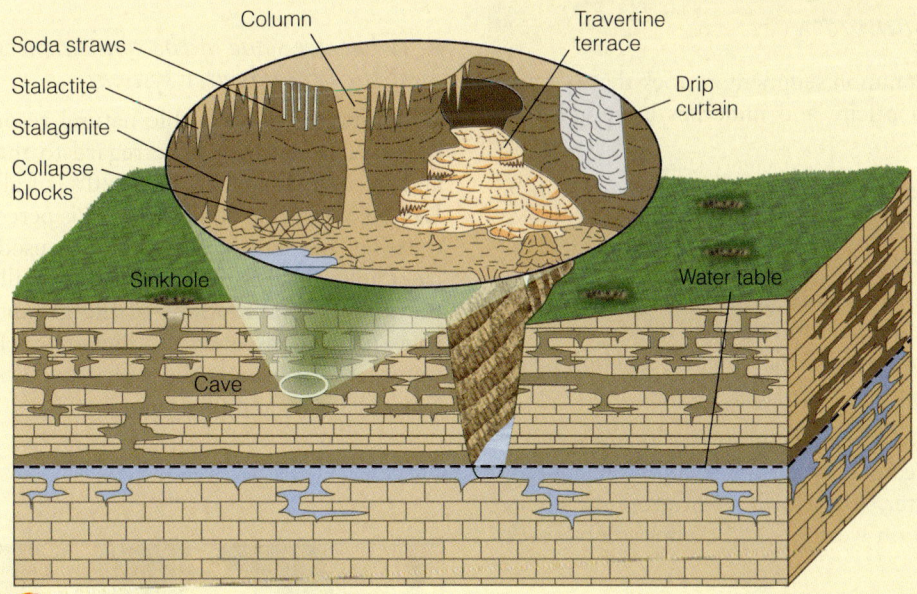

c As the surface streams erode deeper valleys, the water table drops and the abandoned channelways form an interconnecting system of caves and caverns.

▶ **Figure 16.12 Cave Deposits** Stalactites are the icicle-shaped structures hanging from a cave's ceiling, whereas the upward-pointing structures on the floor are stalagmites. Columns result when stalactites and stalagmites meet. All three structures are visible in Buchan Cave, Victoria, Australia.

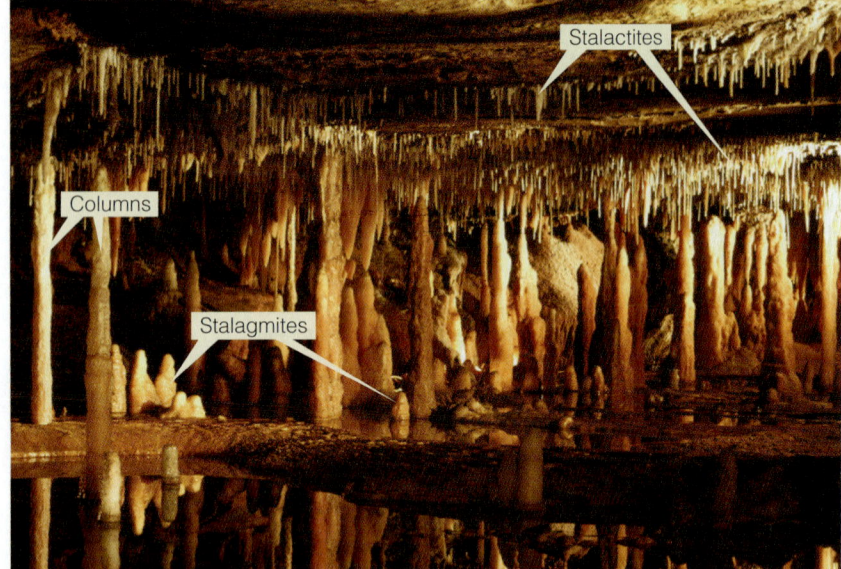

they form a *column*. Groundwater seeping from a crack in a cave's ceiling may form a vertical sheet of rock called a *drip curtain*, and water flowing across a cave's floor may produce *travertine terraces* (▶ Figure 16.11).

Section 16.7 Summary

- Limestone is a common sedimentary rock that is composed primarily of calcite and underlies large areas of Earth's surface.

- Sinkholes are depressions in the ground that form by dissolution of the underlying soluble rocks or when a cave's roof collapses.

- Karst topography develops largely by groundwater erosion in areas underlain by soluble rocks. Features characteristic of karst topography include sinkholes, solution valleys, disappearing streams, springs, and caves.

- Caves are naturally formed subsurface openings that are generally connected to the surface and large enough for a person to enter. They form when groundwater in the zone of saturation weathers and erodes soluble rock such as limestone.

- Common cave deposits include stalactites, which are icicle-shaped structures that hang from a cave's ceiling and form as a result of precipitation of calcite from dripping water; stalagmites, which are upward-growing projections that form when water dripping from a cave's ceiling precipitates a small amount of calcite as it hits the floor; and columns, which form when stalactites and stalagmites merge.

16.8 Modifications of the Groundwater System and Their Effects

■ *What are some of the consequences of modification of the groundwater system?*

Groundwater is a valuable natural resource that is rapidly being exploited with little regard to the effects of overuse and misuse. Currently about 20% of all water used in the United States is groundwater. This percentage is rapidly increasing, and unless this resource is used more wisely, sufficient amounts of clean groundwater will not be available in the future. Modifications of the groundwater system may have many consequences, including (1) lowering of the water table, causing wells to dry up; (2) saltwater incursion; (3) subsidence; and (4) contamination.

What Would You Do?

Your well periodically goes dry and never produces the amount of water you would like or need. Is your well too shallow, is it drilled into Earth materials that have low permeability, or are nearby pumps used for irrigation causing your problems? What would you do to resolve this problem?

Lowering the Water Table

Withdrawing groundwater at a significantly greater rate than it is replaced by either natural or artificial recharge can have serious effects. For example, the High Plains aquifer is one of the most important aquifers in the United States. It underlies more than 450,000 km², including most of Nebraska, large parts of Colorado and Kansas, portions of South Dakota, Wyoming, and New Mexico, as well as the panhandle regions of Oklahoma and Texas, and accounts for approximately 30% of the groundwater used for irrigation in the United States (▶ Figure 16.13).

Irrigation from the High Plains aquifer is largely responsible for the region's agricultural productivity, which includes a significant percentage of the nation's corn, cotton, and wheat, and half of U.S. beef cattle. Large areas of land (about 18 million acres) are currently irrigated with water pumped from the High Plains aquifer. Irrigation is popular because yields from irrigated lands can be triple what they would be without irrigation.

Although the High Plains aquifer has contributed to the high productivity of the region, it cannot continue to provide the quantities of water that it has in the past. In some parts of the High Plains, from 2 to 100 times more water is being pumped annually than is being recharged. Consequently, water is being removed from the aquifer faster than it is being replenished, causing the water table to drop significantly in many areas (▶ Figure 16.13).

What will happen to this region's economy if long-term withdrawal of water from the High Plains aquifer greatly exceeds its recharge rate so that it can no longer supply the quantities of water necessary for irrigation? Solutions range from going back to farming without irrigation to diverting water from other regions such as the Great Lakes. Farming without irrigation would result in greatly decreased yields and higher costs and prices for agricultural products. The diversion of water from elsewhere would cost billions of dollars and the price of agricultural products would still rise.

Another excellent example of what we might call deficit spending with regard to groundwater took place in California during the drought of 1987–1992. During that time, the state's aquifers were overdrawn at the rate of 10 million acre/feet per year (1 acre/foot is the amount of water that covers 1 acre 1 foot deep). In short, during each year of the drought, California was withdrawing 12 km³ of groundwater more than was being replaced. Unfortunately, excessive depletion of the groundwater reservoir has other consequences, such as subsidence, which involves sinking or settling of the ground surface (discussed in a later section).

Water supply problems certainly exist in many areas, but on the positive side, water use in the United States actually declined during the 5 years following 1980 and has remained nearly constant since then, even though the population has increased. This downturn in demand resulted largely from improved techniques in irrigation, more efficient industrial water use, and a general public awareness of water problems coupled with conservation practices. Nevertheless, the rates of withdrawal of groundwater from some

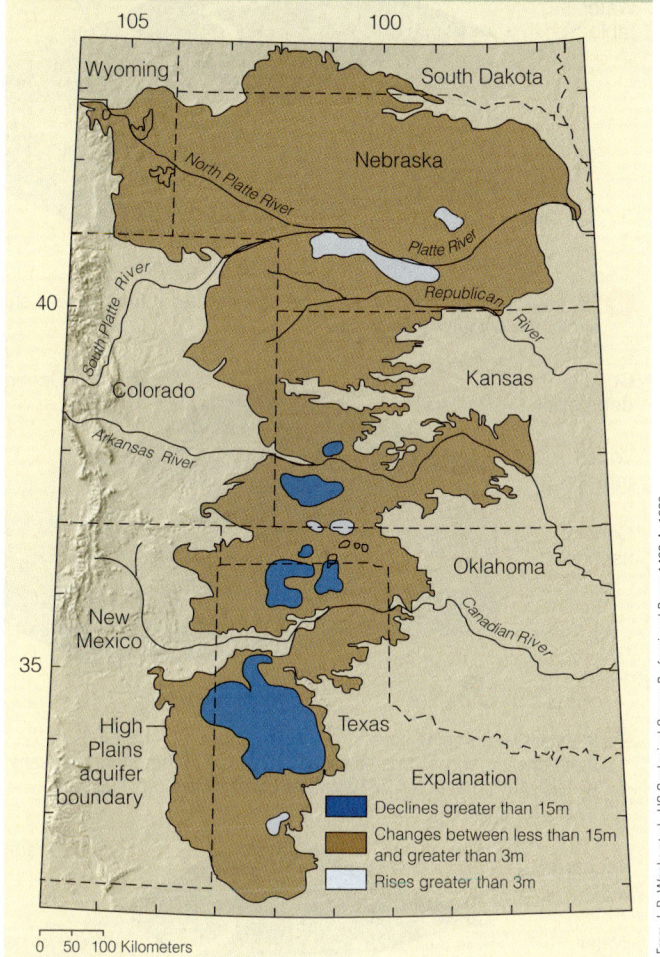

▶ **Figure 16.13 High Plains Aquifer** The geographic extent of the High Plains aquifer and changes in the water level from predevelopment through 1993. Irrigation from the High Plains aquifer is largely responsible for the region's agricultural productivity.

aquifers still exceed their rates of recharge, and population growth in the arid to semiarid Southwest is putting greater demands on an already limited water supply.

Saltwater Incursion

The excessive pumping of groundwater in coastal areas can result in *saltwater incursion* such as occurred on Long Island, New York, during the 1960s. Along coastlines where permeable rocks or sediments are in contact with the ocean, the fresh groundwater, being less dense than seawater, forms a lens-shaped body above the underlying saltwater (▶ Figure 16.14a). The weight of the freshwater exerts pressure on the underlying saltwater. As long as rates of recharge equal rates of withdrawal, the contact between the fresh groundwater and the seawater remains the same. If excessive pumping occurs, however, a deep cone of depression forms in the fresh groundwater (▶ Figure 16.14b). Because some of the pressure from the overlying freshwater has been removed, saltwater forms a *cone of ascension* as it rises to fill the pore space that formerly contained freshwater. When this occurs, wells become contaminated with saltwater and

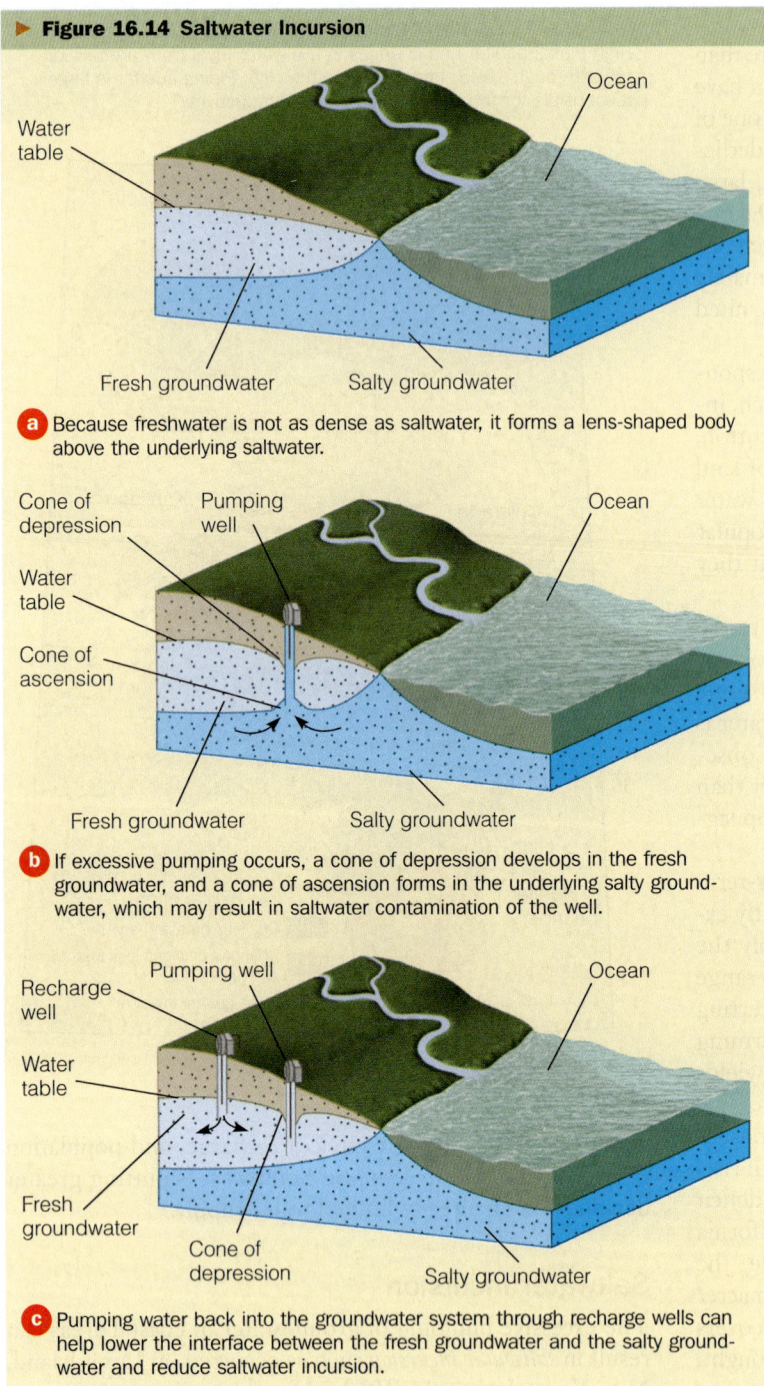

▶ Figure 16.14 Saltwater Incursion

(a) Because freshwater is not as dense as saltwater, it forms a lens-shaped body above the underlying saltwater.

(b) If excessive pumping occurs, a cone of depression develops in the fresh groundwater, and a cone of ascension forms in the underlying salty groundwater, which may result in saltwater contamination of the well.

(c) Pumping water back into the groundwater system through recharge wells can help lower the interface between the fresh groundwater and the salty groundwater and reduce saltwater incursion.

remain contaminated until recharge by freshwater restores the former level of the fresh-groundwater water table.

Saltwater incursion is a major problem in many rapidly growing coastal communities. As the population in these areas grows, greater demand for groundwater creates an even greater imbalance between recharge and withdrawal.

Not only is saltwater incursion a major concern for some coastal communities, it is also becoming a problem in the Salinas Valley, California, which produces fruits and vegetables valued at about $1.7 billion annually. Here, in an area encompassing about 160,000 acres of rich farmland 160 km south of San Francisco, saltwater incursion caused by excessive pumping of several shallow-water aquifers is threatening the groundwater supply used for irrigation. Because of droughts in recent years and increased domestic needs caused by a burgeoning population, this overpumping has resulted in increased seepage of saltwater into the groundwater system such that large portions of some of the aquifers are now too salty even for irrigation. At some locations in the Salinas Valley, seawater has migrated more than 11 km inland during the past 13 years. If this incursion is left unchecked, the farmlands of the valley could become too salty to support most agriculture. Uncontaminated deep aquifers could be tapped to replace the contaminated shallow groundwater, but doing so is too expensive for most farmers and small towns.

To counteract the effects of saltwater incursion, recharge wells are often drilled to pump water back into the groundwater system (▶ Figure 16.14c). Recharge ponds that allow large quantities of fresh surface water to infiltrate the groundwater supply may also be constructed. Both of these methods are successfully used on Long Island, New York, which has had a saltwater incursion problem for several decades.

Subsidence

As excessive amounts of groundwater are withdrawn from poorly consolidated sediments and sedimentary rocks, the water pressure between grains is reduced, and the weight of the overlying materials causes the grains to pack closer together, resulting in *subsidence* of the ground. As more and more groundwater is pumped to meet the increasing needs of agriculture, industry, and population growth, subsidence is becoming more prevalent.

The San Joaquin Valley of California is a major agricultural region that relies largely on groundwater for irrigation. Between 1925 and 1977, groundwater withdrawals in parts of the valley caused subsidence of nearly 9 m (▶ Figure 16.15). Other areas in the United States that have experienced subsidence are New Orleans, Louisiana, and Houston, Texas, both of which have subsided more than 2 m, and Las Vegas, Nevada, where 8.5 m of subsidence has taken place (Table 16.2).

Elsewhere in the world, the tilt of the Leaning Tower of Pisa is partly due to groundwater withdrawal (▶ Figure 16.16). The tower started tilting soon after construction began in 1173 because of differential compaction of the foundation. During the 1960s, the city of Pisa withdrew ever-larger amounts of groundwater, causing the ground to subside further; as a result, the tilt of the tower increased until it was considered in danger of falling over. Strict control of groundwater withdrawal, stabilization of the foundation, and recent renovations have reduced the amount of tilting to about 1 mm per year, thus ensuring that the tower should stand for several more centuries.

▶ **Figure 16.15 Subsidence in the San Joaquin Valley, California** The dates on this power pole dramatically illustrate the amount of subsidence in the San Joaquin Valley, California. Because of groundwater withdrawals and subsequent sediment compaction, the ground subsided nearly 9 m between 1925 and 1977. For a time, surface water use reduced subsidence, but during the drought of 1987–1992 subsidence started again as more groundwater was withdrawn.

▶ **Figure 16.16 The Leaning Tower of Pisa, Italy** The tilting is partly the result of subsidence due to the removal of groundwater. Strict control of groundwater withdrawal, recent stabilization of the foundation, and renovation of the structure itself have ensured that the Leaning Tower will continue leaning for many more centuries.

Table 16.2

Subsidence of Cities and Regions Due Primarily to Groundwater Removal

Location	Maximum Subsidence (m)	Area Affected (km²)
Mexico City, Mexico	8.0	25
Long Beach and Los Angeles, California	9.0	50
Taipei Basin, Taiwan	1.0	100
Shanghai, China	2.6	121
Venice, Italy	0.2	150
New Orleans, Louisiana	2.0	175
London, England	0.3	295
Las Vegas, Nevada	8.5	500
Santa Clara Valley, California	4.0	600
Bangkok, Thailand	1.0	800
Osaka and Tokyo, Japan	4.0	3000
San Joaquin Valley, California	9.0	9000
Houston, Texas	2.7	12,100

Source: Data from R. Dolan and H. G. Goodell, "Sinking Cities," *American Scientist* 74 (1986): 38–47; and J. Whittow, *Disasters: The Anatomy of Environmental Hazards* (Athens: University of Georgia Press, 1979).

A spectacular example of continuing subsidence is taking place in Mexico City, which is built on a former lakebed. As groundwater is removed for the increasing needs of the city's 17.8 million people, the water table has been lowered up to 10 m. As a result, the fine-grained lake deposits are compacting, and Mexico City is slowly and unevenly subsiding. Its opera house has settled more than 3 m, and half of the first floor is now below ground level. Other parts of the city have subsided as much as 7.5 m, creating similar problems for other structures (▶ Figure 16.17). The fact that 72% of the city's water comes from the aquifer beneath the metropolitan area ensures that problems of subsidence will continue.

The extraction of oil can also cause subsidence. Long Beach, California, has subsided 9 m as a result of many decades of oil production. More than $100 million of damage was done to the pumping, transportation, and harbor facilities in this area because of subsidence and encroachment of the sea (▶ Figure 16.18). Once water was pumped back into the oil reservoir, thus stabilizing it, subsidence virtually stopped.

Groundwater Contamination

A major problem facing our society is the safe disposal of the numerous pollutant by-products of an industrialized economy. We are becoming increasingly aware that streams, lakes, and oceans are not unlimited reservoirs for waste and that we must find new, safe ways to dispose of pollutants.

The most common sources of groundwater contamination are sewage, landfills, toxic-waste-disposal sites, and agriculture. Once pollutants get into the groundwater system, they spread wherever groundwater travels, which can make their containment difficult. Furthermore, because groundwater moves so slowly, it takes a long time to cleanse a groundwater reservoir once it has become contaminated.

In many areas, septic tanks are the most common way of disposing of sewage. A septic tank slowly releases sewage into the ground, where it is decomposed by oxidation and microorganisms and filtered by the sediment as it percolates through the zone of aeration. In most situations, by the time the water from the sewage reaches the zone of saturation, it has been cleansed of any impurities and is safe to use (▶ Figure 16.19a). If the water table is close to the surface or if the rocks are very permeable, water entering the zone of saturation may still be contaminated and unfit to use.

Landfills are also potential sources of groundwater contamination (▶ Figure 16.19b). Not only does liquid waste seep into the ground, but rainwater also carries dissolved chemicals and other pollutants down into the groundwater reservoir. Unless the landfill is carefully designed and lined with an impermeable layer such as clay, many toxic compounds such as paints, solvents, cleansers, pesticides, and battery acid will find their way into the groundwater system.

Toxic-waste sites where dangerous chemicals are either buried or pumped underground are an increasing source of groundwater contamination. The United States alone must dispose of several thousand metric tons of hazardous chem-

▶ **Figure 16.17 Subsidence in Mexico City** Excessive withdrawal of groundwater from beneath Mexico City has resulted in subsidence and uneven settling of buildings. The right side of this church (Our Lady of Guadalupe) has settled slightly more than 1 m.

▶ **Figure 16.18 Oil Field Subsidence, Long Beach, California** The withdrawal of petroleum from the Long Beach, California, oil field resulted in up to 9 m of ground subsidence in some areas because of sediment compaction. In this photograph, note that the ground has settled around the well stems (the white "posts"), leaving the wellheads up above the ground. The levee on the left edge of the photo was built to keep seawater in the adjacent marina from flooding the oil field. It was not until water was pumped back into the reservoir to replace the petroleum that was being extracted that ground subsidence finally ceased.

ical waste per year. Unfortunately, much of this waste has been and still is being improperly dumped and is contaminating the surface water, soil, and groundwater.

Examples of indiscriminate dumping of dangerous and toxic chemicals can be found in every state. Perhaps the most famous is Love Canal, near Niagara Falls, New York. During the 1940s, the Hooker Chemical Company dumped approximately 19,000 tons of chemical waste into Love Canal. In 1953 it covered one of the dump sites with dirt and sold it for one dollar to the Niagara Falls Board of Education, which built an elementary school and playground on the site. Heavy rains and snow during the winter of 1976–1977 raised the water table and turned the area into a muddy swamp in the spring of 1977. Mixed with the mud were thousands of toxic, noxious chemicals that formed puddles in the playground, oozed into people's basements, and covered gardens and lawns. Trees, lawns, and gardens began to die, and many of the residents of the area suffered from serious illnesses. The cost of cleaning up the Love Canal site and relocating its residents exceeded $100 million, and the site and neighborhood are now vacant.

■ *What determines groundwater quality?*

Finding groundwater is rather easy because it is present beneath the land surface nearly everywhere, although the depth to the water table varies considerably. But just finding water is not enough. Sufficient amounts in porous and permeable materials must be located if groundwater is to be withdrawn for agricultural, industrial, or domestic use. The availability of groundwater was important in the westward expansion in both Canada and the United States, and now more than one-third of all water for irrigation comes from the groundwater system. More than 90% of the water used for domestic purposes in rural America and the water for a number of large cities come from groundwater, and, as one would expect, quality is more important here than it is for most other purposes.

If we discount contamination by humans from landfills, septic systems, toxic-waste sites, and industrial effluents,

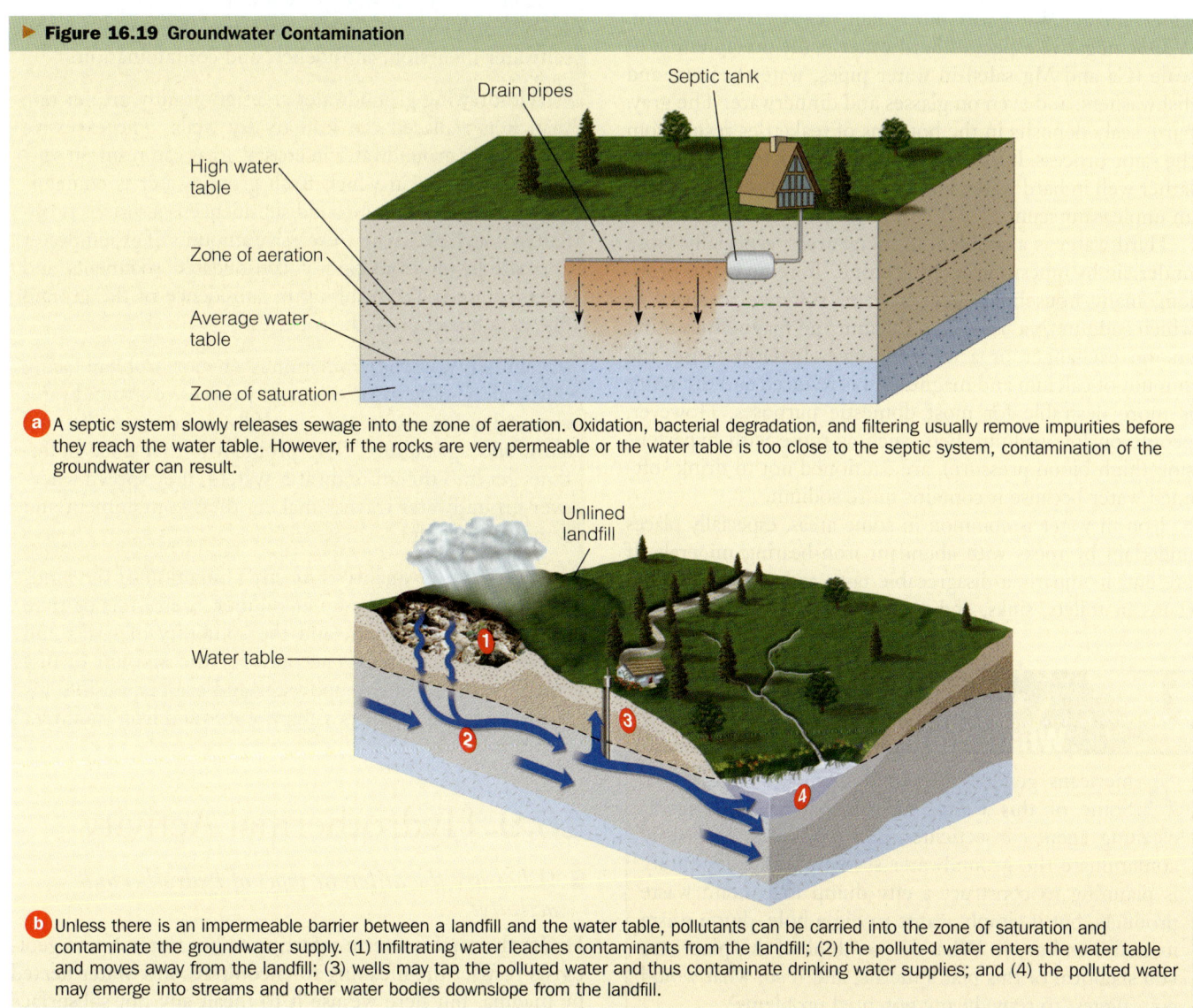

▶ **Figure 16.19 Groundwater Contamination**

a A septic system slowly releases sewage into the zone of aeration. Oxidation, bacterial degradation, and filtering usually remove impurities before they reach the water table. However, if the rocks are very permeable or the water table is too close to the septic system, contamination of the groundwater can result.

b Unless there is an impermeable barrier between a landfill and the water table, pollutants can be carried into the zone of saturation and contaminate the groundwater supply. (1) Infiltrating water leaches contaminants from the landfill; (2) the polluted water enters the water table and moves away from the landfill; (3) wells may tap the polluted water and thus contaminate drinking water supplies; and (4) the polluted water may emerge into streams and other water bodies downslope from the landfill.

groundwater quality is mostly a function of (1) the kinds of materials that make up an aquifer, (2) the residence time of water in an aquifer, and (3) the solubility of rocks and minerals. These factors account for the amount of dissolved materials in groundwater, such as calcium, iron, fluoride, and several others. Most pose no health problems, but some have undesirable effects such as deposition of minerals in water pipes and water heaters and offensive taste or smell, or they may stain clothing and fixtures, or inhibit the effectiveness of detergents.

Everyone has heard of *hard water* and knows that *soft water* is more desirable for most purposes. But just what is hard water, and how does it become hard? Recall from Chapter 6 that some rocks, especially limestone, composed of calcite ($CaCO_3$), and dolostone, composed of dolomite [$CaMg(CO_3)_2$], are readily soluble in water that contains a small amount of carbonic acid. In fact, the amount of dissolved calcium (Ca^{+2}) and magnesium (Mg^{+2}) is what determines whether water is hard or soft. Water with less than 60 milligrams of Ca^{+2} and Mg^{+2} per liter (mg/L) is soft, whereas values from 61 to 120 mg/L indicate moderately hard water, values from 121 to 180 mg/L characterize hard water, and any water with more than 180 mg/L is very hard.

One negative aspect of hard water is the precipitation of scale (Ca and Mg salts) in water pipes, water heaters, and dishwashers, and even on glasses and dinnerware. The gray, hard, scaly deposits in the bottoms of teakettles result from the same process. Furthermore, soaps and detergents do not lather well in hard water, and dirt and soap combine to form an unpleasant scum.

Hard water is a problem in many areas, especially those underlain by limestone and dolostone. To remedy this problem, many households have a water softener, a device in which sodium (Na^+) replaces calcium and magnesium using an ion exchanger or a mineral sieve. In either case, the amount of calcium and magnesium is reduced and the water is more desirable for most domestic purposes. However, people on low-sodium diets, such as those with hypertension (high blood pressure), are cautioned not to drink softened water because it contains more sodium.

Iron in water is common in some areas, especially places underlain by rocks with abundant iron-bearing minerals. If present, it imparts a disagreeable taste and might leave red stains in toilets, sinks, and other plumbing fixtures. Light-colored clothing washed in iron-rich water takes on a rusty color. Even though iron poses no health threat, it is so disagreeable that most people remove it. Water softeners will remove some of it, but in many cases some kind of iron filter must be placed on the incoming water line before it reaches the softener. In any case, the problem is usually easily solved, but of course it adds another expense to household maintenance.

Not all dissolved materials in groundwater are undesirable, at least in small quantities. Fluoride (F^-), for instance, if present in amounts of 1.0–1.5 parts per million (ppm), combines with the calcium phosphate in teeth and makes them more resistant to decay. Too much fluoride—more than 4.0 ppm—though, gives children's teeth a dark, blotchy appearance. Fluoride in natural waters is rare, so not many communities benefit from it, and artificially adding fluorine to water has met with considerable opposition because of possible health risks.

Section 16.8 Summary

- Modifications to the groundwater system can cause serious problems, including lowering of the water table, saltwater incursion, subsidence, and contamination.

- Withdrawing groundwater at a significantly greater rate than it is replaced can lead to dry wells. The excessive pumping of groundwater in coastal areas can result in saltwater incursion, in which fresh groundwater is contaminated with saltwater. Ground subsidence is a serious problem in many areas where excessive amounts of groundwater are withdrawn from poorly consolidated sediments and sedimentary rocks, resulting in subsidence of the ground due to compaction of the rocks.

- Groundwater contamination is a major problem facing our society. The most common sources of groundwater contamination are sewage, landfills, toxic-waste-disposal sites, and agriculture. The problem is that once pollutants get into the groundwater system, they spread wherever groundwater travels, making their containment and removal difficult.

- Groundwater quality is mostly a function of the kinds of materials that make up an aquifer, the residence time of water in an aquifer, and the solubility of rocks and minerals. These factors account for the amount of dissolved materials in groundwater and are responsible for such undesirable effects as hard water and iron staining.

What Would You Do?

Americans generate tremendous amounts of waste. Some of this waste, such as battery acid, paint, cleaning agents, insecticides, and pesticides, can easily contaminate the groundwater system. Your community is planning to construct a city dump to contain waste products, but it simply wants to dig a hole, dump waste in, and then bury it. What do you think of this plan? Are you skeptical of this plan's merits, and if so, what would you suggest to remedy any potential problems?

16.9 Hydrothermal Activity

- *What are the different types of hydrothermal activity?*

Hydrothermal is a term referring to hot water. Some geologists restrict the meaning to encompass only water heated by magma, but here we use it to mean any hot subsurface water and surface activity resulting from its discharge. One manifestation of hydrothermal activity in areas of active or recently active volcanism is the discharge of gases, much of it as steam, at vents known as *fumaroles* (see Figure 5.2). Of more immediate concern here, however, is the groundwater that rises to the surface as *hot springs* or *geysers*. The groundwater may be heated by its proximity to magma or by Earth's geothermal gradient because it circulates deeply.

Hot Springs

A **hot spring** (also called a *thermal spring* or *warm spring*) is any spring in which the water temperature is higher than 37°C, the temperature of the human body (▶ Figure 16.20a). Some hot springs are much hotter, with temperatures up to the boiling point in many instances (▶ Figure 16.20b). Another type of hot spring, called a *mud pot*,

▶ **Figure 16.20 Hot Springs**

a Hot spring in the West Thumb Geyser Basin, Yellowstone National Park, Wyoming.

b The water in this hot spring at Bumpass Hell in Lassen Volcanic National Park, California, is boiling.

c Mud pot at the Sulfur Works, also in Lassen Volcanic National Park.

d The U.S. Park Service warns of the dangers in hydrothermal areas, but some people ignore the warnings and are injured or killed.

results when chemically altered rocks yield clays that bubble as hot water and steam rise through them (▶ Figure 16.20c). Of the approximately 1100 known hot springs in the United States, more than 1000 are in the far West, with the others in the Black Hills of South Dakota, Georgia, the Appalachian region, and the Ouachita region of Arkansas.

At Hot Springs National Park in Arkansas, 47 hot springs with temperatures of 62°C are found on the slopes of Hot Springs Mountain. Hot springs are also common in other parts of the world. One of the most famous is at Bath, England, where shortly after the Roman conquest of Britain in A.D. 43, numerous bathhouses and a temple were built around the hot springs (▶ Figure 16.21).

The heat for most hot springs comes from magma or cooling igneous rocks. The geologically recent igneous activity in the western United States accounts for the large number of hot springs in that region. The water in some hot springs, however, circulates deep into Earth, where it is warmed by the normal increase in temperature, the geothermal gradient. For example, the spring water of Warm Springs, Georgia, is heated in this manner. This hot spring was a health and bathing resort long before the Civil War (1861–1865); later, with the establishment of the Georgia Warm Springs Foundation, it was used to help treat polio victims.

Geysers

Hot springs that intermittently eject hot water and steam with tremendous force are known as **geysers.** The word comes from the Icelandic *geysir*, "to gush" or "to rush forth." One of the most famous geysers in the world is Old Faithful in Yellowstone National Park in Wyoming (▶ Figure 16.22). With a thunderous roar, it erupts a column of hot water and steam every 30 to 90 minutes. Other well-known geyser areas are found in Iceland and New Zealand.

Geysers are the surface expressions of an extensive underground system of interconnected fractures within hot igneous rocks (▶ Figure 16.23). Groundwater percolating down into the network of fractures is heated as it comes in contact with the hot rocks. Because the water near the bottom of the fracture system is under higher pressure

▶ **Figure 16.21 Bath, England** One of the many bathhouses in Bath, England, that were built around hot springs shortly after the Roman conquest in A.D. 43.

▶ **Figure 16.22 Old Faithful Geyser** Old Faithful Geyser in Yellowstone National Park, Wyoming, is one of the world's most famous geysers, erupting faithfully every 30 to 90 minutes and spewing water 32 to 56 m high.

than the water near the top, it must be heated to a higher temperature before it will boil. Thus, when the deeper water is heated to near the boiling point, a slight rise in temperature or a drop in pressure, such as from escaping gas, will instantly change it to steam. The expanding steam quickly pushes the water above it out of the ground and into the air, producing a geyser eruption. After the eruption, relatively cool groundwater starts to seep back into the fracture system, where it heats to near its boiling temperature and the eruption cycle begins again. Such a process explains how geysers can erupt with some regularity.

Hot-spring and geyser water typically contains large quantities of dissolved minerals because most minerals dissolve more rapidly in warm water than in cold water. Because of this high mineral content, some believe the waters of many hot springs have medicinal properties. Numerous spas and bathhouses have been built at hot springs throughout the world to take advantage of these supposed healing properties.

When the highly mineralized water of hot springs or geysers cools at the surface, some of the material in solution is precipitated, forming various types of deposits. The amount and type of precipitated minerals depend on the solubility and composition of the material that the groundwater flows through. If the groundwater contains dissolved calcium carbonate ($CaCO_3$), then *travertine* or *calcareous tufa* (both varieties of limestone) are precipitated. Spectacular examples of hot-spring travertine deposits are found at Pamukhale in Turkey and at Mammoth Hot Springs in Yellowstone National Park (▶ Figure 16.24a). Groundwater containing dissolved silica will, upon reaching the surface, precipitate a soft, white, hydrated mineral called *siliceous sinter* or *geyserite*, which can accumulate around a geyser's opening (▶ Figure 16.24b).

■ *What is geothermal energy and why is it important?*

Geothermal energy is any energy produced from Earth's internal heat. In fact, the term *geothermal* comes from *geo*, "Earth," and *thermal*, "heat." Several forms of internal heat are known, such as hot dry rocks and magma, but so far only hot water and steam are used.

Approximately 1 to 2% of the world's current energy needs could be met by geothermal energy. In those areas where it is plentiful, geothermal energy can supply most, if not all, of the energy needs, sometimes at a fraction of the cost of other types of energy. Some of the countries currently using geothermal energy in one form or another are Iceland, the United States, Mexico, Italy, New Zealand, Japan, the Philippines, and Indonesia.

The city of Rotorua, New Zealand, is world famous for its volcanoes, hot springs, geysers, and geothermal fields. Since the first well was sunk in the 1930s, more than 800 wells have been drilled to tap the hot water and steam. Geothermal energy in Rotorua is used in a variety of ways, including heating homes, commercial buildings, and greenhouses.

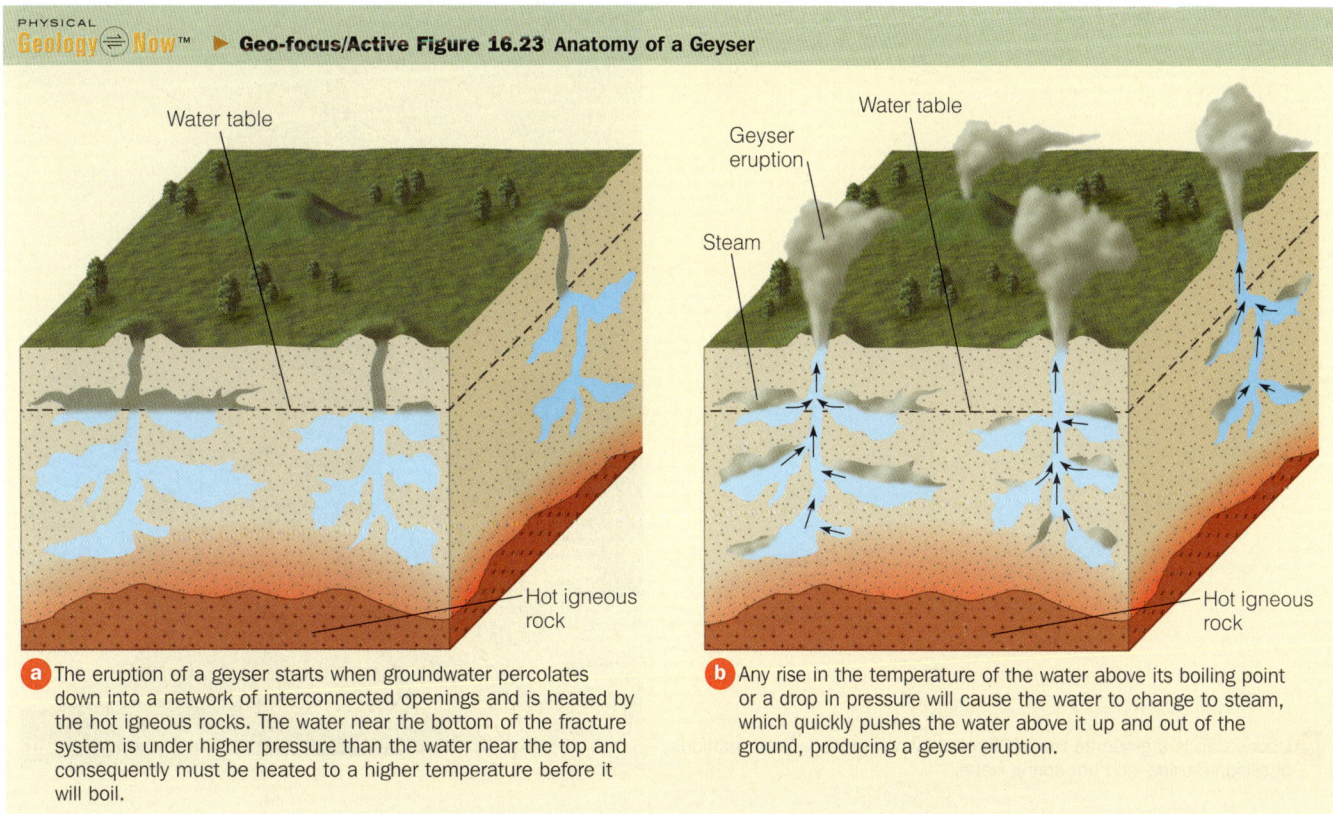

▶ Geo-focus/Active Figure 16.23 Anatomy of a Geyser

a. The eruption of a geyser starts when groundwater percolates down into a network of interconnected openings and is heated by the hot igneous rocks. The water near the bottom of the fracture system is under higher pressure than the water near the top and consequently must be heated to a higher temperature before it will boil.

b. Any rise in the temperature of the water above its boiling point or a drop in pressure will cause the water to change to steam, which quickly pushes the water above it up and out of the ground, producing a geyser eruption.

Figure 16.24 Hot-Spring Deposits in Yellowstone National Park, Wyoming

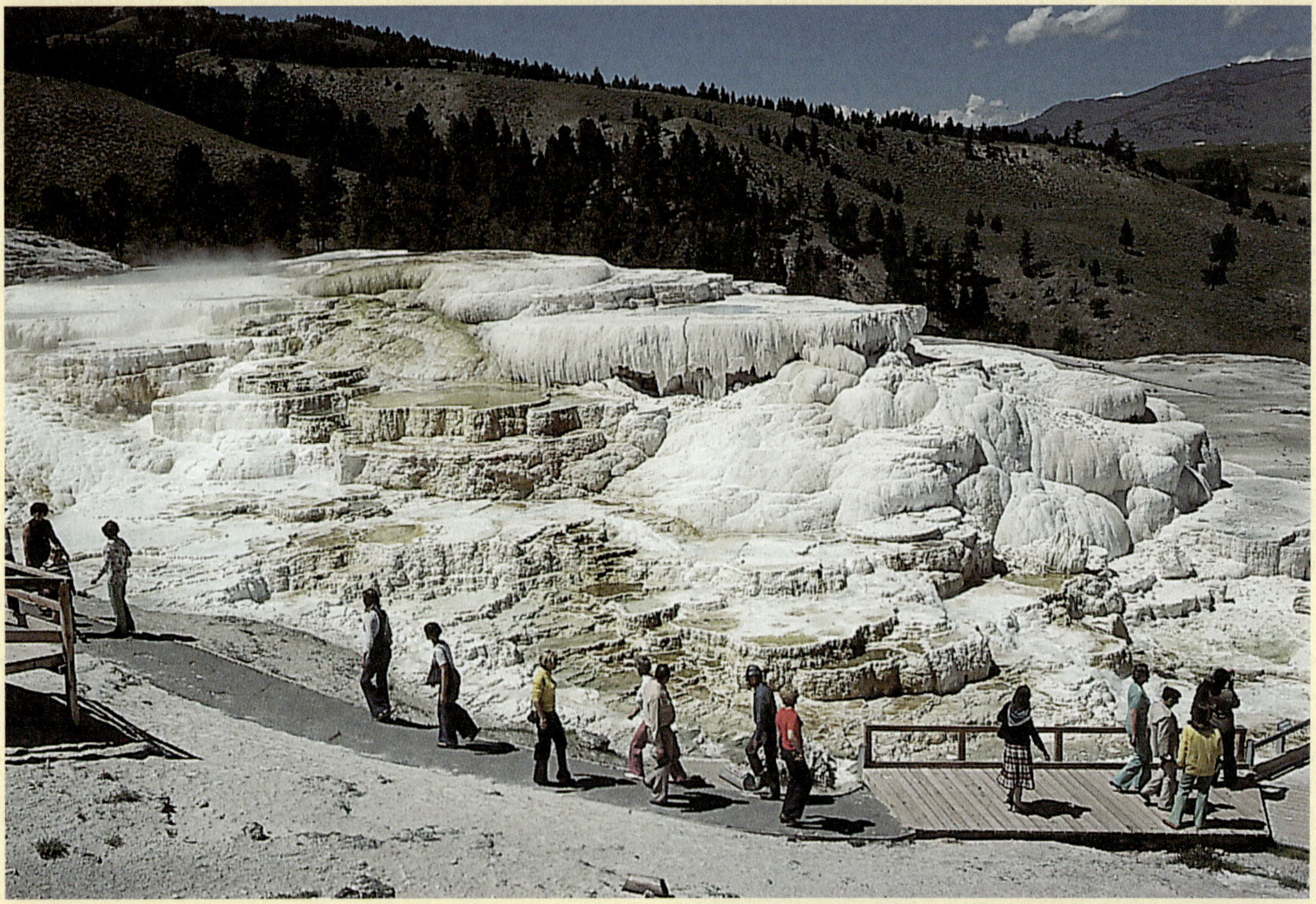

Reed Wicander

a Minerva Terrace formed when calcium-carbonate–rich hot-spring water cooled, precipitating travertine.

Reed Wicander

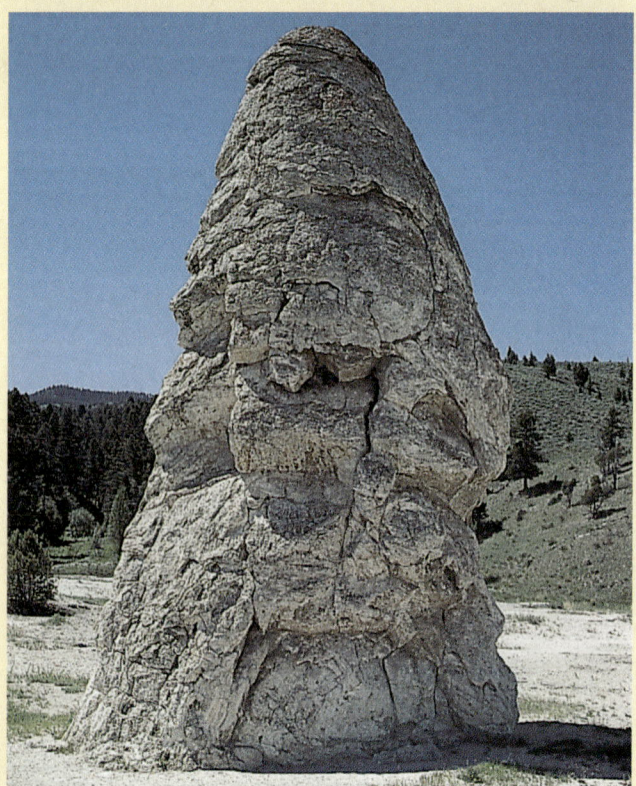

b Liberty Cap is a geyserite mound formed by numerous geyser eruptions of silicon-dioxide–rich hot-spring water.

In the United States, the first commercial geothermal electrical generating plant was built in 1960 at The Geysers, about 120 km north of San Francisco, California. Here, wells were drilled into the numerous near-vertical fractures underlying the region. As pressure on the rising groundwater decreases, the water changes to steam, which is piped directly to electricity-generating turbines and generators (▶ Figure 16.25).

As oil reserves decline, geothermal energy is becoming an attractive alternative, particularly in parts of the western United States, such as the Salton Sea area of southern California, where geothermal exploration and development have begun.

Section 16.9 Summary

• *Hydrothermal* refers to hot water, and we use the term here to include any hot subsurface water and the surface activity resulting from its discharge.

• Manifestations of hydrothermal activity include fumaroles, which are locations for the discharge of gases such as steam; hot springs, which are streams in which the water temperature is higher than 37°C; and geysers, which intermittently eject hot water and steam with tremendous force. Hot-spring and geyser water typically contains large quantities of dissolved minerals, which commonly precipitate as the water cools, producing many colorful and varied structures.

• Geothermal energy is any energy produced from Earth's internal heat. It is a relatively nonpolluting form of energy that is used as a source of heat and for generating electricity.

▶ **Figure 16.25 The Geysers, Sonoma County, California** Steam rising from one of the geothermal power plants at The Geysers in Sonoma County, California. Steam from wells drilled into this geothermal region, about 120 km north of San Francisco, is piped directly to electricity-generating turbines to produce electricity that is distributed throughout the area.

Review Workbook

ESSENTIAL QUESTIONS SUMMARY

16.1 Introduction
■ *Why is it important to study groundwater?*
Groundwater is part of the hydrologic cycle and an important natural resource. As the world's population and industrial development expand, the demand for water, particularly groundwater, will increase.

16.2 Groundwater and the Hydrologic Cycle
■ *What is groundwater?*
Groundwater is all subsurface water trapped in the pores and other open spaces in rocks, sediments, and soil.

16.3 Porosity and Permeability
■ *What is the difference between porosity and permeability?*
Porosity is the percentage of a material's total volume that is pore space. Permeability is the capacity to transmit fluids. Permeability is dependent on porosity, but also on the size of the pores or fractures and their interconnections.

16.4 The Water Table
■ *What is the water table and what does it look like?*
The water table is the surface separating the zone of aeration (in which the pores are filled with air and water) from the underlying zone of saturation (in which the pores are filled with water). The water table is a subdued replica of the overlying land surface in most places.

16.5 Groundwater Movement
■ *How does groundwater move?*
Groundwater moves slowly downward under the influence of gravity through the zone of aeration to the zone of saturation. Some of it moves along the surface of the water table, and the rest moves from areas of high pressure to areas of low pressure. Groundwater velocity varies greatly and depends on various factors. Generally, the average velocity of groundwater is a few centimeters per day.

16.6 Springs, Water Wells, and Artesian Systems
■ *What are springs?*
Springs are found wherever the water table intersects the surface. When percolating water reaches the water or an impermeable layer, it flows laterally, and if this flow intersects the surface, water is discharged as a spring.

■ *What are water wells?*
Water wells are openings made by digging or drilling down into the zone of saturation. When the zone of saturation has been penetrated, water percolates into the well, filling it to the level of the water table.

■ *What is a cone of depression?*
When water is pumped from a well, the water table in the area around the well is lowered, forming a cone of depression. A cone of depression forms because the rate of water withdrawal from the well exceeds the rate of water inflow to the well, thus lowering the water table around the well.

■ *What is an artesian system?*
In an artesian system, confined groundwater builds up high hydrostatic pressure. For an artesian system to develop, three geologic conditions must be met: The aquifer must be confined above and below by aquicludes; the aquifer is usually tilted and exposed at the surface so it can be recharged; and precipitation must be sufficient to keep the aquifer filled.

16.7 Groundwater Erosion and Deposition
■ *What are sinkholes?*
Sinkholes are depressions in the ground formed by the dissolution of the underlying soluble rocks or by the collapse of a cave roof.

■ *What is karst topography?*
Karst topography largely develops by groundwater erosion in many areas underlain by soluble rocks. Sinkholes, along with springs, solution valleys, disappearing streams, and caves, are some of the features that characterize karst topography.

■ *How do caves form and what are the common cave deposits?*
Caves form when groundwater in the zone of saturation weathers and erodes soluble rock such as limestone. Common cave deposits include stalactites, which are icicle-shaped structures hanging from a cave's ceiling and form from the precipitation of calcite by dripping water; stalagmites, which are upward-pointing structures on the floor of caves and form when water dripping from a cave's ceiling precipitates a small amount of calcite when it hits the floor; and columns, which result when stalactites and stalagmites meet.

16.8 Modifications of the Groundwater System and Their Effects
■ *What are some of the consequences of modification of the groundwater system?*
Modifications to the groundwater system can cause serious problems such as lowering of the water table, saltwater incursion, subsidence, and contamination.

■ *What determines groundwater quality?*
Groundwater quality is mostly a function of the kinds of materials that make up an aquifer, the residence time of water in an aquifer, and the solubility of rocks and minerals. These factors account for the amount of dissolved materials in groundwater and are responsible for such undesirable effects as hard water and iron staining. Contamination by humans from landfills, septic systems, toxic-waste sites, and industrial effluents also affects the quality of groundwater.

16.9 Hydrothermal Activity

■ *What are the different types of hydrothermal activity?*
Hydrothermal refers to hot water, typically heated by magma but also resulting from Earth's geothermal gradient as it circulates deeply beneath the surface. Manifestations of hydrothermal activity include fumaroles, which are vents where gases such as steam are discharged; hot springs, which are springs where the water temperature is higher than 37°C; and geysers, which intermittently eject hot water and steam with tremendous force.

■ *What is geothermal energy and why is it important?*
Geothermal energy is energy produced from Earth's internal heat and comes from the steam and hot water trapped within Earth's crust. It is a relatively nonpolluting form of energy that is used as a source of heat and to generate electricity.

ESSENTIAL TERMS TO KNOW

artesian system (p. 503)
cave (p. 507)
cone of depression (p. 503)
geothermal energy (p. 521)
geyser (p. 520)
groundwater (p. 498)

hot spring (p. 519)
hydrothermal (p. 518)
karst topography (p. 506)
permeability (p. 499)
porosity (p. 498)
sinkhole (p. 505)

spring (p. 502)
water table (p. 500)
water well (p. 503)
zone of aeration (p. 500)
zone of saturation (p. 500)

REVIEW QUESTIONS

1. Describe the configurations of the water table beneath a humid area and an arid region. Why are the configurations different?
2. Discuss the role of groundwater in the hydrologic cycle.
3. Why should we be concerned about how fast the groundwater supply is being depleted in some areas?
4. Why isn't geothermal energy a virtually unlimited source of energy?
5. Why does groundwater move so much slower than surface water?
6. Describe three features you might see in an active hydrothermal area. Where in the United States would you go to see such activity?
7. Explain how groundwater weathers and erodes Earth materials.
8. Explain how saltwater incursion takes place and why it is a problem in coastal areas.
9. Explain how some Earth materials can be porous yet not permeable. Give an example.
10. What is the difference between an artesian system and a water well? Is there any difference between the water obtained from an artesian system and the water from a water well?

APPLY YOUR KNOWLEDGE

1. One concern geologists have about burying nuclear waste in present-day arid regions such as Nevada is that the climate may change during the next several thousand years and become more humid, thus allowing more water to percolate through the zone of aeration. Why is this a concern? What would the average rate of groundwater movement have to be during the next 5000 years to reach canisters containing radioactive waste buried at a depth of 400 m?

FIELD QUESTION

1. Withdrawal of large quantities of groundwater in the Las Vegas area has resulted in differential subsidence and damage to roads and buildings. What evidence of subsidence is visible in this photograph?

GEOLOGY IN FOCUS — Arsenic and Old Lace

Many people probably learned that arsenic is a poison from either reading or seeing the play *Arsenic and Old Lace*, written by Joseph Kesselring. In the play, the elderly Brewster sisters poison lonely old men by adding a small amount of arsenic to their homemade elderberry wine.

Arsenic is a naturally occurring toxic element found in the environment, and several types of cancer have been linked to arsenic in water. Arsenic also harms the central and peripheral nervous systems and may cause birth defects and reproductive problems. In fact, because of arsenic's prevalence in the environment and its adverse health effects, Congress included it in the amendments to the Safe Drinking Water Act in 1996. Arsenic gets into the groundwater system mainly as arsenic-bearing minerals dissolve in the natural weathering process of rocks and soils.

A map published in 2001 by the U.S. Geological Survey (USGS) shows the extent and concentration of arsenic in the nation's groundwater supply (▶ Figure 1). The highest concentrations of groundwater arsenic were found throughout the West and in parts of the Midwest and Northeast. Although the map is not intended to provide specific information for individual wells or even a locality within a county, it helps researchers and policymakers identify areas of high concentration so that they can make informed decisions about water use. We should point out, however, that a high degree of local variability in the amount of arsenic in the groundwater can be caused by local geology, type of aquifer, depth of well, and other factors. The only way to learn the arsenic concentration in any well is to have it tested.

What is considered a safe level of arsenic in drinking water? In 2001 the U.S. Environmental Protection Agency (USEPA, or EPA) lowered the maximum level of arsenic permitted in drinking water from 50 μg (50 micrograms) of arsenic per liter to 10 μg of arsenic per liter. This is the same figure used by the World Health Organization.

From the data in Figure 1, additional maps were created. ▶ Figure 2 shows the arsenic concentrations found in at least 25% of groundwater samples per county. Based on these data, approximately 10% of the samples in the USGS study exceed the new standard of 10 μg of arsenic per liter of drinking water.

Public water supply systems that exceed the existing EPA arsenic standard are required to either treat the water to remove the arsenic or find an alternative supply. Although reducing the acceptable level of arsenic in drinking water will surely increase the cost of water to consumers, it will also decrease their exposure to arsenic and the possible adverse health effects associated with this toxic element.

Figure 1
Arsenic concentrations for 31,350 groundwater samples collected from 1973 to 2001.

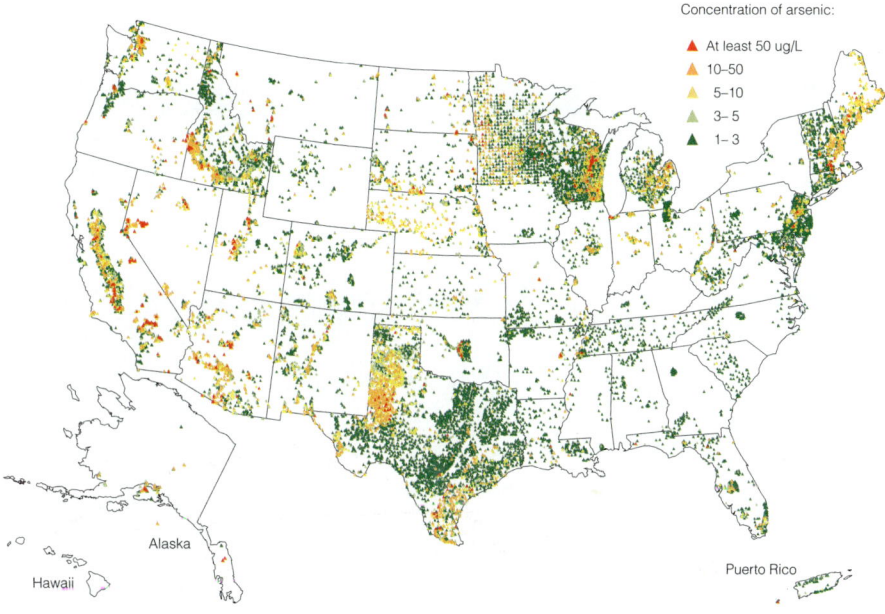

Figure 2
County map showing arsenic concentrations found in at least 25% of groundwater samples per county. The map is based on 31,350 groundwater samples collected between 1973 and 2001.

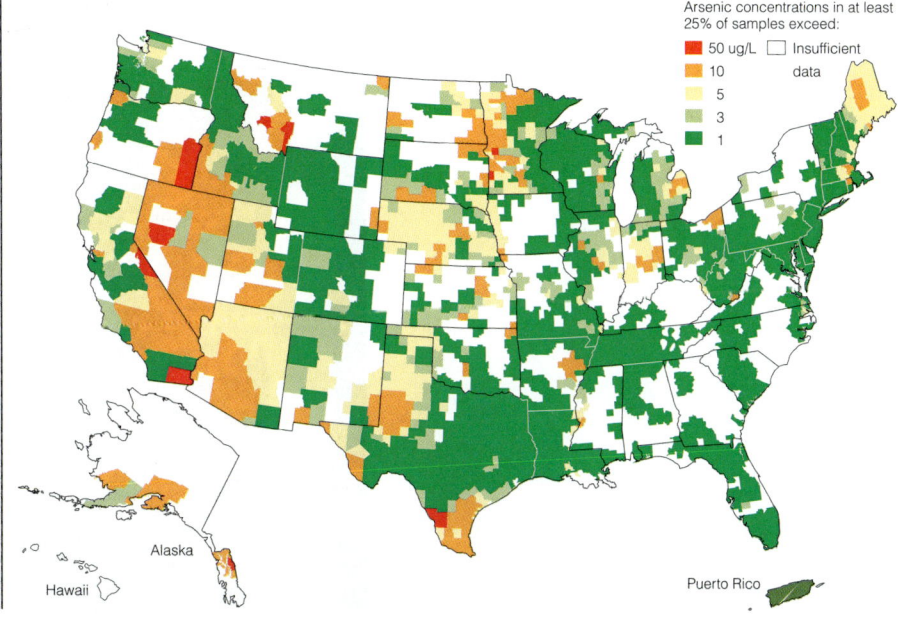

GEOLOGY MATTERS
GEOLOGY IN UNEXPECTED PLACES

Water-Treatment Plants

Your local or regional water-treatment plant may not be the first place you think of when it comes to geology, but surprisingly a fair amount of geology is associated with it. To start off, there is the source of water, which frequently comes from wells drilled below the local water table. The quality of this water depends on the local geology. The water frequently contains high concentrations of iron, calcium, and magnesium; the latter two contribute to the hardness of the water. These elements are usually removed to soften the water.

The first step in softening the water is to aerate it with oxygen to help remove iron, carbon dioxide, and other gases. Removing the carbon dioxide from the water reduces the amount of chemicals needed in subsequent stages of the softening process.

The next step in softening the water is adding lime and sodium hydroxide to precipitate the dissolved calcium and magnesium. This precipitate settles out to form a layer of sludge that is pumped to holding ponds; it can be used later as an agricultural soil conditioner (▶ Figure 1).

Following the removal of calcium and magnesium, carbon dioxide is again added to the water to reduce its pH to about 9, which means it is slightly alkaline. This step stops the softening process and stabilizes the water's chemical composition.

The final process is to pass the treated water through filters to remove fine particles and then add chemicals to ensure that the water is safe for drinking (▶ Figure 2). The water is now ready to be pumped to the plant's storage reservoirs, from which it is distributed to customers.

Figure 1
One of the sludge ponds of the Mt. Pleasant, Michigan, Water Treatment Plant. The sludge produced by the softening process is diverted into a sludge pond. The sludge is reused later as an agricultural soil conditioner.

Figure 2
A worker changes filters in a reverse osmosis water treatment plant in Cape Coral, Florida. Reverse osmosis removes impurities and salt from the water and is one of the treatments used to make it potable.

GEOLOGY MATTERS

Geology AND CULTURAL CONNECTIONS

Dowsing

Finding good drinking water—which generally means finding groundwater—has been a concern of people since time immemorial. It's no wonder that our culture has developed more than one way to locate good, clean water for daily use.

This is not an abstract topic. If you live outside a town or city some time in your life, you'll likely rely on your own well to supply the water your household needs. Imagine that you buy a piece of farmland on which no one has lived in the past. You sketch out plans for a house. But where should you drill a water well, and how deep (and therefore how expensive) will it be?

Most people turn to a local geologist or hydrologist for advice about where to place a well. Such a professional uses all the information you have studied in this chapter to look at the local geology and try to predict where significant amounts of groundwater may lie. Everything that's known about the porosity and permeability of local soil and rock guides the effort. Still, in the end, the proof is in the drilling of the well. Will the drillers hit water, and in quantities large enough to be useful? Or will they continue to drill deeper and deeper without finding enough groundwater to run a household?

No matter how good a geologist is, the day the drilling rig starts work is a tense one for all concerned. Sometimes both geologists and drillers are surprised by large volumes of good-quality groundwater at shallow depths. On other days, a hole may come up "dry" even at great depths, much to the frustration of the homeowner who is paying by the foot for the drilling.

Some people choose to bring a "water witch" or "dowser" to their property before they call in the drillers. Dowsers are men and women who believe they can find groundwater without reference to the local geology (▶ Figure 1). Some dowsers are paid for their efforts; others believe that they have a "gift" for finding water and are happy to help people who need their services.

Dowsing is a simple procedure. The dowser walks over the land with two sticks, a forked willow branch, or two rods (depending on the dowser) (▶ Figure 2). The dowser holds out the sticks or rods. At some point in the survey, the sticks or rods either suddenly cross each other or point downward. The dowser believes that a force coming from the groundwater causes this movement. If you ask the dowser what the force is, or how it affects the sticks or rods, he or she will likely shrug off the question. That kind of attitude can drive a scientist to distraction, but from the dowser's point of view, what matters is that the movement occurs, not why it does so.

Thousands of water wells in North America have been drilled strictly on the advice of dowsers. Some of them have produced abundant and pure groundwater; some have yielded no useful water at all. Unfortunately, no one has kept records on the success rates of dowsers. Still, those who believe in dowsing are unshaken in their confidence when a hole comes up dry. The knowledgeable among them point out that geologists are also unshaken in their beliefs when one of their holes comes up dry.

It is interesting to note that some dowsers seem to recommend the types of places for a well that you yourself would think about after studying this chapter. That is, some dowsers have a strong tendency to recommend drilling on the lowest point of a piece of property or near a stream (where the water table is likely to be highest) or at the base of a prominent hill (where groundwater will be moving in significant volume to lower elevations).

If you ever have to decide on the location of a well, it's wise to consult with everyone you can before the drillers arrive. Your neighbors who have wells on similar land, local geologists, and even experienced dowsers may all have valuable insights for you to consider. In the end, however, the choice of who to believe—what cultural framework you want to accept—will be up to you, just as the bill for the drilling will be all yours!

Figure 1
King Faria, a former dairyman, finds himself in great demand as a dowser during a drought in Marin County, California. Using a weeping willow dowsing rod, Mr. Faria helps residents locate the best place to drill a water well on their property.

Figure 2
Some of the different objects and instruments used by water witches or dowsers in their search for groundwater.

Chapter 17

Glaciers and Glaciation

Nansen Fiord, Greenland
Fractures of glacier, luminous and white float in the fiord like thousands of ceramic plates fragmented against a floor of dark blue water. Greenland is the largest island in the world. More than 80% of its surface is cloaked in ice known as the Greenland ice cap. Most people in Greenland live along the fiords of the western coast where the climate is warmer and the land is not veiled by an immense membrane of ice.
—A. W.

ESSENTIAL QUESTIONS TO ASK

17.1 Introduction
- What was the Little Ice Age, when did it take place, and which areas were most heavily affected?

17.2 The Kinds of Glaciers
- How does a glacier differ from floating sea ice or a permanent snowfield?
- What are the distinguishing features of valley glaciers and continental glaciers?

17.3 Glaciers—Moving Bodies of Ice on Land
- What role do glaciers play in the hydrologic cycle?
- What two mechanisms account for glacial movement?
- Where are present-day glaciers found?

17.4 The Glacial Budget—Accumulation and Wastage
- How are the advance and retreat of a glacier's terminus controlled by its budget?
- What are glacial surges?

17.5 Erosion and Transport by Glaciers
- What are the specific processes that account for erosion by glaciers?
- How do U-shaped glacial troughs form?
- How do arêtes and horns originate?

17.6 Deposits of Glaciers
- How does a recessional moraine form?
- What accounts for the origin of drumlins?
- What kinds of deposits are found in glacial lakes?

17.7 The Ice Age (The Pleistocene Epoch)
- When was the Ice Age, and which areas were most heavily affected by glaciers?
- What are pluvial and proglacial lakes?

17.8 Causes of the Ice Age
- How does the Milankovitch theory account for the onset of glacial episodes?

GEOLOGY MATTERS

GEOLOGY IN UNEXPECTED PLACES:
Evidence of Glaciation in New York City

GEOLOGY IN YOUR LIFE:
A Visit to Cape Cod, Massachusetts

GEOLOGY IN FOCUS:
Glaciers and Global Warming

GEOLOGY AND CULTURAL CONNECTIONS:
The Ice Man

 This icon, which appears throughout the book, indicates an opportunity to explore interactive tutorials, animations, or practice problems available on the Physical GeologyNow website at **http://now.brookscole.com/phygeo6**.

17.1 Introduction

Scientists know that Earth's surface temperatures have increased during the last few decades, although they disagree about how much human activities have contributed to climatic change. Is the temperature increase part of a normal climatic fluctuation, or does the introduction of greenhouse gases to the atmosphere have an adverse effect on climate? Although the evidence for a connection between greenhouse gases, especially carbon dioxide, and global warming is increasing, this question is not fully answered. We know from the geologic record, however, that an Ice Age took place between 1.8 million and 10,000 years ago, and since that time, Earth has experienced several entirely natural climatic changes.

About 6000 years ago, during the Holocene Maximum, temperatures on average were slightly higher than they are now, and some of today's arid regions were much more humid. The Sahara Desert of North Africa had enough precipitation to support lush vegetation, swamps, and lakes. Indeed, Egypt's only arable land today is along the Nile River, but until only a few thousands of years ago much of North Africa was covered by grasslands.

- *What was the Little Ice Age, when did it take place, and which areas were most heavily affected?*

Following the Holocene Maximum was a time of cooler temperatures, but from about A.D. 1000 to 1300, Europe experienced the Medieval Warm Period during which wine grapes grew 480 km farther north than they do today. Then a cooling trend beginning in about A.D. 1300 led to the *Little Ice Age*, which lasted from 1500 to the middle or late 1800s. During the Little Ice Age, glaciers expanded (▶ Figure 17.1), summers were cooler and wetter, winters were colder, and sea ice near Greenland, Iceland, and the Canadian Arctic islands persisted for long periods. Because of the cooler, wetter summers, growing seasons were shorter, which accounted for several widespread famines.

During the coldest part of the Little Ice Age (1680–1730), the growing season in England was about five weeks shorter than it was during the 1900s, and in 1695 Iceland was surrounded by sea ice for much of the year. Occasionally, Eskimos following the southern edge of the sea ice paddled their kayaks as far south as Scotland, and the canals in Holland froze over during some winters. In 1607 the first Frost Fair was held in London, England, on the Thames River, which had begun to freeze over nearly every winter. In the late 1700s, New York Harbor froze over, and 1816 is known as the "year without a summer," when unusually cold temperatures persisted into June and July in New England and northern Europe. (The eruption of Tambora in 1815 contributed to the cold spring and summer of 1816.)

Most of you probably have some idea of what a glacier is and have heard of the Ice Age, although it is doubtful that you know much about the dynamics of glaciers or the Little Ice Age. In any case, *glaciers* are moving bodies of ice on land that are particularly effective at erosion, sediment transport, and deposition. They deeply scour the land they move over, producing easily recognized land-

▶ **Figure 17.1 Glacier Expansion During the Little Ice Age**

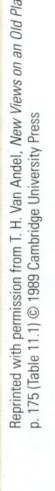

(a) Samuel Birmann (1793–1847) painted this view in Switzerland, titled *Unterer Grindelwald,* in 1826 when the glacier extended onto the valley floor in the foreground.

(b) The same glacier today. Its terminus is hidden behind a rock projection at the lower end of its valley.

forms, and they deposit huge quantities of sediment, much of it important sources of sand and gravel for construction. Glaciers today cover about 10% of the land surface, but during the Ice Age they were much more widespread than they are now.

Unfortunately, our period of recordkeeping is too short to determine whether the last Ice Age and the Little Ice Age are truly events of the past or simply phases of long-term climatic changes and likely to occur again, although the latter case seems most likely. Studying glaciers and their possible causes may help clarify some aspects of long-term climatic change and the debate on global warming. Glaciers are very sensitive to even short-term changes in climate, so scientists closely monitor them to see whether they advance, remain stationary, or retreat.

Section 17.1 Summary

- The Little Ice Age from about 1500 to the middle to late 1800s was a time of cooler weather, especially during summers, advancing glaciers, and persistent sea ice for long periods.

- Climate records are not sufficient to determine whether the Ice Age is over or whether we are in a temporary warm period.

17.2 The Kinds of Glaciers

■ *How does a glacier differ from floating sea ice or a permanent snowfield?*

Geologists define a **glacier** as a moving body of ice on land that flows downslope or outward from an area of accumulation. We will discuss how glaciers flow in a later section. Our definition of a glacier excludes frozen seawater as in the North Polar region, and sea ice that forms yearly adjacent to Greenland and Iceland. Drifting icebergs are not glaciers either, although they may have come from glaciers that flowed into lakes or the sea. The critical points in the definition are *moving* and *on land*. Accordingly, permanent snowfields in high mountains, though on land, are not glaciers because they do not move. All glaciers share several characteristics, but they differ enough in size and location for geologists to define two specific types, valley glaciers and continental glaciers, and several subvarieties.

■ *What are the distinguishing features of valley glaciers and continental glaciers?*

Valley Glaciers

Valley glaciers are confined to mountain valleys where they flow from higher to lower elevations (▶ Figure 17.2), but continental glaciers cover vast areas, they are not confined by the underlying topography, and they flow outward in all directions from areas of snow and ice accumulation. We use the term **valley glacier,** but some geologists prefer the synonyms *alpine glacier* and *mountain glacier.* Valley glaciers commonly have tributaries, just as streams do, thereby forming a network of glaciers in an interconnected system of mountain valleys.

Valley glaciers are common in the mountains of western North America, especially Alaska and Canada, as well as the Andes in South America, the Alps in Europe, the Alps of New Zealand, and the Himalayas in Asia. Even a few of the loftier mountains in Africa, though near the equator, are high enough to support small glaciers. In fact, Australia is the only continent that has no glaciers.

A valley glacier's shape is obviously controlled by the shape of the valley it occupies so it tends to be a long, narrow tongue of moving ice. Where a valley glacier flows from a valley onto a wider plain and spreads out, or where two or more valley glaciers coalesce at the base of a mountain range, they form a more extensive ice cover called a *piedmont glacier* (▶ Figure 17.2b) Valley glaciers that flow into the ocean are called *tidewater glaciers*; they differ from other valley glaciers only in that their terminus is in the sea rather than on land (▶ Figure 17.2c).

Valley glaciers are small compared with the much more extensive continental glaciers, but even so they may be as large as several kilometers across, 200 km long, and hundreds of meters thick. For instance, the Saskatchewan Glacier in Canada is 555 m thick, and some glaciers in Alaska are up to 1500 m thick. Erosion and deposition by valley glaciers were responsible for much of the spectacular scenery in several U.S. and Canadian national parks, notably Yosemite, Glacier, and Banff-Jasper.

Continental Glaciers

Continental glaciers, also known as *ice sheets*, are vast, covering at least 50,000 km^2, and they are unconfined by topography; that is, their shape and movement are not controlled by the underlying landscape. Valley glaciers are long, narrow tongues of ice that conform to the shape of the valley they occupy, and the existing slope determines their direction of flow. In contrast, continental glaciers flow outward in all directions from a central area or areas of accumulation in response to variations in ice thickness.

In Earth's two areas of continental glaciation, Greenland and Antarctica, the ice is more than 3000 m thick and covers all but the highest mountains (▶ Figure 17.3a,b). The continental glacier in Greenland covers about 1,800,000 km^2, and in Antarctica the East and West Antarctic Glaciers merge to form a continuous ice sheet covering more than 12,650,000 km^2. The glaciers in Antarctica flow into the sea, where the buoyant effect of water causes the ice to float in vast *ice shelves;* the Ross Ice Shelf alone covers more than 547,000 km^2 (▶ Figure 17.3a). During March 2000, an iceberg 295 km long and 37 km wide broke off from the Ross Ice Shelf, and later that same year it yielded another iceberg nearly half as large.

▶ Figure 17.2 Valley Glaciers

a A valley glacier in Alaska. Notice the tributaries that unite to form a larger glacier.

b This glacier on Baffin Island in Canada flows from its valley onto a plain, where it spreads out to form a piedmont glacier.

c Wellesley Glacier in Alaska flows into the sea, so it is a tidewater glacier.

Although valley glaciers and continental glaciers are easily differentiated by size and location, geologists also recognize an intermediate variety called an **ice cap.** Ice caps are similar to, but smaller than, continental glaciers, covering less than 50,000 km². The 6000-km² Penny Ice Cap on Baffin Island, Canada (▶ Figure 17.3c), and the Juneau Icefield in Alaska and Canada at about 3900 km² are good examples. Some ice caps form when valley glaciers grow and overtop the divides and passes between adjacent valleys and coalesce to form a continuous ice cover. They also form on fairly flat terrain in Iceland and some of the islands in the Canadian Arctic.

Section 17.2 Summary

- All glaciers are moving bodies of ice on land, but valley glaciers are confined to mountain valleys where they flow from higher to lower elevations, whereas continental glaciers cover vast areas and flow outward in all directions from areas of snow and ice accumulation.

17.3 Glaciers—Moving Bodies of Ice on Land

You already know that a *glacier* is a moving body of ice on land; we use the term **glaciation** to indicate all glacial activity including the origin, expansion, and retreat of glaciers as well as their impact on Earth's surface. Presently glaciers cover nearly 15 million km², or about 10% of Earth's land surface. As a matter of fact, if all glacial ice were in the United States and Canada, it would form a continuous ice cover about 1.5 km thick!

At first glance glaciers appear static. Even briefly visiting a glacier may not dispel this impression because, although glaciers move, they usually do so slowly. Nevertheless, they do move and just like other geologic agents such as running water, glaciers are dynamic systems that continuously adjust to changes. For example, a glacier may flow slower or more rapidly depending on decreased or increased amounts of snow or the absence or presence of water at its base. And glaciers may expand or contract depending on climatic changes.

▶ **Figure 17.3 Continental Glaciers**

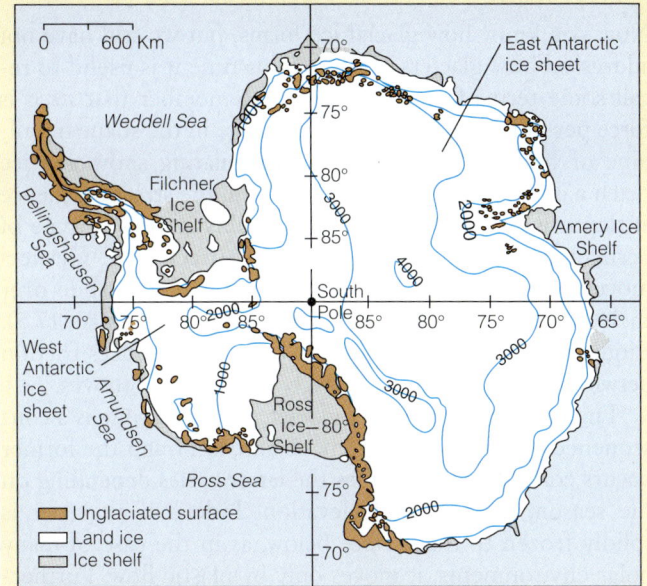

a The West and East Antarctic ice sheets merge to form a nearly continuous ice cover that averages 2160 m thick. The blue lines are lines of equal thickness.

b View of part of the Antarctic ice sheet. Notice the *nunatak*, which is a peak extending above the glacial ice.

c View of the Penny Ice Cap on Baffin Island, Canada. It covers about 6000 km².

Glaciers—Part of the Hydrologic Cycle

■ *What role do glaciers play in the hydrologic cycle?*

In Chapter 1, you learned that one of Earth's systems, the hydrosphere, consists of all surface water, including water frozen in glaciers. So glaciers make up one reservoir in the hydrologic cycle where water is stored for long periods, but even this water eventually returns to its original source, the oceans (see Figure 15.3). Glaciers at high latitudes, as in Alaska, northern Canada, and Scandinavia, flow directly into the oceans where they melt, or icebergs break off (a process known as calving) and drift out to sea where they eventually melt. At lower latitudes or areas remote from the oceans, glaciers flow to lower elevations where they melt and the liquid water enters the groundwater system (another reservoir in the hydrologic cycle) or it returns to the seas by surface runoff.

In addition to melting, glaciers lose water by *sublimation*, when ice changes to water vapor without an intermediate liquid phase. Sublimation is not an exotic process; it occurs in the freezer compartment of a refrigerator. Because of sublimation, the older ice cubes at the bottom of the container are much smaller than the more recently formed ones. In any case, the water vapor so derived enters the atmosphere where it may condense and fall as rain or snow, but in the long run this water also returns to the oceans.

How Do Glaciers Originate and Move?

Ice is a crystalline solid with characteristic physical properties and a specific chemical composition, and thus is a mineral. Accordingly, glacial ice is a type of metamorphic rock, but one that is easily deformed. Glaciers form in any area where more snow falls than melts during the warmer seasons and a net accumulation takes place (▶ Figure 17.4). Freshly fallen snow has about 80% air-filled pore space and 20% solids, but it compacts as it accumulates, partially thaws, and refreezes, converting to a granular type of snow known as **firn**. As more snow accumulates, the firn is buried and further compacted and recrystallized until it is transformed into **glacial ice,** consisting of about 90% solids and 10% air (▶ Figure 17.4).

▶ **Figure 17.4 Glacial Ice**

a The conversion of freshly fallen snow to firn and then to glacial ice.

b This iceberg in Portage Lake in Alaska shows the blue color of glacial ice. The longer wavelengths of white light are absorbed by the ice, but blue (short wavelength) is transmitted into the ice and scattered, accounting for the blue color.

■ *What two mechanisms account for glacial movement?*

Now you know how glacial ice forms, but we still have not addressed how glaciers move. At this time it is useful to recall some terms from Chapter 13. Remember that *stress* is force per unit area and *strain* is a change in the shape or volume or both of solids. When accumulating snow and ice reach a critical thickness of about 40 m, the stress on the ice at depth is great enough to induce **plastic flow,** a type of permanent deformation involving no fracturing. Glaciers move mostly by plastic flow, but they may also slide over their underlying surface by **basal slip** (▶ Figure 17.5). Liquid water facilitates basal slip because it reduces friction between a glacier and the surface over which it moves.

The total movement of a glacier in a given time is a consequence of plastic flow and basal slip, although the former occurs continuously, whereas the latter varies depending on the season, latitude, and elevation. Indeed if a glacier is solidly frozen to the surface below, as in the case of many polar environments, it moves only by plastic flow. Furthermore, basal slip is far more important in valley glaciers as they flow from higher to lower elevations, whereas continental glaciers need no slope for flow. Although glaciers move by plastic flow, the upper 40 m or so of ice behaves like a brittle solid and fractures if subjected to stress. Large crevasses commonly develop in glaciers where they flow over an increase in slope of the underlying surface or where they flow around a corner (▶ Figure 17.6a). In either case, the ice is stretched (subjected to tension) and crevasses open, which extend down to the zone of plastic flow. In some cases, a glacier descends over such a steep precipice

▶ **Figure 17.5 Part of a Glacier Showing Movement by a Combination of Plastic Flow and Basal Slip** Plastic flow involves internal deformation within the ice, whereas basal slip is sliding over the underlying surface. If a glacier is solidly frozen to its bed, it moves only by plastic flow. Notice that the top of the glacier moves farther in a given time than the bottom does.

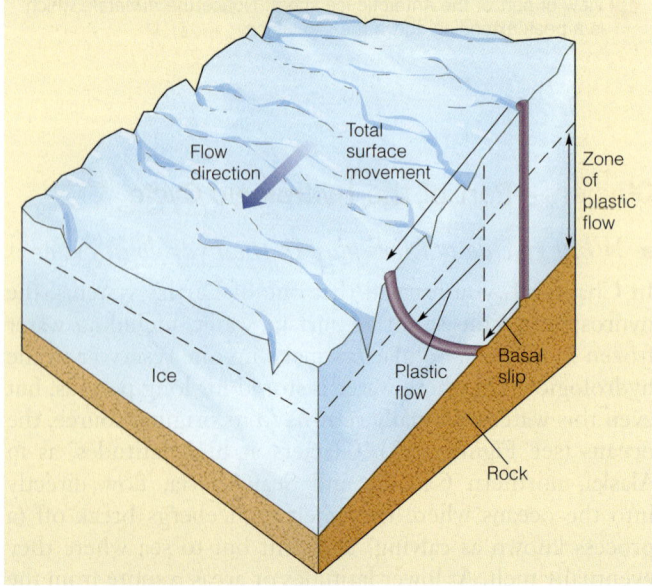

that crevasses break up the ice into a jumble of blocks and spires, and an icefall develops (▶ Figure 17.6c).

Distribution of Glaciers

■ *Where are present-day glaciers found?*

You now have some idea of what glaciers are, how they form, and how they move, but what controls their distribution? As you might suspect, the amount of snowfall and temperature are important factors. Parts of northern Canada are cold enough to support glaciers but receive too little snowfall, whereas some mountain areas in California receive huge amounts of snow but are too warm for glaciers.

Of course temperature varies with elevation and latitude, so we would expect to find glaciers in high mountains and at high latitudes, if these areas receive enough snow.

Many small glaciers are present in the Sierra Nevada of California but only at elevations exceeding 3900 m. In fact, the high mountains in California, Oregon, and Washington all have glaciers because they receive so much snow. Mount Baker in Washington had almost 29 m of snow during the winter of 1998–1999, and average accumulations of 10 m or more are common in many parts of these mountains.

Glaciers are also found in the mountains along the Pacific Coast of Canada, which also receive considerable snowfall and of course they are farther north. Some of the higher peaks in the Rocky Mountains in both the United States and Canada

▶ **Figure 17.6 Crevasses in Upper Parts of Glaciers** Crevasses are common in the upper parts of glaciers when the ice is subjected to tension.

a Crevasses open where the brittle part of a glacier is stretched as it moves over a steeper slope in its valley.

b These crevasses are on the Exit Glacier in Kenai Fjords National Park and Preserve in Alaska.

c The area of chaotic ice on this small glacier on the Jungfrau in Switzerland is an icefall.

also support glaciers; Glacier National Park in Montana and Banff–Jasper National Park in Alberta, Canada, are good examples. At even higher latitudes, as in Alaska, northern Canada, and Scandinavia, glaciers exist at sea level.

> ### Section 17.3 Summary
>
> - Glaciers make up one reservoir in the hydrologic cycle where water is stored for long periods. They are found in high mountains and at high latitudes as in Alaska and Scandinavia.
>
> - Glaciers form where winter snowfall exceeds summer melt and accumulates, thereby converting snow first to firn and then to glacial ice.
>
> - When a body of ice becomes about 40 m thick, pressure causes it to move by plastic flow and, if a slope is present, by basal slip.

17.4 The Glacial Budget— Accumulation and Wastage

■ *How are the advance and retreat of a glacier's terminus controlled by its budget?*

Just as a savings account grows and shrinks as funds are deposited and withdrawn, a glacier expands and contracts in response to accumulation and wastage. We describe a glacier's behavior in terms of a **glacial budget,** which is essentially a balance sheet of accumulation and wastage. For instance, the upper part of a valley glacier is a **zone of accumulation,** where additions exceed losses and the surface is perennially snow covered. In contrast, the lower part of the same glacier is a **zone of wastage,** where losses from melting, sublimation, and calving of icebergs exceed the rate of accumulation (▶ Figure 17.7).

At the end of winter, a glacier's surface is completely covered with the accumulated seasonal snowfall. During the spring and summer, the snow begins to melt, first at lower elevations and then progressively higher up the glacier. The elevation to which snow recedes during a wastage season is the *firn limit* (▶ Figure 17.7). Geologists can easily identify the zones of accumulation and wastage by noting the location of the firn limit.

The firn limit on a glacier may change yearly, but if it does not change or shows only minor fluctuations, the glacier has a balanced budge. That is, additions in the zone of accumulation are exactly balanced by losses in the zone of wastage, and the distal end, or terminus, of the glacier remains stationary (▶ Figure 17.7a). If the firn limit moves up the glacier, indicating a negative budget, the glacier's terminus retreats (▶ Figure 17.7b). If the firn limit moves down the glacier, however, the glacier has a positive budget, additions exceed losses, and its terminus advances (▶ Figure 17.7c).

Even though a glacier may have a negative budget and a retreating terminus, the glacial ice continues to move toward the terminus by plastic flow and basal slip. If a negative budget persists long enough, though, the glacier continues to recede and it thins until it is no longer thick enough to maintain flow. It then ceases moving and becomes a *stagnant glacier*; if wastage continues, the glacier eventually disappears.

Also notice in ▶ Figure 17.7 that a valley glacier with a balanced budget transports and deposits sediment at its terminus as a *terminal moraine*. At a later time the same glacier may have a negative budget, in which case its terminus retreats and perhaps becomes stabilized again if the budget is once more balanced. It then deposits another moraine, but since the glacier's terminus has retreated this one is known as a *recessional moraine*. We will discuss moraines and other glacial landforms later in this chapter.

We used a valley glacier as an example, but the same budget considerations control the flow of ice caps and continental glaciers as well. The entire Antarctic ice sheet is in the zone of accumulation, but it flows into the ocean where wastage occurs.

How Fast Do Glaciers Move?

In general, valley glaciers move more rapidly than continental glaciers, but the rates for both vary from centimeters to tens of meters per day. Valley glaciers moving down steep slopes flow more rapidly than glaciers of comparable size on gentle slopes, assuming that all other variables are the same. The main glacier in a valley glacier system contains a greater volume of ice and thus has a greater discharge and flow velocity than its tributaries (▶ Figure 17.2a). Temperature exerts a seasonal control on valley glaciers because, although plastic flow remains rather constant year-round, basal slip is more important during warmer months when meltwater is abundant.

Flow rates also vary within the ice itself. For example, flow velocity increases downslope in the zone of accumulation until the firn limit is reached; from that point, the

> ### What Would You Do?
>
> Suppose you are a high school earth science teacher trying to explain to students that ice is a mineral and rock, and how a solid like ice can flow like a fluid. Furthermore, you explain that the upper 40 m or so of a glacier is brittle and fractures, whereas ice below that depth simply flows when subjected to stress. Now that your students are thoroughly confused, how will you explain and demonstrate that ice can behave like a solid yet show properties of fluid flow? (*Hint:* Refer to some of the discussion in Chapter 11.)

Geo-focus/Active Figure 17.7 Response of a Hypothetical Glacier to Changes in its Budget

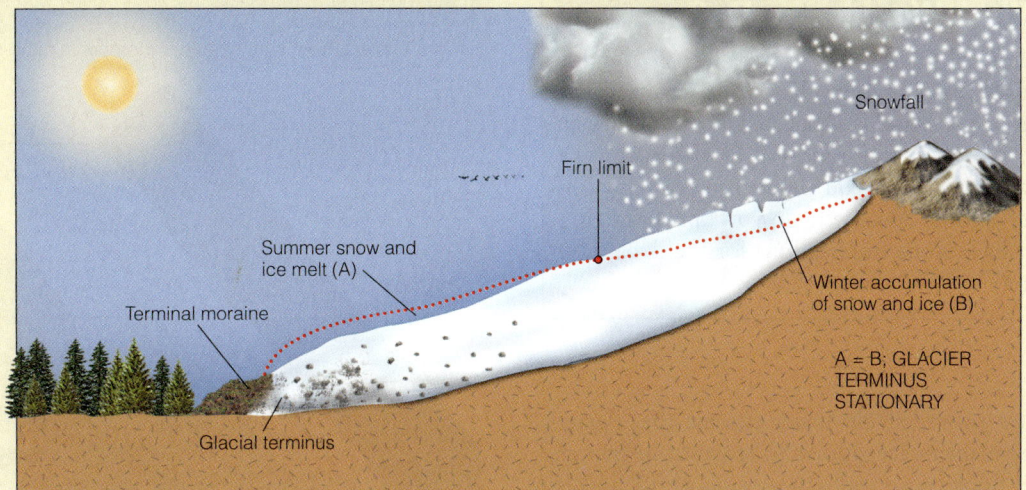

a Winter accumulation (B) and summer snow and ice melt (A) are equal. That is, additions and losses are equal so the glacier's terminus remains stationary. The terminal moraine is deposited at the terminus of a glacier.

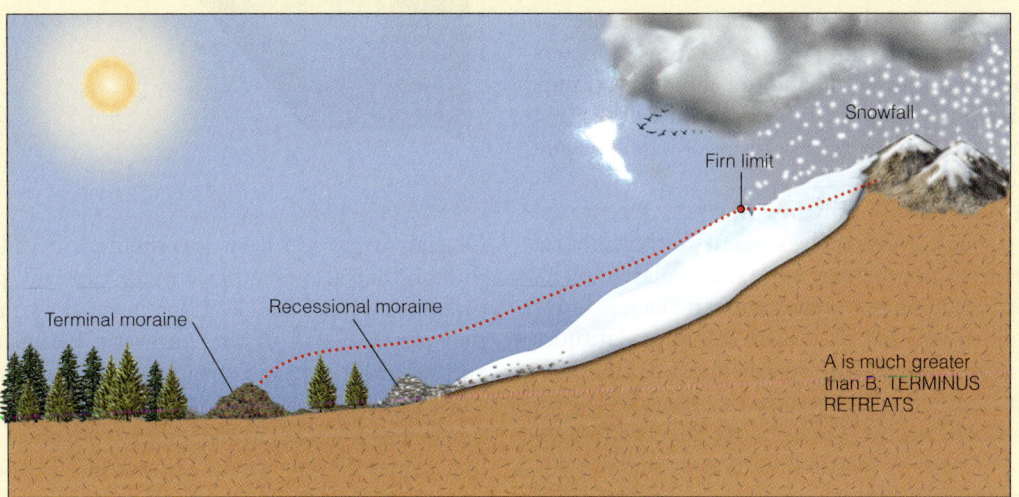

b Summer snow and ice melt (A) are much greater than winter accumulation (B), and the glacier's terminus retreats, although the glacier continues to move by plastic flow and basal slip. The recessional moraine is deposited at the glacier's new terminus.

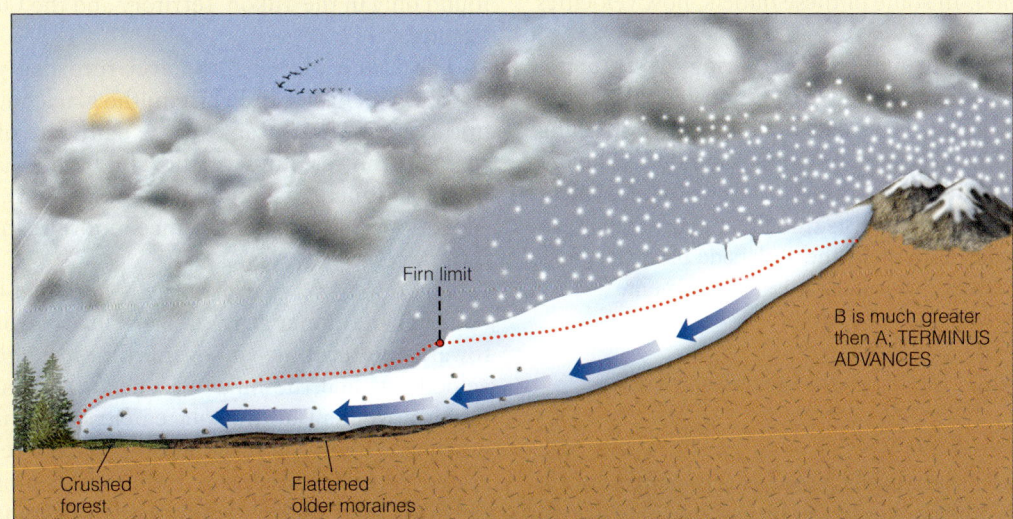

c Winter accumulation (B) is much greater than summer snow and ice melt (A), so the glacier's terminus advances. As it does so, it overrides and modifies its previously deposited moraines.

What Would You Do?

After carefully observing the same valley glacier for several years, you conclude (1) that its terminus has retreated at least 1 km and (2) that debris on its surface has obviously moved several hundred of meters toward the glacier's terminus. Can you think of an explanation for these observations? Also, why do you think studying glaciers might have some implications for the debate on global warming?

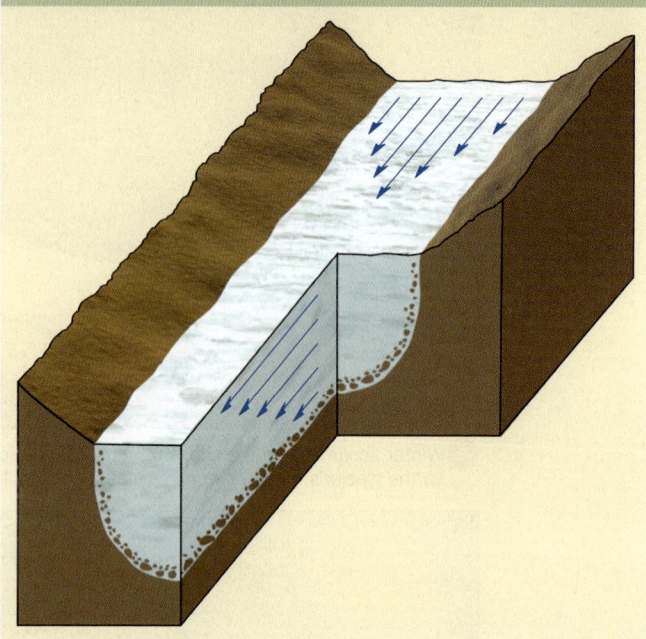

▶ **Figure 17.8 Flow Velocity in a Valley Glacier** Flow velocity in a valley glacier varies both horizontally and vertically. Velocity is greatest at the top center of the glacier because friction with the walls and floor of the trough slows the flow adjacent to these boundaries. The lengths of the arrows in the figure are proportional to velocity.

velocity becomes progressively lower toward the glacier's terminus. Valley glaciers are similar to streams, in that the valley walls and floor cause frictional resistance to flow, so the ice in contact with the walls and floor moves more slowly than the ice some distance away (▶ Figure 17.8).

Notice in ▶ Figure 17.8 that flow velocity in the interior of a glacier increases upward until the top few tens of meters of ice are reached, but little or no additional increase occurs after that point. This upper ice layer constitutes the rigid part of the glacier that is moving as a result of basal slip and plastic flow below.

Continental glaciers ordinarily flow at a rate of centimeters to meters per day. Nevertheless, even a rather modest rate of a meter or so per day has a great cumulative effect after several decades. One reason continental glaciers move comparatively slowly is that they exist at higher latitudes and are frozen to the underlying surface most of the time, which limits the amount of basal slip. But some basal slip does occur even beneath the Antarctic ice sheet, although most of its movement is by plastic flow. Nevertheless, some parts of continental glaciers manage to achieve extremely high flow rates. Near the margins of the Greenland ice sheet, the ice is forced between mountains in what are called *outlet glaciers*. In some of these outlets, flow velocities exceed 100 m per day.

In parts of the continental glacier covering West Antarctica, scientists have identified ice streams in which flow rates are considerably higher than in adjacent glacial ice. Drilling has revealed a 5-m-thick layer of water-saturated sediment beneath these ice streams, which acts to facilitate movement of the ice above. Some geologists think that geothermal heat from subglacial volcanism melts the underside of the ice, thus accounting for the layer of water-saturated sediment.

Glacial Surges

■ *What are glacial surges?*

A **glacial surge** is a short-lived episode of accelerated flow in a glacier during which its surface breaks into a maze of crevasses and its terminus advances noticeably (▶ Figure 17.9).

Glacial surges are best documented in valley glaciers, although they also take place in ice caps and perhaps in continental glaciers. In 1995, for instance, a huge ice shelf at the northern end of the Antarctic Peninsula broke apart and several ice streams from the Antarctic ice sheet surged toward the ocean.

During a surge, a glacier may advance several tens of meters per day for weeks or months and then return to its normal flow rate. Surging glaciers constitute only a tiny proportion of all glaciers, and none of these are in the United States outside of Alaska. Even in Canada they are found only in the Yukon Territory and the Queen Elizabeth Islands.

The fastest glacial surge ever recorded was in 1953 in the Kutiah Glacier in Pakistan; the glacier advanced 12 km in three months, for an average daily rate of about 130 m. In 1986 the terminus of the Hubbard Glacier in Alaska began advancing at about 10 m per day, and in 1993 Alaska's Bering Glacier advanced more than 1.5 km in just three weeks.

The onset of a glacial surge is heralded by a thickened bulge in the upper part of a glacier that begins to move toward the terminus at several times the glacier's normal velocity. When the bulge reaches the terminus, it causes rapid movement and displacement of the terminus by as much as 20 km. Surges are also probably related to accelerated rates of basal slip rather than more rapid plastic flow. One theory holds that thickening in the zone of accumulation with concurrent thinning in the zone of

▶ **Figure 17.9 Hubbard Glacier Surge** During a 1986 surge, Hubbard Glacier in Alaska advanced across Russell Fiord, isolating it from the open sea and raising the water level 23 m above sea level. Environmentalists saved some of the marine mammals trapped in the former bay, but hundreds of seals and porpoises died. The ice dam was eventually breached, but it formed again in 2002.

17.5 Erosion and Transport by Glaciers

Erosion Processes

■ *What are the specific processes that account for erosion by glaciers?*

As moving solids, glaciers erode, transport, and eventually deposit huge quantities of sediment and soil. Indeed, they have the capacity to transport any size of sediment, including boulders the size of houses as well as clay-sized particles. Important processes of erosion include bulldozing, plucking, and abrasion. During the Pleistocene Epoch (Ice Age) glaciers covered much larger areas than they do now and were far more important than their present distribution indicates.

Although *bulldozing* is not a formal geologic term, it is fairly self-explanatory; glaciers shove or push unconsolidated materials in their paths. This effective process was aptly described in 1744 during the Little Ice Age by an observer in Norway:

> When at times [the glacier] pushes forward a great sound is heard, like that of an organ and it pushes in front of it unmeasurable masses of soil, grit and rocks bigger than any house could be, which it then crushes small like sand.*

wastage increases the glacier's slope and accounts for accelerated flow. But another theory holds that pressure on soft sediment beneath a glacier squeezes fluids through the sediment thereby allowing the overlying glacier to slide more effectively.

Section 17.4 Summary

• The behavior of a glacier depends on its glacial budget, which is the relationship between accumulation and wastage. A budget may be balanced, negative, or positive.

• Glaciers move at varying rates depending on slope, discharge, and season. Valley glaciers tend to move faster than continental glaciers. At times glaciers move very rapidly, as much as tens of meters per day during surges.

Plucking, also called *quarrying*, results when glacial ice freezes in the cracks and crevices of a bedrock projection and eventually pulls it loose. One manifestation of plucking is a landform called a *roche moutonnée*, a French term for "rock sheep." As shown in ▶ Figure 17.10, a glacier smooths the "upstream" side of an obstacle, such as a small hill, and plucks pieces of rock from the "downstream" side by repeatedly freezing to and pulling away from the obstacle.

Bedrock over which sediment-laden glacial ice moves is effectively eroded by **abrasion** and develops a **glacial polish,** a smooth surface that glistens in reflected light (▶ Figure 17.11a). Abrasion also yields **glacial striations,** consisting of rather straight scratches rarely more than a few millimeters deep on rock surfaces (▶ Figure 17.11b). Abrasion thoroughly pulverizes rocks, yielding an aggregate of clay- and silt-sized particles that have the consistency of flour—hence, the name *rock flour*.

---

*Quoted in C. Officer and J. Page, *Tales of the Earth* (New York: Oxford University Press, 1993), p. 99.

▶ **Figure 17.10 Origin of a Roche Moutonnée** A roche moutonnée, a French term meaning "sheep rock," is a bedrock projection that was shaped by glacial abrasion and plucking.

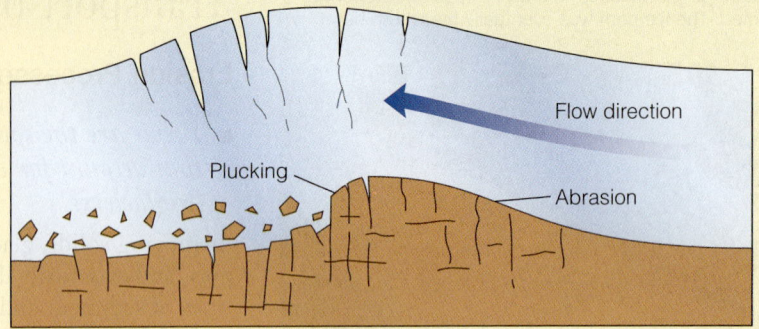

a Origin of a roche moutonnée. A glacier abrades and polishes the "upstream" side of a bedrock projection and shapes its "downstream" side by plucking.

b A roche moutonnée in Montana. **Can you determine the direction that the glacier moved?** *The glacier moved from right to left.*

Rock flour is so common in streams discharging from glaciers that the water has a milky appearance (▶ Figure 17.11c).

Continental glaciers derive sediment from mountains projecting through them, and wind-blown dust settles on their surfaces, but most of their sediment comes from the surface they move over. As a result, most sediment is transported in the lower part of the ice sheet. In contrast, valley glaciers carry sediment in all parts of the ice, but it is concentrated at the base and along the margins (▶ Figure 17.12). Some of the marginal sediment is derived by abrasion and plucking, but much of it is supplied by mass wasting processes, as when soil, sediment, or rock falls or slides onto the glacier's surface.

Erosion by Valley Glaciers

Mountain ranges are scenic to begin with, but when eroded by valley glaciers, they take on a unique appearance of angular ridges and peaks in the midst of broad, smooth valleys with near-vertical walls. The erosional landforms produced by valley glaciers are easily recognized and enable us to appreciate the tremendous erosive power of moving ice. See "Valley Glaciers and Erosion" on pages 544–545, which features U-shaped glacial troughs, fjords, hanging valleys, cirques, arêtes, and horns.

■ *How do U-shaped glacial troughs form?*

U-Shaped Glacial Troughs A **U-shaped glacial trough** is one of the most distinctive features of valley glaciation. Mountain valleys eroded by running water are typically V-shaped in cross section; that is, they have valley walls that descend to a narrow valley bottom (▶ Figure 17.13a). In contrast, valleys scoured by glaciers are deepened, widened, and straightened so that they have very steep or vertical walls but broad, rather flat valley floors; thus they exhibit a U-shaped profile (▶ Figure 17.13c).

Many glacial troughs contain triangular-shaped *truncated spurs*, which are cutoff or truncated ridges that extend into

▶ **Figure 17.11** Glacial Polish

a Glacial polish on quartzite in Michigan.

b Glacial polish and striations, the straight scratches, on basalt at Devil's Postpile National Monument in California.

c The water in this stream in Switzerland is discolored by rock flour, small particles yielded by glacial abrasion.

▶ **Figure 17.12** Sediment Transport by Valley Glaciers

a Debris on the surface of the Mendenhall Glacier in Alaska. The largest boulder is about 2 m across. Notice the icefall in the background. The person left of center provides scale.

b The dark material on this glacier's surface is sediment.

the preglacial valley (▶ Figure 17.13c). Another common feature is a series of steps or rock basins in the valley floor where the glacier eroded rocks of varying resistance; many of the basins now contain small lakes.

During the Pleistocene, when glaciers were more extensive, sea level was as much as 130 m lower than at present, so glaciers flowing into the sea eroded their valleys below present sea level. When the glaciers melted at the end of the Pleistocene, sea level rose and the ocean filled the lower ends of the glacial troughs so that now they are long, steep-walled embayments called **fiords.**

Fiords are restricted to high latitudes where glaciers exist at low elevations, such as Alaska, western Canada, Scandinavia, Greenland, southern New Zealand, and southern Chile. Lower sea level during the Pleistocene was not entirely responsible for the formation of all fiords. Unlike running water,

Valley Glaciers and Erosion

Valley glaciers effectively erode and produce several easily recognized landforms. Where glaciers move through mountain valleys, the valleys are deepened and widened, giving them a distinctive U-shaped profile. The peaks and ridges rising above valley glaciers are also eroded and they become jagged and angular. Much of the spectacular scenery in Grand Teton National Park, Wyoming, Yosemite National Park, California, and Glacier National Park, Montana, resulted from erosion by valley glaciers. In fact, valley glaciers remain active in some of the mountains of western North America, especially in Alaska and Canada.

▶ 1, ▲ 2. Part of southwestern Greenland (1) where valley glaciers merged to form an ice cap. Should these glaciers melt, several landforms like those in Figure 17.13c would be present. The Teton Range in Wyoming (2) acquired its angular peaks and ridges and broadly rounded valleys as a result of erosion by valley glaciers.

▲ 3, ▲ 4, ▶ 5. U-shaped glacial troughs. The glacial trough (3) is in northern Montana, whereas the one in (4) is in southern Germany. The lake is impounded behind a glacial deposit known as an end moraine. The steep-walled glacial trough in Norway (5) extends below sea level, so it is a fiord.

◀ **6.** The bowl-shaped depression on Mount Wheeler in Great Basin National Park, Nevada, is a cirque. It has steep walls on three sides and opens out into a glacial trough.

▶ **7.** Lake Helen on Lassen Peak in Lassen Volcanic National Park, California, is a tarn—that is, a lake in a cirque.

▲ **8.** Bridalveil Falls in Yosemite National Park, California, plunges 190 m from a hanging valley. The valley in the foreground is a huge U-shaped glacial trough.

▲ **9,** ◀ **10.** The Matterhorn (9) in Switzerland is a well-known horn. This view of the Jungfrau (10) in Switzerland shows two small glaciers, a cirque headwall, and an arête.

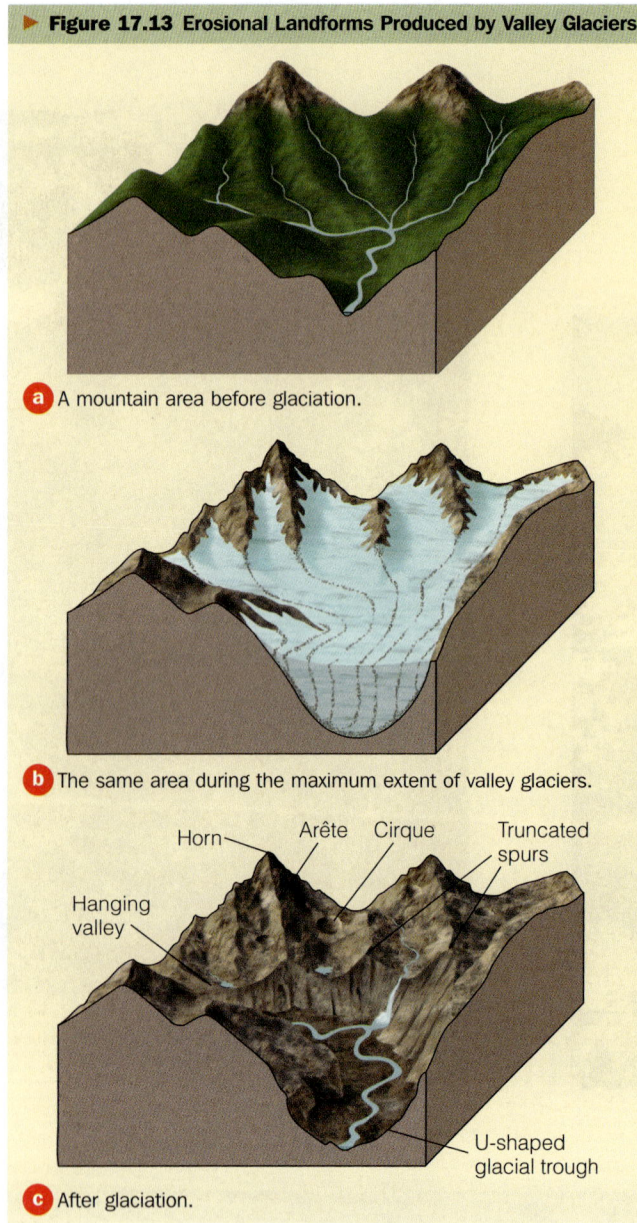

Figure 17.13 Erosional Landforms Produced by Valley Glaciers

a A mountain area before glaciation.

b The same area during the maximum extent of valley glaciers.

c After glaciation.

Labels on figure c: Horn, Arête, Cirque, Truncated spurs, Hanging valley, U-shaped glacial trough

Although not all hanging valleys form by glacial erosion, many do. As ▶ Figure 17.13 shows, the large glacier in the main valley vigorously erodes, whereas the smaller glaciers in tributary valleys are less capable of large-scale erosion. When the glaciers disappear, the smaller tributary valleys remain as hanging valleys.

Cirques, Arêtes, and Horns Perhaps the most spectacular erosional landforms in areas of valley glaciation are at the upper ends of glacial troughs and along the divides that separate adjacent glacial troughs (see "Valley Glaciers and Erosion" on pp. 544–545). Valley glaciers form and move out from steep-walled, bowl-shaped depressions called **cirques** at the upper end of their troughs (▶ Figure 17.13c). Cirques are typically steep-walled on three sides, but one side is open and leads into the glacial trough. Some cirques slope continuously into the glacial trough, but many have a lip or threshold at their lower end (▶ Figure 17.13c).

The details of cirque origin are not fully understood, but they probably form by erosion of a preexisting depression on a mountain side. As snow and ice accumulate in the depression, frost wedging and plucking, combined with glacial erosion, enlarge and transform the head of a steep mountain valley into a typical amphitheater-shaped cirque. Tension in the upper part of the glacier may reduce the erosive power of the ice on the immediate, downslope side of a cirque, leaving a lip or threshold in the valley floor after the ice melts away. Small lakes of meltwater, called *tarns*, often form on the floors of cirques behind such thresholds.

Cirques become wider and are cut deeper into mountainsides by headward erosion as a result of abrasion, plucking, and several mass wasting processes. For example, part of a steep cirque headwall may collapse while frost wedging continues to pry loose rocks that tumble downslope, so a combination of processes erode a small mountainside depression into a large cirque.

■ *How do arêtes and horns originate?*

The fact that cirques expand laterally and by headward erosion accounts for the origin of two other distinctive erosional features: arêtes and horns. **Arêtes**—narrow, serrated ridges—form in two ways. In many cases, cirques form on opposite sides of a ridge, and headward erosion reduces the ridge until only a thin partition of rock remains (▶ Figure 17.13c). The same effect occurs when erosion in two parallel glacial troughs reduces the intervening ridge to a thin spine of rock.

The most majestic of all mountain peaks are **horns,** steep-walled, pyramidal peaks formed by headward erosion of cirques. For a horn to form, a mountain peak must have at least three cirques on its flanks, all of which erode headward (▶ Figure 17.13c). Excellent examples of horns are Mount Assiniboine in the Canadian Rockies, the Grand Teton in Wyoming, and the most famous of all, the Matterhorn in Switzerland.

glaciers can erode a considerable distance below sea level. In fact, a glacier 500 m thick can stay in contact with the seafloor and effectively erode it to a depth of about 450 m before the buoyant effects of water cause the glacial ice to float! The depth of some fiords is impressive; some in Norway and southern Chile are about 1300 m deep.

Hanging Valleys Waterfalls form in several ways, but some of the world's highest and most spectacular are found in recently glaciated areas. Bridalveil Falls in Yosemite National Park, California, plunge from a **hanging valley,** which is a tributary valley whose floor is at a higher level than that of the main valley. Where the two valleys meet, the mouth of the hanging valley is perched far above the main valley's floor (▶ Figure 17.13c). Accordingly, streams flowing through hanging valleys plunge over vertical or steep precipices.

▶ **Figure 17.14 An Ice-Scoured Plain in Northwest Territories of Canada** This low-relief surface is an ice-scoured plain in the Northwest Territories of Canada. Numerous lakes, little or no soil, and extensive bedrock exposures are typical of these areas eroded by continental glaciers.

Continental Glaciers and Erosional Landforms

Areas eroded by continental glaciers tend to be smooth and rounded because these glaciers bevel and abrade high areas that project into the ice. Rather than yielding the sharp, angular landforms typical of valley glaciation, they produce a landscape of subdued topography interrupted by rounded hills because they bury landscapes entirely during their development.

In a large part of Canada, particularly the vast Canadian Shield region, continental glaciers have stripped off the soil and unconsolidated surface sediment, revealing extensive exposures of striated and polished bedrock. These areas have deranged drainage (see Figure 15.23e), numerous lakes and swamps, low relief, extensive bedrock exposures, and little or no soil. They are referred to as *ice-scoured plains* (▶ Figure 17.14). Similar though smaller bedrock exposures are also widespread in the northern United States from Maine through Minnesota.

Section 17.5 Summary

- Glaciers are solids in motion that erode by bulldozing, plucking, and abrasion, and they transport and deposit huge quantities of soil and sediment. Valley glaciers transport sediment in all parts of the ice, whereas continental glaciers carry most of their sediment in the lower part of the ice.

- Erosion of mountains by valley glaciers produces several sharp, angular landforms such as cirques, arêtes, and horns as well as broadly rounded U-shaped glacial troughs, fiords, and hanging valleys.

- Continental glaciers abrade and bevel high areas, forming a smooth, rounded landscape known as an ice-scoured plain.

17.6 Deposits of Glaciers

Glacial Drift

You now know how glaciers erode and transport sediment and soil and that glacial erosion yields several distinctive landforms. But what about glacial deposits? What kinds do geologists recognize, and are they as distinctive as the erosional features? Both valley and continental glaciers deposit sediment as **glacial drift,** a general term for all deposits resulting from glacial activity.

A vast sheet of Pleistocene glacial drift is present in the northern tier of the United States and adjacent parts of Canada. Smaller but similar deposits are found where valley glaciers existed or remain active. The appearance of these deposits may not be as inspiring as some landforms resulting from glacial erosion, but they are important as reservoirs of groundwater, and in many areas they are exploited for their sand and gravel.

One conspicuous aspect of glacial drift is rock fragments of various sizes that were obviously not derived from the underlying bedrock. These **glacial erratics,** as they are called, were derived from some distant source and transported to their present location (▶ Figure 17.15). A good example is the popular decorative stone called puddingstone found in Michigan, Ohio, Illinois, Indiana, Iowa, and Missouri. It consists of quartzite with pieces of red jasper eroded from rock exposures near Sault Sainte Marie in Ontario, Canada. Some erratics are gigantic. For instance, the Madison Boulder in New Hampshire and Daggett Rock in Maine weigh about 4550 and 7270 metric tons, respectively. The glacial erratic shown in ▶ Figure 17.15b is not the world's largest, but it is one of many in a narrow belt of erratics stretching more than 640 km from their source.

As noted, *glacial drift* is a general term, and geologists define two types of drift: till and stratified drift. **Till**

▶ Figure 17.15 Glacial Erratics

a A glacial erratic in the making. This boulder on the surface of the Mendenhall Glacier in Alaska will eventually be deposited far from its source.

b This is the Airdrie erratic near Calgary, Alberta, Canada. It is about 14 m long and 7.6 m high. The world's largest erratic is about one-third again as large as this one.

consists of sediments deposited directly by glacial ice. They are not sorted by particle size or density, and they exhibit no stratification. The till of both valley and continental glaciers is similar, but that of continental glaciers is much more extensive and usually has been transported much farther.

As opposed to till, **stratified drift** is layered—that is, stratified—and it invariably exhibits some degree of sorting by particle size. As a matter of fact, most stratified drift is actually layers of sand and gravel or mixtures thereof that accumulated in braided stream channels. In Chapter 15 we mentioned that streams issuing from melting glaciers are commonly braided because they receive more sediment than they can effectively transport.

Landforms Composed of Till

Landforms composed of till include several types of *moraines* and elongated hills know as *drumlins*.

End Moraines The terminus of any glacier may become stabilized in one position for some period of time, perhaps a few years or even decades. Stabilization of the ice front does not mean that the glacier has ceased to flow, only that it has a balanced budget (▶ Figure 17.7). When an ice front is stationary, flow within the glacier continues, and any sediment transported within or upon the ice is dumped as a pile of rubble at the glacier's terminus (▶ Figures 17.16 and 17.17a). These deposits are **end moraines,** which continue to grow as long as the ice front remains stationary. End moraines of valley glaciers are crescent-shaped ridges of till spanning the valley occupied by the glacier. Those of continental glaciers similarly parallel the ice front but are much more extensive.

Following a period of stabilization, a glacier may advance or retreat, depending on changes in its budget. If it advances, the ice front overrides and modifies its former moraine. If it has a negative budget, though, the ice front retreats toward the zone of accumulation. As the ice front recedes, till is deposited as it is liberated from the melting ice and forms a layer of **ground moraine**. Ground moraine has an irregular, rolling topography, whereas end moraine consists of long ridgelike accumulations of sediment.

■ *How does a recessional moraine form?*

After a glacier has retreated for some time, its terminus may once again stabilize, and it deposits another end moraine. Because the ice front has receded, such moraines are called **recessional moraines** (▶ Figure 17.18). During the Pleistocene Epoch, continental glaciers in the mid-continent region extended as far south as southern Ohio, Indiana, and Illinois. Their outermost end moraines, marking the greatest extent of the glaciers, go by the special name **terminal moraine** (valley glaciers also deposit terminal moraines).

▶ **Figure 17.16 End Moraines**

ⓐ An end moraine that is also a terminal moraine near Mammoth Lakes, California. The glacier is gone, but you can see part of the valley it occupied.

ⓑ Excellent closeup view of an end moraine deposited by the Russell Glacier in western Greenland.

As the glaciers retreated from the positions where their terminal moraines were deposited, they temporarily stopped retreating numerous times and deposited dozens of recessional moraines.

Lateral and Medial Moraines Valley glaciers transport considerable sediment along their margins, much of it abraded and plucked from the valley walls, but a significant amount falls or slides onto the glacier's surface by mass wasting processes. In any case, this sediment is transported and deposited as long ridges of till called **lateral moraines** along the margin of the glacier (▶ Figure 17.17).

Where two lateral moraines merge, as when a tributary glacier flows into a larger glacier, a **medial moraine** forms (▶ Figure 17.17b). A large glacier will often have several dark stripes of sediment on its surface, each of which is a

▶ **Figure 17.17 Types of Moraines** A moraine is a mound or ridge of unstratified till. The types shown here—lateral, medial, and end moraines—are defined by their position.

a Lateral moraine and end moraine deposited by a glacier on Baffin Island in Canada.

b Lateral and medial moraines on the Bernard Glacier in the St. Elias Mountains in Alaska.

medial moraine. Although medial moraines are identified by their position on a valley glacier, they are, in fact, formed from the coalescence of two lateral moraines. One can determine how many tributaries a valley glacier has by the number of its medial moraines (▶ Figure 17.17b).

Drumlins In many areas where continental glaciers deposited till, the till has been reshaped into elongated hills known as **drumlins** (▶ Figure 17.18). Some drumlins are as large as 50 m high and 1 km long, but most are much smaller. From the side, a drumlin looks like an inverted spoon, with the steep end on the side from which the glacial ice advanced and the gently sloping end pointing in the direction of ice movement. Drumlins are rarely found as single, isolated hills; instead, they occur in *drumlin fields* that contain hundreds or thousands of drumlins. Drumlin fields are found in several states and Ontario, Canada, but perhaps the finest example is near Palmyra, New York.

■ *What accounts for the origin of drumlins?*

According to one hypothesis, drumlins form when till beneath a glacier is reshaped into streamlined hills as the ice moves over it by plastic flow. Another hypothesis holds that huge floods of glacial meltwater modify till into drumlins.

Landforms Composed of Stratified Drift

Stratified drift is a type of glacial deposit that exhibits sorting and layering, both indications that it was deposited by running water. Stratified drift is deposited by streams discharging from both valley and continental glaciers, but as one would expect, it is more extensive in areas of continental glaciation.

Outwash Plains and Valley Trains Glaciers discharge meltwater laden with sediment most of the time, except perhaps during the coldest months. This meltwater forms a series of braided streams that radiate out from the front of continental glaciers over a wide region. So much sediment is supplied to these streams that much of it is deposited within their channels as sand and gravel bars. The vast blanket of sediments so formed is an **outwash plain**.

Valley glaciers also discharge large amounts of meltwater and, like continental glaciers, have braided streams extending from them. However, these streams are confined to the lower parts of glacial troughs, and their long, narrow deposits of stratified drift are known as **valley trains**.

Outwash plains, valley trains, and some moraines commonly contain numerous circular to oval depressions, many of which contain small lakes. These depressions are *kettles* that form when a retreating glacier leaves a block of ice that is subsequently partly or wholly buried (▶ Figure 17.18). When the ice block eventually melts, it leaves a depression; if the depression extends below the water table, it becomes the site of a small lake. Some outwash plains have so many kettles that they are called *pitted outwash plains*.

Kames and Eskers **Kames** are conical hills of stratified drift up to 50 m high (▶ Figures 17.18 and 17.19a). Many kames form when a stream deposits sediment in a depression on a glacier's surface; as the ice melts, the deposit is

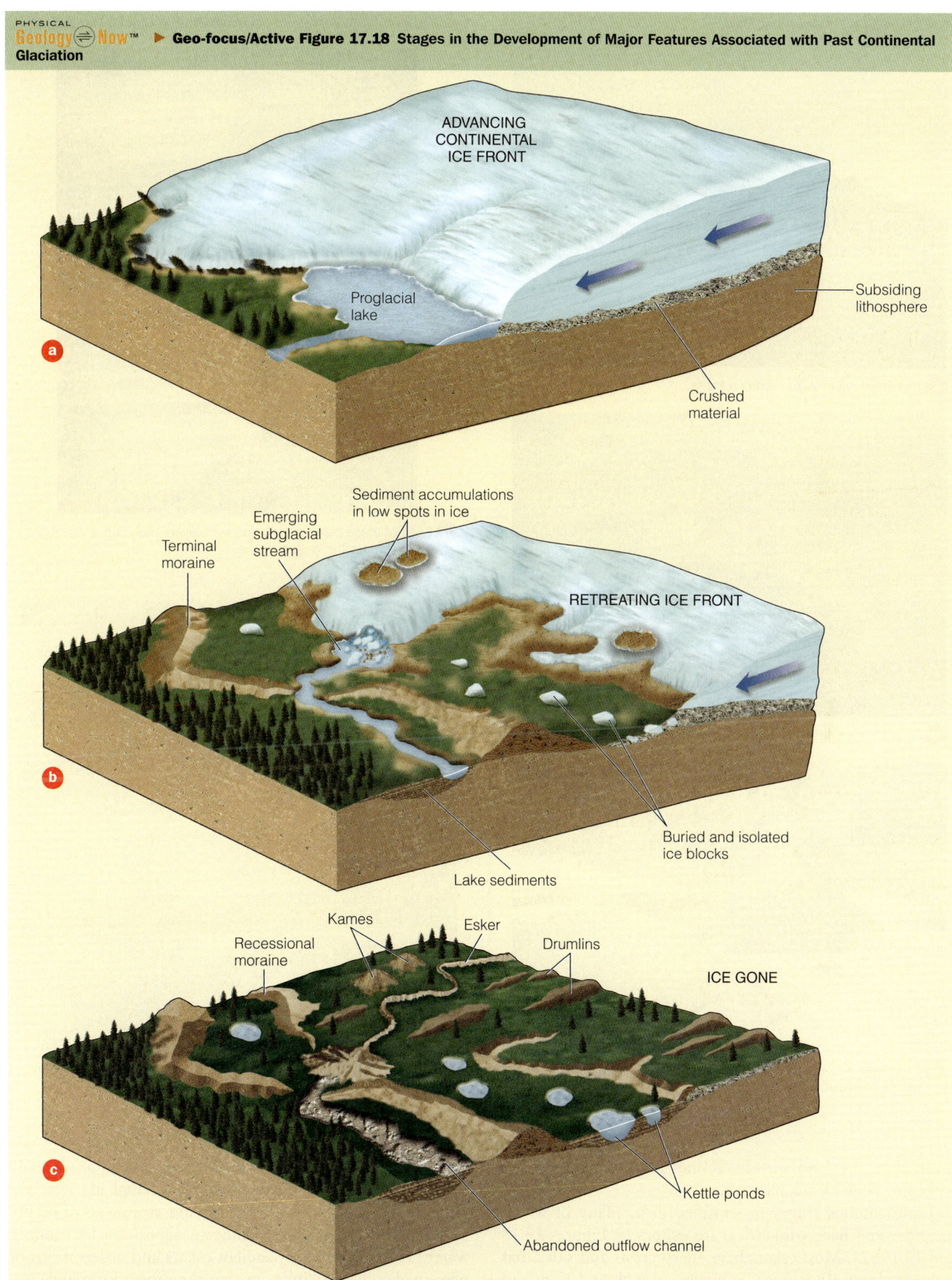

Geo-focus/Active Figure 17.18 Stages in the Development of Major Features Associated with Past Continental Glaciation

(a–c) Moraines, eskers, and drumlins form during glaciation, though eskers and drumlins originate under the ice cover. Kames and kettles develop at the end of glaciation.

▶ Figure 17.19 Kames and Eskers

a This small hill in Wisconsin is a kame.

b An esker near Dahlen, North Dakota.

c This esker was exposed in 1992 from a melting glacier on Bylot Island, Nunavut, Canada.

▶ Figure 17.20 Varves and a Dropstone in Glacial Deposits

a Each of these pairs of dark and light layers make up a varve, an annual deposit in a glacial lake.

b These varves have a dropstone that was probably liberated from floating ice.

lowered to the land surface. Kames also form in cavities within or beneath stagnant ice.

Long sinuous ridges of stratified drift, many of which meander and have tributaries, are **eskers** (▶ Figures 17.18 and 17.19b,c). Most eskers have sharp crests and sides that slope at about 30 degrees. Some are as high as 100 m and can be traced for more than 500 km. Eskers are usually found in areas once covered by continental glaciers and large valley glaciers. The sorting and stratification of the sediments in eskers clearly indicate deposition by running water. The properties of ancient eskers and observations of present-day glaciers indicate that they form in tunnels beneath stagnant ice. Excellent examples of eskers can be seen at Kettle Moraine State Park in Wisconsin and in several other states, but the most extensive eskers in the world are in northern Canada.

Deposits in Glacial Lakes

■ *What kinds of deposits are found in glacial lakes?*

Some lakes in areas of glaciation formed as a result of glaciers scouring out depressions; others occur where a stream's drainage was blocked; and others are the result of water accumulating behind moraines or in kettles. Regardless of how they formed, glacial lakes, like all lakes, are areas of deposition. Sediment may be carried into them and deposited as small deltas, but of special interest are the fine-grained deposits. Mud deposits in glacial lakes are commonly finely laminated (having layers less than 1 cm thick) and consist of alternating light and dark layers. Each light–dark couplet is a *varve* (▶ Figure 17.20a), which represents an annual episode of deposition. The light layer formed during the spring and summer and consists of silt and clay; the dark layer formed during the winter when the smallest particles of clay and organic matter settled from suspension as the lake froze over. The number of varves indicates how many years a glacial lake has existed.

Another distinctive feature of glacial lakes with varves is *dropstones* (▶ Figure 17.20b). These are pieces of gravel, some of boulder size, in otherwise very fine-grained deposits. The presence of varves indicates that currents and turbulence in these lakes were minimal; otherwise, clay and organic matter would not have settled from suspension. How then can we account for dropstones in a low-energy environment? Most of them were probably carried into the lakes by icebergs that eventually melted and released sediment contained in the ice.

Section 17.6 Summary

- Glacial drift applies to all deposits of glaciers, but geologists recognize two types of drift—till and stratified drift. Ridgelike accumulations of till, called moraines, are terminal, recessional, lateral, or medial, depending on their positions.

- Drumlins are also composed of till that was shaped into streamlined hills by glaciers or by floods of glacial meltwater.

- Stratified drift is sediment deposited by meltwater streams as outwash plains adjacent to continental glaciers, or valley trains near valley glaciers. Ridges called eskers and conical hills known as kames are also made up of stratified drift.

- Deposits in glacial lakes consist of dark and light couplets of fine-grained sediment called varves.

What Would You Do?

During a visit to the Pacific Northwest, you notice that a stream has cut a gorge in which you observe the following. In the lower part of the gorge is a bedrock projection with features similar to those shown in Figure 17.10. Overlying this bedrock is a sequence of thin (2–4 mm thick) alternating layers of dark and light very fine-grained materials (silt and clay) with a few boulders measuring 10 to 15 cm across. And in the uppermost part of the gorge is a deposit of mud, sand, and gravel showing no layering or sorting. How would you decipher the geologic history of these deposits?

17.7 The Ice Age (The Pleistocene Epoch)

■ *When was the Ice Age, and which areas were most heavily affected by glaciers?*

You already know that the popular name for the Pleistocene Epoch is the Ice Age, a time from 1.8 million to 10,000 years ago during which glaciers were much more widespread than they are now, especially on the Northern Hemisphere continents. Geologists identify at least four major episodes of Pleistocene glaciation in North America, and six or seven glacial advances and retreats took place in Europe. Now there is evidence from deep-sea cores for more than 20 warm–cold cycles during the Ice Age. Based on the best available evidence, the Pleistocene ended about 10,000 years ago, but scientists do not know whether the present warmer times will persist indefinitely or Earth will enter another glacial interval.

In hindsight, it is hard to believe that so many competent naturalists of the 1800s were skeptical that widespread glaciers existed on the northern continents during the not too distant past. Many naturalists invoked the biblical flood to account for the large boulders throughout Europe that occur far from their sources. Others thought that the boulders were rafted to their present positions by icebergs floating in floodwaters. It was not until 1837 that the Swiss naturalist Louis Agassiz argued convincingly that the displaced boulders, many sedimentary deposits, polished and striated bedrock, and many valleys in Europe resulted from huge ice masses moving over the land.

The onset of glacial conditions really began about 40 million years ago (MYA) when surface ocean waters at high southern latitudes suddenly cooled with the tectonic drift on Antarctica into an isolated polar position. By about 38 MYA, glaciers had formed in Antarctica, but a continuous ice sheet did not develop there until 15 MYA. Following a brief warming trend during the Late Tertiary Period, ice sheets began forming in the Northern Hemisphere, and by 1.8 MYA the Pleistocene Ice Age was under way. At their greatest extent,

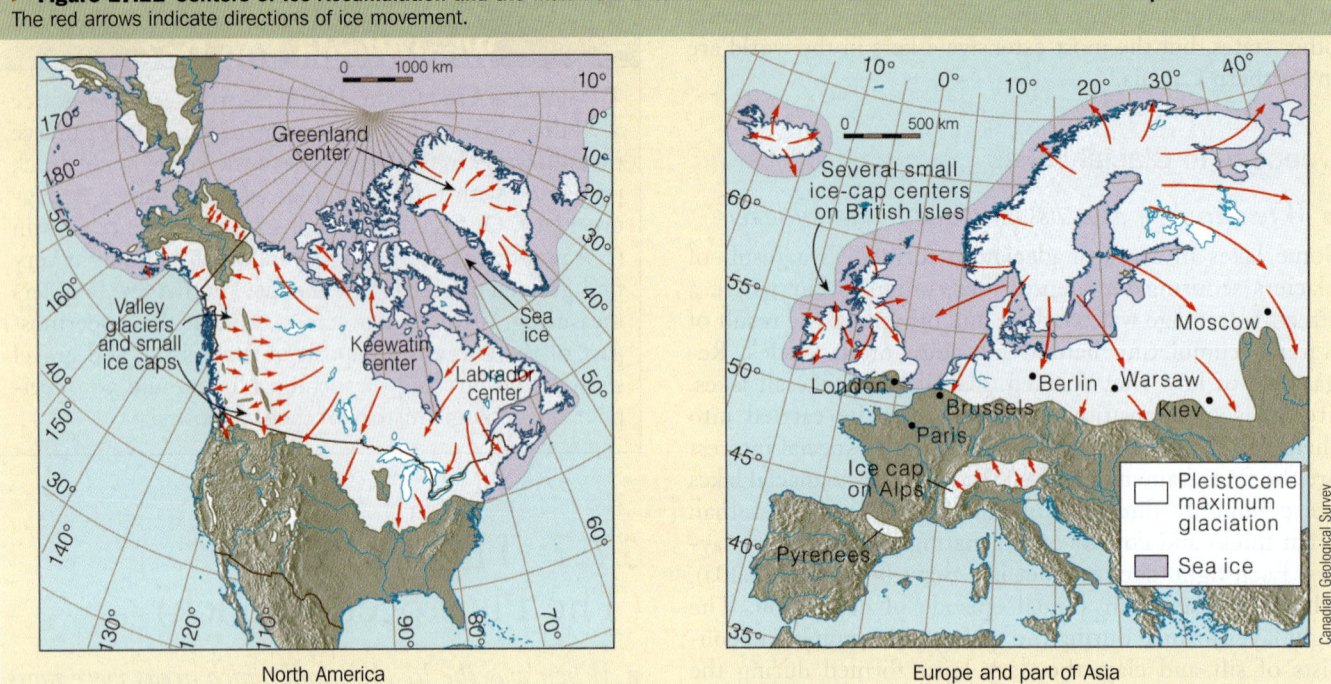

▶ **Figure 17.21 Centers of Ice Accumulation and the Maximum Extent of Pleistocene Glaciers on the Northern Hemisphere Continents** The red arrows indicate directions of ice movement.

North America

Europe and part of Asia

Pleistocene glaciers covered about three times as much of Earth's surface as they do now (▶ Figure 17.21). Large areas of North America were covered by glacial ice, as were Greenland, Scandinavia, Great Britain, Ireland, and a large part of northern Russia. Mountainous areas also experienced an expansion of valley glaciers and the development of ice caps.

Pleistocene Climates

As you would expect, the climatic effects responsible for Ice Age glaciers were worldwide. Contrary to popular belief, though, the world was not as frigid as it is commonly portrayed in cartoons and movies. In addition, the climatic changes responsible for glaciers did not take place in a few days or weeks as depicted in the popular movie *The Day After Tomorrow*. Nevertheless, during times of glacier growth and expansion, those areas in the immediate vicinity of glaciers had short summers and long, wet winters, and fierce cold winds descended from the glaciers.

Areas outside the glaciated regions experienced varied climates. During times of glacial growth, lower ocean temperatures reduced evaporation so that most of the world was drier than it is today, but some areas that are arid now were much wetter. For example, as the cold belts at high latitudes expanded, the temperate, subtropical, and tropical zones were compressed toward the equator, and the rain that now falls on the Mediterranean shifted so that it fell on the Sahara of North Africa, enabling lush forests to grow in what is now desert. California and the arid southwestern United States were also wetter because a high-pressure zone over the northern ice sheet deflected Pacific winter storms south.

Following the Pleistocene, mild temperatures prevailed between 8000 and 6000 years ago. After this warm period, conditions gradually became cooler and moister, favoring the growth of valley glaciers on the Northern Hemisphere continents. Careful studies of the deposits at the margins of present-day glaciers reveal that during the last 6000 years (a time called the *Neoglaciation*), glaciers expanded several times. The last expansion, which took place between 1500 and the middle to late 1800s, was the Little Ice Age (see Section 17.1).

Pluvial and Proglacial Lakes

■ *What are pluvial and proglacial lakes?*

During the Pleistocene, many of the basins in the western United States contained large lakes that formed as a result of more precipitation and overall cooler temperatures (especially during the summer), which lowered the evaporation rate (▶ Figure 17.22). The largest of these *pluvial lakes*, as they are called, was Lake Bonneville, which attained a maximum size of 50,000 km^2 and a depth of at least 335 m. The vast salt deposits of the Bonneville Salt Flats west of Salt Lake City in Utah formed as parts of this ancient lake dried up; Great Salt Lake is the remnant of this once much larger lake.

Another large pluvial lake existed in Death Valley, California, which is now the hottest, driest place in North America. During the Pleistocene, Death Valley received enough rainfall to maintain a lake 145 km long and 178 m deep. When the lake waters evaporated, the dissolved salts were precipitated on the valley floor; some of these evaporite deposits, especially borax, are important mineral resources (see Chapter 18).

In contrast to pluvial lakes, which form far from glaciers, *proglacial lakes* form when meltwater accumulates along the margin of a glacier. In fact, one shoreline of a proglacial lake is the ice front itself. Lake Agassiz, named in honor of the natu-

▶ **Figure 17.22 Pleistocene Lakes in the Western United States** Great Salt Lake in Utah and Pyramid Lake in Nevada are shrunken remnants of extensive ancient lakes. Of all the lakes shown, only Lake Columbia and Lake Missoula were proglacial lakes. When the 600-m-high ice dam impounding Lake Missoula failed, the lake drained through Lake Columbia and scoured out the scablands of eastern Washington.

sheet retreated north during the Late Pleistocene, the ice front periodically stabilized, and numerous recessional moraines were deposited. By about 14,000 years ago, parts of the Lake Michigan and Lake Erie basins were ice free, and glacial meltwater began to form proglacial lakes (▶ Figure 17.23). As the ice sheet continued to retreat—although periodically interrupted by minor readvances of the ice front—the Great Lakes basins were uncovered, and the lakes expanded until they eventually reached their present size and configuration. Currently, the Great Lakes contain nearly 23,000 km³ of water, about 18% of the water in all freshwater lakes worldwide.

Another proglacial lake of interest is glacial Lake Missoula in Montana, which formed when ice dammed the drainage of the Clark Fork River in Idaho (▶ Figure 17.24). The shorelines of this 7800-km² lake are clearly visible on the mountainsides around Missoula, Montana. On several dozen occasions the ice dam collapsed, resulting in huge floods responsible for gravel ridges, called giant ripple marks, as well as the *scablands* of eastern Washington where surface deposits were scoured, thus exposing the underlying bedrock (▶ Figure 17.24).

Glaciation and Changes in Sea Level

More than 70 million km³ of snow and ice covered the continents during the maximum extent of Pleistocene glaciers. The storage of ocean waters in glaciers lowered sea level 130 m and exposed vast areas of the continental shelves, which were quickly covered with vegetation. Indeed, a land bridge existed across the Bering Strait from Alaska to Siberia. The term *bridge* brings to mind a long, narrow span linking two areas, but the Bering land bridge, also known as *Beringia*, was actually an area about 1600 km wide between Siberia and Alaska (▶ Figure 17.25). Native Americans crossed the Bering land bridge, and various animals migrated between the continents; the American bison, for example, migrated from Asia. The British Isles were connected to Europe during the glacial intervals because the shallow floor of the North Sea was above sea level. When the glaciers disappeared, these land bridges were flooded, drowning the plants and forcing the animals to migrate.

ralist Louis Agassiz, was a large proglacial lake covering about 250,000 km², mostly in Manitoba, Saskatchewan, and Ontario, Canada, but extending into North Dakota and Minnesota. It persisted until the glacial ice along its northern margin melted, at which time the lake drained northward into Hudson Bay.

Before the Pleistocene, no large lakes existed in the Great Lakes region, which was then an area of lowlands with broad stream valleys draining to the north. As the glaciers advanced southward, they eroded the valleys more deeply, forming what were to become the Great Lakes basins; four of the five basins were eroded below sea level.

At their greatest extent, the glaciers covered the entire Great Lakes region and extended far to the south. As the ice

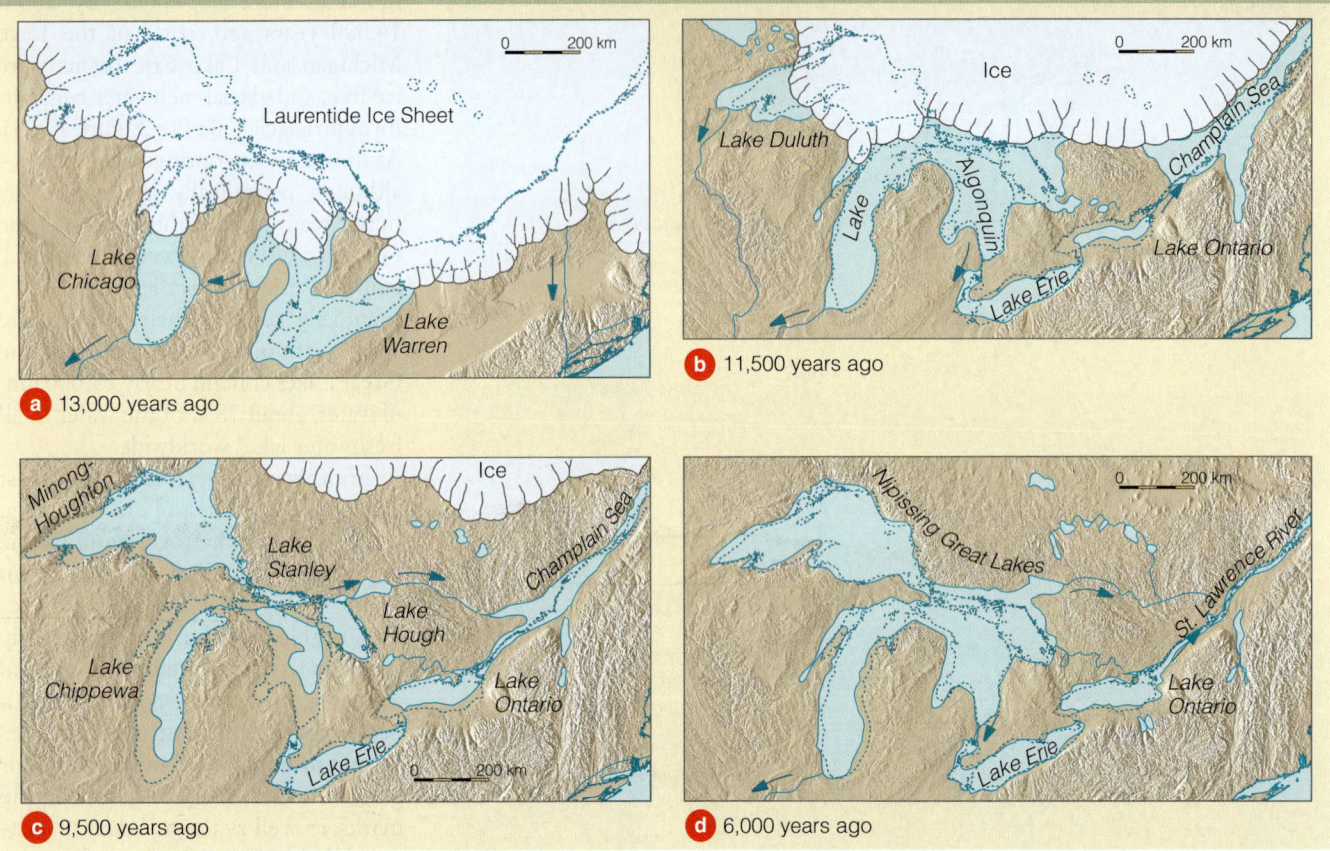

▶ **Figure 17.23 Four Stages in the Evolution of the Great Lakes** As the glacial ice retreated northward, the lake basins began filling with meltwater. The dotted lines indicate the present-day shorelines of the lakes. Source: From V. K. Prest, *Economic Geology Report 1*, 5d, 1970, pp. 90–91 (fig. 7-6), Geological Survey of Canada. Department of Energy, Mines, and Resources, reproduced with permission of Minister of Supply and Services.

Lower sea level during the Pleistocene also affected the base level of most rivers and streams. When sea level dropped, streams eroded deeper valleys as they sought to adjust to a new lower base level (see Chapter 15). Stream channels in coastal areas were extended and deepened along the emergent continental shelves. When sea level rose at the end of the Pleistocene, the lower ends of the valleys along the East Coast of North America were flooded and are now important harbors (see Chapter 19). On the West Coast, two rivers joined just east of present-day San Francisco and flowed to the seashore, which was about 30 km to the west of its present location.

A tremendous quantity of water is still stored on land in glaciers. If these glaciers should melt, sea level would rise about 70 m, flooding all coastal areas of the world where many of the large population centers are located.

Glaciers and Isostasy

In Chapter 11 we discussed the principle of isostasy and noted that Earth's crust responds to an increased load by subsiding, and it rises when a load on it is reduced. When the Pleistocene ice sheets formed and increased in size, their weight caused the crust to slowly subside deeper into the mantle. In some places, the surface was depressed as much as 300 m below preglacial elevations. As the ice sheets disappeared, the down-warped areas gradually rebounded to their former positions. As noted in Chapter 11, parts of Scandinavia are still rebounding at a rate of about 1 m per century (see Figure 11.18). And more than 100 m of isostatic rebound has taken place in northeastern Canada during the last 6000 years.

We noted previously that the Great Lakes evolved as the glaciers retreated to the north. As one would expect, isostatic rebound began as the ice front retreated north, but rebound began first in the southern part of the region because that area was free of ice first. Furthermore, the greatest loading by glaciers, and hence the greatest crustal depression, occurred farther north in Canada in the zones of accumulation. For these reasons, rebound has not been evenly distributed over the entire glaciated area: It increases in magnitude from south to north. As a result of this uneven isostatic rebound, coastal features in the Great Lakes region, such as old shorelines, are now elevated higher above their former levels in the north and thus slope to the south.

Section 17.7 Summary

- Several intervals of widespread glaciation separated by interglacial periods took place during the Pleistocene Epoch. At times of glacial advances, glaciers covered

▶ **Figure 17.24 Glacial Lake Missoula Emptied Rapidly Several Times When the Ice Dam Impounding the Lake Failed**

a This rumpled surface is made up of gravel ridges, the so-called giant ripple marks that formed as glacial Lake Missoula drained across this area near Camas Prairie, Montana.

b Palouse Falls in Washington is in a canyon eroded by the floodwaters from glacial Lake Missoula. See also Figure 17.22.

about 30% of the land surface, especially in the Northern Hemisphere and Antarctica.

• Areas far beyond the Ice Age glaciers were also affected; climate belts were compressed toward the equator, pluvial lakes existed in what are now arid regions, and sea level was 130 m lower than it is now.

• Loading of the crust by glaciers caused isostatic subsidence, and since the Pleistocene several areas have rebounded or continue to show isostatic rebound.

17.8 Causes of the Ice Age

We discussed the conditions necessary for a glacier to form earlier in this chapter: More snow falls than melts during the warm season, thus accounting for a net accumulation of snow and ice over the years. But this really does not address the broader question of what causes ice ages—that is, times of much more extensive glaciation. Actually, we need to

▶ **Figure 17.25 The Bering Land Bridge** During the Pleistocene, sea level was as much as 130 m lower than it is now, and a broad area called the Bering land bridge (Beringia) connected Asia and North America. It was exposed above sea level during times of glacial advances and served as a corridor for the migration of people, animals, and plants.

☐ Bering land bridge

address not only what causes ice ages but also why there have been so few episodes of widespread glaciation in all of Earth history. Only during the Late Proterozoic Eon, the Pennsylvanian and Permian periods, and the Pleistocene Epoch has Earth had glaciers on a grand scale.

For more than a century, scientists have attempted to develop a comprehensive theory explaining all aspects of ice ages, but they have not yet been completely successful. One reason for their lack of success is that the climatic changes responsible for glaciation, the cyclic occurrence of glacial–interglacial episodes, and short-term events such as the Little Ice Age operate on vastly different time scales.

Only a few periods of glaciation are recognized in the geologic record, each separated from the others by long intervals of mild climate. Such long-term climatic changes probably result from slow geographic changes related to plate tectonic activity. Moving plates carry continents to high latitudes where glaciers exist, provided they receive enough precipitation as snow. Plate collisions, the subsequent uplift of vast areas far above sea level, and the changing atmospheric and oceanic circulation patterns caused by the changing shapes and positions of plates also contribute to long-term climatic change.

The Milankovitch Theory

■ *How does the Milankovitch theory account for the onset of glacial episodes?*

Changes in Earth's orbit as a cause of intermediate-term climatic events were first proposed during the mid-1800s, but the idea was made popular during the 1920s by the Serbian astronomer Milutin Milankovitch. He proposed that minor irregularities in Earth's rotation and orbit are sufficient to alter the amount of solar radiation received at any given latitude and hence bring about climate changes. Now called the **Milankovitch theory,** it was initially ignored but has received renewed interest since the 1970s and is now widely accepted.

Milankovitch attributed the onset of the Pleistocene Ice Age to variations in three aspects of Earth's orbit. The first is *orbital eccentricity*, which is the degree to which Earth's orbit around the sun changes over time (▶ Figure 17.26a). When the orbit is nearly circular, both the Northern and Southern Hemispheres have similar contrasts between the seasons. However, if the orbit is more elliptic, hot summers and cold winters will occur in one hemisphere, while warm summers and cool winters will take place in the other hemisphere. Calculations indicate a roughly 100,000-year cycle between times of maximum eccentricity, which corresponds closely to the 20 warm–cold climatic cycles that took place during the Pleistocene.

Milankovitch also pointed out that the angle between Earth's axis and a line perpendicular to the plane of the ecliptic shifts about 1.5 degrees from its current value of 23.5 degrees during a 41,000-year cycle (▶ Figure 17.26b). Although changes in *axial tilt* have little effect on equatorial latitudes, they strongly affect the amount of solar radiation received at high latitudes and the duration of the dark period at and near Earth's poles. Coupled with the third aspect of Earth's orbit, precession of the equinoxes, high latitudes might receive as much as 15% less solar radiation, certainly enough to affect glacial growth and melting.

The last aspect of Earth's orbit that Milankovitch cited is *precession of the equinoxes*, which refers to a change in the time of the equinoxes. At present, the equinoxes take place on about March 21 and September 21 when the Sun is directly over the equator. But as Earth rotates on its axis, it also wobbles as its axial tilt varies 1.5 degrees from its current value, thus changing the time of the equinoxes. Taken alone, the time of the equinoxes has little climatic effect, but changes in Earth's axial tilt also change the times of *aphelion* and *perihelion*, which are, respectively, when Earth is farthest from and closest to the Sun during its orbit (▶ Figure 17.26c). Earth is now at perihelion, closest to the Sun, during Northern Hemisphere winters, but in about 11,000 years perihelion will be in July. Accordingly, Earth will be at aphelion, farthest from the Sun, in January and have colder winters.

Continuous variations in Earth's orbit and axial tilt cause the amount of solar heat received at any latitude to vary slightly through time. The total heat received by the planet changes little, but according to Milankovitch, and now many scientists agree, these changes cause complex climatic variations and provided the triggering mechanism for the glacial–interglacial episodes of the Pleistocene.

Short-Term Climatic Events

Climatic events with durations of several centuries, such as the Little Ice Age, are too short to be accounted for by plate tectonics or Milankovitch cycles. Several hypotheses have been proposed, including variations in solar energy and volcanism.

Variations in solar energy could result from changes within the Sun itself or from anything that would reduce the amount of energy Earth receives from the Sun. The latter could result from the solar system passing through clouds of interstellar dust and gas or from substances in the atmosphere reflecting solar radiation back into space. Records kept over the past 90 years indicate that during this time the

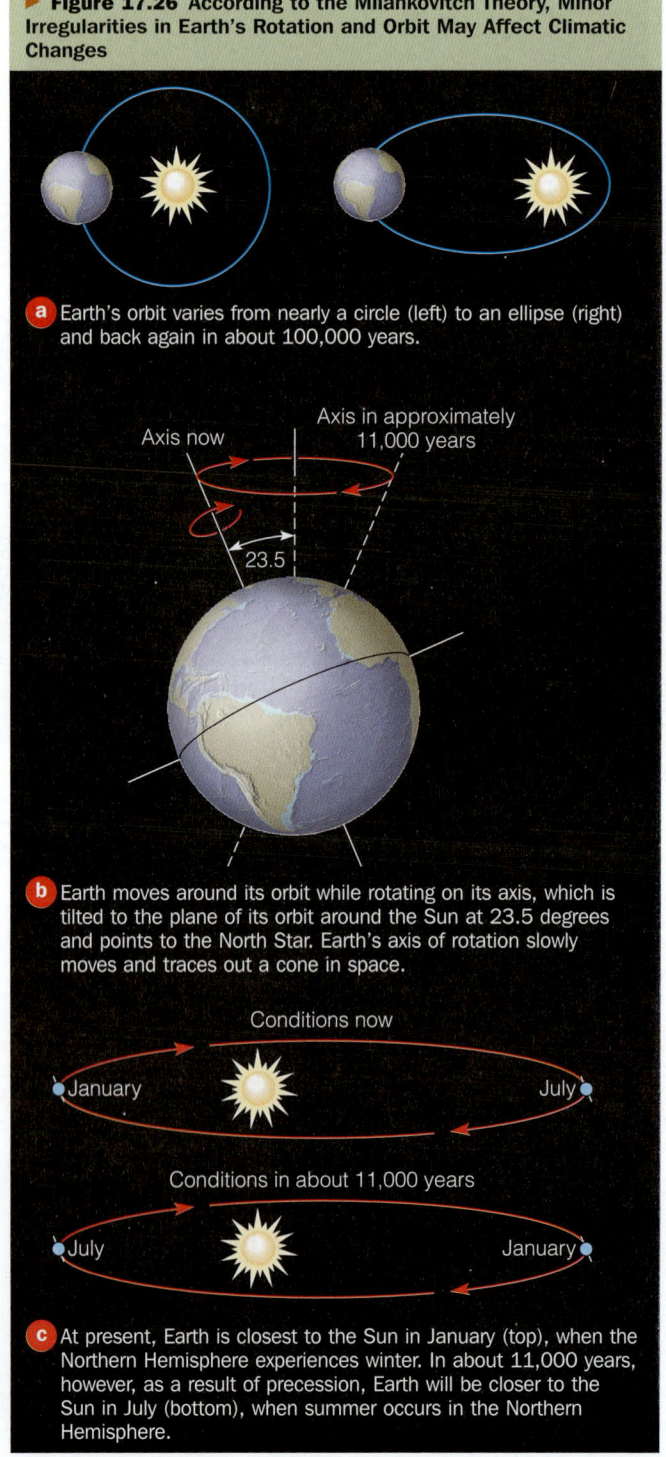

▶ **Figure 17.26 According to the Milankovitch Theory, Minor Irregularities in Earth's Rotation and Orbit May Affect Climatic Changes**

a Earth's orbit varies from nearly a circle (left) to an ellipse (right) and back again in about 100,000 years.

b Earth moves around its orbit while rotating on its axis, which is tilted to the plane of its orbit around the Sun at 23.5 degrees and points to the North Star. Earth's axis of rotation slowly moves and traces out a cone in space.

c At present, Earth is closest to the Sun in January (top), when the Northern Hemisphere experiences winter. In about 11,000 years, however, as a result of precession, Earth will be closer to the Sun in July (bottom), when summer occurs in the Northern Hemisphere.

amount of solar radiation has varied only slightly. Although variations in solar energy may influence short-term climatic events, such a correlation has not been demonstrated.

During large volcanic eruptions, tremendous amounts of ash and gases are spewed into the atmosphere, where they reflect incoming solar radiation and thus reduce atmospheric temperatures. Small droplets of sulfur gases remain in the atmosphere for years and can have a significant effect on climate. Several large-scale volcanic events have occurred, such as the 1815 eruption of Tambora, and are known to have had climatic effects. However, no relationship between periods of volcanic activity and periods of glaciation has yet been established.

Section 17.8 Summary

- Glacial intervals separated by tens or hundreds of millions of years probably result from changing positions of tectonic plates, which cause changes in oceanic and climatic circulation patterns.

- Currently, the Milankovitch theory is widely accepted as the explanation for glacial–interglacial intervals.

- Two proposed causes of short-term climatic changes, such as during the Little Ice Age, are variations in the amount of solar radiation received by Earth and volcanism.

Review Workbook

ESSENTIAL QUESTIONS SUMMARY

17.1 Introduction

■ *What was the Little Ice Age, when did it take place, and which areas were most heavily affected?*

The Little Ice Age was a time of colder winters and short, wet summers during which glaciers advanced and sea ice persisted for long periods of time. It occurred from about 1500 to the middle to late 1800s and had its greatest effect on Northern Europe and Iceland.

17.2 The Kinds of Glaciers

■ *How does a glacier differ from floating sea ice or a permanent snowfield?*

By definition glaciers are moving bodies of ice on land, whereas sea ice is nothing more than frozen seawater. Snowfields are on land but they show no movement.

■ *What are the distinguishing features of valley glaciers and continental glaciers?*

Valley glaciers are long, narrow tongues of ice that move from higher to lower elevations within a valley, whereas continental glaciers, which cover vast areas and flow outward in all directions from an area of accumulation, are unrelated to the underlying topography.

17.3 Glaciers—Moving Bodies of Ice on Land

■ *What role do glaciers play in the hydrologic cycle?*

Glaciers are a reservoir in the hydrologic cycle where water is stored for long periods as it moves from the oceans to land and back to the oceans.

■ *What two mechanisms account for glacial movement?*

If a slope is present glaciers may slide over their underlying surface, a phenomenon called basal slip, but most of their movement is accomplished by plastic flow, a type of deformation that takes places in response to stress.

■ *Where are present-day glaciers found?*

Glaciers exist only where there is sufficient precipitation in the form of snow and where temperatures are low enough so that they do not melt away. These conditions prevail in high mountains, some even near the equator, and at high latitudes as in Alaska, the Canadian Arctic islands, Greenland, and Antarctica.

17.4 The Glacial Budget—Accumulation and Wastage

■ *How are the advance and retreat of a glacier's terminus controlled by its budget?*

If a glacier has a positive budget—that is, gains exceed losses—its terminus advances, but if it has a negative budget—losses exceed gains—the terminus retreats.

■ *What are glacial surges?*

During a glacial surge, accelerated flow in a glacier causes its terminus to advance rapidly and its surface breaks into a maze of crevasses. One theory for the cause of surges holds that water-saturated sediment below a glacier allows it to slide, but another theory holds that a glacier's slope increases due to thickening in the zone of accumulation and thinning in the zone of wastage.

17.5 Erosion and Transport by Glaciers

■ *What are the specific processes that account for erosion by glaciers?*

Glaciers effectively push or bulldoze loose materials in their paths, but they also erode by abrasion—that is, the movement of sediment-laden ice over rock surfaces. In addition, glaciers erode by plucking when ice freezes in or around bedrock projections and pulls them loose.

■ *How do U-shaped glacial troughs form?*

Valleys in mountains typically have V-shaped profiles, but when eroded by glaciers they are deepened and widened so that they have flat or gently rounded valley floors and near-vertical valley walls. At the upper end of a glacial trough, a scoop-shaped depression, or cirque, eroded into a mountain side marks the place where a glacier formed and moved out into a trough.

■ *How do arêtes and horns originate?*

Both are landforms generated by valley glacier erosion. An arête is a serrated ridge between U-shaped glacial troughs or between adjacent cirques, whereas a horn is a pyramid-shaped peak left when headward erosion takes place by at least three glaciers on the same peak.

17.6 Deposits of Glaciers

■ *How does a recessional moraine form?*

Suppose that a glacier reaches its maximum extent and has a balanced budget. Accordingly it deposits a terminal moraine. If it then has a negative budget, its terminus retreats and perhaps becomes stabilized once again if its budget is balanced, in which case another end moraine is deposited but it is a recessional moraine.

■ *What accounts for the origin of drumlins?*
The origin of drumlins is not fully resolved, but according to one theory moving glacial ice shapes these streamlined hills of till. Another theory holds that floods of glacial meltwater are responsible for them.

■ *What kinds of deposits are found in glacial lakes?*
The most distinctive deposits in glacial lakes are varves that consist of couplets of dark and light, laminated, fine-grained sediment. The dark layers form during the winter when small particles of clay and organic matter are deposited. The light layers are made up of silt and clay that form during the warmer months.

17.7 The Ice Age (The Pleistocene Epoch)
■ *When was the Ice Age, and which areas were most heavily affected by glaciers?*
The Ice Age took place between 1.8 million and 10,000 years ago when glaciers covered about 30% of Earth's land surface. Glaciers existed in high mountains everywhere, but the areas most affected by continental glaciers were Antarctica, Greenland, and the Northern Hemisphere continents. For example, much of Canada and the northern tier of U.S. states were covered by glacial ice several times during the Pleistocene.

■ *What are pluvial and proglacial lakes?*
During the Ice Age large pluvial lakes formed in areas that are now arid because precipitation was greater and evaporation rates were lower. There were also many lakes along the margins of glaciers, the so-called proglacial lakes, such as the ancestors of the Great Lakes.

17.8 Causes of the Ice Age
■ *How does the Milankovitch theory account for the onset of glacial episodes?*
Milankovitch claimed that orbital eccentricity, changes in axial tilt, and precession of the equinoxes—that is, irregularities in Earth's rotation and orbit—bring about complex climatic changes that provide the triggering mechanism for glacial episodes.

ESSENTIAL TERMS TO KNOW

abrasion (p. 541)
arête (p. 546)
basal slip (p. 536)
cirque (p. 546)
continental glacier (p. 533)
drumlin (p. 550)
end moraine (p. 548)
esker (p. 552)
fiord (p. 543)
firn (p. 536)
glacial budget (p. 538)
glacial drift (p . 547)
glacial erratic (p. 547)

glacial ice (p. 536)
glacial polish (p. 541)
glacial striation (p. 541)
glacial surge (p. 540)
glaciation (p. 534)
glacier (p. 533)
ground moraine (p. 548)
hanging valley (p. 546)
horn (p. 546)
ice cap (p. 534)
kame (p. 550)
lateral moraine (p. 549)
medial moraine (p. 549)

Milankovitch theory (p. 558)
outwash plain (p. 550)
plastic flow (p. 536)
recessional moraine (p. 548)
stratified drift (p. 548)
terminal moraine (p. 548)
till (p. 547)
U-shaped glacial trough (p. 542)
valley glacier (p. 533)
valley train (p. 550)
zone of accumulation (p. 538)
zone of wastage (p. 538)

REVIEW QUESTIONS

1. How is it possible for glacial ice, which is a solid, to flow?
2. What is the firn limit on a glacier, and how does its position relate to a glacier's budget?
3. What kinds of evidence would indicate that glaciers once covered an ice-free area?
4. A valley glacier has a cross sectional area of 400,000 m² and a flow velocity of 2 m per day. How long will it take for 1 km³ of ice to move past a given point?
5. Explain how glaciers can erode below sea level but streams and rivers cannot.
6. How does glacial ice originate and why do geologists consider it to be rock?
7. What are medial and lateral moraines and how do they form?
8. How do terminal and recessional moraines form?
9. Draw side views of a drumlin and a roche moutonée, indicate the direction of ice movement, and explain how each originates.
10. Give a brief summary of the origin and evolution of the Great Lakes.

APPLY YOUR KNOWLEDGE

1. What evidence from glaciers indicates that global warming is taking place?

FIELD QUESTIONS

1. What glacial features are indicated in the images below?

2. Refer to Figure 17.11b and determine which of two possible directions the glacier moved that abraded this rock surface. How would you determine which of these directions that the glacier actually moved?

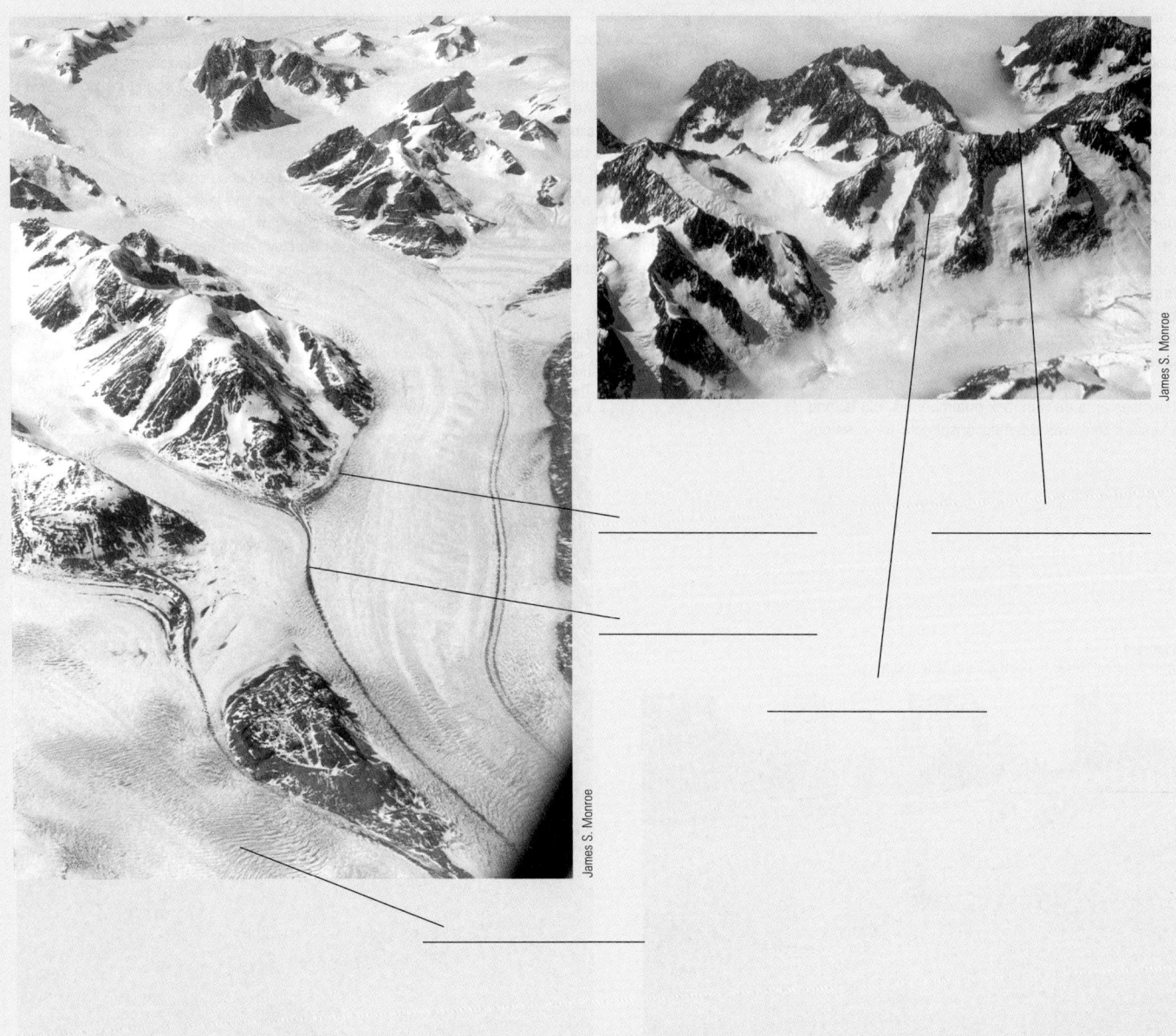

GEOLOGY MATTERS

GEOLOGY IN UNEXPECTED PLACES

Evidence of Glaciation in New York City

So far our discussion of glaciers has focused on Antarctica, Greenland, the Canadian Arctic islands, Alaska, and high mountains around the world. However, you can see the effects of erosion and deposition by continental glaciers in several urban areas, too. Good examples are found in several northern cities, but here we concentrate on the New York City area. As a matter of fact, the area now covered by the city as well as Long Island to the east are excellent places to see evidence of a continental glacier that was there as recently as 12,000 years ago.

If it were not for deposition of huge end moraines, most of Long Island and parts of New York City would be beneath the Atlantic Ocean. The end moraine in some parts of New York City is as high as 150 m, and it extends far to the west. Actually it is part of a system of moraines that extends hundreds of kilometers. In addition to depositing huge moraines in and around the city, the glacier left enormous chunks of ice as it retreated that became partly or completely buried. When these ice blocks melted, they formed kettle lakes locally known as ponds (see Figure 17.18).

That part of New York City designated as Central Park was an area with rock outcroppings, bluffs, and marshes that was deemed unsuitable for development. Nevertheless, when the city acquired the initial 700 acres of land for the park during the mid-1800s, it was necessary to displace about 1600 poor people who resided there. In 1863 the city acquired an additional 143 acres, bringing the park to its present size.

Many of the rocks resting on the surface in Central Park are glacial erratics—that is, boulders unlike the underlying bedrock that were carried far from their sources, much like those in Figure 17.15. You can see glacial polish, striations, and grooves on many of the erratics as well as on the exposed bedrock. In fact, the rock exposure in ▶ Figure 1 is a large roche moutonnée that formed as the "upstream" side of a bedrock projection was abraded and polished, but plucking accounts for the irregular "downstream" side (see Figure 17.10). The orientation of this roche moutonnée and the striations and grooves on the rocks indicate that the glacier moved from northwest to southeast across what is now Central Park.

Today New York Harbor is an important seaport, but during the maximum advance of the glaciers, the shoreline was nearly 130 km to the east where the Hudson River entered the sea. The river eroded a deep canyon on the continental shelf that was eroded farther seaward across the continental margin by turbidity currents. As sea level rose at the end of the Pleistocene, the Hudson Canyon was flooded and now lies below sea level (▶ Figure 2). However, varved deposits indicate that the moraines on Long Island and in New York City formed a dam across the ancestral Hudson River behind which glacial lakes were impounded. Periodically, though, the dam was breached, and as the lakes drained, erosion deepened the area now known as the Verrazano Narrows between Brooklyn and Staten Island, now the site of one of the longest suspension bridges in the world.

Figure 1
A roche moutonnée in Central Park, New York City.

Figure 2
Side-scan radar image of Hudson Canyon off the East Coast of North America. The black line shows sea level, but remember that sea level was far to the east during the Ice Age. The underwater topography has been exaggerated by a factor of 5 relative to the land topography. The area shown here measures about 330 km from west to east.

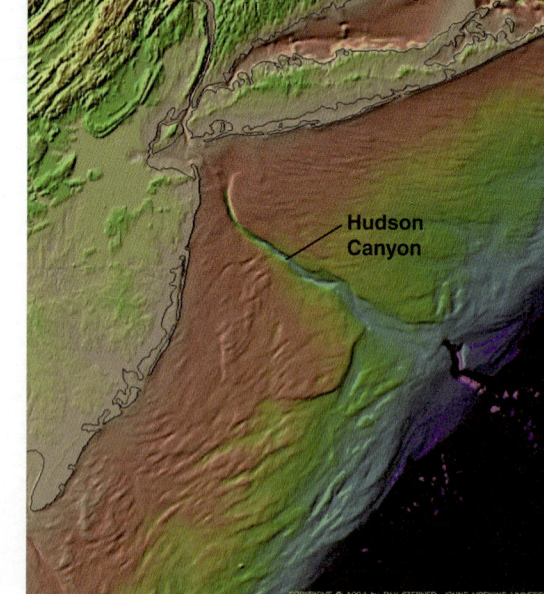

GEOLOGY IN YOUR LIFE

A Visit to Cape Cod, Massachusetts

GEOLOGY MATTERS

Cape Cod, some of which is designated as a National Seashore, is a popular tourist destination of both scenic and historic interest. Despite the persistent myth that the Pilgrims first landed at Plymouth Rock in 1620, their first landfall was actually on Cape Cod near what is now Provincetown (▶ Figure 1a). Nevertheless, the entire area has great symbolic interest as one of the first places in North America that was settled by Europeans. Unfortunately the Native Americans of Cape Cod nearly ceased to exist by 1764, mostly because of diseases.

Cape Cod resembles a large human arm extending into the Atlantic Ocean. It projects 65 km east from the mainland to the "elbow" and then extends another 65 km north where it resembles a half-curled hand (▶ Figure 1a). Cape Cod as well as Nantucket Island, Martha's Vineyard, and the Elizabeth Islands all owe their existence to deposition by Late Pleistocene (Ice Age) glaciers and the subsequent modification of these deposits by waves. The granite bedrock upon which the cape and nearby islands were built lies at depths of 90–150 m.

Between about 23,000 and 16,000 years ago, during the greatest southerly advance of glaciers in this region, the ice front stabilized in the area of present-day Martha's Vineyard and Nantucket Island and deposited a huge terminal moraine (▶ Figure 1b). As the climate became warmer, the glacier's terminus retreated northward. About 14,000 years ago, the ice front stabilized again and deposited a recessional moraine that makes up the Elizabeth Islands and Cape Cod (▶ Figure 1c).

When the continental glacier withdrew entirely from this region, Cape Cod looked different than it does now. On its east side were several projections of the shoreline called headlands and several large embayments, but by 6000 years ago sea level had risen (remember sea level was lower during the Ice Age) and wave activity eroded the headlands and the embayments filled with sediment, thus smoothing the shoreline. In fact, the original shoreline was 1–4 km east of where it is now, but it has eroded and the entire cape has moved west at about 1 m per year. Most of the erosion takes place during storms, so it is episodic rather than continuous.

Other distinctive features of Cape Cod are its numerous circular to oval ponds, which are actually kettles that extend below the water table. A few of these ponds show as dark spots in ▶ Figure 1a. Remember that kettles form when large blocks of glacial ice are partially or completely buried by glacial deposits (see Figure 17.18). When these masses of ice melted, they left depressions measuring up to 0.8 km in diameter. The depressions filled with water when sea level rose following the Ice Age, and the water table also rose.

If you visit Cape Cod, you can see evidence that the cape has and continues to migrate westward. One observation that verifies this is the organic-rich marsh deposits that formed on the cape's landward (west) side. They are now exposed on the beaches in several locations on the seaward side (east) of the cape. So present-day Cape Cod owes its existence to deposition by glaciers and the modification of the glacial deposits by marine processes, another excellent example of the interaction among Earth's systems.

Figure 1
The geologic evolution of Cape Cod, Massachusetts, and nearby areas during the Ice Age. (a) Cape Cod and the nearby islands are made up of mostly end moraines, although the deposits have been modified by waves since they were deposited 14,000 to 23,000 years ago. (b) Position of the glacier when it deposited a terminal moraine that would become Martha's Vineyard and Nantucket Island. (c) Position of the glacier when it deposited a recessional moraine that now forms much of Cape Cod.

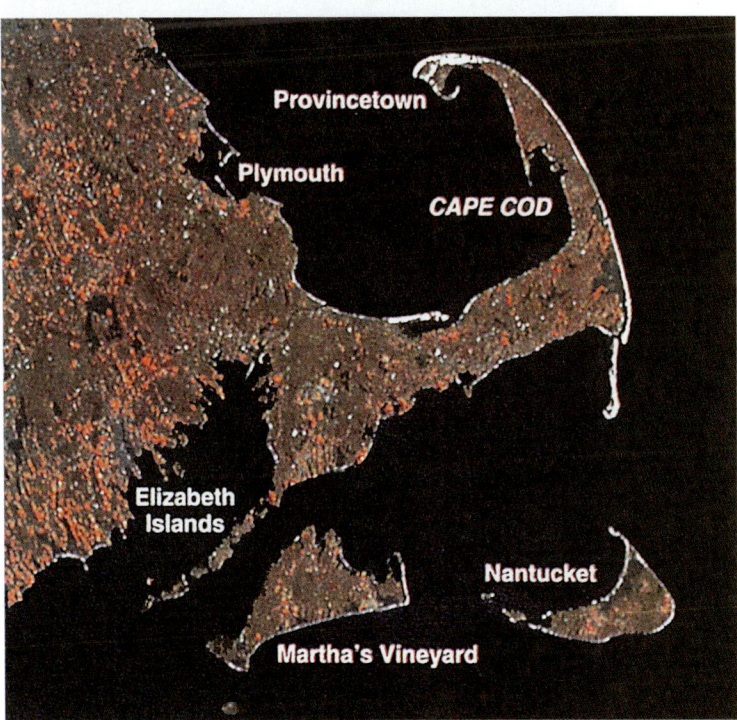
(a)

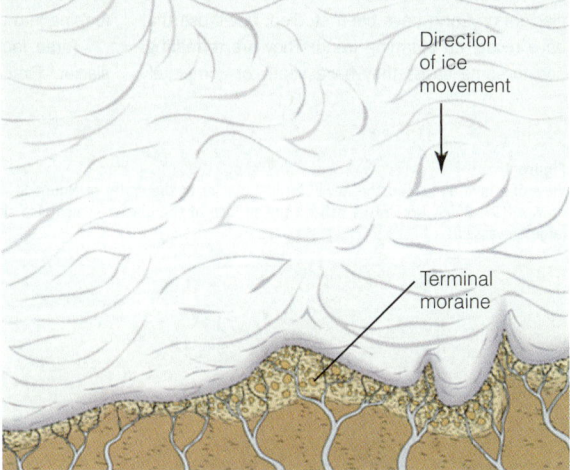
(b)

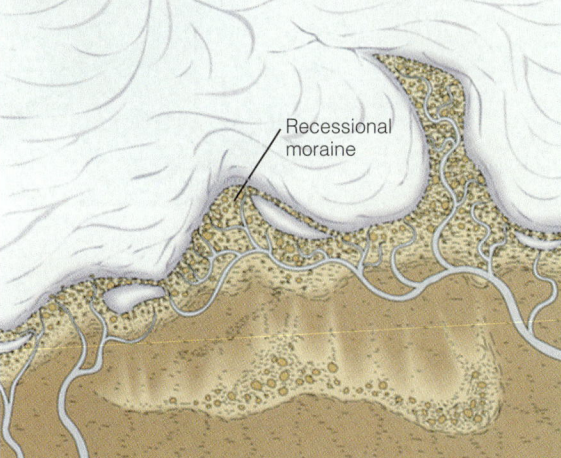
(c)

563

GEOLOGY MATTERS

GEOLOGY IN FOCUS Glaciers and Global Warming

No doubt you have heard of *global warming*, a phenomenon involving warming of Earth's atmosphere during the last 100 years or so. Many scientists think that the concentration of greenhouse gases (carbon dioxide, methane, and nitrous oxide) as a result of human activities, especially the combustion of fossil fuels, is the cause of global warming. There are, however, dissenters who acknowledge that Earth's surface temperatures have increased but attribute the increase to normal climatic variations. Needless to say, the issue has not been resolved.

Whatever the cause of climate change, no one doubts that glaciers are good indicators of short-term variations in climate. According to one estimate there are about 160,000 glaciers outside Antarctica and Greenland, with Alaska alone having several tens of thousands. What do valley glaciers tell us about climate? Remember that the behavior of valley glaciers depends on their glacial budget, which is in turn controlled by temperature and precipitation.

It is true that not very many of the 160,000 or so glaciers on Earth have been studied, but those that have reveal an alarming trend: They are retreating, and in some cases they have nearly or completely disappeared. For example, of the approximately 150 glaciers in Glacier National Park in Montana in 1850, only a few very small ones remain. Many of the valley glaciers in Alaska at lower elevations are much smaller than they were a few decades ago, and so it goes just about everywhere glaciers are studied, including those in the Cascade Range of the Pacific Northwest.

Recall from Chapter 5 that the Cascade Range is made up of several large composite volcanoes and hundreds of smaller volcanic cones and vents (see "The Cascades Volcanoes" on pp. 154–155). All of the higher peaks in the range have glaciers, although some are very small and most are retreating. When Mount St. Helens erupted in May 1980, all of its 12 glaciers were destroyed or at least considerably diminished. But by 1982 the lava dome in the crater had cooled sufficiently for snow to accumulate yearly and it is now as much as 190 m thick. In any case, it is thick enough so that pressure on the snow at depth converts it to glacial ice, and "giant cracks in the ice, called crevasses, and other flow features, indicate that the ice body is transforming into a glacier"* (▶ Figure 1).

Three factors account for the birth of this new glacier. First, the Cascade Range receives huge amounts of snowfall. Second, the crater provides protection for the accumulating snow. And third, rockfalls from the crater walls help insulate the forming glacier. This new glacier formed in little more than 20 years, and it is the only one in the continental United States that is advancing.

The story for the other Cascade Range glaciers is not so comforting. Glacier Peak in Washington, which last erupted in 1880, has more than a dozen glaciers, all of which are retreating, and Whitechuck Glacier will soon be inactive (▶ Figure 2). During the Little Ice Age, Whitechuck Glacier had northern and southern branches, each with a separate accumulation zone, that merged to form a single glacier. It covered about 4.8 km^2, but a rapid episode of retreat began in 1930 and now it covers only about 0.9 km^2 and has thinned to about 35 m thick. In fact, by 2002 the glacier's northern branch was gone, and what had been an ice-filled valley is now filled with rubble. The south branch may persist for a few more decades, but it too is retreating and thinning.

---

*U.S. Forest Service Volcano Review, Summer 2002, contribution by Charlie Anderson, Director of the International Glaciospeleological Survey (http://glaciercaves.com/html/birtho_1.HTM).

Figure 1
View of a lava dome and the newly formed glacier in the crater of Mount St. Helens on April 19, 2005. Notice the ash on the surface of the glacier, which also shows large crevasses.

Figure 2
Whitechuck Glacier on Glacier Peak in Washington State. The south branch of the glacier on the right has a small accumulation area, but the north branch has none.

GEOLOGY MATTERS

Geology AND CULTURAL CONNECTIONS

The Ice Man

In 1991 glacial ice high in a mountain pass between Austria and Italy melted to reveal an unusual scientific treasure: the body of a prehistoric man in his 40s who had died some 5200 years before (▶ Figure 1). The man had evidently been buried in alpine snow that compacted into glacial ice. The ice mummified the man's remains, including near-perfect preservation of his clothing and implements. The man was soon dubbed "The Ice Man" by the press.

Other, more common archaeological finds include human bones and implements buried with people. What is unusual about the Ice Man is that "soft goods" like clothing were preserved. The physical evidence both on and in the Ice Man has provided clues about how this person lived and died well before Greek civilization or even the earliest biblical times.

The Ice Man, who was found at an elevation of 10,500 ft, had a backpack made of wood and fur. His clothing and footwear were made of leather and grasses. He carried an ax made of copper and a bow made of yew wood. Although he clearly perished high in the mountains, he didn't have food with him—or if he did, he had become separated from it before he died.

At first people assumed that the Ice Man had simply died in a mountain storm and been buried in snow. This image of a relatively peaceful death, however, was shattered when scientists began to examine the mummified body using the techniques of modern medicine.

The first discovery made came from a careful inspection of the mummified remains. Doctors and scientists established that the Ice Man had wounds on his hands, wrists, and rib cage. The image of a man falling asleep in a snowstorm and quietly dying of hypothermia quickly evaporated. Next came DNA analyses. The Ice Man's clothing and weapons had the blood of four other people on them, in addition to his own. Taken together with the wounds on his body, it seemed clear that the Ice Man was involved in a bloody fight with other people. Next, using imaging techniques on the mummified remains as a whole, scientists discovered that the Ice Man had an arrowhead buried in his left shoulder. He had been shot, as well as hacked at, in the hours before his death.

High mountain passes are often the boundaries between different cultures and groups of people. Because the Ice Man died in such a place, it seems possible that he was involved in a border clash with people outside his own tribe. He may have done real damage to his enemies (as witnessed by the blood of other people on his weapons), but he, too, was gravely injured. He may have died from blood loss or from a snowstorm that trapped him in his weakened condition at high elevations. But in any event his relatively long life was ended by violence.

Recently scientists have cut open the Ice Man's remains to examine his internal organs. A sample of material from his intestine, just below his stomach, reveals the man's last meal, likely eaten 6 to 8 hours before his death. He had eaten meat, some kind of greens or herb, and a simple kind of bread made of einkorn, one of the few domesticated grains in that part of the world some 5000 years ago. Einkorn is low in gluten, the substance that gives modern bread dough its elastic properties and holds in the tiny pockets of gas that yeast makes. So the "bread" that the Ice Man ate was likely hard, something like rye crisp or a tough cracker. Scientists have even identified charcoal and soot in the Ice Man's gut, likely from the fire that cooked his last meal and baked the einkorn bread.

No one expected that a glacier would give so many clues about what life was like for people in the Alps 5200 years ago. The Ice Man has provided a treasure trove of information for scientists and archaeologists. Tourists now visit the Ice Man in a refrigerated cell in a museum built especially for him in the Italian town of Bolzano. Every year Bolzano celebrates its good fortune in receiving this glacial gift from the prehistoric world.

Figure 1
The Ice Man **(a)** These 5200-year-old remains, found in 1991 by two hikers, were preserved in glacial ice in the mountains of northern Italy. **(b)** Close-up view of The Ice Man in September 2000, when he was thawed so that scientists could take samples.

(a)

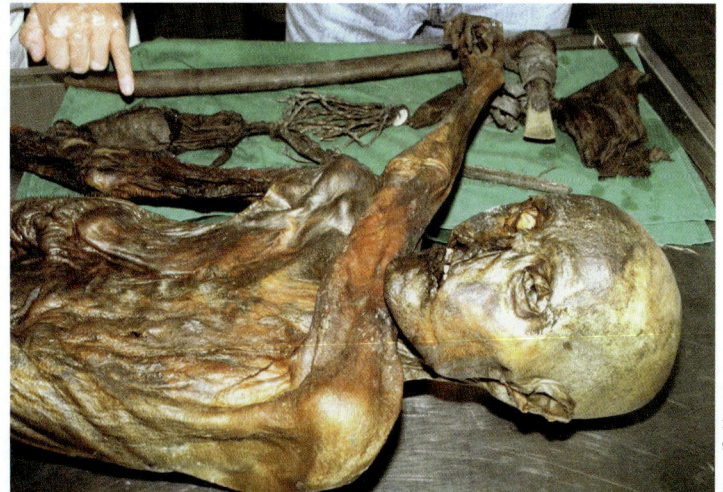
(b)

Chapter 18

The Work of Wind and Deserts

Don Quixote "Tilting at the Windmills"
Like a theatrical drama on a stage that can carry us forward or backward in history, this painted scene of Don Quixote and Sancho Panca takes us back to a time when men on horseback galloped the hillsides, and windmills with thatched rooftops powered the world. This Spanish tilework at the Plaza de Espana in Seville, Spain depicts a scene from Miguel de Cervantes' 1605 novel "Don Quixote de la Mancha."
—A. W.

ESSENTIAL QUESTIONS TO ASK

18.1 Introduction
- What is desertification and why is it a major problem in the world?
- Why is it important to understand how desert processes operate?

18.2 Sediment Transport by Wind
- What is bed load?
- What is suspended load?

18.3 Wind Erosion
- What is abrasion and what type of products are formed by abrasion?
- What is deflation and what features are formed by this process?

18.4 Wind Deposits
- How do dunes form and migrate?
- What are the four major dune types and how do they form?
- What is loess, how does it form, and why is it important?

18.5 Air-Pressure Belts and Global Wind Patterns
- How do Earth's air-pressure belts and the Coriolis effect determine global wind patterns?

18.6 The Distribution of Deserts
- Where are deserts located?

18.7 Characteristics of Deserts
- What are the characteristics of deserts?

18.8 Desert Landforms
- What are the major landforms of deserts and how do they form?

GEOLOGY MATTERS

GEOLOGY IN FOCUS:
Radioactive Waste Disposal—Safe or Sorry?

GEOLOGY IN UNEXPECTED PLACES:
Spiral Jetty

GEOLOGY IN YOUR LIFE:
Windmills and Wind Power

This icon, which appears throughout the book, indicates an opportunity to explore interactive tutorials, animations, or practice problems available on the Physical GeologyNow website at http://now.brookscole.com/phygeo6.

567

18.1 Introduction

■ *What is desertification and why is it a major problem in the world?*

During the past several decades, deserts have been advancing across millions of acres of productive land, destroying rangeland, croplands, and even villages (▶ Figure 18.1). Such expansion, estimated at 70,000 km² per year, has exacted a terrible toll in human suffering. Because of the relentless advance of deserts, hundreds of thousands of people have died of starvation or been forced to migrate as "environmental refugees" from their homelands to camps where the majority are severely malnourished. This expansion of deserts into formerly productive lands is called **desertification** and is a major problem in many countries.

Most regions undergoing desertification lie along the margins of existing deserts where a delicately balanced ecosystem serves as a buffer between the desert on one side and a more humid environment on the other. These regions have limited

▶ **Figure 18.1 Desertification** The Saharan community of El Gedida in western Egypt is slowly being overwhelmed by advancing sand.

potential to adjust to increasing environmental pressures from natural causes as well as human activity. Ordinarily, desert regions expand and contract gradually in response to natural processes such as climatic change, but much recent desertification has been greatly accelerated by human activities.

In many areas, the natural vegetation has been cleared as crop cultivation has expanded into increasingly drier fringes to support the growing population. Because grasses are the dominant natural vegetation in most fringe areas, raising livestock is a common economic activity. However, increasing numbers of livestock in many areas have greatly exceeded the land's ability to support them. Consequently, the vegetation cover that protects the soil has diminished, causing the soil to crumble and be stripped away by wind and water, which results in increased desertification.

One particularly hard-hit area of desertification is the Sahel of Africa (a belt 300–1100 km wide, lying south of the Sahara). Because drought is common in the Sahel, the region can support only a limited population of livestock and humans. Unfortunately, expanding human and animal populations and more intensive agriculture have increased the demands on the lands. Plagued with periodic droughts, this region has suffered tremendously as crops have failed and livestock has overgrazed the natural vegetation, resulting in thousands of deaths, displaced people, and the encroachment of the Sahara. Ironically, drought conditions in this region may be exacerbated by the climatary impact of greater air pollution in the wealthier regions of Europe, thousands of kilometers to the north.

The tragedy of the Sahel and prolonged droughts in other desert fringe areas remind us of the delicate equilibrium of ecosystems in such regions. Once the fragile soil cover has been removed by erosion, it takes centuries for new soil to form (see Chapter 6).

There are many important reasons to study deserts and the processes that are responsible for their formation. First, deserts cover large regions of Earth's surface. More than 40% of Australia is desert, and the Sahara occupies a vast part of northern Africa. Although deserts are generally sparsely populated, some desert regions are experiencing an influx of people, such as Las Vegas, Nevada, the high desert area of southern California, and various locations in Arizona. Many of these places already have problems with increasing population growth and the strains it places on the environment, particularly the need for greater amounts of groundwater (see Chapter 16).

■ *Why is it important to understand how desert processes operate?*

With the current debate about global warming, it is important to understand how desert processes operate and how global climate changes affect various Earth systems and subsystems. By understanding how desertification operates, people can take steps to eliminate or reduce the destruction done, particularly in terms of human suffering.

Learning about the underlying causes of climate change by examining ancient desert regions may provide insight into the possible duration and severity of future climatic changes. This can have important ramifications in decisions about whether burying nuclear waste in a desert, such as Yucca Mountain, Nevada, is as safe as some claim and in our best interests as a society.

As an example, more than 6000 years ago, the Sahara was a fertile savannah supporting a diverse fauna and flora, including humans. Then the climate changed, and the area became a desert. How did this happen? Will this region change back again in the future? These are some of the questions geoscientists hope to answer by studying deserts.

And last, many agents and processes that have shaped deserts do not appear to be limited to our planet. Features found on Mars, especially as seen in images transmitted by the *Opportunity* and *Spirit* rovers, are apparently the result of the same wind-driven processes that operate on Earth.

Section 18.1 Summary

- Desertification is the expansion of deserts into formerly productive lands and is a major problem in many countries. Deserts cover large areas of Earth. Increased population growth in many desert regions is placing a strain on the environment, especially the need to find sufficient quantities of groundwater.

- Understanding how desert processes work and examining ancient desert regimes may provide insights into the cause of climate changes and help in dealing with present-day environmental issues.

18.2 Sediment Transport by Wind

Wind is a turbulent fluid and therefore transports sediment in much the same way as running water. Although wind typically flows at a higher velocity than water, it has a lower density and thus can carry only clay- and silt-sized particles as *suspended load*. Sand and larger particles are moved along the ground as *bed load*.

■ *What is bed load?*

Sediments too large or heavy to be carried in suspension by water or wind are moved as bed load either by *saltation* or by rolling and sliding. As we discussed in Chapter 15, saltation is the process by which a portion of the bed load moves by intermittent bouncing along a streambed. Saltation also occurs on land. Wind starts sand grains rolling and lifts and carries some grains short distances before they fall back to the surface. As the descending sand grains hit the surface, they strike other grains, causing them to bounce along by saltation (▶ Figure 18.2). Wind-tunnel experiments have shown that once sand grains begin moving, they will continue to move, even if the wind drops

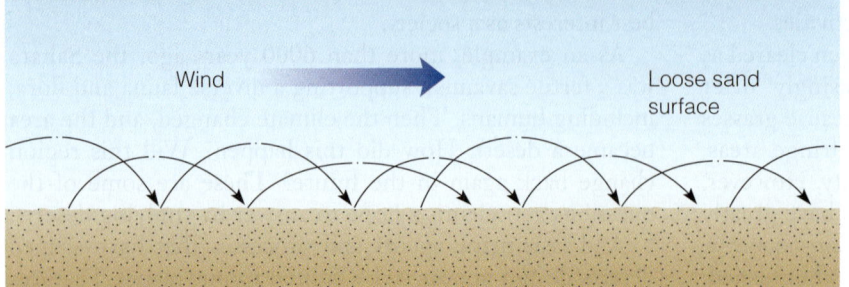

▶ **Figure 18.2 Saltation** Most sand is moved near the ground surface by saltation. Sand grains are picked up by the wind and carried a short distance before falling back to the ground, where they usually hit other grains, causing them to bounce and move in the direction of the wind.

below the speed necessary to start them moving! This happens because once saltation begins, it sets off a chain reaction of collisions between sand grains that keeps the grains in constant motion.

Saltating sand usually moves near the surface, and even when winds are strong, grains are rarely lifted higher than about a meter. If the winds are very strong, these wind-whipped grains can cause extensive abrasion. A car's paint can be removed by sandblasting in a short time, and its windshield will become completely frosted and translucent from pitting.

■ *What is suspended load?*

Silt- and clay-sized particles constitute most of a wind's suspended load. Even though these particles are much smaller and lighter than sand-sized particles, wind usually starts the latter moving first. The reason for this phenomenon is that a very thin layer of motionless air lies next to the ground where the small silt and clay particles remain undisturbed. The larger sand grains, however, stick up into the turbulent air zone, where they can be moved. Unless the stationary air layer is disrupted, the silt and clay particles remain on the ground, providing a smooth surface.

This phenomenon can be observed on a dirt road on a windy day. Unless a vehicle travels over the road, little dust is raised even though it is windy. When a vehicle moves over the road, it breaks the calm boundary layer of air and disturbs the smooth layer of dust, which is picked up by the wind and forms a dust cloud in the vehicle's wake.

In a similar manner, when a sediment layer is disturbed, silt- and clay-sized particles are easily picked up and carried in suspension by the wind, creating clouds of dust or even dust storms. Once these fine particles are lifted into the atmosphere, they may be carried thousands of kilometers from their source. For example, large quantities of fine dust from the southwestern United States were blown east and fell on New England during the dust storms of the 1930s.

Section 18.2 Summary

- Wind is a turbulent fluid and transports sediment in much the same way as running water.

- Sand and larger particles are moved along the ground as bed load either by saltation, in which grains move by intermittent bouncing, or by rolling and sliding.

- Silt- and clay-sized particles make up most of a wind's suspended load. These particles are carried in the wind, sometimes thousands of kilometers.

18.3 Wind Erosion

Although wind action produces many distinctive erosional features and is an extremely efficient sorting agent, running water is responsible for most erosional landforms in arid regions, even though stream channels are typically dry (see Figure 15.2). Wind erodes material in two ways: abrasion and deflation.

■ *What is abrasion and what types of products are formed by abrasion?*

Abrasion involves the impact of saltating sand grains on an object and is analogous to sandblasting. The effects of abrasion are usually minor because sand, the most common agent of abrasion, is rarely carried more than a meter above the surface. Rather than creating major erosional features, wind abrasion typically modifies existing features by etching, pitting, smoothing, or polishing. Nonetheless, wind abrasion can produce many strange-looking and bizarre-shaped features (▶ Figure 18.3).

Ventifacts are a common product of wind abrasion; these are stones whose surfaces have been polished, pitted, grooved, or faceted by the wind (▶ Figures 18.4 and 18.7c). If the wind blows from different directions or if the stone is moved, the ventifact will have multiple facets. Ventifacts are

What Would You Do?

You have been asked to testify before a congressional committee charged with determining whether the National Science Foundation should continue to fund research devoted to the study of climate changes during the Cenozoic Era. Your specialty is desert landforms and the formation of deserts. What arguments would you make to convince the committee to continue funding research on ancient climates?

▶ **Figure 18.3 Wind Abrasion** Wind abrasion has formed these bizarre-shaped structures by eroding the lower part of the exposed limestone in Desierto Libico, Egypt.

most common in deserts, yet they can also form wherever stones are exposed to saltating sand grains, as on beaches in humid regions and some outwash plains in New England.

Yardangs are larger features than ventifacts and also result from wind erosion (▶ Figure 18.5). They are elongated, streamlined ridges that look like an overturned ship's hull. Yardangs are typically found in clusters aligned parallel to the prevailing winds. They probably form by differential erosion in which depressions, parallel to the direction of wind, are carved out of a rock body, leaving sharp, elongated ridges. These ridges may then be further modified by wind abrasion into their characteristic shape. Although yardangs are fairly common desert features, interest in them was renewed when images radioed back from Mars showed that they are also widespread features on the Martian surface.

■ *What is deflation and what features are formed by this process?*

Another important mechanism of wind erosion is **deflation**, which is the removal of loose surface sediment by the wind. Among the characteristic features of deflation in

▶ **Figure 18.4 Ventifacts**

a A ventifact forms when wind-borne particles (1) abrade the surface of a rock, (2) forming a flat surface. If the rock is moved, (3) additional flat surfaces are formed.

b Large ventifacts lying on desert pavement in Death Valley National Monument, California.

▶ **Figure 18.5 Yardang** A profile view of a streamlined yardang in the Roman playa deposits of the Kharga Depression, Egypt. Yardangs form by wind erosion.

many arid and semiarid regions are *deflation hollows,* or *blowouts* (▶ Figure 18.6). These shallow depressions of variable dimensions result from differential erosion of surface materials. Ranging in size from several kilometers in diameter and tens of meters deep to small depressions only a few meters wide and less than a meter deep, deflation hollows are common in the southern Great Plains of the United States.

In many dry regions, the removal of sand-sized and smaller particles by wind leaves a surface of pebbles, cobbles, and boulders. As the wind removes the fine-grained material from the surface, the effects of gravity and occa-

▶ **Figure 18.6 Deflation Hollow** A deflation hollow, the low area, between two sand dunes in Death Valley, California. Deflation hollows result when loose surface sediment is differentially removed by wind.

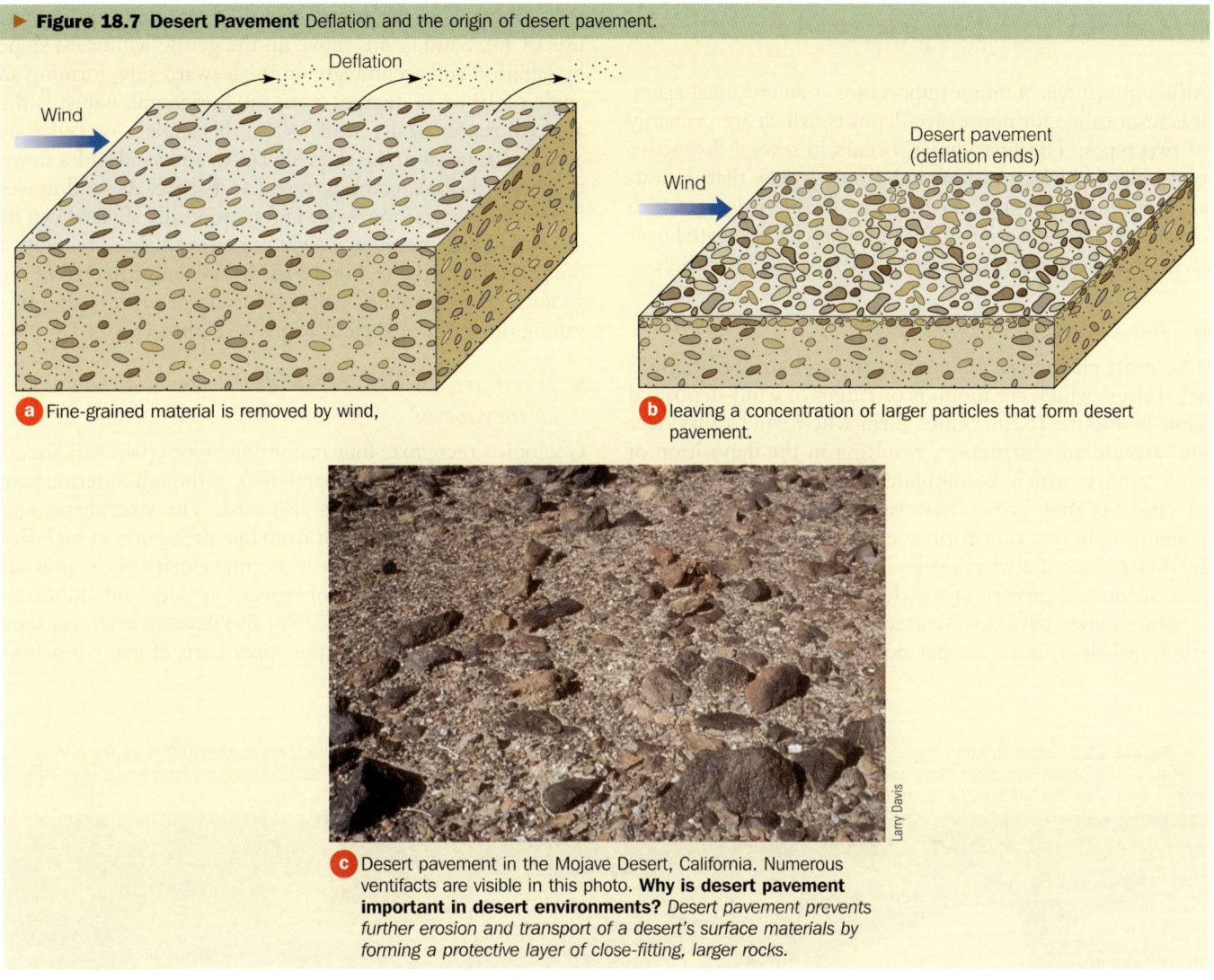

▶ **Figure 18.7 Desert Pavement** Deflation and the origin of desert pavement.

ⓐ Fine-grained material is removed by wind,

ⓑ leaving a concentration of larger particles that form desert pavement.

ⓒ Desert pavement in the Mojave Desert, California. Numerous ventifacts are visible in this photo. **Why is desert pavement important in desert environments?** *Desert pavement prevents further erosion and transport of a desert's surface materials by forming a protective layer of close-fitting, larger rocks.*

sional heavy rain, and even the swelling of clay grains, rearrange the remaining coarse particles into a mosaic of close-fitting rocks called **desert pavement** (▶ Figures 18.4b and 18.7). Once desert pavement forms, it protects the underlying material from further deflation.

What Would You Do?

As an expert in desert processes, you have been assigned the job of teaching the first astronaut crew that will explore Mars all about deserts and their landforms. The reason is that many Martian features display evidence of having formed as a result of wind processes, and many Martian landforms are the same as those found in deserts on Earth. Describe how you would teach the astronauts to recognize wind-formed features and where on Earth you would take the astronauts to show them the types of landforms they may find on Mars.

Section 18.3 Summary

• Wind erodes material by abrasion and deflation. Abrasion involves the impact of saltating sand grains on an object and is analogous to sandblasting. Wind abrasion modifies features by etching, pitting, smoothing, and polishing. Ventifacts are stones whose surfaces have been polished, pitted, grooved, or faceted by abrasion. Ventifacts form wherever stones are exposed to saltating sand grains. Yardangs are larger features than ventifacts and also result from wind erosion. They are elongated, streamlined ridges that probably formed by differential erosion.

• Deflation is the removal of loose surface sediment by wind. It produces such features as deflation hollows, which are shallow depressions resulting from differential erosion of surface materials, and desert pavement, in which fine-grained material is removed, leaving a surface covered by larger particles.

18.4 Wind Deposits

Although wind is of minor importance as an erosional agent, it is responsible for impressive deposits, which are primarily of two types. The first, dunes, occurs in several distinctive types, all of which consist of sand-sized particles that are usually deposited near their source. The second is loess, which consists of layers of wind-blown silt and clay deposited over large areas downwind and commonly far from their source.

■ *How do dunes form and migrate?*

The most characteristic features in sand-covered regions are **dunes,** which are mounds or ridges of wind-deposited sand (▶ Figure 18.8). Dunes form when wind flows over and around an obstruction, resulting in the deposition of sand grains, which accumulate and build up a deposit of sand. As they grow, these sand deposits become self-generating in that they form ever-larger wind barriers that further reduce the wind's velocity, resulting in more sand deposition and growth of the dune.

Most dunes have an asymmetric profile, with a gentle windward slope and a steeper downwind, or leeward, slope that is inclined in the direction of the prevailing wind (▶ Figure 18.9a). Sand grains move up the gentle windward slope by saltation and accumulate on the leeward side, forming an angle of 30 to 34 degrees from the horizontal, which is the angle of repose of dry sand. When this angle is exceeded by accumulating sand, the slope collapses and sand slides down the leeward slope, coming to rest at its base. As sand moves from a dune's windward side and periodically slides down its leeward slope, the dune slowly migrates in the direction of the prevailing wind (▶ Figure 18.9b). When preserved in the geologic record, dunes help geologists determine the prevailing direction of ancient winds (▶ Figure 18.10).

■ *What are the four major dune types and how do they form?*

Geologists recognize four major dune types (barchan, longitudinal, transverse, and parabolic), although intermediate forms and additional types also exist. The size, shape, and arrangement of dunes result from the interaction of such factors as sand supply, the direction and velocity of the prevailing wind, and the amount of vegetation. Although dunes are usually found in deserts, they can also develop wherever sand is abundant, such as along the upper parts of many beaches.

▶ **Figure 18.8 Sand Dunes** Large sand dunes in Death Valley, California. Well-developed ripple marks can be seen on the surface of the dunes. **What is the prevailing wind direction?** *The prevailing wind direction is from left to right, as indicated by the sand dunes in which the gentle windward side is on the left and the steeper leeward slope is on the right (see Chapter 7).*

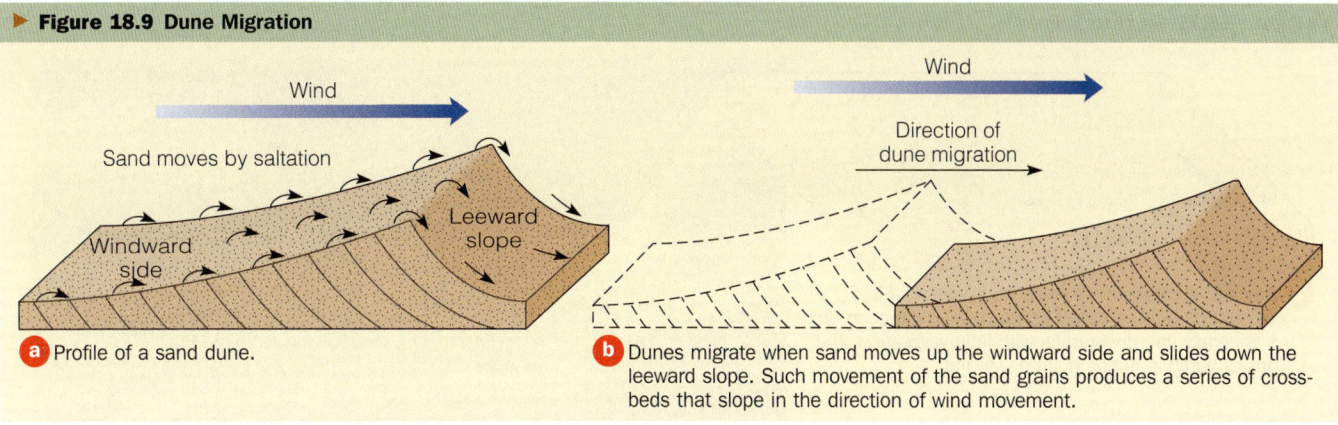

▶ **Figure 18.9 Dune Migration**

a Profile of a sand dune.

b Dunes migrate when sand moves up the windward side and slides down the leeward slope. Such movement of the sand grains produces a series of cross-beds that slope in the direction of wind movement.

Barchan dunes are crescent-shaped dunes whose tips point downwind (▶ Figure 18.11). They form in areas that have a generally flat, dry surface with little vegetation, a limited supply of sand, and a nearly constant wind direction. Most barchans are small, with the largest reaching about 30 m high. Barchans are the most mobile of the major dune types, moving at rates that can exceed 10 m per year.

Longitudinal dunes (also called *seif dunes*) are long, parallel ridges of sand aligned generally parallel to the direction of the prevailing winds; they form where the sand supply is somewhat limited (▶ Figure 18.12). Longitudinal dunes result when winds converge from slightly different directions to produce the prevailing wind. They range in size from about 3 m to more than 100 m high, and some stretch for more than 100 km. Longitudinal dunes are especially well developed in central Australia, where they cover nearly one-fourth of the continent. They also cover extensive areas in Saudi Arabia, Egypt, and Iran.

Transverse dunes form long ridges perpendicular to the prevailing wind direction in areas that have abundant sand and little or no vegetation (▶ Figure 18.13). When viewed from the air, transverse dunes have a wavelike appearance and are therefore sometimes called *sand seas*. The crests of transverse dunes can be as high as 200 m, and the dunes may

▶ **Figure 18.10 Ancient Cross-Bedding** Ancient cross-bedding in sandstone beds in Zion National Park, Utah, helps geologists determine the prevailing direction of the wind that formed these ancient sand dunes. **Did the prevailing direction of wind change during the time the sand forming these sandstones was deposited?** *Yes, as can be seen in the different directions the cross-beds are pointing in the different layers.*

Figure 18.11 Barchan Dunes

a Barchan dunes form in areas that have a limited amount of sand, a nearly constant wind direction, and a generally flat, dry surface with little vegetation. The tips of barchan dunes point downwind.

b An aerial view of several barchan dunes west of the Salton Sea, California. **Can you tell the direction of the prevailing wind?** The prevailing wind direction is from left to right, as indicated by the barchan dune's tips pointing toward the right.

c A ground-level view of several barchan dunes.

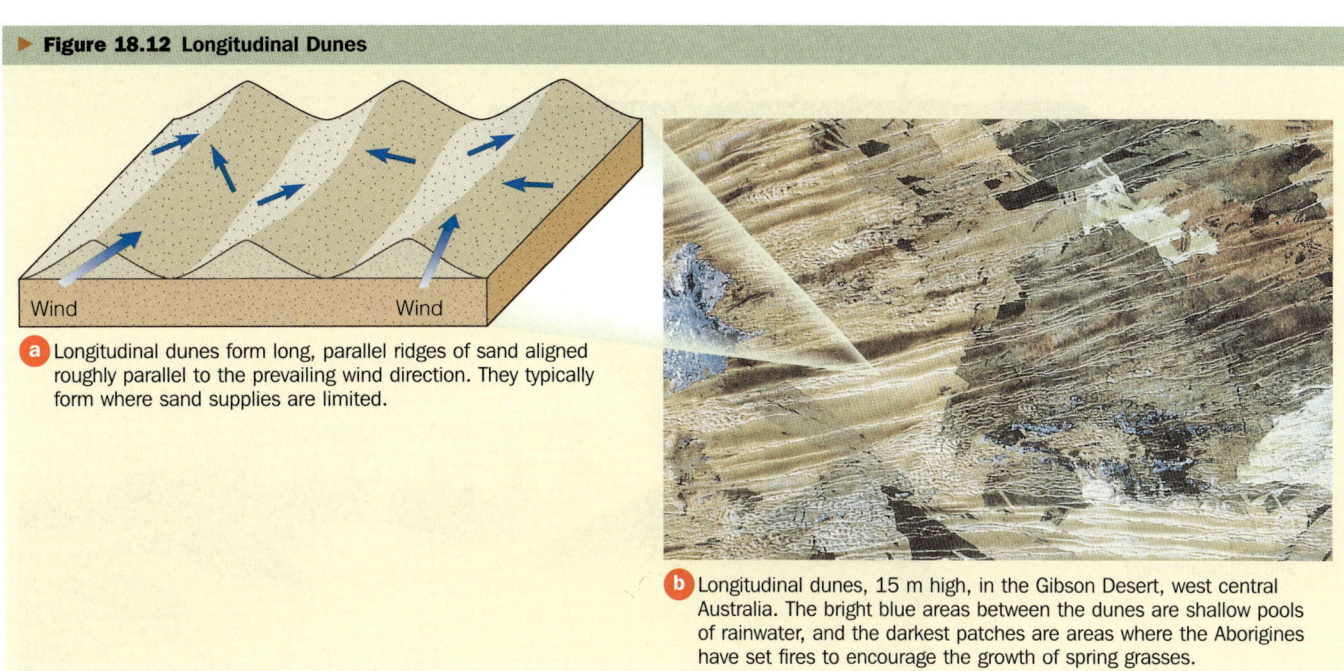

▶ Figure 18.12 Longitudinal Dunes

a Longitudinal dunes form long, parallel ridges of sand aligned roughly parallel to the prevailing wind direction. They typically form where sand supplies are limited.

b Longitudinal dunes, 15 m high, in the Gibson Desert, west central Australia. The bright blue areas between the dunes are shallow pools of rainwater, and the darkest patches are areas where the Aborigines have set fires to encourage the growth of spring grasses.

be as wide as 3 km. Some transverse dunes develop a clearly distinguishable barchan form and may separate into individual barchan dunes along the edges of the dune field where there is less sand. Such intermediate-form dunes are known as *barchanoid dunes*.

Parabolic dunes are most common in coastal areas with abundant sand, strong onshore winds, and a partial cover of vegetation (▶ Figure 18.14). Although parabolic dunes have a crescent shape like barchan dunes, their tips point upwind. Parabolic dunes form when the vegetation cover is broken and deflation produces a deflation hollow or blowout. As the wind transports the sand out of the depression, it builds up on the convex downwind dune crest. The central part of the dune is excavated by the wind, while vegetation holds the ends and sides fairly well in place.

Another type of dune commonly found in the deserts of North Africa and Saudi Arabia is the *star dune*, so named because of its resemblance to a multipointed representation of a star (▶ Figure 18.15). Star dunes are among the tallest in the world, rising, in some cases, more than 100 m above the surrounding desert plain. They consist of pyramidal hills of sand, from which radiate several ridges of sand, and they

▶ Figure 18.13 Transverse Dunes

a Transverse dunes form long ridges of sand that are perpendicular to the prevailing wind direction in areas of little or no vegetation and abundant sand.

b Transverse dunes, Great Sand Dunes National Monument, Colorado. The prevailing wind direction is from lower left to upper right.

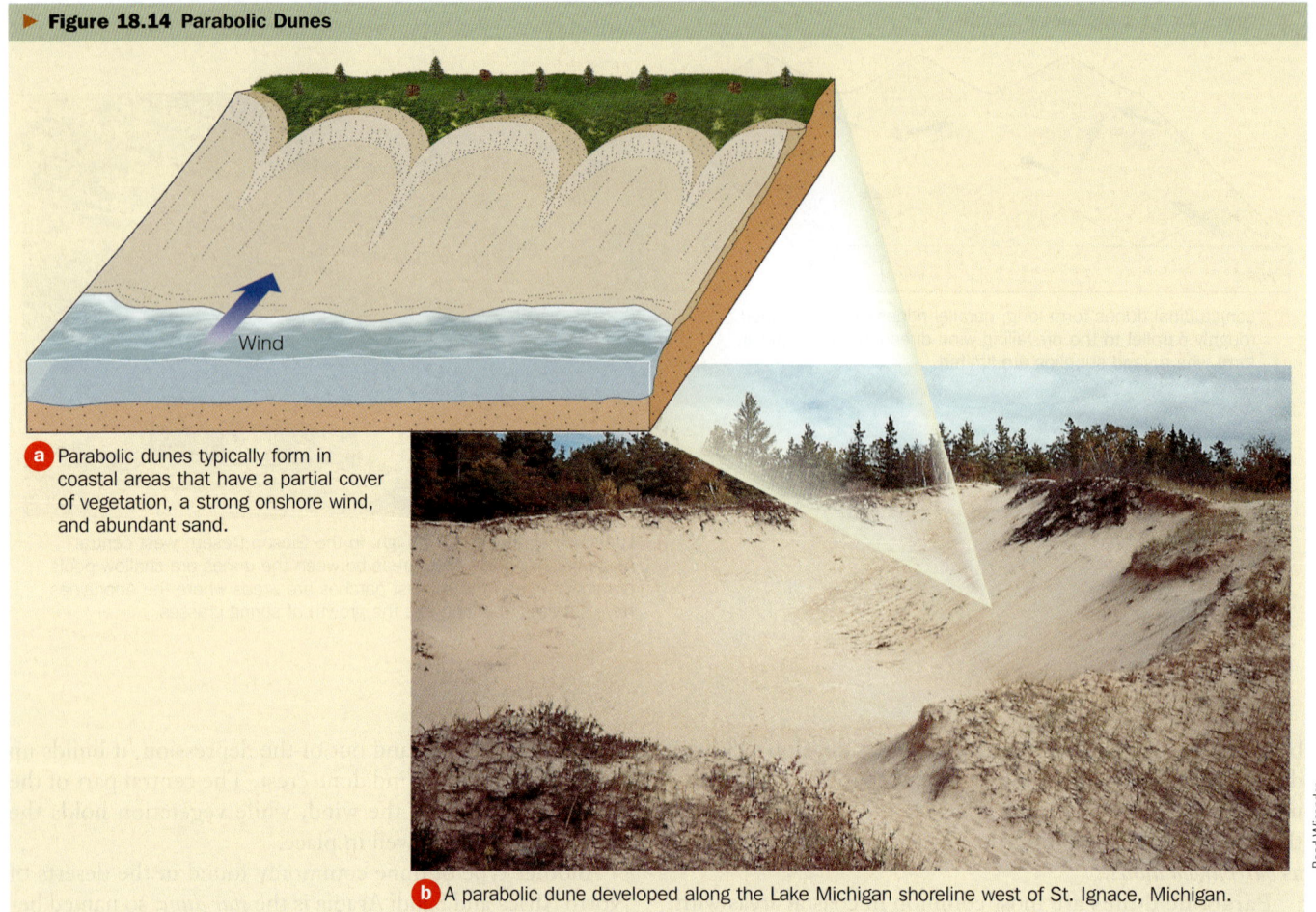

▶ **Figure 18.14 Parabolic Dunes**

a Parabolic dunes typically form in coastal areas that have a partial cover of vegetation, a strong onshore wind, and abundant sand.

b A parabolic dune developed along the Lake Michigan shoreline west of St. Ignace, Michigan.

develop where the wind direction is variable. Star dunes can remain stationary for centuries and have served as desert landmarks for nomadic peoples.

■ *What is loess, how does it form, and why is it important?*

Wind-blown silt and clay deposits composed of angular quartz grains, feldspar, micas, and calcite are known as **loess**. The distribution of loess shows that it is derived from three main sources: deserts, Pleistocene glacial outwash deposits, and the floodplains of rivers and ancient pluvial lake beds in semiarid regions. Loess must be stabilized by moisture and vegetation in order to accumulate. Consequently loess is not found in deserts, even though deserts provide much of its material. Because of its unconsolidated nature, loess is easily eroded, and as a result, eroded loess areas are characterized by steep cliffs and rapid lateral and headward stream erosion (▶ Figure 18.16).

At present, loess deposits cover approximately 10% of Earth's land surface and 30% of the United States. The most extensive and thickest loess deposits are found in northeast China, where accumulations greater than 30 m are common. The extensive deserts in central Asia are the source for this loess. Other important loess deposits are on the North European Plain from Belgium eastward to Ukraine, in Central Asia, and in the Pampas of Argentina. In the United States, loess deposits are found in the Great Plains, the Midwest, the Mississippi River Valley, and eastern Washington.

Loess-derived soils are some of the world's most fertile (▶ Figure 18.16). It is therefore not surprising that the world's major grain-producing regions correspond to large loess deposits, such as the North European Plain, Ukraine, and the Great Plains of North America.

Section 18.4 Summary

● The most characteristic features of sand-covered regions are dunes. Dunes form when wind blows over and around an obstruction, resulting in the deposition of sand grains, which accumulate and build up a deposit of sand. Most dunes have an asymmetric profile with a gentle windward slope and a steeper leeward or downwind slope. Dunes migrate by sand moving up and over the gentle windward slope by saltation and accumulating and sliding down the leeward side.

Figure 18.15 Star Dunes

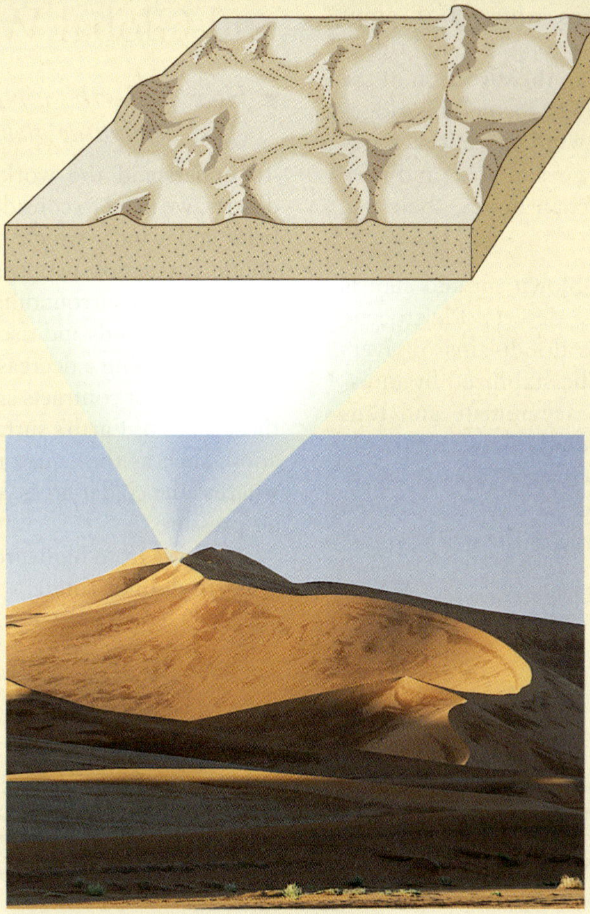

a Star dunes are pyramidal hills of sand that develop where the wind direction is variable.

b A ground-level view of star dunes in Namib-Naukluft Park, Namibia.

Nigel J. Dennis/Photo Researchers, Inc.

Figure 18.16 Terraced Wheat Fields in the Loess Soil at Tangwa Village, China Because of the unconsolidated nature of loess, many farmers live in hillside caves they carved from the loess.

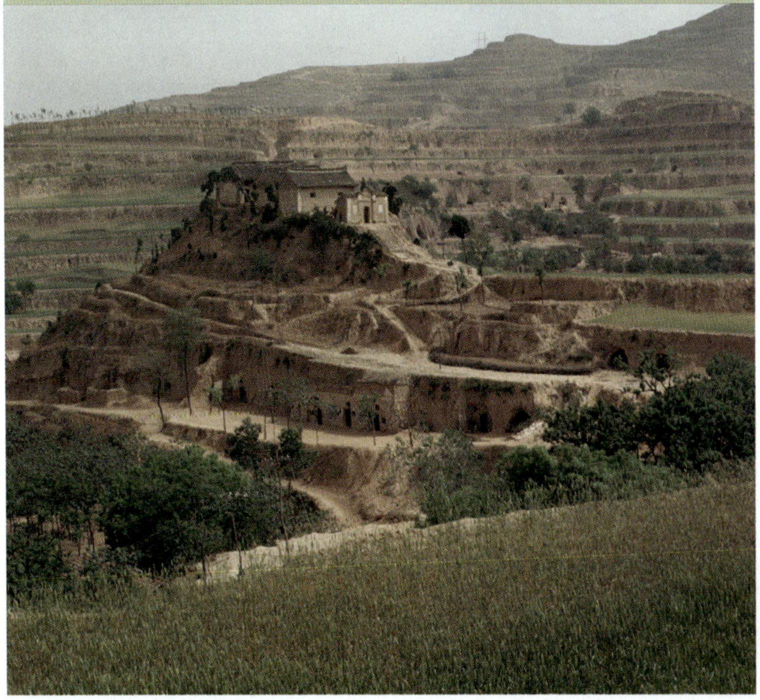

Lowell Georgia/Corbis

- Four major dune types are recognized. Barchan dunes are crescent-shaped dunes whose tips point downwind. Longitudinal dunes are long, parallel ridges of sand aligned generally parallel to the direction of the prevailing winds. Transverse dunes form long ridges perpendicular to the prevailing wind direction. Parabolic dunes are most common in coastal areas and have a crescent shape, but unlike barchan dunes, their tips point upwind.

- Loess, which consists of wind-blown silt and clay, is derived from three main sources: deserts, Pleistocene glacial outwash deposits, and the floodplains of rivers in semiarid regions. Loess must be stabilized by moisture and vegetation in order to accumulate and thus isn't found in deserts, even though deserts provide most of its material. Loess covers approximately 10% of Earth's land surface and 30% of the United States. Loess-derived soils are some of the world's most fertile.

18.5 Air-Pressure Belts and Global Wind Patterns

■ *How do Earth's air-pressure belts and the Coriolis effect determine global wind patterns?*

To understand the work of wind and the distribution of deserts, we need to consider the global pattern of air-pressure belts and winds, which are responsible for Earth's atmospheric circulation patterns. Air pressure is the density of air exerted on its surroundings (that is, its weight). When air is heated, it expands and rises, reducing its mass for a given volume and causing a decrease in air pressure. Conversely, when air is cooled, it contracts and air pressure increases. Therefore those areas of Earth's surface that receive the most solar radiation, such as the equatorial regions, have low air pressure, whereas the colder areas, such as the polar regions, have high air pressure.

Air flows from high-pressure zones to low-pressure zones. If Earth did not rotate, winds would move in a straight line

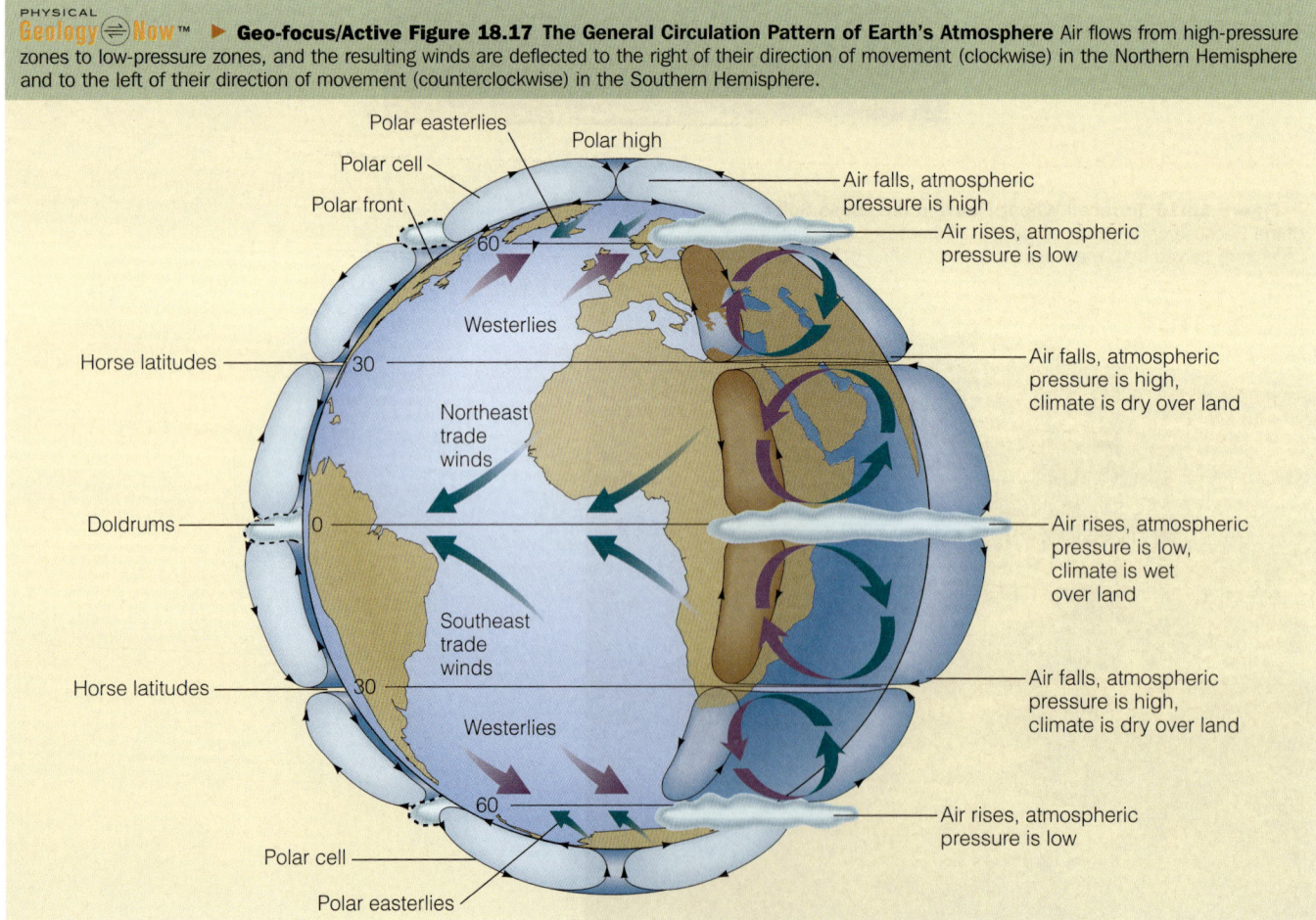

PHYSICAL Geology Now™ ▶ **Geo-focus/Active Figure 18.17 The General Circulation Pattern of Earth's Atmosphere** Air flows from high-pressure zones to low-pressure zones, and the resulting winds are deflected to the right of their direction of movement (clockwise) in the Northern Hemisphere and to the left of their direction of movement (counterclockwise) in the Southern Hemisphere.

from one zone to another. Because Earth rotates, however, winds are deflected to the right of their direction of motion (clockwise) in the Northern Hemisphere and to the left of their direction of motion (counterclockwise) in the Southern Hemisphere. This deflection of air between latitudinal zones resulting from Earth's rotation is known as the **Coriolis effect.** The combination of latitudinal pressure differences and the Coriolis effect produces a worldwide pattern of east-west–oriented wind belts (▶ Figure 18.17).

Earth's equatorial zone receives the most solar energy, which heats the surface air and causes it to rise. As the air rises, it cools and releases moisture that falls as rain in the equatorial region (▶ Figure 18.17). The rising air is now much drier as it moves northward and southward toward each pole. By the time it reaches 20 to 30 degrees north and south latitudes, the air has become cooler and denser and begins to descend. Compression of the atmosphere warms the descending air mass and produces a warm, dry, high-pressure area, providing the perfect conditions for the formation of the low-latitude deserts of the Northern and Southern Hemispheres (▶ Figure 18.18).

Section 18.5 Summary

- Air pressure is the density of air exerted on its surroundings. Those areas of Earth's surface that receive the most solar radiation, such as the equatorial regions, have low air pressure, whereas the colder areas, such as the polar regions, have high air pressure. Air flows from high-pressure zones to low-pressure zones, and is deflected clockwise in the Northern Hemisphere and counterclockwise in the Southern Hemisphere due to the Coriolis effect. This combination of latitudinal pressures and the Coriolis effect produces a worldwide pattern of east-west–oriented wind belts.

18.6 The Distribution of Deserts

■ *Where are deserts located?*

Dry climates occur in the low and middle latitudes, where the loss of water by evaporation may exceed the yearly precipitation (▶ Figure 18.18). Dry climates cover 30% of Earth's land surface and are subdivided into semiarid and arid regions. *Semiarid regions* receive more precipitation than arid regions, yet they are moderately dry. Their soils are usually well developed and fertile and support a natural grass cover. *Arid regions*, generally described as **deserts,** are dry; on average, they receive less than 25 cm of rain per year, have high evaporation rates, typically have poorly developed soils, and are mostly or completely devoid of vegetation.

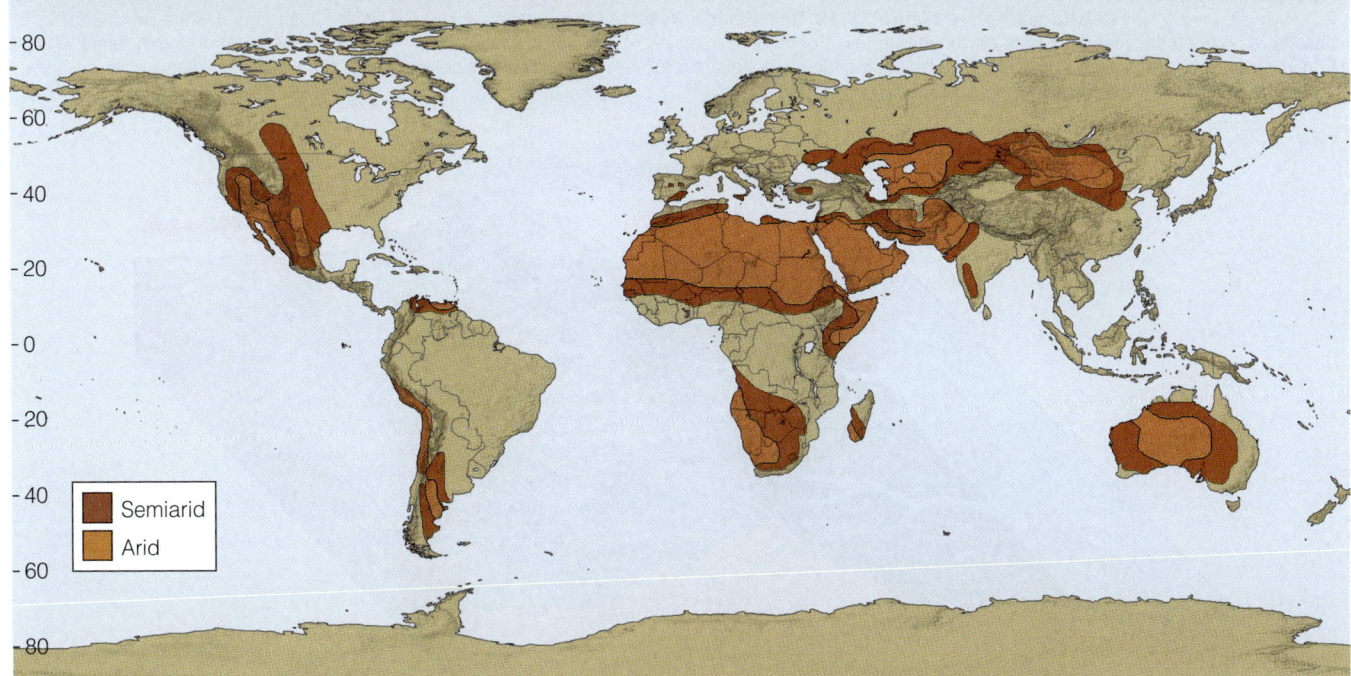

▶ **Figure 18.18 The Distribution of Earth's Arid and Semiarid Regions** Semiarid regions receive more precipitation than arid regions, yet they are still moderately dry. Arid regions, generally described as deserts, are dry and receive less than 25 cm of rain per year. The majority of the world's deserts are located in the dry climates of the low and middle latitudes.

The majority of the world's deserts are in the dry climates of the low and middle latitudes (▶ Figure 18.18). In North America most of the southwestern United States and northern Mexico are characterized by this hot, dry climate, whereas in South America this climate is primarily restricted to the Atacama Desert of coastal Chile and Peru. The Sahara in northern Africa, the Arabian Desert in the Middle East, and the majority of Pakistan and western India form the largest essentially unbroken desert environment in the Northern Hemisphere. More than 40% of Australia is desert, and most of the rest of it is semiarid.

The remaining dry climates of the world are found in the middle and high latitudes, mostly within continental interiors in the Northern Hemisphere (▶ Figure 18.18). Many of these areas are dry because of their remoteness from moist maritime air and the presence of mountain ranges that produce a **rainshadow desert** (▶ Figure 18.19). When moist marine air moves inland and meets a mountain range, it is forced upward. As it rises, it cools, forming clouds and producing precipitation that falls on the windward side of the mountains. The air that descends on the leeward side of the mountain range is much warmer and drier, producing a rainshadow desert.

Three widely separated areas are included within the mid-latitude dry-climate zone (▶ Figure 18.18). The largest is the central part of Eurasia extending from just north of the Black Sea eastward to north-central China. The Gobi Desert in China is the largest desert in this region. The Great Basin area of North America is the second largest mid-latitude dry-climate zone and results from the rainshadow produced by the Sierra Nevada. This region adjoins the southwestern deserts of the United States that formed as a result of the low-latitude subtropical high-pressure zone.

The smallest of the mid-latitude dry-climate areas is the Patagonian region of southern and western Argentina. Its dryness results from the rainshadow effect of the Andes. The remainder of the world's deserts are found in the cold but dry high latitudes, such as Antarctica.

Section 18.6 Summary

• Dry climates cover 30% of Earth's land surface and are subdivided into semiarid and arid regions. Semiarid regions receive more precipitation than arid regions, yet they are moderately dry. Arid regions, generally described as deserts, are dry. They receive less than 25 cm of rain per year, have high evaporation rates, typically have poorly developed soils, and are mostly or completely devoid of vegetation.

• The majority of the world's deserts are in the dry climates of the low and middle latitudes. The remaining dry climates of the world are found in the middle and high latitudes, mostly within continental interiors in the Northern Hemisphere.

18.7 Characteristics of Deserts

■ *What are the characteristics of deserts?*

To people who live in humid regions, deserts may seem stark and inhospitable. Instead of a landscape of rolling hills and gentle slopes with an almost continuous cover of vege-

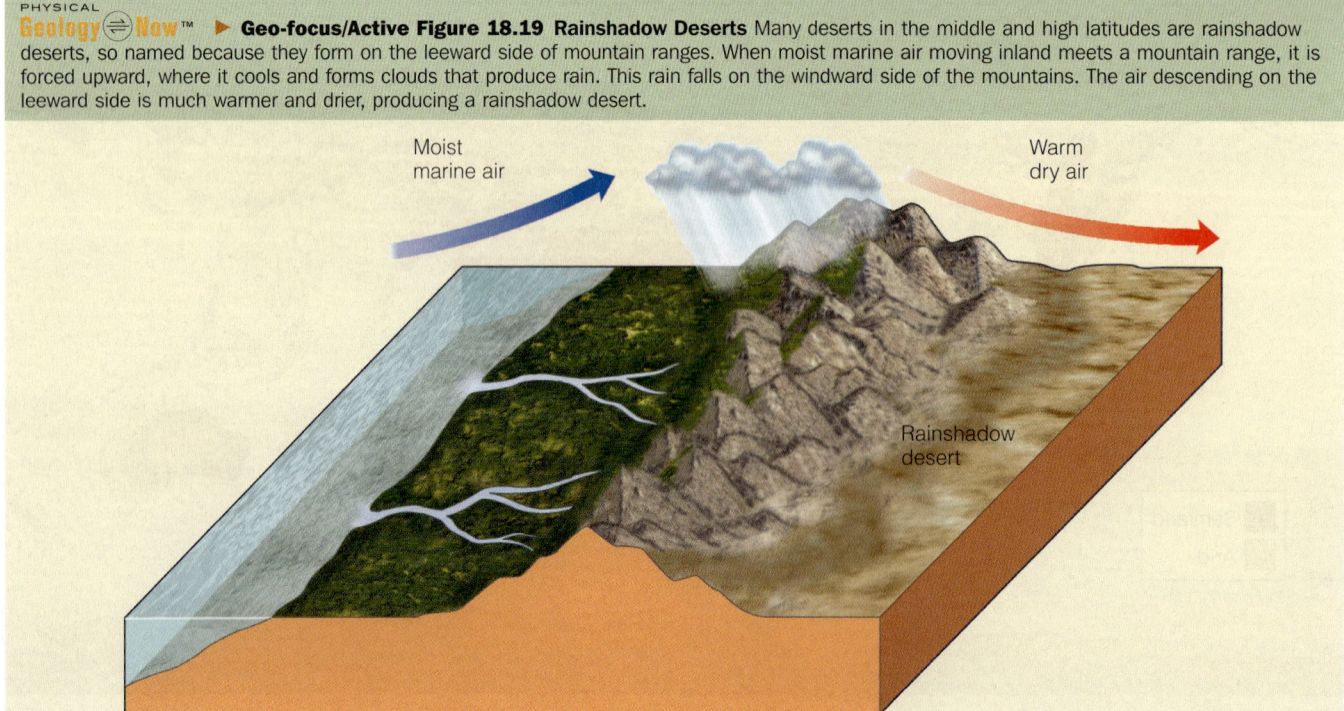

PHYSICAL Geology Now™ ▶ **Geo-focus/Active Figure 18.19 Rainshadow Deserts** Many deserts in the middle and high latitudes are rainshadow deserts, so named because they form on the leeward side of mountain ranges. When moist marine air moving inland meets a mountain range, it is forced upward, where it cools and forms clouds that produce rain. This rain falls on the windward side of the mountains. The air descending on the leeward side is much warmer and drier, producing a rainshadow desert.

tation, deserts are dry, have little vegetation, and consist of nearly continuous rock exposures, desert pavement, or sand dunes. Yet despite the great contrast between deserts and more humid areas, the same geologic processes are at work, only operating under different climatic conditions.

Temperature, Precipitation, and Vegetation

The heat and dryness of deserts are well known. Many of the deserts of the low latitudes have average summer temperatures that range between 32° and 38°C. It is not uncommon for some low-elevation inland deserts to record daytime highs of 46° to 50°C for weeks at a time. The highest temperature ever recorded was 58°C in El Azizia, Libya, on September 13, 1922.

During the winter months when the Sun's angle is lower and there are fewer daylight hours, daytime temperatures average between 10° and 18°C. Winter nighttime lows can be quite cold, with frost and freezing temperatures common in the more poleward deserts. Winter daily temperature fluctuations in low-latitude deserts are among the greatest in the world, ranging between 18° and 35°C. Temperatures have been known to fluctuate from below 0°C to higher than 38°C in a single day!

The dryness of the low-latitude deserts results primarily from the year-round dominance of the subtropical high-pressure belt, whereas the dryness of the mid-latitude deserts is due to their isolation from moist marine winds and the rainshadow effect created by mountain ranges. The dryness of both is further accentuated by their high temperatures.

Although deserts are defined as regions that receive, on average, less than 25 cm of rain per year, the amount of rain that falls each year is unpredictable and unreliable. It is not uncommon for an area to receive more than an entire year's average rainfall in one cloudburst and then to receive very little rain for several years. Thus yearly rainfall averages can be misleading.

Deserts display a wide variety of vegetation (▶ Figure 18.20). Although the driest deserts, or those with large areas of shifting sand, are almost devoid of vegetation, most deserts support at least a sparse plant cover. Compared to the vegetation in humid areas, desert vegetation may appear monotonous. A closer examination, however, reveals an amazing diversity of plants that have evolved the ability to live in the near absence of water.

Desert plants are widely spaced, typically small, and slow growing. Their stems and leaves are usually hard and waxy to minimize water loss by evaporation and protect the plant from sand erosion. Most plants have a widespread, shallow root system to absorb the dew that forms each morning in all but the driest deserts and to help anchor the plant in what little soil there may be. In extreme cases, many plants lie dormant during particularly dry years and spring to life after the first rain shower, with a beautiful profusion of flowers.

Weathering and Soils

Mechanical weathering is dominant in desert regions. Daily temperature fluctuations and frost wedging are the primary forms of mechanical weathering (see Chapter 6). The breakdown of rocks by roots and from salt crystal growth is of minor importance. Some chemical weathering does occur, but its rate is greatly reduced by aridity and the scarcity of organic acids produced by the sparse vegetation. Most chemical weathering takes place during the winter months when more precipitation occurs, particularly in the mid-latitude deserts.

An interesting feature seen in many deserts is a thin, red, brown, or black shiny coating on the surface of many rocks (see "Rock Art for the Ages" on pp. 584–585). This coating, called *rock varnish*, is composed of iron and manganese oxides (▶ Figure 18.21). Because many of the varnished rocks contain little or no iron and manganese oxides, the varnish is thought to result either from wind-blown iron and manganese dust that settles on the ground or from the precipitated waste of microorganisms. Rock varnish can form in moist environments as well; it is simply easier to see in the lightly vegetated desert landscape.

Desert soils, if developed, are usually thin and patchy because the limited rainfall and the resultant scarcity of

▶ **Figure 18.20 Desert Vegetation** Desert vegetation is typically sparse, widely spaced, and characterized by slow growth rates. The vegetation shown here in Organ Pipe National Monument, Arizona, includes saguaro and cholla cacti, paloverde trees, and jojoba bushes and is characteristic of the vegetation found in the Sonoran Desert of North America.

Rock Art for the Ages

Rock art includes rock paintings (where paints made from natural pigments are applied to a rock surface) and petroglyphs (from the Greek *petro*, meaning "rock," and *glyph*, meaning "carving or engraving"), which are the abraded, pecked, incised, or scratched marks made by humans on boulders, cliffs, and cave walls.

Rock art has been found on every continent except Antarctica and is a valuable archaeological resource that provides graphic evidence of the cultural, social, and religious relationships and practices of ancient peoples. The oldest known rock art was made by hunters in western Europe and dates back to the Pleistocene Epoch. Africa has more rock art sites than any other continent. The oldest known African rock art, found in the southern part of the continent, is estimated to be 27,000 years old.

Petroglyphs are a fragile and nonrenewable cultural resource that cannot be replaced if they are damaged or destroyed. A commitment to their preservation is essential so that future generations can study them as well as enjoy their beauty and mystery.

In the arid Southwest and Great Basin of North America where rock art is plentiful, rock paintings and petroglyphs extend back to about 2000 B.C. Here, rock art can be divided into two categories. *Representational art* deals with life-forms such as humans, birds, snakes, and human-like supernatural beings. Rarely exact replicas, they are more or less stylized versions of the beings depicted. *Abstract art*, in contrast, bears no resemblance to any real-life images.

▲ **1.** Various petroglyphs exposed at an outcrop along Cub Creek Road in Dinosaur National Monument, Utah.

◀ **2.** A human-like petroglyph, an example of representational art, exposed at an outcrop along Cub Creek Road in Dinosaur National Monument, Utah. Note the contrast between the fresh exposure of the rock where the upper part of the petroglyph's head has been removed, the weathered brown surface of the rest of the petroglyph, and the black rock varnish coating the rock surface.

Petroglyphs are the most common form of rock art in North America and are made by pecking, incising, or scratching the rock surface with a tool harder than the rock itself. In arid regions, many rock surfaces display a patination, or thin brown or black coating, known as rock varnish (see Figure 18.21). When this coating is broken by pecking, incising, or scratching, the underlying lighter colored natural rock surface provides an excellent contrast for the petroglyphs.

Petroglyphs are especially abundant in the Southwest and Great Basin area, where they occur by the thousands, having been made by Native Americans from many cultures during the past several thousand years. Petroglyphs can be seen in many of the U.S. national parks and monuments, such as Petrified Forest National Park, Arizona; Dinosaur National Monument, Utah; Canyonlands National Park, Utah; and Petroglyph National Monument, New Mexico—to name a few.

▲ **3.** Examples of both representational and abstract art are displayed in these petroglyphs from Arizona.

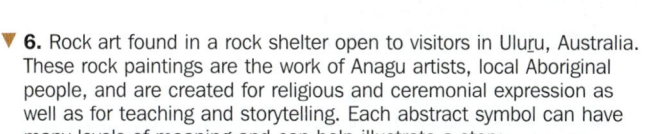

▶ **4.** Several petroglyphs on basalt boulders in Rinconada Canyon, Petroglyph National Monument, Albuquerque, New Mexico.

▼ **6.** Rock art found in a rock shelter open to visitors in Uluru, Australia. These rock paintings are the work of Anangu artists, local Aboriginal people, and are created for religious and ceremonial expression as well as for teaching and storytelling. Each abstract symbol can have many levels of meaning and can help illustrate a story.

◀▲ **5.** Examples of rock art from Tassili-n-Ajjer, Algeria.

◀ **7.** The paints used in this rock art are made from natural mineral substances mixed with water and sometimes with animal fat. Red, yellow, and orange pigments come from iron-stained clays, whereas the white pigments come from either ash or the mineral calcite. The black colors come from charcoal.

▶ **Figure 18.21 Rock Varnish** The shiny black coating on this rock exposed at Castle Valley, Utah, is rock varnish. It is composed of iron and manganese oxides.

vegetation reduce the efficiency of chemical weathering and hence soil formation. Furthermore, the sparseness of the vegetative cover enhances wind and water erosion of what little soil actually forms.

Mass Wasting, Streams, and Groundwater

When traveling through a desert, most people are impressed by such wind-formed features as moving sand, sand dunes, and sand and dust storms. They may also notice the dry washes and dry streambeds. Because of the lack of running water, most people would conclude that wind is the most important erosional agent in deserts. They would be wrong! Running water, even though it occurs infrequently, causes most of the erosion in deserts (see Figure 15.2). The dry conditions and sparse vegetation characteristic of deserts enhance water erosion. If you look closely, you will see the evidence of erosion and transportation by running water nearly everywhere except in areas covered by sand dunes.

Most of a desert's average annual rainfall of 25 cm or less comes in brief, heavy, localized cloudbursts. During these times, considerable erosion takes place because the ground cannot absorb all of the rainwater. With so little vegetation to hinder the flow of water, runoff is rapid, especially on moderately to steeply sloping surfaces, resulting in flash floods and sheetflows. Dry stream channels quickly fill with raging torrents of muddy water and mudflows, which carve out steep-sided gullies and overflow their banks. During these times, a tremendous amount of sediment is rapidly transported and deposited far downstream.

Although water is the major erosive agent in deserts today, it was even more important during the Pleistocene Epoch when these regions were more humid (see Chapter 17). During that time, many of the major topographic features of deserts were forming. Today that topography is being modified by wind and infrequently flowing streams.

Most desert streams are poorly integrated and flow only intermittently. Many of them never reach the sea because the water table is usually far deeper than the channels of most streams, so they cannot draw upon groundwater to replace water lost to evaporation and absorption into the ground. This type of drainage in which a stream's load is deposited within the desert is called *internal drainage* and is common in most arid regions.

Although most deserts have internal drainage, some deserts have permanent through-flowing streams such as the Nile and Niger Rivers in Africa, the Rio Grande and Colorado River in the southwestern United States, and the Indus River in Asia. These streams can flow through desert regions because their headwaters are well outside the desert and water is plentiful enough to offset losses resulting from evaporation and infiltration. Demands for greater amounts of water for agriculture and domestic use from the Colorado River, however, are leading to increased salt concentrations in its lower reaches and causing political problems throughout the region.

Wind

Although running water does most of the erosional work in deserts, wind can also be an effective geologic agent capable of producing a variety of distinctive erosional and depositional features (▶ Figure 18.3). Wind is effective in transporting and depositing unconsolidated sand-, silt-, and dust-sized particles. Contrary to popular belief, most deserts are not sand-covered wastelands but rather vast areas of rock exposures and desert pavement. Sand-covered regions, or sandy deserts, constitute less than 25% of the world's deserts. The sand in these areas has accumulated primarily by the action of wind.

Section 18.7 Summary

- Although deserts have different features than more humid regions, the same geologic processes are at work, only operating under different climatic conditions. Mechanical weathering is dominant in deserts, and soil, if developed, is usually thin and patchy.

- Deserts are known for their hot, dry conditions, yet they contain a wide variety of vegetation that has adapted to these conditions. Rainfall, though sparse, is usually concentrated in cloudbursts. During these times, considerable erosion takes place, and tremendous amounts of sediment are rapidly transported and deposited elsewhere. Most desert streams are poorly integrated, flow only intermittently, and rarely reach the sea. When a stream's load is deposited within the desert, the condition is known as internal drainage.

- Although running water is the dominant erosional agent in deserts, wind produces a variety of erosional and depositional features.

18.8 Desert Landforms

■ *What are the major landforms of deserts and how do they form?*

Because of differences in temperature, precipitation, and wind, as well as the underlying rocks and recent tectonic events, landforms in arid regions vary considerably. Running water, although infrequent in deserts, is responsible for producing and modifying many of the distinctive landforms found there.

After an infrequent and particularly intense rainstorm, excess water not absorbed by the ground may accumulate in low areas and form *playa lakes* (▶ Figure 18.22a). These lakes are temporary, lasting from a few hours to several months. Most of them are shallow and have rapidly shifting boundaries as water flows in or leaves by evaporation and seepage into the ground. The water is often very saline.

When a playa lake evaporates, the dry lakebed is called a **playa,** or *salt pan*, and is characterized by mud cracks and precipitated salt crystals (▶ Figures 18.22b, c). Salts in some playas are thick enough to be mined commercially. For example, borates have been mined in Death Valley, California, for more than 100 years.

Other common features of deserts, particularly in the Basin and Range region of the United States, are alluvial fans and bajadas. **Alluvial fans** form when sediment-laden streams flowing out from the generally straight, steep mountain fronts deposit their load on the relatively flat desert floor. Once beyond the mountain front where no valley walls contain streams, the sediment spreads out laterally, forming a gently sloping and poorly sorted fan-shaped sedimentary deposit (▶ Figure 18.23). Alluvial fans are similar in origin and shape to deltas (see Chapter 15) but are formed entirely on land. Alluvial fans may coalesce to form a *bajada*, a broad alluvial apron that typically has an undulating surface resulting from the overlap of adjacent fans (▶ Figure 18.24).

▶ **Figure 18.22 Playas and Playa Lakes**

a A playa lake formed after a rainstorm filled Croneis Dry Lake, Mojave Desert, California. Playa lakes are ephemeral features, lasting from a few hours to several months.

b A playa is the dry lakebed that remains after the water in a playa lake evaporates. Racetrack Playa, Death Valley, California. The Inyo Mountains can be seen in the background.

c Salt deposits and salt ridges cover the floor of this playa in the Mojave Desert, California. Salt crystals and mud cracks are characteristic features of playas.

▶ **Figure 18.23 Alluvial Fan** An aerial view of an alluvial fan, Death Valley, California. Alluvial fans form when sediment-laden streams flowing out from a mountain deposit their load on the relatively flat desert floor, forming a gently sloping, fan-shaped, sedimentary deposit.

Large alluvial fans and bajadas are frequently important sources of groundwater for domestic and agricultural use. Their outer portions are typically composed of fine-grained sediments suitable for cultivation, and their gentle slopes allow good drainage of water. Many alluvial fans and bajadas are also the sites of large towns and cities, such as San Bernardino, California; Salt Lake City, Utah; and Teheran, Iran.

Most mountains in desert regions, including those of the Basin and Range region, rise abruptly from gently sloping surfaces called pediments. **Pediments** are erosional bedrock surfaces of low relief that slope gently away from mountain bases (▶ Figure 18.25). Most pediments are covered by a thin layer of debris, alluvial fans, or bajadas.

The origin of pediments has been the subject of much controversy. Most geologists agree that they are erosional features developed on bedrock in association with the erosion and retreat of a mountain front (▶ Figure 18.25a). The disagreement concerns how the erosion has occurred. Although not all geologists would agree, it appears that pediments are produced by the combined activities of lateral erosion by streams, sheet flooding, and various weathering processes along the retreating mountain front. Thus pediments grow at the expense of the mountains, and they continue to expand as the mountains are eroded away or partially buried.

Rising conspicuously above the flat plains of many deserts are isolated steep-sided erosional remnants called *inselbergs*, a German word meaning "island mountain." Inselbergs have survived for a longer period of time than other mountains because of their resistance to weathering. Uluṟu (formerly known as Ayers Rock) is an excellent example of an inselberg (see "Uluṟu and Kata Tjuṯa" in Chapter 9 on pp. 276–277).

Other easily recognized erosional remnants common to arid and semiarid regions are mesas and buttes (▶ Figure 18.26). A **mesa** is a broad, flat-topped erosional remnant bounded on all sides by steep slopes. Continued weathering and stream erosion form isolated pillarlike structures known as **buttes.** Buttes and mesas consist of

▶ **Figure 18.24 Bajada** Coalescing alluvial fans form a bajada at the base of these mountains in Death Valley, California.

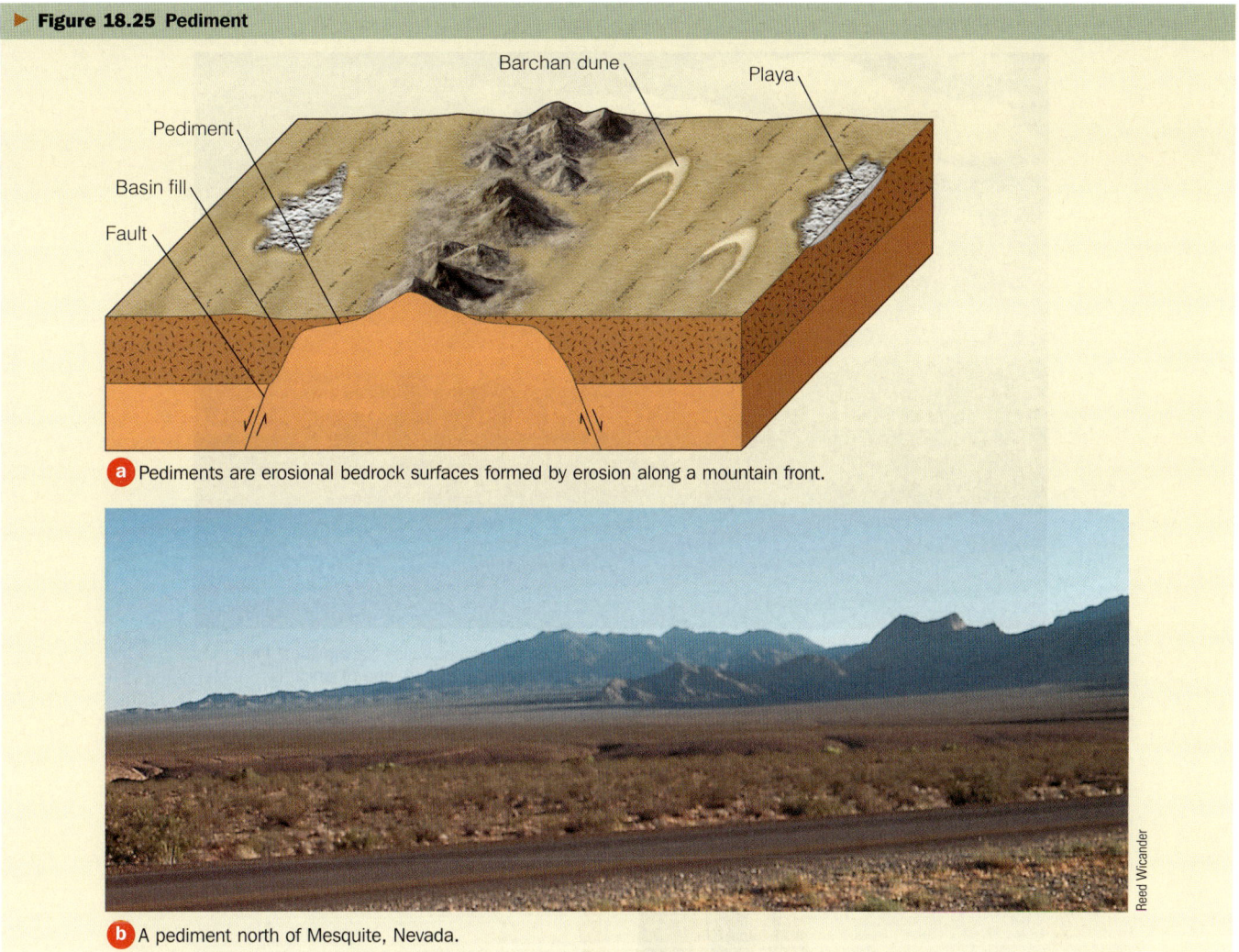

▶ **Figure 18.25 Pediment**

a Pediments are erosional bedrock surfaces formed by erosion along a mountain front.

b A pediment north of Mesquite, Nevada.

relatively easily weathered sedimentary rocks capped by nearly horizontal, resistant rocks such as sandstone, limestone, or basalt. They form when the resistant rock layer is breached, which allows rapid erosion of the less resistant underlying sediment. One of the best-known areas of mesas and buttes in the United States is Monument Valley on the Arizona–Utah border (▶ Figure 18.26).

Section 18.8 Summary

- One type of landform found in deserts is playa lakes, which are temporary lakes that form after an intense rainstorm when the excess water cannot be absorbed. Evaporation of these lakes forms playas, in which the dry lakebed is characterized by mud cracks and precipitated salt crystals.

- Alluvial fans are fan-shaped structures formed by the deposition of sediments at the base of a mountain front. Coalescing alluvial fans form bajadas. Pediments are erosional bedrock surfaces of low relief that slope gently away from mountain bases. Inselbergs are isolated, steep-sided erosional remnants that rise above the flat plains of many deserts. Mesas are broad, flat-topped erosional remnants bounded on all sides by steep slopes. Buttes are similar to mesas but more pillarlike.

Figure 18.26 Buttes and Mesas

a Several mesas and buttes can be seen in this aerial view of Monument Valley Navajo Tribal Park, Navajo Indian Reservation. The mesas are the broad, flat-topped structures, one of which is prominent in the foreground, whereas the buttes are more pillarlike structures.

b Left Mitten Butte and Right Mitten Butte in Monument Valley on the border of Arizona and Utah.

Review Workbook

ESSENTIAL QUESTIONS SUMMARY

18.1 Introduction
■ *What is desertification and why is it a major problem in the world?*
Desertification is the expansion of deserts into formerly productive lands. It destroys croplands and rangelands, causing massive starvation and forcing hundreds of thousands of people from their homelands.

■ *Why is it important to understand how desert processes operate?*
By understanding how desert processes operate and how global climate changes affect Earth's systems and subsystems, people can take steps to reduce the spread of desertification and its effects and to better understand and deal with present-day environmental changes.

18.2 Sediment Transport by Wind
■ *What is bed load?*
Bed load is the material that is too large or heavy to be carried in suspension by water or wind. Particles are moved along the surface by saltation, rolling, or sliding.

■ *What is suspended load?*
Suspended load is the material that can be carried in suspension by water or wind. Silt- and clay-sized particles constitute most of a wind's suspended load, and they can be carried thousands of kilometers.

18.3 Wind Erosion
■ *What is abrasion and what type of products are formed by abrasion?*
Abrasion is the impact of saltating sand grains on an object. Its effect is similar to sandblasting. Ventifacts, which are stones whose surfaces have been polished, pitted, grooved, or faceted by the wind, are common products of wind abrasion. Larger objects include yardangs, which are elongated, streamlined ridges that look like an overturned ship's hull.

■ *What is deflation and what features are formed by this process?*
Deflation is the removal of loose surface material by wind. Deflation hollows resulting from differential erosion of surface material are common features of many deserts, as is desert pavement, which effectively protects the underlying surface from additional deflation.

18.4 Wind Deposits
■ *How do dunes form and migrate?*
Dunes are mounds or ridges of wind-deposited sand that form when wind flows over and around an obstruction, resulting in the deposition of sand grains, which accumulate and build up a deposit of sand. Most dunes have an asymmetric profile, with a gentle windward slope and a steeper downwind, or leeward, slope that is inclined in the direction of the prevailing wind. Dunes migrate by sand moving up and over the gentle windward slope by saltation and accumulating and sliding down the leeward side.

■ *What are the four major dune types and how do they form?*
The four major dune types are barchan, longitudinal, transverse, and parabolic. Barchan dunes are crescent-shaped dunes whose tips point downwind and form in areas that have a generally flat, dry surface with little vegetation, a limited supply of sand, and a nearly constant wind direction. Longitudinal dunes are long, parallel ridges of sand aligned generally parallel to the direction of the prevailing winds. They form where the sand supply is somewhat limited. Transverse dunes form long ridges perpendicular to the prevailing wind direction in areas that have abundant sand and little or no vegetation. Parabolic dunes are most common in coastal areas that have abundant sand, strong onshore winds, and a partial cover of vegetation.

■ *What is loess, how does it form, and why is it important?*
Loess is wind-blown silt and clay deposits composed of angular quartz grains, feldspar, micas, and calcite. Loess is derived from deserts, Pleistocene glacial outwash deposits, and river floodplains in semiarid regions. Loess covers approximately 10% of Earth's land surface and weathers to a rich, productive soil.

18.5 Air-Pressure Belts and Global Wind Patterns
■ *How do Earth's air-pressure belts and the Coriolis effect determine global wind patterns?*
The winds of the major air-pressure belts, oriented east-west, result from the rising and cooling of air. The winds are deflected clockwise in the Northern Hemisphere and counterclockwise in the Southern Hemisphere by the Coriolis effect to produce Earth's global wind patterns.

18.6 The Distribution of Deserts
■ *Where are deserts located?*
Dry climates cover 30% of Earth's land surface and are subdivided into semiarid and arid regions. Semiarid regions receive more precipitation than arid regions, yet are moderately dry. Arid regions, generally described as deserts, are dry and receive less than 25 cm of rain per year. The majority of the world's deserts are located in the low-latitude dry-climate zone between 20 and 30 degrees north and south latitudes. Their dry climate results from a high-pressure belt of descending dry air. The remaining deserts are in the middle latitudes, where their distribution is related to the rainshadow effect, and in the dry polar regions.

18.7 Characteristics of Deserts
■ *What are the characteristics of deserts?*
Deserts are characterized by high temperatures, little precipitation, and a sparse plant cover. Mechanical weathering is the dominant form of weathering and, coupled with slow rates of chemical weathering, results in poorly developed soils. Running water is the major agent of erosion in deserts, with most streams being poorly integrated and flowing intermittently. Wind, though secondary to water as an erosional agent in deserts, is still capable of producing a variety of distinctive erosional and depositional features.

18.8 Desert Landforms
■ *What are the major landforms of deserts and how do they form?*
Important desert landforms include playa lakes, playas, alluvial fans, bajadas, pediments, inselbergs, buttes, and mesas. Playas are dry lakebeds characterized by mud cracks and precipitated salt crystals. When temporarily filled with water following a rainstorm, they are known as playa lakes. Alluvial fans are fan-shaped sedimentary deposits that form when sediment-laden streams flow out from mountain fronts and deposit their load on the relatively flat desert floor. Coalescing alluvial fans form bajadas. Pediments are erosional bedrock surfaces of low relief that slope gently away from mountain bases. Inselbergs are isolated, steep-sided erosional remnants that rise above desert plains. Buttes and mesas are, respectively, pillarlike and flat-topped erosional remnants with steep slopes.

ESSENTIAL TERMS TO KNOW

abrasion (p. 570)
alluvial fan (p. 587)
barchan dune (p. 575)
butte (p. 588)
Coriolis effect (p. 581)
deflation (p. 571)
desert (p. 581)

desert pavement (p. 573)
desertification (p. 568)
dune (p. 574)
loess (p. 578)
longitudinal dune (p. 575)
mesa (p. 588)
parabolic dune (p. 577)

pediment (p. 588)
playa (p. 587)
rainshadow desert (p. 582)
transverse dune (p. 575)
ventifacts (p. 570)

REVIEW QUESTIONS

1. If deserts are dry regions in which mechanical weathering predominates, why are so many of their distinctive landforms the result of running water and not of wind?

2. Using what you now know about deserts, their location, how they form, and the various landforms associated with them, how can you determine where deserts may have existed in the past?

3. Much of the recent desertification has been greatly accelerated by human activity. Can anything be done to slow the process?

4. Why are so many desert rock formations red?

5. Why is desert pavement important in a desert environment?

6. Is it possible for Mars to have the same type of sand dunes as Earth? What would that tell us about the climate and geology of Mars?

7. How do dunes form and migrate? Why is dune migration a problem in some areas?

8. Why are low-latitude deserts so common?

9. What is loess, and why is it important?

10. As noted in the text, many large towns and cities are built on alluvial fans. What are the advantages and disadvantages of such a location?

APPLY YOUR KNOWLEDGE

1. As more people move into arid and semiarid areas such as Las Vegas, Nevada, Phoenix, Arizona, and southern California, an increasing strain is placed on the environment of these areas. What are some of the issues that government entities must face in dealing with these population increases? What are some of the problems that must be dealt with from a geologic perspective?

FIELD QUESTIONS

1. This photograph shows a desert windstorm in progress. What other desert features, landforms, and processes are visible? You should be able to recognize at least three features, landforms, or processes that are characteristic of deserts.

2. What desert landforms are present in this photograph? You should be able to recognize at least three landforms.

GEOLOGY IN FOCUS: Radioactive Waste Disposal—Safe or Sorry?

One problem of the nuclear age is finding safe storage sites for radioactive waste from nuclear power plants, the manufacture of nuclear weapons, and the radioactive by-products of nuclear medicine. Radioactive waste can be grouped into two categories: low-level and high-level waste. Low-level wastes are low enough in radioactivity that, when properly handled, they do not pose a significant environmental threat. Most low-level wastes can be safely buried in controlled dump sites where the geology and groundwater system are well known and careful monitoring is provided.

High-level radioactive waste, such as the spent uranium-fuel assemblies used in nuclear reactors and the material used in nuclear weapons, is extremely dangerous because of large amounts of radioactivity; it therefore presents a major environmental problem. Currently some 77,000 tons of spent nuclear fuel are temporarily stored at 131 sites in 39 states while awaiting shipment to a permanent site in Yucca Mountain, Nevada (▶ Figure 1).

In 1987 Congress amended the Nuclear Waste Policy Act and directed the Department of Energy (DOE) to study only Yucca Mountain as a possible nuclear waste repository. With the passage and signing by President George W. Bush of House Joint Resolution 87 in 2002, the way was cleared for the DOE to prepare an application for a Nuclear Regulatory Commission license to begin constructing a nuclear waste repository at Yucca Mountain. When completed, the repository could begin receiving shipments of spent nuclear fuel by 2010.

Why Yucca Mountain? After more than 20 years of study and $8 billion spent on researching the region, many scientists think Yucca Mountain has the features necessary to isolate high-level nuclear waste from the environment for at least 10,000 years, which is the minimum time the waste will remain dangerous. What makes Yucca Mountain so appealing is its remote location and long distance from a large population center—in this case, Las Vegas, which is about 166 km southeast of Yucca Mountain. It also has a very dry climate, with less than 15 cm of rain per year, and a deep water table that is 500–720 m below the repository. In fact, the radioactive waste will be buried in volcanic tuff at a depth of about 300 m in canisters designed to remain leakproof for at least 300 years.

What then are the concerns and why the opposition to Yucca Mountain? One of the main concerns is whether the climate will change during the next 10,000 years. If the region should become more humid, more water would percolate through the zone of aeration. This would increase the corrosion rate of the canisters and could cause the water table to rise, thereby decreasing the travel time between the repository and the zone of saturation. This area of the country was much more humid during the Ice Age, 1.8 million to 10,000 years ago (see Chapter 17).

Another concern is the seismic activity in the area. It is in fact riddled with faults and has experienced numerous earthquakes during historic time. Nevertheless, based on underground inspections at Yucca Mountain as well as the tunnels at the Nevada Test site nearby, the DOE is convinced that earthquakes pose little danger to the underground repository itself because the disruptive effects of an earthquake are usually confined to the surface. Furthermore, it is required that the facilities, both above and below ground, be designed to withstand any severe earthquake likely to strike the area.

Finally, some people worry about the possibility of sabotage to the facility as well as the problems of transporting high-level nuclear waste to the facility. Being buried 300 m below ground renders it virtually impenetrable to acts of terrorism or sabotage, but the possibility of an accident or terrorist attack on the way to the repository is still a concern to many.

Although it appears that Yucca Mountain meets all the requirements for a safe high-level radioactive waste repository, the site is still controversial and at the time of this writing still embroiled in lawsuits seeking to block its construction and funding battles in Congress.

Figure 1
The location and aerial view of Nevada's Yucca Mountain, the proposed high-level radioactive waste site.

GEOLOGY MATTERS
GEOLOGY IN UNEXPECTED PLACES

Spiral Jetty

Much like the numerous ephemeral playa lakes found throughout Utah and the Southwest, the *Spiral Jetty* is again slowly disappearing from view. Constructed in 1970 by artist Robert Smithson (1938–1973), the *Spiral Jetty* is a large, 5-m-wide, 457-m-long spiral that coils counterclockwise outward from the shoreline of the Great Salt Lake, Utah (▶ Figure 1). Made of black olivine basalt from the nearby shoreline of Rozel Point, the *Spiral Jetty* appears and disappears from view in conjunction with the falling and rising lake levels. Thus its visibility is in harmony with the natural lake-level cycles of the Great Salt Lake. Two years after the spiral earthwork sculpture was created, Smithson (▶ Figure 2) is quoted as saying in a Dia Art Foundation press release: "As an artist, it is interesting to take on the persona of a geological agent and actually become part of that process rather than overcome it."

It was during a dry period in 1970, when the level of the Great Salt Lake was lower than its historic average, that Smithson created his earthwork sculpture. Following its creation, spectators could walk along the spiral; however, rising water levels soon submerged the structure. Its appearance and disappearance are dictated by the rise and fall of lake levels in the Great Salt Lake, which are ultimately dependent on weather cycles. During the past 35 years the *Spiral Jetty* has been submerged several times, the longest period between 1982 and 1992 when the lake level rose more than 5 m. Following a 1999 drought in which the lake level steadily dropped, Smithson's creation reappeared again in 2002. However, its appearance may be somewhat short-lived as rising water levels may soon submerge it again.

Precipitation and temperature are the two factors that determine the level of the Great Salt Lake. The more snow and rain, the more water there is to fill the shallow playa the Great Salt Lake occupies. Because there are no outlets from the lake, however, evaporation is the primary way in which water escapes. Temperature therefore plays a critical role in determining the amount of water that evaporates from the lake. The longer the temperatures remain high, the greater the amount of evaporation that can take place. There is thus a dynamic balance between the amount of water entering the lake from winter's melting snow packs and rain, and how much of that water will evaporate due to spring and summer temperatures. If inflow is high and temperatures remain relatively cool, then the lake level rises. If drought conditions prevail, evaporation is dominant and lake levels fall.

Even though the *Spiral Jetty* is disappearing again, it can still delight and intrigue those who visit it, with its contrasting black contours set beneath the reddish salty waters of the Great Salt Lake, where algae, bacteria, and brine shrimp thrive to give it its rust-colored hue.

Figure 1
Spiral Jetty, the earthwork sculpture created by artist Robert Smithson, can be seen coiling outward from the shore of the Great Salt Lake, Utah. Whether or not this 457-m-long, 5-m-wide spiral can be seen depends on the lake level. During periods of drought, the lake level is low and the sculpture is visible, but during periods of abundant rainfall and snow, the lake level rises, submerging the sculpture from view.

Figure 2
American artist and sculptor Robert Smithson, creator of the earthwork sculpture *Spiral Jetty*, is shown here sitting in a director's chair on November 7, 1969.

GEOLOGY MATTERS

GEOLOGY IN YOUR LIFE

Windmills and Wind Power

Whoosh, whoosh, whoosh. Ah, the gentle sound of a windmill's blades turning in the wind. A pastoral landscape in Holland dominated by a classic Dutch windmill is the image most people associate with windmills (▶ Figure 1), or Don Quixote tilting at windmills in the famous novel *Don Quixote de la Mancha* by Miguel Cervantes. Today, instead of a whoosh, whoosh, whoosh, the sound of modern electricity-generating windmills in a wind farm is more like a woomph, woomph, woomph (▶ Figure 2).

As early as 5000 B.C. people began to harness the power of wind to propel boats along the Nile River. The Chinese used windmills to pump water for irrigating crops as long ago as 200 B.C. Wind power was used in the Middle Ages in Europe, particularly in Holland, where windmills have played an important role in society. Windmills were first used to grind corn, which is where the term *windmill* originally came from. Later windmills were used to drain lakes and marshes from the low-lying districts and to saw timber. Settlers in the United States in the late 19th and early 20th centuries used this technology to pump water and generate electricity in the Great Plains.

With the application of steam power and industrialization in Europe and later the United States, the use of windmills rapidly declined. However, industrialization led to the development of larger and more efficient windmills exclusively designed to generate electricity. Denmark began using such windmills as early as 1890, and other countries soon followed suit. Interest in electricity-generated windmills has always mirrored the price of fossil fuels. When the price of petroleum and coal is low, it is cheaper to use these fuels to generate electricity. When the price of fossil fuels goes up, interest in wind power increases. Today wind farms, which are the locations of groups of windmills, are an important source of electricity, both in the United States and elsewhere around the world, particularly Europe. As advances in wind turbine technology increase their efficiency, the price of wind-generated electricity has decreased significantly, such that wind farms can compete in many areas with traditional fossil fuel–burning power plants.

How do windmills produce electricity? Simply stated, wind turbines (the term commonly used to describe electricity-producing windmills) convert the kinetic energy (the energy of an object due to motion) of the wind to mechanical power—in this case, the generation of electricity. This electricity is sent to the local power grid, where it is distributed throughout the area.

In order to be effective, numerous wind turbines are clustered together on wind farms that are located in areas with relatively strong, steady winds. These farms can range from several turbines to thousands, as in California (▶ Figure 2). In addition to wind farms on land, wind farms offshore are becoming more popular because they are out of sight and people cannot hear the blades turning. In fact, wind installations in the North Sea already provide approximately 20% of Denmark's energy needs. In the United States, wind power production currently accounts for about 9% of the total power produced by all means.

What are the advantages and disadvantages of wind power? First, the wind is a free, renewable energy source, so it cannot be used up. It is also a clean source of energy that doesn't pollute the water or atmosphere or contribute to greenhouse gases. Thus it reduces the consumption of fossil fuels. The land on which windmills are sited can still be used for farming and ranching and as a source of income to the landowner who leases the land to utilities. In addition, wind farms can benefit the local economy of rural areas as well as supply energy to remote areas.

There are some disadvantages. The major one is that the wind doesn't always blow with sufficient strength to be totally reliable, which necessitates backup generation. Furthermore good wind sites are frequently in remote areas, far from the areas where large quantities of electricity are needed, or in coastal areas where land is expensive and local residents do not want large wind turbines as neighbors. The initial start-up costs of a wind farm are usually higher than the costs of a conventional power plant. However, as the cost of wind power has decreased because of better technology, wind-generated electricity can compete favorably with traditional power plants in many areas. The "not in my backyard" opposition to wind farms can make siting a wind farm difficult. The major objection to wind farms is the noise generated by the turbines, although as the windmills are built taller and the turbines are more efficient at noise reduction, that is not the major concern it once was. Of course, aesthetics still plays a role in many people's objection to wind farms, and the fact that some birds are killed by the rotating blades is another issue. Studies have indicated, however, that the impact of wind turbines on bird mortality and injury is less than that of many other structures, such as buildings, power lines, and communication towers.

As the price of fossil fuel continues to rise and our dependency on foreign sources of oil increases, the use of a centuries-old staple, the windmill, albeit modernized, will continue to gain in popularity and use, both locally and regionally.

Figure 1
Five traditional Dutch windmills lined up along a canal at Kinderdiik, the Netherlands.

Figure 2
A windmill farm in the Coachella Valley, California. California has more windmills than any other state in the United States.

Chapter 19
Shorelines and Shoreline Processes

The Coast of Campeche
Like fabric that has been knotted, dyed, then unfurled to reveal an irregular pattern of color, Campeche's coast echoes a tie dye motif of brilliant blue and red. Campeche stretches across the western portion of Mexico's Yucatan peninsula. In this satellite image you can see part of the Terminos Lagoon, which is 72 km long and 20 km wide. The lagoon's entrance is sheltered by a barrier island, Isla Del Carmen. During the Earth's Mesozoic era, the Midwestern United States' shoreline was flooded by a tropical inland sea and resembled the coast of Campeche.
—A. W.

ESSENTIAL QUESTIONS TO ASK

19.1 Introduction
- What is the difference between a coast and a shoreline?
- How does the study of shorelines provide an example of interacting systems?

19.2 Shoreline Processes
- How do the Moon and Sun generate tides?
- What is wave base?
- What is wave refraction, and how does it generate nearshore currents?

19.3 Shoreline Erosion
- What processes account for erosion along shorelines?
- How do sea arches and sea stacks form?

19.4 Deposition along Shorelines
- What are summer beaches and winter beaches?
- What is a tombolo and how does one form?

19.5 The Nearshore Sediment Budget
- What is the nearshore sediment budget?

19.6 The Classification of Coasts
- How do depositional and erosional coasts differ?

19.7 Storm Waves and Coastal Flooding
- What causes the most fatalities during hurricanes?

19.8 Managing Coastal Areas as Sea Level Rises
- What factors influence whether sea level rises or falls?
- What measures do coastal communities use to minimize the effects of shoreline erosion?

GEOLOGY MATTERS

GEOLOGY IN UNEXPECTED PLACES:
Rising Sea Level and the Fate of Venice, Italy

GEOLOGY IN YOUR LIFE:
Giant Killer Waves

GEOLOGY AT WORK:
Erosion and the Cape Hatteras Lighthouse

GEOLOGY IN FOCUS:
Energy from the Oceans

Physical Geology Now — This icon, which appears throughout the book, indicates an opportunity to explore interactive tutorials, animations, or practice problems available on the Physical GeologyNow website at http://now.brookscole.com/phygeo6.

19.1 Introduction

■ *What is the difference between a coast and a shoreline?*

No doubt you know that a **shoreline** is the area of land in contact with the sea or a lake, but we can expand this definition by noting that ocean shorelines include the land between low tide and the highest level on land affected by storm waves. How does a *shoreline* differ from a *coast*? Actually, the terms are commonly used interchangeably, but *coast* is a more inclusive term that includes the shoreline as well as an area of indefinite width both seaward and landward of the shoreline. For instance, in addition to the shoreline area, a coast includes nearshore sandbars and islands, and areas on land such as wind-blown sand dunes, marshes, and cliffs (▶ Figure 19.1).

Our main concern in this chapter is ocean shorelines, or seashores, but waves and nearshore currents are also effective in large lakes; many erosional and depositional features typical of seashores are present along the shorelines of the Great Lakes (▶ Figure 19.2). Waves and nearshore currents are certainly more vigorous along seashores, and even the largest lakes have insignificant tides. Lake Superior has a tidal rise and fall of only about 2.5 cm, whereas tidal fluctuations on seashores may be several meters.

You already know that the hydrosphere consists of all water on Earth, most of which is in the oceans. In this enormous body of water, wave energy is transferred through the water to shorelines where it has a tremendous impact. Accordingly, understanding the geologic processes operating on shorelines is important to many people. Indeed, many centers of commerce and much of Earth's population are concentrated in a narrow band at or near shorelines. In addition, coastal communities such as Myrtle Beach, South Carolina, Fort Lauderdale, Florida, and Padre Island, Texas, depend heavily on tourists visiting their beaches.

Geologists, oceanographers, coastal engineers, and marine biologists, among others, are interested in the dynamics of shorelines. So are elected officials and city planners of coastal communities because they must be familiar with shoreline processes in order to develop policies and zoning regulations that serve the public while protecting the fragile shoreline environment. Understanding shorelines is especially important now because in many areas sea level is rising, so buildings that were far inland are now in peril or have already been destroyed. Furthermore, hurricanes expend much of their energy on shorelines, resulting in extensive coastal flooding, numerous fatalities, and widespread property damage (see Figure 15.1).

■ *How does the study of shorelines provide an example of interacting systems?*

The study of shorelines provides another excellent example of systems interactions—in this case between part of the hydrosphere and the solid Earth. The atmosphere is also involved in transferring energy from wind to water, thereby causing waves, which in turn generate nearshore currents. And, of course, the gravitational attraction of the Moon and Sun on ocean waters is responsible for the rhythmic rise and fall of tides. As dynamic systems, shorelines continuously adjust to any change that takes place, such as increased wave energy or an increase or decrease in sediment supply.

The continents have more than 400,000 km of shoreline, some of it rocky and steep as along North America's West Coast and its northeast coast in Maine and the Maritime Provinces of Canada. Other areas such as much of the U.S. East Coast and Gulf Coast have broad, sandy beaches or long, narrow, offshore islands. Whatever the type of shoreline, though, waves, nearshore currents, and tides bring about change continuously.

▶ **Figure 19.1 U.S. Pacific Coast Shoreline** The shoreline of this part of the U.S. Pacific Coast consists of the area from low tide, about where the waves start to break, to the base of the sea cliffs. The coast, however, extends farther seaward and also includes the sea cliffs as well as the area some distance inland.

▶ **Figure 19.2 Seashores and Lakeshores** Seashores and lakeshores have many similar features because waves and nearshore currents modify them although tides are insignificant in lakes.

a The Pacific coast of the United States.

b This feature called Miner's Castle along the Lake Superior shoreline in Michigan formed by erosion just as similar features do along seashores.

Section 19.1 Summary

- Waves, tides, and nearshore currents continuously modify shorelines of lakes and especially the seashore.

- Scientists and others are interested in shorelines because they illustrate the interactions among Earth's systems, and because shorelines provide the economic base for many communities.

19.2 Shoreline Processes

Anyone who has visited the seashore or a lakeshore is aware that waves expend energy on shorelines. Waves also generate currents in the nearshore zone and these are particularly effective at sediment transport and deposition. Tides also play a role in shoreline evolution. Shoreline processes are restricted to a long, narrow zone at any particular time, but remember that shorelines may migrate landward or seaward depending on changing sea level. Accordingly, during times when sea level rises, the shoreline migrates landward, and as it does, so do the processes that operate on shorelines. If sea level falls with respect to the land, however, just the opposite takes place.

Several biological, chemical, and physical processes operate continuously in the marine realm. Organisms change the local seawater chemistry and contribute their skeletons to nearshore sediments, and seawater temperature and salinity change depending on such factors as evaporation and inflow of freshwater from land. However, the processes most important for shoreline modification are purely physical ones: waves, nearshore currents, and tides. We cannot totally discount some of the other processes, though; offshore reefs constructed by organisms, for example, may protect a shoreline from most of the energy of waves.

Tides

■ *How do the Moon and Sun generate tides?*

Along shorelines the surface of the oceans rises and falls twice daily in response to the gravitational attraction of the Moon and Sun. These regular fluctuations in the ocean's surface, or **tides,** result in most seashores having two daily high tides and two low tides as sea level rises and falls anywhere from a few centimeters to more than 15 m (▶ Figure 19.3). A complete tidal cycle includes *flood tide*, which progressively covers more and more of the shoreline until high tide is reached, followed by *ebb tide* during which the shoreline is once again exposed.

Both the Moon and the Sun have sufficient gravitational attraction to exert tide-generating forces strong enough to deform the solid body of Earth, but they have a much greater influence on the oceans. The Sun is 27 million times more massive than the Moon, but it is 390 times farther away so its tide-generating force is only 46% as strong as that of the Moon. Accordingly, the tides are dominated by the Moon, but the Sun plays an important role as well.

If tides were caused solely by the Moon acting on a spherical, water-covered Earth, the Moon's tide-generating force would cause two bulges on the ocean surface. One bulge would point toward the Moon because of its gravitational attraction, and the other would be on the opposite side of Earth, pointing away from the Moon because of centrifugal force resulting from Earth's rotation, and the fact that the Moon's gravitational attraction is somewhat less (▶ Figure 19.4a). As Earth rotates, the fixed locations of these bulges cause the twice-daily rise and fall of tides at a particular seashore location. The Moon revolves around Earth in 28 days, however, so its position with respect to any latitude changes slightly each day because it takes the Moon about 50 minutes longer each day to return to the same position it was in the previous day. As a result high tide on one day, say at 1:00 P.M., would be followed the next day by high tide at 1:50 P.M.

Of course the combined effects of the Moon and Sun complicate tides; the Moon generates *lunar tidal bulges*, whereas the Sun produces *solar tidal bulges*. When the Moon and Sun are aligned every two weeks, their combined forces generate *spring tides*, which are about 20% higher than average high tides, and the low tides are lower than usual (▶ Figure 19.4b). *Neap tides*, also at two-week intervals, take place because the Moon, Sun, and Earth form a right angle when the Moon is in its first and third quarters. Accordingly, the Sun's tide-generating force cancels some of the Moon's, and high tides are about 20% lower than average, but the low tide is not as low as usual (▶ Figure 19.4c).

The shape of a shoreline also affects the tidal range. Broad, gently sloping

▶ **Figure 19.3 Low and High Tides** Low tide **(a)** and high tide **(b)** in Turnagain Arm, part of Cook Inlet in Alaska. The tidal range here is about 10 m. Turnagain Arm is a huge fiord now being filled by sediment carried in by rivers. Notice the mudflats in **(a)**.

a Low tide.

b High tide.

continental shelves as in the Gulf of Mexico have low tidal ranges, whereas steep, irregular shorelines experience much greater rise and fall of tides. Tidal ranges are greatest in some narrow, funnel-shaped bays and inlets. In the Bay of Fundy in Nova Scotia, the tides rise and fall 16.5 m, and ranges greater than 10 m occur in several other areas.

Tides have an important impact on shorelines because the area of wave attack constantly shifts onshore and offshore as the tides rise and fall. Tidal currents themselves have little modifying effect on shorelines, except in narrow passages where tidal current velocity is great enough to erode and transport sediment. Indeed, if it were not for strong tidal currents, some passageways would be blocked by sediments deposited by nearshore currents. Tides are one potential source of energy.

Waves

You can see **waves,** or oscillations of a water surface, on all bodies of water, but they are most significant along seashores where they are responsible for most of the erosion, transport,

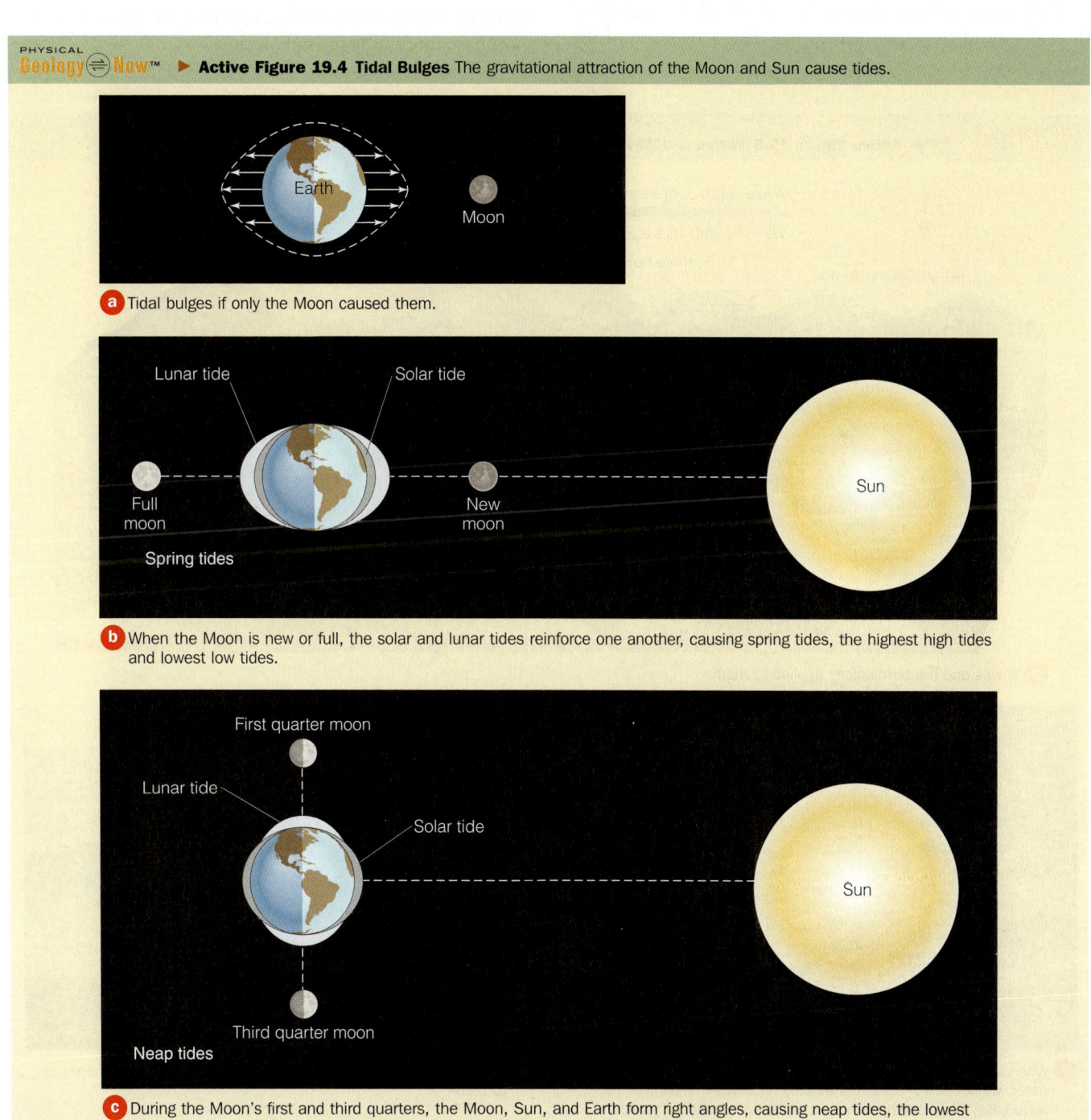

PHYSICAL Geology Now™ ▶ **Active Figure 19.4 Tidal Bulges** The gravitational attraction of the Moon and Sun cause tides.

a Tidal bulges if only the Moon caused them.

b When the Moon is new or full, the solar and lunar tides reinforce one another, causing spring tides, the highest high tides and lowest low tides.

c During the Moon's first and third quarters, the Moon, Sun, and Earth form right angles, causing neap tides, the lowest high tides and highest low tides. Size of the tidal bulges are greatly exaggerated.

and deposition. ▶ Figure 19.5 shows a typical series of waves in deep water and the terminology applied to them. The highest part of a wave is its **crest,** and the low point between crests is the **trough. Wavelength** is the distance between successive wave crests (or troughs), and **wave height** is the vertical distance from trough to crest. You can calculate the speed at which a wave advances, called *celerity* (C), if you know the wavelength (L) and the **wave period** (T), which is the time required for two successive wave crests (or troughs) to pass a given point:

$$C = L/T$$

The speed of wave advance (C) is actually a measure of the velocity of the waveform rather than a measure of the pulse of energy in the water. In fact, the water in waves moves forward and back as a wave passes but has little or no net forward movement. As waves move across a water surface, the water "particles" rotate in circular orbits and transfer energy in the direction of wave advance (▶ Figure 19.5).

■ *What is wave base?*

The diameters of the orbits followed by water particles in waves diminish rapidly with depth, and at a depth of about one-half wavelength ($L/2$), called **wave base,** they are essentially zero (▶ Figure 19.5). At depths exceeding wave base, the water and seafloor, or lake floor, are unaffected by surface waves (▶ Figure 19.6). We will explore the significance of wave base more fully in later sections.

a Waves and the terminology applied to them.

b When swells in deep water move toward shore, the orbital motion of water within them is disrupted as they enter water shallower than wave base. Wavelength decreases while wave height increases, causing the waves to oversteepen and plunge forward as breakers.

c For scale, notice the surfers waiting for this wave that is beginning to break.

Wave Generation Processes such as the displacement of water by landslides, volcanic explosions at sea, and movement of the seafloor by faulting generate waves. Some of these waves are huge and may be devastating to coastal areas; the tsunami of December 26, 2004, that killed more than 220,000 people is an excellent example (see Figure 10.18). Nevertheless, most of the geologic work on shorelines is done by wind-generated waves, especially storm waves. When wind blows over water, the frictional drag of one fluid moving over another transfers some of the energy from the wind to the water, causing the water surface to oscillate.

Sharp-crested, irregular waves of various sizes called *seas* form beneath storms at sea. As these waves move from the area where they form, waves of different heights and lengths are sorted into broad *swells* that have long, rounded crests, and most are about the same size (▶ Figure 19.5b). The harder and longer the wind blows, the larger the waves, but wind strength and duration are not the only factors controlling wave size. No matter how strong or how persistent wind might be, it will never generate large waves on a pond. In fact, the surface of a pond or lake smoothes out quickly once the wind stops blowing. The surface of the ocean is in constant motion, however, and waves as high as 34 m have been recorded during storms.

The reason for the disparity in wave sizes on ponds and lakes and on the oceans is the **fetch,** which is the distance wind blows over a continuous water surface. Fetch is limited on ponds and lakes by their length or width, depending on wind direction, but wind can blow over huge areas in the open oceans. For the largest waves to form, more energy must be transferred from wind to water; hence large waves are generated beneath large storms at sea.

Shallow-Water Waves and Breakers As waves move out from their area of generation, they form swells and lose only a small amount of energy as they travel across the ocean. In deep-water swells, the water surface oscillates and water particles move in orbital paths, with little net displacement of water in the direction of wave advance. When these waves enter progressively shallower water, though, the water is displaced in the direction of wave advance (▶ Figure 19.5).

As deep-water waves enter shallow water, they are transformed from broad, undulating swells into sharp-crested waves. This transformation begins at a water depth of wave base; that is, it begins where wave base intersects the seafloor. At this point, the waves "feel" the bottom, and the orbital motions of water particles within waves are disrupted (▶ Figure 19.5). As they move farther shoreward, the speed of wave advance and wavelength decrease, and wave height increases. In effect, as waves enter shallower water, they become oversteepened; the wave crest advances faster than the wave form, until eventually the crest plunges forward as a **breaker** (▶ Figure 19.5b). Breakers are commonly several times higher than deep-water waves, and when they plunge forward their kinetic energy is expended on the shoreline.

The waves just described are the classic plunging breakers that pound the shorelines of areas with steep offshore slopes, such as on the north shore of Oahu in the Hawaiian Islands (▶ Figure 19.7a). In contrast, shorelines with gentle offshore slopes have spilling breakers, where the waves build up slowly and the wave's crest spills down the front of the wave (▶ Figure 19.7b). Whether the breakers spill or plunge, the water rushes onto the shore, then returns seaward to become part of another breaking wave.

▶ **Figure 19.6 Wave Base and Breakers**

a The waves in this lake have a wavelength of about 2 m, so we can infer that wave base is 1 m deep. Where the waves encounter wave base, they stir up the bottom sediment, thus accounting for the nearshore turbid water.

b These 2-m-high waves at Jenner, California, and the nearshore water are brown because of sediment. If you look closely, you can see the blue water of the open ocean just beyond where the small wave is breaking in the middle distance, which is the position of wave base.

▶ **Figure 19.7 Plunging and Spilling Breakers**

a A plunging breaker on the north shore of Oahu, Hawaii.

b A spilling breaker.

The size of breakers varies enormously. Waves tend to be larger and more energetic in winter because storms are more common during that season. In addition, waves of various sizes merging in the area where they form account for variations in the size of waves breaking on the shore. For instance, when waves with different lengths merge, smaller waves result, whereas merging of waves with the same length forms larger waves. As a result, several small waves might break on a beach followed by a series of larger ones. Surfers commonly take advantage of this phenomenon by swimming out to sea during a relative calm where they wait for a large set of waves to ride toward shore.

Nearshore Currents

We identify the *nearshore zone* as the area extending seaward from the upper limits of the shoreline to just beyond where waves break. It includes a *breaker zone* and a *surf zone*, which is where breaking waves rush forward onto the shore followed by seaward movement of the water as backwash (▶ Figure 19.5). The width of the nearshore zone varies depending on the wavelength of approaching waves because long waves break at a greater depth, and thus farther offshore, than do short waves. Two types of currents are important in the nearshore zone: longshore currents and rip currents.

■ *What is wave refraction, and how does it generate nearshore currents?*

Wave Refraction and Longshore Currents Deep-water waves have long, continuous crests, but rarely are their crests parallel with the shoreline. One part of a wave enters shallow water where it encounters wave base and begins breaking before other parts of the same wave. As a wave begins breaking, its speed diminishes, but the part of the wave in deep water races ahead until it too encounters wave base. The net effect of this oblique approach is that waves bend so that they more nearly parallel the shoreline, a phenomenon known as **wave refraction** (▶ Figure 19.8).

Even though waves are refracted, they still usually strike the shoreline at some angle, causing the water between the breaker zone and the beach to flow parallel to the shoreline. These **longshore currents,** as they are called, are long and narrow and flow in the

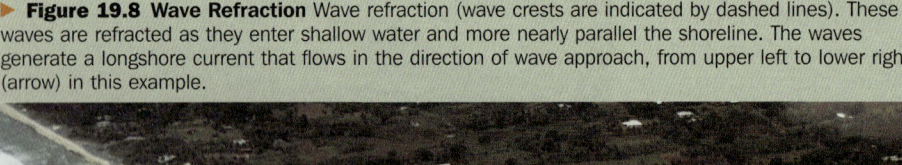

▶ **Figure 19.8 Wave Refraction** Wave refraction (wave crests are indicated by dashed lines). These waves are refracted as they enter shallow water and more nearly parallel the shoreline. The waves generate a longshore current that flows in the direction of wave approach, from upper left to lower right (arrow) in this example.

same general direction as the approaching waves (▶ Figure 19.8). Longshore currents are particularly important because they transport and deposit large quantities of sediment in the nearshore zone.

Rip Currents Waves carry water into the nearshore zone, so there must be a mechanism for mass transfer of water back out to sea. One way that water moves seaward from the nearshore zone is in **rip currents,** which are narrow surface currents that flow out to sea through the breaker zone (▶ Figure 19.9). Surfers commonly take advantage of rip currents for an easy ride out beyond the breaker zone, but rip currents pose a danger to inexperienced swimmers. They may flow at several kilometers per hour, so if a swimmer is caught in one, it is useless to try to swim directly back to shore. Instead, because rip currents are narrow and usually nearly perpendicular to the shore, one can swim parallel to the shoreline for a short distance and then turn shoreward with no difficulty.

Rip currents are circulating cells fed by longshore currents. When waves approach a shoreline, the amount of water builds up until the excess moves out to sea through the breaker zone. Rip currents are fed by nearshore currents that increase in speed from midway between each rip current (▶ Figure 19.9).

Seafloor configuration also plays an important role in determining the location of rip currents. They commonly develop where wave heights are lower than in adjacent areas. Differences in wave height are controlled by variations in water depth. For instance, if waves move over a depression, their heights over the depression tend to be less than in adjacent areas.

▶ **Figure 19.9 Rip Currents**

a Rip currents are fed on each side by currents moving parallel to the shoreline.

b Suspended sediment, indicated by discolored water, is being carried seaward in these rip currents.

c Sign along the California coast warning swimmers of dangerous rip currents, higher than normal (sleeper) waves, and backwash.

What Would You Do?

While swimming at your favorite beach, you notice that after about 20 minutes in the water you are more than 100 m down the beach from where you entered the water. Furthermore, the size of the waves you were swimming in has diminished considerably. You decide to swim back to shore and then walk back to your original starting place, but no matter how hard you swim, you are carried farther and farther offshore. Assuming that you survive this incident, explain what happened and what you did to remedy the situation.

Section 19.2 Summary

- The gravitational attraction of the Moon and Sun causes tides, which is the twice daily rise and fall of sea level along seashores.

- When wind-generated waves enter nearshore water shallower than wave base the wavelength decreases, wave height increases, and the waves become oversteepened and plunge forward as breakers.

- Because waves usually hit a shoreline at some angle, they generate longshore currents that flow parallel to the shoreline.

- Excess water along a shoreline returns to the open sea in rip currents that carry water seaward through the breaker zone.

PHYSICAL GeologyNow™ Log into GeologyNow and select this chapter to work through **Geology Interactive** activities on "High and Low Tides" (click Waves, Tides and Current→High/Low Tides), "Wave Properties" (click Waves, Tides and Current→Wave Properties), "Wave Progression" (click Waves, Tides and Current→Wave Progression) and "Wind-Wave Relationships" (click Waves, Tides and Current→Wind-Wave Relationships).

19.3 Shoreline Erosion

■ *What processes account for erosion along shorelines?*

The impact of water (hydraulic action), abrasion, and the chemical wearing away of rocks (corrosion) are responsible for shoreline erosion.

You already know that waves, nearshore currents, and tides expend energy on shorelines. As a result they are responsible for erosion as well as sediment transport and deposition. In fact, along sea coasts where erosion rather than deposition predominates, beaches are lacking or they are small, discontinuous, and restricted to protected areas, and sea cliffs are commonly present. Sea cliffs are pounded by waves, especially during storms, and the cliffs wear back by the combined effects of hydraulic action, abrasion, and corrosion.

▶ **Figure 19.10 Wave Erosion by Abrasion and Hydraulic Action**

a The rocks in the lower part of this image on a small island in the Irish Sea have been smoothed by abrasion, but the rocks higher up are out of the reach of waves.

b Hydraulic action and abrasion have undercut these sea cliffs near Bodega Bay, California. Erosion was particularly intense during storms of February 1998. Attempts to stabilize the shoreline have failed and the homes have been abandoned.

Corrosion is erosion that involves the wearing away of rocks by chemical processes, especially the solvent action of seawater. The force of waves pounding on coastal rocks, called *hydraulic action*, is particularly effective, especially on unconsolidated materials or highly fractured rock. *Abrasion* involves the grinding action that takes place when waves hurl sand and gravel against coastal materials (▶ Figure 19.10). Remember from Chapter 15 that running water in channels also erodes by hydraulic action and abrasion, and sediment-laden glacial ice abrades rock surfaces (see Chapter 17).

Wave-Cut Platforms

Waves erode sea cliffs, but the rate at which the cliffs erode and retreat landward depends on wave intensity and the resistance of the coastal rocks or sediments. Most sea cliff retreat takes place during storms and occurs most rapidly in sea cliffs composed of sediment. For example, some parts of the White Cliffs of Dover in Great Britain retreat at a rate of more than 100 m per century. By comparison, sea cliffs consisting of dense igneous or metamorphic rocks retreat much more slowly.

Sea cliffs erode mostly as a result of hydraulic action and abrasion at their bases. As a sea cliff is undercut by erosion, the upper part is left unsupported and susceptible to mass wasting processes. Thus sea cliffs retreat little by little, and as they do, they leave a beveled surface known as a **wave-cut platform** that slopes gently seaward (▶ Figure 19.11a). Broad wave-cut platforms are found in many areas, but invariably the water over them is shallow because the abrasive planing action of waves is effective to a depth of only about 10 m. Sediment eroded from sea cliffs is transported seaward until it reaches deeper water at the edge of the wave-cut platform. There it is deposited and forms a seaward extension of the wave-cut platform called a *wave-built platform* (▶ Figure 19.11a). Wave-cut platforms now above sea level are known as **marine terraces** (▶ Figure 19.11b).

Sea Caves, Arches, and Stacks

■ *How do sea arches and sea stacks form?*

Sea cliffs do not retreat uniformly because some of the materials of which they are composed are more resistant to erosion than others. Those parts of a shoreline that consist of resistant materials might form seaward-projecting headlands, whereas less resistant materials erode more rapidly, yielding embayments where beaches may be present. Wave refraction around headlands causes them to erode on both sides so that *sea caves* form; if caves on the opposite sides of a headland join, they form a *sea arch*

(▶ Figure 19.12). Continued erosion causes the span of an arch to collapse, yielding isolated *sea stacks* on wave-cut platforms (▶ Figure 19.12).

In the long run, shoreline processes tend to straighten an initially irregular shoreline. They do so because wave refraction causes waves to expend more energy on headlands and less on embayments. Headlands erode, and some of the sediment yielded by erosion is deposited in the embayments, straightening the shoreline (▶ Figure 19.13).

Section 19.3 Summary

• The combined effects of hydraulic action, abrasion, and corrosion yield gently sloping surfaces along shorelines known as wave-cut platforms. Wave-cut platforms that are now above sea level are marine terraces.

• Other features formed by erosion along rocky shorelines include sea caves, arches, and sea stacks.

▶ **Figure 19.11 Origin of a Wave-Cut Platform**

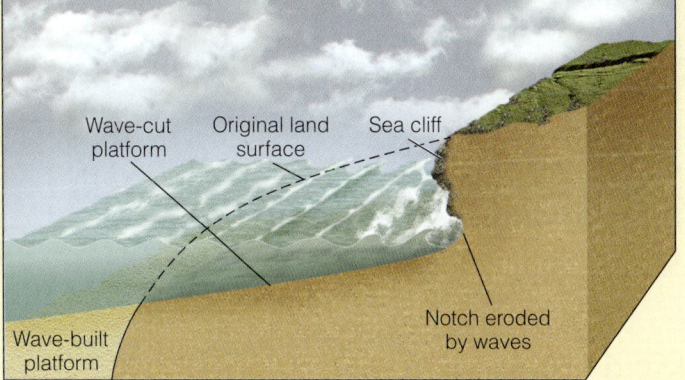

a Wave erosion causes a sea cliff to migrate landward, leaving a gently sloping surface, called a wave-cut platform. A wave-built platform originates by deposition at the seaward margin of the wave-cut platform.

b This gently sloping surface along the coast of California is a marine terrace. Notice the sea stacks rising above the terrace.

Figure 19.12 Erosion of a Headland

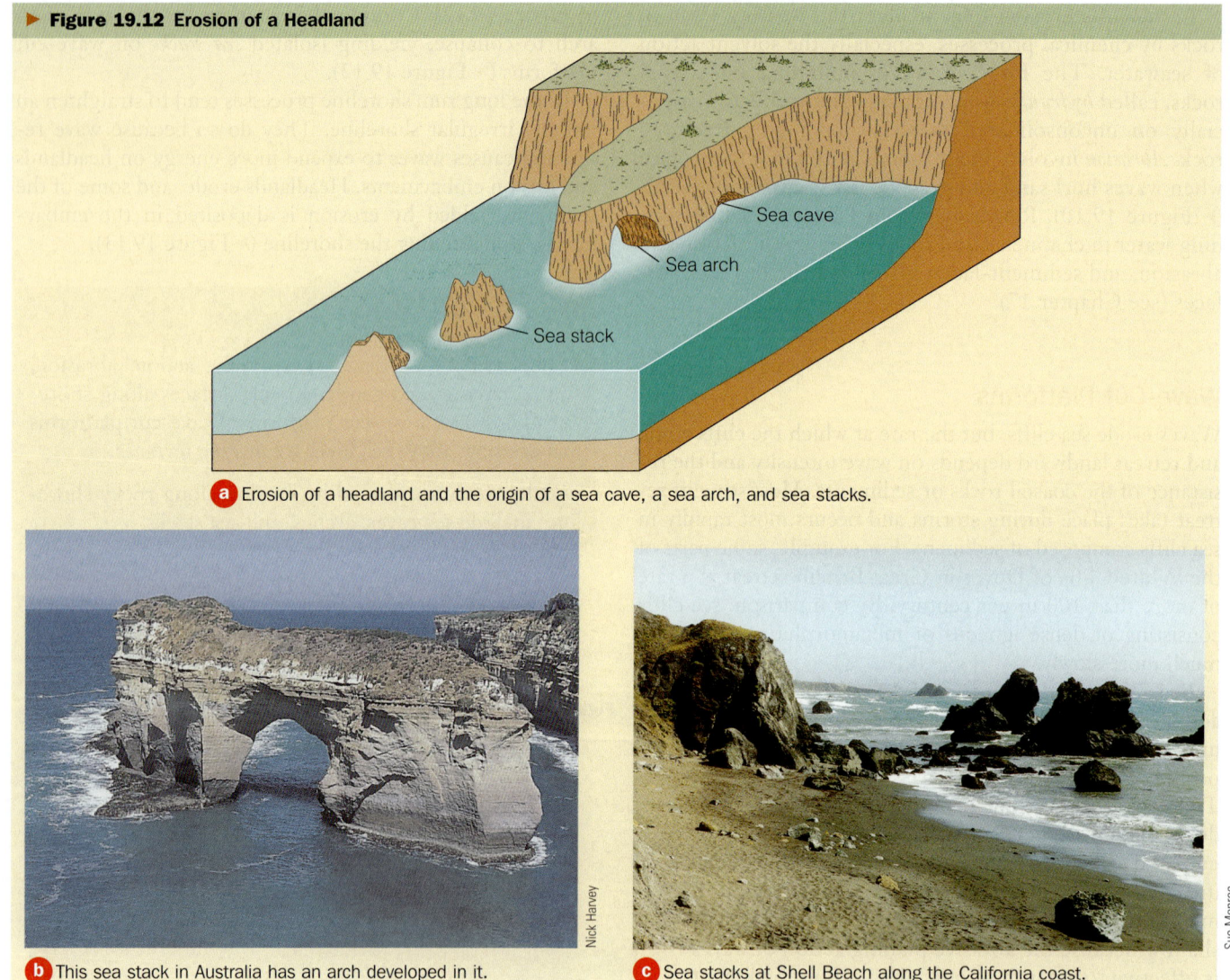

a Erosion of a headland and the origin of a sea cave, a sea arch, and sea stacks.

b This sea stack in Australia has an arch developed in it.

c Sea stacks at Shell Beach along the California coast.

19.4 Deposition along Shorelines

What accounts for deposition along shorelines? You know from previous sections that longshore currents are effective at transporting sediment. In fact, we can think of the area from the breaker zone to the upper limit of wave swash as a "river" that flows along the shoreline. Unlike rivers on land, though, its direction of flow changes if waves approach from a different direction. Nevertheless, the analogy is apt, and just like rivers on land, our longshore current "river's" capacity for transport varies depending on flow velocity and water depth.

Wave refraction and the resulting longshore currents are the primary agents of sediment transport and deposition along shorelines. Tides play a role because as they rise and fall, the position of wave attack shifts onshore and offshore, and rip currents are also important because they transport fine-grained sediment (silt and clay) offshore through the breaker zone.

Beaches

Beaches are the most familiar of all coastal landforms, attracting millions of visitors each year and providing the economic base for many communities. They consist of a long, narrow strip of unconsolidated sediment, commonly sand, and are constantly changing. Depending on shoreline materials and wave intensity, beaches may be discontinuous, existing only as *pocket beaches* in protected areas such as embayments, or they may be continuous for long distances (see "Shoreline Processes and Beaches" on pp. 610–611).

By definition, a **beach** is a deposit of unconsolidated sediment extending landward from low tide to a change in topography such as a line of sand dunes, a sea cliff, or the point where permanent vegetation begins. Typically, a beach has several component parts, including a *backshore* that is usually dry, being covered by water only during storms or exceptionally high tides. The backshore consists of one or more **berms,** platforms composed of sediment deposited by waves. Berms are nearly horizontal or slope gently landward. The sloping area below a berm that is exposed

PHYSICAL Geology Now™ ▶ Geo-focus Figure 19.13 Wave Refraction and Shoreline Straightening

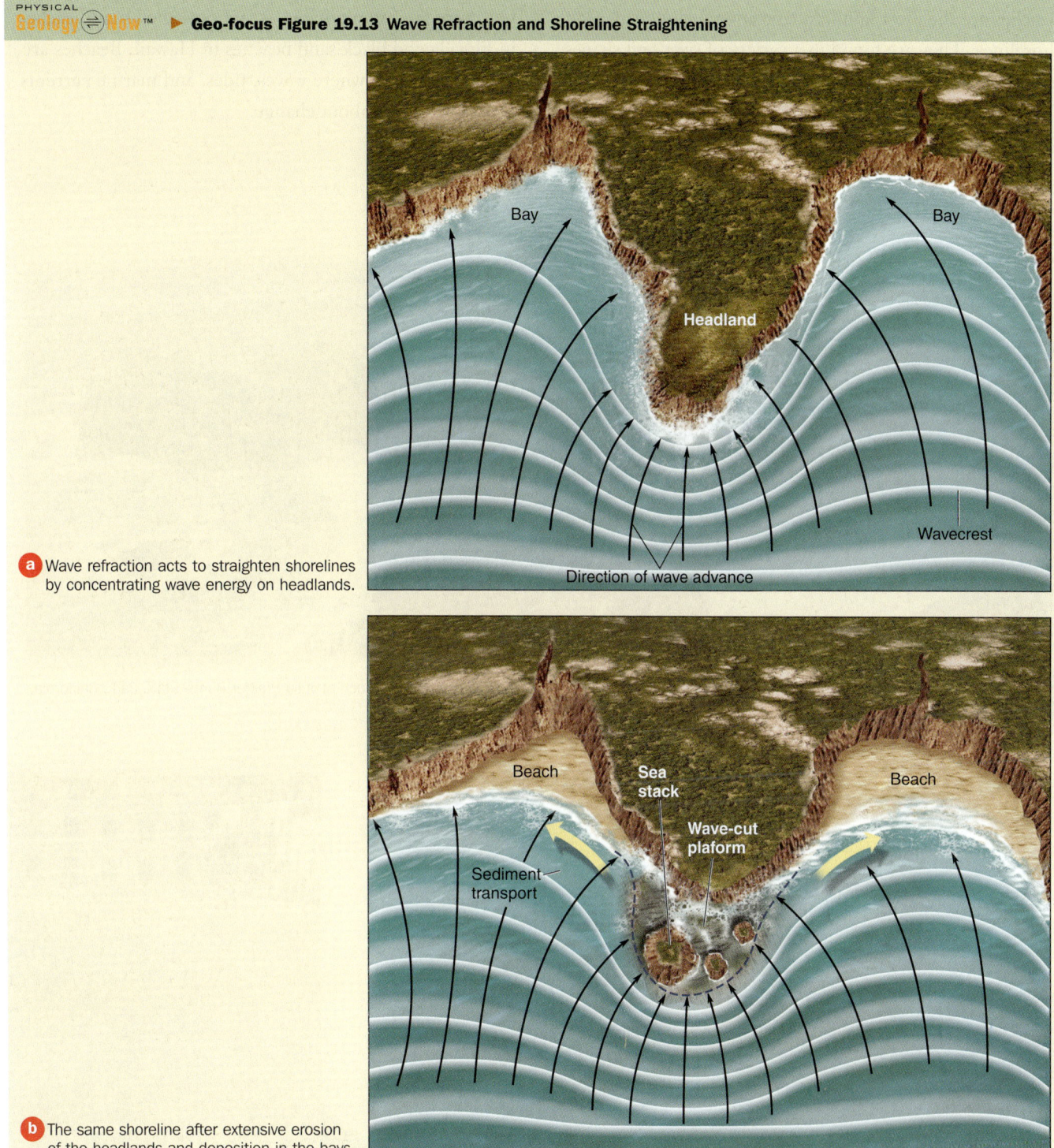

a Wave refraction acts to straighten shorelines by concentrating wave energy on headlands.

b The same shoreline after extensive erosion of the headlands and deposition in the bays.

to wave swash is the *beach face*. The beach face is part of the *foreshore*, an area covered by water during high tide but exposed during low tide.

Some sediment on beaches is derived from weathering and wave erosion of the shoreline, but most of it is transported to the coast by streams and redistributed along the shoreline by longshore currents. **Longshore drift** is the phenomenon involving sand transport along a shoreline by longshore currents (see "Shoreline Processes and Beaches" on pp. 610–611). As noted, waves usually strike beaches at some angle, causing the sand grains to move up the beach face at a similar angle, but as backwash carries the sand grains seaward, they move perpendicular to the long axis of the beach. So individual sand grains move in a zigzag

Shoreline Processes and Beaches

Beaches are the most familiar depositional landforms along shorelines. They are found in a variety of sizes and shapes, with long sandy beaches typical of the East and Gulf Coasts, and smaller, mostly protected beaches along the West Coast. The sand on most beaches is primarily quartz, but there are some notable exceptions: shell sand beaches in Florida and black sand beaches in Hawaii. Beaches are dynamic systems where waves, tides, and marine currents constantly bring about change.

▼ **1.** The Grand Strand of South Carolina, shown here at Myrtle Beach, is 100 km of nearly continuous beach.

▲ **2.** Small pocket beach at Julia Pfeiffer Burns State Park, California.

▼ **3.** Diagram of a beach showing its component parts.

▶ **4.** The backshore area of a pocket beach along the Pacific coast. Note that the berm ends at the rocks on the right and the beach face slopes steeply seaward.

▲ **5.** The origin of beach cusps like these at Mykanos, Greece, is poorly known, but they are very common.

▼ 6, ▶ 7. Longshore currents transport sediment along the shoreline, a phenomenon called longshore drift, between the breaker zone and the upper limit of wave action. These groins (right) at Cape May, New Jersey, interrupt the flow of longshore currents. Sand is trapped on their upcurrent sides, whereas erosion takes place on their downcurrent sides. **Can you tell the direction of longshore drift in this image?** *The longshore currents moved toward the top of the image.*

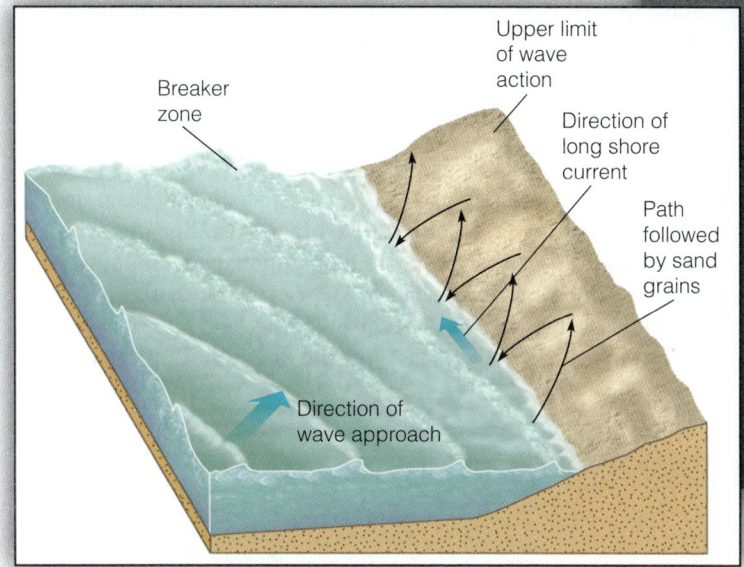

◀ 8. Although most beaches are made up dominantly of quartz sand, there are exceptions. This black sand beach on Maui in Hawaii is made up of small basalt rock fragments and particles of obsidian that formed when lava flowed into the sea.

▶ 9. The beach shown on Oahu in Hawaii is composed mostly of fragmented shells of marine organisms, although some minerals from basalt are also included.

▲ 10. This California beach is made up of quartz and sand- and gravel-sized rock fragments.

pattern in the direction of longshore currents. This movement is not restricted to the beach but extends seaward to the outer edge of the breaker zone.

In an attempt to widen a beach or prevent erosion, shoreline residents build structures known as *groins* that project seaward at right angles from the shoreline. Groins interrupt the flow of longshore currents, causing sand deposition on their upcurrent side, widening the beach at that location. However, erosion inevitably occurs on the downcurrent side of a groin.

Many beaches are sandy, but in areas of particularly vigorous wave activity, they might be gravel covered. Most beach sand is composed of quartz, but other mineral and rock fragments may be present as well. One of the most common accessory minerals in beach sands is magnetite; because of its greater density, magnetite is commonly separated from other minerals and is visible as thin, black layers.

Although quartz is the most common mineral in most beach sands, there are some notable exceptions. The black sand beaches of Hawaii are composed of sand-sized basalt rock fragments, and some beaches are made up of fragmented calcium carbonate shells of marine organisms. In short, beaches are composed of whatever material is available; quartz is common simply because it is abundant in most areas and is the most durable and stable of the common rock-forming minerals (see Chapter 6).

Seasonal Changes in Beaches

Wave energy is dissipated on beaches, so the loose grains composing it are constantly moved by waves. But the overall configuration of a beach remains unchanged as long as equilibrium conditions persist. We can think of the beach profile as consisting of a berm or berms and a beach face shown in ▶ Figure 19.14a as a profile of equilibrium; that is,

▶ Figure 19.14 Seasonal Changes in Beaches

(a) Seasonal changes in beach profiles.

(b) and (c) San Gregorio State Beach, California. These photographs were taken from nearly the same place, but (c) was taken two years after (b). Much of the change from (b) to (c) can be accounted for by beach erosion during 1997–1998 winter storms.

all parts of the beach are adjusted to the prevailing conditions of wave intensity and nearshore currents.

■ *What are summer beaches and winter beaches?*

In many areas, beach profiles change with the seasons; so we recognize *summer beaches* and *winter beaches*, which have adjusted to the conditions prevailing at these times (▶ Figure 19.14). Summer beaches are sand covered and possess a wide berm, a steep beach face, and a smooth offshore profile. During winter, however, when waves more vigorously attack the beach, a winter beach develops, which has little or no berm, a gentle slope, and offshore sand bars paralleling the shoreline (▶ Figure 19.14). Seasonal changes can be so profound that sand moving onshore and offshore yields sand-covered beaches during summer and gravel-covered beaches during winter.

Seasonal changes in beach profiles are related to changing wave intensity. During the winter, energetic storm waves erode the sand from the beach and transport it offshore where it is stored in sand bars (▶ Figure 19.14). The same sand that was eroded from the beach during the winter returns the next summer when it is driven onshore by the more gentle swells that occur during that season. The volume of sand in the system remains more or less constant; it simply moves offshore or onshore depending on the energy of waves.

The terms *winter beach* and *summer beach*, though widely used, are somewhat misleading. A "winter beach" profile can develop at any time of the year if a large storm occurs, and a "summer beach" profile can develop during a prolonged calm period in the winter.

Spits, Baymouth Bars, and Tombolos

Beaches are the most familiar depositional features of coasts, but spits, baymouth bars, and tombolos are common, too. In fact, these features are simply continuations of a beach. A **spit,** for instance, is a fingerlike projection of a beach into a body of water such as a bay, and a **baymouth bar** is a spit that has grown until it completely closes off a bay from the open sea (▶ Figure 19.15). Both are composed of sand or more rarely gravel that was transported and deposited by longshore currents where they weakened as they entered the deeper water of a bay's opening. Some spits are modified by waves so that their free ends are curved; they are called *hooks* or *recurved spits* (▶ Figure 19.15).

■ *What is a tombolo and how does one form?*

A rarer type of spit, a **tombolo,** extends out into the sea and connects a seastack or an island to the mainland. Tombolos develop on the shoreward sides of islands, as shown in ▶ Figure 19.16a. Wave refraction around an island causes converging currents that turn seaward and deposit a sand bar connecting the shore with the island. A similar feature may form when an artificial breakwater is constructed offshore (▶ Figure 19.16c).

Although spits, baymouth bars, and tombolos are most common on irregular seacoasts, many examples of the same features are present in large lakes. Whether along seacoasts or lakeshores, these sand deposits present a continuing problem where bays must be kept open for pleasure boating or commercial shipping. The entrances to these bays must be either regularly dredged or protected.

▶ **Figure 19.15 Spits and Baymouth Bars**

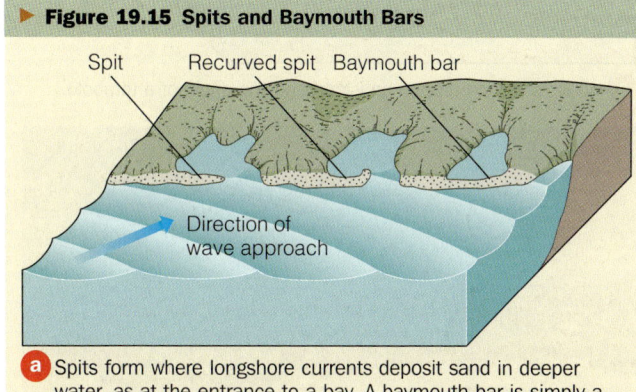

a Spits form where longshore currents deposit sand in deeper water, as at the entrance to a bay. A baymouth bar is simply a spit that has grown until it extends across the mouth of a bay.

b A spit at the mouth of the Russian River near Jenner, California.

c A baymouth bar.

▶ **Figure 19.16** Tombolos and Breakwaters

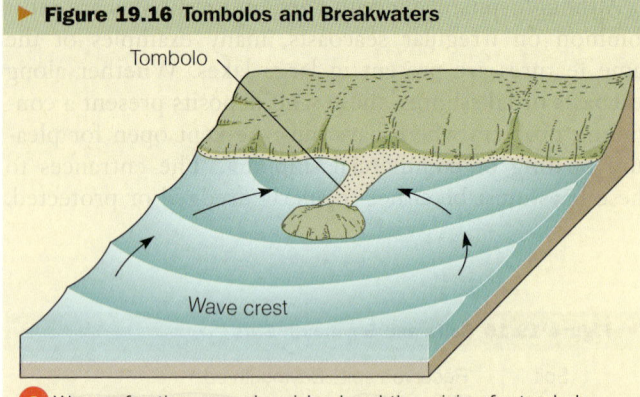

a Wave refraction around an island and the origin of a tombolo.

b A small tombolo along the Pacific coast of the United States.

c Soon after this breakwater was constructed at Oxnard, California, a bulge similar to a tombolo appeared in the beach. Notice the jetty that was constructed to protect the channel.

▶ **Figure 19.17** Barrier Islands

a View from space of the barrier islands along the Gulf Coast of Texas. Notice that a lagoon up to 20 km wide separates the long, narrow barrier islands from the mainland.

b Aerial view of Padre Island on the Texas Gulf Coast. Laguna Madre is on the left, and the Gulf of Mexico is on the right.

The most common way to protect entrances to bays is to build *jetties*, which are structures extending seaward (or lakeward) that protect the bay from deposition by longshore currents (▶ Figure 19.16c).

Barrier Islands

Long, narrow islands composed of sand parallel to the shoreline but separated from the mainland by a lagoon are **barrier islands** (▶ Figure 19.17). The seaward margins of

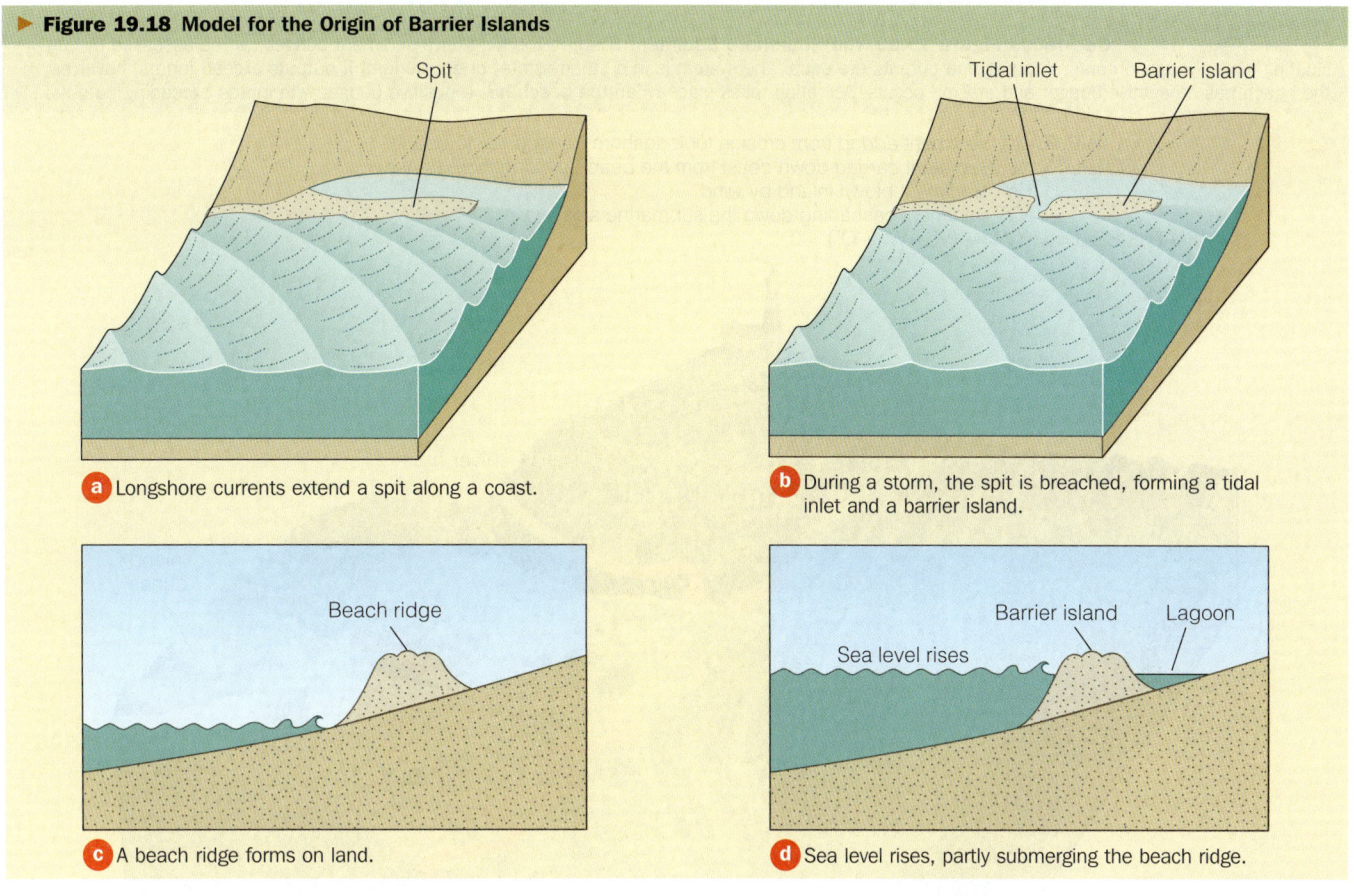

▶ Figure 19.18 Model for the Origin of Barrier Islands
a. Longshore currents extend a spit along a coast.
b. During a storm, the spit is breached, forming a tidal inlet and a barrier island.
c. A beach ridge forms on land.
d. Sea level rises, partly submerging the beach ridge.

barrier islands are smoothed by waves, but their lagoon sides are irregular because during storms waves overtop the island and deposit lobes of sand in the lagoon. Once deposited, these lobes are modified only slightly because they are protected from further wave action. Wind-blown sand dunes are common and are generally the highest part of barrier islands.

The origin of barrier islands has been long debated and is still not completely resolved. It is known that they form on gently sloping continental shelves with abundant sand in areas where both tidal fluctuations and wave-energy levels are low. Although barrier islands are found in many areas, most of them are along the East Coast of the United States from New York to Florida and along the U.S. Gulf Coast. According to one model, barrier islands formed as spits that became detached from the land, whereas another model proposes that they formed as beach ridges on coasts that subsequently subsided (▶ Figure 19.18).

Because sea level is currently rising, most barrier islands are migrating landward. Migration is a natural consequence of evolution of these islands, but it is a problem for the island residents and communities. Barrier islands migrate rather slowly, but the rates for many are rapid enough to cause shoreline problems (discussed in a later section).

Section 19.4 Summary

• A beach is a deposit of sediment along a shoreline derived mostly from sediment transported to the shoreline by running water and redistributed by waves and longshore currents.

• Beaches usually change seasonally because wave intensity, and thus erosion and transport, varies.

• Longshore currents transport and deposit sand as spits, baymouth bars, and tombolos. The origin of barrier islands is not fully understood, but they are sand bodies along shorelines but separated from the shore by a lagoon.

19.5 The Nearshore Sediment Budget

■ *What is the nearshore sediment budget?*

We can think of the gains and losses of sediment in the nearshore zone in terms of a **nearshore sediment budget** (▶ Figure 19.19). If a nearshore system has a balanced budget, sediment is supplied as fast as it is removed, and the volume

Geo-focus Figure 19.19 The Nearshore Sediment Budget The long-term sediment budget can be assessed by considering inputs versus outputs. If inputs and outputs are equal, the system is in a steady state, or equilibrium. If outputs exceed inputs, however, the beach has a negative budget and erosion occurs. Accretion takes place when the beach has a positive budget with inputs exceeding outputs.

INPUT V^+ Sediment added from erosion for longshore transport onto beach
OUTPUTS V^- Sediment carried down-coast from the beach by longshore transport
W^- Sediment blown inland by wind
O^- Sediment cascading down the submarine slope
STABLE BEACH: $(V^+) + (V^- + W^- + O^-) = 0$

of sediment remains more or less constant, although sand may shift offshore and onshore with the changing seasons. A positive budget means gains exceed losses, whereas a negative budget means losses exceed gains. If a negative budget prevails long enough, the nearshore system is depleted and beaches may disappear (▶ Figure 19.19).

Erosion of sea cliffs provides some sediment to beaches, but in most areas probably no more than 5–10% of the total sediment supply comes from this source. There are exceptions, though; almost all the sediment on the beaches of Maine is derived from the erosion of shoreline rocks. Most sediment on typical beaches is transported to the shoreline by streams and then redistributed along the shoreline by longshore drift. Thus longshore drift plays a role in the nearshore sediment budget because it continually moves sediment into and away from beach systems (▶ Figure 19.19).

The primary ways that a nearshore system loses sediment are longshore drift, offshore transport, wind, and deposition offshore. Offshore transport involves mostly fine-grained sediment that is carried seaward, where it eventually settles in deeper water. Wind is an important process because it removes sand from beaches and blows it inland, where it piles up as sand dunes.

If the heads of submarine canyons are nearshore, huge quantities of sand are funneled into them and deposited in deeper water. La Jolla and Scripps submarine canyons off the coast of southern California funnel off an estimated 2 million m³ of sand each year. In most areas, however, submarine canyons are too far offshore to interrupt the flow of sand in the nearshore zone.

If a nearshore system is in equilibrium, its incoming supply of sediment exactly offsets its losses. Such a delicate balance tends to continue unless the system is somehow disrupted. One common change that affects this balance is the construction of dams across the streams that supply sand. Once dams have been built, all sediment from the upper reaches of the drainage system is trapped in reservoirs and thus cannot reach the shoreline. Indeed, about 98% of the suspended load of the Nile River in Egypt is trapped in reservoirs, and as a result the shoreline retreats at about 10 m per year.

> *Section 19.5 Summary*
>
> - The nearshore sediment budget is a balance sheet of gains versus losses in a coastal area. If gains (deposition) equal losses (erosion), a nearshore system is in equilibrium.

What Would You Do?

While visiting a barrier island, you notice some organic-rich silt and mud deposits on the beach. In fact, so much organic matter is present that the sediments are black and have a foul odor. Their presence on the beach does not make sense because you know that sediments like these are deposited in a marsh on the landward (opposite) side of the barrier island. How can you explain your observations?

19.6 The Classification of Coasts

Coasts are difficult to classify because of variations in the factors that control their development. So rather than attempt to categorize all coasts, we note that we have already discussed two types of coasts: those dominated by deposition and those on which erosion is more important. We will also discuss submergent and emergent coasts which owe their distinctive features to changing sea level.

Depositional and Erosional Coasts

Depositional coasts, such as the U.S. East Coast and Gulf Coast, are characterized by an abundance of detrital sediment and wide, sandy beaches, deltas, and barrier islands. In contrast, erosional coasts are steep and irregular and typically lack well-developed beaches except in protected areas (see "Shoreline Processes and Beaches" on pp. 610–611). They are further characterized by sea cliffs, wave-cut platforms, and sea stacks. Many of the beaches along the West Coast of North America fall into this category.

The following section will examine coasts in terms of their changing relationships to sea level. But note that although some coasts, such as those in southern California, are described as emergent (uplifted), these same coasts may be erosional as well. In other words, coasts commonly possess features that allow them to be classified in more than one way.

Submergent and Emergent Coasts

■ *How do depositional and erosional coasts differ?*

If sea level rises with respect to the land or the land subsides, coastal regions are flooded and said to be **submergent** or *drowned* (▶ Figure 19.20). Much of the East Coast of North America from Maine southward through South Carolina was flooded during the rise in sea level following the Pleistocene Epoch, so it is extremely irregular. Recall that during the expansion of glaciers in the Pleistocene, sea level was as much as 130 m lower than it is now and that streams eroded their valleys more deeply and extended across continental shelves. When sea level rose, the lower ends of these valleys were drowned, forming *estuaries* such as Delaware and Chesapeake Bays (▶ Figure 19.20). Estuaries are the seaward ends of river valleys where seawater and freshwater mix.

Submerged coasts are also present at higher latitudes where Pleistocene glaciers flowed into the sea. When sea

▶ **Figure 19.20 Submergent Coasts** Submergent coasts tend to be extremely irregular with estuaries such as Chesapeake and Delaware Bays. They formed when the East Coast of the United States was flooded as sea level rose following the Pleistocene Epoch.

▶ **Figure 19.21 An Emergent Coast in California** Emergent coasts tend to be steep and straighter than submergent coasts. Notice the several sea stacks and the sea arch. Also, a marine terrace is visible in the distance.

Section 19.6 Summary

• Coasts are difficult to classify because the processes responsible for them vary. Nevertheless, geologists categorize coasts as depositional and erosional, and as submergent and emergent.

19.7 Storm Waves and Coastal Flooding

■ *What causes the most fatalities during hurricanes?*

Many people think it is strong wind, but that is incorrect. Actually, coastal flooding caused by storm-generated waves and heavy rainfall is more dangerous. For examples you need only look at the devastation caused by the four hurricanes that hit Florida and nearby areas during the 2004 hurricane season. Another good example is Hurricane Katrina, which hit the Gulf Coast on August 29, 2005, and caused the most costly and perhaps the most deadly natural disaster in the United States (see Figure 15.1).

Coastal flooding during hurricanes results when large waves are driven onshore and by as much as 60 cm of rain in as little as 24 hours. In addition, as a hurricane moves over the ocean, low atmospheric pressure beneath the eye of the storm causes the ocean surface to bulge upward as much as 0.5 m. When the eye reaches the shoreline, the bulge coupled with wind-driven waves piles up in a *storm surge* that may rise several meters above normal high tide and inundate areas far inland (▶ Figure 19.22).

In 1900 Galveston, Texas, was hit by a hurricane and storm waves surged inland, eventually covering the entire barrier island the city was built on. Buildings and other structures near the shoreline were battered to pieces, and "great beams and railway ties were lifted by the [waves] and driven like battering rams into dwellings and business houses"* farther inland. Between 6000 and 8000 people died. In an effort to protect the city from future storms, a huge seawall was constructed (▶ Figure 19.23) and the entire city was elevated to the level of the top of the seawall.

Although Galveston, Texas, has been largely protected from more recent storm surges, the same is not true for several other areas. We have already mentioned the storm surges that hit the Outer Banks of North Carolina in 2003 and Florida in 2004, but one of the most tragic natural disasters in U.S. history occurred when Hurricane Katrina devastated large areas in Louisiana and Mississippi in August 2005. Indeed, Gulfport and Biloxi, Mississippi, were destroyed by a storm surge more than 7 m high (▶ Figure 19.22a), and it caused the levee failures and flooding in New Orleans (see Figure 15.1). The hurricane was responsible for nearly 1400 fatalities, and it caused about $75 billion in damages.

level rose, the lower ends of the glacial troughs were drowned, forming fiords.

Emergent coasts are found where the land has risen with respect to sea level (▶ Figure 19.21). Emergence takes place when water is withdrawn from the oceans, as occurred during the Pleistocene expansion of glaciers. Presently coasts are emerging as a result of isostasy or tectonism. In northeastern Canada and the Scandinavian countries, the coasts are irregular because isostatic rebound is elevating formerly glaciated terrain from beneath the sea.

Coasts that rise in response to tectonism tend to be rather straight because the seafloor topography being exposed as uplift proceeds is smooth. The west coasts of North and South America are rising as a result of plate tectonics. Distinctive features of these coasts include *marine terraces* (Figure 19.11b), which are wave-cut platforms now elevated above sea level. Uplift appears to be episodic rather than continuous, as indicated by the multiple levels of terraces in some areas. In southern California, for example, several terrace levels are present; each probably represents a period of tectonic stability followed by uplift. The highest of these terraces is now about 425 m above sea level.

---

*L. W. Bates, Jr., "Galveston—A City Built upon Sand," *Scientific American*, 95 (1906), p. 64.

Figure 19.22 Hurricane Katrina

a Destruction caused by the storm surge from Hurricane Katrina. This image shows a man in Biloxi, Mississippi, trying to find his house.

b Map of the New Orleans area, including part of the Mississippi River delta. Notice the location of the Chandeleur Islands.

c Images showing the Chandeleur Islands before (left) and after (right) Hurricane Katrina devastated the area.

Another consequence of Hurricane Katrina was its effect on the Chandeleur Island off of Louisiana's southeast coast (▶ Figure 19.22b). These are barrier islands that absorb some of the impact of approaching storms, especially storm surges. When Hurricane Katrina swept across this area, the islands were battered and reduced to small shoals (▶ Figure 19.22c).

Section 19.7 Summary

- Although shorelines are battered by high wind during storms, especially hurricanes, coastal flooding causes most fatalities.

- Storm surges occur during hurricanes because low atmospheric pressure allows the ocean surface to bulge up beneath the eye of the storm, so when the eye makes landfall, the bulge and wind-driven waves pile up in a storm surge that floods low-lying areas.

▶ **Figure 19.23 Seawall Construction** Construction of this seawall to protect Galveston, Texas, from storm waves began in 1902. Notice that the wall is curved to deflect waves upward.

19.8 Managing Coastal Areas as Sea Level Rises

■ *What factors influence whether sea level rises or falls?*

Rise and fall of sea level depends on the volumes of the ocean basins, the temperature of ocean waters, the rate of uplift or subsidence of the continents, and the amount of water frozen in glaciers.

Most of the options available for managing shorelines as sea level rises involve placing restrictions on coastal development and engineering projects designed to protect the shoreline from erosion. The related question, Why is sea level rising?, requires a more lengthy answer. We know that sea level has risen about 12 cm worldwide during the last century, but the absolute rise in a particular area depends on two factors. First, the volume of water in the oceans is increasing because of thermal expansion of near-surface seawater (as water heats up, it expands and occupies more space) and melting glacial ice. Many scientists think that sea level will continue to rise because of global warming resulting from the concentration of greenhouse gases in the atmosphere.

The second factor that controls sea level is the rate of uplift or subsidence of coastal areas. In some parts of the world, uplift of the land takes place rapidly enough that sea level is actually falling with respect to the land.* Unfortunately for many coastal communities, sea level is rising and the land is subsiding, which result in a net sea level change of as much as 30 cm per century. Perhaps such a "slow" rate seems insignificant. After all, it amounts to only a few millimeters per year, but in gently sloping coastal areas a change of only centimeters eventually has widespread effects. And in cities such as Venice, Italy, which is at sea level to begin with, a sea level rise of a few centimeters poses a great risk.

Many of the nearly 300 barrier islands along the East and Gulf Coasts are migrating landward as sea level rises. They do so because during storms waves erode the seaward-facing sides of the islands and deposit sand in their lagoons, resulting in a gradual shift of the entire island complex (▶ Figure 19.24). Landward migration of barrier islands would pose few problems if it were not for communities, resorts, and vacation

---

*Remember from Chapter 7 that rates of seafloor spreading also can cause sea level to rise or fall, but the two factors discussed above are by far the most important for planning.

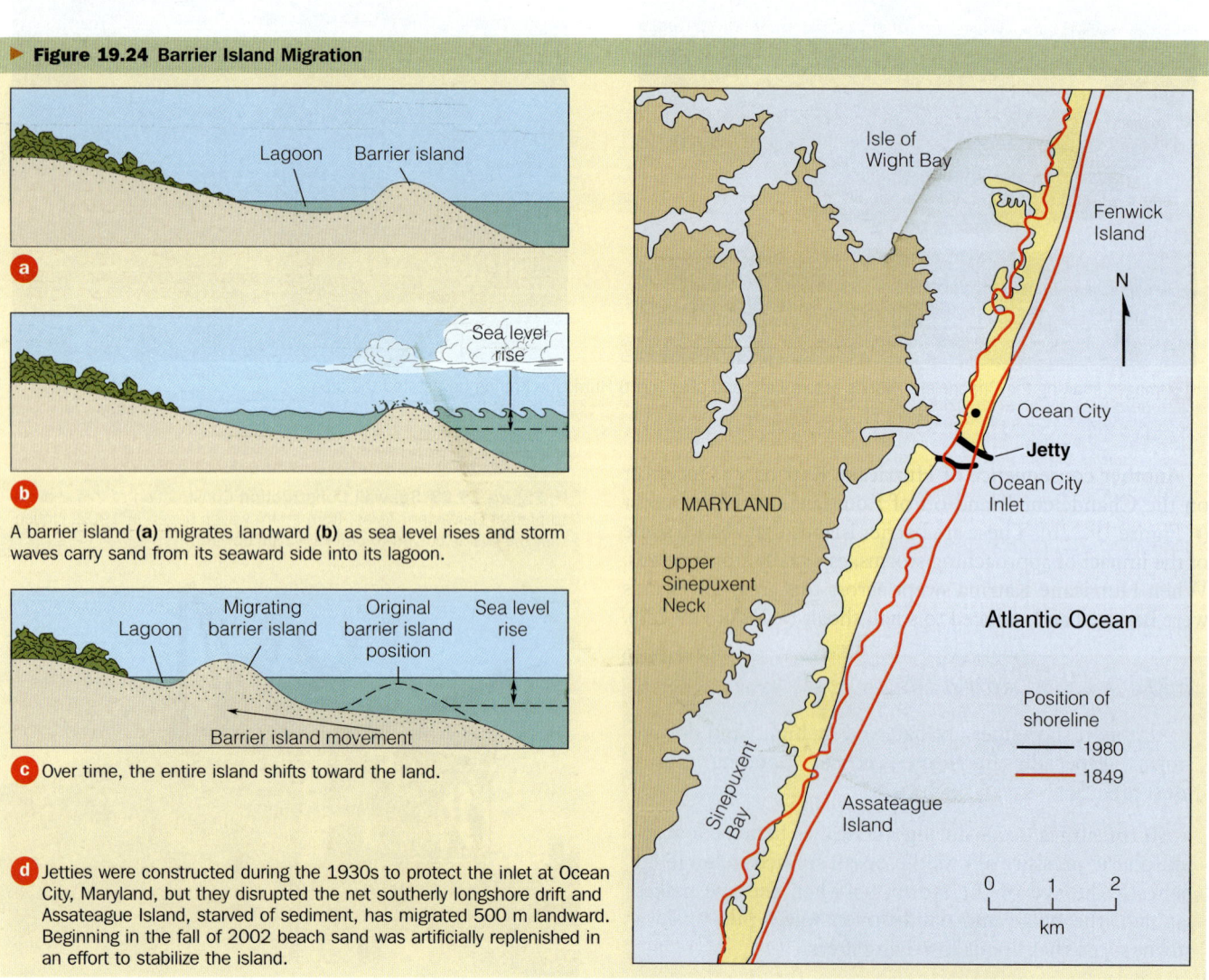

▶ **Figure 19.24 Barrier Island Migration**

(a) A barrier island (a) migrates landward (b) as sea level rises and storm waves carry sand from its seaward side into its lagoon.

(c) Over time, the entire island shifts toward the land.

(d) Jetties were constructed during the 1930s to protect the inlet at Ocean City, Maryland, but they disrupted the net southerly longshore drift and Assateague Island, starved of sediment, has migrated 500 m landward. Beginning in the fall of 2002 beach sand was artificially replenished in an effort to stabilize the island.

homes on them. Moreover, barrier islands are not the only threatened areas. Louisiana's coastal wetlands, an important wildlife habitat and seafood-producing area, are currently being lost at a rate of about 90 km^2 per year. Much of this loss results from sediment compaction, but sea-level rise exacerbates the problem.

Rising sea level also threatens beaches that many communities depend on for revenue. The beach at Miami Beach, Florida, was disappearing at an alarming rate until the Army Corps of Engineers began replacing the eroded beach sand (▶ Figure 19.25). Sea-level rise is a problem in North America, but in some parts of the world it is much more serious. A rise in sea level of only 2 m would inundate large areas of the U.S. East and Gulf Coasts but would cover 20% of the entire country of Bangladesh. Other problems include increased coastal flooding during hurricanes and saltwater incursions into the groundwater system (see Chapter 16).

■ *What measures do coastal communities use to minimize the effects of shoreline erosion?*

We can do nothing directly to prevent sea level from rising, so engineers, scientists, planners, and political leaders must examine what they can do to prevent or at least minimize the effects of shoreline erosion. One measure is to put strict controls on shoreline development. North Carolina, for example, permits large structures to be sited no closer to the shoreline than 60 times the annual erosion rate. A growing awareness of shoreline problems has resulted in similar legislation elsewhere, but some states have virtually no restrictions on coastal development.

Regulating coastal development is commendable, but it has no impact on existing structures and coastal communities. A general retreat from the shoreline is possible but expensive for individual structures and perhaps even for small communities. Of course, it is impractical to move large population centers inland. Such communities as Atlantic City, New Jersey, Miami Beach, Florida, and Galveston, Texas, have opted for one of two strategies to combat coastal erosion.

One strategy, adopted by both Atlantic City and Miami Beach, is to pump sand onto the beaches to replace sand lost to erosion (▶ Figure 19.25). This, too, is expensive as the sand must be replenished periodically because erosion is a continuing process. In many areas, groins are constructed to preserve beaches, but unless additional sand is artificially supplied to the beaches, longshore currents invariably erode sand from the down-current sides of the groins.

Another option is to build protective barriers such as seawalls. Seawalls, like the one at Galveston, Texas, are effective, but they are tremendously expensive to construct and maintain (▶ Figure 19.23). Unfortunately, seawalls retard erosion only in the area directly behind them; Galveston island west of the seawall has been eroded back about 45 m.

Armoring shorelines with seawalls and similar structures has had an unanticipated effect on Oahu in the Hawaiian Islands. Erosion of coastal rocks is the primary supply of sand to many Hawaiian beaches, so beaches maintain their width even as sea level rises. However, seawalls trap sediment on their landward sides, thus diminishing the amount of sand on the beaches. In four areas studied on Oahu, 24% of the beaches either have been lost or are much narrower than they were several decades ago.

▶ **Figure 19.25** U.S. Army Corps of Engineers Miami Beach, Florida, Nourishment Project

a Miami Beach before the beach nourishment project.

b Miami Beach after the U.S. Army Corps of Engineers beach nourishment project.

What Would You Do?

You and other members of the city council of a seaside community are alarmed because your pristine beach has become noticeably narrower each year and tourist revenues have fallen off markedly. What would you do to identify the source of the problem? What remedial action would you suggest?

Seawalls and riprap* (► Figure 19.26a), protect beachfront homes, hotels, and apartment buildings effectively, but they are expensive to build and during large storms are commonly damaged, requiring repairs. Furthermore, they may lead to loss of sand on beaches. North Carolina, South Carolina, Rhode Island, Oregon, and Maine no longer allow construction of seawalls.

The futility of artificially maintaining beaches is aptly demonstrated by the efforts to protect homes on the northeast end of the Isle of Palms, a barrier island off the coast of South Carolina. Following each spring tide, heavy equipment excavates sand from a deposit known as an *ebb tide delta* and constructs a sand berm to protect the houses from the next spring tide (► Figure 19.26b). Two weeks later, the process must be repeated in a never-ending cycle of erosion and artificial replenishment of the beach.

Section 19.8 Summary

- In most coastal areas, sea level is rising with respect to the land. Rising, or falling, sea level depends on how much subsidence or uplift of continents takes place, the volume of the ocean basins, the temperature of sea water, and how much water is frozen in glaciers.

- Some consequences of rising sea level are migration of barrier islands toward their nearby continent, as well as erosion of beaches and sea cliffs, thus threatening structures or communities.

- Protecting shoreline areas with barriers, replenishing eroded beaches, and restricting shoreline development are the methods used by communities threatened by shoreline erosion.

*Riprap is an irregular pile or pile of fitted rocks along a shoreline or riverbank used to prevent erosion.

► **Figure 19.26 Riprap and Sand Berm**

a Riprap made up of large pieces of basalt was piled on this beach to protect a luxury hotel just to the left of this image.

b Heavy equipment builds a berm, an embankment of sand, on the seaward side of beach homes on the Isle of Palms, South Carolina, to protect them from waves. The berm must be rebuilt every two weeks after each spring tide.

Review Workbook

ESSENTIAL QUESTIONS SUMMARY

19.1 Introduction
- *What is the difference between a coast and a shoreline?*

A shoreline is the area from low tide to the highest level on land affected by storm waves, whereas a coast includes the shoreline, nearshore sandbars and islands as well as features onshore such as sand dunes, marshes, and sea cliffs.

- *How does the study of shorelines provide an example of interacting systems?*

Part of the hydrosphere is involved as waves and tides, which are caused by wind and the gravitational attraction of the Moon and Sun, respectively, act on shoreline materials.

19.2 Shoreline Processes
- *How do the Moon and Sun generate tides?*

The gravitational attraction of both the Moon and Sun deform the solid Earth, but their effect on the oceans is even greater, causing the rise and fall of tides twice daily in most places. Depending on the positions of the Moon and Sun, exceptionally high or low tides take place every two weeks.

- *What is wave base?*

As waves enter water shallower than wave base, a depth corresponding to one-half wavelength, the wavelength decreases but the wave height increases. The waves then become oversteepened and plunge forward as breakers.

- *What is wave refraction, and how does it generate nearshore currents?*

Most waves approach a shoreline at some angle, in which case one end of the wave slows while the other races ahead, thereby bringing the wave more nearly parallel with the shoreline. Nevertheless, the waves usually hit the shore at an angle that generates a current along the shore in the same direction as the approaching waves.

19.3 Shoreline Erosion
- *What processes account for erosion along shorelines?*

The impact of water (hydraulic action) coupled with abrasion, the grinding away of rock by water-carrying sediment, erode shorelines.

- *How do sea arches and sea stacks form?*

Wave refraction around a rocky headland may erode the headland to form sea caves on each side. With continued erosion, the caves merge to form a passageway, a sea arch. A collapsed arch leaves isolated rocks or sea stacks projecting above sea level. Oregon is a good place to see arches and stacks because it has a rocky shoreline.

19.4 Deposition along Shorelines
- *What are summer beaches and winter beaches?*

Winter beaches tend to be steeper and made up of coarser materials than summer beaches because waves are more energetic during that season. The sand eroded from beaches during winter is stored in offshore bars until it is driven back shoreward by more gentle waves.

- *What is a tombolo and how does one form?*

A tombolo is a spit that extends more or less perpendicular to the shoreline to an offshore island or rock. Tombolos form because waves are refracted around the offshore obstacle, causing currents to converge and transport sediment seaward.

19.5 The Nearshore Sediment Budget
- *What is the nearshore sediment budget?*

It refers to the balance of sediment inputs and sediment losses in a shoreline system. Most sediment input comes from streams and rivers transporting sediment to the shore, although erosion of seashore rocks contributes, too. Sediment is lost to longshore drift, wind, and offshore sediment transport.

19.6 The Classification of Coasts
- *How do depositional and erosional coasts differ?*

Depositional coasts have an abundance of detrital sediment—especially sand—wide, sandy beaches, deltas, and barrier islands, whereas erosional coasts are steep and irregular, beaches are restricted to protected areas, and sea cliffs, sea stacks, and wave-cut platforms are common.

19.7 Storm Waves and Coastal Flooding
- *What causes the most fatalities during hurricanes?*

Contrary to popular belief, strong wind, although dangerous, is not the main cause of hurricane fatalities. More lives are lost to coastal flooding resulting from storm surge and intense rainfall.

19.8 Managing Coastal Areas as Sea level Rises
- *What factors influence whether sea level rises or falls?*

Several factors are involved. First, uplift or subsidence of the land results in falling and rising in sea level, and so does the volume of the ocean basins, which is affected by the rate of seafloor spreading (see Chapter 7). Another factor controlling sea level is global warming because as near-surface water becomes warmer, it expands and occupies more space.

- *What measures do coastal communities use to minimize the effects of shoreline erosion?*

The only way to completely solve the problem is to move structures farther inland, which is not practical for cities, or to restrict shoreline development. Options for controlling erosion include importing sand to replenish beaches and erecting structures such as seawalls.

ESSENTIAL TERMS TO KNOW

barrier island (p. 614)
baymouth bar (p. 613)
beach (p. 608)
berm (p. 608)
breaker (p. 603)
crest (p. 602)
emergent coast (p. 618)
fetch (p. 603)
longshore current (p. 604)

longshore drift (p. 609)
marine terrace (p. 607)
nearshore sediment budget (p. 615)
rip current (p. 605)
shoreline (p. 598)
spit (p. 613)
submergent coast (p. 617)
tide (p. 600)
tombolo (p. 613)

trough (p. 602)
wave (p. 601)
wave base (p. 602)
wave-cut platform (p. 607)
wave height (p. 602)
wavelength (p. 602)
wave period (p. 602)
wave refraction (p. 604)

REVIEW QUESTIONS

1. What are the important processes operating on shorelines? How effective are they along seacoasts as opposed to lakeshores?

2. Explain how waves break and how they generate longshore currents.

3. How do rising sea level and coastal development, including developments on barrier islands, complicate efforts to control shoreline erosion?

4. Why does an observer at a shoreline location experience two high and two low tides daily?

5. While driving along North America's West Coast, you notice a surface sloping gently seaward with several large masses of rock rising above it. How would you explain the origin of this landform to your children?

6. Why are most beaches made up of quartz sand? Are there other materials that make up beaches? If so, what are they?

7. What are sleeper waves, and why are they dangerous?

8. How does a baymouth bar form?

9. Explain why waves are so small in ponds and lakes but much larger ones develop in the oceans.

APPLY YOUR KNOWLEDGE

1. As the member of a planning committee for a coastal community where shoreline erosion is a continuing problem, what recommendations would you make to remedy or at least mitigate the problem?

2. A hypothetical shoreline has a balanced budget, but dams are built on the rivers that supply sediment and a seawall is built to protect sea cliffs from erosion. Explain what will likely happen to the beaches in this area.

FIELD QUESTION

1. The image below shows two distinctive shoreline features. What are they, are how do they form?

GEOLOGY MATTERS

GEOLOGY IN UNEXPECTED PLACES

Rising Sea Level and the Fate of Venice, Italy

Venice, Italy, is known worldwide for its charm, historic treasures, and streets of water rather than pavement. It is a city built on islands in a lagoon, with a narrow causeway connecting the main island to the mainland. All traffic in the city moves on a maze of canals, except in a small area in and around Venice's port (▶ Figure 1). The problem is that Venice is at sea level but sea level is rising and the city is sinking, creating a huge problem of repeated flooding.

Settlers from the mainland escaping from barbarians sweeping through Italy founded Venice during the fifth century. The entire city rests on pilings that were driven into the muddy sandy sediments to form level foundations for construction. Of course pilings settle and rot, but a much greater concern is the rise in sea level and the increased frequency of flooding. During especially high tides, or what locals call *aqua alta* (high water), water from the Adriatic (part of the Mediterranean) Sea inundates many parts of the city. And even though the flooded areas are usually covered by a meter or less of water, flooding takes place more than 50 times each year.

Three factors account for Venice's current dilemma. One is that sea level is rising as a result of global warming. Whether humans are partly responsible for global warming is irrelevant—the fact is that sea level is rising. The second factor is that the city rests on a foundation of mud and sand that tends to compact under its own weight. And third, withdrawal of groundwater from 20,000 wells in the region has contributed to general subsidence of the land. Pumping ceased by 1970 and now an aqueduct brings water to the area. Since its founding, Venice has subsided about 10 m relative to sea level. So what can be done to prevent flooding? One proposal is that water or carbon dioxide be pumped beneath the area to raise it. This proposal has not been taken seriously, but the Italian government in 2003 approved a colossal project to protect Venice's lagoon from floods.

The solution involves constructing 79 huge hinged gates on the seafloor at the three entrances to the lagoon. During exceptionally high tides, the gates will be raised to form dams that separate the lagoon from the open sea. Current estimates are that the project will take eight years to complete and cost between $2 and $4 billion. Of course environmental groups are concerned about the health of the lagoon if it must be closed off as many as 70 or 80 times a year, not to mention the constant interruptions to shipping and tourism. Others express the concern that in the long term the gates will not be totally effective.

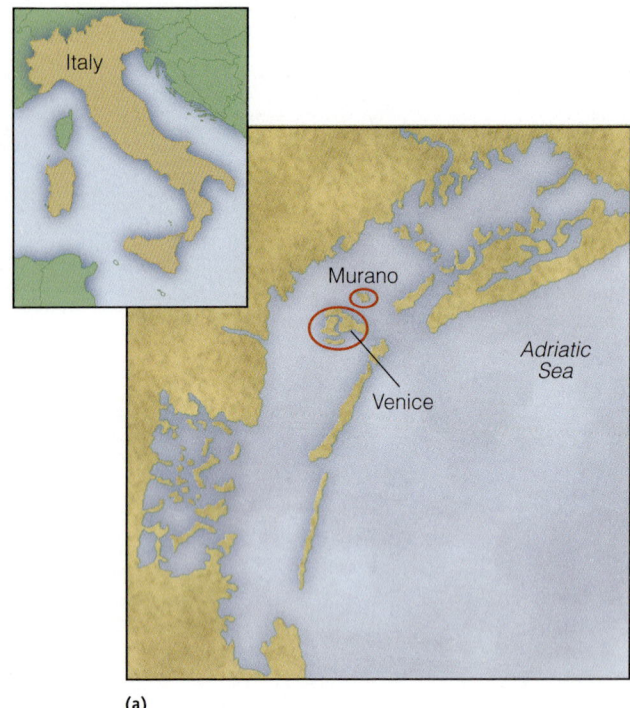

(b)

(c)

Figure 1
(a) Location of Venice, Italy in the Adriatic Sea. **(b)** View of the Grand Canal in Venice. **(c)** Buildings along the canals have waterproof foundations but now the sea commonly overtops the foundations allowing green algae to grow on the porous brickwork of these 14th- and 15th-century structures.

GEOLOGY IN YOUR LIFE

Giant Killer Waves

During the last two decades, more than 200 supertankers and container ships at least 200 m long have been lost at sea, many of them apparently hit by monstrous waves, or what are commonly reported as freak waves. Not to be confused with tsunami, which at sea are rarely more than a meter or so high, these freak waves (actually they are called *rogue waves*) pose a threat to commercial shipping, cruise liners, and pleasure craft. Rogue waves may be as high as 30 m, three to five times higher than average waves, and they arise out of an otherwise comparatively calm sea (▶ Figure 1).

Even the largest ships are in peril when hit by a rogue wave. In 1942 HMS *Queen Mary*, while carrying 15,000 American soldiers, was hit broadside by a rogue wave and nearly capsized (the ship heeled more than 45 degrees before coming upright), and in 1995 the *Queen Elizabeth 2* was hit by a rogue wave 29 m high. More recently, at 6:00 A.M. on April 16, 2005, the Norwegian cruise ship *Dawn* was damaged by a 21-m-high rogue wave that flooded more than 60 cabins and injured four passengers, none of them seriously. Recent research indicates that rogue waves are more common than previously thought; a German research group reported that about ten were detected in a three-week period in 2001.

Encounters with rogue waves are terrifying, they cause considerable damage, and of course they sink ships. For example:

> The Italian Lines' beautiful flagship *Michelangelo* was severely damaged by a rogue wave in a 1966 storm only 1500 kilometers . . . out of New York. Inch-thick glass was shattered on the bridge 24 meters . . . above the waterline, steel railings on the upper decks were ripped away, and the bow was flattened. Many passengers were injured; three died.*

Ships are not the only structures vulnerable to rogue waves. On February 15, 1982, a huge wave crashed into a drilling platform in the North Atlantic about 275 km east of St. John's, Newfoundland. The platform collapsed and sank, taking with it all 84 people on board.

Although scientists have much to learn about where and when to expect rogue waves, they do know something about how they are generated. We mentioned in the section on wave generation that waves with different heights, lengths, and periods may merge to increase or decrease wave height. In some cases, though, many waves approach a single location and merge, a phenomenon called *constructive interference*, and a rogue wave suddenly rises out of the sea. Thus rogue waves form under these exceptional conditions, and because they are made up of waves moving at different rates, they do not last long. They may separate into individual waves or their crests may break, but while intact, however briefly, they can damage or destroy even the largest ships.

Constructive interference is also a threat along shorelines. Many beaches have signs warning about *sleeper waves*, which are much larger than the other waves in a series (see Figure 19.9c). Every year several people in North America, particularly along the West Coast, lose their lives to sleeper waves. These waves are truly dangerous; as more than one website advises, "never turn your back on the ocean."

Another situation for generating rogue waves is when wind-generated waves encounter fast-moving surface currents. As a result rogue waves are more frequent where surface currents are strong, as in the Gulf Stream east of North America and especially off the southeastern coast of Africa in the area of the westerly-flowing Agulhas Current.

What can be done to minimize the impact of rogue waves? Perhaps developing the ability to anticipate rogue waves and giving timely warnings will help, but these waves arise suddenly and dissipate quickly, so reducing encounters with them is not likely. Perhaps the best solution is to design and construct ships that are more likely to survive encounters with rogue waves.

*T. Garrison, *Oceanography: An Invitation to Marine Science*, 4th ed. (Pacific Grove, CA: Wadsworth/Thomson Learning), p. 250.

Figure 1
This image of a rogue wave was taken in 1935 from the liner SS *Washington* as it was crossing the North Atlantic.

GEOLOGY MATTERS
Geology AT WORK

Erosion and the Cape Hatteras Lighthouse

The Outer Banks of North Carolina are a series of barrier islands separated from the mainland by a lagoon up to 48 km wide. Actually, the Outer Banks are made up of several barrier islands and spits extending about 240 km from the Virginia state line to Shackleford Banks near Beaufort, North Carolina (▶ Figure 1a). On September 18, 2003, Hurricane Isabel roared across the Outer Banks, causing wind damage and flooding there and in several adjacent states, particularly Virginia.

Since 1526, more than 600 ships have been lost on the dangerous sand bars adjacent to the Outer Banks, such as Diamond Shoals, an area of very shallow seawater. To reduce ship losses, several lighthouses were built on the islands; the most famous is the Cape Hatteras Lighthouse. It was constructed in 1802, although the current lighthouse was built in 1870. In 1999, however, the National Park Service, which administers Cape Hatteras National Seashore, had the lighthouse moved 500 m inland and 760 m to the southwest (▶ Figure 1b). The reason? The Outer Banks, just like other barrier islands, are migrating toward the mainland as storm waves carry sand from the seaward margins into the lagoon. Indeed, Hurricane Isabel cut a new channel through Hatteras Island, thereby isolating communities to the north and south.

When the Cape Hatteras Lighthouse was built in 1870, it was 457 m from the shoreline, but when it was moved in 1999, it stood only 36.5 m inland. Given that the annual erosion rate is about 3 m per year and all previous attempts to stabilize the shoreline had failed, the Park Service found it prudent to act when it did. The Cape Hatteras Lighthouse, the tallest brick lighthouse in the world, should now be safe for several centuries.

Figure 1
(a) The Outer Banks of North Carolina consist of barrier islands separated from the mainland by a wide lagoon, Pamlico Sound. (b) The Cape Hatteras Lighthouse during its 1999 move 500 m inland. Before it was moved it stood about 36 m from the sea and the annual erosion rate was 3 m/yr.

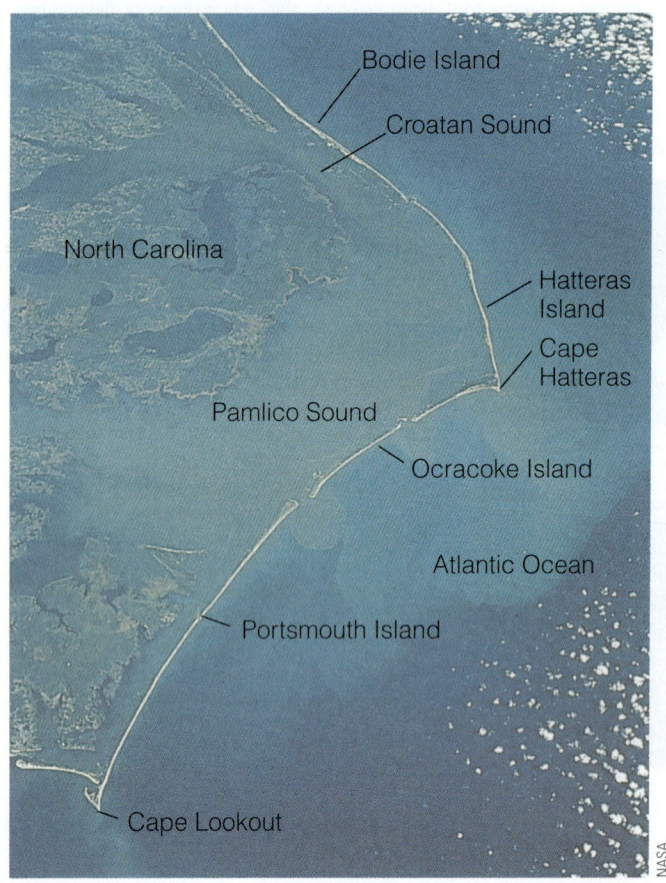
(a)

(b)

GEOLOGY MATTERS

GEOLOGY IN FOCUS Energy from the Oceans

If we could harness the energy of waves, ocean currents, temperature differences in ocean waters, and tides, an almost limitless, largely nonpolluting energy supply would be ensured. Unfortunately, ocean energy is diffuse, meaning that the energy for a given volume of water is small and thus difficult to concentrate and use. Of the several sources of ocean energy, only tides show much promise for the near future.

Ocean thermal energy conversion (OTEC) exploits the temperature difference between surface waters and those at depth to run turbines and generate electricity. The amount of energy from this source is enormous, but a number of practical problems must be solved. For one thing, large quantities of water must be circulated through a power plant, which requires large surface areas devoted to this purpose. Despite several decades of research, only a few have been tested in Hawaii and Japan.

Ocean currents also possess energy that might be tapped to generate electricity. Unfortunately, these currents flow at only a few kilometers per hour at most, whereas hydroelectric power plants on land rely on water moving rapidly from higher to lower elevations. Furthermore, ocean currents cannot be dammed, their energy is diffuse, and any power plant would have to contend with unpredictable changes in flow direction.

Harnessing wave energy to generate electricity is not a new idea, and in fact it is used on a limited scale. Any facility using this form of energy would obviously have to be designed to withstand the effects of storms and saltwater corrosion. No large-scale wave-energy plants are operating, but the Japanese have developed devices to power lighthouses and buoys, and a facility with a capacity to provide power to about 300 homes began operating in Scotland during September 2000.

Tidal power has been used for centuries in some coastal areas to run mills, but its use at present for electrical generation is limited. One limitation is that the tidal range must be at last 5 m, and there must also be a coastal region where water can be stored following high tide. Suitable sites for tidal power plants are limited not only by tidal range but also by location. Many areas along the U.S. Gulf Coast would certainly benefit from a tidal power plant, but the tidal range of generally less than 1 m precludes the possibility of development. However, an area with an appropriate tidal range in some remote area such as southern Chile or the Arctic islands of Canada offers little potential because of the great distance from population centers. Accordingly, in North America only a few areas show much potential for developing tidal energy.

The idea behind tidal power is simple, although putting it into practice is not easy. First, a dam with sluice gates to regulate water flow must be built across the entrance to a bay or estuary. When the water level has risen sufficiently high during flood tide, the sluice gates are closed. Water held on the landward side of the dam is then released and electricity is generated just as it is at a hydroelectric dam. Actually, a tidal power plant can operate during both flood and ebb tides (▶ Figure 1).

The first tidal power-generating facility was constructed in 1966 at the La Rance River estuary in France. In North America, a much smaller tidal power plant has been operating in the Bay of Fundy, Nova Scotia, where the tidal range, the greatest in the world, exceeds 16 m.

Although tidal power shows some promise, it will not solve our energy needs even if developed to its fullest potential. Most analysts think that only 100 to 150 sites worldwide have sufficiently high tidal ranges and the appropriate coastal configuration to exploit this energy resource. This, coupled with the facts that construction costs are high and tidal energy systems can have disastrous effects on the ecology (biosphere) of estuaries, makes it unlikely that tidal energy will ever contribute more than a small percentage of all energy production.

Figure 1
Rising and falling tides produce electricity by spinning turbines connected to generators, just as at hydroelectric plants. This view in the foreground is a cross section showing how water flows into and out of the basin, but the basin would actually be closed off here by land. **(a)** Water flows from ocean to basin during flood tide. **(b)** Basin full. **(c)** Water flows from basin to ocean during ebb tide.

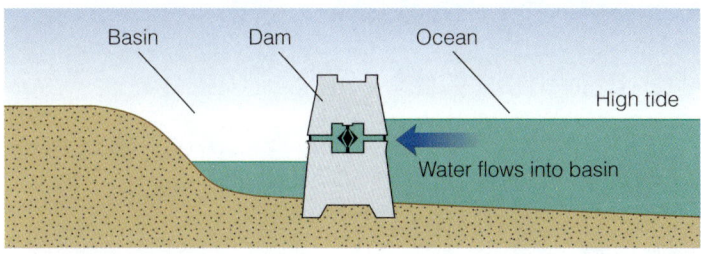

(a)

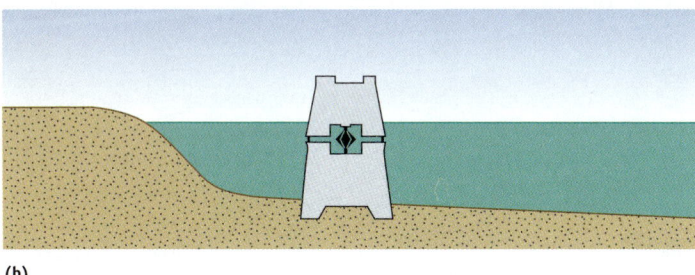

(b)

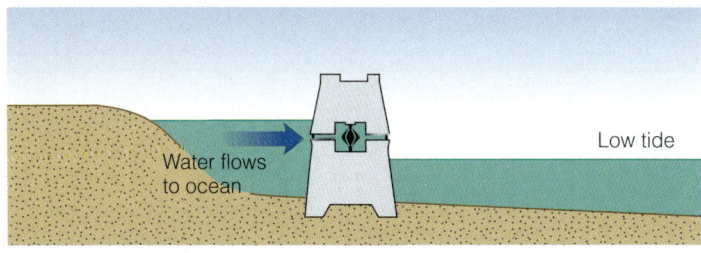

(c)

Chapter 20

Geology and Humanity

ESSENTIAL QUESTIONS TO ASK

20.1 Introduction

20.2 Geology and History
- What were the prerequisites for civilization to begin?
- What role did geology play in trade and conflict?
- How did geology influence science, culture, and international affairs?

20.3 Humans as Geologic Agents
- How seriously have people increased erosion rates worldwide?
- What are the geologic consequences of building dams?
- How does mining compare with the natural movement of sediments?

20.4 Climate Change
- How much have sea levels changed since the end of the last Ice Age?
- What are some examples of natural switches that can cause sudden climate change?
- How does ice shelf collapse occur, and why is it important?

Seeing Earth
Memory can be a mountain in the mind that we climb up to look down at the valley of our past. A photograph can show us a past beyond the valley of our own experience. Ansel Adams saw Nature. His photos read like visual poems of light and shadow that tell us the way Earth once was.

—A. W.

GEOLOGY MATTERS

GEOLOGY IN FOCUS:
Atomic Testing and the Earth System

GEOLOGY AT WORK:
Environmental Geology

GEOLOGY IN YOUR LIFE:
Organic Farming

GEOLOGY IN UNEXPECTED PLACES:
Mines and Flooding

PHYSICAL Geology Now™ This icon, which appears throughout the book, indicates an opportunity to explore interactive tutorials, animations, or practice problems available on the Physical GeologyNow website at **http://now.brookscole.com/phygeo6**.

20.1 Introduction

At the beginning of the 21st century, civilization faces serious problems that have never existed before. During the past 50 years, the world has experienced the greatest population increase and economic growth ever recorded in human history. There are three times more people living today than in 1950, and the average global citizen consumes the energy equivalent of 20 "energy slaves," meaning the power of 20 human equivalents working around the clock, to provide the services that you easily take for granted—from a refrigerator and a car to your word processor and nighttime lighting in the college library (▶ Figure 20.1). This is *five* times the amount of energy an average individual consumed a lifetime ago, despite the increase in the total number of people. Human ingenuity and a booming economic order made all of this possible. Parallel increases in the human impact on the natural world can be seen practically everywhere, and it is not surprising that the natural world is showing signs of strain. The historian J. R. McNeill notes in his book *Something New under the Sun: An Environmental History of the Twentieth-Century World* (p. 17):

> The great modern expansion, while liberating in a fundamental sense, brought disruption with it. The surges in population, production, and energy use affected different regions, nations, classes, and social groups quite unevenly, favoring some and hurting others. Many inequalities widened, and perhaps more wrenching, fortune and misfortune often were reshuffled. Intellectually, politically, and in every other way, adjusting to a world of rapid growth and shifting status was hard to do. Turmoil of every sort abounded. The preferred policy solution after 1950 was yet faster economic growth and rising living standards: if we can all consume more than we are used to, and expect to consume still more in the years to come, it is far easier to accept the anxieties of constant change and the inequalities of the moment. Indeed, we erected new policies, new ideologies, and new institutions predicated on continuous growth. Should this age of exuberance end, or even taper off, we will face another set of wrenching adjustments.

What does all of this have to do with you? You are poised at the edge of a future that will bring with it great uncertainties and changes. You need to have the broad education and skills to adapt accordingly, and in ways that none of us can forecast very well. Both individual and community problem solving will be required on occasion, and in many cases, *geology is going to have to be given serious consideration*. You might think of geology as a "quaint" science that excites museum goers, rock hounds, and grade-school students, but it bears on your life, even directly, in ways far greater than you may appreciate.

▶ **Figure 20.1** The Scale of Human Activity over the Past Half-Century Has Affected the Natural World as Never before in History

The role that geology plays in human affairs, as well as our role in Earth's geologic processes, is usually not apparent, given the urban or semiurban world in which so many of us now live. Before you read this book, were you aware of the connection between geology and economic and political power, or how vitally dependent we are on geologic materials for building our civilization and to make most of the products that we enjoy in our daily lives? In fact, we need to realize that the history and culture of human civilization are deeply rooted in the geologic processes that have shaped the world around us. It is important to appreciate these connections if for no other reason than to acquire a better sense of our "place" in the scheme of things.

Throughout most of humanity's history, people have lived in a world with many frontiers, new places to explore,

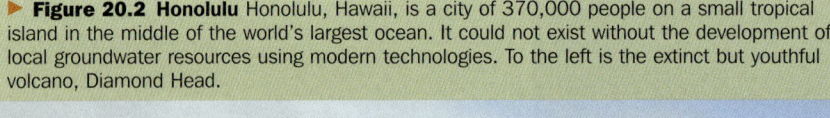

▶ **Figure 20.2 Honolulu** Honolulu, Hawaii, is a city of 370,000 people on a small tropical island in the middle of the world's largest ocean. It could not exist without the development of local groundwater resources using modern technologies. To the left is the extinct but youthful volcano, Diamond Head.

colonize, and develop. Only 100 years ago no human being had yet been to the South Pole, and the bottom of the sea was as mysterious a setting as the surface of Mars. It seemed that our natural resources would never run out because there were still so many unexplored and untapped areas. We could always settle new untamed lands in case other parts of the world became crowded and unfavorable. But during the last 50 years, something new has happened. The surface of the planet has filled up with an advanced, densely populated civilization; the world has become economically globalized; and the result is a unique situation never before seen in Earth's history. Human civilization's effects on this planet produce signs of tremendous changes that rival and even override many of the geologic processes discussed in this book (▶ Figure 20.2).

Human activity alters the makeup of the global atmosphere, causing weather and climate changes. It affects water quality, causes excess erosion and sedimentation, and severely affects the natural biota with which we share Earth. Many vital resources such as petroleum and groundwater are extracted much faster than nature can replenish them, while pollutants are accumulating in quantities that cannot be rapidly dispersed by natural processes. Civilization has even advanced far enough to take notice of these afflictions and help to repair their damage by containing and treating them artificially. Geologists have set up a special branch of earth science, called **environmental geology,** to develop possible solutions to the problems we create for ourselves.

In the short sections that follow, we briefly discuss some of the general points raised in this introduction. By no means can we comprehensively address the fascinating topic of geology and humanity, but we can present some teasers to illustrate that there is, in fact, a very human dimension to the study of Earth.

Section 20.1 Summary

- The spectacular growth in human population and resource consumption during the past 50 years has created problems that require broad learning, including geology, to understand and solve.

20.2 Geology and History

Throughout history, geology has influenced where and how people live. Many great human events, from the gold rushes of the American West to the rise and fall of Mayan civilization, were constrained by geologic phenomena and processes. The historian Will Durant once remarked that "civilization exists by geological consent." Let's see what he means by that statement.

■ *What were the prerequisites for civilization to begin?*

Most of human existence, as recorded in prehistoric rock and cave art, has been nomadic and has taken place in a colder world with a more hostile climate than we face today. It is unlikely that civilization could have arisen during the ice ages for two reasons: *the inability to establish a productive, reliable agricultural base*, and *the inability to carry on maritime trade owing to rapid changes in sea level*. The sudden warming of the world at the start of the Holocene Epoch 10,000–12,000 years ago not only opened up a broader range of latitudes favorable for human habitation but, more important, accompanied significant stabilization of the climate. Agricultural experimentation was under way in the Middle East while a remnant of the vast Canadian ice sheet still clung to the rim of emergent Hudson Bay. When climatic opportunity knocked, people did not waste any time taking advantage of the changes offered them (▶ Figure 20.3).

The earliest centers of urban–agricultural life were sites that combined rich soil with abundant water and just enough flooding to revitalize depleted croplands with fresh nutrients. Even today, 5000 years later, these deltas and floodplains remain home to a large percentage of the human population—an echo of our earliest attempts to become civilized. As people migrated outward from these centers, many settled in humid or subhumid volcanic areas, which also provided fertile soils and farmland. The Campanian Plain, next to the volcano Vesuvius, and the ash-rich soils of Sicily were breadbaskets for early Greeks and Romans. The Romans perfected the development

▶ **Figure 20.3 Wupatki ("Big House") Pueblo, near Flagstaff, Arizona** Wupatki was inhabited by as many as 300 Sinagua people 800 to 900 years ago. This was the first attempt at civilization in this region, and climate change may have contributed to its early demise.

of concrete and set the stage for modern engineering using mortar and brick made from pumice and ash-rich volcanic rocks found throughout their countryside. We owe our best snapshot of Roman life—the preservation of the ancient city of Pompeii—to a powerful volcanic eruption (▶ Figure 20.4). Other ancient communities that give us insights into the lives of our ancestors are preserved in volcanic strata in Greece, El Salvador, New Zealand, and Mexico. We see that the same geologic processes that determine where humanity can thrive can also dictate where it cannot.

▶ **Figure 20.4 Students Pose by a 2000-Year-Old Roman Arch in the Excavated Ruins of Pompeii, Italy** Vesuvius volcano, whose eruption destroyed Pompeii, looks rather friendly and benign in the background.

■ *What role did geology play in trade and conflict?*

The quest for metals to make jewelry, weapons, and tools drew people into international trade spanning thousands of kilometers. The sites of interest were mines in ancient tectonic rifts and convergent plate zones, from Burma and south China to the wilds of Cornwall, England. Geopolitical activity and empire building were oriented around a combination of geography and tectonically associated geologic resources (▶ Figure 20.5). The Romans fought their war with Hannibal in large part for the control of Spanish ore deposits and Mediterranean trade routes, including access to copper from the Island of Cyprus. (Cyprus derives its name, in fact, from *aes cyprium*, the Latin word for "island of copper.") This struggle for the control of geologic resources continues today, just as it did thousands of years ago, and is reflected in such political slogans as "No Blood for Oil" (see Chapter 2).

Geology not only drives the scramble for international power but also has played a key role in politics, economics, arts, and the humanities. The **code of Hammurabi,** written 4500 years ago in ancient Sumeria, "the mother of all civilizations," is the world's first state legal regulation. It dictated the fair use of water for irrigation purposes in the Tigris–Euphrates floodplain. Some violations even carried the death penalty because Sumerian society had evolved a delicate balance and dependence on the *cooperative use* of the hydrologic cycle. Ironically, lack of understanding of local soil chemistry and the consequences of overirrigation led to salinization of the soil in the hot dry climate, and thus contributed to the collapse of Sumeria. The Sumerian pantheon of deities reflected powerful natural forces—rain, sun, flood, and drought—upon which lives depended in a way that is difficult for us to appreciate today.

■ *How did geology influence science, culture, and international affairs?*

Three salient scientific events brought about crises in contemporary religious thinking: the discovery by Copernicus of direct evidence for a Sun-centered solar system in the 15th century, the insights of Charles Darwin and others regarding the evolution of species in the

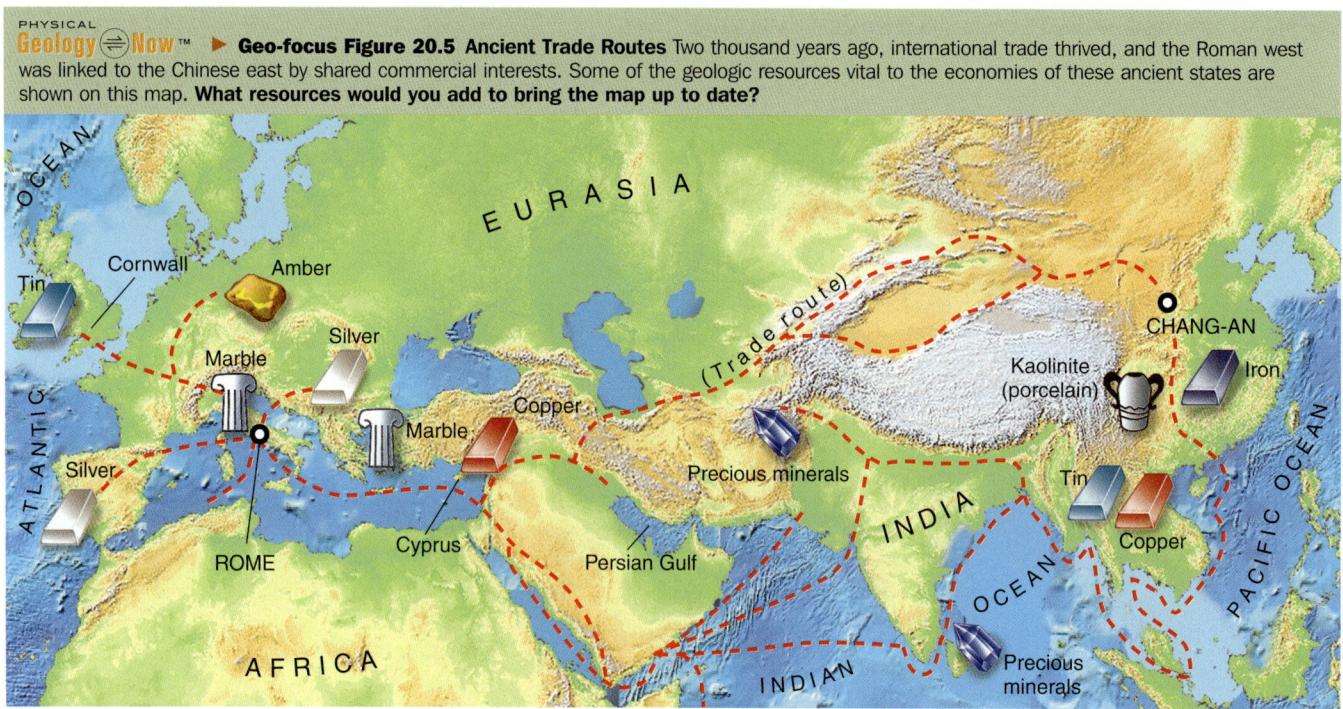

Geo-focus Figure 20.5 Ancient Trade Routes Two thousand years ago, international trade thrived, and the Roman west was linked to the Chinese east by shared commercial interests. Some of the geologic resources vital to the economies of these ancient states are shown on this map. **What resources would you add to bring the map up to date?**

19th century, and the discovery of how the world works, geologically speaking, beginning with Scotsman James Hutton in the late 18th century. These new ideas challenged the two views that humans are at the center of creation and that the world develops along a linear path, from Creation to Apocalypse, rather than via spontaneous, largely cyclical natural processes.

Geology has also rattled Western philosophy. In 1755 a catastrophic earthquake and tsunami destroyed the port of Lisbon, Portugal, killing more than 100,000 people. This disaster, rare for Europe, raised feelings that Nature is indifferent to humanity, contrary to the prevailing notion that people are the central focus of all creation. The earthquake inspired the famous philosopher Francois Voltaire (1694–1778) to write his great work *Candide*, which is a classic commentary on the essence of suffering and the perils of utopian thinking.

Across the ocean in the British colonies, geologic features helped set the stage for the Revolutionary War and the birth of the United States. Colonists living on the Atlantic Seaboard sought to emigrate across the Appalachian mountain barrier in an unconstrained way. But for economic reasons, the British government thought that the Appalachians represented a "natural" western frontier for the colonists and passed the Intolerable Acts to regulate the lives and movements of its American subjects. Important passageways for emigration extended through water gaps in the long anticlines that make up the spine of this range, permitting many would-be settlers to filter westward anyway. The Cumberland Gap was one of the most famous of these natural highways, the "Gateway to Kentucky" popularized by Daniel Boone. It was only a matter of time before the colonists exploited a geographically porous frontier and collided with the financial needs of the Crown.

Geology also played a crucial role in many battles during the U.S. Civil War. For example, U.S. General Grant attempted to neutralize the threat of the Confederate fortress at Vicksburg, which sat on top of a cutbank slope of the Tuscumbia Meander on the strategically vital Mississippi River (▶ Figure 20.6). Grant ordered his engineers to cut a canal across the narrow neck of the meander to provide a safe alternative route for boat traffic. The Tuscumbia Cutoff was a great idea, but flooding destroyed the temporary dam that protected the canal works during construction and filled the artificial channel with overbank sediments. There was no time to spare for removing this detritus, and Vicksburg therefore had to be taken the hard way, at the cost of thousands of lives.

The importance of geologic features and materials in military strategy continues to the present day. **Terrain analysis** is a special study involving the interpretation of the strength and makeup of land surfaces based on careful examination of satellite images and air photography. Whether a feature in enemy possession is a sand dune or a bedrock knob could make all the difference in the success of a military operation.

Geologic features also have a prominent part in the modern environmental movement. The first images from space of the whole Earth, captured by Apollo lunar astronauts in 1968–1972, riveted the public imagination as no science-fiction movie has ever done. Coupled with the birth of modern ecology, these images made Earth suddenly seem both fragile and important to protect. The stunningly beautiful images of glacial Yosemite Valley and other natural landscapes captured by Ansel Adams, Nancy Newhall, and other

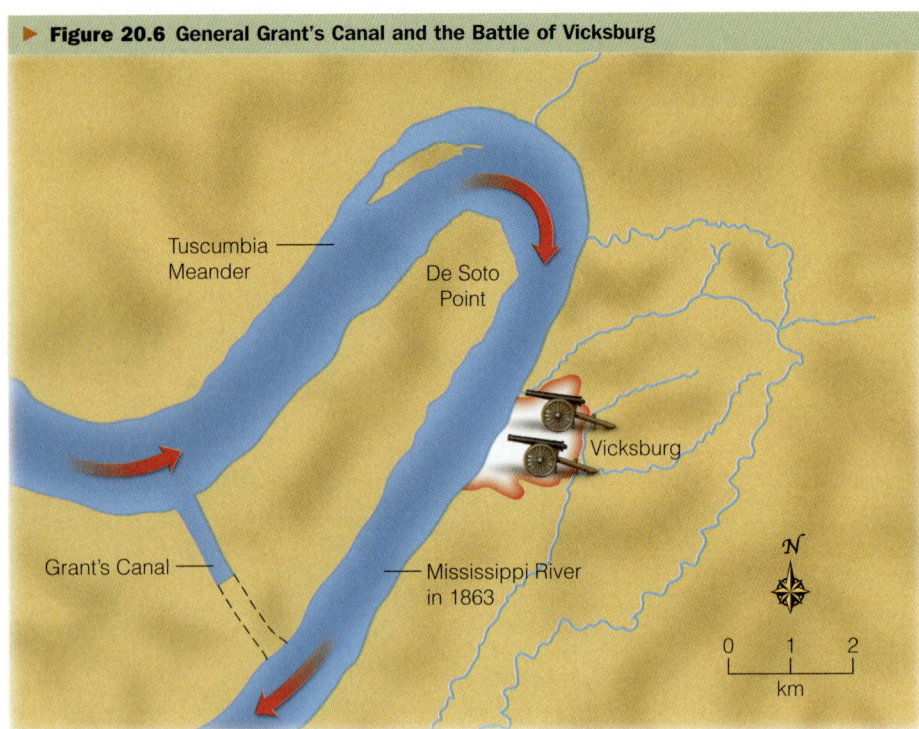

▶ **Figure 20.6 General Grant's Canal and the Battle of Vicksburg**

pioneer landscape photographers rallied people around the cause of conservation. Photographs and films of geologic wonders helped prevent dams from being constructed in Dinosaur National Monument, Colorado and Utah, and in the Grand Canyon, Arizona, in the 1950s and 1960s. They play a key role today in galvanizing the public movement for draining Lake Powell Reservoir in Arizona and Utah to uncover spectacular incised river meanders, natural arches, red rock slot gorges, hoodoos, and other drowned rock wonders.

Although people may not always understand how geology works, the products of its many interrelated processes are widely appreciated and enjoyed.

Finally, the future prospects for civilization, as we will subsequently discuss, are intimately linked to geologic changes. Foremost, perhaps, is the preservation of reasonably stable global climate, fertile soil covers, and freshwater supplies. Just as at the beginning of civilization, the ultimate human concern is *food* (▶ Figure 20.7). Without respecting

▶ **Figure 20.7 Cattle Browse in a French Pasture, Blissfully Unaware That 20,000 Years Ago This Land Could Not Grow Food** Some geologists worry that human-induced climate change could potentially render such land unproductive again.

geologic constraints, we will never be able to achieve and maintain food security for long. But just how big a problem is this today?

> **Section 20.2 Summary**
>
> • A stable world climate and sea level were geological prerequisites for establishing early civilizations. Geology controlled the distribution of resources, especially metallic ores, which governed patterns of trade in the ancient world.
>
> • The need for the cooperative use of geologic resources brought about some important ancient laws. Later, insights about geology and specific geologic events influenced the development of philosophy and religion.
>
> • Landforms related to ordinary geologic processes influenced the course of important military battles, while stimulating efforts to protect and preserve nature elsewhere.
>
> • Most of the influence geology had on early civilizations also affects us today.

20.3 Humans as Geologic Agents

■ *How seriously have people increased erosion rates worldwide?*

People have had an astonishingly large impact on global erosion rates, owing to agriculture, the cutting of forests, urbanization, and other less significant activities.

We can roughly estimate the rate at which continental land areas weather and erode by measuring the volume of the sediment and sedimentary rock found on and around the margins of the continents and dividing that quantity by its geologic age. As you learned in Chapters 2 and 7, sediment is gradually converted to rock and may be subducted and destroyed, so it is no surprise that there is in general much less older sediment than younger around the world. In fact, the amount of surviving sediment decreases exponentially with age. Nevertheless, to produce the amount of sediment thought to have existed during the past half-billion years of Earth history requires a mean erosion rate that lowers the average land surface outside the polar regions by about 25 m per million years, or about 2.5 cm every thousand years. Tectonic uplift roughly matches or slightly exceeds this denudation rate in active mountain belts, whereas the input of meteorite sediment from space during this time has been trivial.

At present, we live under much more erosive conditions than have prevailed during most of the past half-billion years, thanks largely to ice age glacial activity. The erosive denuding of lands is probably on the order of 65 m per million years, averaged over the past 3 million years. But denudation from farmed and grazed lands occurs much faster, as mentioned in Chapter 6, with rates estimated to be 700–1400 m per million years. This amounts to an average of 21 tons of mostly human-generated sediment per person every year. Think about this next time you sink your teeth into an ear of corn or a piece of chicken. Because almost 40% of Earth's land surface is covered with crops and pastureland, this is not a trivial figure geologically speaking—120 billion tons total annually. At this rate, we could fill the Grand Canyon with this material once every 50 years!

By far the most erosion occurs from agricultural land and deforested areas in lesser-developed countries, such as Haiti and Ethiopia (▶ Figure 20.8). High rates of soil erosion in these lands contribute directly to poverty, disease, famine, and natural disasters such as flooding. In 2004 Hurricane Jeanne slammed onto the island of Santo Domingo in the Caribbean rim. More than 3000 people died from landslides, mudflows, and floods in heavily eroded Haiti, but only 18 people perished next door in the Dominican Republic, where the forest and soil cover are much thicker. Given these issues, volunteers for the U.S. Peace Corps are trained as part of their overseas assignments to recognize and prevent erosion.

The world's lands are certainly in no danger of disappearing into the sea. But these erosion rates do threaten to destroy much of the world's fertile supply of soil, which remains the basis of civilization even in the Information Age. Recent research suggests that the overall rate of soil loss is an order of magnitude greater than the rate of natural soil formation in typical agricultural regions. Such rates cannot be sustained for more than a few centuries, except in certain floodplains and deltas. A revolution in agricultural practices is needed if we plan to continue feeding the ever-burgeoning human population.

■ *What are the geologic consequences of building dams?*

Fortunately (or not) most of the sediment leaving agricultural lands moves to adjacent slopes and lands; it does not move all at once to the ocean, and some of it may not even make it that far. As much as 30% of this sediment ends up trapped in the

▶ **Figure 20.8 Deforestation** Charcoal workers cut down trees March 17, 2005, in Miragoane, Haiti. According to reports, charcoal and wood make up to 70% of Haiti's energy resource. Deforestation has become a major problem throughout the country despite efforts of organizations like CARE to plant approximately 800,000 trees a year.

Geo-focus Figure 20.9 Use and Abuse of a River Do the costs outweigh the benefits? Situation **(a)** shows an undammed river, whereas **(b)** shows the same river dammed. **What is needed to improve the situation in (b)?** *Control of deforestation in watersheds, dredging and transport of sediment from reservoir to river downstream, restoration of coastal and riverside wetland habitats, restriction of importation of non-native fishes and other species.*

a Before

b After

reservoirs behind dams, gradually filling up reservoir capacity and reducing our future ability to store surface water and produce hydroelectric power (▶ Figure 20.9). To extend the lifetimes of some reservoirs, expensive and continuous dredging operations are undertaken that extract sediment from reservoir bottoms for transport downstream, where it is dumped back into the rivers below the dams. In other words, along many rivers, people have taken over Nature's role in moving sediment toward the sea.

Loss of sediment downstream from reservoirs has a number of negative consequences, including increasing rates of riverbank erosion as sediment-depleted river waters seek to meet their load capacities (▶ Figure 20.9; see Chapter 15). Given that there are nearly 40,000 dams in the world, this is a serious problem. The ancient Egyptian civilizations that created the great pyramids and thrived for thousands of years were dependent on an annual flooding of the Nile River, which enriched the depleted soil beds with nutrient-rich sediments. The Egyptian people continued to benefit from this cycle after the fall of the pharaohs, through thousands of years, until in 1964 the Aswan High Dam reduced the flooding and trapped the sediment, dropping loads to only 8% of their former level downstream where most Egyptians live and farm. Water tables in important agricultural areas fell as less water infiltrated the floodplain of the lower Nile. Along the coast of the Nile delta, saline groundwater intruded, poisoning wells and water supplies. In return, Egypt got lots of inexpensive electricity to develop industry and some new capacity to irrigate desert lands farther upstream along the Nile. But will the loss of valuable agricultural land downstream from the Aswan High Dam be worth it in the long run?

Depending on who you are, erosion is not necessarily a bad thing. Since the Spanish conquest 500 years ago, cutting of forests, grazing, and European-style farming in higher areas of southern Mexico have formed gullies as deep as 5 m. But native Mixtec farmers downstream have found the sediment influx to be a boon; it allows them to actually extend irrigated traditional farms as much as a thousand years old. Likewise wide, fertile valley floors in northwestern Europe, including some prime wine-grape cultivation, probably owe their existence to soils derived from failed farming efforts in surrounding highlands cleared of forests thousands of years ago.

■ *How does mining compare with the natural movement of sediments?*

As in the case of soil erosion, people have moved soil, sediments, and bedrock at rates and in quantities far exceeding those of natural processes. Most of this is done for the sake of construction, either to excavate foundations or provide raw building materials.

▶ **Figure 20.10 Resources Taken from Earth and Used for Construction** A sand and gravel quarrying operation provides vital raw materials for construction in Southern California.

If fossil fuels are one foot on which industrial civilization stands, metals are the other. Obtaining this valuable material requires the mining and processing of vast quantities of ore-grade minerals (see Chapter 3). Far more abundant than the ore itself is the "worthless" rock rubble that must be excavated in the mining process. By some measures, as much as 100 tons of rock debris is left behind someplace in the world for every gold wedding ring that is manufactured and sold. Construction materials, including gravel for roads, marble for tombstones and countertops, and limestone for cement, make up another large category of solid resources that are taken out of Earth (▶ Figure 20.10). Much of this material is shipped thousands of kilometers before it is processed for commercial products. The total amount of sediment carried to the oceans by all of the world's rivers probably does not exceed 24 billion metric tons per year. In contrast, the annual extraction of rock and mineral products via mining and quarrying is close to 3 trillion tons!

Section 20.3 Summary

- Agriculture, including grazing, seriously accelerates the rate of land degradation, producing far more mobile surface material (alluvium, sediment) than would exist under natural conditions. This is a long-term threat to the global food supply.

- Dams capture much of this sediment, but the silting of reservoirs shortens their service lives, while a decreased supply of sediment downstream triggers bank-side erosion and lowering of the water table.

- Mining is another major way that people move natural materials at Earth's surface.

- Human activities rival natural processes in shaping Earth's modern landscape.

What Would You Do?

What is the price of love? If you buy an engagement ring, you are also "purchasing" 150 tons of waste rock left at a mine entrance somewhere else in the world, much of which will generate stream or groundwater contaminants. A wedding ring involves depositing an additional 100 tons of waste rock. Will you think twice and buy your loved one a paper friendship band instead? What would you recommend we do to reduce the harmful aspects of this tradeoff?

20.4 Climate Change

■ *How much have sea levels changed since the end of the last Ice Age?*

Twenty thousand years ago, sea levels worldwide were about 130 m lower than at present, thanks mostly to the growth of giant ice sheets in North America and Eurasia. The shoreline lay 150 km southeast of where it currently lies along the waterfront at New York City, out in the Atlantic Ocean near the edge of the continental shelf. Likewise, to reach the ice-age Venice Beach in Los Angeles, California, where the slope of once-dry, somewhat steeper land is now submerged, would require going 7 km farther west than today (▶ Figure 20.11).

One hundred twenty thousand years ago, sea level stood several meters higher than at present, owing to the probable collapse and melting of one or several of the ice shelves that presently fringe Antarctica. This seemingly modest rise in sea level was enough to drown much of modern-day Florida, including the Everglades and Miami–Dade County region. The climate was similar to today's, although it may have been somewhat less stable. Could even a slight increase in current temperatures create the same sea-level rise that occurred under natural conditions 100,000 years ago? Where would all the residents of Florida go?

■ *What are some examples of natural switches that can cause sudden climate change?*

We know from the change in the makeup and chemistry of marine sediments that sea level and mean near-surface temperatures tend to change rapidly. Various critical **natural switches,** or sensitive environmental thresholds, make this possible. For example, between 900,000 and 600,000 years ago, the cycle of ice ages suddenly slowed down while continental glaciation, especially in the Northern Hemisphere, greatly intensified. In other words, global ice ages became longer and colder. What accounted for this sudden transition? Some geologists think that a gradual cooling of the southern Pacific Ocean accompanied by expansion of the ice sheets in Antarctica caused the **thermocline**—the immiscible boundary between warm, shallow seawater and colder,

Figure 20.11 U.S. Shoreline Shift from Ice Age

(a) Post–Ice Age shift in the shoreline, northeastern United States (state boundaries indicated).

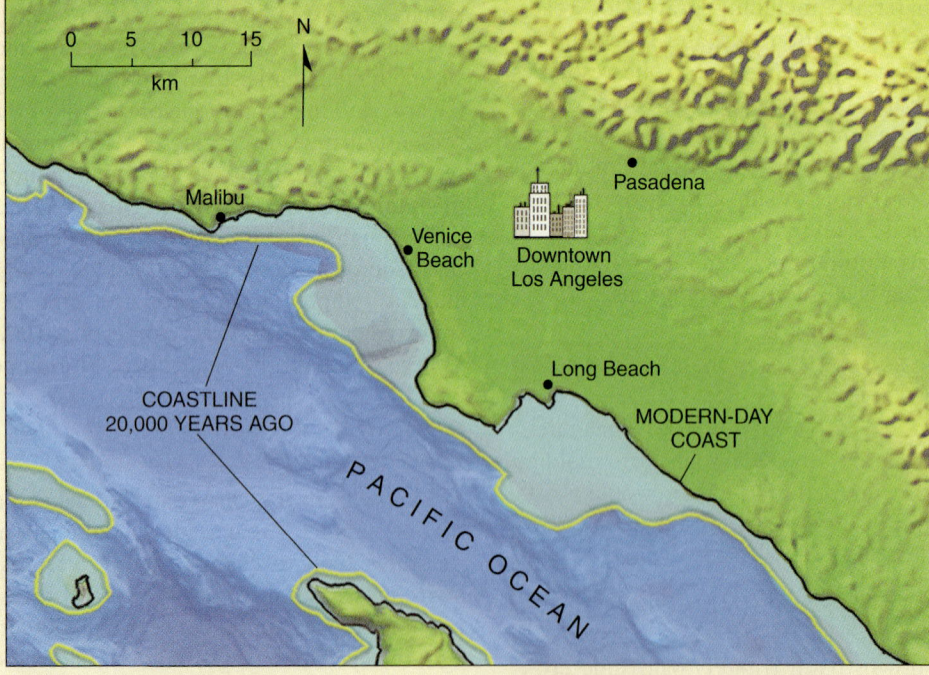

(b) Post–Ice Age shift in the shoreline, Los Angeles, California.

deep water—to migrate upward in the water column. Eventually a critical threshold of shallowness developed whereby trade winds blowing off the Andean coast of northern South America could drive away the warm water layer altogether, allowing frigid deep water to well up and replace it at the ocean surface for the first time in this region (▶ Figure 20.12). At this critical near-equatorial location, the cold water effectively refrigerated the climate worldwide. We now call the seasonal upwelling and the occasional lack of upwelling off the west coast of South America the El Niño–La Niña cycle.

Another well-known natural switch exists in the central Atlantic Ocean. Warm sea temperatures promote evaporation and convection in the atmosphere above the Atlantic between West Africa and Brazil. Hundreds of thunderstorm cells may develop on any given summer day. In fact, the thermal energy may be so great that these cells swirl together and merge into hurricanes. The occurrence of hurricanes depends on the sea-surface temperature. At 26°C virtually no hurricanes can form; the system does not have enough heat. But at just 2°C higher, hurricanes form with seasonal regularity—and the warmer it gets, the more intense the storms become. This is a great concern for people living in low-lying port cities, as shown by the 2005 Hurricane Katrina disaster in New Orleans (see Chapter 19).

Crossing natural thresholds could shift world climate rapidly into a new state with severe repercussions for agriculture and the whole of humanity. There is a critical need to identify and understand these switches well. But even without them, significant climate change is already underway, and there is less and less controversy that human beings play an important role in it. As we are loading the atmosphere with larger amounts of greenhouse carbon dioxide from the burning of fossil fuels, methane from livestock and rice paddies, and other pollutants, the issue becomes a moral as well as a scientific one. Indeed, a lock-step correlation can be seen among the consumption of petroleum (as a proxy for all fossil fuel use), an increased concentration of CO_2 in the atmosphere, and a rise in the annual average temperature in the lower atmosphere since about 1985 (▶ Figure 20.13). Carbon dioxide levels are also rising as virgin forestlands, a natural storehouse for CO_2, are logged away or converted to other types of land use. Atmospheric methane, an even more potent greenhouse gas, is also increasing as populations of livestock and rice paddy cultivation grow. Human beings are now experiencing higher temperatures

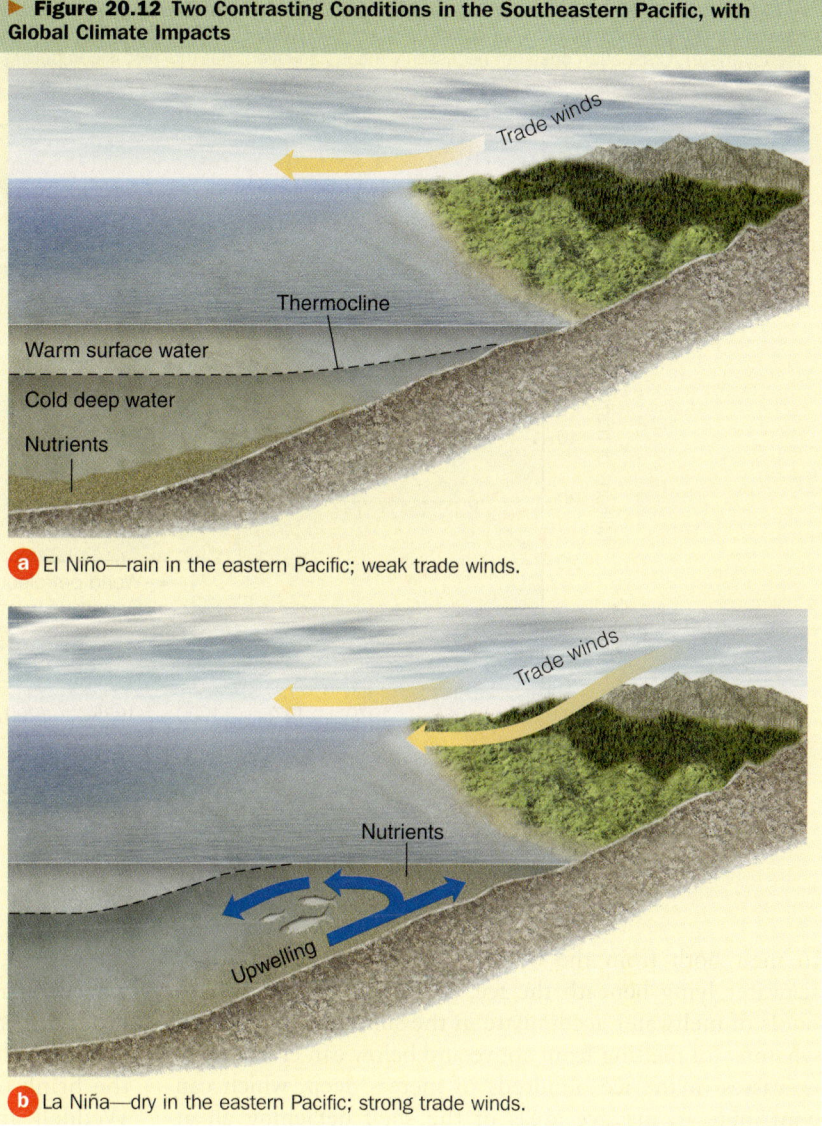

▶ **Figure 20.12** Two Contrasting Conditions in the Southeastern Pacific, with Global Climate Impacts

a El Niño—rain in the eastern Pacific; weak trade winds.

b La Niña—dry in the eastern Pacific; strong trade winds.

than at any time in the past 1000 years, although it is likely that the Holocene Epoch around 8000 years ago was somewhat warmer. Cores from certain sedimentary deposits and from ice sheets indicate that atmospheric CO_2 levels have not been higher in at least 420,000 years, and perhaps in *20 million* years.

■ *How does ice shelf collapse occur, and why is it important?*

The stability of the Antarctic ice mass is a critical factor in this regard. Where the polar ice sheets meet the sea, they are supported in many places by gigantic floating **ice shelves** up to several hundred kilometers wide. The ice shelves act like natural dams to slow the seaward movement of the ice sheets. The shelves normally melt slowly along their seaward edges, infrequently shedding large icebergs into the polar ocean. The rate of ice flow roughly matches the rate of melt-back at the shelf edge, making the geographic size of the ice shelf constant in a perfectly stable climate. If warming occurs, however, the ice shelf will begin

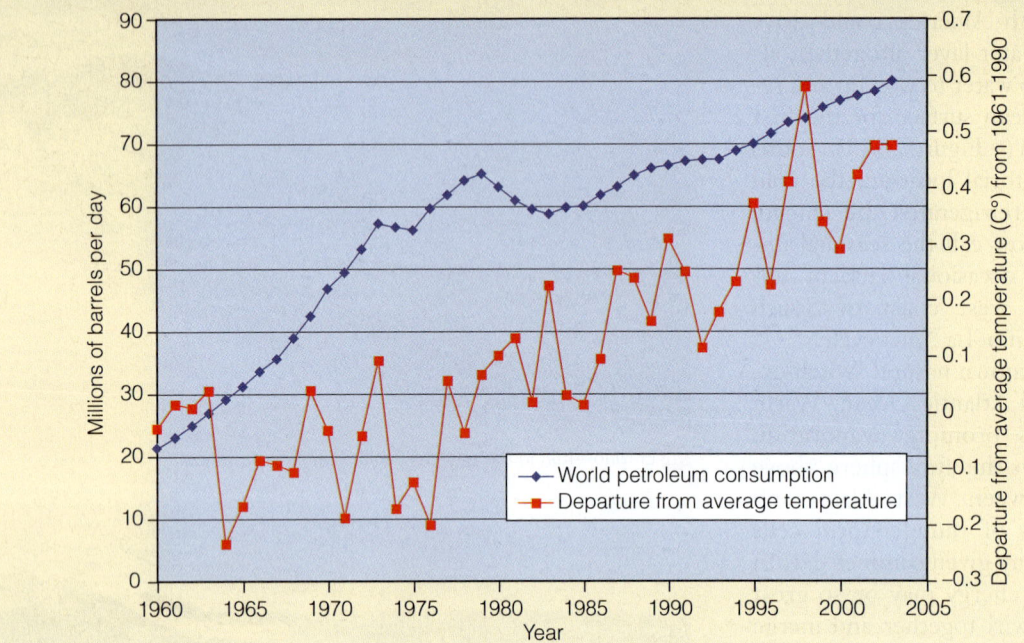

▶ **Figure 20.13 Consumption of Oil Compared with Global Land-Surface Average Annual Temperature, 1950 to Present** For each Qbtu (quadrillion British Thermal Units) of oil we consume, we release 19–20 tons of CO_2 to the atmosphere (added to the roughly 1000 billion tons already there). Correlation does not necessarily mean causation, but this plot is compelling. (Data: U.S. Energy Information Administration and National Center for Atmospheric Research)

to melt both from the bottom, where it meets warmer seawater lying beneath the ice, and from the top, where pools of meltwater accumulate in the summer. The double whammy of melting from above and below can create huge crevasses in the ice hundreds of meters deep, which can contribute to massive areas of the shelf detaching all at once. Such catastrophes destabilize the sluggishly moving ice system all the way into the interior of the adjoining continent (▶ Figure 20.14).

The destruction and melting of an ice shelf in and of itself does nothing to raise sea level because the density of ice is less than that of water. The volume of water added by melting ice merely compensates for the volume of water displaced by the ice when it was floating. The real problem starts with sudden release of a pent up ice sheet on the adjoining land. The suddenly unconstrained glacial ice accelerates its movement, and where it meets the ocean, it creates vast numbers of bergs that are free to drift to warmer latitudes (▶ Figure. 20.14). Sea level will rise as these bergs melt, because the land-born ice never displaced water in its original state. It may take centuries or millennia for sea level to stabilize again as a new ice shelf slowly develops, or the remaining land ice melts away completely.

In 2002 the Larsen B ice shelf abruptly collapsed and disappeared along the northeastern coast of Antarctica's Palmer Peninsula. Fortunately, it was a small shelf, as ice shelves go, but it could portend larger collapses elsewhere in the future on time scales of human interest. Two other ice shelves besides the Larsen B shelf have collapsed in the past decade, and two more appear to be on the brink of doing so. The larger ice shelves filling the Weddell and Ross Seas are even more worrisome and must have disintegrated substantially during the Eemian interglacial 120,000 years ago, which was the brief warm interval with a higher sea level that preceded the most recent ice age.

The connection between human activity and climate change is inescapable, irrespective of one's political stripe. The question is what do we do about it and how fast. Like it or not, we are forced to consider a "climate management" role in modern civilization. New technological efforts are under way to withdraw CO_2 from the atmosphere, perhaps storing it underground in old or abandoned oil fields and salt beds. Growing new forests or plantations to take up CO_2 in living tissues and a switch to alternative energy sources that do not release CO_2 into the atmosphere no doubt would help as well, but in a world with growing numbers of people striving for affluence, to what extent can these "fixes" really work? There is an increasing sense among scientists that humanity is in a race against time. In the end, a profound understanding of geology will be essential for solving the great issues of our day.

▶ **Figure 20.14 Antarctica**

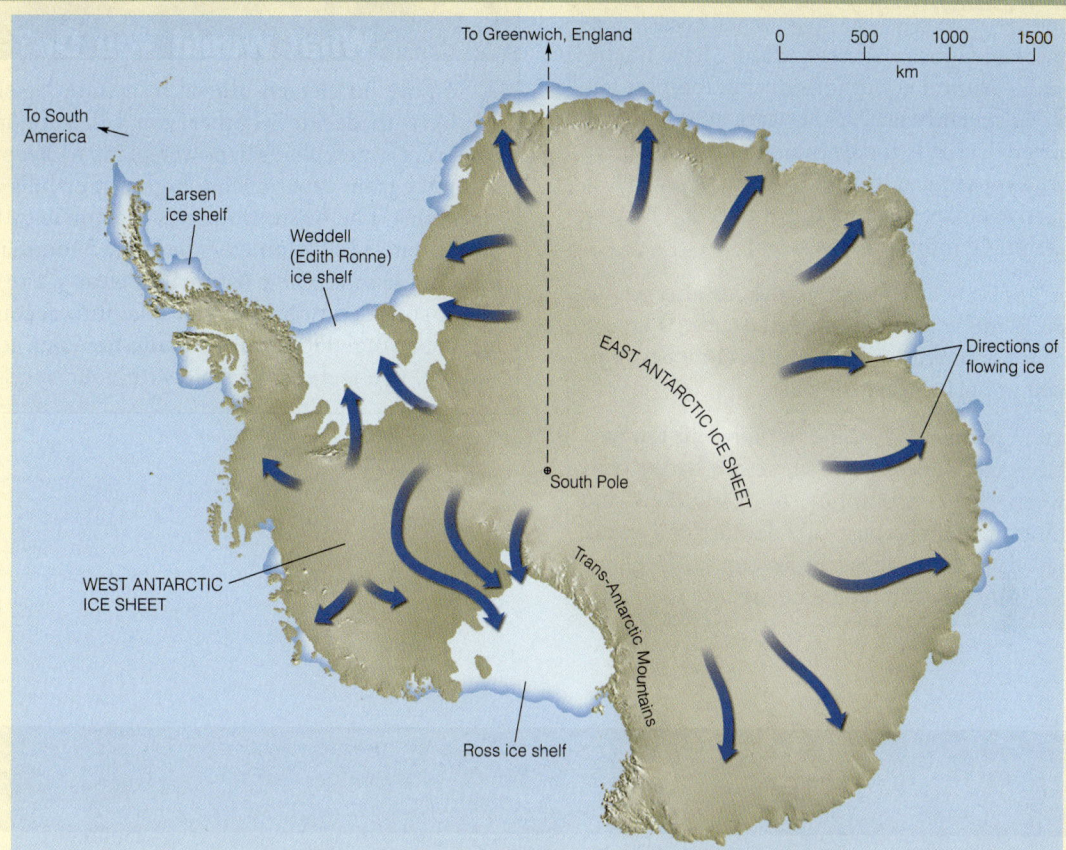

a Antarctic ice shelves, outlined in blue, fringe the continent of Antarctica. Their stability is a major factor in the stability of Earth's climate and sea level. Part of the Larsen ice shelf ("Larsen B") collapsed in 2002.

1 Stable ice shelf.

2 Ice melts from above and below as sea and atmosphere grow warmer.

b Stages in ice-shelf disintegration.

3 Crevasses open, and ice shelf collapses into numerous icebergs. Currents sweep these away to warmer latitudes, where they melt.

Section 20.4 Summary

- Sea level and global mean air temperature have fluctuated wildly under natural conditions during the past few million years. As recently as 20,000 years ago, sea level was 150 m lower than it is today. Sudden changes in sea level and air temperature relate to natural switches, which are sensitive environmental thresholds that may be crossed by seemingly minor changes.

- One major concern at present is the stability of ice shelves fringing the continent of Antarctica. The ice shelves can degrade rapidly, raising sea level catastrophically for coastal populations worldwide.

- To what extent people contribute to global air temperature is controversial, but evidence suggests a strong connection between the consumption of fossil fuels and overall warming during the past two decades. All in all, human beings have become as important as geologic processes in causing environmental change.

What Would You Do?

You are an elected official in a state legislature. You have to decide whether you will support a bill to construct a new nuclear power plant to keep your constituency from experiencing high power bills and rolling blackouts. The waste from this reactor is to be shipped out of state 3200 kilometers to Yucca Mountain, Nevada, for storage in the new national nuclear waste repository. What considerations will you take into account in casting your vote? How will you vote for your constituents, and will you feel comfortable with your decision?

Review Workbook

ESSENTIAL QUESTIONS SUMMARY

20.1 Introduction

20.2 Geology and History

■ *What were the prerequisites for civilization to begin?*
Two main prerequisites were a stable climate and stable sea level. A stable climate was needed for food production, and stable sea level was needed for maritime trade and the gathering of resources from far away.

■ *What role did geology play in trade and conflict?*
The distribution of resources, especially metallic ore deposits, played a key role in the way that trade routes developed, and it influenced the wars that people fought, both to secure resources and to finance conflicts.

■ *How did geology influence science, culture, and international affairs?*
Hutton's insights about the development of Earth came into direct conflict with the religious teachings of his day. Disasters like the 1755 Lisbon earthquake underscored the "indifference" of Nature to humans and had an impact on philosophical thinking. Geography, linked directly to geology, guided the settling of new lands and influenced the resolution of key military battles that forever changed history. Finally, the beauty of geologic features played an essential role in the development of the modern environmental movement.

20.3 Humans as Geologic Agents

■ *How seriously have people increased erosion rates worldwide?*
Human-induced erosion rates, from farming, grazing, and deforestation, now exceed natural erosion rates by roughly 10 to 20 times. This has severe repercussions in some countries hard hit by erosion and contributes to poverty, famine, disease, and exacerbated natural disasters.

■ *What are the geologic consequences of building dams?*
Dams trap sediments in their reservoirs, inducing downstream erosion, a lowering of water tables, and in arid countries, soil salinization. The reservoirs themselves have shortened lifetimes as they fill with sediment. Some require dredging and exportation of sediment to be sustained.

■ *How does mining compare with the natural rates of erosion?*
The amount of material removed by mining annually exceeds all the sediment transported in rivers worldwide by more than a hundred times; 24 billion tons of sediment are transported by rivers worldwide every year, while people move on the order of 3 trillion tons through mining activity.

20.4 Climate Change

■ *How much have sea levels changed since the end of the last Ice Age?*
Sea levels have risen 150 m in the past 20,000 years. This is enough to have shifted the shoreline 150 km inland to the present-day coast of New York City, and 7 km inland along the coastline of Los Angeles.

■ *What are some examples of natural switches that can cause sudden climate change?*
A slight change in sea-surface temperature in the central Atlantic can determine whether there will be hurricanes in that ocean. Likewise, a slight rise in the thermocline in the southeastern Pacific can allow the upwelling of deep, cold Antarctic bottom water, establishing conditions for El Niños and making the world cooler overall.

■ *How does ice shelf collapse occur, and why is it important?*
Ice shelves melt from a combination of warmer air and sea temperatures; they melt from their bases as well as their surfaces. When thinning reaches a critical point, an ice shelf can rapidly disintegrate to icebergs, which melt when currents carry them out to sea. Ice shelf collapse can cause a sudden rise in sea level, which would have a serious impact on coastal populations worldwide. Small ice shelf collapses have occurred recently, and past geologic data suggest that conditions are ripe for much larger collapses, perhaps in a human time frame.

ESSENTIAL TERMS TO KNOW

code of Hammurabi (p. 634)

environmental geology (p. 633)

ice shelves (p. 641)

natural switches (p. 639)

terrain analysis (p. 635)

thermocline (p. 639)

REVIEW QUESTIONS

1. In the quotation in Section 20.1, J. R. McNeill states: "Should this age of exuberance end, or even taper off, we will face another set of wrenching adjustments." What do you suppose he means by this? What adjustments does he have in mind?

2. If climates and sea levels began to change rapidly, what problems would this pose for our modern technological civilization? What prospects do you hold for living a civilized life in a world with an erratic natural environment? What adjustments would we have to make to help civilization survive in such a world?

3. Can you think of examples, other than the ones discussed in your reading, of how geology has influenced history, culture, or current affairs?

4. What measures might be taken to reduce the erosion problem vexing so many parts of the world? Should we be concerned about this problem, even if we don't have it in our own home region?

5. Dams clearly can have negative impacts, but there are many positive ones as well, discussed elsewhere in this book. Under what circumstances would you argue that a dam should be built, even if dredging cannot be done to maintain the reservoir's capacity? Put together a hypothetical example.

6. Mining moves far more material every year than rivers and streams worldwide. But is this necessarily a bad thing? Under what conditions is it bad, and under what conditions would it not really create problems? What benefits do we derive from mining that would be sacrificed if we stopped it altogether?

7. How concerned should we be about climate change based on what we know of the geologic record?

8. If we are in the business of managing the climate, who should be in charge of managing it, and what objectives should be set?

9. How has reading this chapter changed your perspective about the human situation? List the key problems explored in the text, and then write the cause of each problem and some practical steps that can be taken to address it. How would the world be different if we implemented your solutions? Do you think that we can or should try to reshape the world this way?

10. How has your general perspective about geology evolved over the course of the semester?

GEOLOGY MATTERS

GEOLOGY IN FOCUS Atomic Testing and the Earth System

Undoubtedly the greatest curse of the past hundred years has been the introduction of new technologies without a complete understanding of their geologic, social, political, and ecological ramifications. The development of nuclear power and weaponry is perhaps the most Faustian of these technological bargains. Scientists were shocked that half of the 140,000 fatalities from explosion of the Hiroshima bomb in 1945 were due to the aftereffects of radiation sickness. Nevertheless, this did not prevent proposals for the "peaceful applications" of nuclear explosions over the ensuing 20 years, which included the following:

- A proposal to stop hurricanes by detonating nuclear devices in the eyes of these storms while they were still out at sea. (This initiative was endorsed by the National Academy of Science and U.S. Geological Survey.)
- A project supported by the U.S. Army Corps of Engineers, National Academy of Science, and the U.S. Geological Survey to create a corridor across California's Bristol Mountains for Interstate 40, then under construction, by simultaneously detonating 23 nuclear devices with a total yield of 1,830 kilotons. (For comparison, the Hiroshima bomb had a yield of 13 kilotons).
- Proposals by British and American atomic agencies to excavate new fishing harbors in Australia and Alaska through the detonation of undersea nuclear devices, similar to the Bristol Mountains proposal.
- A proposal to detonate underground nuclear devices in Colorado, Mississippi, and New Mexico to increase the permeability of bedrock for natural gas extraction by explosion fracturing. (The Rullison nuclear test in western Colorado actually generated a modest amount of commercial-grade gas, but it was too radioactive to use.)

The testing of hundreds of nuclear bombs at the Nevada Test Site for military as well as "peaceful" purposes during a 30-year period in the late 20th century has contaminated groundwater and soils with plutonium and other potentially deadly radioisotopes across an area larger than the state of Rhode Island (▶ Figure 1). Although much of this radioactivity is confined to a few valley floors and playas, lower-grade wastes are spread well beyond the boundaries of the test site, mostly toward southern Utah and northern Nevada. (The military deliberately avoided testing on those days when the wind blew toward populous southern California.) Quite often the contamination escaped in unexpected ways, in large part because people simply did not think through the implications of what they were doing at the time. For example, in April 1957, engineers in the appropriately labeled "Area 13 Experiment" blew up an atomic bomb with ordinary TNT to investigate how deadly plutonium would disperse through the environment. This project is of interest today because of growing concern over terrorism and the detonation of "dirty" bombs. The explosion so contaminated more than a thousand acres of land that the test engineers had to fence it off "indefinitely," meaning 240,000 years. Unfortunately, subsequent range fires and dust storms spread the contamination beyond the enclosure. More innocuously, natural soil crusts within the enclosure concentrated plutonium in fungal growths and other living matter eagerly consumed by native darkling beetles. Birds in turn ate the beetles and then flew away to spread the contaminants farther through the food chain.

We are moving into an era of increasing use of nuclear power as conventional fossil fuel supplies diminish. Revival of the nuclear weapons program has been proposed through the U.S. Bunker Buster program. But confining nuclear wastes has become a major headache—financially, politically, and environmentally. Geologists must be consulted to ascertain whether the waste can be isolated from groundwater supplies, possible fault displacements, volcanic eruptions, and radical changes in global climate for hundreds of thousands of years. But how can we do this? There will never be a "zero-percent" chance of contaminants escaping. We can never be 100% sure of what we know.

Research suggests that if we were to stop maintaining civilization in the United States beginning tomorrow, mechanical and chemical weathering would reduce all of our buildings, roads, and other structures to unrecognizable heaps of soil and rubble within 10,000 years, except for a few large Pyramid-scale structures, such as Washington's Grand Coulee Dam and Gateway Arch in St. Louis. So we plainly cannot confine nuclear wastes above ground. Storing them in underground vaults, such as at Yucca Mountain, Nevada, also has serious problems. After billions of dollars of research and study, it is now acknowledged that Yucca Mountain will probably leak contaminants to groundwater if this repository is ever used; and no one wants to put future generations at risk from drinking radioactive well water. One promising alternative is to place nuclear wastes in impermeable, self-sealing salt beds deep underground in tectonically quiescent areas. In other words, we could use *ordinary geologic processes* to isolate the contaminants for us. In fact, the Department of Energy has already begun to store deadly transuranic wastes this way at the Waste Isolation Pilot Plant (WIPP) in Carlsbad, New Mexico. But it will take many more sites like WIPP at a cost of hundreds of billions or trillions of dollars to handle all of the high-level and transuranic nuclear waste that we presently have, and that many authorities propose we continue producing in the future.

We seem to learn as we go, jury-rigging solutions along the way. H. G. Wells, esteemed British historian and science-fiction writer, recognized this painful reality when he concluded that human history "is a race between education and catastrophe."

Figure 1
These are not volcanic craters, but pits left from the explosions of underground nuclear bombs at Yucca Flat, Nevada Test Site, Nevada. Sedan Crater, in foreground, resulted from a 104-kiloton hydrogen bomb blast at 200 m beneath the surface, part of a Plowshare Program investigation of using nuclear weapons for excavation. The explosion ejected 8 million tons of rock debris out of the 400 m wide, 100 m deep crater. Four million tons fell back into the crater, the rest drifted as far away as Canada. Other craters resulted from deeper nuclear bomb detonations that did not eject material, but formed when deep fractured rocks collapsed into blast cavities.

GEOLOGY MATTERS

Geology AT WORK

Environmental Geology

How are geologists working to understand and solve some of the problems described in this chapter? The career paths in environmental geology range from cleaning up local soils contaminated by fuel spills around gas stations and estuaries, to evaluating the potential for landslides and searching for new clean water supplies. Environmental consulting firms seek to hire graduating students with backgrounds in the earth sciences to staff their offices and to provide field support. Skills developed on the job include knowledge of basic physical science, including geology, as well as legal expertise and business management.

Environmental geologists are also heavily involved with the preparation of environmental impact statements, which are mandated for most government-subsidized development projects whenever there is potential for negative impact on highly valued natural environment or habitat. For example, in the late 1990s, a proposal for oil and gas drilling along the northern perimeter of Carlsbad Caverns National Park, New Mexico, was denied on the basis of the potential threat to the groundwater, which is vital for the formation of caves and the sustenance of fragile cave biota. Geologic considerations were foremost in this decision.

On a broader scale, geologists who work in government agencies and research universities, such as the U.S. National Oceanographic and Atmospheric Agency or the Lamont-Doherty Earth Observatory of Columbia University, investigate how carbon cycles through Earth system—from atmosphere, to hydrosphere, to biosphere, and back again. Marine geologists examine drill cores taken from remote ocean floors all over the world to decipher the record of past climate change preserved in foraminifera and other marine fossils. Quaternary geologists pursue the same objective by studying the record of past glaciations on land. Soil geologists pitch in too, studying the carbon flux in soils ranging from the tropics to the arctic tundra. All of this is vital for evaluating just how much of an impact we have on global climate. Many of these researchers also work with graduate students, who often contribute to important breakthroughs.

Increasingly, geologists are beginning to work closely with biologists, chemists, and other scientists in integrated research teams because no one person can know it all. The fact is that we arbitrarily divide up our disciplines in academic programs; however, tackling the great issues confronting humanity in the new millennium requires reaching across these divisions to work in combinations that were unknown in the past. The Earth system, after all, is a single, seamlessly operating system. Nature did not create a stand-alone Department of Biology to handle all of its life and a separate Department of Geology to manage Earth's physical substance.

Finally, teaching is an important contribution at all levels of education. Building awareness of how the Earth works, as we hope this book does, is an essential foundation for future problem solving and a better way of living in the world (▶ Figure 1).

Figure 1
(a) Visitors follow the Park service pathway into the main entrance of world-famous Carlsbad Caverns in southeastern New Mexico. (b) A cluster of cave formations in the Big Room of Carlsbad Caverns. One of these formations is called "The Lion's Tail." Can you tell which one?

(a)

(b)

GEOLOGY MATTERS

GEOLOGY
IN YOUR LIFE

Organic Farming

Of the world's 35 billion hectares of land area, about 5 billion hectares (ha) are under cultivation at present, with only a modest amount of highly fertile, arable land left to put under the plow. This is not much more land than was farmed fifty years ago. Yet yields of food crop per hectare have increased, largely keeping pace with the increase in human population. This increase in productivity, the Green Revolution, was largely enabled by two factors: the development of new strains of food crops, and the applications of commercial fertilizers, pesticides, and herbicides to farmlands.

However, there has been a heavy environmental price to pay for this success. Nitrogen from synthetic fertilizers has run off into streams, rivers, and even seas, killing aquatic life and promoting algal blooms through hypoxia. Pesticides and herbicides poison soils and groundwater supplies, in many cases indefinitely. Atrazine and alachlor, two common herbicides widely used in the Midwestern United States, have been implicated in the loss of fertility in male farm workers exposed to these chemicals. Methyl bromide, which is used as a fumigant in strawberry and carrot fields, and for commercial flowers, leaks to the atmosphere where it acts as a powerful agent to destroy life-protecting ozone.

Organic farming is an alternative approach toward food production that uses far fewer synthetic chemicals than conventional modern agriculture, and little or no artificial fertilizer (▶ Figures 1 and 2). Livestock grown on organic farms are range-fed rather than fed processed animal products in feed lots or sheds. Organic farming greatly reduces the environmental impacts of growing food "the modern way," but the common trade off is a drop in yield per hectare and a more severe problem with insect pests, blights, and plant diseases.

Some critics assert that organic farming alone cannot feed the whole world's population, but some organic farms show surprisingly competitive productivity, and the quality and taste of their food products is often more appealing, as many who shop at Farmers Markets and college food co-ops will attest. Non-industrial ways of fertilizing soils include the planting of legumes between crop plants and the use of compost, manure, and mulch. Mulch left on the ground between harvests also has the benefit of inhibiting soil erosion. As for pests and other plant ailments, some plant varieties prove to be naturally insect, fungus, and frost resistant, and can go a long way toward reducing crop loss, even if only partly mixed with other plantings. For example, there are over 5000 kinds of potatoes grown organically by Native Americans in the Andes, where there is rarely a problem with famine, despite great swings in the weather and the presence of many natural pests; but just a few varieties are cultivated industrially in the United States (▶ Figure 3). Although Americans can boast larger potato crops per hectare, these crops are highly vulnerable to the whims of nature, and take far more energy and other natural resources to bring to the table.

Organic food products generally cost more than food grown on large industrial farms. But there is a hidden environmental cost to industrial farming. Some assert that it is wrong for us to be placing this debt on the unborn.

Figure 1
Manually prepared soil beds support organically grown fruits and vegetables at a student organic garden, Pomona College, California.

Figure 2
A student-built tool and seed shelter, called the Dome, lies at the center of the college garden.

Figure 3
Traditional farmers harvest potatoes grown organically in the High Andes. Over 5000 varieties of potatoes are grown in this region, whereas only a few are grown agro-industrially in the United States.

GEOLOGY MATTERS

GEOLOGY IN UNEXPECTED PLACES

Mines and Flooding

One expects flooding to be a problem in places like the bottomlands along rivers, but in fact, flooding is a common problem in mines and quarries wherever excavations penetrate below the water table. A mine or quarry is essentially like a giant well in which the level of flooding rises to the level of the water table in the adjacent wall rock. One of the stimulants of the Industrial Age was the flooding of English coal mines by water seeping into deep tunnels through the walls. To dry out the deep mines so that Britain could obtain more coal for its breweries and other purposes, James Watt (▶ Figure 1) introduced the steam engine (▶ Figure 2), and modern times soon ensued.

A gold boom is currently underway in Nevada, where some mines penetrate as deeply as 300 m beneath the water table. Dewatering the gold mines consumes huge amounts of energy just to operate the pumps alone. A single mine may consume as much energy as needed to power a city of 100,000 people. An enormous amount of water is also removed, lowering the surrounding water table for several kilometers around by as much as 250 m. Springs and streams dry up, depriving wildlife of precious habitat in a dry land. Some researchers estimate that as many as 200 springs may soon disappear in northern Nevada due to mine dewatering operations. Moreover, the remaining groundwater can become highly polluted as rainwater leaches toxic metals from exposed mine and quarry tailings. Then there is the problem of what to do with the already polluted water taken out of the mines. These factors underscore the fact that digging precious metal out of Earth is only a small fraction of the total cost to bring to market many of the products that we take for granted.

Figure 1
James Watt (1736–1819) became a wealthy man after introducing the steam engine in 1773. His rotary-motion steam engine design (1781) largely launched the Industrial Revolution. Watt also introduced the term "horsepower" to measure the strength of his machines.

Figure 2
Steam engine. Historical engraving of Old Bess, a steam engine designed by the British engineer James Watt (1736–1819). It was the first engine to give rotary rather than pumping motion. Watt's engine also drove steam out of a cylinder to be condensed in a separate vessel. Previous engines, based on a design by Newcomen, condensed the steam within the cylinder, which was far less efficient. Watt patented his design in 1769 and by 1772 had entered into a highly profitable partnership with Matthew Boulton in Birmingham. While Boulton supplied capital, Watt continuously refined his design until his retirement in 1800. Old Bess was installed in Watt's Birmingham works in 1778.

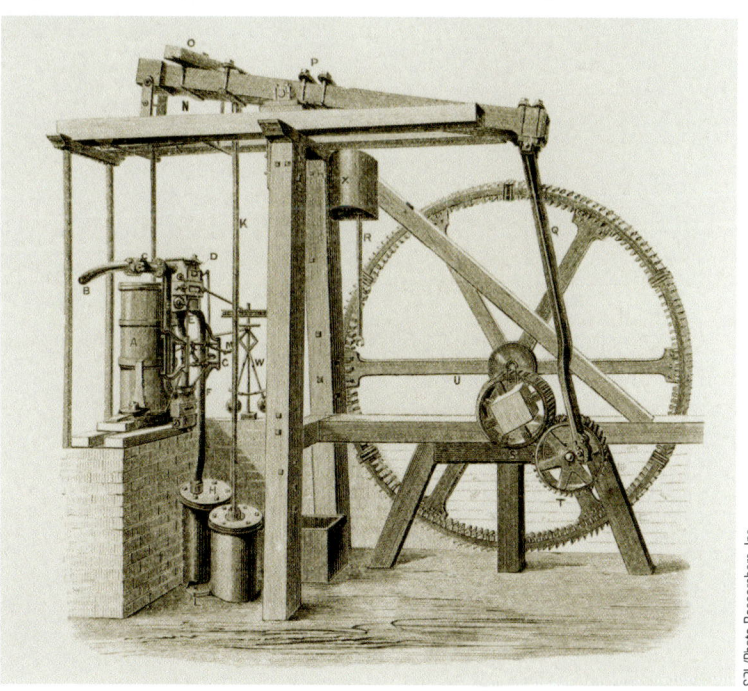

APPENDIX A

ENGLISH–METRIC CONVERSION CHART

	English Unit	Conversion Factor	Metric Unit	Conversion Factor	English Unit
Length	Inches (in.)	2.54	Centimeters (cm)	0.39	Inches (in.)
	Feet (ft)	0.305	Meters (m)	3.28	Feet (ft)
	Miles (mi)	1.61	Kilometers (km)	0.62	Miles (mi)
Area	Square inches (in.^{2})	6.45	Square centimeters (cm^2)	0.16	Square inches (in.^{2})
	Square feet (ft^2)	0.093	Square meters (m^2)	10.8	Square feet (ft^2)
	Square miles (mi^2)	2.59	Square kilometers (km^2)	0.39	Square miles (mi^2)
Volume	Cubic inches (in.^{3})	16.4	Cubic centimeters (cm^3)	0.061	Cubic inches (in.^{3})
	Cubic feet (ft^3)	0.028	Cubic meters (m^3)	35.3	Cubic feet (ft^3)
	Cubic miles (mi^3)	4.17	Cubic kilometers (km^3)	0.24	Cubic miles (mi^3)
Weight	Ounces (oz)	28.3	Grams (g)	0.035	Ounces (oz)
	Pounds (lb)	0.45	Kilograms (kg)	2.20	Pounds (lb)
	Short tons (st)	0.91	Metric tons (t)	1.10	Short tons (st)
Temperature	Degrees Fahrenheit (°F)	$-32° \times 0.56$	Degrees centigrade (Celsius)(°C)	$\times 1.80 + 32°$	Degrees Fahrenheit (°F)

Examples:
10 inches = 25.4 centimeters; 10 centimeters = 3.9 inches
100 square feet = 9.3 square meters; 100 square meters = 1080 square feet
50°F = 10.1°C; 50°C = 122°F

APPENDIX B

PERIODIC TABLE OF THE ELEMENTS

Key:
- 47 — Atomic Number
- **Ag** — Symbol of Element
- silver — Name of Element
- 107.9 — Atomic Mass Number (rounded to three significant figures)

Color legend: Representative Elements | Transition Elements | Inner-Transition Elements | Noble Gases

Period	(1)* I A	(2) II A	(3) III B	(4) IV B	(5) V B	(6) VI B	(7) VII B	(8) VIII B	(9) VIII B
1	1 **H** hydrogen 1.008								
2	3 **Li** lithium 6.941	4 **Be** beryllium 9.012							
3	11 **Na** sodium 22.99	12 **Mg** magnesium 24.31							
4	19 **K** potassium 39.10	20 **Ca** calcium 40.08	21 **Sc** scandium 44.96	22 **Ti** titanium 47.90	23 **V** vanadium 50.94	24 **Cr** chromium 52.00	25 **Mn** manganese 54.94	26 **Fe** iron 55.85	27 **Co** cobalt 58.93
5	37 **Rb** rubidium 85.47	38 **Sr** strontium 87.62	39 **Y** yttrium 88.91	40 **Zr** zirconium 91.22	41 **Nb** niobium 92.91	42 **Mo** molybdenum 95.94	43 **Tc** technetium 98.91	44 **Ru** ruthenium 101.1	45 **Rh** rhodium 102.9
6	55 **Cs** cesium 132.9	56 **Ba** barium 137.3	57 **La** lanthanum 138.9	72 **Hf** hafnium 178.5	73 **Ta** tantalum 180.9	74 **W** tungsten 183.9	75 **Re** rhenium 186.2	76 **Os** osmium 190.2	77 **Ir** iridium 192.2
7	87 **Fr** francium (223)	88 **Ra** radium 226.0	89 **Ac** actinium (227)	104 **Rf** rutherfordium (261)	105 **Db** dubnium (262)	106 **Sg** seaborgium (263)	107 **Bh** bohrium (262)	108 **Hs** hassium (265)	109 **Mt** meitnerium (266)

Lanthanides

58 **Ce** cerium 140.1	59 **Pr** praseodymium 140.9	60 **Nd** neodymium 144.2	61 **Pm** promethium (147)	62 **Sm** samarium 150.4

Actinides

90 **Th** thorium 232.0	91 **Pa** protactinium 231.0	92 **U** uranium 238.0	93 **Np** neptunium 237.0	94 **Pu** plutonium (244)

( ) Indicates mass number of isotope with longest known half-life.

* Number in ( ) heading each column represents the group designation recommended by the American Chemical Society Committee on Nomenclature.

			(13) III A	(14) IV A	(15) V A	(16) VI A	(17) VII A	(18) Noble Gases
								2 **He** helium 4.003
			5 **B** boron 10.81	6 **C** carbon 12.01	7 **N** nitrogen 14.01	8 **O** oxygen 16.00	9 **F** fluorine 19.00	10 **Ne** neon 20.18
(10)	(11) I B	(12) II B	13 **Al** aluminum 26.98	14 **Si** silicon 28.09	15 **P** phosphorus 30.97	16 **S** sulfur 32.06	17 **Cl** chlorine 35.45	18 **Ar** argon 39.95
28 **Ni** nickel 58.71	29 **Cu** copper 63.55	30 **Zn** zinc 65.37	31 **Ga** gallium 69.72	32 **Ge** germanium 72.59	33 **As** arsenic 74.92	34 **Se** selenium 78.96	35 **Br** bromine 79.90	36 **Kr** krypton 83.80
46 **Pd** palladium 106.4	47 **Ag** silver 107.9	48 **Cd** cadmium 112.4	49 **In** indium 114.8	50 **Sn** tin 118.7	51 **Sb** antimony 121.8	52 **Te** tellurium 127.6	53 **I** iodine 126.9	54 **Xe** xenon 131.3
78 **Pt** platinum 195.1	79 **Au** gold 197.0	80 **Hg** mercury 200.6	81 **Tl** thallium 204.4	82 **Pb** lead 207.2	83 **Bi** bismuth 209.0	84 **Po** polonium (210)	85 **At** astatine (210)	86 **Rn** radon (222)
110 **Uun** ununnilium (269)	111 **Uuu** unununium (272)	112 **Uub** ununbium (277)	113	114 **Uuq** ununquadium (289)	115	116 **Uuh** ununhexium (289)	117	118 **Uuo** ununoctium (293)

63 **Eu** europium 152.0	64 **Gd** gadolinium 157.3	65 **Tb** terbium 158.9	66 **Dy** dysprosium 162.5	67 **Ho** holmium 164.9	68 **Er** erbium 167.3	69 **Tm** thulium 168.9	70 **Yb** ytterbium 173.0	71 **Lu** lutetium 175.0
95 **Am** americium (243)	96 **Cm** curium (247)	97 **Bk** berkelium (247)	98 **Cf** californium (251)	99 **Es** einsteinium (254)	100 **Fm** fermium (257)	101 **Md** mendelevium (258)	102 **No** nobelium (255)	103 **Lr** lawrencium (256)

APPENDIX C

MINERAL IDENTIFICATION TABLES

Metallic Luster

Mineral	Chemical Composition	Color	Hardness / Specific Gravity	Other Features	Comments
Chalcopyrite	$CuFeS_2$	Brassy yellow	3.5–4 / 4.1–4.3	Usually massive; greenish black streak; iridescent tarnish	Important source of copper. Mostly in hydrothermal rocks.
Galena	PbS	Lead gray	2.5 / 7.6	Cubic crystals; 3 cleavages at right angles	The ore of lead. Mostly in hydrothermal rocks.
Graphite	C	Black	1–2 / 2.09–2.33	Greasy feel; writes on paper; 1 direction of cleavage	Used for pencil "leads" and dry lubricant. Mostly in metamorphic rocks.
Hematite	Fe_2O_3	Red brown	6 / 4.8–5.3	Usually granular or massive; reddish brown streak	Most important ore of iron. An accessory mineral in many rocks.
Magnetite	Fe_3O_4	Black	5.5–6.5 / 5.2	Strong magnetism	An important ore of iron. An accessory mineral in many rocks.
Pyrite	FeS_2	Brassy yellow	6.5 / 5.0	Cubic and octahedral crystals	Found in some igneous and hydrothermal rocks and in sedimentary rocks associated with coal.

Nonmetallic Luster

Mineral	Chemical Composition	Color	Hardness / Specific Gravity	Other Features	Comments
Anhydrite	$CaSO_4$	White, gray	3.5 / 2.9–3.0	Crystals with 2 cleavages; usually in granular masses	Found in limestones and evaporite deposits. Used as a soil conditioner.
Apatite	$Ca_5(PO_4)_3F$	Blue, green, brown, yellow, white	5 / 3.1–3.2	6-sided crystals; in massive or granular masses	The main constituent of bone and dentine. A source of phosphorus for fertilizer.
Augite	$Ca(Mg,Fe,Al)(Al,Si)_2O_6$	Black, dark green	6 / 3.25–3.55	Short 8-sided crystals; 2 cleavages; cleavages nearly at right angles	The most common pyroxene mineral. Found mostly in mafic igneous rocks.
Barite	$BaSO_4$	Colorless, white, gray	3 / 4.5	Tabular crystals; high specific gravity for a nonmetallic mineral	Commonly found with metal ores and in limestones and hot spring deposits. A source of barium.
Biotite (mica)	$K(Mg,Fe)_3AlSi_3O_{10}(OH)_2$	Black, brown	2.5 / 2.9–3.4	1 cleavage direction; cleaves into thin sheets	Found in felsic and mafic igneous rocks, metamorphic rocks, and some sedimentary rocks.
Calcite	$CaCO_3$	Colorless, white	3 / 2.71	3 cleavages at oblique angles; cleaves into rhombs; reacts with dilute hydrochloric acid	The most common carbonate mineral. Main component of limestone and marble.
Cassiterite	SnO_2	Brown to black	6.5 / 7.0	High specific gravity for a nonmetallic mineral	The main ore of tin.
Chlorite	$(Mg,Fe)_3(Si,Al)_4O_{10}(Mg,Fe)_3(OH)_6$	Green	2 / 2.6–3.4	1 cleavage; occurs in scaly masses	Common in low-grade metamorphic rocks such as slate.
Corundum	Al_2O_3	Gray, blue, pink, brown	9 / 4.0	6-sided crystals and great hardness are distinctive	An accessory mineral in some igneous and metamorphic rocks. Used as a gemstone and for abrasives.

Nonmetallic Luster

Mineral	Chemical Composition	Color	Hardness / Specific Gravity	Other Features	Comments
Dolomite	$CaMg(CO_3)_2$	White, yellow, gray, pink	3.5–4 / 2.85	Cleavage as in calcite; reacts with dilute hydrochloric acid when powdered	The main constituent of dolostone.
Fluorite	CaF_2	Colorless, purple, green, brown	4 / 3.18	4 cleavage directions; cubic and octahedral crystals	Occurs mostly in hydrothermal rocks and in some limestones and dolostones. Used in the manufacture of steel and the preparation of hydrofluoric acid.
Garnet	$Fe_3Al_2(SiO_4)_3$	Dark red	7–7.5 / 4.32	12-sided crystals common; uneven fracture	Found mostly in gneiss and schist. Used as a semiprecious gemstone and for abrasives.
Gypsum	$CaSO_4 \cdot 2H_2O$	Colorless, white	2 / 2.32	Elongate crystals; fibrous and earthy masses	The most common sulfate mineral. Found mostly in evaporite deposits. Used to manufacture plaster of Paris and cements.
Halite	$NaCl$	Colorless, white	3–4 / 2.2	3 cleavages at right angles; cleaves into cubes; cubic crystals; salty taste	Occurs in evaporite deposits. Used as a source of chlorine and in the manufacture of hydrochloric acid, many sodium compounds, and food seasoning.
Hornblende	$NaCa_2(Mg,Fe,Al)_5 (Si,Al)_8O_{22}(OH)_2$	Green, black	6 / 3.0–3.4	Elongate, 6-sided crystals; 2 cleavages intersecting at 56 and 124 degrees	A common rock-forming amphibole mineral in igneous and metamorphic rocks.
Illite	$(Ca,Na,K)(Al,Fe^{+3},Fe^{+2},Mg)_2(Si,Al)_4O_{10}(OH)_2$	White, light gray, buff	1–2 / 2.6–2.9	Earthy masses; particles too small to observe properties	A clay mineral common in soils and clay-rich sedimentary rocks.
Kaolinite	$Al_2Si_4O_{10}(OH)_8$	White	2 / 2.6	Massive; earthy odor; particles too small to observe properties	A common clay mineral. The main ingredient of kaolin clay used for the manufacture of ceramics.
Muscovite (mica)	$KAl_2Si_3O_{10}(OH)_2$	Colorless	2–2.5 / 2.7–2.9	1 direction of cleavage; cleaves into thin sheets	Common in felsic igneous rocks, metamorphic rocks, and some sedimentary rocks. Used as an insulator in electrical appliances.
Olivine	$(Fe,Mg)_2SiO_4$	Olive green	6.5 / 3.3–3.6	Small mineral grains in granular masses; conchoidal fracture	Common in mafic igneous rocks.
Plagioclase feldspars	Varies from $CaAl_2Si_2O_8$ to $NaAlSi_3O_8$	White, gray, brown	6 / 2.56	2 cleavages at right angles	Common in igneous rocks and a variety of metamorphic rocks. Also in some arkoses.
Potassium feldspar — Microcline	$KAlSi_3O_8$	White, pink, green	6 / 2.56	2 cleavages at right angles	Common in felsic igneous rocks, some metamorphic rocks, and arkoses. Used in the manufacture of porcelain.
Potassium feldspar — Orthoclase	$KAlSi_3O_8$	White, pink			
Quartz	SiO_2	Colorless, white, gray, pink, green	7 / 2.67	6-sided crystals; no cleavage; conchoidal fracture	A common rock-forming mineral in all rock groups. Also occurs in varieties known as chert, flint, agate, and chalcedony.
Siderite	$FeCO_3$	Yellow, brown	4 / 3.8–4.0	3 cleavages at oblique angles; cleaves into rhombs	Found mostly in concretions and sedimentary rocks associated with coal.

Nonmetallic Luster

Mineral	Chemical Composition	Color	Hardness / Specific Gravity	Other Features	Comments
Smectite	$(Al,Mg)_8(Si_4O_{10})_3(OH)_{10} \cdot 12H_2O$	Gray, buff, white	1–1.5 / 2.5	Earthy masses; particles too small to observe properties	A clay mineral with the property of swelling and contracting as it absorbs and releases water.
Sphalerite	ZnS	Yellow, brown, black	3.5–4 / 4.0–4.1	6 cleavages; cleaves into dodecahedra	The most important ore of zinc. Commonly found in hydrothermal rocks.
Talc	$Mg_3Si_4O_{10}(OH)_2$	White, green	1 / 2.82	1 cleavage direction; usually in compact masses	Formed by the alteration of magnesium silicates. Mostly in metamorphic rocks. Used in ceramics and cosmetics and as a filler in paints.
Topaz	$Al_2SiO_4(OH,F)$	Colorless, white, yellow, blue	8 / 3.5–3.6	High specific gravity; 1 cleavage direction	Found in pegmatites, granites, and hydrothermal rocks. An important gemstone.
Zircon	Zr_2SiO_4	Brown, gray	7.5 / 3.9–4.7	4-sided, elongate crystals	A common accessory in granitic rocks. An ore of zirconium and used as a gemstone.

APPENDIX D

TOPOGRAPHIC MAPS

Nearly everyone has used a map of one kind or another and is probably aware that a map is a scaled-down version of the area depicted. For a map to be of any use, however, one must understand what is shown on a map and how to read it. A particularly useful type of map for geologists, and people in many other professions, is a *topographic map*, which shows the three-dimensional configuration of Earth's surface on a two-dimensional sheet of paper.

Maps showing relief—differences in elevation in adjacent areas—are actually models of Earth's surface. Such maps are available for some areas, but they are expensive, difficult to carry, and impossible to record data on. Thus paper sheets that show relief by using lines of equal elevation known as *contours* are most commonly used. Topographic maps depict (1) relief, which includes hills, mountains, valleys, canyons, and plains; (2) bodies of water such as rivers, lakes, and swamps; (3) natural features such as forests, grasslands, and glaciers; and (4) various cultural features, including communities, highways, railroads, land boundaries, canals, and power transmission lines.

Topographic maps known as *quadrangles* are published by the U.S. Geological Survey (USGS). The area depicted on a topographic map is identified by referring to the map's name in the upper right and lower right corners, which is usually derived from some prominent geographic feature (Lincoln Creek Quadrangle, Idaho) or community (Mt. Pleasant Quadrangle, Michigan). In addition, most maps have a state outline map along the bottom margin, and shown within the outline is a small black rectangle indicating the part of the state represented by the map.

Contours

Contour lines, or simply contours, are lines of equal elevation used to show topography. Think of contours as the lines formed where imaginary horizontal planes intersect Earth's surface at specific elevations. On maps, contours are brown, and every fifth contour, called an *index contour*, is darker than adjacent ones and labeled with its elevation (▶ Figure D1). Elevations on most USGS topographic maps are in feet, although a few use meters; in either case, the specified elevation is above or below mean sea level. Because contours are defined as lines of equal elevation, they cannot divide or cross one another, although they will converge and appear to join in areas with vertical or overhanging cliffs. Notice in Figure D1 that where contours cross a stream they form a V that points upstream toward higher elevations.

The vertical distance between contours is the *contour interval*. If an area has considerable relief, a large contour interval is used, perhaps 80 or 100 feet, whereas a small interval such as 5, 10, or 20 feet is used in areas with little relief. The values recorded on index contours are always multiples of the map's contour interval, shown at the bottom of the map. For instance, if a map has a contour interval of 10 feet, index contour values such as 3600, 3650, and 3700 feet might be shown (Figure D1). In addition to contours, specific elevations are shown at some places on maps and may be indicated by a small ×, next to which is a number. A specific elevation might also be shown adjacent to the designation BM (benchmark), a place where the elevation and location are precisely known.

Contour spacing depends on slope, so in areas with steep slopes, contours are closely spaced because there is a considerable increase in elevation in a short distance. In contrast, if slopes are gentle, contours are widely spaced (Figure D1). Furthermore, if contour spacing is uniform, the slope angle remains constant, but if spacing changes, the slope angle changes. However, one must be careful in comparing slopes on maps with different contour intervals or different scales.

Topographic features such as hills, valleys, plains, and so on are easily shown by contours. For instance, a hill is shown by a concentric pattern of contours with the highest elevation in the central part of the pattern. All contours must close on themselves, but they may do so beyond the confines of a particular map. A concentric contour pattern also might show a closed depression, but in this case special contours with short bars perpendicular to the contour pointing toward the central part of the depression are used (Figure D1).

Map Scales

All maps are scaled-down versions of the areas shown, so to be of any use they must have a scale. Highway maps, for example, commonly have a scale such as "1 inch equals 10 miles," by which one can readily determine distances. Two types of scales are used on topographic maps. The first and most easily understood is a graphic scale, which is simply a bar subdivided into appropriate units of length

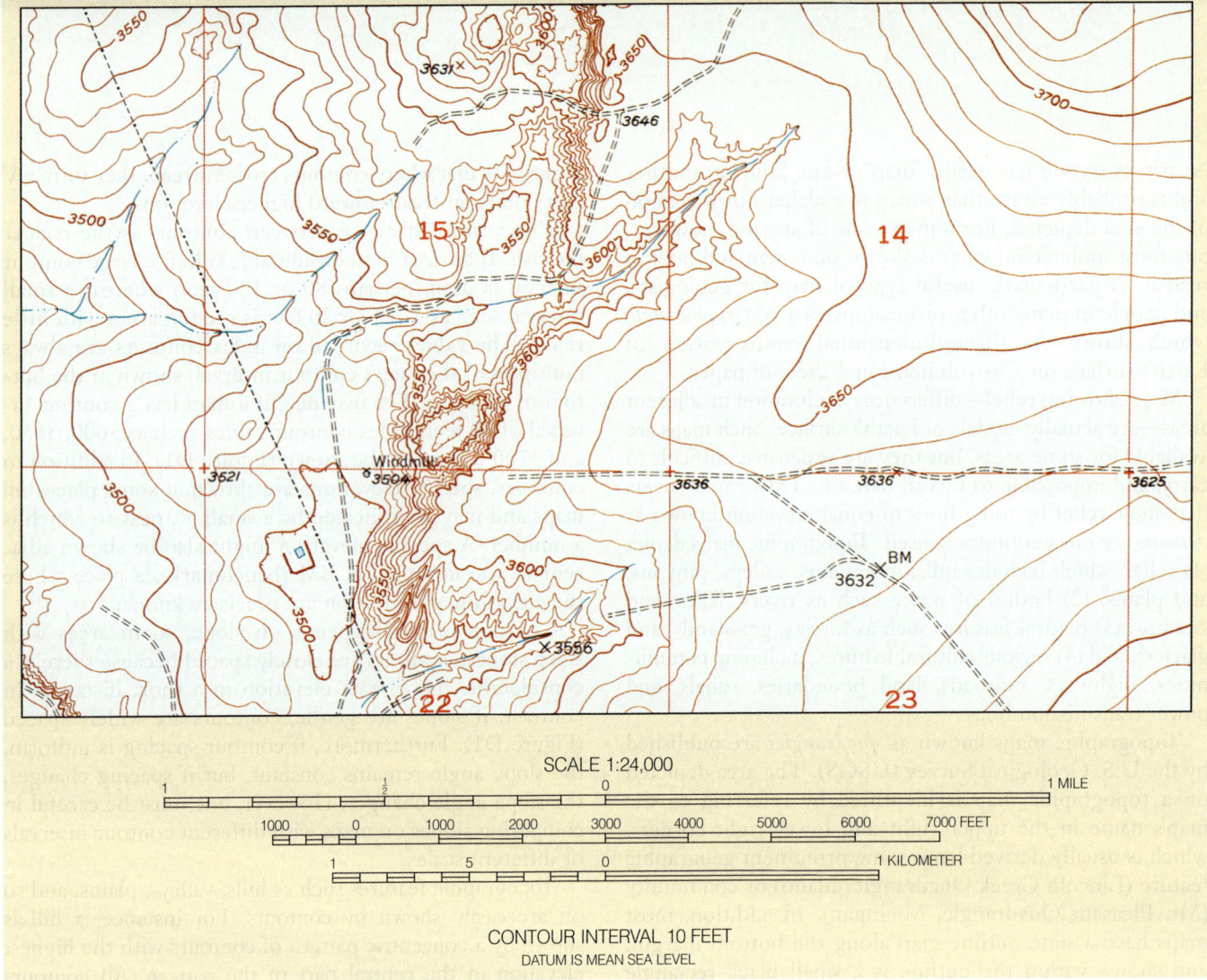

▶ **Figure D1** Part of the Bottomless Lakes Quadrangle, New Mexico, which has a contour interval of 10 feet; every fifth contour is darker and labeled with its elevation. Notice that contours are widely spaced where slopes are gentle and more closely spaced where they are steeper, as in the central part of the map. Hills are shown by contours that close on themselves, whereas depressions are indicated by contours with hachure marks pointing toward the center of the depression. The dashed blue lines on the map represent intermittent streams; notice that where contours cross a stream's channel they form a V that points upstream.

(Figure D1). This scale appears at the bottom center of the map and may show miles, feet, kilometers, or meters. Indeed, graphic scales on USGS topographic maps generally show both English and metric distance units.

A ratio or fractional scale, which represents the degree of reduction of the area depicted, appears above the graphic scale. On a map with a ratio scale of 1:24,000, for instance, the area shown is 1/24,000th the size of the actual land area (Figure D1). Another way to express this relationship is to say that any unit of length on the map equals 24,000 of the same units on the ground. Thus 1 inch on the map equals 24,000 inches on the ground, which is more meaningful if one converts inches to feet, making 1 inch equal to 2000 feet. A few maps have scales of 1:63,360, which converts to 1 inch equals 5280 feet, or 1 inch equals 1 mile.

USGS topographic maps are published in a variety of scales such as 1:50,000, 1:62,500, 1:125,000, and 1:250,000. One should also realize that large-scale maps cover less area than small-scale maps, and the former show much more detail than the latter. For example, a large-scale map (1:24,000) shows more surface features in greater detail than does a small-scale map (1:125,000) for the same area.

Map Locations

Location on topographic maps can be determined in two ways. First, the borders of maps correspond to lines of latitude and longitude. Latitude is measured north and south of the equator in degrees, minutes, and seconds, whereas

▶ **Figure D2** The General Land Office Grid System. Each 36-square-mile township is designated by township and range numbers. Townships are subdivided into sections, which can be further subdivided into quarter sections and quarter-quarter sections.

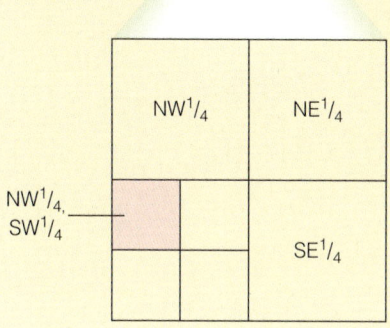

the same units are used to designate longitude east and west of the prime meridian, which passes through Greenwich, England. Maps depicting all areas within the United States are noted in north latitude and west longitude. Latitude and longitude are noted in degrees and minutes at the corners of maps, but usually only minutes and seconds are shown along the margins. Many USGS topographic maps cover $7\frac{1}{2}$ or 15 minutes of latitude and longitude and are thus referred to as $7\frac{1}{2}$- and 15-minute quadrangles.

Beginning in 1812, the General Land Office (now known as the Bureau of Land Management) developed a standardized method for accurately defining the location of property in the United States. This method, known as the General Land Office Grid System, has been used for all states except those along the eastern seaboard (except Florida), parts of Ohio, Tennessee, Kentucky, West Virginia, and Texas.

As new land acquired by the United States was surveyed, the surveyors laid out north–south lines they called *principal meridians* and east–west lines known as *base lines*. These intersecting lines form a set of coordinates for locating specific pieces of property. The basic unit in the General Land Office Grid System is the *township*, an area measuring 6 miles on a side and thus covering 36 square miles (▶ Figure D2). Townships are numbered north and south of base lines and are designated as T.1N., T.1S., and so on. Rows of townships known as *ranges* are numbered east and west of principal meridians—R.2W and R.4E, for example. Note in Figure D2 that each township has a unique designation of township and range numbers.

Townships are subdivided into 36 1-square-mile (640-acre) *sections* numbered from 1 to 36. Because of surveying errors and the adjustments necessary to make a grid system conform to Earth's curved surface, not all sections are exactly 1 mile square. Nevertheless, each section can be further subdivided into half sections and quarter sections designated $NE\frac{1}{4}$, $NW\frac{1}{4}$, $SE\frac{1}{4}$, and $SW\frac{1}{4}$, and each quarter section can be further divided into quarter-quarter sections. To show the complete designation for an area, the smallest unit is noted first (quarter-quarter section) followed by quarter section, section number, township, and range. For example, the area shown in Figure D2 is the $NW\frac{1}{4}$, $SW\frac{1}{4}$, Sec. 34, T.2N., R.3W.

Because only a few principal meridians and base lines were established, they do not appear on most topographic maps. Nevertheless, township and range numbers are printed along the margins of $7\frac{1}{2}$- and 15-minute quadrangles, and a grid consisting of red land boundaries depicts sections. In addition, each section number is shown in red within the map. However, small-scale maps show only township and range.

Where to Obtain Topographic Maps

Many people find topographic maps useful. Land use planners, personnel in various local, state, and federal agencies, as well as engineers and real estate developers might use these maps for a variety of reasons. In addition, hikers, backpackers, and others interested in exploring undeveloped areas commonly use topographic maps because trails are shown by black dashed lines. Furthermore, map users can readily determine their location by interpreting the topographic features depicted by contours, and they can anticipate the type of terrain they will encounter during off-road excursions.

Topographic maps for local areas are available at some sporting goods stores, at National Park Visitor Centers, and from some state geologic surveys. Free index maps showing the names and locations of all quadrangles for each state are available from the USGS to anyone uncertain of which specific map is needed. Any published topographic map can be purchased from two main sources. For maps of areas east of the Mississippi River, write to

Branch of Distribution
U.S. Geological Survey
1200 S. Eads Street
Arlington, Virginia 22202

Maps for areas west of the Mississippi River can be obtained from

Branch of Distribution
U.S. Geological Survey
Box 25286 Federal Center
Denver, Colorado 80225

ANSWERS

ANSWERS TO FIELD QUESTIONS

Chapter 3
1. These fluorite specimens have a glassy luster, but they do not show the brilliant luster of diamond. You could easily scratch fluorite with a knife, but diamond is much too hard for a knife to scratch it.
2. The most obvious places in the rock cycle where minerals form are (1) when magma and lava cool and crystallize and (2) when metamorphism takes place.

Chapter 4
1. The oldest rock in this scene across the alpine lake appears to be the red sandstone, because it is cut by a basalt dike, which in turn is cut by diorite that presumably is related to the small granitic stock making up most of the main mountain face to the right. The basaltic dike flares out into large mass of basaltic rock at the top of the ridge. Is this large mass made up of lava flows, or is this some other intrusive feature; part of a mafic magma chamber almost entirely eroded away perhaps? We can't tell from the available information. We would have to get closer to find out.

    The granitic stock is zoned. Large xenoliths of sandstone exist near the top, and pegmatite dikes cut across the pluton, branching out into the country rock near the ridgeline. A sill and a laccolith extend to the right.

    The gabbro-granite rock in the right foreground shows granite pipes and gabbro pillows. The gabbroic melt must have intruded the granitic magma to create these features. Are these rocks related to the stock across the lake? Possibly, but again, we don't know for sure because we don't see a direct connection. We would have to go around the corner to the right and see if there isn't some evidence that will help us hidden from view.

    What of the geologic history? You can say this much: Red sandstone accumulated in a sedimentary environment. Later tectonic activity impacted the area, causing magmas to form. The earliest magma to penetrate this area was mafic (hence the basaltic intrusion), followed by emplacement of granitic-dioritic melt that cooled and consolidated, developing pegmatite during the final stages of crystallization. Perhaps gabbroic magma invaded the base of the granitic magma to create the pipes we see in the right foreground. Maybe the gabbroic melt even mixed somewhat with the granitic magma to create the diorite zoning that we see. You just can't be sure. Finally, uplift and erosion exposed the scenery shown here, and you walked onto the scene, wondering "how did all of this come to be?"

Chapter 5
1. The elliptical objects represent pillow lava that was erupted on the seafloor.
2. As this ancient lava flow cooled, it contracted, forming columnar joints.
3. Top—cinder cone; Next two—composite volcanoes; Bottom—shield volcano. The composite volcanoes are most likely to erupt explosively because some have lava domes emplaced in their craters.
4. The overlying lava flow baked the pyroclastic materials, providing a small-scale example of contact metamorphism.

Chapter 6
1. The granite in the image on the left shows spheroidal weathering, and the granite in the other image shows the phenomenon of exfoliation (it is a small exfoliation dome).

Chapter 7
1. In this cross section the graded bedding, current ripples, and mud cracks indicate that the youngest layer is the one at the lower right. These layers are overturned, that is, rotated more than 90° from their original horizontal or nearly horizontal position. If you rotate these layers back into their original position, the currents that formed the current ripples flowed from left to right.
2. Deposition of the lowest layer (the mudstone and siltstone with lenses of sandstone). The dinosaur fossils indicate deposition on land. Next, there was a time of erosion, followed by deposition of sandstone, then shale, and then limestone during a marine transgression (see Figure 7.12). Uplift above sea level then occurred and a wind-blown dune deposit accumulated (see Figure 7.19). Finally, erosion yielded the present topography.

Chapter 8

1. First of all, you must collect as much geologic information as you can from what is shown. Don't worry about putting it all together into a history until after you've done your fact-finding:

   - Notice that the metamorphic grade seems to increase toward you across this landscape. There are unmetamorphosed sedimentary rocks on the distant ridge, low-grade metamorphic rocks in the middle hills, and medium-grade rocks (schists and marble) in the foreground. It looks like you are situated near the edge of a metamorphic belt.
   - Notice that stylolites and foliated rocks are in the foreground. There must have been differential pressure acting on these rocks. Moreover, the directions of pressure were from the upper right and lower left corners of the field of view—in other words, perpendicular to the stylolites and foliation. Most of the rocks show the textures of regional metamorphism.
   - The garnet porphyroblasts in the mica schist are not rotated, so you can't say that you have evidence of thrust nappe activity or shear forces during metamorphism. These things may have existed; you just can't tell definitely based on what you see here.
   - Metamorphism transformed several kinds of rocks, as seen in the foreground. Pelitic rock formed the mica schist. In fact, perhaps it is the same rock that makes up the shale in the distant ridge. Mafic rock, likely a basalt flow, transformed into greenschist, and limestone became the marble. This must have been a coastal area with nearby volcanoes originally. In other words, the outcrops show evidence that this area was once an active plate boundary.
   - A dike cuts across the marble in the lower right corner. A small aureole borders it. The facts that this aureole exists and that the dike is unmetamorphosed indicate that the dike intruded after metamorphism. The marble and chlorite schist and other regionally metamorphosed rocks must have risen to a much shallower depth in the crust in order for them to experience contact metamorphism. A fault cuts the dike, so it must be younger than the dike.

Constructing a History A tectonically active continental margin with a variety of sedimentary rocks underwent regional metamorphism. These rocks ejected fluids during metamorphism as garnet porphyroblasts crystallized through heterogeneous metamorphic reactions and stylolites formed in the nearby marble. The presence of greenschist shows that this rock and probably its neighbors belong to the greenschist facies. In other words, peak metamorphic conditions were 200°C to 450°C temperature and 1–9 kb pressure, as shown in Figure 8.20. Uplift and erosion then took place, accompanied by intrusion of mafic magma and later faulting. These rocks are still being eroded, as shown by the stream flowing by, and the hills in the background may be the erosional remnants of the mountain range in which the metamorphism originally took place.

A field geologist would construct a history such as this, and then go about testing it by finding more evidence to support or refute the conclusion. By persistent exploration a more truthful picture eventually emerges!

Chapter 9

1. The tilted beds were deposited first and deposited horizontally. They were then either tilted and then intruded by the batholith, or the batholith intruded them and then the beds and the batholith were tilted. In either case, the area was then uplifted and the surface was eroded, followed by deposition of the next sedimentary unit indicated by the –·–·–·– pattern. The dike was then intruded and the area was again uplifted and eroded. The next sedimentary unit, indicated by the ····· pattern, was deposited, and then either the sill was intruded followed by deposition of the sedimentary unit indicated by the - - - - pattern, or the two sedimentary units (· · · · and - - -) were deposited and the sill was then intruded. Without any more evidence, it is not possible to say which of the two previous scenarios is the correct one. Based on the evidence shown in this diagram, either interpretation is correct.

   All of the tilted sedimentary units must be older than 400 million years because they are intruded by the batholith. The sedimentary unit above the batholith and below the second unconformity is between 400 million and 100 million years old. It is lying unconformably on the batholith, and so is younger than the batholith (400 million years), and it has been intruded by the dike (100 million years), so it must be older than the dike. The sandstone unit (. . . .) above the second unconformity is between 100 million and 25 million years old. It lies above the dike (100 million years) and is intruded by the sill (25 million years). The uppermost sedimentary unit is younger than the underlying sandstone unit because of the principle of superposition. Its age is either younger than 25 million years if the dike intruded immediately after deposition of the sandstone, or it is between 100 million and 25 million years old (but younger than the underlying sandstone), if it was deposited after the sandstone but before the sill intruded.

   The lower unconformity consists of an angular unconformity where the horizontal beds overlie the tilted beds, and a nonconformity where the sedimentary rocks overlie the batholith. The presence of inclusions of the batholith in the overlying sedimentary rock prove the relationship is a nonconformity, and not a recent intrusion.

   The upper unconformity is a disconformity between the two sedimentary units, and a nonconformity where the sedimentary rocks overlie the dike.

2. The unconformity is an angular unconformity.

   The geologic history of the area is that the nearly vertical beds were deposited horizontally, the area was uplifted, tilting the beds to a near vertical position, the area was eroded, then submerged with the present horizontal beds being deposited on the eroded surface of the near vertical beds.

3. Both the principle of inclusions and cross-cutting relationships are evident in this photograph. The dark-colored elliptical-shaped 'blobs' are inclusions. The white colored planar structure cutting across from the upper-left to lower-right side of the photo is a dike.

   The geologic history of this photograph would be the lighter-colored mass intruded into a dark-colored unit, but wasn't able to completely melt it, leaving these dark-colored inclusions. Thus, the inclusions are older than the light-colored igneous body. This was followed by intrusion of the white-colored dike, which is younger than the igneous body and the inclusions, because it cuts through both.

Chapter 11

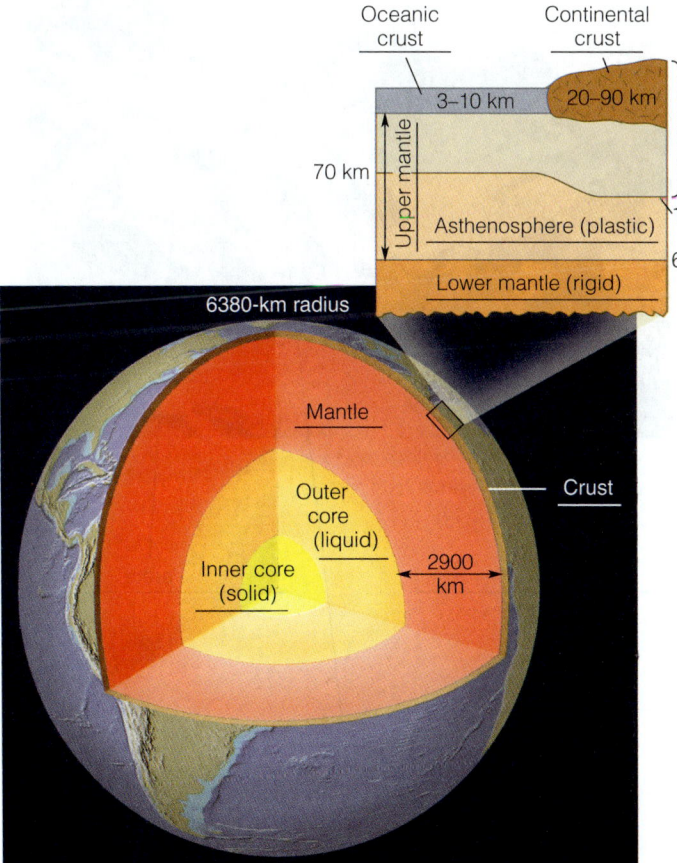

Chapter 12

1. The image in the upper left represents deposition on the upper part of the fan (closer to the source of the turbidity current), whereas the one in the lower right was deposited much farther from the source area.

Chapter 13

1. First, the sedimentary structures indicate that the oldest rock layer is the one in the core of the fold which is overlain by successively younger layers, so the fold is a recumbent anticline. Accordingly, the sedimentary rock layers were deposited, then they were folded, and finally they were cut by a reverse fault (the hanging wall block moved relatively up).

2. The fault is inclined downward (dips) steeply to the left. Look closely at the layers on the left side of the fault, that is, on the hanging wall block. Notice that some of them have been bent so that they curve upward, indicating that this is the side of the fault that moved down. As the hanging wall block moved relatively downward (remember that the footwall block could have moved upward), friction caused the layers to bend as fault movement took place. This is a normal fault because the hanging wall block moved relatively down (or the footwall block moved relatively up).

Chapter 14

1. It looks like the hill's slope has been reduced by grading. The hillside also shows benching to help reduce and stabilize the slope. Culverts have been added to carry away rainwater so as not to saturate the soil during rains, and vegetation has been planted which takes up water and helps stabilize the soil.

Chapter 15

1. The deposit is a point bar that formed on the inner bank of a meander where flow velocity is least. Flow velocity on the opposite side of the river would be highest.

Chapter 16

1. Cracked sidewalks are one feature indicating subsidence. The sidewalk in the background has subsided to a lower level than the sidewalk in the foreground. The section on the right side of the photo looks lower than the left side and the house in the background looks to be tilted.

Chapter 17

1.

2. These striations indicate that the glacier moved either toward the foreground or toward the background. To resolve which of these two directions is the correct one you must identify other glacial features (not shown here) such as cirques and end moraines.

Chapter 18

1. You can see desert pavement in the foreground with many ventifacts visible on the surface. You can also see a pediment gently sloping away from the mountain front, and lastly the white area is either a playa or a sandy region. The blowing wind emphasizes that small particles can be carried by the wind, but larger particles are moved along by saltation.
2. The whitish colored area is a playa. Sloping away from the mountain front are individual alluvial fans, and bajadas.

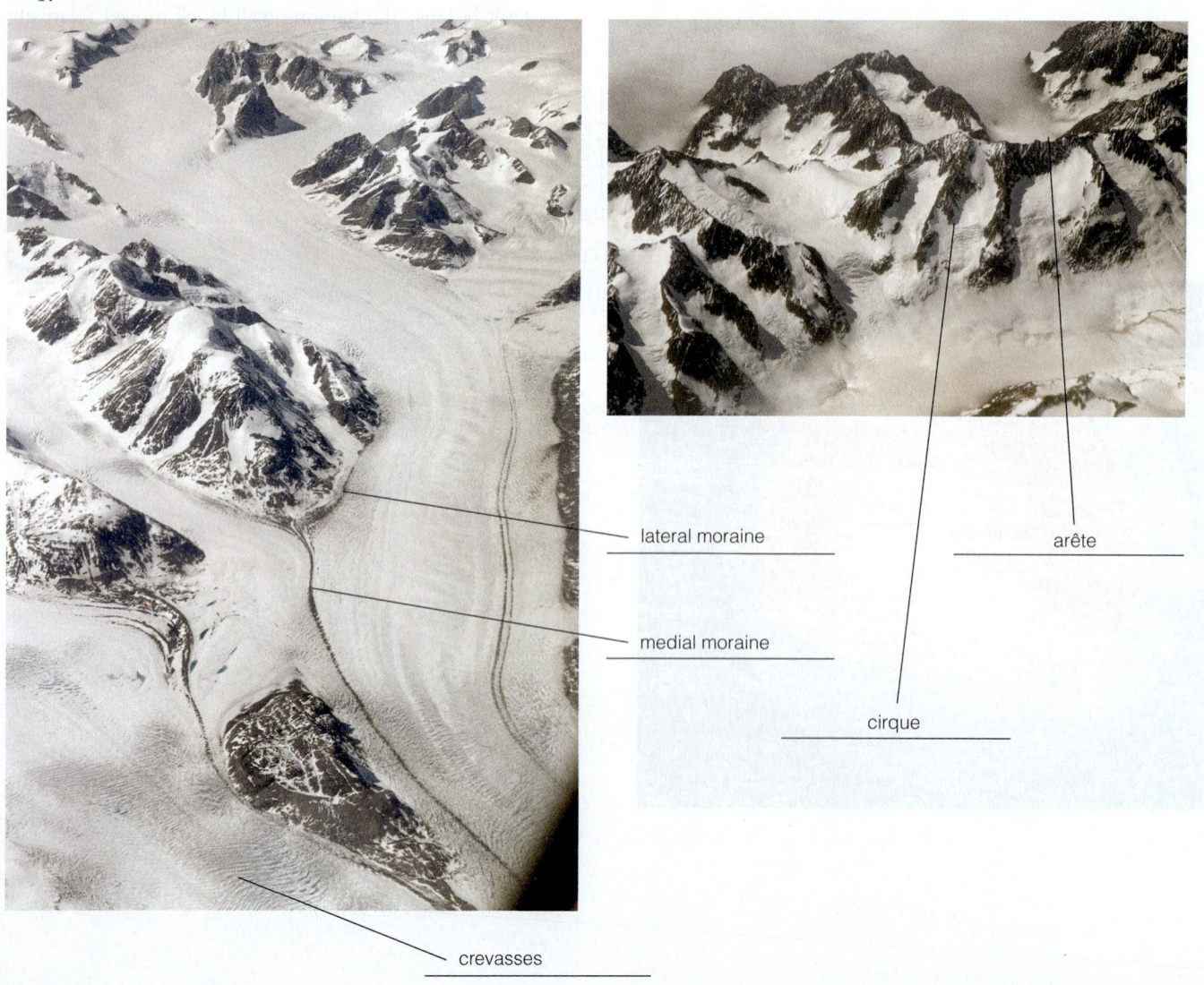

Chapter 19

1. The gently sloping surface in the distance (with buildings on it) is a marine terrace, and the sand bar pointing out to sea toward the rock is a tombolo.
2. A baymouth bar is simply a spit that has grown as a result of deposition by longshore currents until it completely closes off the opening to a bay.
3. Wind intensity, duration, and the distance it blows over a water surface account for large waves. Waves on ponds and lakes are much smaller than those in the oceans because no matter how hard or how long the wind blows it blows over a limited distance.

GLOSSARY

A

aa A lava flow with a surface of rough, angular blocks and fragments.

abrasion The process whereby exposed rock is worn and scraped by the impact of solid particles.

absolute dating The process of assigning ages in years before the present to geologic events. Various radioactive decay-dating techniques yield absolute ages. See also *relative dating*.

abyssal plain A vast flat area on the seafloor adjacent to the continental rises of passive continental margins.

active continental margin A continental margin characterized by volcanism and seismicity at the leading edge of a continental plate where oceanic lithosphere is subducted. See also *passive continental margin*.

alluvial fan A cone-shaped alluvial deposit formed where a stream flows from mountains onto an adjacent lowland.

alluvium A general term for all detrital sediment transported and deposited by running water.

angular unconformity An unconformity below which older rocks dip at a different angle (usually steeper) than the overlying strata. See also *disconformity* and *nonconformity*.

anticline A convex upward fold in which the oldest exposed rocks coincide with the fold axis and all strata dip away from the axis.

aphanitic texture An igneous texture in which individual minerals are too small to be seen without magnification.

arête A narrow, serrated ridge separating two glacial valleys or adjacent cirques.

artesian system A confined aquifer in which high hydrostatic (fluid) pressure builds up causing water to rise above the level of the aquifer.

aseismic ridge A ridge or broad area rising as much as 2 to 3 km above the surrounding seafloor and lacking seismic activity.

ash Pyroclastic material measuring less than 2 mm.

assimilation A process in which magma changes composition as it reacts with country rock.

asthenosphere The part of the mantle that lies below the lithosphere; behaves plastically and flows.

atom The smallest unit of matter that retains the characteristics of an element.

atomic mass number The total number of protons and neutrons in the nucleus of an atom.

atomic number The number of protons in the nucleus of an atom.

aureole A zone surrounding a pluton in which contact metamorphism took place.

B

barchan dune A crescent-shaped dune with the tips of the crescent pointing downwind.

barrier island A long, narrow island composed of sand oriented parallel to a shoreline but separated from the mainland by a lagoon.

basal slip A type of glacial movement in which a glacier slides over its underlying surface.

basalt plateau A large area built up by numerous flat-lying lava flows erupted from fissures.

base level The lowest level to which a stream can erode.

basin The circular fold in which all strata dip toward a central point, and the youngest exposed rocks are in the center.

batholith A discordant, irregularly shaped pluton with a surface area of at least 100 km^2.

baymouth bar A spit that has grown until it closes off a bay from the open sea.

beach A deposit of sediment extending landward from low tide to a change in topography or where permanent vegetation begins.

bed (bedding) A bed is an individual layer of rock, especially sedimentary rock, whereas bedding is the layered arrangement of rocks. See also *stratification*.

bed load The part of a stream's sediment load transported along its bed; consists of sand and gravel.

berm The backshore area of a beach, consisting of a platform composed of sediment deposited by waves. Berms are nearly horizontal or slope gently landward.

Big Bang A model for the evolution of the universe in which a dense, hot state was followed by expansion, cooling, and a less dense state.

biochemical sedimentary rock A sedimentary rock resulting from the chemical processes of organisms.

black smoker A submarine hydrothermal vent that emits a plume of black water colored by dissolved minerals.

body wave An earthquake wave that travels through Earth. Both P- and S-waves are body waves.

bonding The process whereby atoms are joined to other atoms.

Bowen's reaction series A mechanism accounting for the derivation of intermediate and felsic magmas from mafic magma. It has a discontinuous branch of ferromagnesian minerals that change from one to another over specific temperature ranges and a continuous branch of plagioclase feldspars whose composition changes as the temperature decreases.

braided stream A stream with multiple dividing and rejoining channels.

breaker A wave that steepens as it enters shallow water until its crest plunges forward.

butte An isolated, steep-sided, pinnacle-like hill formed by the breaching of a resistant cap rock, which allows rapid erosion of the less resistant underlying rocks.

C

caldera A large, steep-sided circular to oval volcanic depression usually formed by summit collapse resulting from partial draining of the underlying magma chamber.

Canadian shield The exposed part of the North American craton consisting of various ancient rocks.

carbon 14 dating technique An absolute dating method that relies on determining the ratio of C^{14} to C^{12} in an organic substance; useful back to about 70,000 years ago.

carbonate mineral A mineral with the negatively charged carbonate ion $(CO_3)^{-2}$, as in calcite $(CaCO_3)$ and dolomite $[CaMg(CO_3)_2]$.

carbonate rock A rock containing mostly carbonate minerals (such as limestone and dolostone).

Cascade Range A mountain range stretching from northern California through Oregon and Washington and into southern British Columbia, Canada. Made up of about a dozen large composite volcanoes and thousands of smaller volcanic vents.

cave A natural subsurface opening that is generally connected to the surface and is large enough for a person to enter.

cementation The precipitation of minerals as binding material in and around sediment grains, thus converting sediment to sedimentary rock.

chemical sedimentary rock Rock formed of minerals derived from materials dissolved during chemical weathering.

chemical weathering The decomposition of rocks by chemical alteration of parent material.

cinder cone A small, steep-sided volcano composed of pyroclastic materials that accumulate around a vent.

circum-Pacific belt A zone of seismic and volcanic activity that nearly encircles the Pacific Ocean basin.

cirque A steep-walled, bowl-shaped depression formed by erosion at the upper end of a glacial valley.

cleavage The breaking or splitting of mineral crystals along planes of weakness.

Code of Hammurabi Hammurabi, the ruler of Babylon from 1795 to 1750 B.C., developed a body of laws in orderly groups so that people would know what was required of them.

columnar joint Joints in some igneous rocks consisting of six-sided columns that form as a result of shrinkage during cooling.

compaction A method of lithification whereby the pressure exerted by the weight of overlying sediment reduces the amount of pore space and thus the volume of a deposit.

complex movement A combination of different types of mass movements in which one type is not dominant. Most complex movements involve sliding and flowing.

composite volcano A volcano composed of pyroclastic layers, lava flows typically of intermediate composition, and mudflows; also called a *stratovolcano*.

compound A substance resulting from the bonding of two or more different elements, such as water (H_2O) and quartz (SiO_2).

compression Stress resulting when rocks are squeezed by external forces directed toward one another.

concordant An intrusive igneous body with boundaries parallel to the layering in the country rock. See also *discordant*.

cone of depression A cone-shaped depression in the water table around a well, resulting from pumping water from an aquifer faster than it is replenished.

contact (thermal) metamorphism Metamorphism of country rock adjacent to a pluton.

continental accretion The phenomenon whereby continents grow by additions of new material along their margins.

continental–continental plate boundary A convergent plate boundary along which two continental lithospheric plates collide, such as the collision of India with Asia.

continental crust The rocks of continents overlying the upper mantle made up of a wide variety of rocks, but with an overall granitic composition and an average density of about 2.70 g/cm³.

continental drift The theory that the continents were once joined into a single landmass that broke apart with the various fragments (continents) moving with respect to one another.

continental glacier A glacier covering a vast area (at least 50,000 km²) and unconfined by topography. Also called an *ice sheet*.

continental margin The area separating the part of a continent above sea level from the deep seafloor.

continental rise The gently sloping area of the seafloor beyond the base of the continental slope.

continental shelf The area between the shoreline and the continental slope where the seafloor slopes gently seaward.

continental slope The relatively steep area between the shelf–slope break (at an average depth of 135 m) and the more gently sloping continental rise or an oceanic trench.

convergent plate boundary The boundary between two plates that are moving toward one another; three types of convergent plate boundaries are recognized. See also *continental–continental plate boundary*, *oceanic–continental plate boundary*, and *oceanic–oceanic plate boundary*.

core The interior part of Earth, beginning at a depth of about 2900 km; probably composed mostly of iron and nickel; divided into an outer liquid core and an inner solid core.

Coriolis effect The deflection of winds to the right of their direction of motion (clockwise) in the Northern Hemisphere and to the left of their direction of motion (counterclockwise) in the Southern Hemisphere, due to Earth's rotation.

correlation The demonstration of the physical continuity or time equivalency of rock units in different areas.

country rock Any rock that is intruded by and surrounds a pluton.

covalent bond A bond formed by the sharing of electrons between atoms.

crater A circular or oval depression at the summit of a volcano resulting from the extrusion of gases, pyroclastic materials, and lava; connected by a conduit to a magma chamber below Earth's surface.

craton The relatively stable part of a continent; consists of a shield and a platform, a buried extension of a shield; the ancient nucleus of a continent.

creep A type of mass wasting in which soil or rock moves slowly downslope.

crest The highest part of a wave.

cross-bedding Layers in sedimentary rocks deposited at an angle to the surface on which they accumulated.

crust Earth's outermost layer; the upper part of the lithosphere, which is separated from the mantle by the Moho; divided into continental and oceanic crust.

crystal A naturally occurring solid of an element or a compound with a specific internal structure that is manifested externally by planar faces, sharp corners, and straight edges.

crystalline solid A solid in which the constituent atoms are arranged in a regular, three-dimensional framework.

Curie point The temperature at which iron-bearing minerals in a cooling magma or lava attain their magnetism.

D

debris flow A type of mass wasting that involves the flow of a viscous mixture of water, soil, and rocks; much like a mudflow, but at least half the particles are larger than sand.

deflation The removal of loose surface sediment by wind.

deformation Any change in shape or volume, or both, of rocks in response to stress. Deformation involves folding and fracturing.

delta An alluvial deposit formed where a stream flows into a lake or the sea.

density The mass of an object per unit of volume; usually expressed in grams per cubic centimeter (g/cm^3).

depositional environment Any area where sediment is deposited, such as a floodplain or a beach.

desert Any area that receives less than 25 cm of rain per year and has a high evaporation rate.

desertification The change that takes place when fertile or habitable land becomes more arid and desert conditions become more prevalent.

desert pavement A surface mosaic of close-fitting pebbles, cobbles, and boulders found in many dry regions; formed by the removal of sand-sized and smaller particles by wind.

detrital sedimentary rock Rock consisting of the solid particles (detritus) of preexisting rocks.

differential pressure Pressure that is not applied equally to all sides of a rock body.

differential weathering Weathering of rock at different rates, producing an uneven surface.

dike A tabular or sheetlike igneous intrusion.

dip A measure of the maximum angular deviation of an inclined plane from horizontal.

dip-slip fault A fault on which all movement is parallel with the dip of the fault plane. See also *normal fault*, *reverse fault*, and *thrust fault*.

discharge The volume of water in a stream moving past a particular point in a given period of time.

disconformity An unconformity above and below which the strata are parallel. See also *angular unconformity* and *nonconformity*.

discontinuity A boundary across which seismic wave velocity or direction changes abruptly, such as the mantle–core boundary.

discordant Pluton with boundaries that cut across the layering in the country rock. See also *concordant*.

dissolution seam An area in metamorphic rocks where dissolution takes place along parallel or nearly parallel surfaces oriented at right angles to the greatest pressure acting on the rock.

dissolved load The part of a stream's load consisting of ions in solution.

divergent plate boundary The boundary between two plates that are moving apart.

divide A topographically high area that separates adjacent drainage basins.

dome A rather circular geologic structure in which all strata dip away from a central point, and the oldest exposed rocks are at the dome's center.

drainage basin The surface area drained by a stream and its tributaries.

drainage pattern The regional arrangement of channels in a drainage system.

drumlin An elongate hill of till formed by the movement of a continental glacier or by floods.

dune A mound or ridge of wind-deposited sand.

dynamic metamorphism Metamorphism in fault zones where rocks are subjected to high differential pressure.

E

earthflow A mass wasting process involving downslope flow of water-saturated soil.

earthquake Vibrations caused by the sudden release of energy, usually as a result of the displacement of rocks along faults.

echo sounder An instrument used to determine the depth of the oceans by the travel time of sound waves from the source to the seafloor and back.

elastic rebound theory A theory that explains how energy is suddenly released during earthquakes: When rocks are deformed, they store energy and bend; when the inherent strength of the rocks is exceeded, they rupture and release energy, causing earthquakes.

elastic strain A type of deformation in which the material returns to its original shape when stress is relaxed.

electron A negatively charged particle of very little mass that encircles the nucleus of an atom.

electron shell Electrons orbit rapidly around the nuclei of atoms at specific distances known as electron shells.

element A substance composed of atoms that all have the same properties; atoms of one element can change to atoms of another element by radioactive decay, but otherwise they cannot be changed by ordinary chemical means.

emergent coast A coast where the land has risen with respect to sea level.

end moraine A pile or ridge of rubble deposited at the terminus of a glacier. See also *recessional moraine* and *terminal moraine*.

environmental geology A branch of geology that concentrates on developing solutions to environmental problems created by humans.

epicenter The point on Earth's surface directly above the focus of an earthquake.

equilibrium A condition in which minerals and rocks are in balance with their environment, i.e., they are stable under existing conditions.

erosion The removal of weathered materials from their source area.

esker A long, sinuous ridge of stratified drift formed by deposition by running water in tunnels beneath stagnant ice.

evaporite A sedimentary rock that formed by inorganic chemical precipitation of minerals from an evaporating solution (such as rock salt and rock gypsum).

Exclusive Economic Zone An area extending 371 km seaward from the coast of the United States and its territories in which the United States claims all sovereign rights.

exfoliation The process whereby slabs of rock bounded by sheet joints slip or slide off the host rock.

exfoliation dome A large rounded dome of rock resulting from the process of exfoliation.

expansive soil A soil in which the volume increases when water is present.

extrusive Refers to magma and rocks formed from magma that reaches Earth's surface.

F

fault A fracture along which movement has occurred parallel to the fracture surface.

fault plane A fracture surface along which blocks of rock on opposite sides have moved relative to one another.

felsic Magma containing more than 65% silica and considerable sodium, potassium, and aluminum but little calcium, iron, and magnesium.

ferromagnesian silicate A silicate mineral containing iron or magnesium or both.

fetch The distance the wind blows over a continuous water surface.

fiord An arm of the sea extending into a U-shaped glacial trough eroded below sea level.

firn Granular snow formed by partial melting and refreezing of snow.

fission track dating The process of dating samples by counting the number of small linear tracks (fission tracks) that result when a mineral crystal is damaged by rapidly moving alpha particles generated by radioactive decay of uranium.

fissure eruption An eruption in which lava or pyroclastic material is emitted from a long, narrow fissure or group of fissures.

floodplain A low-lying, relatively flat area adjacent to a river or stream, which is partly or completely covered with water when the stream overflows its banks.

focus The place within Earth where an earthquake originates and energy is released.

fold A type of geologic structure in which planar features in rock layers such as bedding and foliation have been bent.

foliated texture A texture of metamorphic rocks in which platy and elongate minerals are arranged in a parallel fashion.

footwall block The block of rock that lies beneath a fault plane. See also *hanging wall block*.

fossil Remains or traces of prehistoric organisms preserved in rocks. See also *body fossil* and *trace fossil*.

fracture A break in a rock resulting from intense applied pressure.

frost action The disaggregation of rocks by repeated freezing and thawing of water in cracks and crevices.

frost wedging The opening and widening of cracks by the repeated freezing and thawing of water.

G

geologic record The record of prehistoric physical and biological events preserved in rocks.

geologic structure Any feature in rocks that results from deformation, such as folds, joints, and faults.

geologic time scale A chart with the designation for the earliest interval of geologic time at the bottom, followed upward by designations for progressively more recent time intervals.

geology The science concerned with the study of Earth; includes studies of Earth materials (minerals and rocks), surface and internal processes, and Earth history.

geothermal energy Energy that comes from the steam and hot water trapped within the crust.

geothermal gradient Earth's temperature increase with depth; it averages 25°C/km near the surface but varies from area to area.

geyser A hot spring that intermittently ejects hot water and steam.

glacial budget The balance between expansion and contraction of a glacier in response to accumulation and wastage.

glacial drift A collective term for all sediment deposited by glaciers or associated processes; includes till deposited by ice and outwash deposited by streams derived from melting ice.

glacial erratic A rock fragment carried some distance from its source by a glacier and usually deposited on bedrock of a different composition.

glacial ice Water in the solid state within a glacier; forms as snow partially melts and refreezes and is compacted so that it is transformed first into firn and then into glacial ice.

glacial polish A smooth, glistening rock surface formed by the movement of a sediment-laden glacier over it.

glacial striation A straight scratch rarely more than a few millimeters deep on a rock caused by the movement of sediment-laden glacial ice.

glacial surge A time of greatly accelerated flow in a glacier. Commonly results in displacement of the glacier's terminus by several kilometers.

glaciation Refers to all aspects of glaciers, including their origin, expansion, and retreat, and their impact on Earth's surface.

glacier A mass of ice on land that moves by plastic flow and basal slip.

***Glossopteris* flora** A Late Paleozoic association of plants found only on the Southern Hemisphere continents and India.

Gondwana One of six major Paleozoic continents; composed of the present-day continents of South America, Africa, Antarctica, Australia, and India and parts of other continents such as southern Europe, Arabia, and Florida.

graded bedding Sedimentary bedding in which an individual bed is characterized by a decrease in grain size from bottom to top.

graded stream A stream that has an equilibrium profile in which a delicate balance exists between gradient, discharge, flow velocity, channel characteristics, and sediment load such that neither significant erosion nor deposition occurs within its channel.

gradient The slope over which a stream flows; expressed in m/km or ft/mi.

gravity The force exerted between any two bodies in the universe as a function of their mass and the distance between their centers of mass.

gravity anomaly (positive and negative) A departure from the expected force of gravity; a gravity anomaly might be positive, indicating a mass excess, or negative, indicating a mass deficiency.

ground moraine The layer of sediment liberated from melting ice as a glacier's terminus retreats.

groundwater Underground water stored in the pore spaces of rock, sediment, or soil.

guide fossil Any fossil that can be used to determine the relative geologic ages of rocks and to correlate rocks of the same relative age in different areas.

guyot A flat-topped seamount of volcanic origin rising more than 1 km above the seafloor.

H

half-life The time required for one-half of the original number of atoms of a radioactive element to decay to a more stable daughter product (e.g., the half-life of potassium 40 is 1.3 billion years).

hanging valley A tributary glacial valley whose floor is at a higher level than that of the main glacial valley.

hanging wall block The block of rock that overlies a fault plane. See also *footwall block*.

hardness A term used to express the resistance of a mineral to abrasion.

heat flow The flow of heat from Earth's interior to its surface.

heterogeneous metamorphic reaction A metamorphic reaction in which the mineral composition of the parent rock changes.

horn A steep-walled, pyramid-shaped peak formed by the headward erosion of at least three cirques.

hot spot A localized zone of melting below the lithosphere.

hot spring A spring in which the water temperature is warmer than the temperature of the human body (37°C).

humus The material in soils derived by bacterial decay of organic matter.

hydraulic action The power of moving water.

hydrologic cycle The continuous recycling of water from the oceans, through the atmosphere, to the continents, and back to the oceans.

hydrolysis The chemical reaction between the hydrogen (H^+) ions and hydroxyl (OH^-) ions of water and a mineral's ions.

hydrothermal A term referring to hot water as in hot springs or geysers.

hypothesis A provisional explanation for observations; subject to continual testing and modification. If well supported by evidence, hypotheses are then generally called *theories*.

I

ice cap A dome-shaped mass of glacial ice covering less than 50,000 km^2.

ice sheet See *continental glacier*.

ice shelf A thick sheet of ice floating on the sea but attached to the land where it is partly fed by glaciers.

igneous rock Any rock formed by cooling and crystallization of magma or by the accumulation and consolidation of pyroclastic materials.

incised meander A deep, meandering canyon cut into bedrock by a stream or river.

infiltration capacity The maximum rate at which soil or sediment absorbs water.

intensity The subjective measure of the kind of damage done by an earthquake, as well as people's reaction to it.

intermediate magma Magma with a silica content between 53% and 65% and an overall composition intermediate between felsic and mafic magmas.

intrusive Refers to magma and rocks that formed within Earth's crust.

intrusive igneous rock See *plutonic rock*.

ion An electrically charged atom produced by adding or removing electrons from the outermost electron shell.

ionic bond A bond that results from the attraction of positively and negatively charged ions.

isograd A line on a map connecting points of equal metamorphic intensity.

isostasy See *principle of isostasy*.

isostatic rebound The phenomenon in which unloading of Earth's crust causes it to rise upward until equilibrium is again attained. See also *principle of isostasy*.

J

joint A fracture along which no movement has occurred or where movement has been perpendicular to the fracture surface.

Jovian planet Any of the four planets (Jupiter, Saturn, Uranus, and Neptune) that resemble Jupiter. All are large and have low mean densities, indicating they are composed mostly of lightweight gases, such as hydrogen and helium, and frozen compounds, such as ammonia and methane. See also *terrestrial planet*.

K

kame Conical hill of stratified drift originally deposited in a depression on a glacier's surface.

karst topography Landscape consisting of numerous caves, sinkholes, and solution valleys developed by groundwater solution of rocks such as limestone and dolostone.

L

laccolith   A concordant pluton with a mushroomlike geometry.

lahar   A mudflow composed of volcanic materials such as ash.

lateral moraine   Ridge of sediment deposited along the margin of a valley glacier.

laterite   A red soil, rich in iron or aluminum or both, that forms in the tropics by intense chemical weathering.

Laurasia   A Late Paleozoic, Northern Hemisphere continent composed of the present-day continents of North America, Greenland, Europe, and Asia.

lava   Magma at Earth's surface.

lava dome   A bulbous, steep-sided structure formed by viscous magma moving upward through a volcanic conduit.

lava tube   A tunnel beneath the solidified surface of a lava flow through which a molten flow continues to move. Also, the hollow space left when the lava within the tube drains away.

lithification   The process of converting sediment into sedimentary rock.

lithosphere   Earth's outer, rigid part consisting of the upper mantle, oceanic crust, and continental crust.

lithostatic pressure   Pressure exerted on rock by the weight of overlying rocks.

loess   Windblown deposit of silt and clay.

longitudinal dune   A long ridge of sand generally parallel to the direction of the prevailing wind.

longshore current   A current between the breaker zone and the beach that flows parallel to the shoreline and is produced by wave refraction.

longshore drift   The movement of sediment along a shoreline by longshore currents.

Love wave (L-wave)   A surface wave in which the individual particles of material move only back and forth in a horizontal plane perpendicular to the direction of wave travel.

low-velocity zone   The zone within the mantle between 100 and 250 km deep where the velocity of both P- and S-waves decreases markedly. It corresponds closely to the asthenosphere.

luster   The appearance of a mineral in reflected light. Luster is metallic and nonmetallic, although the latter has several subcategories.

M

mafic   Magma containing between 45% and 52% silica and proportionately more calcium, iron, and magnesium than intermediate and felsic magmas.

magma   Molten rock material generated within Earth.

magma chamber (reservoir)   A reservoir of magma within the upper mantle or lower crust.

magma mingling   In magma mingling, magmas that come together do not mix and there is a destabilizing exchange of heat that results in explosive eruptions.

magma mixing   The process of mixing magmas of different composition, thereby producing a modified version of the parent magmas.

magnetic anomaly   Any change, such as a change in average strength, of Earth's magnetic field.

magnetic field   The area in which magnetic substances are affected by lines of magnetic force emanating from Earth.

magnetic reversal   The phenomenon in which the north and south magnetic poles are completely reversed.

magnetism   A physical phenomenon resulting from moving electricity and the spin of electrons in some solids in which magnetic substances are attracted toward one another.

magnitude   The total amount of energy released by an earthquake at its source. See also *Richter Magnitude Scale*.

mantle   The thick layer between Earth's crust and core.

marine regression   The withdrawal of the sea from a continent or coastal area, resulting in the emergence of the land as sea level falls or the land rises with respect to sea level.

marine terrace   A wave-cut platform now elevated above sea level.

marine transgression   The invasion of coastal areas or much of a continent by the sea, resulting from a rise in sea level or subsidence of the land.

mass wasting   The downslope movement of material under the influence of gravity.

meandering stream   A stream with a single, sinuous channel with broadly looping curves.

mechanical weathering   Disaggregation of rocks by physical processes that yield smaller pieces retaining the same composition as the parent material.

medial moraine   A moraine formed where two lateral moraines merge.

Mediterranean belt   A zone of seismic and volcanic activity extending westerly from Indonesia through the Himalayas, across Iran and Turkey, and through the Mediterranean region of Europe.

mesa   A broad, flat-topped erosional remnant bounded on all sides by steep slopes.

metamorphic belt   A zone with metamorphic rocks that share the same general age and conditions of formation.

metamorphic facies   A group of metamorphic rocks characterized by particular mineral assemblages formed under the same broad temperature–pressure conditions.

metamorphic fluid   A mobile substance with the properties of liquids and gases at high temperature and under high pressure within Earth; composed mostly of water, carbon dioxide, and dissolved minerals.

metamorphic grade   The degree to which parent rocks have undergone metamorphic change; the higher the grade the greater the change, as in low-, medium-, and high-grade.

metamorphic rock   Any rock altered by high temperature and pressure and the chemical activities of fluids is said to have been metamorphosed (such as slate, gneiss, or marble).

metamorphic zone   The region between lines of equal metamorphic intensity known as isograds.

metamorphism   The phenomenon of changing rocks subjected to heat, pressure, and fluids so that they are in equilibrium with a new set of environmental conditions.

metasomatism   A metamorphic process during which an interaction with hot fluids results in a chemical change in the country rock.

Milankovitch theory   An explanation for the cyclic variations in climate and the onset of ice ages as a result of irregularities in Earth's rotation and orbit.

mineral   A naturally occurring, inorganic, crystalline solid that has characteristic physical properties and a narrowly defined chemical composition.

Modified Mercalli Intensity Scale A scale with values ranging from I to XII used to characterize earthquake intensity based on damage.

Moho See *Mohorovičić discontinuity*.

Mohorovičić discontinuity The boundary between the crust and mantle; also called the Moho.

monocline A bend or flexure in otherwise horizontal or uniformly dipping rock layers.

mud crack A crack in clay-rich sediment that has dried out.

mudflow A flow consisting of mostly clay- and silt-sized particles and more than 30% water; most common in semi-arid and arid environments.

N

native element A mineral composed of a single element (such as gold).

natural levee A ridge of sandy alluvium deposited along the margins of a channel during floods.

natural switch A sensitive aspect of the environment that triggers a rapid climate change, particularly a rapid change in near-surface temperatures.

nearshore sediment budget The balance between additions and losses of sediment in the nearshore zone.

neutron An electrically neutral particle found in the nucleus of an atom.

nonconformity An unconformity in which stratified sedimentary rocks overlie an erosion surface cut into igneous or metamorphic rocks. See also *angular unconformity* and *disconformity*.

nonferromagnesian silicate A silicate mineral that has no iron or magnesium.

nonfoliated texture A metamorphic texture in which there is no discernable preferred orientation of minerals.

normal fault A dip-slip fault on which the hanging wall block has moved downward relative to the footwall block. See also *reverse fault* and *thrust fault*.

nucleus The central part of an atom consisting of protons and neutrons.

nuée ardente A mobile dense cloud of hot pyroclastic materials and gases ejected from a volcano.

O

oblique-slip fault A fault showing both dip-slip and strike-slip movement.

oceanic–continental plate boundary A convergent plate boundary along which oceanic lithosphere and continental lithosphere collide; characterized by subduction of the oceanic plate beneath the continental plate and by volcanism and seismicity.

oceanic crust The crust underlying the ocean basins. It ranges from 5 to 10 km thick, is composed of gabbro and basalt, and has an average density of 3.0 g/cm³.

oceanic–oceanic plate boundary A type of convergent plate boundary along which two oceanic lithospheric plates collide and one is subducted beneath the other.

oceanic ridge A submarine mountain system composed of volcanic rock found in all the oceans.

oceanic trench A long, narrow feature restricted to active continental margins and along which subduction occurs.

ooze Deep-sea sediment composed mostly of shells of marine animals and plants.

ophiolite A sequence of igneous rocks representing a fragment of oceanic lithosphere; composed of peridotite overlain successively by gabbro, sheeted basalt dikes, and pillow lavas.

orogeny The process of forming mountains, especially by folding and thrust faulting; an episode of mountain building.

outwash plain The sediment deposited by the meltwater discharging from a continental glacier's terminus.

oxbow lake A cutoff meander filled with water.

oxidation The reaction of oxygen with other atoms to form oxides or, if water is present, hydroxides.

P

pahoehoe A type of lava flow with a smooth, ropy or billowy surface.

paired metamorphic belts Found at oceanic-continental convergent plate boundaries; consists of a belt of dynamically metamorphosed rocks, an accretionary wedge, and forearc basin rocks adjacent to a belt of island arc rocks changed by contact and regional metamorphism.

paleocurrent The direction of an ancient current as indicated by sedimentary structures such as cross-bedding.

paleomagnetism Remnant magnetism in rocks, studied to determine the intensity and direction of Earth's past magnetic field.

Pangaea The name Alfred Wegener proposed for a supercontinent consisting of all Earth's landmasses that existed at the end of the Paleozoic Era.

parabolic dune A crescent-shaped dune in which the tips point upwind.

parent material The material that is chemically and mechanically weathered to yield sediment and soil.

passive continental margin The trailing edge of a continental plate consisting of a broad continental shelf and a continental slope and rise. A vast, flat abyssal plain is commonly present adjacent to the rise. See also *active continental margin*.

pedalfer A soil formed in humid regions with an organic-rich A horizon and aluminum-rich clays and iron oxides in horizon B.

pediment An erosion surface of low relief gently sloping away from a mountain base.

pedocal A soil characteristic of arid and semiarid regions with a thin A horizon and a calcium carbonate–rich B horizon.

pelagic clay Brown or red deep-sea sediment composed of clay-sized particles.

pelitic rock A clay-rich sedimentary rock, or a metamorphic rock derived from pelitic rock, as in pelitic schist.

permafrost Permanently frozen soil, subsoil, or sediment found mostly in the Arctic but also in some high mountains.

permeability A material's capacity for transmitting fluids.

phaneritic texture Igneous rock texture in which minerals are easily visible without magnification.

pillow lava Bulbous masses of basalt, resembling pillows, formed when lava is rapidly chilled under water.

plastic flow The flow that occurs in response to pressure and causes permanent deformation.

plastic strain The result of stress in which a material cannot recover its original shape and retains the configuration produced by the stress such as folding of rocks.

plate An individual segment of lithosphere that moves over the asthenosphere.

plate tectonic theory The theory that large segments of the outer part of Earth (lithospheric plates) move relative to one another.

platform That part of a craton that lies buried beneath flat-lying or only mildly deformed sedimentary rocks.

playa A dry lakebed found in deserts.

plunging fold A fold with an inclined axis.

pluton An intrusive igneous body that forms when magma cools and crystallizes within the crust (such as a batholith).

plutonic (intrusive igneous) rock Igneous rock that crystallizes from magma intruded into or formed in place within Earth's crust.

point bar The sediment body deposited on the gently sloping side of a meander loop.

polymorphic transformation A reaction during metamorphism which brings about changes in a mineral's shape and size as well as its atomic structure.

porosity The percentage of a material's total volume that is pore space.

porphyritic texture An igneous texture with minerals of markedly different sizes.

porphyroblast Large crystal that grew in the solid state in a metamorphic rock.

pressure release A mechanical weathering process in which rocks that formed under pressure expand on being exposed at the surface.

primary wave See *P-wave*.

principle of cross-cutting relationships A principle used to determine the relative ages of events; holds that an igneous intrusion or fault must be younger than the rocks it intrudes into or cuts.

principle of fossil succession A principle holding that fossils, and especially assemblages of fossils, succeed one another through time in a regular and predictable order.

principle of inclusions A principle holding that inclusions, or fragments, in a rock unit are older than the rock unit itself (for example, granite fragments in a sandstone are older than the sandstone).

principle of isostasy The theoretical concept of Earth's crust "floating" on a denser underlying layer.

principle of lateral continuity A principle holding that sediment layers extend outward in all directions until they terminate.

principle of original horizontality A principle holding that sediment layers are deposited horizontally or very nearly so.

principle of superposition A principle holding that in a vertical sequence of undeformed sedimentary rocks, the relative ages of the rocks can be determined by their position in the sequence—oldest at the bottom followed upward by successively younger layers.

principle of uniformitarianism A principle holding that we can interpret past events by understanding present-day processes; based on the assumption that natural laws have not changed through time.

proton A positively charged particle found in the nucleus of an atom.

P-wave A compressional, or push–pull, wave; the fastest seismic wave and one that can travel through solids, liquids, and gases; also known as a primary wave.

P-wave shadow zone The area between 103 and 143 degrees from an earthquake focus where little P-wave energy is recorded by seismographs.

pyroclastic materials Fragmental substances, such as ash, explosively ejected from a volcano.

pyroclastic (fragmental) texture A fragmental texture characteristic of igneous rocks composed of pyroclastic materials.

pyroclastic sheet deposit Vast, sheetlike deposits of felsic pyroclastic materials erupted from fissures.

Q

quick clay A clay deposit that spontaneously liquefies and flows like water when disturbed.

R

radioactive decay The spontaneous change of an atom to an atom of a different element by emission of a particle from its nucleus (alpha and beta decay) or by electron capture.

rainshadow desert A desert found on the lee side of a mountain range; forms because moist marine air moving inland forms clouds and produces precipitation on the windward side of the mountain range so that the air descending on the leeward side is much warmer and drier.

rapid mass movement Any kind of mass movement involving a visible downslope displacement of material.

Rayleigh wave (R-wave) A surface wave in which the individual particles of material move in an elliptic path within a vertical plane oriented in the direction of wave movement.

recessional moraine An end moraine that forms when a glacier's terminus retreats, then stabilizes and till is deposited. See also *end moraine* and *terminal moraine*.

recrystallization The solid state formation of new crystals in a metamorphic rock; the new crystals may or may not have the same composition as the original crystals but in most cases they are larger than the originals.

reef A moundlike, wave-resistant structure composed of the skeletons of organisms.

reflection The return to the surface of some of a seismic wave's energy when it encounters a boundary separating materials of different density or elasticity.

refraction The change in direction and velocity of a seismic wave when it travels from one material into another of different density and elasticity.

regional metamorphism Metamorphism that occurs over a large area, resulting from high temperatures, tremendous pressures, and the chemical activity of fluids within the crust.

regolith The layer of unconsolidated rock and mineral fragments and soil that covers most of the land surface.

relative dating The process of determining the age of an event relative to other events; involves placing geologic events in their correct chronologic order but involves no consideration of when the events occurred in terms of number of years ago. See also *absolute dating*.

reserve The part of the resource base that can be extracted economically.

resource A concentration of naturally occurring solid, liquid, or gaseous material in or on Earth's crust in such form and amount that economic extraction of a commodity from the concentration is currently or potentially feasible.

retrograde metamorphism A partial reversal of metamorphic grade when heat decreases slowly but enough fluid is present to bring about change.

reverse fault A dip-slip fault in which the hanging wall block has moved upward relative to the footwall block. See also *normal fault* and *thrust fault*.

Richter Magnitude Scale An open-ended scale that measures the amount of energy released during an earthquake.

rill erosion Erosion by running water that scours small channels in the ground.

rip current A narrow surface current that flows out to sea through the breaker zone.

ripple mark Wavelike (undulating) structure produced in granular sediment such as sand by unidirectional wind and water currents or by oscillating wave currents.

rock An aggregate of one or more minerals, as in limestone or granite, or a consolidated aggregate of rock fragments, as in conglomerate; includes rocklike materials such as coal and natural glass.

rock cycle A group of processes through which Earth materials may pass as they are transformed from one rock type to another.

rockfall A common type of extremely rapid mass wasting in which rocks fall through the air.

rock-forming mineral A mineral common in rocks, which is important in their identification and classification.

rock slide Rapid mass movement in which rocks move downslope along a more or less planar surface.

rotated garnet Garnet porphyroblasts that show the effects of rotation as they grew in rocks undergoing deformation by thrust faulting.

rounding The process by which the sharp corners and edges of sedimentary particles are abraded during transport.

runoff The surface flow of streams and rivers.

R-wave See *Rayleigh wave*.

S

salt crystal growth A mechanical weathering process in which rocks are disaggregated by the growth of salt crystals that grow in crevices and pores.

scientific method A logical, orderly approach that involves gathering data, formulating and testing hypotheses, and proposing theories.

seafloor spreading The theory that the seafloor moves away from spreading ridges and is eventually consumed at subduction zones.

seamount A submarine volcanic mountain rising at least 1 km above the seafloor.

secondary wave See *S-wave*.

sediment Loose aggregate of solids derived from preexisting rocks, or solids precipitated from solution by inorganic chemical processes or extracted from solution by organisms.

sedimentary facies Any aspect of a sedimentary rock unit that makes it recognizably different from adjacent sedimentary rocks of the same, or approximately the same, age (e.g., a sandstone facies).

sedimentary rock Any rock composed of sediment (such as sandstone and limestone).

sedimentary structure Any structure in sedimentary rock formed at or shortly after the time of deposition (such as cross-bedding, mud cracks, and animal burrows).

seismic profiling A method in which strong waves generated at an energy source penetrate the layers beneath the seafloor. Some of the energy is reflected from the various layers to the surface, making it possible to determine the nature of the layers.

seismic risk map A map based on the distribution and intensity of past earthquakes; such maps indicate the potential severity of future earthquakes and are useful in planning.

seismic tomography A method of analyzing numerous seismic waves to develop a three-dimensional image of Earth's interior.

seismograph An instrument that detects, records, and measures the various waves produced by an earthquake.

seismology The study of earthquakes.

shear strength The resisting forces that help maintain a slope's stability.

shear stress The result of forces acting parallel to one another but in opposite directions; results in deformation by displacement of adjacent layers along closely spaced planes.

sheet erosion Erosion that is more or less evenly distributed over the surface and removes thin layers of soil.

sheet joint A large fracture more or less parallel to a rock surface resulting from pressure released by expansion of the rock.

shield A vast area of exposed ancient rocks on a continent; the exposed part of a craton.

shield volcano A dome-shaped volcano with a low, rounded profile built up mostly of overlapping basalt lava flows.

shock metamorphism An extreme type of dynamic metamorphism caused by a meteorite impact.

shoreline The area between mean low tide and the highest level on land affected by storm waves.

silica A compound of silicon and oxygen atoms.

silica tetrahedron The basic building block of all silicate minerals, it consists of one silicon atom and four oxygen atoms.

silicate A mineral that contains silica (such as quartz [SiO_2]).

sill A tabular or sheetlike concordant igneous intrusion.

sinkhole A depression in the ground that forms in karst regions by the solution of the underlying carbonate rocks or by the collapse of a cave roof.

slide Mass wasting involving movement of material along one or more surfaces of failure.

slow mass movement Mass movement that advances at an imperceptible rate and is usually only detectable by its effects.

slump Mass wasting that takes place along a curved surface of failure and results in the backward rotation of the slump mass.

soil Regolith consisting of weathered material, water, air, and humus that can support plants.

soil degradation Any process leading to a loss of soil productivity; may involve erosion, chemical pollution, or compaction.

soil horizon A distinct soil layer that differs from other soil layers in texture, structure, composition, and color.

solar nebula theory A theory for the evolution of the solar system from a rotating cloud of gas.

solifluction Mass wasting involving the slow downslope movement of water-saturated surface materials; especially the flow at high elevations or high latitudes where the flow is underlain by frozen soil.

solution A reaction in which the ions of a substance become dissociated in a liquid and the solid substance dissolves.

sorting A term referring to the degree to which all particles in a sediment deposit or sedimentary rock are about the same size.

specific gravity The ratio of a substance's weight, especially a mineral, to the weight of an equal volume of water; for example, the specific gravity of quartz is 2.65.

spheroidal weathering A type of chemical weathering in which corners and sharp edges of rocks weather more rapidly than flat surfaces, thus yielding spherical shapes.

spit A continuation of a beach forming a point of land that projects into a body of water, commonly a bay.

spring A place where groundwater flows or seeps out of the ground.

stock An irregularly shaped discordant pluton with a surface area of less than 100 km^2.

stoping A process in which rising magma detaches and engulfs pieces of the surrounding country rock.

strain Deformation caused by stress. See also *elastic strain* and *plastic strain*.

strata (stratification) Strata (singular *stratum*) are the layers in sedimentary rocks, whereas stratification is the layered aspect of sedimentary rocks. See also *bed* (*bedding*).

stratified drift Glacial deposits that show both sorting and stratification.

stratovolcano See *composite volcano*.

stream terrace An erosional remnant of a floodplain that formed when a stream was flowing at a higher level.

stress The force per unit area applied to a material such as rock.

strike The direction of a line formed by the intersection of a horizontal plane with an inclined plane, such as a rock layer.

strike-slip fault A fault involving horizontal movement so that blocks on opposite sides of a fault plane slide sideways past one another. See *dip-slip fault*.

submarine canyon A steep-walled canyon best developed on the continental slope but some extend well up onto the continental shelves.

submarine fan A cone-shaped sedimentary deposit that accumulates on the continental slope and rise.

submarine hydrothermal vent A crack or fissure in the seafloor through which superheated water issues. See *black smoker*.

submergent coast A coast along which sea level rises with respect to the land or the land subsides.

superposed stream A stream that once flowed on a higher surface and eroded downward into resistant rocks while still maintaining its course.

surface wave Earthquake waves that travel along Earth's surface. Rayleigh (R-) and Love (L-) waves are both surface waves.

suspended load The smallest particles carried by a stream, such as silt and clay, which are kept suspended by fluid turbulence.

S-wave A shear wave that moves material perpendicular to the direction of travel, thereby producing shear stresses in the material it moves through; also known as a secondary wave.

S-wave shadow zone Those areas more than 103 degrees from an earthquake focus where no S-waves are recorded.

syncline A down-arched fold in which the youngest exposed rocks coincide with the fold axis and all strata dip toward the axis.

system A combination of related parts that interact in an organized fashion. Earth systems include the atmosphere, hydrosphere, biosphere, and solid Earth.

T

talus An accumulation of coarse, angular rock fragments at the base of a slope.

tension A type of stress in which forces act in opposite directions but along the same line, thus tending to stretch an object.

terminal moraine An end moraine marking the greatest extent of a glacier. See also *end moraine* and *recessional moraine*.

terrain analysis The analysis of a particular part of Earth's surface to evaluate its effect on some kind of operation, commonly a military operation.

terrane A block of rock with characteristics quite different from those of surrounding rocks. Terranes probably represent seamounts, oceanic rises, and other seafloor features that accreted to continents during orogenies.

terrestrial planet Any of the four innermost planets (Mercury, Venus, Earth, and Mars). They are all small and have high mean densities, indicating that they are composed of rock and metallic elements. See also *Jovian planet*.

theory An explanation for some natural phenomenon that has a large body of supporting evidence. To be considered scientific, a theory must be testable—for example, plate tectonic theory.

thermal convection cell A type of circulation in the asthenosphere during which hot material rises, moves laterally, cools and sinks, and is reheated and continues the cycle.

thermal expansion and contraction Mechanical weathering in which the volume of rock changes in response to heating and cooling.

thermocline The boundary between warm seawater and colder water at depth.

thrust fault A type of reverse fault with a fault plane that dips less than 45 degrees.

tide The regular fluctuation in the sea's surface in response to the gravitational attraction of the Moon and Sun.

till All sediment deposited directly by glacial ice.

tombolo A spit that extends out into the sea or a lake and connects an island to the mainland.

transform fault A fault along which one type of motion is transformed into another; commonly displaces oceanic ridges, but movement on opposite sides of the fault between displaced ridge segments is the opposite of the apparent displacement; on land recognized as a strike-slip fault, such as the San Andreas fault.

transform plate boundary Plate boundary along which plates slide past one another and crust is neither produced nor destroyed; on land recognized as a strike-slip fault.

transverse dune A long ridge of sand perpendicular to the prevailing wind direction.

tree-ring dating The process of determining the age of a tree or wood in structures by counting the number of annual growth rings.

trough The lowest point between wave crests.

tsunami A sea wave that is usually produced by an earthquake but can also be caused by submarine landslides or volcanic eruptions.

turbidity current A sediment–water mixture, denser than normal seawater, that flows downslope to the deep seafloor.

U

unconformity A break in the geologic record represented by an erosion surface separating younger strata from older rocks. See also *angular unconformity, disconformity,* and *nonconformity.*

U-shaped glacial trough A valley with steep or vertical walls and a broad rather flat floor; formed by the movement of a glacier through a stream valley.

V

valley A linear depression bounded by higher areas such as ridges, hills, or mountains.

valley glacier A glacier confined to a mountain valley or to an interconnected system of mountain valleys.

valley train A long, narrow deposit of stratified drift confined within a glacial valley.

vein An irregular, sheetlike intrusion no more than a few centimeters thick that may be concordant or discordant; many veins are composed of quartz or quartz and potassium feldspar.

velocity A measure of the distance traveled per unit of time, as in the flow velocity in a river or stream.

ventifact A stone with a surface has been polished, pitted, grooved, or faceted by wind abrasion.

vesicle A small hole or cavity formed by gas trapped in cooling lava.

viscosity A fluid's resistance to flow.

volcanic A term referring to igneous rocks that formed from magma that reached the surface as lava or pyroclastic materials.

volcanic explosivity index (VEI) A semiquantitative scale for the size of a volcanic eruption based on evaluation of such criteria as volume of material explosively ejected and height of eruption cloud.

volcanic neck An erosional remnant of the material that solidified in a volcanic pipe.

volcanic pipe The conduit connecting the crater of a volcano with an underlying magma chamber.

volcanic (extrusive igneous) rock An igneous rock formed when magma is extruded onto Earth's surface where it cools and crystallizes, or when pyroclastic materials become consolidated.

volcanic tremor Ground motion lasting from minutes to hours resulting from magma moving below the surface, as opposed to the sudden jolts produced by most earthquakes.

volcanism The process whereby magma and its associated gases rise through the crust and are extruded onto the surface or into the atmosphere.

volcano A mountain formed around a vent as a result of the eruption of lava and pyroclastic materials.

W

water table The surface that separates the zone of aeration from the underlying zone of saturation.

water well A well made by digging or drilling into the zone of saturation.

wave An undulation on the surface of a body of water, resulting in the water surface rising and falling.

wave base A depth of about one-half wavelength, where the diameter of the orbits of water in waves is essentially zero; the depth below which water is unaffected by surface waves.

wave-cut platform A beveled surface that slopes gently seaward; formed by the retreat of a sea cliff.

wave height The vertical distance from wave trough to wave crest.

wavelength The distance between successive wave crests or troughs.

wave period The time required for two successive wave crests (or troughs) to pass a given point.

wave refraction The bending of waves so that they more nearly parallel the shoreline.

weathering The physical breakdown and chemical alteration of rocks and minerals at or near Earth's surface.

Z

zone of accumulation The part of a glacier where additions exceed losses and the glacier's surface is perennially covered with snow. Also refers to horizon B in soil where soluble materials leached from horizon A accumulate as irregular masses.

zone of aeration The zone above the water table that contains both water and air within the pore spaces of the rock or soil.

zone of leaching Another name for horizon A of a soil; the area in the upper part of a soil where soluble minerals are removed by downward moving water.

zone of saturation The area below the zone of aeration in which all pore spaces are filled with groundwater.

zone of wastage The part of a glacier where losses from melting, sublimation, and calving of icebergs exceed the rate of accumulation.

INDEX

Note: Italic page numbers refer to illustrations.

A

Aa lava flows, 137, *137*, 160–161
Abrams, Harry N., 141
Abrasion, *138*, 202, 245, 465, 541, 570, 573, 607
    and glaciers, 541
    running water, 465, *465*
    wind erosion, 570
Abrasives, 30, 254
Absolute ages. *See* Numerical ages
Absolute dates determination, for sedimentary rocks, 288
Absolute dating
    definition of, 262
    isotopes and, 281, 284
    methods of, 265–274
    radioactive decay and, 281–282
    refinements in, 262
Abyssal hills, 374, 383
Abyssal plains, 369, 374, 382
Acadian Orogeny, 67
Acasta Gneiss, 413
Accessory minerals, 90
Accretionary wedge, 406
Accurate radiometric dates, 286
Acid rain, 6, 196
Acid runoff, 178, *179*
Acid soils, 178
Acidic solutions, 178
Actinolite, 236, 251
Active continental margins, 367, *368*, 382–383
Active volcanoes, 136
Adirondack Mountains, 404
Aegean Sea, 164
African plate, 156
Aftershocks, 300
Agassiz, Louis, 553–555
Agate, *70–71*
Agricultural revolution, 197
Ahorn, Lukas, 227
Air pressure, 580–581, 591
Airy, George, 349–350
Alaska earthquake (1964), 313, 314, 439, 443
Alaska Peninsula, 112
Alaska pipeline, 441
Alaska Tsunami Warning Center, 333, *333*
Alaska Volcanoes Observatory, 152
Alaskan volcanoes, 152–153
Alberta, Canada, 467
Aleutian Islands, 54, 130, 152–153
Aleutian Trench, Alaska, 415
Algae, 175, 376
Alighieri, Dante, *338*
Alkaline, 178
Alluvial fans, 473, 490, 587, *588*, 589, 591
Alluvium, 467
Alpha decay, 281, *282*
Alpha particle X-ray spectrometer, 69
Alpine glacier, 533
Alpine–Himalayan orogenic belt, 406
Alps, 55, *245*, 342, 390, 405, 533
Aluminum (Al), *8*, 16, 19, 72–73, 125, 183, 191–192, 219
Aluminum ore, 191–192
Alvin, 362, 373, 387
Amalfi Coast, *230–231*
Amazon River, South America, 462, 464
Amber, 73, *73*, 94
American Association of Petroleum Geologists, 35, 218, 228
American Institute of Professional Geologists, 228
American River, California, *86*
American Samoa, 378
Ammonia, 12–13
Ammonites, *216*
Amoco Production Company, 228
Amorphous, 77
Amphibole asbestos, 258
Amphiboles, 81, 120, 178
Amphibolite, 244
Amphibolite facies, 248
Anchorage, Alaska, 140, *443*
Ancient rifting, 52, *52*, 63
Ancient ruins, 421
Ancient Seafloor, San Francisco, 385
Andalusite, 243
Andes Mountains, South America, *53*, 55, 61, 119, *388–389*, 390, 405, 407, *409*, 533
Andesite, 55, 119–120, 122, 126, *154*
Andesitic volcanoes, 413
Andrews, Sarah, 30
Angle of repose, 427
Angle of slope, 187
Angular unconformity, 267, 271, *272*, 274, 292
Anhydrite (CaSO$_4$), 206, *209*
Antarctic ice shelves, *643*
Anthracite, 209, 247
Anticlines, 394, *395*, 395–396, *396*, 397, 419
Antillean island arc, 54
Apatosaurus, 295
Aphanitic rocks, *119*
Aphanitic texture, 115, 126
Aphelion and perihelion, 558
Aphyric, 118
Aplites, 122
Appalachian Mountains, 18, 36–37, 55, 390, 404–405
Appleton, Wisconsin, 494
Aqua alta, 626
Aquifers, 499–500
Arabian Desert, Middle East, 582
Arabian Peninsula, *48*, 49
Arabian plate, 409
Aragonite, 83
Archean Eon, 390
Arches National Park, Utah, *168–169*, 171, 176–177
Archimedes, 350
Arctomys Cave, British Columbia, 507
Arêtes, 546, 560
Argentite (Ag$_2$S), *87*
Argo, 362
Argon, 75, 281
Arid regions, 581
Arkose, 205, *206*
Arsenic, toxic element, 527
Artesian-pressure surface, 503
Artesian system, 503–505, 524
Asbestos, 236, 254, 258
Aseismic ridges, 57, 374, 383
Ash, 19–20, 110, 139, *140*, 150, *157*, 160–161
Ash falls, 139, 287, 422
Ash flow, 139
Ashland city, Tennessee, 475
Asia plate, 409
Assimilation, 107, 126
Asteroid belt, 12
Asteroids, 11–12, 25
Asthenosphere, *17*, *23*, 25–26, 46, 59
    definition of, 13, 16
    lithosphere and, *18*
    low-velocity zone and, 343, 355
    plate boundaries and, *18*
Astronomy, 21
Atbara River, Sudan, 493
Athabaska tar sands, Alberta, 220
Athens, 360
Atlantic City, New Jersey, 621
Atlantic Ocean, 18, 35–38, 54, 151, 306, 360–361, 373
Atlantic Ocean basin, 370
Atlantis, *360*, 360
Atmosphere, 13, 50
    circulation patterns of, 557, 580, *580*
    greenhouse effect and, 9–10, 31
    increase in temperature, 25
    plate tectonic theory and, 18
    shoreline processes, 599
    as subsystem of Earth, 4, 25
Atmospheric pollution, 196
Atmospheric pressure, *241*
Atolls, 377–378, 382
Atomic mass number, 74, 94–95, 281
Atoms, 11, 19, 73–75, *74*, 77, 94–95, 281, 291
Atrazine and alachlor, herbicides, 648
Atrypa, 280
Au Train River, Michigan, 482
Augite crystal, cleavage in, *88*
Augustine volcano, Alaska, *147*
Aureole, 238, 256
Avalanches, 319, 430
Axial tilt, 558
Axial volcano, *151*
Ayers Rock, 276, 588
Azores, 135, 360, 371

B

Back-arc basin, 54, *177*, 407
Back reef deposits, *381*
Background radiation, 11, 13
Backshore, 609
Baffin Island, Canada, 534, *534*, 550
Bahamas, 360
Baja California, 49, 415, 450
Bajada, 587
Balanced rocks, 176, *177*
Balboa Boulevard, Northridge, *317*
Balkan region, 343
Ballard, Robert D., 387
Baltic Sea, 73
Baltis Vallis, *50*
Banda Aceh shore, *320*
Banded iron formation (BIF), *221*, 221–222, 223
Banff-Jasper National Park, Alberta, 538
Bangkok, Thailand, 515
Barchan dunes, 575–577, 580, 591
Barrier islands, 202, *614*, 614–615
Barrier reefs, 377
Barrow, George, 248–249
Basal slip, 536, 538
Basalt (B), 16, *22*, 35, 103, 118–119, *119*, 120, 122, 125–127, *139*, 153, 160, 244, 342–343, 360, 370
Basalt columns, *100–101*
Basalt dikes, 375
Basalt lava flow, *103*, *163*, *182*

677

Basalt plateaus, 149, *150*, 160–161
Basalt porphyry, 115–116
Basaltic lava, *44*
Basaltic magma, 47, *107*
Base level, 480, *482*
Basins, 342, 397, *400*, 419
Bates, L. W., Jr., 618
Batholiths, 122, 126, 156, 266, *293*
Battle of Little Big Horn, Montana, 87
Battle of Shiloh, 262
Bauxite, *8*, *191*, 191–192
Bay of Fundy, Nova Scotia, 601
Bay of Naples, Italy, *134*, 135
Baymouth bar, 613
BBC News, 333
Beach face, 609
Beaches, 202, 610–611
    definition of, 608
    erosion and, 612, 616, 621
    rising sea level and, 621
    seasonal changes in, *612*, 612–614
Bed load, 490, 569, 591
Bedded chert, 208, *209*, 375
Bedding, 213, 223
Bedouin, 7
Bedrock, 37, 103, 124, *183*, *185*, 187, 315, 386, 427, 428, 431, 443, 447, 448, *450*, 485, 541, 547, 588
Beds, 212
Bellini, Giovanni, 7
Ben-Avraham, Zvi, *413*
Benching, 448, *449*
Benioff zones, 52, 304, *305*, 370
Bentonite, *30*, 30
Bering glacier, Alaska, 533
Beringia, 555
Berms, 609
Bermuda, 360
Beryllium, 121
Beta decay, 281, *282*, 286
Big Bang, 10–11, 13, 25–26
Bighorn Mountains, Wyoming, 395, *402*
Bighorn River, Wyoming, 488
Billiard table, 257
Biochemical sedimentary rocks, 206, 210, 223
Biosphere
    plate tectonic theory and, 18
    as subsystem of Earth, 4, 25
    weathering and, 19
Biotite, 82–83, 88, *88*, 111, 120, 122, 178, 201, *241*, 244, 248
Bird's-foot deltas, 472
Birmann, Samuel, 7, 532
Bishop Tuff, California, 150
Bituminous coal, 209, 220–221
Black Canyon, Gunnison River, 485
Black Hills, South Dakota, *87*, 121, 397, 504, 520
Black pearls, 73
Black sand beach, Maui, Hawaii, *611*
Black smoker, *372*, 373, 383
Blake Plateau, 379
Block diagram, *274*
Block faulting, 404
Block ice, 180
Block slide, 434
Blocks, 140, 160
Blowouts, 572
Blue Haze Famine, 136, 166
Blue Mountains, 404
Blue Nile, Ethiopia, 493
Blueberries, 69
Blueschist facies, 240, 406
Blush, 98
Body fossils, 213
Body waves, 307–308, 328–329
Bolivia earthquake (1994), 370
Bolshoi Semiachik Volcano, Kamchatka, *104*
Bombs, 140, 160
Bonding, 73, 75–77, 94–95
Bone Valley Formation, Florida, 384
Boone, Daniel, 635
Bootlegger Cove Clay, 439, *443*
Bottomset beds, 470
Bowen, Norman L., 110
Bowen's reaction series, 110, *111*, 115–116, 120, 125–126, 180

Bowling Green, Kentucky, *510*
Braided stream, 467, 474
Breaker zone, 603–604
Bright Angel shale, *212*
British Isles, 229
Brittle rocks, 392, 418
Broad plateaus, 362
Bronze Age Eruption of Santorini, 164
Brown, G. C., *343*
Bryce Canyon National Park, Utah, 278, *279*
Bullard, Edward, Sir, 36
Bulldozing, 541
Buoys, deep-ocean, *333*, *333*
Burren region, 508
Buried lava flow, *268*
Burrows, *216*
Bush, George W., 333, 593
Bushveld Complex, South Africa, *241*
Buttes, 588, *590*
Bylot Island, Nunavut, *552*
Bypasses, 479

C

Calcareous ooze, 376
Calcite (CaCO$_3$), *22*, 83, *88*, 89, 94, 165, 178, 183, 201, 206, 233
Calcium (Ca), 72, 179, 206, 518
Calcium carbonate (CaCO$_3$), 77, 83, 91, 204, 206, *208*, 215, 222, 288, 376
Calderas, 141, 161
Calgary, Canada, *548*
California beach, *611*
California gold rush (1848–1853), 219
California Institute of Technology, 313
Calving, 535
Camas Prairie, Montana, *557*
Cambrian Period, 284
Campeche Knolls, 365
Canadian shield, 414, 419
Caniapiscau River, Quebec, 226
Canyons, 277, 484
Capacity, measure for total load, 466
Cape Cod, Massachusetts, 564
Cape May, New Jersey, *610*
Capillary fringe, 501
Capone, Al, 385
Carbon-14 dating technique, 285–286, *286*, 287–288, 292
Carbon (C), 74–76, 78, 89, 209, 215, 379
Carbon dioxide (CO$_2$), 9, 10, 29, 75, 107, 115, 136, 140, 178, 253
Carbon isotopes, *75*, 75, 286, 288
Carbon monoxide, 136
Carbonate (CO$_3$), 84, 90, 206, 253
Carbonate ion, 238
Carbonate minerals, 83–84, 94–95
Carbonate rocks, 178, 206, 210, 222–223
Carbonates calcite (CaCO$_3$), 90
Carbonatite, *128*, 128
Carbonic acid, 178, 505
Carboniferous age, 37
Caribbean Sea, 145
Carlsbad Caverns, New Mexico, 178, 507, 647
Carmel formation, *279*
Carnegie Institute, Washington DC, 110
Carnotite, 221
Carrizo Plain, California, *57*
Cascade Range, 119, 144, 152–153, 154–155, 158, 160–161, 204, 479
Cascades Volcano Observatory, Vancouver, 153
Cast, 215, *216*
Castle Valley, Utah, 586
CAT scans. *See* Computerized axial tomography
Cataclastic, 240
Catastrophes, 356
Catskill Delta, New York, 67
Catskill Mountains, New York, 67, *67*
Cave, 507, 512, 524
    deposits, *512*
    formation, 506, *511*
Cavendish, Henry, 337
Cavern, 507
Cementation, 19, *20*, 204, 222–223
Cemeteries, 195
Cenozoic Era, 66, 278, 415
Central Park, New York, 28, *28*, 563

Central rift, *371*
Central Sea, 337, *339*
Centre for Forensic Science, 30
Ceramics, 92, 219
Cervantes, Miguel, 595
Cesium, 121
Chalk, 206, *208*, 243
Challenger Deep, 370
Channel flow, 462
Chaos crags, *147*
Chaos Jumbles, *147*
Chapel of Saint Michel d'Aiguilhe, 129, *129*
Charleston earthquake (1886), 306, *306*, 369
Checkerboard Mesa, 218
Chemical bonds, types of, 75, 94
Chemical deterioration, 188, 190
Chemical sediment, 201
Chemical sedimentary rocks, 204–209, 210, 223
Chemical weathering, 171, *174*, *181*, 193, 431
    climate and, *180*, 180
    decomposition of Earth materials, 175, 178–181
    definition of, 175
    mechanical weathering and, 192
    organisms and, 174–175
    oxidation and, 178
    particle size and, *180*
    processes, 175
    rate of, 179
Chert, 208, *209*, 210
Cheyenne legend, *129*
Chief Keoua, 141
Chief Mountain, *410*
Chippewa River, Ontario, Canada, 465
Chlorine (Cl), 75, 76, 79, 84, 109, 115, 136, 206
Chlorine monoxide molecule (ClO), 166
Chlorite, 236, 248, 251
Chlorofluorocarbons (CFCs), 9, 166, *166*
Chrysotile, specimen of, *258*
Cinder cones, *103*, 141–142, 144, *144–145*, 148, 152, 160–161
Circum-Pacific belt, 151, 161, 304, 307, 406
Cirques, 546
Civil War, 87
Civilization, prerequisites for, 644
Clams, 73, 91, 201, *216*, 218
Clark Fork River, Idaho, 555
Clay, *8*, *8*, 81, 92, 179, 182, 191–192, 202, 219
Clay minerals
    chemical weathering and, 175
    definition of, 81
    formation of, 91, 175, 178
    metamorphic rocks and, 238
    as natural resource, 191
    residual concentrations and, 191
    sheet structure of, 81, 83
    soil and, 182, 183, 188
Clay-rich rocks, 242
Claystone, 205
Cleavage, 77, 88–89, *88*, 95
Cleetwood Cove, *109*
Cleopatra's Needle, New York City, *28*, 28
Climate, 187
    changes in, 288, 289–290, 290, 292
    chemical weathering and, 180
    glaciers and, 553, 563, 564
    mass wasting and, 427
    Pleistocene, 554
    short-term climatic events, 558
    soil formation and, *184*
    weathering and, 427–428
Climate Treaty Meeting, 29
Climatologic evidence, 35, 37
Closed system, 284
Cloud formation and condensation process, 459
Coal, *8*, 8, 37, 208–209, 220–221, 295, 379–382, 435
    beds, 35
    deposits, 35, 38
    layers, 34
    origin of, 210
    swamp flora, 34
Coal miners, 398
Coast, 598, 623
    classification of, 617–618, 623
    depositional, 617
    erosional, 617

Coastal areas, managing, 620
Coastal flood, 459
Coastal plutons, 113
Coastal upwelling, 384
Cobalt, 72, 92
Code of Hammurabi, 634
Coke, 221
Cole, Thomas, 7
Colorado Plateau, 221, 278, *279*
Colorado River, 262, 448
Columbia River basalt, 149, *150*
Columbium, 92
Columbus, Christopher, 361–362
Columnar joints, 138, *138*, 140, 161
Comets, 11–12, 25, 413
Common lava, 125
Compaction, 19, *20*, 203–204, 223
Competence, of stream, 466
Complex movements, 443, *445*, 446, 451
Complex pegmatites, 121
Composite volcanoes, 141, 144, *146*, 156, 161
Compound, 75, 95
Compression, 391, 418–419
Computerized axial tomography (CAT) scans, 340
Comstock Lode, North America, 87
Concordant plutons, 122, 126
Concurrent range zones, 280, *280*
Condie, Kent C., *416*
Conduit, 124
Cone of ascension, 513
Cone of depression, 503, *503*, 505, 524
Conglomerate, 22, 204–205, *205*, 222, 225
Coniferous trees, 73, *73*, 217
Consolidation, 20
Constancy of interfacial angles, 78
Constructive volcanism, 135, 160
Contact metamorphism, 237–238, *239*, 241–242, 255–256
Continental accretion, 390, 409, 414, 419
Continental crust, 16, *23*, 25, 46, 342–343, 351–352, 355, 360, 409, 417
 characteristics of, 16, 232, 342
 composition of, 336, 342
 continental rise and, 365
 mafic magma and, 107
 origin of, 413, *414*
 plates and, 46
 principle of isostasy and, 349, 360
 silica and,
Continental drift, 34–36, 62–64
 evidence for, 36–39, *39*, 39, 63
 fossil evidence and, 38–39
 glacial evidence and, 37
 hypothesis, 35–36
 seafloor spreading and, 43
Continental environments, *202*
Continental fit, *36*, 36
Continental glaciers, 533, *535*, 538
 and erosional landforms, 547
Continental lithosphere, 408
Continental margins, 364–365, *365*, 364–369, 382, 383
 definition of, 369
 types of, 367, 369
Continental nuclei, 413–414
Continental plates, 409
Continental rise, 365, 382–383
Continental shelf, *23*, 202, 365, 369, *379*, 383
Continental slope, 36, 39, 365, 369, 383
Continental–continental plate boundaries, 52, *53*, 54, 64, 406–409, *412*
Continental–continental plate convergence, 418
Continent–continent collision, 409
Continents, 13, 34, 37
 formation and evolution of, 413–417, *418*
Continuing evolution of North America, 415, *417*
Continuous branch, 110
Contour plowing, *190*
Contraction, 173
Convection cells, 16
Convergent plate boundaries, 55, 64, *250*, 304, 390, 406, 410
 copper deposits and, *62*
 definition of, 52
 igneous activity at, 153, 156

types of, 52, *53*, 63
volcanoes at, *151*
Cook Islands, 73, *195*, 195, *380*
Cook, James, 362
Copernicus, 634
Copper, 8, *8*, 72, 92, 191, 219
 convergent plate boundaries and, *62*
 deposits, 61
Coprolite, *216*
Coquina, 206, *208*
Coral reefs, 91, 201, 219, 376
Cordierite, 236
Core, 13, 25–26, 40–41, 50, 342, 355
 composition of, 345–346
 as concentric layer of Earth, 13, 16
 definition of, 13, 344
 density of, 345–346
 discovery of, 345
 as subsystem of Earth, 4, 25
Core–mantle boundary, 345
Coriolis effect, 581
Correlation, 278–281, 291–292
Corrosion, 92, 607
Country rock, 107, 126
Covalent bonding, 75, *76*, 76–77, 95
Crater Lake, Oregon, *109*, 141, *142*, 150, 164
Crater Mountain, California, *143*
Crater of Diamonds State Park, Arkansas, *357*, 357
Craters, 12, 141, 161
Cratons, 413–414, 417–419
Crazy Horse Memorial, *102*, 103
Creep, 441, 443, *445*, 446, 451
Crest, 602
Cretaceous Period, 66, 295
Crevice Cave, Missouri, 288
Crichton, Michael, 297
Cross-bedding, 213, *213*, 218, 223, 227, 291
Cross-dating, 286
Crown Jewels, *97*, 97
Crushed ice, 180
Crust, 13, *13*, 25–26, 50, 342–343, 354–355. *See also* Continental crust; Oceanic crust
 atmosphere and, 9
 as concentric layer of Earth, 13, 16
 definition of, 16
 elements in Earth's, *80*
 fracturing of, 55
 thinning of, 55
Crust–mantle boundary, 342–343, 354
Crystal form, 77, 85, 88, 94–95
Crystal moonbeams, 96
Crystalline rocks, 173, 200
Crystalline solids, 72, 77, 79, 94–95, 109
Crystallization, *20*, 283
Crystals
 gravitational differentiation of, 112, *112*
 shapes of, 233, *233*
Cubic crystals, 78
Cumberland Falls State Resort Park, Kentucky, 482
Cumberland Gap, Kentucky, 487, 635
Cumulates rocks, 118
Curie, Marie, 281
Curie, Pierre, 281
Curie point, 40–41, *42*, 63–64
Current ripple marks, 213, *214*, 225
Custer, George Armstrong, 87
Cut-and-fill method, 448, *449*
Cut bank, 468
Cynognathus, 38, *39*
Cyprus, 62

D

Dacite, *154*
Daggett Rock, Maine weigh, 547
Dahl, Arlene, *339*
Dalradian schists, Scotland, 248
Dams, geologic consequences of building, 637, 639, 644
Dana Point, California, 332
Darwin, Charles, 17, 264, 281, 378, 634
Darwinian evolution, 265
Darwin's hypothesis, 378
Dates, 262
Dating techniques, high-precision, 290

Daughter element, 282
De Buffon, Georges Louis (1707–1788), 263
Dead Horse State Park, Utah, *487*
Dead Sea, 206
Death Valley National Park, California, *459*, 484, 571, *572*, 588
Debris avalanche, 443, *446*
Debris flows, 438, *441*, 446, 451
Deception Island, 151
Dedham Granodiorite, 130
Deep canyons, 362
Deep-diving craft, 360
Deep erosion, 124
Deep focus earthquakes, 303
Deep-ocean basins, 369–373, 374–375, 382
Deep oceans, 382
Deep-Sea Drilling Project, 43, 63, 361
Deep-sea sediments, 45, 364
 cores of, 45
 pillow lavas and, 52, 55, 63
Deep seafloor, *370*, 375–376, *377*, 382
Deep-water waves, 603
Deflation, 571, 573, 591
Deflation hollows, 572
Deforestation, 29, 189, *638*
Deformation
 definition of, 390
 effects of, *390*
 faults, 400–403
 folded rock layers, 394–397
 geologic structures and, 390, 395–397
 joints, 398–400
 mountain building, 404–409, 412–413
 occurrence of, 390–391
 stress and, *392*
 strike and dip, 393
Deformed sedimentary rocks, 413
Delaware and Chesapeake bays, 618
Delicate Arch, *176*
Deltas, 202, 470, 473
Denali National Park, Alaska, *467*
Dendritic drainage, 479, *481*
Density, 89, 95, 340
Density contrast, 107
Denver, Colorado, 295, *295*
Denver Museum of Nature and Science, 295
Deposition
 characteristics of, 202
 glaciers and, 532, 533
 groundwater and, 505–507, 512
 rock cycle and, 19, *20*, 23
 running water and, 467–473, 552
 sedimentary facies and, 210–211
 sedimentary structures and, 213
 shelf–slope break and, 365
 shorelines and, 608–609, 612–615
 surface processes and, 170, 171
 waves and, 602–603
 wind and, 574–578
Depositional coasts, 623
Depositional environment, 201–202, *202*, 222–223
Depressions, 512
Depth, 362
Deranged drainage, 479, *481*
Descartes, René, 11
Desertification, 568, *568*, 569, 591
Deserts, 7, 9, 202, 276, 581, 586, 591
 characteristics of, 582–586, 591
 dune deposit, *213*, 218, 223
 landforms of, 587–588, 591
 pavement, 573
 plants, 583
 processes operation, 569, 591
 soils, 583
Destructive volcanism, 135, 160
Detrital sedimentary rocks, 201, *203*, 204–205, 210, 223
Devil's Backbone, *105*
Devil's Postpile National Monument, California, *138*, 5
Devil's Tower, 129, *129*
Devonian rocks, 67
Diamond-anvil pressure, 345
Diamond (C), 72, 78, 84–85, 88, *88*, 92, 94, 96, 191, 357
Diatomite, 219

Differential erosion, 171
Differential pressures, 239, 242, 255–256
Differential weathering, *171*, 171, 193
Dike-filling breccia, *105*
Dike slices, *104*
Dikes, 47, 123–124, *124*, 125–126, 160, 271
Dilatancy model, 325–326, 328
Dinosaurs
   bones, *216*
   duck-billed, 226
   egg clutches, 297
   fossils, 218
   Jurassic, 226
   plant-eating, 295
Diorite (D), 110, 119–120, 121–122, 126–127
Dioxide (SiO$_2$), 222
Dip-slip faults, 402–403, 410, 418–419
Dip-slip movements, 403
Dipping, 304, *305*, 393, *393*, 395, 397, 419
Direction of a slope, 187
Disappearing streams, 506
Discharge, 464
Disconformity, 267, *271*, 271, 274, 292
Discontinuity, 340, 355
Discontinuous branch, 110
Discordant plutons, 122, 126
Displaced terrane, 409
Dissolution seams, 237, 256
Dissolved load, 466, 490
Divergent plate boundaries, 49, 55, 63–64, 347, 360, 410
   cross section of, *48*
   definition of, 47
   igneous activity and, 153
   volcanoes at, *151*
Divides, 479
Dodson, Janice, 30
Dodson, John, 30
Dolomite [CaMg(CO$_3$)$_2$], 84, *88*, 90, 94, 165, 206
Dolostone, 19, 84, 178, 206–207, 210, 219, *247*, 249, 518
Domes, 397, *400*, 419
Dormant volcanoes, 136
Double Arch, *177*
Double refraction, 89
Downcutting, 484
Downwelling, 384
Dowsing, 529
Doyle, Arthur Conan, Sir, 30
Drainage basins, 447, 479, *480*, 490
Drainage patterns, 479
Drainpipes, 447, *448*, *450*
Drill hole, deepest, 337, 347
Drill Rig Storage, *337*
Drip curtain, 512
Droppings, *216*
Dropstones, 507, 553
Droughts, 7, 9, 31
Drowned coast, 617
Drumlin fields, 550
Drumlins, 548–550, 553, 560
Dry climates, 581, 582
Dry fog, 136
Dry rocks, 109
Du Toit, Alexander, 35–36, 38
Ductile rocks, 392, 418
Dune Migration, 575
Dunes, 574, 578, 591
Dunite rocks, 118
Dunton Pegmatite, Maine, *121*
Durand, Asher Brown, 7, *7*, 25
Dust Bowl, 194, *194*
Dynamic equilibrium, 427
Dynamic metamorphism, 239–240, 242, 255–256
Dynamothermal metamorphism, 241

E

Early exploration, 361
Earth, 12–14, *15*, 50. *See also* Core; Crust; Geologic time; Mantle
   age of, 263–264, 291
   common minerals in, 90
   composition and density of, 336
   concentric layers, 13, 25
   cross section of, *13*
   crust, 342–343, 354
   density of, 337–341, 353
   diameter and radius of, 337
   as dynamic and evolving planet, 13, 16, 25
   formation of, 12–13, 25
   history of, 13, 17–18
   as homogeneous, 341
   internal heat of, 336, 346–347, 354, 390
   internal processes of, 353
   internal structure of, *336*, 337–341
   as internally active planet, 418
   magnetic field of, 39, *40*, 40–41, *42*, 43–44, 63, *352*, 352–353, 354
   mantle, 343–344, 354
   mass, 343
   moon of, 337
   oceans and seas coverage in, 458
   overall density of
   past geography, 34, 62
   place in cosmos, 338–339
   place in solar system, 10–13, 25
   plate tectonic theory and, 18
   plates, *17*, 46
   plates movement, *16*
   simple model of, 46
   size of, 337–341, 353
   structure of, 353
   subsystems of, 4, *5*
   systems interaction, 135
   temperature of, 347
   as terrestrial planet, 337, 458–459, 490
   volcanoes, 151
   warming, 9
   water on, 459–462
Earth Institute at Columbia University, 4
Earthflows, 438, *442*, 446, 451
Earthquakes, 13, 16–17, 19, 329, 410
   average number of, 313
   control, 326, *327*, 329
   deep focus earthquakes, 303
   definition of, 300–301
   destructive effects of, 314–319, *317*, 323, 328
   distance from, *309*
   distribution of, 328
   epicenter of, *303*, 303–304, *310*, 320, 328, 366
   focus of, *303*, 303–304, 328
   frequency and distribution of, 304, 306–307, 328
   in Indian Ocean (2004), 4
   intensity, 311
   intermediate focus earthquakes, 303
   location of, 308–310, 328
   magnitude of, 313, 323
   mass wasting and, 431
   measurement of, 311, 313, 314, 328
   as natural disasters, 4, 7, 25
   occurrence of, 304, 307, 328
   plate boundaries and, *304*
   plate tectonics and, 34
   precursors, *324*, 324–325, 326, 328
   prediction programs, 323–326, 328–329
   preparation for, 325
   research programs, 326
   resistant structures, 331
   shallow-focus earthquakes, 303
   some significant, 300
   study of, 301, 327
Earth's center, 346
Earth's gravitation and rotation, *348*
Earth's principal subsystems, interactions among, 4
East African Rift valley, *48*, 49, *49*, *52*, 165
East Pacific Rise, 151, 153, 370–371
Easter Island, 371
Ebb tide delta, 600, 622
Echo sounding, 362, *362*, 382–383
Economics
   geology and, 7–8
   metamorphic materials and, 253–254
   valuable metamorphic rocks and, 255
EEZ. *See* Exclusive economic zone
Egypt invasion (1798), 7
Eifel Range, Germany, 128
Einstein, Albert, 10
El Azizia, Libya, 583
El Diablo, 300, 336
El Niño, 384
El Salvador, New Zealand, 634
Elastic deformation, 418
Elastic rebound theory, *301*, 301–302, 327, 329
Elastic strain, 419
Elasticity, 308
Electrical resistivity, 281
Electricity, 8
Electromagnetic force, 11
Electron capture, 281, *282*
Electron shell, 73, 95
Electrons, 73, 77, 95, 281
Elements, 73–75, 94–95, 281, 291
   native, 84
   periodic table of, *74*
Emeralds, 72
Emergent coasts, 618
Emperor Seamount–Hawaiian Island chain, 57, *58*, 59, 374
Enamel, 92
Enclaves, 120
End moraines, *544*, 548
Energy, 29
   matter and, 25
   resources, 8, 25, 92, 378–379, 382
   seismic waves and, 332
Engineering and geology, 423
Enstatite, 81
Entrada Sandstone, Jurassic-aged, 176
Environment of deposition, 215, 218
Environmental geology, 632, 647
Environmental issues, geology and, 8–10, 25
Eocene fossil forests, 226
Eocene times, 253
EPA. *See* U.S. Environmental Protection Agency
Epicenter, 329
Equal grade of metamorphism, 248, 255
Equator, 35–37, 40–41, 66, 337
Equilibrium, 233–237, 254, 256
Eratosthenes (C.276–C.195 B.C.), 337, *338*
Eroded anticline, 395, *397*, 418
Eroded laccolith, *129*
Eroded Sandstone, *260–261*
Eroded syncline, 395, *397*, 418
Erosion, 13, *23*, 36, 123, *187*, 193, 267
   on Chief Mountain, *410*
   definition of, 170–171
   measure to minimize the effects of, 621
   of mountain belts, 253
   processes of, 541, 591
   of sea cliffs, 616
   by valley glaciers, 542
   weathering and, 170
Erosional coasts, 623
Erta Ale, Ethiopia, 151, 157
Eruptive fissure, *109*
Eskers, 552
Eskola, Pentii, 248–249
Estes Park, Colorado, *441*
Estuaries, 618
Evacuation maps, 333
Evans, David M., 326
Evaporation, 589
Evaporites, 206–207, *209*, 210, 223
Evapotranspiration, 460
Evolution theory, 17, 264
Ewa Beach, Hawaii, 319, *333*
Excavating trenches, *330*
Exclusive economic zone (EEZ), *378*, 378–383
Exfoliation, 172, 193
Exfoliation domes, 172, *173*, 193
Exotic terrane, 409
Expansion, *174*
Expansive soils, *188*, 188–190, 191–193
Explosions, 430
Extinct volcanoes, 136
Extraterrestrial factors, and plate tectonic theory, 18
Extrusive igneous rocks, 19, *22–23*, 115, 126

F

FAMOUS (French-American Mid-Ocean Undersea forming Study) project, 371
Fault
   deformation and, 400–403
   dip-slip, 402–403, 410, 418–419

earthquakes and, *see* Earthquakes
  normal, 419
  reverse, 419
  strike-slip, 402–403, 410, 419
  thrust, 402, 419
  transform, 55, 64
  uplift and, 395
  wave generation and, 603
Fault breccia, 401, *402*
Fault gouge, 240
Fault plane, 400, 410, 419
Fault scarp, *322*, 400
Faulting, 223
Faults, 332, *373*, 398, 400–403, *402*, 403, *404–405*, 410–411, 418–419, *420*
Federal Bureau of Investigation, 30
Feldspars, 82, 88, 91–92, 109, *131*, 171, 191, 201, 244
Felsic igneous rocks, *120*, 126
Felsic magma, 109–110, *113*, 115, 144
FeO (iron oxide), 344
Ferromagnesian phenocysts, 118
Ferromagnesian silicates, 81–82, 84, 90, 95, 204
Fertile soils, 135
Fertilizers, 84, 92
Fetch, 603
Figuier, Louis, *197*
Fin, *177*
Fire, 131, 315, *319*, 323
Fire stone, 118
Firn, granular type of snow, 536
Firn limit, 538
Fission-track dating, *285*, 285, 287, 292
Fissure eruptions, 149, 161
Fjords, 542
Flake micas, 98
Flash flood, 459, 474, *474*
Flint, 208
Floating continents, 349–352, 354
Floating microorganisms, 91
Flood-frequency curves, 475
Flood of '93, 476
Flood tide, 600
Floodplain, 202, 469, *471*, 474
Floods
  as natural disasters, 4, 7, 25
  controlling, 478–479
  predicting and controlling, 474–479
  types of, 459
  waning stages of, 470
Floodwalls, 479
Floridan aquifer system, 504
Florissant Fossil Beds National Monument, Colorado, *217*
Flow velocity, maximum, *463*
Flows, 435, 446, 451
Flue-gas desulfurization, 196
Fluid activity, 233–237
Fluid-bearing limestone, 237
Fluid flow, 460–461
Fluorapatite [$Ca_5(PO_4)_3F$], 384.
Fluorine, 84, 115
Fluorite (CaF_2), 84, *85*, 88, *88*, 89
Focal depth, 310
Focus, 329
Folded rock layers, 394–397, 418
Folds, 394, 403, 419
  topography and, *396*
  types of, *398*
Foliated metamorphic rocks, 19, *22*, 243, 255
Foliated texture, 243, *244*, 255–256
Foliation, 19
Folium, 243
Footwall block, *401*, 401, 419
Forces, 418
  on fluids, 460
  on solid, 460
  of wind, 13
Forecasting volcanic eruptions, 156–159, 157–159, 161
Foreign rocks, 107
Forensic geology, 30
Forensic palynologists, 30
Forensic palynology, 30
Foreset beds, 470

Foreshore, 609
Forsterite, 79
Fort Lauderdale, Florida, 598
Fort Tejon earthquake (1857), 313
Fossil corals, 218
Fossil *Glossopteris* leaves, *35*
Fossil horses, *226*
Fossil insect, *216*
Fossil succession, 291
Fossilization, 216–217, 226
Fossilized tree stump, *217*
Fossils, 30, 37, 212–215, 218, 223, 226
  animals remains, 38
  dinosaur, 38
  evidence for continental drift, 38
  plant and animal, 39
Foundations, 170
Fractures, 88–89, 179, *179*, 181, *200*, 303, 337, *373*, 419
Fracturing, 245
Fragmental texture, 117
Franciscan Complex, *413*
Freezing and thawing, 192
Fresh water, source of, 459
Fringing reefs, *377*
Frost action, 171–172, 175, 193
Frost heaving, 172
Frost wedging, *172*, 172, 193
Fujiyama, Japan, 144
Fumeroles, *136*, 519, 523
Fungi, 175

G

Gabbro dike cuts, *267*
Gabbro (Gb), 16, 118–119, *119*, 121–122, 125–127, 153, 160, 342–343, 360, 364, 370, 375
Gabbroic (guest) magma, 113
Galápagos Islands, 135
Galápagos Rift, 373
Galena, 85, *88*
Galilean moons, Jupiter's, *51*
Galileo spacecraft, *51*
Gallatin National Forest, Montana, 332
Galveston, Texas, 618, *619*, 621
Ganges-Brahmaputra delta, Bangladesh, 472, *472*
Ganges River, 457
Gansu earthquake (1920), 319
Gardener, Marshall B., 336
Garnets, 85, 241, 243, 248, 251, 254
Garrison, T., *374*
Gas turbine engines, 92
Gases, 25, 73, 161
Gemstones, 30, 72, *73*, 254
Geologic agents, 637, 644
Geologic evidence, 413
Geologic evolution of North America, 414–415
Geologic history of an area, 291
Geologic maps, 393, 403, *422*, 422
Geologic record, 212, 223
Geologic structures, 390, 393, 419
Geologic time, 288, 290, 292
  early concepts of, 263–264, 291
  geologic time scale, 21, 23, *24*, 26, 31, *263*, 287–288, 292
  James Hutton and recognition of, 291
  measurement of
  study of, 262, 291
  uniformitarianism and, 21–24, 26
Geological Survey of India, 345
Geology, 637
  citizens and, 8
  consumers and, 8
  definition of, 6, 8, 25
  economics and, 7–8
  engineering and, 423
  environmental issues and, 8–10, 25
  formulation of theories and, 16–17, 25
  gravestones and, 195
  in Hollywood
  human experience and, 7
  intensity and, *312*
  mass wasting and, *430*
  natural disasters and, 7, 25
  politics and, 7–8

references to, 7
  on science, culture, and international affairs, 634–635, 644
  slope stability and, 430
  specialties of, 6
  topography and, 430, 451
  in trade and conflict, 644
  in unexpected places, *225*
Geometric radioactive decay, *283*
Geosynchronous orbit, 58
Geothermal energy, 521, 523, 525
Geothermal gradient, 346–347, 354–355
Gettysburg National Military Park, Pennsylvania, 196
Geyserite, 521
Geysers, 519–521
Giant crystals, 96
Giant ground sloths, 295
Giant's Causeway, Northern Ireland, *162*
Gibson Desert, Australia, *577*
Gillen, C., 241
Glacial budget, 538
Glacial deposits, 218
Glacial drift, 547, 553
Glacial erratics, 547
Glacial evidence for continental drift, 37, *38*
Glacial ice, 386, 536
Glacial Lakes, deposits in, 553
Glacial marine sediments, *375*
Glacial polish, 541, *542*
Glacial rock sequences, 36, *37*, 39, 63
Glacial striations, 38, 542
Glacial surge, 540–541
Glaciation, 23, 34, 37, 534
Glacier National Park, Montana, *215*, *410*
Glaciers, *138*, 170, 202, 212, 352, 406, 417, 532–534, 538, 559
  in Antarctica and Greenland, 459
  behavior of, 540
  crevasses in, *537*
  deposits of, 547
  distribution of, 537–538
  erosion and transport by, 541–547
  expansion, 532
  global warming and, 565
  kinds of, 533–534, 559
  movement of, 538–540
  National Park, Montana, 538, *544*
Glass, 30, 72, 92
Glauconite, 283
Glaucophane, 251
Glen Canyon Dam, Colorado River, 448
Global climate
  changes, 253, 290
  metamorphism and, 255
Global economy, global warming and, 29
Global ecosystem, 9
Global positioning system (GPS), 158
Global Seismic Hazard Assessment Map, 323, *323*
Global warming, 6, 196
  climate change and, 31
  global economy and, 29
  greenhouse effect and, 9–10, *10*, 25
Glomar Challenger, 43, 362
Glossopteris flora, 34–35, *37*, 38, *39*, 63–64
Glowing cloud, 145
Gneiss, *22*, 244, 247
Gobi Desert, China, 582
Gold alabaster, 99
Gold (Au), *8*, 72, 78, 84, *92*, 94, 109, 191
Gold mining, 86, 92
Goldschmidt, V.M., 248
Goma city, 156
Gondwana continents, 34–37, *37*, *38*, 38, *39*, 63–64
Gondwanaland, 34
Gonzales, F. I., 322
Good Friday earthquake (1964), 439
Goodrich Castle, England, *67*
Gophers, 184
Gorda Ridge, 379
Gorges, 7, 484
Gorridge, Henry, 28
Gossan, 191
Gothic arch, 421

GPS. *See* Global positioning system
Grabens, 405, *406*
Graded bedding, 213, *214*, 223, *225*
Graded stream, 481, 490
Gradient, 462, *462*, 463, 464
Grand Alignment, 356, 356
Grand Canyon, Arizona, 7, 210, *212*, 262, *266*, *273*, 278, 279, 480, 484, *502*
Grand Canyon, Yellowstone River, *167*
Grand Ronde River, Washington, *150*
Grand Teton National Park, Wyoming, 544
Granite, *22*, 102–103, 120, 122, 125–127, 172–173, *193*, 201, 266, 360
   mafic pillows in, *113*
   pipes, 113, *114*
   porphyries, 116, 120
   weathering of, *170*
Granitic continental crust, 342–343
Granitic magma, 125
   origin of, 106, 125
   relationship of basaltic to, *107*
Granitic pegmatites, 120
Granitic plutons, 407
Granitization, 112
Grant, Alan, 297
Granular sugar, 180
Granulite facies, 248
Graphite (C), 84, *89*, 247
Grass Valley, California, *86*
Gravel, *8*, 8, 200, 202–203, 219, 375, 382, *450*
Gravestones, *195*, 421
Graveyards, 253
Gravimeters, *348*, 349
Gravitational compression, 13
Gravitational differentiation, 111, *112*
Gravitational force (*F*), 348, 427
Gravity, 355
   Big Bang and, 11
   catastrophes and, 356
   definition of, 347
   determination of, 347, 348–349, 354
   magnetism and, 352, 354
   planetary alignments and, 356
   surveys of, 370
Gravity anomalies, 349, 355
Graywacke, 385
Great Barrier Reef, Australia, 377
Great Basin National Park, Nevada, 545
Great Dark Spot, *15*
Great Glen fault, *411*
Great Hot Blast, *154*
Great Lakes, *221*, 353, *381*
Great Plains of the United States, 184
Great Red Spot, *15*
Great Salt Lake, Utah, 206
Great Smokey Mountains, 404
Great Star of Africa diamond, 97
Great Wall of China, *296*, 296
Great Wave of Kanagawa, *2–3*
Green Mountains, 404
Green River Formation, Wyoming, 220
Greenhouse effect, 6, 31
   global warming and, 9–10, *10*, 25
   on Venus, 458
Greenhouse gas emissions, 29
Greenland, Iceland, 532
Greenschist facies, 243, 248
Greenstone, 246
Greywacke, 385
Grindelwald, Switzerland, *467*
Grofé, Ferde, 7, 25
Groins, 612
Ground failure, 319, *322*
Ground moraine, 548
Ground shaking, 314–315, *318*, 323
Ground squirrels, 184
Groundmass, 116
Groundwater, 498, 524
   artesian system, 503–504
   contamination of, 516, *517*
   deposition and, 505–512
   erosion and, 505–506
   factors determining quality of, 517–518, 524
   hydrologic cycle and, 498
   modification of, 512–518, 524

movement of, *501*, 501
source of freshwater, 498
springs, 502
velocity of, 501
water wells, 503
withdrawing, 518
Guam, 378
Guide fossils, 278, *280*, 292
Gulf Coast, 219, 417
Gulf of Aden, 66
Gulf of California, 49, 55, *57*, 226, 316, *363*, 370, 415
Gulf of Mexico, 188, 201, 306
Gulf Stream, 384
Gulf War (1990–1991), 8, 66
Gullies, 189
Gully, 484
Gullying, 189
Gunnison River, Colorado, 484
Gutenberg, Beno, 345
Guyots, *374*, 374, 382–383
Gypsum, 8, *84*, 84, 90–91, *96*, 206, 209

H

Haicheng earthquake (1975), 324–325
Half-lives, 281–282, 286, 292
Halides, 84, *84*, 90
Halite (NaCl), 76, 79, 84, *84–85*, 88, *88*, 90–91, 178, 206
Hallet's Cove, Australia, 38
Hanging valleys, 546
Hanging wall block, 401, *401*, 419
Hannibal, Missouri, 479
Hansen, Em, 7, 30
Harbor wave, 315
Hard water, 518
Hardness, 89, 95
Harmonic tremor, 158
Harney Peak Granite, *102*, 103
Hart, Johnny, 7
Hawaii Volcano Observatory, 167
Hawaiian-type eruptions, 141
Hawaiian volcanoes, 110, 141, 148
Headward erosion, 484
Heart Reef, Australia, *358–359*
Heat, 233, 237
   in Earth, *106*
Heat flow, *347*, 347, 355
Heavy rainfall, 424–425, 428, *431*
Heezen, Bruce E., *367*
Heliopolis, great temple of, 28
Helium (He), 11–13, 74
Hells Canyon, Snake River, 484
Hematite (Fe_2O_3), 84, 178, 204, 221
Henry Mountains, southern Utah, 124
Herchenberg volcano, Germany, *124*
Herculaneum city, 135, 151
Hess, Harry, 41–42, 63
Heterogeneous metamorphic reaction, 256
Hiatus, 267, *270*
High Plains Aquifer, *513*
High-pressure metamorphism, 253
Highway excavation, *429*
Hills, 362
Himalaya Mine, California, 72
Himalayan Mountains, *53*, 54–55, 253, 342, 349, 390, 405, 407, 409, *412*, 533
Historical geology, 6, 8, 25
HMS *Beagle*, 378
HMS *Challenger* expeditions (1872–1876), 35, 362
HMS *Titanic*, 362, 387
Hokusai, Katsusika, *2–3*
Hole-in-the-Wall, *177*
Hollister, Charles D., *367*
Holmes, Arthur, 42
Holmes, Sherlock, 30
Homestake Mine, South Dakota, 87
Homo sapiens, 197
Homogeneous accretion theory, *12*
Honolulu, Hawaii, 632
Hoodoos, *177*
Hooks, 613
Horizon A, 182
Horizon B, 183
Horizon C, 183

Horizon O, 182
Horizontal bedding, *381*
Hornblende, *88*, 122, *131*, 244
Hornblende crystals, 244
Horner, John R., *297*, 297
Hornfels, 247
Horns, 560
Horsts, 405, *406*
Hot spots, 57, *58*, 64, 156
Hot Springs National Park, Arkansas, 519–520
Houston, Texas, 514
Huang Ho, China, 479
Hubbard Glacier, Alaska, 541
Hudson River School, 7, 124
Humus, 193
Hurricanes, 110, 618, 641
   Isabel, 628
   Jeanne, 637
   Katrina, 458, *619*, 641
Hutton, James (1726–1797), *131*, 131, 229, 264–265, 267, 281, 291, 634
Hybrid igneous rocks, 126
Hybridization, 113
Hydraulic action, 465, 466, 607
Hydrocarbons, 219–220, 228, 365, 380, 403
Hydroelectric power plants, 494
Hydrofluorocarbons, 29
Hydrogen chloride (HCl) gas, 166
Hydrogen (H), 11–13, 74, 136
Hydrogen sulfide (H_2S), 91, 107, 136
Hydrologic cycle, 459, 490
   glaciers role in, 535
   stages of, 460
Hydrolysis, 175, 178–179, 192–193
Hydrosphere, 598
   on land surface, 458
   plate tectonic theory and, 18
   as subsystem of Earth, 4, 25
Hydrothermal activity, 136, 518, 523, 525
Hydrothermal convection, 347
Hydrothermal vents, 384
Hydroxide limonite, 178
Hydroxides, 178, 204
Hydroxyapatite [$Ca_5(PO_4)_3OH$], 384
Hypocenter, 303
Hypotheses
   definition of, 16–17
   testing, 25–26
Hypothetical area, block diagram of a, *274*

I

Ice Age, 358–359, 385, 553–557
Ice cap, 534
Ice crystals, 382
Ice Jam flood, 459
Ice-scoured plains, 547
Ice sheets, 533
Ice shelf collapse, 641, 645
Ice shelves, 533
Icebergs, 350, *375*
Icelandic dike, *104*
Idaho, southern, 103
Igneous activity, *23*, 390, 406
   convergent plate boundaries and, 153, 156
   divergent plate boundaries and, 153
Igneous intrusions, 122, 123, 125, 126
Igneous rocks, 20, *23*, 26, 54, 102, 126–127, 160
   characteristics of, 115–117, 126
   classification of, 19, 117, 118–122, 126
   common, 126
   composition of, 19
   extrusive, 115
   felsic, *120*
   hand specimens of, *22*
   intermediate, *119*
   intrusive, 115
   mafic, *119*
   major, 103
   mineral content and, 117
   oceanic crust and, 153
   plutonic, 115
   rock cycle and, 19
   various textures of, *116*
   volcanic, 115
Igneous textures, 115

Inactive volcanoes, 136
Incised meanders, 485
Inclusions, 291
Index fossils, 278
Index minerals, 248
India and Asia collision, 253
India earthquake (1993), 315
India earthquake (2001), 34, 306
Indian Ocean, 31, 151, 319, 370, 373, 384
Indian Ocean tsunami (2004), *320*
Indian Ridge, 151, 153, 370
Indiana limestone, 99
Indus River, Asia, 586
Industrialization, 196
Infiltration capacity, 461
Inorganic substances, naturally occurring, 77
Insects, swarms of, 423
Inselbergs, 588
Institute for Exploration, 387
Intensity, of earthquakes, 311, *312*, 328–329
Intermediate focus earthquakes, 303
Intermediate igneous clan, 110
Intermediate igneous rocks, *119*
Intermediate magma, 110, 115, 126, 144
Internal drainage, 586
Interplanetary dust, 11, 25
Interstellar matter, 13, 25
Interstellar space, 11
Intraplate earthquake, *306*
Intraplate volcanism, 156
Intrusive igneous bodies, 54
Intrusive igneous rocks, 19, *22–23*, 115, 104, *123*, 126
Io, moon of Jupiter, *51*, 136
Ionic bonding, 75, *76*, 77, 95
Ions, 75, *79*, 95
Iran earthquake (2003), 34, 300
Iran–Iraq War (1980–1989), 66
Iron (Fe), *8*, 12–13, 16, 25, 72–73, 79–81, 92, 94, 109, 125, 183, 191, 219, 222, 518
Iron oxides, 204
Iron sulfide (FeS$_2$), 215
Island of Cyprus, Mediterranean, 61
Island of Rarotonga, *73*, 195, *195*
Island of Santorini, 151
Island of Staffa, Inner Hebrides, 7
Isograd maps, 248–249, 255–256
Isostasy principle, *556*
Isostatic rebound, *351*, 351–352, *352*, 355
Isotopes, 75, 281, 291
Italian white alabaster, 99

J

J. Paul Getty Museum, Malibu, 232
James, Jesse, 507
Jason Foundation for Education, 387
Jasper, 208
Jeffreys, Harold, 345
John Day Fossil Beds National Monument, Oregon, *200*, 200
Johns Hopkins University, 301
Johnson, Kirk R., 295
Johnston Island, *379*
Johnston, Joe, 297
Johnstown Pennsylvania flood, *458*, 458
Joint Oceanographic Institutions for Deep Earth Sampling (JOIDES) Resolution, 362
Joints, 138, *172*, 398–400, 403, 418–419
Joly, John, 264
Jovian moons, 50
Jovian planets, 11–13, *14–15*, 26
Juan de Fuca plate, 55, *57*, 152, *154*
Juan de Fuca Ridge, 151, *379*
Julia Pfeiffer Burns State Park, California, *610*
Jungfrau, Switzerland, *545*
Jupiter, 11–14, *15*, 356
Jurassic age, 36, *37*, 39, 176, *218*
Jurassic dinosaurs, 226
Jurassic strata, *271*

K

Kaibab limestone, *279*
Kames, 550
Kandinsky, Wassily, 335
Kaolinite, 191
Karren, *508*
Karst topography, 506, *512*, 524
Kata Tjuta, 264, 276–277, *277*
Katmai National Park and Preserve, 152
Kauai, 156
Kelvin, Lord (1824–1907), 264–265, 291
Kermadec–Tonga arc, 54
Kerogen, 219
Kettles, 550
Key beds, 278
Kilamanjaro, Tanzania, 151
Kilauea volcano, *132–133*, 136, 141, 156
Kilobars (kb), 240, *241*
Kimberlite pipes, 344, *344*, 357, *357*
Kinetic energy, 465
Kircher, Anthanasium, 360
Klamath Falls, Oregon, *402*
Kobe earthquake (1995), 304, *305*
Komatiites, 118
Kootenay National Park, British Columbia, Canada, *396*
Kouros, 232
Krafft, M., 141
Krakatau eruption (1883), 135, 318
Kunming, China, *510*
Kutiah Glacier, Pakistan, 540
Kyanite, 248, 254
Kyoto Protocol, *29*, 29

L

La Brea Tar Pits, Los Angeles, *217*, 226
La Conchita, California, 455, *455*
Laccoliths, 124–126
Lacher See, Germany, 128, *128*
Lagoons, 205
Laguna Beach, 434–435, *439*, 455, *455*, 455
Lahars, 144, *146*, 148, *157*, 161
Lake Agassiz, 554
Lake Michigan, St. Ignace, 578
Lake Missoula, Montana, 555
Lake Nyos, 136
Lake Pontchatrain, New Orleans, *458*, 479
Lake Powell Reservoir, Arizona, 636
Lake Superior, 84, 191, 221, 599
Lake waters, 206
Laki fissure, 136, 149
Laminar flow, 460, *461*
Landfills, 516
Landon, Susan M., *228*, 228
Landslides, 4, 7, 25, 323, 406, 426, *431*, *450*
Lapilli, 140, *140*, 160
Larsen ice shelf, *643*
Larson, Gary, 7
Las Vegas, Nevada, 514–515, 569
Lassen Peak, California, 110, 152–153, *154–155*, 160
Lassen Volcanic National Park, California, *136*, 136, *145*, *147*, 152, *174*, *519*, *545*
Lateral accretion, 469
Lateral continuity, 291
Lateral moraines, 549
Laterites, 183, *185–186*, 187, 192–193
Laurasia, 36–37, 63–64
Lava, 91, 94, 102, 110, 122, 126
   behavior of, 125
   blob of, 139
   properties and behavior of, 103–107, 109–113, 115, 125
Lava beds, 145
Lava domes, 120, 141, 144–146, *147*, 156, 160–161
Lava flows, *102*, 135, 137, *139*, 140–141, *144*, 148, 151, 153, 156, 159, 187, 287, 422. *See also* Volcanism and volcanoes
Lava fountains, 141, 153, 159
Lava tubes, 137, *137*, 140, 161, 371
Le Puy, France, 129, *129*
Leaching, 182
Lead, *8*, 92, 96, 191
Leaning Tower of Pisa, Italy, 514, *515*
Lee, J. C., 30
Left-lateral strike-slip fault, 403
Lehman Cave, Nevada, 507
Lehmann, Inge, 345, *345*
Leonardo da Vinci, 7
Lewis and Clark Caverns, Montana, 507
Lewis overthrust, Montana, 403
Lignite, 208, 221
Limb, 395
Limestone, 8, 19, *22*, 84, 178, 192, 196, 204, 206–207, *208*, 210, 219, *225*, *247*, 249, 296, 435, 499, 505, 512, 518
Limestone facies, 210
Limonite, 204
Lincoln Memorial, 253
Lingula, 280
Lion Monument in Lucerne, Switzerland, 227, *227*
Liquefaction, *315*, 315
Liquid petroleum, 220
Liquid water, 72, 94
Liquids, state of matter, 73
Lithification, 19, *20*, *203*, 203–204, 222–223, *268*
Lithium, 96, 109, 121
Lithosphere, 16, *17*, 19, 25–26, 46, 64, 355
   asthenosphere and, *18*
   plate boundaries and, *18*
   as subsystem of Earth, 4, 25
Lithosphere–atmosphere–hydrosphere system, 17
Lithostatic pressures, 239, *241*, 255–256
Little Ice Age, 7, 532–533, 559
Liwei, Yang, 296
Lizard, 297
Local or temporary base levels, 480
Loch Ness, Scotland, *411*
Lodestone, 352
Loess, 578, 580, 591
Loihi seamount, 156, 161
Loma Prieta earthquake (1989), 304, *305*, 311, 315, 319
London obelisk, *28*
Long Beach, California, 515, 516, *516*
Long Island, New York, 513
Long Valley caldera, 150, 158, 167
Long Valley Observatory, Menlo Park, California, 167
Longitudinal dunes, 575, *577*, 580, 591
Longitudinal profile, 483
Longshore currents, 604
Longshore drift, 609
Los City hydrothermal vent field, 373
Louann Salt, 207
Love, A. E. H., 308
Love waves (L-waves), 308, *308*, 309, 328–329
Low-velocity zone, 343–344, 355
Lunar tidal bulge, 600
Luster, 85, 88, 95
Lyell, Charles, 264, 281
Lystrosaurus, 38, *39*

M

Madison Boulder, New Hampshire, 547
Madison Canyon earthquake (1959), 319
Madison River, Montana, *486*
Mafic igneous rocks, *119*, 126
Mafic lava, 150
Mafic magma, 109–110, 115, 125
Mafic magmas, 153, 237
Mafic pillows in granite, *113*
Mafic plutons, 111
Mafic rocks, 122, 244, 247, 248
Magellan spacecraft, *50*
Magma, *23*, 94, 126, 156, 158, 407
   basaltic, 47
   chambers, 113
   chemical composition of, 109–110, 125
   collision of, 112, 126
   cooling of, 40, 42
   crystallization of, 26, 126, *283*
   definition of, 16, 102
   felsic, 109–110, *113*, 144
   formation of, 115
   gabbroic (guest), 113
   generation of, 19–20, 34
   heat of, 110, 125
   igneous rocks formation, 20
   intermediate, 110, 144
   intrusion of, 55
   mafic, 109–110, 237
   Mg-Fe-rich, 109, 125

Magma *(continued)*
 mixing, 113
 origin of minerals and, 91
 properties and behavior of, 103–107, 109–113, 115, 125
 rise of, 107, 125
 role of, 125
 stationary columns of, 57
 viscosity of, 125
 volcano and, 140
 younger rising, *113*
Magma bodies growth, *108*
Magma chamber, 106, 123, 126, 364
Magma mingling, 115, 126
Magma mixing, 126
Magma production, *106*
Magnesium (Mg), 13, 72, 79–81, 94, 109, 125, 206, 518
Magnesium oxides, 12
Magnetic anomalies, 43, *44*, 45, *59*, 353
Magnetic anomaly, 64
Magnetic field, 352, 355
Magnetic poles, 40–41, 63
Magnetic reversals, 41–45, *42*, 59, 63–64
Magnetism, 11, 352, 354, 355
Magnetite (Fe₃O₄), 84, *89*, 221
Magnetometer, 353
Magnitude, 311, 329
Maldives Islands, 31
Mammal and vegetation restoration, *200*
Mammoth Cave, Kentucky, 178, 498, 502, 507
Mammoth Mountain, California, *157*, 158
Mammoths, *295*
Manganese, 72, 191, *376*
Mantle, 13, 25–26, 40–43, 50, 342, 343–344, 354–355
 as concentric layer of Earth, 13, 16
 convection cells, *16*
 definition of, 13
 low-velocity zone in, 343–344
 lower, 13, 16, 25
 as subsystem of Earth, 4, 25
 upper, 16, 25, 54, *364*
Mantle peridotite rocks, 118
Marble, 19, 178, 196, 232, 246–247, *247*
Marianas Trench, Pacific Ocean, 370
Marias Pass, *410*
Marin Headlands terrane, *385*
Marine environments, *202*
Marine regressions, 210–211, *211*, 212, 223
Marine rock sequences, 36, *37*, 39, 63
Marine sediments, 276
Marine terraces, 607, 618
Marine transgressions, 210–211, *211*, 212, 222–223
Mariner 10, 50
Mars, 11–14, *15*, 50, *51*, 54, 288, 337, 356
 dust, *69*
 exploration Rover project, 69
 photomosaic of, *51*
Marshall, James, 86
Mason, James, *339*
Mass movements, 426, 446
 characteristics of, 432
 classification of, 432, 451
Mass spectrometer, 282
Mass wasting, 426–427, 431, 436, 451
 factors that influence, 427
 geology and, *430*
 major factors that cause, 427
 recognizing and minimizing effects of, 446–450, 451
 slope stability and, *430*
 triggering mechanisms of, 430–431, 451
 types of, 431–432, 434–439, 441, 443, 446, 451
Matter, 11, 19, 94
 composition of, 73–77, *77*
 energy and, 25
Matterhorn, Switzerland, *545*
Maui pearls, *73*
Mauna Loa, Hawaii, *51*, 141, *143*, 156
Mausoleums, *421*
Mayon volcano, Philippines, 136, *146*, 156
McNeill, J. R., 632
Meager Mountain, British Columbia, 152, 160
Meandering streams, 467
Meanders, 467

Mechanical weathering, 171, *174*, 193
 chemical weathering and, 192
 definition of, 175
 disaggregation of Earth materials, 171–175
 organisms and, 174–175
Medial moraine, 549
Medicine Lake Highland, California, *103*
Medicine Lake Volcano, California, 152, 154
Mediterranean belts, 151–153, 160–161
Mediterranean–Asiatic belt, 304, 306–307
Megacrystic granites, 115, 120
Megacrysts, 116, 120
Mélange, 54
Melting, 23
Mendelssohn, Felix, 7
Mendenhall Glacier, Alaska, 543
Mendocino fracture, 316
Meramec Caverns, Missouri, 507
Mercury, 12–14, *14*, *50*, 50, 337, 356
Meridiani Planum, 69, *69*
Mesa, 588
Mesosaurus, 38, *39*, 63, *68*
Mesozoic Era, 36, 38, 66–67, 278, 415
Metaconglomerates, 247
Metallic bonding, 75, 76–77, *77*
Metallic core, 14
Metallic resources, 92
Metallurgy, 254
Metamorphic aureoles, *240*
Metamorphic belt, 250, *252*, 256
Metamorphic facies, 248–249, *250*, 255–256
Metamorphic fluids, 236–237, 255–256
Metamorphic grade, 241, 256
Metamorphic materials, economic uses of, 253–254, 255
Metamorphic outgassing, 253
Metamorphic Petrologists, 259
Metamorphic reactions, 233, 253
Metamorphic rocks, 20, *23*, 26, 115, 125, 131, 253, 413
 classification of, 243–247, 255
 composition of, 19
 foliated metamorphic rocks, 19
 hand specimens of, *22*
 metamorphism and, 241
 nonfoliated metamorphic rocks, 19
 rock cycle and, 19
 types of, 247
Metamorphic zones, *248*, 248–249, 255–256
Metamorphism, 20, *23*, 91, 256, 390, 406
 causes of, 233–237, 254
 common type of, 241–242
 contact (thermal), 237, *239*, 255
 definition of, 232, 254
 dynamic, 239–240, 255
 dynamothermal, 241
 environments of, *238*
 equilibrium and, 233–237
 of fluid-bearing limestone
 global climate and, 255
 high-grade, 241
 high-pressure, 253
 low-grade, 241
 medium-grade, 241
 plate tectonics and, 250–253, 254–255
 principal agents of, 233
 on radiometric dating
 regional, 241
 rock cycle and, 19
 of rocks, 237
 shapes of crystals and, *233*
 shock, 240–241, 255
 slab, 253
 types of, 237–242, 255
Metasomatism, 238, *240*, 256
Meteor, 370
Meteorites, 11, 12–13, 25, 128, 413
Meteorologist, 35
Methane, 9, 12–13, 29
Methane hydrates, 378–379, 379–382
Metropolitan Museum of Art, 28, *28*
Mexico City earthquake (1985), 324
Mexico City, Mexico, 515
Mg-Fe-rich magma, 109, 125
MgO (magnesium oxide), 344
Miami Beach, Florida, 621

Micas, 81–82, *88*, 92, 98, *131*
Michigan Basin, 397
Michigan, Ohio, Illinois, 547
Microcline, 82
Microcontinents, 374
Mid-Atlantic Ridge, 35, 43, *48*, *106*, 151, 153, 360, 370–371, *371*, 407
Mid-oceanic ridges, *23*, 43, 160
Middle Ages, 220
Middle Devonian reef, *381*
Migmatites, 245, *246*, 247
Migration, 615
Milankovitch, Milutin, 557–559
Milankovitch theory, *558*, 557–559
Milky Way Galaxy, 11, 13
Million years ago (MYA), *270*
Millipedes, 184
Milne, Lynne, Dr., 30
Mine, deepest, 347
Mineral content, 126
Mineral galena (PbS), 84
Mineral grains, 115
Mineral Information Institute, *8*
Mineral water, 128
Minerals, 26, 30, 72–73, 77, 79, 95, 237
 alteration of, 192
 amphibole group of, 81
 carbonate, 83–84
 chemical bonds and, 75
 chemical composition of, 78–79
 cleavage, 88, *88*, 94
 commodities, *93*
 common rock-silicate, *83*
 composition of, 19, 115
 crystal shapes, *78*
 crystals, 77, 96
 with crystals, *85*
 definition of, 79
 exploring the, 77
 in granite, *91*
 groups, 80–84, 94
 identification of, 85, 88–90
 ions in, 79
 in museum display, *72*
 origin of, 91, 94
 physical properties of, 79
 properties of, 90
 pyroxene, 81
 radicals in, *81*
 resources, 8, 25
 rock-forming minerals, 90, 94
 stability of silicate, 180
Minerals alteration, 171
Miner's Castle, 599
Mines, 337, 649
Ming Dynasty (A.D. 1368–1644), 296
Mining, 639
Minoan eruption of Santorini, *164*
Mission Mountains, 404
Mississippi River, 306, 417, *458*, 466, 470–473, *472*, 477, 494
Mississippian rocks, *271*
Modified Mercalli Intensity Scale, 311, 314, 329
Moenkopi formation, *279*
Moessbauer spectrometer, 69
Moho, 342, 343, 354–355
Mohole project, 343
Mohorovičić, Andrija, 342, *342*, 343, 354
Mohorovičić discontinuity, 343, 354–355
Mohs Hardness Scale, 89
Mojave Desert, California, *587*
Mold, 215, *216*
Molecular solution, *175*
Molecules, 11, 77
Mollo, Terry S., 99
Mollusks, 73
Molokai Island, Hawaii, *104*
Molten rock, 125
Molybdenum, 109
Mona Schist, Marquette, Michigan, *162*
Monoclines, *394*, 394, 395–396, 419
Mont Pelée eruption, 146
Montana earthquake (1959), *322*
Montreal Protocol, 166
Monument valley, Utah, *198–199*
Monuments, 237, 421

Monzonite, 102
Moon, 11–12, *14*, 25, 136, 173, 348, 356
    Earth and, 35
    of Jupiter, 136
    of Neptune, 136
    of Saturn, 136
Moraines, 548
Moskva, 98
Mosley, J., 356
Mosses, 175
Mount Baker, Washington, 156, *537*
Mount Currie Conglomerate, *277*
Mount Erebus, Antarctica, 151
Mount Etna, Italy, 141, 144, 148
Mount Hood, Oregon, 152
Mount Katmai volcano, 15, 112
Mount Mazama, Oregon, *105*
Mount Pelée, Martinique, 145, *147*, 150
Mount Pinatubo, Philippines, 34, 135, 146, 156, *157*, 158
Mount Rainier, Washington, 153
Mount Rushmore, South Dakota, *102*, 103
Mount Shasta, California, 144, *146*, 152
Mount Spurr, Alaska, 152
Mount St. Helens, Washington, 110, *128*, *140*, 145, *147*, 152–153, *155*, 156–159, 226
Mount Tehama, California, *154*
Mount Vesuvius, Italy, *134*, 135, 144, 148, 151, 156
Mountain building, 18–19, *252*, 391, 418
    deformation and, 404–409, 412–413
    metamorphic belt development and, *252*
    plate tectonics and, 405–409
Mountain glacier, 533
Mountain of God, 165
Mountain ranges, 16, 36, 404
Mountain systems, 34, 404, 413
Mountains
    hills and, 13
    origin of, 404–409, *409*, 413, 418
Muav limestone, *212*
Mud, 200, 202–203, 380
Mud cracks, 213, *215*, 218, 223, *225*
Mud flows, *217*, 446
Mud pot, 519
Mud shallows, 360
Mudflows, 135, 422, 435–438, *441*, 451
Mudrocks, 205, 207, *207*
Mudslide, aerial view of, *424–425*
Mudstones, 205, *225*, 233, 296
Mundus Subterraneus (1664), 360
Mural, *226*
Muscovite mica, 83, *88*, 98,
Museum of Natural History, Princeton University, 297
Museum of the Rockies, Bozeman, 297
Musset, A. E., *343*
Mylonites, 239, *242*
Myrtle Beach, South Carolina, 598

N

Nail polish, 98
Namib-Naukluft Park, Namibia, 579
Napoleon Bonaparte, 7
Nappes, 251
Narrow shelf, 369
National Aeronautics and Space Administration (NASA), 69, 296
National Earthquake Information Center, 333
National Oceanic and Atmospheric Administration's (NOAA) National Weather Service, 333
National Theatre, Reykjavik, *100–101*
Native elements, 78, 84, 95
Natural Bridges National Monument, Utah, *213*, 487
Natural disasters, geology and, 4, 7, 25
Natural gas, 8, *8*, 219, 223, 228, 379–382
    traps, *220*
    wells, 379
Natural glass, 117
Natural hazards, 423
Natural levees, 470
Natural resources, weathering and, 92–93, 94
Natural switches, 639
Navajo Sandstone, *218*, 218, 223, *279*

Navajo Tribal Park, *198–199*
Neap tides, 600
Nearshore currents, 623
Nearshore sediment budget, 615, 616, *616*, 617, 623
Nearshore zone, 604
Nebula, center of, 12
Negative gravity anomaly, 349
Negative magnetic anomalies, *353*
Neil, Sam, 297
Nematodes, 184
Neoglaciation, 554
Neon, 75
Neptune, 12–14, *15*
Nests, *216*
Neutral buoyancy position, 107
Neutrons, 11, 73, 77, 95, 281
Nevado del Ruiz, Colombia, 144, 156
New Hampshire, 454
New Orleans, Louisiana, *458*, 514–515
New York obelisk, 28
Newberry Volcano, Oregon, 142, 152, 154
Newfoundland, Quebec, Canada, 92
Newton, Sir Isaac (1642–1727), 16, 337, *347*
Niagara Falls, New York, 517
Nickel, 13, 16, 25, 191
Nicolet, Quebec, Canada, *442*
Niger River delta, Africa, 473, 586
Niigata earthquake (1964), 324
Niigata liquefaction (1964), 315, *315*
Nile River, Egypt, 186, *472*, *472*, 493, 586, 617
Nitrogen, 136, 286
Nitrous oxide, 9, 29
No-till farming, 197
No-till planting, 190
Noble gases, 75
Nodules, 208
Nonconformity, 267, 271, *273*, 274, 292
Nondeposition, 267
Nonferromagnesian silicates, 82–83, 84–85, 90, 95
Nonfoliated metamorphic rocks, 19, *22*, 246–247, 255
Nonfoliated metamorphic texture, 246–247, 255
Nonfoliated texture, 256
Nonmarine rock sequences, 36, *37*, 39, 63
Nonmetallic resources, 92
Nonplunging folds, 396
Nonrenewable energy resources, 8
Nonsaline water, 459
Nordlingen meteorite impact and crater, Germany, *128*
Normal fault, 419
North America
    active volcanoes of, 152–153, 160
    continuing evolution of, 415, 417
    geologic evolution of, 414–415, *416*
    Precambrian evolution of, 418
    Westward drift of, *417*
North American plate, 316
North Pole, *42*, *339*
North Slope of Alaska, 219
Northern Hemisphere, 35, 37
Northridge earthquake (1994), 304, *305*, 313, *317*, 315
Novarupta, Alaska, 152–153, 156
Nuclear force, strong, 11
Nucleus, 11, 73–74, 77, 95
Nuée ardente, 145, *147*, 161
Numerical ages, 22
Nuremberg, southern Germany, 128
Nyiragongo, 151, 156

O

Oahu, Hawaii, *604*, *611*
Oases, artesian springs, 504
Oblique-slip faults, 403, *411*, 419
Obsidian, 117, *120*, 122, 208
Obsidian-speckled cathedral, Nordlingen, 128
Ocean basins, 16, 34, *45*, 360
Ocean Drilling Program, 362
Oceanic circulation, 384
Oceanic crust, *23*, 25, 42–43, 46, 54, 160, 342–343, 355, 360
    age of, *45*
    basaltic, 35
    composition of, *364*

    definition of, 16
    igneous rocks and, 153
    made of, 364
    structure of, 382
Oceanic lithosphere, 408
Oceanic plates, 409
Oceanic ridges, *48*, 370–371, 383
Oceanic slab dewatering
Oceanic trenches, 362, 369–370, 383
Oceanic–continental plate boundaries, 52, *53*, 54, 63–64, 67, 406–407, *409*, 413, 418
Oceanic–oceanic plate boundaries, 52–54, *53*, 63–64, 406–407, *408*, 413
Oceanographic research, 36, 39, *363*
Oceans, 13, 202
    energy from the, 629
    exploration of, 362
    resources from, 378–379, 382
Officer, C., 306
Oil, 8, 228, 379–382
    extraction from shale, *221*
    gas exploration and, 387
    plate tectonics and, 66
    politics and, 66
Oil shale, 219–220, 223
Oil traps, *220*
Oil wells, 379
Old Man of the Mountain, 454, *454*
Old Red Sandstone, 67, *67*
Oldham, R. D., 345
Oldoinyo Lengai volcano, Tanzania, *128*, *165*, 165
Olivine, 79–81, 111, 122, 126, 178, 236
    basalts, 118
    crystals, 124
Olympus Mons, *51*
100-year flood, 478, 490
Ontario, Canada, 550
Ooids, 206, *208*, 218
Oolitic limestones, 206
Ooze, 375–376, 382–383
Open fissures, 447
Ophiolites, 54, *55*, 55, 344, 364, 383
Ophir Creek, Nevada, *441*
Opportunity, Mars Exploration Rover, *69*, 69
Orange Beach, Alabama, *619*
Orbital eccentricity, 558
Ordovician Period, 295
Organisms, 174–175, 187, 192
    activities of, 175, 184
    reefs and, *376–378*, *380–381*, 382
    weathering and, *174*
Original horizontality, 291
Orogenic activity, present-day, 407
Orogenic metamorphic belt, 251, *253*, 255, 406
Orogeny, 405–406, *408*, 418–419
Oroville Dam, California, 478
Orthoclase, 82, 126, 236
Orthoclase (KAlSi$_3$O$_8$), 80, 82, 111, 178
Osaka and Tokyo, Japan, 515
Oshodi Market, Nigeria, *9*
Outlet glaciers, 540
Outwash plain, 550
Oval reef, *381*
Overloading, 428, 430, 451
Overpopulation, 8–9, *9*, 10, 25
Overturned fold, 395
Oxbow lakes, 468
Oxidation, 175, 178, 193
Oxides, 84, 178
Oxygen (O), 17, 19, 73, 79, 222, 288
Oysters, 73, 91, 201
Ozone depletion, 166
Ozone layer (O$_3$), 6, 166, 196
Ozymandias, 25

P

Pacific belts, 152–153, 160, 304
Pacific Ocean, 40, 110, 151, 314, 333, 373, *381*, 379
Pacific Ocean basin, 54, 316, 319, *321*, 378
Pacific Palisades, *435*, 455
Pacific plate, 152, 156, 161, 316
Pacific Rim, 253
Pacific Tsunami Warning System, 319, *321*, *333*, 333
Pacific–Antarctic Ridge, 370

Padre Island, Texas, 598, 614
Page, J., 306
Pahoehoe, *137*, 137–138, 160–161
Paired metamorphic belts, 250, 256
Pakistan earthquake (2005), 34, 306
Paleogeography, 41
Paleomagnetic data, 43
Paleomagnetic studies, 40
Paleomagnetism, 40–41, 63–64
Paleontologic evidence, 35
Paleontologists, 226
Paleontology, 297
Paleoseismology, 330
Paleozoic Era, Late, 37, *39*, 39, 46, 63, *68*, 112, 278, 415
Paleozoic plant fossils, 34
Paleozoic sedimentary rocks, *273*
Palisades of the Hudson River, *52*
Palisades Sill, 124
Palm frond, *217*
Palmyra, New York, 550
Palouse Falls, Washington, *557*
Palynology, 30
Pamukhale, Turkey, 521
Panamint Range, California, 473
Pangaea, 18, 35–36, *39*, 39, 46–47, 62–64, 67, 407, 409
 breakup of, 58
 formation of, 295
Papua New Guinea tsunami (1998), 34
Parabolic dunes, 577, *578*, 580, 591
Paradoxides, 280
Parent element, 282
Parent material, 170, 180–181, 184, 187, *187*, 193
Parent–daughter isotope pairs, long-lived, 284, 286
Parent–daughter ratio, 284
Parícutin, 142, *144*
Parkersburg, West Virginia, 479, *478*
Particle size, 179–180, *180*
Pasha, Khedive Ismail, 28, *28*
Passive continental margins, 367, *368*, 369, 382–383
Pearls, 73, 94
Peat, 208
Pedalfers, 183, *185*, 187, 193
Pediments, 589
Pedocals, 183, *185*, 187, 193
Pedon, 183
Pegmatite (Pg), 120, *121*, 126–127, 191, 588
Pelagic clay, 375, 382–383
Pele, a volcano goddess, 141
Pelitic rocks, 242–243, 247, 249, 256
Pelitic slate, 243
Pennsylvanian age, 36, 37, 38, *39*, 39, 63, 295
Penny Ice Cap, Canada, 535
Perched water table, 502
Perfluorocarbons, 29
Peridotite, 13, 16, 25, *118*, 118, 126, 344, 354, 375
Periodic table of elements, 74
Permafrost, 439, *444*, 451
Permeability, 219, 499–500, 524
Permian age, 37, 38
Permian lava flows, 40
Permian period, *39*
Persian Gulf, 66, 206, 219
Peru earthquake (1970), 319
Peru–Chile Trench, 54, 369–370, 385, 407
Petrified wood, 215
Petroglyphs, 584–585
Petroleum, 8, *8*, 72, 92, 219, 223, 340
Petrologists, 259
pH value, 196
Phaneritic rocks, *119*
Phaneritic texture, 115, 126
Phanerozoic Eon, *416*
Pharaoh Tuthmosis III (1504–1450 B.C.), 28
Phenocrysts, 116
Phoenix, River, 177
Phosphate, 191
Phosphate rock, *8*, 92
Phosphatization, 384,
Phosphorite, 379
Phosphorus, 84

Photons, 11
Photosynthesis, 29
Phyllites, 243
Physical deterioration, 188
Physical geology, 6, 8, 25
Piedmont glacier, 533
Pilgrim Memorial State Park, Plymouth, 130
Pillars of Hercules, 360
Pillow lava, 139–140, *140*, 160–161, 371, 375
 deep-sea sediments and, 52, 55, 63
 flows, 47
Pinnicales, 176
Piston corers, 362
Pitted outwash plains, 550
Placer deposits, 219
Plagioclase crystals, 120, 122, 126, 236
Plagioclase feldspars, 82, *88*, 90, 111, 124, 171, 178–179
Plagioclase minerals, 180
Planetary alignments, 356
Planetesimals, 11–12, *12*, 14
Plasma, state of matter, 73
Plastic flow, 536, 538
Plastic strain, 392, 419
Plate boundaries, 17, 61, 404
 asthenosphere and, *18*
 divergent plate boundaries, 47
 earthquake epicenters and, *304*
 lithosphere and, *18*
 types of, 47, 49, 52, 54–55, 63
Plate divergence, 390
Plate movement, 34, 46, 57, 60, *61*, 64, 391
 average rate of, 59
 determination of, 58–59
 on Earth, 16, 19
Plate tectonics, *411*
 definition of, 20, 47, 62
 driving mechanism of, 59–60, 64
 Earth systems and, 18
 introduction to, 17
 metamorphism and, 250–253, 254–255
 mountain building and, 405–409
 natural resources and, 60–62, 64
 oil and, 66
 plutons and, 153, 156, 160
 politics and, 66
 rock cycle and, 19, *23*
 setting, *154*
 terrestrial planets and, *50–51*
 theory, 17–18, 19, 25–26, 34, 60, 63–64
 as unifying theory, 46–47, 63
 volcanism and, 156
 volcanoes and, 153, 156, 160
Plates, 16, 26, 64
 African plate, 49, 156
 American plate, 49
 arabian plate, 66
 Arabian plate, 409
 Asia plate, 409
 definition of, 46–47
 of Earth, *46*
 Eurasian plate, 49
 European plate, 62
 Juan de Fuca plate, 55, *57*, 152, *154*
 North American plate, 55, *57*, 316
 oceanic Nazca, 54
 Pacific plate, 55, *57*, 57, *58*, 152, 156, 161, 316
 thermal convection cells and, 59
Platforms, 414, 418–419
Platinum (Pt), 84
Plato, 360, *360*
Playa lakes, 587
Pleistocene Epoch, 365, 367
Plinian volcanic eruptions, 148
Pliny the Elder, 135, 141
Pliny the Younger, 135, 148
Plucking, 541
Plumb lines, 349, *350*
Plunging folds, 396–397, *399*, 419
Pluto, 14
Plutonic rocks, 102–106, 112, 115, 122, 125–126
Plutons, 54, 121, 125–126, 156, 413. *See also* Volcanism and volcanoes
 characteristics of, 122–124, 126
 emplacement of, 390

 mafic, 111
 origins of, 122–124, 126
 plate tectonics and, 153, 156, 160
 to volcano, *104–105*
 volcanoes and, 153, 156, 160
Pluvial lakes, 554–555, 560
Plymouth rock, 130, *130*
Pocket beaches, 608
Point bars, 469, *469*, 490
Point Fermin, 434–435, *436*, 436–437, *448*, 455
Polar wandering, 40–41, *41*, 41
Polarizing microscopes, 85
Politics
 geology and, 7–8
 oil and, 66
 plate tectonics and, 66
Pollen, of plants, 30, 38
Pollution, 8, 190
Polymorphic transformation, 236, 256
Pompeii, Italy, *134*, 135, 151, *634*
Popping, 173
Porcelain, 92, 254
Porosity, 498–499, *500*, 500, 524
 in igneous and metamorphic rocks, 499
 and permeability, 498–499
 in sedimentary rocks, 499
 types of, 499
 values of, 499
Porous sediments, 455
Porphyritic rocks, 122
Porphyritic texture, 116, 126
Porphyroblasts, 243, 256
Porphyry, 116
Port Royal earthquake (1692), 319
Portage Lake, Alaska, 536
Positive gravity anomaly, 349
Positive magnetic anomalies, *353*
Potassium feldspars, 82, *88*, 90, 111, 116, 120, 122, 124, 178, 192
Potassium (K), 125, 171, 206, 281
Potassium–argon pair, 285–286
Potential energy, 465
Potholes, manifestation of abrasion, 465
Poulnabrone portal tomb, *509*
Powell, John Wesley, 262
Powell Memorial, 262
Prairie soils, 190
Pratt, J. H., 349–350
Precambrian Era, 278, 295, 415, 417–418
Precambrian evolution, 414, 418
Precambrian metamorphic rocks, *273*
Precession of the equinoxes, 558
Precious metals, 86
Precursors, 324
Pressure, 106, 233, 237
Pressure release, 172–173, 175, 193
Pressure–Temperature (P–T) path, 259
Pressure–Temperature–time (P–T–t) path, 259
Primary wave shadow zones, *345*, 345, 354–355
Primary wave velocity, *343*, 354
Primary waves, 307, *307*, 308–310, 328–329, 340, 342–343, 354
Primum Mobile, 338
Princeton University, 41
Principle of buoyancy, 350
Principle of cross-cutting relationships, 265, 267, 274, 292
Principle of faunal and .oral succession, 266
Principle of fossil succession, 266, 270, 274, 292
Principle of geology, 264
Principle of inclusions, 265, *269*, 272, 274, 292
Principle of isostasy, 349–352, *350*, 352, 354–355, 360
Principle of lateral continuity, 265, 274, 278, 292
Principle of original horizontality, 265, 274, 292, 393
Principle of relative dating, 267, 274, 291, *294*
Principle of superposition, 265, 267, 274, *279*, 292
Principle of uniformitarianism, 264
Prism, 78
Professor Hardwigg, 336–337
Proglacial lakes, 554–555
Prograding delta, *471*
Project FAMOUS (French-American Mid-Ocean Undersea forming Study), 371

Proterozoic Eon, 221–222
Protons, 11, 73, 77, 95, 281
Prudhoe Bay, 441
P–S time interval, 309
Pseudotachylite, 240
Pterosaurs, *295*
Puerto Rico, 378
Pumice deposits, 126
Pyramid Lake, Nevada, *554*
Pyramids and Sphinx, Giza, 493
Pyrite (FeS_2), 85, *85*, 88, 178, 191
Pyritohedron crystals, *78*
Pyroclastic flow, *147*
Pyroclastic layers, 148
Pyroclastic materials, 102, 126, 135, 139, *140*, 140, 144, 148, 150, 153, 160–161, *163*, 181
Pyroclastic rock, 120
Pyroclastic sheet deposits, 149–150, 160–161
Pyroclastic texture, 117, 126
Pyrolite rocks, 118
Pyroxene, 111, 122, 126, 178, 180, 236, 251
Pyroxenite rocks, 118

Q

Quarries, 173
Quarrying, 541
Quartz-bearing limestone, 236
Quartz clocks, 96
Quartz sandstones, 205, *206*, *233*, *247*, 249
Quartz (SiO_2), 8, 72, 77, 78, 79–81, 83, 85, 89, 92, 94, 96, 111, 120–122, 124, 126, *131*, 171, 175, 180, 182, 201, 205, 208, 219, 222, 233–236, 244
Quartzite, 22, 180, 184, *187*, 246–247
Queen Elizabeth II, 97, *97*
Queen's Jewels, 97
Quick clays, 438, *442–443*, 446, 451

R

Racetrack Playa, Death Valley, California, *587*
Radial drainage, 479, *481*
Radicals, 80
Radioactive decay, 11, 13, 281, 292, 347
 definition of, *286*
 for lead 206, *282*
 three types of, 281, *282*
 for uranium 240, *282*
Radioactive isotope pairs, 284–285, 292
Radioactive isotopes, 75
Radioactive parent and stable daughter atoms, *283*
Radioactive waste disposal, 593
Radioactivity, 21, 281, 291
Radiometric dating, 251
 radioactive isotope pairs in, 284
 sources of uncertainty in, 282–284
 techniques, 22
 uncertainty in, 292
Rainfall, 434
Rainshadow Deserts, *582*, 582
Rainwater, 178, 189
Ramage, E. W., 360
Range Province, 342, 405
Rapid mass movements, 431, 451
Rare Earth elements, 165
Rayleigh, Lord, 308
Rayleigh waves (R-waves), *308*, 308–309, 328–329
Recessional moraine form, 538, 548, 560
Recharge, 502
Recrystallization, 232, 246, 256
Rectangular drainage, 479, *481*
Recumbent, 395
Recurrence interval, 475
Recurved spits, 613
Red granite, 28
Red Planet. *See* Mars
Red River, Minnesota, *471*
Red sandstones, 37, 127
Red Sea, 47, *48*, 49, *49*, 62, 66, 151, *380*
Redoubt Volcano, Alaska, 140
Reefs, 204, 383
 evolution of, *380*
 organisms and, 376–378, 380–381, 382
Reflection, 355
Regional metamorphism, 241–242, 256
Regolith, 181, 193, 438

Reid, H. F., 301
Relative dating, 262, 274, 292
 methods of, 265–267, 271–272, 274, 291
 principles of, *275*
 six fundamental principles of, 265, 291
Relative geologic time scale, 262
Relative movement, 401
Relativity theory, 10
Relief, 184
Reserve, 92, 95
Reservoir rock, 219
Residual bonding, 77
Residual concentrations, 191
Resource, 92, 95
Restite, 118
Retrograde metamorphism, 236, 256
Reverse fault, 419
Rhinoceroses, *226*
Rhizoids, 175
Rhyolite, 120, 122, 125–126, *154*
Richter, Charles F., 311, 313
Richter Magnitude Scale, 311, *313*, 313, 314, 328–329
Ridge-push, 60, 64
Rift valley, 55, 342, 484
Rifting, 415
Right-lateral strike-slip fault, 403
Rill erosion, 189, 193
Rills, 189
Rio Grande, *475*, 586
Rip currents, *605*, 605
Ripple marks, 213, *215*, 218, 223
Riprap and Sand Berm, 622
River, definition of, 462
Roadways, 170
Rock bolts, 448, *450*
Rock bursts, 172
Rock chips, 280
Rock cycle, 19–20, *20–21*, 23, 26, 170, 201
Rock deformation, 418
Rock flour, 542
Rock-forming mineral, 95
Rock gypsum, 90, 206–207, *209*, 210
Rock reformation, 391–393
Rock salt, 90, 206–207, *209*, 210
Rock sequences, mountain ranges and, 36
Rock slide, *438*–440, 446, 451, *455*
Rock types, 127
Rock units, correlation of, 278–281, *278–279*, 291
Rock varnish coating, 583, *586*
Rockfalls, 431, *432*–433, 446, 451
Rocks, 26, 30, 73, 90, 95, 122, 210, 237, 434
 alteration of, 192
 conduction of heat, 173
 cumulates, 118
 definition of, 19–20
 dunite, 118
 groups of, 19
 mantle peridotite, 118
 peridotite, *118*, 118
 pyrolite, 118
 pyroxenite, 118
 ultramafic, 118
Rocks alteration, 171
Rocky Mountain Arsenal, 326, *327*
Rocky Mountains, 172, 306, 342, 404, 431, 479
Rogue waves, 627
Roman arch, 421
Roman Sculpture Busts, *230–231*
Romans, 336
Roosevelt, Theodore, 129
Rotated garnets, 251, *251*, 256
Rotorua, New Zealand, 521
Rounding, 202, 223
Roza flow, 149
Rubble, 204
Rubidium, 281, 284, 286
Ruby, 96
Running, 202
Running water, 170, 406, 417, 503
 definition of, 461–462
 deposition by, 467–473
 laminar or turbulent flow in, 462
 power of, 458

velocity of, 463
viscosity in, 460
Runoff, 459
Runways, 170
Rupturing, 325
Russian River, California, 613
Russo-Japanese War (1904–1905), 7

S

Sacramento, California, 479
Sahara Desert, North Africa, 532, 582
Sahel, Africa, 569
Salinas Valley, California, 514
Salinization, 190
Salt crystal growth, *8*, 173, 175, 193
Salt domes, 219, 365
Salt Lake, Utah, *554*, 588
Salt pan, 587
Saltation, 466, 569
Salton Sea, California, 576
Saltwater incursion, 513–514
San Andreas fault, 57, *298*–299, 301, 316–317, *316–317*, 330, 385, 403, *411*, 415, *423*, 423
San Bernardino, California, 588
San Clemente, California, *447*
San Francisco earthquake (1906), 311, *312*, 315, *317*, 523
San Joaquin Valley, California, 514–*515*, *515*
San Lucas submarine canyon, *368*
Sand, 8, *8*, 200, 202–203, 219, 375, 382
Sand seas, 575
Sandstone, 131, 184, 205, 207, 222, *225*, 266
Sandstone arches, 176
Sandstone beds, 393
Sandstone facies, 210
Sandstone Lion, 227
Sandstone varieties, *206*
Sandstones, 296
Santa Clara Valley, California, 515
Santa Cruz Mountains gap, *324*
Santa Monica Freeway, Los Angeles, 423
Santa Monica Mountains, California, *124*
Santo Domingo, Caribbean rim, 637
Sapphires, 72
Saskatchewan glacier, Canada, 533
Satellites, 11, 58–59, 64
Saturn, 12–14, *15*, 356
Sault Sainte Marie, Ontario, 547
Scablands, 555
Scarps, *50*, 447
Scarth, A., 146
Schist, 243
Schistose foliation, 243
Schistosity, 243, *245*
Scientific method, 16–17, 26
Scotia island arc, 54
Scrap, 98
Sea arches, 607, 623
Sea caves, 607
Sea level
 consequences of rising, 622
 factors determining, 620, 623
Sea of Japan, 54
Sea stacks, 607, 623
Seafloor. *See also* Oceanic crust; Oceans
 constitution of, *361*
 fractures, *364*, *373*, 373
 importance of, 360
 methods to study, 382
 methods to study the, 360–362
 resources from, 379
 sampling, *363*, 369
 sediment
 spreading, 41–45, *44–45*, 63–64
 typography, *371*
Seamounts, *374*, 374, 382–383, 409
Seashore, 202
Seawall construction, *619*
Seawater, 22, 206, 222
Secondary wave shadow zones, 345, *346*, 354–355
Secondary waves, *307*, 307–310, 328–329, 340, 343, 351, 354
Sedimentary breccia, 204–205, *205*, 222
Sedimentary facies, 210–212, 212, 222–223
Sedimentary particles, 202

Index

Sedimentary rocks, 20, *23*, 26, 115, 125, 131, 179, 200–201, 222–223, *225*, 408
    absolute dates determination for, *288*
    banded iron formation and, *221*
    classification of, 205–209, *210*
    common cements in, 204, 222
    composition of, 19
    determination of absolute dates for, *288*
    detrital, 204–205
    dolostone and, 84
    fossils and, 37
    hand specimens of, *22*
    history of Earth and, 19
    important resources in, 219–222, 223
    phosphate-rich, 379
    rock cycle and, 19
    sediments and, 203–204
    story of, 212–215, 218, 222
    types of, 204–210, 222
Sedimentary structures, 212–213, *214*, 218, 223
Sediments, 19, *20*, *23*, 170, 223, 380. *See also* Deposition
    definition of, 200–201, 222
    deposition of, 202, 222
    important resources in, 223
    lithification of, 203–204
    origin of, 202
    sedimentary rocks and, 203–204, 222
    sources of, 201, 222
    transport of, 201–202, *202*, 222
    types of, 375–376, 382
Seif dunes, 575
Seismic activity, 407
Seismic data, 13
Seismic discontinuity, 342, *342*
Seismic gaps, 324, *324*
Seismic-moment magnitude scale, 313, 314, 328
Seismic profiles, 281, 369
Seismic profiling, 361, *362*, 383
Seismic risk maps, 323, 326, 329
Seismic sea wave, 315
Seismic stations, 354
Seismic tomography, 340–341, 355
Seismic waves, 303, 307–309, 328, 340, 344
    amplitude of, *314*
    fronts of, *340*
    reflection of, *340*
    refraction of, *340*
    velocities of, *341*
Seismicity, 369, 391. *See also* Earthquakes
Seismogram, 303
Seismographs, *302*, 303, *309*, 310, 319, 329
Seismologists, 304, 309, 313, 314, 328
Seismology, 303–304, 309, 328–329
Self-exciting dynamo, 353
Semiarid regions, 581
Septic tanks, 516
Serpentine, 236
Serpentine asbestos, 258
Shaking, duration of, *314*
Shale facies, 205, 210, *248*
Shales, *221*, 233, 435
Shallow-depth earthquakes, 55
Shallow dike, *105*
Shallow-focus earthquakes, 303
Shallow intrusions, *105*
Shallow-water waves, 603
Shanghai, China, 515
Shatter cones, 240, *242*
Shear strength, 427, 451
Shear stress, 391, 419
Shear waves, 308
Sheep Mountain anticline, Wyoming, *399*
Sheep Rock, *177*, *200*
Sheet erosion, 189, 193, 462
Sheet flow, 462
Sheet joints, 172, *173*–174, 192–193
Sheet lava flows, 371
Sheetlike intrusions, 125
Shelf–slope break, 365
Shell models for atoms, *74*
Shelley, Percy B., 7, *25*
Shells, 210, *223*
Shield volcanoes, 141, *143*, 148, 161
    eruptions of, 141

Shields, 414, 418–419
Shiprock, New Mexico, 124–125, 129, *129*
Shishaldin volcano, Alaska, 152
Shock metamorphism, 240–241, 242, 255–256
Shoreline, 170, 598, 623
    deposits of, 608–609, 623
    erosion, 606–607, 623
    processes of, 599–605, 623
    study of, 598, 623
Siccar Point, Scotland, 267
Sierra Nevada, *173*, 253, 431
Silent City of Death, 146
Silica (SiO_2), 80, 95, 110, 215, 219, 357, 376
Silica tetrahedron, 80–81, 84, 95
Silicates, 12, 80, *82*, 83, 95, 175, 178. *See also* Ferromagnesian silicates; Nonferromagnesian silicates
    mantle-crust, 14
    minerals, 80–83, 94
Siliceous ooze, 376
Siliceous sinter, 521
Silicon dioxide (SiO_2), 77, 91, 201, 204, 210, 215, *216*, 344
Silicon (Si), 12, 16, 19, 76, 79
Sillimanite, 236, 241, 248
Sills, 123, *124*, 124–126
Silly Putty, 351
Silt, 202
Siltstone, 205
Silurian lava flows, 40
Silver (Ag), 78, 84, 96
Silver mines, *87*
Silver sulfide argentite (Ag_2S), 84
Sinkholes, 505, *506*, *512*, 524
Skaergaard Intrusion, Greenland, 111
Skaftafjell National Park, Iceland, *100–101*
Skel, volcano god, 141
Skylight, 137
Skyscrapers, 331
Slab metamorphism, 253
Slab-pull, 60, 64
Slash-and-Burn agriculture, *187*
Slate, 243, *244*, 257, 257
Slaty cleavage, 243
Slides, 432, 446, 451
Slope angle, 186–187, 427, 451
Slope base, *428*–429, 434
Slope direction, 186–187
Slope failure, 427
Slope hill, 434
Slope influence on soil development, *187*
Slope shear strength, 427, *427*, 431, 451
Slope stability, 451
    geology and, *430*
    maps, *447*, 447
    mass wasting and, *430*
    topography and, 430
Slope stabilization, 452
Slow mass movements, 431, 451
Slump block, *437*
Slumps, 432, 434, *434*–435, *437*, 446, 451
Smith, Robert Angus, 196
Smith, William (1769–1839), 266
Smithson, Robert, 594
Smog, 196
Snake River Plain, Idaho, 149, *150*, 464
Sodium chloride (NaCl), 75, *175*, 378
Sodium (Na), 75, 76, 79, 125, 206
Softwater, 518
Soil, 30, 170, 193, 428
    arid regions, *186*
    climate and, 183–184
    composition of, *183*
    its origin, 192
    made of, 187
    parent material and, 182
Soil conservation practices, 190, *190*
Soil degradation, 188–190, 191–193
    chemical, 190
    from erosion, 189
    physical, 190
    types of, 188
Soil development, *187*
Soil erosion, 188–190

Soil formation, 187
    climate and, *184*
    factors in, 183
    rate of, 183
Soil horizons, 182, *183*, 193
Soil organisms, 184
Soil pollution, 190
Soil profiles, 182–183, 187
Solar energy, variations in, 558
Solar nebula, 11
Solar nebula theory, 11–12, 26
Solar system, 10–13, 25
    composition of, 11–12
    diagrammatic representation of, *11*
    lava flow channel, *50*
    origin and history of, 25
Solar tidal bulge, 600
Solids, state of matter, 73
Solifluction, 439, *444*, 446, 451
Solution, 175, 178, 193
Solution valleys, 505
Sonoma County, California, 523
Sorting, 202, 223
Source rock, 219
South Equatorial Current, 384
South Pole, 36–39, *42*, 63
South Rim of the Grand Canyon, 262
Southern California landslides, 455
Southern Hemisphere, 35, 37
Sowbugs, 184
Sparkplugs, 254
Specific gravity, 89, 95
Sphalerite, *88*, 88
Spheroidal weathering, 180, *181*–182, 193
Spielberg, Steven, *297*, 297
Spinosaurus, 297
Spiral Jetty, 594
Spires, 176
Spirit, Mars Exploration Rover, *69*, 69
Spit, 613
Spodumene crystals, 96
Sponges, 376
Spores, of plants, 38
Spreading ridges. *See* Divergent plate boundaries
Spring tides, 600
Springs, *502*, 502, 504, 524
St. Charles County, Missouri, 477
St. Elias Mountains, Alaska, 550
St. Pierre, Martinique, 145, *147*
Stabiae city, 135, 151
Stability of silicate minerals, 180
Stabilizing a hillside, *449*
Stagnant glacier, 538
Stalactites, 507
Stalagmites, 288, *289*–290, 507
Star dunes, 577, 579
Staurolite, 243, 248
Steam (water vapor), 166
Steel, 8
Steinbeck, John, 194
Steno, Nicolas (1638–1686), 77, 265
Stocks, 122, 126
Stone Mountain, Georgia, 172, *173*
Stoping, 107, 126
Storm surges, 618–619
Storm waves, 618, 623
Strain, *391*, 391, 392–393, 393, 418–419, 536
Strait of Gibraltar, *360*, 360
Strata, 212, 223
Stratification, 223
Stratified drift, 553
Stratigraphic traps, 219, 223
Stratosphere, 166
Stratovolcanoes. *See* Composite volcanoes
Stream channels, 202
Stream-dominated deltas, 472
Stream erosion, *428*
Stream gradient, 462
Stream piracy, 484
Stream/river flood, 459
Stream/river monitoring, 475
Stream terraces, 484, *486*
Streamlines, 460
Streams and rivers, 462, 463
Stress, *391*, 391, *392*, 393, 418–419, 536

Stretched-pebble metaconglomerates, 244, *246*
Striations, 83
Strike, 393, *393*, 395, 397, 401, 419
Strike-slip faults, 402–403, 410, 419
Strike-slip movements, 403
Strip-cropping, *190*
Stromboli and Mount Etna, Italy, 157
Strombolian volcanic eruptions, 148
Strontium, 281
Structural basin, 397, 405
Structural dome, 397
Structural traps, 219, 223
Stylolites, *237*, 237
Subbituminous coal, 220
Subduction, 251
Subduction complex, 52
Subduction zones, 360, 407, 413
Sublimation, 535
Submarine canyons, 366–367, *368*, 383
Submarine fans, *366*, 366–367, 369, 383
Submarine hydrothermal vents, 371, *372*, 373, 382–383
Submarine lava flow, 373
Submarine ridges, 362
Submarine volcanism, 153
Submergent coast, 617
Submersibles, 360
Subsoil, 183
Subsurface units, correlation of, 280, 291
Subsurface water, *448*
Sudbury pluton, Ontario, Canada, 241
Suess, Edward, 34–35, 68
Suevite breccia, blocks of, *128*
Suevites, 128
Suez Canal, 28
Sulfates, 80, *84*, 84, 90, 206
Sulfide mineral galena (PbS), *84*
Sulfides, 191
Sulfur, 84, 137, 140
Sulfur dioxide (SO_2), 91, 136–137, 196
Sulfur gases, 115
Sulfur hexafluoride, 29
Sulfur Works, *519*
Sulfuric acid (H_2SO_4), 178, 191, 196
Sumatra earthquake (2004), 314–315
Sumida River, 315
Summer beaches, 613
Summit of Kilauea Volcano, Hawaii, *109*
Sun, 11–13, 25, 73, 166, 336, *339*, 348, 356, 387
Sunderbans, a swamp forest, 457
Sunken city, 436
Superalloys, 92
Supercontinent cycle, 46–47, 63
Superposed streams, 485
Superposition, 291
Supervolcano, 167
Surf zone, 604
Surface waves, 307–308, 328–329
Suspect terrane, 409
Suspended load, 466, 490, 569, 591
Suspended water, 500
Syenite, 102
Sylvite (KCl), 219
Synclines, 394, *395*, 395–396, *396*, 397, 419
Synthetic oil, 221
System, definition of, 4, 25–26

T

T. rex, 297
Table salt, 378
Tabloids, 314
Taconic orogeny, 130
Taipei Basin, Taiwan, 515
Talc, 89, 254
Talus, 172, 193
Tambora eruption, Indonesia, 135–136, 156
Tangshan earthquake (1976), 315, 325
Tangwa Village, China, 579
Tapeats sandstone, *212*
Tar sands, 219–220, *220*
Tarn, 546
Taupo, New Zealand, 167
Taylor, Frank, 35
Tectonite, 245
Teed, Cyrus Reed, 336

Teeth, 223
Teheran, Iran, 588
Tehuacan Valley, Mexico, 285
Tektites, 128
Telegraph Plateau, 370
Temple of Amun at Luxor, Egypt, *421*
Templin Highway, Castaic, California, *267*
Tennant Creek earthquake (1988), 306
Tension, 391, 418–419
Tentative explanations. *See* Hypotheses, definition of
Terminal moraine, 538, 548
Terra, 12
Terrain, 409, 635
Terranes, 409, 413, 419
Terrestrial planets, 12–13, *14–15*, 26, *50–51*, 136
Teton Range, Wyoming, 403–404, *544*
Texture, 115, 122, 126
Thames River, London, 97
Theory, definition of, 16–17, 25–26
Thermal contraction, 173
Thermal convection cells, 42, 59, *60*, 64
Thermal expansion, 173, 175, 193
Thermal spring, 519
Thermocline, 639
Thomasson Partner Associates, 228
Thorium, 230, 284, 286, 288
Thrust fault, 402, 419
Thunder, loud claps of, 430
Tidal flats, 202
Tidal power, 629
Tidal wave, 315
Tide-dominated deltas, 472
Tides, 600
  low and high, *600*
  moon and sun generating, 600, *601*, 606, 623
  on shorelines, 601
Tidewater glaciers, 533
Tigris–Euphrates floodplain, 634
Time, 186–187, 233, 237. *See also* Geologic time
Time–distance graphs, 309–310
Tin, 191
Titan, 136
Toba caldera, Sumatra, 141, 167
Tombolos, 613, *614*, 623
Tombstones, 253
Topaz, 96
Topography
  of Atlantic Ocean basin, *43*
  deep seafloor and, 375
  folds and, *396*
  geology and, 430, 451
  military strategy and, 7, 25
  slope stability and, 430
Topset beds, 470
Topsoil, 182
Torrential rains, 426
Tourmaline, 72, 121, *121*
Tower of London, London, 97
Toxic-waste sites, 516
Trace fossils, 214, *216*
Tracks, *216*
Trade routes, ancient, *635*
Transform faults, 55, 64
Transform plate boundaries, 55, *56–57*, 64, 410
Transition zone, 344
Transitions, 104
Transportation, 19–20, 202
Transverse dunes, 575, 580, 591
Travertine terraces, 206, 512
Tree-ring dating, 286, 287, *287*, 292
Trellis drainage, 479, *481*
Trench, 23
Triassic fault basins, 52, *52*
Triassic period, 38, 46
Triassic rifting, Late, *52*
Triton, 136
Trough, 602
Tsunami, *2–3*, 315, 323, 329, 627
  anatomy of a, *321*
  in Crescent City, California (1964), 318
  fatalities since 1990, 322
  in Hilo, Hawaii (1946), 318–319
  in Indian Ocean (2004), 4, 303, 315, 319, *320*
  killer waves, 315–319

  as natural disasters, 4, 7, 25
  Papua New Guinea (1998), 34
  plate tectonics and, 34
Tuff, 117, 150, 200
Tungstate ion, 238
Tungsten, 109
Turbidity currents, 366–367, *367*, 383
Turbulent flow, 460, *461*
Turkey earthquake (1999), 306, *318*
Turkey-track, 119
Turnagain Heights, 315, 423, 439, 443
Turquoise, 72
Turtle Mountain, Canada, 435, *440*
Turtles, *226*
Tuscumbia cutoff, 635
Twin Falls, Idaho, *150*
20-year flood, 475, 490

U

U-shaped glacial trough, 542, 559
Udachnaya kimberlite pipe, Russia, *344*
Ultimate base level, 480, 484
Ultramafic crust, 413
Ultramafic rocks, 118, 122
Uluru, 264, 276–277, *277*, 588
Unconformities, 266–267, *270*, 274, 291–292
Uniformitarianism, 21–24, 22–23, 26
United States Great Plains, 184
Universal gravitation law, 16, 347
Universal gravitational constant, 347
Universal Studios, Los Angeles, 297
Universe, origin of, 10–13, 25
University of California, 387
University of Rhode Island, 387
University of Western Australia, 30
Unzen volcano, Japan, 124, 145
Uplift, 253. *See also* Mountain building
Upper mantle, 23
Upper Peninsula, Michigan, 248
Upper Permian Dunedoo Formation, Australia, 35
Upwelling, 23, 384
Urals in Russia, 409
Uraninite (UO_2), 221
Uranium (U), 8, 74, 221, 236, 237, 240, 281, 288
Uranium–lead pair, 284, 286
Uranus, 12–14, *15*
Urban flood, 459
U.S. Army, 326
U.S. Coast and Geodetic Survey, 319
U.S. Environmental Protection Agency (EPA), 236, 258
U.S. Fifth Circuit Court of Appeals, 258
U.S. Geological Survey (USGS), 92–93, *132–133*, 158, 259, 303, 325, 326, 333, 362
U.S. Navy, 28
U.S. oil and natural gas production, 378
USGS. *See* U.S. Geological Survey

V

Vaiont Dam disaster, *453*, 453
Vaiont Reservoir area, *453*
Valley glaciers, 533–534, *540*, 544, 546
Valley of the Kings, Egypt, *209*
Valley trains, 550
Valleys, 484–489, *486*, 491
Valuable metamorphic rocks, 255
van der Waals bonding, 75, 77
Varve, *552*
Vast plains, 362
Vegetation, 180, 196, 200, 428, 451
VEI. *See* Volcanic explosivity index
Veins, 123–126
Velociraptors, 297
Velocity, 463–464
Veniaminof volcano, Alaska, 152
Venice Beach, Los Angeles, 639
Venice, Italy, 515, 626
Ventifacts, 570, *571*, 573
Venus, 12–14, *14*, 50, 50, 136, 337, 356
Vermillion Cliffs Wilderness, Arizona, *260–261*
Verne, Jules, 7, 336–337, *339*
Verrazano Narrows, 563
Vertical accretion, 470
Vesicles, 117, 126
Vesicular rocks, 117

Index

Vestmannaeyjar town, 142
Victoria Transvaal diamond, *357*
Village of Rueun, Switzerland, *424–425*
Vinalhaven Island, Maine, *113*
Viscosity, 110, 115, 125–126
Viscous magmas, 156
Vog, 136–137
Volatiles, 107, 115, 120, 125
Volcanic, 126
Volcanic activity, 17, 19
Volcanic arc, 54
Volcanic ash, 126, *145*, 438
Volcanic behavior, 159
Volcanic block, *140*
Volcanic bomb, *140*
Volcanic breccia, *117*, 117
Volcanic deposits, 104
Volcanic domes. *See* Lava domes
Volcanic eruptions, 13, 16, *42*, 125, *148*, 430
    duration of, 156–157
    forecasting, 157–159
    Hawaiian, 148
    as natural disasters, 4, 7, 25
    notable, 135
    plate tectonics and, 34
    Plinian, 148
    size of, 156–157
    Strombolian, 148
    Vulcanian, 148
Volcanic explosivity, 161
Volcanic explosivity index (VEI), 156, *158*, 159
Volcanic gases, 136, *157*, 166
Volcanic hazards, 156–159, *157*, 161
Volcanic igneous rocks, 115, 126
Volcanic island arc, 54, *408*
Volcanic islands, 130, 406
Volcanic landforms, 149–150, 160
Volcanic mudflows, 135, *146*
Volcanic necks, 124, 126, 129, *129*
Volcanic pipes, 124, 126
Volcanic rocks, 102–106, 122, 125
Volcanic sheet intrusions, 125
Volcanic smog (vog), 136
Volcanic tremor, 158, 161
Volcanism and volcanoes, 57, 135, 140, 160–161, 347, 369, 390–391. *See also* Lava flows; Plutons
    in Alaska, 152–153
    at convergent plate boundaries, *151*
    distribution of, 151–152, 160
    at divergent plate boundaries, *151*
    eruptions of, 146, 148
    most dangerous manifestations of, 156, 161
    in North America, 152–153, 160
    plate tectonics and, 153, 156, 160–161
    plutons and, *104–105*, 153, 156, 160
    seismic activity, 407
    shield, 141
    types of, 141–148, 146, 148, *149*, 160
Volcano monitoring, 156–159, 161, 167
Volcano Sapas Mons, *51*
Voltaire, Francois, 635
Voluminous eruptions, 167
Vougioukalakis, G., 164
Voyager spacecraft, *51*
Vulcan, god of fire, 336

Vulcan, Roman deity of fire, 141
Vulcanian volcanic eruptions, 148

W

Wabash River's drainage basin, Ohio River, *478*
Wake Island, 378
Walls retaining, *450*
Walvis Ridge, South Atlantic, 374
Warm Springs, Georgia, 519–520
Wasatch Range, Utah, 404
Washington Monument, 253, *254*
Washington State, 110
Waste disposal, 8
Water. *See* Running water; Ground water
Water content, 428
Water-filled channel, dimension of, 464
Water gaps, 487
Water softeners device, 518
Water table, *500*, 500, 524, 638
    in arid and semiarid regions, 501
    lowering of, 513
Water-treatment plants, 528
Water vapor, 9, 72, 94, 115, 136, 140, 460
Water wells, 503, 505, 524
Watson, John H., Dr., 30
Watt, James, 649
Wave action, *429*
Wave base, 602, 623
Wave-built platform, 607
Wave-cut platforms, 607, *607*
Wave-dominated deltas, 472
Wave erosion by abrasion and hydraulic action, *606*
Wave-formed ripple marks, 213
Wave generation, 603
Wave height, 602
Wave period, 602
Wave rays, 340
Wave refraction, 604, 608, 623
Wavelength, 602
Waves, 202, 417, 601. *See also* Seismic waves
Weak nuclear force, 11
Wear-resistant alloys, 92
Weather phenomena, 423
Weathering, 19–20, 23, 192–193, *193*
    alterations of rocks and minerals, 170
    chemical weathering, 171
    climate and, 427–428
    definition of, 170–171, 191
    differential weathering, *171*, 171
    erosion and, 170
    fractures and, *179*
    freezing and, 171
    of granite, *170*
    mechanical weathering, 171
    natural resources and organisms and, *174*
    thawing and, 171
Wegener, Alfred, 35, *35*, 36, 38, 41, 62, 68
Welded tuff, 150
Well cuttings, 280
Well logs, 280, *281*
Wellesley glacier, Alaska, 534, *534*
Wells, H. G., 646
Westward drift of North America, *417*
White alabaster, *99*

White Nile, Sudan, 493
White Pine Mine, 228
White River Badlands, South Dakota, 226
Whitmore, Paul, 333
Whittow, J., 426
Widespread deposition, *351*
Wildfires, 423
William the Conqueror (1066–1087), 97
Wilson cycle, 46
Wilson, J. Tuzo, 46, 63
Wind, 19, 170, 202, *384*, 417, 586
Wind abrasion, *571*
Wind Cave and Jewel Cave, South Dakota, 507
Wind deposits, 574–580, 591
Wind erosion, 570–573
Wind gap, 487
Wind River Range, Wyoming, 404
Windmills, 595
Windpower, 595
Winston, Stan, 297
Winter beaches, 613
Wizard Island, 142, *142*, 152
Wollastonite ($CaSiO_3$), 233–236
Wood, 215
Woods Hole Oceanographic Institution, 387
World Bank, 4
World War I (1914–1918), 194
World War II (1939–1945), 66, 92, 96–97

X

X-ray beams, 77, 85
Xenoliths, 107, 109

Y

Yangtze River, People's Republic of China, 494
Yardangs, 571, *572*
Yellow River, China, *465*
Yellowjacket Mine, Nevada, 87
Yellowstone caldera, Wyoming, *167*, 167
Yellowstone National Park, Wyoming, 167, 226, 268, *469*, 519, 520, *520*, 522
Yellowstone Parks of deep sea, 387
Yellowstone Tuff, *167*
Yosemite National Park, California, 172, 544, *545*
Young mountain range, 369
Younger komatiites, 118
Younger rising magmas, *113*
Yucatán Peninsula, Mexico, 365
Yucca Mountain, Nevada, 569

Z

Zagros Mountains, Iran, 409
Zagros suture, 66
Zhou Dynasty (770–476 B.C.), 296
Zinc, 8, 88, 191
Zion National Park, Utah, *218*, 278, *279*
Zircons ($ZrSiO_4$), 413
Zone of accumulation, 183, 193, 538
Zone of aeration, 500
Zone of leaching, 182, 193
Zone of saturation, 500
Zone of wastage, 538
Zoned plagioclase feldspars, 111
Zoned plutons, 113, 126